Comprehensive
Clinical
Neurophysiology

Comprehensive Clinical Neurophysiology

KERRY H. LEVIN, M.D.
Staff Neurologist
Department of Neurology
The Cleveland Clinic Foundation
Cleveland, Ohio

HANS O. LÜDERS, M.D., Ph.D.
Professor and Chairman
Department of Neurology
The Cleveland Clinic Foundation
Cleveland, Ohio

W.B. SAUNDERS COMPANY
A Harcourt Health Sciences Company
Philadelphia • London • New York • St. Louis • Sydney • Toronto

BS

W.B. SAUNDERS COMPANY
A Harcourt Health Sciences Company

The Curtis Center
Independence Square West
Philadelphia, Pennsylvania 19106

Library of Congress Cataloging-in-Publication Data

Comprehensive clinical neurophysiology / [edited by] Kerry H. Levin, Hans O. Lüders.

p.   cm.

ISBN 0–7216–7656–1

1. Electrodiagnosis.   2. Nervous system—Disease–Diagnosis.
3. Neurophysiology.   4. Electroencephalography.   5. Evoked
potentials.   6. Electromyography.   I. Levin, Kerry H.   II. Lüders, Hans O.
[DNLM: 1. Nervous System Diseases—diagnosis.   2. Electroencephalography.
3. Electromyography.   4. Evoked Potentials.   5. Nervous System Physiology.
WL 141 C737 2000]
RC386.6.E86 C66 2000

616.8′047547—dc21

DNLM/DLC   99–059398

Acquisitions Editor: Allan Ross
Project Manager: Evelyn Adler
Production Manager: Norman Stellander
Illustration Specialist: Fran Moriarty

COMPREHENSIVE CLINICAL NEUROPHYSIOLOGY   ISBN 0–7216–7656–1

Printed in the United States of America.

Last digit is the print number: 9   8   7   6   5   4   3   2   1

10/29/03

CONTRIBUTORS

Vahe E. Amassian, M.B.
Professor of Physiology and Neurology, SUNY Health
Sciences Center at Brooklyn, Brooklyn, New York
Emerging Applications in Neuromagnetic Stimulation

Karen L. Barnes, Ph.D.
Professor, Department of Physiology and Cell Biology,
College of Medicine and Public Health, Ohio State
University, Columbus, Ohio; Director, Research Programs
Development, The Lerner Research Institute; Staff,
Department of Neurology, The Cleveland Clinic
Foundation, Cleveland, Ohio
*Excitable Cells: The Ionic Basis of Membrane Potentials • Excitable
Membranes: Local Responses and Propagation*

Selim R. Benbadis, M.D.
Associate Professor of Neurology, Departments of
Neurology and Neurosurgery, University of South Florida
College of Medicine, Tampa, Florida
*Focal Nonepileptic Disturbances of Brain Function • Pediatric Epilepsy
Syndromes*

Richard C. Burgess, M.D., Ph.D.
Professional Staff and Head, Section of Neurological
Computing, The Cleveland Clinic Foundation, Cleveland,
Ohio
*Signal Conditioning of Neurophysiological Signals: Amplifiers and
Filters • Clinical Neurophysiological Instrumentation: Electrical Safety*

Thomas F. Collura, Ph.D., P.E.
Lecturer in Clinical Neurophysiology, The Cleveland Clinic
Foundation; President, BrainMaster Technologies, Inc.,
Cleveland, Ohio
Basic Electronics and Physics • Noise, Averaging, and Statistics

Vedran Deletis, M.D., Ph.D.
Director, Division of Intraoperative Neurophysiology,
Institute for Neurology and Neurosurgery, Beth Israel
Medical Center, New York, New York
Emerging Applications in Neuromagnetic Stimulation

Dudley S. Dinner, M.D.
Staff Neurologist, Department of Neurology, The Cleveland
Clinic Foundation, Cleveland, Ohio
Physiology of Sleep

Ronald G. Emerson, M.D.
Professor, Department of Neurology, Columbia University
College of Physicians and Surgeons; Attending, New York
Presbyterian Hospital, New York, New York
Somatosensory Evoked Potentials

Charles M. Epstein, M.D.
Associate Professor of Neurology, Emory University School
of Medicine, Atlanta, Georgia
Visual Evoked Potentials

Mark A. Ferrante, M.D.
EMG Laboratory Co-Director and Associate Professor,
University of South Florida, Tampa, Florida
Plexopathies • Myopathies

Nancy Foldvary, D.O.
Director, Sleep Disorders Center, Department of Neurology,
The Cleveland Clinic Foundation, Cleveland, Ohio
*Normal Electroencephalogram and Benign Variants • Focal Epilepsy
and Surgical Evaluation*

Eric B. Geller, M.D.
Clinical Assistant Professor, New York University Medical
School, New York, New York; Director, Adult
Comprehensive Epilepsy Program, Saint Barnabas Institute
of Neurology and Neurosurgery, Livingston, New Jersey
*Generalized Disturbances of Brain Function: Encephalopathies, Coma,
Degenerative Diseases, and Brain Death*

Mark Hallett, M.D.
Clinical Director, NINDS, National Institutes of Health,
Bethesda, Maryland
Electrodiagnosis in Movement Disorders

Hajo M. Hamer, M.D.
Department of Neurology, University of Marburg, Marburg,
Germany
*Electric Montages and Localization of Potentials in Clinical
Electroencephalography*

C. Michel Harper, M.D.
Associate Professor, Mayo Clinic Foundation; Consultant in
Neurology, Mayo Clinic, Rochester, Minnesota
Intraoperative Monitoring

Richard A. Hrachovy, M.D.
Professor of Neurology, Baylor College of Medicine;
Medical Staff, The Methodist Hospital, St. Luke's Episcopal
Hospital, Texas Children's Hospital; Director, EEG, Ben
Taub General Hospital, Veterans Affairs Medical Center,
Houston, Texas
Development of the Normal Electroencephalogram

Jun Kimura, M.D.
Professor Emeritus, Department of Neurology, Kyoto
University School of Medicine, Kyoto, Japan; Professor,
Division of Clinical Neurophysiology, Department of
Neurology, University of Iowa Hospitals, Iowa City, Iowa
Waveform Analysis and Near- and Far-Field Concepts

George Klem, M.D., R. EEG T.
Supervisor, Section of Epilepsy and Sleep Disorders, The
Cleveland Clinic Foundation, Cleveland, Ohio
Long-Term Electroencephalographic/Video Monitoring

Prakash Kotagal, M.D.
Section of Pediatric Epilepsy and Sleep Disorders,
Department of Neurology, The Cleveland Clinic
Foundation, Cleveland, Ohio
Polysomnography and Multiple Sleep Latency Testing

Richard J. Lederman, M.D., Ph.D.
Associate Professor of Neurology, Ohio State University
College of Medicine, Columbus, Ohio; Staff Neurologist,
The Cleveland Clinic Foundation, Cleveland, Ohio
Nerve Conduction Studies

Heidi Lai, M.D.
Clinical Neurologist, St. Elizabeth Medical Center and
Home Hospital, Lafayette, Indiana
Muscle Physiology and Pathophysiology

A. Arturo Leis, M.D.
Clinical Professor of Neurology, University of Mississippi
Medical Center, Jackson, Mississippi; Electrodiagnostic
Consultant, Mississippi Methodist Rehabilitation Center,
Jackson, Mississippi
*Refractory Period and Collision Technique • Silent Period Studies and
Long Latency Reflexes*

Kerry H. Levin, M.D.
Staff Neurologist, Department of Neurology, The Cleveland
Clinic Foundation, Cleveland, Ohio
*Needle Electrode Examination • Radiculopathy • Disorders of
Neuromuscular Junction Transmission*

Hans O. Lüders, M.D., Ph.D.
Professor and Chairman, Department of Neurology, The
Cleveland Clinic Foundation, Cleveland, Ohio
*Electrode Montages and Localization of Potentials in Clinical
Electroencephalography • Auditory Evoked Potentials*

Paul J. Maccabee, M.D.
Associate Professor of Neurology, State University of New
York, Health Science Center at Brooklyn, Brooklyn, New
York
Emerging Applications in Neuromagnetic Stimulation

Hiroshi Mitsumoto, M.D.
Professor of Neurology, Columbia University; Columbia-
Presbyterian Medical Center, New York, New York
Muscle Physiology and Pathophysiology

Eli M. Mizrahi, M.D.
Head, Section of Neurophysiology; Professor of Neurology
and Pediatrics, Baylor College of Medicine, Houston, Texas
Neonatal Seizures

Harold H. Morris, M.D.
Head, Section of Epilepsy and Sleep Disorders, The
Cleveland Clinic Foundation, Cleveland, Ohio
Long-Term Electroencephalographic/Video Monitoring

Erik P. Pioro, M.D., D.Phil., F.R.C.P.C.
Director, ALS Center, Section of Neuromuscular Diseases,
Department of Neurology, The Cleveland Clinic
Foundation, Cleveland, Ohio
Motor Neuron Disorders

Robert W. Shields, Jr., M.D.
Staff Neurologist, The Cleveland Clinic Foundation,
Cleveland, Ohio
*Quantitative Electromyography and Special Electromyographic
Techniques • Autonomic Nervous System Testing*

Thomas H. Swanson, M.D.
Associate Professor of Psychology, Bowling Green State
University, Bowling Green, Ohio; Director, Comprehensive
Epilepsy Center, St. Vincent Mercy Hospital, Toledo, Ohio
*Synaptic Transmission • Basic Cellular and Synaptic Mechanisms
Underlying the Electroencephalogram*

Kiyohito Terada, M.D.
Department of Neurology, Takeda General Hospital, Ishida-
Mori-Minami 28–1, Fushimi, Kyoto, Japan
Auditory Evoked Potentials

Eric Wassermann, M.D.
Senior Investigator, National Institute of Neurological
Disorders and Stroke, National Institutes of Health,
Bethesda, Maryland
Emerging Applications in Neuromagnetic Stimulation

Asa J. Wilbourn, M.D.
Director, EMG Laboratory, Department of Neurology, The
Cleveland Clinic Foundation, Clinical Professor of
Neurology, Case–Western Reserve University School of
Medicine, Cleveland, Ohio
*Principles of Electrodiagnosis • Mononeuropathies • Plexopathies •
Multiple Mononeuropathies and Polyneuropathies • Myopathies*

Elaine Wyllie, M.D.
Head, Section of Pediatric Epilepsy, The Cleveland Clinic
Foundation, Cleveland, Ohio
Pediatric Epilepsy Syndromes

Ulf Ziemann, M.D.
Assistant Professor of Neurology, Clinic of Neurology,
J. W. Goethe University of Frankfurt, Frankfurt am Main,
Germany
Emerging Applications in Neuromagnetic Stimulation

PREFACE

Since 1992, the Department of Neurology at the Cleveland Clinic has conducted an annual course covering the wide spectrum of clinical neurophysiology. Over the years, as the syllabus material for the course became more complete and as the course faculty refined the teaching approach to the material, the idea for a text of comprehensive clinical neurophysiology developed. All but two of the contributors to this text have participated in the course over the years, and they represent a group of clinical neurophysiologists who are recognized not only for their achievements in their fields but who also have proved themselves as effective communicators.

The text's mission is not to serve as a primer of techniques of clinical neurophysiological testing. Rather, it is meant to serve as a guide to proper electrophysiological testing, a reference for the proper interpretation of clinical electrophysiological data, and a review of the physiological concepts underlying this practice of neurology.

Although the material presented here represents the state of the art of the practice of clinical neurophysiology at this time, the gestation of a project of this size is long enough that by the time it is published, new areas of interest and clinical application will have already developed. A course can be updated on a yearly basis by adding new faculty and new subjects, but an updated text must await a second edition.

KERRY H. LEVIN, M.D.

HANS O. LÜDERS, M.D., Ph.D.

CONTENTS

SECTION I
Basic Neurophysiology 1

1 Basic Electronics and Physics 1
Thomas F. Collura, Ph.D., P.E.

2 Noise, Averaging, and Statistics 11
Thomas F. Collura, PhD, PE

3 Signal Conditioning of Neurophysiological Signals: Amplifiers and Filters 19
Richard C. Burgess, MD, PhD

4 Clinical Neurophysiological Instrumentation: Electrical Safety 31
Richard C. Burgess, MD, PhD

5 Excitable Cells: The Ionic Basis of Membrane Potentials 43
Karen L. Barnes, PhD

6 Excitable Membranes: Local Responses and Propagation 50
Karen L. Barnes, PhD

7 Synaptic Transmission 57
Thomas H. Swanson, MD

8 Waveform Analysis and Near- and Far-Field Concepts 69
Jun Kimura, MD

9 Muscle Physiology and Pathophysiology 81
Heidi Lai, M.D. • Hiroshi Mitsumoto, M.D.

SECTION II
Electromyography 89

10 Nerve Conduction Studies 89
Richard J. Lederman, MD, PhD

11 Refractory Period and Collision Technique 112
A. Arturo Leis, MD

12 Needle Electrode Examination 122
Kerry H. Levin. MD

13 Quantitative Electromyography and Special Electromyographic Techniques 140
Robert W. Shields, Jr., MD

SECTION III
Electrodiagnosis of Neuromuscular Disorders 162

14 Principles of Electrodiagnosis 162
Asa J. Wilbourn, MD

FOCAL DISORDERS 174

15 Mononeuropathies 174
Asa J. Wilbourn, MD

16 Radiculopathy 189
Kerry H. Levin, MD

17 Plexopathies 201
Mark A. Ferrante, MD • Asa J. Wilbourn, MD

GENERALIZED DISORDERS 215

18 Multiple Mononeuropathies and Polyneuropathies 215
Asa J. Wilbourn, MD

19 Motor Neuron Disorders 235
Erik P. Pioro, MD, DPhil, FRCPC

20 Disorders of Neuromuscular Junction Transmission 251
Kerry H. Levin, MD

21 Myopathies 268
Mark A. Ferrante, MD • Asa J. Wilbourn, MD

SECTION IV
Evaluation of Central Influences on Peripheral Function 281

22 Electrodiagnosis in Movement Disorders 281
Mark Hallett, MD

23 Silent Period Studies and Long Latency Reflexes 295
A. Arturo Leis, MD

24 Autonomic Nervous System Testing 307
Robert W. Shields, Jr., MD

25 Emerging Applications in Neuromagnetic Stimulation 325
Paul J. Maccabee, MD • Vahe E. Amassian, MB
Ulf Ziemann, MD • Eric Wassermann, MD
Vedran Deletis, MD, PhD

SECTION V
Electroencephalography 349

26 Basic Cellular and Synaptic Mechanisms Underlying the Electroencephalogram 349
Thomas H. Swanson, MD

27 Electrode Montages and Localization of Potentials in Clinical Electroencephalography 358
Hajo M. Hamer, MD • Hans O. Lüders, MD, PhD

28 Development of the Normal Electroencephalogram 387
Richard A. Hrachovy, MD

29 Normal Electroencephalogram and Benign Variants 414
Nancy Foldvary, DO

30 Long-Term Electroencephalographic/Video Monitoring 433
Harold H. Morris, MD • George Klem, MD, R EEG T

SECTION VI
Electroencephalography in Disturbance of Cerebral Function 438

31 Generalized Disturbances of Brain Function: Encephalopathies, Coma, Degenerative Diseases, and Brain Death 438
Eric B. Geller, MD

32 Focal Nonepileptic Disturbances of Brain Function 457
Selim R. Benbadis, MD

SECTION VII
Electroencephalography in the Epilepsies 468

33 Pediatric Epilepsy Syndromes 468
Selim R. Benbadis, MD • Elaine Wyllie, MD

34 Focal Epilepsy and Surgical Evaluation 481
Nancy Foldvary, DO

35 Neonatal Seizures 497
Eli M. Mizrahi, MD

SECTION VIII
Evoked Potentials and Intraoperative Monitoring 507

36 Visual Evoked Potentials 507
Charles M. Epstein, MD

37 Auditory Evoked Potentials 525
Hans O. Lüders, MD, PhD • Kiyohito Terada, MD

38 Somatosensory Evoked Potentials 543
Ronald G. Emerson, MD

39 Intraoperative Monitoring 565
C. Michel Harper, MD

SECTION IX
Sleep Disorders 589

40 Physiology of Sleep 589
Dudley S. Dinner, MD

41 Polysomnography and Multiple Sleep Latency Testing 597
Prakash Kotagal, MD

Index 611

BASIC NEUROPHYSIOLOGY

Chapter 1

Basic Electronics and Physics

Thomas F. Collura, PhD, PE

DIRECT CURRENT ELECTRICAL CIRCUIT
   CONCEPTS
SERIES AND PARALLEL RESISTANCE
ALTERNATING CURRENT CIRCUIT
   CONCEPTS
LOW-PASS AND HIGH-PASS CIRCUITS

FREQUENCY RESPONSE AND TRANSIENT
   RESPONSE
PRINCIPLES OF SIGNAL MEASUREMENT
DIFFERENTIAL AMPLIFIERS AND THE
   EFFECTS OF INTERFERENCE
APPLICATION TO ELECTROPHYSIOLOGY

Certain basic electrical and physical principles underlie the theory and application of clinical neurophysiology. These principles dictate the manner in which electrical signals behave in physiological tissues and govern the manner in which these signals can be measured. Basic concepts are necessary for the understanding of two main aspects of this discipline. An understanding is required of the nature of the physical, chemical, and biological processes that produce neurophysiological phenomena. Secondly, an understanding of the instrumentation and recording techniques is required to make neurophysiological measurements.

The major challenge of measuring such phenomena rests in the amplification of minute physiological signals. The following discussion emphasizes the fundamental concepts that are common to all areas of clinical neurophysiology, and provides a foundation for understanding neurophysiological processes in specific settings.

DIRECT CURRENT ELECTRICAL CIRCUIT CONCEPTS

From physics, we know that electric current in a conductor consists of the flow of electrons in response to an applied electrical potential. The amount of current depends on the size of the electrical potential and the resistance of the conducting medium. Many conductors are metals, which have a natural abundance of free electrons due to their atomic structure. Current can also flow in solutions or living tissue, by virtue of the ionic processes that produce free electrons, or a loose molecular structure that sheds electrons. Insulators are materials that have very few free electrons and, hence, conduct very little current.

In certain circumstances, such as conduction of muscle or nerve potentials through body fluids, current can also be conveyed by ions in a tissue or solution, rather than by electrons. Figure 1–1 illustrates the basic representation of a constant (direct-current, or DC) voltage source, connected to a resistance, resulting in the flow of current.

Potential can be considered a force that causes the movement of charged particles. Potential exists whenever uneven charges exist in a material, producing a potential difference. Even though the net charge may be zero, whenever charges are unevenly distributed, there will be nonzero potentials in association with the charge distribution. Electrons have negative charge. In a potential field, they will flow from a point of negative charge to a point of positive charge. This

is because they are both repelled by the negative charge and are attracted by the positive charge. Even though current is carried in this way by negatively charged electrons, it is the convention to describe current as flowing from positive to negative. Thus, the current is said to flow in a direction opposite to the electron flow. The flow of current, if constant, can be written as $I = Q/t$, where Q is the quantity of charge, and t is the time in which is transferred. The basic unit of current, the ampere, is defined as one coulomb of charge per second. The coulomb is defined as follows: 1 coulomb = 6.24×10^{18} units of charge, where an electron has one unit of negative charge. Whenever current flows in response to the potential difference, it meets with resistance to the flow, which limits the flow to a particular value. The voltage difference across the resistance is known as the voltage "drop" across it. In general, Ohm's law specifies the

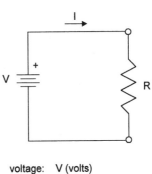

voltage: V (volts)

current: I (amperes)

power: P (watts)

$$V = I \times R$$

$$I = V / R$$

$$P = V \times I = (I \times R) \times I = I^2 \times R = V^2 / R$$

FIGURE 1–1. A DC voltage source, V, connected to a resistor, R, results in a current, I, flowing out of the voltage source and through the resistor. The relationship between voltage and current is given by Ohm's law, $V = I \times R$, and the power is the product of the voltage and the current.

relation between these as $V = I \times R$, where V = voltage in volts, I = current in amperes, and R = resistance in ohms (Ω). Thus, for example, if a 5-volt drop exists across a 30-Ω resistance, the current will be $I = V/R = 5/30 = 0.167$ amperes.

In common terminology, the positively charged terminal of a battery is called the anode. The negatively charged terminal is called the cathode. Since the cathode is negatively charged, it has an abundance of electrons. If a resistance is provided from the anode to the cathode, electrons will flow from the cathode, through the resistance, to the anode, and we say the circuit is closed. By convention, the current is said to flow from the anode, through the resistance, to the cathode. If the resistance is removed or increased to a very large value, we say the circuit is open, and no current flows. If an effectively zero resistance is provided so that current is limited only by the battery itself, we say there is a short circuit.

Energy in electricity is associated with the amount of charge and the size of the electrical potential, so that $E = Q \times V$. Power is the rate of transfer of energy. Since the rate of transfer of the charge is simply the current, electrical power is simply the current times the voltage: $P = E/T = Q \times V/T = (Q/T) * V = I \times V$. The power in an electrical signal, expressed in watts (W), is defined as the product of the voltage and the current. As such, we can have either DC power or AC power.

Example:
If 2 milliamperes pass through a load of 100 Ω, the power is $E = 0.002 \times 100 = 0.2$ W = 200 milliwatts (mW).

Electrical quantities are expressed as voltage, current, resistance, and related values. Often, however, what is important is not the numerical value of some measurement, but its proportionality to another value, or the ratio of the values. Ratios can be expressed either as direct ratios (e.g., 20:1), as a percentage (e.g., 5%), or as decibels, abbreviated as dB.

Power Ratio (dB) = 10 x log10 (P$_2$ / P$_1$)

Given amplitudes A$_1$ and A$_2$:

Amplitude Ratio (dB) = 10 x log10 (P$_2$ / P$_1$)

Amplitude Ratio (dB) = 10 x log10 (A$_2^2$ / A$_1^2$)

Amplitude Ratio (dB) = 20 x log10 (A$_2$ / A$_1$)

Power Ratio	Amplitude Ratio	dB	
100	10	20	
10	3.16	10	
2	1.4	3	
1	1	0	"unity-gain point"
0.5	0.707	-3	"half-power" point
0.1	0.316	-10	
0.01	0.1	-20	

FIGURE 1–2. Power is commonly expressed as a ratio, and this ratio can be expressed in decibels. A decibel is defined as 10 times the log (base 10) of the power ratio. It is therefore equal to 20 times the log of the amplitude (voltage) ratio.

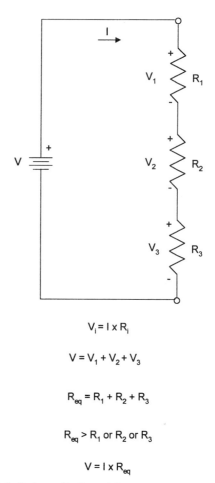

$V_i = I \times R_i$

$V = V_1 + V_2 + V_3$

$R_{eq} = R_1 + R_2 + R_3$

$R_{eq} > R_1$ or R_2 or R_3

$V = I \times R_{eq}$

FIGURE 1–3. Resistors R_1, R_2, and R_3, connected in series, produce a "voltage divider" in which a portion of the input voltage (V_1, V_2, and V_3, respectively) appears across each resistor, and each resistor carries the same current, I. The voltage source sees an equivalent resistance that is equal to the sum of the individual resistances.

The decibel is defined as 10 times the logarithm of the power ratio, or *ratio (dB)* $= 10 \times \log (P_2/P_1)$, where P_2 and P_1 are the power values of the variable of interest. Since $P = I \times V$ and $I = V/R$, we can derive that $P = V^2/R$. Thus, power is proportional to the square of the voltage across a load. The dB ratio can also be expressed in terms of voltage (or energy), as *ratio (dB)* $= 10 \times \log_2 (V_2/V_1) = 20 \times \log (V_2/V_1)$. Thus, a voltage ratio of 10:1 is equivalent to 20 dB (Fig. 1–2).

SERIES AND PARALLEL RESISTANCE

Circuits are drawn by connecting the components with wires that are thought of as having zero resistance. Within a circuit, we think of the current as traveling within loops and voltage as existing between points in the circuit. For example, in Figure 1–3, we think of the voltage, V, as producing a current, I, that travels around the single loop. The current then produces voltage drops across the other components. If a component has high resistance (or impedance), a high voltage will be produced across it. If a component has low resistance (or impedance), a low voltage will be produced across it. It is important to understand that the output of most circuits is defined as a particular voltage across a particular pair of points. This pair of points is considered to be the output of the circuit, and the output impedance of the circuit is the impedance between these two points.

In common practice, electrical circuits are encountered in which a number of circuit elements are present such as resistors, capacitors, and electrical potential sources. These circuit configurations appear often, particularly in situations such as the use of electrodes, and their connections to amplifier inputs. Such circuits may then consist of a single current path, or may contain multiple current paths. The most basic of these are series and parallel circuits.

Resistors are said to be in series when they are connected, successively, end to end, so that a current that flows through one of them must then pass through the next, and so on. Figure 1–3 illustrates the situation with series resistances. Since the total current must flow through each of the resistors, each resistor contributes to the voltage built up. This is also called a "voltage divider," because the voltage across each resistor is a fraction of the total. As a result, an equivalent resistance is produced, which is equal to the sum of the resistances. The equivalent resistance is thus greater than any of the individual resistances. We can then calculate the equivalent resistance as $R_{eq} = R_1 + R_2 + R_3$. For example, if $R_1 = 10K \ \Omega$, $R_2 = 15K \ \Omega$, and $R_3 = 12K \ \Omega$, we have: $R_{eq} = 10K + 15K + 12K = 37K \ \Omega$.

Resistors are said to be in parallel when they are connected together, so that a current can flow through any one of them at a time. Figure 1–4 illustrates what happens with parallel resistances. The current is divided through the resistors, and each carries a share of the total current, resulting in a "current divider." Note that each resistor sees the same voltage, which is the input voltage. As a result, an equivalent resistance is produced that is less than any of the individual resistances. In this case, the net effect is to produce a resistance that is smaller than any one of the individual resistances. Mathematically, we handle parallel resistances by writing: $1/R_{eq} = 1/R_1 + 1/R_2 + 1/R_3$. With R_1, R_2, and R_3 as previously stated, for example, we have: $1/R_{eq} = 1/10K +$

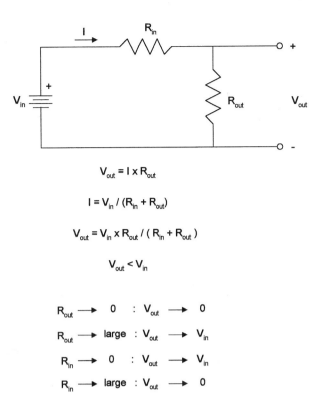

$$V_{out} = I \times R_{out}$$

$$I = V_{in} / (R_{in} + R_{out})$$

$$V_{out} = V_{in} \times R_{out} / (R_{in} + R_{out})$$

$$V_{out} < V_{in}$$

$$R_{out} \longrightarrow 0 \quad : \quad V_{out} \longrightarrow 0$$

$$R_{out} \longrightarrow large \quad : \quad V_{out} \longrightarrow V_{in}$$

$$R_{in} \longrightarrow 0 \quad : \quad V_{out} \longrightarrow V_{in}$$

$$R_{in} \longrightarrow large \quad : \quad V_{out} \longrightarrow 0$$

FIGURE 1–5. A voltage divider produced by a pair of resistors can be viewed as a reducing circuit that produces an output voltage across R_{out} in response to an input voltage, V_{in}, presented through R_{in}. The ratio of the resistors determines the attenuation of the circuit.

$1/15K + 1/12K = 0.0001 + 0.000067 + 0.000083 = 0.00025$. $R_{eq} = 1/0.00025 = 4K \ \Omega$.

When a voltage source is fed into two or more resistors, and the output voltage is taken across a subset of them, a voltage divider results, as shown in Figure 1–5. From the voltage source's point of view, the resistors are in series, so that the voltage is divided between them. However, the output voltage is taken only across R_{out}, so that it sees a smaller voltage. In this case, the smaller R_{out} is, the less voltage will appear at the output. For example, the voltage divider produced by the parallel resistance of an electrode's impedance with the input impedance of an amplifier produces such a voltage drop, which must be minimized, as we shall see later in the chapter.

ALTERNATING CURRENT CIRCUIT CONCEPTS

The previous discussion has described constant, steady electrical current, such as that from a battery through a resistor. Such behavior is called direct current, or DC. Electrical currents often alternate in direction, instead of flowing in one direction. This alternating current, called AC, may occur across a range of frequencies. For example, common household current as provided for residential electrical service operates at 60 cycles or Hertz (Hz); that is, the current alternates from one direction to the other 60 times per second. Physiological signals are often AC in nature. For example, the electroencephalogram (EEG) is typically taken to be an AC signal with a frequency range between about 0.5 and 100 Hz.

AC signals have properties that, in addition to voltage

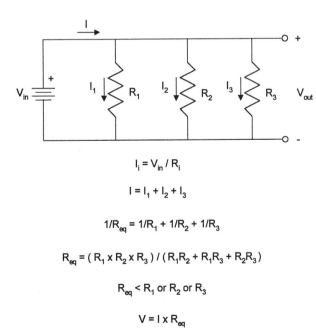

$$I_i = V_{in} / R_i$$

$$I = I_1 + I_2 + I_3$$

$$1/R_{eq} = 1/R_1 + 1/R_2 + 1/R_3$$

$$R_{eq} = (R_1 \times R_2 \times R_3) / (R_1 R_2 + R_1 R_3 + R_2 R_3)$$

$$R_{eq} < R_1 \ or \ R_2 \ or \ R_3$$

$$V = I \times R_{eq}$$

FIGURE 1–4. Resistors R_1, R_2, and R_3, connected in parallel, produce a "current divider" in which a portion of the input current (I_1, I_2, and I_3, respectively) flows through each resistor, but each resistor sees the full input voltage, V. The voltage source sees an equivalent resistance that is less than any of the individual resistances.

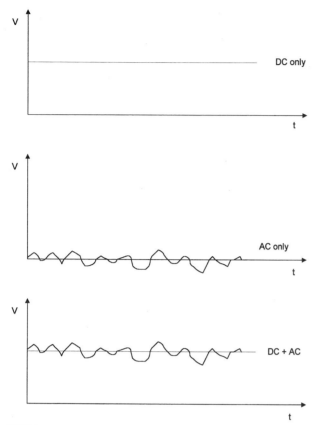

FIGURE 1–6. A signal can generally be thought of as consisting of a DC, or constant bias value, and an AC, or time-varying part. The combination of these two parts produces the entire signal.

and current, must be expressed in terms of frequency. For example, household voltages are generally 110 volts, 60 Hz. If such a voltage is applied to a 50-Ω resistance, the resulting current would be $110/50 = 2.2$ amps, also at 60 Hz. AC signals can be graphically displayed as a sinusoidal waveform (Fig. 1–6). Signals that are more complex can generally be described as combinations of sinusoidal components through the technique of Fourier analysis, which is a mathematical formula for breaking any signal into simpler, sine wave components. The effect of any circuit upon a complex signal can generally be understood by considering, separately, its effect with each sinusoidal frequency present in the input, and then combining the individual outputs to get the total output.

If our systems consisted only of resistive elements, circuits would be simple to describe and understand. However, some electrical devices and systems are not purely resistive, because of capacitance. A capacitor is an electrical device or natural system that stores electrical charge, and produces a resulting electrical potential. Capacitance is measured in farads; 1 farad of capacitance produces 1 volt of potential when 1 coulomb of charge is on it. Capacitance is produced whenever a pair of conductors is separated by an insulator, as in a "parallel-plate" capacitor, or in wires that are closely spaced. In common practice, capacitance is measured in microfarads. The current in a capacitor is proportional to its capacitance, and the rate of change of voltage across it, given by I (amperes) $= C$ (farads) $\times dV/dt$ (volts/second).

A capacitor can be viewed as having an equivalent resistance, called impedance. A capacitor presents no impedance to high-frequency AC signals, but it presents high impedance to low frequencies. It cannot pass DC signals at all. This is

shown in Figure 1–7. The current passing through a capacitor is always associated with the change in the charge distribution across its layers. It is also known as a "displacement" current, because charge must be displaced, for current to flow. DC signals cannot pass across a capacitor indefinitely.

Capacitors have the property that their equivalent resistance, called impedance, is a function of the frequency of the applied voltage. If a DC signal is applied to a capacitor, it will charge up to that voltage and fail to pass more current. Its resistance is thus effectively infinite. We say that a capacitor will not pass DC signals. However, if the applied voltage is AC, the capacitor will allow current to flow, since its alternating nature successively charges and discharges the capacitor.

The equivalent impedance of a capacitor is given by $Zeq = 1/(2 \times \Pi \times f \times C)$, where $\Pi = 3.14159...$, f = the AC frequency in cycles per second, and C = the capacitance in farads. If, for example, a 10-microfarad capacitor is used in a circuit that operates at 60 Hz, the equivalent impedance of the capacitor is Zeq = $1/(2 \times 3.14159 \times 60 \times 10^6)$ = 265 Ω (see Fig. 1–7).

The amount of current (and energy) flowing through the capacitor in this case is exactly as it would be if a 265-Ω resistor were used. If, for example, 24 volts AC were applied, the current would be I = $24/265 = 0.090 = 90$ milliamps AC current.

Although the impedance of a capacitor is expressed in

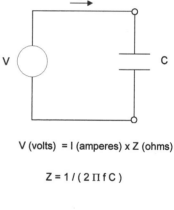

$$V \text{ (volts) } = I \text{ (amperes) } \times Z \text{ (ohms)}$$

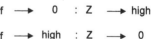

$$Z = 1/(2\Pi f C)$$

$$f \rightarrow 0 : Z \rightarrow high$$
$$f \rightarrow high : Z \rightarrow 0$$

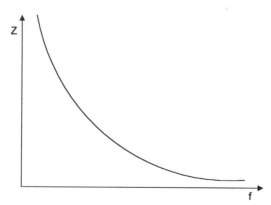

FIGURE 1–7. A capacitor is a circuit element that has an equivalent resistance ("impedance"), Z, that depends on capacitance, C, and frequency, f. The larger the frequency, the smaller the equivalent impedance.

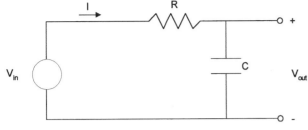

$$V_{out} = V_{in} \times Z_c / (R + Z_c)$$

$$V_{out} = V_{in} / (1 + 2 \Pi f R C)$$

$f \longrightarrow$ large $\quad : V_{out} \longrightarrow 0 \quad$ "blocks highs"

$f \longrightarrow 0 \quad\quad : V_{out} \longrightarrow V_{in} \quad$ "passes DC"

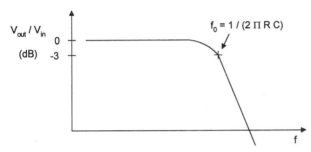

FIGURE 1–8. A "low-pass" filter results when a resistor, *R*, and a capacitor, *C*, are combined as shown in a voltage divider. The output voltage, *V*, produced by current *I* passing through the output capacitor is a function of frequency and is attenuated at high frequencies. Only very low frequencies, or DC voltage, can pass through the circuit.

ohms of resistance to current flow, this is a complex impedance which also introduces so-called phase shifting as well as resistance to current. Its effects are visible in the shapes and sizes of waveforms as they pass through the circuit. Whereas a capacitor freely conducts current at high frequencies, thus presenting a low impedance, an inductor does not conduct current well at high frequencies. Conversely, a capacitor will not pass current at DC (0 Hz), whereas an inductor will. Capacitances and inductances, which can occur by design or may be present as contaminants or "parasitic" elements, will affect the frequency behavior of a circuit, particularly at the high- and low-frequency extremes.

Capacitors can be used in series and parallel, as can resistors. However, the capacitances combine in exactly the opposite manner. That is, capacitors in parallel provide a capacitance that is equal to the sum of the individual capacitances, and capacitors in series provide a capacitance that is less than that of each capacitor alone. Because of this fact, and because the impedance is inversely proportional to the capacitance, the equivalent impedances of series or parallel capacitors behave exactly the same as the equivalent resistances in series or parallel.

LOW-PASS AND HIGH-PASS CIRCUITS

When resistors and capacitors are used together, the combination is known as an RC filter, with either a low-pass or a

high-pass configuration. Such circuits have two main properties, their frequency response and their transient response. The frequency response of a low-pass filter is shown in Figure 1–8, and that of a high-pass filter is shown in Figure 1–9. These represent the simplest forms of filters.

Mathematically, for a simple RC circuit, the cutoff frequency is found to be: $f_c = 1/(2 \cdot \Pi \cdot R \cdot C) = 1/(2 \cdot \Pi \cdot \tau)$. For example, if a 10K resistor is used in a low-pass filter with a 0.15-microfarad capacitor, the cutoff frequency is $f_c = 2 \cdot 3.14159 \cdot 10,000 \cdot 0.15 \cdot 10^6 = 106.1$ Hz.

Since power is the square of the voltage, at the corner frequency, the voltage ratio is 1 over the square root of 2, or 0.707. In dB, this ratio is: $ratio_{corner} = 10 \cdot \log (0.5) = -3.0$. Thus, the output is "3 dB down" at the corner frequency. This is the point on the graph at which the output, which is at full value for frequencies below this point, is now 3 dB below its full value. This frequency is also sometimes called the "3 dB frequency," as seen in Figures 1–8 and 1–9.

FREQUENCY RESPONSE AND TRANSIENT RESPONSE

The frequency response of an RC circuit can be simply described as the change in the impedance of the capacitor with changing frequency, in the setting of a voltage divider. For example, In the low-pass circuit, with the capacitor's high impedance to low frequencies, the voltage divider trans-

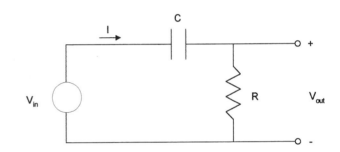

$$V_{out} = V_{in} \times R / (R + Z_c)$$

$$V_{out} = V_{in} \times 2 \Pi f R C / (1 + 2 \Pi f R C)$$

$f \longrightarrow$ large $\quad : V_{out} \longrightarrow V_{in} \quad$ "passes highs"

$f \longrightarrow 0 \quad\quad : V_{out} \longrightarrow 0 \quad$ "blocks DC"

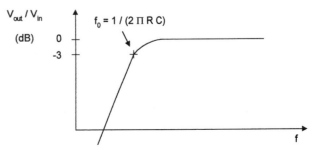

FIGURE 1–9. A "high-pass" filter results when a capacitor, *C*, and a resistor, *R*, are combined as shown in a voltage divider. The output voltage produced by current *I* passing through the output resistor is attenuated at low frequencies and is 0 at DC.

fers most of its input to the output terminals. In effect, the capacitor is more like an "open circuit" and does not reduce the signal. At high frequencies, however, the capacitor conducts readily and presents a low impedance. The output is therefore somewhat "short circuited." Thus, the voltage divider transfers only a fraction of its output to the input; most of the energy is lost across the series resistor. By using this argument and evaluating the actual impedance expression for the capacitor, the exact form of the frequency response can be easily derived. All filter circuits, regardless of their complexity, are understandable in terms of these concepts.

If the output capacitor of a low-pass filter has high capacitance (low impedance), the voltage across it will be low. This is because the capacitor "wants" to conduct current and thus offers low impedance. This may be confusing, since the current flows freely through the capacitor; something that freely conducts current should be associated with a high output. This confusion is resolved by understanding the difference between the series and the parallel impedances encountered in interpreting a circuit. High output is associated with a low series impedance, whereas high output is associated with a high parallel impedance.

A filter is a device whose purpose is to pass specific frequencies while attenuating others. Filters are used to "clean up" signals, for visual suitability or for further processing. Filters generally have low-pass, high-pass, or band-pass characteristics. In addition to the frequency response characteristic, a low-pass or high-pass filter can be understood in terms of its time response, or transient response. These are commonly studied in terms of the step response, which is useful for theoretical, as well as practical, reasons. Instruments often have calibration signals that are essentially step functions,

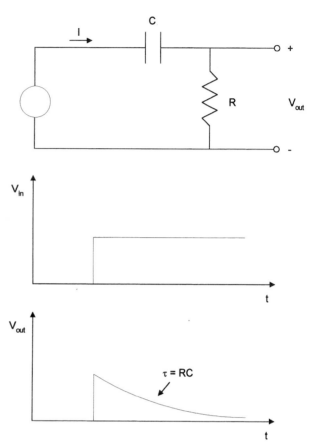

FIGURE 1–11. The high-pass filter has a time-domain response in which the output immediately follows the step input change, then decreases back to zero. The response time is characterized by the time constant, τ, which is the product of R and C.

and the step response is an important way to characterize and calibrate them.

The transient response can be understood from the step response of the circuit, which is the output that occurs if the input voltage is suddenly increased from one level and held at the new level indefinitely. The transient response of a low-pass filter is shown in Figure 1–10. The general form of this response is given as: $V_{output} = V_{input} \times (1 - e^{-t/\tau})$. It can be seen that the output cannot move as quickly as the input, so that it follows with a "lag." However, it is capable of reaching the full output value, given enough time. The transient response of a high-pass filter is shown in Figure 1–11. Although this circuit can immediately produce the full output voltage, it is unable to hold that value, and it will reduce its output to zero, given enough time. This property is sometimes referred to as "droop."

The time response of any low-pass or high-pass filter can be characterized by the "time constant." The time constant of an RC circuit is defined as: τ (seconds) = R (Ω) $\times$ C (farads). If R is in ohms and C is in microfarads, τ is in microseconds. In a simple RC circuit, τ is the time required for the step response to rise from the initial value to 0.63 of the final value when charging or to fall to 0.63 of the initial value when discharging. In each case, the time constant represents the time for the majority of the change to occur. Since the rise or decay of the signal is exponential, the rate of change slows down as time goes by, and the output never actually reaches 100% of its theoretical final, asymptotic value. As another rule of thumb, it takes about three time constants for the output to achieve 95% of its total change.

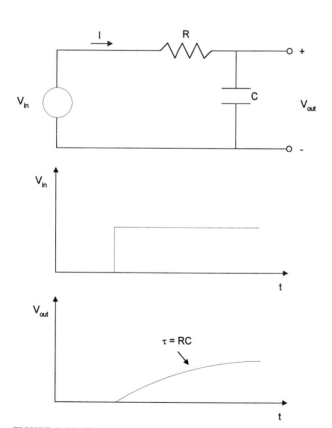

FIGURE 1–10. The low-pass filter has a time-domain response that can be characterized by the "step response." The output will take some time to reach the output value. The response time is characterized by the time constant, τ, which is the product of R and C.

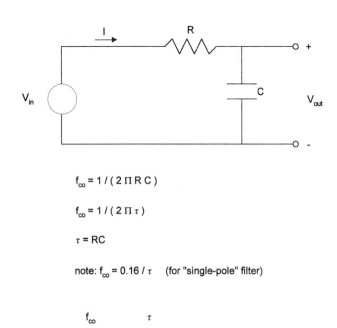

$$f_{co} = 1 / (2 \Pi R C)$$

$$f_{co} = 1 / (2 \Pi \tau)$$

$$\tau = RC$$

note: $f_{co} = 0.16 / \tau$ (for "single-pole" filter)

f_{co}	τ
1 Hz	0.16 sec
0.5 Hz	0.32 sec
0.16 Hz	1 sec
0.02 Hz	8 sec

FIGURE 1–12. The time constant, τ, is defined as the product of R and C for an RC circuit. It is related to the cutoff frequency (for simple filters) by a simple proportional relationship.

A "fast transient" is any signal that requires the circuit to respond faster than its time constant will allow it to follow. Such signals are therefore distorted when they pass through the circuit.

For simple filters, there is a direct, inverse relationship between time constant and cutoff frequency. For example, a time constant of 1 second corresponds to a cutoff frequency of 0.16 Hz, and a time constant of 5 seconds corresponds to a cutoff frequency of 0.03 Hz (Fig. 1–12). Filters also have the important property of phase shifting. Because the filter attenuates the signal differently for different frequency components, it introduces a shift in the phase of the output relative to the input. For a low-pass filter, for example (Fig. 1–13), at the cutoff frequency, the output will be shifted 45 degrees relative to the input, and the phase shift will increase with increasing frequency. The situation for a high-pass filter is diagrammed in Figure 1–14. This phase shifting can also lead to phase distortion of the input signal, in which fast transients appear delayed in comparison to slower components, altering the appearance of the signal. In general, phase shifting is not a problem for signals that are far from the cutoffs, but must be considered for signals that are near the limits.

In a clinical example, certain evoked potentials, such as visual and auditory evoked potentials, may have fast components that are near the limits of the recording system. If these components are to be measured with diagnostic accuracy, it is important to ensure that the bandwidth of the recording system is adequate to avoid distorting these waves.

PRINCIPLES OF SIGNAL MEASUREMENT

An amplifier is a device that provides electrical gain, so that its output is generally larger than its input. Amplifiers can provide voltage gain, current gain, or both. Voltage gain

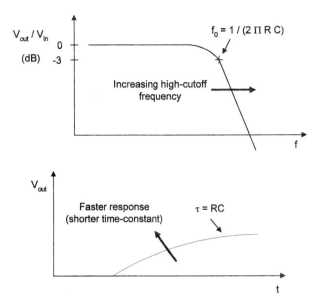

FIGURE 1–13. The relationship between frequency response and time response is shown for a low-pass filter. As the high-cutoff frequency is increased, the time-following response becomes sharper, reflected by a reduction in the time constant.

is used to increase the value of a voltage, for example, to increase the EEG signal from microvolts to volts. Current gain is used to increase the power of the output, for example to drive a strobe light or a speaker. Amplifiers are active devices, and must have a source of external power in order to develop their output. Amplifiers also have characteristics such as input noise, frequency response, common-mode rejection ratio, and harmonic distortion. These and other properties must be considered in the design, analysis, and application of the various types of amplifiers.

When we talk about voltages, it is useful to define a standard reference voltage for discussion. We say that a voltage of zero volts is ground potential, and any circuit element

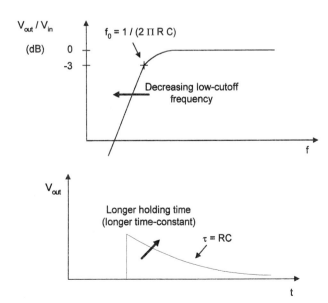

FIGURE 1–14. The relationship between frequency response and time response is shown for a high-pass filter. As the low-cutoff frequency is reduced, the time-following response is able to hold the DC level longer, reflected by an increase in the time constant.

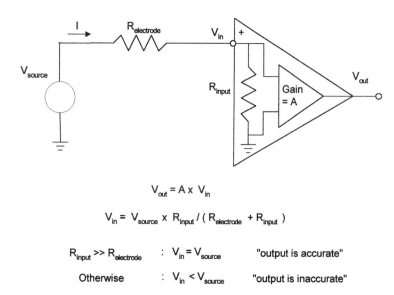

$$V_{out} = A \times V_{in}$$

$$V_{in} = V_{source} \times R_{input} / (R_{electrode} + R_{input})$$

$R_{input} \gg R_{electrode}$: $V_{in} = V_{source}$ "output is accurate"

Otherwise : $V_{in} < V_{source}$ "output is inaccurate"

FIGURE 1–15. An amplifier has a characteristic input resistance, R_{in}, that will interact with the resistance of the source, including the electrode $R_{electrode}$. The input current, I, is produced by the input voltage, V, flowing through both resistances in series. If the amplifier input resistance is much greater than the electrode resistance, then an accurate signal can be recorded. If this is not the case, then the output will not be an accurate reflection of the source voltage; it will also vary with the electrode resistance. The gain of the amplifier is A, so that the output is the product of A and the input voltage, V_{in}.

that is connected to this is said to be grounded. Ground is usually taken to mean literally the potential of the earth, and can be obtained by connecting to one of two things. The first is a cold water pipe, building frame, or other large metal object that is embedded into the earth. While this will usually provide a reliable ground potential, it is not guaranteed. The other, which should be guaranteed, is the ground connection that is to be provided with any three-prong electrical outlet. This plug, which is the extra, round plug in the United States, is connected, according to building and electrical codes, to some large metal object that goes deep into the earth. If this contact does not provide a sound, reliable ground connection, then the plug does not provide adequate safety and is also in violation of governmental ordinances.

Figure 1–15 shows a simple, "monopolar" amplifier. It measures the voltage at its input and produces a proportionally larger output voltage. Such devices are at the heart of all electrophysiological instrumentation. In addition to the obvious property of gain, amplifiers also have the property of having a specific input impedance. The input impedance of an amplifier must be high if it is to accurately measure a voltage from any source. The required magnitude of the input impedance is such that it must be many times larger than the source impedance of the measured entity.

DIFFERENTIAL AMPLIFIERS AND THE EFFECTS OF INTERFERENCE

In clinical electrophysiology, differential amplifiers (Fig. 1–16) are commonly used. These provide an output that is proportional to the difference between two inputs. This is useful because typical electrophysiological sources have, in addition to the (usually small) differential voltage that is desired to be measured, a potentially large common-mode voltage that appears at both inputs. Common-mode voltages include electrical interference, effects of breathing, electrocardiogram (ECG), and so on, but nonetheless appear on the input leads. In order for a differential amplifier to provide an accurate signal, it must have the ability to cancel out common-mode signals. This property is known as the "common-mode rejection ratio," or CMRR. For effective clinical recording, this must be very high (>100 dB) on typical electrophysiological instruments.

Figure 1–16 shows that the CMRR of an electrophysiological amplifier is also dependent on the relative magnitudes of the source (including electrode) resistances compared to the amplifier input resistance. The amplifiers must have a very large input resistance or else the effects of any electrode impedance mismatch will be to degrade the CMRR, allowing (especially 60-Hz) common-mode signals to appear at the output, thus providing a noisy signal. The amplifier operates by amplifying only the difference between the inputs, and not their absolute values. In order to do so, it must effectively subtract the two common-mode inputs. The accuracy with which the amplifier can do this is what is specified by the CMRR.

Note that 60-Hz electrical interference, as caused by electromagnetic induction in the wiring, appears first as an induced current, then causes voltages to appear by virtue of the resistances and impedances in the circuits. Thus, high resistances in critical paths, such as electrode connections, cause susceptibility to interference and should be avoided. Voltage introduced in the leads in this way is often common mode, so that a high CMRR will help to minimize the effects of 60-cycle noise in the environment.

Another source of unexpected circuit components is the existence of stray or parasitic elements. For example, there may be parasitic capacitance in the electrode wires leading from a patient, and there may be parasitic inductance in these wires as well. Although connecting wires ideally have no resistance, inductance, or capacitance, real-world devices are never ideal. Although stray or parasitic values are generally small, they cannot be entirely neglected. They generally appear when the extreme limits of performance are encountered, such as noise, frequency response, or electrical interference. For example, the presence of stray capacitance in the electrodes or wires will cause problems for high-frequency recording, as these signals may be attenuated by stray RC circuits that appear in the system. Often, these situations are dealt with by using superior materials, special insulation, or shielding, or by using specially designed components or instruments.

APPLICATION TO ELECTROPHYSIOLOGY

Figure 1–17 is a simple circuit model of phenomena that occur when a low-level electrophysiological potential, such

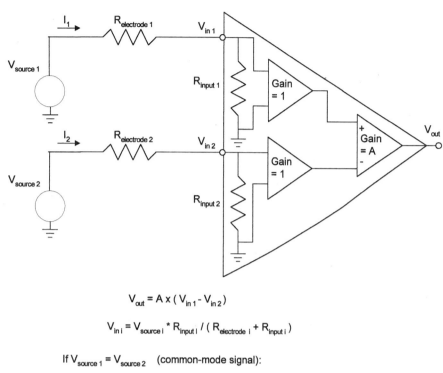

$$V_{out} = A \times (V_{in\,1} - V_{in\,2})$$

$$V_{in\,i} = V_{source\,i} * R_{input\,i} / (R_{electrode\,i} + R_{input\,i})$$

If $V_{source\,1} = V_{source\,2}$ (common-mode signal):

$R_{input\,i} >> R_{electrode\,i}$: $V_{out} = 0$ "good common-mode rejection"

Otherwise   : $V_{out} \neq 0$ "poor common-mode rejection"

FIGURE 1–16. A differential amplifier, such as used in electrophysiology, has two inputs, and its output is the difference between them. There are two voltage sources, $V_{source1}$ and $V_{source2}$, producing two input currents, I_1 and I_2. Thus, two electrodes must be used with resistance, $R_{electrode1}$ and $R_{electrode2}$, plus each input has its own input resistance, R_{input1} and R_{input2}. If the input resistances are much larger than the electrode resistances, then the inputs will be matched, and the amplifier will be able to cancel out any common-mode signal that appears on both inputs. However, if this is not the case, than any electrode resistance mismatch that exists will produce an input difference between the channels, and appear as an error signal on the output.

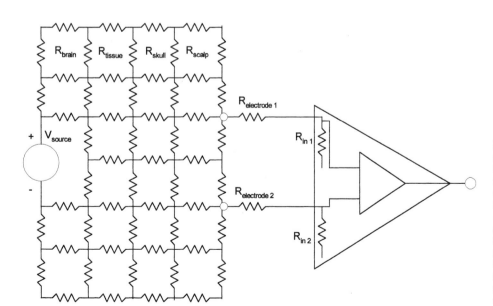

FIGURE 1–17. A biological source, such as the EEG, can be modeled as a voltage source that passes current through a distributed network of resistances (a "volume conductor"), representing the layers of brain (R_{brain}), tissue (R_{tissue}), skull (R_{skull}), and scalp (R_{scalp}). The electrodes pick up the resultant signal on the surface and connect it to an amplifier, through electrode resistances $R_{electrode1}$ and $R_{electrode2}$, across which it is carried to the input resistances R_{in1} and R_{in2}, where it can be measured.

as the electrical field of a brain cell, is measured using electrodes and a sensitive amplifier. The associated events are as follows:

1. Physiological processes displace ions.
2. Local potentials are produced.
3. Currents are conducted throughout the head.
4. Potentials appear on the scalp and the electrodes.
5. Tiny currents are induced in the wires.
6. Potentials appear at the amplifier inputs.
7. Larger potentials appear at the amplifier outputs.
8. Something visible occurs such as pen deflection, meter indication, oscillograph display, computer sample, and so forth.

These basic events, with minor modification, explain how any electrophysiological phenomenon can give rise to its measurable counterpart, for example, muscle activity producing the electromyogram (EMG), eye activity producing the electrooculogram (EOG), or retinal activity producing the electroretinogram (ERG).

REFERENCES

Bueche F: Introduction to Physics for Scientists and Engineers. New York, McGraw-Hill, 1969.

Clark JW: The origin of biopotentials. In Webster JG (ed): Medical Instrumentation Application and Design. Boston, Houghton Mifflin, 1978.

Jackson HW: Introduction to Electric Circuits. Englewood Cliffs, NJ, Prentice-Hall, 1970.

Katz B: Nerve, Muscle, and Synapse. New York, McGraw-Hill, 1966.

Litt B, Fisher RS: EEG engineering principles. In Daly DD, Pedley TA (eds): Current Practice of Clinical Electroencephalography. New York, Raven Press, 1990.

Neuman MR: Biopotential electrodes. In Webster JG (ed): Medical Instrumentation Application and Design. Boston, Houghton Mifflin, 1978.

Nunez PL: Electric Fields of the Brain. New York, Oxford University Press, 1981.

Reitz JR, Milford FJ, Christy RW: Foundations of Electromagnetic Theory. Reading, MA, Addison-Wesley, 1980.

Speckmann J-E, Elger CE: Introduction to the neurophysiological basis of the EEG and DC potentials. In Niedermeyer E, Lopes da Silva F (eds): Electroencephalography. Baltimore, Urban & Schwarzenberg, 1987.

Chapter 2

Noise, Averaging, and Statistics

Thomas F. Collura, PhD, PE

INTRODUCTION
PRINCIPLES OF DIGITIZATION
STATISTICAL PRINCIPLES
PRACTICAL CONSIDERATIONS
EFFECTS OF DATA SAMPLING

CONDITIONS FOR SUCCESSFUL
  AVERAGING
AVERAGING AS FILTERING
ADDITIONAL TECHNIQUES

INTRODUCTION

Signal averaging is a technique that is widely applied to neurophysiological measurements, as a way to measure small signals of interest that are obscured by noise or other background activity. It is particularly applicable to evoked potentials or event-related potentials, including movement-related potentials, cognitive-related potentials, and other signals associated with discrete events. Signal averaging is based on several simple statistical principles, and many commercial computer programs are available to perform this task. However, all such techniques draw from the same fundamental rules in order to maximize the signal-to-noise ratio (SNR) of the signal of interest.[1, 2, 5, 15]

Signal averaging can be applied whenever a desired signal can be obtained in a time-locked fashion. Generally, the response is buried in background activity or noise, and is not normally visible in the raw recording (Fig. 2–1). For example, a typical sensory evoked potential (visual, auditory, somatosensory) is from 1 to 20 μV in amplitude, but is obscured by electroencephalographic signals of 20 to 100 μV, occurring simultaneously.[3] However, because the measured signal component occurs at a specific point in time, it is possible to apply mathematical techniques to determine its properties and separate it from the background activity. Although it is most common to use a stimulus-induced signal that is time-locked to the stimulus, it is also possible to apply averaging techniques to signals that are not time-locked to a stimulus. For example, thought-related potentials, movement-related potentials, or epileptiform spikes can be averaged post hoc by identifying the synchronization times after the fact using electromyography, a button-press signal, or by manually lining up the waveforms.[7, 14] In the following discussion, the model of a stimulus-response paradigm is used; bear in mind that the same techniques are equally applicable to post hoc averaging methods.

Ultimately, whatever the method of determining the synchronizing information, it is possible to exploit the fact that the desired signal has a known time reference, while the background has a random relationship to time. It is common to use repetitive or periodic stimulation to elicit a large number of responses, and to collect them for averaging. However, random or irregular stimuli can also be used, and there are advantages and disadvantages of each technique. The physiological response to the stimulus occurs at a fixed interval of time after the stimulus, whereas the other signal components should not be synchronized. It is therefore possible to use a simple computer program or a dedicated signal averager to separate the response from the other activity.

PRINCIPLES OF DIGITIZATION

Response averages are computed from the raw electroencephalographic signal, which must be acquired in a manner consistent with any good electroencephalographic practice. In particular, the amplifiers, filters, and digital sampling parameters must all be appropriate to the task. Shortcomings in any of these will compromise the measurement. If amplifier gain is too low, the signal will be too small to measure. If it is too high, the amplifiers will saturate, causing distortion in the form of flat lines at the signal extremes. If filters are set too narrowly, critical amplitudes and latencies will be distorted. If filters are set too widely, excess noise or baseline drift will be introduced.

Averaging is used to create an estimate of a desired response by separating it from the obscuring background activity. The estimate is a waveform that is presented as a time-series plot, and viewed exactly as if it had been measured in a noise-free manner to begin with. The waveform is examined to evaluate the amplitude and latency of critical signal components. Both amplitude and latency are generally taken at the waveform maxima or minima (peaks or valleys) by visual inspection or by computer-assisted techniques. It is critical in response averaging to ensure that the waveform peaks provide correct and meaningful information, and are not distorted. Both amplitude and latency distortion are possible, and can be minimized if basic principles are kept in mind.[16]

Latency is measured in milliseconds and interpreted according to established clinical ranges. It can be measured either as absolute latency (from the stimulus onset), or as interwave or interpeak latency, measured between wave maxima or minima. Peaks at common latencies for evoked potentials are identified and labeled according to clinical conventions, such as the P100 and P300 (visual), the N140 to P190 complex (somatosensory), and so on. These identifiers do not indicate the precise latency of the components, but rather are used to specify components that can be seen by an experienced eye, in the context of the entire evoked potential. Ranges for these latencies have been well characterized along with their statistical distributions, and provide valuable clinical information.[3, 6, 8, 9]

Amplitude is measured in microvolts and can be expressed in a variety of ways. These include baseline to peak, peak to peak, or the area under a peak. Choice is usually dictated by clinical conventions and consideration of the shape of the response curve. Amplitudes are less well defined than latencies, and their statistical distributions are also less likely to have a normal distribution. Their absolute values are not as important as comparative values, variations, or left-right asymmetries.

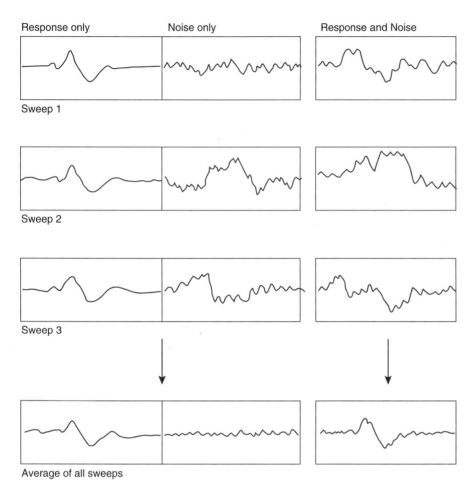

Response only Noise only Response and Noise

Sweep 1

Sweep 2

Sweep 3

Average of all sweeps

FIGURE 2–1. Illustrating the noise reduction achieved by signal averaging. On the left, the idealized event-related potential, or "response," is shown. Going from top to bottom, successive sweeps are shown. The additive noise is shown, for each sweep, as well as the resulting Response + Noise, on the right. When all sweeps are averaged, then the responses are seen to be reinforced, while the noise is reduced in the averaged Response + Noise when compared to any of the raw Response + Noise epochs.

In order to accurately record and analyze the signal, digital techniques are generally used. Therefore, the signal must be sampled to produce a representation of the signal as a series of numbers, distributed evenly in time. Sampling rate is defined by the time delay between successive samples of the signal. If the sampling rate is too low, irrecoverable signal loss will occur as a result of "aliasing," a term that connotes the introduction of error from inadequate signal sampling. The minimum acceptable sampling rate, in order to mathematically represent the signal, is the Nyquist frequency, which is twice the highest frequency found in the raw data. For example, if an electroencephalogram (EEG) is recorded with a high cutoff frequency of 70 Hz, it must be sampled at no less than 140 samples per second, to avoid aliasing error. If sampling is below this frequency, then irrecoverable loss of signal information will occur, and high frequencies can produce apparent energy at lower frequencies, which is considered aliased. To accurately visualize the signal, however, higher sampling rates are advised, with a practical value of 5 to 10 times the highest frequency present. If the sampling rate is higher still, there will be no signal degradation; however, digital storage will be wasted owing to an excess of sampling.

The two basic parameters that define the digitization of the recording are the horizontal and the vertical resolution. The vertical resolution, expressed in microvolts per sample, depends on the full-scale range of the system and the number of bits of analog-to-digital (A/D) conversion. For example, with a 500-μV total full scale and a 12-bit A/D converter, the vertical scale is divided into 4096 quanta, so that the vertical resolution is $500/4096 = 0.122$ μV per quantum. The total number of sample bins (and their values) for typical

A/D bit resolutions are shown in Table 2–1. The operating range can be specified either as a plus-or-minus voltage, or as peak-to-peak. Note that if the full scale is specified as plus-or-minus 250 μV, for example, the peak-to-peak range of the conversion is twice that, or 500 μV. In such a system, the zero point is taken to be a value at mid-range. Thus, in a 12-bit system, the digital values and their microvolt counterparts might be as shown in Table 2–2. This could also be referred to as an 11-bit plus sign, where the sign bit provides the accuracy of a 12-bit converter.

The horizontal resolution, expressed in milliseconds per sample, is determined by the epoch length and the sampling rate of the A/D converter. For example, with a 100-msec epoch containing 1000 samples, the horizontal resolution is $100/1000 = 0.1$ msec per sample, which is equivalent to a

TABLE 2–1	Relationship Between the Number of Bits of Digital Resolution and the Number of Possible Sample Values	
Bits of Resolution[a]	Number of Sample Bins	Possible Sample Values
8	256	0–255
10	1024	0–1023
12	4096	0–4095
16	65,536	0–65,535

[a]For higher bit converters, the output has a larger number of possible values, or bins, providing correspondingly greater accuracy of the A/D conversion.

TABLE 2–2	Example of Sampling with a 12-bit Converter, Mapped to a ± 250-μV Signal Range

Digital Sample (quanta)[a]	Signal Value (μV)
4095	+249.88
4094	+249.76
—	—
2049	+0.12
2048	0.00
2047	−0.12
—	—
1	−249.88
0	−250.00

[a]Each digital sample has a corresponding value to which it is nominally assigned. Signal values between these center values are rounded to the nearest value in the digitizer, resulting in a loss of precision. This loss of accuracy is referred to as digitization error.

sampling rate of 10,000 samples per second. This resolution is also referred to as the dwell time of the measurement. This time dictates the accuracy with which any latencies or critical features can be determined, and also limits the accuracy with which fast signal components can be clearly visualized. This will generally be somewhat higher than the previously described Nyquist frequency.

Accurate sampling also requires that the acquisition of data be performed with appropriate filter settings. A time-limited or "transient" waveform such as an evoked response can be thought of as consisting of a range of frequencies, and the upper and lower ranges depend on the amount of fast (sharp) or slow features that it contains. If the acquisition filter settings are significantly narrower than these, then the waveform will be distorted and its critical amplitudes and latencies will be misrepresented.

As an example of signal averaging, Figure 2–2 shows a 5-second epoch containing a photic stimulator channel, 6 seconds of EEG, and an electrocardiogram (ECG) channel. The photic stimulation, at 1 flash per second, is nearly synchronous with the ECG, but not exactly. In Figure 2–3, 30 epochs are averaged, based on the third stimulus pulse in the epoch. The stimuli are thus reinforced; the third pulse is largest, because the epochs are lined up with it, and the others are slightly smaller, owing to the stimulus jitter. A visual evoked potential is evident on the EEG channels, and the ECG is becoming smaller, as it is not precisely locked to the stimulus. All of these effects are more pronounced in Figure 2–4, which is the result of 118 averages. In Figure 2–5, 60 epochs are averaged, based on the ECG channel; the signals were aligned on the QRS complex, as detected by the computer. The ECG is therefore reinforced in all channels. The other signals are all reduced. This illustrates the precision of even simple signal averaging in emphasizing events synchronized to the averaging marks, and in reducing unrelated signals.

STATISTICAL PRINCIPLES

The underlying principle for separating the response from the background is the statistical concept of expected value of a random variable. A random variable is any measured value that is said to arise from a process that has known statistical properties. Thus, any given sample of an electroencephalographic signal is a random variable. The sequence of EEG samples that comprise a waveform constitute a random process, which is a sequence of random variables. The expected value of a random variable is mathematically defined in terms of its probability density and probability distribution.[5]

The probability distribution, written as $F(x) = p(X < x)$, is a function that gives the probability that the signal will be below a given value, as a function of all possible values. Thus,

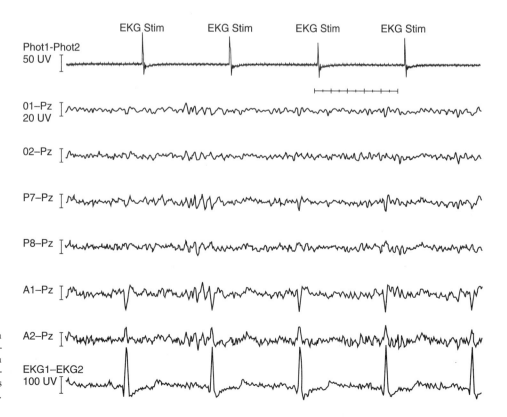

FIGURE 2–2. This 5-second epoch contains a photic stimulator channel, 6 seconds of EEG, and an ECG channel. At 1 flash per second, the photic stimulation is nearly synchronous with the ECG.

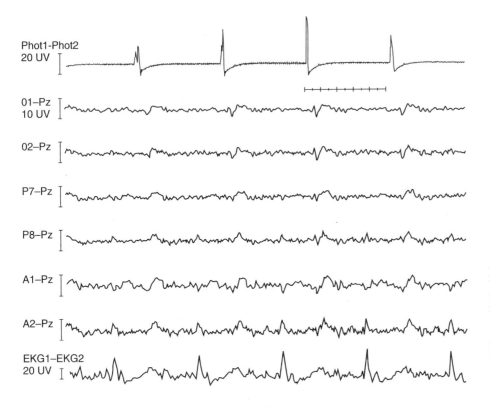

FIGURE 2–3. Averaging of 30 epochs based on the third stimulus pulse in the epoch; the stimuli are thus reinforced, the third pulse is largest, and the others are slightly smaller, owing to the stimulus jitter; a visual evoked potential is evident on the EEG channels, and the ECG is becoming smaller, as it is not precisely locked to the stimulus.

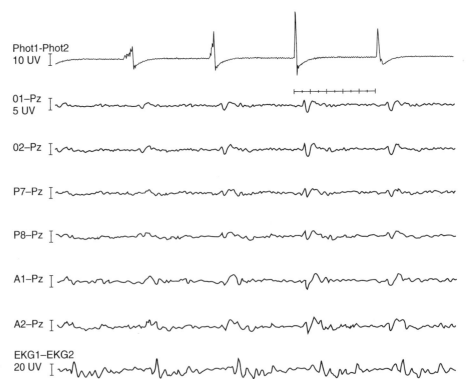

FIGURE 2–4. All of the effects in Figure 2–3 are more pronounced in this display, which is the result of 118 averages.

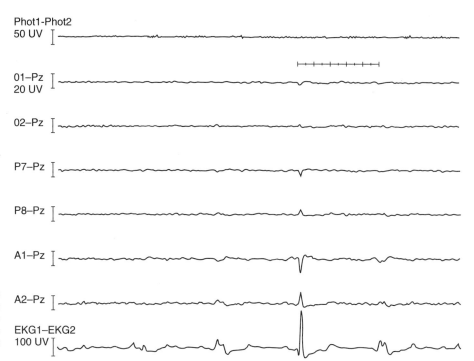

FIGURE 2–5. Averaging of 60 epochs based on the ECG channel; the signals were aligned on the QRS complex, as detected by the computer; the ECG is reinforced in all channels, and the other signals are all reduced.

for an alternating current (AC)-coupled signal, there is a 50% likelihood that the value will be negative and a 50% likelihood that it will be positive. Similarly, there is a decreasing likelihood that it will be greater than a given value as that value increases.

The probability density, written as $f(x) = F'(x)$, is the first derivative of the probability distribution, and thus reflects that likelihood that the signal will take on a particular given value. For continuous signals, $f(a) = 0$ because an infinite number of values are possible and the probability of any given value is infinitesimally small. However, for digitally sampled signals, a finite number of sample values are possible, so that $f(a)$ will have a meaningful and nonzero value for each possible sample value. One could then write $f(x) = p(X = x)$.

The expected value, or mean, of a random variable is defined as:

$$\mu_x = E(X) = \int_{x = -\infty}^{x = \infty} x f(x) \, dx$$

in which $E(X)$ denotes the "expected value" of X, which sums up all possible values and their respective probabilities to arrive at the most likely observed value. This also represents the value that would be expected if many observations were taken, and the results averaged. Therefore, the expected value is the value to which the average recordings are expected to converge, over a large number of observations. Note that the expected value has the same units as the signal itself, so in electroencephalographic studies, this value is expressed in microvolts.

Because the desired signal is time-locked to the stimulus, if a number of samples are averaged, the averaged value will progressively approach that of the signal itself. In contrast, the averaged value of the signal components that are not time-locked will progressively approach a different value, specifically, zero. By letting $n(t)$ be the noise signal, $r(t)$ be

the response signal, and $EEG(t)$ be the measured electroencephalographic signal, this can be written mathematically as:

$$EEG(t) = r(t) + n(t)$$
$$E(r(t)) = r(t)$$
$$E(n(t)) = 0$$
$$E(EEG(t)) = r(t).$$

Thus, the expected value of the electroencephalographic signal is in fact the noise-free response. However, any given sample or epoch of the signal will contain a full amount of the noise, which will obscure the response. By combining many repeated measurements (observations) of the EEG and applying averaging, the noise is expected to go to zero, while the response signal is expected to remain the same.

The preceding analysis represents a simple, ideal case for signal averaging. Signal variance, which is a measure of the expected value of the deviation of the signal from its mean, adds complexity to this simplistic case. Mathematically, $Var(X) = E((X - \mu)^2)$. Variance has the units of signal power and reflects the energy contained in the signal variation. Variance quantifies the contribution of the background noise to the response estimate. It also characterizes any changes or differences in the individual responses that may contribute uncertainty to the measurement. The variance of a signal reflects its power and thus relates to the amplitude of the recording.

If all successive responses are identical,

$$Var(r(t)) = E((r(t) - E(r(t)))^2)$$
$$= E((r(t) - r(t))^2) = 0.$$

Thus, variance in a stable response is zero. This conclusion indicates that, after repeated observations are combined, the result is still expected to look like a single response. For the noise, however,

$$Var(n(t)) = E((n(t) - E(n(t)))^2)$$
$$= E((n(t) - 0)^2)$$
$$= E(n(t)^2)$$

which equals the squared amplitude, or energy, in the noise.

If the background EEG activity and other noise are uncorrelated with the stimuli or the responses, it will tend to cancel away with increasing number of averages. In general, the uncorrelated activity will decrease with the square root of the number of records included in the average. Thus, if $A(s,n)$ is the amplitude of the average of n records of signal s and the records are uncorrelated with the sample times, $E(A(s,n)) = A(s)$.

This result is true for either gaussian (normal) noise that is uncorrelated with the signal or for coherent (sinusoidal) noise whose frequency and phase is unrelated to the stimulus or the response. Gaussian noise is approximated by some natural sources such as background instrumentation or electrophysiological noise, where coherent noise is produced by sources such as 60-cycle power line interference, or oscillating or vibrating devices (motors, etc.).

Similarly, if $A(r,n)$ is an average of n records of a response r and the records are correlated with the sample times, $E(A(r,n)) = A(r)$ so that, whereas the uncorrelated signal is expected to decrease with increasing averages, the response is expected to stay constant, and thus "emerge" from the background signal (see Fig. 2–1). This equation gives the amount of reduction expected in the noise, as responses are combined. For example, if 64 epochs are included in an average, the noise should reduce by a factor of 8, while the responses stay the same, resulting in an enhancement of 8:1 in the SNR. Increasing the number of epochs fourfold to 256 produces an additional twofold ($2\times$) improvement, to an SNR enhancement of 16:1. This result shows that as increasing numbers of epochs are included in an average, the improvement is less and less, owing to the square-root law of the improvement. Refer to Table 2–3.

PRACTICAL CONSIDERATIONS

If the responses to successive stimuli are identical, as more and more averages are taken, the background activity will decrease, allowing the response average to contribute a larger portion to the final average. If more samples are taken, there will come a time when taking more and more samples into the average will no longer have a visible effect on the average. When this happens, the average is said to have converged to a stable estimate. This situation will occur when the responses are constant with time, so that $E(r(t)) = r(t)$. However, if the responses are themselves varying in time, then this relation is not true. Instead, the expected value will be some other signal, which reflects the average of the responses themselves. For example, if visual evoked

potentials are averaged over a 5-minute period of time, it is possible that the responses are changing as a result of factors such as eye movement, habituation, boredom, changes in attention or arousal, drug effects, and so on. Therefore, although a response estimate may be calculated, it may not resemble any of the individual responses themselves.

If the reponses are continually changing owing to persistent and ongoing variations, it is possible that the averaged evoked potential may never converge to a particular value. This situation will occur as a result of any changes in the responses, whether they be in amplitude or in time. For example, if the latency of the wavelets varies across the individual responses, the measured evoked response will again be an average of the individual responses, with some apparent (and erroneous) latency value in between the extremes of the actual latencies. The problem with response variability is exacerbated by the square-root law in SNR improvement, because significantly increasing the number of epochs only makes it harder to gather a large number of identical responses.

EFFECTS OF DATA SAMPLING

When signals are digitized for averaging, some error may be introduced by the fact that the stimuli are not exactly lined up with the sampling times. Thus, the stimuli and their associated responses can have jitter with respect to the sampling intervals. Each response may be offset by up to half the sampling interval, introducing an amount of error in the latency measurement. This is not the same as the error introduced by phase-shifting in the signal acquisition. This problem is effectively avoided by using a fast enough sampling rate. This requirement increases the minimum acceptable sampling rate from the normally accepted Nyquist rate of twice the maximum signal component, to four times the maximum signal component.[12] In practice, the sampled signal should provide a suitable visual representation of the signal; this requirement also dictates a minimum rate in the vicinity of four times the maximum frequency. However, because sampling speed and memory usage have an associated cost, the rate used should be fast enough to be acceptable, but should not be significantly higher. This explains why evoked potential measurements generally use a higher sampling rate than that commonly acceptable for routine electroencephalographic studies, even though the underlying physiological signals are in principle the same. In general electroencephalographic studies, the requirements for latency measurements and reconstruction accuracy are less stringent than those in event-related potential studies.

CONDITIONS FOR SUCCESSFUL AVERAGING

In summary, averaging is effective in separating a desired response from other activity when the following assumptions are satisfied:

Stationarity of Response and Noise. The statistics of the response should be constant in time. In other words, the mean and variance of the response as well as the spectral characteristics and energy of the noise should ideally be the same throughout the recording session. Although averaging will function if this is not true, the reliability of the measurement will not be as controlled and will be more difficult to evaluate.

Linear Superposition of Response. This condition means that if successive responses overlap in time, they must not

TABLE 2–3	Signal-to-noise Improvement Ideally Achievable by a Signal Averager Reducing Noise That Is Perfectly Uncorrelated with the Sampling Epochs

Number of Sweeps	S/N Improvement[a]	Improvement in dB
4	2X	3.0
64	8X	9.0
256	16X	12.0
1024	32X	15.0
4096	64X	18.0

[a]The signal-to-noise (S/N) improvement is stated as the expected ratio of component amplitudes. This is the best improvement possible, using simple signal averaging. The S/N improvement in dB (decibels) is also shown.

have an effect on each other, aside from adding their waveforms in an arithmetic fashion, which is linear. If they interfere, then the addition is nonlinear. Averaging will still work, but results will depend in a complex way on the repetition rate and other parameters. Short-latency responses again provide a good example of signals that are well behaved in this regard, whereas longer latency components are much more susceptible to being altered by higher stimulation rates, and interference by preceding responses. Other effects that provide nonlinearity are facilitation (each response makes the next one stronger) and entrainment (the system resonates with the stimulus frequency, which reinforces the responses).

Independence of Responses from Each Other. Similar to linear superposition, this condition requires that each response has no effect on the others. Even if the responses are separated in time, certain conditions may cause responses to affect each other. For example, an unexpected auditory stimulus may produce an alerting response, which could affect a visual evoked potential taken near the same time. Alternatively, a large number of stimuli administered under certain conditions can produce habituation, fatigue, or other global changes, thus causing changes in later responses. This will, again, not prevent averaging from working, but it will affect the interpretation of the results.

Time-invariance of Response. All individual responses should ideally be identical, so that the average can converge to a meaningful, true average response. Again, if this is not true, averaging will work, but the estimate will not be as meaningful. Figure 2–6 shows an example of a filtering technique that demonstrates the time-variations that can occur in an evoked potential measurement.

Time-locked Response. Each response should have a fixed, well-defined relationship to the stimulus. Variations in latency caused by physiological or instrumentation factors will degrade the quality of the response estimate.

Uncorrelated Noise. The noise to be removed should be uncorrelated with the stimulus and with the response. If the noise has any systematic relationship to the stimulus, it will not be effectively removed. For example, stimulus artifact that occurs at the same time in each epoch will not be averaged away. Similarly, if the stimulus is triggered by a timer that is related to the power line frequency, it will be possible to introduce interference that will not be removed. Physiologically, any associated signal such as electromyogram (EMG), electrooculogram (EOG), or other biological activity that is correlated with the stimulus/response events, will not be removed.

It is not necessary for noise to have a "hard" or causal connection with the stimulus or response in order to be correlated. If, for example, stimuli are applied at a rate of two per second, and there is a standing 60-Hz interference signal, exactly 30 cycles of the interference will fit into each epoch. Therefore, this signal will be included in the average, simply because it happens to line up exactly with the sample epochs. Such noise can be effectively removed by using random or pseudo-random stimulus intervals. This approach is achieved by using sampling intervals that start at times that are not regular, but give the effect of starting the sampling intervals at various delays, thus cancelling out any regular background interference.

AVERAGING AS FILTERING

Because of its mathematical properties, signal averaging does not act identically on all frequency components. In particular, when averaging is applied to a regular stimulus interval, certain frequencies will be removed significantly better than others. With a regular stimulus interval T, the effect of averaging is to produce a filter that selectively

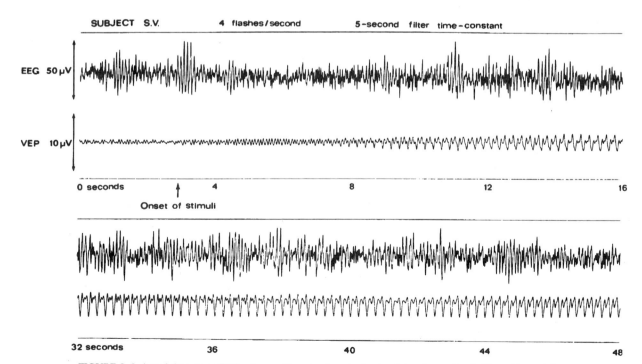

FIGURE 2–6. A real-time trace EEG, along with a visual evoked potential estimate that is filtered in real time. It is clearly evident that the evoked potential is undergoing continual change, with pronounced variations occurring with periods of 5 to 10 seconds. When signal averaging is used over periods of 30 to 60 seconds or more, these variations are lost, and cannot be seen. Reprinted with permission from Collura TF: Real-time filtering for the estimation of steady-state visual evoked brain potentials. IEEE Trans Biomed Eng 37: 650–652, 1990.

emphasizes frequencies equal to the stimulus interval and its integer harmonics. Thus, frequencies equal to $f_n = n/T$ will not be reduced.

At the same time, the averager will remove signals equal to 0.5 times f, and any odd harmonics thereof. Thus, with an interstimulus interval of 0.2 second, for example, averaging will perfectly cancel out sinewaves of 2.5 Hz, 7.5 Hz, 15 Hz, 17.5 Hz, and so on. Sinewaves at 5 Hz, 10 Hz, 15 Hz, 20 Hz, and so on, however, will be minimally attenuated. Averaging in this context, therefore, is equivalent to a comb-filter with stop-bands precisely placed at frequencies dictated by the interstimulus interval. The sharpness of the comb-filtering bands is a function of the number of epochs included in the averaging, so that sharper filtering is achieved with larger numbers of epochs. Note that comb-filtering does not occur when random or pseudo-random interstimulus intervals are used; all frequencies are ideally attenuated an equal amount in this case.

It is possible to apply filtering directly to the raw EEG in an effort to extract event-related potentials. Several such techniques are mentioned in the next section. These can be successful whenever the event-related signals have signal properties that distinguish them from the unwanted signal activity. No technique can violate the basic principles of information theory; it is not possible to create information when it is not originally present. Any technique that enhances the SNR of an event-related potential must do so either by combining many responses in order to produce an estimate, or by using assumptions regarding signal timing or statistical properties.

ADDITIONAL TECHNIQUES

Additional techniques continue to be developed. The dominant thrust is to provide signal estimates in a shorter period of time, particularly by using fewer stimuli and applying more sophisticated analysis techniques. Recent advancements include the use of random stimuli to avoid frequency-dependent following or entrainment effects.[17] Some new techniques provide enhancements or variations on the basic theme of signal averaging, whereas others attempt to estimate single responses based on pattern recognition techniques,[10] or to recover signals in real-time using filtering techniques.[4, 11, 13, 18]

REFERENCES

1. Bendat JS, Piersol AG: Random Data: Analysis and Measurement Procedures. New York, Wiley, 1971.
2. Blum JR, Rosenblatt JI: Probability and Statistics. Philadelphia, WB Saunders, 1972.
3. Chiappa KH: Evoked Potentials in Clinical Medicine. New York, Raven Press, 1990.
4. Collura TF: Real-time filtering for the estimation of steady-state visual evoked brain potentials. IEEE Trans Biomed Eng 37:650–652, 1990.
5. Colton T: Statistics in Medicine. Boston, Little, Brown, 1974.
6. Dustman RE, Beck EC: Long-term stability of visually evoked potentials in man. Science 14:1480–1481, 1963.
7. Furst M, Vlau A: Optimal posterior time domain filter for average evoked potentials. IEEE Trans Biomed Eng 38:827–833, 1991.
8. Grass ER, Johnson E: An Introduction to Evoked Response Signal Averaging. Quincy, MA, Grass Instruments, 1985.
9. Lopes da Silva F: Event-related potentials: Methodology and quantification. In Niedermeyer E, Lopes da Silva F (eds): Electroencephalography (2nd ed.). Baltimore, Urban & Schwarzenberg, 1987, pp 763–772.
10. McGillem CD, Aunon JI: Measurements of signal components of single visually evoked brain potentials. IEEE Trans Biomed Eng 24:232–241, 1977.
11. Mita M: Adaptive analysis of harmonic oscillation for biological signals. Med Biol Eng Comput 26: 379–382, 1988.
12. Nakamura M, Nishida S, Shibasaki H: Deterioration of average evoked potential waveform due to asynchronous averaging and its compensation. IEEE Trans Biomed Eng 38:309–312, 1991.
13. Nakamura M, Shibasaki H, Nishida S, Neshige R: A method for real-time processing to study recovery functions of evoked potentials. IEEE Trans Biomed Eng 37:738–740, 1990.
14. Neshige R, Luders H, Shibasaki H: Recording of movement-related potentials from scalp and cortex in man. Brain 111:719–736, 1988.
15. Papoulis A: Probability, Random Variables, and Stochastic Processes. New York, McGraw-Hill, 1965.
16. Regan D: Human Brain Electrophysiology. New York, Elsevier, 1990.
17. Shi Y, Hecox KE: Nonlinear system identification by m-pulse sequences: Application to brainstem auditory evoked responses. IEEE Trans Biomed Eng 38:834–845, 1991.
18. Thakor NV, Vaz CA, McPherson RW, Hanley DF: Adaptive Fourier series modeling of time-varying evoked-potentials: Study of human somatosensory evoked response to etomidate anesthetic. Electroencephalogr Clin Neurophysiol 80:108–118, 1991.

Chapter 3

Signal Conditioning of Neurophysiological Signals: Amplifiers and Filters

Richard C. Burgess, MD, PhD

INTRODUCTION
AMPLIFIERS
Differential Amplifiers
Accurate Measurement of Signal

FILTERS
Filter Types
Phase Shift
Filter Order
Filter Characteristic
DIGITAL FILTERING

INTRODUCTION

Neurophysiological signals, including electroencephalograms (EEGs) and evoked potentials (EPs), are exquisitely weak. As they progress from the peripheral nerve, spinal cord, or brain to the surface where the recording electrodes sense them, the signals are greatly attenuated—partly by the short-circuiting effect of the cerebrospinal fluid (CSF), which "bridges" electrical generator areas; partly by the high resistance of the skull and vertebral bodies; and partly by the dispersion effect of the scalp and other overlying integument. The scalp electroencephalogram (EEG) represents the summation of the activity of millions of neurons inside the cranial vault, smeared out and blurred during transmission to the surface by these effects caused by the CSF, skull, and scalp. Before the minute signals emanating from neurons and muscles can be employed for any diagnostically useful purpose, they must be "conditioned"; that is, electronically manipulated so that they can be viewed, analyzed, and interpreted. The most fundamental signal-conditioning operations are amplification and filtering. These essential signal-conditioning components are block diagrammed in Figure 3–1.

The signals that are picked up by the sensors (usually metallic electrodes) range from a couple of microvolts (e.g., scalp-recorded brain stem EPs) to a few tens of millivolts (e.g., needle electromyogram [EMG]). The low voltage of these signals, and the correspondingly low current they drive into the recording apparatus, is far too small to move the pen on a strip chart, deflect the beam on an oscilloscope, or even produce a detectable signal at the input to a computer system. Therefore, the signals must be made larger, or amplified. The signal after amplification can then be sent to a variety of devices, either simultaneously or individually, so that it can be viewed for expert interpretation or further refined by subsequent analysis instrumentation. The process of amplification of neurophysiological signals is directly comparable to the process of amplification of the electrical signals generated by a microphone in response to sound waves. These small electrical signals must be amplified so that they can be fed to a loudspeaker for widespread "viewing" or sent to a tape recorder for later "analysis."

Every physiological signal consists of a variety of frequency components. Depending on the purpose of the recording or monitoring, only a relatively restricted "band" of the frequency components is relevant to the diagnostic question

SIGNAL CONDITION COMPONENTS OF ANY ELECTROENCEPHALOGRAPH

FIGURE 3–1. Neurophysiological recording systems sense signals from the body (EEG, EMG, EP, etc.) through metallic electrodes either applied to the skin or inserted into active tissue. The microvolt level signals are made larger by amplifiers, then fed to high-pass filters to eliminate baseline wander and to low-pass filters to remove high-frequency artifacts such as muscle noise. To ensure that the system is properly representing the input signals, the capability to switch a known calibration signal into the circuit in place of the physiological signal is provided.

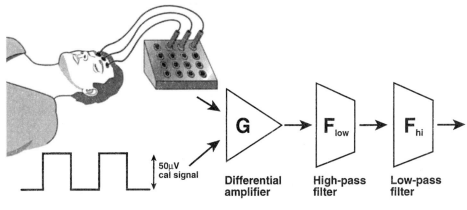

50μV cal signal

G — Differential amplifier

F_{low} — High-pass filter

F_{hi} — Low-pass filter

FILTERING TO SEPARATE SIGNAL FROM NOISE

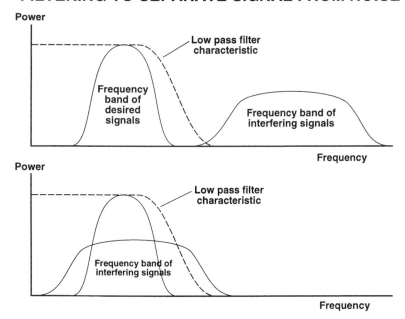

FIGURE 3–2. The signal of interest and the signals to be excluded do not always occupy separable portions of the frequency spectrum.

under study. For example, to the cardiologist, the EMG is unwanted "noise"; whereas to the electroencephalographer, the electrocardiogram (ECG) is noise. Some of the undesirable noise and artifacts have frequency components that are predominantly in a band different from the desired signal. The process of filtering allows investigators to separate signals on the basis of their frequency, attenuating (reducing in amplitude) the unwanted frequency components, or emphasizing the components that are important to them, or all of these, as illustrated in the top portion of Figure 3–2. Filtering is exactly what is done by a radio tuner, which separates stations on the basis of their frequency, so that one hears the station at 104.1 megahertz (MHz), but not the adjacent one at 104.5 MHz. Unfortunately, physiological signals are not subject to the regulations of the Federal Communications Commission (FCC) and so are not as nicely separated as radio stations. There tends to be considerable overlap in frequency between the signals of interest and the unwanted noise signals, as depicted at the bottom of Figure 3–2. Nevertheless, considerable "cleaning up" of the signal can be accomplished by screening out the frequencies that are irrelevant to neurophysiological study and that would otherwise corrupt the desired signal. Naturally, high-quality electrode contact is the foremost prerequisite for high-quality recording. The old adage applied to computers, "garbage in, garbage out," obviously applies to amplifiers as well. No amount of signal conditioning, artifact rejection, or filtering can make up for poor-quality electrode contact.

AMPLIFIERS

The fundamental property of all amplifiers is "gain," the ability to take a signal at the amplifier input and reproduce that signal at the output, but much larger (i.e., *multiplied* by the gain factor). When paper was the primary output medium for neurophysiological signals, the term sensitivity was used frequently, defined as the amount of electrical potential corresponding to a unit deviation on the paper, typically microvolts per millimeter (μV/mm). With the advent of digital recording and paperless electroencephalography, and

with the associated rescaling, zooming, and variable-sized printouts, the meaning of the term sensitivity has become inconsistent. The linear relationship between the input to the amplifier, seen at the electrodes, and the output is shown in Figure 3–3. Amplifier gain (as well as filter attenuation) is measured in a curious engineering unit called a decibel (named in honor of the inventor of the telephone), abbreviated dB. A dB is defined as 20 times the log of the ratio of the output signal/input signal. Therefore, for example, a difference of 20 dB between two signals represents an amplitude ratio of 10:1, whereas a difference of 6 dB represents a ratio of 2:1. Ratios of less than one (e.g., 1:10 or 1:2) are expressed as negative dB, because fractions have a logarithm that is negative. A few typical gain ratios along with their associated dB values are shown in Table 3–1. To increase or decrease the output signal representing a given input, the scale of the amplifier output-to-input relationship is changed (i.e., the slope of the gain curve is altered), as shown in Figure 3–4.

Differential Amplifiers

All amplifiers for neurophysiology are based on the "differential" amplifier, so named because it has two input terminals and is designed to amplify only the *difference* between the signals at the two inputs. Thus, signals that are common to the two inputs will not be amplified. These common mode

TABLE 3–1	Gain or Loss in Decibels
Amplitude Ratio	**Decibels**
0.10	−20 dB
0.50	−6 dB
0.71	−3 dB
1.0	0 dB
10	20 dB
100	40 dB
1000	60 dB
10,000	80 dB

AMPLIFICATION

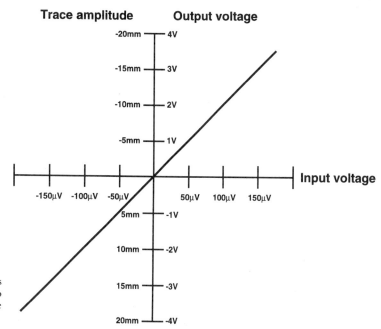

FIGURE 3–3. Within their operating range, amplifiers provide an output that has a direct linear relationship to the input signal. In the example shown, the gain of the amplifier is 20,000 V/V, or 86 dB.

signals will be cancelled out, and only those signals that are different at input 1 and input 2 will remain, as shown in Figure 3–5. This cancellation applies not only to coupled mains interference, but also to distant EMG potentials that affect the two input terminals equally, as illustrated by the examples in Figure 3–6. If the signals at the two inputs to a differential amplifier are respectively $V_{input\ 1}$ and $V_{input\ 2}$, and the gain is G, then the output voltage will simply be:

$$V_{output} = G \times (V_{input\ 1} - V_{input\ 2})$$

Figure 3–7 shows how four entirely different sets of input signals can result in the same appearance at the output of a differential amplifier.

The polarity of the output will be an algebraic function of the polarities of the input signals as given by the preceding equation. The *representation* on a strip chart or display screen

of the polarity of a signal (either positive upward or negative upward) has nothing to do with the amplifier and is purely a matter of convention. Unfortunately, this convention varies between neurophysiological specialties, between signals within the same specialty, and even between laboratories considering the same signals.

Accurate Measurement of the Signal

For an amplifier to faithfully represent at its output an event that has occurred at its input, four key criteria pertain:[5]

Amplitude linearity. Within the working range, the input-to-output relationship must be linear. For example, if the input signal doubles (or halves) in amplitude, then the output signal must also double (or halve). Amplifier saturation, or "clipping," represents an extreme case of

GAIN

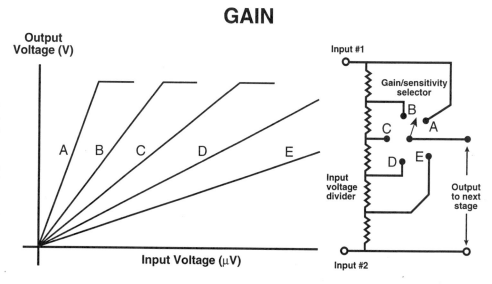

FIGURE 3–4. A resistive divider serves to adjust the gain of an amplifier. As the selector switch is rotated from position A through E, a smaller and smaller proportion of the input signal is available at the output (i.e., the gain is reduced). For gains that are too high to accommodate the full range of the signal, amplifier saturation at the amplifier's maximum output voltage will cause "clipping," as seen at the top of curves A, B, and C.

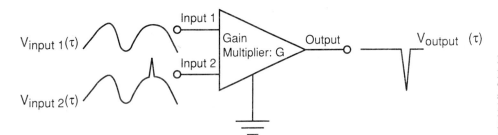

FIGURE 3–5. The differential amplifier cancels out signals which are common to input 1 and input 2, such as the sinusoidal signal shown here. Signal components that differ between the two inputs are multiplied by the gain of the amplifier.

nonlinearity, and is shown at the top of three of the gain curves in Figure 3–4. Two other examples of non-linear input-output relationships are shown in Figure 3–8.

Adequate bandwidth. An amplifier must amplify with an equal gain factor all the frequencies that are presented to its input, within the "meaningful bandwidth," otherwise distortion will occur. This meaningful bandwidth obviously differs for the particular signal under study, but is limited not just to the extremes of the filter cutoff frequencies. An amplifier should generally be capable of response somewhat outside the expected frequency range in order to build in a margin of safety. Excessive bandwidth, however, is also to be avoided. For example, the direct current (DC) component is generally not diagnostically useful in neurophysiology, and amplifying it to the same degree as the alternating current (AC) components would, in most cases, only lead to saturation.

Phase linearity. All of the frequency components that are amplified must be shifted by a constant time.

Low noise. Noise generated within the instrumentation should never be of sufficient amplitude to interfere with the signal of interest. Generally this will be achieved if

the amplifier noise is smaller than any of the other external or biopotential sources of noise.

Noise is defined as *any* unwanted or extraneous signal; that is, as any disturbance that interferes with the information in the signal under study.[3] Noise currents are introduced into amplification systems by (1) capacitive coupling from external electric fields, (2) induction of an electrical current into a conductive loop by an external magnetic field, or (3) artifacts generated within the body or at the interface of the transducer to the body. Modern "op-amp" instrumentation amplifiers[7] are manufactured using high-precision integrated circuits (ICs), which are composed of an array of transistors, capacitors, resistors, and diodes in a single package fabricated using photolithographic techniques on a silicon chip. With close attention to high input impedance, low noise design, low input bias current, and adequate common mode rejection, the amplifier is usually not the limiting factor.[11]

Ordinary 60-Hz mains interference, seen as a common mode signal, can quite easily reach an amplitude several hundred thousand times greater than the physiological signal under study. Differential amplifiers, because they are constructed from real-world components, are not 100% effective in screening out common mode signals. The degree

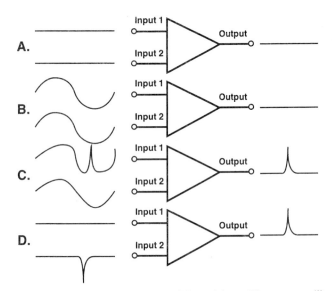

FIGURE 3–6. The polarity of the differential amplifier output will depend on whether the signal is largest at input 1 or input 2. In *A*, both inputs are zero, so the output is zero. In *B*, the two sinusoidal inputs are identical, so the output is zero. In *C*, a spike on input 1 is superimposed on the same sinusoidal inputs seen in *B*. Since the output is input 1 minus input 2, the only signal remaining at the output is the spike. Likewise in *D*, a negative spike has been superimposed on the flat lines seen in *A*. Subtraction of this spike on input 2 from input 1 yields the same output as in *C*.

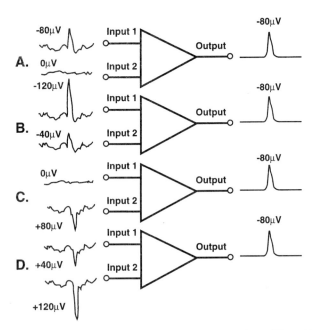

FIGURE 3–7. This illustration employs a differential amplifier with a gain (G) of 1 to show how four different combinations of spikes on input 1 and input 2 can result in the same output. As a result of the subtractive process, only the *relative* difference between input 1 and input 2 is reflected at the output, regardless of the absolute input amplitude.

NON-LINEAR DEVICES

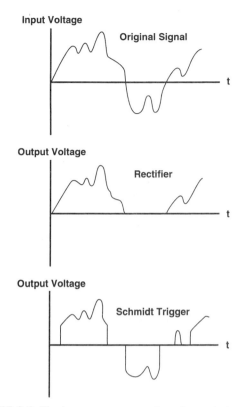

FIGURE 3–8. The input to two types of nonlinear devices is shown in the top trace. The output of a rectifier (diode), which conducts only in one direction, is shown in the middle trace. In the lower trace is the output of back-to-back Schmidt triggers, devices that pass the input to the output only when it exceeds a threshold.

to which they are effective is mathematically expressed as the common mode rejection ratio (CMRR), ordinarily measured in dB. Good neurophysiological amplifiers should have CMRRs greater than 100,000 (100 dB). Even when a high-quality differential amplifier is employed, it is virtually impossible to eliminate conversion of some of the common mode noise to differential noise voltage as a result of electrode impedance mismatch[6] or from unequal magnetic/capacitive coupling of external noise voltage. To minimize conversion of a common noise signal into a differential signal, the amplifier input impedance must be many times higher than any impedance mismatch between the electrodes. Even if the electrode impedances were closely matched, but the amplifier input impedance was rather low (thereby "loading" the source), flow of a significant amount of current into the amplifiers would, in fact, distort the electrical field at the scalp. Hence, high input impedance amplifiers are needed not only for faithful recording, but also so that the measurement system itself does not corrupt the original data being measured.

Measurements of electrical potentials, even in differential amplifiers, are made with respect to a "ground." The ground electrode attached to the patient is not meant to contribute to the actual signal seen at the output of the amplifier, but rather is designed to establish the "zero" level for the measurements that are made at input 1 and input 2 and subsequently subtracted. It is important to recognize that the ground attached to the patient is not ordinarily the same as the ground of the building's power systems (i.e., the green

wire or center D-shaped pin on the electrical receptacle). High-quality instrumentation amplifiers establish a separate "floating" ground, usually called isolated ground (abbreviated as "iso-ground"). This isolated patient ground is electrically separated from the chassis of the equipment and the building ground in order to (1) decrease electrical safety risks and (2) decrease common mode interference.

FILTERS

The most basic signal-processing tool is filtering, designed to make the waveform of interest stand out from the background. Filters are circuits or transformation processes that have a deliberately nonuniform transfer function with respect to frequency; that is, the gain (or loss) varies with the frequency of the signal applied to the input of the filter. The transfer function is the ratio of the output to the input, expressed in relation to the frequency spectrum. Filtering is used in general to (1) reduce external interference, (2) eliminate large-amplitude out-of-band signals that would otherwise block subsequent amplification stages, (3) remove baseline offset, (4) limit the frequency range under study or modify the frequency domain properties of the signal, (5) prevent aliasing during analog-to-digital conversion, and (6) smooth the results of a processed waveform, such as an ensemble average.

Filter Types

Most filters are designed to favor, or "pass," signals in some frequency band, while markedly attenuating, or "rejecting," signals in all others.[9] Filters are ordinarily divided into four categories, as shown in Figure 3–9, based on the band of frequencies they pass (i.e., allow through unchanged) or the band of frequencies they stop (i.e., attenuate to acceptably low levels). These categories are:

Low-pass filter. A filter that allows frequencies from zero, or DC, up to the cutoff frequency to pass through unimpeded, but attenuates all frequencies above the cutoff frequency (Fig. 3–10).

High-pass filter. A filter that allows all frequencies above the cutoff frequency to pass through unimpeded, but attenuates all frequencies below the cutoff frequency. Baseline offset (i.e., a DC potential present in the signal), as well as very slowly varying (i.e., close to DC) artifacts (such as sweat artifact) can be removed using this type of filter (Fig. 3–11).

Band-pass filter. A filter that allows all frequencies in the "passband" between the low cutoff frequency and the high cutoff frequency to pass through unimpeded, but attenuates all frequencies above and below the passband.

Band-stop filter. A filter that attenuates all frequencies within a (usually narrow) "stopband" and that allows all the other frequencies to pass through unimpeded. This type of filter, also called a notch filter when the stopband is very narrow, is most often used to eliminate interference from the 50- or 60-Hz power mains.

Because of the relationship of the type of cutoff frequency to the type of filter, as well as the use of the words "pass" and "stop," the nomenclature is sometimes confusing to the uninitiated. Because a low-pass filter allows frequencies through up to a certain point, it has a *high* cutoff frequency; likewise, a high-pass filter has a *low* cutoff frequency. In common parlance then, low-pass filters are therefore sometimes referred to as "high-frequency filters" and high-pass

TYPICAL FILTERS

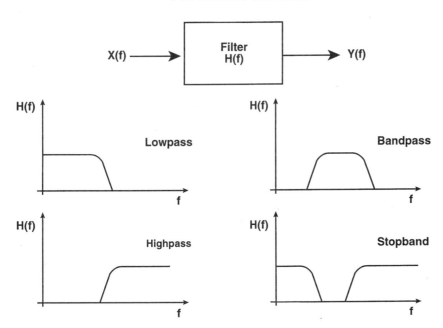

FIGURE 3–9. Filtering consists of taking an input signal, X, and modifying its frequency domain properties according to the transfer function, H, to produce an output function, Y, schematized in the top diagram. Idealized frequency domain representations of the four major types of filter transfer functions are shown. For each of these, the vertical axis shows the relative proportion (H) of the input X, which can be seen at the output Y, as frequency (f) changes along the horizontal axis.

filters as "low-frequency filters." And, because they take a chunk of frequencies somewhere out of the middle of the frequency range of interest (such as 60 Hz), band-stop filters are frequently called notch filters. Filter terminology is further confused by the tendency of some manufacturers to use terms such as high linear filter (HLF) and low linear filter (LLF). These terms originate from a description of the differential equation that expresses the behavior of the filter, and correspondingly the amplitude-versus-log frequency curve, which is a series of straight lines connected by curved transitional regions.

Typical neurophysiological instrumentation filters are

"tunable," meaning that the cutoff frequencies can be adjusted by the operator (or under computer control) to optimally reject the unwanted frequencies, shown as three different cutoff frequencies in Figures 3–10 and 3–11. Fixed frequency filters also have a place, such as antialiasing filters and 60-Hz notch filters. Note that although the frequency cutoff may be controlled for all of the channels together, separate filters (like separate amplifiers) are required for each channel. Filters can be constructed as "analog" filters or as "digital" filters. Analog filters accept as their input a *continuous* voltage versus time and produce at their output a continuous waveform as well. These are the conventional

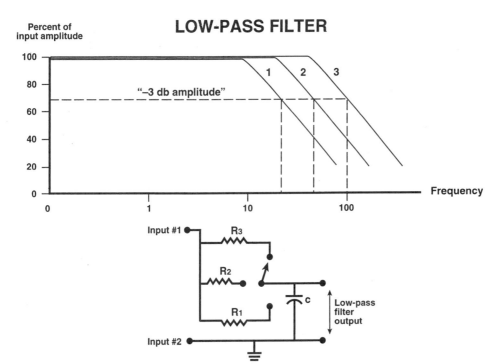

FIGURE 3–10. At higher frequencies, the capacitor shown in the circuit becomes a better conductor, shunting some of the signal to ground. Selection of one of the three frequency cutoffs shown in the graph is done by selecting one of the resistors to change the resistor-capacitor (RC) time constant. In this example R1 > R2 > R3. When the product of R times C is greater, the frequency cutoff (as measured by the 71%, or −3 db, point) is lower.

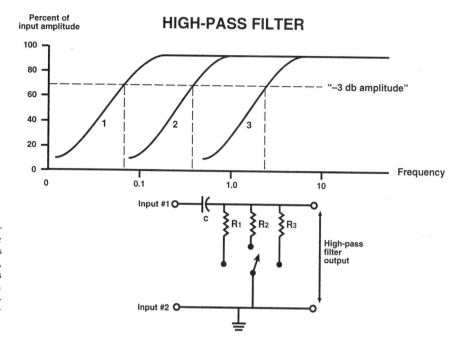

FIGURE 3–11. In the case of the high-pass filter, the better conductivity of the capacitor at higher frequencies allows more of the signal through to the output, resulting in the output amplitude versus frequency graph shown at the top. As in Figure 3–10, R1 > R2 > R3. The −3-db cutoff frequency is lower for a larger RC product.

electronic filters that have always been present in neurophysiology instrumentation. Digital filters, on the other hand, operate on *discrete samples* of the waveform, and so require digitization of the data by an analog-to-digital converter before the filtering operation (Fig. 3–12). Digital filtering is ordinarily carried out by computers, and therefore may not produce an "output" except to a video screen. However the output of a digital filter can be transformed back into an analog voltage by a digital-to-analog converter (shown in the lower right of Fig. 3–12), thus mimicking an analog filter. The effect on a typical EEG signal of progressive decreases in the cutoff frequency of an adjustable low-pass filter is shown in Figure 3–13.

The theoretically perfect filter that would allow passage of the desired frequencies without any alteration or distortion and would completely reject all frequencies outside the desired range does not, of course, exist in either the analog or the digital world. This ideal filter would have a constant gain of exactly one (unity) throughout the passband, then abruptly drop off vertically to a gain of zero outside the passband. A practical filter (i.e., one that can be actually implemented and used) has a gain of approximately one in the passband but with some variation, especially as the cutoff frequency is approached. Around the region of the cutoff frequency is the transition band, where the filter changes from a device that freely passes signal to one that attenuates signal. This gain transition certainly does not occur abruptly but rather deviates from the ideal vertical dropoff with a characteristic "rolloff," or downward slope. This transition band has a finite width in the frequency domain, and the gain does not necessarily vary smoothly in this range, therefore affecting the different frequencies in the transition band in different ways. In the stopband, where it would be preferable if the frequencies were all attenuated to zero, they are not—a certain percentage of the original signal still gets through. In addition, depending on the design of the filter, the reduction in the signal amplitude may not progress monotonically as one moves, further and further into the

DIGITAL FILTERING

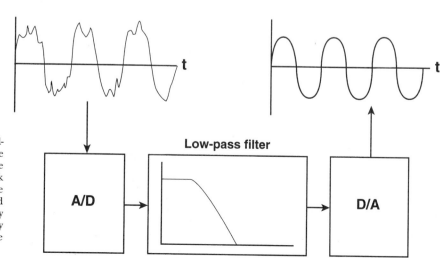

FIGURE 3–12. Digital filters require two additional components: one to transform the analog signal into digital samples, and one to transform the digitally filtered data back into a continuous time domain signal. The characteristics of the filter are determined by the weighting coefficients, and by whether those coefficients are applied only to the raw input data, or recursively to the output data as well.

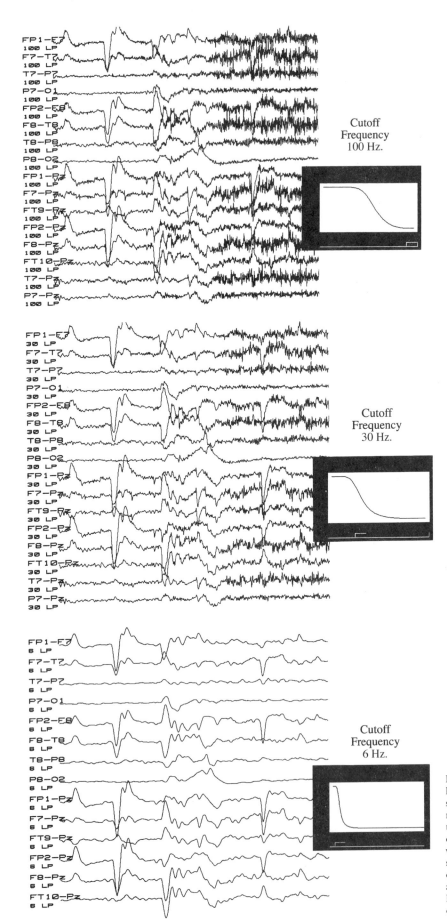

Cutoff
Frequency
100 Hz.

Cutoff
Frequency
30 Hz.

Cutoff
Frequency
6 Hz.

FIGURE 3–13. An EEG signal containing both low frequencies (the eye movements on the left side of the tracing) and high frequencies (the muscle artifact toward the right side) illustrates the effect of various degrees of low-pass filtering on a real signal. The original signal, recorded with an upper frequency cutoff of 100 Hz, is shown at the top. Successive traces illustrate the effect of lowering the cutoff frequency to 30 Hz and 6 Hz. At the lowest cutoff, the EMG activity has been completely attenuated, but the slower eye movements remain.

FILTER TYPES AND ROLLOFF CHARACTERISTICS

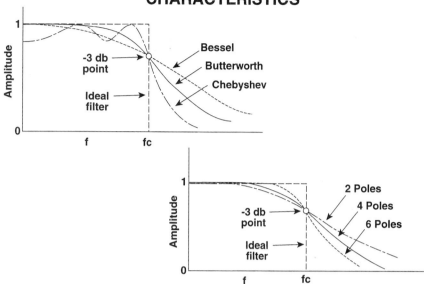

FIGURE 3–14. The design of the filter (both number of poles and filter family type) determines the deviation from ideal in terms of overshoot and phase delay. The bottom panel shows increasingly steep rolloff with greater number of filter poles. The top panel shows how faster rolloff can be achieved at the expense of ripple in the passband by choosing a Chebyshev type of filter.

stopband (i.e., it exhibits "ripple"). Because the ideal filter with flat response, infinite attenuation, and linear phase is simply not mathematically and physically realizable, compromises (usually under the control of the filter designer and not the operator) are required. Selection of the proper filter properties is a multidimensional tradeoff involving cutoff frequency, steepness of the rolloff, variation in the passband, attenuation in the stopband, and phase distortion. Several different filter types and rolloffs are shown in Figure 3–14.

Basic filtering is accomplished through the use of combinations of resistors and capacitors (i.e., RC networks), as seen in Figures 3–10 and 3–11, which rely on the fact that capacitors exhibit a decreasing opposition to the flow of electrons at higher frequencies. Capacitors can be modeled as an open circuit at or near DC, and as a short circuit at high frequency. Analog filters are referred to as either active or passive, depending on whether they are composed of only inactive components such as resistors, capacitors, and inductors, or contain devices requiring external power such as transistors and op-amps. Significant performance advantages usually accrue through the use of active filters, and because neurophysiology instrumentation always incorporates amplifiers (which are active components), it is generally a simple matter for the manufacturer to make the filters active. Active filters are readily combined in functional blocks to build combinations of low-pass, high-pass, and notch filters. For narrow band filters, the steepness of the rolloff (i.e., the degree to which the immediately adjacent frequencies are affected) is expressed by an engineering quantity, the "Q." Hence, high Q 60-Hz notch filters have a very deep notch and little attenuation of the frequencies around 60 Hz. A conventional notch filter constructed from a passive twin-T network of resistors and capacitors has a Q of approximately 0.3, so that much of the important signal will be distorted in the process of eliminating the mains interference. By adding an active component (an op-amp voltage follower), the Q can be easily raised to 50 or more.

Phase Shift

All filters exhibit some delay between the components of the signal seen at the input and the filtered versions of the

same components seen at the output of the filter. (It is possible using digital filters to correct for this delay [i.e., to make it zero[8]], as the computer has access to the data before and after the point undergoing analysis.) Generally speaking, the delay is proportional to the complexity of the filter. Obviously, a delay time can be thought of as a proportion of one cycle of a sine frequency; that is, the "phase shift" (see the illustration in Figure 3–15). If the delay is constant irrespective of the frequency, then the phase shift will be exactly proportional to frequency, called a linear phase characteristic. Filters with a nonlinear characteristic will cause different frequency components of the signal to be delayed by various amounts of time, causing distortion of the resulting output signal (i.e., there will be alterations in the *relative* phase of the components). Phase shift distortion is the effect of time shifting different frequency components by different amounts.[12] For example, in some circumstances, it is possible to shift the spike component of a spike-and-wave complex from between two waves into the center of the wave component. Shifting of the apparent time location of an EP peak as a result of filtering[4] can make a normal EP latency appear abnormal, or vice versa. Phase shift is ordinarily worst near the cutoff frequency.

As previously stated, the "cutoff" frequency is not actually a sharp cutoff, but rather marks the transition from a horizontal gain characteristic to a steep rolloff. Even though it does not represent a highly distinct demarcation between passband and stopband, the cutoff frequency is measured exactly, using a mathematical definition. The cutoff frequency is that frequency at which the amplitude of the signal is "3 db down" from the amplitude of the signal in the passband. Several filters, all with the same −3-dB point are shown in Figure 3–14. If the output of the filter is 1.0 in the passband, then the −3-db point corresponds to the frequency where the signal is reduced to 0.71. (The cutoff frequency of some digital filters is specified by a slightly different method.) Occasionally manufacturers do not use the standard 30% down point for characterizing their filters; rather, they indicate their filter cutoffs by the point that is 20% down. It is also possible to represent the frequency cutoffs by means of the "time constant" defined by the equation $T = 1/2\pi F$. This is most often used for characterization of the low-frequency cutoff because it corresponds to

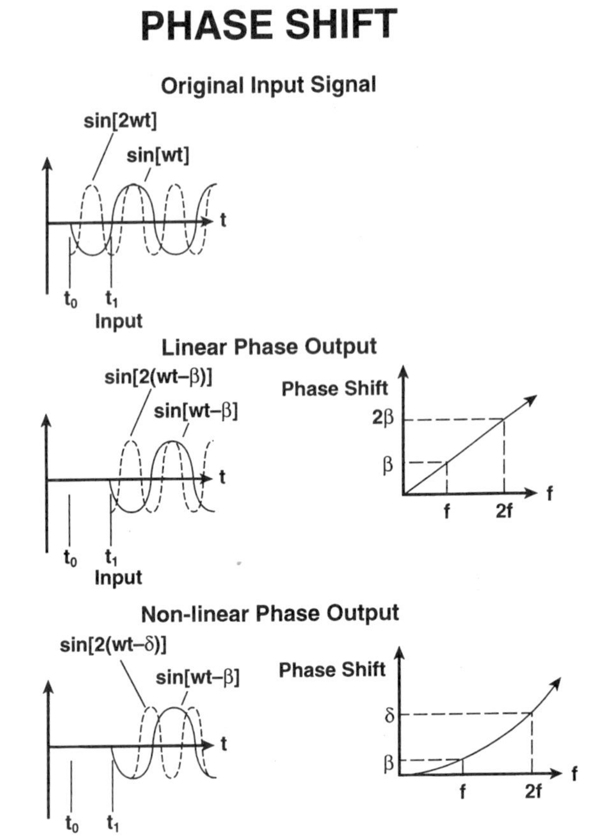

PHASE SHIFT

Original Input Signal

sin[2wt]

sin[wt]

t

t_0 t_1

Input

Linear Phase Output

sin[2(wt−β)]

sin[wt−β]

t

t_0 t_1

Input

Phase Shift

2β

β

f 2f

Non-linear Phase Output

sin[2(wt−δ)]

sin[wt−β]

t

t_0 t_1

Phase Shift

δ

β

f 2f

FIGURE 3–15. Phase distortion causes a relative time delay between components of different frequencies (e.g., the high-frequency spike and the lower frequency wave of a 3-per-second absence discharge, or the high-frequency early components and the lower frequency later components of an evoked potential).

the time in seconds for a step input (such as an artifact or calibration signal) to decay by a factor of 0.37 (i.e., $1 \div e$, the natural log base), illustrated in Figure 3–16. It happens to be easy to calculate from the product of resistance and capacitance used to build a simple passive filter, $T = RC$. Although characterizing filters using measures such as the time constant, 20% attenuation, 50% attenuation, and so on, are theoretically valid methods, it is more sensible to adhere to the standard engineering definitions.

Filter Order

The steepness of the rolloff outside the passband is governed by the "order" of the filter (which is also related to the number of "poles" in the filter). As the number of poles increases, the transition band becomes narrower and the rolloff, steeper (as shown in the lower part of Figure 3–14), approaching the amplitude response of the ideal filter. Unfortunately, the number of poles cannot be increased indefinitely, even if the number of components or complexity of the calculation were not at issue. Very high order filters have such excessive phase shift that they tend to exhibit unstable behavior, manifest as ringing (brief resonation that occurs in response to rapid input transients then dampens out) or oscillation (undampened resonation that builds up and persists). Oscillation at the output of an unstable filter (applied to the same signal as in Figure 3–13) is shown in Figure 3–17. In the case of digital filters, high order filters also take longer to produce an output. At a cutoff frequency

close to zero or Nyquist (the top end), more poles actually increase the width of the transition band.

Filter Characteristic

In addition to the type of filter, cutoff frequency, and rolloff, the filter characteristic or *transfer function* is an important attribute. There are various common filter transfer functions, some seen in the upper portion of Figure 3–14. The Butterworth filter is maximally flat in the passband—that is, it allows all frequencies in the passband through without any change in amplitude (variation in amplitude is called "ripple")—but it produces overshoot in response to a rapid input transient. Choosing a Chebyshev, inverted Chebyshev, or elliptical filter permits a sharper transition without increasing the number of poles. The Chebychev filter characteristic has a controllable equi-ripple passband (i.e., the amount of variation of the frequencies can be kept equal throughout the passband and can be minimized to an acceptable level), and it rolls off faster than the Butterworth. The Elliptical filter falls off faster than any other filter but cannot infinitely attenuate signals at the extreme of the stopband because of ripple in the stopband. The tradeoffs between transition bandwidth and ripple are summarized in Table 3–2. There are other filter characteristics that minimize phase distortion and eliminate overshoot, such as the Bessel, Gaussian, and Paynter filters. Design parameters that change filter characteristics are illustrated in Figure 3–18.

DIGITAL FILTERING

The inclusion of microprocessors as "embedded" components has dramatically altered the capabilities of traditional clinical neurophysiological instruments. Currently, only a small proportion of the electroencephalographs sold are of the paper strip-chart variety. In digital EEG machines,[1] the output of the amplifiers is instead fed directly to an "analog-to-digital convertor," which changes the analog waveform into a series of ones and zeros that can be easily manipulated by a computer. The process is the same as that employed to

TIME CONSTANT

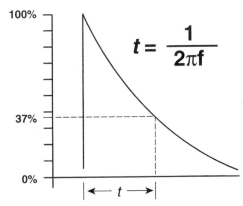

100%

$$t = \frac{1}{2\pi f}$$

37%

0%

$\leftarrow t \rightarrow$

FIGURE 3–16. Time constant is measured as the time that it takes for a sudden impulse to decay from 100% of its value down to 37%. The time constant, t, is inversely proportional to the frequency cutoff, f.

FILTER RINGING

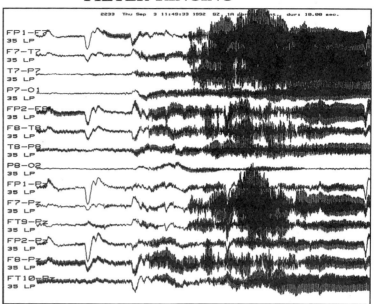

FIGURE 3–17. The same signal used in Figure 3–13 has been applied to a very high order filter. As the filter is activated by the higher frequency EMG activity, it begins to oscillate, generating a false output.

put music onto a compact disc and, if done correctly, conveys all of the information present in the original analog recording.

Analog-to-digital conversion of the raw, amplified signal is done by sampling, a process of measuring the amplitude of the continuous signal at regular intervals, and converting each sample into a digital value. Selection of the appropriate sampling rate is based on considerations of the frequency content of the neurophysiological signal and the data volume capabilities of the computer system. Sampling at points that are too closely spaced will not only result in excessive quantities of digital data, but will also yield correlated and redundant data. On the other hand, insufficiently frequent samples of the EEG will lead to the condition known as aliasing, wherein high-frequency and low-frequency components of the signal become confused, producing distortion. Although Shannon's sampling theorem dictates that the sampling frequency must be at least twice the frequency of the fastest component in the original data (the Nyquist frequency), practical systems employing realistic antialiasing filters should sample at three or four times the Nyquist frequency.[2] In typical EEG applications, low-pass filtering prior to digitization is done at 70 Hz, and analog-to-digital conversion is carried out at 200 Hz. Once the data have been digitized, they can be further conditioned—just as analog data can—but with more flexibility. In particular, posthoc filtering and gain changes can be accomplished.

Averaging is one of the simplest forms of a digital filter.

Typically used in eliciting EPs from a high-noise background, it assumes that the signal of interest is a more or less invariant response to a stimulus, providing a time function that bears the same relationship to the time of each stimulus. The noise, on the other hand, is assumed to be random (i.e., not time-locked to the stimulus), and therefore cancels out as N (the number of trials summed to produce the average) approaches infinity. It can be shown that the improvement in signal-to-noise ratio is proportional to the square root of N.

Key Filter Parameters

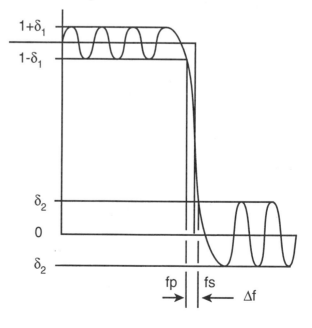

FIGURE 3–18. Transition bands (Δ_f), as well as the passbands and stopbands of complex filters, exhibit nonlinear characteristics. The design of the filter is a tradeoff between passband ripple (δ_1), rolloff, and stopband attenuation.

Filter Characteristic	Sharpness	Tradeoffs
TABLE 3–2 Filter Characteristics		
Chebyshev	Decreases transition region	Introduces ripple into the passband
Inverted Chebyshev	Decreases transition region	Introduces ripple into the stopband (leakage)
Elliptical	Sharpest transition region	Introduces ripple into both passband and stopband

Digital filters carry out a form of averaging on adjacent data points. Digital filtering is roughly akin to a sliding average, but with specific components chosen to effect the desired waveshaping. The number of points included in the weighted average determines the number of "taps" of the filter. Specification of the weighting (a multiplier or coefficient) applied to each tap determines how various frequencies are emphasized or deemphasized. Digital filters are constructed using simple operators: delays, multiplies, and adds. Digital filters offer (1) sharper rolloffs, (2) linear or zero phase, (3) no necessity for calibration, (4) perfect stability with time or temperature, (5) independence from power supply variations or component value tolerances, and (6) the facility for adaptation to the data, as the parameters are controlled in software.[10]

Digital filters are implemented in one of two ways: Finite impulse response (FIR) or infinite impulse response (IIR). FIR filters are so named because an impulse fed into an FIR filter generates no more nonzero terms than the order of the filter. The IIR filter introduces some measure of phase shift distortion and therefore has the potential for infinite ringing when forced by an impulse. All digital filters introduce some amount of time shifting of a processed signal. It is possible to design a class of FIR filters that is phase shift distortionless. However, the disadvantage of FIR filters is that they are not as efficient as IIR filters and, hence, take longer to process the signals.

FIR filters (1) have linear phase because of constant delay, (2) do not accumulate roundoff error, because every output is a function of only N multiplications and accumulations; and (3) are implemented using design by windows, or through computer-aided design (CAD) employing the Remez iteration. IIR filters (1) are inherently unstable and can also oscillate as a result of roundoff errors; (2) are not linear phase, but can be made fairly linear within the pass-band; (3) require far fewer coefficients than the FIR for the same filter performance; and (4) can be implemented by the methods of invariant transformation, bilinear transformation, or using CAD through chains or biquads. Table 3–3 contrasts the critical properties of FIR and IIR filters.

Where all of the data are available for retrospective refiltering (e.g., after an EP sweep), adaptive techniques such as Weiner filtering, Kalman filtering, and other time-varying filtering techniques may provide even better performance. These techniques are extensions of standard digital filtering methods modified by a priori knowledge of the expected waveform's signal components. Digital filtering can provide a continuous output, albeit delayed by a few sampling intervals, and can be used to provide alteration of the frequency domain characteristics of a signal in a controllable way.

TABLE 3–3	Digital Filter Design Tradeoffs
FIR Filter Properties	**IIR Filter Properties**
Always stable	Potentially unstable
Can have linear phase	Nonlinear phase
Can be adaptive	Feedback
Low roundoff noise	More efficient than FIR filters
Easy to understand, design, and implement	Subtle design issues

FIR, finite impulse response; IIR, infinite impulse response.

REFERENCES

1. Barlow JS: Computerized clinical electroencephalography in perspective. IEEE Trans Biomed Eng 26:377–391, 1979.
2. Bendat JS, Piersol AG: Random data: Analysis and measurement procedure. New York, Wiley, 1971.
3. Brittenham D: Recognition and reduction of physiological artifacts. Am J EEG Technol 14:158–165, 1974.
4. Desmedt JE, Brunko E, Debecke J, Carmeliet J: The system bandpass required to avoid distortion of early components when averaging somatosensory evoked potentials. Electroencephalogr Clin Neurophysiol 37:407–410, 1974.
5. Geddes LA, Baker LE: Principles of Applied Biomedical Instrumentation. New York, 1968, Wiley, p 446.
6. Gordon M: Artifacts created by imbalanced electrode impedance. Am J EEG Technol 20:149–160, 1980.
7. Graeme JG, Tobey GE, Huelsman LP: Operational Amplifiers, Design and Application. New York, McGraw-Hill, 1971.
8. Green JV, Nelson AV, Michael D: Digital zero-phase-shift filtering of short latency somatosensory evoked potentials. Electroencephalogr Clin Neurophysiol 63:384–388, 1986.
9. Gussow M: Schaum's Outline of Theory and Problems of Basic Electricity. New York, McGraw-Hill, 1983, pp 243–249.
10. Macabee P, Hassan N, Cracco R, Schiff J: Short latency somatosensory and spinal evoked potentials: Power spectra and comparison between high pass analog and digital filter. Electroencephalogr Clin Neurophysiol 65:177–187, 1986.
11. Motchenbacher CD, Fktchen FC: Low-Noise Electronic Design. New York, Wiley, 1973.
12. Seaba PJ: Understanding frequency response. Am J EEG Technol 15:20–41, 1975.

Chapter 4

Clinical Neurophysiological Instrumentation: Electrical Safety

Richard C. Burgess, MD, PhD

INTRODUCTION
ELECTRICAL POWER SERVICE
ELECTRICAL SHOCK
GROUNDING
PATIENT ELECTRICAL SAFETY HAZARDS
Sources of Hazardous Electrical Current

GUIDELINES FOR SAFE OPERATION OF
  CLINICAL NEUROPHYSIOLOGICAL
  EQUIPMENT
Hospital Electrical Safety Guidelines
CONCLUSION

INTRODUCTION

In the neurophysiology laboratory, patients are at risk primarily because of the electrodes that have been attached to them. The designated connections between the subject and the recording equipment are the *recording electrodes* and the *stimulating electrodes* (if any). However, a number of other possible connections between the patient and sources of electrical current also exist. These may include both obvious (and intentional) connections as well as unrecognized current pathways. For example, there may be low resistance pathways through electrodes connected to the patient from other equipment (such as electrocardiogram [ECG] monitors) or through indwelling catheters carrying conductive intravenous (IV) solutions ("long lines" to the heart or great vessels are especially dangerous because of their intimate relationship to the vulnerable cardiac ventricular tissue); or higher resistance pathways (but nevertheless dangerous) through casual contact with metal objects such as bed frames, other furniture, wall fixtures, or even personal devices such as radios and hair dryers.

Because of these connections to the patient during neurophysiological recordings (e.g., electroencephalogram [EEG], electromyogram [EMG], evoked potential [EP], polysomnogram [PSG], etc.), the patient may sustain an electrical shock as a result of accidental leakage current, or because of component failure within the instrumentation that brings the patient connections into contact with dangerous voltages. Controlled electrical stimulation of nervous system tissue is frequently employed in somatosensory-evoked potentials (SEPs) and electromyography. During neurophysiological testing, which includes stimulation, the patient may also be subject to accidental spread of the purposefully introduced stimulation current or secondary to possible component failure. Unpleasant or dangerous electrical shock to a patient occurs when (1) one portion of the patient is connected to a *source* of electrical current, and (2) that current follows a pathway through the patient and exits elsewhere to a *sink*. The usual electrical sink is one or another unexpected "grounds." Understanding these unexpected connections and following a few simple rules of practice will keep the patient out of harm's way.

ELECTRICAL POWER SERVICE

The 110- to 120-V (220- to 230-V in many countries) alternating current (AC) that comes to the electrical outlets in the wall in the laboratory or hospital room is distributed from circuit breaker panels, much like those in residential home wiring. Each receptacle has three openings (Fig. 4–1). The short slot, connected to the black wire, is the "hot" conductor and the long slot, connected to the white wire, is the "neutral" conductor. These two conductors allow a device plugged into the receptacle to complete a circuit, with the energy coming into the device from the short slot/black wire and returning to the receptacle through the long slot/ white wire. The additional D-shaped hole is connected to the green ground wire and provides a path to the "ground." Farther upstream, at some distance from the patient room, the green ground wire and the neutral white wire are bonded to a true ground, that is, one with a low-resistance connection to the earth. Normally the power cord to a hospital device connects the D-shaped pin on a hospital plug to the case of the medical equipment. The purpose of the green wire is to harmlessly conduct to ground any stray current that reaches the case of the equipment and that might otherwise be conducted to the patient. Faults in the power-line system, either in the hot, neutral, or ground conductors, can pose an electrical danger to the patient, owing to the changes in current described in Table 4–1. More often, however, danger to the patient comes not from power-line problems, but from faults in the apparatus or from ground loop currents. The presence of conductive fluids in the health care environment increases the chance for electrical malfunction.

ELECTRICAL SHOCK

Because of the electrodes attached to the skin and the consequent low-resistance connections to electrical equipment, patients are much more likely to be exposed to unusual electrical currents than are nonpatients. In addition, indwelling catheters provide direct routes to the heart for passage of current which on the skin surface would not even be detectable, but conducted in this fashion can cause lethal arrythmias.

The minimum current detectable at the skin is approximately 0.5 milliamperes (mA).[2] When electrical current passes through the body, in addition to producing an unpleasant sensation, the current may cause tetanic muscle contraction, resulting in an inability to let go of the electrical conductor, or may produce paralysis. At higher current levels, pain will be induced, muscles tonically contracted, and

POWER DISTRIBUTION SYSTEM

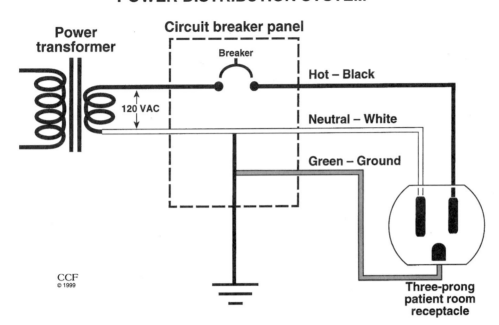

FIGURE 4–1. The AC power from the electric utility company is stepped down by a large power transformer (in the basement or on a utility pole) then routed to a circuit breaker panel. The power is delivered from the circuit breaker panel to each receptacle by wires with three conductors. The mains power neutral, which completes the power circuit back to the source, is bonded to the mains power ground at the service entrance panel. (Copyright © Cleveland Clinic Foundation.)

finally tissue destroyed. Although this "macroshock" hazard is of concern to linesmen and others who work with electrical power, it is the risk of "microshock" that is of greatest concern to medical patients. When applied directly to the heart, a current of as little as 20 μA has been shown to produce ventricular fibrillation in dogs,[5] as shown in Figure 4–2. For humans the threshold is about 180 μA. In cases of death by electrocution, most subjects are killed by the passage of electrical current through the heart, resulting in ventricular fibrillation. If current is applied to the skin, fibrillation can still be produced. This would only occur if a minimum of 100 mA passed through the skin, and a portion of it flowed through the heart. Current flow through the heart from both skin electrode attachments and more invasive connections is illustrated in Figure 4–3.

For electricity to flow, a complete circuit is required (i.e., from the "hot" back to the neutral). Because the ground wire is also ultimately (e.g., at the circuit breaker box) connected to the neutral, the circuit may also be completed from hot to ground. Electricity may complete its circuit through any ground, and will take the path of least resistance. In a hospital room, it is difficult to avoid having the patient grounded at least transiently. The side rails of the bed or a metal wall plate may be grounded. Contact by the patient with the sink plumbing fixtures, either directly or through the hands of a health care worker, also creates a low-resistance ground.

Since the patient is ordinarily grounded or partially grounded (i.e., there is a finite resistance to ground), any electrical potential with which he or she comes in contact will be conducted through the patient's body to ground, subjecting him or her to the risk of shock. If this patient is grounded through a very low-resistance pathway, such as the indwelling line shown in Figure 4–4, he or she is at extreme risk for microshock. As it happens, human susceptibility to shock is maximal at 60 Hz, exactly the power-line frequency.

Although high voltages are dangerous, it is not the absolute voltage level but rather the current passing through the body that is dangerous. Thus, a 110-V shock sustained by a person standing on dry ground and insulated by normal clothing may scarcely be felt, because it is less than 1 mA. On the other hand, a person standing barefoot on wet ground who touches a 110-V hot conductor with his or her arm may be killed, because the current is likely to be more than 100 mA and will pass through the heart.

GROUNDING

Understanding the concept of electrical ground and the differences between various electrical points called "ground" is essential to a satisfactory understanding of electrical safety concerns.

Earth Ground. Our planet is composed of material which, because of the presence of water and salts, is conductive. A metal rod with sufficient surface area driven into the earth will have a relatively low resistance contact with the surrounding soil. Likewise, the metal structures of a building, such as I-beams or water pipes that reach into the ground, also are in good contact with the soil. Because they run through the earth for considerable distances, cold water (supply) pipes are frequently considered to be the most practical high-quality earth grounds. Because of its size, the earth, and consequently earth ground connections, have a limitless capacity for absorbing currents from these connections.

Mains Power Ground. The third contact (i.e., the D-

TABLE 4–1	Powerline Faults	
Fault Condition	**Effect**	
Reversal of hot and neutral	Leakage current may increase	
Interruption of neutral	Function taken over by ground, causing excessive current on ground conductor	
Excessive current through ground conductor	Ground terminal no longer at 0, V	
Two "grounds" at different potentials	Current flow between two instruments, perhaps through patient	

VENTRICULAR FIBRILLATION FROM CURRENT CONDUCTED DIRECTLY TO THE HEART

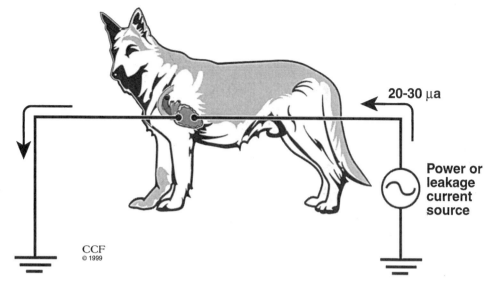

20-30 μa

Power or leakage current source

CCF
© 1999

FIGURE 4–2. Electrical current passing through the heart at a current as low as 20 μA has been shown to produce ventricular fibrillation in dogs. In humans, the threshold is about 10 times higher. (Copyright © Cleveland Clinic Foundation.)

shaped pin in the United States, shown in Fig. 4–1) on the electrical power plug is routed from the power receptacle on the wall, through the building wiring and circuit panels, and eventually attached to earth ground. It therefore serves as a sink for any stray currents introduced into the electrical ground pin.

Mains Power Neutral. The longer of the two rectangular slots in a mains power receptacle in the United States (referring again to Fig. 4–1) is electrical "neutral," that is, the return path for the 110-V power for the equipment plugged into the wall.[3] Eventually (usually somewhere in the building's basement) the neutral is connected to earth ground. En route from the wall receptacle to the basement, however, the neutral is kept separate from the mains power ground so that the latter is not carrying any current during normal operation.

Chassis Ground. The enclosure of electrical equipment, if fashioned of a conductive material, is usually connected to mains power ground → earth ground. This is done in order to provide a path for stray currents (resulting from failures occurring inside the enclosure) harmlessly to earth ground rather than accidentally through the patient. It also serves to shield components inside the enclosure or other equipment outside from interference, as shown in Figure 4–5. For equipment located in patient areas, connectors (such as BNCs) should ordinarily be insulated from the chassis so that circuit ground does not employ chassis ground as a signal pathway.

Circuit Ground. The electrical circuitry inside the equipment will normally "reference" power and signal lines to "circuit ground." This may be the 0-V tap on the equipment power supply, or it may be chassis ground, depending on the application, or both. In medical instrumentation, very often,

SURFACE VS. INVASIVE CONNECTIONS: CURRENT DENSITY

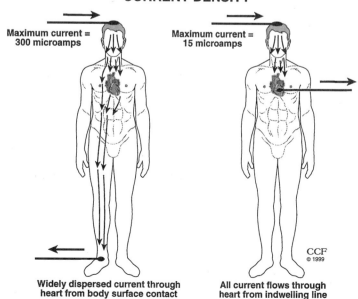

Maximum current = 300 microamps

Maximum current = 15 microamps

Widely dispersed current through heart from body surface contact

All current flows through heart from indwelling line

CCF
© 1999

FIGURE 4–3. Any electrical current passing through the body may pose a high risk, as some portion of it may pass through the heart. The proportion directed into the vulnerable cardiac tissue is much higher, as shown on the right, when the current exits the heart through an indwelling long line. (Copyright © Cleveland Clinic Foundation.)

MICROSHOCK HAZARD IN PATIENTS WITH INDWELLING LINES

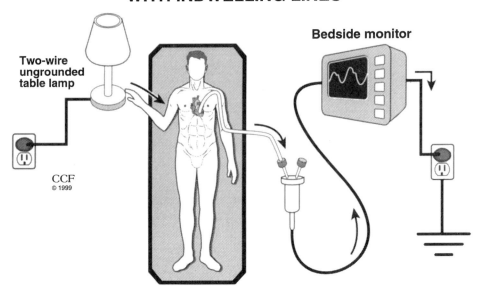

FIGURE 4–4. Any mains-powered electrical device will "leak" current, which may enter the patient through direct contact (such as touching the metal lamp base) or indirect pickup (such as capacitive coupling from nearby wiring). If there is a pathway to the heart providing a highly conductive route (such as the Swan-Ganz catheter shown), a higher proportion of the leakage current will pass through the heart. (Copyright © Cleveland Clinic Foundation.)

there are several different circuit grounds that are isolated from each other (e.g., digital circuit ground, analog output circuit ground, isolated input circuit ground, stimulator ground, etc.). Signals flowing from one stage of a piece of equipment to another do so through signal lines that complete their circuit back to the originating stage through one of these circuit grounds.

Patient Iso(lated) Ground. One type of circuit ground (which is specialized to medical instrumentation and illustrated in Fig. 4–6) is engineered to maximize the safety of patients by remaining separated (i.e., "isolated" by an extremely high impedance) from any other circuit ground, earth ground, or chassis ground. Differential amplifiers used in medical instrumentation applications (such as those for

EPs) need a ground connection to the patient in order to reject common mode signals. Stimulators, such as the ones used in SEP applications, employ patient "grounds" to decrease stimulus artifact at the recording site. In both of these cases, these grounds are kept separate and isolated from all other grounds, as well as from each other (i.e., no current should flow from these circuits in or out of any other circuit or ground).

Ground loops occur when a galvanic connection exists between two different grounds and current accidentally introduced into one ground completes the circuit by flowing through another ground. Unfortunately, completion of a circuit between the two grounds at different potentials is often carried out by the patient.

CHASSIS LEAKAGE CURRENT

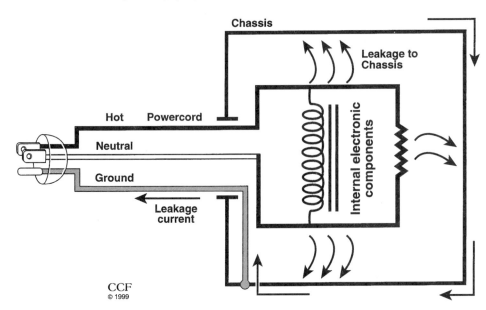

FIGURE 4–5. All of the components inside an electrical instrument have the potential to "leak" some current to the surrounding metal case, depending on the capacitance between them and the case. This leakage current is normally drained away harmlessly by the green ground wire. (Copyright © Cleveland Clinic Foundation.)

PATIENT ISOLATION

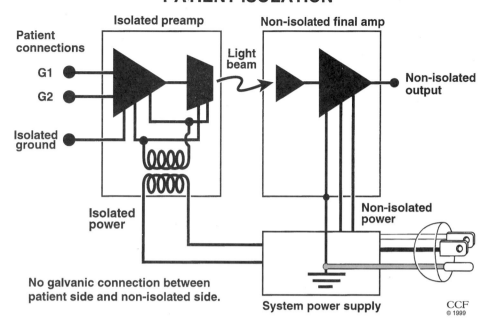

FIGURE 4–6. A medical grade isolation amplifier has no galvanic connection between the preamplifier and the subsequent stages of processing. The preamplifier receives its power from a transformer where the primary and secondary are galvanically isolated, and the signal is conveyed from the preamplifier to the subsequent stage through a light beam across an insulated gap. (Copyright © Cleveland Clinic Foundation.)

Patient isolation by means of isolated power supplies, optically or transformer-coupled isolation amplifiers (Fig. 4–6), double insulation, and so forth, are attempts to prevent the occurrence of current flow to or from *any* ground connection by any equipment located near the patient.

Chassis *leakage current* refers to the small amount of current that flows from the equipment inside the chassis to the metal enclosure,[4] usually as a result of capacitive or inductive coupling from the power supply. Medical instrumentation must follow strict guidelines to keep this leakage current to an acceptably safe level (less than 100 μA), established in the United States by Underwriters Laboratories.[1] Safe current limits are listed in Table 4–2. In addition, as shown in Figure 4–5, any current that does manage to leak to the chassis will be carried harmlessly to earth ground through the mains power ground.

PATIENT ELECTRICAL SAFETY HAZARDS

Shock to a patient does not generally come from an exposed 110-V conductor (e.g., a frayed extension cord), but rather through leakage current from the chassis of electrically powered equipment. The proximity of the electrical conductors inside an electrical device (ranging from vacuum cleaners to table lamps to ventilators) to the metal case allows capacitive coupling of some current, called leakage current, to the case. If a patient comes in contact with a metal case, and he or she provides a better pathway to ground than any other conductor, then the leakage current will be conducted through the patient's body to ground. An example of leakage from a two-wire (i.e., ungrounded) table lamp through the patient to a venous pressure monitor is illustrated in Figure 4–4. Incorporating a low-resistance connection between the metal case of electrically powered equipment and ground substantially decreases this hazard because the leakage current will preferentially travel through the green wire, as shown in the upper portion of Figure 4–7.

Macroshock Hazards. Macroshock refers to the electrical shock sustained by a person who comes into skin contact with a current-carrying conductor and completes a circuit, allowing the electricity to pass through his or her body. Many people have experienced the highly unpleasant macroshock sensation that occurs when one accidentally touches an exposed lightbulb contact or a malfunctioning kitchen appliance delivering 110-V, 60-Hz current to the fingers. The effects of macroshock are a function of the current magnitude, pathway, and duration. These effects are summarized in Table 4–3. If the macroshock results in ventricular fibrillation or paralysis, it may be fatal. The risk of ventricular fibrillation is a function of body weight,[6] as graphed in Figure 4–8, and therefore children may be more easily killed by electrical shocks.

Microshock: Indwelling Lines. Patients who have electrodes affixed to the skin or inserted into muscles, as is often the case in the neurophysiology laboratory, are at high risk because of the low-resistance pathway to the body established by these electrodes. Even more dangerous, however, are those patients with indwelling lines such as IVs, central venous catheters, pacemaker leads, and so forth. Placement into the bloodstream (or worse still, directly into a cardiac chamber) provides a low-resistance pathway to the heart, and therefore places the patient at a much higher risk of microshock. As outlined on Table 4–4, the impedances provided by a metal wire into the heart or a tube of conducting fluid into the central circulation are orders of magnitude lower than the intact skin. These connections to patients are dangerous because a mere 150 to 200 μA applied directly to

TABLE 4–2	Safe Current Limits for Medical Instrumentation
Condition	**Current**
Chassis leakage under either normal or fault condition	100 μA
Patient electrodes from electrode input to ground	50 μA
Electrically sensitive patients (neonates or adults with indwelling lines). Maximum current for electrically sensitive patients (neonates or adults with indwelling lines) when 120-V AC applied in series with patient leads.	10 μA

AC, alternating current.

MICROSHOCK FROM CHASSIS LEAKAGE CURRENT

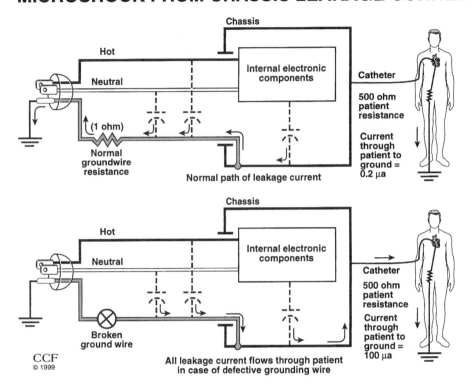

FIGURE 4–7. Capacitance from the power conductors in the power cord to the green conductor, as well as from the internal components to the instrument case, causes a certain amount of leakage current to ground—about 100 μA in this example. Because the resistance to ground is 500 times lower through the green ground wire than through the patient, all but 0.2 μA returns harmlessly through the building wiring. In the lower panel, however, the green ground wire has been fractured (causing no noticeable change in the instrument's operation) so that the 100-μA current has no way to flow to ground except through the patient. Because this is a pressure monitor measuring through a right heart catheter, the current is directed into the patient entirely through the heart tissue, exiting through the right leg ECG lead. (Copyright © Cleveland Clinic Foundation.)

the human heart can induce ventricular fibrillation. Often the patients who require portable EEGs have some form of heart disease, rendering them more susceptible to ventricular fibrillation. In Figure 4–9, one of these patients receives a shock directly to the heart through a pacemaker lead when a technician unknowingly becomes part of a completed circuit.

Sources of Hazardous Electrical Current

Chassis or Frame Leakage Current

The major source of potentially lethal currents from medical equipment is leakage from the 110-V power transformer primary winding. The leakage current reaches the case of the medical equipment as a result of capacitive coupling from the transformer to metal parts in contact with the case, as illustrated in Figure 4–5. Electrical equipment (not just medical equipment) is designed so that this leakage current flows from the instrument case to ground through the three-wire power cord provided with the device. General purpose electrical equipment, such as that found in the home, can be expected to produce up to 500 mA of leakage current. However, hospital medical equipment must have a leakage below 100 mA. Shock to patients results both from hardware

faults (i.e., abnormal malfunctions such as fracturing of a ground lead in a power cord, as illustrated in Figure 4–7), as well as from improper use and configuration of normally functioning equipment.

Inductively Coupled Leakage Currents

If a patient is connected by means of skin electrodes across a 1-V AC source, Ohm's law dictates that approximately 100 μA will flow through the subject as a result of the approximately 10,000-Ω summated electrode connections. Such a source may result from a ground loop (Fig. 4–10), that is, current flowing from a ground of one potential to a ground of a lower potential. Another source of ground loop currents occurs when two parts of a patient's body are connected to ground through wires that are long or widely separated, or both. Under this condition, the wires and the patient form a loop, which becomes the secondary of a transformer (Fig. 4–11). The primary of the transformer is the magnetic field generated by the power-line wiring in the walls and ceiling. The primary induces a current in the secondary, and because the patient is part of that secondary, a hazardous current may result.

The ground loop shown in Figure 4–11 can be eliminated by either (1) making certain that the patient is grounded at only one point so that ground loop currents do not flow through the patient's body, or (2) eliminating the "galvanic" connection of one of the monitors to ground (i.e., providing only an "isoground" connection, a very high-impedance connection to ground).

Capacitively Coupled Leakage Currents

Far more common than inductively coupled leakage currents are leakage currents that result from capacitive coupling of AC sources to the patient or to equipment connected to the patient. Ordinarily, patients are grounded by direct connection to a grounded electrode, or by inadvertent connection to a grounded bed or other metal object within

TABLE 4–3	Danger from Electricity: Macroshock
Current	**Physiological Effect**
0.5–1.0 mA	Threshold of sensation
1.0–5.0 mA	Threshold of pain
10–20 mA	Let-go current (muscle contraction)
50 mA	Mechanical injury, syncope
100–500 mA	Ventricular fibrillation
5–10 A	Sustained myocardial contraction, burns, paralysis

Fibrillation Threshold as a Function of Body Weight

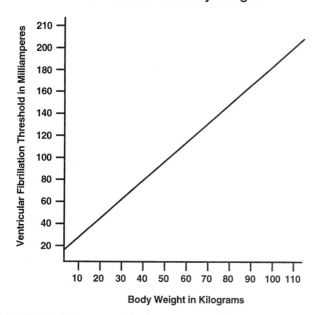

FIGURE 4–8. Because of the lower current-diffusing capacity of a smaller amount of body tissue, younger, tinier patients can develop ventricular fibrillation at a lower current level.

TABLE 4–4	Body Contact Impedances	
Route of Contact		**Impedance**
Intact skin		5000–10,000 Ω
Point to point through internal tissue volume		100 Ω
Central venous pressure line		800 Ω
Swan-Ganz catheter		500 Ω
Transvenous pacemaker lead		1 Ω

native pathway for the current to flow (i.e., through the shield rather than through the patient).

In the typical polysomnography suite, the patient, who is grounded, is in a monitoring room, while the monitoring equipment is in an adjacent room. The cabling from the patient travels through the walls, in close proximity to the electrical mains wiring. Capacitive coupling of the 60-Hz AC mains current into the patient-connected cable occurs over the length of the cable. The current induced in the cable passes through the patient and returns to ground through the ground electrode. To protect against this kind of leakage current, the electrode leads may be shielded.

Figure 4–7 shows a piece of electrical equipment connected to patient monitoring electrodes. The 60-Hz wiring inside the equipment case (especially the equipment power supply) generates an electrical field that is capacitively coupled to the case or electrode leads, or both. This induced current then flows through the patient-connected ground or lead wire into the body, and then returns to ground. Ordinarily the current induced into metal components of electrical equipment is preferentially returned to ground through the green ground wire rather than through the patient, because the connection to the building ground through the green wire is a lower impedance pathway than that through the patient. However, a broken ground conductor in the equipment power cord (schematized in the lower portion of Figure 4–7), which will not cause a noticeable change in

their reach. Electrical mains wiring in the ceiling is capacitively coupled to the patient through the air, inducing a current in the patient, which then returns through the patient to ground. Typical capacitance between the wiring and the patient may be on the order of 3 pfd. This situation can be prevented by (1) shielding the utility wiring, or (2) shielding the patient. Normally, the utility wiring is enclosed in grounded metal conduit. Such shielding provides an alter-

MICROSHOCK THROUGH HEALTHCARE PERSONNEL

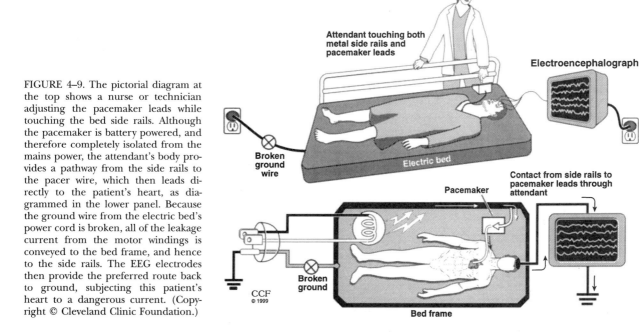

FIGURE 4–9. The pictorial diagram at the top shows a nurse or technician adjusting the pacemaker leads while touching the bed side rails. Although the pacemaker is battery powered, and therefore completely isolated from the mains power, the attendant's body provides a pathway from the side rails to the pacer wire, which then leads directly to the patient's heart, as diagrammed in the lower panel. Because the ground wire from the electric bed's power cord is broken, all of the leakage current from the motor windings is conveyed to the bed frame, and hence to the side rails. The EEG electrodes then provide the preferred route back to ground, subjecting this patient's heart to a dangerous current. (Copyright © Cleveland Clinic Foundation.)

NON-EQUIPOTENTIAL GROUNDS

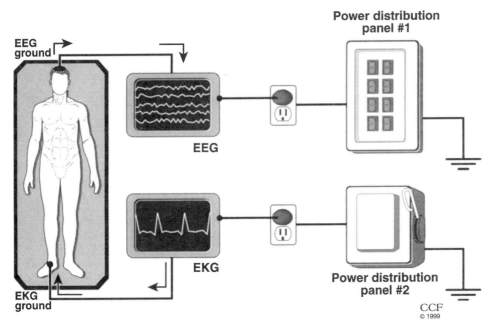

FIGURE 4–10. When a portable EEG is performed on a patient in a crowded hospital room, the EEG machine may be plugged into a receptacle at some distance from the patient—in this case powered by a branch circuit from a new circuit breaker panel. The ECG monitor is already connected to the patient, but is powered from a second circuit, installed many years before when the hospital was built. Because of differences in the wiring and the amounts of leakage current carried by the mains power grounds from panel 1 and panel 2, the two grounds connected to the patient end up a volt or two apart. The patient conducts the current from one ground to the other. (Copyright © Cleveland Clinic Foundation.)

equipment operation, will make the patient the only pathway for return of these stray leakage currents to ground.

Nonequipotential Grounds

When multiple ground points are accessible in a patient room or intensive care area, there are no guarantees that all of these ground points will be at the same potential, unless a very low-resistance connection exists between all of them. This is the reason that plugging all patient-connected monitors into the same outlet cluster is advised in intensive care

units. If this is not done, and a potential difference of only 30-mV exists between the various ground points, this can produce sufficient current through a patient to cause death by electrocution. These subtle ground potential differences are difficult to detect but may be fatal if undiscovered.

Metal equipment cases, or any other grounded surfaces, within the patient room are supposed to be grounded through the green wire. Unfortunately, the metal cases of different electrical instruments within the patient room may not all be grounded to the same point, and therefore may

GROUND LOOP CURRENTS

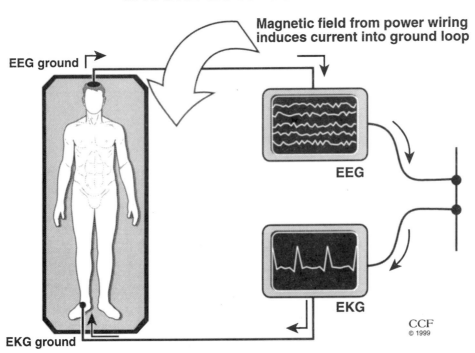

FIGURE 4–11. An alternating magnetic field passing through a conductive loop of wire induces a current in the wire. If, owing to the positioning of the EEG and the ECG cable, the patient happens to be a conductive part of that loop, current will pass through him or her, perhaps causing injury in a susceptible patient. (Copyright © Cleveland Clinic Foundation.)

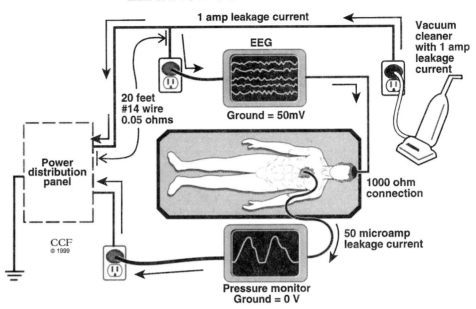

MICROSHOCK FROM REMOTE LEAKAGE CURRENT SOURCE

FIGURE 4–12. Because of the large leakage current (1 A is not unusual) generated by motorized cleaning equipment, there will be a significant voltage drop along the mains ground wire as it returns from the vacuum cleaner to the circuit breaker panel. The EEG is plugged into a receptacle along the way, which has a ground pin 50 mV above true ground owing to this voltage drop. The central venous pressure monitor is connected to the patient's heart through the column of conductive IV fluid. With the 1000-Ω EEG electrodes connected to the patient's head, the 50-mV voltage difference produces a current of 50-μA right through the patient's heart—well above the allowable limit. (Copyright © Cleveland Clinic Foundation.)

have varying degrees of conductivity to earth. Leakage current flowing from the case of one instrument to ground may put the case of that instrument at a different electrical potential than the case of piece of electrical equipment in the same room that has a different amount of leakage current or a different connection to ground at a different potential. Should the patient come into contact with these two grounds an electrical current can flow through the patient from one ground to the other.

In hospitals that have undergone renovation over a period of time, it is not unusual to find more than one electrical service within the same patient care area, as illustrated in Figure 4–10. A new service may have been added to handle the additional power demanded since the original construction. This second utility service may come from a different part of the hospital or have a different route to the basement, and the electrical potential of its ground conduction may therefore differ from the potential of the ground conductor of the older service by 1 V or more.

Current Sources Remote from the Patient

Especially insidious is the effect of a significant ground fault current flowing in a remote device. For example, referring to Figure 4–12, a large vacuum cleaner or floor buffer being used in a hallway outside a patient's room may be plugged into the same circuit as the electrical monitor inside the patient's room. These devices with powerful motors have notoriously high leakage current, even up to 1 A in this example. Although this leakage current is insufficient to trip the 20-A circuit breaker, the 1-A current flowing through 20 feet of number 14 ground wire is sufficient to produce a voltage drop of 50 mV. Monitoring equipment plugged into a receptacle at one end of the room, while the patient is grounded (perhaps via the bed) by a receptacle at the other end of the room, may produce enough current flow through a patient instrumented with indwelling catheters to cause a fatality. Service outlets within the patient room should be grouped to provide a single ground point, and no extension of the branch circuit outside the patient room should be permitted.

High-Current Faults

In addition to the low leakage currents, which may be introduced into a patient by an inappropriate wiring configuration, very high currents—though sufficient to trip a circuit breaker—can still place the patient at risk owing to the time factor. Circuit breakers are electromechanical devices that take on the order of 100 msec (as detailed in Table 4–5) to trip, sufficient time to expose a patient to a lethal current during the "vulnerable period" of the cardiac cycle.

Because all fault conditions cannot be anticipated, and because patients may come in contact with sources of dangerously high electrical current, monitoring equipment can include second order shock protection, especially in patients instrumented with indwelling lines. Current-limiters (devices that switch into a high resistance state when a large current is encountered, as diagrammed in Fig. 4–13) can be provided to guard against unexpected fault conditions.

Stimulators

All other stimulation devices, including defibrillators and electrocautery units, should have their outputs galvanically isolated from ground. If their large-amplitude stimulus pulses are not isolated, it is possible for current to flow through the neurophysiological recording electrodes to ground, producing burns or muscle stimulation at those sites. This would also normally be prevented by proper isolation of the recording equipment from ground.

TABLE 4–5	Over-Current Protection[3]	
Fault Current Level		**Time to Open**[a]
25 A		500 sec
40 A		30 sec
125 A		2 msec

[a]For a nominal 20-A circuit breaker.

CURRENT LIMITING DEVICES

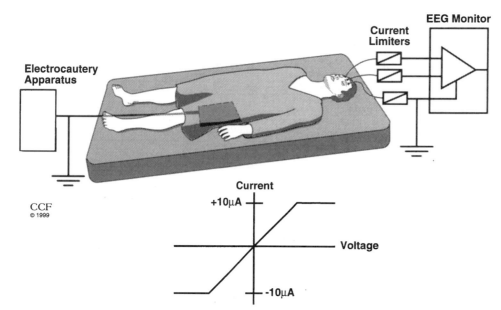

FIGURE 4–13. Where the patient may be subjected to very high currents, such as electrocautery in the operating room, current limiters can be used in series with the electrode leads. These devices have a characteristic, shown in the lower portion of the figure, such that the low voltages generated by physiological sources will induce appropriate currents into the amplifiers, but that increased voltages will not result in increased current conducted through the patient into the amplifiers. (Copyright © Cleveland Clinic Foundation.)

GUIDELINES FOR SAFE OPERATION OF CLINICAL NEUROPHYSIOLOGICAL EQUIPMENT

Coupling of electromagnetic sources to biological sinks cannot be completely eliminated (because of capacitive coupling and other high impedance connections); thus, no environment will be completely absent of danger to the patient. By following certain guidelines, however, the risk can be reduced to a negligible level.

Hospital Electrical Safety Guidelines

Power Receptacles. Use hospital-approved three-prong plugs, receptacles, and power cords. It should go without saying that the use of adaptors, "cheator cords," and even regular extension cords is taboo. This will ensure good contact with the mains power ground to route stray currents to earth ground in the event of an equipment fault condition. Make sure that all receptacles have been properly tested for adequate grounding and physical retention of the power cord.

Power Circuit. Make sure that the circuit is dedicated for patient use only. Avoid utilizing circuits serving extraneous equipment. Make sure that the power circuit is properly protected by a circuit breaker.

Single Power Source. Connect all of the patient-associated equipment in a room to one cluster of power receptacles. Medical "headwalls" with banks of receptacles all located together near the head of the bed are specifically designed to avoid accidentally inducing current into the patient from ground loops.

Patient Area. Remove devices with two-wire cords such as radios, hair dryers, and so on. Where possible, avoid the use of metal enclosures and other metal contact areas within reach of the bedside for electrically susceptible patients. Do not forget that medical personnel themselves may serve as connections from the patient to ground (see Fig. 4–9), for example, when a nurse is simultaneously touching a faucet and the patient.

Grounding. Do not connect the patient to earth ground.

In particularly vulnerable patients (i.e., those with indwelling catheters), all recording and stimulating leads (including the ground lead) must be individually isolated. Leakage through these cables may not be in excess of 10 μA. Use only equipment that has an isoground connection to the patient, and do not run jumpers or alligator-clipped leads to water pipes.

Equipment Condition. All patient care area electrical equipment should be periodically tested for leakage current. Any unusual appearance or function should prompt immediate removal of equipment from service until it has been rechecked for safe electrical function. This includes equipment that has been wet, equipment that has been physically abused, equipment that generates a tingling sensation when touched, equipment that is unusually hot or generates unusual odors, equipment with obviously damaged power cables, or equipment that suddenly develops an unusual interference pattern.

Equipment Operation. Always turn the equipment on before hooking up the patient and disconnect the equipment from the patient before turning the equipment off. Do not use extension cords. Employ battery-operated equipment where possible.

Patient. Contact between the patient and a metal bed or furniture, plumbing, metal architectural elements, cases of other hospital machines, or other mains-powered devices should be avoided. The technologist should take care to avoid providing an electrical pathway through his or her body, as shown in Figure 4–9. Current limiters in the ground leads of equipment which may be subject to substantial currents (e.g., in the operating room) should be provided. Because many of the susceptible patients are comatose or unable to communicate, remember that they will be unable to signify the subjective sensation of a shock.

Electrical Stimulation. When stimulating a patient, do not exceed recommended intensity or duration of stimulation. The stimulus delivery subsystem, very often separately packaged at the patient bedside, should be entirely isolated from building ground. Recording electrodes also should not be connected to building ground, but only through isoground. Ground electrodes and stimulus electrodes should be placed on the same limb. Stimulating electrodes should be placed close together, and the cardiac area should not be within the stimulating field.

Equipment Electrical Safety Testing. Make certain that the equipment has been checked for compliance with the applicable safety standards by the hospital (or shared medical services) biomedical engineering team. A sticker will be affixed to the equipment attesting to its safety or lack thereof. In most medical centers, a clinical engineering or otherwise authorized department carries out testing at regular intervals to determine electrical safety performance of all hospital equipment. The following tests for equipment grounding and leakage current should be carried out:

1. Visually inspect the power cord and plug for signs of damage or breakage.
2. Measure the resistance of the ground pin and the plug of the instrument chassis; the resistance should be less than one-tenth of an ohm.
3. Visually inspect all wall receptacles to make sure that they are secure and that the wiring is not exposed.
4. Measure the resistance between the wall receptacle grounds and a known adequate ground (e.g., a cold water pipe or grounding bus).
5. Measure the force required to withdraw a ground pin from wall receptacles; the contact tension should be greater than 10 ounces.
6. Measure the chassis leakage current under conditions of normal polarity, reverse polarity, with the equipment on, and with the equipment off; the leakage current should be less than 100 μA.
7. Measure the leakage current from each patient termi-

nal to ground; the current that can flow through all patient connections should be less than 20 μA.
8. Measure the leakage current from every hot terminal to each patient terminal; the leakage current should be less than 20 μA.

CONCLUSION

Although not frequently considered, electrical safety risk is one of the disadvantages of the increasingly instrumented patient. Not only are patients connected to multiple kinds of instruments, but this equipment has often been used or abused by many individuals in a variety of different locations. Proper and safe operation remains the first responsibility of the clinical neurophysiology laboratory.

REFERENCES

1. Association for the Advancement of Medical Instrumentation: AAMI/ANSI ESL—1985, Safe Current Limits for Electromedical Apparatus. Arlington, VA, AAMI, 1988.
2. Cromwell L, Weibell F, Pfeiffer E, Usselman L: Biomedical Instrumentation and Measurements. Englewood Cliffs, NJ, Prentice-Hall, 1973, pp 371–387.
3. Holt M: Neutrals and grounds in the 1999 code. Electrical Contractor 64:113–116, 1999.
4. Seaba P: Electrical safety. Am J EEG Technol 20:1–13, 1980.
5. Starmer C, McIntosh H, Whalen R: Electrical hazards and cardiovascular function. N Eng J Med 284:181–186, 1971.
6. Whitaker: Electric Shock As It Pertains to the Electric Fence. Kilpatrick Associates, 1971.

Chapter 5

Excitable Cells: The Ionic Basis of Membrane Potentials

Karen L. Barnes, PhD

MEMBRANE POTENTIAL CHARACTERISTICS
NONGATED ION CHANNELS GENERATE
  THE MEMBRANE POTENTIAL
SELECTIVE PERMEABILITY OF ION
  CHANNELS
DIFFUSION THROUGH NONGATED
  CHANNELS PRODUCES
  CONCENTRATION GRADIENTS
IONIC CONCENTRATION DIFFERENCES
  PRODUCE VOLTAGE GRADIENTS
EQUILIBRIUM POTENTIAL IN GLIAL CELLS

NEURONS: MULTIPLE IONS CONTRIBUTE TO
  THE RESTING POTENTIAL
VOLTAGE-GATED CHANNELS AND THE
  ACTION POTENTIAL
MODEL FOR VOLTAGE-GATED SODIUM
  CHANNELS
VOLTAGE-GATED POTASSIUM CHANNELS
CONTRIBUTION OF VOLTAGE-GATED
  SODIUM AND POTASSIUM CHANNELS
  TO THE ACTION POTENTIAL WAVEFORM
SUMMARY

All cells have an electrical voltage difference across their membranes, the *membrane potential,* which provides their source of energy, like a battery. Several kinds of pores in the membrane, called ion channels, are responsible for maintaining and altering the membrane potential. This chapter first discusses the mechanisms by which ion channels that are always open (nongated channels) generate the membrane potential, and how alterations in ionic flows through these channels modulate this potential. Comprehensive reviews are found in Kandel, Schwartz, and Jessell,[7] Kandel,[6] and Knowles and Lüders.[8]

Information is transmitted within glial cells and neurons in the nervous system by means of changes in the membrane potential, and between neurons by all-or-none electrical impulses that release chemical messages. Because glial cells contain only nongated ion channels, their membrane potential can vary only passively, as changes in the ionic concentrations of the extracellular fluid (ECF) or cell interior generate local *electrotonic potentials.* In contrast, neurons are "excitable" cells, because in addition to supporting these local electronic potentials, they have voltage-sensitive gated ion channels that enable them to propagate all-or-none *action potentials* along the membrane of their dendrites and axons and modulate the membrane potential of other cells at synaptic junctions. The action potential is a very efficient, noise-resistant mechanism for propagating information within and between cells. This chapter also examines the properties of the voltage-gated ion channels that contribute to generation and propagation of the action potential.[6, 8]

MEMBRANE POTENTIAL CHARACTERISTICS

In the nervous system, the electrical potential gradient between the interior of a neuron or glial cell and the surrounding ECF ranges between 60 and 100 millivolts (mV). The inside of the cell is negative with respect to the ECF. This potential gradient reflects a separation of positive and negative charges across the membrane: there are more positively charged ions on the outside of the membrane and more negative charges on the inner surface. Figure 5–1 shows that the membrane potential can be measured directly

by finding the difference between the voltages at an intracellular electrode and at a reference electrode outside the cell, where the potential is arbitrarily defined as zero. The membrane potential of a quiescent neuron or a glial cell is called the *resting potential.* If an inward current pulse is given through an intracellular electrode, it produces an increase in the charge separation across the membrane (*hyperpolarization*) because the inside of the cell has become *more* negative relative to the ECF (Fig. 5–2). In contrast, outward current pulses generate a decrease in the charge separation across the membrane, or *depolarization,* as the inside of a cell becomes *less* negative relative to the ECF. In neurons, if a depolarization is large enough to reach the cell's *threshold* level, an action potential is triggered.

NONGATED ION CHANNELS GENERATE THE MEMBRANE POTENTIAL

The membrane potential of a cell depends on its ability to maintain concentrations of ions within its cytoplasm that are different from those found in the ECF. Two forces maintain these concentration differences:

1. Ions cannot cross the cell membrane freely.
2. The channels in the membrane that allow ions to move between the cell interior and the ECF are selectively permeable to only certain ions.

Because the cell membrane is a nonpolar lipid bilayer through which polarized ions cannot pass, it insulates the cytoplasm from the ECF very effectively. However, the membrane contains complex protein structures that span the membrane and surround an aqueous pore: the ion channels. Although there are several types of ion channels in the membrane that are always open (nongated), the potassium (K^+) and sodium (Na^+) nongated channels are primarily responsible for establishing and maintaining the membrane potential.

SELECTIVE PERMEABILITY OF ION CHANNELS

The most important feature of the ion channels that generate and sustain the membrane potential in both neurons

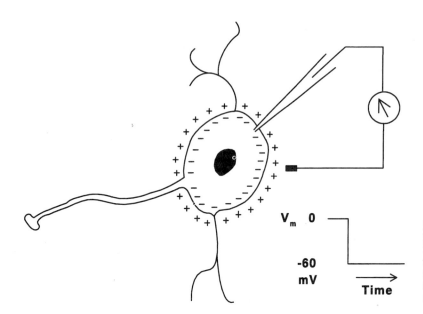

FIGURE 5–1. Intracellular recording of the membrane potential (V_m) from the cell body of a neuron. The resting potential of -60 mV is measured by a voltmeter connected between the intracellular electrode and a reference electrode in the extracellular fluid surrounding the neuron, where the voltage is defined as zero.

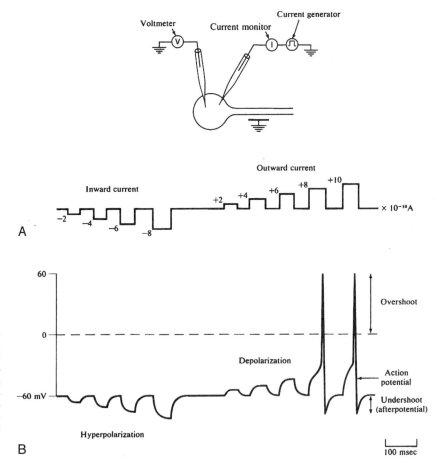

FIGURE 5–2. Intracellular stimulation. *A,* The membrane potential of a neuron is recorded while current pulses are injected into the cell body. *B,* Inward current pulses (left side) hyperpolarize the membrane; outward current pulses depolarize the membrane. If the depolarization reaches the neuron's threshold potential, an all-or-none action potential is triggered. (From Kandel ER, Schwartz JH, Jessell TM: Principles of Neural Science, 3rd ed. New York, Elsevier, 1991, p 101.)

and glial cells is their *selective permeability,* which allows only one species or family of ions to pass through their pores. This selectivity for ions is determined by both the physical size and the structure of the pore, as well as by the amino acid composition of the proteins that constitute the channel walls. First, since ions differ in physical size, in part due to the number of water molecules they carry in their hydration shell, the diameter of the channel pore determines the maximum diameter of the ions that can traverse it. Second, the interior of the channel has a net charge, determined by the positive or negative charge of the hydrophilic (polar) amino acid residues of the proteins that line the pore, which makes the channel selective for either negative or positive ions, respectively. Although most channels allow passage of only one species of ion, some channels are selective for a family of ions, such as all small positive ions.[3]

TABLE 5–1	Intracellular and Extracellular Concentrations in Millimoles and Hydration Number of Ions at the Resting Membrane Potential of Mammalian Excitable Cells

Ion Species	Intracellular Concentration (mM)	Extracellular Concentration (mM)	Hydration Number
K^+	155	4	2.9
Na^+	12	145	4.5
Cl^-	4.2	123	2.9
Organic Anions	140	—	—

DIFFUSION THROUGH NONGATED CHANNELS PRODUCES CONCENTRATION GRADIENTS

The *kinetic energy* of an ion causes it to undergo random motion *(diffusion),* which disperses each species evenly within the cell interior or the ECF (Fig. 5–3). The rate of migration of an ionic species is a function of its *mobility,* or capacity to move through its fluid environment. The mobility of an ion is determined by its functional size, which depends on the number of water molecules it carries in its shell, indicated by its *hydration number.* Although elemental Na^+ ions are physically smaller than K^+ ions, in the brain Na^+ carries more water molecules in its hydration shell than K^+, which reduces the mobility of Na^+ in aqueous solutions and makes the ion too large to traverse K^+ channels (Table 5–1).[9]

Table 5–1 shows that in mammalian excitable cells none of the major ionic species are in equal concentrations on each side of the membrane. A key factor in the development of concentration gradients across the membrane is the pool of negatively charged amino acids and proteins trapped within the cell because they are too large to traverse any ion channels (organic anions A^-; see Fig. 5–3). These anions attract the positively charged extracellular ions (cations; K^+, Na^+) into the cell and drive chloride (Cl^-) ions out. Because most of the open channels in the resting neuron are selective

for K^+, this ion is rapidly drawn into the cell. This influx of K^+ establishes a voltage gradient across the membrane, which then tends to force Na^+ out of the cell.

IONIC CONCENTRATION DIFFERENCES PRODUCE VOLTAGE GRADIENTS

The ionic concentration differences across the cell membrane result in two forces that act on each ion. First, a *diffusional force* produced by the concentration gradient tends to move ions from the compartment with the higher concentration to the side with the lower concentration. Thus, K^+ tends to diffuse outward, while both Na^+ and Cl^- are attracted into the cell (Fig. 5–4). However, the rate at which the diffusion force moves an ion across the membrane depends on the relative permeability of the membrane to that species, which is proportional to the fraction of all the membrane channels that are selective for the ion.

The simplest example of how ionic concentration gradients generate the membrane potential is the glial cell, because all its open nongated channels are selective for K^+ (Fig. 5–5). The concentration gradient drives K^+ ions from the higher density inside the cell out into the lower concentration of the ECF, and it produces a net transfer of positive charge to the outside of the membrane. Because the trapped

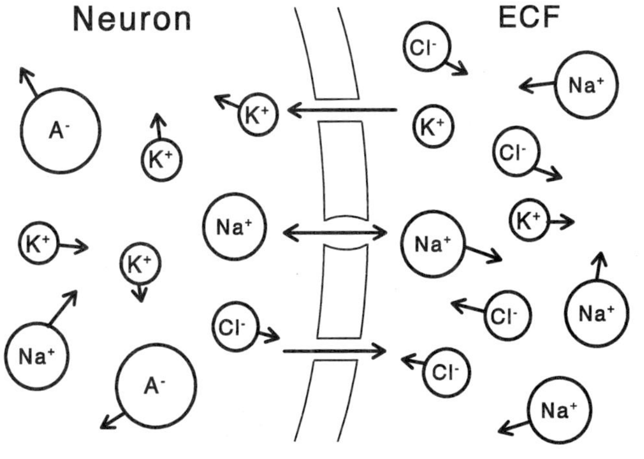

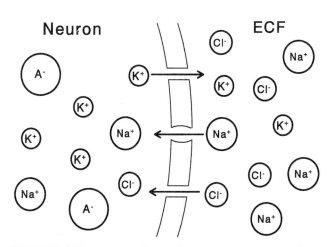

FIGURE 5–3. Diffusion of ions within the cytoplasm of a neuron and in the ECF. The functional size of each ion is indicated by the size of the circle; its kinetic energy is represented by an arrow. The large anions (A^-) trapped inside the membrane attract Na^+ and K^+ into the cell, and drive Cl^- into the ECF.

FIGURE 5–4. The concentration gradient tends to move an ion from the region where its concentration is higher across the membrane to the area with a lower concentration: K^+ is driven out of the neuron, whereas both Na^+ and Cl^- will diffuse into the cell.

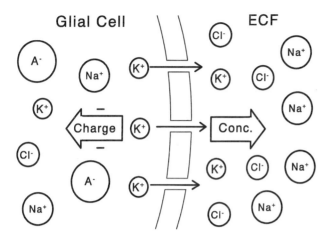

FIGURE 5-5. Glial cell membranes have only functional K⁺ channels. The concentration gradient that drives K⁺ out of the cell produces a voltage gradient attracting positive charges into the cell. The glial cell reaches equilibrium when the inward force of the charge gradient is equal and opposite to the outward force of the concentration gradient.

organic anions (A⁻) generate a net negative charge inside the cell, the separation of charge establishes a voltage gradient across the membrane. This voltage gradient exerts the second force on each ion: the *electrostatic force*. The charge separation attracts positively charged K⁺ and Na⁺ ions to the negative intracellular charge, and negatively charged ions to the extracellular side of the membrane.[6, 7]

EQUILIBRIUM POTENTIAL IN GLIAL CELLS

In glial cells, K⁺ ions continuously diffuse through nongated channels out of the cell, down their concentration gradient. However, because the loss of K⁺ increases the positive charge on the outside of the cell as well as the negativity inside the cell, the outward flow of K⁺ ions decreases. As shown in Figure 5-5, the outward movement of K⁺ ions will stop when the inward force of the voltage gradient that attracts K⁺ to the inside of the membrane balances the outward force of the concentration gradient that pulls K⁺ into the extracellular compartment. At this point the glial cell has reached *equilibrium,* when the diffusional force is equal and opposite to the electrostatic force. When the cell is in equilibrium, equal numbers of K⁺ ions move in both directions through the nongated K⁺ channels.

If we know the concentrations of an ionic species inside and outside the cell, we can calculate the equilibrium voltage at which the diffusional and electrical forces on that ion will be symmetrical. The *Nernst equation* uses basic principles of thermodynamics to convert the concentration gradient of an ionic species into its equivalent voltage gradient, or *equilibrium potential* (E_{ion}), in volts:

$$E_{ion} = \frac{RT}{ZF} \ln \frac{[ion]_o}{[ion]_i} V$$

where R is the universal gas constant, T is temperature in degrees Kelvin, Z is the valence of the ion, F is the Faraday constant, and $[ion]_o$ and $[ion]_i$ are the concentrations of the ion outside and inside the cell, respectively.[9] If we convert this expression from natural to base 10 logarithms and from volts to millivolts, at 37°C, the constant equals 62. Substituting the concentrations for K⁺ from Table 5-1, in a glial cell

at a temperature of 37°C, the equilibrium potential for K⁺ is given by

$$E_K = 62 \ln \frac{[K^+]_o}{[K^+]_i} mV = 62 \ln \frac{4}{155} mV = -98 \, mV$$

We find that a voltage gradient of −98 mV across the cell membrane would exactly balance the diffusion gradient produced by these concentrations of K⁺. This calculated equilibrium potential is very close to the membrane potential of −100 mV actually measured in glial cells. Because this equilibrium depends only on the temperature of the cell, the ionic concentrations on each side of the membrane, and passive ionic diffusion through nongated channels, it does not require metabolic energy for its maintenance. If the concentration gradient for K⁺ is reduced by increasing the amount of extracellular K⁺, the membrane potential will decrease, depolarizing the cell. Because only about 0.0005% of the ions inside the cell actually move into the ECF to establish the membrane potential, there is no measurable change in the ratio of the ionic concentrations.

NEURONS: MULTIPLE IONS CONTRIBUTE TO THE RESTING POTENTIAL

Although the majority of nongated channels that are open in the resting neuron are selective for K⁺, some Na⁺ and Cl⁻ channels are also open. We can consider the neuron as a glial cell with some Na⁺ channels added to the membrane. The Nernst equation shows us that if only Na⁺ channels were open in the membrane of a neuron, the equilibrium potential for Na⁺ would be

$$E_{Na^+} = 62 \ln \frac{[Na^+]_o}{[Na^+]_i} mV = 62 \ln \frac{145}{12} mV = +67 \, mV$$

If the membrane potential of a neuron were set at the equilibrium potential for K⁺ (−98 mV), both the concentration gradient for Na⁺ and the strong electrostatic force of the membrane potential (nearly 170 mV) would move Na⁺ ions into the cell, making the membrane potential *more positive* (Fig. 5-6). But because the K⁺ channels are also open, the concentration gradient for K⁺ tends to move posi-

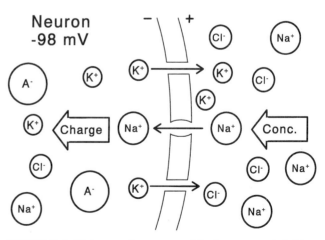

FIGURE 5-6. If the membrane potential of a neuron is held at −98 mV, both the concentration gradient for Na⁺ and the 170-mV difference between the equilibrium potential for Na⁺ (+67 mV) and the membrane potential will drive Na⁺ into the cell.

tive charge out of the cell, making the membrane potential *more negative.* In order to determine where the membrane potential will stabilize, we must be able to calculate the net effect of the concentration gradients for Na$^+$, K$^+$, and Cl$^-$ on the membrane potential, which depends on the relative permeabilities of the membrane for these ions.

While studying the squid giant axon more than 40 years ago, Hodgkin and Katz[5] found that the *Goldman equation*[1] can predict the resting membrane potential of a neuron (V_m) as a function of the concentration gradients of Na$^+$, K$^+$, and Cl$^-$ and their relative permeabilities P_{Na}, P_K, and P_{Cl}:

$$V_m = \frac{RT}{F} \ln \frac{P_K [K^+]_o + P_{Na} [Na^+]_o + P_{Cl} [Cl^-]_i}{P_K [K^+]_i + P_{Na} [Na^+]_i + P_{Cl} [Cl^-]_o}$$

Very simply, ions with higher concentration gradients or permeabilities, or both, have a greater effect on the membrane potential. Although this equation looks complicated, it is really just the sum of the Nernst equations for each individual ion, with each concentration multiplied by the membrane permeability for that ion. Let us look at a few examples. First, because glial cell membranes are almost impermeable to Na$^+$ and Cl$^-$, the terms for these ions equal zero. Thus, the equation reduces to the Nernst equation for K$^+$, which we know predicts the membrane potential for glial cells as -98 mV. Second, because in most neurons the resting membrane is about 25 times more permeable to K$^+$ than to Na$^+$, the membrane potential should be closer to -98 mV than to $+67$ mV. In addition, the neuronal membrane permeability to Cl$^-$ is extremely high. Therefore, the concentration gradient for Cl$^-$ quickly adapts to the existing membrane potential, and its contribution to the resting potential is negligible. If the Goldman equation is solved for K$^+$ and Na$^+$ with their usual concentrations inside and outside the neuron, a resting membrane potential of about -60 mV is predicted (Fig. 5–7). This resting potential is very close to values obtained during intracellular recordings from a wide range of neurons.

At the neuron's resting potential of -60 mV, both K$^+$ and Na$^+$ ions are far from their equilibrium potential. Thus their concentration gradients generate an outward diffusion of K$^+$ and inward flow of Na$^+$, which slowly depletes the cell of K$^+$ and adds excess Na$^+$ ions. Although there is no net change in charge separation to alter the resting potential, the concentration gradients of both K$^+$ and Na$^+$ will slowly decrease,

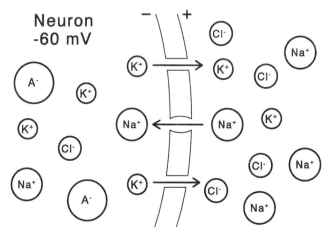

FIGURE 5–7. Solving the Goldman equation for the concentrations of K$^+$ and Na$^+$ inside and outside the neuron reveals that the diffusion forces are balanced in a neuron with a membrane potential of -60 mV.

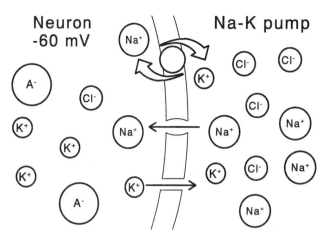

FIGURE 5–8. The Na$^+$-K$^+$ ATPase pump maintains the ionic concentration gradients in neurons required for a stable resting potential.

and would ultimately abolish the membrane potential unless the cell is able to restore the ionic concentration gradients. However, Figure 5–8 shows that neuronal membranes contain an Na$^+$-K$^+$ ATPase pump that uses energy from adenosine triphosphate to transport Na$^+$ ions out of the cell and K$^+$ ions back inside, against their concentration gradients. Because this metabolically dependent pump generates active flows of Na$^+$ and K$^+$ ions that are equal and opposite to their passive flows, it maintains the ionic balance of the neuron required for a stable membrane potential.

VOLTAGE-GATED CHANNELS AND THE ACTION POTENTIAL

Since the permeabilities of the nongated ion channels that generate the resting potential remain constant, regardless of the level of the membrane potential, the steady ionic flows keep the membrane potential stable. However, the membrane also contains gated channels that are selective for Na$^+$ or K$^+$ ions and have permeabilities that change dramatically as the transmembrane voltage gradient varies. If the membrane is either at the resting potential or hyperpolarized, the "gates" in the pores of these channels are closed. When a brief depolarizing potential from an excitatory synaptic input depolarizes the membrane to the threshold for these channels, the gates open. Because the voltage-gated channels for Na$^+$ and K$^+$ differ in their voltage thresholds and have a range of times for opening and closing, the permeability changes to each ion summated over the membrane have different time courses. Let us examine the mechanisms by which these voltage-gated ion channels enable the neuron to generate and transmit action potentials along its processes.[6, 7]

The Goldman equation, which helped us find the resting potential for a neuron, also predicts that if the relative permeability of the membrane to an ionic species increases, that ion will move down its electrochemical gradient, and the membrane potential will shift toward the Nernst potential for the ion. Therefore, if the permeability of voltage-gated Na$^+$ channels increases, Na$^+$ will enter the cell and depolarize the membrane. However, if the permeability of K$^+$ channels increases, K$^+$ will move out of the cell, hyperpolarizing the membrane. Because the permeabilities of these gated channels both are altered by the voltage across the membrane and produce changes in membrane voltage, the membrane depolarizes rapidly by a feedback process. Hodgkin and Huxley[4] derived an elegant set of differential equations that describe the time and voltage dependencies of the gates in these Na⁺ and K⁺ channels and predict the ionic flows that underlie the changes in membrane voltage that comprise the action potential. Although the actual structures of these channels are still unknown, the equations suggest the following models for voltage-gated Na⁺ and K⁺ channels.

MODEL FOR VOLTAGE-GATED SODIUM CHANNELS

Voltage-gated Na⁺ channels are modeled as having three discrete states. When the neuron is at its resting potential, these channels are in their *resting state,* with the gates closed so that Na⁺ cannot pass through. If the membrane is depolarized by excitatory synaptic inputs, the channel shifts to the *active state,* and the gates open. As Na⁺ flows rapidly into the cell and depolarizes the membrane further, this feedback cycle quickly drives the membrane toward the Nernst potential for Na⁺ (+67 mV). However, after about 1 msec in the active state, the Na⁺ channel closes and shifts to the *inactivated state.* This state differs from the resting state because the channel cannot reopen as long as the membrane remains depolarized. After the membrane has returned to the resting potential for a few milliseconds, the channel reverts to the resting state and is ready to open again if the membrane is depolarized.

The present model for the voltage-gated Na⁺ channel is based on the assumption that the channel has two internal gates, one that activates the channel and the other that inactivates it (Fig. 5–9). Because closing either gate blocks passage of ions through the channel, both gates must be open for Na⁺ to cross the membrane. In the resting state, the inactivating gate is open, but the activating gate is closed, so Na⁺ cannot cross. Depolarization of the membrane opens the activating gate, allowing Na⁺ ions to diffuse down their concentration gradient into the cell and further depolarize the membrane. After 1 msec, the inactivating gate closes and blocks the channel. When other ionic flows cause the neuron to repolarize, the voltage-sensitive activating gate also closes. However, the channel remains inactivated until the inactivating gate reopens several milliseconds later and returns the cell to the resting state. Only then can a depolarizing stimulus reactivate the voltage-gated Na⁺ channel.

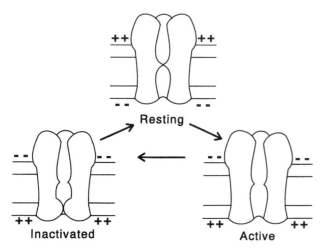

FIGURE 5–9. A model for the three states of the voltage-gated Na⁺ channel.

VOLTAGE-GATED POTASSIUM CHANNELS

Unlike the Na⁺ channel, the voltage-gated K⁺ channel appears to have only two states and is modeled with a single gate, which is closed at the resting potential. Like the voltage-gated Na⁺ channel, depolarization of the membrane opens the gate. However, the K⁺ gate opens much more slowly than the Na⁺ channel, only after a delay of about a millisecond. Then K⁺ ions flow out of the cell, down their concentration gradient, and start to reverse the depolarization generated by the rapid influx of Na⁺. When the K⁺ outflow returns the membrane to the resting potential, the K⁺ gate closes.

CONTRIBUTION OF VOLTAGE-GATED SODIUM AND POTASSIUM CHANNELS TO THE ACTION POTENTIAL WAVEFORM

The membrane voltage changes that form the action potential depend on the characteristics of the voltage-gated Na⁺ and K⁺ channels described earlier. Figure 5–10 shows the time courses of the action potential and of the Na⁺ and K⁺ currents that generate it.[2] That voltage-gated Na⁺ channels are responsible for the explosive depolarization of the membrane potential that forms the action potential "spike" has been verified by documenting that removing Na⁺ ions from the ECF or blocking the channels pharmacologically prevents the spike. Any input that depolarizes the membrane will cause the voltage-sensitive Na⁺ channels with the lowest thresholds to open. Na⁺ ions enter the cell and depolarize the neuron further, which opens Na⁺ channels with higher thresholds. This positive feedback process quickly opens all the voltage-gated Na⁺ channels, and the membrane potential actually reverses, or *overshoots,* approaching the Nernst potential for Na⁺ (+67 mV). After about a millisecond of this rapid depolarization phase, the Na⁺ channels begin to inactivate and close, while at the same time the voltage-gated K⁺ channels begin to open. As the membrane permeability to Na⁺ decreases and that to K⁺ increases, the net current flow begins to shift from inward Na⁺ to an outward K⁺ flux, which drives the membrane potential back toward the resting potential. Thus, the voltage-gated K⁺ channels are responsible for the repolarization of the membrane potential following the "spike," because they enable K⁺ ions to leave the neuron.

In many neurons, the membrane potential actually becomes more negative than the resting potential for several milliseconds, producing an *afterhyperpolarization.* Two factors contribute to this hyperpolarization. First, some of the K⁺ channels do not close for several milliseconds after the membrane potential returns to its resting voltage, and the continuing outflow of K⁺ shifts the membrane to a more negative voltage than the resting potential. Second, because the Na⁺ channels are inactivated at this time, the membrane permeability to Na⁺ is lower than in the resting state, which causes the influx of Na⁺ to be less than normal. However, when the K⁺ channels close and the Na⁺ channels are released from inactivation, the membrane finally returns to the resting potential.

These properties of voltage-gated ion channels also determine several other important features of neuronal behavior:
1. The *threshold voltage* at which depolarization of the membrane produces an action potential is that voltage at which the inward Na⁺ current that depolarizes the membrane becomes greater than the steady state outward K⁺ current that repolarizes the membrane, and it

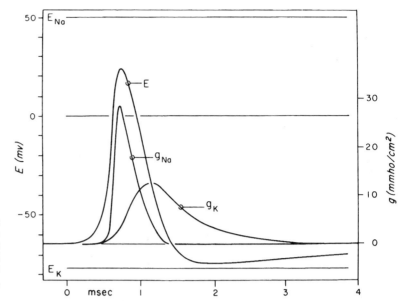

FIGURE 5–10. Diagram of the time course of the action potential *(E)* superimposed on the Na$^+$ (g_{Na}) and K$^+$ (g_K) currents that produce it. (From Hille B: Ionic basis of resting and action potentials. In Kandel ER (ed): Cellular Biology of Neurons, 2nd ed, vol I. Bethesda, American Physiological Society, 1977, p 119.)

sets off the explosive positive feedback process that opens all the Na$^+$ channels.

2. Slowly rising depolarizations have a higher threshold voltage for generating an action potential than rapid depolarizations, because during the depolarization some of the K$^+$ channels begin to open, offsetting the effect of the Na$^+$ channel openings until a greater level of depolarization is achieved.

3. There is a period of several milliseconds after an action potential during which a second spike cannot be evoked, regardless of the magnitude of the depolarizing stimulus. This *absolute refractory period* is the time when the voltage-gated Na$^+$ channels are in the inactivated state and cannot reopen until the inactivating gates have closed. The absolute refractory period is followed by the *relative refractory period,* another several milliseconds when the threshold for evoking an action potential is much higher than normal. This period corresponds to the time when the voltage-gated K$^+$ channels remain open and keep the membrane hyperpolarized.

SUMMARY

Two mechanisms are responsible for the generation of the resting potential: (1) The selective permeability of the membrane to ions due to the characteristics of nongated ion channels, and (2) the concentration gradients for K$^+$ and Na$^+$ across the resting membrane. The resting potential of glial cells is determined solely by nongated K$^+$ channels, because no other ion channels are open at the resting potential. The Nernst equation for each ion predicts the membrane potential at which the voltage gradient is equal and opposite to the concentration gradient for that ionic species. The Goldman equation indicates how the relative membrane permeabilities of each ion determine the contribution of each ionic species to the resting potential. Because the membrane of a neuron is much more permeable to K$^+$ than to Na$^+$ in the resting state, its resting potential is close to the Nernst potential for K$^+$ and far from the Nernst potential

for Na$^+$. The nongated channels allow K$^+$ ions to diffuse out of the cell and Na$^+$ to enter the neuron, down their respective concentration gradients. However, these passive flows are counterbalanced by the metabolic Na$^+$-K$^+$ pump, which restores the ionic balances. In contrast, the membrane permeability for Cl$^-$ ions is so high that Cl$^-$ distributes passively according to the membrane potential and has little influence on the resting potential.

The action potential is generated by voltage-gated Na$^+$ and K$^+$ channels whose permeability changes with the membrane voltage. Depolarizing inputs open Na$^+$ channels, generating an explosive influx of Na$^+$ that depolarizes and then reverses the membrane potential. After about a millisecond, the Na$^+$ channels inactivate and close, while the K$^+$ channels finally begin to open in response to the depolarization. The outward K$^+$ current repolarizes the membrane and usually produces an afterhyperpolarization. This negative afterpotential slowly returns to the resting level after several milliseconds, when the Na$^+$ channels shift from inactivation to the resting state and restore the resting level of Na$^+$ influx.

REFERENCES

1. Goldman DE: Potential, impedance, and rectification in membranes. J Gen Physiol 27:37, 1943.
2. Hille B: Ionic basis of resting and action potentials. In Kandel ER (ed): Cellular Biology of Neurons, 2nd ed, vol I. Bethesda, American Physiological Society, 1977, p 99.
3. Hille B: Ionic Channels of Excitable Membranes, 2nd ed. Sunderland, MA, Sinauer Associates, 1992.
4. Hodgkin AL, Huxley AF: A quantitative description of membrane current and its application to conduction and excitation in nerve. J Physiol 117:500, 1952.
5. Hodgkin AL, Katz B: The effects of sodium ions on the electrical activity of the giant axon of the squid. J Physiol 108:37, 1949.
6. Kandel ER: Cellular Basis of Behavior. San Francisco, W.H. Freeman and Co, 1976, pp 95–163.
7. Kandel ER, Schwartz JH, Jessell TM: Principles of Neural Science, 3rd ed. New York, Elsevier, 1991, pp 66–234.
8. Knowles WD, Lüders HO: Normal neurophysiology: The science of excitable cells. In Wyllie E (ed): The Treatment of Epilepsy: Principles and Practice. Malvern, PA, Lea & Febiger, 1993, p 71.
9. Miles FA: Excitable Cells. London, W. Heinemann, 1969, pp 12–70.

Chapter 6

Excitable Membranes: Local Responses and Propagation

Karen L. Barnes, PhD

INTEGRATION OF SYNAPTIC INPUTS: CABLE
   PROPERTIES
LOCAL ELECTROTONIC POTENTIALS
TIME CONSTANT
SPACE CONSTANT
SIGNAL AMPLIFICATION
FEED-FORWARD AND FEEDBACK
   INHIBITION

ACTIVE PROPERTIES OF DENDRITES
NEURONAL MECHANISMS OF SYNAPTIC
   PLASTICITY: THE CELLULAR BASES OF
   MEMORY?
SUMMARY

Neurons convey information both along their membranes and across synaptic junctions with neurons or effector cells primarily by means of electrochemical signals. This capacity to transmit information depends upon the neuron's ability to establish and renew the high-energy resting state that serves as its battery, as was discussed in Chapter 5. Afferent (sensory) inputs, which are located primarily on dendrites, evoke graded, locally decrementing changes in the postsynaptic membrane potential. As the postsynaptic potential spreads along the membrane, it is integrated with graded local membrane potential changes produced by other synaptic inputs and may be amplified or filtered as it travels across dendritic branch points and into the cell body. When the integrated sum of all postsynaptic potentials produces a depolarization of the axon initial segment that exceeds its threshold level, action potentials are generated and propagated to the axon terminals, where the information is transferred to other neurons or effector cells.

The information-processing capacities of neurons have traditionally been modeled by the functions of a passive electrical network that integrates and filters the analog signals from synaptically generated postsynaptic potentials and transforms the result into the all-or-none output of action potentials.[6, 8] However, it is now clear that neurons process information by both passive and active processes. The *passive mechanisms* involve the local electrical properties of the membrane, modeled as resistances and capacitances, that enable both spatial and temporal summation of the electrotonic changes in postsynaptic membrane potential that are evoked by synaptic inputs from afferent terminals. *Active mechanisms* include propagation of the action potential along the axon, signal amplification or attenuation, and synaptic transmission. However, because generation of an action potential at the axon initial segment generally requires passive summation of multiple synaptic inputs as well as signal amplification before the membrane depolarization reaches threshold, the passive and active processes are closely interdependent.[6]

INTEGRATION OF SYNAPTIC INPUTS: CABLE PROPERTIES

The passive properties of the membrane result from its structure as an insulating lipid bilayer, which contains aqueous pores and separates two conductive media, the cytoplasm and the extracellular fluid (ECF). The *cable properties* of the resting membrane can be modeled as an electrical circuit with both *resistance* (R_m), which represents the restriction on the movement of ions produced by the membrane and its open ion channels, and *capacitance* (C_m), which indicates the amount of electrical charge, q, that can be stored on the membrane (Fig. 6–1). The conductive pathways on either side of the membrane through the cytoplasm and the ECF have very low resistances to current flow compared to the transmembrane resistance.

LOCAL ELECTROTONIC POTENTIALS

Because the relationship of membrane resistance to transmembrane current and voltage follows Ohm's law, we can determine R_m experimentally by injecting graded hyperpolarizing current pulses into the cell and calculating the ratio of the change in membrane potential (V_m) to the input current: $\Delta V_m/I_m = R_m$. Figure 6–2A shows that V_m changes much more slowly than the rectangular current pulse. Because the membrane capacitance and resistance are connected in parallel across the membrane (Fig. 6–2B), the initial inward negative current alters or even reverses the charge on the capacitive membrane elements (C_m). As an increasing portion of the current shifts to the resistive ion channels, V_m slowly rises as the ionic charges on each side of the lipid bilayer are altered. C_m is larger in neurons with greater surface area and is inversely related to V_m by the charge, q, on the membrane.

Synaptic inputs to the membrane produce a local *electrotonic potential* (ETP), which decreases in size with distance from the stimulus (Fig. 6–3). Inhibitory synapses *hyperpolarize* the membrane and generate a negative ETP. Excitatory synapses *depolarize* the membrane and evoke a positive ETP. Although both hyperpolarizing and depolarizing inputs produce local ETPs, only depolarizing stimuli can reduce the membrane potential to the *threshold voltage* for generating an all-or-none *action potential*. Figure 6–3 shows that as the distance from a stimulation site increases, the change in membrane potential evoked by current injection decays quickly.[7] Thus, the graded ETPs that are generated by synaptic inputs extend only a few millimeters along the membrane. Because the threshold for evoking an action potential is very high in the dendrites and cell body, depolarizing ETPs must reach

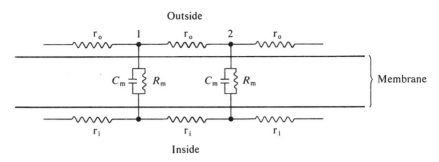

FIGURE 6–1. The electrical equivalent circuit of the neuronal membrane. Membrane resistance (R_m) and capacitance (C_m) are connected in parallel across the membrane; r_o = extracellular fluid resistance; r_i = intracellular cytoplasm resistance. (Adapted from Kandel ER: Cellular Basis of Behavior. San Francisco, W.H. Freeman and Co., 1976, p 119.)

the axon initial segment before the cell will fire an action potential that propagates along the axon. However, neurons have both spatial and temporal properties that enable multiple synaptic inputs to summate in order to depolarize the axon to its threshold: the *time constant* and the *space constant*.

TIME CONSTANT

The time constant, τ, is the product of R_m times C_m and reflects how long it takes to charge or discharge the membrane capacitance. τ denotes the time required for a prolonged stimulus to change the membrane voltage by 63% of the maximum amplitude produced by constant stimulation. Figure 6–4 shows that if a 10-mV depolarizing input takes 100 msec to depolarize the membrane potential from −60 mV to −53.7 mV in a neuron, the cell is considered to have a time constant of 100 msec. Dendrites and cell bodies with larger diameters and greater membrane surface have higher

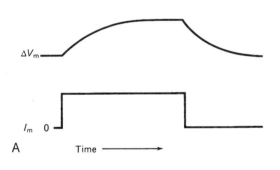

A

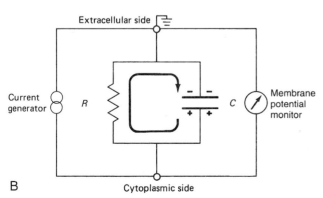

B

FIGURE 6–2. The effect of a rectangular current pulse on membrane potential. *A*, The time required to alter the charge of the membrane capacitance C_m slows the rise of and fall of the membrane potential (ΔV_m) in response to a rectangular current pulse (I_m). *B*, At the end of the current pulse, V_m falls gradually as C_m discharges through R_m. (Adapted from Kandel ER, Schwartz JH, Jessell TM: Principles of Neural Science, 3rd ed. New York, Elsevier, 1991, p 96.)

capacitance, charge more slowly, and have a larger value of τ than smaller neurons. Thus in larger cells, successive input signals are more likely to produce *temporal summation* and depolarize the axon initial segment sufficiently to generate an action potential than in a smaller cell with a short τ (Fig. 6–5).

SPACE CONSTANT

The influence of synaptic inputs that are spatially separated along the dendritic tree or soma on the membrane potential of a neuron is amplified by the mechanism of *spatial summation*. The graded ETP produced by a synaptic input is maximal at the synapse. Its amplitude decays with increasing distance along the membrane from the synapse, because a portion of the intracellular current leaks out through open ion channels in the membrane. The rate of the decay in the ETP with distance can be expressed as a function of the space constant, λ, the distance over which a synaptically evoked ETP decays to 37% of its original amplitude (Fig. 6–6). Typical neuronal space constants range from 100 to 1000 μm, depending on what fraction of the total synaptic current leaks out through the ion channels in the membrane compared to the current that flows along the inside of the dendrite or soma membrane.

The space constant, λ, is determined by the square root of the ratio of the membrane resistance, R_m, to the axial resistance to current flow inside the dendrite or soma, R_a:

$$\lambda = \sqrt{\frac{R_m}{R_a}}$$

This equation indicates that dendrites with larger diameter (smaller R_a) or higher membrane resistance (R_m), or both, have a larger space constant, so that synaptically generated ETPs extend farther along their membrane toward the soma and the axon initial segment.

Synapses on the dendritic tree obviously vary in their influence on the membrane potential at the soma and axon initial segment, depending on their distance from the cell body. Inhibitory synapses on distal dendrites generally have relatively little influence on the output of the cell. The probability that a neuron will fire depends on the combination of the spatial density and temporal pattern of synaptic inputs to the neuron and the passive electrotonic properties, τ and λ, of its membranes, all of which determine the likelihood that a given pattern of inputs will generate an action potential.

The magnitudes of τ and λ also play an important role in determining the conduction velocity of unmyelinated axons. Sequential regeneration of the action potential along the axonal membrane depends on the spread of current from previously activated regions of the membrane into quiescent areas. As the value of λ increases, electrotonic depolarizing

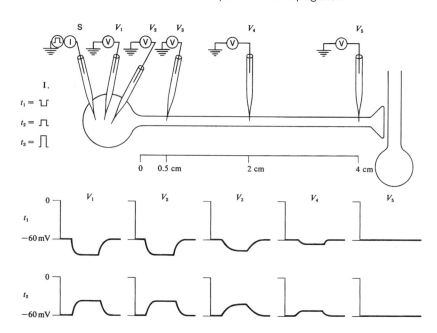

FIGURE 6–3. Propagation of graded local electrotonic potentials (ETPs) along the axonal membrane. The magnitude of both hyperpolarizing (t_1) and depolarizing (t_2) ETPs decreases exponentially with the distance of the recording electrode from the stimulus. (Adapted from Kandel ER: Cellular Basis of Behavior. San Francisco, W.H. Freeman and Co., 1976, p 124.)

currents extend farther along the axon, and the action potential will propagate more quickly. As in the case of dendrites, axons with a larger diameter have a lower R_a and thus a larger value of λ.

SIGNAL AMPLIFICATION

Thus far the discussion has considered only the classic properties of neuronal membranes as passive signal integrators, but their gated ion channels enable neurons to actively modify signal transmission both within a cell and at synaptic junctions. Signal transfer between cells in the brain is primarily mediated by chemical transmission at synapses from axon terminals to other neurons. However, not all signals are transferred from one cell to the next unaltered: some are attenuated, and others may be amplified. Signal amplification results from both the active properties of the

individual neuron and the characteristics of its synaptic inputs. At the level of the neuronal membrane, amplification takes place primarily by activation of calcium (Ca^{2+}) channels such as those associated with the NMDA ionotropic excitatory amino acid receptor. For example, Figure 6–7 shows that release of L-glutamate from a presynaptic terminal in the hippocampus will evoke a local depolarizing ETP in a dendritic spine on the postsynaptic neuron. This initial depolarization of the postsynaptic membrane causes the release of magnesium (Mg^{2+}) ions that block Ca^{2+} flux through Ca^{2+} channels associated with NMDA receptors at the resting potential. Another stimulus that releases L-glutamate close in time or space to the initial stimulus will then permit Ca^{2+} ions to enter the dendrite through the NMDA receptor channel and will produce a greater depolarization of the membrane than did the first stimulus. Thus, the effect of the L-glutamate released by the second stimulus is amplified by the influx of Ca^{2+}, and it increases the likelihood that an action potential will be generated. These mech-

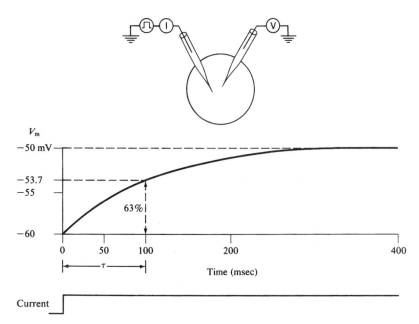

FIGURE 6–4. Time constant. A constant 10-mV depolarizing stimulus decreases the membrane potential from −60 mV to −53.7 mV in 100 msec in a neuron with $\tau = 100$ msec. (From Kandel ER: Cellular Basis of Behavior. San Francisco, W.H. Freeman and Co., 1976, p 122.)

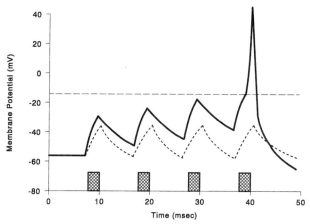

FIGURE 6–5. Equal synaptic inputs *(hatched bars)* to neurons with different time constants produce different postsynaptic ETPs. In a neuron with a short time constant *(dashed trace)*, each ETP decays back to the resting potential. In a neuron with a long time constant *(solid trace)*, the prolonged ETPs summate, so that the voltage reaches threshold and triggers an action potential.

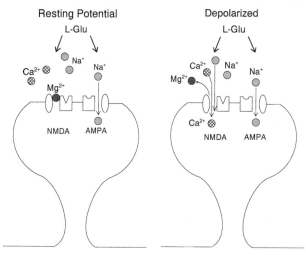

FIGURE 6–7. Signal amplification by NMDA receptor Ca^{2+} channels. *Left:* L-glutamate (L-Glu) released from a presynaptic terminal depolarizes a postsynaptic dendritic spine, releasing Mg^{2+} that blocks NMDA receptor Ca^{2+} channels at the resting potential. *Right:* A second input drives Ca^{2+} into the dendrite through the Ca^{2+} channels and evokes a larger depolarization than the first stimulus.

anisms are discussed in more detail in the section on synaptic plasticity later in this chapter.

Another mechanism that amplifies neuronal signals utilizes voltage-gated Ca^{2+} channels in dendrites to increase the transmitter-induced membrane depolarization produced by a given excitatory input. Membrane regions at dendritic branch points have particularly high concentrations of these channels. The depolarization induced by synaptic inputs to the distal portions of a dendrite activates these channels at proximal branch points, permitting Ca^{2+} currents to flow into the dendrite and augment the original stimulus-induced depolarization. Thus, the activation of voltage-gated Ca^{2+} channels enhances the capacity of distal excitatory dendritic inputs to transmit their information to the soma and axon initial segment, and to elicit action potentials.[18, 20, 21]

Furthermore, the output of a single axon can also be amplified if that axon has multiple branches. When axonal branches form excitatory synapses on many different postsynaptic neurons, the impact of each action potential in the presynaptic cell on activity in a pathway will be magnified in proportion to the number of postsynaptic cells it activates

(Fig. 6–8A). Alternatively, if the branches of the primary cell axon produce inhibition of neighboring cells, they will enhance the signal-to-noise ratio of the excitatory input of the main axon to its postsynaptic neuronal target. This form of signal amplification is known as "surround inhibition" (Fig. 6–8B).[10, 14, 17]

FEED-FORWARD AND FEEDBACK INHIBITION

The final output produced by a neuron depends on the relative balance between excitatory processes and various inhibitory mechanisms that act on the cell. Some inhibitory mechanisms are due to intrinsic properties of the neuronal membrane, whereas others depend on inhibitory synaptic inputs to the cell. However, most inhibitory processes involve potassium (K^+) or chloride (Cl^-) conductances. Mecha-

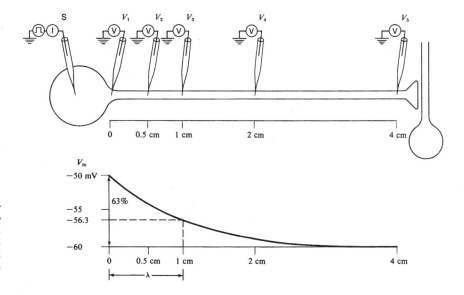

FIGURE 6–6. Space constant. A 10-mV depolarization decays from -60 mV to -53.7 mV 1000 μm from the stimulation site in a neuron with $\lambda = 1000$ μm. (From Kandel ER: Cellular Basis of Behavior. San Francisco, W.H. Freeman and Co., 1976, p 122.)

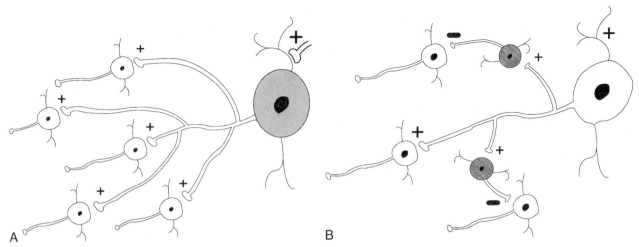

FIGURE 6–8. Signal amplification mediated by multiple axonal branches. *A,* Action potentials evoked in a neuron with multiple axonal branches (shaded cell body) are transmitted to five postsynaptic cells. *B,* Surround inhibition: Axonal branches of a primary excitatory neuron activate inhibitory interneurons (hatched cell bodies), which inhibit adjacent pathways, enhancing the signal-to-noise ratio of the primary excitatory pathway.

nisms that are intrinsic to the cell membrane include voltage-gated K^+ channels or Ca^{2+}-activated K^+ channels, both of which are opened by high-frequency action potential trains. Because of the concentration gradient for K^+, these channels generate an outward K^+ current, which hyperpolarizes the membrane potential. Furthermore, under some conditions excitatory inputs to a neuron can actually produce synaptic inhibition of the same pathway:

Feed-forward inhibition occurs when multiple axonal branches of a presynaptic neuron that activates a postsynaptic cell simultaneously activate inhibitory interneurons that produce inhibitory postsynaptic potentials (IPSPs) in the same postsynaptic cell (Fig. 6–9A).[5] These IPSPs can occur either simultaneously with or closely following the excitatory postsynaptic potentials (EPSPs) generated in the postsynaptic cell by the output from the presynaptic cell. Because this inhibition of the postsynaptic neu-

ron is produced by graded local inhibitory postsynaptic potentials that hyperpolarize the membrane and make the cell less responsive to excitatory synaptic inputs, this inhibitory mechanism does not require the postsynaptic neuron to fire action potentials.

Feedback inhibition requires that the axon of a presynaptic excitatory neuron that activates a postsynaptic cell send branches to inhibitory feedback interneurons (Fig. 6–9B).[1, 2, 9, 16] These interneurons send their axons back to form inhibitory synapses on the presynaptic cell and generate IPSPs that hyperpolarize the membrane potential of the presynaptic cell. Thus, the postsynaptic neuron is excited only briefly, until the presynaptic membrane is hyperpolarized. This feedback inhibition pathway contains at least two synapses, and both the primary neuron and the feedback interneurons must fire action potentials to generate the inhibition.

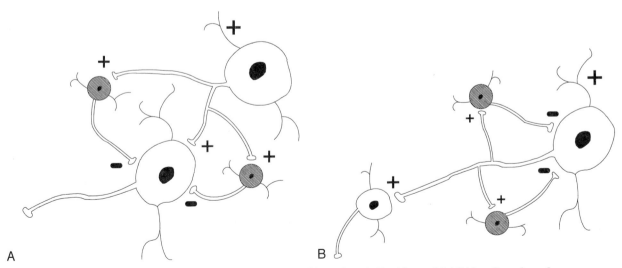

FIGURE 6–9. Inhibition mediated by multiple axonal branches. *A,* Feed-forward inhibition: Branches of a presynaptic excitatory neuron activate inhibitory interneurons *(hatched cell bodies)* that send axons to the postsynaptic excitatory neuron, terminating excitation in the pathway. *B,* Feedback inhibition: Branches of a presynaptic excitatory neuron activate inhibitory interneurons *(hatched cell bodies),* which send axons back to the presynaptic neuron and terminate its excitation of the postsynaptic cell.

ACTIVE PROPERTIES OF DENDRITES

The development of high-speed fluorescence imaging and patch-clamp recording from dendrites has now revealed that, at least in certain neurons, dendritic membranes actively modify the transmission of membrane potential changes, primarily via activation of voltage-gated Na^+ and Ca^{2+} channels, or by Ca^{2+} influx through NMDA receptors. In addition to traveling orthodromically along the axon, in CA1 hippocampal or neocortical pyramidal cells, action potentials are propagated retrogradely from the axon into the soma and dendrites. Recording simultaneously from pairs of synaptically coupled neocortical neurons, Markram and coworkers[13] discovered that action potentials that back-propagated into the dendrites of the postsynaptic neurons and coincided with EPSPs synaptically evoked by the presynaptic neuron substantially increased the amplitude of the dendritic EPSPs. Because the EPSP potentiation was abolished by NMDA receptor blockade, the synaptic enhancement was probably mediated by Ca^{2+} influx through NMDA receptors on the dendritic spines, as shown earlier in Figure 6–7. These back-propagated action potentials may serve to provide feedback to dendritic regions that receive synaptic inputs that the cell has generated an output spike.[11] In contrast, cerebellar Purkinje cell dendrites do not appear to support back-propagated action potentials. Recent findings on active dendritic properties are reviewed by Johnston and associates.[6]

NEURONAL MECHANISMS OF SYNAPTIC PLASTICITY: THE CELLULAR BASES OF MEMORY?

In addition to integrating multiple excitatory and inhibitory synaptic inputs when signals are transmitted from presynaptic terminals to postsynaptic dendrites, some neuronal pathways can also undergo sustained modulation of the strength of signals that are transmitted from one cell to the next. Although Ramon y Cajal proposed more than 100 years ago that information is stored in the brain as changes in synaptic strength, the cellular and biochemical substrates for "memory" are only now beginning to be uncovered. A central hypothesis is that activity-dependent changes in synaptic strength, or functional synaptic plasticity, are essential for the encoding of memory. The form of experimental synaptic plasticity that has been studied most extensively and is the most convincing model for memory is long-term potentiation (LTP). This enhancement of the strength of excitatory synaptic connections between neurons that lasts for hours, days, or longer was first identified at perforant path synapses on dentate granule cells in the hippocampus in 1973.[4] Although most studies of LTP have examined hippocampal pathways, this synaptic strengthening process has been identified in other brain regions, including neocortex and cerebellum. In the last several years a process has been identified in some brain pathways that appears to be the converse of LTP because it generates activity-dependent sustained weakening of the strength of synaptic excitation, designated as long-term depression (LTD). A comprehensive review of work on activity-dependent synaptic plasticity prior to 1993 is provided by Bliss and Collingridge;[4] an extensive review of recent work is provided by Wang and coworkers.[19]

Initial studies of LTP determined that activity-dependent synaptic potentiation can appear in susceptible neurons within milliseconds after a single brief train of high-frequency afferent stimulus pulses.[4] The frequency, intensity, and patterning of the stimuli, as well as the number of afferent fibers they activate, determine how long the potenti-

ation of synaptic responsiveness lasts. At least three types of short-term facilitation have been identified: synaptic facilitation lasts for less than a second, post-tetanic potentiation (PTP) continues for seconds, and short-term potentiation (STP) persists for minutes. In contrast, LTP is sustained for days, weeks, or longer. Synaptic efficacy can also be enhanced by low-frequency, low-intensity afferent stimulation if the postsynaptic membrane is adequately depolarized during the stimulus.[4] The mechanisms that underlie LTP appear to include changes in the properties of either the presynaptic or the postsynaptic neuron, or in both cells: after stimulation the sensitivity of the postsynaptic cell to the neurotransmitter may be increased, whereas the presynaptic terminal may also release an increased amount of neurotransmitter. These alterations in neuronal function increase the strength of the synaptic connection between the cells, which is usually measured by the magnitude of the EPSP produced in the postsynaptic neuron.[12] Furthermore, morphological changes in dendritic branching and synaptic junctions have been described in hippocampus, neocortex, and cerebellum after induction of LTP; these changes resemble the structural modifications that follow behavioral training.[19] Studies using inhibitors of gene transcription or protein synthesis first documented a transient activation of genes that encode factors required for protein synthesis. New techniques for producing mouse strains with specific gene deletions or augmentation have strongly reinforced LTP as the leading candidate for the mechanisms that underlie consolidation of long-term memory.

Although L-glutamate is the primary neurotransmitter at all three primary hippocampal synapses, two different mechanisms for synaptic potentiation have been identified in these pathways.[15] At synapses on dentate granule cells and CA1 pyramidal cells, L-glutamate released from presynaptic terminals opens Ca^{2+} channels associated with postsynaptic NMDA receptors, producing an influx of Ca^{2+} that depolarizes the postsynaptic cell. Although blockade of NMDA receptors prevents induction of LTP at these synapses, subsequent upregulation of AMPA receptors in the postsynaptic dendritic spines may contribute to the stabilization of LTP. In contrast, synaptic potentiation at mossy fiber inputs to CA3 neurons appears to be independent of the state of the postsynaptic neuron,[15] because LTP can be produced at these synapses both during blockade of NMDA receptors and when either activation by L-glutamate or depolarization of the postsynaptic CA3 cell is prevented. Instead, LTP appears to require a rise in Ca^{2+} within the presynaptic mossy fiber terminal, which generates a sustained increase in the release of L-glutamate from the terminal.

The concept of synaptic plasticity implies that modulation of synaptic strength should be bi-directional. Recent studies have documented the existence in hippocampal pathways of sustained activity-dependent depression of synaptic strength, or LTD, and suggest that LTD may be as important as LTP in both information storage processes and brain development.[3, 12, 19] LTD can also be induced in neocortex, striatum, nucleus accumbens, and cerebellum, but the required stimulation characteristics differ. The optimal pattern for producing LTD in hippocampus is continuous, low-frequency stimulation for 10 minutes, whereas in neocortex, striatum, and accumbens the high-frequency tetanic bursts that evoke LTP are most effective. However, in all regions modest depolarization and an increase in intracellular Ca^{2+} in dendrites of the postsynaptic neuron must be paired with afferent stimulation for successful induction of LTD.[3] Studies in neocortex have revealed that gradually increasing the level of depolarization of the postsynaptic neuron produces a continuum of synaptic plasticity, evoking depression of EPSPs at a threshold level, followed by no alteration, and then synaptic potentiation at levels required for generation of LTP.[3] This bipolar modulation of synaptic efficacy seems ideal to produce the adaptation of behavior that characterizes both learning and memory.

SUMMARY

Neurons have both passive and active mechanisms to process information by integration, filtration, and amplification of synaptic inputs to produce an output pattern of action potentials. They can integrate changes in membrane potential evoked by synaptic inputs in both time (temporal summation) and space (spatial summation). The degree to which input signals are integrated depends on the time and space constants of the neuron. The time constant, τ, reflects how long it takes to charge the membrane capacitance, and it is the product of the resistance and capacitance of the neuron. The space constant, λ, is determined by the resistance of the membrane to leakage of current out of the cell, expressed as the ratio of the transmembrane resistance to the axial resistance of the cell.

Signal amplification occurs primarily by means of voltage-gated Ca^{2+} channels on the postsynaptic membrane, such as those associated with postsynaptic NMDA receptors on dendritic spines or other voltage-sensitive Ca^{2+} channels, which are concentrated at branch points of the dendrites. Axonal branches can also amplify signals, either directly by increasing the number of excited postsynaptic neurons, or indirectly by inhibiting adjacent neuronal pathways. The output of a neuron depends on the relative balance between excitatory inputs and inhibitory mechanisms. Intrinsic K^+ membrane currents are activated by high-frequency action potential trains. Inhibitory synaptic inputs can be either feed-forward or feedback, and they generally operate by increasing K^+ or Cl^- conductances.

New imaging and electrophysiological techniques have revealed that dendrites actively modulate graded membrane potential changes, and in some cases support back-propagation of action potentials, which may provide feedback to dendrites about axonal output. Activity-dependent synaptic plasticity is thought to provide the mechanism for encoding information storage in neurons. Long-term potentiation is the most convincing model of memory consolidation. Simultaneous depolarization of presynaptic terminals and postsynaptic neurons, along with the entry of Ca^{2+} into the neurons, produces lasting increases in the strength of synaptic connections. The recent discovery of long-term depression of synaptic efficacy confirms the bi-directional character of synaptic plasticity.

REFERENCES

1. Andersen P: Organization of hippocampal neurons and their interconnections. In Isaacson RL, Pribram KH (eds): The Hippocampus. New York, Plenum Press, 1975, p 155.
2. Andersen P, Eccles JC, Loyning Y: Pathway of postsynaptic inhibition in the hippocampus. J Neurophysiol 27:609, 1964.
3. Artola A, Singer W: Long-term depression of excitatory synaptic transmission and its relationship to long-term potentiation. TINS 16:480, 1993.
4. Bliss TVP, Collingridge GL: A synaptic model of memory: Long-term potentiation in the hippocampus. Nature 361:31, 1993.
5. Buzsàki G: Feed-forward inhibition in the hippocampal formation. Prog Neurobiol 22:131, 1984.
6. Johnston D, Magee JC, Colbert CM, Christie BR: Active properties of neuronal dendrites. Annu Rev Neurosci 19:165, 1996.
7. Kandel ER: Cellular Basis of Behavior. San Francisco, W.H. Freeman and Co, 1976.
8. Kandel ER, Schwartz JH, Jessell TM: Principles of Neural Science, 3rd ed. New York, Elsevier, 1991.
9. Knowles WD, Schwartzkroin PA: Local circuit synaptic interactions in hippocampal brain slices. J Neurosci 1:318, 1981.
10. Knowles WD, Traub RD, Wong RKS, Miles R: Properties of neural networks: Experimentation and modeling of the epileptic hippocampal slice. TINS 8:73, 1985.
11. Magee JC, Johnston D: A synaptically controlled, associative signal for Hebbian plasticity in hippocampal neurons. Science 275:209, 1997.
12. Malenka RC, Nicoll RA: NMDA-receptor-dependent synaptic plasticity: Multiple forms and mechanisms. TINS 12:521, 1993.
13. Markram H, Lübke J, Frotscher M, Sakmann B: Regulation of synaptic efficacy by coincidence of postsynaptic APa and EPSPs. Science 275:213, 1997.
14. Miles R, Wong RKS: Excitatory synaptic interaction between CA3 neurons in the guinea pig hippocampus. J Physiol 373:397, 1986.
15. Nicoll RA, Malenka RC: Contrasting properties of two forms of long-term potentiation in the hippocampus. Nature 377:115, 1995.
16. Schwartzkroin PA, Knowles WD: Local interactions in hippocampus. TINS 6:88, 1983.
17. Traub RD, Knowles WD, Miles R, Wong RKS: Models of the cellular mechanism underlying propagation of epileptiform activity in the CA2–CA3 region of the hippocampal slice. Neuroscience 21:457, 1987.
18. Traub RD, Llinàs R: Hippocampal pyramidal cells: Significance of dendritic ionic conductances for neuronal function and epileptogenesis. J Neurophysiol 42:476, 1979.
19. Wang J-H, Ko GYP, Kelly PT: Cellular and molecular bases of memory: Synaptic and neuronal plasticity. J Clin Neurophysiol 14:264, 1997.
20. Wong RK, Prince DA: Participation of calcium spikes during intrinsic burst firing in hippocampal neurons. Brain Res 159:385, 1978.
21. Wong RK, Prince DA: Dendritic mechanisms underlying penicillin-induced epileptiform activity. Science 204:1228, 1979.

Chapter 7

Synaptic Transmission

Thomas H. Swanson, MD

PHYSIOLOGY OF SYNAPTIC TRANSMISSION
INTRODUCTION
GAP JUNCTIONS
CHEMICAL SYNAPSES
Synaptic Vesicles
Neurotransmitter Receptors
Ion Channels
NEUROMUSCULAR JUNCTION
FREQUENCY-DEPENDENT CHANGES IN SYNAPTIC TRANSMISSION
CONCLUSION

NEUROTRANSMITTER RECEPTOR SYSTEMS
INTRODUCTION
CLASSIC NEUROTRANSMITTERS
Catecholamines
Serotonin
Acetylcholine
Gamma-Aminobutyric Acid
Glutamate
Adenosine Triphosphate
Neuropeptides
OTHER SIGNALING MOLECULES
Adenosine
Nitric Oxide
CONCLUSION

Physiology of Synaptic Transmission

INTRODUCTION

The synapse is the basic structural and functional building block of neural circuits. The biology of the synapse is complex, and trying to understand synaptic circuitry is an overwhelming task. It has been estimated that the brain contains up to 60 trillion synapses.[57] There is thus a tremendous capacity to form very complicated microcircuits, about which we know very little. However, it is important to understand the principles of synaptic transmission because of the role of the synapse in disease, learning, memory, development, and in the generation of clinically relevant electrophysiological signals. The principal function of the synapse is to facilitate cell-to-cell communication, both among neurons and between neurons and muscle cells. More important from a therapeutic standpoint, the synapse serves as a critical regulatory site during information transfer and is a prime target for drugs. Important synaptically centered mechanisms have evolved to both limit, and facilitate, information transfer at the synapse.

An increasing number of modulatory substances and processes have been identified that both inhibit and enhance synaptic transmission. Many of these regulatory processes are initiated by synaptic activity, and occur in the normal brain. Some of these processes are pathological and initiated by cell loss or other biochemical perturbations. The purpose of this chapter is to outline some of the basic principles of synaptic transmission and review the normal biochemical messenger molecules that participate in neurotransmission. This material provides a framework of basic information from which an understanding of clinically applicable electrophysiology can be built.

GAP JUNCTIONS

This type of cell-to-cell communication is included here because, historically, gap junctions were known as electrical synapses. At gap junctions, cells are electrically coupled so that ions flow between the two cells without entering the extracellular space. Gap junctions allow the passage of not only ions, but other relatively small molecules such as second messengers and metabolites. The channels themselves are large, on the order of about 1.6 nm, with conductances around 150 pS.

Gap junction channels consist of hexamers of subunits from a family of proteins called connexins. For every gap junction, each cell contains a hexamer of connexins surrounding a central pore. Different types of connexins are expressed in different tissues, each with N- and C-terminal cytoplasmic domains and four α-helical transmembrane spanning regions.[41] Like other ion channels, gap junctions undergo conformational changes, particularly in response to altered calcium concentration, which alters ion and metabolite flow through the channel.

The advantage of gap junctions is that signaling occurs quickly, with no synaptic delay. Gap junctions are prevalent in invertebrate nervous systems where they mediate fast reflex activity. In humans, gap junctions exist between astrocytes, where they mediate K^+ buffering. They are also present in the retina, inferior olive, vestibular nuclei, nucleus of the trigeminal nerve, and the reticular nucleus of the thalamus. In general, gap junctions provide a mechanism by which populations of cells can be synchronously activated. The mechanisms of their opening and closing are incompletely understood (for reviews, see Hall[19] and Hille[24]).

CHEMICAL SYNAPSES

Chemical synapses are more prevalent and widespread than gap junctions. Most of the communication between neurons in the brain takes place through such synapses.

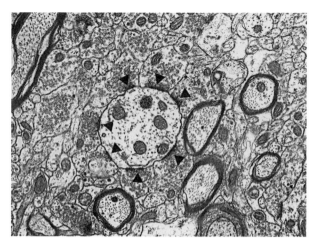

FIGURE 7–1. Electron micrograph of an excitatory synapse in the normal rat dentate hilus. This 25-K magnification demonstrates typical excitatory synapses *(arrowheads)* of dentate granule cell axon terminals onto a hilar neuron dendrite. Note the dense postsynaptic basement membrane, round synaptic vesicles filling the terminals, and large active zone. (Courtesy of Evelyn Dean and Robert S. Sloviter.)

Synapses occur in regions where two neurons come into close physical proximity, usually with 20 to 50 nm. Each synapse contains a presynaptic and a postsynaptic element; the presynaptic elements are specialized regions of the axon that contain neurotransmitters, whereas the postsynaptic elements are specialized regions of axons, dendrites, or cell bodies where transmitter receptors are located. Ample evidence for presynaptic receptors exists, but the details of their function are incomplete compared to postsynaptic receptors.

Synapses have been identified and classified in several ways. Gray[17] described two basic types of synapses based on ultrastructure, each associated usually, but not always, with either excitation or inhibition. Accordingly, type 1 synapses have asymmetrical membrane appositions, dense postsynaptic basement membranes, prominent presynaptic dense projections, round synaptic vesicles, a large postsynaptic active zone, and a wide synaptic cleft. Type 1 synapses are usually excitatory and often associated with small, round, clear vesicles.[17] Type 2 synapses are associated with flattened synaptic vesicles, less obvious dense projections, small postsynaptic active zones, a narrow synaptic cleft, and a modest basement membrane, and are usually inhibitory. Figure 7–1 is an electron micrograph demonstrating several excitatory synapses in the rat hippocampus.

As synapses have both a pre- and a postsynaptic element, they are also labeled according to these elements, for example axo-somatic or axo-dendritic. Although strict rules for nomenclature are lacking, the presynaptic component is usually named first, as in the neuromuscular junction (NMJ).

Within the presynaptic apparatus, neurotransmitter substances are contained in vesicles that contain either a single neurotransmitter substance or, depending on the synapse, multiple transmitter substances.[20] Neurotransmitters are defined as chemical substances that are released from nerve terminals in response to rapid rises in presynaptic calcium concentration, and which then interact directly with the postsynaptic membrane, causing either a metabolic or electrical change in the postsynaptic cell.[54] As such, neurotransmitters differ from other chemical messengers in the brain, such as adenosine and nitric oxide, which are released from cells and may act at nonsynaptic sites. These other messenger molecules have been called neuromodulators, but this term is confusing, as it has been used to describe many different

substances that are not thought to be "primary" neurotransmitters. Neuropeptides, for example, which are G-protein–coupled, have also been called neuromodulators, but have a different mechanism of action than do nitric oxide and adenosine. Most of these modulator substances act through G-proteins, mediate slower actions, and act at sites distant from the site of their release. Thus, the distinction between neurotransmitters and neuromodulators is blurred; there is really a spectrum of actions ranging from rapid, discrete, and localizable (classic neurotransmitters) to the slow, diffuse actions termed neuromodulation.

Synaptic Vesicles

The presynaptic vesicle's main function is to store neurotransmitters in a readily releasable pool, although some contain neurotransmitter synthetic enzymes. Transmitter release is initiated when the nerve terminal is depolarized by an action potential, calcium (Ca^{2+}) enters the nerve terminal through rapidly activating, voltage-gated Ca^{2+} channels in the terminal membrane, concentrated near the apposing postsynaptic membrane.[9] Ca^{2+} entry leads to a highly localized transient increase in nerve terminal Ca^{2+}, which quickly dissipates by diffusion and active transport, allowing repetitive, rapid fluctuations in terminal Ca^{2+}.[9, 42] Increased Ca^{2+} concentration, through a poorly understood mechanism, causes synaptic vesicle fusion to the presynaptic membrane and exocytosis of the transmitter substances contained within the vesicle.[5] Figure 7–2 schematically depicts the events involved with Ca^{2+} influx into the terminal.

Calcium entry into the presynaptic terminal is a critical step in the synaptic transmission process, which is subject to regulation by a variety of neuromodulators and drugs. The specific types of Ca^{2+} channels involved in neurotransmitter release are still being defined. Calcium channels are grouped

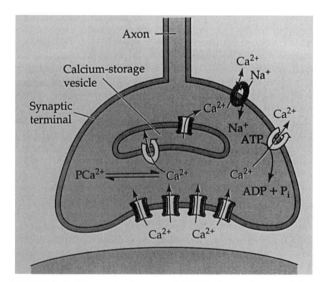

FIGURE 7–2. Schematic diagram of Ca^{2+} influx and removal from the cytosol following an action potential in the presynaptic nerve terminal. Ca^{2+} enters through the voltage-gated Ca^{2+} channels at the active zone and diffuses into the terminal where it binds proteins (P), is taken up into the Ca^{2+} storage compartment, and is actively transported out of the cell. As the cytosolic concentration falls, the storage compartment slowly releases Ca^{2+} to be transported out of the cell by a Ca^{2+} ATPase and an Na^+/Ca^{2+} exchange pump. At high concentrations of cytosolic Ca^{2+} (0.5 μM), mitochondria also take up Ca^{2+}. (Reprinted with permission from Hall Z: The nerve terminal. In Hall Z (ed): An Introduction to Molecular Neurobiology. Sunderland, MA, Sinauer, 1992, pp 148–168.)

into three categories, which vary both in their persistence and in the voltage at which they are activated: I_T ("transient," low threshold), I_L ("long-lasting," high threshold), and I_N ("neither," transient, high threshold).[66] It is the I_N channels classically, but also possibly the I_L channels, that appear to participate in presynaptic transmitter release.[66]

Because Ca^{2+} is critical in triggering transmitter release, and sustained high Ca^{2+} concentration is lethal to the cell, it is tightly regulated in the nerve terminal. Several mechanisms act to decrease terminal Ca^{2+} levels, including rapidly inactivating Ca^{2+} channel kinetics and intracellular sequestration of Ca^{2+} by organelles. The concentration of Ca^{2+} at the site of vesicle fusion, a fraction of a micrometer from the Ca^{2+} channel, has been estimated at tens or even hundreds of micromolar; that of the rest of the cell is kept at submicromolar levels (circa 10^{-7} M). The delay between Ca^{2+} entry and the beginning of transmitter release is less than 1 msec at a rapidly acting synapse. Considering the rapidity of nerve transmission necessary for complex integration of brain activity, the nerve terminal must be finely tuned to allow repetitive, rapid fluctuations in terminal Ca^{2+}.[9, 42] Many drugs that cause "inhibition" by blocking neurotransmitter release are thought to work by decreasing terminal Ca^{2+} levels, but often the precise mechanisms of this reduced terminal Ca^{2+} are unknown.

The function of the vesicle is to store, and in some cases synthesize transmitter, as well as release packages, or "quanta" of transmitter at the appropriate time. Vesicles actively take up transmitters which, with the exception of glutamate, glycine, and neuropeptides, are synthesized in the nerve terminal by cytosolic enzymes with energy supplied by terminal mitochondria. Many neurons are specialized to manufacture and release transmitters such as acetylcholine (ACh), gamma-aminobutyric acid (GABA), dopamine, serotonin, and norepinephrine. The enzymes for neurotransmitter synthesis may be present either in the cytoplasm of the nerve terminal, or in the synaptic vesicles themselves (as for some enzymes involved in catecholamine synthesis).[20]

The mechanisms of vesicle fusion and release are complex, and only partially understood. Vesicle fusion and release require the interaction of several specialized proteins. The vesicle appears to be linked to the presynaptic membrane by actin and other filaments via a protein called synapsin I, a 35-nm long phosphorylated protein found only in nerve terminals.[54] This attachment serves to help concentrate the vesicles near the terminal active zone. The vesicle membranes themselves contain many integral proteins which serve to anchor the vesicle to the region of the terminal and participate in vesicle exocytosis. Four proteins are common to all vesicle membranes: synaptophysin, SV2, synaptotagmin, and synaptobrevin, and although little is known of their individual functions, they are thought to be important in vesicle fusion and exocytosis.[9, 42, 56] Recent evidence has demonstrated that synaptophysin can form hexameric ion channels which resemble gap junctions.[42] This ionophore property of synaptophysin suggests it may play a critical role in exocytosis. The initial event in exocytosis may be the formation of a fusion pore, linking the cytoplasm of the vesicle to the extracellular space. Many proteins in nerve terminals and synaptic vesicle membranes can bind Ca^{2+} (e.g., calmodulin), but the molecular site at which Ca^{2+} acts to trigger release is still uncertain.[5, 20] No doubt the mechanism of vesicle fusion will become clear as more is known about these vesicle proteins.

Neurotransmitter Receptors

Once the transmitter is released into the synaptic cleft, it diffuses to the postsynaptic membrane and binds to one of two general classes of specific cell surface transmitter receptors. The receptors may be directly linked to ion channels, in which case they are called ionotropic receptors; the channels to which they bind are called ligand-gated ion channels. Alternatively, the receptors may be linked to guanosine triphosphate (GTP)-binding regulatory proteins (G-proteins), in which case they are called metabotropic receptors. Ionotropic receptors act quickly to open ion channels and produce transmembrane currents, an effect that lasts only a few milliseconds. G-protein–linked responses are slower, taking hundreds of milliseconds to occur, and last from seconds to days. In addition to opening or closing ion channels, G-proteins alter metabolism, growth, cytoskeletal structure, and gene expression. These effects can be produced directly by the G-protein or via changes in the intracellular concentration of second messenger molecules.

Ionotropic Receptors

Although relatively few neurotransmitters act through ionotropic receptors, they are nonetheless the most important receptor class for "day-to-day" signaling in both the central and peripheral nervous systems. Both glutamate, the major excitatory transmitter in the central nervous system, and GABA, the major inhibitory transmitter, bind to ionotropic receptors which are responsible for the majority of synaptic information flow in the brain. With few exceptions, all other transmitter receptors serve to modulate glutamatergic and GABAergic neurotransmission. In addition to glutamate and GABA, there are also ionotropic receptors for ACh (predominately in the peripheral nervous system), serotonin, glycine, and adenosine triphosphate (ATP).

The ionotropic receptors for ACh, GABA, glycine, and serotonin all belong to a superfamily that appears to have evolved from a common ancestral receptor.[44] Each receptor includes several homologous subunits (probably five, some of which may be identical), and each subunit has four putative transmembrane regions. The ionotropic receptors for glutamate and ATP also consist of subunits, but they display no apparent sequence homology with the others. It is the subunit composition that gives each receptor its unique ion channel kinetics. Recent evidence suggests, at least in the brain, that the subunits of ionotropic receptors of a given transmitter may be dissimilar in different brain regions. The arrangement of these subunits may become altered and be responsible for pathological conditions such as epilepsy.

Ionotropic receptors produce fast synaptic potentials. Several properties of these receptor systems ensure reliable neurotransmission. The receptors are located within 1 μm of the site of release, and the concentration of transmitter in the synaptic cleft is high, approaching 1 mM. In order to limit the duration of receptor activation, affinity of the receptor for the transmitter is low (10 to 100 μM). Several other systems ensure rapid termination of transmitter action. For example, at the NMJ, any ACh not bound to an ACh receptor is rapidly hydrolyzed by acetylcholinesterase, an extremely fast enzyme. Adenosine is deactivated by adenosine deaminase, although the products inosine and hypoxanthine also weakly activate adenosine receptors. Other mechanisms for removal of transmitter include simple diffusion, desensitization, and reuptake.

G-protein–Coupled Receptors

The second major class of transmitter receptors are those linked to G-proteins. All transmitters, including those substances that bind to ionotropic receptors, have specific G-protein–coupled receptors. The molecular biology of G-protein receptors is complex; they comprise a superfamily of at least 100 different proteins that share considerable sequence homology and structure, all with seven membrane spanning domains. Figure 7–3 demonstrates the typical structure of a G-protein–coupled receptor. Among members of this family are catecholamines muscurinic cholinergic, rhodopsin, *N*-methyl-D-aspartate (NMDA), GABA_B, and adenosine receptors.

The function of G-protein–coupled receptors is different from the ionotropic receptors. Whereas ionotropic receptors have evolved to efficiently transfer information from neuron to neuron, G-protein receptors have evolved to modify the function and even structure of neurons that express them.[51] G-proteins regulate a multitude of different effector systems, including cAMP, cGMP, protein kinase C and A, other protein kinases, calcium ions, diacylglycerol, inositol triphosphate, arachidonic acid, and nitric oxide. A single receptor may be linked to more than one G-protein which, in turn, may be linked to multiple effector systems. Built into this system is a tremendous capacity for diversity; there may be as many as 1000 different protein kinases in a single cell. This explains why one transmitter has different effects on different cell populations. The ultimate intracellular effect is a function of the G-protein and the G-protein effector systems with which the receptor is associated. Conversely, several different receptors may be linked to the same type of G-protein in a given cell, and thus have the same effects. This combination of divergence and convergence is typical of G-protein–mediated effects, in the nervous system as elsewhere.[25]

Functionally and mechanistically, slow synaptic potentials mediated via G-protein–coupled receptors are intermediate between classical fast synaptic potentials and classical hormone actions. It is no coincidence that many of the neurotransmitters and neuropeptides that act through G-protein–coupled receptors also function as hormones. Although some slow synaptic potentials produce a direct effect, for example, the inhibition produced by opening of potassium channels by GABA_B receptors, others produce their action through a more convoluted route. For example, inhibition of the M current, a voltage-dependent potassium current, produces slow excitatory postsynaptic potentials (EPSPs) in

sympathetic neurons. The slow EPSP allows these neurons to switch from phasic to nearly tonic firing upon maintained depolarization.[2] Many of the ultimate actions that follow activation of G-protein–coupled receptors have yet to be worked out.

Unlike transmitters that activate ionotropic receptors, those acting through G-protein–coupled receptors have much higher affinity, and might act at longer distances from their site of release. In the autonomic nervous system, for example, the nerve terminals are often located micrometers from the postsynaptic targets on end organs. In the heart, muscarinic receptors are located diffusely on atrial cells, rather than being concentrated near nerve terminals.[21] In sympathetic ganglia, there is evidence that peptides can diffuse to cells away from the actual synaptic contacts and can act for a period of several minutes before the action is finally terminated.[26]

Ion Channels

Whether a receptor is G-protein–linked or ion channel coupled, the final common event in neurotransmission is the opening of ion channels. The diversity and complexity of individual ion channels are large, but knowledge of basic ion channel function is important when trying to understand how electrophysiological signals are generated.

Ligand-gated Ion Channels

Binding of the transmitter to the receptor results in an alteration of the channel which, in turn, leads to either a decrease or an increase in the conductance of a specific ion. For ionotropic receptors, the specific ion that flows through the channel of a given receptor is predictable. Every receptor alters the conductance of the same ion each time it is activated. This is because the receptor protein itself is part of the ion channel, and binding of the transmitter changes the conformation of the receptor protein in such a way as to alter ion flow directly. This ion flow induces either a depolar-

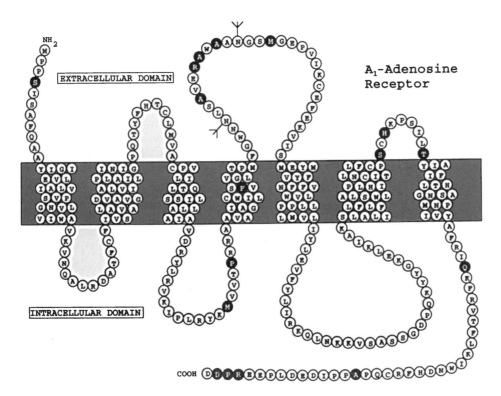

FIGURE 7–3. The adenosine A_1 receptor sequence. This receptor has seven transmembrane spanning domains and is typical of the superfamily of G-protein–coupled receptors, which include the catecholamine, muscarinic, rhodopsin, metabotropic glutamate, and GABA_B receptors.

TABLE 7–1	Common Neurotransmitter Actions	
Response	**Neurotransmitter**	**Receptor**
$\uparrow$ g Na$^+$, $\uparrow$ gK$^+$	Glutamate	Quisqualate/Kainate
$\uparrow$ g Na$^+$,	Glutamate	N-Methyl-D-aspartate
$\uparrow$ gK$^+$, $\uparrow$ g Ca^{2+}	Acetylcholine	Nicotinic
$\uparrow$ g Cl$^-$, $\uparrow$ gK$^+$r	Gamma-aminobutyric acid	GABA$_A$
	Acetylcholine	M$_2$
	Norepinephrine	Alpha$_2$
	Serotonin	5-HT$_{1A}$
	Gamma-aminobutyric acid	GABA
	Dopamine	D$_2$
	Adenosine	A$_1$
	Somatostatin	
	Enkephalins	Mu, delta
$\downarrow$ g AHP	Acetylcholine	Muscarinic
	Norepinephrine	Beta$_1$
	Serotonin	(?)
	Histamine	H$_2$
$\downarrow$ g M	Acetylcholine	Muscarinic
$\downarrow$ g K, I	Acetylcholine	Muscarinic
	Norepinephrine	Alpha$_1$
	Serotonin	(?)
$\downarrow$ g A	Norepinephrine	Alpha$_1$
$\downarrow$ g Ca^{2+}	Multiple transmitters	

A, A current; AHP, afterhyperpolarizing potential; Cl, chloride; g, conductance; K, potassium; M, M current; Na, sodium.
From Shepherd G (ed): Synaptic Organization of the Brain, 4th ed. New York, Oxford University Press, 1998. Copyright © 1998 by Oxford University Press, Inc. Used by permission of Oxford University Press, Inc.

ization or hyperpolarization of the membrane, which has a discrete time course governed by the kinetics of the channel and the physical properties of the membrane. Table 7–1 summarizes the action of some common neurotransmitters on ion channels.

The kinetics of an ion channel can be modified by changes in membrane potential, ligand, and other protein binding, or by the presence of ions in the ion channel. Ion channel kinetics are studied using a technique called patch clamping. In this technique, a microelectrode is placed on the service of a cell, and a piece of the cell membrane is sucked up onto the pipette. The pipette can then be left with the cell attached (cell attached mode) or withdrawn from the cell, breaking off a patch of membrane. This patch contains isolated ion channels, and the potential across these channels can be recorded in response to various drug and ion treatments (current clamp mode). In addition, changes in current flow can be recorded if constant voltage is maintained (voltage clamp mode). In this way, the ions, and potentials at which they flow, can be studied. Figure 7–4 schematically depicts a typically patch-clamp configuration and an example of the types of recording obtained.

The duration of the ligand/receptor interaction governs the duration of opening of the channel. Many drugs act to alter the interaction of the transmitter with the receptor, and thus influence the kinetics of channel opening. This may result in an increase in the mean opening time of the channel or an increase in the frequency of opening. For example, benzodiazepines bind to a modulatory site on the GABA$_A$ receptor and increase the probability of channel opening. Barbiturates also bind to a site on the GABA$_A$ receptor, but act to increase the channel opening time.

The current flow across the postsynaptic membrane changes the potential across the membrane. This potential decays with time and distance down the membrane as a function of the intrinsic capacitance and resistance of the membrane. The resistance capacitance (RC) circuit characteristics of the membrane impart a certain time constant to the decay of the potential. Likewise, a space constant for given pieces of membrane is governed by integral membrane components. This decay allows for temporal and spatial sum-

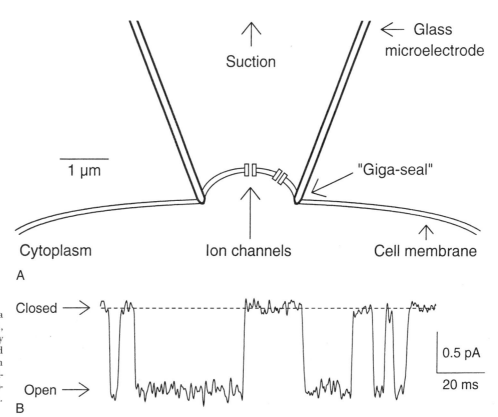

FIGURE 7–4. A is a drawing of a "cell attached" patch preparation, typical of the type used to study neuronal ion channel kinetics and gating. B is an example of ion channel gating for an L-type calcium channel from a smooth muscle cell. (Courtesy of Dr. Carlos A. Obejero-Paz.)

mation of other inputs to add or detract from the first voltage change. Because of these membrane properties, inputs onto a single neuron or group of neurons will be affected by neighboring inputs. Neurotransmitters will, therefore, cause varied effects depending on the combination of transmitters affecting the synapse at a given time.

Voltage-Activated Ion Channels

In addition to ligand-gated ion channels, a large, diverse number of voltage-activated ion channels exist on neurons, although some ionic currents are activated both by voltage and transmitters. Table 7–2 summarizes some of the well-known voltage-sensitive ionic currents that have been identified electrophysiologically. Molecular cloning techniques have led to the identification of even more ionic currents. Thus, over a dozen K^+ channels, one or two Na^+, and an increasingly large number of Ca^{2+} channels exist on neurons. However, the details of which channels occur on which neurons remains unclear. It is important to understand ion channels because as their function becomes known, drugs are being developed that alter their properties, and thus affect synaptic transmission.

One example of a clinically useful drug that works directly on ion channels is 3-4-diaminopyridine (DAP), used in the treatment of Lambert-Eaton myesthenic syndrome (LEMS). This compound antagonizes DAP-sensitive, presynaptic K^+ channels at the NMJ,[35] likely of the delayed rectifier type. Such K^+ channel blockade results in a delay of repolarization of the action potential. The net effect of this action is to increase the duration of terminal depolarization, which in turn prolongs the opening time of the voltage-sensitive calcium channels responsible for terminal calcium entry. This increase in terminal calcium allows greater numbers of vesicles to undergo fusion and release of transmitter. In the case of LEMS, more ACh is released per nerve action potential, which clinically improves muscle contraction. As research in the area of ion channels continues to blossom, new therapeutic agents are sure to emerge.

Neuromuscular Junction

The NMJ has proven to be a critical synapse, not only because it is the site of transduction of brain impulses into muscle contraction, but because of its accessibility to controlled research. Historically, ACh acting at this synapse was the first compound to be identified as a neurotransmitter. The binding of agonist to nicotinic receptors at the NMJ opens Na^+ and K^+ channels, leading to depolarization of the muscle cell, which in turn facilitates calcium entry and initiates contraction. Many disease processes affect the number and arrangement of NMJs, such as myasthenia gravis, motor neuropathies, and radiculopathies.

Understanding how synapses are formed gives us insight into how they might be modified. Recent evidence has shown that the motor nerve itself participates in NMJ formation. Agrin, contained in motor axons, is released, and binds to receptors at the postsynaptic muscle membrane.[53] Through a process that involves a number of proteins, including dystroglycan, MuSK, and rapsyn, agrin induces the formation of the postsynaptic NMJ apparatus. As more is known about these proteins, we may begin to understand how synapses are formed. Therapeutically, the ability to manipulate synapse formation has potential applications in virtually all areas of neurologic disease.

FREQUENCY-DEPENDENT CHANGES IN SYNAPTIC TRANSMISSION

Some excitatory synapses, particularly in the hippocampus, exhibit a use-dependent change in efficacy of neurotransmission. Repetitive stimulation of afferent fibers to these synapses results in long-lasting (hours to weeks) enhancement of synaptic transmission. This phenomenon is called long-term potentiation (LTP), and was originally described in the dentate gyrus in 1966.[32] See the review by Madison and associates.[34] Recently LTP has become more thoroughly studied, as it is thought to underlie plastic changes associated with learning and memory. The mechanisms of LTP are debated, but there appear to be both presynaptic and postsynaptic changes that account for the enhanced synaptic efficacy. A minimum "threshold" stimulus strength is necessary to induce LTP, suggesting that a certain number of input fibers need to be activated simultaneously.[36] The higher the frequency of stimulation, the lower the required stimulus strength needed to induce this change.[7] Thus, LTP demonstrates the importance of spatial and temporal integration of inputs discussed earlier.

In addition to LTP, simple increases in firing of a neuron may increase the amount of transmitter released, a process

TABLE 7–2	Neuronal Ionic Currents	
Current	**Description**	**Function**
Na^+		
$g_{Na,t}$	Transient; rapidly inactivating	Action potentials
$g_{Na,p}$	Persistent; noninactivating	Enhances depolarizations; contributes to steady state firing
Ca^{2+}		
g_T, low threshold	"Transient"; rapidly inactivating; threshold negative to −65 mV	Underlies rhythmic burst firing
g_L, high threshold	"Long lasting"; slowly inactivating; threshold around −20 mV	Underlies Ca^{2+} spikes that are prominent in dendrites; involved in synaptic transmission
g_N	"Neither"; rapidly inactivating; threshold around −20 mV	Underlies Ca^{2+} spikes that are prominent in dendrites; involved in synaptic transmission
K^+		
g_K	Activated by strong depolarization	Repolarization of action potential
g_C	Activated by increases in $(Ca^{2+})_i$	Action potential repolarization and interspike interval
g_{AHP}	Slow after hyperpolarization; sensitive to increases in $(Ca^{2+})_i$ Transient, inactivating	Slow adaptation of action potential discharge
g_A	Transient, inactivating	Delayed onset of firing; interspike interval; action potential repolarization
g_M	"Muscarine" sensitive; activated by depolarization	Contributes to spike frequency adaptation
g_Q, g_h	"Queer"; activated by hyperpolarization/ mixed cation current	Prevents strong hyperpolarization
$g_{K,leak}$	Contributes to neuronal resting "leak" conductance	Helps determine resting membrane potential

Ca, calcium; g, conductance; K, potassium; Na, sodium.
From Synaptic Organization of the Brain, 4th ed, edited by Gordon Shepherd. Copyright © 1998 by Oxford University Press, Inc. Used by permission of Oxford University Press, Inc.

termed facilitation. Facilitation appears to be a function of residual increase in terminal Ca^{2+} left over from the first stimulation. Because of the rapidity of Ca^{2+} removal in the terminal, this second stimulation must be within milliseconds of the first. Conversely, a second stimulation may result in inhibition, not facilitation of the response. In this case, activation of an inhibitory interneuron which "feedsback" on the first to inhibit firing causes this apparent inhibition. This feedback inhibitory system operates strongly at the normal dentate granule cell/perforant path synapse. In epilepsy, several alterations of this feedback inhibition have been identified that may play a role in seizure initiation and propagation.[63]

Stimulation frequency may also affect the type of transmitter released. In the submandibular gland, parasympathetic stimulation at 2 Hz leads to the release of both ACh and vasoactive intestinal polypeptide (VIP).[53] However, at 10 Hz stimulation, only VIP is released.[53] VIP may then alter the subsequent release of ACh at lower frequencies. In this way, the frequency of firing at certain synapses is critical for integration of inputs.

CONCLUSION

The synapse is a prime target for potential therapeutic agents. As we learn more about the molecules and mechanisms involved in vesicle packaging, release, transmitter elimination, and activation of ion channels, more therapeutic options will emerge. For example, presynaptic modulation of neurotransmitter release is a common mechanism by which drugs ultimately alter neuronal excitability. Part of the effect of virtually all anticonvulsant drugs can be attributed to alteration of transmitter release. Accordingly, phenytoin and carbamazepine both block fast Na^+ channels, which reduces action potential propagation and thus the amount of transmitter released at nerve terminals. Adenosine, a potent endogenous anticonvulsant, inhibits neurotransmitter release, partly by inhibiting axon potential propagation down the axon[64] as well as by blocking voltage-dependent Ca^{2+} channels. New agents that interact with the adenosine inhibitory system will undoubtedly become valuable for pharmacotherapy, particularly in epilepsy, where drugs with different mechanisms of action used in combination offer the best chance of seizure control in many patients.

Neurotransmitter Receptor Systems

INTRODUCTION

All neurotransmitters bind to postsynaptic receptors, and the principal ones are directly ion channel linked. Discrete functions of transmitter systems result from specific distribution of the receptor subtypes and the axons that release the transmitter. An understanding of this heterogeneity is important to understand the varied effects of the different transmitters, and drug interactions with transmitter receptors. This section reviews the major transmitter systems with emphasis on the type and distribution of their respective receptors. Table 7–3 summarizes the effects of anticonvulsants on brain neurotransmitter systems.

CLASSIC NEUROTRANSMITTERS

Catecholamines

The catecholamines include epinephrine, norepinephrine (NE), and dopamine. NE-containing cell bodies are found in the sympathetic ganglia and in the brain stem lateral tegmental nuclei and locus ceruleus of the brain stem. These brain stem NE-containing cell bodies project highly branched axons to widespread areas of the central nervous system. NE exerts its effect through adrenergic receptor subtypes, which include alpha$_{1,2}$ and beta$_{1,2}$. Each receptor is an integral membrane glycoprotein with seven membrane spanning regions and a molecular weight of 64 to 80 kD.

NE has diverse effects in the brain, predominately inhibitory, and is an anticonvulsant in most animal models of epilepsy. For example, depletion of NE facilitates amygdala kindling in rats[10] and worsens audiogenic seizures in genetically epilepsy-prone rats.[27]

The adrenergic receptors are all G-protein linked, and nine subtypes have been cloned: three alpha$_1$ receptor subtypes, three alpha$_2$ receptor subtypes, and three beta receptor subtypes. Pharmacological definition of the receptor subtypes conflicts with the cloning data; four alpha$_2$ receptors

have been defined pharmacologically. As is typical with receptor nomenclature, reconciling the traditional pharmacological properties with molecular cloning data is ongoing. As members of the G-protein–coupled family of receptors, the adrenergic receptors have seven transmembrane spanning domains with marked sequence homology (see Fig. 7–3). For a current review of adrenergic receptors the reader is referred to Hein and Kobilka.[22]

Dopamine is contained in neurons of the retina, olfactory bulb, diencephalon, periaquaductal gray, dorsal motor nucleus of the vagus, nucleus of the solitary tract, ventral tegmental area, and substantia nigra. Dopamine binds to several different G-protein–coupled receptors. These are termed D_1 through D_5, but have also been called $D_{1a,b}$ and $D_{2a,b,c}$. Classification of these receptors has classically relied on whether they increase or decrease adenylyl cyclase activity; the D_1 family stimulates, whereas the D_2 family inhibit the enzyme.

Only a few drugs used for neurological disorders target dopamine receptors. The most effective anti-Parkinson drugs are those that activate D_2 receptors. Anticonvulsant drug interaction with dopamine receptors is limited. Phenytoin reduces dopamine neurotransmission, which may be the mechanism of oral facial dyskinesia with this drug.[8] Carbamazepine decreases release and uptake of NE. Decreasing NE synthesis rate, or blocking receptors, decreases the effect of phenobarbital.

Serotonin

Serotonin (5HT)-containing neurons are found in the brain stem, mostly within the raphe nucleus. Axons of these neurons are widely distributed in the forebrain, cerebellum, and spinal cord. The K_m of tryptophan hydroxylase, the rate-limiting enzyme in 5HT synthesis, is 50 to 120 μM, whereas the brain concentration of tryptophan is about 30 μM. Thus, increasing dietary tryptophan can increase brain serotonin. Serotonin receptor pharmacology and molecular biology are rapidly changing fields. Three separate families of serotonin receptors have been identified: 5HT$_1$ (5HT$_{1a,1b,1d}$), negatively

TABLE 7-3	**Effect of Antiepileptic Drugs on Brain Neurotransmitters**

| | | **Effect on Neurotransmitters** | |
| | | **Plasma Concentration** | |
Drug	**Epilepsy Types**	*Therapeutic*	*Nontherapeutic*
Phenytoin	GTC, P	↑ 5HT levels (?) ↓ ACh release ↓ Adenosine uptake (?) ↑ GABA action (?)	↓ NE reuptake ↓ NE release ↓ Glutamate release
Carbamazepine	GTC, P	↑ 5HT release (?) ↓ NE release Block adenosine receptors	↑ ACh (?) ↓ Glutamate release ↓ NE reuptake
Valproic acid	Ab, GTC	↑ GABA levels* ↓ Aspartate (?)	
Phenobarbital	GTC, P	↑ GABA action* ↓ Glutamate action (?) ↑ Aspartate levels (?)	↓ NE release ↓ 5HT synthesis ↓ ACh release
Benzodiazepines	SE (diazepam) Ab (clonazepam)	↑ GABA action*	↓ Adenosine uptake ↓ 5HT synthesis ↓ ACh release ↑ ACh levels ↓ NE synthesis ↓ Dopamine synthesis
Ethosuximide	Ab	↑ Dopamine action	↑ GABA levels

Ab, absence; GTC, generalized tonic-clonic; P, partial; SE, status epilepticus; (?) indicates that this has not been well studied or effect is controversial; *indicates good evidence linking effect to mechanism of action.
Reprinted with permission from Browning RA: Overview of neurotransmission: Relationship to the action of antiepileptic drugs. In Fairgold CL, Fromm GH (eds): Drugs for Control of Epilepsy: Action on Neuronal Networks Involved in Seizure Disorders. Boca Raton, FL, CRC Press, 1992, pp 23–56.

coupled to adenylate cyclase through a G_I protein; $5HT_2$ series, linked to phospholipase C and phosphoinositide second messenger systems through a G-protein; and $5HT_3$, an ionotropic receptor permeable to cations, including Ca^{2+} (for review see Peroutka[45]).

The $5HT_1$ receptor appears to mediate the action of ergots and sumatriptan in headache treatment. Some anticonvulsants can also interact with the 5HT system. Phenytoin increases the synthesis rate of brain serotonin and cerebrospinal fluid levels of 5-HIAA, mostly at toxic levels. Carbamazepine may decrease 5HT synthesis in hippocampus. Lastly, 5HT uptake blockers can enhance the action of phenobarbital, carbamazepine, and phenytoin. A defect in the serotonin receptor system has been identified in the genetically epilepsy-prone rat, and pharmacological potentiation of 5HT can block seizures in this model.[27]

Acetylcholine

In addition to lower motor neurons, ACh is contained in neurons found in small nuclear groups of the ventral forebrain, the septal nuclei, the diagonal band of Broca, the basal nucleus of Meynert, some tegmental brain stem nuclei, and in interneurons of the striatum. There are two types of brain ACh receptors, nicotinic and muscarinic, both with several subtypes.

The NMJ nicotinic ACh receptor is an integral membrane glycoprotein with four types of subunits. Two alpha, one beta, one gamma, and one delta subunit form a pentamer 80 to 90-Å in diameter, which requires the binding of two molecules of ACh on the alpha subunits. Neuronal ACh receptors are formed from distinct but homologous subunits. This pentameric structure is typical of ligand-gated ion channels such as the $GABA_A$ receptor (see later discussion). Nicotinic receptors are found primarily in peripheral autonomic ganglia and other neural crest–derived tissues, and on skeletal muscle, although some areas of brain, such as the optic tectum, contain nicotinic receptors.

Muscurinic ACh receptors are the predominate ACh re-

ceptor in brain. These receptors are G-protein linked, and are pharmacologically grouped into M_1, M_2, and M_3 types, although five separate genes for muscurinic receptors have been cloned. Some of the effects of activation of muscurinic ACh receptors are inhibition of adenylyl cyclase, stimulation of phosphoinositol hydrolysis, and activation of a K^+ channel. The net effect of such receptor occupation thus depends on the individual processes with which the G-protein is coupled. Both pre- and postsynaptic receptors exist; postsynaptic sites may cause inhibition or excitation, whereas the presynaptic sites reduce transmitter release, leading to inhibition. Postsynaptic excitation is thought to be mediated by closing of voltage-gated K^+ channels (the M current), and antagonizing Ca^{2+}-dependent K^+ channels.

Some anticonvulsants interact with the ACh system. Phenytoin increases K^+-evoked ACh release.[49] Carbamazepine may increase brain ACh, particularly in the striatum.[55] Phenobarbital can decrease the rate of release and synthesis of ACh at high concentrations. Diazepam may increase brain ACh levels. For a review of these and other interactions see Browning.[8] Lastly, tacrine, used to treat dementia, is a cholinesterase inhibitor, enhancing synaptic ACh levels in brain.

Gamma-Aminobutyric Acid

GABA is the main inhibitory neurotransmitter in the nervous system. Two types of GABA receptors exist: the $GABA_A$ and $GABA_B$ receptors. The $GABA_A$ receptor is a heterooligomeric synaptic protein complex that constitutes a gated Cl^- channel, and is thus an ionotropic receptor. The $GABA_A$ receptor belongs to the superfamily of ligand-gated ion channels which includes the nicotinic acetylcholine receptors, glycine receptors, and the $5HT_3$ receptor. Like other members of this family, it is a pentameric structure, formed by the association of various combinations of subunits, of which there are at least three different families: alpha$_{1-6}$, beta$_{1-3}$ and gamma$_{1-3}$. These subunits coalesce to form a variety of functional $GABA_A$ receptor subtypes with differing sensitivities to modulatory substances such as steroids and benzodiazepines. There appears to be regional brain heterogeneity in $GABA_A$ receptor subunit composition. Substances known to modulate this receptor include barbiturates, which increase channel opening time, and benzodiazepines, which increase the probability of channel opening and increase GABA binding. Pharmacological activity at the $GABA_A$ receptor has been classically explored using muscimol as an agonist and biculculine as an antagonist. The reader is referred to Lüddens and colleagues[33] for further review of the $GABA_A$ receptor.

The $GABA_A$ receptor has been the target for several experimental anticonvulsant drugs, as many currently used drugs appear to interact with the $GABA_A$ receptor complex. Most notably the effects of barbiturates and benzodiazepines are mediated through the $GABA_A$ receptor. In addition, valproic acid may increase brain and nerve terminal GABA concentration.[15] Tiagabine is a GABA uptake blocker that is thought to enhance synaptic GABA concentrations.

The $GABA_B$ receptor is G-protein–linked and found on both presynaptic terminals and postsynaptic dendrites and cell bodies. Activation of postsynaptic $GABA_B$ receptors leads to an increase in potassium conductance. The overall effect of GABA on many neurons, as for example on hippocampal pyramidal cells, is early $GABA_A$-mediated inhibition (tens of milliseconds) followed by a later $GABA_B$ mediated inhibitory response (hundreds of milliseconds).[13, 65] $GABA_B$ receptors are bicuculline insensitive, activated by baclofen, and antagonized by phaclofen and saclofen. Presynaptic $GABA_B$ receptors inhibit transmitter release, leading to depression of both EPSPs and inhibitory postsynaptic potentials (IPSPs) in the postsynaptic cell.[14, 29] There is some evidence that the pre- and postsynaptic receptors are pharmacologically distinct.[14]

Glutamate

The main excitatory neurotransmitter in the central nervous system is glutamate. This excitatory amino acid binds to multiple receptor types, which differ with respect to activation and inactivation time courses, desensitization kinetics, ion permeability, and conductance properties. Selective agonists and antagonists have been used to differentiate between the different glutamate receptor types. These receptor subtypes are grouped into three main classes: NMDA (N-methyl-D-aspartate), non-NMDA, and metabotropic. In addition to mediating excitatory transmission, glutamate receptors are also critically involved in neuronal plasticity (processes such as learning and memory). These receptors also mediate the excitotoxic damage, thought to play a role in many neurological diseases.

Non-NMDA Glutamate Receptors

Two distinct non-NMDA receptor complexes have been identified via molecular cloning techniques, the AMPA (alpha amino-3-hydroxy-5-methyl-4-isoxazole propionic acid) receptor and the kainate receptor. Because both are activated by the same agonists, and specific pharmacological criteria for separating the action of glutamate on these two receptors are incomplete, they are designated as non-NMDA receptors, although some authors refer to them as separate receptor classes. The AMPA receptor is found in the majority of fast excitatory synapses and activates channels with fast kinetics (onset and offset desensitization time course in the order of milliseconds) with high Na^+ and low Ca^{2+} permeability. Four genes encode the AMPA receptor termed Glu_{R1-4}, and no significant sequence homology exists between the AMPA or kainate receptors and other ionotropic receptors. These AMPA receptor subunits can aggregate into homomeric and heteromeric receptor configurations with unique biophysical properties. For example, the permeability

to Ca^{2+} is imparted by the Glu_{R1} and Glu_{R3} subunits, but prevented by the Glu_{R2} subunit.[18] The nomenclature and distribution of these receptors in evolving, and the reader is referred to Bettler and Mulle[6] for review.

The kainate receptor subunits are coded by five distinct genes, but the biophysical properties of this receptor are less well understood. Few specific agonists and antagonists capable of differentiating the two receptors exist, and most "fast" actions of glutamate are thought to be mediated via AMPA receptors.

NMDA Glutamate Receptors

This receptor is also activated by glutamate, but has vastly different properties from the AMPA type receptor. NMDA channels have much slower kinetics, high Ca^{2+} permeability, large single channel conductance, and require glycine as a co-agonist. The receptor channel is tonically blocked by Mg^{2+} in a voltage-dependent manner; depolarization releases Mg^{2+} from the pore, allowing ions to pass. Glutamate dissociates slowly from the NMDA receptor, with time constants measured in tens to hundreds of milliseconds. Like the AMPA and kainate receptor, the NMDA receptor is composed of multimeric heteromers of subunits. Two families have been identified: the widely distributed $NMDA_{R1}$, and $NMDA_{R2}$, which exists in a more limited distribution in the central nervous system.[37, 39]

Calcium entry via the NMDA channel triggers trophic developmental actions, movement of neuronal growth cones, and activity-dependent resetting of synaptic strength termed long-term potentiation (LTP, see later discussion). Calcium entry also leads to cell destruction, and is responsible for the "excitotoxic" effects of glutamate.[18] The NMDA receptor also contains binding sites for polyamines and phencyclidine (PCP), which modulate receptor activation. Competitive antagonists at the NMDA receptor disrupt memory and motor performance, but also act as anticonvulsants. Several NMDA antagonist compounds, most notably MK-801, have shown promise as anticonvulsants, but their toxicity limits therapeutic usefulness. Noncompetitive antagonists disrupt memory and motor performance, and may cause PCP-like psychosis.

Metabotropic Glutamate Receptors

The aforementioned ionotropic glutamate receptors mediate fast excitatory synaptic transmission. Another, more recently discovered class of glutamate receptors was initially shown to activate phospholipase C, and subsequently shown to inhibit glutamate release at the presynaptic terminal. L-2-amino-4-phoshponobutyrate (L-AP4), a glutamate analog, was the first agonist described at this site. Metabotropic glutamate receptors (mGluRs), as these receptors have come to be known, are linked to G-proteins and adenylyl cyclase, and exert long-lasting intracellular effects. The molecular biology of this receptor is complex, and still being defined. Eight genes coding mGluRs have been identified, and many more mRNAs are produced from these genes via alternative splicing. For a more comprehensive review of this evolving topic, the reader is referred to Pin and Duvoisin.[48]

Adenosine Triphosphate

Purine receptors are grouped into those that respond to adenosine, termed P_1 receptors, and those that respond to ATP, termed P_2 receptors. Adenosine receptors are discussed later under other signalling molecules. ATP satisfies many of the criteria for a classical neurotransmitter: it is stored in synaptic vesicles, released from nerve terminals in a Ca^{2+}-dependent manner (usually co-released with another transmitter), binds to extracellular ATP receptors, and is rapidly removed from the synaptic cleft by degradative enzymes; and exogenous application of ATP mimics the effects of nerve stimulation in many preparations. Until recently, there have been no specific antagonists for the various subtypes of ATP receptors, a factor that has hampered unequivocal demonstration of the transmitter role of ATP.

ATP binds to two functionally distinct cell surface receptors: the ligand-gated ionotropic ATP receptor (P_{2X}) and the G-protein–coupled receptor (P_{2Y}). Three P_{2Y} receptor subtypes have been cloned with the typical seven transmembrane spanning domains, closely resembling the thrombin receptor (P_{2Z}, P_{2T}, and P_{2U}). The nomenclature of the ATP receptor, like that of the dopamine and serotonin receptors, undergoes almost annual revision.[1] However, each of the subtypes is linked to either an ion channel or a G-protein. The ionotropic receptor shares little homology with other ionotropic receptors such as the $5HT_3$, GABA, AMPA, or nicotinic receptors. ATP acts through the P_{2X} receptor to open cation selective channels with rapid kinetics (a few milliseconds), and strong desensitization.[62]

Neuropeptides

A variety of peptides function as neurotransmitters and neuromodulators, notably vasoactive intestinal polypeptide, substance P, enkephalin, endorphin, neuropeptide Y, cholecystokinin, and somatostatin. Most of these peptides exist in several brain regions. Often, a neuron will contain both a small-molecule neurotransmitter, and one or more neuropeptides. The actions of neuropeptides are mediated by G-protein–coupled receptors in most cases. Thus, neuropeptide actions are relatively slow, but not necessarily different from those of classical neurotransmitters. The physiological significance of most of these peptide and nonpeptide neuromodulatory systems is unknown.

OTHER SIGNALING MOLECULES

Adenosine

Adenosine is a purine nucleoside with diverse biological functions. An integral component in the regulation of nucleic acid synthesis[28] and energy metabolism,[61] adenosine also acts as an extracellular signal to influence the function of numerous tissues including brain. In the brain, adenosine modulates synaptic transmission[60] by blocking neurotransmitter release[46, 50] and influences cerebral blood flow.[46, 47] In addition, adenosine analogs block experimentally induced seizures in animals[4] and act as an endogenous anticonvulsant in humans,[11, 12] whereas methylxanthine adenosine antagonists promote seizure activity.[58] Adenosine also may protect the brain against ischemic damage.[52]

One source of adenosine is ATP, which is co-released with several neurotransmitters,[40] and is presumably broken down to adenosine in the extracellular space. Intra- and extracellular enzymes, and pre- and postsynaptic nucleoside transporters tightly regulate tissue levels of adenosine. However, in contrast to classical neurotransmitters, which are discretely released only after neuronal activation, adenosine is released by all cells under basal conditions.[23, 68] After intracellular accumulation, a facilitated diffusion system transports adenosine to the extracellular space.[30] Large fluctuations in extracellular adenosine concentration can occur transiently,[31, 68] although basal levels are in the nanomolar range and quite tightly regulated.[23, 30, 68] Following increased tissue activity, or decreased blood flow, there is rapid production of adenosine from cellular ATP stores. During seizures[12] or hypoxic ische-

mia,[16, 31] extracellular adenosine levels rise to more than 1 μM.

Because adenosine is released by all cells, the specificity of its action occurs at the receptor level. Four specific, G-protein–coupled, cell surface adenosine receptors have been identified. Based on their pharmacology and their action on the cAMP regulatory system, adenosine receptors are grouped into A_1, A_2, and A_3 receptor subtypes.[59] A_1 and A_3 adenosine receptors are inhibitory, whereas A_{2A} and A_{2B} adenosine receptors stimulate adenylyl cyclase activity.[59] However, owing to the complexity imparted by G-protein coupling, activation of adenosine receptors in various brain regions is unpredictable. Nonetheless, it is clear that adenosine acts as a strong inhibitory substance in almost all brain regions.

Adenosine receptor mediated inhibition may involve pre- and postsynaptic sites of action. The postsynaptic action of adenosine is probably mediated by an increase in K^+ conductance, although decreases in Ca^{2+} and Cl^- conductance have also been described. Presynaptically, inhibition of transmitter release by adenosine is likely mediated through a decrease in Ca^{2+} conductance. It may hyperpolarize the presynaptic terminal by enhancing K^+ conductance, or decrease action potential amplitude along the axon.

Nitric Oxide

Nitric oxide (NO) is a unique intercellular messenger because it is both lipid and water soluble. Thus, the diffusion barriers that restrict the action of other chemicals do not hinder movement of NO, although its short half life of a few seconds limits its action. The nitric oxide synthases (NOS), enzymes that manufacture NO from L-arginine, are heme-containing cytochrome p450-like enzymes which require multiple co-factors for full activity. NOS exist in constitutive and inducible forms. Three human genes encode three separate types of NOS: the neuronal form located on chromosome 12, the endothelial form on chromosome 7, and the inducible form on chromosome 17.[67]

NOS-containing neurons are found in the cerebellum, olfactory bulb, cortex, hippocampus, posterior pituitary gland, and in autonomic fibers of the retina. Increase in intracellular Ca^{2+}, for example by NMDA receptor activation, can activate constitutive NOS leading to NO production. The role of NO in the brain is not fully understood, but it appears to participate in LTP,[43] arousal,[3] and pain perception,[38] in addition to its role as an inhibitor of platelet aggregation and vasodilatation.[67] The role of NO in epilepsy is unclear.

CONCLUSION

Many current therapies target neurotransmitter receptors and associated proteins involved in synaptic transmission. Emerging therapies will undoubtedly become more selective in the target neurons they affect as more is known about signal transduction and ionic currents. One promising approach is vector-based gene therapy. As regional defects in receptors and other functional proteins become known, replacement or augmentation of these substances will be accomplished by inserting the parent gene into neighboring cells. Likewise, focally "knocking out" certain harmful gene products may be accomplished by adding antisense to inhibit selective mRNA to target cells. Both viral and synthetic polymer-based vector systems are proving valuable in this regard. Understanding the process of synaptic transmission and neurotransmitter receptor systems will become even more critical to the rational application of these new therapeutic technologies.

REFERENCES

1. Abbracchio MP, Cattabeni F, Fredholm BB, Williams M: Purinoceptor nomenclature: A status report. Drug Dev Res 28:207–213, 1993.
2. Adams PR, Jones SW, Pennefather P, et al: Slow synaptic transmission in frog sympathetic ganglia. J Exp Biol 124:259–285, 1986.
3. Bagetta G, Iannone M, Del Luca C, Nistico G: N-nitro-L-arginine methyl ester of the electrocortical arousal response in rats. Br J Pharmacol 108:296–297, 1993.
4. Barraco RA, Swanson TH, Phillis JW, Berman RF: Anticonvulsant effects of adenosine analogs on amygdala-kindled seizures in rats. Neurosci Lett 46:317–322, 1984.
5. Bennett MK, Scheller RH: A molecular description of synaptic vesicle membrane trafficking. Annu Rev Biochem 63:63–100, 1994.
6. Bettler B, Mulle C: Review: Neurotransmitter receptors II AMPA and kainate receptors. Neuropharmacology 34(2):123–139, 1995.
7. Bliss TVP, Lomo T: Long-lasting potentiation of synaptic transmission in the dentate area of the anaesthetized rabbit following stimulation of the perforant path. J Physiol 232:331–356, 1973.
8. Browning RA: Overview of neurotransmittion: Relationship to the action of antiepileptic drugs. In Faingold CL, Fromm GH (eds): Drugs for Control of Epilepsy: Action on Neuronal Networks Involved in Seizure Disorders. Boca Raton, FL, CRC Press, 1992, pp 23–56.
9. Burgoyne RD, Morgan A: Ca^{2+} and secretory-vesicle dynamics. TINS 18(4):191–196, 1995.
10. Corcoran ME, Mason ST: Role of forebrain catecholamines in amygdaloid kindling, Brain Res 190:473–484, 1980.
11. Dragunow M, Goddard GV, Laverty R: Is adenosine an endogenous anticonvulsant? Epilepsia 26:480–487, 1985.
12. During MJ, Spencer DD: Adenosine: A potential mediator of seizure arrest and postictal refractoriness. Ann Neurol 32:618–624, 1992.
13. Dutar P, Nicoll RA: A physiological role for GABA_b receptors in the central nervous system. Nature 332:156–158, 1988.
14. Dutar P, Nicoll RA: Pre- and postsynaptic GABA_b receptors in the hippocampus have different pharmacological properties. Neuron 1:585–591, 1988.
15. Faingold CL, Browning RA: Mechanisms of anticonvulsant drug action II. Drugs primarily used for absence epilepsy. Eur Pediatr 146:8–17, 1987.
16. Fredholm BB, Dunwiddie TV, Bergman B, Lindstrom K: Levels of adenosine and adenine nucleotides in slices of rat hippocampus. Brain Res 295:127–136, 1984.
17. Gray EG: Axo-somatic and axo-dendritic synapses of cerebral cortex: An electron-microscope study. J Anat 93:420–433, 1959.
18. Greenamyre JT, Porter RHP: Anatomy and physiology of glutamate in the CNS. Neurology 44(suppl 8):S7–S13, 1994.
19. Hall Z: Ion channels. In Hall Z (ed): An Introduction to Molecular Neurobiology. Sunderland, MA, Sinauer, 1992, pp 81–118.
20. Hall Z: The nerve terminal. In Hall Z (ed): An Introduction to Molecular Neurobiology. Sunderland, MA, Sinauer, 1992, pp 148–178.
21. Hartzell HC: Distribution of muscarinic acetylcholine receptors and presynaptic nerve terminals in amphibian heart. J Cell Biol 86:6–20, 1980.
22. Hein L, Kobilka BK: Adrenergic receptor signal transduction and regulation. Neuropharmacology 34(4):357–366, 1995.
23. Hertz L: Nucleoside transport in cells: Kinetics and inhibitor effects. In Phillis JW (ed): Adenosine and Adenine Nucleotides as Regulators of Cellular Function. Boca Raton, FL, CRC Press, 1991, pp 85–107.
24. Hille B: Ionic Channels of Excitable Membranes, 2nd ed. Sunderland, MA, Sinauer, 1992, p 607.
25. Hille B: Modulation of ion-channel function by G-protein-coupled receptors. TINS 12:531–536, 1994.
26. Jan LY, Jan YN: Peptidergic transmission in sympathetic ganglia of the frog. J Physiol 327:219–246, 1982.
27. Jobe PC, Mishra PK, Dailey JW: Genetically epilepsy-prone rats: Actions of antiepileptic drugs and monoaminergic neurotransmitters. In Faingold CL, Fromm GH (eds): Drugs for Control of Epilepsy. Boca Raton, FL, CRC Press, 1992, pp 253–275.
28. Kornberg A: Aspects of DNA replication. Cold Spring Harbor Symp Quant Biol 43:1–9, 1979.
29. Lanthorn TH, Cotman CW: Baclofen selectively inhibits excitatory synaptic transmission in the hippocampus. Brain Res 255:171–178, 1981.
30. Linden J: Purinergic systems. In Seigel GJ, Agranoff BW, Albers BW, Molinoff PB (eds): Basic Neurochemistry. New York, Raven Press, 1994, pp 401–416.
31. Lloyd HGE, Lindström K, Fredholm BB: Intracellular formation and release of adenosine from rat hippocampal slices evoked by electrical stimulation or energy depletion. Neurochem Int 23:173–185, 1993.
32. Lomo T: Frequency potentiation of excitatory synaptic activity in the dentate area of the hippocampal formation. Acta Physiol Scand 68(277):128, 1966.
33. Lüddens H, Korpi ER, Seeburg PH: GABA_A/benzodiazepine receptor heterogeneity: Neurophysiological implications. Neuropharmacology 34(3):245–254, 1995.
34. Madison DV, Malenka RC, Nicoll RA: Mechanisms underlying long-term potentiation of synaptic transmission. Annu Rev Neurosci 14:379–397, 1991.
35. McCormick DA: Membrane properties and neurotransmitter actions. In Shepherd GM, (ed): The Synaptic Organization of the Brain. New York, Oxford University Press, 1990, pp 32–66.
36. McNaughton BL, Douglas RM, Goddard GV: Synaptic enhancement in fascia dentata: Cooperativity among coactive afferents. Brain Res 157:277–293, 1978.
37. Monyer H, Sprengel R, Schoepfer R: Heteromeric NMDA receptors: Molecular and functional distinction of subtypes. Science 256:1217–1221, 1992.
38. Moore PK, Babbedge RC, Wallace P, et al: 7-Nitro indazole, an inhibitor of nitric oxide synthase, exhibits anti-nociceptive activity in the mouse without increasing blood pressure. Br J Pharmacol 108:296–297, 1993.
39. Moriyoshi K, Masu M, Ishii T: Molecular cloning and characterization of the rat NMDA receptor. Nature 354:31–37, 1991.
40. Mosqueda-Garcia R, Tseng C-J, Appalsamy M, et al: Cardiovascular excitatory effects of adenosine in the nucleus of the solitary tract. Hypertension 18:494–502, 1991.
41. Nagy JI, Yamamoto T, Shiosaka S, et al: Immunohistochemical localization of gap junction protein in rat CNS: A preliminary account. In Hertzberg EL, Johnson RG (eds): Gap Junctions. New York, Alan R. Liss, 1988, pp 375–389.
42. O'Connor V, Augustine GJ, Betz H: Synaptic vesicle exocytosis: Molecules and models. Cell 76:785–787, 1994.
43. O'Dell TJ, Hawkins RD, Kandel ER, Arancio O: Tests of the role of two diffusable substances in long-term potentiation: Evidence for nitric oxide as a possible early retrograde messenger. Proc Natl Acad Sci USA 88:1285–1298, 1991.
44. Ortels MO, Lunt GG: Evolutionary history of the ligand-gated ion-channel superfamily of receptors. Trends Neurosci 18:121–127, 1995.
45. Peroutka SJ: 5-Hydroxytryptamine receptor subtypes and the pharmacology of migraine. Neurology 43(suppl 3):S34–S38, 1993.
46. Phillis JW, Wu PH: The role of adenosine and its nucleotides in central synaptic transmission. Prog Neurobiol 16:187–239, 1981.
47. Phillis JW, Wu PH: Roles of Adenosine and Adenine Nucleotides in the Central Nervous System. New York, Raven Press, 1983, pp 219–235.
48. Pin JP, Duvoisin R: Review: Neurotransmitter receptors I. The metabotropic glutamate receptors: Structure and functions. Neuropharmacology 34(1):1–26, 1995.
49. Pincus JH, Kiss A: Phenytoin reduces early acetylcholine release after depolarization. Brain Res 397:103–108, 1986.
50. Ribeiro JA: Purinergic regulation of transmitter release. In Phillis JW (ed): Adenosine and Adenine Nucleotides as Regulators of Cellular Function. Boca Raton, FL, CRC Press, 1991, pp 155–167.
51. Ross EM: G proteins and receptors in neuronal signaling. In Hall Z (ed): An Introduction to Molecular Neurobiology. Sunderland, MA, Sinauer, 1992, pp 181–206.
52. Rudolphi KA: Manipulation of purinergic tone as a mechanism for controlling ischemic brain damage. In Phillis JW (ed): Aden-

osine and Adenine Nucleotides as Regulators of Cellular Function. Boca Raton, FL, CRC Press, 1991, pp 423–436.

53. Ruegg MA, Bixby JL: Agrin orchestrates synaptic differentiation at the vertebrate neuromuscular junction. TINS 21(1):22–27, 1998.

54. Scheller RH, Hall Z: Chemical messengers at synapses. In Hall Z (ed): An Introduction to Molecular Neurobiology. Sunderland, MA, Sinauer, 1992, pp 119–147.

55. Schmutz M: Carbamazepine. In Frey HH, Janz D (eds): Antiepileptic Drugs: Handbook of Experimental Pharmacology. Berlin, Springer-Verlag, 1985, pp 479–495.

56. Schweizer FE, Betz H, Augustine GJ: From vesicle docking to endocytosis: Intermediate reactions of excocytosis. Neuron 14:689–696, 1995.

57. Shepherd GM, Koch C: Introduction to synaptic circuits. In Shepherd GM (ed): The Synaptic Organization of the Brain. New York, Oxford University Press, 1990, pp 3–31.

58. Snyder SH, Katims JJ, Annau Z, et al: Adenosine receptors and behavioral actions of methylxanthines. Proc Natl Acad Sci USA 78:3260–3264, 1981.

59. Stiles GL: Adenosine receptors. J Biol Chem 267:6451–6454, 1992.

60. Stone TW: Adenosine as a neuroactive compound in the central nervous system. In Phillis JW (ed): Adenosine and Adenine Nucleotides as Regulators of Cellular Function. Boca Raton, FL, CRC Press, 1991, pp 329–338.

61. Stryer L: Metabolism: Basic concepts and design. In Stryer L (ed): Biochemistry. San Francisco, WH Freeman, 1981, pp 235–254.

62. Surprenant A, Buell G, North RA: P_{2X} receptors bring new structure to ligand-gated ion channels. TINS 18:224–229, 1995.

63. Swanson TH: The pathophysiology of human mesial temporal lobe epilepsy. J Clin Neurophysiol 12(1):2–22, 1995.

64. Swanson TH, Krahl SE, Liu YZ, et al: Evidence for physiologically active axonal adenosine receptors in the rat corpus callosum. Brain Res 784:188–198, 1998.

65. Thompson SM, Gahwiler BH: Comparison of the actions of baclofen at pre- and postsynaptic receptors in the rat hippocampus *in vitro*. J Physiol 451:329–345, 1992.

66. Tsien RW, Lipscombe D, Madison DV, et al: Multiple types of neuronal calcium channels and their selective modulation. TINS 11:431–438, 1988.

67. Vallance P, Moncada S: Nitric oxide—from mediator to medicines. Royal Coll Phys Lond 28(3):209–219, 1994.

68. White TD, Hoehn K: Adenosine and adenine nucleotides in tissues and perfusates. In Phillis JW (ed): Adenosine and Adenine Nucleotides as Regulators of Cell Function. Boca Raton, FL, CRC Press, 1991, pp 109–118.

Chapter 8

Waveform Analysis and Near- and Far-Field Concepts

Jun Kimura, MD

GENERATION AND PROPAGATION OF
   ACTION POTENTIAL
VOLUME CONDUCTION
CURRENT PATHWAYS
PHYSIOLOGICAL SOURCES

NEAR-FIELD RECORDING
TEMPORAL DISPERSION AND PHASE
   CANCELLATION
FAR-FIELD RECORDING

GENERATION AND PROPAGATION OF ACTION POTENTIAL

Externally applied current can excite nerves or muscles. These excitable tissues generate action potentials when depolarized to the threshold. At rest they are polarized with inside negativity. To depolarize a nerve, one must attract more positive charges inside the axon and more negative ones outside. This can be attained by capacitative current directed outward from the axon. Thus, an externally applied current sink or *cathode* can generate an artificial action potential.

A pair of electrodes placed on the surface of a nerve or muscle at rest register no difference of potential between them. If, in the tissue activated at one end, the propagating action potential reaches the nearest electrode (G1), then G1 becomes negative relative to the distant electrode (G2). This results in an upward deflection of the tracing according to the convention of clinical electrophysiology (Fig. 8–1, *top*). With further passage of the action potential, the trace returns to the baseline at the point where the depolarized zone affects G1 and G2 equally. When the action potential moves farther away from G1 and toward G2, G2 becomes negative relative to G1 (or G1 becomes positive relative to G2). Therefore, the trace now shows a downward deflection. It then returns to the baseline as the nerve activity becomes too distant to affect the electrical field near the recording electrodes. This produces a diphasic action potential. In the same situation, permanent depolarization induced by damage at one end results in monophasic rather than diphasic recording (Fig. 8–1, *bottom*). The preceding discussion dealt with the action potential recorded directly with no external conduction medium intervening between the pickup electrodes and the nerve or muscle.

VOLUME CONDUCTION

During clinical study, connective tissue and interstitial fluid act as a volume conductor surrounding the generator sources.[5, 9, 17] Here, an electrical field spreads instantaneously from a source represented as a dipole; that is, a pair of positive and negative charges. In a volume conductor, currents move into an infinite number of pathways between the positive and negative ends of the dipole, with the greatest densities (the number of charges passing through a unit area per unit time) along the straight path.

Analysis of potential distribution in a volume conductor requires knowledge of the pathways of current that generates potential gradients.

CURRENT PATHWAYS

Current path is described as current lines connecting both current source and sink of the dipole. The more current lines penetrate a unit area (or, the higher the current density), the steeper is the potential gradient. In other words, isoelectric lines (points of identical unit potentials), which transect the current lines, become more crowded. Current pathways are predicted by a simple law that *current tends to flow along the path of minimal resistance.*

If the current flows in an infinite homogeneous volume conductor, a potential at a given point can be calculated mathematically; the potential is inversely related to the square of the distance from the dipole and proportional to the cosine of the angle subtended by the point and the orientation of the dipole. By contrast, the human body provides a notoriously inhomogeneous and finite-shaped conductor. As a consequence, the analysis of current lines becomes extremely complicated and hardly predictable. Various tissues in the body have their own resistance (Table 8–1).[35] For instance, the skull has a resistivity (resistance per unit length) approximately 100 times higher than that of the brain. Brain potentials are more apt to be confined within the brain than to penetrate the skull, thus having more diffuse scalp distribution than those recorded directly from the brain.

The volume conductor has another characteristic of importance for determining the potential. *Capacitance* is the ability of the conductor to store charges. If current flows in a volume, a portion is used to store positive and negative charges of equal quantity. If capacitance is high, it takes longer to generate potential difference at distance. If low, the same current produces steeply rising potential. In the presence of a fixed capacitance, a quick build-up of current gives rise to a volume-conducted potential at more distant locations than a slow development of the same current. Source currents with highly synchronized onset tend to generate greater volume-conducted potentials than those with gradual onset. Thus, capacitance serves much like a resistance in a situation with changing current. The term *impedance* is used to denote the total resistance acting against the changing current, including the resistance per se and the capacitance.

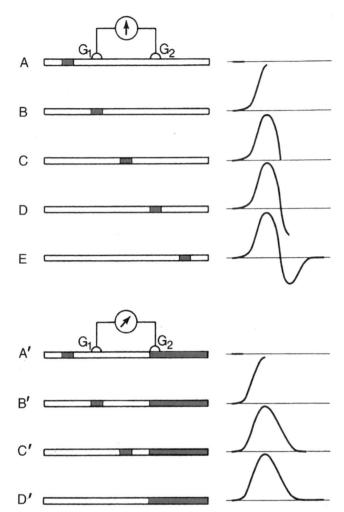

FIGURE 8–1. Diphasic *(top)* and monophasic recording *(bottom)* of an action potential represented by the shaded area. As the impulse propagates from left to right in the top series, the two electrodes see no potential difference in *A, C,* and *E.* Relative to the reference electrode (G2), the active electrode (G1) becomes negative in *B,* and positive in *D,* resulting in a diphasic potential. In the bottom, the darkened area on the right indicates a killed end with permanent depolarization, making G1 positive relative to G2 in *A', C',* and *D'.* In *B',* G1 and G2 see no potential difference, causing upward deflection from the positive baseline to 0 potential.

PHYSIOLOGICAL SOURCES

Physiological sources of current comprise those generated by synaptic potentials and action potentials. Action potentials are a propagating source of current with front positivity. They are always drawing current lines starting from this front current source and ending at the sink just behind, thus producing currents distributed widely. Potentials associated with these currents can be recorded from distance in a volume conductor. Such potential fields are called *open fields.* Some synaptic potentials in central nuclei make a spherical

dipolar sink-source. The central current sink is juxtaposed with the source surrounding it. In this case, current lines have a short path confined within the nucleus. As a result, no volume-conducted potentials are recorded at a distance. Such a configuration is termed a *closed field* distribution.

The current flow decreases in proportion to the square of the distance from the generator source. Thus, the effect of the dipole gives rise to voltage difference between an active recording electrode in the area of high current density and a reference electrode at a distance. Whether the electrode records positive or negative potentials depends on its spatial orientation to the opposing charges of the dipole. For example, in an idealized homogenous volume the active electrode located at a point equidistant from the positive and negative charges registers no potential, provided the reference electrode is indifferent. The factors that determine the amplitude of the recorded potential include charge density (the net charge per unit area), surface areas of the dipole, and its proximity to the recording electrode.

Moving the recording electrode short distances away from the muscle fibers results in obvious reduction in amplitude. Additionally, the rise time, or the duration of the positive-to-negative rising phase, becomes greater. The rise time gives an important clue in determining the proximity to the generator source. Amplitude may not serve for this purpose, because it may decrease with smaller muscle fibers or lower fiber density.

The specific potential recorded under a particular set of conditions depends not only on the location of the recording

TABLE 8–1	Volume Conduction in Inhomogeneous Media	
	Medium	**Resistivity (Ω * cm)**
	Seawater	20
	CSF	64
	Blood	150
	Spinal cord	180–1200
	Cortex	230–350
	White matter	650
	Bone	16,000
	Skull	20,000

From Nunez PL: An overview of electromyogenetic theory. In Nunez PL (ed): Electric Fields of the Brain. New York, Oxford University Press, 1981.

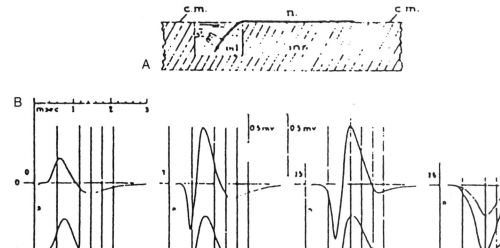

FIGURE 8–2. Volume-conducted potentials were recorded at various points after stimulating the frog sciatic nerve placed on saline-extended wax paper. The reference electrode is placed at a distance. The recording electrodes put in line and far from the nerve terminal register positive stationary peaks (two superimposed traces at the right upper corner). (From Lorente de Nó RA: A study of nerve physiology. Studies from the Rockefeller Institute, 132, 384–477, 1947.)

electrodes relative to the active tissue at any instant in time but also on the physical characteristics of the volume conductor.[8, 21, 25, 33, 37] The near-field potential (NFP) and far-field potential (FFP) distinguish two different manifestations of the volume-conducted field.[19] The NFP represents recording of a potential change near the source, usually by a pair of closely spaced electrodes over the propagating impulses. In contrast, the FFP implies the detection of a voltage step that takes place at a certain point along the signal pathways, usually recorded by a pair of widely separated electrodes located far from the traveling volleys.

NEAR-FIELD RECORDING

Analyzing the waveform plays an important role in the assessment of a nerve or muscle action potential. A sequence of potential changes arises as the two sufficiently close wave fronts travel in the volume conductor from left to right. This results in a positive-negative-positive triphasic wave as the moving fronts of depolarization and repolarization approach, then reach, and finally pass beyond the point of the recording electrode. Figure 8–2 depicts an example of a propagating action potential recorded by series of electrodes placed along the nerve at various depth from the source.[32]

The theory of solid angle pertains to the analysis of an action potential recorded through a volume conductor. The resting transmembrane potential consists of a series of dipoles arranged with positive charges on the outer surface and negative charges on the inner surface. The solid angle subtended by an object equals the area of its surface divided by the squared distance from a specific point to the surface.[2, 18] Thus, it increases in proportion to the size of the polarized membrane viewed by the electrode and decreases with the distance between the electrode and the membrane.

Solid angle approximation (Figs. 8–3 and 8–4) closely predicts the potential derived from a dipole layer. The propagating action potential, visualized as a positively charged wave front, represents depolarization at the cross-section of the nerve at which the transmembrane potential reverses.[37] A negatively charged wave front follows, signaling repolarization of the activated zone.

Thus, an orthodromic sensory action potential from a deeply situated nerve gives rise to a triphasic waveform in surface recording. The potentials originating in the region near the electrode, however, lack the initial positivity, in the absence of an approaching volley. A compound muscle action potential therefore appears as a negative-positive diphasic waveform when recorded with the electrode near the end-plate region where the volley initiates. In contrast, a pair of electrodes placed away from the activated muscle registers a positive-negative diphasic potential, indicating that the impulse approaches but does not reach the recording site. Similarly, a pair of electrodes placed at the site of injury during spinal cord monitoring registers a positive-negative diphasic potential. This indicates, as implied by the term *killed-end effect*, that the impulse fails to propagate across the recording site.

The number of triphasic potentials generated by individual muscle fibers summate to give rise to a motor unit potential in electromyography. The waveform of the recorded potential varies with the location of the recording tip relative to the source of the muscle potential.[3] Thus, the same motor unit shows multiple profiles depending on the site of the exploring needle.

TEMPORAL DISPERSION AND PHASE CANCELLATION

Temporal dispersion increases with more proximal stimulation in proportion to the distance to the recording site[24] because the impulses of slow conducting fibers lag increasingly behind those of fast conducting fibers over a long conduction path.[4, 29] With increasing distance between stimulating and pick-up electrodes, the recorded potentials become smaller in amplitude, and contrary to the common belief, the area under the waveform also diminishes.

Thus, for example, stimulation proximally in the axilla or Erb's point may give rise to a small or inconsistent digital potential in contrast to a large response elicited by stimulation at the wrist or palm.[22, 36, 43] For the same number of conducting fibers activated by the stimulus, the sizes of both sensory and muscle action potentials change almost linearly with the length of the nerve segment, but not to the same degree.[24]

Physiological temporal dispersion affects the sensory action potential more than the muscle response, perhaps because of the difference in duration of individual unit dis-

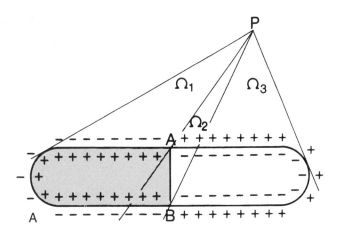

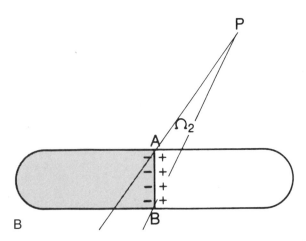

FIGURE 8–3. Potential recorded at P from a cell with active (dark area) and inactive region. In A, total solid angle consists of Ω_1, Ω_2, and Ω_3. Potential at P subtending solid angles Ω_1 and Ω_3 equals zero as, in each, the nearer and farther membranes form a set of dipoles of equal magnitude but opposite polarity. In Ω_2, however, cancellation fails because these two dipoles show the same polarity at the site of depolarization. In B, charges of the nearer and farther membranes subtending solid angle Ω_2 are placed on the axial section through a cylindrical cell. A dipole sheet equal in area to the cross-section then represents the onset of depolarization traveling along the cell from left to right with positive poles in advance. (Adapted from Patton HD: Special properties of nerve trunks and tracts. In Ruch, HW, Patton HD, Woodbury JW, Towe AL (eds): Neurophysiology, ed 2. Philadelphia, WB Saunders, 1969, pp 73–94.)

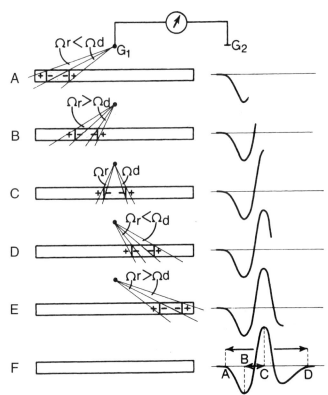

FIGURE 8–4. Triphasic potential characterized by amplitude, duration (A–D), and rise time (B–E). A pair of wave fronts of opposite polarity represents depolarization and repolarization. The action potential travels from left to right in a volume conductor with the recording electrode (G1) near the active region and reference electrode (G2) on a remote inactive point.

As shown in A, G1 initially sees the positivity of the first dipole, which subtends a greater solid angle (Ωd) than the second dipole of negative front (Ωr). In B, this relationship reverses with gradual diminution of Ωd compared with Ωr, as the active region approaches G1. In C, the maximal negativity signals the arrival of the impulse directly under G1, which now sees only negative ends of the dipoles. In D, the negativity declines as G1 begins to register the positive end of the second dipole. In E, the polarity reverses again as Ωr exceeds Ωd. In F, the trace returns to the baseline when the active region moves further away. The last positive phase, though smaller in amplitude, lasts longer than the first, indicating a slower time course of repolarization.

charges between nerve and muscle (Fig. 8–5). In support of this view, the duration change of the sensory potential, expressed as a percentage of the respective baseline values, far exceeds that of the muscle response.[24] With short-duration diphasic sensory spikes, a slight latency difference could line up the positive peaks of the fast fibers with the negative peaks of the slow fibers, cancelling both (Fig. 8–6). In motor unit potentials of longer duration, however, the same latency shift would superimpose peaks of opposite polarity only partially, resulting in little cancellation (Fig. 8–7).

Mathematical models[7, 38] allow estimation of conduction velocity distribution in a nerve bundle based on a detailed model of the compound muscle action potential as a weighted sum of asynchronous single fiber action potentials. In one study,[12] nerve conduction velocities of the large myelinated axons that contributed to the surface recorded response varied by as much as 25 m/sec between fast and slow sensory fibers, but over a much narrower range of 11 m/sec for motor fibers. This observation, although not universally accepted,[11] would also, at least in part, explain the different

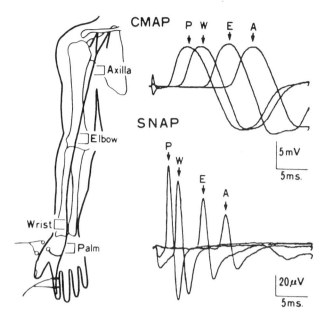

FIGURE 8–5. Simultaneous recordings of compound muscle action potentials (CMAP) from the thenar eminence and antidromic sensory nerve action potentials (SNAP) from index finger after stimulation of the median nerve at palm, wrist, elbow, and axilla. A series of stimuli from the wrist to the axilla elicited nearly the same CMAP but progressively smaller SNAP. (From Kimura J, Machida M, Ishida T, et al: Relation between size of compound sensory or muscle action potentials and length of nerve segment. Neurology 36:647–652, 1986, with permission.)

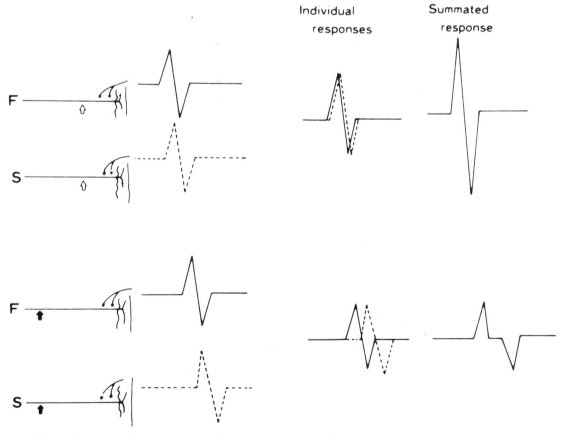

FIGURE 8–6. A model for phase cancellation between fast (F) and slow (S) conducting sensory fibers. With distal stimulation, two unit discharges summate in phase to produce a sensory action potential twice as large. With proximal stimulation, a delay of the slow fiber causes phase cancellation between the negative peak of the fast fiber and positive peak of the slow fiber, resulting in a 50% reduction in size of the summated response. (From Kimura J, Machida M, Ishida T, et al: Relation between size of compound sensory or muscle action potentials and length of nerve segment. Neurology 36:647–652, 1986, with permission.)

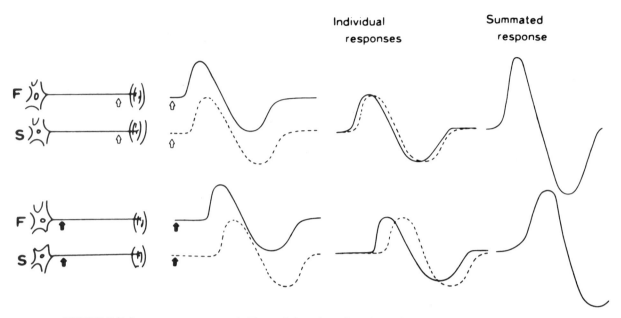

FIGURE 8–7. Same arrangements as in Figure 8–6 to show the relationship between fast (F) and slow (S) conducting motor fibers. With distal stimulation, two unit discharges representing motor unit potentials summate to produce a muscle action potential twice as large. With proximal stimulation, long duration motor unit potentials still superimpose nearly in phase despite the same latency shift of the slow motor fiber as the sensory fiber shown in Figure 8–6. Thus, a physiological temporal dispersion alters the size of the muscle action potential only minimally, if at all. (From Kimura J, Machida M, Ishida T, et al: Relation between size of compound sensory or muscle action potentials and length of nerve segment. Neurology 36:647–652, 1986, with permission.)

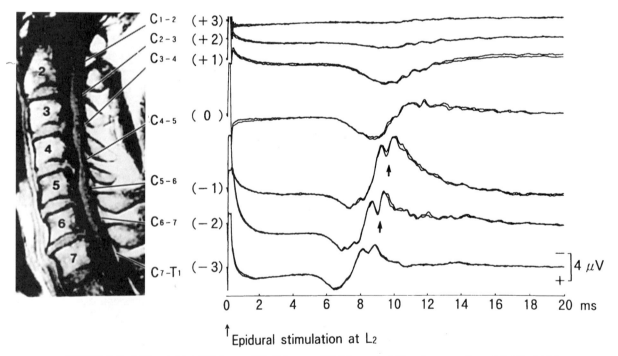

FIGURE 8–8. A T1-weighted MR image (TR 400 msec; TE 13 msec) (left) and a recording of SSEPs (right) obtained from a 65-year-old patient with cervical myelopathy. The SSEPs were recorded unipolarly from the ligamentum flavum of C7–T1 through C12 after epidural stimulation at L2. Note the progressive increase in size of the negative component (arrows pointing up) from C7–T1 (−3) through C5–6 (−1) with the abrupt reduction at C4–5 (0) followed by a monophasic positive wave at C3–4 (+1). The negative wave doubled in amplitude and quadrupled in area at −1 compared to −3. The 0 level corresponded to the level of the spinal cord showing the most prominent compression on the MR image. (From Tani T, Ushida T, Yamamoto H, et al. Waveform analysis of spinal somatosensory evoked potential: Paradoxically enhanced negative peaks immediately caudal to the site of conduction block. Electroencephalogr Clin Neurophysiol 108:325–330, 1998.)

effect of temporal dispersion on sensory and motor fibers for a given length of nerve segment.[36]

For the reasons stated above, a major reduction in size of the compound sensory action potential can result solely from physiologic phase cancellation. In contrast, the same degree of temporal dispersion has little effect on compound muscle action potential. However, this may not hold if the latency difference between fast and slow conducting motor fibers increases substantially as predicted by computer-simulation, with a broader spectrum of motor nerve conduction velocities.[30]

These observations support the view that temporal dispersion can effectively reduce the area of diphasic or triphasic evoked potentials recorded in a bipolar derivation. Thus, in a demyelinating neuropathy, a substantial phase cancellation of the muscle action potential could give rise to a false impression of motor conduction block. Sustained reduction in size of the compound muscle action potential, therefore, does not necessarily imply a prolonged neurapraxia. The loss of area under the waveform seen in the absence of conduction block implies a duration-dependent phase cancellation of unit discharges within the compound action potential. An awareness of this possibility helps properly analyze dispersed action potentials in identifying various patterns of neuropathic processes.[24]

Referential derivation of a monophasic waveform conserves the area irrespective of stimulus sites and circumvents the ambiguity. This type of recording, however, may register a stationary far-field potential generated by the propagating impulse crossing the partition of the volume conductor.[26] Such a steady potential could, in turn, distort the waveform of the near-field activity.

A conduction block ranks first among abnormalities associated with spinal cord compression causing clinical weakness and sensory loss. Similar to the studies of the peripheral nerve, understanding of the waveform changes would maximize the diagnostic value of spinal somatosensory evoked potential (SSEP). The precise site of conduction block can be demonstrated by an abrupt reduction in the amplitude of the evoked potential. In addition, the computer model predicts that this change is accompanied by an increased negative wave caudally and an enhanced monophasic positive wave rostrally, making SSEP assessment of a localized lesion considerably easier.[41] The apparently paradoxical enhancement of negative peak probably results from the loss of physiological phase cancellation.

In the presence of a conduction block, the leading dipole stops traveling when it reaches the site of involvement, although the trailing dipole continues to propagate until it arrives at the same point (see Fig. 8–4). If the block involves only the fast conducting fibers, the absence of their terminal-positive phases paradoxically enhances the negative peak of the spinal SSEPs because the negative phases of the slower fibers escape the physiological phase cancellation. At the points immediately preceding the block, the identical mechanism enhances the spinal SSEP even when all the fibers sustain a conduction block. Thus, SSEPs would consist of positive-negative diphasic waves with enhanced negativity at points immediately preceding the block, a diphasic wave with reduced negativity at the point of the block, and initial-positive waves alone or abolition of any wave at points beyond the block.

Figure 8–8 shows spinal SSEPs recorded unipolarly from the ligamentum flavum at multiple levels after epidural stimulation of the cauda equina in a patient with cervical spondylotic myelopathy.[40] In our experience, this combination of an abrupt reduction in amplitude of the negative peak at one level, augmentation of the negative peaks in the leads closely caudal to that level, and monophasic positive waves at

more rostral levels constitutes a typical pattern of waveform changes, indicating a complete focal conduction block. The mechanism underlying such alterations in waveform can be analyzed with the help of a simplified mathematical model. Paradoxical increase of amplitude in a partial conduction block also can best be explained by the same underlying mechanism.

FAR-FIELD RECORDING

A bipolar derivation, used in conventional nerve conduction studies, registers primarily, though not exclusively, NFP from the axonal volley along the course of the nerve. In contrast, a monopolar or referential montage preferentially detects FFP, although it may also register NFP, if the impulse passes near the active (G1) or indifferent (G2) electrode. The far-field recording has gained popularity in the study of evoked potentials for detection of a voltage source generated at a distance.[6, 10, 33]

Earlier studies on short latency auditory evoked potentials suggested that neural discharges from the brain stem may account for FFPs.[19] This assumption led to the common belief that stationary peaks of cerebral evoked potentials generally originated from fixed neural generators, such as those which occur at relay nuclei. Subsequent animal experiments,[21, 35] however, emphasized the role of a synchronized volley of action potentials within afferent fiber tracts as the source of FFPs. Besides, the initial positive peaks of the scalp recorded median (P9 and P11) and tibial (P17 and P24) somatosensory evoked potential (SEP) are generated before the propagating sensory nerve action potentials reach the second order neurons in the dorsal column.[6, 10, 20, 27, 28, 33] These peaks, therefore, must result from axonal volleys of the first order afferents.[27, 28]

Hence, two types of peaks can occur in far-field recording: a volume-conducted potential representing a fixed neural activity such as synaptic discharges, and a stationary peak generated by the axonal volley propagating through the volume conductor. Short sequential segments of the brain stem pathways may summate in far-field recording, resulting in successive peaks of the recorded potentials.[1, 21, 42] This mechanism by itself, however, does not account for FFPs derived from the propagating volleys in SEPs of the median or tibial nerve.

We postulated that stationary far-field activity can appear from a moving source at certain fixed points in time when the impulse traverses the boundary of the heterogeneous volume conductor. This hypothesis was confirmed by latency comparison between the bipolarly recorded traveling source and the referentially recorded stationary peaks.[15, 25, 28, 31] In fact, the initial positive peaks of the median (P9) and tibial SEPs (P17) arise when the propagating volleys enter the shoulder and pelvic girdles.[15, 16, 25] In addition, change in the position of shoulder girdle slightly but significantly alters the latency of P9.[10] Similarly, the subsequent positive peaks of the median and tibial SEPs develop as the impulses reach the cervical cord (P11) and conus medullaris (P24), and finally traverse the foramen magnum (P14 and P30).

To gain further insights into the origin of FFPs, we tested our hypothesis using the palm and digits as compartments of a volume conductor. In referential recording of antidromic radial sensory potentials (Figs. 8–9 and 8–10), the digital electrodes detected two stationary FFPs, PI–NI and PII–NII.[23] When compared to a bipolar recording of the traveling source, PI occurred with the arrival of the propagating sensory impulse at the wrist, and PII, at the base of the digit. In addition, we found that the higher the stimulus

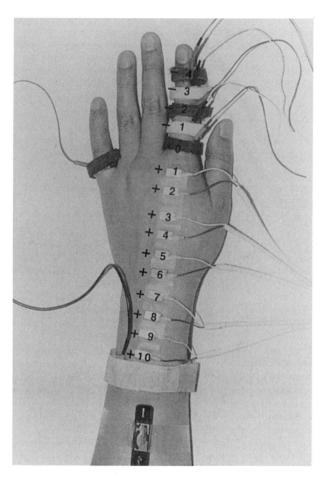

FIGURE 8–9. Stimulation of the radial nerve 10 cm proximal to the styloid process of the radius and serial recording of antidromic sensory potentials in 1.5-cm increments along the length of the radial nerve. The "0" level was at the base of the second digit where the volume conductor changes abruptly. In most hands, " + 6" was near the distal crease of the wrist, where another, less obvious, transition of volume conductor geometry takes place. The ring electrode around the fifth digit is an indifferent lead for referential recordings.

intensity, the greater the amplitude of the stationary peak (Fig. 8–11), in proportion to the propagating volleys passing the boundary of the volume conductor.[23] All these findings are consistent with the notion that FFPs are generated by a propagating action potential composed of a leading and trailing dipole, as stated later in this discussion.[13, 14, 22]

Following stimulation of the radial nerve at the forearm, no field change is registered from the digit while the antidromically propagating sensory nerve action potential is contained within the forearm. With the passage of the positive front of the leading dipole across the wrist, the palm becomes suddenly positive as the current begins to circulate in this larger volume. The base of the index finger becomes positive nearly instantaneously by this circulating current, albeit some potential gradient. The current circulating from the positive to the negative end of the dipole does not penetrate into the digit, which, being a constricted medium, has a high resistance. In the absence of current flow, there is no gradient of potential. Thus, the positivity at the base is also registered simultaneously at the tip of the digit (PI) without potential decay as though the volume on the far side is connected by wire. As the nerve volley travels farther distally, the current continues to circulate primarily within the palm compartment, changing the voltage of the digit very little. With the arrival of the positive front of the leading dipole at the base of the index finger, the entire volume of the digit suddenly becomes positive (PII) by the same mechanism as postulated for PI. Both PI and PII reverse to NI and NII, presumably reflecting the arrival of the negative front of the trailing dipole at the respective boundary.

For PI generated at the wrist, the trailing dipole is contained in a relatively restricted volume of the forearm when the leading dipole crosses into a large volume of the palm.

For PII generated at the metacarpophalangeal junction, the leading dipole enters a small cylindrical volume of the digit while the trailing dipole is still located in a large flat volume of the palm. In both instances, one of the dipoles is trapped in a smaller volume, creating an increased electrical field density per unit volume compared to the other dipole in a larger volume. Thus, the two wave fronts have a different dipole moment, thereby generating a potential difference at the boundary. This theory also accounts for the positive polarity of FFPs, if recorded with G1 at far side, whether the propagating impulses enter or exit the constriction. The computer model of Stegman and associates[39] indeed predicts that, in the case of a boundary constriction, all points on the far side begin to go positive when the generator approaches the boundary. For the impulse moving from a small to large volume, Cunningham and coworkers[8] believe that the chamber approached also detects a shallow positive potential initially, only to revert suddenly to negativity as the generator crosses the partition.

To draw an analogy, an oncoming train (axonal volley) becomes simultaneously visible (FFP) to all bystanders at a distance (series of recording electrodes) as it emerges from a tunnel (partition of volume conductor), whereas the same bystanders see the train pass by at different times (NFP), depending on their position along the railroad.[23] The designation junctional or boundary potential specifies the source of the voltage step by location and differentiates this type of FFP from fixed neural generators (Fig. 8–12). A pair of electrodes only a short distance apart best detects such a stationary potential, so long as they are placed across the partition in question. This observation calls for reassessment of the commonly used dichotomy, equating a referential recording with FFP and a bipolar recording with NFP.

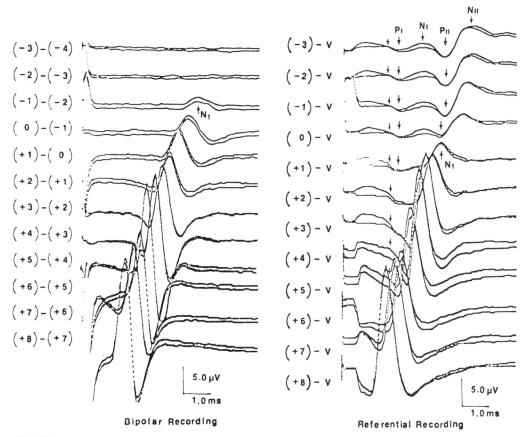

Bipolar Recording

Referential Recording

FIGURE 8–10. Sensory nerve potentials across the hand and along the second digit in a normal subject recorded antidromically after stimulation of the superficial sensory branch of the radial nerve 10 cm proximal to the styloid process of the radius. The site of recording is indicated (Fig. 8–9). In a bipolar recording *(left)*, the initial negative peaks, N1 *(arrow pointing up)*, showed a progressive increase in latency and reduction in amplitude distally and no response was recorded beyond −1. In a referential recording *(right)*, biphasic peaks PI–NI and PII–NII *(arrows pointing down)* showed greater amplitude distally, with a stationary latency irrespective of the recording sites along the digit. The onset of PI extended proximally to the recording electrodes near the wrist *(small arrows pointing down)*, whereas PII first appeared at the base of the digit.

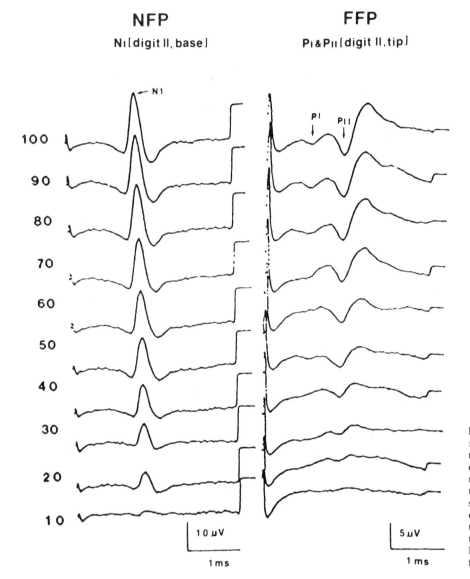

FIGURE 8–11. The far-field potential (FFP; *right*) recorded referentially with G1 at the tip of the second digit and G2 at the fifth digit, and near-field potential (NFP; *left*) registered bipolarly with G2 at the base of the digit and G1 1 cm proximally, after stimulation of the radial nerve. With reduction of stimulus from a maximal *(top)* to a threshold intensity *(bottom)* in 10 steps, the amplitude of far-field potential, PI and PII, declined in proportion to that of near-field potential, N1.

FAR-FIELD POTENTIAL

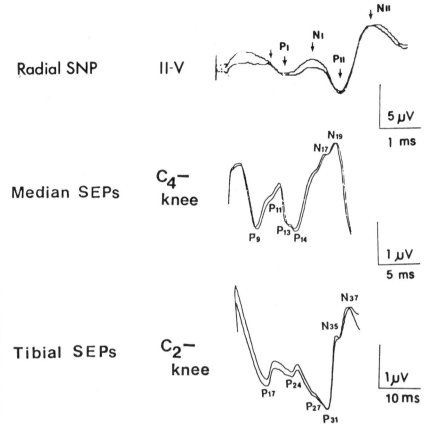

FIGURE 8–12. Scalp recorded somatosensory evoked potentials (SEPs) using a noncephalic reference after stimulation of the median nerve at the wrist *(middle)* and tibial nerve at the ankle *(bottom)*. Both median and tibial SEPs consist of four positive peaks initially and two negative peaks thereafter, all within the first 20 and 40 msec following the stimulus, respectively. For comparison, the top tracing shows far-field potentials, PI–NI and PII–NII, recorded from digit II referenced to digit V, after stimulation of the radial sensory fibers at the forearm.

Clinical studies of cerebral evoked potential exploit far-field recording in the evaluation of subcortical pathways not otherwise accessible. The stationary peaks recorded by this means may not exclusively relate to a specific neural generator, thus, providing no direct measurement of neural activities responsible for sensory transmission. In fact, it should be noted that certain abnormalities of somatosensory and other evoked potentials could result from changes affecting the surrounding tissue and not necessarily from lesions involving the sensory pathways per se. Nonetheless, junctional potentials render clinically useful information, disclosing the arrival of the impulse at a given anatomical landmark forming a partition of the volume conductor.

REFERENCES

1. Arezzo JC, Vaughan HG Jr: The contribution of afferent fiber tracts to the somatosensory evoked potentials. In Bodis-Wollner I (ed): Evoked Potentials. Ann NY Acad Sci 388:679–682, 1982.
2. Brown BH: Theoretical and experimental waveform analysis of human compound nerve action potentials using surface electrodes. Med. Biol Eng 6:375, 1968.
3. Buchthal F, Guld C, Rosenfalck P: Volume conduction of the spike of the motor unit potential investigated with a new type of multielectrode. Acta Physiol Scand 38:331–354, 1957.
4. Buchthal F, Rosenfalck A: Evoked action potentials and conduction velocity in human sensory nerves. Brain Res 3:1–122, 1966.
5. Clark J, Plonsey R: The extracellular potential field of the single active nerve fiber in a volume conductor. Biophys J 8:842–864, 1968.
6. Cracco RQ, Cracco JB: Somatosensory evoked potential in man: Far field potentials. Electroencephalogr Clin Neurophysiol 41:460–466, 1976.
7. Cummins KL, Perkel DH, Dorfman LJ: Nerve fiber conduction-velocity distributions. I. Estimation based on the single-fiber and compound action potentials. Electroencephalogr Clin Neurophysiol 46:634–646, 1979.
8. Cunningham K, Halliday AM, Jones SJ: Stationary peaks caused by abrupt changes in volume conductor dimensions: Potential field modelling [abstract]. Electroencephalogr Clin Neurophysiol 61:S100, 1985.
9. Delisa JA, Brozovich FV: AAEE Minimonograph No. 10, Volume Conduction in electromyography. Rochester, MN, American Association of Electromyography and Electrodiagnosis, 1979.
10. Desmedt JE, Huy NT, Carmeliet J: Unexpected latency shifts of the stationary P9 somatosensory evoked potential far field with changes in shoulder position. Electroencephalogr Clin Neurophysiol 56:623–627, 1983.
11. Dominique J, Shahani BT, Young RR: Conduction velocity in different diameter ulnar sensory and motor nerve fibers. Electroencephalogr Clin Neurophysiol 50:239P–245P, 1980.
12. Dorfman LJ: The distribution of conduction velocities (DCV) in peripheral nerves: A review. Muscle Nerve 7:2–11, 1984.
13. Dumitru D, Jewett DL: Far-field potentials. Muscle Nerve 16:237–254, 1993.
14. Dumitru D, King JG: Far-field potential production by quadrupole generators in cylindrical volume conductors. Electroencephalogr Clin Neurophysiol 88:421–431, 1993.
15. Eisen A, Odusote K, Bozek C, Hoirch M: Far-field potentials from peripheral nerve: Generated at sites of muscle mass change. Neurology 36:815–818, 1986.
16. Frith RW, Benstead TJ, Daube JR: Stationary waves recorded at the shoulder after median nerve stimulation. Neurology 36:1458–1464, 1986.
17. Gath I, Stalberg E: On the volume conduction in human skeletal muscle: In situ measurements. Electroencephalogr Clin Neurophysiol 43:106–110, 1977.

18. Hodgkin AL: The Conduction of the Nervous Impulse. The Sherrington Lectures, Vol 7. Liverpool, Liverpool University Press, 1965.
19. Jewett DL, Williston JS: Auditory-evoked far fields averaged from the scalp of humans. Brain 94:681–696, 1971.
20. Jones SJ: Short latency potentials recorded from the neck and scalp following median stimulation in man. Electroencephalogr Clin Neurophysiol 43:853–863, 1977.
21. Kaji R, Tanaka R, Kawaguchi S, et al.: Origin of short-latency somatosensory evoked potentials to median nerve stimulation in the cat: Comparison of the recording montages and effect of laminectomy. Brain 109:443–468, 1986.
22. Kimura J: Electrodiagnosis in Diseases of Nerve and Muscle: Principles and Practices. Philadelphia, FA Davis, 1989.
23. Kimura J, Kimura A, Ishida T, et al.: What determines the latency and the amplitude of stationary peaks in far-field recordings? Ann Neurol 19:479–486, 1986.
24. Kimura J, Machida M, Ishida T, et al: Relation between size of compound sensory or muscle action potentials and length of nerve segment. Neurology 36:647–652, 1986.
25. Kimura J, Mitsudome A, Beck DO, et al.: Field distributions of antidromically activated digital nerve potentials: model for far-field recording. Neurology (Cleveland) 33:1164–1169, 1983.
26. Kimura J, Mitsudome A, Yamada T, Dickins QS: Stationary peaks from a moving source in far-field recording. Electroencephalogr Clin Neurophysiol 58:351–361, 1984.
27. Kimura J, Yamada T: Short-latency somatosensory evoked potentials following median nerve stimulation. Ann NY Acad Sci 388:689–694, 1982.
28. Kimura J, Yamada T, Shivapour E, Dickins QS: Neural pathways of somatosensory evoked potentials: Clinical implication. In Buser PA, Cobb WA, Okuma T (eds): Kyoto Symposium (EEG Suppl 36), Amsterdam, Elsevier, 1982, pp 328–335.
29. Lambert EH: Diagnostic value of electrical stimulation of motor nerves. Electroenceph Clin Neurophysiol (suppl 22):9–16, 1962.
30. Lee R, Ashby P, White D, Aguayo A: Analysis of motor conduction velocity in human median nerve by computer simulation of compound action potentials. Electroencephalogr Clin Neurophysiol 39:225–237, 1975.
31. Lin JT, Phillips LH II, Daube JR: Far-field potentials recorded from peripheral nerves. Electroencephalogr Clin Neurophysiol 50:174, 1980.
32. Lorente de Nó RA: A study of nerve physiology. Studies from the Rockefeller Institute, 132, 384–477, 1947.
33. Luders H, Dinner SD, Lesser RP, Klem G: Origin of far field subcortical evoked potentials to posterior tibial and median nerve stimulation. Arch Neurol 40:93–97, 1983.
34. Nakanishi T: Origin of action potential recorded by fluid electrodes. Electroencephalogr Clin Neurophysiol 55:114–115, 1983.
35. Nunez PL:, An overview of electromyognetic theory. In Nunez PL (ed): Electric Fields of the Brain. New York, Oxford University Press, 1981, pp 42–74.
36. Olney RK, Miller RG: Pseudo-conduction block in normal nerves. Muscle Nerve 6:530, 1983.
37. Patton HD: Special properties of nerve trunks and tracts. In Ruch HD, Patton HD, Woodbury JW, Towe AL (eds): Neurophysiology, 2nd ed. Philadelphia, WB Saunders, 1965, pp 73–94.
38. Stegeman DF, De Weerd JPC, Notermans SLH: Modelling compound action potentials of peripheral nerves in situ. III. Nerve propagation in the refractory period. Electroencephalogr Clin Neurophysiol 55:668–679, 1983.
39. Stegeman D, Van Oosteron A, Colon E: Simulation of far field stationary potentials due to changes in the volume conductor [abstract]. Electroencephalogr Clin Neurophysiol 61:S228, 1985.
40. Tani T, Ushida T, Yamamoto H, Kimura J: Waveform analysis of spinal somatosensory evoked potential: Paradoxically enhanced negative peaks immediately caudal to the site of conduction block. Electroencephalogr Clin Neurophysiol 108:325–330, 1998.
41. Tani T, Ushida T, Yamamoto H, Okuhara Y: Waveform changes due to conduction block and their underlying mechanism in spinal somatosensory evoked potential: A computer simulation. J Neurosurg 86:303–310, 1997.
42. Vaughan HG Jr: The neural origins of human event-related potentials. In Bodis-Wollner I (ed): Evoked Potentials. Ann NY Acad Sci 388:125–138, 1982.
43. Wiechers D, Fatehi M: Changes in the evoked potential area of normal median nerves with computer techniques. Muscle Nerve 6:532, 1983.

Chapter 9

Muscle Physiology and Pathophysiology

Heidi Lai, MD • Hiroshi Mitsumoto, MD

NEURAL CONTROL
Motor Control System
Concept of the Final Common Path
Feedback System of Muscle Contraction
Anterior Horn Cells
MOTOR UNIT
Physiological Classification
Histochemical Classification of Muscle
 Fiber Type
MUSCLE FIBER
Development and Structure
Subcellular Structure
MUSCLE METABOLISM

STRUCTURAL ABNORMALITIES OF MUSCLE
 FIBERS
Muscle Fiber Atrophy
Muscle Fiber Hypertrophy
Muscle Fiber Degeneration and
 Necrosis
Vacuolar Degeneration
Structural Fiber Changes in Congenital
 Myopathies
Mitochondrial Abnormalities
Fiber Type Abnormality
Ringbinden

The human body has 434 skeletal muscles composed of approximately 250 million muscle fibers; this muscle tissue constitutes 40% to 45% of the total body mass. Therefore, any physiological and biochemical changes in the muscles may affect the entire body significantly. The intricate relationships between structure, biochemistry, physiology, and function in skeletal muscle are among the best understood in humans. Involvement of skeletal muscle disease has brought insight to a number of congenital and genetic disorders. For example, McArdle's disease, the first identified hereditary metabolic error in humans, is a myophosphorylase deficiency myopathy. The concept and the genetic defects of mitochondrial cytopathies were also initially elucidated in muscle disease. The investigation of inclusion body myositis has provided some insight into mechanisms of neurodegeneration. In this chapter, we review skeletal muscle structure, biochemistry, physiology, and function along with structural abnormalities seen in the principal diseases of skeletal muscle.

NEURAL CONTROL

Motor Control System

To execute a voluntary skillful muscle movement, the motor control system implements and executes planning, transmitting, and "fine tuning" commands. The motor control system has three distinct features. First, the motor neurons and their fiber tracts are somatotopically organized. Second, the motor system has proprioceptive inputs for feedback control to adjust fine movements and posture, and feedforward control to prepare rapid movements. Third, the motor system is organized hierarchically and in parallel: the cortical motor neurons, limbic system, and brain stem control the interneuron system and alpha motor neurons of the brain stem and spinal cord.

Concept of the Final Common Path

In 1947, Sherrington showed that virtually all reflexes involve an integrated activation of synergist muscles and inhibition of antagonist muscles, largely coordinated by the spinal interneurons. The spinal interneurons, located in the anterior horn of the spinal cord, constitute the major motor control system and form intricate neuronal circuits with the corticospinal and brain stem tracts and the limbic motor control system. Eventually, all interneuronal paths converge on the lower motor neurons that innervate the skeletal muscles. Sherrington called this final activation system of the lower motor neurons the *final common path*.

Feedback System of Muscle Contraction

The feedback system for muscle contraction is a complex network of afferent and efferent pathways. Through it, the central nervous system modulates muscle contraction by adjusting the sensitivity of afferent muscle spindle receptors and overall muscle tone. Muscle spindles are the principal kinetic receptors, providing the central nervous system with information concerning muscle length and the degree of muscle contraction. The spindles differ from ordinary (extrafusal) muscle fibers in that they are encapsulated by connective tissues and contain the intrafusal (or spindle) muscle fibers. These spindle fibers are classified as one of two types: nuclear bag fibers (bag 1 or bag 2, depending on their length) or nuclear chain fibers. Nuclear bag fibers contain numerous large nuclei closely packed in a bag in the central portion of each intrafusal fiber, whereas nuclear chain fibers contain a single row of central nuclei. These intrafusal fibers attach to connective tissues, muscle tendons, and extrafusal muscle fibers to convey afferent information from extrafusal fibers and also from intrafusal fibers.

Sensory innervation of the intrafusal fibers occurs through primary and secondary endings. The primary endings are located at both the nuclear bag and nuclear chain fibers and give rise to the largest size myelinated group Ia nerve fibers. These endings provide afferent signals concerning only the rate of change in the length of the intrafusal spindle fibers (dynamic changes). In contrast, the secondary endings are found at the nuclear chain fibers and give rise to slightly smaller myelinated group II fibers. These endings provide afferent signals concerning the absolute length of the intrafusal spindle fibers (static changes). Muscle spindle function is best demonstrated with a simple muscle stretch reflex. When a muscle stretch is induced by tapping a tendon, the stretch stimulates the primary endings of Ia afferent nerve fibers and triggers a monosynaptic myotatic reflex on the large anterior horn cells, resulting in agonist muscle contraction. Gamma motor neurons, which innervate the intrafusal muscle fibers, change the length of the intrafusal fibers so that the afferent sensitivity is altered: the shorter the intrafusal muscle fibers, the more sensitive they are to changes in length.

Golgi tendon organs are encapsulated stretch receptors situated at the junction between the muscle and its tendon. Structurally, afferent receptors and extrafusal fibers are meshed together, so that when muscle contraction stimulates the tendon organs, afferent signals are produced. The Ib afferent nerve fibers originate from the tendon organs to provide (1) polysynaptic inhibition on the alpha motor neurons innervating the agonist muscles, and (2) polysynaptic facilitation on the motor neurons controlling the antagonist muscles (Fig. 9–1).

Anterior Horn Cells

Skeletal muscle fiber contraction is controlled and modulated by the activation of the alpha, beta, and gamma motor neurons and the interneurons. Alpha motor neurons are among the largest in the central nervous system and innervate the extrafusal muscle fibers. These neurons are excited monosynaptically by the Ia afferent and pyramidal tract fibers. Gamma motor neurons (fusimotor neurons) are the smallest anterior horn cells and innervate the intrafusal muscle fibers only. Beta motor neurons (skeletofusimotor neurons) are moderate-sized and innervate both extrafusal and intrafusal muscle fibers, enhancing the sensitivity of intrafusal fibers. Of the remaining small anterior horn cells, the majority are interneurons. Renshaw cells are unique feedback interneurons that provide collateral inhibitory synaptic actions on their own alpha motor neurons.

MOTOR UNIT

Physiological Classification

The motor unit is the smallest functional unit of the motor control system and consists of an alpha motor neuron, its

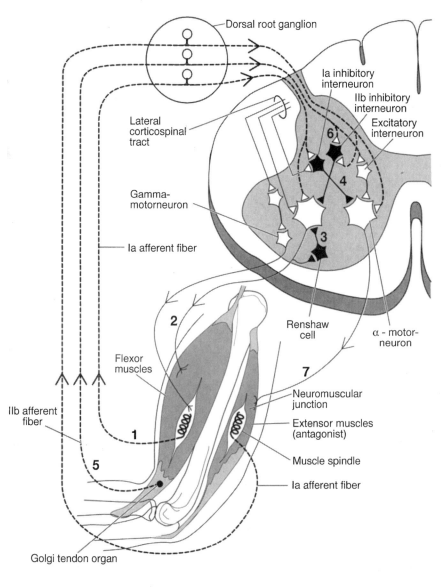

FIGURE 9–1. A simplified diagrammatic presentation of the interneurons and alpha motor neurons. The Ia afferent and corticospinal tract fibers transmit monosynaptic excitatory signals to alpha motor neurons, resulting in direct muscle contractions. All muscle contractions are coordinated by the interneuron system. When a flexor muscle is stretched, as shown, the stretch stimulates intrafusal spindle fibers, sending excitatory impulses through Ia afferent fibers (1), which stimulates the contraction of the same muscle (2). The Ia afferent fibers simultaneously stimulate Ia inhibitory interneurons, which inhibit the alpha motor neurons of antagonist extensor muscles (3), so that flexor muscle contraction can be carried out without resistance. The excitation of alpha motor neurons sends collateral signals to Renshaw cells that regulate their own excitation (4). At the same time, the mechanical contraction of the flexor muscle stretches the Golgi tendon organ, which then sends IIb inhibitory signals (5). This signal stimulates the IIb inhibitory interneurons (6), which exert polysynaptic inhibitory inputs on the original alpha motor neurons and polysynaptic excitatory inputs on alpha motor neurons controlling antagonist extensor muscles (6); the antagonist contraction (7) inhibits further flexor contraction and restores the flexor muscle to the original position. In another mechanism, flexor muscle contraction shortens intrafusal spindle muscle fibers, which stimulates another feedback system, signaling gamma motor neurons to contract the intrafusal spindle fibers so that the sensitivity of the spindle fibers to stretch is restored. The corticospinal tract fibers also stimulate polysynaptic excitatory interneurons to stimulate gamma motor neurons so that the sensitivity to stretch is increased, causing hyperreflexia. (From Mitsumoto H, Chad DA, Pioro EP: Amyotrophic Lateral Sclerosis. Philadelphia, FA Davis, 1997, pp 40–41.)

axon, and all the extrafusal muscle fibers it innervates. One motor neuron innervates multiple muscle fibers. The innervation ratio (the number of muscle fibers innervated by one motor neuron) depends on the size of the muscle. For example, the innervation ratio of the external ocular muscles is approximately 10 muscle fibers to 1 motor neuron, that of the hand muscles is approximately 100 muscle fibers to 1 motor neuron, and that of gastrocnemius is about 2000 muscle fibers to 1 motor neuron.

Physiologically, motor units are classified as belonging to one of three categories according to both the time the fibers require to reach the peak force during a twitch and the degree to which they fatigue: (1) slow nonfatigable, (2) fast nonfatigable, and (3) fast fatigable. The differences among these three types of motor units are shown in Table 9–1. The three types of units vary in the force they generate. Fast fatigable units produce 100 times more force than slow nonfatigable units. The differences in force are based on the innervation ratio and the cross-sectional areas of individual muscle fibers. Fast fatigable muscles have the largest innervation ratio and largest cross-sectional area, whereas slow nonfatigable muscles have the lowest ratio and cross-sectional area. Resistance to fatigue also correlates with the oxidative capacity of a muscle.

Histochemical Classification of Muscle Fiber Type

Muscle fibers can also be defined on the basis of their enzyme histochemistry. Muscle fiber type classification by enzyme histochemistry roughly corresponds to the physiological motor unit classification, and is highly useful in the interpretation of muscle biopsy specimens. The lower motor neuron determines the physiological and thus histochemical property. The fiber types—type I, type IIA, and type IIB—are differentiated by routine adenosine triphosphatase (ATPase) enzyme histochemistry. In type I fibers, ATPase is activated at an acidic pH whereas in type II fibers, it is activated at an alkaline pH. The histological features of these muscle fiber types are shown in Table 9–2. Fast twitch fibers (type II fibers) contain large amounts of stored glycogen and rely on anaerobic metabolism to provide the immediate energy required for muscle activation. They stain darkly with glycolytic enzyme stain. Type II fibers have short endurance as their aerobic metabolic capacity is low. Type I muscle fibers are slow twitch fibers and contain significant amounts of myoglobin and mitochondria, which are required for aerobic metabolism. They rely on aerobic metabolism for ATP production.

Animal muscles are often composed of either predominantly type I or type II fibers. Visually, the muscles consisting of type I fibers appear dark and informally are referred to as "red muscle"; those consisting of type II fibers are termed "white muscle." However in human muscle, such distinctions are less relevant because almost all muscles contain both type I and II fibers. The proportion of fiber types varies from muscle to muscle. Generally, in humans the phasic muscles of the extremities contain more type II fibers whereas the truncal postural muscles have more type I fibers. Within a single muscle, women tend to have more type I fibers than II fibers.

MUSCLE FIBER

In this section, we describe the development, and cellular and subcellular structure and function of the muscle fibers. The sliding theory of myofilaments and excitation-contraction coupling are also reviewed.

Development and Structure

Skeletal muscle develops from the embryonic somites. Individual muscle stem cells (myoblasts) fuse to form myotubes and initiate synthesis of the contractile filaments. Satellite cells are probably derived from the myoblasts and are situated beneath the basal lamina of skeletal muscle fibers. The proliferation of satellite cells (muscle fiber regeneration) is triggered by acute muscle injury. The fusion of numerous embryonic mononucleated myoblasts produces the multinuclear fibers (myocytes) that are characteristic of skeletal muscle. In all skeletal muscle fibers, the nuclei are oval and usually located at the cell periphery, just below the cell membrane. Only about 3% of normal muscle fibers have nuclei at or near the center of the fiber. Each myofiber is covered by the cell (sarcoplasmic) membrane, or *sarcolemma*. A delicate layer of collagen fibers, the *endomysium*, lies parallel to the muscle fibers, and loosely binds neighboring muscle fibers to each other. These adjacent groups of myofibers are gathered into fascicles by a connective tissue, the *perimysium*. Finally, a dense fibrous sheath, the *epimysium* or *deep fascia*, surrounds the entire muscle. The connective tissue, including tendon, is important in the mechanical transmission of the force generated by contracting muscle. The myofibers at each end of the muscle are attached to the tendons by complex interdigitation and tight junctions.

Subcellular Structure
Dystrophin-Glycoprotein Complex

Dystrophin is a large rod-shaped cytoskeletal protein found at the inner surface of the sarcolemma that provides

TABLE 9–2	Histochemical Characteristics of Muscle Fibers		
Characteristic	**Type I**	**Type IIA**	**Type IIB**
ATPase pH 9.4	Light	Dark	Dark
ATPase pH 4.6	Dark	Pale	Intermediate
ATPase pH 4.3	Dark	Pale	Pale
Succinic dehydrogenase	Dark	Intermediate	Light
NADH-TR	Dark	Intermediate	Light
Cytochrome oxidase	Dark	Intermediate	Light
PAS	Light	Dark	Intermediate
Phosphorylase	Pale	Dark	Dark
Lipid	Intermediate	Light	Light
Metabolism	Oxidative	Oxidative and glycolytic	Glycolytic
Corresponding physiological type	Slow nonfatigable	Fast nonfatigable	Fast fatigable

ATPase, adenosine triphosphatase; NADH-TR, NADH-tetrazolium reductase; PAS, periodic acid-Schiff

TABLE 9–1	Physiological Classification of Motor Units		
Characteristic	**Slow Nonfatigable**	**Fast Nonfatigable**	**Fast Fatigable**
Contraction time	Longer	Shorter	Shorter
Tetanic contraction	Easier	More difficult	More difficult
Fatigue	Resistant	Resistant	Easier
Force generation	Smaller	Intermediate	Larger

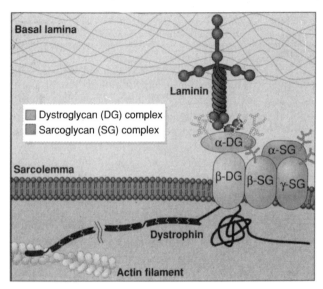

FIGURE 9–2. The structural relationship of dystrophin and the dystrophin-glycoprotein complex. (Adapted from Worton R: Muscular dystrophies: Diseases of the dystrophin-glycoprotein complex. Science 270:755–756, 1995.)

structural support to the muscle membrane (Fig. 9–2). The dystrophin-glycoprotein complex is composed of dystrophin and at least eight different glycoproteins associated with dystrophin. Two of these glycoproteins span the sarcolemma (the dystroglycans), five are found within the sarcolemma (the sarcoglycans), and one is at the inner edge of the sarcolemma (utrophin). The dystrophin-glycoprotein complex provides a bridge between the inner cytoskeleton and the extracellular matrix (laminin and the basal lamina). Therefore, dystrophin and its associated proteins are essential in maintaining the structural integrity of the muscle fiber. For example, if dystrophin is absent, as in Duchenne muscular dystrophy, or deficient, as in Becker muscular dystrophy, the sarcolemma becomes structurally unstable, resulting in excess Ca^{2+} influx and muscle necrosis.

Ion Channels

Electrically excitable cells are exquisitely controlled by ion concentration. Macromolecular protein tunnels spanning the lipid bilayer of the cell membrane, the ion channels are usually classified according to the type of ion (Na^+, K^+, Ca^{2+}, or Cl^-) they allow to pass. They may be gated by extracellular ligands, changes in transmembrane voltage, or intracellular second messengers. Abnormalities in these ion channels can produce a number of diseases (Table 9–3).

Neuromuscular Junction

In skeletal muscle, the site of neuromuscular transmission is the neuromuscular junction. The junction region is apposed by an enlarged lower motor neuron axon terminal and is covered by basement membrane derived from both Schwann cells and myofibers. The postsynaptic region consists of junctional folds made of specialized sarcoplasmic membrane. Nicotinic acetylcholine receptors are concentrated at the peaks of the junctional folds, apposing the axon terminal. Acetylcholinesterase is anchored by the collagen matrix of the basal lamina at the junctional folds. Membrane-bound synaptic vesicles containing acetylcholine are located in the axon terminal axoplasm.

When the action potential travels along the motor nerve axon and reaches the axon terminal, the depolarization in-

duces the opening of voltage-gated calcium channels. With the influx of Ca^{2+} ions, synaptic vesicles fuse with the presynaptic membrane to release acetylcholine into the synaptic cleft by an exocytotic mechanism. Acetylcholine then binds to the nicotinic acetylcholine receptors at the postsynaptic membrane to open the nonselective cation channels of the acetylcholine receptor; Na^+, K^+, and to a lesser extent, Ca^{2+} and Mg^{2+}, ions then flow into the fiber. This ion flow induces a voltage change to produce a localized electrical end plate potential. If this end plate potential reaches sufficient amplitude, it then generates an action potential that triggers muscle contraction. Excess acetylcholine molecules undergo hydrolysis by acetylcholineasterase. When the muscle is at rest, a small amount of acetylcholine is released spontaneously and very slowly, producing miniature end plate potentials (MEPPs). These MEPPs are insufficient to produce an action potential. Details of neuromuscular transmission are discussed in Chapter 20.

Transverse Tubules

The transverse tubules (T tubules) are membrane structures that extend deep into the fiber from its surface (sarcoplasmic membrane or sarcolemma). The T tubules transmit the initial depolarization at the motor end plate throughout the muscle fiber. The T tubules are located near the junction of the A and I bands (see later discussion) and are oriented perpendicular to myofibrils. The T tubules are apposed on each side by the terminal cisterns of the sarcoplasmic reticulum, which lie parallel to the myofibrils (Fig. 9–3). This sarcoplasmic reticulum-T-tubule-sarcoplasmic reticulum complex is known as the *triad* and is important in converting electrical signals (membrane action potentials) to chemical signals (Ca^{2+} ion release).

Sarcoplasmic Reticulum

The sarcoplasmic reticulum irregularly surrounds each myofibril. It enables the uptake, storage, and release of Ca^{2+} ions, which are critical in controlling myofibril contraction and relaxation. The sarcoplasmic reticulum consists of longitudinal elements that are closely associated with the myofibrils. At the ends of each longitudinal element are terminal cisternae (lateral sacs) in which Ca^{2+} ions are stored. Skeletal muscle sarcoplasmic reticulum contains ryanodine R1 receptors located at the terminal cisterns; in response to sarcoplasmic reticulum depolarization at these receptors, Ca^{2+} ions

TABLE 9–3	Ion Channel Disorders in Skeletal Muscles		
Channel	**Location**	**Genetic Disorders**	**Acquired Disorders**
Sodium	Sarcoplasmic membrane	Hyperkalemic periodic paralysis Paramyotonia	Myasthenia gravis
Chloride	Sarcoplasmic membrane	Myotonia congenita (Thomsen, Becker)	None known
Calcium	Sarcoplasmic reticulum	Malignant hyperthermia Central core disease	None known
Calcium	Sarcoplasmic membrane	Hypokalemic periodic paralysis Episodic ataxia	None known
Potassium	Sarcoplasmic membrane	Episodic ataxia + myokymia	Isaacs' syndrome

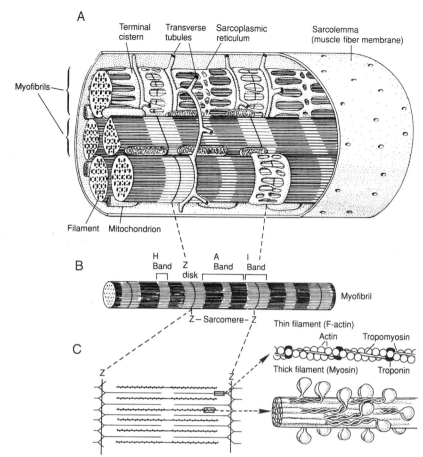

FIGURE 9–3. Alternating light and dark bands within the myofibrils give skeletal muscle its characteristic striated appearance. *A,* Three-dimensional reconstruction of a sector of muscle fiber showing the relationships of the membrane and tubular system to the myofibrils. *B,* Individual myofibril showing light and dark bands. Individual sarcomeres are separated by thin Z disks. The dark bands correspond to regions of overlap of thin and thick filaments. *C,* Schematic cross-section of an individual sarcomere. The thin filaments are composed principally of polymerized actin whereas the thick filaments are made up of arrays of myosin molecules. The myosin molecule includes a stem and a globular double-head that protrudes from the stem. (Reproduced from Kandel ER, Schwartz JH, Jessell TM: Principles of Neural Science, 3rd ed. New York, Elsevier, 1991, p 549, with permission.)

are passively released into the region of the overlapping thick and thin filaments. These released Ca^{2+} ions enable excitation-contraction coupling—they bind to troponin and allow the interdigitation between actin and myosin (see later discussion). When the membrane depolarization ceases, the sarcoplasmic reticulum acts as a Ca^{2+} reservoir and actively transports the Ca^{2+} ions back into the longitudinal cistern, which terminates the contractile activity.

Myofilament and Myofibril Structure

Muscle fibers are composed of numerous myofibrils; the myofibrils are oriented parallel to the muscle fibers. Each myofibril is formed by a bundle of myofilaments and is surrounded by organelles, such as glycogen granules, mitochondria, T tubules, and the sarcoplasmic reticulum. In the longitudinal view, the myofibril is divided into sarcomeres, which are the smallest contractile units. The sarcomere is the portion of myofibril between two regions on the myofibril called the Z bands (or Z disks), which are composed of actinin. Structurally, the sarcomere consists of two distinct filaments, the thick and thin filaments, which are arranged parallel to each other in bundles (see Fig. 9–3). The thin filaments are composed of actin and are isotropic upon exposure to polarized light. The portion of these filaments that does not overlap with the thick filaments is termed the I band. The thick filaments are composed of myosin, are anisotropic upon exposure to polarized light, and are termed the A band. The I band is attached to each end of the Z band whereas the A band lies at the center of the sarcomere. The A band is overlaped by actin except at the central zone, which is composed solely of myosin without overlap from actin (H band) (see Fig. 9–3).

Contractile Proteins

Skeletal myofilaments consist of at least four major contractile proteins: actin, tropomyosin, troponin, and myosin. *Actin* is a long filamentous (F-actin) polymer composed of two strands of globular (G-actin) monomers twisted around each other in a double helix. Each G-actin monomer has a binding site for myosin. *Tropomyosin* is a long thin molecule containing two polypeptide chains in the form of a double helix. *Troponin* has three subunits: TnT (binds to tropomyosin), TnC (binds Ca^{2+} ions), and TnI (inhibits the actin-myosin interaction). Thin filaments consist of three proteins: actin, tropomyosin, and troponin. Each tropomyosin molecule spans seven G-actin molecules and has one troponin complex bound to its surface. The thick filaments are composed of the fourth protein, *myosin,* which is a complex molecule consisting of two functionally different regions. The long straight portion, consisting of light meromyosin, forms the "tail" region. The remainder of the molecule, consisting of heavy meromyosin, forms both a globular head portion (S1 region) and "chain" portion (S2 region). The S1 region contains an actin-binding site and ATP-binding site. The S2 region functions as a flexible hinge between the head and tail regions.

Excitation-Contraction Coupling

Sarcomeres and the I band shorten during contraction because the thick and thin filaments slide past each other during contraction, not because the absolute length of the filaments changes. The sliding filament theory, first proposed by Hugh Huxley and Andrew Huxley in 1954, is outlined next.

When the muscles are at rest, ATP binds to the ATPase

site on the myosin molecule and is hydrolyzed very slowly; however, myosin cannot bind actin at this point because the actin-binding sites on the myosin head are blocked by the troponin-tropomyosin complex of the F-actin filament.

After an action potential is generated, the T tubules transmit this initial depolarization into the fiber. The depolarization then penetrates the triad complex to initiate signal transduction to release Ca^{2+} ions through ryanodine receptors. At this point, Ca^{2+} ions become available at the junction of the thick and thin filaments, they bind to the troponin TnC subunit. The binding causes a conformational change among the three troponin subunits that allows the tropomyosin molecule to move deeper into the groove of the actin helix. This change exposes the myosin-binding sites of the G-actin monomers, and actin is then free to attach to the head of the myosin molecule.

When the hydrolysis products (ADP and Pi) are released, the myosin head can bend, which pulls the actin past the myosin filament. Consequently, the thin filament of the sarcomere is drawn farther into the A band. When a new ATP molecule binds to myosin, the myosin head detaches from the thin filament. The bound ATP is subsequently hydrolyzed by the free myosin head, which prepares the head for the next interaction with the thin filaments. When Ca^{2+} ions are actively transported back into the sarcoplasmic reticulum, the original configuration of troponin and tropomyosin is restored, and the G-actin active attachment site is again covered.

Summary of Excitation-Contraction Coupling

The sequence of events in excitation-contraction coupling is as follows: (1) the action potential enters the T tubule, (2) the T tubule propagates the action potential inward, (3) the action potential penetrates the trial complex, (4) Ca^{2+} is released from the ryanodine R1 receptor into the intracellular space surrounding the myofilaments, (5) Ca^{2+} ions bind to the TnC subunits on the thin filaments and produce a conformational change, (6) myosin hydrolyzes ATP to prepare the head for attachment at the actin attachment site, (7) muscle contraction occurs by sliding between actin and myosin filaments, (8) a binding of the new ATP to myosin dissociates the actin and becomes ready for the next cycle of actin attachment, and (9) contraction ceases when Ca^{2+} is removed from the sarcoplasm to the sarcoplasmic reticulum by active transport.

MUSCLE METABOLISM

In this section, we review muscle metabolism briefly; a useful detailed review can be found in the Engel and Franzini-Armstrong monograph. For skeletal muscle contraction, the contractile mechanism requires ATP. Two important metabolic pathways in skeletal muscle contraction are the glycolytic (anaerobic) and oxidative (aerobic) pathways. Which pathway is used depends on the level of exercise activity. At maximal exercise level, glucose provides the energy for muscle contraction. Two molecules of ATP are produced for every molecule of glucose consumed. The source of glucose may be circulating blood glucose or its storage form, glycogen, in muscle cells.

The end product of glycolysis, pyruvic acid (pyruvate), can be broken down to lactic acid. The accumulation of lactic acid is the largest contributor to oxygen debt. Lactic acid and pyruvic acid move into the mitochondria where they enter the Krebs cycle and are used in oxidative metabolism. The inner mitochondrial membrane contains the enzymes

and carriers for electron transfer and phosphorylation of ADP to ATP. Because the Krebs cycle leads to complete breakdown of pyruvic acid, substrates become available for terminal electron transport and oxidative phosphorylation. At the completion of cyclic reduction and oxidation, 36 molecules of ATP are produced, in addition to the 2 molecules of ATP generated from glycolysis.

The muscles contain significant amounts of myoglobin, which is best used during aerobic metabolism. Myoglobin is a globular heme protein and gives muscles their characteristic red color. Myoglobin binds only one oxygen molecule, and this binding is not modified by the sarcoplasmic pH; thus, myoglobin enhances the diffusion of oxygen through exercising muscle tissue by binding and releasing oxygen molecules as they move down their concentration gradient inside the sarcoplasm.

Glucose is the preferred energy source for muscle contraction at higher levels of exercise, but fats also provide energy for muscle contraction at lower exercise levels. Fat provides energy in the form of free fatty acids when metabolized inside the mitochondria. Free fatty acids are transported into the mitochondria across their membrane in the form of acyl-carnitine by the enzyme carnitine palmitoyltransferase. During the oxidation of fatty acids in the Krebs cycle, coenzyme A is linked to an acetyl group to form acetyl coenzyme A, which is the main precursor of β-oxidation in the Krebs cycle.

The "second-wind" phenomenon described in patients with McArdle's disease is a unique example of metabolic compensation. When these patients begin to exercise, they experience an initial transient period of progressive fatigue and weakness in the exercised muscle because they are unable to metabolize muscle glycogen. However, if they continue to exercise at a low level, because they become able to use glucose and free fatty acids effectively, the exercise becomes easier than when it was initiated.

Protein can serve as an energy source for muscle contraction, especially during starvation or during heavy exercise. Proteins are broken down into amino acids to provide energy for muscle contraction and metabolized in mitochondria.

STRUCTURAL ABNORMALITIES OF MUSCLE FIBERS

A number of disease states result in alteration of muscle fiber structures. We briefly summarize typical changes that occur in the more common muscle tissue pathologies.

Muscle Fiber Atrophy

Muscle fiber atrophy can occur in many conditions. Denervation typically leads to fiber atrophy, and the atrophied fibers are variably angular in shape on cross-section. Denervation always involves both type I and type II muscle fibers. In the early stages of denervation, scattered individual atrophic fibers are seen. As the denervation process continues, small groups of atrophic fibers develop.

Primary muscle disease can reduce muscle fiber size, but the atrophied fibers typically are round on cross-section and irregular in size. In type II fiber atrophy, the diameter of only type II fibers is reduced; such a reduction can be caused by *disuse, cachexia, excess corticosteroids,* or *collagen vascular disease.* Type I fibers can also undergo selective atrophy, which is often seen in *myotonic dystrophy* or other congenital myopathies.

Muscle Fiber Hypertrophy

Muscle can become hypertrophied if the work of a muscle is increased, which increases the muscle fiber diameter. Such an increase is a familiar result of muscle-strengthening exercises. However, in conditions such as *myotonia congenita*, the muscle fiber diameter is also increased, resulting in hypertrophy. In various other myopathies, such as *Duchenne muscular dystrophy*, a mixture of small atrophied fibers and hypertrophic fibers is seen.

Muscle Fiber Degeneration and Necrosis

Muscle fibers undergoing degeneration have either a very pale cytoplasm or a homogeneous eosinophilic cytoplasm that has lost its normal striation. Large swollen hyalinated fibers are also often characteristic of muscle fiber degeneration in Duchenne muscular dystrophy. When necrosis occurs, the cytoplasm is often invaded by macrophages or replaced by necrotic debris. Regenerating myoblasts that are derived from satellite cells may often be seen in necrotic fibers; these fibers that have both degeneration and regeneration are sometimes called "degen/regen" fibers. Regenerating fibers stain dark blue with hematoxyline because of their increased ribonucleic acid content. Segmental necrosis may separate a portion of a muscle fiber from the portion in which the neuromuscular junction is located, resulting in denervation in the detached portion of the muscle fiber.

Vacuolar Degeneration

Upon histological examination, vacuoles in myofibers appear as empty spaces, but in vivo they most likely had contained normal or abnormal material and fluid. Two types of vacuoles are seen in muscle degeneration. Autophagic vacuoles are membrane bound and are commonly filled with cellular debris. Rimmed vacuoles also contain debris but in addition contain multilaminated membranous structures; typically, they are seen in inclusion body myositis.

Glycogen storage diseases can cause vacuolar changes. These vacuoles stain positively with celloidin–processed periodic acid–Schiff stain or periodic acid–Schiff staining of alcohol-fixed tissue. Lipid storage disorders can also produce vacuolar changes. These vacuoles stain positively with oil red-O and Sudan black B. A unique type of vacuole occurs in *hypokalemic periodic paralysis*. These vacuoles are caused by proliferation, degeneration, and autophagic destruction of the sarcoplasmic reticulum and membrane organelles associated with the T tubule system.

Structural Fiber Changes in Congenital Myopathies

Structural changes occur in several congenital myopathies. Nemaline, or rod body, myopathy is caused by an abnormal accumulation of actinin of the Z bands. In cytoplasmic body myopathy, individual small round cytoplasmic bodies develop; they consist of a tangle of irregularly oriented fine filaments and are surrounded by a halo of radiating myofilaments. In centronuclear myopathy, central nuclei appear in the majority of muscle fibers. The central nucleus is surrounded by a halo devoid of normal organelles. Central core disease primarily affects the high-oxidative, low-glycolytic type I fibers and is characterized by a central core in the muscle fiber in which oxidative enzyme activity and mitochondria are deficient. Multicore (minicore) fibers have multiple small cores similar to the central cores.

Mitochondrial Abnormalities

Mitochondria stain darkly with the oxidative stains cytochrome C oxidase and modified Gomori trichrome. Muscle fibers with increased numbers of mitochondria have a highly basophilic subsarcolemmal cytoplasm because of accumulated oxidative enzymes; they stain bright red with Gomori stain, hence the name "ragged red fibers." These mitochondrial abnormalities are associated with several mitochondrial cytopathologies: *mitochondrial myopathy, encephalopathy, lactic acidosis and stroke-like episodes (MELAS) syndrome, Kearns-Sayre syndrome (oculocraniosomatic syndrome), myoclonic epilepsy with ragged red fibers (MERRF),* and *chronic progressive external ophthalmoplegia (CPEO).*

Fiber Type Abnormality

Fiber type grouping is seen in chronic neurogenic atrophy occurring after reinnervation. The normal random distribution of fiber types is altered such that increased numbers of fibers of a single type appear immediately adjacent to each other. *Fiber type disproportion* is characterized by a reduction of at least 12% in the mean diameter of the fiber type in question. Type I fiber disproportion occurs in myotonic dystrophy and in various congenital myopathies.

Ringbinden

The *ringbinden* (annular fibers) is a peripheral band of myofibrils that encircles a central longitudinal core of normally oriented fibers. This band apparently is an artifact that occurs because of a difference in contractility between the external and internal myofibrils. Ringbinden is nonspecific but is often found in myotonic dystrophy and in various other myopathies.

REFERENCES

Ackerman MJ, Clapham DE: Ion channels—basic science and clinical disease. N Engl J Med 336:1575–1585, 1997.

Brown RH: Dystrophin-associated proteins and the muscular dystrophies. Ann Rev Med 48:457–466, 1997.

Engel A, Franzini-Armstrong C (eds): Myology, 2nd ed., Vol. 1, New York, McGraw-Hill, Inc., 1994.

Ghez C: Muscles: Effectors of the motor systems. In Kandel ER, Schwartz JH, Jessell TM (eds): Principles of Neural Science. 3rd ed., New York, Elsevier, 1991, pp 548–563.

Griggs RC, Mendell JR, Miller RG: Evaluation and Treatment of Myopathies. Philadelphia, FA Davis, 1995.

Mitsumoto H, David CA, Pioro EP: Functional neuroanatomy of the motor system In Mitsumoto H, David CA, Pioro EP (eds): Amyotrophic Lateral Sclerosis. Philadelphia, FA Davis, 1998, pp 34–44.

Worton R: Muscular dystrophies: Diseases of the dystrophin-glycoprotein complex. Science 270:755–756, 1995.

II ELECTROMYOGRAPHY

Chapter 10

Nerve Conduction Studies

Richard J. Lederman, MD, PhD

ANATOMY AND PHYSIOLOGY
Terminology Used for Nerve Fibers
Nerve Fiber Size and Function
Muscle Location and Size
GENERAL PRINCIPLES OF ELECTRICAL
  NERVE STIMULATION
Stimulating Electrodes
Technique
Caveats in Interpreting the Effects of
  Nerve Stimulation
RECORDING OF ACTION POTENTIALS
  FROM PERIPHERAL NERVE
Extracellular Recording from Multiple
  Nerve Fibers
Factors Influencing Response Amplitude
RECORDING OF ACTION POTENTIALS
  FROM MUSCLE
Difference Between Nerve and Muscle
  Recording
Factors Influencing Response Amplitude
BASIC FEATURES OF NERVE CONDUCTION
  STUDIES
Techniques of Stimulation and
  Recording
Motor Nerve Conduction Studies:
  Median Nerve

Other Motor Nerve Conduction Studies
Sensory Nerve Conduction Studies:
  Median Nerve
Other Sensory Nerve Conduction Studies
Median and Ulnar Palmar (Mixed
  Sensorimotor) Studies
Late Responses: H-Reflexes, F-Waves
FACTORS INFLUENCING THE RECORDED
  RESPONSE
Stimulus Intensity
Filter Settings
Recording Electrode Position
Body Temperature
Age
Height
Anomalous Pathways
CRANIAL NERVE CONDUCTION STUDIES
Facial Motor Conduction
The Blink Reflex
Techniques
Clinical Applications
Facial Nerve Lesions
Trigeminal Nerve Lesions
Central Nervous System Lesions
The Jaw Jerk
CONCLUSION

ANATOMY AND PHYSIOLOGY

The peripheral nervous system, strictly speaking, is defined as those portions of nerve fibers that are supported by Schwann cells and, at least for all but the smallest nerve fibers, covered by an investment of "peripheral" myelin derived from these Schwann cells. The Schwann cell investment begins within a short distance of the central nervous system (CNS) and continues to the termination of the peripheral nerve axon, either as a free nerve ending or specialized receptor for afferent fibers or at an effector cell such as a muscle fiber or glandular cell for efferent fibers. Functionally and clinically, this separation of the peripheral neuron is arbitrary, for the sensory nerve cell continues on into the CNS to terminate in either the spinal cord or the brain. The motor neuron has its cell body within the CNS, and its synaptic relationship to the muscle fiber (or other organ in the case of autonomic efferents) is not only functionally important but also critical for interpretation of the clinical neurophysiological studies.

A peripheral nerve consists of axons of widely differing diameters from less than 1 μ to as large as 20 to 22 μ. These belong to the somatic as well as to the autonomic nervous system. Although the autonomic nervous system has obvious and profound clinical importance, the present discussion largely ignores this population. Furthermore, although smaller, unmyelinated axons are also of crucial importance both physiologically and clinically, the ensuing discussion largely ignores these as well, for reasons that will become apparent.

Terminology Used for Nerve Fibers

Terminology used for the different fiber groups within peripheral nerve unfortunately evolved without consensus, some based on size or diameter, others on physiological parameters. These differing systems have stayed with us. Thus, we are left with designations employing Roman letters and numbers as well as Greek letters. Not only is there overlap and redundancy, but the terminologies used are often mixed.

The most common system uses Roman letters A, B, and C for fibers differing in diameter from the largest (A) to the smallest (C).[33] A-fibers range in size from about 1 μ in diameter to about 20 μ and consist of both afferent and efferent axons. B-fibers are efferent, myelinated preganglionic autonomic axons and fall in the lower end of the size range for A-fibers. C-fibers are unmyelinated, ranging in

TABLE 10-1	Terminology of Nerve Fibers			
Designation	Fiber Size (μ)	Myelin	Modality	Direction of Transmission
A	1–17	+	Motor, sensory	Efferent, afferent
-alpha	6–17	+	Cutaneous sensory	Afferent
-delta	1–6	+	Cutaneous sensory	Afferent
B	1–3	+	Autonomic	Efferent
C	0.3–1.3	−	Sympathetic	Efferent
			Somatosensory	Afferent
I	12–20	+	Sensory (muscle)	Afferent
II	6–12	+	Sensory (muscle)	Afferent
III	1–6	+	Sensory (muscle)	Afferent
IV	0.3–1.3	−	Sympathetic	Efferent
			Somatosensory	Afferent

diameter from about 0.3 μ to as large as 1.3 μ, and represent either efferent postganglionic sympathetic or afferent somatosensory fibers.

Roman numerals I through IV have been applied primarily to nerves supplying muscles, comprising primarily afferent fibers.[33] Groups I, II, and III all fall within the A-fiber range: I—12 to 20 μ, II—6 to 12 μ, and III—1 to 6 μ in diameter. Group IV is generally equivalent to the C-fiber population. Greek letter designations alpha, beta, gamma, and delta were derived from individual peaks or elevations of the compound action potential recorded in early studies of nerve trunks in the 1930s, with alpha representing the fastest and delta the slowest fibers within the A group (the elevation associated with the unmyelinated fibers was recognized only later as technological advances allowed, and these fibers are generally referred to as C-fibers in all systems). To make matters worse, lowercase Roman letters are sometimes added to indicate anatomical or functional groupings (e.g., the Ia and Ib afferents deriving from muscle spindles and Golgi tendon organs, respectively, subserving the tendon reflex).

Despite this potential for confusion, clinical neurophysiology generally focuses on specific designations for certain populations. Afferent cutaneous fibers are represented by A-alpha and A-delta (myelinated groups) and C-fibers. Sensory fibers from muscle are classified as groups I and II afferents, representing the primary and secondary endings from muscle spindles. Motor nerve axons are generally characterized

functionally as either alpha or gamma within the A-fiber range, supplying either extrafusal or intrafusal (muscle spindle) fibers, respectively. These anatomical and functional features are summarized in Table 10–1.

Nerve Fiber Size and Function

There is a direct relationship between fiber size and conduction velocity in mammalian myelinated nerve fibers throughout the range that has been studied[15] (Fig. 10–1). The exact ratio may vary from one species to another, but a rough approximation of human fiber diameter in microns can be obtained by dividing the observed nerve conduction velocity in meters per second by 6. Thus, a conduction velocity of 60 m/sec corresponds to a fiber of about 10 μ in external diameter, including both axon and investing myelin.

The course of peripheral nerves is remarkably consistent among individuals, although some differences, particularly for certain nerves, exist. These variations can have important implications for electrodiagnostic studies as well as for clinical manifestations. Some of these implications are pointed out later in this chapter. Nonetheless, the pathway followed by a peripheral nerve from its origin (or termination, as the case may be) centrally to its most peripheral branch can generally be reliably predicted. The peripheral nerve lies farthest away from the surface at its most central point and is often, but not always, most superficial at the periphery.

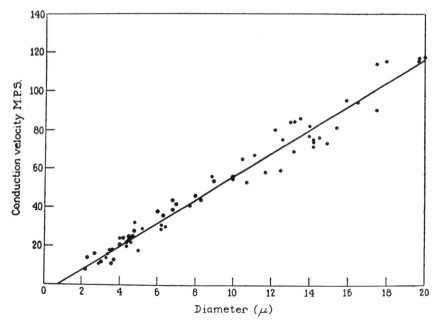

FIGURE 10–1. The relationship between conduction velocity and mammalian myelinated nerve fiber diameter. (From Gasser HS: The classification of nerve fibers. Ohio J Sci 41:154, 1941, with permission.)

The depth of the nerve from the skin surface, however, may vary considerably, and some peripheral nerves, such as the ulnar nerve in the intercondylar groove and the peroneal nerve at the fibular neck, are just beneath the skin and barely protected. The distance of the nerve from the surface has important implications regarding the ability to study the nerve in the electrodiagnostic laboratory, both for ease of stimulation and for ease of recording with surface electrodes.

In passing from the spinal cord (or brain) to the periphery, nerve fibers course through numerous passages and are in contact with many tissues, including bone, blood vessels, muscles, and connective elements. Nerves cross joints and therefore must be able to accommodate the range of motion dictated at that joint. This has major implications specifically in nerve conduction studies when carrying out and interpreting studies of the ulnar nerve in the region of the elbow.

We have focused so far on the nerve fiber distribution of peripheral nerve and on its relationship to other structures along its course, but it also has an internal structure that is important and relevant to nerve conduction studies. Axons within the peripheral nerve are grouped together in bundles or fascicles, including fibers of various sizes, both myelinated and unmyelinated, and including both afferent and efferent fibers in muscle nerves. Nerves carrying motor fibers contain the broadest range of axons, including the largest in humans, the group I afferents. Purely cutaneous (sensory) nerves lack these largest fibers and are composed primarily of A-alpha and -delta sensory fibers, as well as both somatosensory and autonomic C-fibers.

The axons within a fascicle are embedded in connective tissue, the endoneurium, and each fascicle is surrounded by a sleeve of additional connective tissue, the perineurium. The individual fascicles are, themselves, surrounded and separated by a third layer of connective tissue, the epineurium (Fig. 10–2). These connective tissue investments are not, in themselves, of critical importance in the execution and interpretation of clinical electrodiagnosis, but they do have some relevance in influencing the stimulation and recording from the nerve fibers.

There has been considerable controversy regarding the fascicular content of peripheral nerves and whether the axonal makeup of the fascicle remains constant or changes dramatically along the course of the nerve (see Stewart[42] for a discussion of these issues). In general, the segregation of fibers destined for a target begins quite proximally and becomes increasingly specific more distally. This means that a focal proximal injury to a large nerve trunk may predominantly affect fibers destined for a single or restricted group of muscles supplied by that nerve, suggesting that fascicular segregation has already taken place at the proximal level.

The motor nerves terminate in small branches, which supply individual muscle fibers. The majority of these neuromuscular junctions are clustered around the belly of the muscle, an area called the motor point or end plate region. Some muscles may have multiple motor points. A single motor nerve axon may supply only a few muscle fibers or several hundred, depending on the particular muscle. The motor unit is composed of the motor neuron and all of the muscle fibers it supplies. Muscles vary widely, not only in innervation ratio (number of muscle fibers supplied by a single motor neuron) or size of the motor unit, but also in the number of motor units within the muscle, which can range from 100 to 200 for individual muscles of the hand to about 1000 for large antigravity leg muscles.[38] These factors are of considerable importance in clinical electrodiagnosis, not only for the needle examination (see Chap. 12) but also for motor nerve conduction studies.

Muscle Location and Size

The anatomical position of the muscle also has a bearing on our ability to study it electrodiagnostically. The ideal muscle would (1) be anatomically discrete or isolated, (2) be just beneath the skin surface, (3) have all its fibers running in parallel from proximal to distal connections, and (4) be supplied by a single nerve readily accessible to electrical stimulation at two or more points along its course. Some muscles that we are able to study come close to this ideal, such as the extensor digitorum brevis (which, unfortunately, has other problems, including susceptibility to the effects of local trauma, aging, and anatomical variability). The abductor digiti minimi and abductor pollicis brevis, while fulfilling many of the desirable features, lie on and between other muscles, the activation of which may "contaminate" the response of the primary target. Muscles that are deeply placed are difficult to study with surface recording, as are muscles that are tightly packed together (e.g., the forearm flexors or the interossei and lumbricals). The more proximal and larger muscles may be very accessible themselves, while the nerves supplying them are not. Large and particularly flat, thin muscles, or complex ones such as the trapezius, are also readily accessible, but surface electrodes can record from only a small portion, leading to inadequate standardization of the technique and irreproducible results.

GENERAL PRINCIPLES OF ELECTRICAL NERVE STIMULATION

Stimulating Electrodes

Activation of a peripheral nerve is accomplished by depolarizing the surface membrane of each nerve fiber sufficiently to produce a propagated action potential. This requires the accumulation of negative charges at the surface of the nerve, initiating the sequence of ionic permeability changes associated with the excitation process. Stimulating

Cross-section of Peripheral Nerve

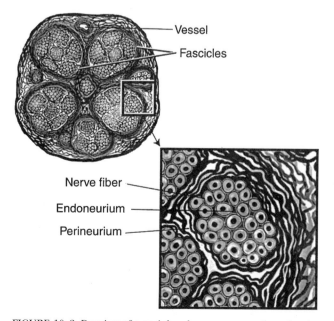

FIGURE 10–2. Drawing of a peripheral nerve cross-section, showing fascicles and connective tissue sheaths.

electrodes are usually placed on the skin surface overlying the nerve, and sufficient current is passed to produce the required transmembrane and longitudinal (both intra- and extra-axonal) ionic movements to achieve depolarization and impulse propagation. It is to be remembered that an all-or-none action potential is produced in each fiber that is adequately depolarized, and that there are many thousands of fibers of widely differing thickness, both myelinated and unmyelinated. These fibers will have different thresholds for stimulation that depend on·size, as well as position, within the nerve (see later discussion). Clinical nerve conduction studies usually aim to stimulate all of the larger fiber components by applying a "supramaximal stimulus."

Technique

Most clinical electrodiagnostic laboratories utilize surface electrodes for nerve conduction studies. Various forms are available, and the choice is often determined by anatomical factors at the stimulation site, as well as convenience (Fig. 10–3). For stimulation of nerves in the upper or lower extremities, the cathode and the anode are commonly composed of metal disks or pods applied to the surface overlying the nerve, either taped to the skin or positioned over the nerve as part of a specially constructed hand-held stimulator. The cathode and anode must be close enough together to produce focused current flow through the underlying tissues, sufficient to depolarize the nerve, but far enough separated and adequately isolated from each other electrically to prevent shunting of current flow superficially. Usually, a distance of 2 to 3 cm is practical and effective. On occasion, stimulation can be carried out via a needle (cathode) placed in the vicinity of the nerve, with the anode being either a second needle appropriately placed to facilitate current flow through the nerve, or a surface electrode on the opposite side of the limb.

For stimulating nerves in the digits, most laboratories use rings or clips applied to the surface of the digit, usually over adjacent phalanges. In addition to the type of electrode used and the interelectrode distance, other stimulus variables include the intensity (current), voltage, pulse duration and waveform (monophasic or biphasic), and repetition rate. Skin resistance is minimized by cleansing the skin and applying a conducting paste or jelly, taking care that surface interelectrode conduction is not facilitated by intervening conducting media. Pulse durations are generally in the range of 0.05 to 0.1 msec for routine upper limb studies and from

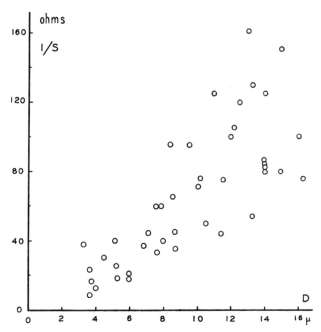

FIGURE 10–4. The relationship between threshold for stimulation and nerve fiber diameter. Ordinate: reciprocal of threshhold stimulus strength; abscissa: fiber diameter. (From Tasaki I: Nervous Transmission. Springfield, IL, Charles C Thomas, 1953, p 123, with permission.)

0.1 msec to 1 msec for the lower limb, depending on the nerve being studied, the electrode type and position, body habitus, and the health of the nervous system. Most systems utilize alternating or biphasic current for stimulation.

Axons within the peripheral nerve have differing thresholds for electrical stimulation. Because propagation of the action potential requires longitudinal current (ion) flow within the axoplasm beneath the surface membrane, as well as just outside the nerve fiber, and because the resistance to internal current flow is inversely related to fiber diameter, the largest axons have the lowest threshold and are the first to be activated as the stimulus current is increased (Fig. 10–4). Although position of the axon within the nerve trunk is a less important determinant, axons lying closest to the stimulating cathode are more readily activated, based on the

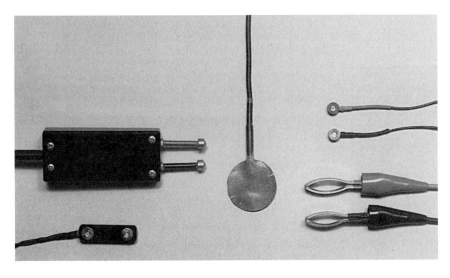

FIGURE 10–3. Electrodes commonly used for nerve conduction studies. *Upper left:* hand-held stimulator; *lower left:* bar stimulator; *center:* ground electrode; *upper right:* surface disk electrodes; *lower right:* clip electrodes.

electrical field that is generated, than those placed farther away.

Issues of electrical safety have been discussed in Chapter 4. Suffice it to say here that adequate grounding is essential for stimulation studies and that certain situations such as the presence of an indwelling cardiac pacemaker or defibrillator may preclude safe study or mandate special precautions.

Caveats in Interpreting the Effects of Nerve Stimulation

It must be remembered that the stimulating pulses are being applied to peripheral nerves passing in the vicinity of the electrodes. The initiated action potential will be conducted bi-directionally, both centrifugally and centripetally from the point of stimulation (or both orthodromically and antidromically for motor and sensory fibers), although recorded in only one direction. It should also be recognized that simultaneous activation of all fibers within a peripheral nerve is a highly unphysiological event, virtually never occurring in the course of normal function. This means that one cannot automatically make assumptions regarding clinical function on the basis of responses to electrical stimulation. The simplest and most apparent example of this might be the "normal" compound muscle action potential (CMAP) in response to stimulation of a nerve supplying a muscle paralyzed by a CNS process or even by a very recent (few days) proximal peripheral nerve lesion, before axonal and neuromuscular junction failure. Another and very important example might be the normal compound sensory nerve action potential (SNAP) elicited in a patient with total sensory loss in the distribution of that nerve, caused by a process proximal to the dorsal root ganglion, such as nerve root avulsion. More subtle differences may be seen when, for instance, relatively normal sensation or muscle strength is preserved in a patient with markedly dispersed, low-amplitude sensory or motor responses associated with demyelination and asynchronous conduction between the stimulating and recording sites. Slow or asynchronous conduction, in the absence of profound conduction block, may be quite consistent with adequate or even normal sensory and motor function, as may be seen, for example, in some inherited demyelinating polyneuropathies.

RECORDING OF ACTION POTENTIALS FROM PERIPHERAL NERVE

Extracellular Recording from Multiple Nerve Fibers

Information is transmitted along an individual nerve fiber by the propagation of an action potential with an amplitude of approximately 70 to 90 mV (as measured across the axon membrane) and a duration of about 0.5 to 1 msec. This potential difference across the surface membrane is a reflection of the flow of charged ions across the very high resistance surface membrane (transiently rendered selectively permeable) and the concomitant local current flows longitudinally along the axon both internally and externally in a local circuit. The relationship of the current flow, resistance, and potential difference is defined by Ohm's law, $E = IR$. How is it, then, that what we record clinically is so different in amplitude, duration, and waveform?

It must first be recognized that we are measuring the potential difference produced by local circuits of current flowing in an extracellular medium surrounding the nerve fiber. Although these tiny ionic currents produce a large

action potential when measured across the high resistance of the axolemma, the potential difference created extracellularly, in an environment of relatively very low resistance, is, according to Ohm's law, also very low. Even immediately outside the membrane itself the recorded potential difference falls dramatically by some two to three orders of magnitude from the 70- to 90-mV transmembrane potential (Fig. 10–5). This drop, and the further attenuation as the recording electrode moves some distance from the nerve fiber, along with the expected changes in configuration and duration, is defined by the specific features of the recording arrangement and, most importantly, by the principles of volume conduction, which have already been discussed in Chapter 8.

It is also a property of electrical conduction in a volume conductor that the recorded potential difference will be the summation of the local current flows, and resultant potential differences, of all the electrically active units (fibers) within the vicinity of the recording electrode. Thus, if there is sufficient synchronization, the principles of spatial and temporal summation will permit the recording of a "compound" action potential of measurable amplitude, even at some distance from the generator or source. The influence of dis-

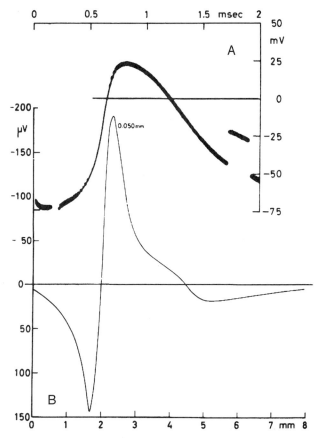

FIGURE 10–5. Relationship between intracellular and extracellular recording of the action potential. *Upper trace:* A recorded intracellular action potential from a frog muscle fiber. The ordinate A shows the amplitude scale in millivolts (mV). *Lower trace:* A calculated action potential recorded just outside the surface of the fiber. The ordinate B shows the amplitude scale in microvolts (μV). The intracellularly recorded action potential of approximately 90 mV would drop to less than 350 μV with extracellular recording and the configuration reflects the results of volume conduction. (From Rosenfalck P: Intra- and extracellular potential fields of active nerve and muscle fibres: A physical–mathematical analysis of different models. Acta Physiol Scand (suppl 321): 118, 1969.)

tance is of great importance, however, not only theoretically, but also practically for clinical electrophysiology. The amplitude of a response in a volume conductor will be inversely proportional to the square of the distance from the source of the potential generation.

The degree of synchronization is also of major importance, since the measured compound action potential is a summation of individual fiber contributions. If the positive and negative phases of the individual extracellular action potentials are not sufficiently synchronized, the positives and negatives may, in fact, cancel each other out, leading to a net zero potential difference as recorded at a distant site. This phenomenon of phase cancellation is also a most important principle underlying the interpretation of the compound action potential.

Despite the profound influence of distance of the recording site from the source of an electrical potential difference, most clinical recording of compound nerve (and muscle) action potentials is accomplished using surface electrodes. These are most commonly applied to the skin overlying the nerve, which is, at that point, sufficiently close to the surface to provide a measurable compound action potential. Sometimes, however, we may prefer to use needle electrodes to allow closer approximation to the nerve itself, especially when the nerve is more deeply situated, either as the result of its normal course or because of excessive tissue (e.g., adipose tissue or edema) between the nerve and the skin. However, the practical advantages of comfort, time, and convenience associated with surface recording usually prevail.

Because stimulation of nerve fibers produces bi-directional propagation of action potentials, the possibilities of recording orthodromically (in the natural or physiological direction) or antidromically (the opposite direction) exists for sensory nerve studies. In general, stimulating proximally and recording antidromically conducted potentials distally (e.g., over the digits) is associated with a higher amplitude response, since the nerve fibers are closer to the surface distally than proximally. However, there is more likely to be contamination from much larger amplitude CMAPs, which are also elicited by the proximal stimulus and are recorded in the volume conductor, even though the muscle fibers per se are no longer present beyond the midportion of the proximal phalanx of the digit.

At times, the summated or compound nerve action potentials are of insufficient amplitude at the surface to allow recordings or to separate them from the "noise" inherent in the recording arrangement. Averaging techniques (i.e., the computerized separation of evoked electrical signals from the random variations of the baseline) may allow differentiation of a response at a fixed interval after the applied stimulus. The number of repetitions required to separate the true propagated compound nerve action potential from the noise of the system may be anywhere from a few (10 to 20 or less) to 1000, or more. In most clinical situations, however, the elicitation of a pathological, tiny amplitude response by averaging is no more meaningful than obtaining a null response.

Factors Influencing Response Amplitude

As previously stated, the extracellular recording involves the detection of potential differences produced just outside the nerve fiber by current flow associated with the propagating action potential. The amplitude of the recorded action potentials is proportional to the magnitude of the current flow, as dictated by Ohm's law. The relationship of current flow, potential difference, and resistance holds within the internal environment of the axon, as well. For any given

potential difference generated by the action potential (which is approximately the same for nerve fibers of all sizes), the intracellular current flow will be larger in magnitude for a fiber of larger internal diameter, since the internal resistance to ionic flow within a larger axon is lower than that in a smaller fiber (Fig. 10–6). Since the extracellular current flow must, of necessity, be equal (and opposite in direction) to that inside the cell, a larger individual fiber will generate a higher amplitude volume-conducted action potential than the smaller fiber. Therefore, the summated or compound action potential of a group of simultaneously activated large fibers will be much larger than that produced by a comparable number of small fibers. This means that the contribution of larger fibers to the overall volume-conducted compound nerve action potential is much greater than that of small fibers. Indeed, the smallest fibers have little if any impact on the recorded response, even though, in cutaneous nerves, they may be far more numerous than the larger ones. There are other factors, as well, including the broad range of conduction velocities, producing desynchronization and phase cancellation, and the fact that clinical nerve conduction recordings generally focus on the earliest, and therefore, fastest-conducting portion of the overall response. The component comprising the smallest fibers would have a markedly delayed latency, if visible at all.

Although the absolute number of fibers conducting simultaneously past the point of recording is the major variable determining the amplitude of a response, the density of nerve fibers (number of fibers per unit volume) is also important. A thick nerve trunk that has been partially depleted of large fibers may, because of the distribution of field potentials dictated by the geometry of the recording system (including the size of the electrode and the distance from the nerve), produce a smaller recorded response than a normally constituted, but smaller, peripheral nerve trunk. Factors producing decreased absolute numbers, as well as diminished density of fibers, are considered later in this discussion.

The depth of the peripheral nerve trunk from the surface recording site is usually the most important distance factor in determining response amplitude. A fiber closer to the skin surface will, theoretically, have a greater influence on the recorded amplitude than one of comparable size either in the center of the nerve trunk or on the deeper surface of the nerve. Physiologically, this rarely makes a difference, but clinically it could if, for instance, fibers already segregated

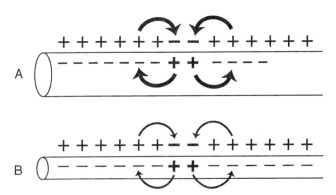

FIGURE 10–6. Potential changes and current flow outside and within the axon following electrical stimulation of A, a large fiber, and B, a small fiber. Although the potential changes across the membrane are identical in the two fibers of different size, the thicker arrows in A indicate a larger longitudinal current flow because of lower internal resistance within the larger axon, producing a larger amplitude response recorded extracellularly.

for one specific branch were more superficially placed in the nerve and were more, or less, affected by a disease process than those more deeply situated.

Finally, the issue of synchrony has been mentioned previously but is an extremely important one to emphasize. Because the volume-conducted response is a summation of individual fiber action potentials, simultaneous arrival of action potentials is critical for a recording. As a theoretical example, let us look at Figure 10–7. In this figure, we are attempting to record a compound nerve action potential from 10 large myelinated nerve fibers, each of which is producing a propagated action potential of 0.5 μV amplitude and 1 msec duration at the recording site. Assume, because of the variable demyelination of the nerve fibers between the site of stimulation and recording, the action potential of each individual fiber arrives at the recording site at 1-msec intervals over 10 msec. This amplitude of 0.5 μV will be insufficient to be detected above the level of noise of the recording system, and therefore there will be "no response." If, however, all 10 individual action potentials are propagated at the same speed and arrive simultaneously at the recording site, a response of 5 μV in amplitude will be detected. Lesser degrees of synchrony may produce a lower, but still recordable, compound nerve action potential.

RECORDING OF ACTION POTENTIALS FROM MUSCLE

Difference Between Nerve and Muscle Recording

The recording of CMAPs in the clinical electrodiagnostic laboratory has many features in common with the recording of nerve action potentials, as described in the previous section. Extracellular electrodes are used to record the summated responses of electrically excitable membranes, which have produced propagated action potentials resulting from the transmembrane and longitudinal current flow associated with ionic movement. The recorded CMAP is subject to the same principles of electrical fields, involving factors of distance, number of active fibers, direction of propagation of the action currents, number and density of fibers, and synchrony of activation. There are several very important differences, however, between the recording of CMAPs and the previously described recording of compound nerve action potentials. Although muscle fibers can be directly activated by electrical current applied externally, clinical re-

cording of CMAPs is carried out by stimulating the motor nerve fibers supplying a muscle. Although motor nerves contain many afferent fibers, including the large group I fibers from muscle spindles, the responses we are currently discussing are those elicited by direct activation from the alpha (group A) motor axons. The motor nerve action potentials arrive synchronously at the end plate region of the muscle, and after the time required for neuromuscular transmission, the muscle fiber action potentials are produced. As has been previously discussed, each motor nerve axon excites a group of muscle fibers comprising a motor unit, and with the supramaximal nerve stimulus commonly employed, virtually all muscle fibers are activated, more or less simultaneously. Individual muscle fiber action potentials have a somewhat different configuration from those of nerve. Although the amplitude of the transmembrane potential change is similar to that of nerve, the duration is somewhat longer, there is a gradual tailing off of the negative phase, and the speed of propagation along the muscle fiber is slow (about 3 m/sec) compared to that of the large motor and sensory nerve fibers.

Factors Influencing Response Amplitude

The amplitude of the surface recorded compound muscle action is very much larger than that of the compound nerve action potential. The reasons for this should be fairly obvious from the earlier discussion. Most importantly, the muscle from which we are recording with surface electrodes is much closer to the surface than the nerve, and both the total number of muscle fibers and the density are greater than in nerve, producing a summated potential of two to three orders of magnitude greater than those recorded from nerve. The differences between the individual nerve and muscle fiber action potential configuration, however, have little or no impact on the compound recorded response.

Unlike recording of nerve action potentials, the surface electrode utilized will often be considerably smaller than the muscle from which the recording is derived, and therefore, the recorded response will reflect only a portion of the activated muscle. Larger recording surfaces can be used, but these are then less selective and are more likely to record electrical changes from adjacent muscles. In the usual situation, the recording electrode is placed over the end plate region of the muscle, usually in the center or "belly" of that muscle. Since this is the site of origin of the muscle action potentials, an ideally placed electrode will produce a sharply defined negative takeoff, making it easy to determine an accurate latency. This distal latency is, of course, the result of the time it takes the nerve action potential to arrive at the neuromuscular junction from the point of nerve stimulation, including the slower conduction along the thin nerve terminals, the time of the electrochemical events at the neuromuscular junction, and the initiation of the propagated muscle action potential arising from the end plate potentials. The interpretation of the amplitude of compound muscle action potential is more complex than that of nerve, since it is determined by the number of excitable motor nerve axons, the adequacy of neuromuscular transmission, and the number of muscle fibers subsequently activated. In most cases, however, it is the number of motor nerve axons that is the major determining factor of the CMAP amplitude for a given set of recording parameters (e.g., distance from the source, size of recording electrode).

The position of the reference or "indifferent" electrode is also important in determining both amplitude and configuration of the recorded compound response. In most cases, the second recording electrode is placed over the tendon of the muscle, where it is relatively inactive. A refer-

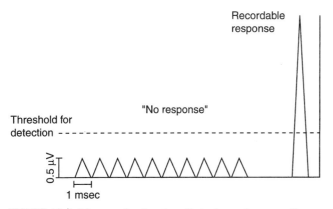

FIGURE 10–7. Diagram showing the effect of asynchrony or dispersion on the amplitude of a response to electrical stimulation. See text for explanation.

ence electrode placed over the active muscle will produce more of a bipolar response as the propagated potential approaches and passes the second recording electrode, possibly truncating the amplitude, as well as the duration.

It is possible to record compound muscle action potentials with needle electrodes beneath the surface, and there are some situations in which this is preferred. These electrodes may be subcutaneous or intramuscular; they may be useful when attempting to record from a deeper muscle. The much smaller recording surface of the needle tip, however, means that a much smaller proportion of the muscle will be sampled and the overall amplitude will be much more variable. The inevitable movement of the tip of the needle as the muscle is activated also adds to this variability, making amplitude measurements far less accurate and predictable. However, latency may be quite constant, and, particularly for large deep muscles of the trunk (such as the serratus anterior or the supraspinatus), this approach can be extremely useful. In most cases, a concentric needle electrode would be used for this purpose.

BASIC FEATURES OF NERVE CONDUCTION STUDIES

The remainder of this chapter deals with the usual methods of carrying out nerve conduction studies in the clinical electrodiagnostic laboratory and the clinically important factors in interpreting the results. The discussion reviews the key features of such studies and the techniques utilized. Since this is not meant to be a manual of nerve conduction studies, the method of performing one motor and one sensory nerve conduction study is reviewed in some detail, using the median nerve as an example of each. We also look briefly at some of the other commonly performed sensory and motor nerve studies, primarily to review the specific features of those studies that differentiate them from others. Next, we look briefly at a mixed sensory and motor nerve conduction study and review the principles and technique of recording late responses, the H-reflex and the F-response. We then review the factors that influence the interpretation of the recorded responses, utilizing the basic principles with which this chapter was introduced and reviewing how an understanding of these principles allows us to differentiate the effects of disease from the artifacts produced by technical factors and physiological variables. In the final section, we consider the commonly performed cranial nerve conduction studies, involving trigeminal and facial nerves.

Techniques of Stimulation and Recording

The clinical nerve conduction study begins with the stimulation of a peripheral nerve. Usually, the cathode and anode are placed longitudinally over the nerve, the cathode being closer to the recording site, as this is the point at which activation will take place, and reversing the poles risks having the propagated action potential blocked by anodal current or hyperpolarization between the site of stimulation and the site of recording. Usually, the stimulus intensity and duration are set at the lowest levels initially. Many, if not most, modern stimulators have fixed stimulus durations beginning with 0.05 msec. The intensity is gradually increased until a maximal response is produced. As we have seen, the recorded response reflects activation of the largest fibers in the nerve, and once they are all stimulated, no significant further increase in the compound action potential is seen. For most purposes, the stimulus used for the study is increased about 20% to 25% above the level producing a maximal response,

to ensure that all large fibers are activated. The positioning of both the cathode and the anode may be varied to find the point at which the threshold is lowest, the point at which the response is maximal, and the point at which the stimulus artifact is minimized. Sometimes, this requires moving the anode away from the nerve on one side or the other, or it may require adjustment of the position of the ground electrode, which is usually placed between the stimulation and recording sites. Stimulus artifact is generally more of a problem when recording sensory nerve action potentials, primarily because of the small amplitude of these responses. Care must be taken to make efficient contact between the stimulating electrode and the skin surface, and to avoid shunting of stimulus current externally, thus requiring much higher stimulus intensities. It should be mentioned that if a maximal stimulus cannot be achieved with the shortest duration, a situation that may occur in the face of nerve disease or when the nerve to be stimulated is at a greater than usual distance from the surface, then the duration can be gradually increased, sometimes to as much as 1 msec.

The recording electrodes for sensory or mixed nerve action potentials are placed over the nerve at a specified distance from the stimulating cathode and for motor nerve conduction studies over the belly of the muscle, as outlined earlier. Once again, the positioning of these electrodes may be modified by trial and error to find the point of maximal response, both in terms of amplitude and in terms of waveform. Most laboratories have specified positions for various nerve conduction studies, based on anatomical landmarks, including distances between stimulus and recording points and the sites of the active and "reference" recording electrodes. These may be modified depending on the body habitus and other variables, but any modification could influence the responses.

For the calculation of maximal conduction velocity, two stimulus points along an intermediate portion of the limb are utilized (see Fig. 10–9). This is critical for motor nerve conduction studies since the conduction time to the recording point includes contributions from slow terminal nerve branches and the neuromuscular transmission time. With the sensory nerve conduction study, the distal latency of the response reflects a true nerve conduction time, but maximal conduction velocity calculation still requires two points of stimulation.

The latency of the recorded response is measured from the peak of the stimulus artifact to the onset of the negative motor or sensory compound action potential or, in some laboratories, to the peak of the sensory nerve action potential, which, because of problems with stimulus artifact and stability of the baseline, may be a more constant and reliable parameter. The amplitude of the compound muscle action potential is measured from the baseline to the peak of the negative phase, although the area under the curve may more accurately reflect the actual number of active units. Most laboratories establish their own normal values, primarily for amplitude. The amplitude of sensory nerve action potentials may be measured from baseline to peak of negative phase, or peak to peak, depending on the specific nerve being studied and the particular laboratory. As long as the normal values are available and consistent, both measurement techniques are acceptable (see Figs. 10–8 and 10–11). Conduction velocity is determined by the distance traveled, generally between two points of stimulation and the time required for the impulse to negotiate that distance. This represents, therefore, a calculated parameter and is highly dependent on the accuracy of the measurements of latency and distance. Most modern electromyogram (EMG) machines provide very accurate latency determinations, but the distance must be provided by surface measurement. In some situa-

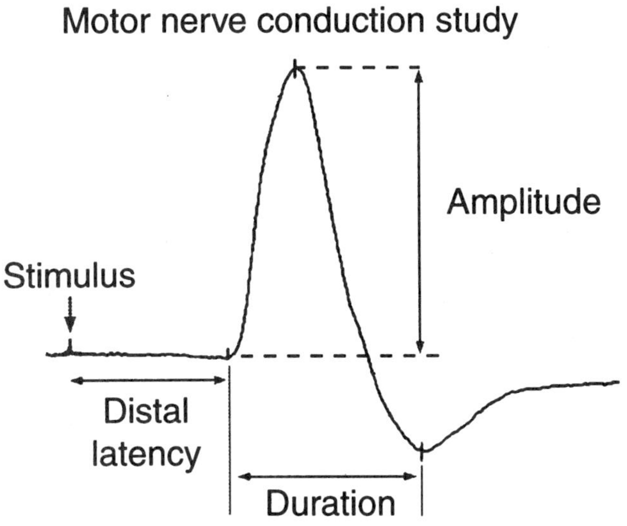

Motor nerve conduction study

Amplitude

Stimulus

Distal latency

Duration

FIGURE 10–8. A typical compound muscle action potential (CMAP) recorded at the surface in response to electrical stimulation of the motor nerve. Distal latency, from the stimulus to the onset of the negative response, amplitude, and response duration are noted.

tions, this is very simple and easily determined, for instance, measuring between two stimulation points on the forearm. There are situations, however, in which this measurement is much more difficult and subject to error, for instance, in determining the distance between stimulation points above and below the elbow. Since this nerve trunk must negotiate a joint that has a wide range of movement, the measured distance may vary significantly depending on the angle of the elbow at which the measurement is taken. Again, it is critical that for any given laboratory the normal values be determined for a given position of the elbow, and consistency of technique is critical for determining what is normal and what is not.

Motor Nerve Conduction Studies: Median Nerve

The motor nerve conduction study is carried out by stimulating a motor or mixed sensorimotor nerve and recording the compound action potential produced by the muscle it supplies (Fig. 10–8). For most routine purposes, the recording is made utilizing a surface disc placed over the belly of the muscle, ideally at the motor point or end plate region (G1), with the reference, or G2, electrode placed more distally over the tendon or other relatively inactive site. The typical arrangement is illustrated in Figure 10–9, using the median nerve.

The G1 recording electrode for the median motor nerve conduction study is placed over the center of the thenar pad, which usually corresponds to the end plate region of

the abductor pollicis brevis (APB). The adjacent muscles (the opponens policis laterally and the medial head of the flexor pollicis brevis [FPB] medially) may also contribute to the recorded CMAP, but fortunately these all share median innervation. The underlying lateral portion of the FPB and the adductor pollicis, both supplied by ulnar fibers, contribute relatively little to the CMAP recorded from this point and are rarely a significant factor even with median nerve lesions.

Placement over the source of the CMAP at the end plate region ensures that the takeoff of the CMAP will be abrupt and smooth, without an initial positive "dip." The appearance of such a dip should signal a need to relocate the G1 electrode, indicating that the potential source does not underlie the electrode. This can occur if the stimulus is being delivered to sufficient ulnar nerve fibers, either because of incorrect placement of the stimulating electrode or, as discussed later, if there is an anomalous communication, ulnar nerve fibers being temporarily housed in the median nerve.

The reference, or G2, electrode is usually placed 3 to 5 cm distally at the tendon insertion of the APB distal to the metacarpophalangeal joint of the thumb. The stimulating electrodes are placed longitudinally over the median nerve at the wrist, usually just medial to the prominent palmaris longus tendon. The cathode is distal to the anode, as is demonstrated in Figure 10–9.

The stimulus duration is initially 0.05 msec, and the stimulus intensity is gradually increased from its lowest level, until a maximum CMAP with sharp negative takeoff is elicited. The stimulus intensity is further increased beyond the maximum by about 20% to 25% to a "supramaximal" level for the clinical recording (Fig. 10–10A). Any deviation of the CMAP from normal (in either amplitude or configuration) requires repositioning of stimulating and recording electrodes before assuming that an abnormality exists. The distal amplitude and duration of the CMAP are noted, recalling that the distal latency includes the terminal nerve branch conduction time and electrochemical events at the neuromuscular junction.

To determine the motor nerve conduction velocity, the same procedure is repeated with stimulation of the median nerve more proximally, usually in our laboratory just proximal to the elbow crease on the ventromedial aspect of the forearm. Some trial and error is usually required to find the point at which the lowest threshold response can be obtained. The resulting CMAP should be virtually identical in configuration to that obtained with more distal stimulation, and the amplitude should also be the same or close (usually within 20% to 25%; Fig. 10–10B). The appearance of an initial positive dip again indicates either contamination with nonmedian fiber stimulation or an anomalous pathway. Motor nerve conduction velocity is calculated by dividing the distance between proximal and distal cathodal stimulus points by the latency difference between the responses to proximal and distal stimulation.

In some situations, even more proximal stimulation of median fibers may be required, including axillary or supra-

FIGURE 10–9. Stimulus and recording arrangement for a median motor nerve conduction study. The cathodal pole of the stimulator for stimulation both at the wrist and at the elbow is distal to the anode. In this and all subsequent figures, the darker stimulator prong represents the cathode. The ground electrode is on the dorsum of the hand.

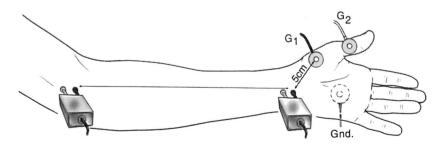

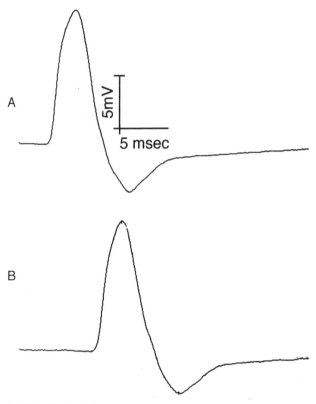

FIGURE 10–10. Typical CMAP recorded from the abductor pollicis brevis in response to median nerve stimulation at the wrist (A) and at the elbow (B). Median nerve conduction velocity can be calculated by dividing the distance between the proximal and distal stimulating electrodes (as in Fig. 10–9) by the difference in measured distal latency between responses A and B.

clavicular sites. At proximal, less accessible sites of stimulation, the stimulus intensity required generally is considerably higher, there is greater likelihood of stimulating nonmedian nerve fibers leading to CMAP contamination by responses of nonmedian innervated adjacent muscles, and CMAP amplitudes are reduced as incomplete median nerve activation and temporal dispersion become increasingly greater factors. In the absence of contamination by nonmedian stimulation, the CMAP configuration should be similar or identical to that obtained with distal stimulation.

Other Motor Nerve Conduction Studies

For ulnar motor nerve conduction studies, the recording electrodes may be placed over the hypothenar pad (abductor digiti minimi) or, less commonly, over the first dorsal interosseous with equally good results. Each recording site has some advantages and the two together may provide useful diagnostic information. The ulnar nerve is routinely stimulated at the wrist and proximal to the elbow, just above the medial epicondyle. Because of the frequency with which ulnar neuropathy occurs in the region of the elbow, stimulation below the intercondylar groove is also frequently carried out, particularly if there is either a significant amplitude drop between distal and proximal stimulation or significant slowing of conduction velocity when calculated for the distance between the proximal elbow stimulus and the wrist.

There has been considerable controversy regarding the ideal position of the elbow for ulnar nerve conduction studies. Some laboratories prefer a fully extended elbow whereas others favor variable degrees of flexion. The problem relates

to the measurement of the ulnar nerve length, which can only be estimated, because of the redundancy of the nerve at the elbow, and because the surface measurement changes with elbow position. It is critical that the normative data be determined for a specific elbow configuration and all studies be done in that position. As with the median nerve conduction study, the ulnar nerve can also be stimulated in the axilla and at Erb's point when attempting to localize a more proximal lesion. The ulnar nerve fibers within the lower trunk or medial cord of the brachial plexus cannot be reliably stimulated with sufficient precision and accuracy of localization with surface electrodes to allow determination of conduction velocity across the so-called thoracic outlet. In my opinion, therefore, attempts to relate such measurements to pathological compression at that site are fruitless at best and misleading at worst.

A number of other motor nerve conduction studies are a part of the repertoire of many electrodiagnostic laboratories, some utilized routinely or frequently, others only occasionally. In the upper extremity, radial motor studies can be very helpful, obviously in cases of suspected radial nerve injury, but also in investigating suspected lesions of the brachial plexus. Recording is made over the extensor indicis proprius or the extensor digitorum communis, with stimulation at the elbow, at the spiral groove, the axilla, and in the supraclavicular space. When abnormal responses are generated, or when infrequently performed nerve conduction studies are required, the contralateral limb should always be used as a control.

The musculocutaneous nerve can be stimulated in the axilla with recording over the biceps muscle. The axillary nerve can be stimulated in the supraclavicular fossa with recording over the deltoid. The spinal accessory nerve can be stimulated as it emerges from the posterior border of the sternocleidomastoid muscle, with recording over the upper trapezius. For all of these, the potential for spread of the stimulus to nerve fibers other than those intended, and the difficulties in achieving supramaximal stimulation, produce an element of unreliability, particularly regarding amplitude of the response.

In the lower extremity, common peroneal and posterior tibial motor nerve conduction studies would be part of the routine for most laboratories. Stimulation of the peroneal nerve occurs at the ankle, and proximally above and just below the fibular neck, recording over the foot muscle extensor digitorum brevis. A useful addition to this would be recording over the anterior tibialis muscle with stimulation above the fibular neck. Anterior tibialis recording can provide important information, because the extensor digitorum brevis is liable to local trauma and is not directly affected in cases of foot drop. The relatively short conduction distance to the anterior tibialis can produce high and at times inaccurate conduction velocity values.

The posterior tibial nerve can be stimulated at the ankle or behind the knee, recording the response in the foot over the abductor hallucis muscle. The femoral motor response can be recorded over the quadriceps, stimulating the femoral nerve at the inguinal ligament. For further details and specific technical aspects of any of these additional conduction studies, several authoritative sources are available.[19, 29, 34]

Sensory Nerve Conduction Studies: Median Nerve

Compound SNAPs can be at least theoretically recorded from any peripheral nerve in the body that contains sensory fibers. A characteristic SNAP is shown in Figure 10–11, demonstrating the usual parameters measured. Because there are

Sensory nerve conduction study

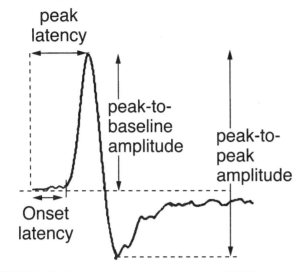

FIGURE 10–11. A sensory nerve action potential (SNAP). Latency may be measured to the onset of the negative phase or to the negative peak. Amplitude may be measured from baseline to negative peak or from negative to positive peak.

no specific features that allow us to recognize an evoked compound nerve action potential as being derived from sensory axons, we must rely on the stimulus and recording arrangement to identify the response as "sensory." In the physiological state, of course, a series of impulses traveling centripetally would be presumed to be derived from sensory axons, but we have already indicated that responses to electrical stimulation of nerve are carried bi-directionally. Identification of a response as being from sensory fibers must, therefore, be accomplished either by studying only "pure" sensory nerves or by limiting the stimulus or the response to sensory fibers. Strictly speaking, there is probably no such thing as a "pure" sensory nerve. Even cutaneous nerves, such as the sural, contain autonomic efferent and afferent fibers and may, in a small percentage of humans, contain motor fibers.[1] Recall, however, that the compound nerve action potentials recorded with surface electrodes basically reflect activity in only the larger fibers, and hence, recordings from a sensory (e.g., cutaneous) nerve truly are "pure" sensory responses. A similar argument can be applied to other cutaneous nerves such as the superficial radial, medial and lateral antebrachial cutaneous, and posterior cutaneous nerves in the upper limb, and the lateral femoral cutaneous, saphenous, and distal superficial peroneal nerves in the leg. Strictly sensory fiber action potentials can be isolated even in mixed motor and sensory nerves by applying the stimulus to, or recording from, a portion of the nerve in which there are no motor fibers; in the commonly studied median and ulnar nerves, this would be at the level of the digits, distal to the termination of all motor axons.

We can illustrate the "typical" SNAP study by utilizing median sensory nerve conduction as an example. The median sensory fibers can be isolated by stimulating them distally at the digital level, recording the response at the wrist or even more proximally, as the impulses travel orthodromically. An alternative is to reverse the process and record from the sensory fibers in the digits, stimulating at the wrist or more proximally. In this situation, we are recording impulses carried antidromically. A stimulus applied to the median nerve at the wrist or in the forearm necessarily activates

some motor as well as sensory fibers, and both transmit impulses bi-directionally. Even though there may be some "contamination" of the recorded response over the digital nerves by the much larger CMAPs produced by the proximal median nerve stimulus, the recorded motor response will be greatly attenuated by the distance of the digital electrodes from the potential generator site.

We use the antidromic median sensory responses in this example. The stimulating and recording arrangements are illustrated in Figure 10–12. The stimulating electrodes are placed over the median nerve at the wrist, the cathode more distally since the antidromic response will be recorded. Usually, a hand-held stimulator with fixed interelectrode distance is used for convenience, as the position can be more easily modified by trial and error to minimize threshold and maximize the amplitude and waveform of the response. The cathode is placed at a fixed distance (13 cm in our laboratory) from the proximal recording electrode, usually utilizing a ring or clip on the index or middle finger (or less frequently the thumb), with the second recording electrode placed more distally, usually 3 cm apart. This interelectrode distance will have some effect on the configuration and amplitude of the response, but if standardized and consistently utilized, the normal response characteristics can be determined. A ground electrode is fixed to a convenient location, usually on the dorsum of the hand.

The stimulus is applied at the wrist, beginning with subthreshold intensity and a duration usually of 0.05 msec. The stimulus current is gradually increased until a response is obtained and is further increased until a maximum amplitude is achieved. The current is then increased to a supramaximal level, usually about 20% above that eliciting a maximal response. The purpose of this latter increase is to ensure stability and reproducibility of the response, abolishing the effect of slight movements or alterations of threshold. An excessive increase, however, is to be avoided as it is unnecessarily uncomfortable for the subject, aggravates the problem of stimulus artifact, and can alter the response and the apparent conduction time by stimulating the nerve at some distance from the point just beneath the cathode.

The recorded SNAP (Fig. 10–13) is analyzed with respect to latency, amplitude, duration, and waveform. Latency may be measured from the stimulus artifact to the onset of the negative potential, signaling the arrival of impulses carried by the largest and hence fastest-conducting fibers, or to the peak of the negative potential, indicating that the bulk of

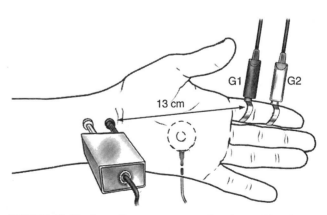

FIGURE 10–12. Recording arrangement for the antidromic recording of median sensory responses. Once again, the stimulus cathode is more distal. The distance between the cathode and G1 electrode is generally fixed at 13 cm in our laboratory. If proximal stimulation is desired, the proximal electrode would be placed as in Figure 10–9.

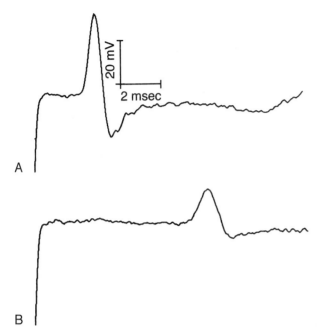

FIGURE 10–13. Median SNAP recorded at the index finger in response to median nerve stimulation at the wrist (A) and at the elbow (B). Note the drop in amplitude and increase in response duration with more proximal stimulation. See text for details.

the largest fibers have depolarized. The amplitude may be measured from baseline to negative peak, or from negative peak to positive nadir as shown in Figure 10–11 (while both will vary with interelectrode distance, the latter will be even more dependent on relative electrode placement than the former). The configuration of the response is also of importance. Synchronous arrival of the individual action potentials produces a relatively narrow (short duration) potential with short rise time and smooth contour. Asynchronous arrival, produced potentially by widely disparate conduction times even in the relatively short distance of travel (generally 13 cm), generates a broader, dispersed, multipeaked, and lower amplitude response and may indicate a pathological state.

Although the conduction time (i.e., either onset or peak latency) can be taken as an indicator of conduction velocity (by dividing the distance between stimulating cathode proximally and recording electrode by the response latency) measured in milliseconds, it is customary in our laboratory and generally more accurate to determine nerve conduction velocity by performing an additional procedure of stimulating the median nerve more proximally (e.g., at the elbow) and calculating the conduction velocity using the distance between the two stimulus points and the difference in latencies (either peak or onset) between responses to proximal and distal stimulation. This obviates several inherent inaccuracies encountered when calculating the terminal conduction velocity from the distal latency value, including a shorter distance over which the velocity is calculated, and a slower velocity measurement reflecting small terminal nerve fiber diameters. The SNAP elicited by proximal stimulation (e.g., upper forearm) is lower in amplitude and somewhat broader in duration than that produced by more distal stimulation (e.g., at the wrist; see Figure 10–13). Factors that contribute to this effect include dispersion of the waveform that occurs secondary to the differences in individual fiber conduction velocity magnified by the greater distance of travel, and phase cancellation of arriving nerve fiber action potentials. This discrepancy between responses to distal and proximal

stimuli is much more prominent for sensory than for motor nerve conduction studies, partly because of the somewhat broader range of sensory fiber size in the nerves being studied, but mostly because of the relative amplitudes of the responses, a small decrease in synchrony having a greater impact on the response measured in microvolts than one measured in millivolts.

Other Sensory Nerve Conduction Studies

Ulnar SNAPs are produced by stimulating the ulnar nerve at the wrist and recording, antidromically, at a predetermined distance utilizing ring or clip electrodes on the little finger. As with the median nerve, the procedure may be reversed for orthodromic recording. An ulnar sensory conduction velocity can be obtained by stimulating more proximally, either below or above the elbow.

A radial SNAP can be recorded by stimulating the superficial radial nerve in the distal forearm, where it can often be palpated, and recording on the thumb with the ring or clip electrodes. The standard distance between cathode and proximal recording electrode in our laboratory is 10 cm.

Medial and lateral antebrachial cutaneous responses can generally be obtained antidromically, by surface recordings, stimulating above the elbow and recording from the forearm. These studies are particularly useful in evaluating suspected lesions of the brachial plexus and in cases of multiple mononeuropathies.

In the lower extremity, the sural sensory response is generally recorded antidromically at the level of the lateral malleolus, following stimulation at a standard distance (14 cm in our laboratory) or at one of several more proximal sites. The reference electrode is placed along the lateral aspect of the foot several centimeters distal to the G1 electrode. On occasion, it may be useful to record distal to the lateral malleolus, using the contralateral foot as a control. To calculate the sural sensory conduction velocity, we measure the latency difference between stimulation sites at 21 cm and 10 cm above the recording electrodes, for a conduction distance of 11 cm.

The superficial peroneal sensory response can be obtained in most normal subjects by recording over the nerve (or one of its branches) at the level of the distal fibula and stimulating proximally along the anterior border of the fibula.[25] As with the sural nerve study, this can be extremely useful in polyneuropathies (and is not infrequently recordable even when a sural response cannot be obtained), and in differentiating lumbar plexopathy from L5 root lesions.

Techniques for recording saphenous and lateral femoral cutaneous sensory responses are described and can be useful in specific clinical situations. Because of anatomical variability and technical difficulties, particularly in all but the most slender young patients, contralateral studies are mandatory for comparison. The inability to obtain a response, even with averaging techniques, cannot automatically be presumed to be pathological.

Median and Ulnar Palmar (Mixed Sensorimotor) Studies

A variation of the median sensory nerve conduction study, utilized specifically in cases of suspected carpal tunnel syndrome (CTS), is the median palmar study. The rationale, as first suggested by Buchthal and Rosenfalck,[8] is that the routine median sensory study, either orthodromic or antidromic, includes the entire segment of median nerve between the wrist and the digits, whereas the site of pathology in CTS is in the proximal palm. Excluding the segment of

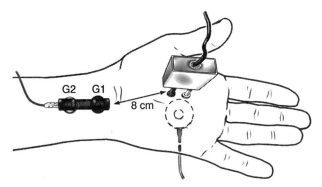

FIGURE 10–14. Technique for the median (mixed) palmar response, stimulating median nerve at the midpalm and recording from the median nerve at the wrist.

nerve from midpalm to digit reduces the "dilution" factor by eliminating a portion of the nerve with presumably normal conduction. The technique involves percutaneous stimulation of the median nerve in the palm and orthodromic recording at the wrist, generally 8 cm proximal to the stimulus site[12, 40] (Fig. 10–14). The resulting mixed (but mainly sensory) nerve action potential (Fig. 10–15), usually much higher in amplitude than in the routine antidromic or orthodromic studies, can be compared to the ipsilateral ulnar palmar response, stimulating the ulnar fibers in the midpalm and recording over the ulnar nerve at the wrist, 8 cm proximally. In most normal cases, the latency difference will be 0.2 msec or less between the two. Many laboratories, including our own, prefer a greater latency difference (≥ 0.4 msec) before considering the median palmar study abnormal.

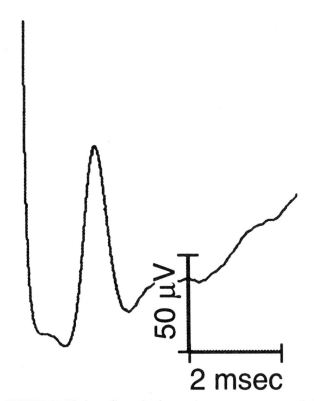

FIGURE 10–15. A median mixed sensorimotor response recorded at the wrist with palmar stimulation. See text for further details.

Late Responses: H Reflexes, F Waves

Recording of so-called late responses is commonly part of the routine electrodiagnostic study. They are particularly useful in spinal radiculopathies and in polyneuropathies, particularly segmental demyelinating types, as these responses depend on conduction along the proximal segments of the peripheral nerve. Although H reflexes and F waves represent indirectly activated compound muscle action potentials and the latency of their responses is roughly similar, they represent very different processes, which should readily be distinguishable despite their common features.

The H reflex, named after Hoffmann, who first described it in 1918, is a true reflex response, roughly equivalent to the myotatic or tendon reflex. In normal adults it is reliably elicited only from the tibial nerve, recording over the triceps surae (gastrocnemius-soleus) muscle complex, and less consistently, from the homologous flexor muscles of the forearm, stimulating the median nerve. The H reflex represents the CMAP elicited by monosynaptically activated motor neurons in response to stimulation of the largest afferent (Ia) fibers in the tibial nerve. Like the tendon reflex, therefore, it can serve as a measure of the excitability of the motor neuron pool and may be more widely elicitable in patients with upper motor neuron disorders.

The tibial nerve is stimulated in the popliteal fossa, using the standard bipolar electrode arrangement. Because the response is a true reflex, the fibers being stimulated are the large afferents; hence, the cathode in this case is proximally placed to avoid anodal block of centripetal conduction. The recording electrodes are surface disks. The active (G1) electrode is placed over the gastrocnemius-soleus, with the G2 electrode more distally on the anterior surface near the ankle (Fig. 10–16).

As always, the stimulus intensity is gradually increased, until an initial response is seen. Stimulus duration for the H reflex is usually lengthened (1.0 msec) to maximize the effectiveness of exciting the large afferent fibers. The earliest response seen, usually at a latency of about 30 msec (range: 25 to 35 msec, depending on exact electrode placement and length of limb), is the result of discharge of a small number of motor neurons at the S1 level of the spinal cord activated by the afferent stimulus. As the stimulus intensity is increased, a greater number of afferent fibers activate a larger percentage of the available motor neuron pool, producing an increased amplitude response. At the same time, the increasing stimulus begins to activate the somewhat smaller alpha motor nerve axons in the mixed tibial nerve, thus producing a direct (M) response of the gastrocnemius-soleus muscle complex with a correspondingly much shorter latency. As the stimulus intensity is further increased, the

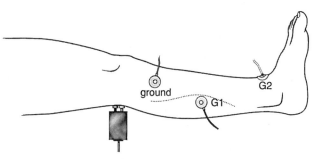

FIGURE 10–16. The stimulation and recording arrangement for the H reflex. In this case, the CMAP is recorded from the medial gastrocnemius. Note particularly the proximal placement of the cathode in this case.

H reflex reaches a peak and then begins to diminish while the direct or (M) response continues to increase to its maximal amplitude (Fig. 10–17). The standard explanation for this phenomenon is "collision" of the orthodromically traveling reflex impulses and the antidromic motor nerve action potentials elicited by the stimulus. There is some controversy as to whether this mechanism adequately explains the complete suppression of the H reflex as the stimulus intensity increases. An alternative explanation implicates inhibitory processes within the spinal cord.[14]

In general, there is a close correlation between the ability to elicit an H reflex and the clinical presence of an Achilles tendon reflex. However, one may clearly be present without the other.[45] Age is a well-recognized factor influencing both reflexes. Body habitus and mechanical factors (e.g., flexion contracture at the knee or local edema at the stimulation or recording site) are likely to interfere with the elicitation of an H reflex. Technique (positioning, type of reflex hammer)

is important in determining the success in eliciting a tendon jerk. It must be emphasized that the H reflex and the Achilles tendon reflex are not identical. The most obvious difference is that the stretch produced by the reflex hammer, activating the muscle spindle, is bypassed with electrical stimulation. It is less obvious that the electrical stimulus simultaneously activates both the Ia spindle afferents and the Ib Golgi tendon organ afferents, whereas the latter are activated in the tendon reflex only by the contraction of the muscle. This may have only theoretical implications clinically.

The F wave, so named because it was initially recognized in the foot muscle, is a late response recorded over a muscle belly after motor nerve trunk stimulation. The response is thought to result from back-firing of the motor neuron upon arrival of the antidromic impulse at the spinal cord.[43] The response can be produced in virtually any nerve containing motor fibers, using maximal or supramaximal stimulus intensities for the M response. It is characteristic of much lower amplitude than the maximal H reflex, representing antidromic activation of only a small number of the motor neurons (usually 10% or less), and the latency is both slightly longer than the H reflex (antidromic conduction velocity along the alpha motor axons is slightly slower than orthodromic conduction along a similar length of group I afferent fibers) and more variable. The technique for eliciting the F wave is identical to that of routine motor nerve conduction studies, aside from the change in the sweep speed to allow visualization of this long latency response. Because of the variability in both amplitude and latency, clinical studies usually include a series of F waves (at least 10). The shortest reproducible latency is considered the most useful measurement (Fig. 10–18). There is controversy as to whether minimum latency or mean latency is the more relevant measure;[14] our laboratory utilizes the former.

FACTORS INFLUENCING THE RECORDED RESPONSE

We can now use the information provided earlier to help interpret the recorded responses, looking at various factors, both technical and pathological, that may influence the response amplitude, configuration, and latency.

Stimulus Intensity

The smallest recordable response will be obtained as the stimulus intensity (or voltage) is gradually increased, as previously described. Although each nerve fiber responds with an all-or-none action potential, the compound or summated response will grow in amplitude as the individual fiber threshold for activation is reached, until no additional fibers are available (Fig. 10–19). There is an orderly and predictable recruitment of nerve fibers, based primarily on axonal diameter, with the largest fibers having the lowest threshold, for reasons elucidated earlier. Although position of the axon within the peripheral nerve trunk may play a role in determining the specific sequence of axonal activation, ultimately all nerve fibers within the electrical field will be activated and additional stimulus strength will not produce a larger amplitude response. As has been previously suggested, the stimulus strength producing this maximal response is ordinarily further increased by 20% to 25% in clinical electrodiagnostic studies to ensure stability. Increasing the stimulus intensity further, beyond this supramaximal level, has several potentially deleterious effects. It will likely increase patient discomfort and will generally have an adverse effect on cooperation. It runs the risk of stimulating

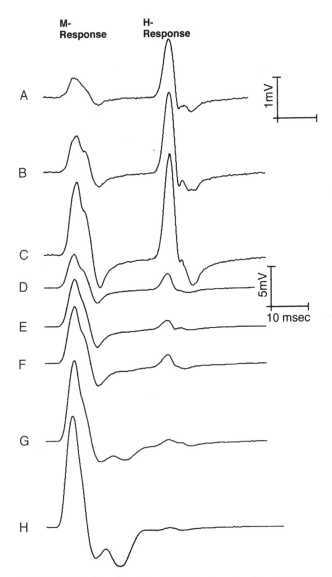

FIGURE 10–17. The H reflex. Stimulus intensity is gradually increased from *A* through *H*. The amplitude of the H reflex increases from *A* to *C* and then progressively declines at higher stimulus strength as the M response continues to increase in amplitude. Note that the amplitude calibration for responses *A* through *C* is 2 mV, but it is 5 mV for responses *D* through *H*. The H reflex in trace *D* is only slightly lower than that in trace *C*.

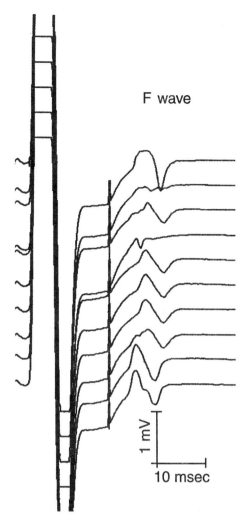

F wave

FIGURE 10–18. The F response. Note the variable amplitude and configuration of the F wave. In this series, the latency is fairly constant. The much larger M response is off the screen at this recording sensitivity to facilitate visualization of the F response.

distant and unwanted axons (e.g., those in an adjacent peripheral nerve trunk). It may increase the stimulus artifact, which may interfere with the ability to recognize a sharp and clean response takeoff. Finally, it may actually shorten the response latency, by producing activation of the axons distal to the expected (and measured) point of stimulation directly beneath the cathode, thereby increasing apparent conduction velocity. This can be a particularly troublesome problem when attempting to stimulate a nerve at a deep and proximal site such as Erb's point. Actual activation of nerve fibers distal to the intended location may account for some of the inordinately high "normal" conduction velocity values reported in the literature.

Filter Settings

The primary purpose of filters in the electrophysiological laboratory is to enhance the recorded response by rejecting unwanted signals or "noise," both high and low frequencies. In a sense, the filters provide a response similar to what an averaging technique would yield. Maximal use of filters allows accurate and reproducible determination of amplitude, latency, and waveform with a minimum of distortion.[16] Inappropriate filter settings, however, can have a pro-

nounced effect on each of these properties. Because the SNAP is so much smaller in amplitude and less easily distinguished from spontaneous fluctuations in the baseline, filter settings can have a particularly profound effect on the sensory nerve recordings. The result of excessive filtering of high-frequency responses on the SNAP is shown in Figure 10–20. While the relatively unfiltered response is cosmetically unattractive because of the high-frequency fluctuations, the amplitude is progressively attenuated and the peak latency delayed as the higher frequency signals are increasingly rejected. At the other end of the spectrum, reducing the lower frequencies of the response can also attenuate the SNAP amplitude, while shortening the peak latency slightly (Fig. 10–21).

Recording Electrode Position

The relationship between response amplitude and distance from the source has been discussed previously. In the course of placing the recording electrodes, the site of maximal amplitude response has generally been identified by trial and error; this usually corresponds to the shortest distance between the potential source and the skin surface. Although the nerve or muscle may be less than 1 mm from the surface, recording from more deeply situated structures may be required. The depth may vary with the amount of subcutane-

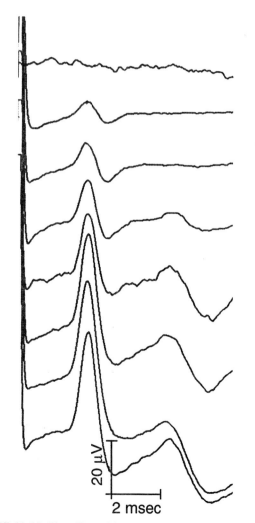

FIGURE 10–19. The effect of increasing stimulus strength on the recorded SNAP, from subthreshold to supramaximal.

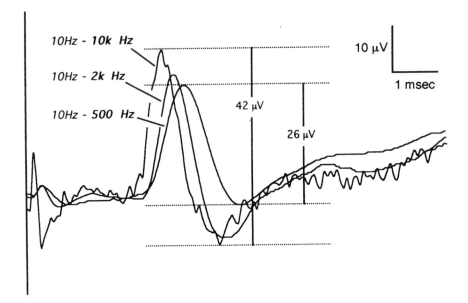

FIGURE 10–20. The effect of changes in high-frequency filter settings from 10,000 Hz to 500 Hz, recording median SNAP. Progressive reduction in high-frequency components leads to slower rise time and reduced amplitude along with a modest increase in peak latency. (From Gitter AJ, Stolov WC: Muscle Nerve 18:805, 1995, with permission.)

FIGURE 10–21. The effect of altering low-frequency filter from 3 to 500 Hz on the SNAP. The response amplitude again diminishes as the low frequencies are progressively filtered out and there is a slight decrease in peak latency. (From Gitter AJ, Stolov WC: Muscle Nerve 18:806, 1995, with permission.)

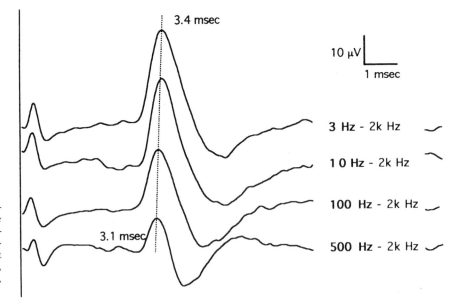

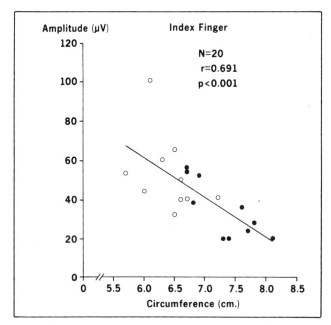

FIGURE 10–22. Relationship between the median digital SNAP and digit circumference in women (open circles) and men (filled circles). (From Bolton, CF: AAEM Minimonograph #17. American Association of Electromyography and Electrodiagnosis, Rochester, MN, 1981, p 6, with permission.)

ous tissue or muscle overlying the specific nerve or muscle being studied. Obesity and peripheral edema are obvious variables. Sensory responses recorded antidromically at the digits are generally lower in amplitude in men than in women, and in either sex, there is an inverse correlation between digit circumference and response amplitude, presumably related to the distance from the nerve to the surface[5] (Fig. 10–22).

The relative position of the active and reference recording electrodes also influences the response obtained. In most cases, both electrodes (G1 and G2) contribute, since the reference electrode (G2) is rarely in a truly "indifferent" position. The interelectrode distance and the proximity of each electrode to the potential source determine the actual configuration of the recorded response (Fig. 10–23). As it is

EFFECT OF DISTANCE BETWEEN G1 AND G2

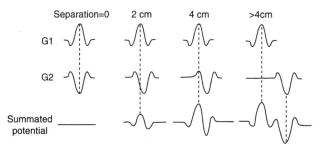

FIGURE 10–23. The effect on the recorded compound action potential of increasing separation between recording electrodes G1 and G2, as measured by a differential amplifier in which the negative of the response at G2 is added to the response at G1. Since both recording electrodes are theoretically over active sites, the recorded response at G1 and G2 will be equal and opposite. At zero separation, the potential difference between G1 and G2 will be zero. At increasing distances, the summated potential differences will vary as shown.

generally the later portions of the response that are most influenced by interelectrode distance, the latency and peak amplitude are not significantly altered if the interelectrode distance is adequate. If either recording electrode is too close to the stimulating electrodes, shock artifact may be a serious and potentially uncontrollable problem. This may influence most prominently the accurate identification of the response takeoff, which is necessary for accurate determination of latency.

Body Temperature

Response amplitude of both SNAPs and CMAPs increases with decreasing temperature (Fig. 10–24). The most important determining factor appears to be the delay in inactivation of the increased sodium conductance with colder ambient temperature, allowing the axon or muscle transmembrane potential to approximate more closely the equilibrium potential for sodium.[13] At the same time, conduction velocity tends to be directly related to body temperature over a wide range; this combination of increased amplitude and increased latency or reduced nerve conduction velocity is characteristic of the cold limb.

Age

Age is also a factor in both response amplitude and conduction velocity.[7, 39] There may be several reasons for this, including fiber attrition for both nerve and muscle, with relatively greater loss of larger axons particularly, and perhaps effects of aging on the nerve and muscle membranes themselves. The decrement in amplitude and conduction velocity may be relatively small per decade, but for SNAPs or mixed sensorimotor responses, particularly, this can result in the inability to record a response (e.g., sural, plantar nerve) in older subjects, even with averaging techniques. Interpreta-

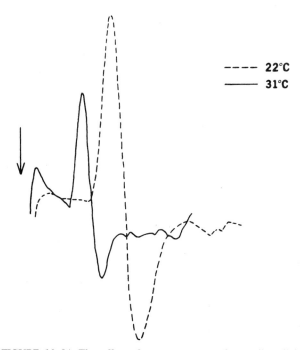

FIGURE 10–24. The effect of temperature on the median digital SNAP. Both the latency and the amplitude are increased at lower temperatures. See text for explanation. (From Bolton, CF: AAEM Minimonograph #17. American Association of Electromyography and Electrodiagnosis, 1981, p 5, with permission.)

tion of both amplitude and nerve conduction velocity must reflect these changes, as determined by normative date for different ages (Table 10–2).

Height

Several studies have suggested that nerve conduction velocity is inversely correlated with height.[9, 35, 46] The mechanism of this relationship is unclear. It is recognized that nerve conduction velocities are generally higher proximally than distally, perhaps a reflection of reduced nerve fiber diameter distally. The suggestion has been made that nerve fiber diameter begins to decrease at a set distance from the cell body, and in taller subjects, the portion of the nerve available for stimulation and recording may be of smaller caliber. Histological studies have not tended to support this hypothesis, however, and a physiological study suggested that longer axons tend to be thinner along their entire length.[46] An alternative explanation may relate to internodal length as it varies with distance along a nerve.[9] Whatever the mechanism, the observed relationship between height and nerve conduction velocity can readily be demonstrated, and failure to take this into account may lead to misinterpretation of nerve conduction velocity data in some individuals. The influence of height on response amplitude recorded from nerves is less clear and less consistent.

Anomalous Pathways

A final factor influencing response amplitude and configuration recorded in clinical electrodiagnostic studies is the anomalous or aberrant pathway. The most important and most common of these is the median-to-ulnar crossover, generally known as the Martin-Gruber anastomosis, named after two of its earliest describers.[17] In this anomaly, ulnar motor fibers derived from the lower trunk and medial cord of the brachial plexus find their way into the median nerve temporarily, where they segregate in most cases with fibers of the anterior interosseous branch. At some point in the forearm, they cross to the ulnar nerve from the anterior interosseous branch and continue on to their proper target muscles, most commonly the first dorsal interosseous, less frequently the hypothenar group, and least often the adductor pollicis (or the deep head of the flexor pollicis brevis). The result of this anomaly in the otherwise normal individual is that stimulation of the median nerve at the elbow may produce a CMAP recorded over the thenar muscle larger in amplitude than that produced by stimulating the median nerve at the wrist—just the opposite of what might be expected. Furthermore, the CMAP may have a somewhat different configuration. This would be due to the presence in the median nerve, at the proximal stimulation site, of ulnar

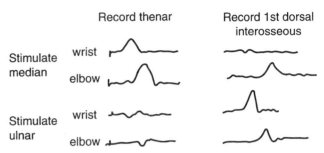

FIGURE 10–25. The effect of median-ulnar crossover (Martin-Gruber anomaly) on the responses to median and ulnar nerve stimulation, proximal and distal, as recorded over the thenar and first dorsal interosseous muscles. See text for explanation.

fibers destined for the first dorsal interosseous or thenar muscles, contributing to the response. More importantly, when recording either first dorsal interosseous or hypothenar, the response to ulnar nerve stimulation proximally may be considerably lower in amplitude than that obtained by stimulating the ulnar nerve at the wrist, again, because some of the ulnar fibers present at the more distal site are contained within the median nerve in the proximal forearm. This may give the erroneous appearance of an ulnar conduction block at the elbow. The correct interpretation is confirmed by stimulating the median nerve and recording directly from the first dorsal interosseous or the hypothenar muscles. When the larger response from elbow stimulation is subtracted from the smaller or absent response from wrist stimulation, the difference approximates the observed discrepancy in the routine study (Fig. 10–25).

In a pathological state, such as median neuropathy at or distal to the wrist (carpal tunnel syndrome), the results can be quite confusing in the face of this anomaly. When stimulating the median nerve proximally, the contained ulnar fibers, bypassing the carpal tunnel via the ulnar nerve at the wrist, produce a thenar-recorded response of "normal" latency. Because the median fibers may produce a lower amplitude, more dispersed, and longer latency response, the ulnar-derived response may in fact predominate. Furthermore, because the thenar recording electrode is not overlying the motor point for the ulnar-innervated contributors to the response, an initial positive deflection or dip may precede the main negative response. With median nerve stimulation at the wrist, the ulnar component would no longer be present, thus producing a "purely" negative, albeit longer latency, and perhaps lower amplitude and more dispersed response owing to the median nerve lesion within the carpal tunnel. Once again, the true nature of the situation will be revealed by recording first dorsal interosseous and hypothenar responses to the proximal median nerve stimulation.

Other communications between median and ulnar nerves may occur more distally in the hand itself, producing unexpected and sometimes confusing results. These are generally referred to as the Riche-Cannieu anomaly; they are apparently much less common than the Martin-Gruber anastomosis and of less potential significance to the electromyographer.[17]

In the lower extremity, the most important anomalous pathway is the presence of an accessory deep peroneal nerve, a branch of the superficial peroneal nerve supplying a portion of the extensor digitorum brevis, instead of the expected deep branch. This can be suspected when the response recorded over the extensor digitorum brevis follow-

TABLE 10–2	Median Sensory Conduction: Normative Data at Different Ages	
Decade	Amplitude (μV)[a]	Peak Latency (msec)[a]
1	20–70 (40)	2.0–2.9 (2.4)
2	20–85 (42)	2.3–3.2 (2.8)
3	20–75 (46)	2.4–3.3 (2.8)
4	15–80 (44)	2.3–3.4 (2.8)
5	15–65 (36)	2.4–3.5 (2.8)
6	15–70 (28)	2.6–3.6 (3.1)
7	10–50 (26)	2.8–3.7 (3.2)
8	10–40 (20)	2.8–3.8 (3.2)

[a] Parenthetical numbers represent the mean.

ing proximal (common) peroneal stimulation is higher in amplitude than that following distal stimulation anterior to the lateral malleolus. The anomalous branch can be found behind or posterior to the lateral malleolus, and stimulation at that site will generally produce the "missing" portion of the extensor digitorum brevis response, thereby solving what otherwise might have been a puzzling discrepancy.

CRANIAL NERVE CONDUCTION STUDIES

The facial, trigeminal, and spinal accessory nerve trunks are accessible to percutaneous electrical stimulation. These nerve conduction studies, when added to data from the needle electrode examination of cranial innervated muscles, contribute to the precise diagnosis of a number of cranial neuropathies. Spinal accessory nerve conduction studies are conceptually no different from these studies in the extremities and are not discussed here.

Facial Motor Nerve Conduction

The facial motor nerve trunk first becomes accessible to percutaneous stimulation as it exits the stylomastoid foramen. A direct evoked motor response can be obtained by placing the cathode of the stimulating electrode just anterior to and below the lower tip of the mastoid, at the tip of the ear lobe. Anatomical variability requires some adjustment of the electrode position for maximal depolarization of the nerve trunk, and it is often necessary to rotate the anode to a position above or slightly forward and above that of the cathode. Care should be taken that the anode is not placed over the masseter muscle, as direct muscle stimulation will produce a volume-conducted artifact at the recording electrodes, obscuring the latency measurement (Fig. 10–26). The active (G1) recording electrode is placed over the nasalis

muscle, with the reference electrode (G2) placed over the contralateral nasalis. The amplitude should be > 1.8 mV and the initial latency < 3.4 msec, with a side-to-side latency difference < 0.6 msec.

The direct facial motor response is useful primarily in the diagnosis of axon loss lesions. With an acute axon loss lesion distal to the site of stimulation, conduction failure is identifiable immediately, as impulses travel through the lesion. With proximal lesions, the amplitude does not begin to fall until wallerian degeneration and neuromuscular transmission failure begin, usually delayed 3 to 4 days from the onset of axon loss.[10] Because the maximal drop in facial CMAP amplitude may not occur until day 8 after injury, direct facial motor conduction studies are not useful in defining the extent and axon loss character of a peripheral facial nerve lesion in the immediate period.

Facial motor conduction studies can be useful in the diagnosis of generalized peripheral polyneuropathies. Latency prolongation can be marked in primary demyelinating polyneuropathies such as Guillain-Barré syndrome and Charcot-Marie-Tooth type I disease. In severe chronic inflammatory demyelinating polyneuropathy (CIDP), facial motor latency prolongation may aid in the diagnosis when extremity conduction studies are unelicitable.

The Blink Reflex

This multisynaptic reflex enters the brain stem via trigeminal sensory afferents and exits the brain stem via facial motor nerve fibers. Assessment of this reflex complements the clinical evaluation of the corneal reflex and can provide valuable information regarding localization and prognosis of trigeminal and facial nerve lesions. Kugelberg was the first to study the blink reflex electrophysiologically, identifying two responses: an early contraction of orbicularis oculi ipsilateral to the side of stimulation, and a late bilateral contraction.[23] The early, or R1, response is not clinically visible, but the

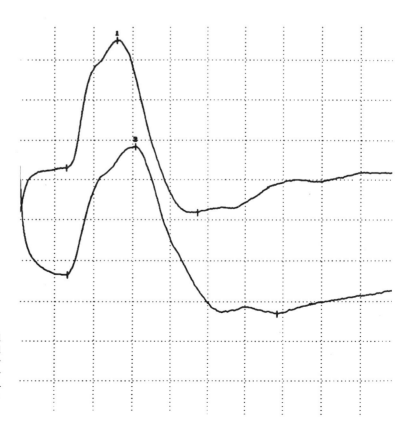

FIGURE 10–26. The upper tracing represents a facial motor CMAP after correct stimulation of the facial nerve trunk at the stylomastoid foramen. The lower CMAP was recorded after the anode of the stimulator was rotated to a position overlying the masseter muscle. Horizontal division = 2 msec, vertical scale = 1 mV.

later R2 response is synchronous with the blink occurring after electrical stimulation of the supraorbital nerve. This reflex is also inducible by mechanical stimulation of the face away from the trigeminal nerve, or by noise and light, and thus is thought to be initiated through exteroceptive (cutaneous) rather than proprioceptive fibers.[37] Although the corneal and blink reflexes share features in common, including brain stem pathways, they cannot be equated. The corneal reflex is carried through smaller afferent nerve fibers, is not preceded by an early (R1) response, and is associated with longer latency and duration than the R2 of the blink reflex.[30]

The afferent limb of the blink reflex is carried through trigeminal nerve fibers, composed mainly of medium-myelinated (A-beta) cutaneous fibers. The efferent limb is carried through facial motor nerve fibers. The R1 response is mediated through an oligosynaptic pathway in the pons, including the main trigeminal sensory nucleus and the ipsilateral facial motor nucleus. The R2 response is mediated through a polysynaptic pathway in the pons and medulla, including the spinal trigeminal nucleus. Projections from the spinal trigeminal nucleus are bilateral, ascending into the pons and synapsing in bilateral facial motor nuclei.[22, 32] Thus, the normal blink reflex is composed of an R1 response ipsilateral to the side of stimulation, and bilateral R2 responses (Fig. 10–27).

Techniques

Tin disk recording electrodes are applied to the skin overlying the orbicularis oculi muscles bilaterally, requiring two-

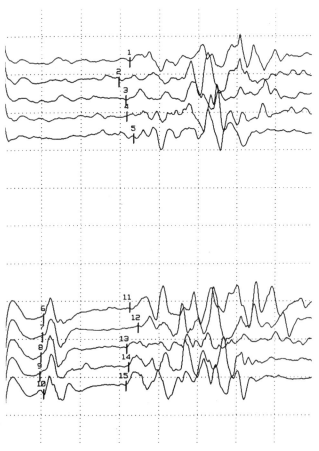

FIGURE 10–27. Normal blink reflexes. The lower tracings were recorded over the left orbicularis oculi muscle after left supraorbital nerve stimulation, and show R1 and R2 responses. The upper tracings were recorded simultaneously over the right orbicularis oculi muscle, and show R2 responses. Horizontal division = 10 msec, vertical division = 500 μV.

channel simultaneous recording capability. Trigeminal nerve trunk stimulation is performed most reliably over the supraorbital nerve at the supraorbital notch in the V_1 distribution. The cathode of the stimulating electrode should be placed at the notch since evoked impulses are being directed orthodromically toward the brain stem, and the anode should be placed away from the contralateral notch to avoid bilateral stimulation. Stimuli should be delivered no faster than every 5 to 7 seconds to avoid habituation of the responses, and minimal current is required to produce the reflex. Difficult-to-obtain R1 responses can be elicited with use of paired stimuli about 5 msec apart, within the period of facilitation that lasts 1 to 9 msec.[6] Although stimulation in the V_2 and V_3 distributions can be accomplished, responses are not reliably obtained in normal subjects.

Only latency measurements of the R1 and R2 responses are routinely made, since amplitudes are too variable. The R1 latency should be < 13 msec, ipsilateral R2, < 41 msec, and contralateral R2, < 44 msec. The latency difference between sides for the R1 should be < 1.2 msec, and for R2, < 8 msec.[41] By 2 years of age, the R1 latency should reach adult values, but the R2 can be absent in children less than 2 years old, and it does not reach adult latency values until ages 5 to 6 years.[2, 11]

Clinical Applications

Blink reflexes are abnormal when there is significant axon loss, conduction block, or conduction velocity slowing along the afferent or efferent limb. Lesions of the trigeminal nerve, the gasserian ganglion, brain stem, and facial nerve can produce loss of the response or prolongation of latency. R1 latency abnormalities are more frequent than R2 changes. An abnormality of only R1 suggests a lesion in either the trigeminal or facial nerve on the side of stimulation. An abnormality limited to one or both R2 responses suggests a lesion in the spinal tract or spinal trigeminal nucleus. Depending upon the specific pattern of abnormalities, facial and trigeminal lesions can be distinguished (Fig. 10–28). In generalized demyelinating polyneuropathic processes, such as Charcot-Marie-Tooth disease type I, and acute and chronic acquired segmental demyelinating polyneuropathies, marked R1 latency prolongation can be seen and can aid in the diagnosis when extremity responses are absent.[20]

Facial Nerve Lesions

Most axon loss facial nerve lesions can be identified on direct stimulation of the facial nerve trunk at the stylomastoid foramen, but isolated conduction block lesions proximal to the point of stimulation are missed. Since the blink reflex travels through the whole length of the facial nerve, it will identify a proximal pure conduction block lesion as long as no normally conducting facial nerve fibers participate in the response. The blink reflex will also identify a proximal axon loss lesion that is fresh and severe immediately, whereas 3 to 4 days must elapse before direct facial nerve trunk stimulation will show first signs of transmission failure and wallerian degeneration.[10]

With recent-onset Bell's palsy, the R1 and R2 responses are always abnormal. In patients with a good prognosis for recovery, the abnormal latencies tend to move toward normal when studied serially over several months after onset of symptoms.[21] In patients with old peripheral facial nerve lesions, aberrant regeneration can be demonstrated by the blink reflex, even when the clinical evidence for facial synkinesis is equivocal. This is especially valuable when trying to separate an old peripheral facial nerve lesion from a CNS lesion. To record synkinesis, the second set of recording

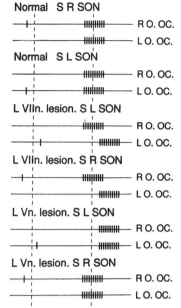
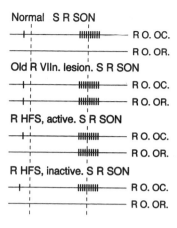

Normal S R SON
———— R O. OC.
———— L O. OC.

Normal S L SON
———— R O. OC.
———— L O. OC.

L VIIn. lesion. S L SON
———— R O. OC.
———— L O. OC.

L VIIn. lesion. S R SON
———— R O. OC.
———— L O. OC.

L Vn. lesion. S L SON
———— R O. OC.
———— L O. OC.

L Vn. lesion. S R SON
———— R O. OC.
———— L O. OC.

Normal S R SON
———— R O. OC.
———— R O. OR.

Old R VIIn. lesion. S R SON
———— R O. OC.
———— R O. OR.

R HFS, active. S R SON
———— R O. OC.
———— R O. OR.

R HFS, inactive. S R SON
———— R O. OC.
———— R O. OR.

S R SON - Stimulate R Supraorbital Nerve
S L SON - Stimulate L Supraorbital Nerve
R O. OC. - R Orbicularis Oculi
R O. OR - R Orbicularis Oris
HFS - Hemifacial Spasm

FIGURE 10–28. Blink reflex patterns in different conditions. Above each set of schematically rendered blink reflexes is listed the clinical situation and the nerve trunk stimulated. To the right of each recording is the muscle from which the recording was made. Vertical dotted lines indicate the upper limit of normal latencies for the R1 and R2 responses. L VII n., left facial nerve lesion; L V n., left trigeminal nerve lesion.

electrodes is placed over the orbicularis oris, so that simultaneous recording of the oculi and oris muscles takes place on the side of stimulation (see Fig. 10–28).

Hemifacial spasm, a disorder characterized by brief, sporadic, stereotyped unilateral contractions of some or many facial innervated muscles, usually results from compression of the facial nerve by an aberrant arterial loop near the cerebello-pontine angle. Focal demyelination at the compression site leading to ectopic impulse propagation and ephaptic transmission between facial nerve fibers are the likely pathogenic mechanisms.[24, 27] Some evidence also suggests that pathological impulse generation may result from facial motor nucleus hyperactivity.[26, 44] Routine direct facial nerve conduction is normal, as are the latencies of the R1 and R2 responses. Patients with hemifacial spasm demonstrate synkinesis in the orbicularis oris muscles during the blink reflex. These responses are inconstant, compared to the fixed, constant synkinesis seen in aberrant regeneration of the facial nerve. They can be recorded shortly after a spasm but not during long quiet periods, and the responses elicited can alternate among an R1, R2, or both (see Fig. 10–28). Surgical removal of the offending arterial compression can alleviate the hemifacial spasm immediately, a phenomenon that can be documented intraoperatively by mandibular and zygomatic facial motor conduction studies, as blink reflexes are suppressed by general anesthesia[28] (see Chap. 39).

Trigeminal Nerve Lesions

With a lesion of the trigeminal nerve, all latencies are prolonged bilaterally when the involved trigeminal pathway is stimulated, and all responses are lost when the lesion is severe. Trigeminal neuralgia and atypical facial pain predictably show no abnormalities of the blink reflex. In sensory neuropathies related to connective tissue diseases, such as scleroderma, Sjögren's syndrome, and mixed connective tissue disease, the R1 or R1 and R2 responses may be affected, whereas the R2 responses are seldom affected alone.[18]

Central Nervous System Lesions

CNS lesions produce abnormalities of the blink reflex. In multiple sclerosis, R1 abnormalities are more frequent than

R2 changes and can be seen in the absence of clinical signs of brain stem involvement. Isolated abnormalities of the R2 response have been noted in patients with signs of disease in the medulla. Blink reflex abnormalities are also seen in malignancies and vascular lesions affecting the brain stem. In the setting of coma resulting from supratentorial cerebral trauma, loss of the R1 responses bilaterally is a bad prognostic sign. Loss of the R2 response suggests additional lesions in the brain stem.[36] In patients with hemispheric stroke and hemiparesis, abnormalities of the R1 and R2 responses can be seen for several weeks.[4]

The Jaw Jerk

The jaw jerk (masseter reflex) is mediated through a monosynaptic reflex arc and is elicited by tapping the chin. Both the afferent and efferent limbs travel through the mandibular branch of the trigeminal nerve. Stretch of muscle spindles in the masseter muscles sets off this proprioceptive reflex. Cell bodies of the Ia afferent fibers reside in the mesencephalic nucleus of the trigeminal nerve in the midbrain. Fibers synapse on the trigeminal motor nucleus in the pons to activate the efferent limb of the reflex arc.

Two sets of tin disk recording electrodes are applied to the skin overlying the masseter muscles bilaterally, the reference electrodes placed over the mandibles. A specially adapted reflex hammer is connected to the EMG machine and triggers a sweep of the oscilloscope when the hammer strikes. The normal latency is between 6 and 10.5 msec, with a side-to-side latency difference < 1 msec.[41] The most common abnormality is unilateral absence of the jaw jerk. Amplitude measurements are not reliable.

In lesions of the mandibular nerve, usually caused by infiltrative malignancy, the jaw jerk is absent on the side of the lesion, and the needle electrode examination shows denervation of the masseter. This is also the pattern in pontine lesions. In mesencephalic lesions, the jaw jerk is abnormal while the masseter EMG is normal.[31]

The masseter reflex may have differential diagnostic value in separating pure sensory ganglionopathies from sensory axonopathies. Patients with subacute sensory neuronopathies, presumably on an autoimmune basis, have absent deep tendon reflexes, blink reflexes, and H reflexes, but intact jaw jerk responses. It is postulated that the immune attack occurs at peripheral ganglia not protected by blood-nerve barriers but spares the mesencephalic nucleus because of the presence of a blood-brain barrier.[3]

CONCLUSION

The proper interpretation of nerve conduction studies requires an understanding of the many factors that may influence the recorded results. Only when one has comfortably mastered the range of possible results in normal subjects can one begin to feel confident in interpreting the effects of disease. Nerve conduction studies, when combined with the needle electrode examination, provide a powerful tool for elucidating the presence of, and mechanism operative in, peripheral neuromuscular disease. Subsequent chapters in this textbook, and many additional sources, will greatly amplify this statement. Failure to understand the many physiological and technical factors—other than disease—that may influence the interpretation of nerve conduction studies may render the technique less useful or even misleading, a situation that, at times, can be more dangerous than no information at all.

ACKNOWLEDGMENTS

It is a pleasure to thank Nancy Heim, AMI, of the Cleveland Clinic Foundation Department of Medical Illustrations for her expert help with the figures and Dr. Kerry Levin for his contribution to the section on cranial nerve studies.

REFERENCES

1. Amoiridis G, Schöls L, Ameridis N, et al: Motor fibers in the sural nerve of humans. Neurology 49:1725–1728, 1997.
2. Anday EK, Cohen ME, Hoffman HS: The blink reflex: Maturation and modification in the neonate. Dev Med Child Neurol 32:142–150, 1990.
3. Auger RG: Role of the masseter reflex in the assessment of subacute sensory neuropathy. Muscle Nerve 21:800–801, 1998.
4. Berardelli A, Accornero N, Cruccu G, et al: The orbicularis oculi response after hemispheral damage. J Neurol Neurosurg Psychiatry 46:837–843, 1983.
5. Bolton CF: AAEE Minimonograph #17: Factors affecting the amplitude of human sensory compound action potentials. Rochester, MN, AAEE, 1981.
6. Broggi G, Caraceni T, Negri S: An analysis of a trigemino-facial reflex in normal humans. Confin Neurol 35:263–270, 1973.
7. Buchthal F, Rosenfalck A: Evoked action potentials and conduction velocity in human sensory nerves. Brain Res 3:1–122, 1966.
8. Buchthal F, Rosenfalck A: Sensory conduction from digit to palm and from palm to wrist in the carpal tunnel syndrome. J Neurol Neurosurg Psychiatry 34:243–252, 1971.
9. Caruso G, Massini R, Crisci C, et al: The relationship between electrophysiologic findings, upper limb growth and histological features of median and ulnar nerves in man. Brain 115:1925–1945, 1992.
10. Chaudhry V, Cornblath DR: Wallerian degeneration in human nerves: Serial electrophysiological studies. Muscle Nerve 15:687–693, 1992.
11. Clay SA, Ramseyer JC: The orbicularis oculi reflex: Pathologic studies in childhood. Neurology 27:892–895, 1977.
12. Daube JR: Percutaneous palmar median nerve stimulation for carpal tunnel syndrome [abstract]. Electroencephalogr Clin Neurophysiol 43:139–140, 1977.
13. Denys EH: AAEM Minimonograph #14: The influence of temperature in clinical neurophysiology. Muscle Nerve 14:795–811, 1991.
14. Fisher MA: AAEM Minimonograph #13: H reflexes and F waves: Physiology and clinical applications. Muscle Nerve 15:1223–1233, 1992.
15. Gasser HS: The classification of nerve fibers. Ohio J Sci 41:145–159, 1941.
16. Gitter AJ, Stolov WC: AAEM Minimonograph #16: Instrumentation and measurement in electrodiagnostic medicine—Part I. Muscle Nerve 18:799–811, 1995.
17. Gutmann L: AAEM Minimonograph #2: Important anomalous innervations of the extremities. Muscle Nerve 16:339–347, 1993.
18. Hagen NA, Stevens JC, Michet CJ Jr: Trigeminal sensory neuropathy associated with connective tissue diseases. Neurology 40:891–896, 1990.
19. Hammer K: Nerve Conduction Studies. Springfield, IL, Charles C Thomas, 1982.
20. Kimura J: An evaluation of the facial and trigeminal nerves in polyneuropathy: Electrodiagnostic study in Charcot-Marie-Tooth disease, Guillain-Barré syndrome, and diabetic neuropathy. Neurology 21:745–752, 1971.
21. Kimura J: Giron LT Jr, Young SM: Electrophysiological study of Bell palsy: Electrically elicited blink reflex in assessment of prognosis. Arch Otolaryngol 102:140–143, 1976.
22. Kimura J, Lyon LW: Orbicularis oculi reflex in Wallenberg syndrome: Alteration of the late reflex by lesions of the spinal tract and nucleus of the trigeminal nerve. J Neurol Neurosurg Psychiatry 35:228–233, 1972.
23. Kugelberg E: Facial reflexes. Brain 75:385–396, 1952.
24. Kumagami H: Neuropathological findings of hemifacial spasm and trigeminal neuralgia. Arch Otolaryngol 99:160–164, 1974.
25. Levin KH, Stevens JC, Daube JR: Superficial peroneal nerve conduction studies for electromyographic diagnosis. Muscle Nerve 9:322–326, 1986.
26. Møller AR: Interaction between the blink reflex and the abnormal muscle response in patients with hemifacial spasm: Results of intraoperative recordings. J Neurol Sci 101:114–123, 1991.
27. Nielsen VK: Pathophysiology of hemifacial spasm: I. Ephaptic transmission and ectopic excitation. Neurology 34:418–426, 1984.
28. Nielsen VK, Jannetta PJ: Pathophysiology of hemifacial spasm: III. Effect of microsurgical decompression of the facial nerve. Neurology 34:891–897, 1984.
29. Oh SJ: Clinical Electromyography: Nerve Conduction Studies, 2nd ed. Baltimore, Williams & Wilkins, 1993.
30. Ongerboer de Visser BW: Neurophysiological examination of the trigeminal, facial, hypoglossal, and spinal accessory nerves in cranial neuropathies and brain stem disorders. In Brown WF, Bolton CF (eds): Clinical Electromyography, 2nd ed. Boston, Butterworth-Heinemann, 1993, pp 61–93.
31. Ongerboer de Visser BW, Goor C: Jaw reflexes and masseter electromyograms in mesencephalic and pontine lesions: An electrodiagnostic study. J Neurol Neurosurg Psychiatry 39:90–92, 1976.
32. Ongerboer de Visser BW, Kuypers HGJM: Late blink reflex changes in lateral medullary lesions. An electrophysiological and neuroanatomical study of Wallenberg's syndrome. Brain 101:285–294, 1978.
33. Patton HD: Special properties of nerve trunks and tracts. In Ruch T, Patton HD (eds): Physiology and Biophysics, 2nd ed. vol IV. Philadelphia, WB Saunders, 1982, pp 101–127.
34. Preston DC, Shapiro BE: Electromyography and Neuromuscular Disorders: Clinical-Electrophysiologic Correlations. Boston, Butterworth-Heinemann, 1998.
35. Rivner MH, Swift TR, Crout BO, et al: Toward more rational nerve conduction interpretations: The effect of height. Muscle Nerve 13:232–239, 1990.
36. Schmalohr D, Linke DB: The blink reflex in cerebral coma: Correlations to clinical findings and outcome. Electromyogr Clin Neurophysiol 28:233–244, 1988.
37. Shahani B: The human blink reflex. J Neurol Neurosurg Psychiatry 33:792–800, 1970.
38. Slawnych MP, Laszlo CA, Hershler C: A review of techniques employed to estimate the number of motor units in a muscle. Muscle Nerve 13:1050–1064, 1990.
39. Stetson DS, Albers JW, Silverstein BA, et al: Effects of age, sex, and anthropomorphic factors on nerve conduction measures. Muscle Nerve 15:1095–1104, 1992.
40. Stevens JC: AAEM Minimonograph #26: The electrodiagnosis of carpal tunnel syndrome. Muscle Nerve 20:1477–1486, 1997.
41. Stevens JC, Smith BE: Cranial reflexes. In Daube JR (ed): Clinical Neurophysiology, Philadelphia, FA Davis, 1996, pp 321–335.
42. Stewart JD: Focal Peripheral Neuropathies, 2nd ed. New York, Raven, 1993, pp 6–8.

43. Thorne J: Central responses to electrical activation of the peripheral nerves supplying the intrinsic hand muscles. J Neurol Neurosurg Psychiatry 28:482–495, 1965.

44. Valls-Sole J, Tolosa ES: Blink reflex excitability cycle in hemifacial spasm. Neurology 39:1061–1066, 1989.

45. Weintraub JR, Madalin K, Wong M, et al: Achilles tendon reflex and the H-response: Their correlation in 400 limbs [abstract]. Muscle Nerve 11:972, 1988.

46. Zwarts MJ, Guecher A: The relation between conduction velocity and axonal length. Muscle Nerve 18:1244–1249, 1995.

Chapter 11

Refractory Period and Collision Technique

A. Arturo Leis, MD

INTRODUCTION
DEFINITION
TECHNIQUES
Paired Stimuli
Trains of Stimuli
Supernormal and Subnormal Periods
Temperature Effects
CLINICAL UTILITY
COLLISION PHENOMENON
Axillary Stimulation and Collision
    Technique

Martin-Gruber Anastomosis and
    Collision Technique
Refractory Period and Collision
    Technique
Collision Technique to Block Fast or Slow
    Fibers

INTRODUCTION

This chapter reviews the physiology and technique of the *refractory period* and *collision phenomenon*. Although these techniques are uncommonly used in practice, for some conditions they may be more sensitive in demonstrating a disorder than routine nerve conduction studies. The techniques can be used both clinically and experimentally to evaluate the variations in excitability of peripheral nerves caused by subtle and conspicuous abnormalities in the central and peripheral nervous systems.

DEFINITION

In 1899, Gotch and Burch investigated the electrical response in rat sciatic nerve to two stimuli occurring in rapid succession.[17] They found an interval at which a second stimulus ceased to evoke any perceptible electrical response. They termed this the *critical interval*, and noted that "every stimulus of such intensity as that here employed, if it follows a predecessor at a period of less than the critical interval, is ineffective; every stimulus of similar intensity succeeding a predecessor at a period of greater duration than this critical interval, is effective." They further noted "since a second excitatory disturbance cannot be readily evoked until a certain time has elapsed, then nerve excitability must be lowered during nerve activity, and every excitatory disturbance is thus associated with a period of diminished excitability."[17]

The "critical interval" described by Gotch and Burch[17] is now known as the *refractory period* of an axon, defined as the time during which the ability of the axon to conduct an action potential is altered following initiation and propagation of a nerve impulse. The refractory period is divided into two phases: an *absolute refractory period* (ARP), corresponding to a short period immediately following propagation of an action potential when the axon is completely unexcitable, and a *relative refractory period* (RRP), when the intensity of the stimulus required to excite propagation of a second action potential is increased above normal. During the ARP, which typically lasts about 0.7 to 1.0 msec in human nerve,[16] even a high-intensity stimulus will not produce prop-

agation of a second action potential. The reason is that immediately after the action potential is initiated, the voltage-dependent sodium (Na^+) channels are inactivated.[18, 33] During the RRP, which lasts about 2.5 to 3.5 msec,[16] the *amplitude* of the second response is lower, and the *conduction velocity* in response to the second stimulus is slower. The cause of the RRP is twofold: (1) during this time there is residual inactivation of some of the Na^+ channels, and (2) there is opening of the potassium (K^+) channels, causing a state of hyperpolarization that makes it more difficult to initiate another action potential.[18, 33]

Refractoriness of an axon influences two properties of an axon. First, *the ARP and RRP set an upper limit on the maximum rate of repetitive firing of an axon*. Although it is sometimes assumed that the capacity of nerve fibers to conduct trains of impulses at high frequencies greatly exceeds that required during physiological activity, in animal studies, discharge frequencies may reach 400 to 800 Hz in afferent fibers in response to end-organ stimulation.[23, 24] In human radial nerve, discharge frequencies of 330 Hz have been recorded in single mechanoreceptor fibers in response to skin stimulation.[20] Such findings suggest that in animals and humans, the peak frequency of physiological firing may indeed be limited by the refractory period. As stated by Gilliatt and Willison,

> . . . It is clear that an action potential is propagated with a reduced velocity in the relative refractory period; the effect of this is to cause the second of two impulses to lag increasingly behind the first until the end of the refractory period is reached, at which point the velocity of the second impulse will have recovered to equal that of the first. This must have the effect of limiting the frequency of a train of impulses travelling over a long length of nerve, regardless of the frequency of initiation.[16]

Gilliatt and Willison also postulated that since physiological activity can involve discharge frequencies close to the refractory period, one might expect significant disturbances of function if the refractory period were prolonged by disease.[16] It is now well established that in demyelination of the peripheral nervous system, there is a prolongation of the refractory period and impaired conduction of trains of impulses.[13, 34, 36] Moreover, in conditions usually associated with demyelination of the cental nervous system (e.g., multiple sclerosis), the RRP of peripheral nerves is also significantly prolonged,[22] supporting the contention that peripheral myelinated fibers are involved. Under such conditions, the peak frequency of physiological firing in peripheral nerves is more likely to be limited by the refractory period.

The second property influenced by refractoriness of an axon is the direction of conduction of an action potential. *The refractory membrane behind an advancing action potential prevents the action potential from reversing its direction.* Under normal physiological conditions, axons conduct in only one direction; physiological stimuli occur at one end of the axon, and impulses are conducted in an orthodromic direction because the refractory membrane behind the action potential prevents reversal of direction. Under artificial stimulation conditions, axons within a mixed peripheral nerve that are directly activated by an electrical stimulus carry two volleys of impulses, an ascending volley consisting of orthodromic sensory and antidromic motor impulses, and a descending volley containing orthodromic motor and antidromic sensory impulses. The refractory periods in motor and sensory fibers prevent the action potentials in each of these volleys from reversing direction.

Refractory periods occur in individual axons and in peripheral motor and sensory nerves following rapid repetitive stimulation. The time course of both the absolute and relative refractory periods in individual axons is similar to those found in peripheral motor and sensory nerves, although the mechanisms generating the latter may be more complex.

The refractory period of a peripheral nerve is easily demonstrated using paired stimuli.[11] When the interstimulus interval of a pair of stimuli is brief, less than 0.5 to 0.8 msec, there is *no* conducted response to the second stimulus no matter how great the stimulus intensity, because the second stimulus occurs during the nerve ARP. The shortest interval between two stimuli at which a propagated response to the second stimulus can be recorded is the duration of the ARP.[16] This is also known as the refractory period of transmission.[36] As the interstimulus interval increases, a response to the second stimulus can be recorded from the nerve, but the response elicited by the second stimulus is conducted *more slowly* and the response is of *lower amplitude* than the response to the first stimulus. During this time, the nerve is in the RRP. As the interstimulus interval increases further, the latency and amplitude of the second response becomes equal to the latency and amplitude of the first response. The shortest interstimulus interval at which equal latencies and amplitudes occur, when the excitability and conductivity of the nerve return to normal, defines the duration of the RRP.[1, 37] The recovery of normal excitability in human nerves is usually complete by 2.5 to 3.5 msec.[16] Both the ARP and the RRP vary inversely with the conduction velocity (or diameter) of the nerve fibers.[39, 40]

Following the RRP, there is a period of supernormal excitability known as the *supernormal period,* in which a lower intensity second stimulus produces a normally conducted response.[9, 10, 16, 44–46] During this period, the thresholds of individual nerve fibers are reduced by 10% to 20%, so that a submaximal stimulus depolarizes a higher proportion of fibers than it would otherwise. In humans, supernormal excitability is maximal 5 to 8 msec following the first stimulus, and the threshold of excitability returns to normal levels within 10 to 20 msec.[16]

The cause of the supernormal period remains unclear, but a correspondence between the supernormal period and the presence of a *depolarizing afterpotential* has been observed in animal models.[10] This has prompted speculation that the supernormal period may be the result of extracellular K^+ accumulation following an action potential. An increase in

extracellular K^+ would be expected to depolarize the axon and thereby move the membrane potential closer to threshold.[10, 44–46] The supernormal period may also be the result of capacitative discharge of the internodal axon membrane. The depolarization associated with this capacitative current may account for the excitability changes observed during the supernormal period in myelinated axons.[10] The magnitude of the depolarizing afterpotential is altered by activity of the metabolically active sodium-potassium pump, as well as the capacitance across the nerve membrane. Disease conditions that alter the capacitive properties of the nerve membrane, such as demyelination, can be expected to alter the supernormal period.

Following the supernormal period, there is a longer period of subnormal excitability known as the *subnormal period,* in which a slightly higher intensity second stimulus is needed to produce a normally conducted response.[45, 46] During this period, the thresholds of individual nerve fibers are increased by 5% to 13%, so that the same submaximal second stimulus depolarizes a lower proportion of fibers than it would otherwise. In humans, maximal depression can be seen 38 to 60 msec following the first stimulus, and the threshold of excitability returns to normal levels 120 to 200 msec after the conditioning pulse.[45, 46] The subnormal period reflects a relative hyperpolarization of the membrane potential, which seems to result from an electrogenic ion pump activity that depends on internal Na^+ and external K^+ concentrations. Because less K^+ is pumped in than Na^+ is pumped out, a net deficit of positive ions is produced on the inner side of the membrane and a corresponding hyperpolarization results.

TECHNIQUES

Paired Stimuli

A mixed or sensory nerve is stimulated with paired stimuli of *supramaximal* intensity.[1, 11, 37, 49] The ARP and RRP can be determined using orthodromic or antidromic stimulation techniques. Readily testable nerves include the median, ulnar, radial, and sural nerves. For median and ulnar nerves, ring electrodes can be placed on the digits to record the response following wrist stimulation (sensory antidromic technique) or electrodes can be placed over the respective nerve at the wrist to record the response following digital or midpalmar stimulation (sensory orthodromic technique). The refractory period can be recorded from a mixed nerve at any point along its course, or the mixed nerve can be stimulated at any point along its course to elicit a refractory period. The proximal refractory period can also be compared to the distal refractory period using dual channel recording.[15]

The clinical utility of paired stimulation is limited in the evaluation of the refractory period of motor nerves to a muscle, because the response to a second stimulus falls within the compound muscle action potential of the first response. The responses can be separated with computer separation[41] or by blocking the effects of the first stimulus without affecting that of the second[30, 32] (see collision phenomenon, described later). Refractory periods from *single* motor nerve fibers can be measured, however, using paired stimuli.[5, 6, 8]

Because various machines used in electrophysiological studies may differ in delivery of paired stimuli and adjustment of interstimulus intervals, the practitioner is referred to instruction manuals to program the appropriate settings.

Absolute Refractory Period (ARP). To determine the ARP, the interstimulus interval is increased from 0.5 msec in intervals of 0.05 to 0.1 msec, until a second response is elicited (Fig. 11–1). The ARP is the shortest interstimulus interval at which a propagated response to the second stimulus can be recorded.

Relative Refractory Period (RRP)

INTERSTIMULUS INTERVAL MEASUREMENTS. To determine the RRP, the interstimulus interval is increased from 2.5 msec in intervals of 0.1 msec, and the interpeak interval of the response is measured. When the interpeak interval equals the interstimulus interval (i.e., when the response to the second stimulus is elicited as quickly as the response to the first stimulus), the nerve is out of the RRP and conduction has returned to normal (Fig. 11–2). The duration of the "true" RRP is calculated by subtracting the ARP from the measured RRP (true RRP = measured RRP − ARP).

AMPLITUDE MEASUREMENTS. When the response to the second stimulus is equal in amplitude to that of the first, the nerve is out of the RRP.[4] The RRP using amplitude criteria may be up to 1.4 times longer than the RRP using latency.[47]

For commonly assessed nerves (median, ulnar, sural, radial) over short distances, the stimulus artifact of the second stimulus may alter the latency measurement and amplitude from the first stimulus, making comparisons of interstimulus interval or amplitude difficult. This problem can be over-

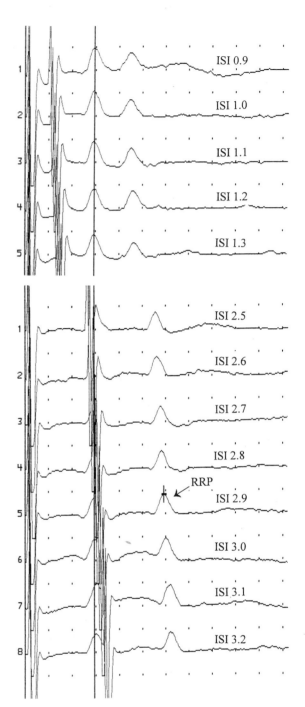

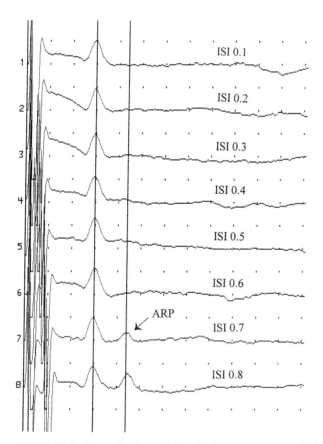

FIGURE 11–1. Determination of the absolute refractory period (ARP). Stimulating the index finger and recording over the median nerve at the wrist. At interstimulus intervals (ISIs) of 0.1 to 0.6 msec, only the response to the first stimulus is recorded, at a peak latency of 2.98 msec. No conducted response is recorded from the second stimulus. At an ISI of 0.7 msec, a small amplitude potential is recorded in response to the second stimulus, with a latency of 4.40 msec following the second stimulus. A small proportion of the axons in the nerve are conducting, and at a slower velocity than in response to the first stimulus. The amplitude of second response increases slightly with an interstimulus interval of 0.8 msec. Calibration: vertical, 1 division = 20 μV; horizontal, 1 division = 1 msec.

FIGURE 11–2. Determination of the relative refractory period (RRP), stimulating the index finger and recording over the median nerve at the wrist. The response to the first stimulus, determined from the top five traces, remains constant at a latency of 2.98 msec. As the interstimulus interval (ISI) increases, the latency of the response to the second stimulus decreases. At the termination of the RRP, the latency of the second response equals or closely approximates that of the first response, and the interpeak interval equals the interstimulus interval. Calibration: vertical, 1 division = 20 μV; horizontal, 1 division = 1 msec.

come by obtaining latency or amplitude measurements of the first response before they are altered by artifact from the second stimulus (the latency and amplitude of the first response will remain constant since the same supramaximal intensity is used throughout the study; see Fig. 11–2). If longer segments are used, for example, between the wrist and elbow or axilla, stimulus artifact of the second stimulus does not alter the latency measurement or amplitude of the first stimulus. Over long segments, one can also compare conduction velocities of nerves in the refractory period.[15, 16]

Trains of Stimuli

Trains of stimuli delivered at frequencies from 100 to 500 Hz can provide information on changes in peripheral nerve function.[11, 51, 52] The transmission of frequent stimuli in single nerve fibers depends on (1) the absolute and relative refractory period, and (2) the conduction velocity of the nerve fiber.[39, 40, 51] In fibers with a conduction velocity of 12 m/sec, the upper transmissible frequency is 350 Hz, whereas in fibers with a conduction velocity of 50 m/sec, trains up to 700 Hz may be conducted without any decrement.[39, 40] In normal human sensory nerves, Tackmann and colleagues[51] found changes in peripheral nerve function (prolongation of latency, decrease of amplitude) at stimulus frequencies of 100 to 200 Hz. They attributed the prolongation of latency and decrease of amplitude at relatively low frequencies to the *subnormal phase* during the recovery cycle of nerve fibers following trains of frequent stimuli.

When using trains of frequent stimuli, the amplitudes or latencies of the last action potential of a train are compared with those obtained from the first action potential of a train and expressed as ratios or percentage values. The ratios in diseased nerves can be compared with those of normal nerves. The stimulus rate at which amplitude falls or latency increases is lower in demyelinating neuropathies, compressive neuropathies (e.g., carpal tunnel syndrome),[34] and alcoholic neuropathies.[52] Failure of transmission of a train of impulses may demonstrate abnormalities not appreciated using paired stimuli,[36] although the high-frequency trains may be difficult for patients to tolerate.[44]

Supernormal and Subnormal Periods

Two stimuli of independently variable intensity are needed to demonstrate the supernormal and subnormal periods.[16] The first stimulus (S1) is usually supramaximal whereas the voltage of the second stimulus (S2) is submaximal and adjusted to give the smallest recognizable response. The stimulating voltage is noted and the procedure repeated for a different interstimulus interval. Gilliatt and Willison[16] found that the voltage of S2 required to produce a minimal response fell as the interstimulus interval was increased, the resting value being reached after 2.5 to 3.5 msec. The recovery of normal excitability was followed by the supernormal period in which the nerve threshold fell below the resting value. Supernormal excitability was maximal 5 to 8 msec after S1; thereafter it declined, with the threshold returning to its resting level within 10 to 20 msec.[16] Stohr and colleagues[45, 46] found that the period of supernormal excitability was terminated by a subsequent subnormal period, which ended 120 to 200 msec after the conditioning S1 stimulus. Maximal depression was seen 38 to 60 msec following S1, with the maximal increase in threshold ratio being 5% to 13%.

To demonstrate the supernormal or subnormal periods, the second stimulus must be *submaximal* in intensity. If a supramaximal stimulus is used, the changes in threshold that define these periods of heightened or diminished excitability would go undetected, as all fibers would be activated by the stimulus.

Temperature Effects

Alterations in temperature have a profound effect on the refractory period.[11, 36] Buchthal and Rosenflack[11] found that

with decreasing temperature the ARP increased 0.2 msec/°C between 32 and 22°C and 0.05 msec/°C between 36 and 32°C. As compared to the temperature dependence of conduction velocity between 32 and 22°C, the ARP varied almost twice as much. This larger dependence on temperature of the refractory period than of conduction time has also been shown in isolated whole nerves and in single fibers.[11] Because repeated stimulation of the digital nerves causes reflex sweating and cooling of the hand, which can prolong the refractory period, the limb should be warmed, and temperature maintained constant during the study.[16, 36] Obviously, control of temperature is necessary before changes in the refractory period can be interpreted.

CLINICAL UTILITY

The RRP is more sensitive than routine measures of nerve conduction in the detection of axonal disorders influencing nerve conduction; *the RRP may be abnormal when other measures of conduction are normal.* Alderson and Petajan[1] showed that the RRP was prolonged in clinical and subclinical alcoholic neuropathy, even when routine nerve conduction studies were normal in the asymptomatic subjects. Tackmann and colleagues[52] applied trains of frequent stimuli to patients with alcoholic polyneuropathy and evaluated the latency and amplitude of the tenth response in each train. They found that abnormal function could be demonstrated even in nerves with normal conduction velocities and amplitudes on conventional studies. In uremic patients with no clinically manifest neuropathy, the RRP to paired stimuli is prolonged.[50] In patients with chronic renal failure being hemodialyzed, the RRP decreases after dialysis to become normal.[37] A membrane abnormality due to uremic poisoning is assumed to cause the reversible prolongation in the RRP. The RRP can also be used to demonstrate impairment of peripheral nerve function at a very early stage of diabetic polyneuropathy, in many instances when sensory nerve conduction studies are not yet affected.[43, 48] The duration of the refractory period can be lengthened by at least 50% following short periods of ischemia.[16] Medications that alter K+ concentrations, such as rubidium, can prolong the RRP.[4] The RRP is also prolonged in axonal, demyelinating, and compressive neuropathies, including carpal tunnel syndrome,[15, 49] and Guillain-Barré syndrome.[34] As mentioned previously, in conditions usually associated with demyelination of the cental nervous system (e.g., multiple sclerosis), the RRP of peripheral nerves is significantly prolonged,[22] although not to the degree demonstrated in axonal and demyelinating neuropathies. Metabolic factors that prolong the RRP include hyperkalemia, which may contribute to a prolonged RRP in uremia and the improvement after dialysis.[37] The motor nerve fiber refractory period is prolonged in motor neuron disease[7] and other neuropathies.[6] The refractory period is *shortened* in hypokalemia, probably as a result of increased resting membrane potential.[37] The RRP has also been reported to shorten with chronic administration of the organophosphate, trichlorphon.[3] The RRP increases slightly with age.[47]

In animal studies, the RRP appears to be very consistent in nerves taken from animals of a given age, strain, and size. This allows small alterations in nerve excitability owing to neurotoxicity to be readily detected using the RRP.[2] Data in animals have demonstrated the value of the RRP when comparing the responses of normal nerves with the responses of nerves exposed to various neurotoxic agents. For example, the RRP is prolonged in phenol-induced peripheral neuropathy, shifted in diisopropylfluorophosphate

(DFP) neurotoxicity, and unchanged in acrylamide toxicity. These observations are consistent with the views that phenol induces progressive axonal degeneration after just a single dose,[2, 42] DFP produces a delayed damage to peripheral nerves,[38] and acrylamide induces a neuropathy that begins at the nerve terminals and only later affects the axons.[28] The RRP thus appears to remain normal after exposure to neurotoxic agents, such as acrylamide, that are initially toxic to parts of the nerve other than the axon; it changes in a manner consistent with the neurotoxicity and is valuable as an index of the extent and type of injury induced by agents that affect peripheral axons. The ARP is generally not as sensitive a measure of nerve pathology,[44] although it may be prolonged in some cases of demyelinating neuropathy[44] and compressive neuropathy, such as carpal tunnel syndrome,[49] and in uremia.[50]

There is evidence that alterations in supernormality may be a subtle index of nerve dysfunction.[44] An impairment of supernormality has been shown in patients with multiple sclerosis. Eisen and colleagues[14] found that 20 of 20 control subjects showed a supernormal response when a test stimulus was presented 6 msec after a conditioning stimulus. However, 17 of 40 multiple sclerosis patients (43%) showed no supernormal response. Routine sensory conduction studies were normal in both groups; thus, the only detectable conduction abnormality was in the supernormal period. An increase in the duration of the supernormal period can also reflect pathological conditions that are not detected by routine nerve conduction studies. For example, local ischemia can extend the supernormal period to more than 40 msec, twice the period seen in control subjects.[46] These changes in the supernormal period can occur despite normal routine conduction studies.[46] A brief period of ischemia has also been reported to abolish the supernormal period.[16]

COLLISION PHENOMENON

The term collision, when used with reference to nerve conduction studies, refers to the interaction of two action potentials propagated toward each other from opposite directions on the same nerve fiber so that the refractory periods of the two potentials prevent propagation past each other.[29, 31] Hopf[21] is generally credited with introducing the contemporary collision method, which makes use of paired supramaximal stimuli at different interstimulus intervals (ISIs). This method has been adapted and modified by several groups of investigators.[12, 19, 25, 26, 29] The collision technique provides a simple, noninvasive way to produce a *physiological nerve block*. In routine nerve conduction studies, collision is most useful during axillary stimulation of median or ulnar nerves to block unwanted nerve stimulation, and during the Martin-Gruber anastomosis, to identify unwanted impulses transmitted via the communicating fibers. It can also be used to assess the refractory period and in the assessment of fast versus slow fiber conduction. In addition, as discussed in Chapter 23, collision of antidromic with orthodromic impulses accounts for the first part of the mixed nerve silent period.[35]

Axillary Stimulation and Collision Technique

Stimulation to the median or ulnar nerves at the axilla tends to activate both nerves as they lie in close proximity.[29, 31] The current intended for the median nerve may spread to the ulnar nerve, so that the electrodes placed on the thenar eminence record not only median but also ulnar innervated muscle potentials. Measured latency may then reflect normal ulnar conduction even if the median nerve conducts more slowly, as in carpal tunnel syndrome. In the case of carpal tunnel syndrome, a stimulus at the elbow activates only the median nerve, which results in a prolonged latency of the compound muscle action potential (CMAP) response. The calculated conduction time between the axilla and the elbow would then suggest an erroneously fast conduction velocity in this segment. In extreme cases, the latency of the median response after stimulation at the elbow may exceed that of the ulnar component elicited with shocks at the axilla (Fig. 11–3).

A physiological nerve block with collision allows selective recording of the median or ulnar component despite coactivation of both nerves proximally.[29, 31] In studying the median nerve, a distal stimulus is delivered to the ulnar nerve at the wrist. The antidromic impulse from the wrist collides with the orthodromic impulse from the axilla in the ulnar nerve, so that only the median impulse can reach the muscle (see Fig. 11–3). The ulnar response induced by the distal stimulus occurs much earlier and does not obscure the median CMAP under study. Delivering the stimulus at the wrist a few milliseconds before the proximal stimulation accomplishes an even greater separation between the median and ulnar CMAP; however, the time interval between the two stimuli should not exceed the conduction time between the distal and proximal points of stimulation. Otherwise, the antidromic impulse from the distal stimulus passes the proximal site of stimulation without collision.[29]

Martin-Gruber Anastomosis and Collision Technique

The same principles apply for the use of a distal stimulus to selectively block unwanted orthodromic impulses traveling through an anastomotic branch to the ulnar nerve (Fig. 11–4). Delivering the distal stimulus to the ulnar nerve at the wrist a few milliseconds before the proximal stimulus to the median nerve at the elbow usually achieves satisfactory separation of the median and ulnar CMAP responses. The interstimulus interval, however, must not exceed the conduction time between the two stimulus sites, otherwise the orthodromic impulse produced by proximal stimulation would escape antidromic collision.[29]

If the Martin-Gruber anastomosis accompanies a median mononeuropathy at the wrist (carpal tunnel syndrome), stimulation of the median nerve at the elbow may evoke two temporally dispersed potentials, a normal ulnar component and a delayed median component. The latency of the initial ulnar response may erroneously suggest normal conduction in median fibers. In contrast, stimulation of the median nerve at the wrist evokes a delayed CMAP without an ulnar component. The discrepancy between proximal and distal stimulation leads to an unreasonably fast conduction velocity from the elbow to the wrist[12, 27] (see Fig. 11–4).

In cases of severe carpal tunnel syndrome and the Martin-Gruber anastomosis, the latency of the median response after stimulation at the wrist may exceed that of the anomalous response elicited with elbow stimulation. The collision technique thus helps not only in accurately calculating conduction velocity, but also in characterizing the anomalous response. For example, if the distal stimulus is applied to the median nerve at the wrist, only the orthodromic impulses traveling through an anastomotic branch to the ulnar nerve would bypass the antidromic impulses and escape collision.[29, 31]

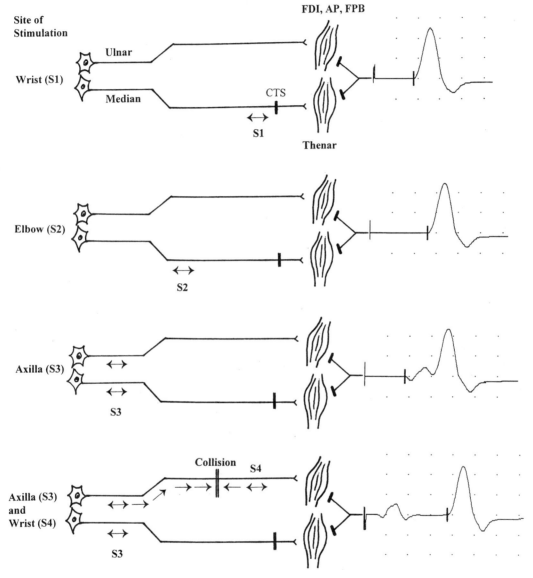

FIGURE 11–3. A 33-year-old woman with carpal tunnel syndrome (CTS). Stimulation to the median nerve at the wrist (S1) or elbow (S2) elicited a compound muscle action potential (CMAP) with increased latency in the thenar eminence. Spread of axillary stimulation (S3) to the ulnar nerve (*third tracing from top*) activated ulnar-innervated thenar muscles with much shorter latency. Another stimulus (S4) applied to the ulnar nerve at the wrist (*bottom tracing*) collided out the descending ulnar impulses. Delaying the S3 stimulus by 4 msec relative to S4 accomplished a greater separation between the ulnar and median CMAP. Calibration: vertical, 1 division = 5 mV; horizontal, 1 division = 5 msec.

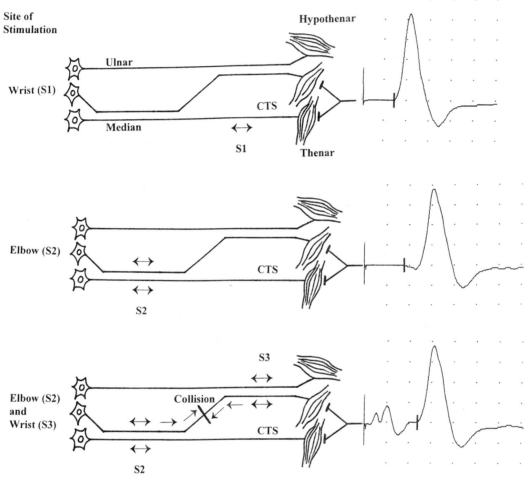

FIGURE 11–4. A 43-year-old woman with carpal tunnel syndrome (CTS) and Martin-Gruber anastomosis. Stimulation to the median nerve at the wrist (S1, *top tracing*) elicited a compound muscle action potential (CMAP) with increased latency in the thenar eminence. Stimulation at the elbow (S2) spread to the ulnar nerve through the Martin-Gruber anastomosis (*middle tracing*), producing a CMAP with shorter latency than expected. This gave rise to a conduction velocity of 100 m/sec in the wrist-to-elbow segment. Another stimulus (S3) applied to the ulnar nerve at the wrist (*bottom tracing*) blocked the impulses transmitted through the communicating fibers. In the bottom tracing, there was no overlap of the CMAP responses despite simultaneous delivery of S3 and S2.

Refractory Period and Collision Technique

A single distal stimulus, combined with paired proximal stimuli, can be used to assess the refractory period in motor fibers.[29, 30] In this arrangement, the antidromic impulse from the distal shock at the wrist, S1, is eliminated by collision with the orthodromic impulse generated by the first of the paired shocks, S2, in Figure 11–5. The orthodromic impulse of the second proximal stimulus, S3, will propagate distally along the motor fibers cleared of antidromic activity. The amplitude and latency of the CMAP from S3 depends solely on the neural excitability in motor fibers after the passage of the conditioning stimulus, S2. Changing the S2–S3 interval defines the range of the absolute refractory period by demonstrating the moment of earliest recovery of the test response amplitude. During the relative refractory period that ensues, the test response shows reduced amplitude and increased latency. The onset of stable latencies and amplitudes identifies the end point of the relative refractory period.

A double collision technique for measurement of the motor nerve refractory period distribution in human peripheral nerves has been described by Ingram and colleagues.[25] The method is computer assisted and takes advantage of subtraction and cross-correlation techniques. Using this technique, Ingram and colleagues[25] showed that the motor nerve refractory period distribution was less dispersed than previously supposed.

Collision Technique to Block Fast or Slow Fibers

The principle of collision can be used to quantitatively assess conduction in fast versus slow fibers, and their respective contribution to the CMAP.[19, 25, 26] Paired shocks of supramaximal intensity are delivered to a proximal and distal site of the same nerve. Shocks applied simultaneously cause collision to occur in all fibers. With increasing delays of the proximal shock, however, the interstimulus interval eventually exceeds the earliest conduction time between the two stimulus sites. At this point, orthodromic impulses in the fastest fibers escape collision, since the fastest antidromic impulses have already passed the proximal site of stimulation. With greater intervals between the two stimuli, the slower fibers also escape collision. Measurement of the minimal interstimulus interval sufficient to produce a maximum CMAP provides an indirect assessment of conduction in the slowest fibers.[29]

To *directly* determine the latency and amplitude of the slowest conducting fibers, collision in the fast-conducting fibers is needed. To demonstrate the technique, two stimuli, S(A1) and S(A2), are delivered through proximal electrodes placed over the ulnar nerve at the axilla and another shock, S(W), is delivered through distal electrodes placed over the same nerve at the wrist (Fig. 11–6). The CMAP is recorded by surface electrodes placed over the abductor digiti minimi muscle. The first axillary stimulus, S(A1), precedes the wrist

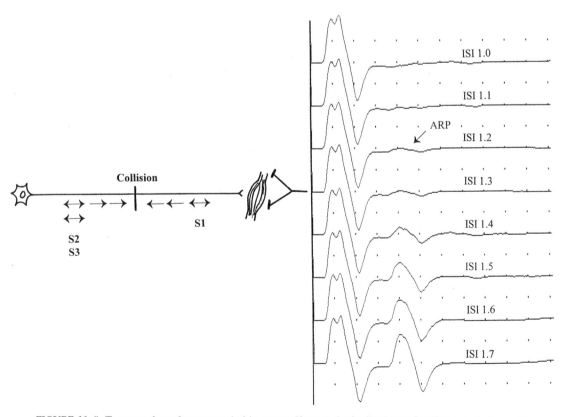

FIGURE 11–5. To assess the refractory period in motor fibers, a single distal stimulus (S1) is combined with paired proximal stimuli (S2, S3). Compound muscle action potentials (CMAPs) are recorded by surface electrodes placed over the abductor digiti minimi after stimulation to the ulnar nerve. The diagram shows the collision between antidromic motor impulses following a stimulus at the wrist (S1) and orthodromic impulses following the first stimulus at the axilla (S2). The wrist stimulus (S1) was delivered 6.0 msec before the axillary stimulus (S2). The amplitude and latency of the CMAP from the second of the paired axillary shocks (S3) depends solely on the neural excitability in motor fibers after passage of the S2 impulses. Changing the interstimulus interval (ISI) defines the absolute refractory period (ARP) by demonstrating the moment of earliest recovery of the S3 response.

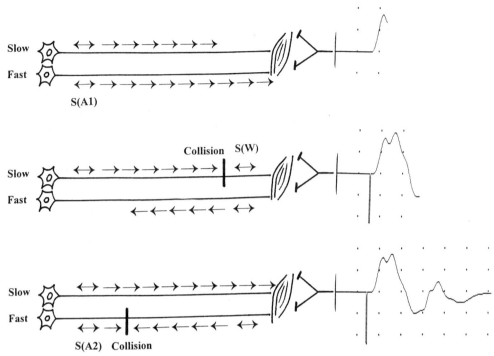

FIGURE 11–6. Compound muscle action potentials (CMAPs) recorded by surface electrode placed over the abductor digiti minimi after stimulation of the ulnar nerve. The diagrams show orthodromic and antidromic motor impulses generated by three stimuli, S(A1), S(W), and S(A2), delivered at the axilla, wrist, and axilla, respectively. In the top tracing, orthodromic S(A1) impulses in fast-conducting fibers escape collision, while impulses in slow-conduction fibers "lag behind." In the middle tracing, S(W), delivered 7.0 msec after S(A1), elicits antidromic impulses that collide with the "lagging" orthodromic impulses in slow fibers from S(A1); there is no collision in fast-conducting fibers. The CMAP produced is a summation of S(A1) impulses in fast-conducting fibers and S(W) impulses in both fast- and slow-conducting fibers. In the last tracing, orthodromic impulses from S(A2), delivered 11.0 msec after S(A1) and 4.0 msec after S(W), collide with antidromic impulses in fast-conducting fibers from S(W). There is no collision in slow-conducting fibers; the orthodromic impulses from S(A2) propagate along the slow fibers to elicit the second CMAP.

stimulus, S(W), by 6 to 8 msec. Under these conditions, collision takes place near the wrist site and only in the slow fibers, sparing the antidromic activity from the S(W) in the fast fibers. The impulse of the subsequent proximal stimulus, S(A2), is adjusted so that it collides with the antidromic activity in fast fibers from S(W). In this way, the CMAP elicited by S(A2) corresponds to the remaining slow-conducting fibers that selectively transmit the orthodromic impulses.[29]

A computer-based collision technique for direct measurement of the motor nerve conduction velocity distribution in human peripheral nerves has been described by Ingram and colleagues.[26] The increased sensitivity of this technique permits accurate measurement of the slowest 1% of motor nerve fibers.

REFERENCES

1. Alderson MK, Petajan JH: Relative refractory period: A measure to detect early neuropathy in alcoholics. Muscle Nerve 10:323–328, 1987.
2. Anderson RJ: Relative refractory period as a measure of peripheral nerve neurotoxicity. Toxicol Appl Pharmacol 71:391–397, 1983.
3. Averbook BJ, Anderson RJ: Electrophysiologic changes associated with chronic administration of organophosphates. Arch Toxicol 52:167–172, 1983.
4. Betts RP, Paschalis C, Jarratt JA, Jenner FA: Nerve fibre refractory period in patients treated with rubidium and lithium. J Neurol Neurosurg Psychiatry 41:791–793, 1978.
5. Borg J: Axonal refractory period of single short toe extensor motor units in man. J Neurol Neurosurg Psychiatry 43:917–924, 1980.
6. Borg J: Axonal refractory period of single short toe extensor motor units in neuropathies and neuromuscular diseases. J Neurol Neurosurg Psychiatry 44:1136–1140, 1981.
7. Borg J: Conduction velocity and refractory period of single motor nerve fibers in motor neuron disease. J Neurol Neurosurg Psychiatry 47:349–353, 1984.
8. Borg J: Refractory period of single motor nerve fibers in man. J Neurol Neurosurg Psychiatry 47:344–348, 1984.
9. Bostock H, Grafe P: Activity dependent excitability changes in normal and demyelinated rat spinal root axons. J Physiol 365:239–257, 1985.
10. Bowe CM, Kocsis JD, Waxman SG: The association of the supernormal period and the depolarizing afterpotential in myelinated frog and rat sciatic nerve. Neuroscience 21:585–593, 1987.
11. Buchthal F, Rosenflack A: Evoked action potentials and conduction velocity in human sensory nerves. Brain Res 3:1–122, 1966.
12. Buchthal F, Rosenflack A, Trojaborg W: Electrophysiological findings in entrapment of the median nerve at wrist and elbow. J Neurol Neurosurg Psychiatry 37:340–360, 1974.
13. Davis FA: Impairment of repetitive impulse conduction in experimentally demyelinated and pressure-injured nerves. J Neurol Neurosurg Psychiatry 35:537–544, 1972.
14. Eisen A, Paty D, Hoirch M: Altered supernormality in multiple sclerosis peripheral nerve. Muscle Nerve 5:411–414, 1982.
15. Gilliatt RW, Meer J: The refractory period in the carpal tunnel syndrome. Muscle Nerve 11:974, 1988.
16. Gilliatt RW, Willison RG: The refractory and supernormal periods of the human median nerve. J Neurol Neurosurg Psychiatry 26:136–147, 1963.
17. Gotch F, Burch GJ: The electrical response of nerve to two stimuli. J Physiol (Lond) 24:410–426, 1899.

18. Guyton AC: Membrane potentials and action potentials. In Guyton AC (ed): Textbook of Medical Physiology, 8th ed. Philadelphia, WB Saunders, 1991, pp 51–66.

19. Hakamada S, Kumagai T, Watanabe K, et al.: The conduction velocity of slower and the fast fibers in infancy and childhood. J Neurol Neurosurg Psychiatry 45:851–853, 1982.

20. Hensel H, Boman KKA: Afferent impulses in cutaneous sensory nerves in human subjects. J Neurophysiol 23:564–578, 1960.

21. Hopf HC: Untersuchungen uber die Unterschiede in der Leitgeschwindigkeit motorische Nervenfasern beim Menschen. Dtsch Z Nervenheilk 183:579–588, 1962.

22. Hopf HC, Eysholdt M: Impaired refractory periods of peripheral sensory nerves in multiple sclerosis. Ann Neurol 4:499–501, 1978.

23. Hunt CC: On the nature of vibration receptors in the hind limb of the cat. J Physiol 155:175–186, 1961.

24. Hunt CC, McIntyre: Characteristics of reponses from receptors from the flexor longus digitorum muscle and the adjoining interosseous region of the cat. J Physiol 153:74–87, 1960.

25. Ingram DA, Davis GR, Swash M: The double collision technique: A new method for measurement of the motor nerve refractory period distribution in man. Electroenceph Clin Neurophysiol 66:225–234, 1987.

26. Ingram DA, Davis GR, Swash M: Motor nerve conduction velocity distributions in man: Results of a new computer-based collision technique. Electroenceph Clin Neurophysiol 66:235–243, 1987.

27. Iyer V, Fenichel GM: Normal median nerve proximal latency in carpal tunnel syndrome: A clue to coexisting Martin-Gruber anastomosis. J Neurol Neurosurg Psychiatry 39:449–452, 1976.

28. Jennekens FGI, Veldman H, Schotman P, Gispen WH: Sequence of motor nerve terminal involvement in acrylamide neuropathy. Acta Neuropathol (Berlin) 46:57–63, 1979.

29. Kimura J: Collision technique: Physiologic block of nerve impulses in studies of motor nerve conduction velocity. Neurology 26:680–682, 1976.

30. Kimura J: A method for estimating the refractory period of motor fibers in the human peripheral nerve. J Neurol Sci 28:485–490, 1976.

31. Kimura J: Facts, fallacies, and fancies of nerve stimulation techniques. In Kimura J (ed): Electrodiagnosis in Diseases of Nerve and Muscle: Principles and Practice, 2nd ed. Philadelphia, FA Davis, 1989, pp 139–166.

32. Kimura J, Yamada T, Rodnitzky RL: Refractory period of human motor nerve fibers. J Neurol Neurosurg Psychiatry 41:784–790, 1978.

33. Koester J: Voltage gated ion channels and the generation of the action potential. In Kandel ER, Schwartz JH, Jessel TM (eds): Principles of Neural Science, 3rd ed. Norwalk, CT Appleton & Lange, 1991, pp 104–122.

34. Lehmann HJ, Tackmann W: Neurographic analysis of trains of frequent electric stimuli in the diagnosis of peripheral nerve disease. Eur Neurol 12:293–308, 1974.

35. Leis AA: Conduction abnormalities detected by silent period testing. Electroenceph Clin Neurophysiol 93:444–449, 1994.

36. Low PA, McLeod JG: Refractory period, conduction of trains of impulses, and effect of temperature on conduction in chronic hypertrophic neuropathy. J Neurol Neurosurg Psychiatry 40:434–447, 1977.

37. Lowitzsch K, Gohring U, Hecking E, Kohler H: Refractory period, sensory conduction velocity and visual evoked potentials before and after haemodialysis. J Neurol Neurosurg Psychiatry 44:121–128, 1981.

38. Lowndes HE, Baker T, Riker WF: Motor nerve dysfunction in delayed DFP neuropathy. Eur J Pharmacol 29:66–73, 1974.

39. Paintal AS: Conduction in mammalian nerve fibers. In Desmedt JE (ed): New Developments in Electromyography and Clinical Neurophysiology, vol. 2, Basel, Karger, 1973, pp 19–41.

40. Paintal AS: Conduction properties of normal peripheral mammalian axons. In Waxman SG (ed): Physiology and Pathobiology of Axons. New York, Raven, 1978, pp 131–144.

41. Reitter B, Johannsen S: Neuromuscular reaction to paired stimuli. Muscle Nerve 5:593–603, 1982.

42. Schaumburg HH, Byck R, Weller R: The effect of phenol on peripheral nerve. J Neuropathol Exp Neurol 29:615–630, 1970.

43. Schutt P, Muche H, Lehmann HJ: Refractory period impairment in sural nerves of diabetics. J Neurol 229:113–119, 1983.

44. Shefner JM, Dawson DM: The use of sensory action potentials in the diagnosis of peripheral nerve disease. Arch Neurol 47:341–348, 1990.

45. Stohr M: Activity dependent variations in threshold and conduction velocity of human sensory fibers. J Neurol Sci 49:47–54, 1981.

46. Stohr M, Gilliatt RW, Willison RG: Supernormal excitability of human sensory fibers after ischemia. Muscle Nerve 4:73–75, 1981.

47. Tackmann W, Lehmann HJ: Refractory period in human sensory nerve fibers. Eur Neurol 12:277–292, 1974.

48. Tackmann W, Lehmann HJ: Conduction of electrically elicited impulses in peripheral nerves of diabetic patients. Eur Neurol 19:20–29, 1980.

49. Tackmann W, Lehmann HJ: Relative refractory period of median nerve sensory fibers in the carpal tunnel syndrome. Eur Neurol 12:309–316, 1974.

50. Tackmann W, Ullerich D, Cremer W, Lehmann HJ: Nerve conduction studies during the relative refractory period in sural nerves of patients with uremia. Eur Neurol 12:331–339, 1974.

51. Tackmann W, Ullerich D, Lehmann HJ: Transmission of frequent impulse series in human sensory nerve fibers. Eur Neurol 12:261–276, 1974.

52. Tackmann W, Ullerich D, Lehmann HJ: Transmission of frequent impulse series in sensory nerves of patients with alcoholic polyneuropathy. Eur Neurol 12:317–330, 1974.

Chapter 12

Needle Electrode Examination

Kerry H. Levin, MD

ANATOMY OF THE MOTOR UNIT
MOTOR UNIT ACTION POTENTIAL
Amplitude
Duration and Configuration
Variation
NEEDLE ELECTRODE EXAMINATION IN
    CONTRACTING MUSCLE
Factors Affecting Morphology in the
    Normal State
Motor Unit Recruitment
NEEDLE ELECTRODE EXAMINATION IN
    RESTING MUSCLE
Normal Insertional Activity
Abnormal Insertional Activity
Normal Spontaneous Activity
Abnormal Spontaneous Activity
Grouped Repetitive Discharges

NEEDLE ELECTRODE EXAMINATION IN
    SPECIFIC DISEASE STATES
Acute Neurogenic Disorders
Chronic Progressive Neurogenic
    Disorders
Myopathy and Defects of
    Neuromuscular Transmission
Case Studies
SAFETY ISSUES REGARDING THE NEEDLE
    ELECTRODE EXAMINATION

When combined with nerve conduction studies, the needle electrode examination (NEE) constitutes the basic technical procedure for the electrodiagnosis of most neuromuscular disorders. Whereas nerve conduction studies measure the excitability of motor and sensory nerve trunks and yield information regarding axon loss and the excitability of muscle fibers, the NEE provides information regarding the integrity of the motor unit. A motor unit is comprised of a single anterior horn cell, its axon process, terminal branches, associated neuromuscular junctions, and muscle fibers. Supramaximal stimulation of a motor nerve trunk activates almost all of the axons present and results in depolarization of most of the muscle fibers in the associated muscle belly, producing a compound muscle action potential summating most of the motor units in that muscle. In contrast, the NEE displays a voluntary contraction of a muscle belly as multiple, separate motor unit discharges. Putting the information from both electrodiagnostic techniques together provides a powerful tool in the diagnosis of diseases of the motor unit.

The NEE is carried out by advancing a bipolar wire electrode through the skin surface and into a muscle belly. The electrode has both an active and a referential recording site, and is connected through a differential amlifier to an audio and visual display system. Recordings are made during states of muscle rest and activation. In muscle at rest, the stationary needle electrode records discharges described as spontaneous activity, and the advancing needle records discharges described as insertional activity. When the muscle is voluntarily activated, the needle electrode records groups of individual motor unit action potentials (MUAPs).

ANATOMY OF THE MOTOR UNIT

There are three main types of motor neurons. Alpha motor neurons have large cell bodies (40 to 80 μ), are associated with fast-conducting axons, and innervate skeletal muscle fibers. Gamma motor neurons have smaller cell bodies (20 to 40 μ), are associated with more slowly conducting axons, and innervate intrafusal muscle fibers in the muscle spindle complexes, organs involved in motor control through the perception of muscle stretch. Beta motor neurons innervate both intrafusal and skeletal muscle fibers.[24] The NEE records only skeletal muscle action potentials.

Information about the structure of the motor unit has come from histological, histochemical, and electrophysiological techniques. Determination of the numbers of muscle fibers in motor units has come mainly from the tedious counting of large motor axons and muscle fibers in cross-sectional histological preparations. Glycogen depletion studies have demonstrated the spatial interrelation of motor units (mosaic pattern) and have shown that muscle fibers belonging to a given motor unit share the same histochemical attributes. Information regarding the size of motor units has come from electrophysiological studies with multilead electrodes.

Based on morphological studies in human muscles, a single motor neuron may innervate as few as 10 muscle fibers in extraocular muscles or as many as 2000 muscle fibers in the gastrocnemius muscles.[20] The innervation ratio (number of muscle fibers innervated by a single motor neuron) tends to be large in muscles where little dexterity is required and small in muscles where fine motor control is required. The number of motor units in a muscle depends on the size of the muscle and the innervation ratio in that muscle, such that there are about 100 motor units in the first dorsal interosseus of the hand, 1000 in the rectus femoris (where the innervation ratio is similar to that in the first dorsal interosseus), and 3000 in the external rectus extraocular muscle.

Skeletal muscle is organized into fascicles consisting of 20 to 60 muscle fibers each. The muscle fibers belonging to a single motor unit can be distributed over as many as 100 fascicles, and they are distributed randomly over a circular or oval region that can reach about 20% to 30% of the muscle's cross-sectional area, with a diameter of 5 to 10 mm, on average. A few muscle fibers belonging to the same motor unit may be contiguous, but in general there is a mosaic pattern of as many as 20 to 50 overlapping motor units in the same muscle volume. A slightly higher density of muscle fibers belonging to the same motor unit is seen at the center of the motor unit.[12]

Research techniques have yielded differing data regarding the number of overlapping motor units in a muscle volume. Histochemical studies indicate some 30 overlapping motor units in the same territory, whereas electrophysiological counts estimate about 10 separate motor units, the difference owing to the fact that electrophysiological studies select for only the earliest-recruiting type I motor units, and cannot adequately assess type II motor units in the same area.[10, 11, 13]

MOTOR UNIT ACTION POTENTIAL

The size and shape of the MUAP derived from a particular motor unit will vary with the recording characteristics. The recorded MUAP is highly dependent upon the characteristics of the needle electrode. The size of the active recording surface dictates the recording distance in the muscle tissue. A so-called single fiber needle electrode has an active recording surface of 25 μ in diameter, with a semicircular recording radius of 250 to 300 μ. Only one or two muscle fibers belonging to the same motor unit would likely be in the immediate recording territory of this needle electrode.[49] In contrast, the standard concentric needle electrode used in clinical electromyographic (EMG) practice has an active recording surface of 150 × 600 μ, and a recording diameter of about 10 mm in muscle tissue. As many as 8 to 20 muscle fibers belonging to the same motor unit may contribute to the recorded MUAP with this electrode. Figure 12–1 provides a schematic comparison of different needle electrodes.

The electrical fields generated by propagating muscle fiber action potentials in a motor unit are summed spatially and temporally in the volume conductor by the recording needle electrode. How much of the electrical field reaches the electrode depends on filtering, which occurs in the physiological tissues and in the recording apparatus. Muscle tissue acts as a low-pass filter, letting through distant low-frequency components of the motor unit under study as well as adjoining motor units not within the immediate recording range. The high-frequency components that constitute the main negative spike potential of the MUAP discharge penetrate poorly through contiguous muscle fibers not belonging to

the motor unit under study. This filtering effect can diminish the main spike amplitude by 90% in distances of 200 to 500 μ.[10] Thus, the MUAP derives its main negative amplitude from the two to three muscle fiber action potentials within 500 μ of the electrode, while the initial and terminal low-amplitude components of the MUAP are derived from the lower frequency contributions of many muscle fiber action potentials farther removed from the electrode.

Amplitude

MUAP amplitude is defined as the height of the main negative spike potential in the discharge, measured from the preceding positive peak. It is primarily derived from the muscle fiber action potentials of the motor unit residing within 500 μ of the active recording surface, and can range from 50 μV to 15 or more mV. In a reinnervating process where muscle fiber grouping occurs due to collateral sprouting of surviving axons, the number of participating muscle fibers close to the electrode is increased and the MUAP amplitude is higher. As the beveled active recording surface of the electrode is moved in the muscle, a different subset of the muscle fibers in the motor unit will be within the recording distance, yielding a different MUAP configuration and amplitude.

Duration and Configuration

MUAP duration, measured in milliseconds, is the time elapsed from the discharge's first deviation from the baseline to its return to the baseline (Fig. 12–2). Duration is related to the time required for all muscle fiber action potentials within the recording distance of the electrode to reach its recording surface. The onset of the MUAP waveform represents the arrival of the fastest muscle fiber action potential. The difference in arrival time of muscle fiber action potentials relates to a number of variables of impulse conduction, including the length of individual terminal nerve branches, the distance of individual neuromuscular junctions from the recording electrode (a factor that correlates with the width of a muscle belly's end plate zone), the diameter of individual muscle fibers, and the conduction velocity along individual muscle fibers (Fig. 12–3).

In normal motor units, the difference between the fastest and slowest arriving muscle fiber action potentials at the recording electrode is minimal, and the electrical summation

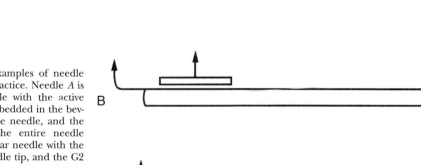

FIGURE 12–1. Schematic examples of needle electrodes used in clinical practice. Needle *A* is a standard concentric needle with the active (G1) recording electrode embedded in the beveled surface of the tip of the needle, and the reference (G2) electrode the entire needle shaft. Needle *B* is a monopolar needle with the G1 occupying the entire needle tip, and the G2 applied separately on the skin surface. Needle *C* is a single fiber needle. The G2 is the cut end of a fine wire embedded in the side port of the shaft.

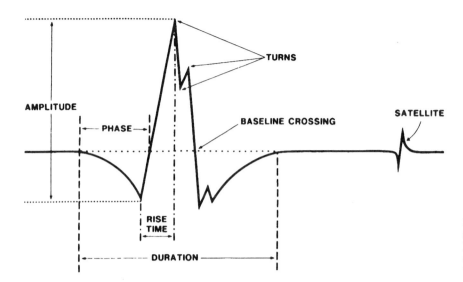

FIGURE 12–2. Components of a motor unit action potential. (From Daube JR: AAEM Minimonograph #11. Needle examination in electromyography. Muscle Nerve 14:685–700, 1991, with permission.)

in the volume conductor produces a classical triphasic waveform. In axon loss disorders, reinnervation by collateral sprouting from surviving axons leads to a number of changes, including lengthening and suboptimal myelination of newly formed terminal nerve branches, widening of the end plate zone of a motor unit, and heterogeneity of muscle fiber diameters, that leads to altered impulse conduction along muscle fibers. All these changes lead to desynchronization of the arrival of muscle fiber action potentials. There is loss of the normal MUAP integration, resulting in a longer duration waveform that has lost its classical triphasic appearance. Individual muscle fiber action potentials may appear in the MUAP configuration as increased turns. Further desynchronization of the MUAP is seen when individual muscle fiber action potentials do not summate in the volume conductor with other action potentials, giving rise to a polyphasic MUAP. A markedly slowed muscle fiber action potential may not be summated in the main MUAP discharge, creating a separate satellite potential. As more muscle fibers are incorporated into the reinnervating motor unit, its territory increases, fiber grouping occurs, and the MUAP increases in amplitude and duration.

In myopathic disorders, several neurophysiological events occur that alter the MUAP. Muscle fibers are lost from the motor unit, decreasing the number of muscle fiber action potentials close to and at a distance from the recording electrode. This results in MUAPs of smaller amplitude and

duration. Muscle fibers lost from the motor unit are reinnervated by collateral sprouting in a fashion identical to that seen in neurogenic disorders. This results in desynchronization of the MUAP and an increase in turns and polyphasia.

The number of phases constituting a motor unit is operationally defined as one higher than the number of baseline crossings in the MUAP waveform. A classical triphasic MUAP has two baseline crossings. Increased polyphasia is arbitrarily defined as five or more phases. Turns and phases have similar significance. However, defining the normal limit for the number of turns is complicated by the tendency for slight changes in the orientation of the recording electrode to alter the internal configuration of some MUAPs.

Variation

With stable positioning of the needle electrode and a steady, minimal to moderate rate of motor unit activation, a normal MUAP will have an identical appearance from one firing to the next. In disease states where there is intermittent failure of impulse transmission along terminal nerve branches or across neuromuscular junctions, affected muscle fibers will not be activated along with the other muscle fibers in the motor unit. The result is a MUAP that changes in shape from firing to firing, depending on the specific muscle fiber action potentials that dropped out or returned to contribute to the summated response at that moment. On the

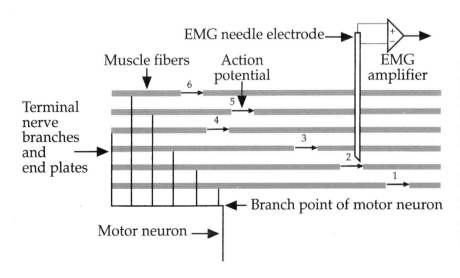

FIGURE 12–3. Schematic representation of motor unit action potential generation and recording by a concentric needle electrode. Individual muscle fiber action potentials numbered 1 through 6 arrive at the recording electrode at different times, depending on factors such as terminal nerve branch length, distance between the muscle fiber's neuromuscular junction and the recording electrode, diameter of the individual muscle fiber, and the conduction velocity of muscle fiber action potentials along individuals muscle fibers. (From Ball RD: Basics of needle electromyography. An AAEE Workshop. American Association of Electrodiagnostic Medicine, Rochester, MN, 1985, with permission.)

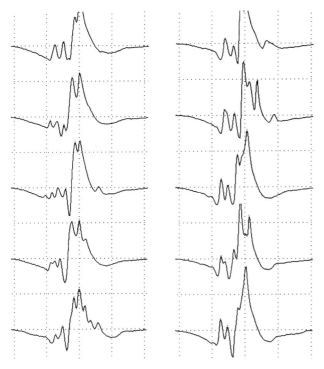

FIGURE 12–4. Five sequential firings of two different motor unit action potentials, demonstrating moment-to-moment variation in configuration.

EMG monitor, this variability in MUAP morphology is identified as moment-to-moment loss or return of turns, phases, satellite potentials, or amplitude (Fig. 12–4). This phenomenon is described as moment-to-moment motor unit variation or motor unit instability. MUAP instability is most easily appreciated during a minimal voluntary activation of motor units, using the trigger delay mode of the EMG machine, at an expanded sweep speed of 2 to 5 msec per division.

MUAP instability is a classical feature of defects of neuromuscular junction transmission such as myasthenia gravis, Lambert-Eaton myasthenic syndrome, and botulism, owing to the diminished safety margin of synaptic impulse transmission and the resulting loss of muscle fiber action potential generation. Momentary rest of the synapse allows for a successful subsequent impulse transmission and action potential generation. Finding MUAP instability in this setting adds great support for the diagnosis of a defect of neuromuscular junction transmission, and is especially useful when repetitive stimulation studies are normal or equivocal. However, in routine clinical EMG practice MUAP instability is more commonly seen in states of reinnervation after motor axon loss. During reinnervation, the newly formed neuromuscular junctions are immature and impulse transmission across their synapses often fails, producing MUAP instability in those motor units. This is likely to be seen in reinnervating traumatic or compressive nerve injuries, recovering axonal polyneuropathies, and active denervating-reinnervating processes such as progressive motor neuron disease (amyotrophic lateral sclerosis). Finding MUAP instability adds support for the diagnosis of recent denervation and ongoing reinnervation.

NEEDLE ELECTRODE EXAMINATION IN CONTRACTING MUSCLE

Performing the NEE is a multifaceted process that includes assessing the order in which new MUAPs appear on the EMG monitor in response to increasing voluntary contraction, counting the total number of MUAPs on the monitor at a given level of muscle contraction, and analyzing the range of sizes and configurations of MUAPs seen at minimal through maximal levels of contraction. The NEE is a complicated process requiring the observation and synthesis of many types of information occurring simultaneously during a muscle contraction. Successful mastery of the NEE requires pattern recognition skills and knowledge of the normal range under a number of different situations. In the normal state, a number of variables influence the appearance of the MUAP (Table 12–1). The ability to assess motor unit recruitment during a voluntary contraction is useful in distinguishing between normal and abnormal patterns.

Factors Affecting Morphology in the Normal State

Type of Needle Electrode

As noted earlier in this chapter, the needle's active recording surface size and its relationship with the reference electrode determine the recording distance in the muscle tissue and the number of muscle fiber action potentials that will contribute in the volume conductor to the summated MUAP. The standard concentric needle electrode has an oval active recording surface on the beveled edge of the needle tip. The referential electrode is the needle electrode shaft. Such a needle configuration incorporates action potentials from 8 to 20 muscle fibers in a single motor unit over a semicircular recording diameter of about 10 mm. Monopolar electrodes consist only of an active recording surface that is located at the bared tip of a Teflon-covered needle. A referential electrode must be applied on the adjoining skin surface, usually in the form of a 10-mm tin disc. This recording configuration produces a somewhat larger recording area, and results in MUAPs that are 10% to 15% larger in amplitude and duration. Monopolar needle electrodes are in declining use since the introduction of single-use disposable concentric needles.

Equipment Filter Settings

Changing the filter settings on the recording amplifier of the EMG machine will alter the size of recorded MUAPs. For evaluation of MUAPs, the general practice is to use a low filter setting of 20 to 50 Hz and a high filter setting of 10 KHz or higher. Increasing the low filter setting will reject more low-frequency components usually derived from distant muscle fiber action potentials, and will result in shortening the duration of the discharge by shrinking the size of the initial and terminal phases of the MUAP. This is the objective in single fiber EMG, when only the muscle fiber action potentials closest to the recording surface are of interest, and low-frequency contamination from outlying muscle fiber action potentials is to be avoided. Lowering the high-frequency filter will reject more of the high-frequency compo-

TABLE 12–1	Factors that Alter the Appearance of the MUAP in the Normal State

Type of needle electrode used
Filter settings of the recording equipment
Gain settings of the recording equipment
Temperature of the muscle
Specific muscle under study
Age of the patient
Needle location
Level of voluntary contraction

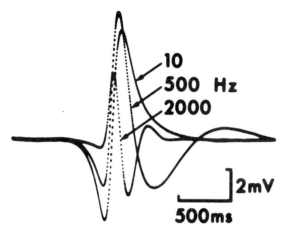

FIGURE 12–5. Effect of filtering on the shape and amplitude of the single fiber action potential. (From Stalberg EV, Trontelj JV: Single Fiber Electromyography, 2nd ed. New York, Raven Press, 1994, p 10, with permission.)

nents of the MUAP, reducing the amplitude of the negative spike potential and somewhat compressing the MUAP duration (Fig. 12–5).

Equipment Gain Settings

The display size of the NEE recording can be increased or decreased by changing the gain setting. The usual setting for the recording of MUAPs during voluntary muscle contraction is 200 μV per division on the vertical scale of the EMG monitor, compared with 50 μV per division for the recording of spontaneous and insertional activity when the muscle is at rest. Change in gain setting shrinks or expands the size of a waveform in both the vertical and horizontal dimensions. The units of measurement on the vertical scale of the monitor automatically change to conform to the gain setting chosen, so that the actual measured amplitude does not change from one gain setting to another. However, the monitor's horizontal unit of measurement remains real time, and is only altered by changes in the sweep speed of the baseline. The effect of increasing the sensitivity of the gain setting is to increase the duration of recorded discharges at any given sweep speed; the effect of decreasing the sensitivity is to shorten the duration. Thus, the measured duration of a MUAP is entirely dependent on the gain setting. The daunting task of remembering the normal range of duration of MUAPs at many gain settings can be simplified by selecting one or two gain settings at which all duration measurements will be made.

Effect of Temperature

The effect of temperature change on MUAP size has been noted in a number of reports, but there is no unanimity of opinion. Some researchers have noted reduction of amplitude with cooling of muscle, whereas others have noted increase in amplitude.[18] Cooling produces delayed inactivation of sodium channels in nerve and muscle membranes, increasing the size of the propagated nerve and muscle fiber action potentials. This can result in increased size of MUAPs in the same way that cooling produces increased amplitude and duration of evoked sensory nerve action potentials and compound motor action potentials. In general EMG practice, significant cooling of an examined muscle is likely to lead to spuriously increased MUAP amplitude and duration measurements (Fig. 12–6). This is especially likely to occur in distal muscles of the arm and leg, where the MUAP changes may be misinterpreted as representing a neurogenic process.

Muscle Specificity

The characteristics of MUAP size and configuration vary from muscle to muscle, each with its own morphological "signature." In part this specificity is related to the muscle's innervation ratio, the number of muscle fibers innervated by a single motor neuron. It is also likely related to the way the muscle's end plate zone is laid out in the muscle belly. All muscles have a varied population of larger and smaller MUAPs, a percentage of which has increased numbers of turns and phases. Muscles such as the gluteus maximus, iliacus, orbicularis oculi, orbicularis oris, frontalis, and paraspinalis tend to have MUAPs with the shortest mean duration, and an increased number of phases. Muscles such as the vastus lateralis, deltoid, tibialis anterior, and triceps have MUAPs with the longest mean duration. Normative data for the mean duration of MUAPs in many muscles are published[34] (Table 12–2). These measurements were made at lower gain and low-frequency settings than are routinely used in clinical practice today, and are therefore somewhat larger than would be expected at currently used settings.

Effect of Age

MUAPs change in size as part of the natural process of aging. MUAPs are smallest in the neonate and increase in both duration and amplitude throughout life.[12] In the neonate, muscle fiber diameters are about 25% of the normal adult size, and the proportionate reduction in the electrical fields generated by neonatal propagated action potentials likely explains the reduced MUAP size. Aging factors in the elderly, such as senescence of motor neurons and subtle collateral sprouting, may in part explain why MUAPs continue to increase in size.

Needle Location

The standard concentric needle electrode records only a percentage of the muscle fiber action potentials in a motor unit under study. Turning the beveled active recording surface of the needle will change the configuration of the

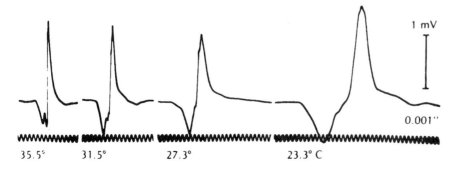

35.5° 31.5° 27.3° 23.3° C

1 mV

0.001"

FIGURE 12–6. Effect of muscle temperature on the motor unit action potential. (From Daube JR: Minimonograph #11. Needle examination in electromyography. Muscle Nerve 14:685–700, 1991, with permission.)

TABLE 12-2	Mean MUAP Duration in Milliseconds

Age	Frontalis	Deltoid	Biceps	Triceps	FDI	Rectus Femoris	Vastus Lateralis	Tibialis Anterior	Gastrocnemius
0	4.1	7.8	7.7	9.0	7.7	8.7	9.7	9.5	7.2
10	4.8	9.3	9.1	10.6	9.1	10.3	11.5	11.2	8.6
20	5.3	10.2	10.0	11.6	10.0	11.3	12.6	12.3	9.4
40	5.5	11.3	11.1	12.2	11.1	12.6	14.0	13.6	10.4
60	5.7	12.1	11.9	12.6	11.9	13.5	15.0	14.7	11.2
80	5.9	13.0	12.8	12.8	12.8	14.4	16.1	15.7	12.0

Data modified from Ludin,[33] with permission, according to data (partly unpublished) from Buchtal F: Einführung in die Elektromyographie. Munich, Urban and Schwarzenberg, 1958. A low filter setting of 16 Hz and a gain setting of 50 to 100 μV per division were used for the recording of these potentials.

MUAP because of changes in the geographical distribution and number of muscle fibers close to the new electrode position.

Where in the muscle the needle electrode is introduced can also alter MUAP morphology. Needle placement within the end plate zone itself can minimize the initial positive phase of the MUAP and may decrease the MUAP duration. Excessive intervening inactive muscle or fibroadipose tissue will attenuate volume-conducted signals and diminish or change the MUAP appearance. Measurement of MUAP size should be made only after assuring that the waveforms reliably represent the true MUAP population. Optimal MUAP recording is achieved when the rise time of the main negative spike potential (from preceding positive peak to negative spike peak) is less than 500 μsec. Continual adjustment of the needle electrode placement is required to maximize the main negative spike potential amplitude.

Effect of Level of Muscle Contraction on Motor Unit Action Potential Population Activated

MUAPs can be designated as type I or type II, as is the case for the motor neurons and muscle fibers from which they are derived. In general, type I motor units are associated with oxidative metabolism, are resistant to fatigue, have lower twitch speed, generate less tetanic force, and have smaller fiber diameters. Type I MUAPs tend to be activated early and tend to predominate during minimal and moderate muscle contraction. Type I MUAPs tend to be smaller than type II MUAPs, which are activated during strong muscle contraction, and have larger fiber diameters, higher twitch speed, and generate more tetanic force, but fatigue easily. Thus, MUAPs in the same muscle will differ in size depending on the level of muscle contraction used during their activation.

Motor Unit Recruitment

Muscle power increases as more muscle fibers contract. Muscle power can increase with an increase in the frequency of firing of motor units already activated, with activation of resting motor units, or both. Motor unit recruitment refers to the orderly increase in the frequency of firing of activated motor units, and the calling up of additional motor units as voluntary muscle contraction increases. In all muscles there is a mixed population of small, weak motor units and larger, stronger motor units. The Henneman size principle refers to the orderly successive activation of motor units during an increasing voluntary muscle contraction, such that small, weak type I motor units are activated first in an early contraction, and sequentially larger, stronger motor units are called up to deliver a smooth increase in muscle power.[23]

The analysis of motor unit recruitment is an important part of the NEE in clinical practice. When a patient is asked to give a minimal voluntary contraction of muscle, a single MUAP may begin to fire in its typical semirhythmic pattern at a frequency of around 5 Hz. When slightly more contraction is requested, the MUAP increases its firing frequency to 10 Hz. With more contraction, a second MUAP is recruited in, at a somewhat slower initial firing frequency than that already achieved by the first. This process continues until an interference pattern of MUAP activation is seen during a maximal voluntary contraction.

Analysis of recruitment on the EMG monitor requires two measurements: an estimate of the number of different MUAPs appearing on the monitor at the same time, and an estimate of the frequency of firing of the fastest MUAPs. The MUAP firing frequency can be measured by a simple calculation. If the sweep speed of the baseline is set at 10 msec per division and there are 10 divisions along the baseline, a MUAP appearing in the same position on the monitor with each successive sweep is firing once per 100 msec, or at a frequency of 10 Hz. If the same MUAP appears twice with each sweep, it has a firing frequency of 20 Hz. A MUAP that appears once per sweep, but shifts to the left with each sweep, is firing at a frequency between 10 and 20 Hz.

The rule of fives is a simplified counting technique that can be employed in the EMG laboratory to measure the adequacy of motor unit recruitment.[17] The first MUAP begins firing at 5 Hz, and a new MUAP is recruited in when the previous MUAP reaches a firing frequency of 10 Hz. Each MUAP gradually increases its firing frequency by 5 Hz, at which time another MUAP is recruited in. When the firing frequency of the fastest MUAP on the screen is divided by the number of different MUAPs firing on the screen, that number should be less than or equal to five. For example, if the fastest MUAP is firing at 20 Hz, there should be at least four different MUAPs on the screen. If the ratio exceeds five, there are too few MUAPs for the frequency of firing of the fastest.

Each muscle has a characteristic MUAP recruitment frequency and recruitment interval. Recruitment frequency refers to the firing frequency of a first MUAP when a second MUAP is recruited at a stable firing rate. Recruitment interval refers to the time elapsed between two firings of the first MUAP when the second MUAP has just been recruited, and is equal to the inverse of the recruitment frequency. Facial muscles have high recruitment frequencies (20 to 30 Hz), whereas most extremity muscles have recruitment frequencies in the 10- to 12-Hz range.[42, 43]

Neurogenic Recruitment

The pattern of reduced recruitment is only seen when there is reduced availability of motor units. This pattern is seen almost exclusively in neurogenic disorders. The term reduced recruitment indicates a reduced number of motor units available for activation on demand. On-the EMG monitor, one sees a pattern of too few MUAPs for the amount of muscle contraction produced, and those MUAPs present are firing at an excessive frequency.

In our laboratory we have established a subjective grading system for the analysis of reduced recruitment. The system is predicated on the observation that it is rare to observe a MUAP that fires faster than 10 to 15 Hz in a normal muscle, as the monitor quickly fills with different MUAPs, obscuring the firing pattern of any one MUAP. A reduced recruitment pattern is reported when a MUAP is identified with a frequency of 15 to 20 Hz or higher. A subjective analysis of the total number of different MUAPs on the monitor is then performed. If three to four other MUAPs are seen, the recruitment is said to be mildly reduced; if two to three other MUAPs are seen, the pattern is said to be moderately reduced; if one to two other MUAPs are seen, the pattern is said to be markedly reduced. If only one MUAP is present, and firing at 20 Hz or higher, a single motor unit firing pattern with reduced recruitment is reported. In extreme neurogenic situations a single MUAP can fire with a frequency of 40 to 50 Hz, a pattern seen only when there is profound motor unit drop out.

Axon loss produces reduction of the available recruitable motor unit pool. Loss of motor neuron cell bodies can occur owing to myelopathy and motor neuron diseases. Peripheral motor axon loss can occur due to traumatic transection, infarction, infiltration, and degeneration. In all these mechanisms, interruption of the motor axonal connection with muscle fibers excludes motor units from the recruitment process.

Conduction block along motor nerve fibers can also inactivate motor units from the recruitment process, without axon loss. When enough demyelination occurs along a motor axon, impulse transmission is prevented, although the underlying axon is intact. Acute and chronic demyelinating disorders, acute and chronic nerve trunk compression and entrapment, and occasionally ischemia can produce demyelinating conduction block. In a neurogenic process in which weakness is present for longer than 1 week, the finding of reduced recruitment in a muscle that has a normal compound muscle action potential (CMAP) to motor nerve stimulation is strong indirect evidence of a conduction block lesion proximal to the site of motor nerve stimulation.

In any significant acute neurogenic disorder, the first pathological EMG feature is likely to be a reduced recruitment pattern of motor unit firing. As recruitment is a measure of the nervous system's ability to respond physiologically to the need for increased voluntary muscle contraction, it will immediately reflect any change in the availability of the motor unit pool. Reduced recruitment can be observed in the abductor pollicis brevis muscle moments after a partial transection of the median nerve trunk, but the median CMAP amplitude will not begin to fall for 3 days, fibrillation potentials will not develop for 2 to 3 weeks, and reinnervation of denervated muscle fibers will not be appreciated for weeks or months. In a patient with severe weakness in the setting of rapidly progressive acute inflammatory demyelinating peripheral polyneuropathy (Guillain-Barré syndrome), the earliest feature is often a reduced recruitment pattern of motor unit firing, sometimes preceding by days the classical nerve conduction study abnormalities.

Myopathic Recruitment

In myopathy, individual muscle fibers drop out of the motor unit, without affecting the integrity of the motor unit's connection with the central nervous system. The result is a normal motor unit pool, but with less contractile force per motor unit. When more muscle contraction is required, the response is the recruitment of more motor units. The re-

cruitment pattern seen in severe myopathy is an activation of too many MUAPs for the degree of muscle contraction achieved, while MUAP firing frequencies are normal. In this scenario the ratio of firing frequency to number of MUAPs is less than five, a pattern described as early or increased recruitment.

Nonspecific Recruitment Patterns

Reduced and increased recruitment patterns represent the only pathognomonic patterns of recruitment. More commonly, nonspecific recruitment abnormalities are seen, characterized by MUAPs firing at reduced numbers and at submaximal frequencies. This pattern is seen in a wide range of states, including central disorders of motor unit control (stroke, myelopathy, spasticity, rigidity, and extrapyramidal disorders), mechanical disorders (joint immobilization and positioning), and reduced voluntary effort (pain and malingering). Patients with clear neurogenic or myopathic states may not demonstrate pathognomonic recruitment patterns because of superimposed conditions as listed earlier, or because profound weakness prevents sufficient motor unit firing to activate the recruitment process.

In light of the preceding comments, it is clear that a standard rule regarding the degree of neurogenic or myopathic damage required to produce of given degree of recruitment abnormality cannot be established. The percentage of motor unit drop out required to produce a reduced recruitment pattern depends on a number of factors, including the patient's ability to give a full effort, the expected recruitment frequency of the muscle under normal circumstances, and the electromyographer's facility in recruitment analysis. In a neurogenic process, given a full effort by the patient, an obvious reduced recruitment pattern should be appreciated with a loss of 50% or more of the motor unit pool. In established axon loss processes, when examining a muscle from which a CMAP can be obtained by nerve conduction studies, a reduced recruitment pattern would be expected when 50% or more of the normal CMAP amplitude is lost. In a myopathic process, the identification of an increased recruitment pattern is even more subjective than identification of reduced recruitment, because it requires a clinical judgment regarding the degree of muscle contraction and the relative increase in MUAP numbers. Therefore, most electromyographers reserve this designation for marked changes.

NEEDLE ELECTRODE EXAMINATION IN RESTING MUSCLE

The NEE in resting muscle records muscle fiber discharges in two settings. Discharges that are induced by needle electrode movement through muscle, and then die out, are termed insertional activity. Discharges that occur without being triggered by needle movement, and continue indefinitely are termed spontaneous activity. For the examination of insertional and spontaneous activity, the gain setting should be 50 μV per division, the low filter setting 20 to 50 Hz, and the high filter setting 10 KHz or higher.

Normal Insertional Activity

Movement of the needle electrode through resting muscle tissue mechanically depolarizes contacted muscle fibers, triggering propagated muscle fiber action potentials. Normal insertional activity is characterized by a flurry of muscle fiber action potentials that commences with the needle movement and ends less than 200 μsec after cessation of the movement.

At the usual recording settings, a normal response exceeds the vertical boundaries of the monitor.

Reduced or absent insertional activity can be seen in a normal individual when the needle electrode is advanced through electrically silent tissue, such as adipose or fibrous tissue. Increased insertional activity can be seen as a variant of normal. Most commonly this activity mimics fasciculation potentials in firing pattern and morphology, except that it always follows needle movement, and resolves within 10 sec of onset. The activity often begins with a burst of potentials with interpotential intervals of 50 to 100 msec, and then wanes. This form of insertional activity tends to be seen in healthy, muscular young men. At times this activity may be accompanied by trains of positive waves, spontaneous fasciculation potentials, and cramp discharges, a condition described as benign fasciculation cramp syndrome.

Abnormal Insertional Activity

The most commonly encountered type of abnormal insertional activity takes the form of trains of potentials with either positive waveform or brief sharp spike morphology (Fig. 12–7). These potentials are identical to the abnormal spontaneous activity known as fibrillation potentials, except that they are induced by needle movement and die out within seconds. These potentials tend to be seen 1 to 3 weeks after onset of axon transection and signify a recent onset axon loss process.[56] When not accompanied by frank fibrillation potentials, the acute process can usually be dated at less than 3 weeks age, although in some patients the onset of spontaneous fibrillation potentials is delayed up to 6 weeks. As the acute axon loss ages, fibrillation potentials become more numerous and the abnormal insertional activity dies out. A midway point is often seen, during which prominent increased insertional activity is accompanied by mild fibrillation, indicating an evolving axon loss process of 2 to 8 weeks in age. Occasionally increased insertional activity is seen in a widespread distribution, never associated with fibrillation

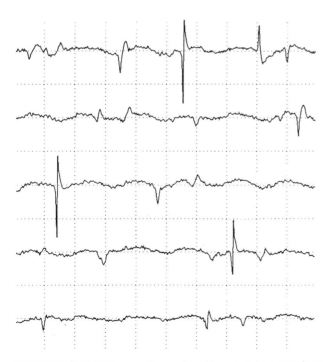

FIGURE 12–7. Brief sharp spikes and positive waves in a muscle at rest. Vertical division = 50 μV, horizontal division = 10 msec.

TABLE 12–3	Types of Insertional Activity

States of Increased Insertional Activity
Variants of normal
Trains of positive waves and brief sharp spikes (early denervation potentials)

States of Decreased Insertional Activity
Needle movement through electrically inactive tissues (necrotic muscle, fibrosis, adipose)
Physiological contracture
  McArdle's disease (myophosphorylase deficiency)
  Phosphofructokinase deficiency
  Paramyotonia congenita
Episodes of periodic paralysis

or pathological MUAP changes. This represents a variant of normal.

Abnormal decreased insertional activity is seen in a number of settings. Advancing the needle electrode through fibrotic, atrophied muscle yields reduced insertional activity. Muscle of normal electrical excitability can become unexcitable, paradoxically when the muscle develops an involuntary contraction, termed a physiological contracture. Physiological contracture occurs in conditions including McArdle's disease (myophosphorylase deficiency), phosphofructokinase deficiency, and paramyotonia congenita. Reduced or absent insertional activity is also seen during flaccid episodes of periodic paralysis. A summary of the types of insertional activity can be found in Table 12–3.

Normal Spontaneous Activity

End plate noise is one of two forms of normal spontaneous activity. It is seen when the needle electrode is in the end plate zone of the muscle, and it is composed of small irregularly occurring potentials of 0.5 to 2 msec duration and 10 to 40 μV amplitude.[5] End plate noise is composed of miniature end plate potentials, the result of subthreshold depolarizations of muscle fibers occurring during the steady state release of single vesicles of acetylcholine from the nerve terminals of neuromuscular junctions (see Chap. 20). Such potentials are too small to record except when the needle electrode is adjacent to the neuromuscular junctions and, as a result, have no initial positivity. At the usual EMG settings, end plate noise is usually manifested by a widening of the baseline width with an associated hissing sound. The patient may complain of increased discomfort, a common finding when the needle electrode is in the end plate region.

End plate spikes are often seen in close association with end plate noise. They also discharge irregularly, but have longer duration (3 to 4 msec) and higher amplitude (100 to 200 μV). One school of thought suggests that these discharges arise from needle electrode irritation of distal nerve fiber axons or the nerve terminals themselves, producing impulse transmission and propagated muscle fiber action potentials.[51] A second theory suggests that end plate spikes arise from the intrafusal muscle fiber action potentials in muscle spindles innervated by gamma motor neurons, a less likely possibility considering end plate spikes and noise are often seen together, and end plate noise is clearly localized at the muscle end plate.[41] End plate spike waveforms are biphasic with an initial negativity, reflecting the fact that they are recorded by the needle electrode at the point of their generation in the end plate region (Fig. 12–8).

Abnormal Spontaneous Activity

Discharges occurring in a muscle at rest without needle electrode movement include fibrillation potentials, complex

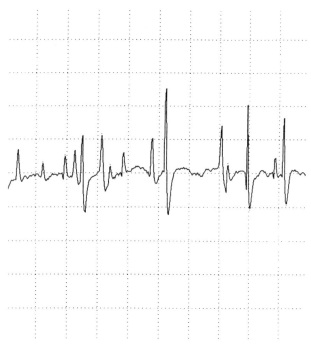

FIGURE 12–8. End plate spikes in a muscle at rest. Note the initial negative deflection of these biphasic potentials. Vertical division = 50 µV, horizontal division = 10 msec.

repetitive discharges, myotonic discharges, fasciculation potentials, and grouped repetitive discharges (Table 12–4). They are usually associated with pathological states.

Fibrillation Potentials

These discharges are arguably the most important form of abnormal spontaneous activity to be identified during routine needle electromyography. They represent the propagated muscle fiber action potentials resulting from spontaneous threshold depolarizations of single muscle fibers, and develop after the neuromuscular junction of a muscle fiber is lost or degenerates. They represent the most obvious electrodiagnostic feature of active denervation, and are seen in axon loss lesions as well as necrotizing myopathies, where a segment of the muscle fiber becomes disconnected from the portion of the muscle fiber containing the neuromuscular junction.

Fibrillation potentials have two distinct waveform morphol-

TABLE 12–4	Forms of Spontaneous Activity

Normal Spontaneous Activity
End plate noise
End plate spikes
Fasciculation potentials in some situations (benign fasciculation cramp syndrome)

Abnormal Spontaneous Activity
Fibrillation potentials
Complex repetitive discharges
Myotonic discharges
Fasciculation potentials
Grouped repetitive discharges
  Myokymia
  Neuromyotonia
  Hemifacial spasm
  Cramp discharges
  Multiple discharges (multiplets and tetany)
  Tremor

ogies. They may appear as brief sharp spikes, and are either biphasic or triphasic with an initial positivity. They range in amplitude from 20 to 200 µV, and in duration from 1 to 5 msec. Or, they may have a positive wave morphology, with similar amplitude, but with duration ranging from 10 to 30 msec (see Fig. 12–7). The size of fibrillation potentials is likely related to the diameter of the muscle fiber generating them.[29] Fibrillation potentials are distinguished from other discharges with similar morphology on the basis of their very regular firing frequency (Table 12–5). However, in some patients early fibrillation may demonstrate an irregular firing pattern.

Fibrillation potentials result from changes in the muscle fiber membrane potential, leading to spontaneous threshold depolarization and action potential propagation. After 4 to 6 days, a denervated muscle fiber's resting membrane potential begins to drift up from −80 mV toward −60 mV, the threshold level for depolarization of the muscle fiber membrane. Spontaneous oscillations of the membrane potential eventually reach the threshold level and a depolarization occurs. During repolarization the membrane reaches a hyperpolarized state from which the drift up begins again, repeating itself at a regular interval.[52] Fibrillation potentials decrease in numbers with decreasing temperature.[9]

The timing between the onset of nerve or muscle injury and the first appearance of fibrillation potentials in the NEE depends on several factors. With direct muscle injury, fibrillation can develop within 7 days.[40] With motor axon transection, histological evidence of wallerian degeneration is first seen after days 2 to 3,[19] and true fibrillation may not appear until after days 14 to 21, depending on the distance between the site of nerve injury and the muscle being examined.[52, 53] For example, in the setting of an axon loss cervical radiculopathy, fibrillation potentials are likely to develop in paraspinal muscles some days before they appear in denervated muscles of the arm.

The resolution of fibrillation potentials is dependent on the establishment of new neuromuscular junctions. In the case of muscle injury, fibrillation potentials decrease within several months, and are no longer seen at 11 months.[40] For neurogenic lesions, the resolution of fibrillation potentials is related to the severity of the axon loss process, the efficiency with which surviving motor nerve terminal branches can produce collateral nerve sprouts to denervated muscle fibers, and (in regard to reinnervation by axonal regeneration from the lesion site) the distance between the site of the nerve injury and the muscle to be reinnervated. For example, paraspinal muscles can lose their fibrillation potentials within 6 weeks of the onset of radiculopathy, whereas extremity muscle fibrillation potentials can persist for months. In some severe denervating processes such as poliomyelitis and traumatic brachial plexopathy, fibrillation potentials can persist in some areas indefinitely.

Complex Repetitive Discharges

Complex repetitive discharges (CRDs) are trains of discharges firing spontaneously in a regular, lock-step rhythm. Each train begins and ends abruptly, without variation in

TABLE 12–5	Discharges Sharing Positive Wave or Brief Sharp Spike Morphology

Increased insertional activity (denervation potentials)
End plate spikes
Fibrillation potentials
Fasciculation potentials
Myotonic discharges

frequency of discharge firing or change in waveform morphology. Each discharge is comprised of a nearly identical group of as many as 10 muscle fiber action potentials (Fig. 12–9). Discharges fire at rates between about 10 and 50 Hz, and sound like the idling of an outboard boat motor. They range in amplitude from 50 to 500 μV. CRDs continue in spite of nerve block and curare, indicating that these discharges are generated in the muscle membrane.[50]

Single fiber EMG studies indicate a jitter value of less than 5 μsec between potentials in a single discharge, suggesting that they result from the ephaptic activation of adjacent muscle fibers, rather than from the activation of individual muscle fibers by transmission across their neuromuscular junctions.[54] CRDs may occur as the result of a spontaneously discharging pacemaker muscle fiber that ephaptically activates adjoining muscle fibers, which in turn reactivate the pacemaker fiber, creating a closed loop of continuously activating fibers.

In clinical EMG practice, the finding of CRDs usually indicates a chronic neuromuscular lesion. They are seen in a wide range of myopathic and neurogenic disorders, including dystrophies, polymyositis, motor neuron disease, and chronic polyneuropathies. They do not always indicate a pathological disorder, as they can be seen in proximal muscles such as the iliacus, rectus femoris, and glutei in otherwise healthy elderly individuals.

Myotonic Discharges

Myotonic discharges are prolonged trains of muscle fiber action potentials firing with waxing and waning frequency and amplitude, giving them the sound of a "dive bomber" or a decelerating and accelerating motor. Although they are traditionally described as a form of spontaneous activity, they require a triggering event, such as external muscle tap, initial voluntary muscle contraction, or needle electrode movement. Within a given train the potentials wax and wane in frequency between about 10 and 100 Hz (Fig. 12–10). When initiated by needle electrode movement, they tend to have a positive wave appearance identical to insertional positive

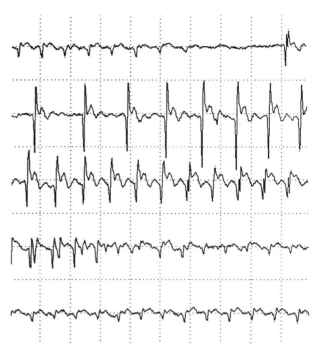

FIGURE 12–10. A train of waning myotonic discharges. Note the marked amplitude decline over time, while the firing rate varies from about 20 to 40 Hz. Vertical division = 100 μV, horizontal division = 50 msec.

waves. When initiated by muscle contraction, they tend to have a brief sharp spike morphology identical to fibrillation potentials.

The pathophysiology of myotonic discharges in humans has not been firmly established, but there is information from several animal models. Hereditary myotonia in goats is related to decreased muscle membrane chloride conductance, and this is also likely to be the case in patients with myotonia congenita, one of the human hereditary myotonias. Chloride conductance in normal muscle membrane stabilizes the membrane potential by shunting depolarizing current.[8] Loss of chloride conductance has the opposite effect, raising the membrane resistance and decreasing the amount of current required to depolarize the membrane. After a normal membrane depolarization, potassium channel opening leads to hyperpolarization. A slight depolarization occurs as the potassium conductance returns to the resting level, but in the unstable membrane with increased resistance caused by reduced chloride conductance, that slight depolarization can trigger another threshold depolarization of the membrane, producing another muscle fiber action potential that restarts the cycle.

Myotonic discharges are a classical feature of the hereditary myotonias: myotonic dystrophy, myotonia congenita, and paramyotonia congenita. In these disorders, myotonic discharges are associated with stiffness of muscle, often clinically manifested by an involuntary delay in the relaxation of a contracted muscle. In myotonic dystrophy, the discharges are not often seen until the second decade of life, even in patients with neonatal clinical manifestations. When they do appear, they are most prominent in distal muscles of the extremities, corresponding to the distribution of myopathic motor unit potential changes in this disorder. In the other two hereditary myotonias, myotonic discharges are seen in a more widespread distribution. In paramyotonia congenita, the stiffness, or paramyotonia, that develops in the setting of continuous exercise or exposure to cold is related to an

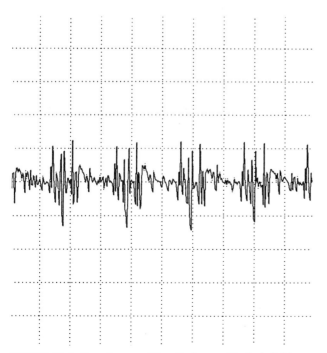

FIGURE 12–9. Complex repetitive discharges. Vertical division = 200 μV, horizontal division = 50 msec.

electromyographically silent physiological contracture during which myotonic discharges and normal motor unit potential activation are inhibited. The characteristics of myotonic discharges in the hereditary myotonias are reviewed in Table 12–6.

Myotonic discharges are also seen in a number of other myopathic conditions not associated with clinical myotonia. Muscle disorders in which myotonic discharges may be seen include hyperkalemic periodic paralysis, polymyositis, oculopharyngeal dystrophy, and any chronic end-stage myopathy. The presence of myotonic discharges in children should raise the suspicion of disorders such as centronuclear myopathy and glycogen storage diseases. The presence of myotonic discharges isolated to paraspinal muscles and proximal trunk muscles in the adult with respiratory insufficiency should raise the suspicion of acid maltase deficiency. Myotonic discharges are also seen in toxic myopathies associated with vacuolar changes and degradation of the muscle membrane. Incriminated agents include clofibrate, lovastatin, gemfibrozil, chloroquine and hydroxychloroquine, and cyclosporine.[16]

Myotonic discharges are also seen in neuropathies. In this setting they represent marked chronicity of the neurogenic process. They can be seen in chronic radiculopathy and peripheral polyneuropathies. In these disorders, the trains of myotonic discharges tend to brief and low in amplitude. Table 12–7 reviews the disorders in which myotonic discharges can be seen.

Fasciculation Potentials

Fasciculation potentials are characterized by the sporadic, involuntary activation of a motor unit or portion of a motor unit. They resemble MUAPs and can be of any size and shape. Fasciculation potentials are typified by the totally random occurrence of nonrepeating MUAPs, each of unique configuration. Large fasciculation potentials near the skin surface are associated with a small, visible muscle twitch within a muscle belly that does not cause movement of a joint. The generator of fasciculation potentials has not been established. Because they are the hallmark of amyotrophic lateral sclerosis, they have traditionally been thought to originate at or near the anterior horn cell. However, several reports suggest they are more likely to originate in the terminal nerve branches[46] or in the muscle itself.[48] One theory suggests that a fasciculation potential arises from a single spontaneously depolarizing nerve terminal whose action potential travels antidromically to a more proximal nerve branch point, at which time it travels orthodromically back down the arborization of terminal nerve branches to depolarize a fraction of the motor unit[33] (Fig. 12–11).

TABLE 12–7	Causes of Myotonic Discharges

Hereditary Myotonias and Inborn Errors of Metabolism
Myotonic dystrophy
Myotonia congenita
Paramyotonia congenita
Oculopharyngeal dystrophy
Centronuclear myopathy
Infantile glycogen storage diseases (Pompe's disease)
Hyperkalemic periodic paralysis
Adult acid maltase deficiency

Acquired Myotonic Discharges
Chronic radiculopathies
Chronic axonal mononeuropathies and polyneuropathies
Polymyositis
End-stage myopathies
Toxic myopathies
  Clofibrate
  Lovastatin
  Gemfibrozil
  Chloroquine
  Hydroxychloroquine
  Cyclosporine

The presence of fasciculation potentials is most closely associated with disorders of the motor unit. Fasciculation potentials can be prominent in disorders of the motor neuron such as amyotrophic lateral sclerosis, spinal muscular atrophy, X-linked bulbospinal muscular atrophy (Kennedy's syndrome), and Creutzfeldt-Jakob disease. They are associated with disorders of peripheral motor nerves, such as chronic entrapment neuropathies, peripheral polyneuropathies, and radiculopathies.

Fasciculation potentials are also seen in metabolic derangements. Endogenous states such as thyrotoxicosis and hypocalcemic tetany can be associated with fasciculation potentials. Certain drugs that act at the neuromuscular junction can produce fasciculation potentials. Acetylcholinesterase inhibitor medications such as pyrodostigmine, used in the treatment of myasthenia gravis, produce fasciculations in periocular muscles and elsewhere, especially at higher doses. Cholinesterase inhibitors such as succinylcholine, used in general anesthesia, and organophosphate insecticides such as malathion and parathion, with antipseudocholinesterase activity, all have the potential to produce both clinical and EMG manifestations of fasciculation.

Fasciculation potentials may also be seen in normal muscle. These discharges likely originate in the muscle itself.[48] They are commonly seen in the gastrocnemius muscles after prolonged running, and can be part of the benign fascicula-

TABLE 12–6	Electrodiagnostic Findings in the Myotonias			
	Myotonic Dystrophy	**Dominant Myotonia Congenita (Thomsen)**	**Recessive Myotonia Congenita (Becker)**	**Paramyotonia Congenita**
Effect of repetitive stimulation	Mild decrement	Mild to marked decrement	Marked decrement	Little or no decrement
Distribution of myotonia	Distal predominance	Generalized	Generalized	Limited in distribution and intensity
Response to exercise	Myotonia subsides	Myotonia subsides	Myotonia subsides and MUAPs fail to activate	Myotonia subsides, MUAPs fail to activate, and onset of physiological contracture
Response to cold	Transient increase in myotonia	Transient increase in myotonia	Transient increase in myotonia	Myotonia subsides, MUAPs fail to activate, and onset of physiological contracture
Muscle unit action potential (MUAP) morphology	Distal predominance of myopathic changes	Normal	Mild myopathic changes	Normal

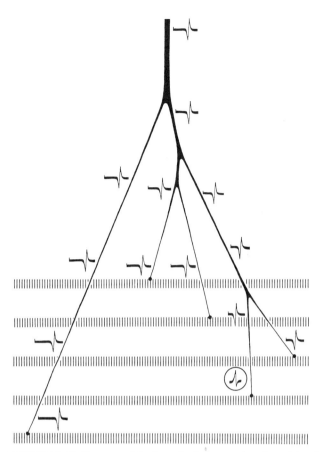

FIGURE 12–11. Diagram of the hypothesis concerning the origin of fasciculations in the nerve twigs. The ringed action potential with negative onset from the baseline is the point of departure of the excitation. From here the excitation runs antidromically to the outer twigs. The length of the baseline before the other action potentials indicates the arrival time of the excitation at different points in the motor unit. (From Ludin HP: Electromyography in Practice. New York, Thieme-Stratton, 1980, p 60, with permission).

tion cramp syndrome. The causes of fasciculation potentials are reviewed in Table 12–8.

Grouped Repetitive Discharges

Grouped repetitive discharges are examples of abnormal spontaneous activity, but they differ from the abnormal spontaneous activity described earlier in that they are composed

TABLE 12–8	Causes of Fasciculation Potentials

Neurogenic Causes
Motor neuron disease (amyotrophic lateral sclerosis)
X-linked bulbospinal muscular atrophy (Kennedy's syndrome)
Creutzfeldt-Jakob disease
Chronic mononeuropathies and polyneuropathies
Radiculopathies

Cholinesterase Inhibitors
Pyridostigmine
Succinylcholine
Organophosphates (malathion, parathion)

Myogenic Causes
Normal variant, increased with exercise
Benign fasciculation cramp syndrome

of repeating MUAPs rather than muscle fiber action potentials. In many cases the exact sites of origin of these discharges, and the underlying pathophysiological differences between one type of grouped repetitive discharge and another are not clear.

Myokymia

Myokymic discharges are classically described as repeating bursts of MUAPs. Within a burst there may be 2 to 10 potentials with intervals ranging from 10 to 20 msec between potentials. Bursts can repeat at regular or semiregular intervals of 100 msec to 10 seconds (Fig. 12–12). From one burst to another there is often some variation in the number of potentials within a burst and the interpotential intervals. Myokymic discharges often manifest themselves clinically as an undulating or rippling subcutaneous movement of muscle. On the EMG monitor, the sound of myokymia resembles a repeating sputter or the locked-step march of soldiers.

A single myokymic burst is thought to represent several repetitive firings of MUAPs. The only recent neurophysiological study of these discharges focused on the neuromyotonic discharges in Isaacs' syndrome, described later in this chapter.[48] In a later publication by the same authors, discharges closely resembling those of Isaacs' syndrome were identified in the thenar muscles of patients with carpal tunnel syndrome.[49] Within a single myokymic burst, the basic discharge consisted of one or several muscle fiber action potentials with jitter exceeding 5 μsec, suggesting that they arose from transmission across separate neuromuscular junctions, and were not ephaptically generated. The basic discharge was repeated several times at irregular intervals, and some spike components in the repeated discharges within a burst were either dropped from the prior discharge, or appeared for the first time. The authors proposed that depolarization in this condition originates in distal nerve fibers, and that there may be more than one trigger point along an individual nerve fiber. The pattern of the repeating bursts would then depend on the recovery cycles of the individual trigger

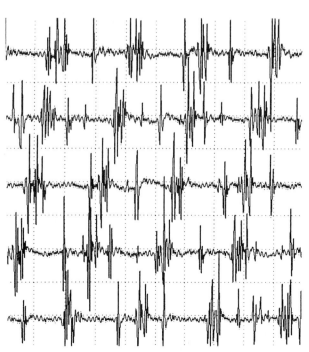

FIGURE 12–12. A train of myokymic discharges. Vertical division = 50 μV, horizontal division = 100 msec.

points and collision between orthodromic and antidromic impulses.

Information regarding the generator of myokymia has come from other sources. Curare blocks the activity, but not general and spinal anesthesia,[26, 32] eliminating muscle and the central nervous system as sites of origin. Peripheral nerve blockade effectively suppresses the activity in some cases. Activity is reduced with pressure cuff–induced ischemia, but becomes exaggerated in the postischemic period.

Focal myokymia refers to the expression of myokymic discharges in restricted groups of muscles in one area of the body. In the extremities this is most often caused by chronic compression of nerve trunks. Classical examples include carpal tunnel syndrome, chronic radiculopathy, and radiation plexopathy,[2] in which radiation-induced vasculopathy produces progressive constrictive fibrosis. Facial myokymia is associated with a group of disorders, including brain stem tumors, multiple sclerosis, syringobulbia,[45] basilar invagination,[15] Guillain-Barré syndrome, and old Bell's palsy.

Generalized myokymia has diverse causes, and may be acquired or hereditary. Table 12–9 lists acquired forms of myokymia. Acquired generalized myokymia has been reported in association with thymoma, Guillain-Barré syndrome,[2] and metabolic derangements related to uremia, thyrotoxicosis, gold toxicity,[14] penicillamine therapy,[44] and timber rattlesnake envenomation.[4] Isaacs' syndrome, a sporadic disorder of generalized grouped repetitive discharges that have been variably described as myokymia and neuromyotonia, is discussed later.

Hereditary forms of generalized myokymia have been recognized. An autosomal dominant hereditary disorder of generalized myokymia producing generalized stiffness as the major clinical feature has been described.[3] In another hereditary disorder, episodic ataxia and myokymia syndrome, paroxysmal attacks of truncal ataxia, tremulousness, and dysarthria are the most prominent features, but close inspection often uncovers subtle clinical evidence of myokymia.[7] Associated symptoms include sensation of stiffness or limpness, jerking movements, titubation, and carpopedal spasms. This disorder occurs as the result of point mutations in the potassium channel gene *KCNA1* located on chromosome 12p13.[6] The potassium channel is also implicated in the acquired form of generalized myokymia known as Isaacs' syndrome (see later discussion).

Neuromyotonia

The term neuromyotonia was first used by Mertens and Zschoche,[35] and it is often used in association with terms such as Isaacs' syndrome, generalized myokymia, continuous muscle fiber activity, and neurotonic discharges, creating

confusion regarding the nature of this form of abnormal spontaneous activity. The term has been applied to cases of generalized muscle stiffness and delayed relaxation, with associated myokymia-like activity.[3] Some authors define neuromyotonic discharges by their tendency to show interpotential intervals less than 10 msec within a burst, differentiating this activity from classical myokymic discharges with intervals greater than 10 msec. Neuromyotonia may be continuous for many seconds, or may occur in bursts. Bursts are composed of repeating motor unit discharges, as described earlier for myokymic discharges. Stalberg has noted that the same fibers involved in bursting can also be activated volitionally.[48] Bursting can be eliminated by cooling[3] and exaggerated with warming.[55] It is most likely that there are no significant pathophysiological distinctions between myokymia and neuromyotonia. Differences in the length of bursting discharges and the interpotential interval may represent differing degrees of expression of the same pathophysiological process.[28]

The term applied to most cases of acquired generalized myokymia is Isaacs' syndrome, first described in 1961.[27] The disorder may be relatively mild, but in severe cases chronic muscle hyperactivity can result in hyperhydrosis, joint contractures, and reduction of muscle stretch reflexes. Antibodies to one or more human brain voltage-gated potassium channels have been found in the sera of patients with Isaacs' syndrome.[22, 36] The effect of the antibodies mimics the effect of potassium channel blocking agents such as 4-aminopyridine, which prolongs nerve membrane depolarization by preventing nerve repolarization. Voltage-gated potassium channels are concentrated in the paranodal and terminal regions of myelinated axons, the likely site of anti–potassium channel antibody action in the provocation of motor nerve membrane hyperexcitability.

The term neurotonic discharge has been used to describe the repeating MUAP firing pattern seen intraoperatively during manipulation and mechanical insult to peripheral nerve trunks.[21] The component discharges have a simple motor unit morphology, may be continuous or broken up into sputtering bursts, and eventually stop spontaneously after cessation of the mechanical insult.

Hemifacial Spasm

This condition is characterized by involuntary bursting of motor units in selected unilateral facial muscles, in the absence of other electrodiagnostic evidence of facial mononeuropathy. The motor unit bursting may occur spontaneously, or may be triggered by voluntary facial movement, especially a blink. There are two described patterns of motor unit activity. The first pattern, designated as the blink burst by Willison, is characterized by a short burst of motor units in a few muscles with a small blink, and a burst in a wider distribution of muscles with a large blink.[57] Blink bursts are composed of 2 to 40 potentials with interpotential intervals of 2.5 to 5 msec. The second pattern is characterized by a prolonged contraction of facial muscles, not as easily triggered by blink or electrical stimulation.[25] The action potentials seen with prolonged contraction fire at a slower frequency, with interpotential intervals of 20 to 50 msec.

The site of origin of this activity is most likely a zone of demyelination along the peripheral facial nerve trunk. Electrophysiological studies favor a combination of ectopic impulse generation and ephaptic transmission.[37] In several cases, intracranial peripheral facial nerve trunk compression by an arterial loop has been identified as the etiology of the abnormal activity, which has resolved after decompression of the nerve trunk.[38]

TABLE 12–9	Causes of Acquired Myokymia

Focal Myokymia
Facial myokymia
  Multiple sclerosis
  Guillain-Barré syndrome
  Brain stem mass lesions
Limb myokymia
  Radiation-induced brachial and lumbar plexopathy
  Chronic focal compressive neuropathies
  Other chronic polyneuropathies

Generalized Myokymia
Isaacs' syndrome
Timber rattlesnake envenomation
Gold therapy
Penicillamine therapy
Thymoma

Cramp Discharges

Cramps are sustained, involuntary, often painful contractions of a muscle, lasting seconds to minutes. They occur in normal states as the result of the sudden stretch of a muscle before and after exertion. They can also result from metabolic derangements occurring in the setting of hyponatremia, hypocalcemia, uremia, hypothyroidism, hyperthyroidism, hypomagnesemia, hypophosphatemia, hypernatremia, hypercalcemia, ischemia, and pregnancy.[31, 47] They are seen in numerous myopathic disorders, including inflammatory myopathies, disorders of glycogenolysis, glycolysis, and fatty acid oxidation. Muscle cramps occur commonly in chronic, active motor axon loss processes, such as amyotrophic lateral sclerosis, X-linked bulbospinal muscular atrophy, radiculopathy, and peripheral neuropathy. They are composed of multiple MUAPs firing synchronously and repetitively at rates of 40 to 60 Hz, and occasionally at even higher frequencies.[39] As a cramp progresses, it may recruit in new territories and more motor units. As a cramp wanes, the trains become discontinuous, and finally stop, often followed by fasciculation potentials.

Many cramp discharges are thought to be of peripheral nerve origin.[31] Local anesthetics applied to peripheral nerve can sometimes stop a cramp; general and spinal anesthesia and diazepam are ineffective.

Multiple Discharges

MUAPs firing in quick succession in doublets, triplets, or multiplets are examples of multiple discharges. The individual MUAPs are separated by intervals of 4 to 20 msec, and a multiplet may repeat itself at a frequency of 5 to 25 Hz. This activity may be clinically manifested as twitching and cramping of muscle, or carpopedal spasm. A general term applied to this clinical picture is tetany. This activity is usually the result of a metabolic derangement, such as hypocalcemia, hypomagnesemia, hypocarbia from hyperventilation, hypothyroidism, and uremia.[31] Tetany in the absence of hypocalcemia is sometimes referred to as spasmophilia. This activity can be enhanced by systemic alkalosis and local ischemia.[30]

Studies indicate a peripheral nerve origin of this activity. Curare blocks most but not all discharges, and peripheral nerve trunk block is not effective, suggesting the generator may be in distal or terminal nerve fibers.[47]

Tremor

Tremor consists of rhythmically or semiregularly firing groups of motor units at rest or during voluntary activation of muscle. Usually of central nervous system origin, the discharges are complex, tend to vary in configuration, and represent the unorderly activation of normal MUAPs. Unlike other grouped repetitive discharges, these are associated with easily seen clinical muscle activation that causes movement across at least one joint.

NEEDLE ELECTRODE EXAMINATION IN SPECIFIC DISEASE STATES

The NEE is a valuable tool in the distinguishing neurogenic and myopathic processes. It can also provide information about the age, activity, and chronicity of a neuromuscular disorder. A summary of these findings appears in Table 12–10.

Acute Neurogenic Disorders

Acute neurogenic processes include traumatic mononeuropathies, acute inflammatory and ischemic disorders, and acute demyelinating polyneuropathies such as Guillain-Barré syndrome. In a neurogenic process that interrupts impulse transmission along motor nerve fibers, the immediate finding is a reduced recruitment pattern of MUAP activation. No increased insertional activity or fibrillation is expected at this stage. There is no change in the size and morphology of MUAPs at this stage, unless the acute process is superimposed on a chronic neurogenic process. When there is severe motor unit drop out and resulting severely reduced recruitment, the remaining few MUAPs firing may be type II motor units, giving the appearance of larger MUAPs than would be expected at the same level of muscle contraction in a normal muscle.

Within 3 weeks after an acute axon loss process, increased insertional activity in the form of brief sharp spikes and positive waves or frank fibrillation may have developed, although some patients may not develop clear fibrillation potentials for up to 6 weeks after the onset of a severe axon loss process. In lesions where there are significant numbers of surviving motor axons, early collateral sprouting from surviving terminal nerve branches may be seen by about 6 weeks after an acute lesion. Expected MUAP changes include satellite potentials, increased turns, and moment-to-moment variation in MUAP shape. As denervated muscle fibers are reinnervated, fibrillation potentials are lost. The reinnervating phase may persist for months.

The NEE pattern in the chronic phase of an acute neurogenic process depends on the degree of axon loss that has occurred. In disorders where there is ongoing motor axon loss, features of ongoing reinnervation coexist with active fibrillation potentials. In processes where reinnervation is nearly complete, MUAPs evolve from a polyphasic appearance with MUAP variation to one in which duration and amplitude are increased. Reduced recruitment often persists indefinitely if nerve regeneration is not complete. In very severe lesions where reinnervation is incomplete because remaining motor nerve fibers are unable to sprout nerve branches to all denervated muscle fibers, subpopulations of small, polyphasic MUAPs can be seen with persistent marked reduced recruitment patterns. Fibrillation potentials may persist indefinitely in patchy areas of muscle even years after the acute injury.

Chronic Progressive Neurogenic Disorders

Amyotrophic lateral sclerosis typifies this group of disorders. In some patients fasciculation potentials are the first manifestation of lower motor neuron involvement. Early in the course of disease, reinnervation of muscle fibers keeps pace with denervation, so few if any fibrillation potentials are seen. MUAPs may have somewhat increased duration and amplitude.

In the middle phase of disease, motor unit drop out has occurred to the point that neurogenic MUAP changes are obvious, with associated reduced recruitment. In some patients little fibrillation is seen, but if the degree of motor nerve fiber loss is accelerated, fibrillation potentials will be evident, and there will be early features of reinnervation, including polyphasia, satellite potentials, and MUAP variation.

In the late phase of disease, fasciculation potentials are lost in severely affected muscles. As heavily reinnervated motor units drop out, severely reduced recruitment develops, and remaining MUAPs may be small or complex. In end-stage neuromuscular disorders of neurogenic or myopathic etiology, what remains in muscle is an indistinct pattern of MUAP changes sharing fragmentary features of both

TABLE 12–10	Needle Electrode Examination in Specific Disease States

Condition	Insertional Activity	Spontaneous Activity	Voluntary Activity
Neurogenic			
Immediate	Normal	Normal	Reduced recruitment
Acute	Increased	Fibrillation	Reduced recruitment
Chronic and active	Increased	Fibrillation	Reduced recruitment, MUAPs: polyphasic, enlarged, unstable
Remote		(Myotonic discharges, complex repetitive discharges)	Reduced recruitment, MUAPs: enlarged if reinnervation successful
Motor Neuron Disease			
Early	(Increased)	Fasciculation (fibrillation)	(MUAPs: enlarged)
Midstage	Increased Reduced	Fasciculation, fibrillation	Reduced recruitment, MUAPs: polyphasic, enlarged, unstable
Advanced	Increased or reduced	Fibrillation	Reduced recruitment, MUAPs: enlarged or small, polyphasic, unstable
Myopathy			
Necrotizing	Increased	Fibrillation (myotonic discharges)	(Early recruitment) MUAPs: small, polyphasic
Slow or old	(Increased)	(Fibrillation, complex repetitive discharges, myotonic discharges)	(Early recruitment) MUAPs: small (and enlarged), polyphasic
Neuromuscular junction defects			
Mild/Early	Normal	Normal	MUAPs: unstable, polyphasic
Advanced	(Increased in proximal muscles)	(Fibrillation in proximal muscles)	(Early recruitment) MUAPs: small, polyphasic, unstable

axon loss and myopathy. Muscle is often atrophic, with little or no electrical activity, although islands of dense fibrillation potentials may be seen.

Myopathy and Defects of Neuromuscular Transmission

In necrotizing myopathies such as polymyositis, disease of only a few weeks' duration are manifested only by fibrillation potentials in proximal muscles and paraspinal muscles, without significant MUAP changes. As portions of muscle fibers segmentally separated from their neuromuscular junctions by the necrotizing process are reinnervated, MUAP changes occur, characterized by satellite potentials, increased turns,

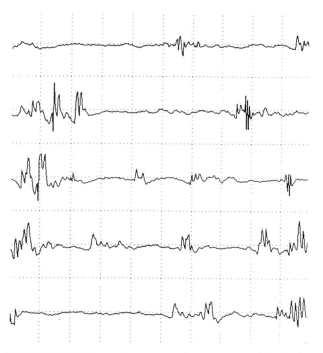

FIGURE ■ 12–13. Myopathic motor unit action potentials. Vertical division = 200 μV, horizontal division = 10 msec.

and polyphasia. MUAP changes of reduced duration and amplitude also occur when large numbers of muscle fibers are lost from the motor unit (Fig. 12–13).

In slowly progressive, non-necrotizing myopathies, such as some forms of muscular dystrophy, few if any fibrillation potentials are seen. MUAP changes are characterized more by reduction of duration and amplitude, than by polyphasia. In infants and young children, this diagnosis is difficult to make because the normal MUAP size is small compared to adults.

In patients with mild defects of neuromuscular junction transmission or with symptoms limited to ocular or bulbar muscles, repetitive stimulation studies may be normal or equivocal. The NEE examination provides an important tool for the support of the diagnosis because many muscles can be analyzed and subtle fatigability of individual muscle fibers can be seen. MUAP variation from one firing to the next is a classical feature of these disorders. Mild to moderate changes in MUAP morphology can be seen, characterized by increased numbers of turns or phases and reduction of amplitude and duration. In markedly weak muscles, a declining amplitude and fullness of the interference pattern during maximal voluntary contraction is the equivalent of severe decrement of the CMAP during repetitive stimulation. In advanced myasthenia gravis, the chronic autoimmune destructive process leaves the neuromuscular junctions scarred, disconnecting the muscle fibers from their nerve terminals, and leading to the development of fibrillation potentials. In botulism, the degree of transmission failure is often profound, resulting in marked reduction in MUAP size. Fibrillation potentials are common in botulism, the result of functional denervation of muscle fibers when acetylcholine transmission across the synapse ceases completely.

Case Studies

CASE STUDY 1

A 45-year-old man has an 8-month history of progressive weakness, characterized by difficulty arising from chairs, climbing stairs, and using his upper arms. The NEE shows fibrillation in the biceps brachii, triceps, deltoid, infraspinatus, iliacus, gluteus medius, and low cervical paraspinal muscles, associated with MUAP changes characterized by short duration, low amplitude, and polyphasia.

The NEE is typical of an active necrotizing myopathy affecting primarily proximal muscles. In this patient's age group, the most likely diagnosis is an inflammatory myopathy such as polymyositis, although the NEE pattern does not allow precise diagnosis. Other diagnostic possibilities include inflammatory myopathies associated with underlying connective tissue diseases. Less likely possible diagnoses include inclusion body myositis and toxic myopathy.

CASE STUDY 2

A 72-year-old woman has a 14-month history of slowly progressive weakness, starting in the legs and subsequently moving into the arms. She cannot dorsiflex her feet or open jars. The NEE shows prominent numbers of fibrillation potentials in distal and proximal leg muscles, in hand and forearm muscles, and in thoracic paraspinal muscles. Throughout the legs and in the arms below the elbows, there are prominent MUAP changes characterized by increased duration and amplitude. In both distal and proximal muscles of all extremities and paraspinal muscles, another population of MUAPs is characterized by polyphasia with increased duration and motor unit instability.

The description of this NEE supports a neurogenic process that is both chronic and active, manifested by both active denervation in the form of fibrillation potentials, and active reinnervation in the form of motor unit instability. Although no mention is made of fasciculation potentials, the pattern is compatible with a progressive motor neuron disease such as amyotrophic lateral sclerosis. This pattern could also be seen in the setting of a chronic progressive axon loss polyradiculopathy, or a disorder of the spinal cord affecting the anterior horns.

CASE STUDY 3

A 22-year-old man has a 1-week history of generalized weakness making ambulation, arising from a chair, and climbing stairs difficult. The nerve conduction studies are inconclusive. The NEE shows no evidence of fibrillation potentials and no changes in MUAP morphology. There is severely reduced recruitment of motor unit activation in most muscles.

This EMG pattern supports a severe neurogenic disorder of recent onset. Because the study was performed only 1 week after onset of symptoms, an acute motor axon loss process cannot be distinguished from an acute demyelinating polyneuropathy with proximal conduction blocks not appreciated on the nerve conduction studies. Although this case study represents a common presentation of Guillain-Barré syndrome, an acute axon loss polyradiculopathy or the fulminant presentation of a generalized mononeuropathy multiplex could also have this appearance.

CASE STUDY 4

A 68-year-old woman has had slowly progressive weakness for at least the past 5 years, characterized by heaviness of the legs, frequent falls, and weakness of the hands. There is recent onset of symptoms of dysphagia. The NEE shows a moderate number of fibrillation potentials in a generalized

distribution, although most prominent in the quadriceps muscles, flexor pollicis longus, and flexor carpi radialis. MUAP changes follow the same distribution, and are characterized by short duration, low amplitude, and polyphasia. There is a subpopulation of long duration, high-amplitude MUAPs in the most severely involved muscles.

This is the classical NEE pattern of inclusion body myositis, showing predilection for the quadriceps and forearm finger flexor muscles. Mixed populations of myopathic and neurogenic-appearing MUAPs are seen in muscles with advanced changes.

CASE STUDY 5

A 65-year-old man has a 1-year history of progressive proximal muscle weakness, head dropping, and dysphagia, with a tendency for symptoms to worsen through the day. Routine nerve conduction studies are normal. The NEE is normal in the extremities except for an increased number of turns in MUAPs in the deltoid and trapezius muscles, associated with motor unit instability. In cervical paraspinal muscles, there are modest numbers of fibrillation potentials associated with MUAP changes like those seen in the shoulder girdle muscles.

In some patients with myasthenia gravis, weakness is centered in proximal limb muscles, and the typical pattern of ocular and bulbar muscle fatigue is not prominent. Repetitive stimulation studies of proximal nerve trunks may be normal or inconclusive, but careful NEE of the weakest muscles will demonstrate typical findings. In the setting of marked scarring of the motor end plates in severely affected muscles, there is a disconnection of the muscle fiber from its nerve fiber, leading to the development of fibrillation potentials.

CASE STUDY 6

A 55-year-old woman has experienced muscle aching and weakness of the legs for 2 months. She was recently diagnosed with hypercholesterolemia and a lipid-reducing agent was prescribed. The NEE shows small numbers of fibrillation potentials in predominately proximal muscles and paraspinal muscles. No definite MUAP changes are seen.

In early or indolent necrotizing myopathies, typical myopathic MUAP changes may not be apparent. The only EMG feature may be increased insertional activity or modest numbers of fibrillation potentials in the most proximal muscles.

SAFETY ISSUES REGARDING THE NEEDLE ELECTRODE EXAMINATION

Aside from the local pain associated with needle electrode penetration of a muscle, the NEE is an extremely well-tolerated procedure. However, there are some contraindications to the NEE, and occasional complications arise. In most routine clinical settings informed consent is not sought before performing an EMG study.

As with any medical procedure involving needle insertion, care should be taken to prevent the spread of communicable diseases. Single-use latex gloves should be worn at all times by the electromyographer. It is ideal to employ single-use needle electrodes to minimize the risk of inadequate resterilization and to minimize the handling of used needles by staff. For nondisposable electrodes (such as single fiber electrodes), reuse requires initial cleaning of the electrode, followed by soaking in a 1:100 dilution of sodium hypochlorite (to neutralize human immunodeficiency viruses), and finally sterilization by either autoclaving or gas.

Patients with hereditary and acquired states of reduced coagulation are at risk of developing bleeding after needle penetration. In this regard, the greatest risk lies in the development of a compartment syndrome from uncontrolled bleeding in a deep muscle. An increased risk of bleeding is said to occur with a platelet count below 50,000/mm³, or a prothrombin time greater than 1.5 to 2 times the control value.[1] A limited NEE can be performed safely in these patients using a fine-gauge electrode, selecting only superficial muscles, and exercising care during needle advancement in the tissue.

In general, antibiotic prophylaxis is not instituted in patients with cardiac valvular disease. In patients with marked lymphedema related to mastectomy and radiation therapy, use of antibiotic prophylaxis may be advised.

REFERENCES

1. AAEM Professional Standards Committee: Guidelines in Electrodiagnostic Medicine. Rochester, MN, American Association of Electrodiagnostic Medicine, 1984, pp 1–12.
2. Albers JW, Allen AA, Bastron JD, et al: Limb myokymia. Muscle Nerve 4:494–504, 1981.
3. Auger R, Daube JR, Gomez MR, et al: Hereditary form of sustained muscle activity of peripheral nerve origin causing generalized myokymia and muscle stiffness. Ann Neurol 15:13–21, 1984.
4. Brick JF, Gutmann L, Brick J, et al: Timber rattlesnake venom-induced myokymia: Evidence for peripheral nerve origin. Neurology 37:1545–1546, 1987.
5. Brown WF: The Physiological and Technical Basis of Electromyography. Boston, Butterworth, 1984, pp 317–368.
6. Browne DL, Gancher ST, Nutt JG, et al: Episodic ataxia/myokymia syndrome is associated with point mutations in the human potassium channel gene, KCNA-1. Nat Genet 8:136–140, 1994.
7. Brunt ERP, Van Weerden TW: Familial paroxysmal kinesigenic ataxia and continuous myokymia. Brain 113:1361–1382, 1990.
8. Bryant SH: The electrophysiology of myotonia, with a review of the congenital myotonia of goats. In Desmedt JE (ed): New Developments in Electromyography and Clinical Neurophysiology, vol 1. Basel, Karger, 1973, pp 420–450.
9. Buchtal F: Fibrillations: Clinical electrophysiology. In Culp WJ, Ochoa J (eds): Abnormal Nerves and Muscle Generators. New York, Oxford University Press, 1982, pp 632–662.
10. Buchtal F, Guld C, Rosenfalck P: Volume conduction of the spike of the motor unit potential investigated with a new type of multielectrode. Acta Physiol Scand 38:331–354, 1957.
11. Buchtal F, Guld C, Rosenfalck P: Multielectrode study of the territory of a motor unit. Acta Physiol Scand 39:83–104, 1957.
12. Buchtal F, Schmalbruch H: Motor unit of mammalian muscle. Physiol Rev 60:90–142, 1980.
13. Burke RE, Levine DN, Tsairis P, et al: Physiological types and histochemical profiles in motor units of the cat gastrocnemius. J Physiol 234:723–748, 1973.
14. Caldron PH, Wilbourn AJ; Gold neurotoxicity and myokymia [letter]. J Rheumatol 15:528, 1988.
15. Chalk CH, Litchy WJ, Ebersold MJ, Kispert DB: Facial myokymia and unilateral basilar invagination. Neurology 38:1811–1812, 1988.
16. Costigan DA: Acquired myotonia, weakness, and vacuolar myopathy secondary to cyclosporine. Muscle Nerve 12:761, 1989.
17. Daube JR: EMG in motor neuron disease. Minimonograph #18, American Association of Electrodiagnostic Medicine, Rochester, MN, 1981.
18. Daube JR: AAEM Minimonograph #11. Needle examination in electromyography. Muscle Nerve 14:685–700, 1991.
19. Dyck PJ, Karnes J, Lais A, et al: Pathologic alterations of the peripheral nervous system of humans. In Dyck PJ, Thomas PK, Lambert EH, Bunge R (eds): Peripheral Neuropathy. Philadelphia, WB Saunders, 1984, pp 760–870.
20. Feinstein B, Lindegard B, Nyman E, et al: Morphologic studies of motor units in normal human muscles. Acta Anat 23:127–142, 1955.
21. Harner SG, Daube JR, Ebersold MJ: Electrophysiologic monitoring of facial nerve during temporal bone surgery. Laryngoscope 96:65–69, 1986.
22. Hart IK, Waters C, Vincent A, et al: Autoantibodies detected to expressed K⁺ channels are implicated in neuromyotonia. Ann Neurol 41:238–246, 1997.
23. Henneman E, Olsen CB: Relation between structure and function in the design of skeletal muscle. J Neurophysiol 28:581–598, 1965.
24. Henneman E, Somjen G, Carpenter DO: Functional significance in cell size in spinal motoneurons. J Neurophysiol 28:560–580, 1965.
25. Hjorth RJ, Willison RG: The electromyogram in facial myokymia and hemifacial spasm. J Neurol Sci 20:117–126, 1973.
26. Irani PF, Purohit AV, Wadia NH: The syndrome of continuous muscle fiber activity. Evidence to suggest proximal neurogenic causation. Acta Neurol Scand 55:273–288, 1977.
27. Isaacs H: A syndrome of continuous muscle-fibre activity. J Neurol Neurosurg Psychiatry 24:319–325, 1961.
28. Jamieson PW, Katirji MB: Idiopathic generalized myokymia. Muscle Nerve 17:42–51, 1994.
29. Kraft GH: Fibrillation potential amplitude and muscle atrophy following peripheral nerve injury. Muscle Nerve 13:814–821, 1990.
30. Kugelberg E: Activation of human nerves by hyperventilation and hypocalcemia. J Neurol Neurosurg Psychiatry 60:153–164, 1948.
31. Layzer RB: Motor unit hyperactivity states. In Vinken PJ, Bruyn GW (eds): Handbook of Clinical Neurology, vol 41, part 2. Amsterdam, North Holland, 1979, pp 295–316.
32. Lublin FD, Tsairis P, Streletz LJ, et al: Myokymia and impaired muscular relaxation with continuous motor unit activity. J Neurol Neurosurg Psychiatry 42:557–562, 1979.
33. Ludin HP: Electromyography in Practice. New York, Thieme-Stratton, 1980, p 60.
34. Ludin HP: Electromyography in Practice. New York, Thieme-Stratton, 1980, pp 124–127.
35. Mertens HG, Zschoche S: Neuromyotonie. Klin Wochenschr 43:917–925, 1965.
36. Newsom-Davis J, Mills KR: Immunological associations of acquired neuromyotonia (Isaacs' syndrome). Report of five cases and literature review. Brain 116:453–469, 1993.
37. Nielsen VK: Pathophysiology of hemifacial spasm I. Ephaptic transmission and ectopic excitation. Neurology 34:418–426, 1984.
38. Nielsen VK, Jannetta PJ: Pathophysiology of hemifacial spasm III. Effects of facial nerve decompression. Neurology 34:891–897, 1984.
39. Norris SH, Gasteiger EL, Chatfield PO: An electromyographic study of induced and spontaneous muscle cramps. Electroencephalogr Clin Neurophysiol 9:139–147, 1957.
40. Partanen JV, Danner R: Fibrillation potentials after muscle injury in humans. Muscle Nerve 5:S70–S73, 1982.
41. Partanen JV, Nousiainen U: End plate spikes in electromyography are fusimotor unit potentials. Neurology 33:1039–1043, 1983.
42. Petajan JH: Clinical electromyographic studies of diseases of the motor unit. Electroencephalogr Clin Neurophysiol 36:395–401, 1974.
43. Petajan JH: Motor unit frequency control in normal man. In Desmedt JE (ed): Progress in Clinical Neurophysiology, vol 9. Basel, Karger, 1981, pp 184–200.
44. Reeback J, Benton S, Swash M, Schwartz MS: Penicillamine-induced neuromyotonia. Br Med J 2:1464–1465, 1979.
45. Riaz G, Campbell WW, Carr J, Ghatak N: Facial myokymia in syringobulbia. Arch Neurol 47:472–474, 1990.
46. Roth G: The origin of fasciculations. Ann Neurol 12:542–547, 1982.

47. Rowland LP: Cramps, spasms, and muscle stiffness. Rev Neurol (Paris) 141:261–273, 1985.
48. Stalberg EV, Trontelj JV: Abnormal discharges generated within the motor unit as observed with single fiber electromyography. In Culp WJ, Ochoa J (eds): Abnormal Nerves and Muscles as Impulse Generators. New York, Oxford University Press, 1982, pp 443–474.
49. Stalberg EV, Trontelj JV: Single Fiber Electromyography, 2nd ed. New York, Raven Press, 1994, pp 198–199.
50. Stoehr M: Low frequency bizarre discharges. Electromyogr Clin Neurophysiol 18:147–156, 1978.
51. Stohr M: Benign fibrillation potentials in normal muscle and their correlation with end plate and denervation potentials. J Neurol Neurosurg Psychiatry 40:765–768, 1977.
52. Thesleff S: Physiological effects of denervation of muscle. Ann NY Acad Sci 228:89–103, 1974.
53. Thesleff S: Fibrillation in denervated mammalian muscle. In Culp WJ, Ochoa J (eds): Abnormal Nerve and Muscle as Impulse Generators. New York, Oxford University Press, 1982, pp 678–694.
54. Trontelj J, Stalberg EV: Bizarre repetitive discharges recorded with single fiber EMG. J Neurol Neurosurg Psychiatry 46:310–316, 1983.
55. Warmolts JR, Mendell JR: Neurotonia: Impulse-induced repetitive discharges in motor nerves in peripheral neuropathy. Ann Neurol 7:24–250, 1980.
56. Wiechers DO, Stow R, Johnson EW: Electromyographic insertional activity mechanically provoked in the biceps brachii. Arch Phys Med Rehabil 58:573–578, 1977.
57. Willison RG: Spontaneous discharges in motor nerve fibers. In Culp WJ, Ochoa J (eds): Abnormal Nerves and Muscles as Impulse Generators. New York, Oxford University Press, 1982, pp 383–392.

Chapter 13

Quantitative Electromyography and Special Electromyographic Techniques

Robert W. Shields, Jr., MD

INTRODUCTION
SINGLE FIBER ELECTROMYOGRAPHY
MACRO ELECTROMYOGRAPHY
SCANNING ELECTROMYOGRAPHY

QUANTITATIVE METHODS FOR MOTOR UNIT
  POTENTIAL MEASUREMENT
INTERFERENCE PATTERN ANALYSIS
MOTOR UNIT NUMBER ESTIMATION

INTRODUCTION

Quantitation is an integral part of the routine clinical electrodiagnostic evaluation, especially with regard to the nerve conduction studies (NCS). Amplitudes, latencies, and conduction velocities are routinely measured and quantitated for precise analysis and clinical application. In contrast to the NCS, the needle electrode examination (NEE) has traditionally remained more subjective and qualitative. Although quantitative methods for the NEE have been described and utilized since the early 1950s,[11, 12, 14] these early methods were very time-consuming and tedious and never gained wide acceptance as a standard component to the clinical electrodiagnostic evaluation. With the advent of computerization of electrodiagnostic units and the subsequent digitalization of the motor unit potential (MUP) and interference pattern (IP), interest has been rekindled in the quantitation of the NEE. Subsequently, a wide variety of quantitative methods has been devised to provide more precise assessment of the MUP and IP. The term *quantitative electromyography* (EMG) has traditionally been applied to those methods. In addition, various quantitative special techniques such as single fiber EMG[140] and macro EMG[115] have been devised to provide additional information regarding the microphysiology of the motor unit, including the assessment of neuromuscular transmission. More recently, techniques have been developed to assess the number of motor

units within a muscle.[19, 26, 74] These various quantitative EMG methods and special techniques have been regarded as complementary methods that supplement the information obtained on the routine NCS and NEE. Some of these techniques have remained primarily investigational, whereas others have developed into very useful diagnostic clinical tools. This chapter describes many of these important quantitative methods and special techniques.

SINGLE FIBER ELECTROMYOGRAPHY

Single fiber EMG (SF EMG) is a special technique first described by Ekstedt and Stålberg in 1963.[33] This technique is capable of recording action potentials from single muscle fibers during voluntary activation of human muscle. SF EMG has evolved over the years into a very important special technique of great value in the understanding of the physiology of muscle fibers and motor units.[36, 140] SF EMG has also been applied to the study of a wide range of neuromuscular disorders, but it has been of particular clinical value in the study of disorders of neuromuscular transmission.[37, 58, 106, 130]

SF EMG requires a specialized needle electrode that utilizes a 25-μm diameter wire as the recording electrode, embedded in a small circular resin port on the side of the electrode shaft, approximately 7.5 mm from its tip[140] (Fig. 13–1). The shaft of the electrode serves as the reference

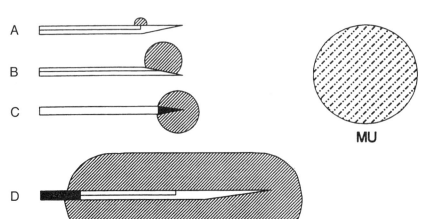

MU

FIGURE 13–1. Different types of electromyographic electrodes illustrating their recording uptake areas (shaded) compared to the average normal motor unit territory (MU). *A.* SF EMG electrode. *B.* Concentric needle electrode. *C.* Monopolar electrode. *D.* Macro EMG electrode. (Reprinted from Stålberg E, Nandedkar SD, Sanders DB, et al: Quantitative motor unit potential analysis. J Clin Neurophysiol 13:401–22, 1996, with permission from Lippincott-Williams & Wilkins Publishers.)

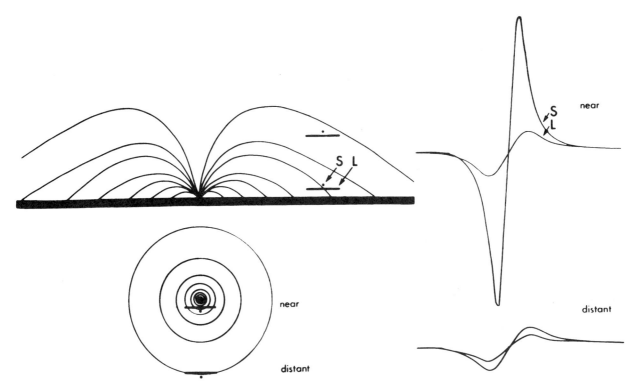

FIGURE 13–2. The electrical field generated by a single muscle fiber action potential and the corresponding potentials measured from a small (S) and a large (L) electrode surface. The large electrode shunts the isopotential lines at short fiber electrode distances, but less so at longer distances. Thus, at a short distance, the large electrode records a lower amplitude than does the small electrode. At distant sites, both electrodes record low amplitude responses. (Reprinted from Stålberg E, Trontelj JV: Single Fiber Electromyography in Healthy and Diseased Muscle, 2nd ed. New York, Raven Press, 1994, with permission from Lippincott-Williams & Wilkins Publishers.)

electrode. The small recording surface of 25 μm is needed in order to selectively record maximum amplitudes from the isopotential lines generated by a muscle fiber action potential[35, 36, 140] (Fig. 13–2). In addition to the small size of the recording electrode, it is essential that the electrode be placed in close proximity to the muscle fiber as the amplitude of the potential attenuates dramatically when it is moved only 100 μm away from the muscle fiber[36, 140] (Fig. 13–2). Because single muscle fiber action potentials are in the range of a few milliseconds in duration, a relatively high setting for the low-pass (high-frequency) filter is needed to avoid waveform distortions.[140] By convention, the low-pass filter is typically set at 8 to 10 KHz. In addition, it is necessary to use a relatively high setting for the high-pass (low-frequency) filter of approximately 500 Hz in order to filter out low-frequency waveforms, which are volume conducted from nearby muscle fibers and which would otherwise contribute to excessive noise and interference.[140] In addition to the special electrode and filter settings, a delay line with a threshold trigger is needed to display the single muscle fiber action potentials during successive discharges. Sweep speeds are typically set at 2 or 5 msec per division.

Standard SF EMG is performed in muscles during modest levels of uniform voluntary contraction. The SF EMG electrode is randomly placed in the muscle with small adjustments made in its position until a single muscle fiber action potential is recorded. The experienced single fiber electromyographer utilizes visual cues from the potentials on the visual display and the acoustical information that is amplified to the loudspeaker of the EMG unit, to make fine adjustments in the placement of the electrode to optimize the recordings. This is especially true in the course of per-

forming jitter analysis when one is seeking to record from two single muscle fibers belonging to the same motor unit, a potential pair. The trained ear can immediately recognize the double potential as a typical cracked echo-like sound, which has been likened to the sound of castanets.[140] Once a single muscle fiber action potential is recorded, small adjustments are made in the positioning of the electrode, until the largest possible amplitude of the potential is recorded. Criteria for identifying a single muscle fiber action potential include a peak-to-peak amplitude greater than 200 μV, a rise time of less than 300 μsec, and a stable waveform during successive discharges.[31] Typically, the spike rise time is between 67 and 200 μsec (median, 112 μsec),[36] depending on the proximity of the recording electrode to the muscle fiber. The total spike duration of a typical single muscle fiber action potential ranges from 265 to 800 μsec (median, 470 μsec) and the spike voltage ranges from 0.7 to 25.2 mV (median, 5.6 mV).[31]

There are two principle SF EMG parameters, fiber density and jitter. Fiber density is an index of the number of muscle fibers that comprise the motor unit.[46, 138, 140] Fiber density is determined by measuring 20 different single muscle fiber action potentials. Typically, four or five measurements are made in four or more sites within the muscle. The technique is to advance the SF EMG electrode into a muscle during weak voluntary contraction to record a single muscle fiber action potential. During weak uniform contraction, numerous MUPs will be discharging successively at relatively regular rates. When a single muscle fiber action potential is recorded, it will discharge each time the entire motor unit discharges. Fine adjustments of the single fiber electrode are made until the single muscle fiber action potential demonstrates a maximum amplitude. On occasion, additional time-locked single muscle fiber action potentials belonging to the same motor unit will be recorded simultaneously. When the triggering muscle fiber action potential is optimized, that potential plus all of the other time-locked potentials that can be recorded with each successive discharge of the MUP are then counted.[138, 140] This represents one recording of a fiber density determination. It has been estimated that the recording radius of the SF EMG electrode is approximately 300 μm[140] (Fig. 13–3). Thus, fiber density measurements record all of the muscle fibers belonging to the same motor unit within a 300-μm radius. A total of 20 fiber density measurements are made at different sites within the muscle. The total number of single fiber muscle action potentials recorded at these 20 different sites is added and divided by 20 to determine the mean fiber density.[138, 140] Fiber density values vary from muscle to muscle and tend to increase with age, particularly for muscles in the lower extremity.[7, 47, 140] Approximately 60% to 70% of the time, a fiber density measurement will record only a single muscle fiber action potential.[138] This is because, in normal muscle, approximately 76% of all muscle fibers belonging to a motor unit, are separated from other muscle fibers of that motor unit.[5] Another parameter that can be derived from the fiber density recordings is the mean of the interspike intervals (MISIs). This represents the total duration of the potentials that are time-locked to the triggering potential, divided by the number of intervals (number of components minus one)

within the potential complex.[140] The MISI may be increased in early reinnervation and in myopathies, particularly muscular dystrophies and polymyositis.[140]

The other principle SF EMG parameter is jitter. Jitter refers to the variability of the interpotential interval (IPI) between two single muscle fiber action potentials belonging to the same motor unit during successive discharges.[129, 140] The strategy for recording jitter measurements is somewhat different from that for fiber density measurements (see Fig. 13–3). Therefore, it is desirable that fiber density measurements and jitter measurements be made at two different times during the SF EMG examination. For jitter measurements, at least two time-locked single muscle fiber action potentials belonging to the same motor unit, a potential pair, must be recorded.[129] For this technique, the triggering single muscle fiber action potential does not necessarily need to be of maximum amplitude. Instead, the SF EMG electrode is adjusted in such a way as to be able to record from two or more time-locked single fiber muscle action potentials simultaneously (see Fig. 13–3). The triggering potential will serve to initiate the sweep of the display with the other action potential recorded in relationship to the triggering potential. A minimum of 50 stable successive discharges are recorded, and the IPI between the triggering potential and the second potential is recorded for each discharge (Fig. 13–4). The variability of the IPI represents the physiological jitter.[129, 140] There are at least five anatomical and physiological factors that contribute to the IPI. These include the

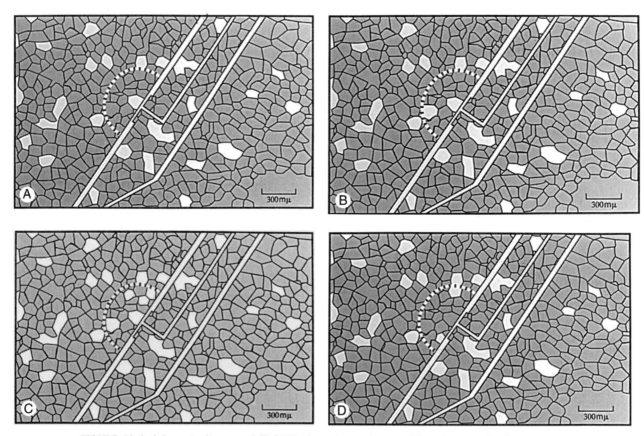

FIGURE 13–3. Schematic diagram of SF EMG electrode uptake area. The white muscle fibers represent muscle fibers belonging to a single motor unit. A. The single fiber electrode is positioned to record a maximum amplitude from a single muscle fiber. In this case, only a single muscle fiber action potential is recorded. B. In this setting, the single fiber electrode records two muscle fiber action potentials belonging to the same motor unit; that is, a fiber density measurement of two. C. In this setting, the SF EMG electrode records five muscle fibers belonging to the same motor unit; that is, a fiber density measurement of five. D. Strategy for recording jitter. Note that the single fiber electrode is positioned to record from two muscle fibers belonging to the same motor unit, although neither fiber may disclose a maximum amplitude.

conduction velocity.[36] If the duration of the IPI is 1 msec or less, and if the firing rate of the motor unit under study is relatively uniform, then most of the variability of the IPI is related to variation of neuromuscular transmission.[140] On the other hand, if the IPI is in the order of 10 msec or longer, and the discharge rate is variable, then the jitter may be influenced to a greater extent by variation of conduction velocity along the distal axon branch and variation of muscle fiber conduction velocity.[140] Thus, it is not desirable to record jitter with IPIs of greater than 4 msec.[140] Controlling for the variability of MUP discharges is more problematic. The mechanism of increased jitter with variable MUP discharge rates involves variability of muscle fiber conduction velocity owing to the velocity recovery function (VRF)[113, 140] (Fig. 13–5). A reduced propagation velocity is seen for an impulse occurring after an interval of 3 to 10 msec, with reduction of approximately 20% for the shortest intervals.[113] A supernormal velocity is found at intervals between 6 and 1000 msec, with a peak at approximately 10 msec intervals[80, 113] (see Fig. 13–5). Thus, nonuniform intervals of motor unit discharging place the muscle fiber conduction velocity along this variable curve, and result in greater variability of the IPI due to changing muscle fiber conduction velocities.[140] This phenomenon can be assessed by utilizing the mean sorted data difference (MSD) calculation, discussed later.

As previously noted, variability of neuromuscular transmission time is the major contributing factor to the jitter. The mechanism of variable neuromuscular transmission time relates to the size and slope of the excitatory postsynaptic potential (EPP) generated at the neuromuscular junction. In normal neuromuscular junctions, the EPP is always above threshold and a muscle fiber action potential is always generated. However, even in the normal neuromuscular junction, the slope of the EPP may vary to some extent, and this results in a slight variation in neuromuscular transmission time (Fig. 13–6). In disorders of neuromuscular transmission, particularly in myasthenia gravis, the amplitude of the EPP will sometimes fall below threshold and neuromuscular transmission block will occur. In addition, the slope of the EPP may be more variable and will contribute to increased variability of neuromuscular transmission time and subsequently increased jitter (see Fig. 13–6).

Most modern computer-based EMG units have single fiber EMG programs that facilitate recording and collecting the SF EMG data, as well as calculation of the jitter. Although standard deviation around a mean IPI value could be used to express variability of the IPI, slow trends that are frequently superimposed on the single fiber recordings can distort the accuracy of the standard deviation as an index of jitter.[140] These slow trends include variable degrees of slowing of propagation velocity along the two muscle fibers during continuous discharges, as well as alterations in potential shape that may result from slight movement or displacement of the electrode during the recordings.[101, 140] The preferred way of expressing variability of the IPI is by calculating the mean value of consecutive differences (MCD).[129] In a Gaussian distribution, MCD equals 1.13 standard deviation.[129] The MCD value tends to measure variability over short periods of the recording and is refractory to being influenced by slow trends.[129, 140] The equation for calculating MCD[32, 129] is:

$$MCD = \frac{(IPI_1 - IPI_2) + (IPI_2 - IPI_3) + .. + (IPI_{n-1} - IPI_n)}{n-1}$$

Another method for measuring variability of the IPI is the calculation of the MSD.[140] This is particularly useful in assessing the impact of variation of motor unit firing rates on the jitter as measured by the MCD.[140] In this method, the IPIs are sorted in increasing order according to the preceding

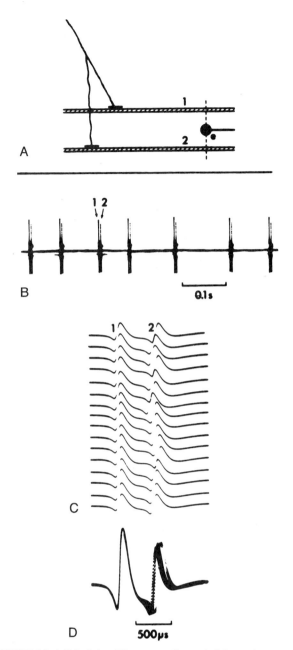

FIGURE 13–4. Principle of jitter recordings. A. Schematic presentation of the recording conditions for jitter measurement in voluntarily activated muscle. The SF EMG electrode is positioned to record from a potential pair (muscle fibers 1 and 2) innervated by the same motor axon. B. Potential pair firing at a low degree of voluntary effort. C. The potential pair recorded at a faster sweep speed utilizing a sweep trigger to position the first potential. The second potential occurs at a varying interval. D. Superimposed successive discharges of the potential pair illustrating the variability of the interpotential interval (the neuromuscular jitter). (Reprinted from Stålberg E, Trontelj JV: Single Fiber Electromyography in Healthy and Diseased Muscle, 2nd ed. New York, Raven Press, 1994, with permission from Lippincott-Williams & Wilkins Publishers.)

length of the distal axon branch from the motor axon to the end plate, conduction velocity along the distal axon branch, neuromuscular transmission, distance of the end plate to the recording electrode, and conduction velocity of the muscle fiber.[36] Of these five factors, only three may contribute to variability of the IPI: conduction velocity along the distal axon branch, neuromuscular transmission, and muscle fiber

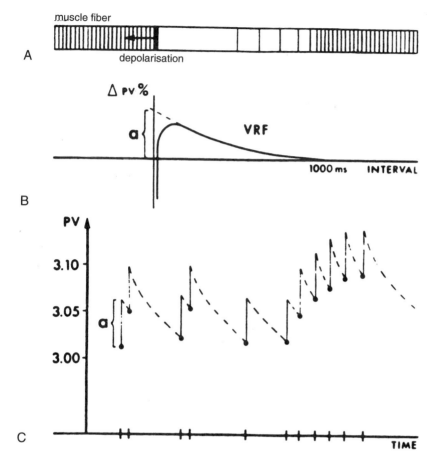

FIGURE 13–5. Propagation velocity recovery function in a muscle fiber following the passage of an impulse. The muscle fiber is initially refractory, after which it propagates a succeeding impulse at subnormal and later supernormal velocity (A). This change in propagation velocity can be illustrated using double pulse stimulation (B). The increment typically varies between 0 and 15% in normal muscle. The remaining effect of preceding impulses is added to subsequent impulses, and at irregular activation rates (C), the propagation velocity for successive discharges varies. (Reprinted from Stålberg E, Trontelj JV: Single Fiber Electromyography in Healthy and Diseased Muscle, 2nd ed. New York, Raven Press, 1994, with permission from Lippincott Williams & Wilkins Publishers.)

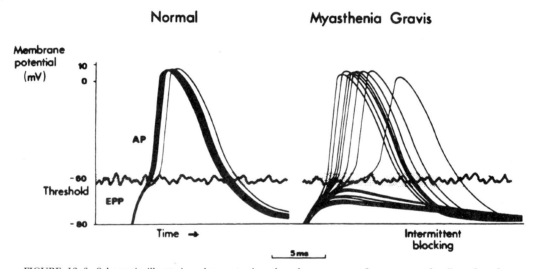

FIGURE 13–6. Schematic illustration demonstrating the phenomenon of neuromuscular jitter based on intracellular recording from human intercostal muscle fiber. In the normal muscle fiber the variation in neuromuscular transmission time is caused by a fluctuation of the membrane potential at which the muscle fiber becomes depolarized with a suprathreshold end plate potential (EPP), as well as to the variation of the EPP amplitude and slope. In the muscle fiber from the myasthenia gravis patient, the EPPs are abnormally low in amplitude and less steep. Some of the EPPs fail to reach threshold and neuromuscular block occurs. The EPPs that reach threshold result in increased variability of neuromuscular transmission time. (Original diagram of the myasthenia gravis recording is reproduced from D. Elmqvist, et al: An electrophysiological investigation of neuromuscular transmission in myasthenia gravis. J Physiol 174:417–434, 1964, with permission from Cambridge University Press. The modified figure in its entirety is reproduced from E. Stålberg and J. V. Trontelj: Single Fiber Electromyography, Studies in Healthy and Diseased Muscle, 2nd ed., 1994, with permission from Lippincott Williams & Wilkins Publishers.)

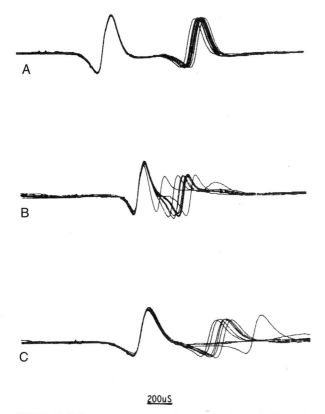

FIGURE 13–7. Jitter measurements in myasthenia gravis illustrating *(A)* normal jitter (MCD = 28 μsec), *(B)* increased jitter (MCD = 88 μsec), *(C)* increased jitter with blocking (MCD = 244 μsec). (Reprinted from Sanders DB, Howard JF: AAEM Minimonograph #25: Single-fiber electromyography in myasthenia gravis. Muscle Nerve 9: 809–819, 1986, with permission from John Wiley & Sons Publishers.)

interdischarge interval of the motor unit. The MCD is then calculated on this sorted data and the ratio of the MCD to the MSD is determined. If this ratio exceeds 1.25, then variability from irregularity of the interdischarge intervals is contributing to the jitter measurement as calculated by the

MCD, and in such a case, the MSD value should be used instead of the MCD value.[140] In most modern EMG units, SF EMG programs automatically calculate the MSD and report the MCD/MSD ratio for each potential pair measured.

In normal muscles, jitter tends to vary from muscle to muscle and generally tends to increase with age, particularly in muscles of the lower extremity.[7, 47] Jitter may also be influenced by other factors, especially temperature. Between 32 and 36°C each degree of cooling may add 2 to 3 μsec of jitter.[129] However, when a muscle is very cool, the impact is even greater with each degree of cooling below 32°C adding 7.5 μsec to the jitter.[129] Thus, it is important to maintain the muscle at a normal warm physiological temperature for accurate jitter measurements.

Rarely in normal subjects, but much more frequently in patients with neuromuscular disease, impulse blocking may be observed. Blocking refers to the failure of one or more time-locked single muscle fiber action potentials to fire in the triggered discharge[140] (Fig. 13–7). This typically occurs in disorders of neuromuscular transmission and may be observed when jitter measurements exceed 100 μsec MCD.[140] On occasion, exceptionally low jitter, in the vicinity of 5 μsec MCD is observed in normal subjects.[34] This usually implies that the recording was taken from a split fiber.[34, 55, 129] Split fibers and the resulting very low jitter measurements are much more commonly encountered in muscle disease, particularly muscular dystrophies.[55] Clearly, these low jitter values should not be used in the determination of the mean jitter. Another unusual pattern of jitter is referred to as "bimodal jitter" or the "flip-flop phenomenon"[146] (Fig. 13–8). This occurs when the jitter is distributed around two separate means. This type of recording may be encountered in 1 out of every 20 jitter measurements in normal muscle.[146] When bimodal jitter is encountered, jitter measurements may be taken from one of the two distributions or the potential may be discarded and not analyzed at all.[146] Failure to recognize bimodal jitter may result in incorrect measurement of excessive jitter in an otherwise normal potential pair. The precise mechanism of the bimodal jitter phenomenon is unclear, but in normal muscle with a short IPI of 1 msec or less it may be related to subdepolarization of the second fiber by the electrical field of the first fiber, resulting in altered propagation velocity for some of the discharges.[140] It is also possible that the action potential of the first fiber

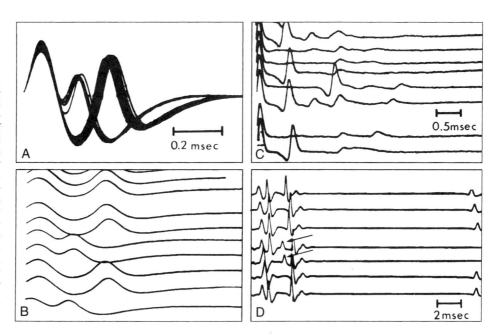

FIGURE 13–8. Bimodal jitter in different neuromuscular disorders. *A* and *B.* The same potential pair from a patient with myasthenia gravis. Note the distribution of jitter around two separate means *(A). C.* Recordings from a patient with muscular dystrophy. Latency change is 1 msec and there are blockings of the jumping potential. *D.* Recordings from a patient with progressive spinal muscular atrophy. The last spike component occasionally appears 16 msec earlier *(arrows).* (Reprinted from Thiele B, Stålberg E: The bimodal jitter: a single fibre electromyographic finding. J Neurol Neurosurg Psychiatry 37:403–411, 1974, with permission from the BMJ Publishing Group.)

summates with the end plate potential of the second fiber.[140] In neuromuscular disorders, particularly with reinnervation, bimodal jitter can occur with IPIs of up to 16 msec. In this setting, the mechanism of the bimodal jitter may be caused by uncertain impulse conduction in nerve twigs, or the existence of two motor end plates[140, 146] (see Fig. 13–8). Another normal phenomenon that may be misinterpreted as a potential pair with excessive jitter or blocking is the recording of a false double potential. This occurs when a triggering potential is followed by a positive monophasic potential which may show excessive jitter and blocking.[140] This does not represent a true potential pair, as the second potential is generated by the triggering muscle fiber as a result of damage to the membrane by the single fiber electrode.[140] Obviously, these types of recordings do not represent potential pairs and should not be measured. Normal values for both fiber density and jitter measurements have been published.[140] As noted earlier, these values vary from muscle to muscle and frequently are influenced by age. More recently, a multicenter collaborative study reported normal reference values for fiber density and jitter measurements in numerous muscles of subjects spanning seven to nine decades of age[7, 47] (Table 13–1).

The SF EMG examination may be abnormal if the fiber density is increased beyond the normal range, or if the jitter is abnormal. Jitter may be abnormal if the mean jitter exceeds the upper limit for that particular muscle, or if more than 10% of the potential pairs have increased jitter or blocking.[140] Typically, a complete SF EMG examination consists of a fiber density measurement (usually performed at the beginning of the study independent of the jitter measurements) and a jitter analysis, which typically encompasses 20 potential pairs. Clearly, if three or more of the initial potential pairs examined disclose excessive jitter or blocking, or

both, a conclusion could be drawn that the SF EMG examination is abnormal.

In addition to the previously outlined method, SF EMG measurements can also be obtained via electrical stimulation of motor axons rather than with voluntary activation.[140, 141, 152, 155, 156] This technique has been termed stimulation SF EMG.[140] The obvious advantage of the electrical stimulation technique is that it can be accomplished in patients who are otherwise unable to provide a sustained uniform voluntary contraction of the muscle under study. Thus, this technique could be utilized in children, adult patients who are unable to cooperate, and in patients who have rather severe weakness. There are other inherent advantages to this technique, including the possibility of recording from numerous end plates during a signal stimulation and the option of varying the rate of stimulation. In addition, the rate of discharge is uniform, eliminating considerations of irregular discharge rates contributing to the jitter. However, there are numerous potential pitfalls and drawbacks regarding this technique, as discussed next.

The technique requires the placement of a monopolar needle electrode several millimeters proximal to the apparent motor point region of the muscle[140] (Fig. 13–9). This will function as the cathode. The motor point may be identified by using a surface stimulator and finding the region of the muscle that has the lowest stimulation threshold for twitch. The anode may be a similar monopolar electrode inserted subcutaneously 15 to 25 mm laterally from the site of the cathode.[140] Alternatively, a surface electrode can also be used as an anode. When both electrodes are in place, the stimulator may be adjusted to elicit twitches in a small portion of the muscle, typically causing a fine jerking of the stimulating needle.[140] This implies that only small muscular branches are being stimulated. A stronger, forceful twitch evoked by

TABLE 13–1	Single Fiber EMG Reference Values

Jitter Values (μsec): Mean MCD (Mean Consecutive Difference)
95% Upper Confidence Limit of Normal/95% Upper Confidence Limit of Normal for Single Fiber Pairs

Muscle	10 yr	20 yr	30 yr	40 yr	50 yr	60 yr	70 yr	80 yr	90 yr
Frontalis	33.6/49.7	33.9/50.1	34.4/51.3	35.5/53.5	37.3/57.5	40.0/63.9	43.8/74.1		
Obicularis oculi	39.8/54.6	39.8/54.7	40.0/54.7	40.4/54.8	40.9/55.0	41.8/55.3	43.0/55.8		
Obicularis oris	34.7/52.5	34.7/52.7	34.9/53.2	35.3/54.1	36.0/55.7	37.0/58.2	38.3/61.8	40.2/67.0	42.5/74.2
Tongue	32.8/48.6	33.0/49.0	33.6/50.2	34.8/52.5	36.8/56.3	39.8/62.0	44.0/70.0		
Sterncleidomas	29.1/45.4	29.3/45.8	29.8/46.8	30.8/48.8	32.5/52.4	34.9/58.2	38.4/62.3		
Deltoid	32.9/44.4	32.9/44.5	32.9/44.5	32.9/44.6	33.0/44.8	33.0/45.1	33.1/45.6	33.2/46.1	33.3/46.9
Biceps	29.5/45.2	29.6/45.2	29.6/45.4	29.8/45.7	30.1/46.2	30.5/46.9	31.0/48.0		
Ext dig comm	34.9/50.0	34.9/50.1	35.1/50.5	35.4/51.3	35.9/52.5	36.6/54.4	37.7/57.2	39.1/61.1	40.9/66.5
Abd digiti V	44.4/63.5	44.7/64.0	45.2/65.5	46.4/68.6	48.2/73.9	51.0/82.7	54.8/96.6		
Quadriceps	35.9/47.9	36.0/48.0	36.5/48.2	37.5/48.5	39.0/49.1	41.3/50.0	44.6/51.2		
Ant tibialis	49.4/80.0	49.3/79.8	49.2/79.3	48.9/78.3	48.5/76.8	47.9/74.5	47.0/71.4	45.8/67.5	44.3/62.9

Fiber Density Values: Mean Fiber Density 95% Upper Confidence Limit of Normal

Muscle	10 yr	20 yr	30 yr	40 yr	50 yr	60 yr	70 yr	80 yr	90 yr
Frontalis	1.67	1.67	1.68	1.69	1.70	1.73	1.76		
Tongue	1.78	1.78	1.78	1.78	1.78	1.79	1.79		
Sterncleidomas	1.89	1.89	1.90	1.92	1.96	2.01	2.08		
Deltoid	1.56	1.56	1.57	1.57	1.58	1.59	1.60	1.62	1.65
Biceps	1.52	1.52	1.53	1.54	1.57	1.60	1.65	1.72	1.80
Ext dig comm	1.77	1.78	1.80	1.83	1.90	1.99	2.12	2.29	2.51
Abd digiti V	1.99	2.00	2.03	2.08	2.16	2.28	2.46		
Quadriceps	1.93	1.94	1.96	1.99	2.05	2.14	2.26	2.43	
Ant tibialis	1.94	1.94	1.96	1.98	2.02	2.07	2.15	2.26	
Soleus	1.56	1.56	1.56	1.57	1.59	1.62	1.66	1.71	

Recommended criteria for an abnormal study: Jitter is abnormal if either: (1) value for mean MCD of 20 fiber pairs greater than the 95% upper confidence limit; or (2) jitter values in more than 10% of pairs is greater than the 95% upper confidence limit for action potential pairs. Fiber density is abnormal if mean value of 20 observations is greater than 95% of upper confidence limit.
Reprinted from Bromberg MD, Scott DM, Ad Hoc Committee of the AAEM Single Fiber Special Interest Group: Single fiber EMG reference values: Reformatted in tabular form. Muscle Nerve 17.820–821, 1994, with permission from John Wiley & Sons Publishers.

Stimulation

+ –

SFEMG Recording

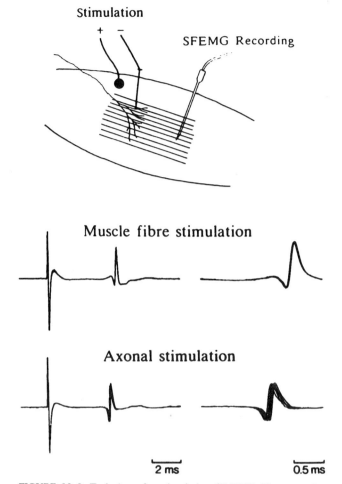

Muscle fibre stimulation

Axonal stimulation

2 ms   0.5 ms

FIGURE 13–9. Technique for stimulation SF EMG. Upper tracing shows low jitter resulting from direct muscle fiber stimulation. Lower tracing discloses normal jitter with axonal stimulation. (Reprinted from Stålberg E, Trontelj JV: Single Fiber Electromyography in Healthy and Diseased Muscle, 2nd ed. New York, Raven Press, 1994, with permission from Lippincott-Williams & Wilkins Publishers.)

stimulation should be avoided, as this tends to indicate a major motor branch is being activated and would result in the discharge of an excessive number of motor units, which would interfere with discrete single muscle fiber recordings.[140] Amplitudes of 0.5 to 10 mA for constant current stimulators are usually sufficient to provide a suprathreshold stimulus.[140] A standard SF EMG electrode is then placed in the muscle distal to the stimulation site, typically 20 mm away. The position of this electrode is adjusted to record the single muscle fiber action potentials that are being generated by the intramuscular stimulation. Further adjustments in both stimulating and recording electrodes are sometimes required to provide optimal recordings. The frequency of stimulation may be varied, but most recordings begin at low frequencies of 2 to 3 Hz. Once the recording is optimized, the stimulus rate may be increased to 10 Hz, which is close to the firing rates at which voluntarily activated jitter is usually measured.[140] It is critical that the stimulation intensity be properly adjusted to provide an accurate jitter measurement. As the strength of the stimulus is increased, there will be a progressive reduction in both jitter as well as latency from the stimulus to the single muscle fiber action potential. The stimulus intensity therefore must be increased to a level at which no further latency shifts or reductions in jitter are detected.[140] Often, several single muscle fiber action poten-

tials can be recorded at a single time, and the jitter measurements may be measured from each recording. However, it is essential that the intensity of the stimulation be adjusted so that each single muscle fiber action potential demonstrates no further alteration in latency or jitter measurement.

Because jitter is measured from the stimulus artifact to the muscle fiber action potential, the jitter measurements for the electrical stimulation technique differ from that of voluntarily activated SF EMG. Jitter measurements from electrical stimulation are approximately 5 μsec less than those derived from voluntary activation.[140] Normal values for individual muscle fibers typically range between 5 and 40 μsec.[140] However, if the variation in latency is very small, less than 4 μsec, one must suspect that direct muscle fiber stimulation has occurred; that is, that the stimulating electrode is directly stimulating a muscle fiber which would, of course, produce a single muscle fiber potential with extremely low jitter.[140, 152] Failure to recognize direct muscle fiber stimulation would clearly result in significant underestimation of the mean jitter. It is recommended that jitter measurements of 4 μsec or less be discarded.[140] Normal values for the extensor digitorum communis (EDC) and orbicularis oculi have been obtained for electrical stimulation techniques.[140, 151] For other muscles, multiplying the normal value for jitter for voluntary activation by a conversion factor of 0.8 approximates the theoretical relationship between voluntarily activated jitter analysis and that from electrical stimulation.[101]

SF EMG has been applied to a wide range of neuromuscular disorders. One of the most important applications has been in the diagnosis of disorders of neuromuscular transmission, in particular, myasthenia gravis.[99, 101] In myasthenia gravis, fiber density is usually normal but the mean jitter is increased, frequently with blocking, and there is often a significant increase in the percentage of abnormal potential pairs (i.e., pairs with excessive jitter or blocking, or both). In one study, 86% of all patients with myasthenia gravis had abnormalities on the SF EMG examination of the EDC.[99] However, it has been observed that a higher diagnostic yield may be obtained when the study of the frontalis or other facial muscle is incorporated in the examination.[101]

SF EMG is a much more sensitive method for the detection of the neuromuscular transmission defect in myasthenia gravis than repetitive nerve stimulation (RNS). In a large series of 550 myasthenia gravis patients who had RNS recording from a shoulder and hand muscle, RNS was abnormal at one site in 76% of patients with generalized disease and in 48% of patients with ocular disease.[101] However, SF EMG of the EDC was abnormal in 89% of patients with generalized disease and in 60% with ocular disease.[101] When facial muscles were added, SF EMG was abnormal in 99% of patients with generalized disease and 97% of patients with ocular disease. Thus, it has been recommended that if SF EMG of a limb muscle is normal, examination of the frontalis or orbicularis oculi muscle should be done before concluding that the SF EMG examination is indeed negative.[101] In serial SF EMG studies in myasthenia gravis, it has been observed that a fall of 10% in mean jitter often correlates with clinical improvement, and conversely, an increase of 10% in mean jitter is usually associated with worsening of the condition.[99] SF EMG tends to remain abnormal even in patients during clinical remission.[99]

Using stimulated SF EMG recordings in myasthenia gravis, it is possible to demonstrate increasing jitter with higher rates of sustained stimulation.[59, 154] However, this is not uniformly present in all potential pairs.[154] Although fiber density is usually normal in most patients with myasthenia gravis, in one study of 71 patients, the fiber density was significantly increased.[57] This was not related to the severity or duration of the disease, but was associated with treatment with anticholinesterase drugs.[57] It was suggested that the increased fiber density resulted from denervation and reinnervation caused by severe involvement of the motor endplates,[57] but a direct neurotoxic effect of anticholinesterase drugs on motor end plates may also be a factor.[38]

SF EMG has also been a valuable aid in the diagnosis of Lambert-Eaton myasthenic syndrome (LEMS).[107, 140] The SF EMG findings are similar to those seen in myasthenia gravis except that in LEMS, jitter tends to be more significantly increased.[107, 140] These findings occur not only in clinically weak muscles, but may be quite prominent in muscles that do not appear to be involved clinically. Utilizing stimulation SF EMG, a reduction of jitter and blocking may be observed as the stimulation rate is increased, particularly when the rate exceeds 10 Hz.[98, 140, 154] When a number of end plates deviate from these patterns and in fact show patterns typical of myasthenia gravis, it may indicate an overlap syndrome,[106] in which antibodies to both calcium channels and acetylcholine receptors are noted.[95] This combined pattern, however, must be interpreted with great caution, as jitter and blocking may also decrease with higher rates of stimulation in some end plates in patients with myasthenia gravis.[154] Fiber density in LEMS may be mildly increased, possibly owing to type II fiber grouping, secondary to the type II fiber atrophy that has been noted in this condition, as well as to functional denervation of the post-synaptic endplate due to severe presynaptic blockade.[140]

SF EMG has also been used to study the neuromuscular transmission abnormality seen in botulism.[104, 140] In botulism, jitter values are usually very high and blocking is prominent.[104] With higher rates of voluntary activation, improvement in these SF EMG abnormalities may be noted similar to that seen in LEMS.[140] Similar SF EMG findings have been observed in patients treated with botulinum toxin injection.[100] These findings are observed in muscles distant from the site of the injection and may be observed for up to 8 months following the injection.[140] Fiber density in botulism tends to increase, again most likely owing to denervated motor end plates from presynaptic blockade.[140]

SF EMG has also been used to study a wide range of neurogenic disorders. One of the most important and early SF EMG observations was made in autogenous facial muscle transplants in humans.[50] The SF EMG findings disclosed increased jitter during the first 3 months after transplantation, followed by a subsequent increase in fiber density.[50] These observations suggest that newly formed neuromuscular junctions created by reinnervation may function imperfectly during the first 3 months of reinnervation, and may result in excessive jitter or blocking. These findings have been used to explain many of the SF EMG findings noted in disorders that produce denervation and reinnervation.

Reinnervation may occur from distal sprouting of surviving axons, which innervate adjacent muscle fibers, or from regeneration of motor axons from a point of proximal injury or transection. The process of reinnervation results in remodeling of MUPs, oftentimes with increased duration and amplitude, and in some occasions, increased phases (polyphasic MUPs). The reinnervation process increases the number of muscle fibers that may belong to a single motor unit, and thus results in an increase in the fiber density[105, 108–110, 116, 120, 140, 145] (Fig. 13–10). Increased fiber density following reinnervation correlates with histological findings of fiber type grouping.[105] Once a collateral branch or axon reinnervates muscle, the nerve terminal matures over a period of 1 to 3 months. During that time, jitter and blocking may occur.[50, 108, 120] By 6 months, the neuromuscular junction is mature and jitter may normalize and blocking is no longer observed.[50, 108] However, in some circumstances, imperfect neuromuscular transmission may persist indefinitely.[108] A special type of blocking is sometimes encountered when collateral sprouting results in numerous neuromuscular junctions supported by a single motor axon branch.[137] During the early stages of this type of reinnervation, blocking of transmission at an axonal branch point may be manifested by simultaneous blocking of two or more muscle fiber action potentials[137, 140] (Fig. 13–11). This pattern of blocking, termed *concomitant blocking* or *neurogenic blocking*, is differentiated from

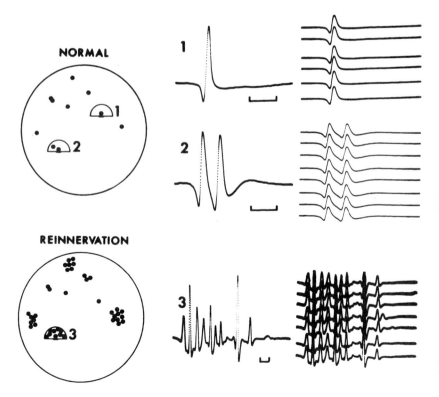

FIGURE 13–10. SF EMG recordings illustrating fiber density measurements in normal and reinnervated muscle. The black dots indicate muscle fibers belonging to the same motor unit, and the semicircular region illustrates the recording territory of the SF EMG electrode. In normal muscle (1 and 2), only one or two muscle fibers are recorded during the fiber density measurement. However, with reinnervation (3), some recordings may disclose numerous muscle fibers, resulting in an increased fiber density. (Reprinted from Stålberg E, Trontelj JV: Single Fiber Electromyography in Healthy and Diseased Muscle, 2nd ed. New York, Raven Press, 1994, with permission from Lippincott-Williams & Wilkins Publishers.)

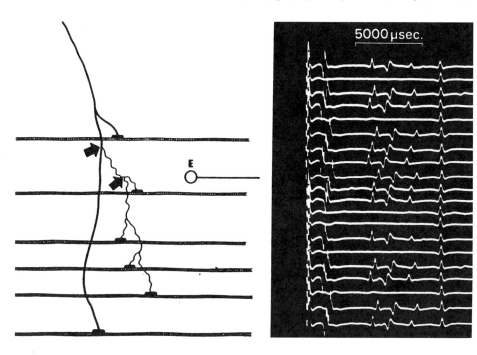

FIGURE 13–11. Concomitant (neurogenic) blocking. Reinnervation has resulted in six muscle fibers being innervated by the same motor unit. Note that the middle four single muscle fiber action potentials block intermittently together, and show a large common jitter in relation to the other two potentials. The origin of the block and the high jitter is situated in the nerve twig common to the four blocking muscle fibers noted by the two arrows. (Reprinted from Stålberg E, Trontelj JV: Single Fiber Electromyography in Healthy and Diseased muscle, 2nd ed. New York, Raven Press, 1994, with permission from Lippincott-Williams & Wilkins Publishers.)

blocking of a single muscle fiber potential that is typically encountered in neuromuscular transmission failure.

Jitter may also be increased during active or acute denervation, as it has been observed within days after onset of Guillain-Barré syndrome or following nerve injury.[71, 101] Increased fiber density and jitter have been observed in a wide variety of disorders that cause denervation and reinnervation. These disorders include anterior horn cell disease, including amyotrophic lateral sclerosis (ALS),[109, 116, 135] motor radiculopathies, and disorders of peripheral motor fibers including a wide variety of polyneuropathies.[64, 110, 147]

By observing the extent of the jitter and the degree of increased fiber density, one can characterize the nature of the denervating-reinnervating condition[140] (Fig. 13–12). For example, in patients with high jitter, but only modest elevations of fiber density, one might suppose that the denervation-reinnervation process is relatively rapid, such that the process does not permit adequate reinnervation for increased fiber density to occur.[140] On the other hand, a disorder characterized by very high fiber density with modest elevations of jitter, would imply a slowly progressive disorder in which reinnervation is allowed to occur in a very complete

fashion.[140] Disorders with relatively modest elevations of jitter and fiber density could imply a slow process with poor reinnervation, whereas disorders with rather high fiber density and high jitter might imply acceleration of the denervation process in a previously slow denervation-reinnervation disorder.[140]

SF EMG findings in myopathies depend largely on the nature of the histopathological changes of the muscle disease. In those muscle disorders that are associated with muscle cell degeneration or muscle cell necrosis, such as progressive muscular dystrophies and polymyositis/dermatomyositis and related disorders, fiber density and jitter are frequently increased to a mild or moderate degree.[101, 114] This is believed to occur due to "myopathic denervation" of individual muscle fibers caused by segmental necrosis of muscle fibers resulting in isolation of part of the muscle fiber from its nerve supply.[23] Collateral sprouting and reinnervation of these isolated muscle fiber segments may result in a regional increase in fiber density and jitter.[53, 114] In those myopathies that typically do not result in myonecrosis or muscle cell degeneration, fiber density and jitter may be normal or only minimally elevated.[140] Other possible explanations for

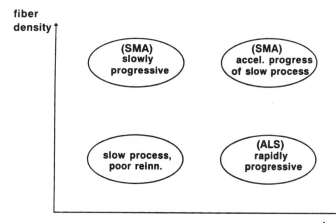

FIGURE 13–12. Schematic illustration of the relationship of fiber density to jitter in disorders causing denervation with reinnervation. (Reprinted from Stålberg E, Trontelj JV: Single Fiber Electromyography in Healthy and Diseased muscle, 2nd ed. New York, Raven Press, 1994, with permission from Lippincott-Williams & Wilkins Publishers.)

increased jitter in myopathy include enhanced VRF, altered propagation velocity of muscle fibers owing to degenerative changes, a primary disturbance of the motor end plate, and uncertain transmission in regenerating muscle fibers.[140] Other findings seen in myopathy include potential pairs with very low jitter, 5 μsec or less, due to split muscle fibers.[55] Other mechanisms have been proposed for the increased fiber density noted in myopathies, including packing of muscle fibers due to atrophy, and ephaptic recruitment of muscle fibers from other motor units.[101]

SF EMG has also been applied as a research tool to explore the nature of various electromyographical findings and phenomena, including fibrillation potentials,[128] fasciculation potentials,[135] complex repetitive discharges,[153] myotonia,[140] and neuromyotonia.[140] SF EMG has also been used to study various reflexes, including the axon reflex,[139, 157] F wave response,[103, 149] H reflex,[60, 148, 150] and blink reflex.[158]

MACRO ELECTROMYOGRAPHY

Macro electromyography (macro EMG) is a special technique developed by Stålberg that provides a quantitative measure of the motor unit by assessing the summated potentials from nearly all the muscle fibers belonging to the motor unit.[115] Macro EMG is a modified SF EMG technique. The macro EMG electrode is similar to the SF EMG electrode, as it also utilizes a 25 μm recording surface located 7.5 mm from the tip of the electrode. However, the shaft of the electrode is insulated to within 15 mm of the tip.[115] Recordings are made via two channels: one utilizing the SF EMG signal, as in a standard SF EMG recording, and the other utilizing the distal shaft of the electrode as a recording electrode with a remote surface skin electrode as the reference electrode[115] (Fig. 13–13). The technique involves examination of voluntarily activated muscle. An SF EMG recording is made and the time-locked macro MUP is recorded from the shaft of the electrode. The design of the electrode permits recording from essentially all of the muscle fibers of the motor unit that are firing in near synchrony and contributing to the MUP.[115, 118] The macro MUP is averaged during successive discharges and the final potential is then analyzed regarding its various characteristics, especially amplitude and rectified area. Filter settings for the single fiber channel are identical to those used in routine SF EMG, but for the macro channel, the filter settings are 8 Hz for the low-frequency filter and 8 KHz for the high-frequency filter.[115] The sweep speed for the macro channel is usually set at 10 msec per division. Typically, over 100 consecutive macro MUP discharges are collected and averaged on line. The determination of a baseline to calculate area under the macro MUP waveform is achieved by utilizing a 60-msec signal window with the triggering potential at its midpoint.[115] The 10 msec before and after this 60-msec period are then averaged to determine the baseline. The area under this segment of the tracing is calculated and the height of the potential from baseline to positive peak is noted for the amplitude.[115] Amplitude and area of macro MUPs display a linear relationship.[118] Because amplitude is a more easily derived parameter, it is the preferred parameter for defining normal values and detecting abnormalities. A complete macro EMG examination of a muscle encompasses 20 different macro MUPs. Macro MUP amplitudes and areas vary from muscle to muscle and tend to increase with age.[118, 132] Age-stratified normal values for several muscles have been reported in the literature and may be defined by the median macro MUP amplitude or the number of individual macro MUPs that fall outside the normal individual range.[118, 132] Thus, a macro

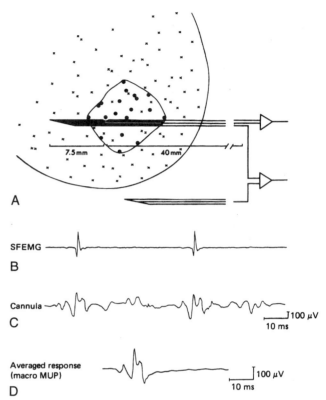

FIGURE 13–13. (A) Principle for macro EMG recording. Channel (B) represents the SF EMG tracing. A single muscle fiber action potential is used to average the time-locked motor unit potential recorded from the canula of the macro EMG electrode in (C). The average macro MUP is noted in (D). (Reprinted from Stålberg E: Single fiber EMG, macro EMG, and scanning EMG. New ways of looking at the motor unit. Crit Rev Clin Neurobiol 2:125–167, 1986, with permission from Karger.)

EMG study is considered abnormal if either the median amplitude value exceeds the normal median range, or if two or more macro MUPs out of 20 studied fall outside the individual macro MUP normal range.

Macro MUPs differ considerably from MUPs recorded with standard concentric or monopolar electrodes. Several factors contribute to this difference. Although the standard concentric electrode may have a pick-up radius of approximately 1 mm,[63] the spike of the recorded MUP from this electrode may originate from a relatively small number of fibers.[92, 115] Furthermore, the MUP is derived from a relatively small proportion of the number of motor fibers that contribute to the motor unit, which may be distributed in an oval area with a diameter approximating 5 to 10 mm.[134] However, the macro MUP is derived from a much larger number of the muscle fibers belonging to the motor unit. Computer simulation studies have suggested that the area of uptake for the macro electrode constitutes a 4-mm diameter cylinder of approximately 15 mm in length.[54, 93, 94] This may result in a MUP that represents 75% of the territory of the motor unit. Because the macro recording canula also functions as a low-pass filter, it tends to lessen the contribution of muscle fibers adjacent to the canula, with regard to their contribution to the spike portion of the waveform.[54, 115] Thus, the macro MUP tends to remain relatively stable with regard to morphology despite the insertion and movement of the macro EMG electrode in different portions of the motor unit territory.[93, 94] Peaks within the macro MUP tracing appear to represent temporal summation of various subgroups of muscle fibers innervated by small nerve branches from a single motor unit.[93, 119]

Macro EMG has been applied to a wide range of neuromuscular disorders. Typically, macro MUPs are of reduced amplitude and area in myopathies.[56, 115, 118, 119] These findings appear to contradict the SF EMG findings of increased fiber density noted in some myopathies, especially muscular dystrophies and polymyositis/dermatomyositis. This apparent inconsistency can be understood when one examines the various mechanisms of increased fiber density seen in myopathies. Fiber density in these disorders can be increased owing to new muscle fiber production from satellite cells, fiber splitting, and reinnervation of muscle fibers that have undergone segmental necrosis[55, 56] These mechanisms can result in a regional increase in fiber density without an overall increase in the number of muscle fibers within the motor unit. In neuropathic disorders causing denervation with reinnervation, macro EMG discloses MUPs of increased amplitude and area.[115, 117-120, 145] In rapidly progressing disorders, as in some cases of ALS, macro EMG may reveal MUPs that are of normal amplitude or only mildly increased in amplitude.[22, 117, 118, 124] In this setting, SF EMG discloses high jitter with modest increase in fiber density, indicating that denervation is occurring too rapidly to permit stable mature neuromuscular junctions to develop in the course of reinnervation.[135] In more slowly progressive disorders, jitter may be increased to a lesser extent but fiber density is higher and macro EMG MUPs are of increased amplitude and area.[120]

SCANNING ELECTROMYOGRAPHY

Scanning EMG is a special quantitative technique developed by Stålberg and Antoni that enables an assessment of the electrical and physical territory of a motor unit.[122] This technique enables a physical description of the generators of the MUP.[122] The physical arrangement of these generators is responsible for the sometimes dramatic changes in mor-phology of MUPs recorded with concentric or monopolar needle electrodes when the recording electrode is moved to different sites within the muscle. The technique involves the use of a single fiber electrode to serve as a trigger to time-lock a MUP discharge.[119, 122] On a second channel, a standard concentric electrode records the same motor unit that is being triggered by the single fiber electrode. The concentric electrode is connected to a computer-controlled step motor, which advances the electrode at 50 μm increments, up to a total distance of 20 mm.[119] Multiple tracings of the MUP are recorded for each location. The tracings are then displayed in a cascade of sequential tracings, which represent the anatomical distribution of the various muscle fibers contributing to the MUP (Fig. 13–14).

A variety of scanning EMG parameters have been defined. Scan length represents the distance along which a motor unit's spike component does not fall below 50 μV in amplitude.[119] Silent areas represent a portion of the scan where the spike amplitude drops below an amplitude of 50 μV.[119] A fraction is defined as an area of the scan that is separated by silent areas or a zone defined by maximal discrete spiked components, extending along the time line.[119] Polyphasic potentials may also be noted and are expressed as a percentage of the total scan. Lastly, the maximal duration is defined as the time spanned by all of the spikes belonging to the complete scanning EMG signal.[119]

Scanning EMG tracings vary from muscle to muscle, but most scanning MUPs demonstrate one to four fractions and a total length of 3 to 10 mm.[119] Typically, there is only a single silent area, although, occasionally none is noted.[119] The fractions within the motor unit are believed to represent groups of individual muscle fibers innervated from a common nerve branch.[119] It has been postulated that these various fractions contribute to the individual spikes that may be observed in a macro MUP, and that these fractions contribute to the MUP recorded from standard concentric or monopolar electrodes.

Scanning EMG has been used primarily as a research tool, helping to elucidate the microphysiology of the motor unit

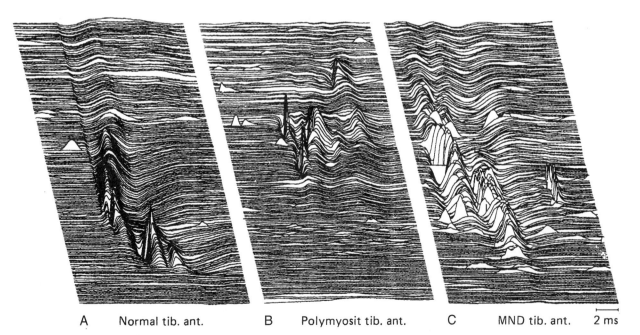

A   Normal tib. ant.   B   Polymyosit tib. ant.   C   MND tib. ant.   2 ms

FIGURE 13–14. Scanning EMG tracings in the tibialis anterior muscles from (A) normal control, (B) a patient with polymyositis, and (C) a patient with motor neuron disease. (Reprinted from Stålberg E: Single fiber EMG, macro EMG, and scanning EMG. New ways of looking at the motor unit. Crit Rev Clin Neurobiol 2:125–267, 1986, with permission from Karger.)

in various disorders, including myopathies and processes in which there is denervation with reinnervation. In neurogenic lesions, the number of fractions is typically unchanged, but each fraction may be increased in amplitude and duration.[119, 124] The scan length may be unchanged or only slightly increased and fewer numbers of silent areas are typically found.[119, 124] Interestingly, the extent of the motor unit is essentially unchanged.[127] In myopathies, there tends to be an increase in the number of fractions, with each fraction being somewhat lower in amplitude and shorter in duration than normal fractions.[119, 124]

QUANTITATIVE METHODS FOR MOTOR UNIT POTENTIAL MEASUREMENT

The traditional method of quantitative EMG involved the precise measurement of various aspects of the MUP that were recorded during the routine concentric needle electrode examination. Such a method was developed by Buchthal and colleagues.[11, 12, 14] In this method, standard needle electrodes were used to measure numerous MUPs in a single muscle during weak sustained voluntary contraction. MUPs with rise times of less than 500 μsec, preferably less than 100 to 200 μsec, were collected and photographed and various parameters of the MUP were then measured. These parameters included duration, amplitude, peak rise time, and the number of phases.[11, 12, 14] Other parameters may also be analyzed, including baseline crossings, turns, and the presence or absence of satellite potentials.[121] (See Chap. 12.)

Duration of a potential is measured from its onset to its termination or return to baseline. The amplitude is typically measured peak to peak, but may be measured from baseline to negative peak. The rise time is defined as the time from the initial negative deflection to the peak of the negative deflection. A baseline crossing is defined as a change in voltage from positive to negative or from negative to positive. The number of total phases is defined as the number of baseline crossings plus one, whereas a turn simply indicates a change in direction of the potential without baseline crossing. When area is assessed, it is usually measured under the negative components of the potential.

Buchthal and colleagues found that the duration of the MUP appeared to be the most clinically relevant parameter. Reduced duration of the MUP was observed in some myopathies, whereas increased duration MUPs were frequently noted in neuropathies.[11] This particular method has proven effective in the diagnosis of various neuromuscular diseases.[13] However, the method is inherently time-consuming and tedious and thus was never widely applied as a routine clinical EMG technique. In addition, Buchthal's technique required a collection of MUPs during weak or minimal activation, clearly indicating that type I MUPs were preferentially being analyzed.

With the capability of digitalizing MUP waveforms came the opportunity to develop new parameters to analyze the MUP. Thickness is one such parameter, assessed by computing the area-to-amplitude ratio of the MUP.[84] A reduction in thickness of the MUP corresponds to reduced duration, thin MUPs noted on subjective assessment of MUPs, and is a sensitive means of detecting myopathic disorders.[84] Surprisingly, MUP thickness is not significantly increased in neuropathic disorders. However, using a scatter plot of MUP amplitude versus thickness resulted in a new parameter termed the *size index*.[112] Size index appears to be more sensitive than amplitude, duration, or thickness in detecting neurogenic disorders.[112, 133] Jiggle of the MUP has been introduced as another quantitative parameter that assesses variability of the

MUP during successive discharges.[136] Variability of the MUP may occur in disorders of neuromuscular transmission as well as in disorders resulting in ongoing denervation with reinnervation. The jiggle is caused by the variable arrival time of the individual muscle fiber action potentials that are contributing to the MUP. Another novel MUP parameter has been devised to quantitate the irregularity of the MUP waveform. This involves computing the total amplitude change of the MUP, termed the MUP length, and dividing it by the peak-to-peak amplitude.[161] The higher the value of this ratio, the more complex or irregular the MUP waveform. These newer parameters are interesting methods for quantitating various aspects of the MUP but have yet to be established as valuable parameters for clinical diagnosis.

Computer programs have been developed to facilitate the recording and analysis of MUPs during standard NEE. These programs typically incorporate some form of template matching to identify individual MUPs during successive discharges.[1, 2, 121, 124] A delay line is used to assist in the recording of individual MUPs. Successive discharges are then averaged and stored as a digitalized MUP signal, which may be automatically analyzed for various parameters. As in Buchthal's technique, it is customary to record and analyze at least 20 MUPs in a given muscle. Although these programs clearly facilitate the collection and analysis of the individual MUPs, the process still remains relatively time-consuming and the analysis is restricted to the motor units that are recruited earliest during mild voluntary contraction, primarily type I motor units.

With the advent of more sophisticated computer applications, a different approach to MUP analysis has been initiated. This new method, termed *decomposition analysis*, involves identifying individual MUPs from an IP. One such program is the automated decompensation EMG program (ADEMG), developed by McGill and Dorfman, which utilizes spike trigger averaging of a 10-second epoch of IP.[76–78] Spikes within the IP are identified and filtered and then used to trigger and identify MUPs within the original IP tracing.[77, 79] The computer program summates an average of the individual MUPs, eliminating the superimposition of other MUPs and noise that typically occurs in an IP recording. This technique may identify up to 15 MUPs in a single IP epoch and provide information regarding amplitude, duration and other parameters of each MUP.[77, 78] ADEMG has clearly demonstrated clinical utility,[29] but it remains time-consuming and there is concern that MUPs of different motor units may appear similar and be misclassified.[133] Another technique termed *precision decomposition* was introduced to provide more precise measurements.[67, 142] This technique involves a multi-channel recording utilizing an electrode with four different recording surfaces capable of recording three different bipolar potentials from the same motor unit.[67, 142] Precision decomposition provides a unique MUP characterization making it possible to identify MUPs at even stronger contractions and, thus, permitting a more complete decomposition of the IP. However, this method is too intricate and time-consuming for practical application.

In an effort to develop a practical automated technique for decomposition of the IP that would be applicable for routine use in a busy EMG laboratory, a new technique termed *multi-MUP analysis* was developed[133] (Fig. 13–15). This technique was developed independently at two different centers.[4, 84, 131] It uses a conventional needle electrode recording from a muscle generating a slight to moderate contraction ranging between 5% and 30% of maximum force.[133] An epoch of 5 to 10 seconds is recorded for analysis. Typically, two or three skin insertions are made and different sites are sampled with each insertion. Approximately 30 MUPs may be collected with a goal of obtaining 20 suitable

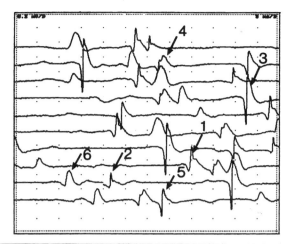

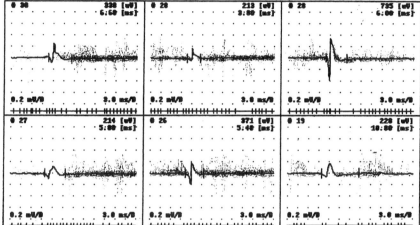

FIGURE 13–15. Multi-MUP analysis. Top tracing indicates free-running EMG recording. Bottom tracings indicate results of matching. Six different classes are found. Individual MUPs from the classes are indicated in the free-running panel. Matching MUPs are superimposed and other MUPs are seen as spurious noise. (Reprinted from Stålberg E, Nandedkar SD, Sanders DB, et al: Quantitative motor unit potential analysis. J Clin Neurophysiol 13:401–422, 1996, with permission from Lippincott-Williams & Wilkins Publishers.)

MUPs for analysis.[133] The method involves the identification of signals with a short rise time and an amplitude exceeding 50 µV. Template matching is performed and the six MUPs with the most matches are then identified and averaged. The individual discharges as well as the average MUP are then displayed in a superimposed fashion. This technique can be accomplished rapidly, usually within 5 to 10 minutes, and has been demonstrated to be reproducible, independent of amplifier gain, and not biased toward the largest MUPs.[133] However, there is a bias toward faster firing MUPs, as these would have a greater chance of being recognized. Furthermore, averaging of the MUPs may result in some alteration in their morphology. The MUPs measured by this technique disclose shorter durations and higher amplitudes compared to MUPs measured with the manual method.[131] This may be due to the ability of the multi-MUP technique to record MUPs at higher levels of voluntary contraction.[4] An abnormal study is defined as one in which 3 or more MUPs out of 20 disclose abnormal parameters.[125] This and other automatic computer-based methods of MUP analysis have yet to be applied as practical clinical tools in diagnostic electromyography.

INTERFERENCE PATTERN ANALYSIS

Various quantitative methods have been employed to study the IP. The IP is generated by an orderly recruitment of motor units during increasing force of contraction.[102] Initially, small motor units are recruited, primarily type I motor units, which have fewer muscle fibers, may generate smaller MUPs and are fatigue resistant. As force of contraction increases, larger motor units are recruited, type II motor units, which produce larger MUPs, begin firing at higher rates, and may be susceptible to fatigue. The orderly recruitment of motor units based upon their size has been termed the *size principle*.[52] As more and more motor units are recruited with increasing force of contraction, the superimposition of the MUPs prevents identifying discrete MUPs. The term *IP* was coined to describe this phenomenon. With increasing force of contraction, the amplitude of the IP increases, due primarily to recruitment of large type II motor units.[102] Although it is possible that the summation of two MUPs may contribute to a large spike within the IP, simulation studies indicate that this is a relatively infrequent occurrence, and that this mechanism is not the primary reason for the increasing amplitude of the IP at a complete or full contraction.[89]

Semiquantitative measures of the IP are commonly determined by experienced electromyographers during the course of the routine NEE. When the IP is reduced, such that individual MUPs are easily identified, the IP is termed *discrete*.[102] With full activation, a complete IP is often described as full or dense, whereas patterns that are intermediate are characterized as incomplete or reduced.[102] These descriptions are usually made during graded muscle contraction. When the muscle is being contracted at full effort, an assessment of the amplitude of the full or dense IP is often made as well. The IP in diseases relates not only to the individual properties of the MUP, but also to the firing rate and recruitment of the MUPs.[30] In neuropathies, with denervation and reinnervation, MUPs of increased amplitude and duration may contribute to a reduced IP and result in a higher amplitude. In muscle disease, the IP may be of lower amplitude, and the interference may become dense more rapidly during incomplete muscle contraction owing to smaller MUP sizes, a finding termed *early recruitment*.

One of the most common and widely used quantitative methods of IP analysis has been termed *turns amplitude analysis* or *turns analysis*.[102] This technique was originally described by Willison.[159, 160] Turns analysis incorporates an assessment of "turns" defined as a change in signal direction of a certain amplitude (Fig. 13–16). Willison utilized a threshold of 100 µV of amplitude to define a turn.[159] The number of turns within a certain defined epoch of time in the IP is an index of the number and firing rate of the MUPs contributing to the IP as well as the number of phases of the individual MUPs.[79] The number of turns increases with increasing force of contraction until the force reaches approximately 50% of maximum.[102] Beyond that level of contraction, only a modest increase in number of turns is noted, primarily due to interference of MUP waveforms superimposed upon one another.[89] However, the amplitude of the turns will increase with graded force of contraction throughout all levels of force.[102] Initially, turns analysis was measured at fixed contractile force. Using these methods, myopathies were characterized by increased numbers of turns with a reduction in the mean amplitude of the turn, whereas neuropathies produced higher mean amplitudes with no significant change in the number of turns.[51, 159, 160] Subsequent studies described higher mean amplitude and reduced turns in neurogenic disorders.[44] The ratio of the number of turns to the mean amplitude was also used as a valuable parameter.[43] This ratio was decreased in neuropathy,[44] primarily because of a decrease in the number of turns, and was increased in myopathies.[43] In some studies, the diagnostic yield of turns analysis for myopathy was similar to that of MUP analysis.[43]

Subsequently, it became apparent that varying force of muscular contraction expressed as the percentage of maximum force provided a more reliable index of force of muscle contraction.[42] The greatest diagnostic yield was found at approximately 30% maximum force of contraction.[44] Unfortunately, measurement of force is somewhat cumbersome, technically demanding, and eliminates the possibility of examining muscles whose force cannot be isolated from synergists. These technical difficulties led to alternative methods of quantitating turns analysis that did not require the measurement of the force of muscle contraction.

Stålberg and colleagues developed a computer-based turns amplitude method utilizing the concept of a "cloud" formed by the scatter plot of the number of turns versus the mean amplitude at varying degrees of muscle contraction.[123, 126] This method does not require precise force measurements. IPs are recorded at different sites within the muscle at three to five levels of muscle contraction, ranging from minimum to maximum at each site. The scatter plot of the number of turns versus mean amplitude is then created and a cloud is formed in such a way as to encompass more than 90% of all the data points from a normal reference population.[126] A study is considered abnormal if more than 10% of data points fall outside the normal cloud. Typically, 20 epochs are analyzed, thus, more than two epochs outside the cloud define an abnormal study. In patients with muscle disease, the amplitude of the IP epochs is reduced, causing some of the points to fall below the cloud, whereas in neuropathy, the amplitude of the IP epochs is increased, resulting in some of the points falling above the cloud.[126] This technique was found to be more sensitive than standard MUP analysis for the detection of myopathic and neuropathic disorders.[96]

Nandedkar and colleagues developed a computer simulation to investigate the IP.[89, 90] and proposed new quantitative measures to be applied to turns amplitude analysis.[87, 88, 90] Their technique was termed Expert Quantitative Interference Pattern (EQUIP) analysis.[87] The measures proposed for EQUIP were designed to reflect the very measures used by electromyographers in assessing the IP more subjectively. The activity parameter is an index of the density of the IP, the number of small segments (NSS) is an index of the complexity or number of MUPs comprising the IP and the upper centile amplitude (UCA) was used to measure the amplitude of the IP signal.[87] Later, an alternative measure for UCA was proposed, termed the *envelop amplitude* (ENAMP).[85] This was determined by measuring the amplitude of the IP envelope after excluding the largest eight positive and negative turns. With this technique, the NSS and the ENAMP are measured against increasing activity. Normal clouds were constructed for these measures using the same guidelines as turns versus amplitude clouds. In patients with muscle disease, the points for some of the epochs tend to fall below the lower margin of the amplitude-activity cloud and above the upper margin of the NSS-activity cloud, whereas patients with neuropathy have data points above the upper margin of the amplitude-activity cloud and below the lower margin of the NSS-activity cloud (Fig. 13–17). More than 2 out of 20 points outside the normal cloud would define an

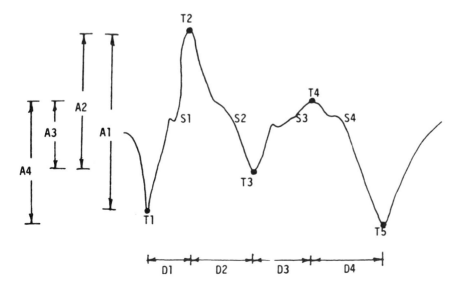

FIGURE 13–16. Schematic of an electromyographic signal illustrating five turns (T1–T5) which define four segments (S1–S4) whose durations (D1–D4) and amplitudes (A1–A4) are indicated. (Reprinted from Nandedkar SD, Sanders DB, Stålberg EV: Automatic analysis of the electromyographic, interference pattern. Part I: development of quantitative features. Muscle Nerve 9:431–439, 1986, with permission from John Wiley & Sons Publishers.)

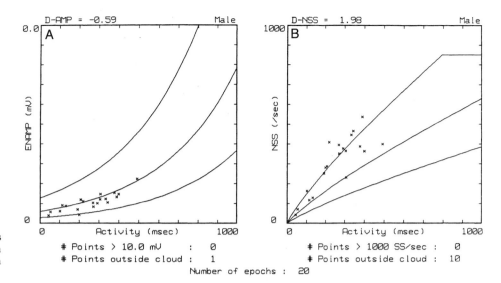

FIGURE 13–17. EQUIP turns analysis findings in a patient with myopathy *(A)* and in a patient with neuropathy *(B)*.

abnormal study. The shape of the various clouds is dependent on numerous variables, including the type of electrode used, the presence of muscle fatigue, gender, age, and the muscle in question.[102] The shape may also be affected by the degree of voluntary muscle contraction.[91] The diagnostic value of the EQUIP analysis has been contrasted with the subjective assessment of the IP by an experienced electromyographer and found to provide comparable results.[48] EQUIP results were found to be superior to the more conventional IP analysis using number of turns and mean amplitude ratios.[48] In another study, the EQUIP analysis appeared superior to the subjective assessment of the IP.[16]

Another important technique of IP analysis has been described by Fuglsang-Frederiksen and colleagues.[40, 41] These authors recorded IPs during a gradual increase in force from 0 to 100% over a 10-second epoch. The IP was measured at three different sites and the number of turns, mean amplitude of the turns and the turns:amplitude ratio were measured every 100 msec. A plot of the turns:amplitude ratio versus amplitude resulted in the determination of the peak ratio of turns:amplitude.[41] A study was abnormal if two or more sites disclosed a peak ratio more than two standard deviations from a control population. This method was abnormal in 82% of patients with myopathy and in 36% of patients with neuropathy.[41] An additional parameter, the time interval between turns, was added to this method.[68] In muscle disease, the time interval is decreased, whereas in neuropathy, it is increased. Combining the peak ratio of turns:amplitude with time interval analysis between turns, permitted all 25 subjects with myopathy and 21 subjects with neuropathy to be properly identified.[69]

Spectral analysis has also been applied to the quantitation of the IP.[30, 97, 102] Fast Fourier transformation of the IP can represent the pattern as a summation of various sine waves which can be plotted as the amount of power at different frequencies, the so-called power spectrum.[30, 97, 102] The power spectrum is an index of the number of MUPs, their firing rate, and their morphology.[30] The typical power spectrum of the IP ranges from 10 to 2000 Hz with a peak at approximately 100 to 200 Hz.[30] The low-frequency region of the spectrum, less than 50 Hz, appears to be related to the firing rate,[65] whereas the higher frequencies relate to the MUP morphology.[70] In myopathies, the power spectrum is shifted to higher frequencies, whereas in neuropathies, the shift is toward lower frequencies.[97] In one study, interference patterns were recorded at 10% and 30% maximum force and power was assessed at four frequencies, 140 Hz, 1400 Hz,

2800 Hz, and 4200 Hz.[97] With increasing force, there was a relative shift of power to lower frequencies. The greatest diagnostic yield was found by assessing the mean power frequency and relative power at 1400 Hz. There was an increase in the mean power frequency and the amount of power at 1400 Hz in myopathy, whereas there was a reduction in the mean power frequency and amount of power at 1400 Hz in neuropathy.[97] This method diagnosed 65% of myopathy patients and 73% of neuropathy patients. Power spectrum analysis has not been widely used in clinical diagnosis, but has been used as an investigational tool for the study of muscle fatigue.[49, 61, 62]

MOTOR UNIT NUMBER ESTIMATION

Estimating the number of motor units within a given muscle provides valuable information regarding the presence and progression of neuropathic disorders in which motor axons or motor neurons are lost. Indirect estimates of the number of motor units within a muscle can be derived from the standard NEE.[20] It is customary for an experienced electromyographer to make a semiquantitative assessment of the number of MUPs that are firing during increasing levels of voluntary contraction. This assessment is clearly a subjective one, as the electromyographer must estimate the number of MUPs firing and correlate that with the degree of voluntary effort that the patient is providing. When there is a reduced number of motor unit potentials (i.e., a reduction in the interference pattern with increased voluntary contraction, particularly when the MUPs are noted to be firing rapidly), the electromyographer may conclude that the number of MUPs in that muscle is reduced.[20] A more quantitative method of estimating motor unit numbers relies on measuring the firing rate of the first motor unit recruited during voluntary contraction and then assessing its firing rate at the time the second MUP is recruited.[20] Typically, the first MUP will be firing at a rate of 7 to 16 Hz at the time the second MUP begins firing.[20] This relationship can also be assessed by determining the ratio of the firing rate to the number of MUPs that are firing.[20] Normally, this ratio is less than 5. A ratio greater than 5 indicates a loss of motor units.[20] This analysis is helpful as it provides evidence regarding reduced MUPs without the need to assess the amount of voluntary activation and effort that the patient may be providing. Nevertheless, these methods are semiquantitative and relatively crude measures of the number of motor units.

The first method devised to more precisely count the number of motor units in a muscle was published by McComas and colleagues in 1971.[75] Since that time, numerous other methods have been developed to improve the precision and reliability of motor unit counting.[19, 26, 74] These techniques or methods can collectively be referred to as motor unit number estimation (MUNE). Most of these methods share a common basic approach, which is to determine the average size of a single MUP and to divide that into the size of the compound muscle action potential (CMAP) obtained from supramaximal stimulation of the appropriate motor nerve.[26] The exception to this is the statistical method introduced by Daube.[19]

The first method introduced for MUNE by McComas and colleagues has been referred to as the "manual incremental method."[72-75] In this method, successive electrical stimuli, manually adjusted, are delivered while recording a CMAP from the muscle in question. Because motor axons are stimulated in an all or none fashion, very fine incremental adjustments in stimulation intensity result in a stepwise increment of the CMAP. Each increment of the CMAP is regarded as the addition of a single motor unit. The total CMAP amplitude is then divided by the average motor unit action potential (MUAP) size to derive the number of motor units within the muscle (Fig. 13–18). This technique was initially applied to the extensor digitorum brevis muscle and then later to intrinsic hand muscles.[8] This continues to represent the most simple and straightforward method for MUNE. However, there are certain difficulties with this technique; chief among them is the problem of alternation.[26, 74] Alternation occurs when the thresholds of electrical stimulation of two or more motor axons overlap. In this circumstance, the steps or alternations represent various combinations of individual MUPs that may result in more steps or increments, thus, resulting in an increased number of the motor axons counted and an underestimation of the individual mean single MUP size. This would result in an overestimation of the number of motor units within the muscle.[9] The number of alternations or extra steps that may be measured is equal to $2^n - 1$ where n represents the number of axons that have similar or overlapping thresholds.[9, 26] The incremental technique also becomes somewhat unreliable when MUPs are small.[19] Clearly, the inability to record small steps results in underestimation of the number of motor units within a muscle. This incremental method has been modified to minimize the difficulty with alternation, by using automated computer measurement of the templates of different single MUPs and utilizing an automated method of incrementing the stimulus strength.[27, 45, 74]

An important modification of the original incremental method is a method termed *multiple-point stimulation* (MPS).[24, 26] This technique is performed by stimulation of the distal motor nerve over an area of 50- to 100-mm proximal to the motor point or recording site of the surface electrode. At each site of stimulation, only the first single reproducible MUP free of alternation is recorded. Typically, 10 or more MUPs may be recorded. This technique does not appear to have a bias toward small or large motor units and does eliminate the problem of alternation.[26] The average MUP size is then divided into the maximum CMAP to determine the number of motor units (Fig. 13–19). The major disadvantage of this technique is that it can only be applied to muscles that have access to 100 mm of their motor nerve for surface stimulation. Nevertheless, this method has been successfully applied to measurements from intrinsic hand muscles, foot muscles, and even proximal upper extremity muscles.[26]

Another method for MUNE utilizes single MUAPs derived

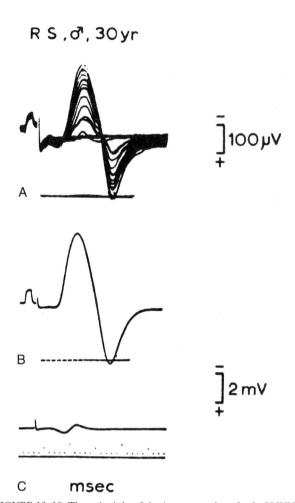

R S , ♂, 30yr

]100μV

A

]2mV

B

C   msec

FIGURE 13–18. The principle of the incremental method of MUNE. *(A)* Eleven increments are noted in the response of the extensor digitorum brevis muscle to stimuli of increasing intensity. *(B)* Maximum CMAP recorded from the extensor digitorum brevis. *(C)* The small contribution of the dorsal interossei to the maximum CMAP. (Reprinted from McComas AJ, Fawcett PRW, Campbell MJ, et al: Electrophysiological estimation of the number of motor units within a human muscle. J Neurol Neurosurg Psychiatry 34:121–131, 1971, with permission from BMJ Publishing Group.)

from F responses.[25, 144] When studying the thenar muscles, it may require 200 to 300 successive stimuli to collect an adequate number of F responses to identify at least 10 separate and unique single MUPs. When comparing the F wave method to the MPS method, it appears that single MUPs are similar in size distribution indicating that the F wave method does not have a particular bias to large or small MUPs.[26] A computer-based algorithm has been applied to this method to identify single MUPs from the F response.[144] This method also eliminates the problem of alternation. Its major disadvantage is the concern that in various disease states there is often an increased chance of an F response occurring, which could result in observing combinations of two or more single MUPs, which may be confused for a single MUP.[26] These considerations are not as significant in normal control populations.

Another method of MUNE utilizes spike-triggered averaging.[10, 26, 66] This technique requires a two-channel recording in which the first channel identifies a single MUP utilizing a delay line and trigger device, with a second channel recording the MUP from a surface electrode placed over the

S - MUAPs

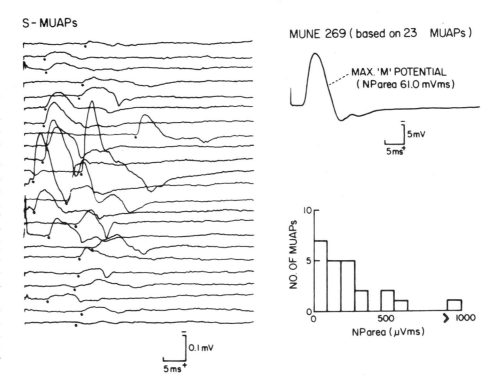

MUNE 269 (based on 23 MUAPs)

MAX. 'M' POTENTIAL
(NP area 61.0 mVms)

5mV
5ms

NO. OF MUAPs

NParea (μVms)

0.1 mV
5ms

FIGURE 13–19. Tracings from multiple point stimulation MUNE. Twenty-three surface detected motor unit potentials (S-MUAPs) from the thenar group are illustrated. The varying latencies reflect the different stimulation sites along the course of the median nerve. The onset of each S-MUAP is indicated by a dot. Each response was then aligned and the mean S-MUAP size was derived from all of the onset-aligned S-MUAPs. The negative peak area of the maximum "M" potential was then divided by the negative peak area of the mean S-MUAP to obtain the motor unit number estimate. (Reprinted from Doherty T, Simmons Z, O'Connell B, et al: Methods for estimating the numbers of motor units in human muscles. J Clin Neurophysiol 12:565–584, 1995, with permission from Lippincott Williams & Wilkins Publishers.)

motor point of the muscle. Thus, the size of the single MUP recorded on channel one is measured on channel two by the surface electrode. Typically, 10 or 20 single MUPs are recorded, and the average size of a single MUP is then derived and divided into the CMAP measured from the same surface recording electrode following supramaximal stimulation of the motor nerve. The needle electrodes used to measure the MUPs can be concentric or monopolar electrodes, or even a macro EMG electrode.[21] Results obtained from spike-triggered averaging techniques and MPS have been comparable.[26] In some ways, this is surprising as it is very likely that the spike-triggered averaging techniques will preferentially be biased toward recording smaller motor units which are activated first during modest voluntary muscle contraction.[10, 82] However, these comparisons have been made primarily in intrinsic hand muscles. Observations in more proximally situated muscles need to be made to further assess reliability of spike-triggered averaging techniques.[26] Another modification of the spike-triggered averaging method is to utilize macro EMG, which may represent a more reliable method of measuring the size of the MUP.[21] Furthermore, a technique involving microstimulation with an intramuscular electrode for collecting single MUPs has also been reported and would tend to reduce the possibility of motor unit size bias from voluntary activation.[81] Another modification of the spike-triggered averaging technique is to use computer algorithms designed for signal decomposition of the interference pattern to derive MUP size.[143] This would permit identifying and recording single MUPs from stronger contractions that may reflect a more representative pool of motor units.

Another approach to MUNE is the statistical method developed by Daube.[18, 19] This method utilizes stimulation of the motor nerve and recording CMAPs. However, the method relies not on the identification of single MUPs via incremental technique, but rather on the known relationship between the variance of multiple measures of step functions and the size of individual steps when these steps have a

Poisson distribution (Fig. 13–20). In a pure Poisson statistic the size of a series of measurements is a multiple of the size of a single component.[19] Thus, this method assesses the variance of the CMAP response assuming a Poisson distribution which, in turn, permits an estimate of the average size of a single MUP.[19] The technique utilizes standard recording electrodes that are used for routine motor nerve conduction studies. A sequence of submaximal stimuli are given. Typically, numerous series of 30 stimuli are delivered until the standard deviation of the result is less than 10%. Initially, a "scan" of the CMAP is obtained utilizing 30 stimuli with increasing intensities to identify any unusually large steps in the CMAP with increasing stimulation. In this way the optimal stimulus intensities can be selected. If all steps are relatively similar and small, then the stimulus intensity that produces a CMAP, approximating 5% to 10% of the maximal response, is usually used. If larger steps are noted, then ranges must increase by 15%. Ten groups of 30 responses are recorded at each stimulus intensity. The statistical method has the advantage of being able to measure MUP size when MUPs are either small or large.[19] However, this method is at a disadvantage when there are numerous MUPs in the CMAP, as the distribution shifts from a Poisson to a more normal distribution, resulting in errors of approximately 10% in the MUNE.[19]

All of the aforementioned techniques have shown similar results and, indeed, this is a form of validation in the absence of an absolute standard.[74] At this point in time, there is no one ideal method for MUNE. Rather, each of the various methods has certain advantages or disadvantages in particular clinical settings.

At this time, MUNE remains primarily an investigational tool. However, it has been successfully applied to various clinical disorders,[74] especially motor neuron diseases. MUNE studies in ALS have noted a rather rapid rate of loss of motor units in the early stages of the disease and a somewhat slower rate of motor unit loss later.[17, 72] The rate of loss of most units in ALS appears greater than the changes noted

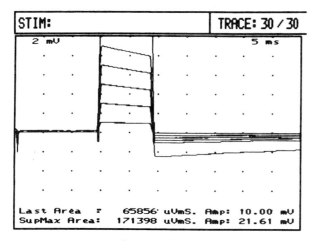

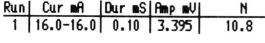

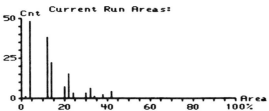

FIGURE 13–20. Histogram *(below)* of the number of occurrences of a series of electronic waveforms *(superimposed above)* shows a stepwise distribution with decreasing numbers of responses at higher levels. This pattern is that of a Poisson distribution. Variance estimate of the motor unit number estimate (MUNE) in the simulated signal with 11 steps was 10.8. (Reprinted from Daube JR: Estimating the number of motor units in a muscle. J Clin Neurophysiol 12:585–594, 1995, with permission from Lippincott-Williams & Wilkins Publishers.)

by other measures of function, including contractile strength.[6, 39] MUNE has also been used to study the effects of aging on the motor system. MUNE shows a clear decline in the number of motor units with advancing age.[8, 15, 28] Most of the decline occurs after the seventh decade, and the degree of change is variable from muscle to muscle.[111]

REFERENCES

1. Andreassen S: Methods for computer-aided measurement of motor unit parameters. Electroencephalogr Clin Neurophysiol Suppl 39:13–20, 1987.
2. Bergmans J: Computer assisted on line measurement of motor unit parameters in human electromyography. Electromyography 11:161–181, 1971.
3. Bertorini TE, Stålberg E, Yuson CP, et al: Single-fiber electromyography in neuromuscular disorders: Correlation of muscle histochemistry, single-fiber electromyography, and clinical findings. Muscle Nerve 17:345–353, 1994.
4. Bischoff C, Stålberg E, Falck B, et al: Reference values of motor unit action potentials obtained with multi-MUAP analysis. Muscle Nerve 17:842–851, 1994.
5. Brandstater ME, Lambert EH: Motor unit anatomy. Type and spatial arrangement of muscle fibers. In Desmedt JE (ed): New Developments in Electromyography and Clinical Neurophysiology. Basel, Karger, 1973, pp 14–22.
6. Bromberg MB, Forshaw DA, Nau HL, et al: Motor unit number estimation, isometric strength and electromyographic measures in amyotrophic lateral sclerosis. Muscle Nerve 16:1213–1219, 1993.
7. Bromberg MB, Scott DM, Ad Hoc Committee of the AAEM Single Fiber Special Interest Group: Single fiber EMG reference values: Reformatted in tabular form. Muscle Nerve 17:820–821, 1994.
8. Brown WF: A method for estimating the number of motor units in thenar muscles and the changes in motor unit count with ageing. J Neurol Neurosurg Psychiatry 35:845–852, 1972.
9. Brown WF, Milner-Brown HS: Some electrical properties of motor units and their effects on the methods of estimating motor unit numbers. J Neurol Neurosurg Psychiatry 39:249–257, 1976.
10. Brown WF, Strong MJ, Snow R: Methods for estimating numbers of motor units in biceps-brachialis muscles and losses of motor units with aging. Muscle Nerve 11:423–432, 1988.
11. Buchthal F: An Introduction to Electromyography. Copenhagen, Scandinavian University Books, 1957.
12. Buchthal F, Guld C, Rosenfalck P: Action potential parameters in normal human muscle and their dependence on physical variables. Acta Physiol Scand 32:200–218, 1954.
13. Buchthal F, Kamieniecka Z: The diagnostic yield of quantified electromyography and quantified muscle biopsy in neuromuscular disorders. Muscle Nerve 5:265–280, 1982.
14. Buchthal F, Pinelli P, Rosenfalck P: Action potential parameters in normal human muscle and their physiological determinants. Acta Physiol Scand 32:219–229, 1954.
15. Campbell MJ, McComas AJ, Petito F: Physiological changes in ageing muscles. J Neurol Neurosurg Psychiatry 36:174–182, 1973.
16. Cao J, Sanders DB: Multivariate discriminant analysis of the electromyographic interference pattern: A statistical approach to discrimination among controls, myopathies, and neuropathies. Med Biol Eng Comput 34:369–374, 1996.
17. Dantes M, McComas A: The extent and time course of motorneuron involvement in amyotrophic lateral sclerosis. Muscle Nerve 14:416–421, 1991.
18. Daube JR: Statistical estimates of number of motor units in the thenar and foot muscles in patients with amyotrophic lateral sclerosis or the residual of poliomyelitis. Muscle Nerve 11:957–958, 1988.
19. Daube JR: Estimating the number of motor units in a muscle. J Clin Neurophysiol 12:585–594, 1995.
20. Daube JR: Assessing the motor unit with needle electromyography. In Daube JR (ed): Clinical Neurophysiology. Philadelphia, FA Davis, 1996, pp 257–281.
21. de Koning P, Wieneke GH, van der Most van Spijk D, et al: Estimation of the number of motor units based on macro-EMG. J Neurol Neurosurg Psychiatry 51:403–411, 1988.
22. Dengler R, Konstanzer A, Kuther G, et al: Amyotrophic lateral sclerosis: Macro-EMG and twitch forces of single motor units. Muscle Nerve 13:545–550, 1990.
23. Desmedt JE, Borenstein S: Relationship of spontaneous fibrillation potentials to muscle fibre segmentation in human muscular dystrophy. Nature 258:531–534, 1975.
24. Doherty TJ, Brown WF: The estimated numbers and relative sizes of thenar motor units as selected by multiple point stimulation in young and older adults. Muscle Nerve 16:355–366, 1993.
25. Doherty TJ, Komori T, Stashuk DW, et al: Physiological properties of single thenar motor units in the F-response of younger and older adults. Muscle Nerve 17:860–872, 1994.
26. Doherty T, Simmons Z, O'Connell B, et al: Methods for estimating the numbers of motor units in human muscles. J Clin Neurophysiol 12:565–584, 1995.
27. Doherty TJ, Stashuk DW, Brown WF: Determinants of mean motor unit size: Impact on estimates of motor unit number. Muscle Nerve 16:1326–1331, 1993.
28. Doherty TJ, Vandervoort AA, Brown WF: Effects of ageing on the motor unit: A brief review. Can J Appl Physiol 18:331–358, 1993.
29. Dorfman L, Howard J, McGill K: Clinical studies using automatic decomposition electromyography (ADEMG) in needle and surface EMG. In Desmedt JE (ed): Computer Aided Electromyography and Expert systems. Clinical Neurophysiology Updates. Amsterdam, Elsevier, 1989, pp 189–204.
30. Dorfman LJ, McGill KC: AAEE Minimonograph #29: Automatic quantitative electromyography. Muscle Nerve 11:804–818, 1988.
31. Ekstedt J: Human single muscle fibre action potentials. Acta Physiol Scand 61 (suppl 226):1–96, 1964.

32. Ekstedt J, Nilsson G, Stålberg E: Calculation of the electromyographic jitter. J Neurol Neurosurg Psychiatry 37:526–539, 1974.
33. Ekstedt J, Stålberg E: A method of recording extracellular action potentials of single muscle fibres and measuring their propagation velocity in voluntarily activated human muscle. Bull Am Assoc EMG Electrodiagn 10:16, 1963.
34. Ekstedt J, Stålberg E: Abnormal connections between skeletal muscle fibers. Electroencephalogr Clin Neurophysiol 27:607–609, 1969.
35. Ekstedt J, Stålberg E: How the size of the needle electrode leading-off surface influences the shape of the single muscle fibre action potential in electromyography. Comput Programs Biomed 3:204–212, 1973.
36. Ekstedt J, Stålberg E: Single fibre electromyography for the study of the microphysiology of the human muscle. In: Desmedt JE (ed): New Developments in Electromyography and Clinical Neurophysiology. Basel, Karger, 1973, pp 89–112.
37. Ekstedt J, Stålberg E: Single muscle fibre electromyography in myasthenia gravis. In: Kunze, K and Desmedt JE (eds): Studies in Neuromuscular Diseases. Basel, Karger, 1975, pp 157–161.
38. Engel AG, Lambert EH, Santa T: Study of long-term anticholinesterase therapy. Effects on neuromuscular transmission and on motor end-plate fine structure. Neurology 23:1273–1281, 1973.
39. Felice KJ: A longitudinal study comparing thenar motor unit number estimates to other quantitative tests in patients with amyotrophic lateral sclerosis. Muscle Nerve 20:179–185, 1997.
40. Fuglsang-Frederiksen A, Dahl K, Lo Monaco M: Electrical muscle activity during a gradual increase in force in patients with neuromuscular diseases. Electroencephalogr Clin Neurophysiol 57:320–329, 1984.
41. Fuglsang-Frederiksen A, LoMonaco M, Dahl K: Turns analysis (peak ratio) in EMG using the mean amplitude as a substitute of force measurement. Electroencephalogr Clin Neurophysiol 60:225–227, 1985.
42. Fuglsang-Frederiksen A, Månsson A: Analysis of electrical activity of normal muscle in man at different degrees of voluntary effort. J Neurol Neurosurg, Psychiatry 38:683–694, 1975.
43. Fuglsang-Frederiksen A, Scheel U, Buchthal F: Diagnostic yield of analysis of the pattern of electrical activity and of individual motor unit potentials in myopathy. J Neurol Neurosurg Psychiatry 39:742–750, 1976.
44. Fuglsang-Frederiksen A, Scheel U, Buchthal F: Diagnostic yield of the analysis of the pattern of electrical activity of muscle and of individual motor unit potentials in neurogenic involvement. J Neurol Neurosurg Psychiatry 40:544–554, 1977.
45. Galea V, deBruin H, Cavasin R, et al: The numbers and relative sizes of motor units estimated by computer. Muscle Nerve 14:1123–1130, 1991.
46. Gath I, Stålberg E: On the measurement of fibre density in human muscles. Electroencephalogr Clin Neurophysiol 54:699–706, 1982.
47. Gilchrist JM, Ad Hoc Committee of the AAEM Special Interest Group on Single Fiber EMG: Single fiber EMG reference values: A collaborative effort. Muscle Nerve 15:151–161, 1992.
48. Gilchrist JM, Nandedkar SD, Stewart CS, et al: Automatic analysis of the electromyographic interference pattern using the turns:amplitude ratio. Electroencephalogr Clin Neurophysiol 70:534–540, 1988.
49. Hägg GM: Interpretation of EMG spectral alterations and alteration indexes at sustained contraction. J Appl Physiol 73:1211–1217, 1992.
50. Hakelius L, Stålberg E: Electromyographical studies of free autogenous muscle transplants in man. Scand J Plast Reconstr Surg 8:211–219, 1974.
51. Hayward M, Willison RG: Automatic analysis of the electromyogram in patients with chronic partial denervation. J Neurol Sci 33:415–423, 1977.
52. Henneman E, Clamann HP, Gillies JD, et al: Rank order of motorneurons within a pool: Law of combination. J Neurophysiol 37:1338–1349, 1974.
53. Henriksson KG, Stålberg E: The terminal innervation pattern in polymyositis: A histochemical and SFEMG study. Muscle Nerve 1:3–13, 1978.
54. Hilton-Brown P, Nandedkar SD, Stålberg EV: Simulation of fibre density in single-fiber electromyography and its relationship to macro-EMG. Med Biol Eng Comput 23:541–546, 1985.
55. Hilton-Brown P, Stålberg E: The motor unit in muscular dystrophy, a single fibre EMG and scanning EMG study. J Neurol Neurosurg Psychiatry 46:981–995, 1983.
56. Hilton-Brown P, Stålberg E: Motor unit size in muscular dystrophy, a macro EMG and scanning EMG study. J Neurol Neurosurg Psychiatry 46:996–1005, 1983.
57. Hilton-Brown P, Stålberg EV, Osterman P-O: Signs of reinnervation in myasthenia gravis. Muscle Nerve 5:215–221, 1982.
58. Howard JF, Sanders DB: Serial single-fiber EMG studies in myasthenic patients treated with corticosteroids and plasma exchange therapy. Muscle Nerve 4:254, 1981.
59. Jabre JF, Chirico-Post J, Weiner M: Stimulation SFEMG in myasthenia gravis. Muscle Nerve 12:38–42, 1989.
60. Jabre JF, Stålberg EV: Single-fiber EMG study of the flexor carpi radialis H reflex. Muscle Nerve 12:523–527, 1989.
61. Jurell KC: Surface EMG and fatigue. Phys Med Rehabil Clin N Am 9:933–947, 1998.
62. Kadefors R, Petersen I, Broman H: Spectral analysis of events in the electromyogram. In: Desmedt JE (ed): New Developments in Electromyography and Clinical Neurophysiology. Basel, Karger, 1973, pp 628–637.
63. King JC, Dumitru D, Nandedkar S: Concentric and single fiber electrode spatial recording characteristics. Muscle Nerve 20:1525–1533, 1990.
64. Konishi T, Nishitani H, Motomura S: Single fiber electromyography in chronic renal failure. Muscle Nerve 5:458–461, 1982.
65. Lago P, Jones NB: Effect of motor-unit firing time statistics on EMG spectra. Med Biol Eng Comput 15:648–655, 1977.
66. Lee RG, Ashby P, White DG, et al: Analysis of motor conduction velocity in the human median nerve by computer simulation of compound muscle action potentials. Electroencephalogr Clin Neurophysiol 39:225–237, 1975.
67. LeFever RS, De Luca CJ: A procedure for decomposing the myoelectric signal into its constituent action potentials. I: Technique, theory and implementation. IEEE Trans Biomed Eng 29:149–157, 1982.
68. Liguori R, Dahl K, Fuglsang-Frederiksen A: Turns-amplitude analysis of the electromyographic recruitment pattern disregarding force measurement. I. Method and reference values in healthy subjects. Muscle Nerve 15:1314–1318, 1992.
69. Liguori R, Dahl K, Fuglsang-Frederiksen A, et al: Turns-amplitude analysis of the electromyographic interference pattern disregarding force measurement. II. Findings in patients with neuromuscular disorders. Muscle Nerve 15:1319–1324, 1992.
70. Lindstrom LH, Magnusson RI: Interpretation of myoelectric power spectra: A model and its applications. Proc IEEE 65:653–662, 1977.
71. Massey JM, Sanders DB: Single-fiber EMG demonstrates reinnervation dynamics after nerve injury. Neurology 41:1150–1151, 1991.
72. McComas AJ: Invited review. Motor unit estimation: methods, results, and present status. Muscle Nerve 14:585–597, 1991.
73. McComas AJ: Motor-unit estimation: the beginning. J Clin Neurophysiol 12:560–564, 1995.
74. McComas AJ: Motor unit estimation: anxieties and achievements. Muscle Nerve 18:369–379, 1995.
75. McComas AJ, Fawcett PRW, Campbell MJ, et al: Electrophysiological estimation of the number of motor units within a human muscle. J Neurol Neurosurg Psychiatry 34:121–131, 1971.
76. McGill KC, Cummins KL, Dorfman LJ: Automatic decomposition of the clinical electromyogram. IEEE Trans Biomed Eng 32:470–477, 1985.
77. McGill KC, Dorfman LJ: Automatic decomposition electromyography (ADEMG): Validation and normative data in brachial biceps. Electroencephalogr Clin Neurophysiol 61:453–461, 1985.
78. McGill K, Dorfman L: Automatic decomposition electromyography (ADEMG), methodologic and technical considerations. In Desmedt JE (ed): Computer Aided Electromyography and Expert Systems. Clinical Neurophysiology Updates. Basel, Karger, 1989, pp 91–101.
79. McGill KC, Lau K, Dorfman LJ: A comparison of terms analysis and motor unit analysis in electromyography. Electroencephalogr Clin Neurophysiol 81:8–17, 1991.
80. Mihelin M, Trontelj JV, Stålberg E: Muscle fiber recovery functions studied in the double pulse stimulation. Muscle Nerve 14:739–747, 1991.

81. Milner-Brown HS, Brown WF: New methods of estimating the number of motor units in a muscle. J Neurol Neurosurg Psychiatry 39:258–265, 1976.
82. Milner-Brown HS, Stein RB, Yemm R: The orderly recruitment of human motor units during voluntary isometric contractions. J Physiol (Lond) 230:359–370, 1973.
83. Nandedkar SD, Barkhaus PE, Charles A: Multi-motor unit action potential analysis. Muscle Nerve 18:1155–1166, 1995.
84. Nandedkar SD, Barkhaus PE, Sanders DB, et al: Analysis of the amplitude and area of concentric needle EMG motor unit action potentials. Electroencephalogr Clin Neurophysiol 69:561–567, 1988.
85. Nandedkar SD, Sanders DB: Measurement of the amplitude of the EMG envelope. Muscle Nerve 13:933–938, 1990.
86. Nandedkar SD, Sanders DB, Stålberg EV: Selectivity of electromyographic recording techniques: A simulation study. Med Biol Eng Comput 23:536–540, 1985.
87. Nandedkar SD, Sanders DB, Stålberg EV: Automatic analysis of the electromyographic interference pattern. Part I: Development of quantitative features. Muscle Nerve 9:431–439, 1986.
88. Nandedkar SD, Sanders DB, Stålberg EV: Automatic analysis of the electromyographic interference pattern. Part II: Findings in control subjects and in some neuromuscular diseases. Muscle Nerve 9:491–500, 1986.
89. Nandedkar SD, Sanders DB, Stålberg EV: Simulation and analysis of the electromyographic interference pattern in normal muscle. Part I: Turns and amplitude measurements. Muscle Nerve 9:423–430, 1986.
90. Nandedkar SD, Sanders DB, Stålberg EV: Simulation and analysis of the electromyographic interference pattern in normal muscle. Part II: Activity, upper centile amplitude and number of small segments. Muscle Nerve 9:486–490, 1986.
91. Nandedkar SD, Sanders DB, Stålberg EV: On the shape of the normal turns-amplitude cloud. Muscle Nerve 14:8–13, 1991.
92. Nandedkar SD, Sanders DB, Stålberg EV, et al: Simulation of concentric needle EMG motor unit action potentials. Muscle Nerve 2:151–159, 1988.
93. Nandedkar S, Stålberg E: Simulation of macro EMG motor unit potentials. Electroencephalogr Clin Neurophysiol 56:52–62, 1983.
94. Nandedkar SD, Stålberg E, Kim YI, et al: Use of signal representation to identify abnormal motor unit potentials in macro EMG. IEEE Trans Biomed Eng 31:220–227, 1984.
95. Newsom-Davis J, Leys K, Vincent A, et al: Immunological evidence for the co-existence of the Lambert-Eaton myasthenic syndrome and myasthenia gravis in two patients. J Neurol Neurosurg Psychiatry 54:452–453, 1991.
96. Nirkko AC, Rosler KM, Hess CW: Sensitivity and specificity of needle electromyography: A prospective study comparing automated interference pattern analysis with single motor unit potential analysis. Electroencephalogr Clin Neurophysiol 97:1–10, 1995.
97. Ronager J, Christensen H, Fuglsang-Frederiksen A: Power spectrum analysis of the EMG pattern in normal and diseased muscles. J Neurol Sci 94:283–294, 1989.
98. Sanders DB: The effect of firing rate on neuromuscular jitter in Lambert-Eaton myasthenic syndrome. Muscle Nerve 15:256–258, 1992.
99. Sanders DB, Howard JF: AAEM Minimonograph #25: Single-fiber electromyography in myasthenia gravis. Muscle Nerve 9:809–819, 1986.
100. Sanders DB, Massey EW, Buckley EG: Botulinum toxin for blepharospasm: Single-fiber EMG studies. Neurology 36:545–547, 1986.
101. Sanders DB, Stålberg EV: AAEM Minimonograph #25: Single-fiber electromyography. Muscle Nerve 19:1069–1083, 1996.
102. Sanders DB, Stålberg EV, Nandedkar SD: Analysis of the electromyographic interference pattern. J Clin Neurophysiol 13:385–400, 1996.
103. Schiller HH, Stålberg E: F responses studied with single fibre EMG in normal subjects and spastic patients. J Neurol Neurosurg Psychiatry 41:45–53, 1978.
104. Schiller HH, Stålberg E: Human botulism studied with single-fiber electromyography. Arch Neurol 35:346–349, 1978.
105. Schwartz MS, Moosa A, Dubowitz V: Correlation of single fibre EMG and muscle histochemistry using an open biopsy recording technique. J Neurol Sci 31:369–378, 1977.
106. Schwartz MS, Stålberg E: Myasthenia gravis with features of the myasthenic syndrome: An investigation with electrophysiologic methods including single-fibre electromyography. Neurology 25:80–84, 1975.
107. Schwartz MS, Stålberg E: Myasthenic syndrome studied with single fiber electromyography. Arch Neurol 32:815–817, 1975.
108. Schwartz MS, Stålberg E, Schiller HH, et al: The reinnervated motor unit in man. A single fibre EMG multielectrode investigation. J Neurol Sci 27:303–312, 1976.
109. Shields RW: Single fiber electromyography in the differential diagnosis of myopathic limb girdle syndromes and chronic spinal muscular atrophy. Muscle Nerve 7:265–272, 1984.
110. Shields RW: Single fiber electromyography is a sensitive indicator of axonal degeneration in diabetes. Neurology 37:1394–1397, 1987.
111. Sica RE, McComas AJ, Upton AR, et al: Motor unit estimation in small muscles of the hand. J Neurol Neurosurg Psychiatry 37:55–67, 1974.
112. Sonoo M, Stålberg E: The ability of MUP parameters to discriminate between normal and neurogenic MUPs in concentric EMG: Analysis of the MUP "thickness" and the proposal of "size index." Electroencephalogr Clin Neurophysiol 89:291–303, 1993.
113. Stålberg E: Propagation velocity in human single muscle fibers in situ. Acta Physiol Scand 70(suppl 287):1–112, 1966.
114. Stålberg E: Electrogenesis in human dystrophic muscle. In Rowland LP (ed): Pathogenesis of Human Muscular Dystrophies. Amsterdam, Excerpta Medica, 1977, pp 570–587.
115. Stålberg E: Macro EMG, a new recording technique. J Neurol Neurosurg Psychiatry 43:475–482, 1980.
116. Stålberg E: Electrophysical studies of reinnervation in ALS. In Rowland LP (ed): Human Motor Neuron Diseases. New York, Raven Press, 1982, pp 47–59.
117. Stålberg E: Macroelectromyography in reinnervation. Muscle Nerve 5:S135–S138, 1982.
118. Stålberg E: Macro EMG. Muscle Nerve 6:619–630, 1983.
119. Stålberg E: Single fiber EMG, macro EMG, and scanning EMG. New ways of looking at the motor unit. Crit Rev Clin Neurobiol 2:125–167, 1986.
120. Stålberg E: Use of single fiber EMG and macro EMG in study of reinnervation. Muscle Nerve 13:804–813, 1990.
121. Stålberg E, Andreassen S, Falck B, et al: Quantitative analysis of individual motor unit potentials: A proposition for standardized terminology and criteria for measurement. J Clin Neurophysiol 3:313–348, 1986.
122. Stålberg E, Antoni L: Electrophysiological cross section of the motor unit. J Neurol Neurosurg Psychiatry 43:469–474, 1980.
123. Stålberg E, Antoni L: Microprocessors in the analysis of the motor unit and the neuromuscular transmission. In Yamazuchi N, Fujizawa K (eds): Proceedings of the Conference on EEG and EMG Data Processing. Amsterdam, Elsevier, 1981, pp 295–313.
124. Stålberg E, Antoni L: Computer-aided EMG analysis. In Desmedt JE (ed): Progress in Clinical Neurophysiology, vol 10. Basel, Karger, 1983, pp 186–234.
125. Stålberg E, Bischoff C, Falck B: Outliers, a way to detect abnormality in quantitative EMG. Muscle Nerve 17:392–399, 1994.
126. Stålberg E, Chu J, Bril V, et al: Automatic analysis of the EMG interference pattern. Electroencephalogr Clin Neurophysiol 56:672–681, 1983.
127. Stålberg E, Dioszeghy P: Scanning EMG in normal muscle and in neuromuscular disorders. Electroencephalogr Clin Neurophysiol 81:403–416, 1991.
128. Stålberg E, Ekstedt J: Single fibre EMG and microphysiology of the motor unit in normal and diseased human muscle. In Desmedt JE (ed): New Developments in Electromyography and Clinical Neurophysiology, vol 1. Basel, Karger, 1973, pp 113–129.
129. Stålberg E, Ekstedt J, Broman A: The electromyographic jitter in normal human muscles. Electroencephalogr Clin Neurophysiol 31:429–438, 1971.
130. Stålberg E, Ekstedt J, Broman A: Neuromuscular transmission in myasthenia gravis studied with single fibre electromyography. J Neurol Neurosurg Psychiatry 37:540–547, 1974.
131. Stålberg E, Falck B, Sonoo M, et al: Multi-MUP EMG analysis—a two year experience in daily clinical work. Electroencephalogr Clin Neurophysiol 97:145–154, 1995.

132. Stålberg E, Fawcett PR: Macro EMG changes in healthy subjects of different ages. J Neurol Neurosurg Psychiatry 45:870–878, 1982.
133. Stålberg E, Nandedkar SD, Sanders DB, et al: Quantitative motor unit potential analysis. J Clin Neurophysiol 13:401–422, 1996.
134. Stålberg E, Schwartz MS, Thiele B, et al: The normal motor unit in man. J Neurol Sci 27:291–301, 1976.
135. Stålberg E, Schwartz MS, Trontelj JV: Single fibre electromyography in various processes affecting the anterior horn cell. J Neurol Sci 24:403–415, 1975.
136. Stålberg E, Sonoo M: Assessment of variability in the shape of the motor unit action potential, the "jiggle," at consecutive discharges. Muscle Nerve 17:1135–1144, 1994.
137. Stålberg E, Thiele B: Transmission block in terminal nerve twigs: A single fibre electromyographic finding in man. J Neurol Neurosurg Psychiatry 35:52–59, 1972.
138. Stålberg E, Thiele B: Motor unit fibre density in the extensor digitorum communis muscle: Single fibre electromyographic study in normal subjects of different ages. J Neurol Neurosurg Psychiatry 38:874–880, 1975.
139. Stålberg E, Trontelj JV: Demonstration of axon reflexes in human motor nerve fibres. J Neurol Neurosurg Psychiatry 33:571–579, 1970.
140. Stålberg E, Trontelj JV: Single Fiber Electromyography in Healthy and Diseased Muscle, 2nd ed. New York, Raven Press, 1994.
141. Stålberg E, Trontelj JV, Mihelin M: Electrical microstimulation with single-fiber electromyography: A useful method to study the physiology of the motor unit. J Clin Neurophysiol 9:105–119, 1992.
142. Stashuk D, DeLuca CJ: Update on the decomposition and analysis of EMG signals. In Desmedt JE (ed): Computer Aided Electromyography and Expert systems. Clinical Neurophysiology Updates (vol 2). Basel, Karger, 1989, pp 39–54.
143. Stashuk DW, Doherty TJ, Brown WF: EMG signal decomposition applied to motor unit estimates. Muscle Nerve 15:1191, 1992.
144. Stashuk DW, Doherty TJ, Kassam A, et al: Motor unit number estimates based on automated analysis of F responses. Muscle Nerve 17:881–890, 1994.
145. Tackmann W, Vogel P: Fibre density, amplitudes of macro-EMG motor unit potentials and conventional EMG recordings from the anterior tibial muscle in patients with amyotrophic lateral sclerosis. A study on 51 cases. J Neurol 235:149–154, 1988.
146. Thiele B, Stålberg E: The bimodal jitter: A single fibre electromyographic finding. J Neurol Neurosurg Psychiatry 37:403–411, 1974.
147. Thiele B, Stålberg E: Single fibre EMG findings in polyneuropathies of different aetiology. J Neurol Neurosurg Psychiatry 38:881–887, 1975.
148. Trontelj JV: H-reflex of single motoneurones in man. Nature 220:1043–1044, 1968.
149. Trontelj JV: A study of the F response by single fibre electromyography. In Desmedt JE (ed): New Developments in Electromyography and Clinical Neurophysiology (vol 3). Basel, Karger, 1973, pp 318–322.
150. Trontelj JV: A study of the H-reflex by single fibre EMG. J Neurol Neurosurg Psychiatry 36:951–959, 1973.
151. Trontelj JV, Khuraibet A, Mihelin M: The jitter in stimulated orbicularis oculi muscle: Technique and normal values. J Neurol Neurosurg Psychiatry 51:814–819, 1988.
152. Trontelj JV, Mihelin M, Fernandez JM, et al: Axonal stimulation for end-plate jitter studies. J Neurol Neurosurg Psychiatry 49:677–685, 1986.
153. Trontelj J, Stålberg E: Bizarre repetitive discharges recorded with single fibre EMG. J Neurol Neurosurg Psychiatry 46:310–316, 1983.
154. Trontelj JV, Stålberg E: Single motor end-plates in myasthenia gravis and LEMS at different firing rates. Muscle Nerve 14:226–232, 1991.
155. Trontelj JV, Stålberg E: Jitter measurements by axonal microstimulation. Guidelines and technical notes. Electroencephalogr Clin Neurophysiol 85:30–37, 1992.
156. Trontelj JV, Stålberg E, Mihelin M, et al: Jitter of the stimulated motor axon. Muscle Nerve 15:449–454, 1992.
157. Trontelj JV, Trontelj M: F-responses of human facial muscles. A single motorneurone study. J Neurol Sci 20:211–222, 1973.
158. Trontelj MA, Trontelj JV: Reflex arc of the first component of the human blink reflex: A single motorneurone study. J Neurol Neurosurg Psychiatry 41:538–547, 1978.
159. Willison RG: A method of measuring motor unit activity in human muscle. J Physiol (Lond) 168:35–36P, 1963.
160. Willison RG: Analysis of electrical activity in healthy and dystrophic muscle in man. J Neurol Neurosurg Psychiatry 27:386–394, 1964.
161. Zalewska E, Hausmanowa-Petrusewicz I: Evaluation of MUAP shape irregularity—a new concept of quantification. IEEE Trans Biomed Eng 42:616–620, 1995.

III ELECTRODIAGNOSIS OF NEUROMUSCULAR DISORDERS

Chapter 14

Principles of Electrodiagnosis

Asa J. Wilbourn, MD

INTRODUCTION
HISTORICAL ASPECTS
OVERVIEW
THE REFERRING PHYSICIAN
THE ELECTRODIAGNOSTIC
   CONSULTANT—REQUIREMENTS
THE ELECTRODIAGNOSTIC
   EXAMINATION—GENERAL ASPECTS
NERVE PATHOPHYSIOLOGY
Axon Loss

Focal Demyelination
Mixed (Axon Loss Plus Demyelinating)
   Lesions
THE ELECTRODIAGNOSTIC EXAMINATION
   REPORT
THE ELECTRODIAGNOSTIC EXAMINATION:
   WHAT TO EXPECT

INTRODUCTION

For more than 50 years, the electrodiagnostic (EDX) examination has been used to complement the clinical neurological examination in the evaluation of the peripheral neuromuscular system. Although several different techniques are used to accomplish this, all rely on one of two basic methods: (1) recording the electrical activity of muscle, generated spontaneously, volitionally, or reflexively, by means of needle recording electrodes, and (2) recording the electrical activity generated by nerves or muscles following electrical stimulation of nerve fibers, using either surface or needle recording electrodes.[21]

All the procedures that collectively compose the EDX examination can be grouped under one of three headings: (1) nerve conduction studies (NCS); (2) needle electrode examination (NEE), or "electromyography"; and (3) "special procedures"; the latter are a potpourri of techniques, nearly all variations of the NCS (Table 14–1). The two basic components of the EDX examination—the NCS and the NEE—are complementary; a limitation of one frequently is countered by a benefit of the other.[19] The advantages and disadvantages of each are shown in Table 14–2.

NCS are of three basic types: motor, sensory, and mixed. They are used to evaluate motor, sensory, and mixed nerves. Motor NCS are used to assess motor nerves and sensory NCS to assess sensory nerves. However, mixed NCS have some major limitations and relatively few benefits for assessing mixed nerves (e.g., median, ulnar, peroneal, etc.). Consequently, these nerves are also assessed by motor and sensory NCS which, for technical reasons, are performed on the motor and sensory fibers of a mixed nerve sequentially, rather than simultaneously. Thus, unlike the motor and sensory NCS, mixed NCS are not an essential component of the basic EDX examination; they are performed mainly to assess the nerves in the hands and feet.[5, 21]

The principal function of NCS is to evaluate the peripheral nerve fibers from their cell bodies of origin (anterior horn of spinal cord for motor; dorsal root ganglia [DRG] for sensory) distally. Two physiological features are determined by the NCS: (1) the ability of the assessed axons to conduct impulses between the stimulating and recording sites; and (2) the speed of impulse transmission along those axons that are capable of conducting impulses. An interrelated function of the NCS is to determine, whenever a nerve fiber lesion is demonstrated, the type of nerve pathophysiology present.

The sensory and mixed NCS assess nerve fibers directly, because they are obtained by stimulating and recording at different points along the nerve; they yield sensory and mixed nerve action potentials (abbreviated, respectively, SNAPs and MNAPs). In contrast, the motor NCS assessment of motor nerve fibers is indirect, because the end point is a compound muscle action potential (CMAP), rather than a motor nerve action potential. Because neuromuscular junctions and muscle fibers are indispensable structures in the

TABLE 14–1	Components of the Electrodiagnostic Examination

1. Nerve Conduction Studies (NCS)[a]
   Motor
   Sensory
   Mixed
2. Needle Electrode Examination (NEE)[a]
3. Special Procedures
   H reflex
   F waves
   Repetitive stimulation
   Blink reflex
   Single fiber EMG

[a]The NCS and NEE, as a unit, often are considered the "basic portion" of the EDX study.

TABLE 14-2	Value and Limitations of the Two Major Components of the Electrodiagnostic Examination

Nerve Conduction Studies	Needle Electrode Examination
Advantages	**Advantages**
Require only passive patient cooperation	Permits widespread examination of motor portion of the peripheral nervous system
Usually produce minimal discomfort	Very sensitive for detecting motor axon loss
Assess some peripheral sensory fibers	Optimal procedure for detecting and localizing motor axon loss lesions
Optimal procedure for detecting and localizing demyelinating lesions	Sensitive for some muscle fiber disorders
Disadvantages	**Disadvantages**
Evaluate only a limited portion of the peripheral nervous system	Requires active patient cooperation
Relatively insensitive to axon loss, especially motor	Produces moderate discomfort
Insensitive for detecting muscle fiber disorders	Does not evaluate sensory nerve fibers
Assess only large myelinated fibers	Assesses motor fibers mainly for axon loss
	Insensitive for detecting demyelinating lesions along motor fibers
	Assesses only large myelinated fibers

Modified from Wilbourn AJ: How can electromyography help you? Postgrad Med 73:187–195, 1983; used with permission.

acquisition of motor NCS responses, the latter can be used to assess them as well; in practical terms, however, the motor NCS are of substantial benefit only for detecting defects in neuromuscular transmission, and not muscle disorders.[21]

The NEE, unlike the NCS, does not assess the sensory portion of the peripheral nervous system (PNS). It does, however, evaluate the motor unit, a function it shares with the motor NCS. Nonetheless, the mechanisms by which each does so are sufficiently different that neither is redundant. In this regard, the NEE is somewhat more versatile than the motor NCS, because it is equally helpful with disorders affecting PNS elements, neuromuscular junctions, and muscle fibers. Concerning PNS lesions, it is highly sensitive to axon loss. Conversely, it is quite insensitive to demyelination, being capable of registering only substantial demyelinating conduction block.[19, 21]

The special procedures consist of several different techniques, each of whose application, in general, is more limited than that of the basic NCS and NEE. Thus, they may assess only some cranial nerves (e.g., blink reflex) or mainly the proximal elements of limb nerves (e.g., F waves, H reflex), or just one portion of the motor unit (e.g., neuromuscular transmission).[1, 11, 21] Although most are variations of the NCS, one of them, single fiber electromyography (EMG), constitutes an entire subspecialty of the NEE.

HISTORICAL ASPECTS

In regard to the history of the NCS and NEE, the latter is the older of the two studies. A 1943 article by Weddell, Feinstein, and Pattle generally is considered the publication that pioneered the clinical use of the NEE,[18] which subsequently was referred to as "clinical electromyography." A 1948 article by Hodes, Larrabee, and German is credited with having initiated a sustained interest in the clinical application of NCS.[10] The first three books published in English that focused on the clinical EDX examination—by Marinacci (1955), Licht (1956), and Buchthal (1957)—dealt solely with the NEE, or "electromyography."[3, 14, 16] Only in the second edition of Licht's textbook, published in 1961, did a chapter on NCS (authored by Gilliatt) first appear in a textbook devoted solely to electrodiagnosis.[15] Even though by the 1960s NCS and NEE were considered two components of the same diagnostic procedure, no suitable name for the combination was forthcoming. Many EDX consultants simply expanded the term "clinical electromyography," using it as a collective title for all the procedures performed in the EDX laboratory. However, this proved confusing, as others continued to use it solely in its original sense, that is, as an alternate name for the NEE.[1, 21] Rather recently, the old term "electrodiagnosis" has been appropriated to serve as the umbrella title. It originally referred only to the "classic" stimulation studies (e.g., chronaxy and strength-duration curve determinations) that are no longer performed.[14] Although highly nonspecific, this term has gained widespread acceptance and will be used throughout this chapter.

In a field as large as neuromuscular diagnosis, and one that spans five decades, a multitude of contributions have been made by a great number of persons. Nonetheless, when the early history of the specialty is reviewed, three names—Buchthal, Gilliatt, and Lambert—are encountered with such regularity that they have been designated by some as the "fathers of electrodiagnosis."[6] This is probably justified, considering that so many of the major advances in the field were made by them, or in their EDX laboratories, or by physicians who they trained.

Buchthal, of Denmark, wrote on neuromuscular physiology in the 1930s. Over the next 40-plus years, he contributed numerous publications on the basic and clinical features of the EDX examination, the majority concerned with various aspects of the NEE findings, particularly the characteristics of the motor unit action potential (MUAP). Buchthal also provided the first detailed analysis of the SNAP.[4] In the late 1950s and early 1960s he was, according to Gilliatt (personal communication), the main moving force in having electrodiagnosis become a discipline separate from electroencephalography.

Gilliatt, of England, first adapted sensory nerve NCS to clinical use.[9] He also wrote the initial article suggesting that polyneuropathies could be divided into two basic types, axon loss and demyelinating, that could be distinguished by their EDX features. In the 1970s, Gilliatt directed a team of researchers that provided an immense amount of information regarding the nature of focal peripheral nerve injuries. He had the unique distinction of being one of the authors of both the 1953 article in which carpal tunnel syndrome (CTS) was first well described, and the 1970 report in which true neurogenic thoracic outlet syndrome (cervical rib and band syndrome) was rediscovered and redefined.[7]

Lambert, of the United States, undertook a very methodical evaluation of motor nerve NCS and their clinical application, following the initial report by Hodes and associates. He formulated guidelines (dubbed "the Lambert criteria") for the EDX confirmation of amyotrophic lateral sclerosis (ALS) that have been used for more than three decades. Lambert also extensively studied the electrophysiological assessment of neuromuscular transmission disorders, in the process of which he described the disorder that bears his name (Lambert-Eaton syndrome).[12]

It is noteworthy that only one of these men, Gilliatt, was a neurologist, reflecting the initial antipathy of members of that profession to electrodiagnosis[2]; in those early times, as Fowler noted, "the majority of practitioners came from a background of physical medicine."[6] Buchthal was a clinical neurophysiologist, whereas Lambert was a respiratory physiologist who had already achieved considerable fame in aviation physiology before he turned to clinical neurophysiology, a discipline in which he was self-taught.[2] Fortunately, he had a superb teacher.

OVERVIEW

The EDX examination augments the clinical neurophysiological examination. It is performed in most EDX laboratories primarily to detect or exclude abnormalities of the peripheral neuromuscular system. It accomplishes this by evaluating (1) the motor unit, which includes the anterior horn cells (AHCs) and the muscle fibers they innervate via the motor nerves and neuromuscular junctions; and (2) the sensory portion of the PNS, principally at and distal to the DRG, which are situated on the most distal aspect of the primary (dorsal) sensory roots.[21]

The results of the EDX examination must be interpreted by specially trained physicians, formerly called electromyographers but now known as EDX consultants. However, both the NCS portion of the EDX examination, as well as the majority of the special procedures, can be performed by well-trained technologists. Nonetheless, unlike most other clinical electrophysiological tests, one portion of the EDX examination, the NEE, cannot be delegated: instead, it must be performed by the EDX consultant. This is because, as Lambert pointed out over 40 years ago, the NEE is a dynamic process that cannot be scripted in advance; moreover, the findings on the NEE are best interpreted at the moment they are obtained, by persons who have extensive knowledge of the peripheral neuromuscular system, the disorders that affect it, and the electrical activity that is generated by it.[13] As has been noted: "Very few nonphysicians attempting to perform EDX studies are likely to possess such requisite knowledge."[21]

The EDX examination also differs considerably from most other clinical electrophysiological procedures in its relative lack of standardization among those who perform it. This absence of uniformity permeates all aspects of the EDX examination, particularly the NCS. For each portion of the EDX examination, it extends to (1) the specific equipment used (e.g., surface versus needle recording electrodes during the NCS; concentric versus monopolar electrodes during the NEE); (2) the particular techniques employed (e.g., orthodromic versus antidromic sensory NCS), as well as to a great number of details regarding each of them. Even if the equipment used and the techniques employed by two EDX consultants are identical, the specific NCS performed, and the particular muscles assessed on NEE, for the same disorder can vary enormously, as these selections mirror the approach to diagnosis embraced by the individual EDX consultants.

One explanation for this overall variability is that among the clinical electrophysiological procedures, the EDX examination is unique in regard to the sheer number of sites on the body from which electrical activity can be recorded. Many different nerves, and segments of those nerves, can be assessed during the NCS, and an even greater number of muscles can be evaluated during the NEE. It is because so many choices are available that two of the integral components of the EDX examination are its formulation before it is even begun, and its modification as it progresses.

One likely cause for the differences in techniques used is that the majority of them, at least in North America, were first devised in one of the relatively few training EDX laboratories functioning in the 1960s, such as Edward Lambert's at the Mayo Clinic and Ernie Johnson's at Ohio State Univer-

sity. The early trainees at these institutions then passed these techniques on to those they trained who, in turn, instilled them in their trainees. (Even today, it is possible to learn where many of a particular EDX consultant's techniques originated, simply by asking the following question: "What distances do you use for your basic median and ulnar sensory NCS?" If the answer is "13 and 11 cm," the techniques came directly, or indirectly, from Lambert's laboratory; if "14 and 14 cm," they came from Johnson's.) Moreover, over the years, new techniques have been devised and refined in various ways by individual EDX physicians.

The principal reason for the differences in the approach to diagnosis is that there are major differences in the underlying philosophy, among EDX consultants, concerning the purpose of the EDX examination. In the simplest terms, some believe it is a "dependent procedure," with its sole function being to support the referring clinical diagnosis, whereas others consider it to be an "independent procedure," performed to establish the actual cause of the patient's complaints, regardless of the referring diagnosis (Table 14–3). As will be obvious in this and subsequent chapters, we consider the EDX examination to be principally an independent procedure.

THE REFERRING PHYSICIAN

To achieve maximal yield, few, if any, clinical electrophysiological techniques require more input from the referring physician than does the EDX examination. The greater the knowledge the referring physicians possess regarding the particular neuromuscular disorders they suspect are present, as well as the EDX examination in general (particularly its limitations), the more likely they are to refer patients at the correct time and for the appropriate reasons.[1, 11, 17]

One important point is that the EDX examination cannot be used as a general screening procedure for peripheral neuromuscular disorders,[6] much as a creatine kinase determination is obtained to detect evidence of muscle disease. Requests to study all four limbs, or to evaluate for multiple generalized disorders (e.g., "rule out ALS, peripheral neuropathy, myasthenia gravis, and myopathy"), reflect ignorance on the referring physician's part regarding the necessary limitations placed on the EDX examination by time constraints, expense, and patient discomfort. So many methods are available for assessing so many nerves and muscles that each EDX examination must, to some extent, be tailored to each patient. Because this depends on the reason for referral, EDX examinations are most beneficial when they are performed after, rather than prior to, clinical examinations by physicians familiar with neuromuscular diseases. Unfortunately, many physicians consider the EDX examination to be a "cheap neurological examination," even though it is neither inexpensive, nor a substitute for the clinical evaluation.

Referring physicians have certain obligations to both their patients and the EDX consultants. First, they should always briefly describe the EDX examination to their patients in advance, including the fact that some unavoidable discomfort will likely be experienced, and then refer only those who are willing to be studied. Second, without prior discussion with their EDX consultants, they should not seek to restrict the extent of the EDX examination in any manner (e.g., by assuring the patient that only one, or certain, limbs will be assessed; that only NCS will be performed; or that only a limited number of muscles, or only muscles of one limb, will be evaluated on the NEE). Even EDX consultants cannot provide such advance guarantees without potentially

TABLE 14-3	Differing Philosophies of Physicians Who Perform Electrodiagnostic Examinations	
	Philosophy 1—Dependent	**Philosophy 2—Independent**
Reason for performing study	To confirm clinical diagnosis by demonstrating findings consistent with it	To confirm clinical diagnosis if correct; to determine actual diagnosis if not correct
Criteria for establishing normal and abnormal diagnosis	Often variable; same finding may be interpreted differently from one study to another, depending on whether it supports the clinical diagnosis	Fixed; same finding has same significance, regardless of clinical diagnosis; EDX diagnosis made whenever preestablished criteria are met
Advantages	Minimal assessment required; high correlation between clinical and EDX diagnoses	False-positive and misleading diagnoses rare; correct diagnosis can be made, even when clinical diagnosis is incorrect or misleading
Disadvantages	False-positive and misleading diagnoses whenever the clinical diagnosis is incorrect or misleading	More extensive, time-consuming assessments often required; false-negative and indeterminate diagnoses more common

EDX, electrodiagnostic.
From Wilbourn AJ, Ferrante MA: Clinical electromyography. In Joynt RJ, Griggs RC (eds.): Clinical Neurology, vol. 1. Philadelphia, Lippincott-Raven, 1997, pp 1–76; used with permission.

compromising the subsequent EDX examinations, because the findings may not be as anticipated. Third, they must be familiar with the time limitations inherent to EDX examinations, particularly in regard to abrupt-onset lesions, in which the assessments can be performed either too soon or too late after onset (i.e., before various EDX abnormalities have had time to fully develop, or so late that many of them have resolved in the interim). Fourth, they should be aware of the relatively few conditions that cause EDX examinations to be contraindicated. Two of the most important of these are (1) automatic cardiac internal defibrillators (ACIDs); NCS cannot be performed on patients with these devices unless they are temporarily deactivated by the patient's cardiologist[21]; and (2) bleeding tendencies, caused by either disease or anticoagulants; only very limited NEEs, typically restricted to small, distal muscles, can be performed on these patients, for fear of causing massive subcutaneous or intramuscular bleeding. Both of these are relative contraindications.[1, 11] Thus, patients with ACIDs cannot be assessed in the EDX laboratory for CTS, which is diagnosed by NCS, but they can be evaluated for a cervical radiculopathy, which is detected by the NEE. Similarly, patients with bleeding disorders can be assessed for those focal and generalized PNS disorders that cause demyelination and are diagnosed by NCS (e.g., CTS), but not for AHC disorders, radiculopathies, and myopathies, all of which are diagnosed principally by the NEE. Another contraindication to EDX examinations is inaccessibility of the body regions to be evaluated. Denuded skin areas, open or recent wounds (surgical or traumatic), bandages, casts, multiple lines for monitoring or blood vessel access, external orthopedic hardware, limb deformities (particularly contractures), limb movement disorders, and gross obesity are just some of the factors that can prevent EDX examinations from being performed, or can seriously compromise their value. Conversely, referring physicians should know how few restrictions there are regarding diet and medication for patients who undergo EDX examinations. Patients can eat, drink, and take (with just two exceptions) all of their drugs prior to their visits to the EDX laboratory, without imperiling the results. The two medication restrictions are (1) anticholinesterase drugs for myasthenia gravis—if patients are referred for EDX confirmation of this disorder, then this medicine should be temporarily discontinued, if possible, for at least 6 hours, and preferably 24 hours, before they are evaluated; and (2) anticoagulants—if the patients need to undergo a NEE, then this medication must be temporarily stopped far enough in advance that their coagulation profile is normal at the time of the EDX assessment. Noteworthy is the fact that analgesics should not be withheld.

THE ELECTRODIAGNOSTIC CONSULTANT—REQUIREMENTS

Physicians performing EDX examinations must possess a basic fund of knowledge regarding the anatomy of the neuromuscular system, the disorders that affect it, and the EDX changes those disorders produce, if the results obtained are to have any clinical value. A certain familiarity with the gross anatomy of the PNS is necessary for performing NCS and for the recognition and localization of root, plexus, and focal or multifocal peripheral nerve lesions. Any physician who does not know, for example, how to distinguish a lower trunk brachial plexopathy from an ulnar neuropathy with the clinical examination is equally unlikely to be able to do so with the EDX examination. Similarly, the gross anatomy of voluntary muscles must be appreciated for a meaningful NEE to be performed. Concerning neuromuscular diseases, the EDX consultant must possess a vital core of information. While it does not have to be encyclopedic, it must include a practical understanding of how the various peripheral neuromuscular disorders—focal, multifocal, and generalized—appear clinically at onset, how they change over time and, most importantly, how they present on the EDX examination. (Their manifestations result from alterations in neurophysiology that they produce.)

Many EDX skills tend not to be acquired until a certain level of expertise is achieved, based mainly on the number and variety of patients studied. Only with experience does the EDX examination come to be used fully as an extension of the neurological examination. One obvious sign of competency is for the EDX examination to be focused on the patient's complaints, rather than being carried out in a rote manner. Consider the patient referred with the diagnosis of "median neuropathy," because he or she complains of dense sensory loss affecting just the middle finger. Probably most inexperienced EDX consultants would perform only the basic median sensory NCS, even though, in most EDX laboratories, it assesses the digital nerves of the index, rather than the middle, finger. In contrast, most experienced EDX consultants would stimulate and record from the symptomatic finger during median sensory NCS. Another skill acquired through experience is judging how significant a finding, or combination of findings, is in the overall EDX assessment. Just as with the clinical examination and most other laboratory diagnostic procedures, the results of the EDX examination frequently do not fit neatly in one of two categories, labeled "normal" and "abnormal," respectively. There is a space between these extremes—a gray area where the findings on many EDX examinations most comfortably can reside. Inexperienced EDX consultants have a tendency to "over-call" normal variants and trivial abnormalities, particularly if more than one is present (Table 14–4). Another skill EDX consultants acquire with experience is the ability to correlate the results of the NCS and the NEE, being able to recognize pathological changes and distinguish them from technical artifacts. One example of this aptitude concerns the interrelationship between the motor NCS and the NEE. Although they are different EDX procedures, they both assess the motor unit, so certain correlations can be established for them. Thus, if during motor NCS a very low-amplitude CMAP is found, then the NEE of the recorded muscle should reveal the cause, as follows: (1) if severe denervation, then a markedly reduced number of MUAPs firing, on maximal effort, at faster than their basal firing rate (i.e., reduced MUAP recruitment), as well as fibrillation potentials and chronic neurogenic MUAP changes, depending on lesion duration; (2) if an unusual myopathy, then MUAPs firing out of proportion to effort (i.e., early MUAP recruitment), with the MUAPs being of decreased amplitude, decreased duration, and highly polyphasic; or (3) if anomalous innervation, then normal MUAP recruitment, normal-appearing MUAPs, and no fibrillation potentials. Conversely, the CMAP amplitude can be used as a check on the NEE findings. Thus, if substantially reduced recruitment of MUAPs of normal configuration is seen on a NEE of a particular muscle, and an axon loss lesion is the suspected cause, then a CMAP recorded from that muscle should be very low in amplitude. If it is not, then there are only a limited number of possibilities, as follows: (1) there has been some error in the interpretation of the NEE findings, or in the performance of the motor NCS; (2) the nerve supplying the muscle in question is affected by a severe but incomplete focal axon loss lesion located proximal to the stimulation site, but the NCS have been performed too early (before the CMAP amplitude has had time to decrease); or (3) the nerve supplying the muscle is affected by a severe but incomplete demyelinating conduction block lesion located proximal to the stimulation site.[20]

THE ELECTRODIAGNOSTIC EXAMINATION—GENERAL ASPECTS

EDX examinations generally are performed in rooms with low noise levels and subdued lighting, which are well away from high electrical sources (e.g., generators). Most often patients are studied while lying on an examination table, because this facilitates obtaining adequate muscle relaxation; however, a few procedures require that patients be sitting, and some EDX consultants prefer to perform the complete examination with patients in that position. Because cool limbs can create several misleading alterations in both the NCS and NEE results, the extremities to be assessed are maintained at normal temperature with blankets or various heating devices; if initially they are cool, then they are warmed to the appropriate temperature, most often with warm water or special heat lamps.[5, 21]

The EMG machines used by the early investigators all had to be constructed by or for them, because none was commercially available. Currently, however, several different manufacturers produce such equipment. The same EMG machine is used for all portions of the EDX examination; however, for both the NCS and for most of the special procedures, it is supplemented with an electrical nerve stimulator (Fig. 14–1). The EMG machine and the patient are linked by insulated wires. These conduct electrical impulses to the patient for nerve stimulation, and they transmit electrical activity from the patient's muscles and nerves to the EMG machine for recording, where it is amplified and then displayed on a screen (as well as, during the NEE, played over a loudspeaker).

Electrodes serve as the interfaces between the patient and the wires from the EMG machine. A wide variety of these is available. For the NCS, needle or surface electrodes can be used for both stimulating and recording. In North America, most EDX consultants employ surface electrodes throughout. These usually consist of metal discs of various sizes and shapes, which are affixed to the skin with adhesive tape over the nerve or muscle to be recorded from, after being coated with an electrolyte gel to reduce skin impedance. Some newer recording electrodes consist of single use, self-adhesive, thin, metallic strips. Ring electrodes are used to record from, or to stimulate, the digital nerves in the fingers. The typical nerve stimulator is a hand-held device with two prongs, each approximately 4 cm long and separated by a fixed distance of 3 cm, connected to the EMG machine by a cable (see Fig. 14–1).

For the NEE, typically either concentric or monopolar needle electrodes are used; each type is manufactured by several different companies that provide them in different lengths and diameters. Generally, only a single type is used in a given laboratory, with the one selected most often depending on the training biases of the EDX consultants using them. Both types of needle recording electrodes can be used multiple times if they are sterilized between patients. Formerly, this was the standard practice in most laboratories. However, since the appearance of the AIDS epidemic in the 1980s, disposable needle electrodes are being used exclusively in most laboratories. In addition to the electrodes used for stimulating and recording during all portions of the EDX examination, a "grounding" electrode must be used to shunt unwanted electrical artifact.[5, 11, 21]

Similar to so many aspects of the EDX examination, there is no standardized approach followed in regard to the specific limb or limbs assessed. As would be expected, when symptoms are localized to a single limb, it is always evaluated. However, if homologous limbs (e.g., both upper extremities) are symptomatic, either one or both limbs may be assessed, depending principally on whether abnormalities are found in the initial limb studied (which customarily is the most symptomatic limb). If abnormalities are not found in the initially studied limb, assessment of the other limb rarely yields a positive result. Whenever a generalized disorder of any type is suspected, two limbs are usually assessed; most

TABLE 14–4	Normal Variants and Insignificant Abnormalities Often Found on Nerve Conduction Studies and Needle Electrode Examination

Nerve Conduction Studies
Bilaterally unelicitable:
  Plantar responses in patients > 45–50 years of age
  Sural and superficial peroneal SNAPs in patients > 60 years of age
Unilateral or bilateral low-amplitude or unelicitable peroneal CMAPs, recording EDB
Bilateral borderline low-amplitude median and ulnar (particularly orthodromic) SNAPs in patients with large hands
Low-amplitude median CMAPs in patients with hand deformities resulting from arthritis

Needle Electrode Examination
Fibrillation potentials, MUAP loss, and chronic neurogenic MUAP changes in EDB muscle; unilateral or bilateral
Fasciculation potentials in the AH muscle; usually bilateral
Complex repetitive discharges in the iliacus muscle; usually bilateral

AH, abductor hallucis; CMAP, compound muscle action potential; EDB, extensor digitorum brevis; MUAP, motor unit action potential; SNAP, sensory nerve action potential.

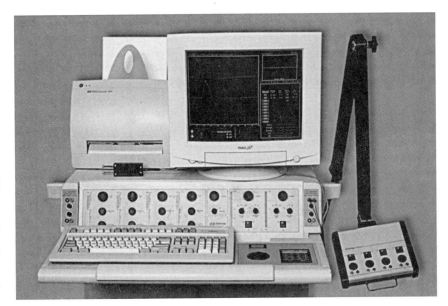

FIGURE 14–1. One of the many EMG machines currently available for performing electrodiagnostic (EDX) studies. The same EMG machine is used for all portions of the EDX examination: nerve conduction studies (NCS), needle electrode examination (NEE), and the special procedures. The major difference is that a nerve stimulator (shown resting in front of the printer, left-hand side) is required for the NCS and some of the special procedures.

often this consists of one upper extremity and the ipsilateral lower extremity. In some laboratories, however, the two limbs are on opposite sides (e.g., right arm and left leg), although such an approach has no apparent benefit. Certainly, it is quite illogical in these situations to study homologous limbs (e.g., both lower extremities), as any abnormalities present usually are bilaterally symmetrical. Two exceptions include early ALS and multifocal motor neuropathies, in which the abnormalities typically are quite asymmetrical. In many patients with suspected generalized disorders, a third limb must be included as well, typically for comparison purposes. Thus, in the presence of a rather severe generalized polyneuropathy, the only means of determining whether some atypical changes in one limb are due to the underlying disorder, or to a superimposed process, is to assess the contralateral limb to determine whether the findings are symmetrical. Multilimb (three- and four-limb) assessments generally are performed only if the EDX consultant considers them necessary for diagnosis, and not because they were requested by the referring physician. A patient referred for a multilimb "rule out everything" study nearly always can derive far more benefit from a thorough clinical examination by a physician experienced in diagnosing neurological disorders than by undergoing an unguided, very uncomfortable, time-consuming, expensive laboratory procedure. Ideally, only after the findings on the neurological examination suggest that an EDX examination is indicated is one scheduled, and with a specific disorder in mind.

Unless unusual circumstances supervene, both components of the standard EDX examination (NCS, NEE) are performed in every instance, even though the electrical changes found may be restricted to just one of them. With CTS, for example, the findings are usually limited to the NCS, with the NEE playing little direct role in the diagnosis. Conversely, with radiculopathies, the NEE is the critical portion of the EDX examination (along with, to a limited extent, the H wave test with lumbosacral lesions); the NCS typically contribute very little directly to the diagnosis, at least with isolated root lesions. Nonetheless, with the majority of neuromuscular disorders, both the NCS and NEE are critical for diagnosis, and even when just one of them is the mainstay procedure, often the other provides useful information in either a positive or negative fashion. Thus, if a severe CTS has been diagnosed using the NCS, then detecting fibrillation potentials in one of the median nerve–innervated the-

nar muscles on the NEE can be helpful in management, because it indicates that what customarily is a focal demyelinating process has converted to axon loss, and thus favors early surgery. Similarly, although the NEE may demonstrate that a patient's foot drop is the result of an L5 radiculopathy, a peroneal motor NCS, recording the tibialis anterior (TA) muscle will provide the most accurate information obtainable regarding the type of pathophysiology of the lesion, and therefore its prognosis.

The sequencing of the various portions of the EDX examination is determined by the individual EDX consultant. While some consultants prefer to perform the NEE before the NCS, the majority tend to perform the NCS first.[1, 5, 21] Regardless, if any special procedures are required, they usually are performed during or immediately after the NCS, because most of them are variations of the NCS. In our laboratory, NCS are always performed before the NEE; not only are they considered by most patients to be less uncomfortable than the NEE and, therefore, better tolerated, but they serve as a screening test to help guide the NEE.

Each EDX consultant also determines which NCS are performed on a given limb and exactly what muscles are sampled on NEE; there are no standardized guidelines in this regard. Nonetheless, a common approach is to perform, at a minimum, a "general survey" on the symptomatic limb(s); this consists of certain "basic" NCS and a stock set of muscles studied on NEE. Both motor and sensory NCS usually are performed on every patient, and often on each limb. Regarding the NEE, at least six to eight muscles are sampled in each limb, including muscles situated in the proximal, mid-, and distal portion of the limb, and innervated by different roots and peripheral nerves. The basic EDX examination can then be supplemented with additional NCS and NEE, depending on the referral diagnosis and what is found as the assessment progresses. Frequently, with focal nerve lesions, at least some of these are performed on the contralateral, uninvolved limb for comparison purposes. The basic upper extremity and lower extremity examinations performed in the EDX laboratory at the Cleveland Clinic are shown in Tables 14–5 and 14–6.

NERVE PATHOPHYSIOLOGY

Virtually all the diagnostic changes seen on the various portions of the EDX examination are the result of altered

TABLE 14-5	Nerve Conduction Studies Performed during the Basic Electrodiagnostic Examination

Upper Extremity	Lower Extremity
Sensory	**Sensory**
Median–D2	Sural
Ulnar–D5	
Radial–thumb base	
Motor	**Motor**
Median (thenar)	Peroneal (EDB)
Ulnar (hypothenar)	Tibial (AH)

( ), recording site; AH, abductor hallucis; D, digit; EDB, extensor digitorum brevis.

physiological properties of nerve fibers, neuromuscular junctions, or muscle fibers. In most, if not all, EDX laboratories, the bulk of abnormalities encountered result from primary lesions of nerve fibers. Consequently, those performing EDX examinations must be cognizant of the changes produced by the various types of nerve pathophysiology, as well as the clinical manifestations (if any) of such changes. A critical fact is that virtually all components of the EDX examination assess only large myelinated nerve fibers; thinly myelinated and unmyelinated axons are not studied.

Regardless of the specific cause (compression, traction, radiation, thermal, etc.) of a focal nerve injury, the large myelinated axons respond in a very limited manner. Pathologically, either axon loss or focal demyelination occurs. Pathophysiologically, these are manifested on NCS as conduction failure, conduction block, or conduction slowing.

Axon Loss

If the lesion is severe enough, it kills axons at the injury site and, subsequently, the entire distal segment of nerve undergoes wallerian degeneration. Any amount (1% to 100%) of the axons at the lesion site can be killed. If enough fibers are affected by this process, then on NCS conduction failure is soon obvious: the degenerated axons no longer contribute to the motor or sensory NCS responses and, therefore, the amplitudes of the latter are correspondingly decreased.[21] There are, however, two notable exceptions.

The first of these is that, for varying periods of time after a focal nerve injury is sustained, the axon segments distal to the lesion site can still conduct impulses, even though they are actively undergoing wallerian degeneration. For motor fibers, the CMAP amplitudes on stimulating and recording distal to the lesion are not affected at all for the first 2 to 3

TABLE 14-6	Muscles Sampled on the Needle Electrode Examination during the Basic Electrodiagnostic Examination

Upper Extremity	Lower Extremity
First dorsal interosseus	Tibialis anterior
Extensor indicis proprius	Medial gastrocnemius
Triceps	Flexor digitorum longus/posterior tibialis
Deltoid	
Biceps	Abductor hallucis
Pronator teres	Extensor digitorum brevis
Flexor pollicis longus	Vastus lateralis/rectus femoris
Cervical paraspinals	Gluteus medius/maximus
	Lumbosacral paraspinals

days after onset; however, after that, they begin to fall rapidly, reaching approximately 15% of their preinjury value at day 5, and their nadir at day 7. For sensory fibers, the SNAP amplitudes are unaffected for the first 5 to 6 days post-onset, and then they progressively drop, reaching their nadir at 10 to 11 days after lesion occurrence[21] (Fig. 14–2). Noteworthy is that mixed NCS responses follow the same time course of change as do the sensory NCS responses. Hence, there is nothing intrinsically different about the "time to conduction failure" between the motor and sensory axons. Instead, as Gilliatt and Hjorth have shown, the CMAPs become unelicitable several days before the MNAPs and SNAPs because of differences in recording techniques, and because transmission failure occurs along intramuscular nerve fibers and across the neuromuscular junctions before it does along nerve trunk fibers.[8] In contrast, SNAPs and MNAPs are recorded directly from the nerve trunk fibers themselves, proximal to terminal nerve branches and neuromuscular junctions.[21] As long as some of the fibers of the nerve segment distal to the lesion are capable of conducting impulses, then if the injured nerve is stimulated proximally and distally to the lesion site, the findings are those of a conduction block—that is, the responses on distal stimulation are higher in amplitude than are those on proximal stimulation. This type of conduction block, termed an "axon noncontinuity" conduction block, is encountered much less frequently than is the far more common demyelinating conduction block, because it is a transient phenomenon, present only during the first 6 to 10 days following lesion onset (Fig. 14–3E). After that time, no matter where the nerve is stimulated along its course, the results are those characteristic of conduction failure: uniform low amplitude or unelicitable NCS responses (Fig. 14–3F).

In sharp contrast to the effects it has on the NCS amplitudes, conduction failure causes little, if any, alteration in the latencies and conduction velocities (CVs), both of which are rate measurements; that is, they measure the speed of impulse transmission along those fibers that are capable of conducting impulses. With partial axon loss lesions, the conduction rate is being determined along the uninjured fibers and, therefore, usually is within the normal range. Occasionally, when very severe but incomplete axon loss has occurred, the fastest conducting axons in the nerve are among those that degenerate and, as a result, the distal latencies and CVs are determined along the slightly slower conducting fibers that survive; typically, the degree of conduction slowing seen in these situations is relatively mild, with the decrease being less than approximately 20%; thus, if the normal CV were 50 m/sec, with severe axon loss lesions causing very low NCS amplitudes, the CV could drop to 40 m/sec. With complete axon loss, all the nerve fibers distal to the lesion undergo wallerian degeneration, soon manifested as conduction failure, so no responses can be elicited and neither latencies nor CVs can be determined. Although it is encountered quite infrequently, there is one important exception to the rule that axon loss lesions do not cause substantial conduction slowing. This is seen whenever NCS are performed years after a total axon loss lesion has occurred, followed by progressive proximal-to-distal regeneration of the axons. Conduction is substantially, permanently slowed along these reconstituted segments, because the regenerated axons are narrower than normal, the myelin coat is thinner, and internodal distances are shorter.[10]

The second exception concerns the fact that, for the peripheral nerve axons to undergo wallerian degeneration following a focal nerve injury, they must be separated from their cell bodies; the latter, for the motor fibers, are the AHCs located in the anterior horns of the spinal cord, but for the sensory fibers they are the unipolar cells situated in

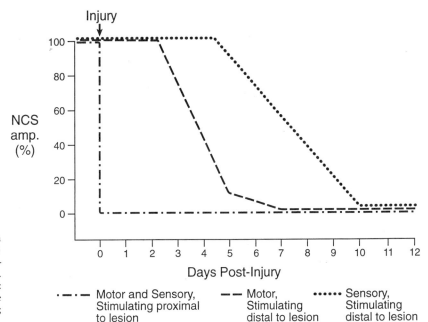

FIGURE 14–2. The time course of changes in the compound muscle action potential (CMAP) and sensory nerve action potential (SNAP) amplitudes following a severe axon loss lesion. (From Wilbourn AJ: AAEE Case Report #12: Common peroneal mononeuropathy at the fibular head. Muscle Nerve 9:825–836, 1986; used with permission.)

the DRG. Because of this anatomical arrangement, axon loss lesions (if severe enough) located at any point along the motor fibers, from the spinal cord distally, will affect the CMAP amplitudes. However, only those axon loss lesions along sensory fibers that are located at or distal to the DRG will affect the SNAP amplitudes. Specifically, lesions at the proximal root level, such as root avulsions, have no effect on the sensory NCS amplitudes, even though substantial, if not total, loss of preganglionic sensory fibers has occurred (Fig. 14–4). In these instances, the sensory axons proceeding centrally from the DRG, into the spinal cord, are those separated from their cell bodies, so it is the central portions of the sensory fibers that degenerate.[21]

The severity of axon loss also determines the changes seen on the EDX examination. Minimal amounts of axon loss affecting either motor or sensory fibers typically have no appreciable effect on the SNAP or CMAP amplitudes. The fact that the responses are decreased by a few microvolts (sensory) or a portion of a millivolt (motor), usually is well within the normal limits of variability. However, minimal amounts of axon loss do register on the NEE, in that a modest number of fibrillation potentials can be detected in the muscles innervated by the damaged motor axons. Thus, fibrillation potentials are the most sensitive EDX indicator of axon loss. Whenever axon loss is more substantial (i.e., when the lesions would be described as being of at least moderate severity), the NCS amplitudes begin to decrease, because of conduction failure along the affected axons. Characteristically, at this stage of severity, only the SNAP amplitudes are affected enough that they are low in amplitude, either absolutely (as judged by EDX laboratory normal values) or relatively (as judged by comparing the SNAP elicited on the symptomatic limb to the corresponding SNAP obtained on the contralateral, normal limb, and having the former be 50% or lower in amplitude). Generally, only when the axon loss has been rather severe, so much so as to reduce the CMAP amplitudes by 50% or more, are the CMAP amplitudes considered low, in either absolute or relative terms[20] (Table 14–7; see also Fig. 14–4). It is unclear why the SNAP amplitudes seem to be more "sensitive" to axon loss than the CMAP amplitudes, that is, why they show substantially greater decreases in amplitude with the same degree of axon loss. Nonetheless, this is a well-founded empirical fact that

proves very helpful during the EDX examination, because it means that the SNAP amplitudes are the most sensitive NCS component to axon loss.

Just as axon loss lesions have predictable effects on the NCS, they have similar effects on the NEE. Mild axon loss, as noted, is manifested solely on the NEE, where it produces some fibrillation potentials in the minimally denervated muscles. When the axon loss is more severe, the number (or density) of the fibrillation potentials usually increases, and the MUAPs elicited by voluntary contraction of the muscle being assessed begin to show changes, initially in their firing pattern, and then ultimately in their internal and external configurations. The instant an axon loss lesion of greater than moderate severity (but still incomplete) is sustained, the MUAP firing pattern seen on maximal effort in the partially denervated muscle is altered: the MUAPs being generated by the surviving motor units are noted to fire in decreased numbers, at faster than their basal firing rates.[1, 5, 13] (See Chap. 12.)

Denervated muscle fibers can become reinnervated by one of two processes: (1) regeneration of the axons from the lesion site, or (2) collateral sprouting from regenerated or surviving axons. If muscle fiber reinnervation is affected by means of proximal-to-distal regeneration, the original motor units may be reconstituted. If this occurs, so-called *reinnervational* MUAPs, or *nascent* MUAPs, are the earliest findings. These MUAPs are low in amplitude, markedly prolonged in duration, and highly polyphasic; typically, they fire at slow to moderate rates and usually in an unsustained fashion. As these motor units mature with the passing of time, the MUAPs they generate gradually undergo alterations in their configuration, ultimately regaining their preinjury appearance. Conversely, if reinnervation is the result of collateral sprouting, then each functioning motor unit contains more than the usual number of muscle fibers. These now larger-than-normal motor units generate MUAPs that are of increased duration and sometimes of increased amplitude as well; such MUAP alterations are referred to as "chronic neurogenic MUAP changes," and often are permanent. Generally, following a motor axon loss lesion, several months (approximately 4 to 6 months, at minimum) must pass before chronic MUAP changes are seen on NEE of the once partially denervated muscles. Regardless of whether muscle

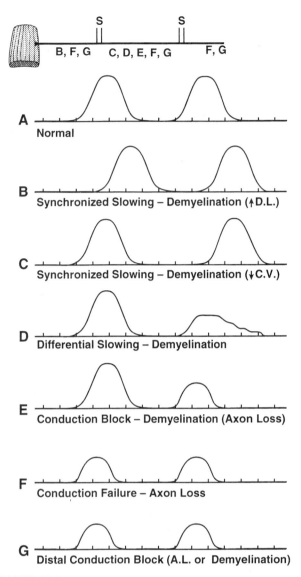

FIGURE 14–3. The motor nerve conduction study changes seen with different types of focal pathophysiology located along various nerve segments both proximal and distal to the nerve stimulation sites. The uppermost drawing illustrates where lesions are situated to produce the abnormalities seen. (From Wilbourn AJ, Ferrante MA: Clinical electromyography. In Joynt RG, Griggs RC (eds.): Clinical Neurology, vol. 1. Philadelphia, Lippincott-Raven, 1997, pp 1–78; used with permission.)

fiber reinnervation occurs by proximodistal axon regeneration or by collateral sprouting, as the number of innervated muscle fibers increases, the number of fibrillation potentials present decreases accordingly.[21]

It is noteworthy that collateral sprouting causes changes not only on the NEE, but also on the motor NCS. Whenever the innervation ratio of the muscle used for recording during motor NCS has been altered considerably by collateral sprouting, the CMAP amplitudes that muscle generates no longer are accurate indicators of the number of motor axons innervating it, as they are in the normal situation. Instead, they reflect the total number of muscle fibers responding to the stimulus, regardless of whether those muscle fibers are innervated by their original axons or by axons that adapted them after they lost their original nerve supply. As a result, with incomplete axon loss lesions, either chronic and ongoing in nature, or static and of at least several months' dura-

tion, the CMAP amplitudes recorded from the affected muscles may be misleading, in that they may be relatively preserved in the presence of significant, permanent motor nerve fiber loss. The latter becomes evident when a NEE is performed on the recorded muscle and demonstrates reduced MUAP recruitment and chronic neurogenic MUAPs.[21]

Focal Demyelination

A focal injury of a myelinated nerve fiber that is not severe enough to cause wallerian degeneration can still compromise the physiological properties of the nerve at the lesion site and thereby impair conduction across it. Two separate processes can result along individual myelinated axons: (1) conduction slowing and (2) conduction block.

Focal demyelinating slowing involving individual axons comprising peripheral nerves has several different NCS presentations, depending on both the number of axons affected and whether, among those so affected, the conduction slowing is to the same or different degrees (i.e., is uniform or nonuniform). Focal slowing across the lesion site is said to be "synchronized," or "uniform," whenever conduction along essentially all the large myelinated fibers is slowed, and to essentially the same degree. In these situations, on NCS the latencies and CVs are altered, but not the amplitudes or durations of the responses. Thus, the only evidence that a

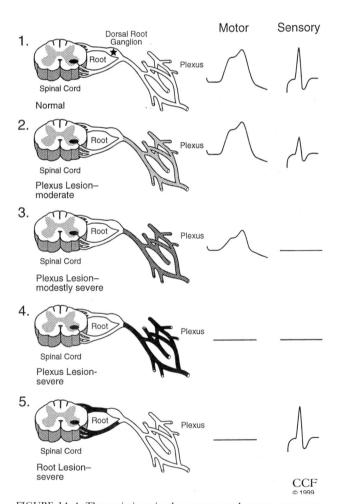

FIGURE 14–4. The variations in the sensory and motor nerve conduction responses elicited following proximal axon loss lesions, depending on the severity of the lesion and on whether preganglionic fibers or ganglion/postganglionic fibers are injured. (Star in topmost illustration identifies the dorsal root ganglion.) (Copyright © 1999 Cleveland Clinic Foundation.)

| TABLE 14–7 | Relative Sensitivity of Various Components of the Electrodiagnostic Examination to Axon Loss and the Order of Appearance of the Abnormalities That Result |

	Amount of Axon Loss					If Abnormal	
	None	Mild	Moderate	Moderately Severe	Severe	First Appearance	Fully Developed
Nerve Conduction Studies							
SNAP amplitudes	Normal	Normal	Decreased	Absent	Absent	5 days	9–11 days
CMAP amplitudes	Normal	Normal	Normal	Reduced	Absent	2–3 days	7 days
Needle Electrode Examination							
Fibrillation potentials	None	Some	Many	Abundant	Abundant	± 14 days	21–35 days
MUAP recruitment	Normal	Normal	Normal	Reduced	None fire	Onset	Onset

CMAP, compound muscle action potential; MUAP, motor unit action potential; SNAP, sensory nerve action potential.
Modified from Wilbourn AJ, Ferrante MA: Clinical electromyography. In Joynt RJ, Griggs RC (eds.): Clinical Neurology, vol. 1. Philadelphia, Lippincott-Raven, 1997, pp 1–76; used with permission.

nerve lesion is present is the fact the SNAPs or CMAPs are delayed in time (see Fig. 14–3B and C). This is a rather common finding with mild CTS and is seen with some ulnar neuropathies at the elbow. When only some of the larger myelinated fibers at the lesion site, not including the fastest conducting ones, are affected by demyelinating conduction slowing, then the only abnormalities seen on the NCS are increases in the duration of the CMAPs and SNAPs; that is, they become prolonged, with a corresponding decrease in their amplitudes if the dispersion is severe enough. The latencies and CVs, determined by the conduction rate along the fastest-conducting fibers, are not affected. This type of demyelinating conduction slowing is referred to as "differential," "desynchronized," or "nonuniform" slowing (see Fig. 14–3D). In some cases, all the fibers are affected by focal demyelinating slowing, but some are affected even more than others. This produces a combination of the two processes described earlier: the distal latencies and CVs are slowed but, in addition, the CMAPs and SNAPs are dispersed, and often low in amplitude.[21]

The NEE is not affected by any type of demyelinating conduction slowing; specifically, fibrillation potentials are not seen, because none of the axons degenerates, and both the configuration and the firing pattern of the MUAPs are unaltered, because all the motor nerve impulses ultimately are reaching their destination.

Focal demyelinating conduction block, similar to axon loss, can affect any percentage of the axons at the lesion site. Because the nerve impulses cannot traverse the lesion site, and therefore cannot reach their intended target, the amplitudes of the NCS responses on stimulation proximal to the lesion site are lower than the amplitudes on stimulation distal to it; the greater number of fibers affected, the lower the amplitudes on proximal stimulation. Conduction block resulting from focal demyelination has an NCS presentation identical to that caused by early axon degeneration (see Fig. 14–3E). For this reason, if conduction block is present along a nerve within the first few days after injury, the underlying pathophysiology (axon loss versus focal demyelination) cannot be determined. To make this distinction, the NCS must be repeated a few days later. At that point, the responses obtained on distal stimulation will have decreased considerably in amplitude if the underlying process is axon loss (see Fig. 14–3F), whereas they will be unchanged if it is focal demyelination.[21]

The NEE findings with demyelinating conduction block are very different from those seen with demyelinating conduction slowing which, as noted, produces no abnormalities. Instead, they are quite similar to those observed with axon loss. Thus, the MUAP recruitment is reduced to an identical degree whenever the same number of motor axons is affected, regardless of whether the responsible process is demyelinating conduction block or conduction failure caused by axon loss. Also, although fibrillation potentials are not seen with absolutely "pure" demyelinating conduction block lesions, such are rarely encountered. Far more often, whenever a focal nerve lesion has been severe enough to produce demyelinating conduction block along most of the motor fibers, it has been severe enough to kill at least a few of them. Hence, fibrillation potentials, sometimes in rather abundant numbers, very commonly are seen with what appears to be, on motor NCS, a focal demyelinating conduction block disorder. Because their NEE presentations are very similar, severe axon loss lesions and severe conduction block lesions can be confused with one another in the EDX laboratory. This most commonly occurs when patients are studied within a few weeks after an abrupt-onset lesion that resulted in predominantly demyelinating conduction block. On NEE of the clinically weak muscles, fibrillation potentials and reduced MUAP recruitment are seen. Because of the presence of the fibrillation potentials, these changes are attributed to a severe axon loss lesion. In fact, they are caused by two separate processes coexisting at the lesion site: (1) axon loss affecting relatively few fibers and causing the fibrillation potentials; and (2) demyelinating conduction block affecting most of the remaining fibers and causing a marked reduction in the number of MUAPs and their neurogenic firing pattern (as well as the clinical weakness noted in the affected muscles). This type of error can result in the patient being given a much worse prognosis than is indicated. It can be avoided if motor NCS are performed, while recording from the weak muscle(s). If these reveal CMAPs of normal or near-normal amplitude when the nerve is stimulated distal to the lesion site, then demyelinating conduction block, rather than conduction failure, is the predominant underlying pathophysiology.[21]

Mixed (Axon Loss Plus Demyelinating) Lesions

In many instances, only one type of pathophysiology results from a focal nerve lesion (e.g., only demyelinating conduction slowing may be found with CTS, or only axon loss with an acute, traumatic nerve injury). However, mixed lesions also occur, in which both axon loss and conduction failure along with one or more types of focal demyelination coexist, or various presentations of focal demyelination (e.g., conduction slowing and conduction block) occur together. In my experience, mixed lesions of this nature most often are seen with ulnar neuropathies at the elbow, although they are relatively common with many abrupt-onset nerve injuries caused by traction or compression, particularly those involving the brachial plexus. (With the latter, the typical combination is conduction failure and demyelinating conduction block; demyelinating conduction slowing is distinctly rare). The fact that more than one type of pathophysiology is present can be appreciated if the nerve is one that can be assessed with NCS, and if it can be stimulated both proximal and distal to the lesion site. Even in those situations in which the nerve can be stimulated only distal to the lesion site (e.g., facial neuropathies, brachial plexopathies involving the proximal trunks), the predominant pathophysiology of the lesion can still be inferred, depending on the amplitude of the SNAPs and CMAPs, and the number of MUAPs seen on NEE. An example is the study of a patient with Bell's palsy 3 weeks after onset of weakness. A facial CMAP is recorded from the paralyzed muscles, which is equal in amplitude to the response on the normal side, and NEE of the muscles shows severely reduced recruitment of MUAPs, indicating that the lesion is almost solely demyelinating conduction block in nature. In such an instance, some fibrillation potentials are likely to be seen as well on NEE, but they do not support the presence of significant axon loss.

THE ELECTRODIAGNOSTIC EXAMINATION REPORT

The EDX report is the principal link between the EDX consultant and the referring physician. Regardless of the abilities of the examiner and the thoroughness and quality of the examination, if the results are not transmitted to the referring physician in a readily understood manner, the assessment has been of no appreciable value. An important point is that, while brevity is often considered a virtue, it can be counterproductive when it is the guiding principle in the preparation of the EDX report. The results of all components of the EDX examination should be integrated into one or more meaningful conclusions and reported as such. Some EDX impressions describe the results of the NCS and NEE separately (e.g., "NEE is normal; NCS shows evidence of . . ."), as if they were two completely separate tests, performed sequentially but otherwise unrelated to one another. In fact, however, they are simply two procedures that assess the same peripheral neuromuscular system, but in somewhat different ways.

All the pertinent data acquired during the EDX examination should be provided in the report, even though the referring physician may only be interested in the diagnosis. If this information is conveyed via preprinted sheets that are completed during the study, then these must provide sufficient space so that all specific details are adequately covered. I have seen NEE forms that contained only a single line for the description of the findings in homologous muscles (e.g., the left and right tibialis anterior muscles). This implied that either (1) both tibialis anterior muscles of the patient would never be examined during one study, or (2) if both were assessed, the findings would always be identical on the two sides, neither of which seems probable. Similarly, on many such forms, the specific muscles are not defined. Thus, the four hamstring muscles are lumped under a single heading, "hamstrings." Such an approach ignores the fact that frequently the findings in these muscles vary considerably, depending on the particular lesion present.

If the EDX examination is incomplete or unsatisfactory for any reason, the referring physician should be alerted to this fact by a statement heading the EDX impression, such as, "Limited EDX Examination—Nerve Conduction Studies Only." The reasons for this can be provided later in the report. Because the EDX examination can be performed on several different body regions, it is helpful to begin the narrative impression by identifying exactly what segments were assessed, for example, "EDX examination of the right upper extremity, and limited nerve conduction studies of the left upper extremity. . . ." If the EDX examination is normal, it is advisable to state not only this fact in the impression, but also that no evidence was found of the specific disorder for which the patient was referred for study, for example, "Normal EDX examination of the right upper extremity. No evidence of an ulnar nerve lesion is seen." Adding such a phrase assures everyone that the EDX assessment was focused on the referring clinical diagnosis; this can prevent questions subsequently being raised on this point. If a PNS lesion is present, the exact nerves involved, the nature of the pathophysiology, its severity, and its chronicity should be reported, if these can be ascertained. The EDX examination identifies abnormalities of the neuromuscular system and not clinical syndromes. For example, CTS is a clinical diagnosis, and an EDX report should describe the results as "showing evidence of a right median neuropathy, at or distal to the wrist, consistent with CTS. . . ." If multiple abnormalities are found during the EDX examination, there are several methods of reporting them, for example, the generalized before the focal, the definite before the more indefinite, those most likely to be responsible for the patient's referral versus those likely to be coincidental or asymptomatic. Sometimes statements appear in the EDX impression regarding the prognosis of various disorders demonstrated on the EDX examination, particularly focal traumatic nerve lesions. This generally is not advisable. A possible exception is the traumatic lesion of recent onset, the underlying pathophysiology of which has been demonstrated on motor NCS to be demyelinating conduction block. It is never prudent to make optimistic predictions regarding the ultimate outcome of severe axon loss lesions assessed soon after their onset. Even if some encouraging changes are seen on the EDX examination (e.g., some "reinnervational" MUAPs in a very weak muscle), conclusions regarding the long-term results should be guarded, and the need for subsequent repeat follow-up EDX examinations must be stressed. If follow-up EDX examinations should be obtained for any reason (e.g., the current examination was performed too near the onset of symptoms for fibrillation potentials to have developed fully in the affected muscles; to determine whether a lesion is progressing; to establish whether reinnervation is occurring), this fact should be explicitly stated at the end of the formal EDX report, with a recommendation concerning when the next assessment should be performed.

THE ELECTRODIAGNOSTIC EXAMINATION: WHAT TO EXPECT

The goal of every EDX examination performed is to provide useful, reliable information to the referring clinician. How often this occurs depends on more than the skills of the EDX consultant, the equipment used, and the cooperation of the patient. The specific information sought, and a multitude of patient complications, can render the results of the EDX study unhelpful. Regarding the former, certain body regions simply cannot be assessed well in the EDX laboratory, primarily because of their inaccessibility for NCS and NEE (Table 14–8). Extremes of age and many confound-

TABLE 14-8	Various Body Regions and Their Accessibility to the Electrodiagnostic Examination

Upper extremity, distal to elbow	Best
Lower extremity, distal to knee	↓
Shoulder girdle/proximal upper extremity	↓
Hip girdle/proximal lower extremity	↓
Head/neck	↓
Trunk	Worst

From: Wilbourn AJ, Ferrante MA: Clinical electromyography. In Joynt RJ, Griggs RC (eds.): Clinical Neurology, vol. 1. Philadelphia, Lippincott-Raven, 1997, pp 1–76; used with permission.

ing factors are in the latter category. Infants and young children cannot cooperate for the EDX study, often necessitating the use of conscious sedation and thereby markedly increasing the time expenditure and expense. In addition, technical factors require the use of special stimulating and recording electrodes, and there are physiological differences, such as significantly slower CVs and the absence of some NCS responses that are easily elicitable later in life. Conversely, in many presumably normal patients older than age 60 years, the lower extremity NCS reveal that the sural and superficial peroneal sensory SNAPs, the H responses, and the mixed plantar responses are all bilaterally unelicitable. Such findings can seriously compromise EDX assessments for polyneuropathies and several focal neurogenic lesions (e.g., peroneal neuropathies, sciatic neuropathies, sacral plexopathies, and lumbar intraspinal canal lesions). Moreover, the older the patient, the more likely that residuals from prior neuromuscular disorders (e.g., poliomyelitis, radiculopathies, or incidental peripheral nerve damage resulting from remote trauma or surgeries) will confound the EDX results. In regard to peripheral nerve fiber disorders, the EDX examination is limited in its ability to detect focal lesions superimposed on regional or generalized ones, to identify two (or more) lesions affecting the same nerve fibers, and to identify recent-onset lesions in the presence of subacute ones. When the preceding factors are taken into consideration, it is apparent that the EDX examination is most likely to yield definite results whenever it is performed on an otherwise healthy nonobese adult, under the age of 60, with no prior significant neuromuscular problems, in whom electrical evidence is sought of a single focal or generalized neuromuscular disorder. The greater the deviation from this model, the less likely the results obtained will be conclusive. Although it may not be at all apparent to the referring physician, there is a marked difference, for example, between performing a unilateral lumbosacral radiculopathy assessment on a 30-year-old athlete and on a 75-year-old obese diabetic patient who has undergone prior lumbar laminectomies and an ipsilateral total hip replacement. This is not to imply that the latter patient should not be referred to the EDX laboratory. Rather, it indicates that all concerned

should appreciate that the EDX results may be less than conclusive and, therefore, suboptimal. Based on such cases, referring physicians may develop the inaccurate impression that little benefit derives from EDX studies. In fact, the EDX examination is the best overall laboratory procedure for assessing the peripheral neuromuscular system. It can, for example, demonstrate the type of nerve pathophysiology present, something neither the clinical examination nor any other laboratory study can do with certainty. Nonetheless, like every other laboratory diagnostic test, it has its limitations.

REFERENCES

1. Aminoff MJ: Electromyography in Clinical Practice, 3rd ed. New York, Churchill Livingstone, 1998.
2. Braddom RL: AAEM: The first five years (1992 presidential address). Muscle Nerve 15:118–123, 1992.
3. Buchthal F: An Introduction to Electromyography. Copenhagen, Scandinavian University Books, 1957.
4. Buchthal F: Publications. Muscle Nerve 1:442–449, 1978.
5. Dumitri D: Electrodiagnostic Medicine. Philadelphia, Hanley and Belfus, 1995.
6. Fowler CJ: Electromyography and nerve conduction. In Osselton JW (ed.): Clinical Neurophysiology. Oxford, Butterworth Heinemann, 1995, pp 45–49.
7. Gilliatt RW: Publications. Muscle Nerve 15:126–129, 1992.
8. Gilliatt RW, Hjorth RJ: Nerve conduction during Wallerian degeneration in the baboon. J Neurol Neurosurg Psychiatry 35:335–341, 1972.
9. Gilliatt RW, Sears TA: Sensory nerve action potentials in patients with peripheral nerve lesions. J Neurol Neurosurg Psychiatry 21:109–118, 1958.
10. Hodes R, Larrabee MG, German W: The human electromyogram in response to nerve stimulation and the conduction velocity of motor nerves. Arch Neurol Psychiatry 60:340–365, 1948.
11. Kimura J: Electrodiagnosis in Diseases of Nerve and Muscle. Principles and Practice, 2nd ed. Philadelphia, FA Davis, 1989.
12. Lambert E: Bibliography. Muscle Nerve (Special Lambert Symposium) 5:S166–S172, 1982.
13. Lambert EH: Electromyography and electrical stimulation of nerve and muscle. In Mayo Clinic staff (eds.): Clinical Examinations in Neurology. Philadelphia, WB Saunders, 1956, pp 287–317.
14. Licht S (ed.): Electrodiagnosis and Electromyography. New Haven, Elizabeth Licht, 1956.
15. Licht S (ed.): Electrodiagnosis and Electromyography, 2nd ed. New Haven, Elizabeth Licht, 1961.
16. Marinacci AA: Clinical Electromyography. Los Angeles, San Lucas Press, 1955.
17. Preston DC, Shapiro BE: Electromyography and Neuromuscular Disorders. Boston, Butterworth-Heinemann, 1998.
18. Weddell G, Feinstein B, Pattle RE: The clinical application of electromyography. Lancet 1:236–239, 1943.
19. Wilbourn AJ: How can electromyography help you? Postgrad Med 73:187–195, 1983.
20. Wilbourn AJ: Pitfalls in nerve conduction studies. In Syllabus, 1995 AAEM Course A: Nerve Conduction Studies—Fundamentals. Rochester, American Association of Electrodiagnostic Medicine, 1995, pp 39–51.
21. Wilbourn AJ, Ferrante MA: Clinical electromyography. In Joynt RJ, Griggs RC (eds.): Clinical Neurology, vol. 1 (looseleaf). Philadelphia, Lippincott-Raven, 1997, pp 1–78.

FOCAL DISORDERS

Chapter 15

Mononeuropathies

Asa J. Wilbourn, MD

INTRODUCTION
PATHOPHYSIOLOGY
ELECTRODIAGNOSTIC AND CLINICAL
   NERVE LESION CLASSIFICATIONS:
   CORRELATIONS
CRANIAL MONONEUROPATHIES
Facial Mononeuropathies

Spinal Accessory Neuropathies
Trigeminal Neuropathies
Hypoglossal Neuropathies
LIMB MONONEUROPATHIES
Traumatic Focal Nerve Lesions
Upper Extremity Nerve Lesions
Lower Extremity Nerve Lesions

INTRODUCTION

One of the very first clinical uses of the electrodiagnostic (EDX) assessment—at that time consisting only of the needle electrode examination (NEE)—was to evaluate patients with traumatic peripheral nerve lesions. As early as 1944, the electroencephalographer Herbert Jasper had devised a portable, clinical electromyograph machine to assess Canadian military personnel with traumatic peripheral nerve injuries.[13] However, because the NEE detects essentially only one type of pathophysiology, axon loss, it is of little value in demonstrating the presence of some compressive and entrapment mononeuropathies, in which focal demyelination is the sole or very predominant pathophysiology. Carpal tunnel syndrome (CTS), for example, by far the most common nontraumatic mononeuropathy encountered in humans, results from demyelinating conduction slowing until late in its course, when axon loss supervenes. Also, the NEE frequently provides unreliable information regarding the severity of axon loss, compared to the amplitudes of the compound muscle action potentials (CMAPs) obtained by recording from clinically weak muscles during the motor nerve conduction studies (NCS). For these reasons, the full potential of the EDX examination in assessing patients with mononeuropathies and multiple mononeuropathies was not realized until NCS came into widespread use in the 1960s. Using these procedures, focal lesions causing only demyelinating conduction slowing could be detected, and focal lesions resulting from demyelinating conduction block could be recognized, and distinguished from those caused by axon loss. Moreover, the severity of motor axon loss lesions could be reliably estimated if a CMAP could be recorded from the weak muscle, and if the lesion was not so chronic that substantial amounts of collateral sprouting would have occurred in the interim. (See Chap. 14.)

An appreciation of the underlying pathophysiology that affects large myelinated peripheral nerve fibers is vital for proper interpretation of the EDX findings, because the pathophysiology determines which components of the EDX examination will be abnormal and in what manner the abnormality will be manifested (e.g., amplitude decreased; latency prolonged). A detailed review of this pathophysiology appears in Chapter 14 of this text. A brief review follows here.

PATHOPHYSIOLOGY

The ability of large myelinated nerve fibers to respond to focal insult, regardless of the exact nature of the latter, is quite limited. If the process is severe enough, it kills a variable number of axons at the lesion site. Conversely, if the insult is less severe, then only the myelin at the lesion site may be affected, resulting in various conduction abnormalities, as follows: (1) demyelinating synchronized conduction slowing, (2) demyelinating desynchronized conduction slowing, or (3) demyelinating conduction block. Any combination of these responses can occur, but very often only one type of pathophysiology predominates. In regard to focal nerve lesions, by far the most underlying pathophysiology, overall, is conduction failure caused by axon loss. In many instances, this is the only detectable type present. Demyelinating conduction slowing is the next most common type of pathophysiology encountered, principally because it is the characteristic underlying abnormality found with CTS, and CTS is a common entity. However, focal demyelinating conduction slowing is seen frequently *only* with CTS and with some ulnar neuropathies along the elbow segment (UN-ES). Demyelinating conduction block occurs less often than the other types. For the most part, it is confined to abrupt-onset lesions caused by mild traction and compression.

The critical difference between focal demyelinating and focal axon loss lesions, for the EDX consultant, is that the focal demyelinating lesions remain focal, affecting only the segment of myelinated axons at the lesion site, whereas focal axon loss lesions, in contrast, always have more widespread effects. This is because they produce wallerian degeneration, resulting in conduction failure, which involves the entire segment of the nerve fiber distal to the lesion. As a result of these differences, demonstrating a focal demyelinating conduction abnormality on NCS requires that the stimulating and recording points straddle the lesion site; that is, one point must be proximal to the lesion, while the other is distal to it, so that either the impulses fail to traverse it, or they do so in an abnormal fashion. If both the stimulating and recording points are distal to the lesion, then conduction is normal, because it is being assessed along a normal segment of nerve. In contrast, with axon loss lesions, once the distal stump fibers have degenerated sufficiently that they can no longer conduct impulses (e.g., by 7 to 11 days postinjury), then low-amplitude or unelicitable NCS re-

174

sponses will be recorded distally, regardless of whether the nerve is stimulated proximal or distal to the lesion site.[27]

Frequently, the type of underlying pathophysiology, or at least the predominant type of pathophysiology, can be predicted, based on symptomatology as well as lesion duration and location. Thus, if clinical weakness or sensory deficits implicating large myelinated fibers are present, then either axon loss with conduction failure, or demyelinating conduction block, or a combination of the two is the responsible pathophysiology. The lesion in these instances cannot be causing pure demyelinating conduction slowing, as neither clinical weakness nor fixed large fiber sensory deficits result from slowing alone. In fact, slowing causes no symptoms, although the focal demyelination responsible for it can, through a separate process, produce sensory symptoms. All but advanced CTS is manifested by positive sensory symptoms (e.g., paresthesias, and less often, pain), rather than by negative motor or sensory symptoms (e.g., weakness, static sensory deficits). In these instances, neither axon loss nor demyelinating conduction block is seen. Conversely, radial nerve lesions at the spiral groove, common peroneal nerve lesions at the fibular head, and ulnar nerve lesions in the hand characteristically present with weakness. Hence, their underlying pathophysiology is axon loss, demyelinating conduction block, or a combination of both. The focal mononeuropathy with the greatest repertoire of underlying pathophysiology is UN-ES. Virtually any type of pathophysiology, or combination of pathophysiology, can be seen with these lesions, although axon loss is the most prevalent overall.[27]

Regarding lesion duration, focal demyelination conduction block characteristically is seen with acute-onset lesions, and usually has resolved by approximately 6 weeks. There are two notable exceptions: multifocal motor neuropathy and radiation-induced plexopathies. Both develop demyelinating conduction blocks in a subacute or gradual fashion that can persist indefinitely. In contrast, focal demyelinating conduction slowing mainly occurs only with chronic lesions, a major exception being the occasional UN-ES of relatively abrupt onset (e.g., one that is first apparent in the postoperative period). Axon loss with conduction failure is by far the most common type of pathophysiology seen, not only with acute lesions (usually resulting from trauma), but also with chronic compressive and entrapment lesions; the only major exceptions to the latter are, as noted, CTS until late in its course, and approximately half of all UN-ES.

Regarding lesion location, any peripheral nerve can sustain a focal injury at any point along its course as a result of acute trauma. Most other etiologies, however, tend to affect the peripheral nervous system (PNS) at relatively predictable locations. Even ischemia and neoplastic compression frequently are found at particular sites along peripheral nerve fibers. Thus, in our experience, primary nerve tumors tend to occur along the posterior tibial nerve near the popliteal fossa, and along the common peroneal nerve near the sciatic notch (where it is nominally the lateral component of the sciatic nerve). Most chronic compressive and entrapment neuropathies, by their nature, appear at predictable, specific locations along nerve fibers, for example, true neurological thoracic outlet syndrome, at the distal T1 anterior primary ramus (APR) or the proximal lower trunk; median nerve, at the wrist (e.g., CTS); ulnar nerve, along the elbow segment or in the hand; radial nerve, at the spiral groove; peroneal nerve, at the fibular head.

ELECTRODIAGNOSTIC AND CLINICAL NERVE LESION CLASSIFICATIONS: CORRELATIONS

Some physicians have attempted to correlate the EDX findings with the clinical classifications of peripheral nerve fiber lesions. While it is tempting to do so, they are not comparable in many respects. The most commonly used clinical classification was provided by Seddon over 50 years ago.[20] With it, three types of nerve fiber lesions are recognized: neurapraxia, axonotmesis, and neurotmesis. Neurapraxia includes demyelinating conduction block, but it also includes the very short-lived conduction block caused by trauma, which presumably is the result of ischemia. Both axonotmesis and neurotmesis are axon loss lesions. They differ from each other based on the degree of injury to the supporting structures of the nerve (e.g., the endoneurium, perineurium, and epineurium) at the lesion site. However, these supporting elements are not assessed by the EDX examination. Hence, the EDX consultant is incapable of distinguishing axonotmesis from neurotmesis—they both register simply as axon loss lesions. Demyelinating conduction slowing, synchronized or desynchronized, is not included in any of the clinical classifications of focal peripheral nerve lesions because it causes no clinical symptoms (Table 15–1).

In the following section, the EDX evaluation of the various mononeuropathies is discussed. Multiple mononeuropathies are reviewed in Chapter 18.

TABLE 15–1	Pathological, Clinical, and Electrodiagnostic Classifications of Severity of Focal Nerve Damage at the Lesion Site[a]					
Pathology at Lesion Site						
Axon	Intact	Intact	Killed	Killed	Killed	Killed
Myelin	Mild damage	More severe damage	Degenerates	Degenerates	Degenerates	Degenerates
Supporting structures						
Endoneurium	Intact	Intact	Intact	Damaged	Damaged	Damaged
Perineurium	Intact	Intact	Intact	Intact	Damaged	Damaged
Epineurium	Intact	Intact	Intact	Intact	Intact	Damaged
Clinical Classifications						
Seddon's	—	Neurapraxia	Axonotmesis	Neurotmesis	Neurotmesis	Neurotmesis
Sunderland's	—	First-degree injury	Second-degree injury	Third-degree injury	Fourth-degree injury	Fifth-degree injury
Electrodiagnostic (EDX) Findings	Focal slowing	Conduction block	Axon loss	Axon loss	Axon loss	Axon loss

[a]Note that (1) the mildest type of nerve abnormality detectible on EDX examination, focal conduction slowing caused by demyelination, has no clinical counterpart, because it produces no symptoms; (2) the EDX examination cannot distinguish the various grades of axon loss lesions from one another because it does not assess the connective tissue supporting structures of the nerves at the lesion site.
From Seddon H: Three types of nerve injury. Brain 66:237–286, 1943, and Sunderland S: Nerve and Nerve Injuries. Edinburgh, Churchill Livingstone, 1968.

CRANIAL MONONEUROPATHIES

For practical purposes, only 2 of the 12 cranial nerves can be assessed adequately in the EDX laboratory: the facial (VII) and the spinal accessory (XI). A limited assessment of the trigeminal (III) and hypoglossal (XII) nerves can also be accomplished. That only suboptimal evaluation is possible for most of the cranial nerves results from lack of accessibility, for both NCS and NEE, of those nerves and the muscles they innervate.[27]

Facial Mononeuropathies

Cranial nerve VII can be assessed with blink reflexes, facial motor NCS, and NEE of many of the facial muscles.

The most common cause for facial nerve evaluation in the EDX laboratory is Bell's palsy. In approximately 90% of instances, the underlying pathophysiology is demyelinating conduction block. In such patients, because the responsible lesion is focal demyelinating in type and proximal to the facial nerve stimulation point (which is at the stylomastoid foramen), the facial CMAP obtained is normal in amplitude and essentially equal to the corresponding response obtained on the contralateral, normal side. The blink reflex, in contrast, is abnormal, because the impulses are blocked at the lesion site. Thus, the R1 and R2 responses on the affected side are unelicitable or very low in amplitude, while the R2 response on the unaffected side is normal. The preserved R2 response indicates that the trigeminal component of the ipsilateral blink reflex is normal. The NEE reveals either absence of motor unit action potentials (MUAPs) on activation or, more often, a substantially decreased number of MUAPs, firing at faster than their basal firing rate of 10 Hz. These MUAP changes are referred to as "reduced MUAP recruitment." Some fibrillation potentials usually are present as well, reflecting the mild, inconsequential amount of axon loss that has occurred. With recovery, ipsilateral R1 and R2 responses can be elicited during the blink reflex. Initially, they are prolonged in latency because the lesion, as it goes from the demyelinating conduction block stage to normal, passes through a demyelinating conduction slowing phase. Concurrently, on NEE, progressively more MUAPs can be activated in the facial muscles.[5, 11, 27]

In approximately 10% of patients with Bell's palsy, the responsible pathophysiology is axon loss. (Typically, this is severe in degree, so that, if reinnervation does occur, often there is significant misdirection of regenerating axons, as discussed later.) With recent-onset lesions, the facial CMAP on the affected side is very low in amplitude or unelicitable, as are the ipsilateral R1 and R2 components of the blink response. On NEE, after 3 weeks, fibrillation potentials usually are noted in abundant numbers in the affected muscles, whereas no, or only a very few, MUAPs are seen, showing reduced recruitment.[11, 27]

Exactly when after onset of facial paralysis the facial CMAP amplitude reaches its nadir, because of conduction failure, is not as clear-cut as with most other focal nerve lesions. This is because, at least conceivably, the first few days of facial weakness may be caused by demyelinating conduction block, and only after that time does axon loss supervene. Consequently, it is prudent to postpone performing facial motor NCS to obtain definitive information regarding the type of pathophysiology operative until the facial weakness is of at least 10 days' duration.

If reinnervation subsequently occurs, the facial CMAP initially is low in amplitude, dispersed, and prolonged in latency, while both the ipsilateral R1 and R2 responses of the blink reflex are either unelicitable or, if detectable, pro-

longed in latency and dispersed. As time passes, both the facial motor NCS and the blink reflex may become more normal in appearance, although frequently they never attain their preinjury status. On NEE, a permanent loss of MUAPs on maximal effort may be obvious and, with lesions of at least 6 months' duration, the MUAPs that are seen typically show chronic neurogenic changes. A common finding is the synkinetic firing of MUAPs in the upper face with voluntary activation of lower facial muscles, and vice versa. These abnormalities are caused by misdirection of regenerating axons. The fact that the regenerating fibers can reach muscles they initially did not innervate indicates that at least some of the supporting structures of the axon, as well as the axon itself, were damaged at the lesion site; thus, a neurotmesis lesion occurred. When this process is substantial, more MUAPs may fire synkinetically in a facial muscle than can be activated volitionally, and the blink reflex may be elicited from muscles other than the orbicularis oculus. As time passes, fibrillation potentials progressively disappear, either because the facial muscles are reinnervated, or because they degenerate.[11, 27] Unlike most other muscles, however, increased mechanical resistance to needle insertion is not encountered with severe degeneration of facial muscles, simply because they lack sufficient bulk for this to be noticeable.

Two types of abnormal facial movements, typically unilateral, can be detected with the EDX examination: facial myokymia and hemifacial spasm. Facial myokymia, usually caused by an intrinsic abnormality of the facial nucleus in the brain stem (characteristically, pontine gliomas or multiple sclerosis) or by cranial nerve VII demyelination within the brain stem or at the root entry zone, often causes the ipsilateral facial CMAP amplitude to be relatively low, presumably because at the time of nerve stimulation some of the muscle fibers are in various stages of contraction. The blink reflex is normal, whereas on NEE myokymic discharges are constantly present, to varying degrees, in all the facial nerve–innervated muscles. With hemifacial spasm, the facial CMAP is normal, and the standard blink reflex is normal. However, the blink reflex "spreads," and can be recorded from muscles other than the orbicularis oculus. On NEE, paroxysmal bursts of large sinusoidal-like waveforms appear intermittently, which vary in duration. Very short bursts of hemifacial spasm, just as very brief cramp potentials, may be asymptomatic, but prolonged bursts are associated with marked facial distortions. Between bursts, which are totally unpredictable in their appearance, the facial MUAPs usually appear normal. It is noteworthy that the lack of myokymic discharges in facial muscles virtually eliminates the possibility of facial myokymia whereas, in contrast, the lack of bursts of hemifacial spasm most definitely does not exclude the diagnosis—it may be simply that no bursts appeared during the NEE.[27]

Spinal Accessory Neuropathies

Cranial nerve XI can be studied with both a spinal accessory NCS, recording upper trapezius, and an NEE of the two muscles that it innervates, the trapezius and sternocleidomastoid. A pertinent point is that during NEE of the upper trapezius, it is often difficult to obtain a full MUAP activation pattern. Hence, the MUAPs frequently fire in decreased numbers at a slow-to-moderate rate.

The most common cause of spinal accessory neuropathies is a minor surgical procedure in the posterior triangle of the neck, frequently performed under local anesthesia, to remove a lymph node or "lump." In these situations, the sternocleidomastoid muscle already has been innervated by the spinal accessory nerve before the latter tracks posteroinferiorly in the posterior triangle, so only the trapezius branch of the nerve is injured. With less common lesions of the nerve, however, both muscles may be affected.

Spinal accessory mononeuropathies almost always are axon loss in type and severe, if not total, in degree. Consequently, the CMAP response recorded during the spinal accessory motor NCS is very low in amplitude (characteristically less than 10% of the corresponding response obtained on the contralateral, normal side), if elicitable at all. With recent-onset lesions, NEE of the upper trapezius typically reveals rather abundant fibrillation potentials and severe, or total, loss of MUAPs; if any are present, they are of normal configuration and demonstrate reduced recruitment on maximal activation. Depending on the etiology, the ipsilateral sternocleidomastoid may or may not show similar NEE changes. If reinnervation is occurring, then an EDX examination performed a few months after onset typically reveals that the CMAP response is low in amplitude, dispersed, and prolonged in latency, and that fibrillation potentials are present in modest numbers in the affected muscles, along with a significantly decreased number of MUAPs, with many often showing "reinnervational" MUAP changes. With lesions that are of 4 to 6 months' or greater duration, chronic neurogenic MUAP changes become evident. However, if recovery is poor, the CMAP amplitude remains very low, and even though fibrillation potentials gradually disappear from the affected muscles, severely reduced MUAP recruitment persists.[27]

Trigeminal Neuropathies

Some of the motor fibers of the trigeminal nerve can be directly assessed by NEE of the masseter muscle, whereas some of the sensory fibers can be indirectly assessed by means of the blink reflex. With certain trigeminal nerve lesions, the ipsilateral R1 and both the ipsilateral and contralateral R2 responses may be uniformly prolonged, or even unelicitable when the ipsilateral trigeminal nerve is stimulated. Conversely, the R2 response on the side of the lesion can be recorded with contralateral trigeminal nerve stimulation. It is noteworthy that trigeminal neuralgia typically does not cause any detectable abnormalities of the blink reflex, because the first division of the trigeminal nerve, the one assessed during the blink reflex, is not affected by this disorder, whereas the second and third divisions, which are involved, cannot be studied.[11] Also, the blink reflex usually is not altered by the various atypical facial pain syndromes.

Hypoglossal Neuropathies

The EDX assessment of cranial nerve XII typically is quite suboptimal. Only the NEE can be performed, and not only can one tongue muscle not be separated from another (e.g., genioglossus from superior longitudinal muscle), but often the evaluation is unsatisfactory for several other reasons. These include the fact that the MUAPs of the tongue (1) usually cannot be inactivated by the patient on request, (2) are much smaller than those of the limb muscles, and (3) are essentially identical in appearance to the biphasic spike form of fibrillation potential. Consequently, it can be difficult to detect fibrillation potentials, or even fasciculation potentials, in the tongue, and it is the search for these that usually initiates the NEE of this muscle. In our experience, the only two situations, both uncommonly encountered, in which NEE of the tongue is informative are the following: (1) when tongue denervation is very severe, and fibrillation potentials are the only electrical activity present, and therefore, readily recognized; and (2) when the MUAPs in the tongue show marked chronic neurogenic changes, so that

they are approaching the size of the MUAPs seen in normal limb muscles.[27]

LIMB MONONEUROPATHIES

Traumatic Focal Nerve Lesions

This category of injury can affect any peripheral nerve at any point along its course. Most of these lesions cause axon loss. Soon after their onset, however, a moderate number of them have variable amounts of associated demyelinating conduction block. To obtain the most accurate data regarding the extent and severity of these lesions, as many sensory and motor NCS that feasibly can be performed to assess them should be done. Thus, with acute-onset median nerve lesions, not only can information be derived by recording from the second digit (e.g., the index finger or D2) while stimulating the median nerve at the wrist, but also while recording from the other median nerve–innervated digits (i.e., D1, D3, D4). Similarly, with traumatic ulnar neuropathies, the ulnar motor NCS, recording first dorsal interosseous (FDI), often is very helpful when performed in addition to the standard ulnar motor NCS, recording hypothenar, as is the dorsum ulnar sensory NCS.

With all suspected mononeuropathies and multiple mononeuropathies, traumatic and otherwise, the EDX examination cannot be terminated as soon as abnormalities are demonstrated in the distribution of the affected peripheral nerve(s). Rather, additional studies must be performed to prove that neighboring nerves are not involved.[11, 27] Thus, as William Litchy (personal communication) has put it so succinctly, "The abnormal must be surrounded by normal."

Upper Extremity Nerve Lesions

Overall, focal neuropathies far more often involve upper extremity nerves, compared to lower extremity ones. The most common ones are discussed next.

Median Neuropathies

The median nerve originates from the C6 through T1 roots; its fibers traverse all three trunks and two of the three cords of the brachial plexus before its medial and lateral heads come together in the axilla to form the median nerve, immediately superficial to the axillary artery. The median sensory fibers supplying the digits arise from the C6 and C7 dorsal root ganglia (DRG), whereas the median motor fibers that supply the lateral thenar muscles originate in anterior horn cells (AHCs) in the C8 and T1 segments of the spinal cord. These fibers do not become contiguous until the lateral and medial heads fuse to form the median nerve, at the most distal cord level of the brachial plexus. An important point, based on the anatomy just described, is that if both the median motor and sensory fibers supplying the hand are affected on NCS, and yet the other NCS of the limb are normal, then the lesion definitely involves the median nerve proper, and not the trunks or cords of the brachial plexus.

Chronic Median Neuropathies at the Wrist (Carpal Tunnel Syndrome). Chronic entrapment of the median nerve beneath the transverse carpal ligament is the most common human entrapment by far. Fortunately for EDX consultants, it is also the one for which the EDX examination is the most sensitive.

Median motor NCS, median sensory NCS, and a number of "single use" NCS (devoted solely to CTS detection) are useful for diagnosis. NEE of the median nerve–innervated thenar muscles is also beneficial in those instances in which axon loss may have supervened.[1]

The very predominant type of pathophysiology found with CTS until late in its course is demyelinating conduction slowing. It is accompanied occasionally by some demyelinating conduction block,[16] but rarely is the latter the sole kind of pathophysiology present. Only with advanced cases of CTS is axon loss found. Why CTS presents so consistently with conduction slowing is unknown; few other entrapments are so predictable in this regard. Nonetheless, this is a godsend for EDX physicians, because it allows accurate localization of even mild lesions.

The initial EDX finding used for localization was a prolonged median motor distal latency. This was originally reported by Simpson, in 1956, in a publication that was the first to describe the value of motor NCS in detecting and localizing focal entrapment neuropathies.[21] This article had a profound effect, both by providing the means of objectively diagnosing a very common (as was becoming apparent) entity, and by markedly expanding the value of the EDX examination for clinicians. Within a few years, it was appreciated that peak latencies of the median sensory nerve action potentials (SNAPs) were even more sensitive to CTS. These were introduced by Gilliatt and Sears, in 1958, who used orthodromically derived SNAPs for this purpose[7]; Campbell, in 1962, described the antidromic technique.[2]

Unfortunately, both the "routine" median motor and sensory latencies have proven to be of increasingly suboptimal sensitivity over the past two decades, as CTS has been recognized clinically at a much earlier stage, owing mainly to heightened public awareness. Consequently, using solely these techniques, in many EDX laboratories the incidence of false-negative EDX examinations with CTS has increased from approximately 12%, as reported in a detailed paper by Thomas and associates in 1967[24] to approximately 40% at present. As a result, additional NCS are now performed in most EDX laboratories to detect CTS in the relatively large percentage of patients who have only electrically mild disease (although their symptoms may be quite disabling). These are "single use" NCS, performed solely for demonstrating CTS. They are therefore *not* substitutes for the basic median sensory and motor NCS, which are "multiuse" studies, done to identify focal lesions at any point along the median nerve fibers, from the C6 and C7 DRG, and the C8 and T1 AHCs, distally. A considerable number of "single use" NCS are available (Table 15–2). The majority of these are versions of palmer NCS, in which the conduction along the mixed nerve fibers between the wrist and palm (i.e., points immediately proximal and distal to the carpal tunnel) is determined. Because these eliminate most of the normal segment of nerve from consideration, they can detect milder degrees of conduction slowing.[11, 19] None of the "single use" NCS has been shown to be substantially more sensitive than another for diagnosing CTS. The technique we use, orthodromic median and ulnar palmar NCS, first described by Eklund in 1975[6] and later modified by Daube in 1977,[3] has proven to be a very sensitive procedure (Fig. 15–1).

The typical sequence of NCS events seen in patients under the age of 60 years with progressively more advanced CTS is rather predictable, and allows for CTS to be graded in regard to severity (Table 15–3). Very severe lesions, in which both the median motor and sensory NCS responses are unelicitable, can no longer be localized to the carpal tunnel segment of the nerve by the standard NCS, because a focal conduction abnormality cannot be demonstrated. However, using a needle recording electrode, a prolonged motor distal latency sometimes can be documented. Obviously, this occurs only if at least a few motor axons supplying the abductor

TABLE 15–2	Selected Electrodiagnostic Techniques Used to Detect Carpal Tunnel Syndrome[a]

Diagnostic NCS:	Results Compared To:
1. Median motor DL	a. Laboratory normal values b. Contralateral median motor DL c. Ipsilateral ulnar motor DL d. Forearm median motor CV
2. Median sensory PL (D1, D2, D3, D4)	a. Normal laboratory values b. Contralateral medial sensory PL c. Ipsilateral ulnar sensory PL
3. Median sensory amp	Ipsilateral ulnar sensory amp
4. Median sensory PL (D1)	Radial sensory PL (D1)
5. Median sensory PL (D4)	Ulnar sensory PL (D4)
6. Median palm-to-digit latency (D1, D2, D3, D4)	a. Median palm-to-wrist motor DL b. Median palm-to-wrist sensory PL
7. Median palm-to-wrist latency	a. Normal laboratory values b. Contralateral median palmar latency c. Ipsilateral ulnar palmar latency d. Median forearm CV
8. Digit (1 or 3)-to-palm median sensory PL	Median palm-to-wrist sensory PL (calculated)
9. Median motor residual latency, carpal tunnel segment	Entire median motor residual latency
10. Median palmar cutaneous PL	Median sensory PL (D1)

Other Techniques
11. Segmental motor latency ("inching" to APB; second lumbrical)
12. Sensory segmental latency ("inching" to D2)
13. Seeking evidence of motor neurapraxia
14. Using terminal latency index
15. Performing NCS during or after "stress" maneuvers
16. Using "repeater" F waves
17. Measuring refractory period

[a]Several variations are available for many of these techniques.
NCS, Nerve conduction study; amp, amplitude; APB, abductor pollicis brevis; CV, conduction velocity; D, digit; DL, distal latency; PL, peak latency.
From Wilbourn AJ: Electrodiagnosis with entrapment neuropathies. In Syllabus: 1992 AAEM Plenary Session I: Entrapment Neuropathies. Rochester, MN, American Association of Electrodiagnostic Medicine, 1992, pp 23–33. Based on references 4, 10, 15 to 19, 22, and 26.

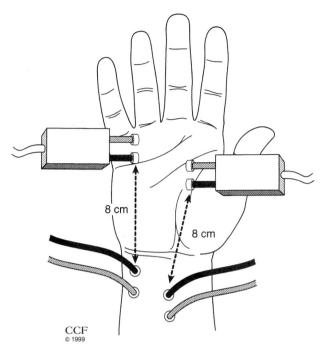

FIGURE 15–1. One of the many procedures designated "palmar nerve conduction studies" that are used to diagnose carpal tunnel syndrome. The median palmar latency can be compared to (1) laboratory normal values, (2) ipsilateral ulnar palmar latency, or (3) contralateral median palmar latency. (Copyright © Cleveland Clinic Foundation.)

in patients in this age group, it is particularly important to perform the basic median motor and sensory NCS, and not just one of the "special use" NCS.

The NEE of the lateral thenar muscles with CTS generally is of value only with the more advanced cases (modestly severe, severe, and very severe lesions). Only in these instances are fibrillation potentials, MUAP loss, and, depending on the duration of the lesion, chronic neurogenic MUAP changes, likely to be found. Many EDX consultants, independently, have endorsed the concept that whenever fibrillation potentials are detected in the APB or OP with CTS, the patient should be treated surgically. Although the idea is reasonable in its underlying premise—that demyelinating conduction slowing has now converted to axon loss with conduction failure, and therefore the process is approaching end stage—this assumption apparently never has been proven with appropriate controlled studies.

Two facts regarding the EDX demonstration of CTS should be appreciated. First, no EDX laboratory currently can be considered up-to-date if CTS is excluded solely by performing the basic median sensory and motor NCS—at least one of the "single use" NCS for CTS must be employed. Second, because CTS is such a common entity, it often coexists with other disorders. In these situations, its presence sometimes can be a hindrance. Thus, when assessing for brachial plexopathies, low-amplitude median SNAPs resulting from moderate or greater CTS compromise the value of some of the major sensory NCS available for assessing the upper and middle trunk and the lateral cord of the brachial plexus. Similarly, a mild ipsilateral asymptomatic electrical CTS in the presence of a C7 radiculopathy, which is actually responsible for the patient's symptoms, can be quite misleading. Also, with many polyneuropathies, particularly in the elderly, the diagnosis rests in large part on finding sensory NCS abnormalities in the upper extremities; the presence of a CTS in these circumstances renders useless the median sensory NCS for this purpose.

EDX studies are important in the assessment of patients with clinically suspected CTS for several reasons: (1) to confirm the diagnosis in questionable or atypical clinical cases; (2) to facilitate early diagnosis and subsequent surgical treatment; (3) to verify the diagnosis prior to surgery; (4) to establish a quantitative baseline in unquestionable clinical cases so that if surgery is postponed, subsequent changes, or lack of same, can be appreciated; (5) to detect unsuspected polyneuropathy in patients thought to have only CTS (in these instances, the diagnosis of a generalized disorder depends on finding abnormalities of other nerves assessed, e.g., the ulnar motor NCS, ulnar sensory NCS, and radial sensory NCS, rather than upon the median NCS); and (6) to detect the actual cause of the patient's symptoms and signs, such as a coexisting cervical radiculopathy, if they are not caused by CTS.[26] We have seen many patients over the

pollicis brevis (APB) or opponens pollicis (OP) are still viable. If virtually all have degenerated, as demonstrated by total absence of MUAPs on NEE of these lateral thenar muscles, then localization to the wrist becomes impossible. Under these circumstances, localization must be accomplished by NEE, by demonstrating that the median nerve–innervated muscles in the forearm—pronator teres (PT), flexor carpi radialis (FCR), and flexor pollicis longus (FPL)—are normal. Unfortunately, because almost all the median nerve branches in the forearm arise from the median nerve trunk near the elbow, the best localization possible under these circumstances is that the lesion is situated somewhere between the proximal forearm and the palm.[27]

While CTS in many patients over the age of 60 years follows the same progression in EDX changes as previously described, in many others it does not. Thus, in patients of advanced age occasionally the median motor distal latencies are abnormal in the presence of normal median palmar and sensory latencies. Conversely, sometimes only the sensory latencies are abnormal. For this reason, if CTS is suspected

TABLE 15–3	Grading of Carpal Tunnel Syndrome, Based on Nerve Conduction Study Changes Alone		
Severity Grade	**Median Palmar Latency**	**Median SNAP**	**Median CMAP**
1. Mild	Prolonged, relative or absolute	Normal	Normal
2. Modest	—	Prolonged latency	Normal
3. Moderate	—	Prolonged latency; normal amplitude	Prolonged latency
	—	Prolonged latency; low amplitude	Prolonged latency
4. Modestly severe	—	Unelicitable	Prolonged latency; normal amplitude
5. Severe	—	Unelicitable	Prolonged latency; low amplitude
6. Very severe	—	Unelicitable	Unelicitable (prolonged latency, using needle recording electrode)

years whose transverse carpal ligaments were needlessly sectioned because their surgeons considered EDX studies superfluous. Included were patients with (1) thenar wasting caused by motor neuron disease and other cervical intraspinal canal lesions, as well as true neurological thoracic outlet syndrome (TOS); (2) one or more "numb" median nerve–innervated fingers, caused by multiple sclerosis, cervical intraspinal canal lesions, vasculitis, early radiation plexopathy, and pure sensory polyneuropathy; and (3) painful hands and fingers, resulting from lesions (particularly iatrogenic) of the cervical roots, brachial plexus, and median nerve. Moreover, probably every high-volume EDX laboratory has encountered hysterics and malingerers whose claim of having an organic disorder (e.g., CTS) had been bolstered by the dreaded combination of a readily available scalpel and poor surgical judgment.[26]

Brief comments are indicated regarding "acute CTS." This is a misnomer; CTS is caused by chronic or, rarely, subacute, compression of the median nerve beneath the transverse carpal ligament. The median nerve can be injured acutely in the carpal tunnel, just as it can be injured anywhere along its course. However, the resulting pathophysiology is either axon loss causing conduction failure, or demyelinating conduction block, and not demyelinating conduction slowing. Consequently, the clinical and EDX features are quite different from CTS as well. The former is manifested as nonfluctuating weakness and sensory loss in a median nerve distribution in the hand, usually without paresthesias. The latter consists of very-low-amplitude (or unelicitable) median CMAPs and SNAPs, with normal or near-normal latencies on NCS, along with fibrillation potentials and severe or total MUAP loss in the APB and OP muscles on NEE.[8]

Anterior Interosseous Neuropathies. The anterior interosseous nerve arises from the median nerve trunk near the elbow. Lesions of this nerve are uncommon and typically affect its proximal portion, resulting in all branches—that is, those to the FPL, flexor digitorum profundus, (FDP-1,2), and pronator quadratus (PQ)—being severely involved. A major exception is neuralgic amyotrophy, in which only one or two of the branches may be substantially affected.

The muscles innervated by the anterior interosseous nerve are too small, too deep, or both, to be satisfactory assessed with surface recording electrodes. Although needle recording electrodes can serve this purpose, they characteristically provide no useful information. This is because nearly all anterior interosseous neuropathies are axon loss in type, and very severe in degree electrically. If they were not, then clinical weakness of the muscles innervated by the nerve would not be present, and it is unlikely that many patients would be referred to the EDX laboratory. Because of the underlying pathophysiology, with complete axon loss lesions no responses will be obtained on recording with needle electrodes in the FPL or PQ, whereas with incomplete lesions the recorded responses will simply reflect the conduction rate along the surviving fibers, which is inconsequential. The amplitudes of the responses obtained with needle recording electrodes are the product of the few viable MUAPs near the needle tip of the recording electrode; they are not a semiquantitative measure of the total number of nerve fibers responding to the stimulus, as they are when surface recording electrodes are used.[27]

With all but very chronic anterior interosseous neuropathies, abundant fibrillation potentials and severe, if not total, loss of MUAPs characteristically are seen on NEE of the affected muscles, two of which, the FPL and the PQ, are easily assessed. With very chronic, very severe lesions in which recovery has been poor or negligible, usually only a few chronic neurogenic MUAPs are present, which show reduced recruitment, or the muscles are completely

atrophic. With the latter, it is often helpful to assess the corresponding muscle in the contralateral normal limb, to verify that the inability to detect the muscle in the affected limb is not the result of faulty NEE sampling technique.[27]

Proximal Main Trunk Median Nerve Lesions. Most of these lesions occur near the elbow, and all are either controversial or quite uncommon. Pronator syndrome is a highly debated entity attributed to compression of the median nerve in the proximal forearm, where it passes between the two heads of the PT muscle. Most EDX consultants have found no changes in patients referred with this diagnosis. The ligament of Struthers syndrome is a rare disorder that affects the median nerve in the distal arm. It typically causes severe axon loss, usually along all fascicles equally. Unfortunately, generally there is no focal demyelinating component, so precise localization cannot be achieved by NCS. As a result, the lesion, based on the EDX examination, can only be localized to some site along the arm segment (i.e., between the axilla and the elbow). Injuries resulting from needles and cannulas are, in many EDX laboratories, the most common cause for median nerve lesions near the elbow. Nearly all of these are axon loss in type. They vary substantially in degree, but often they are severe. Different nerve fibers can be affected, and in various combinations. These include (1) the motor axons supplying the PT, FCR, and other median nerve–innervated forearm muscles; (2) the motor and sensory axons supplying the hand; and (3) the anterior interosseous nerve. It is noteworthy that clinical complaints of pain in the hand and median nerve–innervated fingers may appear disproportionately severe, compared to the amount of axon loss present. It is important to remember, in this regard, that over a century ago Weir Mitchell and co-workers pointed out that the nerve injuries most likely to result in causalgia were incomplete axon loss lesions of the median nerve, tibial nerve, and brachial plexus[14] (i.e., partial injuries of the nerves that provide the main sensory supply to the hand and foot).

Ulnar Neuropathies

The ulnar nerve fibers originate from AHCs in the C8 and T1 segments of the spinal cord, and the C8 DRG. Because the ulnar motor and sensory axons are contiguous along nearly their entire length, lesions at different points along their course—C8 root; lower trunk or medial cord of brachial plexus; ulnar nerve in the arm, at the elbow, or in the forearm—can be mistaken for one another. As a result, it is important on EDX examination to assess axons of other upper extremity nerves that originate from the same roots and traverse the same portions of the brachial plexus. Thus, the NCS should include the median motor, recording thenar, and the medial antebrachial cutaneous, while during the NEE, the APB and FPL (C8–T1, median nerve–innervated) as well as the extensor indicis proprius (EIP) and extensor pollicis brevis (EPB) (C8–radial nerve–innervated) muscles should be sampled.

Of all the peripheral nerves, none can be assessed in the EDX laboratory more thoroughly than the ulnar. Motor NCS can be performed while stimulating supraclavicularly in the axilla, above and below the elbow, and at the wrist. Thus, any potential focal lesion proximal to the wrist can be bracketed by stimulation points. Moreover, several muscles can serve as recording sites, including the hypothenar (standard recording site), as well as the first dorsal interosseous (FDI). Additional recording sites for motor NCS include the adductor pollicis (AdP) muscle in the hand, and the flexor carpi ulnaris (FCU) FDP-3,4 in the forearm. Ulnar SNAPs can be recorded from the fifth digit (D5; standard site), as well as from the fourth finger (D4) and the dorsum of the hand. Essentially all the ulnar nerve–innervated muscles can be

TABLE 15–4	Nerve Conduction Study Changes with an Ulnar Neuropathy at the Left Elbow[a]

Nerve Stimulated	Stimulation Site	Recording Site	Amplitude (µV/mV)		Latency (mS)		Conduction Velocity (M/S)		F Waves	
			L	R	Peak	Distal	L	R	L	R
Median (S)	Wrist	D2	20		3.0			—	—	—
Ulnar (S)	Wrist	D5	NR[b]	16	—	2.9	—	—	—	—
Radial (S)	Dist. forearm	Thumb base	21		2.4		—		—	—
Dorsal ulnar (s)	Dist. forearm	Dorsum hand	4[b]	10	2.4	2.6	—		—	—
Median (M)	Wrist	Thenar	8.0		3.1		—		—	
	Elbow		7.4		8.4		56		26	
Ulnar (M)	Wrist	Hypothenar	5.5	9.9	2.5	2.4	—	—	36	30
	Below elbow	Hypothenar	4.8	—	7.0	—	55	—	—	—
	Above elbow	Hypothenar	4.3	9.3	10.2	8.2	42[b]	55	—	—
Ulnar	Wrist	1st DI	3.0[b]	12.0	3.5	3.3	—	—	—	—
	Below elbow		2.5[b]	—	8.1	—	54	—	—	—
	Above elbow		2.0[b]	11.4	11.6	9.1	41[b]	56	—	—

D, digit; 1st DI, first dorsal interosseous; L, left; R, right; NR, no response.
[a]The unelicitable left ulnar sensory responses and low-amplitude ulnar motor responses (recording first dorsal interosseous) indicate that axon loss has occurred, but definite localization by nerve conduction studies depends on detecting slowing of motor conduction velocity along the elbow segment caused by focal demyelination.
[b]Abnormal response

assessed during the NEE.[25] Nonetheless, both the detection and localization of focal lesions involving this nerve frequently are suboptimal, as will be discussed next.

Ulnar Neuropathies Along the Elbow Segment. Although lesions of the ulnar nerve at this location are relatively common, the exact site of involvement, at the ulnar groove or immediately distal to the groove, in the cubital tunnel, is debated.[4] For several reasons, the EDX assessment of UN-ES tends to be much less satisfactory overall than does the assessment for CTS. The major causes are the pathophysiology of the lesion and various anatomical factors.

The pathophysiology with UN-ES, unlike that with CTS, is strikingly variable. The most common underlying pathophysiology is axon loss with conduction failure, occurring either alone or with various types of focal demyelination present along the surviving fibers. Synchronized conduction slowing is probably the next most common pathophysiology found, followed by demyelinating conduction block and desynchronized conduction slowing. Many combinations are seen, with some fibers affected by one process, and others affected by another[12] (Tables 15–4 and 15–5).

An important point regarding localization is that only those lesions producing focal demyelination (or axon loss lesions studied within the first week of onset, before the distal stump has degenerated enough that it can no longer conduct impulses) can be localized satisfactorily by ulnar motor NCS. Lesions causing solely axon loss, as happens with at least half of all UN-ES, cannot be localized well by the NCS. Nonetheless, they can be detected, if of at least moderate severity, because they affect the amplitudes of the ulnar SNAPs and, with greater severity, the CMAPs. With the typical moderate to mildly severe axon loss UN-ES, the ulnar SNAP, recording D5, is unelicitable, whereas the ulnar CMAPs, recording both hypothenar and FDI, are either normal or slightly low in amplitude at all stimulation sites. The ulnar motor CVs across the elbow segment are normal, because they are being determined along the surviving, unaffected axons. Unfortunately, none of these changes permits localization. If the amplitude of the dorsum ulnar SNAP also is affected, then the lesion is located somewhere at or proximal to the division between the distal and mid-third of the forearm, where that nerve arises from the main trunk of the

TABLE 15–5	Needle Electrode Examination Findings That Accompany the Nerve Conduction Studies Reported in Table 15–4[a]

Muscle	Spontaneous Activity (FIBs)	Motor Unit Action Potentials			
		Recruitment	Duration	Amplitude	Polyphasic
Left					
First dorsal interosseous	2+	↓ ↓ ↓ #, rapidly	Most ↑	Normal	Many +
Abductor digiti minimi	1+	↓ # rapidly	Most ↑	Normal	Some + +
Flexor carpi ulnaris	0	Normal	Normal	Normal	None
Flexor digitorum profundus (3, 4)	0	Normal	Normal	Normal	Few +
Extensor indicis proprius	0	Normal	Normal	Normal	None
Flexor pollicis longus	0	Normal	Normal	Normal	None
Pronator teres	0	Normal	Normal	Normal	None
Biceps	0	Normal	Normal	Normal	None
Triceps	0	Normal	Normal	Normal	None
Abductor pollicis brevis	0	Normal	Normal	Normal	None
Right					
First dorsal interosseous	0	Normal	Normal	Normal	None

FIBs, fibrillation potentials.
[a]Even though the lesion is at the elbow, the forearm ulnar nerve–innervated muscles appear normal. The chronic neurogenic motor unit action potential changes present in the left ulnar nerve–innervated hand muscles indicate that the lesion is of more than 5 to 6 months' duration.

ulnar nerve; thus, an ulnar nerve lesion in the hand or wrist is excluded, but little else. If fibrillation potentials, MUAP loss, or chronic neurogenic MUAP changes are found in the FCU and FDP-3,4 muscles, as well as the ulnar nerve–innervated hand muscles, then the lesion can be localized to some point at or proximal to the elbow, because the ulnar motor branches supplying these forearm muscles arise within the cubital tunnel.[4, 25, 26] Unfortunately, with many severe axon loss UN-ES, these two muscles appear completely normal on NEE. Because this appears to defy all logic, EDX consultants have scurried to find some explanation for it; the most commonly mentioned (albeit quite unsatisfactory), is that, for unconvincing reasons, the particular nerve fascicles supplying them tend to be spared with UN-ES.[23]

Even when UN-ES result from focal demyelination, problems still exist. Because of the substantial redundancy of the ulnar nerve along the elbow segment, markedly variable ulnar motor CVs can be calculated in a normal limb, depending on the degree of elbow flexion present at the time the NCS is performed. These variations result from the disparity between the actual length of the ulnar nerve between the two stimulation points, and that length as determined by surface measurements. When the elbow is extended, the measured length is shorter than the actual length, so the calculated CV is considerably slower than when the measurements are made with the elbow flexed. Much debate has occurred, over the years, regarding the limb position (e.g., elbow fully extended, flexed 90 degrees, flexed 135 degrees, flexed 60 degrees, etc.) most appropriate to use whenever these motor NCS are performed.[4, 25] Regardless of the exact position chosen, it is imperative that all ulnar motor NCS, including those performed on asymptomatic people to obtain normal values, be obtained with exactly the same elbow positioning.[26] An additional problem is that demyelinating conduction block lesions at the elbow, in which the CMAPs are low amplitude or unelicitable on stimulating proximal to the elbow compared to stimulating distal to it, can readily be confused with the ulnar motor NCS changes manifested by median-to-ulnar nerve communications in the forearm ("median-to-ulnar crossovers"), particularly when the FDI muscle is used as the recorded muscle. Typically, patients with this nerve anomaly mistakenly are thought to have an UN-ES, rather than the reverse.[26]

Three final statements regarding the EDX examination with UN-ES are necessary. First, a widespread and very misleading concept is that all UN-ES result from conduction slowing and can, therefore, be localized by ulnar motor NCS; unfortunately, the majority do not manifest demyelinating conduction slowing and, consequently, they cannot be localized by ulnar motor CVs across the ES. Second, mild UN-ES—those causing only intermittent paresthesias or pain in the ulnar nerve–innervated fingers—frequently have no EDX accompaniment; that is, the EDX examinations with them are completely normal, regardless of how extensive they are. Third, clinically severe UN-ES—those manifested as a fixed sensory deficit with marked wasting and weakness in an ulnar nerve distribution—very often cannot be localized well, because they consist exclusively of conduction failure resulting from axon loss, and have no detectable focal demyelinating component to them.[25]

Ulnar Neuropathies in the Hand. These lesions are much less common than UN-ES. Most involve only the deep ulnar (motor) terminal branch of the ulnar nerve, either immediately distal to Guyon's canal, at the base of the hypothenar eminence, or farther distally, in the palm. Unlike with UN-ES, almost the sole types of pathophysiology seen with these lesions are conduction failure or, much less often, demyelinating conduction block. Typically, the ulnar SNAPs are normal, whereas the ulnar CMAP, recording FDI, is al-

most always very low in amplitude or unelicitable. Occasionally, with lesions at or immediately distal to Guyon's canal, the motor fibers supplying the hypothenar muscles are as damaged as those supplying the more distal muscles; in these instances, the ulnar CMAP elicited from the hypothenar muscles is as affected as that recorded from the FDI. Far more often, however, with ulnar nerve lesions in this region the motor branches supplying the hypothenar muscle are much less involved and, consequently, the CMAP recorded from them is only slightly low in amplitude, or just in the low-to-normal range. This disassociation of the ulnar CMAP amplitudes when recording from different hand muscles, as well as the fact that the NEE typically reveals prominent MUAP loss and often (depending on the pathophysiology and duration of the lesion) fibrillation potentials and chronic neurogenic MUAP changes in the FDI and AdP, but not the hypothenar, or other C8-innervated, muscles, allow correct localization.[25, 27] With some incomplete demyelinating conduction block lesions, the motor distal latencies measured along the unblocked axons are prolonged. Unfortunately, this coexisting conduction slowing occurs in only a minority of instances, so it is seldom helpful for localization.

Radial Neuropathies

The radial nerve fibers originate from the C5 through C8 AHCs and from the C6 and C7 DRG. Both the radial nerve proper and its two terminal branches (posterior interosseous nerve, superficial radial nerve) can sustain focal nerve lesions.

Radial motor NCS can be obtained while using either the proximal or distal extensor forearm muscles as recording sites. The principal difference between the two is that the distal one records activation almost solely of muscles innervated by the posterior interosseous nerve, whereas the proximal site records impulses generated primarily by muscles innervated by the main trunk of the radial nerve (e.g., brachioradialis [BR], extensor carpi radialis [ECR]) prior to where it ends, by dividing into its two terminal branches. Superficial radial sensory responses can be recorded from the thumb as well as from the dorsal aspect of the base of the thumb, at the "anatomical snuffbox," but the SNAPs recorded from the latter are considerably higher in amplitude; consequently, it is the preferred radial sensory NCS. On NEE, nearly every muscle innervated by the radial nerve is readily assessed. For NEE localization purposes, the radial nerve and its motor offspring can be considered as three separate segments: (1) that proximal to the spiral groove, from which arise the motor branches to the triceps and anconeus; (2) that distal to the spiral groove, from which arise the motor branches to the BR and ECR; and (3) the posterior interosseous nerve, the source of most of the motor branches to the extensor muscles of the forearm (Table 15–6).

TABLE 15–6	Muscles That Can Be Assessed When a Radial Nerve or Posterior Interosseous Nerve Lesion Is Suspected

Radial Nerve	Posterior Interosseous Nerve
Proximal to spiral groove	Supinator
Triceps	Extensor digitorum communis
Anconeus	Extensor digiti quinti
Distal to spiral groove	Extensor carpi ulnaris
Brachioradialis	Abductor pollicis longus
Extensor carpi radialis (longus)	Extensor pollicis longus
	Extensor pollicis brevis
	Extensor indicis proprius

Radial Neuropathies at the Spiral Groove. Of all focal lesions involving the radial nerve, these are the most common. Characteristically, they present with weakness of the extensor forearm muscles (e.g., wrist drop and finger drop) so, predictably, the underlying pathophysiology is either axon loss with conduction failure, or demyelinating conduction block (Fig. 15–2). The latter is readily demonstrated on radial motor NCS, regardless of whether the recording site is the proximal or distal extensor forearm: the CMAPs are normal on stimulation at the elbow and immediately distal to the spiral groove, while very low in amplitude or unelicitable on stimulations proximal to the spiral groove and in the axilla (as well as supraclavicularly, if the nerve fibers are stimulated at that site). The superficial radial sensory SNAP is normal (Figure 15–2B and C). NEE reveals a marked drop out of MUAPs, and usually some fibrillation potentials as well, in all the muscles innervated by branches arising from the radial nerve distal to the spiral groove. In contrast, with axon loss lesions, the radial CMAPs are uniformly low in amplitude or unelicitable at all stimulation points and the superficial radial sensory SNAP typically is unelicitable (Fig. 15–2D and E). With axon loss lesions, the NEE changes are similar to those seen with substantial demyelinating conduction block lesions at the same site, except that fibrillation

potentials usually are more abundant. It is noteworthy that, unlike the median and ulnar nerves, the radial nerve has many motor branches arising from it at fairly regular intervals along its course. Consequently, focal axon loss lesions of this nerve can be localized by NEE nearly as well as demyelinating conduction block lesions can be localized by motor NCS. Combined lesions, in which both axon loss with conduction failure and demyelinating conduction block coexist, sometimes occur. With them, the CMAPs elicited by stimulating distal to the spiral groove are low in amplitude, whereas those obtained on stimulating proximal to it are appreciably lower, or unelicitable; the superficial radial sensory SNAP usually is unelicitable, but sometimes it is just low in amplitude. Radial motor CVs along the arm segment can be calculated, but they are devoid of value. In normal limbs they frequently are spuriously fast (up to or greater than 90 m/sec) because the actual length of the nerve assessed, as it winds around the mid-humerus, is much shorter than its length as measured on the surface of the limb. (As would be expected, the larger the diameter of the arm, the greater the error of the measured as opposed to the actual distance, and the faster the CV appears.) Conversely, with prominent demyelinating conduction block, the CV may be spuriously slow along the spiral groove segment because the onset of

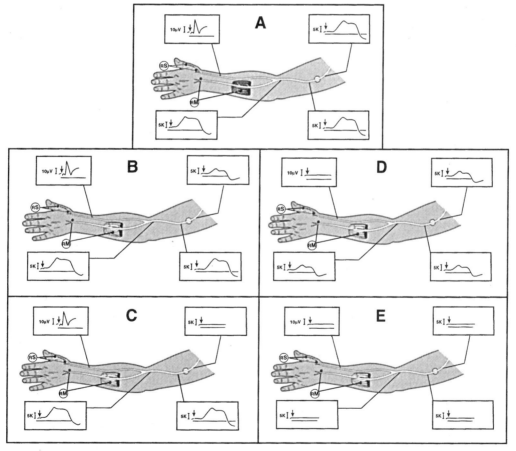

FIGURE 15–2. Radial motor and sensory nerve conduction studies (NCS). In each segment the radial sensory response is in the upper left rectangle, whereas the radial motor responses, elicited by stimulation of the nerve at different points along its course, occupy the remaining three rectangles. The lesion site (spiral groove) appears in all diagrams as a small circle. *A,* Normal motor and sensory NCS. *B* and *C,* Partial and complete conduction block. *D* and *E,* Partial and complete conduction failure, caused by axon loss. Conduction block lesions can be localized precisely to the spiral groove by stimulating immediately above and below the site of injury. Axon loss lesions cannot be localized by radial NCS because the responsible lesion may be situated anywhere along the radial nerve fibers, from the axilla to the elbow; these must be localized by needle electrode examination. (From Dawson DM, Hallett M, Wilbourn AJ (eds.): Entrapment Neuropathies, 3rd ed. Philadelphia, Lippincott-Raven, 1999, pp 204–205. Used with permission.)

the CMAPs, on stimulations above and below the spiral groove, are not being determined for the same fibers. In any case, motor conduction slowing, regardless of severity, cannot account for clinical weakness.[4, 5]

Posterior Interosseous Neuropathies. Two distinctly different syndromes are attributed to lesions of the posterior interosseous nerve, and EDX studies characteristically are abnormal with only one of them. The motor, or supinator, syndrome is almost always pure axon loss in type and severe in degree electrically. The superficial radial sensory SNAP is normal with this entity, as usually are the radial motor CMAPs, recording proximal extensor forearm. In contrast, the radial motor CMAPs, recording distal extensor forearm, uniformly are very low in amplitude, or unelicitable. The NEE reveals fibrillation potentials and severe MUAP loss in all the muscles innervated by the posterior interosseous nerve; however, muscles innervated by the main trunk of the radial nerve, particularly including the BR and ECR, are normal. The radial tunnel, or resistant tennis elbow, syndrome is a highly controversial disorder which, despite the fact that a "pure" motor nerve reputedly is affected, is manifested almost solely as pain. In most EDX laboratories, no abnormalities can be detected in patients considered to have this disorder.[4, 27]

Superficial Radial Sensory Neuropathies. Most lesions of the superficial radial sensory nerve occur at or proximal to the wrist, and almost invariably they are axon loss in type. Consequently, the superficial radial sensory SNAP typically is very low in amplitude with a normal latency, or unelicitable. In contrast, the radial motor CMAPs, as well as NEE of the radial and posterior interosseous nerve–innervated muscles, is normal.[4, 27]

Musculocutaneous Neuropathies

The musculocutaneous nerve fibers derive from the C5 and C6 AHCs and the C6 DRG. A musculocutaneous motor NCS response, recording biceps and brachialis, can be performed, while stimulating both in the axilla and supraclavicularly (where the musculocutaneous nerve fibers are a component of the upper trunk). The lateral antebrachial cutaneous nerve, the terminal component of the musculocutaneous nerve (i.e., the "cutaneous" portion of the musculocutaneous nerve), can be assessed with NCS in the proximal and mid aspects of the anterolateral forearm. On NEE, both the biceps and brachialis are readily sampled.

Because most musculocutaneous neuropathies present with weakness of elbow flexion, the underlying pathophysiology characteristically is either axon loss with conduction failure or, much less often, demyelinating conduction block. The latter usually can be demonstrated on motor NCS: the CMAP response is of normal amplitude on axilla stimulation, while low in amplitude or unelicitable on supraclavicular stimulation. The lateral antebrachial cutaneous NCS response is normal, because the segment of nerve assessed is distal to the lesion. With the much more common axon loss lesions, the musculocutaneous CMAPs are very low in amplitude or unelicitable at both stimulation points, and the lateral antebrachial cutaneous SNAP is unelicitable. With both demyelinating conduction block and axon loss with conduction failure lesions, the results of NEE of the biceps and brachialis are very similar, showing marked dropout of MUAPs and fibrillation potentials (with the latter more abundant if axon loss is the predominant or sole process). In contrast, other muscles innervated by the upper trunk or lateral cord of the brachial plexus (e.g., deltoid, BR, PT, and FCR) appear normal on NEE.[4, 23, 27]

Axillary Neuropathies

The axillary nerve fibers originate from the C5 and C6 AHCs in the spinal cord and the C5 DRG. The principal clinical findings with axillary neuropathies are weakness and wasting of the deltoid and teres minor muscles, and sensory loss overlying a small area of the lateral deltoid. Given the clinical presentation, the underlying pathophysiology is either severe axon loss with conduction failure or, much less commonly, demyelinating conduction block. Axillary motor NCS can be performed by recording from the deltoid muscle while stimulating the axillary nerve fibers supraclavicularly, where they are a component of the upper trunk. A major limitation is that focal axillary neuropathies almost always are located between the supraclavicular stimulation point and the recording site situated over the lateral deltoid muscle. Consequently, severe axon loss lesions cannot be distinguished from severe demyelinating conduction block lesions, because both produce identical abnormalities. These consist of, on NCS, very low amplitude or unelicitable axillary motor CMAPs and, on NEE, significantly reduced MUAP recruitment or no MUAPs, and fibrillation potentials (although with axon loss lesions, the fibrillation potentials may be more abundant). No method for assessing the axillary sensory fibers has been devised. This is unfortunate, because the presence or absence of an axillary SNAP with severe axillary neuropathies could be used to separate axon loss with conduction failure lesions from demyelinating conduction block lesions. Also, because the fibers it would assess derive from the C5 DRG, it would be helpful in the evaluation of upper trunk brachial plexopathies. A practical point is that more than one head of the deltoid should be sampled on NEE, because sometimes the different heads show different changes (e.g., the lateral head may be completely denervated, whereas the posterior head may be severely, but not completely, denervated).[4, 27]

Suprascapular Neuropathies

The suprascapular nerve fibers arise from AHCs in the C5 and C6 spinal cord segments. As with most motor nerves, lesions of the suprascapular nerve are manifested by weakness and often wasting. The expected pathophysiology typically is axon loss with conduction failure, although occasionally demyelinating conduction block is responsible. Lesions can be located at various points along the suprascapular nerve, the two most common being (1) at or near the suprascapular notch, and therefore usually affecting the axons to both the supraspinatus and infraspinatus muscles, and (2) at the spinoglenoid notch, thereby involving just the motor branches to the infraspinatus (Fig. 15–3).

Although techniques for performing motor NCS on the suprascapular nerve have been described, it is difficult to activate the nerve fibers to the supra- and infraspinatus muscles alone, without co-activating nerve fibers supplying other shoulder girdle muscles. As a result, the CMAP amplitudes obtained often are of questionable value (i.e., a low-amplitude response may be elicited from the supraspinatus muscle, even when NEE of that muscle demonstrates no voluntary MUAPs). NEE of both the supraspinatus and infraspinatus muscles should be performed on every patient suspected of having a suprascapular neuropathy. Fibrillation potentials, a substantial loss of MUAPs, and sometimes chronic neurogenic MUAP changes, are seen in the affected muscles. Some notable causes for involvement of solely the infraspinatus muscle (supraspinatus muscle spared) include neuralgic amyotrophy and ganglion cysts situated at the spinoglenoid notch.[4, 27]

Long Thoracic Neuropathies

The long thoracic nerve fibers arise from AHCs in the C5, C6, and C7 segments of the spinal cord. Lesions involving this nerve nearly always are axon loss in type, and because they are manifested as weakness and wasting, they almost

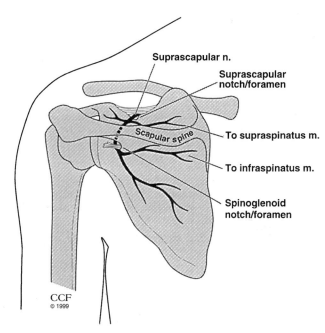

Suprascapular n.

Suprascapular notch/foramen

Scapular spine

To supraspinatus m.

To infraspinatus m.

Spinoglenoid notch/foramen

CCF
© 1999

FIGURE 15–3. Posterior view of the distribution of the suprascapular nerve, showing the two principal sites of injury: suprascapular notch and spinoglenoid notch. (Copyright © Cleveland Clinic Foundation.)

invariably are severe in degree electrically. Because the serratus anterior muscle is composed of numerous strips, rather than being a single entity, it cannot be assessed with a motor NCS. However, NEE can be performed on various segments of the serratus anterior muscle. Sometimes it is helpful to sample more than one segment (e.g., both the C6 and C7 portions) because the findings at the two locations may differ. With lesions of relatively recent onset, fibrillation potentials usually are rather abundant, whereas either no, or only a few, MUAPs are seen on maximal effort. Conversely, with lesions of several months' duration, typically severe MUAP loss and chronic neurogenic MUAP changes affecting the remaining MUAPs are evident. With very severe lesions, the muscle strips may be completely atrophic, so that NEE reveals electrical silence. In these situations, to verify that the needle recording electrode has been inserted in the proper place to assess the serratus anterior muscle, typically the serratus anterior on the contralateral, uninvolved side, is sampled.[27]

Lower Extremity Nerve Lesions

Focal mononeuropathies involving lower extremity nerves are relatively infrequent; the common peroneal nerve is most often affected. The major lower extremity focal nerve lesions are reviewed next.

Peroneal Neuropathies

The axons composing the common peroneal nerve arise from AHCs located in the L4, L5, and S1 spinal cord segments and the L5 DRG. Weakness of foot dorsiflexion ("foot drop") and foot eversion are almost invariable accompaniments of common peroneal neuropathies, most of which occur at the fibular head (FH).

Peroneal motor NCS can be performed, recording from both the anterior and lateral compartments of the leg, labeled tibialis anterior (TA) recordings, and from the extensor digitorum brevis (EDB) muscle on the dorsum of the

foot. Although the latter is the "standard" peroneal motor NCS, the former is generally much more helpful in assessing patients with symptomatic peroneal neuropathies. This is because the muscle (i.e., the TA) whose dysfunction is directly responsible for the patient's major symptom, foot drop, is serving as the recorded muscle. A superficial peroneal sensory NCS can be done, stimulating and recording over the distal aspect of the leg. All muscles innervated, either directly or indirectly, by the common peroneal nerve, are accessible for NEE. These include the short head of the biceps femoris (BF-SH), the only muscle actually innervated by the common peroneal nerve proper, and those innervated by the two terminal divisions of the common peroneal nerve: the deep peroneal and superficial peroneal nerves.[4, 9]

Common Peroneal Neuropathies at the Fibular Head. These are the most frequently encountered focal nerve lesions of the lower extremity. The underlying pathophysiology is either axon loss with conduction failure, or demyelinating conduction block, or a combination of the two (Fig. 15–4). Demyelinating conduction slowing seldom is seen and typically is helpful only when it is found in a patient who is being assessed soon after a demyelinating conduction block lesion, and the symptoms resulting from it, have resolved. With "pure" demyelinating conduction block lesions, the CMAPs obtained on below-FH stimulation are normal, whereas those elicited by stimulation in the popliteal fossa are low in amplitude or unelicitable. The superficial peroneal sensory SNAP is normal (Fig. 15–4, C1 and C2). NEE reveals prominent MUAP loss and a variable number of fibrillation potentials in the muscles of the anterior compartment (e.g., TA; extensor hallucis [EH]), and the lateral compartment (e.g., peroneus longus [PL] and peroneus brevis [PB]) of the leg, as well as the EDB muscle. With axon loss lesions, in contrast, the peroneal motor CMAPs at all stimulation points are uniformly low in amplitude or unelicitable, and the superficial peroneal sensory SNAP usually is unelicitable (Fig. 15–4, B1 and B2). The NEE with axon loss lesions reveals nearly the same findings, in an identical distribution, as with demyelinating conduction block lesions; the only significant difference is that the amount of fibrillation potentials present in the affected muscles usually is greater. Regardless of the type of pathophysiology at the FH, NEE of the BF-SH is normal, because the motor branches supplying that muscle arise from the common peroneal nerve component of the sciatic nerve, in approximately the mid-portion of the thigh. Combined demyelinating conduction block and axon loss with conduction failure lesions sometimes occur. With these, the CMAP amplitude is low on below-FH stimulation, but substantially lower on popliteal fossa stimulation, while the superficial peroneal SNAP generally is unelicitable (Fig. 15–4D). On NEE, the dropout of MUAPs is severe, reflecting the summation of both processes, and fibrillation potentials typically are seen in substantial numbers.

It is noteworthy that only those lesions causing focal demyelination, either alone or coexisting with substantial axon loss, can be localized (by the NCS) to the FH. Those common peroneal neuropathies at the FH that are causing only axon loss must be localized by the NEE. The fact that the BF-SH appears normal, while all the muscles in the anterior and lateral compartments of the leg show obvious neurogenic changes, indicates that the lesion is located at some point between the mid-thigh (i.e., where the motor branches supplying the BF-SH arise) and the FH. No more precise EDX localization is possible with solely axon loss common peroneal nerve lesions at this level.

In addition to common peroneal nerve–innervated muscles, it is very important to assess the tibialis posterior (TP)

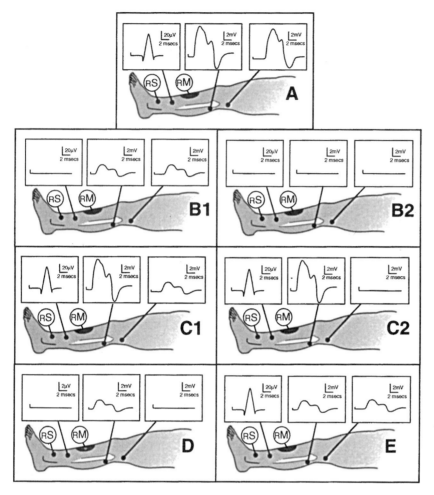

FIGURE 15–4. Peroneal motor and sensory nerve conduction studies (NCS). In each diagram the sensory response is in the rectangle on the left (labeled RS), whereas the motor responses are in the middle and right rectangles (labeled RM). *A*, Normal. *B1* and *B2*, Partial and complete (respectively) conduction failure due to axon loss. *C1* and *C2*, Partial and complete (respectively) conduction block; most often, these are caused by focal demyelination, but they can be caused by axon loss if the studies are performed within the first 5 days or so after onset, while the distal stump fibers are still capable of conducting impulses. *D*, Mixed (axon loss-conduction failure and focal demyelinating) lesion. *E*, Pattern seen with both severe L5 radiculopathy and with proximal deep peroneal neuropathy. Note that *C* and *D* can be localized by nerve conduction studies alone, whereas *B* and *E* require needle electrode examination for localization. (From Dawson DM, Hallett M, Wilbourn AJ (eds.): Entrapment Neuropathies, 3rd ed. Philadelphia, Lippincott-Raven, 1999, pp 280–281. Used with permission.)

and flexor digitorum longus (FDL) muscles, among others, during the NEE in patients with foot drop. These muscles, although innervated by the L5 root and sciatic nerve, receive their terminal innervation from the tibial nerve. Hence, if they also show abnormalities, the lesion does not involve just the common peroneal nerve; rather it is situated at some point proximal to the popliteal fossa, where the sciatic nerve divides to form the common peroneal and tibial nerves.[4, 9]

Proximal Deep Peroneal Neuropathies. These lesions, by definition, affect the very proximal portion of the deep peroneal nerve, usually at the FH. They are often mistaken clinically for common peroneal neuropathies, and frequently they have the same etiology (e.g., leg crossing). However, foot eversion is not affected, nor is there a sensory disturbance in the distribution of the superficial peroneal nerve. The underlying pathophysiology is either axon loss with conduction failure, or demyelinating conduction block. The motor NCS changes can be very similar to those seen with common peroneal neuropathies at the FH. Nonetheless, even if total axon loss is present, CMAPs that are 2 to 3 mV in amplitude can be elicited while recording over the anterior compartment muscles (TA, EH), due to the electrical activity generated by the lateral compartment muscles (PL, PB) reaching the recording electrode by volume conduction; also, the superficial peroneal SNAP is normal, because the superficial peroneal nerve fibers are not affected (Fig. 15–4*E*). On NEE, fibrillation potentials, MUAP loss and, depending on the severity and duration of the axon loss, chronic neurogenic MUAP changes, are restricted to the EDB and the anterior compartment muscles. Specifically, the

PL and PB muscles in the lateral compartment of the leg appear normal or near normal on NEE. Axon loss lesions involving the proximal deep peroneal nerve may be mistaken for severe axon loss L5 radiculopathies on NCS, because the peroneal CMAP and SNAP findings may be identical. Consequently, distinguishing one of these lesions from another customarily depends solely on the NEE.[27]

Distal Deep Peroneal Neuropathies. During NEE of the lower extremity, neurogenic abnormalities indicative of acute and chronic axon degeneration—fibrillation potentials, MUAP loss, chronic neurogenic MUAP changes—commonly are found in isolation in the EDB muscle. Typically, in these instances, the CMAPs recorded from that muscle during peroneal motor NCS are low in amplitude, but the distal latency is normal. Occasionally, however, the CMAPs not only are low in amplitude, but the distal latency is substantially prolonged (e.g., greater than 6.5 msec). Thus, along the distal deep peroneal fibers supplying the EDB, there is a significant amount of axon loss, as well as focal demyelinating slowing along the surviving fibers. When the latter situation occurs, the patient is reported to have the anterior tarsal tunnel syndrome (of Marinacci). In our experience, this disorder, similar to isolated neurogenic changes in the EDB not accompanied by focal slowing, is asymptomatic. Moreover, it probably has the same etiology: compression of the distal deep peroneal nerve at the ankle by shoe wear.[27]

Superficial Peroneal Neuropathies. Lesions located in the proximal leg that involve this nerve produce, on NCS, slightly low-amplitude peroneal CMAPs, recording TA, and unelicitable superficial peroneal sensory SNAPs, along with, on NEE, fibrillation potentials and MUAP loss in the PL and PB muscles. Most focal disorders affecting only this nerve, however, occur in the more distal portion of the leg, and consequently involve only the sensory fibers. Hence, the only abnormality found on the EDX examination is a low-amplitude or unelicitable superficial peroneal SNAP. The NEE of the PL and PB, specifically, is normal.[27]

Tibial Neuropathies

The tibial nerve fibers arise from AHCs in the L5, S1, and S2 segments of the spinal cord and the S1 and S2 DRG. Isolated lesions of this nerve in the leg are quite uncommon, and most cause axon loss with conduction failure. The NCS available for assessing the tibial nerve include the posterior tibial motor NCS, recording both abductor hallucis (AH) and abductor digitorum quenti pedis (ADQP), the direct (M) response of the H-wave test as well as the H response proper, the sural NCS, and medial and lateral plantar NCS. NEE can be performed on tibial-innervated muscles in the thigh (semitendenosis, semimembranosis, BF–long head), the leg (e.g., FDL, gastrocnemii) and the foot (e.g., AH, ADQP). Unlike the situation with the majority of focal peripheral nerve lesions already described, most tibial neuropathies affect the nerve distal to the origin of the main sensory nerve (i.e., the sural), which occurs in the proximal portion of the popliteal fossa. Consequently, the sural NCS usually are normal with these axon loss tibial neuropathies. As a result, the latter are readily mistaken, based on NCS results alone, for severe S1–2 intraspinal canal lesions. Thus, axon loss lesions of the tibial nerve situated in the popliteal fossa present, on NCS, with low-amplitude motor CMAPs, particularly those recorded from the intrinsic foot muscles, a normal sural response, and unelicitable plantar NCS responses. On NEE, fibrillation potentials and MUAP loss are found in all the muscles innervated by the tibial nerve distal to the lesion site, regardless of whether their root innervation is predominantly L5 or S1–2.[27]

The most discussed focal abnormalities of the tibial nerve are entrapment neuropathies affecting its terminal branches: the medial plantar, lateral plantar, and calcaneal. These lesions, referred to collectively as (medial) tarsal tunnel syndrome, initially were thought to cause demyelinating conduction slowing. Depending on the particular terminal nerve(s) involved, they could thus present with prolonged posterior tibial motor distal latencies, recording AH, ADQP, or both; unelicitable, or present but prolonged in latency, medial or lateral plantar responses, or both; and an essentially normal NEE of the tibial nerve–innervated intrinsic foot muscles. However, in most patients referred to the EDX laboratory with a diagnosis of tarsal tunnel syndrome, either no abnormalities are seen, or there is evidence of an S1 radiculopathy or a generalized polyneuropathy. Moreover, when changes are restricted to the distal tibial nerve, the findings usually are indicative of axon loss, rather than of demyelinating conduction slowing. Thus, the typical results are for one or both of the plantar responses to be unelicitable; one or both of the posterior tibial CMAPs to be unelicitable or low in amplitude with essentially normal distal latencies; and for fibrillation potentials, reduced MUAP recruitment, and chronic neurogenic MUAP changes to be seen on NEE in the AH, ADQP, or both. Unfortunately, this combination of EDX changes does not permit satisfactory localization. The responsible lesion could lie anywhere along the tibial nerve axons, from the proximal leg area distally (i.e., distal to the origin of the sural nerve and to the origin of the motor branches supplying the tibial nerve–innervated leg muscles). To compound the problem, the typical patient has undergone several failed tarsal tunnel syndrome surgeries before the EDX examination is performed. As a result, almost all the EDX changes could be residuals of one or more of these operations, rather than evidence of an ongoing lesion.[27]

Sciatic Neuropathies

The sciatic nerve fibers arise from AHCs situated in the L4 through S3 segments of the spinal cord, and the L5 and S1 DRG. Most sciatic neuropathies occur at or near the sciatic notch. The majority are incomplete axon loss lesions and affect the common peroneal fibers more than the tibial. On rare occasions, however, lesions at this level may result at least partially from demyelinating conduction block. The sciatic nerve can be damaged in the mid-portion of the thigh (e.g., with mild mid-shaft fractures of the femur). These characteristically are axon loss in type and very severe in degree electrically, often with total involvement of both the common peroneal and tibial components.

All the NCS performed to assess the common peroneal and tibial nerves can be used to assess the sciatic nerve. The most important of these are the common peroneal motor, recording TA, the M component of the H-wave test, the superficial peroneal sensory NCS, and the sural NCS.

With high sciatic nerve lesions, the peroneal motor CMAPs, recording TA, and the superficial peroneal SNAPs usually are the most severely involved of the four. Characteristically, in all cases, only the amplitudes of these NCS are affected. On NEE, fibrillation potentials, MUAP loss and, with lesions of several months' duration, chronic neurogenic MUAP changes usually are seen in all the sciatic nerve–innervated muscles in the thigh, leg, and foot. However, with some chronic or very slowly progressive lesions, the NEE abnormalities sometimes are minimal or undetectable in the hamstrings. Moreover, with incomplete axon loss lesions near the sciatic notch that affect the common peroneal component more than the tibial component, a characteristic pattern of muscle involvement is noted on NEE. The most abnormal muscles are those innervated by the common peroneal nerve, regardless of their root innervation. The next most severely affected are those innervated by the L5 root, via the tibial nerve; the least affected muscles are those innervated by the S1 root, via the tibial nerve (Table 15–7). The NCS changes seen with sciatic neuropathies, as well as most of the NEE changes, also occur with axon loss sacral plexopathies. The only difference, demonstrable on NEE, is that sacral plexopathies usually (but not always) involve muscles innervated by the superior and inferior gluteal nerves (e.g., the tensor fascia lata and the glutei). Consequently, if these muscles show substantial abnormalities on NEE the lesion very likely affects the sacral plexus, rather than the high sciatic nerve. However, an identical combination of

TABLE 15–7	Common Pattern of Muscle Involvement Seen with Incomplete High Sciatic Neuropathies

Most affected: L4, L5, S1/common peroneal nerve-innervated
  Biceps femoris, short head
  Tibialis anterior
  Extensor hallucis
  Peroneus longus
  Extensor digitorum brevis
Intermediate affected: L5/tibial nerve-innervated
  Semitendinosis
  Semimembranosis
  Flexor digitorum longus
  Tibialis posterior
Least affected: S1/tibial nerve-innervated
  Biceps femoris, long head
  Medial gastrocnemius
  Abductor hallucis

changes can be seen with gluteal compartment syndrome, which yields simultaneously occurring high sciatic and gluteal neuropathies.[27]

One reputed sciatic neuropathy, labeled the piriformis syndrome, produces no detectable changes on the EDX examination. It is in part for this reason that many physicians doubt its existence.

Femoral Neuropathies

The femoral nerve fibers arise from AHCs in the L2 through L4 segments of the spinal cord and the L4 DRG. The femoral motor axons can be assessed with a femoral motor NCS, recording quadriceps. However, there is no reliable NCS for assessing any of the sensory fibers of the femoral nerve, such as the saphenous. On NEE, all heads of the quadriceps muscle can be assessed. The iliacus, innervated directly by branches from the lumbar plexus, and at least one of the thigh adductor muscles (e.g., adductor longus, adductor magnus) innervated by the obturator nerve, should be sampled. These help to distinguish a femoral neuropathy from a lumbar plexopathy or an L2–4 radiculopathy. Most femoral neuropathies are located at or proximal to the inguinal ligament, are axon loss in type, and are rather severe in degree electrically (i.e., severe enough to cause quadriceps weakness). Therefore, the typical changes seen are low-amplitude or unelicitable femoral CMAPs, recording quadriceps, along with fibrillation potentials, reduced MUAP recruitment and, depending on lesion duration, chronic neurogenic MUAP changes. Occasionally, the underlying pathophysiology with a femoral neuropathy is demyelinating conduction block. These can be difficult to distinguish from axon loss lesions, because typically the lesion is at or immediately distal to the inguinal ligament, and stimulating distal to it can be problematic.

Obesity is a major obstacle in the assessment of quadriceps weakness in the EDX laboratory. This renders the femoral motor NCS technically difficult to perform. As a result, the femoral CMAP amplitudes may not be reliable (e.g., they may be unelicitable, and yet NEE of the various quadriceps muscles may demonstrate a considerable number of MUAPs). In any case, if either the iliacus or one of the thigh adductor muscles shows abnormalities similar to those seen in the quadriceps, the responsible lesion is not simply a femoral neuropathy.

Obturator Neuropathies

The obturator nerve fibers derive from the L2 through L4 segments of the spinal cord. There is no NCS available for assessing the obturator nerve. However, one or more of the thigh adductor muscles can be sampled on NEE. Isolated obturator neuropathies are quite rare, with the causes mainly being vasculitis and neoplasms. Far more often, the obturator axons are compromised at the root or plexus level, and the femoral nerve–innervated quadriceps show abnormalities on NEE, as well as the obturator nerve–innervated thigh adductor muscles.

REFERENCES

1. Aminoff MJ: Electromyography in Clinical Practice, 3rd ed. New York, Churchill Livingstone, 1998.
2. Campbell EDR: The carpal tunnel syndrome: Investigation and assessment of treatment. Proc R Soc Med 55:401–405, 1962.
3. Daube JR: Percutaneous palmar median nerve stimulation for carpal tunnel syndrome. Electroencephalogr Clin Neurophysiol 243:139–140, 1977.
4. Dawson DM, Hallett M, Wilbourn AJ: Entrapment Neuropathies, 3rd ed. Philadelphia, Lippincott-Raven, 1999.
5. Dumitru D: Electrodiagnostic Medicine. Philadelphia, Hanley and Belfus, 1995.
6. Eklund G: A new electrodiagnostic procedure for measuring sensory nerve conduction across the carpal tunnel. Ups J Med Sci 80:63–64, 1975.
7. Gilliatt RW, Sears TA: Sensory nerve action potentials in patients with peripheral nerve lesions. J Neurol Neurosurg Psychiatry 21:109–118, 1958.
8. Holmlund T, Wilbourn AJ: Acute median neuropathy at the wrist is not carpal tunnel syndrome. Muscle Nerve 16:1092, 1993.
9. Katirji B: Electromyography in Clinical Practice. St. Louis, Mosby, 1998.
10. Kimura J: The carpal tunnel syndrome: Localization of conduction abnormalities within the distal segment of the median nerve. Brain 102:619–635, 1979.
11. Kimura J: Electrodiagnosis in Diseases of Nerve and Muscle: Principles and Practice. Philadelphia, FA Davis, 1983.
12. Magnotta JA, Wilbourn AJ: Ulnar neuropathy at the elbow: Electrodiagnostic findings in 241 limbs of 200 patients. Muscle Nerve 17:111, 1994.
13. Marinacci AA: Clinical Electromyography. Los Angeles, San Lucas Press, 1955.
14. Mitchell SW, Morehouse GR, Keen WW: Gunshot Wounds and Other Injuries of Nerves. Philadelphia, JB Lippincott, 1864.
15. Nathan PA, Keniston RC, Meadows KD, Lockwood RS: Predictive value of nerve conduction measurements at the carpal tunnel. Muscle Nerve 6:1377–1382, 1993.
16. Pease WS, Cunningham ML, Walsh WE, Johnson EW: Determining neurapraxia in carpal tunnel syndrome. Am J Phys Med Rehabil 67:117–119, 1988.
17. Preston DC: The median–ulnar latency differences studies are comparable in mild carpal tunnel syndrome. Muscle Nerve 17:1469–1471, 1994.
18. Preston DC, Shapiro BE: Electromyography and Neuromuscular Disorders. Boston, Butterworth-Heinemann, 1998.
19. Rivner MH: Carpal tunnel syndrome: A critique of "newer" nerve conduction techniques. In Syllabus: 1991 AAEM Course D: Focal Peripheral Neuropathies: Selected Topics. Rochester, MN, American Association of Electrodiagnostic Medicine, 1991, pp 19–24.
20. Seddon HJ: Three types of nerve injury. Brain 66:238–286, 1943.
21. Simpson JA: Electrical signs in the diagnosis of carpal tunnel and related syndromes. J Neurol Neurosurg Psychiatry 19:275–280, 1956.
22. Stevens JC: AAEE Minimonograph #26: The electrodiagnosis of carpal tunnel syndrome. Arch Neurol 16:635–641, 1967.
23. Stewart JD: Focal Peripheral Neuropathies, 2nd ed. New York, Raven Press, 1993.
24. Thomas JE, Lambert EH, Cseuz KA: Electrodiagnostic aspects of the carpal tunnel syndrome. Arch Neurol 16:635–641, 1967.
25. Wilbourn AJ: Ulnar neuropathy. In Syllabus: Course A: Basic Electrophysiologic Testing in Mononeuropathy. Rochester, MN, American Association of Electromyography and Electrodiagnosis, 1985, pp 27–36.
26. Wilbourn AJ: Electrodiagnosis with entrapment neuropathies. In Syllabus: 1992 AAEM Plenary Session 1: Entrapment Neuropathies. Rochester, MN, American Association of Electrodiagnostic Medicine, 1992, pp 23–33.
27. Wilbourn AJ, Ferrante MA: Clinical electromyography. In Joynt RJ, Griggs RC (eds.): Clinical Neurology, vol. 1 (looseleaf). Philadelphia, Lippincott-Raven, 1997, pp 1–78.

Chapter 16

Radiculopathy

Kerry H. Levin, MD

ANATOMY AND PATHOPHYSIOLOGY
ELECTRODIAGNOSIS
Nerve Conduction Studies
Needle Electrode Examination

SPECIFIC DISORDERS
Individual Root Lesions
Polyradiculopathies

ANATOMY AND PATHOPHYSIOLOGY

In all, there are 31 pairs of spinal nerve roots: 8 cervical, 12 thoracic, 5 lumbar, 5 sacral, and 1 coccygeal. Each spinal nerve root is comprised of a dorsal (somatic sensory) root and a ventral (somatic motor) root, which join in the intraspinal region, just proximal to the intervertebral foramen (Fig. 16–1). In the extraspinal region, just distal to the intervertebral foramen, the nerve root divides in two, forming a small posterior primary ramus that supplies innervation to the paraspinal muscles and skin of the neck and trunk, and a large anterior primary ramus that supplies innervation to the limbs and trunk, including intercostal and abdominal wall muscles. Cell bodies of the motor nerve fibers reside in the anterior horns of the spinal cord, whereas those of the sensory nerve fibers reside in the dorsal root ganglia (DRG). DRG are, in general, located within the intervertebral foramina and are therefore not strictly speaking intraspinal. However, at the lumbar and sacral levels, there is a tendency for DRG to reside proximal to the intervertebral foramina, in intraspinal locations. About 3% of L3 and L4 DRG are intraspinal, about 11% to 38% of L5 DRG are intraspinal, and about 71% of S1 DRG are intraspinal, according to recent cadaver, radiographic, and magnetic resonance imaging (MRI) studies.[15, 18]

Nerve roots are numbered according to their segmental location in the spinal cord, whereas intervertebral foramina are numbered according to the two vertebral bodies that frame the intervertebral foramen from above and below. A cervical root exits above the vertebral body of the same number, such that the C3 root exits the spinal canal via the C2–3 intervertebral foramen. Because there are only seven cervical vertebrae, the C8 root exits through the C7–T1 intervertebral foramen. As a result of this incongruity, all thoracic, lumbar, and sacral roots exit below the vertebral body of the same number.

Nerve root fibers are vulnerable to the same types of injury as other peripheral nerves: entrapment, compression, infiltration, necrosis, and transection. Damage may result in focal demyelination leading to conduction block or conduction velocity slowing along nerve root fibers. Axon loss at the root level results in wallerian degeneration along the whole course of affected nerve fibers. Both conduction block and axon loss produce symptoms and neurological deficits if a sufficient number of nerve fibers are affected. Conduction velocity slowing alone is insufficient to produce weakness or significant sensory loss, although sensory modalities requiring timed vollies of impulse transmission along their pathways, such as vibration and proprioception, can be altered.

Axon loss lesions occurring within the intraspinal canal affect both sensory and motor root fibers and produce both sensory and motor symptoms, but will only produce peripheral wallerian degeneration along the motor fibers, as long as the DRG are distal to the site of nerve root damage. Peripheral sensory axons do not degenerate when they remain connected to their cell bodies.

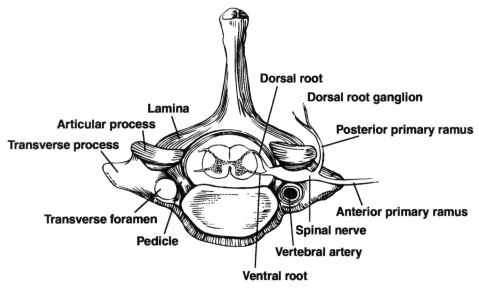

FIGURE 16–1. Cross-sectional view of the spine and neural elements at the cervical level.

ELECTRODIAGNOSIS

Several factors reduce the sensitivity of electrodiagnosis in radiculopathy. First, most radiculopathies are caused by compression from disk protrusion or spondylosis and result in damage to only a fraction of nerve root fibers, producing limited motor and sensory deficits. Second, in the acute setting, radiculopathy manifests itself most commonly by symptoms of pain and alteration of sensory perception. Sensory radiculopathy can be reliably localized segmentally by electrodiagnostic techniques only rarely. This is the case because symptoms of pain and parasthesia are mediated through C-type sensory fibers that are too small to be studied by routine electrodiagnostic techniques, and because the peripheral processes of sensory root fibers remain intact with intraspinal lesions, so sensory nerve action potentials (SNAPs) remain normal. Third, the intraspinal location of most lesions makes it impossible to perform direct nerve conduction studies (NCS) on the nerve root proximal to the damaged segment, preventing the diagnosis of conduction block and focal conduction velocity slowing along the damaged segment of the root.

However, when intraspinal lesions result in significant motor axon loss, electrodiagnosis is both sensitive and specific in its ability to define the features of the motor radiculopathy. Carefully performed NCS and needle electrode examination (NEE) can give valuable information regarding the root level of involvement, the age of the lesion, the severity of damage, and the degree of activity of the pathological process.

Nerve Conduction Studies

Routine Studies

Antidromic sensory NCS performed along peripheral nerve trunks are characteristically normal in radiculopathy. The SNAP distal latency and nerve conduction velocity are never involved in radiculopathy. The SNAP amplitude may be abnormal if DRG are affected in the pathological process. In pathological processes that infiltrate or extend from the intraspinal space into the intervertebral foramen, such as malignancy, infection, and meningiomas, DRG are damaged and wallerian degeneration along sensory axons occurs, resulting in SNAP amplitude loss. When DRG reside in an intraspinal location, as mentioned earlier, they are vulnerable to compression by disk protrusion and spondylosis. For this reason, L5 radiculopathy can uncommonly be associated with loss of the superficial peroneal SNAP.[27] In our experience, S1 radiculopathy has not been associated with sural SNAP amplitude loss. Although S1 DRG are even more commonly intraspinal than L5 DRG, their intraspinal location is caudal to the L5–S1 disk space where most compressive S1 radiculopathies occur. When nerve root damage occurs distal to the intervertebral foramen, SNAP amplitude will be affected.[27, 45]

Motor NCS are relatively insensitive in the diagnosis of motor radiculopathy for several reasons. First, most radiculopathies interrupt only a fraction of the total number of motor root fibers, whereas loss of close to 50% of motor axons in a nerve trunk is required to reliably establish a significant reduction in the compound muscle action potential (CMAP) amplitude, when compared with the same response on the uninvolved side.[7] Second, to identify an abnormality of CMAP amplitude in a motor radiculopathy, the muscle belly from which the CMAP is generated must be in the myotome of the injured root. For example, a severe C8 radiculopathy would be expected to produce some change in the ulnar CMAP amplitude, recording from either the abductor digiti minimi or the first dorsal interosseus. In the C5 myotome, the musculocutaneous and axillary nerve trunks can be stimulated to assess CMAPs from the biceps and deltoid muscles, respectively. However, in the C6 and C7 myotomes there are no easily and reliably accessible muscle bellies from which to assess the CMAP amplitude.

Late Responses

Late responses are evoked motor potentials that can be used to measure the travel time of propagated nerve action potentials from a distal point of electrical stimulation along a peripheral nerve trunk, proximally to the spinal cord, and then back down the limb to a muscle belly innervated by the same peripheral nerve trunk. Theoretically, they make possible the assessment of conduction through the damaged segment of a nerve root, but there are a number of limitations. First, because the traditional measurement is latency, the sensitivity is low because even severe slowing over a short segment will not diagnostically prolong the total latency. Second, as long as a few nerve fibers conduct normally through a damaged segment, a normal "shortest" latency will be recorded, even in the presence of severe nerve root damage. Finally, late responses such as F waves are of limited value in the diagnosis of radiculopathy because they are not recorded along sensory nerve fibers, and are useless in the assessment of sensory symptoms.

The F wave was first described by McDougal and Magladery in 1950, and was so named because they were originally recorded from foot muscles.[30] The F wave is a motor response often recorded from a muscle belly after stimulation of the peripheral nerve trunk innervating the muscle. It is thought to arise from the "backfiring" of motor neurons as impulses arrive antidromically from a peripheral site of nerve trunk stimulation.[32] The F wave occurs after the CMAP, but as the point of nerve trunk stimulation is moved more proximally, the CMAP latency lengthens and the F wave latency shortens, indicating that the impulse eliciting the F wave travels away from the recording electrodes toward the spinal cord before returning to activate distal muscles.[19] Traditionally, the shortest latency of at least eight consecutive discharges is measured. The F amplitude and duration and the range of F wave latencies have not been of clinical value in the diagnosis of focal radiculopathy.

The H reflex is a monosynaptic spinal reflex first described by Hoffmann in 1918.[17] Traditionally, this response is obtained by stimulating the tibial nerve at low voltage in the popliteal fossa to preferentially activate Ia sensory afferents, whose fibers terminate directly on motor neurons in the same spinal cord segment, completing a reflex arc that ends in an H wave recorded over a distally recorded muscle in the tibial nerve distribution, such as the soleus muscle. When elicited from the tibial nerve, the H reflex is the electrophysiological equivalent of the Achilles tendon muscle stretch reflex. The H reflex can be elicited from other nerve trunks under certain circumstances,[12, 38] especially when corticospinal tract disease results in loss of the normal central inhibitory influence on motor neuron pools. However, only the tibial H reflex has been found to be useful in routine clinical practice.

The H reflex is an extremely sensitive test for the assessment of the integrity of the tibial/S1 sensory pathway, including the intraspinal course of the S1 root. It is markedly reduced in amplitude or absent in axon loss lesions affecting the S1 root, the tibial division of the sciatic nerve, and the posterior tibial nerve proximal to the soleus muscle. In our laboratory, the H amplitude is a more reliable measurement than H latency in the electrodiagnosis of S1 radiculopathy.[43] We define an H reflex amplitude as abnormal if it is less than 1 mV, or less than 50% of the H amplitude on the uninvolved side. Other authors have claimed superior sensitivity of the H latency in the diagnosis of unilateral S1 root lesions.[6, 20, 37]

The H reflex is likely to show an abnormality with any disturbance of conduction through the tibial/S1 pathway. This sensitivity is associated with reduced specificity, resulting in a number of clinical limitations. First, the response is not reliably present in normal subjects after the age of 60 years, although normal responses have been identified at all ages. Second, whereas unilateral absence of the H reflex is clearly abnormal at any age, bilateral absence of H responses is often of uncertain clinical significance. Technical factors and generalized neuropathic processes can affect the H reflex. Possible causes include obesity and inadequate penetration of the stimulus in the popliteal fossa, prior lumbar spine surgery, and peripheral polyneuropathy, especially in the setting of diabetes. Bilateral absence of the H reflex may be the earliest electrodiagnostic feature of acute demyelinating peripheral polyneuropathy (Guillain-Barré syndrome). Third, abnormalities anywhere along the tibial/S1 sensory or motor pathway will alter the H response, including posterior tibial mononeuropathies proximal to the branch point of the nerve to the soleus and gastrocnemius muscles. Thus, an H reflex abnormality is insufficient by itself to confirm the presence of an S1 radiculopathy.

Somatosensory Evoked Responses

Theoretically, somatosensory evoked potentials (SEPs) should be a valuable tool in the assessment of conduction abnormalities along sensory fibers at the root level. Electrical stimuli are delivered on the skin surface to a mixed sensory and motor nerve trunk, a sensory nerve trunk, or the skin in a specific nerve or root distribution. Responses are recorded over the spine and scalp, and latencies are measured to assess the conduction time along large diameter sensory fibers across various segments of the peripheral and central conduction pathways primarily subserving proprioception and vibratory sense.

Unfortunately, a number of limitations diminish the value of this technique. First, amplitude measurements are too variable in normal individuals to have clinical significance; thus, the assessment of partial axon loss lesions and partial conduction block is not reliable. Second, focal slowing in the root segment is diluted by normal conduction along the rest of the sensory pathway. Third, nerve trunk stimulation often simultaneously activates nerve fibers belonging to more than one root segment, masking the abnormality in the abnormal root in question.[44]

Given those limitations, SEPs performed after nerve trunk stimulation are of little diagnostic value. Peroneal nerve stimulation is normal in patients with lumbosacral radiculopathy.[3] In patients with signs of cervical radiculopathy, with or without signs of myelopathy, nerve trunk stimulation may yield abnormal SEPs.[10, 11, 41] However, in these studies the abnormal SEP results contributed little to electrodiagnosis, because the patients in general also demonstrated clear electromyographic features of cervical radiculopathy.

SEPs may also be derived by stimulation of the skin in specific dermatomal distributions. Studies have examined the value of dermatomal stimulation between the first and second toes for the L5 root and over the lateral edge of the foot for the S1 root. Aminoff and colleagues found that only 25% of patients with clinically definite L5 and S1 radiculopathy showed significant scalp-recorded abnormalities, whereas the NEE was diagnostic in 75%.[2, 3]

Cutaneous sensory nerves have more specific and isolated root innervations, and thus SEPs derived from cutaneous nerve stimulation have a potential diagnostic advantage. Scalp-recorded cutaneous SEPs were abnormal in 57% of 28 cases of cervical and lumbosacral radiculopathy in one report, based on findings of abnormal amplitude and waveform configuration.[9] Using the same technique Seyal and colleagues found only 20% of their patients had abnormal scalp-recorded recordings, although the number of abnormal cases increased to about 50% when spine-recorded SEP latency or response size was measured.[36] Despite these results, the overall correlation was poor between the SEP abnormality and the clinical localization of the sensory radicular symptoms.

In summary, SEPs do not appear to have either the specificity or sensitivity of other electrodiagnostic techniques, such as the NEE, to recommend them at this time for the routine diagnosis of radiculopathy.

Needle Electrode Examination

General Concepts

Although the NEE assesses only the motor component of radiculopathy, it is the most specific and sensitive of the electrodiagnostic tests for radiculopathy. In many cases the NEE can provide information regarding the root level of involvement, the degree of axon loss present, the degree of ongoing motor axon loss, and the chronicity of the process. Several general comments can be made. An electromyogram (EMG) examination should be considered an electrodiagnostic medical consultation. The patient usually presents with symptoms, a specific clinical diagnosis having not yet been made. The role of the electrodiagnostic medical consultant is to perform a study that adequately evaluates the likely diagnostic possibilities based on the patient's symptoms, and not simply the single diagnostic possibility offered by the referring physician. In our laboratory, patients with arm or leg pain receive a general NEE survey that samples all major root and nerve trunk distributions in the limb in question. If abnormalities are identified, the examination is modified to focus on the cause for the abnormality. If there is a symptom in a specific region of the limb, such as the shoulder girdle or posterior thigh, muscles in that region are also examined. Table 16–1 outlines the screening NCS for nonspecific arm and leg symptoms. Tables 16–2 and 16–3 outline the screening NEE for nonspecific arm and leg symptoms, respectively.

The localization of a nerve root lesion requires the identi-

TABLE 16–1	Screening Nerve Conduction Studies

Arm Pain Sensory
Distal amplitude and latency:
  Median
  Ulnar
  Radial

Motor
Distal latency, distal and proximal amplitudes, conduction velocity, and F latency:
  Median (recording from thenar eminence)
  Ulnar (recording from hypothenar eminence)

Leg Pain Sensory
Distal amplitude and latency:
  Sural
  Tibial H reflex

Motor
Distal latency, distal and proximal amplitudes, conduction velocity, and F latency:
  Posterior tibial (recording from abductor hallucis)
  Peroneal (recording from extensor digitorum brevis)

TABLE 16-2	Screening Needle Electrode Survey for Arm Pain	

Muscle	Root Level	Nerve Trunk
First dorsal interosseus	C8	Ulnar
Flexor pollicis longus	C8	Anterior interosseus (median)
Extensor indicis proprius	C8	Posterior interosseus (radial)
Pronator teres	C6–7	Median
Triceps	C6–7	Radial
Biceps	C5–6	Musculocutaneous
Deltoid	C5–6	Axillary
C7 paraspinal	Overlap	

fication of neurogenic abnormalities in a distribution of muscles that shares the same root innervation. The abnormalities may include increased insertional activity in the form of positive waves or sharp spikes, abnormal spontaneous activity in the form of fibrillation potentials, reduced (neurogenic) recruitment of motor unit activation, and features of chronic neurogenic change in motor unit morphology such as increased duration, increased amplitude, and polyphasia.

The timing of the NEE is important. In acute radiculopathy, fibrillation is the abnormality most likely to confirm the presence of a motor radiculopathy. Fibrillation seldom develops before 2 weeks have elapsed from the onset of symptoms, and in some patients may not appear until 4 to 6 weeks after the onset of symptoms. The most efficient use of the EMG is to delay the performance of the NEE for at least 3 weeks after the onset of motor symptoms.

Root Localization by Needle Electrode Examination

The choice of muscles for the NEE must be tailored to the clinical question and specific symptoms, but must be comprehensive enough to maximize diagnostic certainly. The particular muscles showing neurogenic changes in the myotome in question will vary from case to case because most root lesions are partial, and not all muscles in the myotome will be affected equally. During the NEE, the more muscles identified as abnormal in the myotome, the more secure the electrodiagnosis. To make a reliable diagnosis of a single root lesion, at least two muscles in that myotome should be found with neurogenic changes, and they should not share the same peripheral nerve innervation. Where possible, involvement of proximal and distal muscles should be sought to increase the certainty of the diagnosis and exclude peripheral mononeuropathy as the cause for the abnormalities. To complete the NEE in an individual with an identified single root lesion, muscles in the myotomes

TABLE 16-3	Screening Needle Electrode Survey for Leg Pain	

Muscle	Root Level	Nerve Trunk
Abductor hallucis	S1	Posterior tibial
Medial gastrocnemius	S1	Posterior tibial
Biceps femoris (short head)	S1	Peroneal
Extensor digitorum brevis	L5 (S1)	Peroneal
Tibialis anterior	L5 (L4)	Peroneal
Tibialis posterior	L5	Posterior tibial
Gluteus medius	L5	Superior gluteal
Rectus femoris	L2,3,4	Femoral
S1 paraspinal	Overlap	

training the involved root level should be examined to verify that those myotomes are normal. For example, the biceps and first dorsal interosseus muscles should be normal in a patient with a C7 radiculopathy.

Paraspinal muscle involvement should always be sought, as it adds important support for the diagnosis of an intraspinal lesion, and rules out plexopathy and peripheral mononeuropathy as the cause of extremity muscle involvement. However, a number of factors reduce their value. First, paraspinal muscle fibrillation can be seen not only in disorders of the root, but also in processes affecting anterior horn cells and in muscle disorders such as necrotizing myopathy. Second, paraspinal muscle involvement cannot precisely localize the segmental level of root damage because the segmental innervation of paraspinal muscles can overlap by as much as four to six segments.[13] Third, clear evidence of paraspinal denervation with cervical and lumbosacral radiculopathies is seen in only about 50% of cases.[5, 25] Likely causes include the overlapping segmental innervation of paraspinal muscles and the tendency for muscles close to the site of the nerve lesion to reinnervate sooner and more completely than muscles at greater distance from the point where nerve regeneration must begin. Finally, in paraspinal muscles that are close to a prior laminectomy site, fibrillation may persist indefinitely owing to iatrogenic denervation. In routine practice, we do not examine paraspinal muscles in areas of prior surgery.

Anatomical, clinical, and electromyographic myotomal charts are used to correlate the pattern of EMG abnormalities in a limb with a specific root level. Anatomical charts have been derived by tracing root and peripheral nerve innervations of muscles from cadaver studies. Clinical charts have been derived by correlating the distribution of clinical muscle weakness in patients with specific traumatic lesions. Although these charts are useful, they are not entirely applicable to the NEE. Muscles are chosen for the NEE because of specific attributes of root innervation and accessibility. Some muscles, such as the anconeus, pronator teres, and brachioradialis, are not easily isolated in the clinical examination, but are easily isolated by the NEE, and are important in root localization. Thus, electromyographically derived myotomal charts are useful in the electrodiagnosis of radiculopathy.[5, 25] Figure 16–2 is a representative myotomal chart derived clinically; Figures 16–3 and 16–4 are electromyographically derived myotomal charts.

Defining an Acute Radiculopathy

In an axon loss radiculopathy, determining the age of the lesion requires combining information about the duration of the symptoms with NEE attributes of both active and chronic motor axon loss. When motor unit potentials are of normal configuration and size, the presence of abnormal insertional or spontaneous activity in the form of trains of brief sharp spikes or positive waves indicates recent motor axon loss. Abnormal insertional activity alone suggests that the process may be only several weeks old. The presence of spontaneous activity in the form of fibrillation potentials indicates a process of at least 3 weeks' duration.

As already stated, electrodiagnostic testing for radiculopathy is most valuable when significant axon loss has occurred. However, a prominent conduction block lesion at the root level can be inferred under certain circumstances. When examining a muscle whose CMAP is of normal amplitude, the presence of a reduced recruitment pattern of motor unit potential activation in the absence of fibrillation suggests conduction block. If this pattern is seen in multiple muscles of a specific myotome, a diagnosis of radiculopathy can be made. This diagnostic strategy is not reliable if the onset of weakness is less than 4 weeks prior to the electrodiagnostic

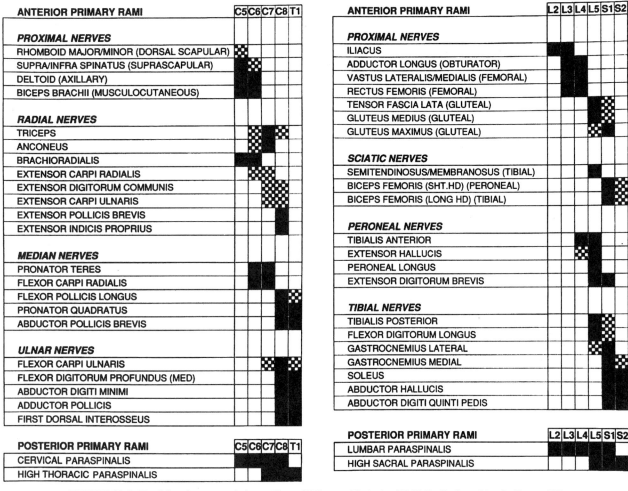

FIGURE 16–2. Traditional myotomal chart. (From Wilbourn AJ, Aminoff MF: Radiculopathies. In Brown WF, Bolton CF [eds]: Clinical Electromyography, 2nd ed. Boston, Butterworth-Heinemann, 1993, p 192, with permission.)

study, since an acute axon loss lesion may not clearly manifest fibrillation potentials for 3 or more weeks after onset of symptoms.

Defining a Chronic Radiculopathy

The diagnosis of a chronic and active or a chronic and remote root lesion is based on the observation of neurogenic motor unit action potential (MUAP) changes, in the presence or absence of evidence of fibrillation potentials. In the early stages of reinnervation of denervated muscle fibers, between 6 and 26 weeks after nerve root injury, collateral sprouting from surviving nerve fiber terminals gives rise to MUAPs that are polyphasic and show moment-to-moment variation in morphology (see Chapter 12). As more time elapses, reinnervation becomes more complete, moment-to-moment MUAP variation resolves, and MUAPs develop the characteristic features of increased duration and amplitude, the typical electrodiagnostic features of a chronic lesion. An NEE demonstrating these chronic neurogenic MUAP changes without fibrillation indicates the residuals of a remote lesion. These MUAP changes are permanent, reflecting the histopathological changes in the reinnervated muscle, and remain unchanged unless the motor unit is injured again. After a significant motor axon loss process has occurred, MUAPs never return to their preinjury morphology.

Chronic lesions can be classified into a chronic and active category if there are both fibrillation potentials and chronic neurogenic motor unit potential changes. In root distributions where the myotome includes muscles in both distal and proximal regions of a limb (especially the L5 and S1, and perhaps the C5 root distributions), the presence of a chronic and ongoing axon loss process can be even more clearly defined when fibrillation potentials are seen in both distal and proximal muscles in the root distribution. In lesions where fibrillation potentials are seen in distal muscles only, the presence of an ongoing axon loss process is less certain. Some inactive but severe axon loss processes never fully reinnervate, especially in muscles farthest from the injury site, leaving some muscle fibers denervated indefinitely. The NEE findings at progressive stages of axon loss radiculopathy are summarized in Table 16–4.

Defining the Severity of a Radiculopathy

The severity of an axon loss process can be assessed during the NEE by the degree of motor unit loss in the root distribution. This is determined by a subjective measurement of the degree of reduced recruitment of motor unit potential activation. Although the degree of reduced recruitment of motor unit potential activation in a muscle correlates with the degree of neurogenic weakness, it does not equate with axon loss unless the CMAP elicited from the same muscle is also reduced in amplitude, since reduced recruitment of motor unit activation is also a feature of a pure demyelinating conduction block lesion along a nerve trunk. Thus,

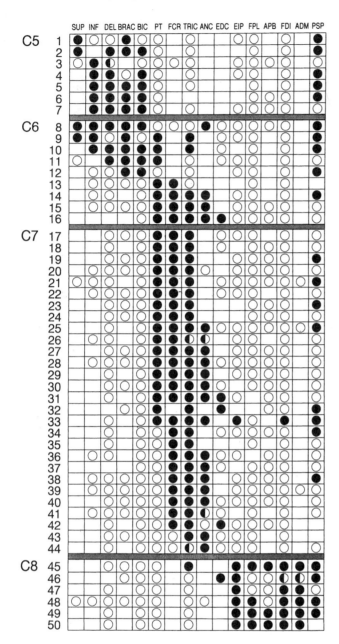

FIGURE 16–3. Needle electrode examination results grouped by the surgically defined root level of involvement. (*Closed circle:* positive waves or fibrillation potentials, with or without neurogenic recruitment and motor unit changes; *half-closed circle:* neurogenic recruitment changes only; *open circle:* normal examination. SUP, supraspinatus; INF, infraspinatus; DEL, deltoid; BRAC, brachioradialis; BIC, biceps; PT, pronator teres; FCR, flexor carpi radialis; TRIC, triceps; ANC, anconeus; EDC, extensor digitorum communis; EIP, extensor indicis proprius; FPL, flexor pollicis longus; APB, abductor pollicis brevis; FDI, first dorsal interosseus; ADM, abductor digiti minimi; PSP, paraspinal muscle.) (From Levin KH, Maggiano HJ, Wilbourn AJ: Cervical radiculopathies: Comparing surgical and EMG localization of single-root lesions. Neurology 46:1022–1025, 1996, with permission.)

defining the severity of an axon loss radiculopathy requires evaluation of both the CMAPs in the myotome in question (when possible) and the degree of reduced recruitment of MUAP activation. Measuring the number of fibrillation potentials present in a muscle is highly subjective and does not correlate as well with the degree of axon loss.

SPECIFIC DISORDERS

Individual Root Lesions

Individual root lesions can be diagnosed with precision in many cases by a comprehensive NEE. By establishing familiarity with the electrical presentation of single root lesions, it is often possible to identify radicular patterns when confronted with more complex combinations of root lesions in polyradiculopathies.

Cervical Radiculopathies

The most complete clinical study of specific cervical root lesions was carried out by Yoss and associates.[42] According to that study, and one by Marinacci,[31] clinical and radiographic evidence of radiculopathy occurs at the C7 level 70% of the time, at the C6 level 19% to 25% of the time, at the C8 level 4% to 10% of the time, and at the C5 level 2% of the time. The following NEE data on individual cervical radiculopathies comes from a study of isolated single root lesions based on confirmed surgical localization[25] (Fig. 16–3).

C5 radiculopathy produces a rather stereotyped pattern of muscle involvement, affecting the spinati, biceps, deltoid, and brachioradialis with about equal frequency, but not all of them together in each patient. In our experience, the pronator teres is never involved in C5 radiculopathy.[25] Because the rhomboid major muscle is said to have prominent C5 innervation, it should be examined in unclear cases. The upper trapezius, with its prominent C4 innervation, is spared in C5 radiculopathy. NCS are not likely to be helpful, al-

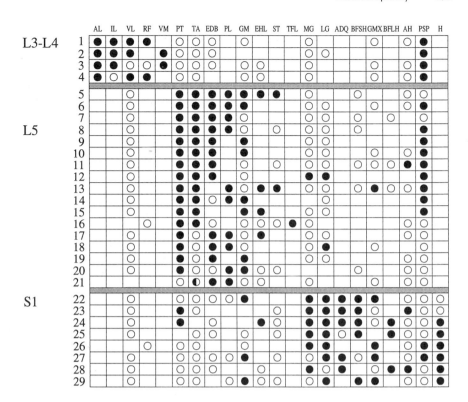

FIGURE 16–4. Needle electrode examination results grouped by the surgically defined root level of involvement. (*Closed circle:* positive waves or fibrillation potentials, with or without neurogenic recruitment and motor unit changes; *half-closed circle:* neurogenic recruitment changes only; *open circle:* normal examination.)

Adductor longus = AL
Iliacus = IL
Vastus lateralis = VL
Rectus femoris = RF
Vastus medialis = VM
Flexor digitorum longus /
Tibialis posterior = PT
Tibialis anterior = TA
Extensor digitorum brevis = EDB
Peroneus longus = PL
Gluteus medius = GM
Extensor hallucis longus = EHL

Semitendinosus = ST
Tensor fascia lata = TFL
Medialis gastrocnemius = MG
Lateralis gastrocnemius = LG
Abductor digiti quinti = ADQ
Biceps femoris (short head) = BFSH
Gluteus maximus = GMX
Biceps femoris (long head) = BFLH
Abductor hallucis = AH
Paraspinal = PSP
H- Reflex = H

though severe lesions may be associated with axillary and musculocutaneous CMAP amplitude loss.

C7 radiculopathy produces a rather stereotyped pattern of muscle involvement, affecting particularly the triceps, but also the anconeus, flexor carpi radialis, and pronator teres. We found no case of isolated C7 radiculopathy where the triceps was not involved.[25] Although traditionally considered to be a C7-innervated structure, the extensor carpi radialis has not shown reliable C7 predominance in our experience,

and is not a part of our routine survey for radiculopathy. An important part of the clinical diagnosis of C7 radiculopathy rests on the finding of a diminished triceps deep tendon reflex, but in the Yoss study[42] and our own,[25] the reflex was abnormal in only 65% and 68% of patients, respectively. There are no reliably performed motor NCS that can be used to generate CMAPs from C7-innervated muscles.

With C6 radiculopathy there is no single characteristic pattern of muscle involvement. Rather, two patterns are dis-

TABLE 16–4	Findings in the Needle Electrode Examination at Progressive Stages of Axon Loss Radiculopathy						
	Recruit	Insertion	PSP	FIB	Poly/Var	Neur	MTP/CRD
<3 weeks	+ +	+ / + +	+				
3–6 weeks	+ +	+ +	+ +	+ + +			
6–26 weeks	+ +	+	+ / −	+ +	+ + +		
Chronic/active	+ +		+ / −	+	+ +	+ +	
Chronic/remote	+ / + +					+ + +	+

RECRUIT, neurogenic recruitment of myotomal motor units; INSERTION, abnormal insertional activity in myotomal muscles; PSP, paraspinal fibrillation; FIB, fibrillation potentials in myotomal muscles; POLY/VAR, polyphasic motor unit potential changes/motor unit potential variation; NEUR, neurogenic motor unit potential changes (increased duration and amplitude); MTP/CRD: myotonic discharges/complex repetitive discharges; + / −, equivocal amount; +, mild amount; + +, moderate amount; + + +, greatest amount.

cernible: the first, very similar to the C5 pattern, with additional involvement of triceps and pronator teres in some; and the second, similar to the C7 pattern. The pronator teres is abnormal in 80% of patients with C6 radiculopathy, but is also abnormal in 60% of the cases of C7 radiculopathy.[25] The triceps is abnormal in over half the cases of C6 radiculopathy. Thus, significant electromyographic overlap occurs between C5 and C6 radiculopathy, and between C6 and C7 radiculopathy. There are no reliably performed motor NCS that can be used to generate CMAPs from C6-innervated muscles.

C8 radiculopathy produces a stereotyped pattern of muscle involvement, including the ulnar-innervated muscles, extensor indicis proprius, and flexor pollicis longus. Abductor pollicis brevis is involved less often, and to a lesser degree than other muscles. Of all the root lesions, C8 radiculopathy is the most clearly identified by NEE because of the limited myotomal overlap. NCS are not likely to be helpful, although severe lesions may be associated with ulnar (recording from the abductor digiti minimi or first dorsal interosseous) CMAP amplitude loss.

T1 radiculopathy is the most uncommon isolated root lesion affecting the arm. Although all C8 muscles of the hand are said to have T1 contributions, the abductor pollicis brevis muscle appears to be predominately T1 innervated.[28] We have recognized a single case of T1 radiculopathy with neuroimaging and intraoperative confirmation of isolated T1 root compression. The EMG picture showed chronic and active denervation limited to the abductor pollicis brevis.[29]

Lumbosacral Radiculopathies

According to one large study, lumbar disk herniation leading to electromyographically determined motor radiculopathy occurs at the L4–5, L5–S1, and L3–4 levels 55%, 43%, and 2% of the time, respectively.[21] At lumbosacral levels, the anatomical localization of the site of root injury, the identification of single root lesions and the accuracy of electrodiagnosis are less successful than at the cervical levels. First, there is the issue of the longer intraspinal course of most lumbosacral roots. All lumbar and sacral spinal nerve roots are constituted at the T12–L1 vertebral level, where the spinal cord ends as the conus medullaris. The roots then course down the canal as the cauda equina, until they exit at their respective neural foramina. Depending on the nature and location of intraspinal compression, roots may be injured at any disk level, from the L1–2 level to the level of their exit into the intervertebral foramen. For example, the L5 roots can be compressed by a central disk protrusion at the L2–3 or L3–4 level, a lateral disk protrusion at the L4–5 level, or foraminal stenosis at the L5–S1 level. Thus, the electrodiagnostic localization of a specific root lesion does not specify the vertebral level of damage. Second, because of the presence of multiple spinal nerve roots in the cauda equina, the likelihood of multiple, bilateral radiculopathies increases. This occurrence reduces electrodiagnostic accuracy and introduces possible confusion with other disorders, such as peripheral polyneuropathy and motor neuron disease. Thus, the identification of a lumbosacral radiculopathy requires at least a limited evaluation of the contralateral side for evidence of concurrent lesions. The following NEE data on individual lumbosacral radicolopathies come from a study of isolated single root lesions based on confirmed surgical localization[5] (see Fig. 16–4).

S1 radiculopathy produces specific abnormalities in the NEE and NCS. There is a stereotyped pattern of muscle involvement, including the gastrocnemius muscles, short head of the biceps femoris (BFSH), and abductor hallucis. In our study, evidence of paraspinal denervation was seen in only 25% of patients, owing to the significant overlap of

paraspinal segmental innervation.[5] The gastrocnemius muscles are often difficult to voluntarily activate, making the assessment of motor unit potential recruitment and morphological changes incomplete. Therefore, the identification of abnormalities in proximal muscles such as the BFSH, long head of the biceps femoris, and gluteus maximus is crucial for the confirmation of an S1 radiculopathy, eliminating the possibility of more distal peripheral mononeuropathies. The BFSH is especially useful in this regard, where it is abnormal in over 80% of surgically proven S1 radiculopathies. In our experience, the BFSH never shows significant involvement in L5 radiculopathy, although some texts have described this muscle as having predominant L5 innervation. As noted earlier, in chronic axon loss states, the identification of fibrillation potentials in proximal muscles supports the presence of ongoing axon loss, whereas the presence of fibrillation potentials in distal muscles only is of uncertain significance, as they could represent the failure of reinnervation in a remote lesion. The H reflex amplitude was less than 50% of the normal side in about 75% of patients in our study.[5] Comparison of our data with other studies has been complicated by the tendency in other studies to use different parameters of the H reflex for measurement (i.e., latency), and because of the lack of surgical confirmation of isolated S1 root involvement in other studies. Motor NCS are not likely to be helpful in the diagnosis of S1 radiculopathy, although severe lesions may be associated with posterior tibial (recording the abductor hallucis) CMAP amplitude loss.

With L5 radiculopathy, the NEE shows involvement of the flexor digitorum longus, posterior tibialis, and tibialis anterior muscles in over 75% of surgically proven cases of L5 radiculopathy. Fifty percent of our patients with L5 radiculopathy demonstrated paraspinal fibrillation potentials. NEE of the posterior tibialis or flexor digitorum longus is critical, as they are the only L5-innervated muscles below the knee not innervated by the peroneal nerve. Abnormalities in either of these muscles exclude the diagnosis of peroneal mononeuropathy. To verify the presence of an L5 radiculopathy, abnormalities should be sought in proximal L5 muscles such as the semitendinosus, tensor fascia lata, and gluteus medius, in order to eliminate the diagnoses of sciatic and peroneal mononeuropathies. This is especially true in elderly individuals whose superficial peroneal sensory responses are absent because of age, and in whom peroneal and sciatic mononeuropathy cannot be as easily excluded.

L5 radiculopathy may not produce significant abnormalities of NCS, unless the degree of axon loss is severe, in which case the peroneal motor response to the extensor digitorum brevis and tibialis anterior muscles may be reduced in amplitude. Rarely, the extensor digitorum brevis is primarily innervated by the S1 root, in which case its CMAP will be paradoxically normal in the face of a reduced amplitude response recording over the tibialis anterior. Uncommonly, the superficial peroneal sensory response is absent or reduced in amplitude, the result of compression of L5 DRG when they reside within the intraspinal canal.[27]

Chronic L5 radiculopathy is diagnosed in the presence of neurogenic motor unit potential changes in distal and proximal L5-innervated muscles. In this setting, the presence of fibrillation potentials in the proximal muscles suggests a chronic and active radiculopathy.

L2, L3, and L4 root lesions cannot be reliably distinguished from each other because of the overlap of innervation of the anterior thigh muscles. The problem in reliable localization is compounded by the absence of proximal and distal muscles to examine, as well as the low incidence of L2, 3, and 4 lesions in general. We routinely examine the rectus femoris, vastus lateralis, iliacus, and adductor longus in patients with a clinical question of upper lumbar radiculopathy.

As the adductor longus is the only muscle not innervated by the femoral nerve, its evaluation is critical for the differentiation of femoral mononeuropathy and L2, 3, and 4 radiculopathy. In our study, all four patients in this category demonstrated paraspinal fibrillation potentials.

Special Radiculopathies

Two disorders have traditionally been categorized as types of brachial plexopathy, but in fact are more likely to represent extraspinal radiculopathies affecting the anterior primary rami. Neurogenic thoracic outlet syndrome, long considered a type of lower trunk brachial plexopathy, produces most severe damage in the abductor pollicis brevis muscle and the medial antebrachial cutaneous SNAP, both sharing principally T1 root innervation.[28] In most cases, lower trunk and C8 structures are affected to a much lesser extent. Median sternotomy brachial plexopathy, a condition that occurs in the course of coronary artery bypass graft and cardiac valve repair procedures, and also considered a type of lower trunk brachial plexopathy, produces most severe damage in the ulnar and C8 root distribution, with little involvement of T1-innervated structures. These two lesions show distributions of involvement that approach being mutually exclusive: the abductor pollicis brevis and the medial antebrachial cutaneous response with neurogenic thoracic outlet syndrome, and C8 muscles and the ulnar sensory

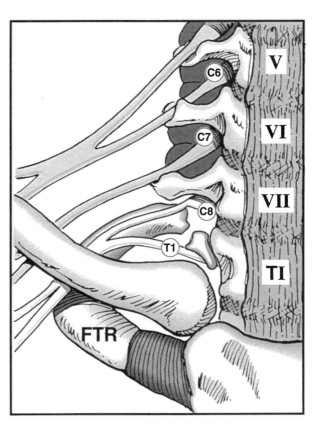

©CCF
1997

FIGURE 16–6. Diagram depicting the anatomical relationship between the C8 nerve root and fracture of the first rib near the costotransverse articulation in a patient who has undergone median sternotomy. Abbreviations as in Figure 16–5. (Copyright © Cleveland Clinic Foundation.)

response with median sternotomy brachial plexopathy. However, the nerve fibers innervating all these structures travel together in the lower trunk of the brachial plexus. Therefore, neurogenic thoracic outlet syndrome and median sternotomy brachial plexopathy more likely represent, respectively, extraspinal T1 and C8 root lesions proximal to the formation of the lower trunk, as diagrammed in Figures 16–5 and 16–6.

CASE STUDY 1

Forearm Aching and Numbness in the Hand

At the time of diagnosis, this 32-year-old man was a neurology resident in our laboratory having routine NCS performed as part of technique practice. The median CMAP amplitude was 7.6 mV, compared with 19 mV on the normal side. Median sensory responses were normal, the ulnar CMAP was 15 mV, and the ulnar SNAP amplitude was 12 μV, compared with 12 mV and 40 μV, respectively, on the unaffected side. The medial antebrachial cutaneous sensory response was 5 μV, compared with 15 μV on the unaffected side. The NEE showed neurogenic motor unit changes in the abductor pollicis brevis to the greatest extent, with lesser changes in the ulnar-innervated hand muscles.

In retrospect, this man had noticed nonspecific forearm

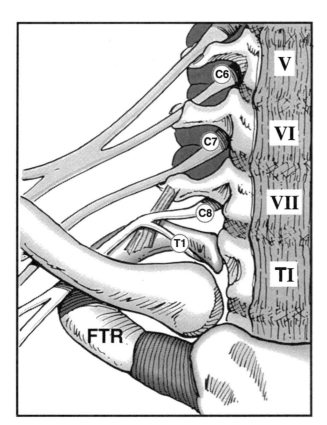

©CCF
1997

FIGURE 16–5. Diagram depicting the likely anatomical relationship between the T1 and C8 nerve roots and the offending ligamentous band in neurogenic thoracic outlet syndrome, showing entrapment of the T1 and C8 nerve trunks. Roman numerals indicate vertebral body levels, circled numbers indicate root levels, and FTR indicates first thoracic rib. (Copyright © Cleveland Clinic Foundation.)

aching and numbness of the fourth and fifth digits for years, seemingly nonprogressive. The electrodiagnostic examination showed substantial involvement in the T1 postganglionic sensory and motor nerve fibers, with C8/ulnar involvement that was not fully appreciated until compared with results on the opposite side. An elongated C7 transverse process was identified on chest radiography, and he subsequently underwent section of a fibrous band that extended from the C7 transverse process to the first thoracic rib.

Polyradiculopathies

The term *polyradiculopathy* indicates damage to multiple root segments simultaneously or in progressive order, occurring in a single limb, or more frequently bilaterally, and sometimes diffusely. The causes are diverse, and at times unclear. In some neurological disorders, polyradiculopathy coexists with lesions in distal peripheral nerves, or lesions in the central nervous system, or both. A brief description of the most prominent causes of polyradiculopathy follows, and Table 16–5 lists causes of polyradiculopathy and their differential diagnosis.

Compressive Polyradiculopathies

Spondylosis of the spine is often multifocal, and multiple roots may suffer compressive damage concurrently. This is especially true at the lumbosacral level, where spondylosis causes lumbar canal stenosis and multilevel neural foraminal stenoses. In our laboratory, we see few elderly patients with single lumbosacral root lesions, but many more with multiple simultaneous radiculopathies, often showing a combination of active and more chronic features. Lumbar canal stenosis exerts compressive effects on the cauda equina, resulting in the potential for multiple root involvement. It may present clinically with weakness in a single root distribution, in several distributions, or as chronic progressive weakness of the legs in a diffuse distribution. Alternatively, lumbar canal stenosis may present as intermittent progressive fatigability and aching of the legs elicited by walking or exercise, a symptom complex known as intermittent neurogenic claudication. The EMG picture of lumbar canal stenosis is extremely variable, spanning the gamut from no abnormality to multilevel, symmetrical motor axon loss.

Regardless of the cause of the lumbosacral polyradiculopathy, electrodiagnostic specificity is hampered when the NEE abnormalities are bilateral and confluent. In the chronic state, the NEE changes are usually most prominent in distal muscles of the myotome, shading to normal in more proximal muscles. When chronic motor axon loss spans the L5 and S1 distributions symmetrically, the electrical picture resembles the confluent changes seen in peripheral polyneuropathy. This is especially true in the elderly, in whom physiological loss of sural and superficial peroneal sensory responses can prevent the clear distinction between axon loss peripheral polyneuropathy and a chronic or active pattern of bilateral L5 and S1 radiculopathies.

When the process is chronic and active, the EMG pattern may be difficult to distinguish from early to mid-stage progressive motor neuron disease (amyotrophic lateral sclerosis, ALS) or progressive necrotizing myelopathies, until serial studies are performed over time. Mid-stage ALS is less likely to produce a picture in which contiguous muscles of the same root or an adjoining root level show markedly different degrees of involvement. ALS at this stage is more likely than active radiculopathy to show a significant distal to proximal gradient of muscle involvement in a limb.

Diabetic Polyradiculopathies

Radiculopathies can be among the most disabling of diabetic peripheral nervous system complications. They occur at the thoracic, lumbar, and sacral levels, and have been rarely reported at cervical levels.[33] About 25% occur in the absence of underlying peripheral polyneuropathy.[4, 24] Thoracic radiculopathies occur either unilaterally or bilaterally. They are clinically characterized by cutaneous pain and dys-

TABLE 16–5	Differential Diagnosis of Polyradiculopathies			
		Polyradiculopathy	Polyneuropathy	Myelopathy
Disorders with True Root Involvement				
Arachnoiditis		+		
Inflammatory polyneuropathy		+	+	
Diabetes		+	+	
HNPP		+	+	
Adrenal insufficiency[1]		+	+	
Procainamide polyradiculoneuropathy[13]		+	+	
Spondylosis		+		+
Radiation		+		+
Vascular malformation (conus medullaris)		+		+
Malignant invasion		+	+	+
Sarcoidosis		+	+	+
Lyme disease		+	+	+
Viral infection (HZ, CMV, HSV, EBV)		+	+	+
Mycoplasma infection		+	+	+
Vasculitis		+	+	+
Angiotropic lymphoma		+	+	+
Disorders Mimicking Root Involvement				
Porphyric polyneuropathy			+	
Alpha-lipoprotein deficiency			+	+
X-Linked bulbospinal neuronopathy			+	+
Motor neuron disease				+
Juvenile monomelic amyotrophy				+
Spinal cord infarction				+
Multiple sclerosis				+
Syringomyelia				+

HNPP, hereditary neuropathy with tendency to pressure palsy; HZ, herpes zoster; CMV, cytomegalovirus; HSV, herpes simplex virus; EBV, Epstein-Barr virus.

esthesia in the posterior and anterior aspects of the torso in the distributions of the involved roots, and there may be weakness and bulging of the abdominal wall from denervation of rectus abdominis muscles. Thoracic radiculopathies can be confused clinically with intra-abdominal disorders. The NEE shows evidence of denervation in thoracic paraspinal muscles as well as in associated rectus abdominis muscles.

Diabetic lumbosacral radiculopathies may occur at any segmental level, but the L3–4 levels are especially vulnerable. In a study by Levin and associates, 15 of 16 cases of diabetic lumbosacral radiculopathy included the L3–4 level, and 5 of the 15 were limited to that distribution.[24] L5 root involvement occurred in 10. S1 root involvement occurred in 7, all but one of which in the presence of L5 root involvement. In only one case did L5 and S1 root involvement occur in the absence of L3–4 root involvement. Bilateral involvement occurred in 11 cases. These data support the clinical observation that diabetic lumbosacral radiculopathy usually begins at the L3–4 level, and often spreads over weeks and months to involve contiguous root levels, and eventually the contralateral side.

Other Causes of Polyradiculopathy

Processes that infiltrate or compress nerve roots lead to polyradiculopathy. Infectious causes include Lyme disease, tuberculosis, syphilis, and fungal infections. In the setting of human immunodeficiency virus (HIV) infection, cytomegalovirus (CMV) can produce a characteristic cauda equina syndrome that includes progressive paraparesis, sensory loss, and sphincter dysfunction. Arachnoiditis of the cauda equina can occur as a result of intrathecal drugs (radiocontrast agents, anesthetics, amphotericin B), surgical procedures, and spinal subarachnoid hemorrhage.

Malignancy produces polyradiculopathy by compression and invasion. Malignancies with a predilection for bone are especially likely to cause polyradiculopathy, myelopathy, or both, because of their tendency to spread into contiguous regions. Malignant cells may also gain entry into the intraspinal canal by hematogenous spread.

CASE STUDY 2

Subacutely Progressive Weakness in a Man with HIV Infection

A 45-year-old man not previously known to be infected with HIV developed right leg pain that was unrelated to position or activity. The neurological examination after 3 weeks of symptoms showed only a reduced right ankle deep tendon reflex, without weakness, sensory impairment, or straight leg raising sign. An MRI scan of the lumbosacral spine without contrast enhancement was normal. An extensive EMG examination showed absence of the H reflexes bilaterally, without evidence of acute or chronic motor axon loss in any lumbosacral root distribution on the right or left. Over the subsequent week, he developed progressive bilateral leg weakness, and was admitted to the hospital with acute urinary retention. A repeat MRI scan with contrast showed marked enhancement of the cauda equina. A repeat EMG examination showed severe active motor axon loss in all lumbosacral root distributions in both legs, with preservation of the SNAPs. A cerebrospinal fluid (CSF) examination showed 200 polymorphonuclear cells per μl, with a protein concentration of 500 mg/dl. Cowdry type I inclusions in polymorphonuclear cells of the CSF supported the diagnosis of CMV polyradiculopathy, subsequently confirmed by antibody testing.

Myelopathy

Disease processes that exert the brunt of damage within the spinal cord can produce clinical and electrodiagnostic patterns of polyradiculopathy, mainly by involvement of anterior horn cells. In the case of spondylotic compressive myelopathy, anterior horn cell loss can occur at levels removed from the actual compressive lesion, giving rise to a clinical and EMG picture of bilateral single or multiroot motor axon loss at the levels apart from direct cord or root compression. This is thought to occur on the basis of anterior horn cell hypoxia from venous congestion resulting from the compressive myelopathy at a distance from the involved segmental levels.[39]

Intramedullary spinal cord disorders that cause loss of anterior horn cells yield clinical pictures of lower motor neuron weakness that can mimic cauda equina syndromes and lumbosacral polyradiculopathy. These disorders often demonstrate coexisting corticospinal tract features. At times the upper motor neuron deficits mask the lower motor neuron features, such as in cases of thoracic spinal cord and conus medullaris infarction caused by anterior spinal artery occlusion or arteriovenous malformation, but electrodiagnostic studies will identify motor axon loss in the thoracic (paraspinal muscle and rectus abdominis) and lumbar root distributions corresponding to the infarcted segments.[23] Lesions in the conus medullaris show a predominance of lower motor neuron involvement, including denervation of sphincter muscles. Lesions in this region often present a confusing clinical picture combining myelopathic sensory features, sphincter dysfunction, and radicular or cauda equina patterns of weakness.[26] Other causes of myelopathy producing anterior horn cell loss include radiation myelopathy, syringomyelia, metastatic and primary malignancies, demyelinating disease, herpes simplex and other viral infections, Lyme disease, and structural compressive disorders. Focal and evolving motor neuron disorders affecting lumbar or cervical segments can also mimic the clinical and electrodiagnostic patterns of polyradiculopathies. Juvenile amyotrophy of the upper extremity has been considered an isolated self-limited form of lower motor neuron disease,[16] although some of these cases may be related to venous occlusion (G Suarez, Mayo Clinic, Rochester, MN, unpublished report). Lesions confined to the spinal cord produce profound motor axon loss pictures sparing sensory nerve action potential responses, despite the presence of marked clinical sensory impairment.

Polyradiculoneuropathies

The diagnosis of polyradiculoneuropathy indicates the presence of coexisting features of polyradiculopathy and peripheral polyneuropathy. The clinical presentation includes weakness in both proximal and distal distributions of the legs (and in some disorders the arms as well), often in a distal to proximal gradient, with associated features of sensory axon loss. Electrodiagnostically, the picture is characterized by loss of sensory responses, combined with features of motor axon loss, or demyelinating conduction block in multiple root distributions, or both. The leading causes include acute inflammatory demyelinating polyradiculoneuropathy (Guillain-Barré syndrome), chronic inflammatory demyelinating polyradiculoneuropathy (CIDP), and diabetes. We now recognize both demyelinating and pure axon loss forms of inflammatory polyradiculoneuropathy.[14] Other causes include systemic vasculitides, CMV polyradiculoneuropathy in the setting of HIV infection, and angiotropic large-cell lymphoma.[26]

Uncommonly, hereditary disorders produce a picture of polyradiculoneuropathy. Hereditary neuropathy with tendency to pressure palsy (HNPP) has been reported to present in this fashion.[22] Alpha-lipoprotein deficiency (Tangier disease) produces motor and sensory neuronopathy in a progressive segmental pattern that often affects the upper extremities first, and mimics a pattern of polyradiculopathy.[8, 35] Porphyric polyneuropathy is characterized by marked proximal and distal weakness, sometimes asymmetrical and affecting the arms more than the legs. Sensory responses are variably affected, but the clinical presentation is predominantly motor.[40]

REFERENCES

 1. Abbas DH, Schlagenhauff RE, Strong HE: Polyradiculoneuropathy in Addison's disease, case report and review of literature. Neurology 27:494–495, 1977.
 2. Aminoff MJ, Goodin DS, Barbaro NM, et al: Dermatomal somatosensory evoked potentials in unilateral lumbosacral radiculopathy. Ann Neurol 17:171–176, 1985.
 3. Aminoff MJ, Goodin DS, Parry GJ, et al: Electrophysiological evaluation of lumbosacral radiculopathy; electromyography, late responses and somatosensory evoked potentials. Neurology (NY) 35:1514–1518, 1985.
 4. Bastron JA, Thomas JE: Diabetic polyradiculopathy, clinical and electromyographic findings in 105 patients. Mayo Clin Proc 56:725–732, 1981.
 5. Bodner RA, KH Levin, AJ Wilbourn: Lumbosacral radiculopathies: Comparison of surgical and EMG localization. Muscle Nerve 18:1071, 1995.
 6. Braddom RI, Johnson EW: Standardization of H reflex and diagnostic use in S1 radiculopathy. Arch Phys Med Rehabil 55:161, 1974.
 7. Bromberg M, Jaros L: Symmetry of normal motor and sensory nerve conduction measurements. Muscle Nerve 21:498–503, 1998.
 8. Case Record of the Massachusetts General Hospital 16-1996. N Eng J Med 334:1389–1394, 1996.
 9. Eisen A, Hoirch M, Moll A: Evaluation of radiculopathies by segmental stimulation and somatosensory evoked potentials. Can J Neurol Sci 10:178–182, 1983.
10. El Negamy E, Sedgwick EM: Delayed cervical somatosensory potentials in cervical spondylosis. J Neurol Neurosurg Psychiatry 42:238–241, 1979.
11. Ganes T: Somatosensory conduction times and peripheral, cervical and cortical evoked potentials in patients with cervical spondylosis. J Neurol Neursurg Psychiatry 43:683–689, 1980.
12. Garcia HA, Fisher MA, Gilai A: H reflex analysis of segmental reflex excitability in flexor and extensor muscles. Neurology (NY) 29:984, 1979.
13. Gough J, Koepke G: Electromyographic determination of motor root levels in erector spinae muscles. Arch Phys Med Rehabil 47:9–11, 1966.
14. Griffin JW, Li CY, Ho TW, et al: Pathology of the motor-sensory axonal Guillain-Barré syndrome. Ann Neurol 39:17–28, 1996.
15. Hamanishi C, Tanaka S: Dorsal root ganglia in the lumbosacral region observed from the axial views of MRI. Spine 18:1753–1756, 1993.
16. Hirayama K, Tamanaga M, Kitano K, et al: Focal cervical poliopathy causing juvenile muscular atrophy of the distal upper extremity: A pathological study. J Neurol Neurosurg Psychiatry 50:285–290, 1987.
17. Hoffmann P: Untersuchungen über die eigenreflexe (sehnenreflexe) menschlicher Muskelm. Berlin, Springer, 1922.
18. Kikuchi S, Sato K, Konno S, Hasue M: Anatomic and radiographic study of dorsal root ganglia. Spine 19:6–11, 1994.
19. Kimura J: Electrodiagnosis in Diseases of the Nerve and Muscle: Principles and Practice. Philadelphia, FA Davis, 1984, p 353.
20. Kimura J: Electrodiagnosis in Diseases of the Nerve and Muscle: Principles and Practice. Philadelphia, FA Davis, 1984, p 384.
21. Knuttson B: Comparative value of electromyographic, myelographic, and clinical neurological examination in the diagnosis of lumbar root compression syndrome. Acta Orthoped Scand 49(suppl):1–135, 1961.
22. Le Forestier N, LeGuern E, Coullin P, et al: Recurrent polyradiculoneuropathy with the 17p11.2 deletion. Muscle Nerve 20:1184–1186, 1997.
23. Levin KH, Daube JR: Spinal cord infarction—Another cause of "lumbosacral polyradiculopathy." Neurology 34:389–390, 1984.
24. Levin KH, Wilbourn AJ: Diabetic radiculopathy without peripheral neuropathy. Muscle Nerve 14:889, 1991.
25. Levin KH, Maggiano HJ, Wilbourn AJ: Cervical radiculopathies: Comparison of surgical and EMG localization of single-root lesions. Neurology 46:1022–1025, 1996.
26. Levin KH, Lutz G: Angiotropic large cell lymphoma with peripheral nerve and skeletal muscle involvement: Early diagnosis and treatment. Neurology 47:1009–1011, 1996.
27. Levin KH: L5 radiculopathy with reduced superficial peroneal sensory responses: Intraspinal and extraspinal causes. Muscle Nerve 21:3–7, 1998.
28. Levin KH, Wilbourn AJ, Maggiano HJ: Cervical rib and median sternotomy related brachial plexopathies: A reassessment. Neurology 50:1407–1413, 1998.
29. Levin KH: Neurological manifestations of compressive radiculopathy of the first thoracic root. Neurology 53:1149–1151, 1999.
30. Magladery JW, McDougal DB Jr: Electrophysiological studies of nerve and reflex activity in normal man. 1. Identification of certain reflexes in the electromyogram and the conduction velocity of peripheral nerve fibres. Bull Johns Hopkins Hosp 86:265, 1950.
31. Marinacci AA: A correlation between operative findings in cervical herniated disc with the electromyograms and opaquae myelograms. Electromyography 6:5–20, 1966.
32. Mayer RF, Feldman RG: Observations on the nature of the F wave in man. Neurology (Minneap) 17:147, 1967.
33. Riley D, Shields RW: Diabetic amyotrophy with upper extremity involvement [abstract]. Neurology 34(suppl 1):216, 1984.
34. Sahenk Z, Mendell JR, Rossio JL, Hurtubise P: Polyradiculopathy accompanying procainamide-induced lupus erythematosus: Evidence for drug-induced enhanced sensitization to peripheral nerve myelin. Ann Neurol 1:378–384, 1977.
35. Schmalbruch H, Stender S, Boysen G: Abnormalities in spinal neurons and dorsal root ganglion cells in Tangier disease presenting with a syringomyelia-like syndrome. J Neuropathol Exp Neurol 46:533–543, 1987.
36. Seyal M, Sandhu LS, Mack YP: Spinal segmental somatosensory evoked potentials in lumbosacral radiculopathies. Neurology 39:801–805, 1989.
37. Shahani BT: Late responses and the "silent period." In Aminoff M (ed): Electrodiagnosis in Clinical Neurology, 2nd ed. New York, Churchill Livingstone, 1986, pp 333–345.
38. Stanley EF: Reflexes evoked in human thenar muscles during voluntary activity and their conduction pathways. J Neurol Neurosurg Psychiatry 41:1016, 1978.
39. Stark RJ, Kennard C, Swash M: Hand wasting in spondylotic high cord compression: An electromyographic study. Ann Neurol 9:58–62, 1981.
40. Thomas PK: Classification and electrodiagnosis of hereditary neuropathies. In Brown WF, Bolton CF (eds): Clinical Electromyography, 2nd ed. Boston, Butterworth-Heinemann, 1993, pp 391–425.
41. Yiannikas C, Shahani BT, Young RR: Short-latency somatosensory evoked potentials from radial, median, ulnar, and peroneal nerve stimulation in the assessment of cervical spondylosis. Comparison with conventional electromyography. Arch Neurol 43:1264–1271, 1986.
42. Yoss RE, Corbin KB, MacCarty CS, Love JG: Significance of symptoms and signs in localization of involved root in cervical disk protrusion. Neurology 7:673–683, 1957.
43. Weintraub JR, Madalin K, Wong M, et al: Achilles tendon reflex and the H response. Their correlation in 400 limbs. Muscle Nerve 11:972, 1988.
44. Wilbourn AJ, Aminoff MJ: Radiculopathies. In Brown WF, Bolton CF (eds): Clinical Electromyography, 2nd ed. Boston, Butterworth-Heinemann, 1993, pp 177–209.
45. Wiltse LL, Guyer RD, Spencer CW, et al: Alar transverse process impingement of the L5 spinal nerve: The far-out syndrome. Spine 9:31–41, 1984.

Chapter 17

Plexopathies

Mark A. Ferrante, MD • Asa J. Wilbourn, MD

INTRODUCTION
PATHOPHYSIOLOGY
VALUE OF VARIOUS COMPONENTS
THE BRACHIAL PLEXUS
Pertinent Anatomy
Electrodiagnostic Assessment
Concluding Comments: Brachial Plexus
   Assessment

THE LUMBOSACRAL PLEXUS
Pertinent Anatomy
Electrodiagnostic Assessment
Concluding Comments: Lumbosacral
   Plexus Assessment

INTRODUCTION

Plexuses are intermediary neural elements situated between the mixed spinal nerves (as they emerge from the intervertebral foramina) and the main nerve trunks. The term *plexus*, a synonym for braid, aptly reflects the intermingling and rearrangement of axons that occur within these portions of the peripheral nervous system (PNS). For optimal assessment of these intricate anatomical structures, the clinical examination must be supplemented with ancillary laboratory procedures. The most beneficial of these, overall, is the electrodiagnostic (EDX) examination. It often can localize plexus lesions and reveal their underlying pathophysiology and severity, thereby providing information that has diagnostic and prognostic value for the referring clinicians. The EDX assessment of plexus lesions, however, is a much more challenging process than is the evaluation of most other portions of the peripheral neuromuscular system.[1, 4, 11] In part, this results simply from how proximally situated the plexuses are; for example, using percutaneous activation, it is quite difficult to stimulate proximally to much of the brachial plexus and all of the lumbosacral plexus. Moreover, the internal anatomy of the plexuses is quite complex. This reflects the fact that many thousands of axons traverse them, while following intermingling routes.[12] Consequently, to accurately assess plexus lesions, the EDX consultant must possess a thorough understanding of (1) plexus anatomy, (2) the EDX manifestations of the various underlying pathophysiologies, and (3) the plexus elements assessed by each nerve conduction study (NCS)—that is, each compound sensory nerve action potential (SNAP) and compound muscle action potential (CMAP)—and by every muscle sampled during the needle electrode examination (NEE). The first and third topics are the focus of this chapter; the second was reviewed in Chapter 14 and, hence, is only briefly addressed here.

PATHOPHYSIOLOGY

Among patients with plexus lesions who are referred to the EDX laboratory, the three types of nerve pathophysiology—demyelinating conduction slowing, demyelinating conduction block, and axon loss–induced conduction failure—are not observed equally. The explanation for this is straightforward. With lesions producing demyelinating conduction slowing, all of the nerve action potentials (NAPs) are transmitted through the lesion site and ultimately reach their destination; for that reason, these characteristically are asymptomatic in regard to negative manifestations (e.g., motor or sensory deficits). Consequently, lesions producing only this type of pathophysiology are rarely encountered in the EDX laboratory. When demyelinating conduction block or axon loss is produced, however, the NAPs are unable to traverse or to extend beyond the lesion site. As a result, clinical symptoms ensue, the magnitudes of which are proportional to the number of affected axons. Whenever a substantial number are affected, patients frequently are referred for EDX assessment. Although demyelinating conduction block and axon degeneration cause similar symptoms, the former typically is observed only early in the course of abrupt-onset plexus lesions and, in general, is accompanied by varying degrees of axon loss. Consequently, the overwhelming majority of plexopathies studied in the EDX laboratory are secondary to axon loss–induced conduction failure. There are two important exceptions, however, concerning demyelinating conduction block causing relatively rapidly reversible lesions, with recovery occurring typically within a few weeks: both multifocal motor neuropathy and radiation plexopathy result in typical conduction blocks that persist indefinitely, and often eventually progress to become axon loss lesions.[14, 16]

VALUE OF VARIOUS COMPONENTS

Each component of the EDX examination—the sensory NCS, motor NCS, and NEE—has a valuable role to play in plexopathy assessment. Regarding the NCS, the sensory NCS responses generally are the most helpful. As Gilliatt pointed out four decades ago, when they were just being introduced clinically, the SNAPs are affected (i.e., they become low in amplitude or unelicitable) by axon loss postganglionic plexus lesions, but not by preganglionic axon loss lesions of the roots of the plexus.[2, 7] In contrast, the CMAPs are equally affected by axon loss lesions of the same severity, regardless of whether they are located extraforaminally within the intraspinal canal. Moreover, for any given incomplete axon loss trunk lesion, the SNAP amplitudes customarily are more affected than the CMAP amplitudes. Thus, with an incom-

The opinions and assertions contained herein are the private views of the authors and are not to be construed as the official policy or position of the U.S. Government, the Department of Defense, or the Department of the Air Force.

201

plete axon loss lesion of mild to moderate severity, the SNAP amplitudes will be low in amplitude while the CMAP amplitudes may still be within the normal range.[14, 16] For this reason, the SNAP amplitudes are the most sensitive NCS indicator of axon loss. Consequently, they are very helpful in both detecting and localizing axon loss plexopathies, as well as in distinguishing them from radiculopathies, the PNS disorder with which plexopathies most often are confused clinically. When a reliable sensory NCS is not available for assessing a plexus element (i.e., the C5 component of the upper trunk of the brachial plexus; the lumbar plexus), or when the sensory NCS that assesses a particular plexus element is affected by a coexisting process (e.g., carpal tunnel syndrome [CTS] with upper and middle trunk brachial plexopathies; polyneuropathies and advanced age with sacral plexopathies), the utility of the EDX examination in plexus evaluation is seriously compromised.

The motor NCSs also have unique roles in plexus assessment. The CMAP amplitudes can be used to directly demonstrate conduction block lesions along the more distal brachial plexus, with stimulations supraclavicularly and in the axilla. Of more widespread application is the fact that CMAP amplitudes recorded from weak muscles with stimulation distal to a focal lesion can be used to (1) distinguish demyelinating conduction block from axon loss/conduction failure (if the CMAP recorded from a clinically weak muscle 7 days or more after lesion onset is normal or nearly normal in amplitude, demyelinating conduction block is responsible; whereas if it is low in amplitude or unelicitable, axon loss is responsible); and (2) determine the severity of axon loss plexopathies (by comparing the CMAP amplitude recorded from an affected muscle with that obtained on the contralateral limb).

Finally, the NEE aids in establishing the vertical and longitudinal extent of all plexus lesions. It is vital with axon loss plexopathies for (1) detecting mild lesions; (2) establishing whether the lesion is complete or severe but incomplete; and (3) demonstrating early reinnervation.[13–15]

THE BRACHIAL PLEXUS

Pertinent Anatomy

The brachial plexus is located between the base of the neck and the axilla and in anatomical textbooks consists of several elements—five anterior primary rami (APR), three *trunks* (upper, middle, and lower), six *divisions* (three anterior and three posterior), three *cords* (lateral, posterior, and medial), and three to five *terminal nerves*[3] (Fig. 17–1). The APR are synonymously referred to as the "roots" of the brachial plexus by most anatomists. In the surgical literature, however, the term *root* in this context encompasses the primary dorsal and ventral roots, the mixed spinal nerve (MSN) they form, and the APR continuous with it (i.e., the APR and all of the neural structures situated between it and the spinal cord).[12] Thus, surgeons who deal with lesions of the brachial plexus view it as beginning within the cervical intraspinal canal, from the C5 to T1 spinal cord segments. As a result, to them root avulsion injuries resulting from violent traction are just as much brachial plexopathies as are trunk and cord lesions.[12] In this chapter, the surgical view of the plexus serves as the basis of discussion.

The APR are located beneath the sternocleidomastoid muscle, between the anterior and middle scalene muscles, in the anteroinferior aspect of the posterior triangle of the neck. Nerves derived directly from the APR include those to the longus colli and scalene muscles (via C5 through C8) and the long thoracic nerve to the serratus anterior muscle (via C5 through C7). In addition, axons originating from the C5 APR contribute to the formation of the phrenic and dorsal scapular nerves. Trunk formation occurs near the lateral edge of the anterior scalene muscle—the C5 and C6 APR unite to form the upper trunk (UT), the C7 APR continues as the middle trunk (MT), and the C8 and T1 APR join to form the lower trunk (LT). These plexus elements are named for their anatomical relationship to each other.

Only two nerves, the subclavian and suprascapular, are derived from this level of the brachial plexus, both via the

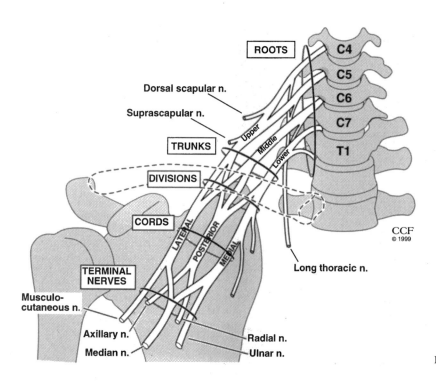

FIGURE 17–1. The brachial plexus.

UT. Each trunk then divides into an anterior and a posterior division; this occurs just distal to the lateral border of the anterior scalene muscle for the UT and MT, and at the level of the first rib for the LT. The three posterior divisions join to form the posterior cord (PC), the anterior divisions derived from the UT and MT unite to form the lateral cord (LC), and the anterior division of the LT continues as the medial cord (MC). The cords are the longest of the brachial plexus elements; they are named for their relationship to the second portion of the axillary artery (the structure which they surround), and are located in the axilla, deep to the pectoralis minor muscle, and about an inch below the coracoid process. Typically, the LC contains axons derived from the C5 through C7 spinal cord segments, the PC from the C5 through C8 spinal cord segments, and the MC from the C8 and T1 spinal cord segments.

The lateral and medial pectoral nerves, which originate from the proximal aspects of the LC and MC, respectively, fuse and innervate the pectoralis muscles.[3, 12] The terminal nerves number three (median, ulnar, and radial) to five (if the musculocutaneous and axillary are also included), depending on the reference source cited.[12] They are given off in the axilla, distal to the pectoralis minor muscle and, according to Narakas, their origin and first "few centimeters" are considered to be the terminal nerve elements of the brachial plexus.[9]

Electrodiagnostic Assessment

Brachial plexopathies are seen with some frequency in EDX laboratories, a reflection of the structures the brachial plexus supplies, and its anatomical susceptibility to traumatic injuries. Its location between two highly mobile structures, the neck and arm, significantly increases its susceptibility to traction-induced injury, the most common insult it sustains. In addition, its large size, superficial location (in part), and proximity to several disease- or trauma-susceptible structures (e.g., apex of lung, axillary lymph nodes, major blood vessels) render it vulnerable to many other injurious processes, the overwhelming majority of which cause axon loss.[12]

The brachial plexus contains both sensory and motor axons and, therefore, all components of the EDX examination have utility and are performed. As discussed in Chapter 14, the sensory NCS responses are very sensitive to axon loss lesions occurring at or distal to the dorsal root ganglia (DRG). Because the brachial plexus essentially is a postganglionic structure, SNAP abnormalities are always sought. In addition, whenever several sensory NCSs are performed on a limb, the pattern of SNAP abnormalities that result can be useful in determining exactly where the lesion is situated (e.g., UT; PC). This is possible because the paths taken through the brachial plexus by the sensory axons subserving each of the routine and nonstandard upper extremity sensory NCSs have been well established[6] (Table 17–1). In our EDX laboratories, the "routine" sensory NCSs include the median, recording index finger (Med-D2), the ulnar, recording fifth finger (Uln-D5), and the radial, recording base of the thumb (Rad-BT). To adequately assess the brachial plexus, however, several "nonstandard" sensory NCSs must be incorporated, including the lateral and medial antebrachial cutaneous (LABC, MABC), recording forearm, and the median, recording thumb and middle fingers (Med-D1, Med-D3). As the information obtained from the dorsal ulnar cutaneous (DUC) NCS, recording dorsum of the hand, is redundant to that provided by the Uln-D5 NCS, it typically is not performed.[6] The incidence of the various SNAP abnormalities associated with trunk and cord lesions is shown in Table 17–1 and Figure 17–2.

In our EDX laboratories, the routine motor NCSs are the

TABLE 17–1	Sensory Nerve Action Potential Domains of the Brachial Plexus and the Frequency of Associated SNAP Abnormalities

Upper Trunk	Middle Trunk	Lower Trunk
Lateral antebrachial cutaneous (100%)	Median, recording D2 (80%)	Ulnar, recording D5 (100%)
Median, recording D1 (100%)	Median, recording D3 (80%)	Medial antebrachial cutaneous (100%)
Radial, recording thumb base (60%)	Radial, recording thumb base (40%)	Median, recording D3 (15%)
Median, recording D2 (20%)		Dorsal ulnar cutaneous (100%)[a]
Median, recording D3 (10%)		

Lateral Cord	Posterior Cord	Medial Cord
Lateral antebrachial cutaneous (100%)	Radial, recording thumb base (100%)	Ulnar, recording D5 (100%)
Median, recording D1 (100%)		Medial antebrachial cutaneous (100%)
Median, recording D2 (100%)		Median, recording D3 (15%)
Median, recording D3 (85%)		

D, digit.
[a]This response typically provides information redundant to that obtained with the ulnar, recording D5.
Data compiled from Ferrante MA, Wilbourn AJ: The utility of various sensory nerve conduction responses in assessing brachial plexopathies. Muscle Nerve 18:879–889, 1995.

median, recording abductor pollicis brevis (Med-APB), and the ulnar, recording abductor digiti minimi (Uln-ADM). Because these assess only the MC, LT, and the C8 and T1 elements, nonstandard motor NCSs often are required, including the musculocutaneous, recording biceps (Musc-Bic), the axillary, recording deltoid (Ax-Delt), and the radial, recording from the extensor aspects of the proximal and distal forearm (Rad-Prox, Rad-Dist, respectively). The CMAP domains for all of the trunk and cord elements of the brachial plexus are well known (Table 17–2).

A clinically relevant classification system for categorizing brachial plexopathies, employed by most surgeons, groups these brachial plexus lesions into supraclavicular and infraclavicular plexopathies. This reflects the anatomical fact that, with the arm resting at the side of the body, the divisions lie behind the middle one-third of the clavicle and, hence, are *retroclavicular*, whereas the roots and trunks are *supraclavicular* and the cords and terminal nerves are *infraclavicular*. Because lesions affecting the divisions rarely occur in isolation, a retroclavicular plexopathy category seldom is necessary. The supraclavicular plexopathies are further divided into upper, middle, and lower plexus lesions. These categories differ in terms of their incidence, severity, and prognosis. For this reason, the EDX assessment of this structure is approached using the same anatomical divisions.[12]

Supraclavicular Plexus

The supraclavicular plexus consists of the C5 through T1 primary roots, APR and MSNs, as well as all three trunks. It extends from the spinal cord proximally to the clavicle distally.

Upper Plexus

The term *upper plexus* encompasses the UT, and the C5 and C6 primary roots, MSNs, and APR. This portion of the brachial plexus is solely or principally damaged more often

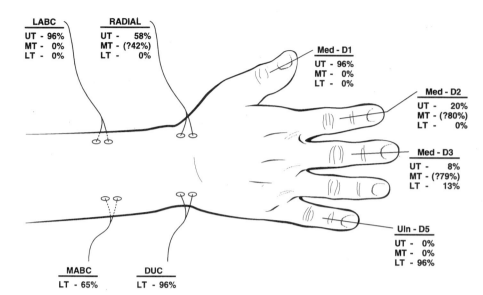

FIGURE 17–2. The incidence of sensory NCS abnormalities with various brachial plexus trunk lesions. (From Ferrante MA, Wilbourn AJ: The utility of various sensory nerve conduction responses in assessing brachial plexopathies. Muscle Nerve 18:879–889, 1995; used with permission.)

than any other region, yet it is the least well evaluated by the routine NCS.

Sensory NCS. None of the available sensory NCSs assesses the C5 component of the upper plexus.[12, 14] The Rad-BT SNAP assesses the C6 component about 60% of the time, the Med-D2 SNAP assesses it about 20% of the time, while the Uln-D5 SNAP does not assess the upper plexus at all. Consequently, adequate evaluation of the upper plexus requires the addition of nonstandard sensory NCSs, specifically, the LABC and Med-D1 SNAPs, because the sensory axons they assess, derived from the C6 DRG, traverse the upper plexus 100% of the time. Thus, they are not affected by middle plexus or lower plexus lesions. With most axon loss upper plexus lesions both of these sensory NCSs are affected, and to similar degrees. Hence, either one may be used for screening purposes.[6] The LABC NCS is preferred, however, if the Med-D1 possibly is compromised by CTS. We perform both whenever an upper plexus lesion is definitely present. As previously noted, lesions affecting only the preganglionic elements of the sensory fibers—that is, root avulsion injuries—do not show abnormalities of any sensory NCSs, including these; unfortunately, many lesions of the upper plexus, especially those resulting from vehicular accidents, are of this nature.

Motor NCS. Both the Musc-Bic and the Ax-Delt NCSs assess the upper plexus, including both its C5 and C6 components. In addition, the Rad-Prox NCS can be used to assess the C6 component of the upper plexus. In general, whenever the deltoid or biceps muscle is clinically weak or shows motor unit axon potential (MUAP) dropout on NEE, then the motor NCS that employs the affected muscle as the recorded muscle is performed, since CMAP amplitude decrements are useful indicators of the amount of motor axon loss present. In this setting, if the distal stimulation site yields a normal or near-normal CMAP, evidence of a demyelinating conduction block lesion is sought by comparison to more proximally elicited responses (e.g., via supraclavicular stimulation).

Needle Electrode Examination. Many muscles innervated by the upper plexus are available for assessment during the NEE (Table 17–3). Once it is apparent that the upper plexus is involved, it is important to determine on NEE (1) the proximal extent of the lesion, by sampling muscles innervated by axons that arise from the very proximal upper trunk (i.e., spinati) or the C5, C6, or C7 APR (i.e., rhomboids, serratus anterior), and (2) whether the middle plexus is also damaged; this requires determining if muscles innervated by the C7 root (e.g., triceps, pronator teres [PT], flexor carpi

TABLE 17–2	Compound Muscle Action Potential Domains of the Brachial Plexus

Upper Trunk	Middle Trunk	Lower Trunk
Musculocutaneous (biceps) Axillary (deltoid)		Ulnar (abductor digiti minimi) Ulnar (first dorsal interosseous) Median (abductor pollicis brevis) Radial (extensor indicis proprius)

Lateral Cord	Posterior Cord	Medial Cord
Musculocutaneous (biceps)	Axillary (deltoid) Radial (extensor digitorum communis) Radial (extensor indicis proprius)	Ulnar (abductor digiti minimi) Ulnar (first dorsal interosseous) Median (abductor pollicis brevis)

TABLE 17–3	Frequently Studied Muscles within the Upper Plexus Muscle Domain

Distal Upper Trunk	Proximal Upper Trunk[a]	Probable Intraspinal Canal[b]
Biceps Deltoid Teres minor Brachioradialis *Triceps *Extensor carpi radialis *Flexor carpi radialis *Pronator teres	All distal UT muscles Supraspinatus Infraspinatus	All proximal UT muscles Rhomboideus major Rhomboideus minor Levator scapulae Serratus anterior Cervical paraspinals

UT, upper trunk.
Note: Muscles with * are also in the domain of the middle plexus.
[a]The suprascapular nerve derives from the proximal aspect of the upper trunk.
[b]The dorsal scapular and long thoracic nerves derive from the anterior primary rami level of the brachial plexus.

TABLE 17–4	Nerve Conduction Study Findings Seen with a Right Upper Trunk Axon Loss Brachial Plexus Lesion Sustained in an Automobile Accident

Nerve Stimulated	Stimulation Site	Recording Site	Amplitude μV/mV		Latency (msec) Peak/Distal		Conduction Velocity (m/s)		F Waves	
			R	L	R	L	R	L	R	L
Median (S)	Wrist	D1	NR[a]	32	—	2.4				
Median (S)	Wrist	D2	24	40	2.6	2.6				
Median (S)	Wrist	D3	38	40	2.7	2.6				
Radial (S)	Wrist	Thumb base	10[a]	32	2.1	1.9				
Ulnar (S)	Wrist	D5	30	32	2.2	2.1				
LABC (S)	Elbow	Forearm	NR[a]	20	—	2.1				
Median (M)	Wrist	Thenar	12	11	2.8	2.9	54	55	27	27
	Elbow		11	10	7.8	8.0				
Ulnar (M)	Wrist	Hypothenar	13	12.5	2.4	2.5	56	55		
	Elbow		12	12	7.0	7.2				
Musculocutaneous (M)	Axilla	Biceps	1.5[a]	5.0	2.4	2.2				
Axillary	Supraclavicular	Deltoid	2.0[a]	10.0	4.0	3.8				
Radial (M)	Elbow	Proximal extensor forearm	5.0	8.0	3.1	3.0				

D, digit; LABC, lateral antebrachial cutaneous; NR, no response.
Note that, of the three median sensory NCSs performed, only the one recording from D1 is definitely abnormal. In this patient, the right radial sensory NCS is low in amplitude, indicating that many of the axons assessed traveled through the upper trunk; however, in other patients with similar lesions, this NCS is normal, so it is not a reliable NCS for assessing the upper plexus. As typically occurs with incomplete axon loss lesions, the sensory NCS responses (median D1, LABC) that assess the affected plexus element are more severely affected than the motor NCS (musculocutaneous; axillary) that do.
[a]Abnormal response.

radialis [FCR]) are involved and, if so, how severely (as they also receive C6 innervation).

Common Entities Affecting the Upper Plexus. High-energy closed trauma, such as that frequently sustained during vehicular accidents, particularly motorcycle mishaps, typically affects the upper plexus (Tables 17–4 and 17–5). Other disorders that commonly involve this portion of the brachial plexus include obstetrical palsy, classic postanesthesia paralysis, rucksack palsy, and the burner syndrome. Whereas high-energy and obstetrical injuries usually cause substantial axon loss, the burner syndrome is characterized by mild axon loss—so mild that the LABC and Med-D1 NCS amplitudes are not affected. Demyelinating conduction block and axon loss often coexist initially with the less severe high-energy and obstetrical lesions,

TABLE 17–5	Needle Electrode Examination Findings that Accompany the Nerve Conduction Studies Reported in Table 17–4

Muscle	Spontaneous Activity	Motor Unit Action Potentials			
		Recruitment	Duration	Amplitude	Polyphasic
Right	**Fibs**				
First dorsal interosseous	None	Nor	Nor	Nor	Nor
Extensor indicis proprius	None	Nor	Nor	Nor	Nor
Flexor policis longus	None	Nor	Nor	Nor	Nor
Pronator teres	1 +	↓ # Rap	Nor	Nor	Some 1 +
Triceps	1 +	↓ # Rap	Nor	Nor	Nor
Brachioradialis	2 +	↓ ↓ # Rap	Nor	Nor	Some 1 +
Biceps	2 +	↓ ↓ ↓ # Rap	Nor	Nor	Many 2 +
Brachialis	2 +	↓ ↓ ↓ # Rap	Nor	Nor	Many 2 +
Deltoid	2 +	↓ ↓ ↓ # Rap	Nor	Nor	Most 2 +
Infraspinatus	3 +	None	—	—	—
Supraspinatus	3 +	↓ ↓ ↓ # Rap	Nor	Nor	Most 2 +
Serratus anterior	None	Nor	Nor	Nor	Nor
Rhomboid major	None	Nor	Nor	Nor	Nor
Trapezius	None	↓ # Mod	Nor	Nor	—
C6 paraspinal	None	—	—	—	—
Left					
Biceps	None	Nor	Nor	Nor	Nor
Deltoid	None	Nor	Nor	Nor	Nor

Fibs, fibrillation potentials; Mod, moderate; Nor, normal; Rap, rapid.
The results are those of an acute axon loss lesion of just a few months duration (since chronic neurogenic motor unit potential changes have not yet developed in the partially denervated muscles). Although the abnormalities present in the right triceps and pronator teres (C6,C7-innervated muscles) are equally consistent with an upper plexus lesion and a combined upper and middle plexus lesion, the normal median D3 response is more consistent with the former.

whereas demyelinating conduction block is the predominant type of pathophysiology with classic postoperative paralysis.[12]

Middle Plexus

The term *middle plexus* encompasses the MT, and the C7 APR, MSN, and primary root. Lesions involving this portion of the supraclavicular plexus are rarely seen in isolation. Instead, they coexist with upper plexus lesions, lower plexus lesions, or both.[12]

Sensory NCS. Of the routine sensory NCSs, the Med-D2 and Rad-BT NCSs assess the middle plexus approximately 80% and 40% of the time, respectively. Among the nonstandard sensory NCSs, the Med-D3 assesses the middle plexus approximately 80% of the time. The Med-D1 and LABC SNAPs can be used to determine the coexistence of an upper plexus lesion, and the Uln-D5 and MABC SNAPs, a lower plexus lesion, as none of the cell bodies of origin of the sensory axons subserving these four sensory NCSs is located within the C7 DRG.[6] With avulsion injuries of the C7 sensory root, all of the SNAPs are affected.

Motor NCSs. There is no satisfactory motor NCS for assessing solely the middle plexus elements, because none of the muscles that receive their major innervation from the C7 root (e.g., triceps) can serve as the recorded muscle during motor NCS—the techniques for doing so have not been devised.[14]

Needle Electrode Examination. Because the muscle domain of the middle plexus overlaps with those of the upper plexus and, to a much lesser extent, the lower plexus, it is challenging to determine, by NEE alone, whether a lesion is confined to the middle plexus. However, when the "shared" muscles (i.e., those contained within the muscle domain of both the middle plexus and either the upper or lower plexus) are affected while the "unshared" muscles (i.e., those belonging to either the upper or lower plexus muscle domains, but not that of the middle plexus) are spared, a middle plexus lesion is probable.

Lower Plexus

The term *lower plexus* encompasses the LT and the C8 and T1 APR, MSNs, and primary roots. Among supraclavicular plexus lesions, those of the lower plexus are less frequent than those of the upper plexus. Similar to those of the upper plexus, however, they can occur in isolation or with coexistent damage to the middle plexus, or to both the middle and upper plexuses. Although the routine NCSs assess the lower plexus better than any other supraclavicular region, nonstandard NCSs occasionally are still important in its evaluation.

Sensory NCSs. Of the three routine sensory NCSs, the sensory axons subserving the Uln-D5 study traverse portions of the lower plexus, whereas those assessed by the Med-D2 and Rad-BT NCSs do not. Among the nonstandard studies, the MABC SNAP is the most helpful for assessing the lower plexus; it complements the Uln-D5 study. Even though the sensory axons assessed by both the Uln-D5 NCS and the MABC NCS traverse the LT,[6] the Uln-D5 SNAP assesses axons that originate principally in the C8 DRG, whereas the MABC SNAP assesses those derived primarily from the T1 DRG. This is important, because two brachial plexopathies labeled LT lesions—postmedian sternotomy (PMS) lesions and true neurological thoracic outlet syndrome (N-TOS)—have quite dissimilar clinical and EDX features, as they actually result from damage to axons at the APR, rather than the LT, level. Thus, PMS lesions affect almost solely the C8 APR and, therefore, the Uln-D5 SNAP amplitudes, whereas true N-TOS injures the T1 APR solely or predominantly, and thereby has a much greater effect on the MABC SNAP amplitude.[8] With all other LT plexopathies, these two SNAPs usually are more equally involved.

Motor NCSs. Both of the routine motor NCSs (Med-APB, Uln-ADM) assess the lower plexus, including the LT and the C8 and T1 APR, MSN, and primary roots. However, just as with the sensory NCSs that assess the lower plexus, the motor NCSs that assess it preferentially evaluate either C8- or T1-derived axons; the Uln-ADM CMAP, the C8 axons; and the Med-APB CMAP, the T1 axons.[8] The Rad-Dist CMAP also assesses a portion of the lower plexus, because the radial nerve–innervated muscles in the distal extensor forearm, which serve as the recorded muscles, receive substantial innervation from the C8 root. This NCS may aid in differentiating a lesion of the LT from one affecting the MC (discussed later).

Needle Electrode Examination. Many muscles are likely to show NEE abnormalities with lower plexus lesions, including all those innervated by the C8 (T1) roots via the ulnar nerve, the (C8) T1 roots via the median nerve, and the C8 root via the radial nerve (see Table 17–6). When denervated, the radial nerve–innervated muscles (e.g., extensor indicis proprius [EIP]; extensor pollicis brevis [EPB]) differentiate LT from MC lesions. Although their involvement excludes an MC lesion, the converse—that is, that their lack of involvement verifies an MC lesion—is not accurate, because a partial LT lesion, one that spared the C8/radial nerve component, cannot be excluded. Of note is that the lateral thenar muscles (i.e., APB and opponens pollicis [OP] muscles), sometimes appear to be innervated solely by the T1 (as opposed to C8 plus T1) root. As a result, with many PMS lesions (mainly C8 APR lesions) they show no abnormalities and, as noted, the Med-APB, which uses them as the recorded muscles, is normal. Conversely, with true N-TOS (T1 APR lesions) they characteristically are the most severely denervated muscles and, as a result, the Med-APB CMAP is always much lower in amplitude than the Uln-ADM CMAP.[6, 12, 15]

Common Entities Affecting the Lower Plexus. Sole or preferential involvement of this portion of the supraclavicular plexus is observed with certain brachial plexopathies, including true N-TOS; postoperatively, with disputed N-TOS and PMS lesions; neoplastic lesions; and sometimes closed traction injuries due to trauma or obstetrical paralysis[6, 14, 15] (Tables 17–7 and 17–8). With the latter two etiologies, C8 and T1 root avulsions are common.

Combined Lesions (UT/MT, MT/LT, UT/MT/LT)

Early traumatic and postanesthesia paralysis plexopathies, as well as late radiation and neoplastic plexopathies, commonly present as diffuse supraclavicular plexus lesions. Depending on the etiology, the infraclavicular plexus elements may also be involved (e.g., radiation plexopathies usually begin in the infraclavicular plexus, especially the lateral cord, and then spread to become diffuse).[12] It may be impossible to differentiate a widespread supraclavicular lesion from a widespread infraclavicular one by EDX examination, unless the pectoral muscles reveal NEE abnormalities. Because the pectoral nerves arise so proximally from the lateral and medial cords, their involvement suggests a supraclavicular process. Fortunately, the history is frequently also useful in making this distinction.

TABLE 17–6	**Frequently Studied Muscles Within the Lower Plexus Muscle Domain**

Radial Nerve	Ulnar Nerve	Median Nerve
Extensor indicis proprius	First dorsal interosseous	Abductor pollicis brevis
Extensor pollicis brevis	Abductor digiti minimi	Opponens pollicis
Extensor digitorum communis	Flexor digitorum profundus-4,5	Flexor pollicis longus
	Flexor carpi ulnaris	

TABLE 17–7	Nerve Conduction Study Findings Seen with a Severe Axon Loss, Right Lower Trunk Brachial Plexopathy Caused by Apical Lung Carcinoma

Nerve Stimulated	Stimulation Site	Recording Site	Amplitude µV/mV		Latency (msec) Peak/Distal		Conduction Velocity (m/s)		F Waves	
			R	L	R	L	R	L	R	L
Median (S)	Wrist	D1	20	2.1	3.0	2.8				
Median (S)	Wrist	D2	29	28	3.0	3.0				
Median (S)	Wrist	D3	21	27	3.1	3.0				
Radial (S)	Wrist	Thumb base	35	38	2.1	2.2				
Ulnar (S)	Wrist	D5	NRa	15	—	2.7				
MABC (S)	Arm	Forearm	NRa	18	—	2.4				
Median (M)	Wrist	Thenar	0.8a	9.5	3.3	2.8	50	59	33a	28.5
	Elbow		0.6a	8.0	9.0	7.8				
Ulnar (M)	Wrist	Hypothenar	1.5a	10.6	2.9	2.3	48a	58	34a	28
	Elbow		1.3a	10.1	9.5	6.8				
Ulnar (M)	Wrist	1st DI	NRa	12.0	—	1.9				
Radial (M)	Wrist	Distal extensor forearm	2	7	2.1	1.6				
Radial (M)	Elbow	Proximal extensor forearm	7.2	9	1.7	1.6				

D, digit; 1st DI, first dorsal interosseous; MABC, medial antebrachial cutaneous; NR, no response.
a Abnormal responses.

Infraclavicular Plexus

The infraclavicular plexus consists of the cords and terminal nerves. It extends from the clavicle to the distal axilla. Because of the intimate relationship of the infraclavicular plexus elements with the axillary blood vessels, primary vascular damage commonly causes secondary plexopathies.

Lateral Cord

This portion of the infraclavicular plexus contains axons derived from the C5 through C7 APR and the C6 and C7 DRG, via the anterior divisions of the UT and MT.

Sensory NCSs. Of the routine NCSs, only the Med-D2 assesses the LC. Consequently, additional nonstandard studies must be added for optimal evaluation. The sensory axons subserving the LABC and Med-D1 NCS traverse this plexus element probably 100% of the time, and those subserving the Med-D3 NCS approximately 85% of the time. Lesions of the LC must be distinguished from those of the UT. The Med-D2, Med-D3, and Rad-BT NCSs are all useful in this regard. The Med-D2 and Med-D3 SNAPs are always affected with LC lesions, but seldom with UT lesions. The sensory axons subserving the Rad-BT SNAP traverse the UT about 60% of the time, but do not traverse the LC. Hence, when this NCS is abnormal, a UT lesion is likely. Unfortunately, the converse—that is, that when it is normal an LC lesion has been identified—is inaccurate for two reasons: (1) a

TABLE 17–8	Needle Electrode Examination Findings that Accompany the Nerve Conduction Study Results Shown In Table 17–7

		Motor Unit Action Potentials			
Muscle	Spontaneous Activity	Recruitment	Duration	Amplitude	Polyphasic
Right	**Fibs**				
First dorsal interosseous	2 +	None	—	—	—
Abductor pollicis brevis	3 +	↓ ↓ ↓ ↓ # Rap	All ↑	Nor	Most 1 +
Abductor digitorum minimi	2 +	↓ ↓ ↓ # Rap	All ↑	Nor	Most 1 +
Flexor pollicis longus	2 +	↓ ↓ ↓ # Rap	All ↑	Nor	Most 1 +
Flexor carpi ulnaris	2 +	↓ ↓ # Rap	All ↑	Nor	All 2 +
Extensor indicis proprius	2 +	↓ ↓ ↓ # Rap	All ↑	Nor	Some 1 +
Extensor pollicis brevis	2 +	↓ ↓ # Rap	All ↑	Nor	Most 1 +
Extensor digitorum communis	1 +	↓ # Rap	Most ↑	Nor	Some 1 +
Brachioradialis	None	Nor	Nor	Nor	Nor
Triceps	None	Nor	Nor	Nor	Nor
Biceps	None	Nor	Nor	Nor	Nor
Deltoid	None	Nor	Nor	Nor	Nor
Pronator teres	Nor	Nor	Nor	Nor	Nor
C8 paraspinals	None	—	—	—	—
T1 paraspinals	—	—	—	—	—
Left					
First dorsal interosseous	None	Nor	Nor	Nor	Nor
Abductor pollicis brevis	None	Nor	Nor	Nor	Nor
Extensor indicis proprius	None	Nor	Nor	Nor	Nor

Fibs, fibrillation potentials; Nor, normal; Rap, rapid.
Note that the lesion appears to be quite active (based on the density of fibrillation potentials present) and chronic (judging by the chronic neurogenic motor unit potential changes present).

partial UT lesion may spare this NCS, and (2) the sensory axons being assessed are derived from the C7 DRG, and therefore traverse the middle trunk about 40% of the time.[6]

Motor NCS. The routine motor NCSs do not assess the LC. Consequently, the nonstandard Musc-BC study must be used. If abnormal, an Ax-Delt NCS is added to help exclude the possibility of a UT lesion.

Needle Electrode Examination. Muscles likely to show abnormalities with LC lesions include those innervated by the musculocutaneous nerve (biceps, brachialis) and the lateral head of the median nerve (PT, FCR). The deltoid and teres minor are unaffected, as are the spinati, the rhomboids, and the serratus anterior muscles. The NEE of some of the latter muscles helps differentiate a lesion of the LC from one involving the UT. To differentiate an LC lesion from a median nerve lesion, the NEE should include muscles innervated by the median nerve via its medial head (e.g., the flexor pollicis longus [FPL] and APB).

Posterior Cord

This portion of the infraclavicular plexus contains axons derived from the C5 through C8 AHCs and the C5 through C7 DRG, via all three posterior divisions.

Sensory NCSs. Only one of the routine sensory NCSs, the Rad-BT SNAP, assesses the PC; among the nonstandard sensory NCSs, the posterior antebrachial cutaneous NCS also assesses this element, but it is rarely required. The radial sensory axons evaluated by the Rad-BT SNAP originate from the C6 DRG, the C7 DRG, or both, and traverse the PC 100% of the time as, anatomically, this is the only cord element to which they have access. The routine Med-D2 and Uln-D5 responses are normal with lesions localized to the PC. The C5 DRG-derived sensory axons that traverse the PC (e.g., those in the axillary nerve) cannot, unfortunately, be assessed by any sensory NCS.

Motor NCSs. Neither of the routine motor NCSs assesses the PC. For this reason, it is necessary to perform nonstandard motor NCSs: the Rad-Prox, the Rad-Dist, and the Ax-Delt.

Needle Electrode Examination. Muscles that may show abnormalities with PC lesions include those innervated by the radial, axillary, and thoracodorsal nerves. Conversely, muscles innervated by the median, ulnar, musculocutaneous, and suprascapular nerves are not affected. For this reason, some of the muscles innervated by these nerves are sampled on the NEE; if abnormal, they indicate that the lesion is not confined to the PC.

Medial Cord

This portion of the infraclavicular plexus contains axons derived from the C8 and T1 APR and DRG, via the LT and lower anterior division. It is, therefore, fairly well assessed by the routine NCS.

Sensory NCSs. Only one of the routine sensory NCSs, the Uln-D5, is useful in MC assessment. Among the nonstandard studies, the MABC NCS is most useful. Anatomically, the sensory axons of the Uln-D5 and MABC NCSs traverse the MC 100% of the time, whereas those of the Med-D3 traverse this structure about 15% of the time.[3]

Motor NCSs. The motor axons subserving both of the routine motor studies (Med-APB, Uln-ADM) traverse the MC, but they also traverse the LT. Hence, neither of these NCSs is capable of differentiating between lesions at these two sites. Consequently, whenever an MC lesion is suspected, the Rad-Dist NCS may be added, because the motor axons it assesses traverse the PC, but not the MC; thus, when it is abnormal, the lesion is likely proximal to the MC and, consequently, is probably affecting the LT. Unfortunately, the converse reasoning—that is, that an MC lesion is probable whenever the Rad-Dist is normal—is incorrect, because a partial LT lesion may spare the axons this NCS assesses.

Needle Electrode Examination. Muscles likely to show abnormalities with MC lesions include those innervated by the ulnar nerve and the medial head of the median nerve. Radial nerve–innervated muscles are never abnormal with lesions localized to the MC.

Terminal Nerves

The terminal nerves are the most distal elements of the brachial plexus. Because they extend only a "few centime-

TABLE 17–9	Nerve Conduction Study Findings Seen with a Right Infraclavicular Plexopathy Involving Predominantly the Lateral Cord, Caused by Remote Radiation for Ipsilateral Breast Carcinoma

Nerve Stimulated	Stimulation Site	Recording Site	Amplitude µV/mV		Latency (msec) Peak/Distal		Conduction Velocity (m/s)		F Waves	
			R	L	R	L	R	L	R	L
Median (S)	Wrist	D1	4[a]	20	2.6	2.5				
Median (S)	Wrist	D2	6[a]	27	2.7	2.5				
Median (S)	Wrist	D3	6[a]	28	2.6	2.6				
Radial (S)	Wrist	Thumb base	13	26	2.0	2.0				
Ulnar (S)	Wrist	D5	21	20	2.1	2.1				
LABC (S)	Elbow	Forearm	16	10		2.0				
MABC (S)	Arm	Forearm	10	14	2.4	2.3				
Median (M)	Wrist	Thenar	9	10	3.0	3.1	57	56	26	27
	Elbow			9.5	7.1	7.0				
Ulnar (M)	Wrist	Hypothenar	12	13	2.5	2.4				
	Elbow		11	12	7.3	7.2	57	57	25	25.5
	Axilla		10		11.5		59			
	Supraclavicular		9		13.7		60			
Musculocutaneous	Axilla	Biceps	6.0	5.5	2.4	2.2	58	59		
	Supraclavicular		1.5	4.8	4.5	4.6				
Axillary	Supraclavicular	Deltoid	9	11	4.0	3.9				
Radial	Elbow	Proximal extensor forearm	5.0	7.5	3.0	3.2				

D, digit; LABC, lateral antebrachial cutaneous; MABC, medial antebrachial cutaneous.
Note that all the right median sensory NCS responses are equally affected, along with the right lateral antebrachial cutaneous NCS response and, minimally, the right radial sensory response. A conduction block is present along the right musculocutaneous nerve fibers between the supraclavicular and axilla stimulation points; similar conduction blocks are not demonstrable along the axillary or proximal ulnar motor fibers.
[a] Abnormal response.

TABLE 17–10	Needle Electrode Examination Findings that Accompany the Nerve Conduction Studies Shown in Table 17–9

Muscle	Spontaneous Activity			Motor Unit Action Potentials			
	Fibs	Fas	MKD	Recruitment	Duration	Amplitude	Polyphasic
Right							
First dorsal interosseous	None	None	None	Nor	Nor	Nor	Nor
Abductor pollicis brevis	None	None	None	Nor	Nor	Nor	Nor
Extensor digitorum communis	1 +	None	None	Nor	Nor	Nor	Nor
Extensor indicis proprius	None	1 +	None	Nor	Nor	Nor	Nor
Flexor pollicis longus	None	None	None	Nor	Nor	Nor	Nor
Triceps	None	1 +	None	Nor	Nor	Nor	Nor
Brachioradialis	None	1 +	None	Nor	Nor	Nor	Nor
Biceps	1 +	2 +	0–1 +	↓ ↓ ↓ # Rap	All ↑	Nor	All 1 +
Brachialis	1 +	2 +	None	↓ ↓ ↓ # Rap	All ↑	Nor	All 1 +
Pronator teres	2 +	1 +	1 +	↓ ↓ # Rap	Most ↑	Nor	Most 1 +
Flexor carpi radialis	1 +	None	None	↓ ↓ # Rap	All ↑	Nor	All 1 +
Deltoid	1 +	1 +	None	Nor	Nor	Nor	Nor
Infraspinatus	None	1 +	Nor	Nor	Nor	Nor	Nor
Left							
Biceps	None	None	None	Nor	Nor	Nor	Nor
Pronator teres	None	None	None	Nor	Nor	Nor	Nor

Fibs, fibrillation potentials; Fas, fasciculation potentials; MKD, myokymic discharges; Nor, normal; Rap, rapid.
Note that the abnormalities localize principally to the lateral cord, although some changes are present in muscles innervated by the posterior cord. Even though the lesion is due to radiation, fibrillation potentials and fasciculation potentials are much more abundant than myokymic discharges.

ters" before they become the most proximal portions of the major peripheral nerve trunks, and because there are no sensory or motor branches derived from these first few centimeters, it can be impossible to determine, by either clinical or EDX examination, whether a specific lesion involves the infraclavicular plexus or the proximal peripheral nerve.[9] This is especially the case if only a single terminal nerve element is involved. Differentiating abnormalities involving the terminal nerve from those involving the cords also can be challenging.

To localize terminal nerve lesions by the EDX examination, all abnormalities must be found in only the distribution of one or more terminal nerves. Thus, with an isolated terminal median nerve lesion, on NCS only the median SNAP and CMAP amplitudes are affected, whereas on NEE only the median nerve–innervated muscles show abnormalities. Because the sensory and motor axons subserving the routine median NCS do not converge until the lateral and medial heads of the median nerve fuse (at the most distal aspects of the LC and MC), it would be impossible for a more proximal lesion to generate this pattern of abnormalities without also producing abnormalities of other brachial plexus elements.

In contrast, the ulnar sensory and motor axons lie adjacent to each other within the MC, LT, and the C8 MSN and APR, in addition to the ulnar nerve. Consequently, EDX abnormalities confined to these axons are insufficient for diagnosing an ulnar nerve lesion. Instead, additional assessments of the MC and LT elements must be performed, and yield normal results. In this setting, the MABC NCS is added, as its involvement indicates a lesion situated proximal to the origin of the ulnar nerve. Also, on motor NCS, if the Med-APB is abnormal, the lesion is proximal to the ulnar nerve. The Rad-Dist NCS can be added, if necessary. Because the NEE is the most sensitive EDX component for identifying motor axon loss, the C8/radial nerve–innervated muscles, as well as the C8, T1/median nerve–innervated FPL, must be sampled.

With radial nerve lesions, in addition to having the EDX abnormalities confined to the sensory and motor axons of the radial nerve, more proximal plexus elements (e.g., PC, MT) must be shown to be spared. To help exclude a PC lesion, an Ax-Delt NCS and NEE of at least one axillary nerve–innervated muscle (e.g., deltoid) are performed. In addition, C7/median nerve–innervated muscles (e.g., PT, FCR) are added to assess the MT. When abnormal, these signify that the lesion is proximal to the origin of the radial nerve.[14]

An even lesser number of NCSs and muscles accessible for NEE exist for the remaining terminal nerve elements. With musculocutaneous nerve lesions, the NCS abnormalities are limited to the LABC and Musc-Bic motor NCSs, whereas NEE abnormalities are found only in the biceps and brachialis muscles. With axillary nerve lesions, there is no available sensory NCS. However, the Ax-Delt NCS can be performed, and both muscles innervated by the nerve (deltoid and teres minor) are accessible for NEE. Other muscles innervated by the UT (e.g., biceps, infraspinatus), the radial nerve (e.g., triceps, EIP), or both (brachioradialis), should appear normal on NEE.

Common Entities Affecting the Cords and Terminal Nerves. These brachial plexus elements may be injured in a variety of ways, most commonly by trauma and iatrogenic insults. Traumatic causes include fracture/dislocation of the humeral head, axillary compression, and blunt or penetrating injuries to the axillary region (e.g., gunshot wounds, stab wounds). Iatrogenic causes include radiation, shoulder operations, axillary arteriograms, axillary brachial plexus blocks, and reduction of humeral head dislocations. Whenever injuries of the axillary blood vessels occur, subsequent hematoma or pseudoaneurysm formation may secondarily injure the infraclavicular plexus.[12] Noteworthy is that simultaneous involvement of two or more terminal nerves can be difficult to distinguish from a cord lesion (Tables 17–9 and 17–10).

Concluding Comments: Brachial Plexus Assessment

1. All three components of the EDX study (sensory NCS, motor NCS, and NEE) must be performed. There are no abbreviated "screening" evaluations capable of adequately assessing this large PNS structure. Almost al-

ways, nonstandard NCSs and additional muscles on NEE must be incorporated into the EDX study for optimal assessment.

2. Whenever the patient is known to have sensory loss or weakness, NCSs and NEEs of the involved portion of the limb are incorporated into the EDX examination.

3. Whenever only a single brachial plexus element is thought to be affected, additional EDX studies are added to assess the immediately proximal, distal, and adjacent brachial plexus elements, thereby ensuring that the boundaries of the lesion are accurately delineated.

4. Whenever a demyelinating conduction block lesion is suspected (e.g., normal CMAP amplitude recorded from a muscle which reveals a neurogenic MUAP recruitment pattern on NEE), motor nerve stimulation at more proximal sites (e.g., supraclavicular stimulation) should be performed to localize the lesion, if possible.

5. Low-amplitude or unelicitable CMAPs in the presence of SNAPs of normal or near-normal amplitude, when found with a proximal lesion, have only two possible causes, as follows: (1) an intraspinal canal lesion (most commonly an avulsion); or (2) the NCSs were performed between days 4 and 7 after lesion onset, at a time when wallerian degeneration has substantially affected conduction along the motor axons, but has not yet begun to alter, or to appreciably alter, conduction along the peripheral sensory axons.

6. Finally, it is neither practical nor necessary to perform extensive EDX examinations on every patient referred with the diagnosis of brachial plexopathy, particularly when the referring physicians have had little experience with these PNS lesions. In these situations, all that is required is for all the major plexus elements to have been assessed with a sensory NCS and with the NEE. In our EDX laboratories, this simply means that a Med-D1 or an LABC NCS is added to the routine NCS, and a few additional muscles are sampled on NEE; for example, brachioradialis and infraspinatus (and deltoid, if it is not included in the basic survey). It is not necessary to perform additional motor NCSs, unless clinical weakness is present, or the sensory NCS or NEE suggests a brachial plexus lesion is present. Regarding the NEE, although it is the most sensitive component of the EDX study for identifying motor axon loss, it cannot distinguish between a brachial plexopathy and a radiculopathy that is sparing the paraspinal muscles. Consequently, it can never be utilized in isolation to "screen" the brachial plexus.

THE LUMBOSACRAL PLEXUS

The *lumbosacral plexus* is a collective term applied to two separate components of the PNS: the lumbar plexus and the sacral plexus. These differ in several respects—including their origins, surrounding anatomical structures, courses, and terminations. They innervate the lower abdominal region and almost the entire pelvic girdle and lower extremities. Lesions affecting these plexuses are much less common than are those involving the brachial plexus, a consequence of their less compact nature, their location distant from highly mobile structures, and the protection they are afforded by neighboring structures—the lumbar plexus is formed within the psoas major muscle, while the sacral plexus, which is surrounded by soft tissue structures (e.g., the colon), lies within the bony pelvis.[3, 5, 10] Combined lesions of these two plexuses (i.e., actual "lumbosacral" plexopa-

thies) are relatively uncommon and, hence, this subsection addresses them individually.

Pertinent Anatomy

Unlike the infrastructure of the brachial plexus, these two plexuses lack trunks and cords. Instead, their APR give rise to superior and inferior branches which, in turn, yield anterior and posterior divisions whose fibers then intermingle and eventually terminate as named peripheral nerves.

The lumbar plexus is situated within the posterior aspect of the upper and middle portions of the psoas major muscle and is contained by the psoas fascia, as are the nerves that arise from it: the obturator, femoral, and lateral femoral cutaneous. It originates from the L1, L2, L3, and a portion of the L4 APR. Except for the L3 APR, each of these divides into superior and inferior branches. The superior branch of the L1 APR terminates by dividing into the iliohypogastric and ilioinguinal nerves (which may also receive a contribution from the T12 APR), while its inferior branch joins the superior branch of the L2 APR to form the genitofemoral nerve. The inferior branch of the L2 APR, the unbranched L3 APR, and the superior branch of the L4 APR divide into anterior and posterior divisions. The anterior divisions fuse to form the obturator nerve while the posterior divisions join to form the femoral nerve. The lateral femoral cutaneous nerve arises from branches derived from the posterior divisions of the L2 and L3 APR. The inferior branch of the L4 APR, along with the L5 APR, is termed the *lumbosacral trunk*, a sacral plexus element (discussed later). The cutaneous domain of the lumbar plexus includes the pubic symphysis, portions of the external genitalia, and the anterolateral, anterior, and medial thigh regions. It innervates anterior and medial thigh muscles (quadriceps and thigh adductors, respectively) as well as the iliacus and psoas (Fig. 17–3).[3, 5, 10]

The sacral plexus is located on the anterior surface of the piriformis muscle, within the posterior aspect of the true pelvis. In addition to the inferior branch of the L4 APR, it includes axons from the L5, S1, S2, and S3 APR; each of the latter divides into anterior and posterior divisions. Branches originating from the divisions ultimately form several peripheral nerve trunks, including (1) the superior gluteal nerve, derived from the posterior divisions of L4 through S1; (2) the inferior gluteal nerve, derived from the posterior divisions of L5 through S2; (3) the posterior femoral cutaneous nerve, derived from the posterior divisions of S1 and S2, as well as the anterior divisions of S2 and S3; and (4) the sciatic nerve, whose tibial nerve component results from fusion of fibers from the anterior divisions of L4 though S1, and whose common peroneal component is formed by axons from the posterior divisions of L4 through S2. The sciatic nerve proper is formed when the tibial and common peroneal nerves become enclosed within a common connective tissue sheath; throughout its course in the thigh, its tibial and common peroneal nerve components remain internally distinct entities. The cutaneous domain of the sacral plexus includes the remainder of the external genitalia, the gluteal region, the posterior thigh, and all of the leg and foot except the medial leg strip supplied by the saphenous vein (derived from the lumbar plexus). The sacral plexus innervates some of the pelvic muscles, the glutei and tensor fascia lata, the hamstrings, and all of the leg and foot muscles.[3, 10]

Electrodiagnostic Assessment

Disorders affecting the lumbar and sacral plexuses are almost always axon loss in type and often quite severe. Unfortunately, each component of the EDX examination has

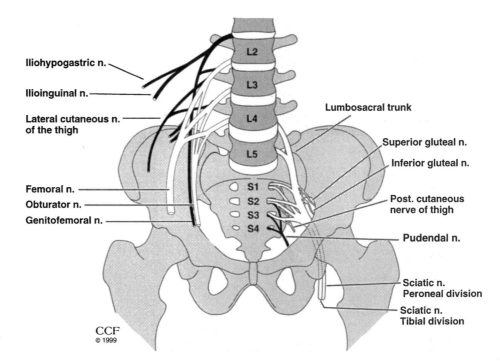

Iliohypogastric n.

Ilioinguinal n.

Lateral cutaneous n. of the thigh

Lumbosacral trunk

Superior gluteal n.

Inferior gluteal n.

Femoral n.

Obturator n.

Genitofemoral n.

Post. cutaneous nerve of thigh

Pudendal n.

Sciatic n. Peroneal division

Sciatic n. Tibial division

L2
L3
L4
L5
S1
S2
S3
S4

CCF
© 1999

FIGURE 17–3. The lumbar and the sacral plexuses.

unique pitfalls. As a result, unlike the situation with brachial plexopathies, the EDX examination is only of definite diagnostic value with some sacral plexopathies and a few lumbar plexopathies. The principle difficulties lie with the unavailability of consistently obtainable sensory NCS.

Sensory NCSs

As already discussed, sensory NCSs play a major role in the EDX recognition of plexopathies, because low-amplitude or unelicitable SNAPs, when found with a proximal axon loss lesion, usually signify that the disorder is ganglionic or postganglionic. Thus, they differentiate plexus lesions from root and anterior horn cell (AHC) lesions. Although there are many sensory NCSs for assessing the brachial plexus, there are only a few available for evaluating the lumbar plexus and sacral plexus. Unfortunately, even these have serious limitations; for example, whereas techniques have been reported for eliciting saphenous and lateral femoral cutaneous SNAPs to assess the lumbar plexus, these responses tend to be low in amplitude or sometimes unelicitable, even among young, healthy individuals who have a thin habitus; consequently, they seldom can be recorded in even the contralateral normal limb among the patients in whom they are sought.

Only two lower extremity sensory NCSs are deemed reliable for plexus assessment, at least in patients under 60 years of age: (1) the sural (routinely performed in most EDX laboratories); and (2) the superficial peroneal (frequently performed in many EDX laboratories). Neither of these, however, assesses the lumbar plexus. Moreover, both of them have suboptimal utility, because sacral plexopathies sometimes are bilateral (reflecting the high incidence of neoplasms spreading bilaterally from more central pelvic structures) and they tend to affect older individuals. Unfortunately, a substantial percentage of normal individuals over the age of 60 years have bilaterally unelicitable sural and superficial peroneal sensory SNAPs, and this is the age group in which sacral plexopathies are most commonly found. Consequently, in these patients it can be impossible to distinguish between age-related and pathological absence of the SNAPs. Additionally, the tendency of these lesions to be bilateral

means that side-to-side amplitude comparisons between homologous SNAPs cannot be made and, consequently, the ability to identify a response as *relatively* abnormal is lost (see Chapter 14). Finally, associated disorders, such as diabetic or chemotherapy-induced polyneuropathy, can render the lower extremity SNAPs unelicitable, as can—owing to the superficial location of the nerves—incidental trauma (e.g., vein stripping to treat varicose veins).

In spite of these limitations, whenever a lumbar plexopathy is suspected clinically, an attempt should be made to elicit saphenous SNAPs bilaterally. Usually they cannot be obtained on either side. Rarely, however, one can be elicited on the contralateral (normal) limb, but not the symptomatic limb, making a lumbar plexopathy a more likely possibility. Similarly, all patients with possible sacral plexopathies should have sural and superficial peroneal sensory NCSs performed bilaterally. In patients over 60 years of age, bilaterally unelicitable SNAPs must be discounted, but the same findings in patients under age 60 years, as well as unilaterally abnormal SNAPs at any age, are significant. An important point is that the sural and superficial peroneal sensory SNAPs are complementary, rather than redundant, in regard to sacral plexus assessment. This is because the superficial peroneal sensory fibers derive principally from the L5 DRG, whereas the sural fibers originate mainly from the S1 DRG. While total sacral plexopathies affect both SNAPs, partial lesions commonly affect only one, or only one substantially; for example, the superficial peroneal sensory SNAP may be low in amplitude or unelicitable, while the sural SNAP is normal.

Motor NCSs

A femoral NCS is available for assessing portions of the motor lumbar plexus. However, a low-amplitude or unelicitable CMAP does not distinguish lesions of the L2–L4 AHC, L2–L4 root, lumbar plexus, or femoral nerve from one another. In addition, it cannot always be elicited in obese individuals. Consequently, the EDX identification, and then only tentative, of lumbar plexopathies relies heavily upon the NEE (discussed later). Nonetheless, femoral motor NCSs should be performed on all patients with possible lumbar plexus lesions. The responses obtained can be compared to laboratory normal values and, in the presence of unilateral symptoms, with the results recorded on the contralateral normal limb.

Regarding sacral plexus assessment, the peroneal NCS, recording extensor digitorum brevis, and the posterior tibial NCS, recording abductor hallucis, are the basic lower extremity motor NCSs. With possible sacral plexopathies, the amplitudes of two other motor NCSs are very helpful: (1) peroneal motor, recording tibialis anterior (TA), and (2) the direct M response obtained during the H response test. With axon loss lesions, their CMAP amplitudes demonstrate the amount of axon loss sustained by the anterior, lateral, and posterior leg muscles; in contrast, the CMAPs of the basic lower extremity NCS only reveal the amount of axon loss in various intrinsic foot muscles. When these motor NCSs are performed bilaterally, it is possible to estimate the amount of motor axon loss affecting the L5/common peroneal fibers (recording TA) and the S1/tibial fiber (recording gastrocnemius/soleus).

Needle Electrode Examination

The NEE with lumbar and sacral plexopathies reveals abnormalities in either myotomal distributions (Table 17–11) or multiple nerve distributions, depending on whether proximal or more distal portions of the plexuses are affected.

The NEE is vital for separating lesions involving the lumbar plexus from those affecting the L2–L4 roots more proximally and the femoral nerve more distally. The NEE abnormalities (fibrillation potentials, MUAP loss, chronic neurogenic MUAPs) will be limited to the quadriceps muscles with a femoral neuropathy, whereas they should also be found in the thigh adductor muscles and possibly the iliacus with radiculopathies and lumbar plexopathies. Unfortunately, the NEE is less helpful in differentiating between the latter two possibilities. If fibrillation potentials are present in the mid and low paraspinal muscles ipsilateral to the symptomatic limb, a root lesion is the likely cause. The converse reasoning, however—that is, that the absence of fibrillation potentials in these paraspinal muscles indicates that a lumbar plexopathy is present—is fallacious, because paraspinal fibrillation potentials cannot be found in many patients with L2–L4 radiculopathies. Hence, in this situation it is not possible to determine with certainty whether the lesion localizes to the lumbar plexus or to the lumbar root(s). The only instances, seldom encountered, in which a lumbar plexopathy can be unequivocally diagnosed is when there is a unilateral abnormality of the saphenous SNAP; this usually occurs only with young, lean patients. If myokymic potentials are detected in the anteromedial thigh musculature, a history of previous pelvic irradiation should be sought; this uncommon type of spontaneous activity is frequently observed with radiation-induced plexopathies, which usually involve the lumbar, rather than the sacral plexus.

Lesions of the sacral plexus are easily confused with both radiculopathies (L5, S1) and proximal sciatic mononeuropathies. The NEE is critical for distinguishing L5, S1 radiculopathies from sacral plexopathies whenever the lower extremity sensory NCS responses are bilaterally unobtainable (usually in elderly patients); fibrillation potentials in the appropriate paraspinal muscles place the lesion at the root level. Unfortunately, identical to the situation encountered with L2–L4/lumbar plexus lesions, the lack of paraspinal muscle abnormalities on NEE does not, by default, diagnose a plexopathy. Instead, the EDX results remain indeterminate. A similar problem is confronted when attempting to distinguish sacral plexopathies from high sciatic mononeuropathies. Because the NCS features of these two disorders may be identical, differentiation rests solely on the NEE of limb muscles innervated by the sacral plexus but not by the sciatic nerve: the glutei and tensor fascia lata muscles innervated by the gluteal nerves. If changes are found in these muscles as well as the sciatic nerve–innervated muscles, a sacral plexopathy is likely. Unfortunately, the proximal location of these muscles makes them less likely to show EDX evidence of denervation and more likely to undergo early reinnervation. In either case, they may appear normal with sacral plexus lesions, resulting in an erroneous diagnosis of a proximal sciatic mononeuropathy. In addition, lesions that simultaneously affect both the gluteal and sciatic nerves (e.g., gluteal compartment syndrome) mimic sacral plexus lesions in their clinical and EDX presentations; these can only be distinguished by history or neuroimaging studies, not by EDX examination.

Common Entities Affecting the Lumbar and Sacral Plexus. Trauma, which plays such a prominent role with brachial plexopathies, is responsible for only a small portion of lumbar and sacral plexopathies. Often, unstable pelvic fractures and injury to other pelvic structures resulting from these, accompany such plexopathies. In contrast, neoplasms produce the majority of lumbosacral lesions. Most of these are the result of direct extension from nearby structures (e.g., rectum, cervix), but some are caused by metastases, especially from breast carcinoma. Hemorrhage and hematoma formation in patients who are anticoagulated or who have clotting disorders affect the lumbar plexus more often than the sacral plexus, whereas maternal paralysis involves the lumbosacral trunk component of the sacral plexus[5, 10, 13, 15] (Tables 17–12 and 17–13).

Concluding Comments: Lumbosacral Plexus Assessment

1. When evaluating for suspected lumbar plexopathies, femoral motor NCS should always be performed bilaterally. If the femoral CMAP amplitude is significantly

TABLE 17–11	Lumbosacral Myotomes		
L2, L3	**L4**	**L5**	**S1**
Iliacus	Rectus femoris	Extensor hallucis	Abductor hallucis
Rectus femoris	Vastus lateralis	Tibialis anterior	Abductor digiti quinti pedis
Vastus lateralis	Vastus medialis	Peroneus longus	Gastrocnemius, medial head
Vastus medialis	Adductor longus	Extensor digitorum brevis	Gastrocnemius, lateral head
Adductor longus	Adductor magnus[a]	Tibialis posterior	Flexor digitorum longus
Adductor magnus[a]	Tibialis anterior	Flexor digitorum longus	Biceps femoris, short head
		Semitendinosus	Biceps femoris, long head
		Tensor fascia lata[a]	Gluteus maximus[a]
		Gluteus medius[a]	

[a] These muscles are innervated by the sacral plexus, but not by the sciatic nerve. To distinguish the two, these muscles must be assessed on needle electrode examination.

TABLE 17–12	Nerve Conduction Study Findings Seen with a Right Sacral Plexopathy Caused by Rectal Carcinoma

Nerve Stimulated	Stimulation Site	Recording Site	Amplitude μV/mV		Latency (msec) Peak/Distal		Conduction Velocity (m/s)	
			R	L	R	L	R	L
Sural (S)	Distal posterior leg	Lateral ankle	NRᵃ	12	—	4.0		
Superior peroneal sensory	Distal anterior leg	Anterior ankle	NRᵃ	16	—	3.8		
Tibial (M)	Ankle	Abductor hallucis	4.0ᵃ	10.7	4.9	4.5	41	44
	Popliteal fossa		2.1	7.0	16	15.7		
Peroneal (M)	Ankle	Extensor digitorum brevis	NRᵃ	5.0	—	4.0	—	43
	Popliteal fossa			4.4		16.0		
Peroneal (M)	Below FH	Tibialis anterior	1.1ᵃ	6.8	3.3	3.0	50	52
	Popliteal fossa		0.7ᵃ	6.6	5.5	5.2		
Femoral (M)	Inguinal ligament	Rectus femoris	6	7	2.1	2.2		
Tibial:	Popliteal fossa	Gastrocnemius/ soleus						
(1) H response			NRᵃ	3.9	5.0	4.8	NR	31.3
(2) M response			3.7ᵃ	11.9				

FH, femoral head; NR, no response.
Note that the sensory NCSs are more affected than the motor, and the peroneal motor NCS more than the tibial motor NCS.
ᵃ Abnormal response.

reduced, the axon loss lesion responsible is much more likely to be affecting the femoral nerve or the lumbar plexus, as individual radiculopathies typically do not produce enough motor axon loss to significantly alter the femoral CMAP amplitude. In addition, saphenous NCS should be attempted, even though the effort is very likely to be futile. On NEE, paraspinal muscles as well as limb muscles innervated by both the obturator and femoral nerves should be sampled. If fibrillation potentials are detected in both nerve distributions but not in the mid-lumbar paraspinal muscles, then a tentative, but only a tentative, diagnosis of a lumbar plexopathy is possible.

2. When evaluating for suspected sacral plexopathy, some additional NCSs should be performed: peroneal motor, recording TA; superficial peroneal sensory; and the H wave test to elicit an M wave. The NEE should include muscles innervated by the common peroneal and the tibial nerves, the superior and inferior gluteal nerves, as well as the paraspinal muscles. Sampling the latter is

TABLE 17–13	Needle Electrode Examination Findings that Accompany the Nerve Conduction Studies Shown in Table 17–12

Muscle	Spontaneous Activity	Motor Unit Action Potentials			
		Recruitment	Duration	Amplitude	Polyphasic
Right	**Fibs**				
Tibialis anterior	3 +	None	—	—	—
Extensor hallucis	3 +	None	—	—	—
Peroneal longus	3 +	↓ ↓ ↓ # Rap	Nor	Nor	None
Extensor digitorum brevis	3 +	None	—	—	—
Tibialis posterior	2 +	↓ ↓ ↓ # Rap	Nor	Nor	None
Flexor digitorum longus	2 +	↓ ↓ ↓ # Rap	Nor	Nor	None
Medial gastrocnemius	1 +	↓ ↓ # Rap	Nor	Nor	None
Abductor hallucis	1 +	↓ ↓ # Rap	Some ↑	Nor	Some 1 +
Vastus lateralis	None	Nor	Nor	Nor	None
Rectus femoris	None	Nor	Nor	Nor	None
Biceps femoris (SH)	3 +	None	—	—	—
Biceps femoris (LH)	1 +	↓ # Rap	Nor	Nor	Some 1 +
Semitendinosus	2 +	↓ # Rap	Nor	Nor	Many 2 +
Gluteus medius	2 +	↓ ↓ # Rap	Nor	Nor	Many 2 +
Gluteus maximus	1 +	↓ # Rap	Nor	Nor	Many 1 +
Tensor fascia lata	2 +	↓ ↓ ↓ # Rap	Nor	Nor	Most 2 +
L5 paraspinals	None	—			
S1 paraspinal	None	—			
Left					
Tibialis anterior	None	Nor	Nor	Nor	None
Medial gastrocnemius	None	Nor	Nor	Nor	None
Tibialis posterior	None	Nor	Nor	Nor	None
Vastus lateralis	None	Nor	Nor	Nor	None

Fibs, fibrillation potentials; Nor, normal; Rap, rapid.
Note that the abnormalities are most severe in the common peroneal and, to a lesser extent, the L5 distribution. The fact that the glutei and tensor fascia lata also show severe changes localizes the lesion to the sacral plexus, as opposed to the proximal sciatic nerve.

particularly important if, for any reason, the sural and superficial peroneal sensory SNAPs are absent bilaterally.

3. Sometimes the segmental nature of the sacral plexus is reflected by its associated NCS abnormalities. When the L5 components are involved, the superficial peroneal SNAP and the peroneal CMAPs usually are most affected, whereas when the S1 components are involved, the sural SNAP and tibial CMAP tend to be the most affected.

REFERENCES

1. Aminoff MJ: Electromyography in Clinical Practice. New York, Churchill-Livingstone, 1998.
2. Bonney G, Gilliatt RW: Sensory nerve conduction after traction lesion of the brachial plexus. Proc R Soc Med 51:365–367, 1958.
3. Clemente CD: Gray's Anatomy, 30th American ed. Philadelphia, Lea and Febiger, 1985.
4. Daube JR: An electromyographer's review of plexopathy. In Syllabus: Neuromuscular Diseases As Seen by the Electromyographer. Rochester, MN, American Association of Electromyography and Electrodiagnosis, 1979, pp 19–31.
5. Donaghy M: Lumbosacral plexus lesions. In Dyck PJ, Thomas PK (eds): Peripheral Neuropathy, 3rd ed, vol 2. Philadelphia, WB Saunders, 1993, pp 951–959.
6. Ferrante MA, Wilbourn AJ: The utility of various sensory nerve conduction responses in assessing brachial plexopathies. Muscle Nerve 18:879–889, 1995.
7. Gilliatt RW, Sears TA: Sensory nerve action potentials in patients with peripheral nerve lesions. J Neurol Neurosurg Psychiatry 11:109–118, 1958.
8. Levin KH, Wilbourn AJ, Maggiano HJ: Cervical rib and median sternotomy-related brachial plexopathies: A reassessment. Neurology 50:1407–1413, 1998.
9. Narakas AO: Traumatic brachial plexus lesions. In Dyck PJ, Thomas PK, Lambert EH et al (eds): Peripheral Neuropathy, 2nd ed. Philadelphia, WB Saunders, 1984, pp 1394–1409.
10. Stevens JC: Lumbosacral plexus lesions. In Dyck PJ, Thomas PK, Lambert EH et al (eds): Peripheral Neuropathy, 2nd ed, vol 2. Philadelphia, WB Saunders, 1984, pp 1425–1434.
11. Wilbourn AJ: Electrodiagnosis of plexopathies. Neurol Clin 3:511–529, 1985.
12. Wilbourn AJ: Brachial plexus disorders. In Dyck PJ, Thomas PK (eds): Peripheral Neuropathy, 3rd ed. Philadelphia, WB Saunders, 1993, pp 911–950.
13. Wilbourn AJ: Plexus injuries. In Evans RW (ed): Neurology and Trauma. Philadelphia, WB Saunders, 1996, pp 375–400.
14. Wilbourn AJ: Assessment of the brachial plexus and the phrenic nerve. In Johnson EW, Pease WS (eds): Practical Electromyography, 3rd ed. Baltimore, William and Wilkins, 1997, pp 273–310.
15. Wilbourn AJ, Ferrante MA: Plexopathies. In Pourmand R (ed): Neuromuscular Disease; Expert Clinicians' Views. (In press)
16. Wilbourn AJ, Ferrante MA: Clinical electromyography. In Joynt RJ, Griggs RC (eds): Clinical Neurology, vol 1. Philadelphia, Lippincott-Raven, 1997, pp 1–76.

Chapter 18

Multiple Mononeuropathies and Polyneuropathies

Asa J. Wilbourn, MD

MULTIPLE MONONEUROPATHIES
Surgical and Nonsurgical Acute Trauma
Tourniquet Paralysis
Mononeuropathy Multiplex
Ischemic Monomelic Neuropathy
Neuralgic Amyotrophy
Multifocal Motor Neuropathy
POLYNEUROPATHIES
Historical Aspects

Classification of Polyneuropathies
Utility of Various EDX Components
An Approach to Diagnosis
Axon Loss Polyneuropathies
Demyelinating Polyneuropathies
Confounding Factors
Diabetic Polyneuropathy
Conclusions

In this chapter, disorders that compromise more than one peripheral nerve, often simultaneously, are reviewed. How frequently the entities included in this broad category are assessed in individual electrodiagnostic (EDX) laboratories varies. As a group, however, polyneuropathies are probably the most common generalized neuromuscular disorders referred for EDX evaluation.

MULTIPLE MONONEUROPATHIES

Patients with two or more mononeuropathies are encountered with some frequency in the typical EDX laboratory. Probably the most common situation is the presence of bilateral median neuropathy distal to the wrist, or carpal tunnel syndrome (CTS); whenever median nerve conduction studies (NCSs) demonstrate CTS in one limb of a patient, a similar lesion can be found in the contralateral limb in the majority of instances.[26] A combination of mononeuropathies less frequently seen is CTS coexisting with an ipsilateral ulnar neuropathy along the elbow segment (UN-ES). Bilateral CTS and bilateral UN-ES are particularly likely to be encountered in diabetic patients.[30]

In this portion of this chapter, a variety of disorders that often cause multiple mononeuropathies are discussed.

Surgical and Nonsurgical Acute Trauma

Two or more peripheral nerves can be injured in a limb when surgical procedures are performed upon it, as well as when it sustains violent trauma; long-bone fractures may accompany the latter. These peripheral nervous system (PNS) lesions typically occur in the arm and elbow segments of the upper extremity, and in the leg and knee segments of the lower extremity. In most cases, axon loss is the predominant, if not the sole, pathophysiology present, and usually the injury sustained by one (if not more) of the peripheral nerves is severe in degree electrically. Focal demyelinating conduction blocks occasionally are present, always limited to

a small proportion of the patients assessed soon after their injuries occur, since "neurapractic" lesions, as these are labeled, generally resolve after three to six weeks. Demyelinating conduction slowing is rarely seen following acute trauma, and it is almost never the only, or even the predominant, type of process present. Based on the likely underlying pathophysiology (i.e., axon loss/conduction failure or, much less likely, demyelinating conduction block), on NCS the amplitudes of the responses are the critical components. Consequently, if the EDX assessment of these patients is limited by time, then it is far more important to obtain nerve conduction amplitudes from more nerves, and branches of nerves, in the affected limb (and often in the uninvolved limb, as well, for comparison purposes) than it is to stimulate motor nerves at two points along their course, seeking slowed conduction velocities (CVs) through the sites of injury.

The main tasks of electromyographers in these situations are to determine the pathophysiology, extent, and severity of the lesions and to pinpoint their location as accurately as possible. In many patients who sustain these injuries, the referring physicians are not aware of all the nerves that are damaged. Thus, a mild radial neuropathy in the arm, coexisting with severe ulnar and median neuropathies, may escape clinical detection. For this reason, the EDX examination performed on the affected limb should be extensive; for example, even if the clinical examination suggests that only the median and ulnar nerves are involved, at least two or three muscles innervated by the distal radial nerve (i.e., the posterior interosseous nerve) should be assessed, and a radial sensory NCS should be performed. Also, because only some of the axons of a particular peripheral nerve may sustain damage, thorough assessments of each affected, and possibly affected, nerve will ensure that the EDX examination accurately reflects the extent and severity of the lesion. Thus, with high median nerve injuries, digital sensory nerve action potentials (SNAPs) from the index finger, or digit two (D2), may suggest that the axon loss is complete for sensory fibers whereas, if additional median SNAPs are recorded from D1 and D3, they may reveal low amplitude responses, thus demonstrating that the nerve lesion, for the sensory fibers at least, is incomplete. Similarly, if the needle electrode examination (NEE) is restricted to only one or two muscles innervated by each injured or possibly injured nerve, the findings may be quite misleading.

Satisfactorily localizing these combined nerve injuries may prove difficult for several reasons. First, because these lesions characteristically are axon loss in type, accurate localization depends primarily on the NEE findings. Thus, motor nerve fiber involvement is required. Moreover, localization is critically dependent upon the site of origin of the motor branches from the affected nerve, in relationship to the point of injury along it. Unfortunately, the anatomy in these situations often is not propitious for precise EDX localization. Thus, because no motor branches arise from the arm segments compared to the forearm segments of either the median or ulnar nerves, it is impossible to determine whether axon loss lesions along those nerve segments are situated near the elbow or are several centimeters proximal, near the axilla. In contrast, the radial nerve has branches arising in the midarm, immediately proximal to the spiral groove, that supply the triceps and anconeus muscles. Hence, if the radial nerve is affected along with the median and ulnar nerves, then more precise localization can be achieved. For similar reasons, EDX studies alone cannot distinguish severe axon loss peroneal and tibial mononeuropathies, caused by high fractures of the tibia and fibula, from a distal sciatic mononeuropathy, i.e., one located in the more distal leg, distal to the origin of the motor branches supplying the hamstring muscles.

As time passes, the EDX abnormalities present in the distribution of less involved nerves may resolve completely. In addition, the muscles closest to the injury site of the affected nerves will progressively reinnervate, via collateral sprouting. Hence, if EDX examinations are performed many months after onset, the extent of the nerve damage easily is underestimated, whereas the actual lesion site may erroneously be located more distally. This latter phenomenon is referred to as "apparent distal displacement of the lesion with time." Although most often it is a problem with radiculopathies, it can be seen with single and multiple mononeuropathies as well. "Apparent proximal displacement of the lesion" also occurs; most often this is due to simultaneous damage of a nerve trunk and some motor branches arising from it at the lesion site. Thus, blunt trauma to the spiral groove area in the midarm may injure not only the radial nerve trunk but also, independently, the radial motor supply to the triceps and anconeus. As a result, the responsible lesion is readily mislocalized, on the NEE, to some point proximal to the spiral groove.

Another problem encountered in localization—again, over which the electromyographer has no control—is when two lesions occur at different points along the same nerve fibers. In these instances, it is important to remember that the sequence of the focal nerve injuries along the nerve determines whether the less severe lesion can be detected. If it is proximal to the more severe one, then both can be identified on EDX examination. Conversely, if it is distal, then the electrical abnormalities that are produced by the more severe one obscure its presence.

Tourniquet Paralysis

This unilateral PNS disorder occurs when surgical procedures are performed on the more distal portions of the lower or upper extremities. With lower extremity lesions, although both the femoral and sciatic nerves may be damaged by tourniquets about the thigh, far more often only the latter are affected. Also, almost invariably, axon loss is the sole pathophysiological process detected. In contrast, when

upper extremity lesions result from tourniquets about the arm, in the majority of instances all three of the major long nerves of the limb—radial, median, and ulnar—are compromised at the midarm level. Of those patients who do not experience "tri-nerve" involvement, usually just the radial nerve is injured. In either case, the characteristic pathophysiology usually is demyelinating conduction block.

Regarding the EDX examination with upper extremity tourniquet paralysis, four features merit comment.

1. For confirmation of these demyelinating conduction blocks, the affected nerves must be stimulated proximal and distal to them (i.e., in the axilla and at multiple points in the arm), while recording radial, median, and ulnar compound muscle action potentials (CMAPs) from extensor forearm, thenar, and hypothenar muscles, respectively.
2. The type of pathophysiology present can be suspected, however, when just the routine upper extremity motor NCSs are done, based on the discrepancy between the relatively preserved CMAPs recorded from the weak muscles with distal stimulation, compared with the degree of clinical weakness and the drop out of motor unit action potentials (MUAPs) noted on NEE of the same muscles.
3. As with most focal, very predominantly demyelinating, conduction block lesions, at least a minimal amount of motor axon loss occurs, manifested by fibrillation potentials in the weak muscles. These may be interpreted, incorrectly, as indicating that the nerve fiber damage is primarily axon loss in type. However, the fact that normal/near normal CMAP amplitudes are recorded from the very weak muscles on distal stimulation excludes axon loss as the cause.
4. Unlike the focal demyelinating nerve lesions produced by acute trauma ("neurapraxia," already discussed), those lesions responsible for upper extremity tourniquet paralysis can persist for many months (up to one year). Thus, based on clinical findings alone, they are readily mistaken for those caused by axon loss, because of the persistent clinical weakness and the muscle atrophy of disuse that results.[29]

Mononeuropathy Multiplex

This syndrome results from focal damage of two or more peripheral nerves. The etiology of mononeuropathy multiplex (MNM) varies. Vasculitis is a common cause; nonetheless, MNM can occur with many other disorders, including diabetes mellitus, human immunodeficiency virus (HIV) disease, Lyme disease, leprosy, infiltrating neoplasms, and cryoglobulinemia. Most often, these multifocal PNS lesions develop abruptly, sequentially, and in different limbs. They are almost invariably axon loss in type, but their severity is quite variable, ranging from extremely severe (total) to mild in degree. Large nerve trunks, rather than small branches, more often are affected. Weakness and especially pain are the major symptoms. The peroneal, ulnar, and median nerves are most frequently involved, and usually the lesions along them are not at the typical compression/entrapment sites. Thus, with ulnar neuropathies, they commonly are in the distal axilla or proximal arm.

Whenever it is apparent that two or more individual peripheral nerve trunks have sustained nontraumatic focal axon loss damage in a step-wise manner, the diagnosis of "nonconfluent MNM" generally is obvious, especially if the patient has a medical disorder known to be associated with MNM. However, in some instances multiple peripheral nerves are involved in a limb, including relatively small cutaneous (sensory) nerves. As a result, the extensive overlapping of motor and sensory nerve abnormalities conceals the true nature of the process (i.e., coexisting involvement of many individual peripheral nerves) from both the clinician and the electromyographer. Whenever "confluent MNM" affects most of the nerves in the distal portions of both lower extremities, in a near-symmetrical manner, the misdiagnosis of an axon loss "polyneuropathy" is readily made.

The EDX examinations with suspected MNM, either of the nonconfluent or confluent type, generally must be quite extensive. Nonconfluent MNM should be suspected whenever nontraumatic focal axon loss lesions, usually of abrupt onset and often of varying ages and severity, involve two or more peripheral nerves at other than the usual entrapment sites. Confluent MNM is much more difficult to identify with any certainty by EDX examination, but it should be suspected whenever evidence of a somewhat asymmetrical, axon loss "polyneuropathy" is encountered, particularly if (1) the patient has a known medical condition that predisposes to MNM; (2) on extensive EDX examination, either the level or severity of axon loss along the individual nerves in corresponding limbs is dissimilar; or (3) on EDX examination, abnormalities are not detected in an upper extremity and yet bilateral lower extremity denervation is severe and extends proximally to or beyond the knee levels.[3, 24, 27, 28]

Ischemic Monomelic Neuropathy

This is a special type of MNM in which all the nerves in the more distal portion of a single limb are injured simultaneously and equally, because of a transient compromise of the major arterial blood flow to the limb. Unlike the situation with typical MNM, the neurological abnormalities in the affected limb are not suggestive of focal lesions involving a discrete segment along each of the injured nerves. Rather, regardless of the particular nerve involved, the changes invariably are most severe distally and shade proximally. Characteristically, no abnormalities are seen proximal to the midarm or midleg level.

Ischemic monomelic neuropathy (IMN) occurs mostly in older adults with vascular disorders of various types and, excluding the rare patient with bilateral lower extremity lesions due to a saddle embolus, is always unilateral. Although it may affect either the upper or lower extremity, involvement of the latter is more frequent. Almost all upper extremity IMNs are iatrogenic in nature, caused by the placement of arteriovenous shunts in the arm to treat renal failure resulting from diabetes mellitus. Lower extremity lesions are much more variable in their causation. Many are iatrogenic, resulting from cannulation of the superficial femoral artery for cardiopulmonary bypass or for intra-aortic balloon pump use. Others, however, are not iatrogenic; they are produced by aortoiliac emboli, acute superficial femoral artery thrombosis, and iliofemoral thrombosis.

Regardless of cause, the main symptom of IMN is burning pain, most often experienced in the hand and fingers or the foot and toes. Sensory and motor deficits usually are prominent only in the hand and foot, although changes may be found, in a graded fashion, up to the midarm and midleg levels. At any given level of limb involvement, the sensory abnormalities about the circumference of the limb usually are essentially equal, regardless of the particular nerve supplying the area.

On EDX examination, IMN presents as a rather severe axon loss polyneuropathy affecting a single limb. Thus, using upper extremity IMN as an example, on NCS the digital SNAPs and the CMAPs recorded from intrinsic hand muscles generally are unelicitable (sensory) or low in amplitude (motor). Reflecting the prominent distal-to-proximal gradient of abnormalities, however, the radial SNAP, recording base of

thumb, usually is less severely affected, whereas the medial antebrachial cutaneous and lateral antebrachial cutaneous SNAPs typically are normal. Similarly, the amplitudes of the CMAPs recorded from forearm muscles—i.e., radial motor, recording extensor indicis proprius; ulnar motor, recording flexor carpi ulnaris—are much less affected than the median and ulnar CMAPs recorded from intrinsic hand muscles. However, it is during the NEE that the characteristic distal-to-proximal shading of axon loss is most obvious. Without exception, the most severely denervated muscles, i.e., those having the most abundant fibrillation potentials and showing the most profound drop out of MUAPs, are the intrinsic hand muscles, which are affected equally, notwithstanding their innervation (i.e., median or ulnar). In sharp contrast, muscles in the proximal forearm usually are normal. Intermediate between these two extremes are muscles sampled in the distal forearm, such as the flexor pollicis longus, extensor indicis proprius, and flexor carpi ulnaris; regardless of their peripheral nerve innervation, muscles at the same forearm level show essentially the same degree of NEE changes.

A number of confounding factors can compromise the diagnosis of IMN, to the point that in some cases a definite diagnosis cannot be provided. These include the following.

1. *Advanced age.* The majority of patients with IMN are elderly and often have unelicitable lower extremity SNAPs bilaterally. This renders a diagnosis of lower extremity IMN problematic, at best.
2. *Generalized polyneuropathy.* An inordinate number of patients with suspected IMN have generalized polyneuropathies, usually diabetic in origin. When these are mild, the superimposed distal axon loss resulting from IMN is easily detected, if extensive NCS and NEE are performed on homologous limbs (with the contralateral limb assessed for comparison purposes). However, if they are axon loss in type and severe in degree electrically, then it becomes impossible to identify a superimposed IMN. As would be expected, polyneuropathies more often cause difficulties in the recognition of lower extremity IMN, as opposed to upper extremity IMN. Nonetheless, when of sufficient severity, they can obliterate evidence of even upper extremity IMN.
3. *Coexisting mononeuropathies or the residuals of such lesions.* These are a greater problem with upper than with lower extremity IMN and include both CTS and UNES, disorders that are encountered with high frequency in patients with upper extremity IMN, nearly all of whom are diabetic.
4. *Lesions of long duration.* This most often relates to lower extremity IMN being confused with a focal sciatic nerve infarction. With recent onset lesions, the distribution of the EDX abnormalities in the limb readily distinguishes between these two possibilities (shading of changes with IMN, abrupt demarcation of abnormalities with sciatic nerve infarcts); however, with chronic lesions (i.e, those of greater than six or more months' duration), a distal-to-proximal gradient of findings may be seen with sciatic nerve infarcts, due to progressive proximal-to-distal regeneration. Consequently, a point in time can be reached at which the EDX examination is no longer capable of differentiating between these two entities.[27, 28, 31]

Neuralgic Amyotrophy (Parsonage-Turner Syndrome)

This upper extremity PNS disorder most likely is a special type of MNM that is immune-mediated. Neuralgic amyotrophy (NA) usually is acquired and affects adults, men more than women. Most often it is unilateral, and repeat bouts are uncommon. NA characteristically has three components: (1) antecedent event; (2) pain; and (3) weakness, sometimes with wasting. Most antecedent events, or "triggers," precede the onset of the pain phase anywhere from a few hours to three to four weeks. Nearly all the generally acknowledged triggers are in the medical realm. These include any medical illness (e.g., bacterial pneumonia), any surgical procedure (e.g., herniorrhaphy); childbirth; and any even minimally invasive diagnostic or therapeutic procedure (e.g., arteriography). Whether excessive arm use alone can trigger NA is debated; we believe it does occasionally. The pain of NA usually is abrupt in onset, has a marked propensity to begin while the patient is sleeping, and reaches maximal intensity rapidly, at which time it is very severe in degree. Most often it is located over the lateral deltoid. It typically persists for 5 to 10 days, and then gradually resolves, often being replaced by a persistent dull ache. The weakness (and wasting) associated with NA can result from involvement of one, two, or several peripheral nerves derived from the brachial plexus, as well as some extraplexal nerves (e.g., spinal accessory; phrenic; laryngeal). NA has a marked predilection to affect "pure," or predominantly pure, motor nerves, e.g., long thoracic; suprascapular; anterior interosseous; motor branches to pronator teres; posterior interosseous; axillary; and spinal accessory. It commonly causes severe denervation of one muscle innervated by an involved nerve, while relatively or completely sparing another. The only sensory nerve it attacks with any frequency is the lateral antebrachial cutaneous (sometimes without involving the motor axons supplying the biceps and brachialis). The characteristic type of pathology responsible for NA is axon loss.

The EDX examination can be extremely helpful in the detection and recognition of NA. However, the assessments must be extensive (usually bilateral), and more than the routine upper extremity NCSs are mandatory, since the latter are abnormal in only about 15% of NA cases; this is because the main median and ulnar nerve trunks are involved so infrequently. The positive yield with NCS increases appreciably if lateral antebrachial cutaneous, axillary motor, and musculocutaneous motor NCSs are performed, especially when the deltoid or biceps muscles are weak. An extensive NEE can demonstrate the nerves and nerve branches involved, as well as the severity of the lesion for each affected muscle. Certain patterns seen on NEE are suggestive of NA, e.g., severe denervation of infraspinatus with supraspinatus spared; severe denervation of pronator teres, with all other C6, C7/median nerve–innervated muscles spared.

Because NA has a predilection for motor nerves, and many muscles supplying the shoulder girdle and proximal arm muscles are predominantly or solely motor nerves, it is easily mistaken for a brachial plexus lesion, specifically one affecting the upper trunk, particularly whenever the suprascapular, axillary, and musculocutaneous nerves are all simultaneously involved. However, on EDX examination, the median D1 SNAP will not be affected, even though the lateral antebrachial cutaneous SNAP is; such dissociated sensory NCS findings are almost never encountered with axon loss, upper trunk brachial plexopathies. Moreover, on NEE, whereas the supraspinatus, infraspinatus, deltoid, and biceps will show severe changes, the brachioradialis, another muscle innervated by the upper trunk, will be unaffected.[8, 27, 32]

Multifocal Motor Neuropathy

This is a rare, immune-mediated PNS disorder that was first described less than 20 years ago. Its hallmark is persistent, focal demyelinating conduction block lesions affecting two or more peripheral nerves. Characteristically, major nerve trunks are involved, as opposed to small, terminal nerve branches. The proper place of multifocal motor neuropathy (MMN) in the nosology of PNS disorders is unclear at this time. Some investigators consider it to be a variant of chronic inflammatory demyelinating polyneuropathy; if this is the case, then it is a most unusual variant.

Multifocal motor neuropathy tends to affect men more than women, and it most often occurs in patients in their third to fifth decade of age. The classic clinical presentation of MMN consists of slowly progressive, asymmetrical, painless weakness of various limb muscles, characteristically in peripheral nerve distributions. Muscle wasting, fasciculations, and cramps may also occur. Typically: (1) upper extremity nerves are affected more often than lower extremity ones; (2) distal muscles are involved more often than proximal ones (e.g., with upper extremity involvement, muscles innervated by the median, the ulnar, or both nerves are usually abnormal); and (3) in many patients, increased levels of IgM anti-GM1 antibodies are found.

It is important for electromyographers to recognize MMN, because clinically it is sometimes confused with amyotrophic lateral sclerosis (ALS). However, in contrast with the latter, it is a treatable disorder. On EDX examination, the characteristic finding with this disorder is seen on the motor NCS to be prominent demyelinating conduction blocks situated along nerves at other than their usual entrapment points, for example, in the upper extremities such focal abnormalities frequently are found along the forearm segments of the median and ulnar nerves (i.e., distal to the below-elbow stimulation point along the ulnar nerve). Lesions situated along more proximal segments of limb nerves sometimes can be identified by stimulating proximal to the usual proximal stimulation points. This is far more easily accomplished in the upper, compared with the lower, extremity: the median, ulnar, and radial nerves can all be stimulated in the axilla and supraclavicularly. Sometimes conduction slowing and differential slowing (causing dispersed responses) accompany the conduction blocks, but it is important to remember that in these cases it is the conduction blocks that are responsible for the patient's primary symptom of weakness.[5]

Similar to radiation-induced brachial plexopathy, the other disorder caused by indefinitely persistent demyelinating conduction blocks, motor axon loss gradually is superimposed. As this occurs, the CMAPs obtained on stimulating distal to the lesion progressively decrease in amplitude. In contrast to the abnormal motor NCS, the SNAPs characteristically are normal, even when the sensory nerve impulses traverse the nerve segment where the motor abnormalities are situated.

On NEE, the findings in the affected muscles with MMN can be very similar to those seen with ALS, including (1) marked drop out of MUAPs (always present); (2) fibrillation potentials, fasciculation potentials, and prominent chronic neurogenic MUAPs (usually present); and cramp potentials and myokymic discharges (sometimes present). Consequently, it is vital during the NEE to determine whether the abnormalities are present in a segment/root distribution or in peripheral nerve distributions. This generally means that the NEE must be extensive, e.g., whenever MMN involves hand muscles, not only must forearm muscles innervated by the median and ulnar nerves be assessed but also the C8 radial nerve–innervated muscles, e.g., extensor indicis proprius and extensor pollicis brevis. Inadequate NEEs are one of the major reasons why MMN is mistaken for ALS in the EDX laboratory. This disorder should be suspected whenever:

1. On clinical examination, the affected muscles have been substantially weak for months and yet they are not atrophic.
2. On EDX examination, whenever conduction blocks are

demonstrated along the motor (but not the sensory) component of nerve trunks, at nonentrapment sites.

3. On EDX examination with proximal lesions, whenever the amplitudes of the CMAPs recorded from weak muscles are relatively preserved, although NEE of the same muscles reveals substantial drop out of MUAPs, commensurate with the clinical weakness present. In these instances, if motor conduction blocks are not detected during the standard motor NCS, then more proximal stimulations should be done, if possible. Sometimes axon loss is so severe in the distribution of the most affected peripheral nerve that a motor conduction block along it cannot be sought, because the CMAP amplitudes elicited even on distal stimulation are very low. Whenever this occurs, proximal stimulations should be performed along less affected nerves, in search of demyelinating conduction blocks.[10, 16, 23, 27, 28]

POLYNEUROPATHIES

The nomenclature concerning this class of PNS disorders is confusing. The term "peripheral neuropathy" is preferred by many neurologists. However, the members of most other specialties, including neurosurgery and orthopedic surgery, consider this label to be a synonym for any type of peripheral nerve lesion, including isolated focal disorders, such as entrapment or traumatic mononeuropathies. Electromyographers unaware of this fact may perform completely inappropriate multilimb assessments on patients who were actually referred for evaluation of a single nerve. Another term sometimes used is "generalized neuropathy." The problem with this name is that often the process, although presumably generalized, is not severe enough in the upper extremities to produce either clinical or EDX manifestations. For example, in one third of a series of patients with diabetic polyneuropathy, the EDX abnormalities were restricted to the lower extremities. Finally, some of the disorders included in this category of peripheral nerve disorders are not "polyneuropathies" at all. Rather, they are polyradiculoneuropathies. Consequently, they often lack many of the features so characteristic of most polyneuropathies, particularly their nerve length dependency aspect, i.e., the most distal nerve fibers are affected earliest, and most severely.[28] In this chapter, the term polyneuropathy is used, in spite of its limitations.

Historical Aspects

The EDX examination has been of quite variable value in the elucidation of the various peripheral neuromuscular disorders. With some, such as entrapment neuropathies and ALS, it has played a significant role. With none, however, has it had more impact than with polyneuropathies. Due almost solely to its NCS component, the EDX examination was responsible for major advances occurring in the recognition and classification of polyneuropathies. It did this principally by demonstrating fundamental differences in their underlying pathophysiology.[1, 28] How relatively small the contribution of the NEE component was to this endeavor becomes obvious when the first book concerned with clinical electrodiagnosis, written in 1955 by Marinacci and dealing only with the NEE, is perused. In its 174 pages of text, with entire chapters devoted to such topics as "Peripheral Nerve Injuries," "Root Compression Syndromes," and "Anterior Poliomyelitis," less than one page concerns polyneuropathies, and the majority of that consists of two brief case reports. The actual discussion occupies a single paragraph: "The EMG characteristic of polyneuropathy is the finding of

denervated activity in a large degree in the periphery of the extremities, and in a lesser degree in the proximal portion of the extremities and the trunk. This is true in alcoholic neuritis, diabetic neuritis, lead neuritis, etc. Polyneuritis, when not fully developed, can present a difficult problem in diagnosis. Some of these cases are even diagnosed as multiple sclerosis."[17]

The fundamental reason for the NCS (specifically, the motor NCS) being so beneficial, compared with the NEE, is that demyelination can be recognized on NCS, whereas only axon loss can be detected during the NEE, with the single exception of substantial conduction block.

In 1956, Lambert reported that motor nerve CVs were substantially slowed with some chronic polyneuropathies, though normal with others.[15] At that time, the only known cause of substantial reduction of impulse propagation along peripheral nerves was regeneration following axon loss. Consequently, as Gilliatt later noted in 1969, "... it seemed astonishing to all of us that peripheral nerves which were still capable of conducting impulses ... should do so in such an abnormal fashion."[12] The reason for the marked slowing was discovered by Kaeser and Lambert.[14] In a 1962 publication that had profound implications, they showed that axon loss (induced with thallium) and demyelination (induced with diphtheria toxin) had markedly different effects on the motor NCS performed on experimental animals. Axon loss, causing conduction failure, affected almost solely the amplitudes of the CMAPs; the latencies and CVs, as long as they could be obtained (i.e., as long as CMAPs were elicitable), usually remained within the normal range, or showed only moderate slowing. In contrast, demyelination caused strikingly different changes, in various combinations, of all portions of the motor NCS: the amplitudes and duration of the responses, as well as the latencies and CVs. One of the most impressive features was the prominent conduction slowing present, as manifested by the two components of the motor NCS that measure the rate of conduction: latencies and CVs.[14]

That substantial slowing of motor conduction is one of the characteristic features of demyelination was soon confirmed by other investigators,[18, 21] and, in 1966, Gilliatt published a landmark article in which he extended the experimental work of Kaeser, Lambert, and others to the clinical population.[13] He suggested that the results of the motor NCS could be used to predict the underlying pathophysiology of polyneuropathies, which appeared to be either axon loss or demyelination. Moreover, he proposed the first EDX classification of polyneuropathies, based on their motor NCS findings (and, therefore, on their known or presumed pathophysiology).[13] Over the next several years, Gilliatt's supposition regarding the relationship of the motor NCS findings and the underlying pathophysiology of nerves in patients with polyneuropathies was confirmed by several different investigators.

One of the most convincing publications on the topic was by McLeod and coworkers in 1973.[19] They performed motor NCS and sural nerve biopsies on 81 patients with polyneuropathies. Using the biopsy material, they divided their patients' polyneuropathies into two groups: axon degeneration and segmental demyelination. When they correlated the relationship of the motor CVs to the type of pathology present, they found that of the 49 patients with axon degeneration, the motor CVs were normal in one, while slowed to varying degrees in the remainder: mild in 44, moderate in three, and gross in one. (In the last patient, even though only axon degeneration was present on teased fiber preparations, electron microscopy revealed definite evidence of demyelination, so the biopsy had been misclassified.) In the 32 patients with segmental demyelination, the motor CVs were slowed in all: mild in six, moderate in four, and gross in the remaining 22. McLeod and coworkers also added a number of other polyneuropathies to Gilliatt's original classification of polyneuropathies.[19]

Since the publication of Gilliatt's article, other NCS techniques, including sensory NCS and late responses (H waves and F waves), have been added to the EDX armamentarium, increasing the capability for detecting and characterizing polyneuropathies, as well as for distinguishing them from other disorders with which they may be clinically confused.[28]

As electromyographers have gained further experience in the assessment of demyelinating polyneuropathies, it has become obvious that the amplitudes and the durations of the CMAPs and SNAPs often are more important in their recognition than the CVs and distal latencies. This is because demyelination can result in conduction blocks and differential slowing, as well as synchronized slowing. An important point is that, whenever clinical weakness results from demyelination, the cause is conduction block, not conduction slowing.[5, 28] In general, conduction block is the dominant pathophysiology with the acquired acute-onset disorders, e.g., acute inflammatory demyelinating polyneuropathy (AIDP), whereas conduction slowing typically predominates with slowly progressive disorders, either familial or acquired, e.g., heredity motor and sensory neuropathy, type I (HMSN, type I), and chronic inflammatory demyelinating polyneuropathy (CIDP).

The EDX examination can provide solutions to several pertinent questions regarding definite or possible polyneuropathies, only some of which can be answered by the clinical evaluation alone.[25] First, are the clinical changes present actually due to a polyneuropathy, or is some other peripheral neuromuscular disorder responsible for them? For example, many elderly patients with bilateral sensory disturbance in their feet mistakenly are assumed to have a polyneuropathy, when the actual cause for their symptoms is lumbar canal stenosis involving the L5 and S1 roots bilaterally.[30] Second, if a polyneuropathy is present, what size nerve fibers are involved? The key point here is that virtually the entire EDX examination assesses only large myelinated fibers. Hence, in a patient with bilateral burning dysesthesia of the feet and a completely normal lower extremity EDX examination, one diagnostic possibility is "small fiber" polyneuropathy. (Nonetheless, in most patients with symptoms of polyneuropathy, large fiber involvement is demonstrable on EDX examination.) Third, if a polyneuropathy is present, are motor fibers, sensory fibers, or both involved? The EDX examination is much more sensitive for detecting abnormalities of the large myelinated motor and sensory fibers than is the clinical assessment. Thus, in many diabetic patients, both the symptoms and the neurological examination suggest that solely sensory fibers are affected; for this reason, these disorders often are referred to as "diabetic sensory polyneuritis." On EDX examination, however, almost invariably abnormalities are found not only during the sensory NCS, but also during the motor NCS and during the NEE of the intrinsic foot muscles.[30] Fourth, if a polyneuropathy is present, and if it is demyelinating in nature, is its etiology familial or acquired? The EDX examination cannot answer this question if only uniform diffuse generalized conduction slowing is present. In many cases, however, it can demonstrate that the demyelination is nonuniform along the individual nerves and it is producing other than synchronized conduction slowing (e.g., differential slowing or conduction block), two findings almost characteristic of an acquired disorder. Fifth, are the EDX abnormalities in homologous limbs symmetrical? This question becomes important with the uncommonly encountered asymmetrical polyneuropathies, almost all of which are acquired disorders. Sixth, if a large fiber polyneuropathy is present, regardless of type, what is its severity and what are

its characteristics over time? The first part of this question is germane whenever single EDX examinations are performed on patients with polyneuropathies, whereas the second part is significant when multiple assessments are performed over a period of several months. Serial EDX examinations can provide objective information regarding whether a particular polyneuropathy has become more severe, has improved, has resolved, or remains static; in many patients, this question cannot be answered satisfactorily by the results of serial clinical examinations alone.

Classification of Polyneuropathies

As with headaches, polyneuropathies have a multitude of causes. In a review article, Donofrio and Albers list over 100 of the known etiologies for polyneuropathies.[9] To pare this list of diagnostic possibilities down to manageable levels, various classifications for polyneuropathies are used, based on clinical, pathophysiological, and pathological criteria. One of the principal functions of the electromyographer, after determining that a polyneuropathy is present, is to help characterize it in such a manner that an etiological diagnosis can be established, if possible, so that the most appropriate treatment can be initiated.[1, 6, 10, 25]

In the EDX laboratory, the linchpin for classifying polyneuropathies is deducing their underlying pathophysiology—axon loss or demyelination—based on their EDX characteristics.

How these two types of pathophysiological processes affect the various components of the EDX examination is discussed in detail below. A summary of these changes is as follows. Axon loss lesions principally alter the NCS amplitudes, producing low amplitude or unelicitable responses, with little to no associated conduction slowing of the latencies and CVs. On NEE, they cause fibrillation potentials, a drop out of MUAPs, and chronic neurogenic MUAPs. In contrast, demyelinating lesions may affect any component of the NCS, including the amplitudes, durations, latencies, and CVs, in various combinations, depending upon whether the demyelination is producing conduction block, differential slowing, or synchronized slowing. Unless axon loss coexists, demyelination typically has little effect on the NEE. Essentially the only abnormalities seen on NEE with "pure" demyelinating polyneuropathies are drop out of MUAPs and some disintegration of MUAPs, and these findings are limited to those polyneuropathies in which demyelinating conduction blocks (as opposed to conduction slowing) are manifest.[28]

It is noteworthy that not all types of polyneuropathies studied in the EDX laboratory can be divided into axon loss or demyelinating types. Some lack distinctive features of either process, usually because they are too mild to classify, whereas others appear to share components of both. With the latter, nonetheless, often one of the two basic types of pathology predominates. Moreover, it is now appreciated that Guillain-Barré syndrome, which has been considered the prototype of AIDP for decades, sometimes is caused by axon loss, not demyelination. This is the explanation for most cases of "atypical" Guillain-Barré syndrome in which substantial deficits persist.

Utility of Various EDX Components

All aspects of the EDX examination are helpful in polyneuropathy assessments, including the NCS, NEE, and the late responses, but overall the NCSs, particularly the motor NCS, are of most value. (However, if patients are shown to have some other peripheral nervous system cause for their symptoms instead of a polyneuropathy, then often the NEE component becomes of critical importance in diagnosis.) With axon loss polyneuropathies, the sensory NCSs often are more helpful than the motor NCSs in detecting early or mild disorders, because the SNAP amplitudes are more sensitive to axon loss than are the CMAP amplitudes. In contrast, the motor NCSs generally are more helpful than the sensory NCSs in identifying demyelinating polyneuropathies, because the SNAPs often appear to be too sensitive to demyelination, frequently becoming unelicitable whenever relatively moderate degrees of differential slowing or conduction block are present. (Unelicitable NCS responses are equally consistent with axon loss and demyelination.) Demyelinating conduction blocks are particularly difficult to detect with sensory NCS, because the SNAP amplitudes, unlike the CMAP amplitudes, decrease substantially as the distance between the stimulation and recording points increases, owing to phase cancellation.[10, 28]

Although typically only the distal one half to one third of peripheral nerves can be assessed by NCS,[5] fortunately those are the segments affected by the majority of polyneuropathies. With most axon loss polyneuropathies and most familial demyelinating polyneuropathies, the presentation of the NCS abnormalities is very consistent. These have been designated the "Electrophysiological Rules Regarding Polyneuropathies." Most polyneuropathies comply with at least five such rules (Table 18–1):

1. The lower extremity NCSs are involved earlier and more severely than their sensory and motor NCS counterparts in the upper extremity. Thus, the sural and superficial peroneal sensory SNAPs may be low in amplitude at a time when the median and ulnar SNAPs are of normal amplitude. With more severe lesions, the lower extremity SNAPs typically become unelicitable at a point when the distal upper extremity SNAPs are low in amplitude but still present.

2. In any limb, the sensory NCS responses generally are involved earlier, and then more severely, than the motor NCS responses. A necessary caveat here is that the sensory and motor nerve fibers must be assessed at approximately the same limb level. Thus, the superficial peroneal sensory SNAPs will be affected before the peroneal motor CMAPs, recording extensor digitorum brevis (EDB).

3. In any limb, NCSs of the same type (e.g., sensory, motor, mixed) will be affected to the same degree; again, with the proviso that nerve segments are evaluated at approximately the same limb level. Thus, if the forearm median motor CV is mildly slow, then the ulnar motor forearm CV in the same limb is likely to be slightly slow as well. Conversely, NCSs assessing nerve fibers of the distal limb will be more severely affected than NCSs of the same type assessing proximal nerve fibers. Thus, with many rather severe axon loss polyneuropathies, the median and ulnar SNAPs, stimulating or recording from the digits, will be low in amplitude, whereas the radial SNAP, recording thumb base, will be normal.

4. The results of motor and sensory NCSs performed on one limb will be very similar to those performed on the homologous nerves in the contralateral limb.

5. In any particular limb, NCSs that assess the more distal segments of a nerve will be more abnormal than those that assess the more proximal segments, if the underlying pathophysiology is axon loss. Thus, the peroneal motor amplitude, recording EDB, will be more severely affected than the peroneal motor amplitude, recording tibialis anterior (TA). Conversely, NCSs that assess the various segments will be affected approximately equally if the pathophysiology is demyelinating conduction slowing.

These "rules" impart a predictability factor to most generalized polyneuropathies that is not found with most acquired

TABLE 18–1	Electrophysiological Rules Regarding Polyneuropathies

1. The LE NCSs are affected earlier, and later more severely, than the UE NCSs, when nerves of the same type (sensory; motor) are assessed at the same relative limb levels.
2. In any limb, LE or UE, the sensory NCSs are affected earlier, and later more severely, than the motor NCSs, when nerves are assessed at the same limb level.
3. In any limb, NCSs of the same type (sensory; motor) are affected equally when assessed at the same limb level; in contrast, NCSs of the same type are not affected the same when they assess nerves at different limb levels—those that assess the more distal nerves are solely or more severely involved.
4. In either the LEs or UEs, NCSs that assess homologous nerves are affected equally.
5. NCSs that assess the more distal segments of a limb nerve are either more affected (axon loss) or equally affected (demyelination), compared with those that assess more proximal segments.

NCSs, nerve conduction studies; LE, lower extremity; UE, upper extremity.

demyelinating polyneuropathies and with a few atypical familial axon loss and demyelinating polyneuropathies.[28]

Compared with the NCS, the NEE is of secondary importance in the assessment of polyneuropathies. Nonetheless, it can provide certain useful information, mostly regarding axon loss polyneuropathies, that the NCS and the late responses are not capable of providing. These consist of the following. (1) It can show involvement of motor fibers, by demonstrating fibrillation potentials in the intrinsic foot muscles, before this is apparent on clinical examination. (2) It can verify that motor axon loss follows a distal-to-proximal gradient, by demonstrating that fibrillation potentials (and sometimes MUAP changes as well) are restricted to the intrinsic foot muscles, or are more severe in them, compared with the distal and particularly the proximal leg muscles. (3) It can reveal the degree of activity of the denervating process, based on the presence and density of fibrillation potentials. With relatively recent-onset axon loss polyneuropathies, these changes typically are restricted to the intrinsic foot muscles. However, with chronic advanced disorders in which the intrinsic foot muscles are completely atrophic, they are found, along with chronic neurogenic MUAP changes, in more proximal muscles of the limb, specifically, distal leg muscles such as the exterior hallucis, flexor digitorum longus, and possibly the TA and gastrocnemii. (4) It can establish the duration of the polyneuropathy, in general terms, by the presence and distribution of chronic neurogenic MUAP changes in the more distal lower extremity muscles.[28] Thus, as Albers has observed, the NEE is useful for detecting axon loss polyneuropathies, as well as for defining their distribution and chronicity.[2]

An Approach to Diagnosis

For optimal polyneuropathy assessment, at least one upper extremity and the ipsilateral lower extremity should be studied. Sensory and motor NCSs, late responses, and an NEE should be performed. Sometimes a third limb, usually the contralateral lower extremity, must be assessed as well, and rarely some evaluation of the remaining limb is necessary; these latter situations most often arise when confounding factors are present, several of which are discussed later.

In the majority of EDX laboratories, the NCSs performed consist of the standard lower and upper extremity motor and sensory NCSs: for motor NCSs, these are the peroneal and posterior tibial, recording intrinsic foot muscles (i.e., EDB and abductor hallucis [AH]), and the median and ulnar, recording intrinsic hand muscles (i.e., thenar and hypothenar muscles); for sensory NCSs, these are the sural, median (recording/stimulating D2), and ulnar (recording/ stimulating D5).[1, 9, 28] To the standard sensory NCS, we always add radial NCS, recording base of thumb, and often the superficial peroneal NCS as well. Some reports have suggested that the most sensitive NCSs for detecting polyneuropathies are the median and lateral plantar responses. Although these are mixed nerves rather than pure sensory nerves, they are situated more distally in the lower extremity than are the sural and superficial peroneal sensory nerves and are therefore more likely to become abnormal with early disease. A major limiting factor to their use is the fact that often they are unelicitable in normal patients who are more than 45 years of age. In regard to late responses, we have found the amplitude of the H response to be extremely sensitive with both axon loss and demyelinating polyneuropathies. In contrast, while great claims are made regarding the F waves in detecting early Guillain-Barré syndrome (GBS), we have found them to be no more sensitive than the H response with that disorder at that time, and far less sensitive with most other polyneuropathies.

The value of the NEE with polyneuropathies depends upon the underlying pathophysiology. Whenever axon loss occurs, which it does, by definition, with all axon loss polyneuropathies as well as with almost all acquired, acute-onset demyelinating polyneuropathies, the NEE may be very informative. Because the majority of polyneuropathies involve the most distal lower extremity muscles first, some of the intrinsic foot muscles, e.g., the EDB, AH, and abductor digiti quinti pedis (ADQP), should always be assessed during the NEE. There is some controversy regarding this point. In one recently published textbook, for example, the authors state that these muscles "... are best avoided in the EMG evaluation of polyneuropathy." One reason given for this opinion is that they show evidence of denervation and reinnervation so frequently that they are of no benefit in this regard.[25] We strongly disagree with this view. Whereas the EDB muscle often shows active and chronic denervation in isolation, presumably caused by shoe wear, we have never found the tibial nerve–innervated intrinsic foot muscles (e.g., the AH and ADQP) to contain fibrillation potentials in normal persons, even though fasciculation potentials often can be recorded from the AH.[9, 28] Consequently, if fibrillation potentials are present in the AH or ADQP, as well as the EDB, they should not be dismissed lightly.

One important point: if the lower extremity NEE is limited to muscles proximal to the ankle, then a considerable number of mild axon loss polyneuropathies will be overlooked.[30] On a more fundamental basis, both the EDB and AH are used as recorded muscles during the standard lower extremity NCS. A basic rule regarding the EDX examination is that any muscle used as the recorded muscle for motor NCS should always be available for NEE assessment. Otherwise, if the CMAP recorded from it is of low amplitude or unelicitable, the cause cannot be determined.[28]

Although it may be argued that in many patients with polyneuropathies the NEE of the lower extremity can be restricted to the distal muscles, most electromyographers do more extensive muscle sampling. In our EDX laboratory, for example, we perform a general survey of the lower extremity on every patient. This is mandatory if fibrillation potentials or MUAP changes are found in the intrinsic foot muscles because, with most polyneuropathies at least, it is important to demonstrate a proximal-distal gradient of abnormalities so as to obtain an acceptable level of diagnostic certainty. Consequently, if such NEE abnormalities are detected in the leg muscles as well as the intrinsic foot muscles, then it is important to sample more proximal lower extremity muscles, particularly the hamstrings and glutei, because they are in-

nervated by the same roots (L5 and S1) as are the distal leg muscles, and frequently the critical distinction is between an axon loss polyneuropathy and a lumbar intraspinal canal lesion involving the L5 and S1 roots bilaterally.

Generally, NEE of the upper extremity can be restricted to the more distal muscles (e.g., intrinsic hand and distal forearm muscles). However, with severe axon loss polyneuropathies, even those muscles are abnormal, so more proximal limb muscles must be sampled and a shading of changes demonstrated. Also, with a few unusual axon loss polyneuropathies (e.g., dapsone neuropathy), the distal upper extremity muscles are preferentially involved. Moreover, with porphyric polyneuropathy and with AIDP, the NEE changes in the upper as well as the lower extremity frequently are limited to, or are most prominent in, muscles other than the distal limb muscles.

Concerning paraspinal muscles, at least the low lumbar/ high sacral region muscles should be assessed in all patients with suspected polyneuropathies, if possible (i.e., if lumbar laminectomies have not been performed). This stems from the general rule that the paraspinal muscles should be sampled in all patients with suspected generalized neuromuscular disorders, including polyneuropathies. Additional motor and sensory NCSs often are helpful in these assessments. Thus, in many patients with axon loss polyneuropathies, the CMAPs, recording intrinsic foot muscles, are unelicitable. By recording from the TA, while stimulating the common peroneal nerve at the fibular head, and from the gastrocnemii (GTN) muscles, while stimulating the tibial nerve in the popliteal fossa (as part of the H wave test), the degree of denervation present in the leg muscles can be ascertained. If marked asymmetry in the amplitudes of these two motor responses (e.g., peroneal/TA amplitude of 4 mV with a tibial/GTN of 0.5 mV) is present, it is very likely that, on subsequent NEE, the findings will be those of bilateral lumbosacral radiculopathies rather than an axon loss polyneuropathy. Also, forearm sensory NCSs (e.g., lateral and medial antebrachial cutaneous) can be of particular benefit when "pure" sensory polyneuropathies are present.

No attempt is made in this chapter to discuss the EDX features of every known etiology of polyneuropathy. Rather, the general features of the two basic types are discussed, and a few disorders are reviewed in greater detail.

Axon Loss Polyneuropathies

An atrocious, albeit often used, synonym for this type of polyneuropathy is "axonal" polyneuropathy. This term does not correspond to its counterpart, "demyelinating"; one concerns anatomy, the other pathology. Consequently, to be consistent, the terms used should be either "axonal" and "myelinative," or "axon loss" and "demyelinating." Throughout this chapter, the latter pair are employed.

Axon loss polyneuropathies are encountered far more often in the EDX laboratory than are demyelinating polyneuropathies. The etiologies are numerous (Table 18–2). Most of the polyneuropathies of unknown cause also fall into this general category. The majority of axon loss polyneuropathies are chronic and slowly progressive in nature, and mixed, sensorimotor in type. Most are symmetrical, but during the EDX examination, particularly the NEE, some leeway must be allowed for sampling differences. Thus, with a moderately severe axon loss polyneuropathy, detecting 1+ fibrillation potentials in one medial gastrocnemius muscle and 2+ in the corresponding muscle in the contralateral limb does not seriously compromise the diagnosis of polyneuropathy.

With relatively few exceptions, axon loss polyneuropathies affect the longest axons initially. Therefore, they produce a distal-to-proximal gradient of abnormalities, beginning in

TABLE 18–2	Some Axon Loss Sensorimotor Polyneuropathies

Familial

Hereditary motor-sensory neuropathy type II (Charcot-Marie-Tooth, Type II)
Porphyria
Amyloidosis

Acquired

Arsenic
Ethyl alcohol
Metronidazole
Amyloidosis
B$_{12}$ deficiency
Guillain-Barré syndrome (axon loss type)

the lower extremities, consistent with a "dying-back" mechanism. Consequently, most comply with the electrophysiological rules already enumerated, and as a result the EDX abnormalities appear in a predictable pattern as they evolve. The earliest changes consist of low amplitude, lower extremity sensory NCS responses, unelicitable H responses, and fibrillation potentials on NEE of the intrinsic foot muscles. Any one, two, or all three of this trio may be present on the initial EDX examination, depending upon the particular patient. In regard to the lower extremity SNAPs, often a discrepancy is apparent in the degree of involvement of sural and superficial peroneal SNAP amplitudes, with the former usually being the more severely affected. This most likely reflects the fact that the sural NCS assesses a slightly more distal segment of nerve than does the superficial peroneal sensory NCS, a critical point in regard to the time of onset of abnormalities with disease progression.

As the polyneuropathy evolves, sufficient denervation occurs in the intrinsic foot muscles that the peroneal and posterior tibial CMAPs, recording EDB and AH, respectively, become low in amplitude, and MUAP drop out becomes evident in those muscles on NEE. Next, fibrillation potentials are detected on NEE of the leg muscles. These exhibit a distal-to-proximal gradient, being found first, and later more severely, in the muscles near the ankle (e.g., FDL; EHL), compared with those assessed nearer the knee, (e.g., TA, GTN). Eventually, the axon loss is sufficient in these leg muscles to cause CMAPs recorded from them (from the TA during peroneal motor NCS; from the GTN as the M component of the H response) to be low in amplitude and eventually unelicitable. At the same time, NEE of them reveals substantial drop out of MUAPs, as well as fibrillation potentials. Generally, no EDX abnormalities are detected in the upper extremity with the typical mixed sensorimotor axon loss polyneuropathy until substantial denervation is present in the proximal leg (sometimes even the distal thigh) muscles.

Characteristically, the first EDX change seen in the upper extremities is a decrease in the median and ulnar SNAP amplitudes. The polyneuropathy usually is more severe before the radial sensory SNAP, recording base of thumb, becomes abnormal, and upper extremity denervation must be substantially greater before the medial and lateral antebrachial cutaneous SNAPs are affected. Only after the median and ulnar SNAPs are very low in amplitude, or unelicitable, are fibrillation potentials likely to appear in the intrinsic hand muscles, and only with much further progression of axon loss do the median and ulnar CMAPs, recorded from the thenar and hypothenar muscles, respectively, become low in amplitude. Presumably, fibrillation potentials consistently are seen in the distal upper extremity muscles

only after digital SNAP abnormalities are obvious, because the digits are situated more distally along the limb than are the intrinsic hand muscles.

In marked contrast to the motor and sensory NCS amplitudes, the motor and sensory distal peak latencies and CVs generally show little alteration with axon loss polyneuropathies, regardless of severity. The major exception, of course, is when the axon loss is so severe that CMAPs and SNAPs cannot be elicited and, therefore, neither latencies nor CVs can be obtained. Typically, the latencies and CVs remain within the normal range—although often they slow slightly, thus moving from the mid to lower range of normal—until the denervating process is rather advanced. When this occurs, as judged by the unelicitable SNAPs and low amplitude CMAPs, the motor CVs and distal latencies often are definitely, but mildly, slow. Only very rarely is the slowing sufficient to put them in the demyelinating range, however (see later). Presumably, this mild slowing results from random loss of the fastest conducting axons, leaving the slower ones for NCS assessment (Table 18–3*A,B*).

With the majority of axon loss polyneuropathies, both motor and sensory fibers are affected, and to approximately equal degrees (as well as this can be judged). However, in some patients, either the motor or the sensory axons are substantially more affected than the other, e.g., if only a modest amount of active and chronic denervation is found in the intrinsic foot muscles at a time when the SNAPs are unelicitable in the lower extremity and low in amplitude in the upper extremity, the diagnosis of a sensorimotor polyneuropathy with predominant sensory involvement is justified. Moreover, occasionally axon loss polyneuropathies detectably involve solely the sensory or the motor fibers, not both. "Pure" motor polyneuropathies that exhibit a prominent proximo-distal gradient of change are quite rare. (The term "motor neuropathy" often is used as a label for anterior horn cell disorders, or predominantly motor polyneuropathies.) We have encountered only three in our EDX laboratory, while performing over 65,000 examinations. In all three, substantial denervation and nerve fiber loss were present in the intrinsic foot and hand muscles. The CMAPs, recording intrinsic hand and foot muscles, were very low in amplitude, whereas those recordings from more proximal muscles were normal. All the SNAPs and the H responses were normal. On NEE, fibrillation potentials and marked MUAP drop out were evident in the distal muscles, shading proximally. In all three patients, the polyneuropathies were acquired and paraneoplastic in type, being remote effects of neoplasms, specifically leukemia or lymphomas.

In comparison to pure motor polyneuropathies, "pure" sensory polyneuropathies are relatively common. The only components of the EDX examination abnormal with these polyneuropathies are the SNAPs, the H responses, and occasionally the blink reflex; the motor NCS, F waves, and NEE remain normal. A major disadvantage in the EDX assessment of pure sensory polyneuropathies is that, for sensory axons, no procedure equivalent to the NEE is available that allows loss of individual fibers to be detected; because a counterpart to fibrillation potentials is lacking, usually it is not possible to determine the site of axon loss damage along sensory fibers. This is a serious deficiency in regard to localization, because nearly all pure sensory polyneuropathies affect solely the SNAP amplitudes and not the peak latencies (or sensory CVs, if they are performed), and frequently they are of sufficient severity that all the sensory NCSs in the upper extremity, including the lateral and medial antebrachial cutaneous, are compromised. Consequently, although widespread sensory NCS abnormalities may be obvious, determining where the responsible lesions lie along the sensory fibers, from their distal terminations up to and including their

TABLE 18–3 (A)	The Nerve Conduction Studies Seen with a Modestly Severe Axon Loss Polyneuropathy

| | | | Amplitude | | Latency (msec) | | Conduction Velocity (m/s) | | F waves (mS) |
| | | | uV/mV | | Peak/Distal | | | | |
Nerve Stimulated	Stimulation Site	Recording Site	R	L	R	L	R	L	
Median (S)	Wrist	D2	NR*		—				
Ulnar (S)	Wrist	D5	NR*		—				
Radial (S)	Distal forearm	Thumb base	6*		2.6				
Sural (S)	Dist. post. leg	Lat. ankle	NR*		—				
SPS (S)	Dist. ant. leg	Ant. ankle	NR*		—				
Median (M)	Wrist	Thenar	9.7		3.4		52		34.9
	Elbow		8.2		8.9				
Ulnar (M)	Wrist	Hypothenar	6.6		2.6		51		34.6
	Elbow		5.0		9.1				
Peroneal (M)	Ankle	Ext. dig. brev.	NR*		—		—		
	Pop. fossa		NR*		—				
Tibial (M)	Ankle	Abd. hal.	NR*		—		—		
	Pop. fossa		NR*		—				
Peroneal (M)	Dist. fib. head	Tib. ant.	2.3*		3.9		35*		
	Pop. fossa		1.5*		6.7				
Tibial	Pop. fossa	Gastroc./soleus							
(1) H response			NR*						
(2) M response			3.0*		5.5				

*Abnormal response.
Note that the motor NCS responses recording intrinsic foot muscles are unelicitable, whereas the peroneal and tibial motor responses, recording from more proximal muscles (i.e., tibial anterior and gastrocnemius, respectively, in the leg), are present but low in amplitude. In contrast, the median and ulnar motor NCSs are normal; only the sensory NCSs are abnormal in the upper extremity.
S, sensory; M, motor; D, digit; NR, no response; SPS, superficial peroneal sensory.

TABLE 18–3 (B)	Needle Electrode Examination Findings That Accompany Table 18–3(A)

| Muscle | Spont. Activity | Motor Unit Action Potentials | | | |
Right	Fibs	Recruitment	Duration	Amplitude	Polyphasic
1st dorsal interosseous	0	NOR	NOR	NOR	—
Abd. pollicis brevis	0	NOR	NOR	NOR	—
Ext. indicis proprius	0	NOR	NOR	NOR	—
Abd. dig. minimi	0	NOR	NOR	NOR	—
Ext. dig. brevis	Atrophic	↓↓↓↓ #RAP	All ↑	NOR	
Abd. hallucis	Atrophic	None fire	—	—	
Ext. hallucis	2+	↓↓↓ #RAP	All ↑	NOR	Most 2+
Flex. dig. longus	2+	↓↓↓ #RAP	All ↑	NOR	Some 1+
Tib. anterior	1+	↓↓ #RAP	All ↑	NOR	—
Med. gastrocnemius	1+	↓↓ #RAP	All ↑	NOR	—
Rectus femoris	0	NOR	NOR	NOR	Some 1+
Biceps femoris (SH)	0	NOR	NOR	NOR	—
Semitendinosus	0	NOR	NOR	NOR	—
Glut. maximus	0	NOR	NOR	NOR	—
High sacral PSP	0	—	—	—	—

Note the distal-to-proximal gradation of neurogenic abnormalities present in the lower extremity. The fact that fibrillation potentials and motor unit potential loss are seen in the tibialis anterior and medial gastrocnemius muscles could be predicted, after compound muscle action potentials of low amplitude were recorded from those muscles during the nerve conduction studies.
Fibs, fibrillation potentials; NOR, normal; RAP, rapidly; SH, short head; PSP, paraspinal muscles.

TABLE 18–4	Some "Pure" Sensory Polyneuropathies (Axon Loss Type)

Familial

Hereditary sensory and autonomic neuropathy (HSAN), types I–V
Spinocerebellar degeneration (especially Friedreich's ataxia)
Kennedy's disease (one component)

Acquired

Cisplatin
Nitrous oxide
Pyridoxine
Paraneoplastic (Denny-Brown's)
Sjögren's syndrome
Idiopathic sensory neuropathy

origins in the dorsal root ganglia, often is impossible. As a result, generalized sensory neuronopathies frequently cannot be differentiated from generalized sensory polyneuropathies by EDX examination.[28]

There are two types of pure sensory polyneuropathies: familial and acquired (Table 18–4). Familial causes include the various types of hereditary sensory and autonomic neuropathy (HSAN), and the spinocerebellar degenerations, particularly Friedreich's ataxia. A pure sensory polyneuropathy is also a component of the PNS abnormalities that occur with X-linked recessive bulbospinal neuronopathy (Kennedy's disease). The acquired cause most often mentioned is the paraneoplastic sensory neuronopathy first described by Denny-Brown. However, a far more commonly encountered acquired pure sensory polyneuropathy, in our EDX laboratory at least, is idiopathic progressive sensory polyneuropathy, which most often affects elderly women and is manifested principally by pain and ataxia.[20] With the familial types, the general rules regarding the EDX changes with polyneuropathies, noted earlier, typically are followed. Hence, the abnormalities usually are most severe in the

lower extremity and are symmetrical. In contrast, with the acquired disorders, the abnormalities frequently begin in the upper extremities and often are quite asymmetrical. The familial pure sensory polyneuropathy that occurs with Kennedy's disease is quite atypical, in that it behaves as though it is an acquired disorder, often affecting upper extremity SNAPs more severely than lower extremity ones, and manifesting striking side-to-side asymmetries.[11]

When assessing for pure sensory polyneuropathies, it is important to perform as many sensory NCSs as feasible. These can be restricted to ipsilateral upper and lower extremities, if the patient's sensory complaints are symmetrical. If they are not, then at least some SNAPs should be obtained in the corresponding limb(s) (Table 18–5). H responses always should be performed because they are so sensitive to these disorders, and if sensory complaints are prominent in the face, as they sometimes are, blink reflexes may be informative.

Demyelinating Polyneuropathies

Demyelinating polyneuropathies are far less common than are axon loss polyneuropathies. However, like the latter, they may be either familial or acquired in nature; these two general types often, but not always, have distinct EDX presentations (Table 18–6).

The familial demyelinating polyneuropathies all share two common characteristics: (1) demyelinating synchronized conduction slowing is their most conspicuous, and consistent, underlying pathophysiology (Fig. 18–1A,B); and (2) they generally comply with the electrophysiological rules regarding polyneuropathies (listed in Table 18–1). Thus, both the NCS and NEE with familial demyelinating polyneuropathies have a certain predictability about them that is not found with the acquired demyelinating polyneuropathies.

In regard to the sensory NCS, those in the lower extremity (e.g., sural SNAP; superficial peroneal sensory SNAP) almost invariably are unelicitable, whereas those in the upper ex-

TABLE 18–5	The Nerve Conduction Studies with a "Pure" Sensory Polyneuropathy

Nerve Stimulated	Stimulation Site	Recording Site	Amplitude uV/mV R	L	Latency (msec) Peak/Distal R	L	Conduction Velocity (m/s) R	L	F waves (mS)
Median (S)	Wrist	D2	NR*	4*	—	2.8			
Median (S)	Wrist	D1	NR*	6*	—	3.0			
Ulnar (S)	Wrist	D5	NR*	5*	—	2.8			
Radial (S)	Dist. forearm	Thumb base	NR*	7*	—	2.4			
LAC (S)	Elbow	Forearm	6*	15	2.5	2.6			
MAC (S)	Arm	Forearm	5*	16	2.7	2.5			
Sural (S)	Dist. post. leg	Lat. ankle	NR*	NR*	—	—			
SPS (S)	Dist. ant. leg	Ant. ankle	NR*	NR*	—	—			
Median (M)	Wrist	Thenar	9.0		3.4		58		31
	Elbow		8.0		8.4				
Ulnar (M)	Wrist	Hypothenar	11.0		2.7		60		30
	Elbow		10.1		7.3				
Peroneal (M)	Ankle	Ext. dig. brev.	5.5		4.7		43		
	Pop. fossa		4.2		14.7				
Post. tib. (M)	Ankle	Abd. hal.	10.7		4.6		44		54
	Pop. fossa		7.2		14.5				
Tibial	Pop. fossa	Gastroc./soleus							
(1) H response			NR*		—				
(2) M response			11.5		4.9				

*Abnormal response.

Needle electrode examination of several muscles in both the right upper and lower extremities, including hand and foot muscles, revealed no abnormalities.
  Ordinarily, the NCS would be confined to one side of the body; sensory NCS were performed on the contralateral (left) side in this elderly woman because her hand symptoms (pain, ataxia) were so asymmetric, right > left. The asymmetric sensory NCS responses in the upper extremities are suggestive of an acquired disorder, or Kennedy's disease.
S, sensory; M, motor; D, digit; NR, no response; LAC, lateral antebrachial cutaneous; SPS, superficial peroneal sensory.

TABLE 18–6	Some Demyelinating Polyneuropathies

Familial

Hereditary motor sensory neuropathy, type I (Charcot-Marie-Tooth, type I)
Hereditary motor sensory neuropathy, type III (Dejerine-Sottas disease)
Hereditary motor sensory neuropathy, type IV (Refsum's disease)

Acquired

Acute inflammatory demyelinating neuropathy—includes most Guillain-Barré syndromes
Chronic inflammatory demyelinating neuropathy (CIDP)
CIDP variants (e.g., MGUS (Monoclonal Gammopathy of Unknown Significance))

tremity are either unelicitable or very low in amplitude, dispersed, and markedly prolonged in peak latency. In some patients, one upper extremity SNAP may be absent (e.g., ulnar), while another (e.g., median) is present, albeit quite abnormal. In these instances, the same pattern of abnormalities typically is found in the contralateral limb.

Concerning motor NCSs, those in the lower extremity, recorded from the intrinsic foot muscles (peroneal motor, recording EDB; tibial motor, recording AH) are unelicitable in some patients. In other patients, however, they are present but often abnormally low in amplitude and dispersed. Nonetheless, the low amplitude, dispersed motor CMAPs are very similar in appearance at all stimulation points (e.g., ankle, below fibular head, and popliteal fossa for peroneal) indicating that differential slowing, although present along motor fibers, is focal and confined to the very distal nerve segments (i.e., the segments situated between the distal stimulation points and the recording site) (Fig. 18–1A). Some textbooks have provided erroneous information on this point, stating that markedly dispersed responses indicate an acquired demyelinating polyneuropathy.

The most striking finding on the motor NCS, however, is the uniform, prominent conduction slowing present.[1, 6] An important question can be posed: "How slow must the motor CVs be to be in the demyelinating range?" Gilliatt, in the initial clinical publication on the subject, suggested that the motor CVs should be "reduced by more than 40%," but he did not state whether this measure is from the mean of the motor CV for that nerve, or from the lower limits of normal (LLN).[13] Currently, although a general consensus is still somewhat elusive, slowing of at least 20% to 30%, measured from the LLN, is considered sufficient.[2, 6, 10, 28] In our EDX

laboratory, this means that the motor CVs in the lower extremity should be decreased to at least 28 to 30 m/s (from an LLN of 40 m/s), whereas those in the upper extremity should be reduced to 35 to 38 m/s, from an LLN of 50 m/s. A very common error in EDX interpretation is for mild generalized motor conduction slowing to be considered indicative of a "demyelinating" polyneuropathy.[1, 28]

The conduction slowing generally is remarkably uniform along the entire length of the motor axons. Consequently, if the motor CV, for example, is half the LLN along the forearm segment of the ulnar nerve, it will also be half the LLN along the arm segment; and the motor distal latency, reflecting conduction along the terminal segment of the nerve, will be approximately twice the upper limit of normal (Fig. 18–1B).[6] Although the major nerves in any given limb share the same degree of slowing, as they do with the nerves of the contralateral limb, the normal motor CV differential of approximately 10 m/s between the upper and lower extremity nerves is maintained. In regard to late responses, H responses almost always are unelicitable, whereas F waves are unelicitable in some limbs, while present, but markedly prolonged in latency, in others. On NEE, abnormalities are always most severe in the intrinsic foot muscles. Often, fibrillation potentials are restricted to those muscles, where they are found in only minimal to modest numbers and accompanied by substantial drop out of MUAPs. Chronic neurogenic MUAPs usually are more widely distributed, frequently being detected as far proximally in the lower extremity as the quadriceps and hamstring muscles, as well as in the intrinsic hand muscles. Nonetheless, a definite distal-proximal gradient of these MUAP changes is obvious in the lower extremities (Table 18–7A,B).

Among the various causes for familial demyelinating polyneuropathies, relatively little variation is seen in the EDX features they produce. A major exception is Dejerine-Sottas disease, in which the electrical abnormalities typically are noticeably more severe. Thus, (1) the upper extremity SNAPs usually are unelicitable; (2) compared with HMSN, type 1, the motor CVs usually are 10 to 15 m/s slower (e.g., in the 20s in the upper extremity and in the teens in the lower extremity, compared with CVs in the 30s in the upper extremity and in the 20s in the lower extremity with HMSN, type I); and (3) active denervation is more prominent and more extensive in the lower extremities, resulting in the peroneal and posterior tibial CMAPs, recording intrinsic foot muscles, usually being unelicitable.[10, 28]

The acquired demyelinating polyneuropathies, AIDP and CIDP, are misnamed, in that they are not polyneuropathies at all but, rather, polyradiculoneuropathies. This accounts

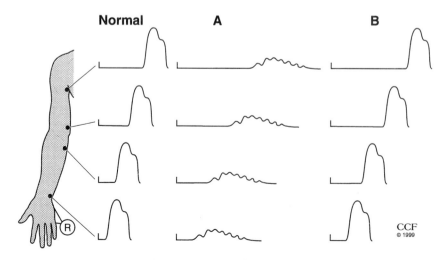

FIGURE 18–1. The uniform conduction slowing seen with familial demyelinating polyneuropathies and some acquired chronic inflammatory demyelinating polyneuropathies is illustrated using the standard ulnar motor nerve conduction study. Dispersed, irregular responses, sometimes of low amplitude (A), are commonly seen, rather than normal-appearing responses (B). The fact that all the responses in the A column share a common configuration indicates that the cause for their altered appearance is located along the terminal segment of the nerve, distal to the most distal stimulation point.

CCF
© 1999

TABLE 18-7 (A)	The Nerve Conduction Studies Seen with a Familial Demyelinating Polyneuropathy

Nerve Stimulated	Stimulation Site	Recording Site	Amplitude uV/mV		Latency (msec) Peak/Distal		Conduction Velocity (m/s)		F waves (mS)
			R	L	R	L	R	L	
Median (S)	Wrist	D2	5 (disp)*		6.0*				
Ulnar (S)	Wrist	D5	6 (disp)*		5.8*				
Radial (S)	Distal forearm	Thumb base	10 (disp)*		4.9*				
Sural (S)	Dist. post. leg	Lat. ankle	NR*		—				
SPS (S)	Dist. ant. leg	Ant. ankle	NR*		—				
Median (M)	Wrist	Thenar	5 (disp)*		8.9*		27*		50*
	Elbow		4 (disp)*		20.0*				
Ulnar (M)	Wrist	Hypothenar	5 (disp)*		8.1*				
	Elbow		3.5 (disp)*		18.6*		29*		NR*
	Axilla		2 (disp)*		23.6*		31*		
	Supraclavicular		NR*		—		—		NR
Peroneal (M)	Ankle	Ext. dig. brev.	0.8*		9.5*		26*		
	Pop. fossa		0.3*		24.6*				
Tibial (M)	Ankle	Abd. hal.	NR*		—		—		
	Pop. fossa		NR*		—		—		
Peroneal (M)	Dist. fib. head	Tib. ant.	4.7		5.8*		24*		
	Pop. fossa		3.4		9.9*				
Tibial (1) H response	Pop. fossa	Gastroc./soleus	NR*						
(2) M response			4.7		7.2				

*Abnormal response.
Note that nerves of the same type in the same limb are affected equally, and that conduction slowing is uniform along the nerve.
S, sensory; M, motor; D, digit; NR, no response; disp, dispersed; SPS, superficial peroneal sensory.

TABLE 18-7 (B)	Needle Electrode Examination Findings That Accompany Table 18-7 (A)

Muscle Right	Spont. Activity Fibs	Motor Unit Action Potentials Recruitment	Duration	Amplitude	Polyphasic
Ext. dig. brevis	2+	↓↓↓↓	All ↑↑	All ↑	1+
Abd. hallucis	1+	None fire	—	—	—
Ext. hallucis	0-1+	↓↓↓	All ↑↑	Fibs	—
Flex. dig. longus	0-1+	↓↓↓	All ↑↑	NOR	—
Tib. anterior	0	↓	All ↑	NOR	—
Med. gastrocnemius	0	↓	All ↑	NOR	—
Rectus femoris	0	NOR	Some ↑	NOR	—
Biceps femoris (SH)	0	NOR	Some ↑	NOR	—
Glut. maximus	0	NOR	NOR	NOR	—
High sacral PSP	0	—	—	—	—
1st dorsal interosseus	0	NOR	Many ↑	NOR	—
Abd. pol. brevis	0	NOR	Many ↑	NOR	—
Ext. ind. proprius	0	NOR	NOR	NOR	—
Flex. pol. longus	0	NOR	NOR	NOR	—
Flex. carpi ulnaris	0	NOR	NOR	NOR	—
Triceps	0	NOR	NOR	NOR	—
Biceps	0	NOR	NOR	NOR	—

Note that the abnormalities are most severe in the intrinsic foot muscles, which are the only muscles that contain a significant amount of fibrillation potentials. The chronic neurogenic motor unit potential changes are more widespread than is common in an axon loss polyneuropathy, being present in thigh and hand muscles; nonetheless, a definite gradation of abnormalities in the limbs is evident.
Fibs, fibrillation potentials; NOR, normal; SH, short head; PSP, paraspinal muscles.

for at least one, if not both, of the two major dissimilarities they have with familial demyelinating polyneuropathies. (1) They manifest a much greater repertoire of demyelinating NCS patterns. Whereas familial demyelinating polyneuropathies are essentially limited to demyelinating synchronized slowing, the pathophysiology of acquired demyelinating polyneuropathies often is demyelinating conduction block or differential slowing. Synchronized conduction slowing may be undetectable with them (particularly with AIDP) and, when present, it frequently coexists with conduction block and differential conduction slowing. (2) They may not comply with any or all of the Electrophysiological Rules Regarding Polyneuropathies, on either the NCS or the NEE.

It must be emphasized that not all acquired demyelinating neuropathies, particularly many cases of CIDP, transgress the polyneuropathy rules. When they do not do so, it can be impossible, by EDX studies alone, to differentiate a familial from an acquired demyelinating polyneuropathy. However, almost all AIDPs and many CIDPs manifest EDX features that are incompatible with a familial demyelinating polyneuropathy. These include the following.

1. Concerning sensory NCS, the sural SNAP may be normal, whereas the median SNAP, and usually the ulnar SNAP as well, are low in amplitude or unelicitable. This combination of SNAP findings, so strikingly different from what is seen with familial demyelinating neuropathies, was first reported at an international meeting in 1977 and first described in the literature in 1980.[22, 33] Nonetheless, Albers and various coworkers have assessed this phenomenon most extensively and have emphasized its diagnostic importance in various publications.[1, 2, 4, 9]

2. In any given limb, the abnormalities of the motor NCS may be more severe than those of the corresponding sensory NCS.

3. The NCS abnormalities along a given nerve may be very nonuniform (e.g., in regard to an upper extremity motor nerve, the distal latency may be very prolonged, whereas the forearm CV may be normal, or vice versa) (Fig. 18–2C,D).

4. Motor NCS may reveal progressive conduction blocks or progressive differential slowing along a nerve (Fig. 18–3E,F). Thus, using the ulnar nerve as an example, a response of 12 mV may be obtained on wrist stimulation, which decreases to 6 mV on below-elbow stimulation and to 3 mV on mid-arm stimulation and becomes unelicitable on supraclavicular stimulation.

5. NCS performed on nerves of the same type (motor, sensory) in the same limb and at the same limb level may yield very different results (e.g., the forearm motor CV may be markedly slow for the median, while normal for the ipsilateral ulnar nerve).

6. Homologous nerves in contralateral limbs may show very different EDX features (e.g, substantial demyelinating synchronized slowing may be present along the forearm nerve segment of the right median nerve, whereas a prominent conduction block may be affecting the same nerve segment of the left median nerve).

7. NEE abnormalities in a given limb often are not found in a distal-proximal gradient, and fibrillation potentials frequently are rather abundant in some muscles, including midlimb and proximal limb muscles.

8. NEE abnormalities may be more substantial in muscles of the upper extremity than of the lower extremity (Table 18–8).

Just as the degree of conduction slowing necessary to be considered indicative of demyelination has been debated, so has the amount of CMAP amplitude (or area) decrease on proximal compared with distal stimulation that is required for designation of conduction block. Percentages of from 20% to 60% have appeared in various publications.[7] We use the conservative figure of 50%. One subject that has been the topic of far more discussion in the AIDP, or GBS, literature than it appears to merit concerns distinguishing conduction block from differential conduction slowing. Granted that the distinction may be difficult when the CMAP responses are quite low in amplitude and dispersed, but it is also moot, in the sense that both processes are due to demyelination, and that is the fact being sought.[6, 7] Moreover, a simple method of ascertaining which of the two is operative is to determine the clinical strength of the recorded muscle. If it is weak, then a substantial amount of demyelinating conduction block is present.

Whereas the diagnosis of the familial demyelinating polyneuropathies rests almost solely on the demonstration of marked conduction slowing along multiple nerves, the diagnostic criteria for the acquired demyelinating polyneuropathies are much broader, and far more often changes are seen that, although consistent with the diagnosis, are not diagnostic of such. This is particularly the case with AIDP; for example, with many early AIDPs, the principal NCS finding is generalized low amplitude CMAPs. Although these could result from distal demyelinating conduction blocks along all the major motor nerves, they equally could be due to sub-

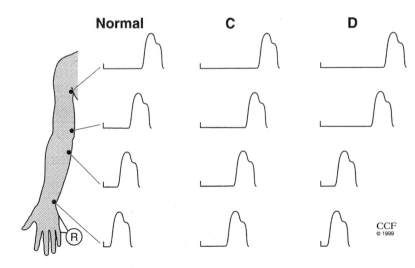

FIGURE 18–2. The segmental conduction slowing seen so often with acquired demyelinating polyneuropathies is illustrated, using the standard ulnar motor nerve conduction study. In C, the slowing is more pronounced between the distal stimulation point and the recording point, whereas in D the slowing is located along the elbow segment, at an entrapment site.

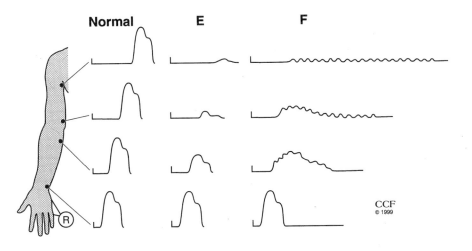

Normal   **E**   **F**

CCF
© 1999

FIGURE 18–3. Two patterns of nerve conduction study changes commonly seen with acute inflammatory demyelinating polyneuropathies are illustrated using the standard ulnar motor nerve conduction study. In *E*, progressive conduction blocks are present along the nerve, causing the responses to be lower in amplitude in a step-wise fashion. In *F*, progressive differential slowing along the nerve is present. A noteworthy point is that the recorded muscles (hypothenar) will be clinically weak only with *E*.

TABLE 18–8 (*A*)	**Nerve Conduction Studies Seen With an Acquired Demyelinating Polyneuropathy**							
			Amplitude		Latency (msec)		Conduction Velocity (m/s)	
			uV/mV		Peak/Distal			
Nerve Stimulated	**Stimulation Site**	**Recording Site**	R	L	R	L	R	L
Median (S)	Wrist	D2	NR*		—			
Ulnar (S)	Wrist	D5	3*		2.5			
Radial (S)	Dist. forearm	Thumb base	25		2.4			
Sural (S)	Dist. post. leg	Lat. ankle	10		6.5*			
SPS (S)	Dist. ant. leg	Ant. ankle	NR*		—			
Median (M)	Wrist	Thenar	4.0*		5.1		53	
	Elbow		3.1*		10.0			
Ulnar (M)	Wrist	Hypothenar	5.5*		3.0*			
	Below elbow		4.3*		7.3*		55	
	Elbow		3.0*		11.3*		37*	
	Axilla		2.1*				44	
Peroneal (M)	Ankle	Ext. dig. brev.	3.1*		4.0*			
	Pop. fossa		2.5*		15.4*		40	
Tibial (M)	Ankle	Abd. hal.	0.9*		4.8*		41	
	Pop. fossa		0.4*		14.8			
Peroneal (M)	Dist. fib. head	Tib. ant.	2.1*		3.1		48	
	Pop. fossa		1.8*		5.7			
Tibial								
(1) H response	Pop. fossa	Gastroc./soleus	NR*					
(2) M response			2.4*		6.7			

*Abnormal response.
The nerve conduction studies (NCSs) seen in a 45-year-old patient with an acute inflammatory demyelinating polyneuropathy (Guillain-Barré syndrome) of 12 days' duration. Note the marked variation from one NCS result to another.
S, sensory; M, motor; D, digit; NR, no response; SPS, superficial peroneal sensory.

TABLE 18-8 (B)	Needle Electrode Examination Findings That Accompany Table 18-8 (A)

Muscle	Spont. Activity	Motor Unit Action Potentials			
Right	Fibs	Recruitment	Duration	Amplitude	Polyphasic
Ext. dig. brevis	0	↓↓ #Mod. RAP	NOR	NOR	—
Abd. hallucis	2+	↓↓↓ #RAP	Most ↑	NOR	—
Tib. anterior	1+	↓↓ #RAP	NOR	NOR	—
Med. gastrocnemius	0	↓↓ #Mod. RAP	NOR	NOR	—
Rect. femoris	0	↓↓ #RAP	NOR	NOR	—
Biceps femoris (SH)	0	↓ #Mod. RAP	NOR	NOR	—
Glut. maximus	0	↓↓ #Mod.	NOR	NOR	—
High sacral PSP	2+	—	—	—	—
1st dors. interosseus	0-1+	↓ #Mod. RAP	NOR	NOR	—
Abd. pol. brevis	1+	↓↓ #RAP	NOR	NOR	—
Ext. ind. proprius	0	↓ #RAP	NOR	NOR	—
Flex. pol. longus	0	↓ #Mod. RAP	NOR	NOR	—
Pro. teres	2+	↓↓ #RAP	NOR	NOR	—
Biceps	0	↓ #Mod.	NOR	NOR	—
Triceps	1+	↓ #Mod. RAP	NOR	NOR	—
Deltoid	0	↓ #Mod.	NOR	NOR	—

Fibs, fibrillation potentials; NOR, normal; RAP, rapid; SH, short head; PSP, paraspinal muscles.

stantial axon loss along the same nerves. Hence, in themselves, they are not diagnostic (Fig. 18-4G). (In contrast, if the CMAPs are not only low amplitude but also dispersed at all stimulation points, this suggests that demyelinating differential slowing is affecting the distal segment of the nerve, i.e., that portion between the distal stimulation point and the recording site [Fig. 18-4H].) Nonetheless, although demonstrating definitive evidence of focal demyelination along multiple nerves is the optimal method for identifying acquired demyelinating polyneuropathies, at times the presence of particular combinations of nonspecific findings can be diagnostic. Thus, if generalized low amplitude CMAPs are found in a patient who also has unelicitable median and ulnar SNAPs in the presence of a normal sural SNAP, the diagnosis is extremely likely to be an acquired demyelinating polyneuropathy, specifically, AIDP. This is so even though neither of these findings alone can be considered unequivocal evidence of demyelination. For example, in regard to the sensory NCS changes, there is no way to determine that the unelicitable upper extremity SNAPs result from demyelination as opposed to axon loss, and certainly a normal sural NCS cannot be considered an abnormal finding in itself.

Several observations are pertinent in regard to AIDP. *First*, this is the type of acquired demyelinating polyneuropathy in which the most variation of NCS and NEE changes occurs. *Second*, because clinical weakness is essentially an indispensable component of this disorder, demyelinating conduction blocks along motor fibers invariably are present although, depending upon their location, they may or may not be demonstrable by NCS. The routine motor and sensory NCSs do not assess the proximal one half to one third or so of the main long nerve trunks of the limbs.[5] While conduction along these segments of the upper extremity nerves can be assessed to some extent with axillary and supraclavicular stimulation, determining conduction along the corresponding nerve segments in the lower extremity is quite difficult. Moreover, evaluating conduction along the most proximal portions of both upper extremity and lower extremity nerves is a major technical problem and usually is not done in most EDX laboratories. Thus, very proximal conduction blocks along motor fibers typically cannot be directly demonstrated, whereas very distal conduction blocks along motor fibers, while obvious on NCS, are nondiagnostic. *Third*, the EDX findings with this disorder are most likely to be minimal or nondiagnostic when it is in its early stages (i.e., at exactly the time when definite diagnosis is most critical, so that appropriate therapy can be initiated). Frequently, during this period, the motor distal latencies and CVs, as well as the

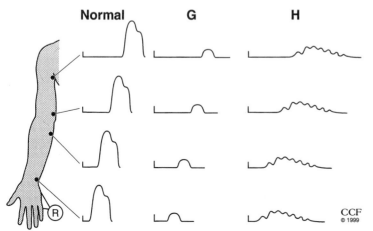

FIGURE 18-4. Two additional nerve conduction study patterns seen with acquired inflammatory demyelinating polyneuropathies are illustrated using the standard ulnar motor nerve conduction study. In *G*, all the responses are uniformly quite low in amplitude, but no conduction slowing is present. This pattern is seen most often during the early stages of this disorder. Although it may be due to distal demyelinating conduction blocks, it may equally be due to distal axon loss; hence, it is a nonspecific presentation. In *H*, the configurations of all the responses are uniformly altered by differential slowing. The fact that all the responses are of the same configuration indicates that this process is affecting the terminal portion of the nerve, between the distal stimulation point and the recording point.

sensory NCS, are normal. The motor NCS amplitudes may be uniformly normal or uniformly low, but in either case, for reasons noted earlier, they are nondiagnostic. The H response is usually unelicitable, but this is a nondiagnostic finding in itself, particularly in patients over the age of 60 years. The F waves may be slow (due to demyelinating synchronized slowing), but they are just as likely to be unelicitable (due to demyelinating conduction block or to differential slowing). NEE performed during this early period, similar to the NCS, yields nonspecific findings. In clinically weak muscles, nonspecific MUAP changes usually are seen, caused by disintegration of the motor unit, resulting from distal blocks along the terminal nerve fibers. Since these changes are equally consistent with disorders of terminal nerve fibers, neuromuscular junctions, and muscle fibers, they are nonspecific in nature and unhelpful for diagnosis. *Fourth,* a key point is that substantial slowing of motor CVs is not a characteristic finding with AIDP at any stage. Moreover, it is least likely to be encountered during the early stages of AIDP, at a time when the diagnosis is unclear. Conversely, if it is present, usually it is found in the convalescence stage, after the demyelination that initially caused conduction blocks (and clinical weakness) has resolved enough that conduction, albeit slow, has been restored along the damaged nerve segments.[5, 10] *Fifth,* although in the past AIDP and GBS have been considered synonymous terms, it is now apparent that GBS is merely a constellation of symptoms, which can be caused not only by widespread demyelination but also by widespread axon loss.[28] *Sixth,* superimposed focal nerve lesions at the usual entrapment sites (e.g., median nerve at the carpal tunnel) occur with some frequency; these can compromise the value of the NCSs that reveal them for diagnosing AIDP.

The EDX features of CIDP, compared with those of AIDP, tend to be more uniform, and marked demyelinating conduction slowing usually is evident. Whereas certain features often seen with AIDP, such as low amplitude or unelicitable upper extremity SNAPs in the presence of normal lower extremity SNAPs, are found in many patients with CIDP, in many others they are not; frequently, all the SNAPs are unelicitable.[10] Whether conduction blocks occur with CIDP has been the topic of much debate. Nonetheless, generalized, severe distal motor conduction blocks, so often seen with AIDP, are rarely, if ever, encountered with CIDP. The NEE findings with CIDP, with few exceptions, are far less

impressive than with AIDP. Fibrillation potentials often are not detected in any muscle, and, while chronic neurogenic MUAPs sometimes are seen in various muscles (most often the more distal lower extremity muscles) in certain patients with very chronic lesions, in other patients no MUAP abnormalities are found. Thus, in marked contrast to the EDX evaluations with AIDP, a completely normal NEE is found with some frequency when CIDP cases are assessed.

The fact that the NCS changes are so predictable with familial demyelinating polyneuropathies, while quite the opposite with acquired demyelinating polyneuropathies, has significant implications in regard to the extent of EDX assessment required to reasonably exclude the members of each of these generalized demyelinating disorders. The results of a single sensory NCS can virtually eliminate the possibility of a familial demyelinating polyneuropathy being present, specifically, if a sural response is well within the normal range for age. In contrast, the results of motor and sensory NCS performed on many upper and lower extremity nerves can be normal, and yet an acquired demyelinating polyneuropathy has not been completely excluded.

Confounding Factors

Nearly all the most vexing confounding factors in the EDX diagnosis of polyneuropathies are directly related to the fact that the majority of these disorders occur in elderly patients. Unfortunately for electromyographers, in the diagnosis of no other peripheral neuromuscular disorder does advanced age have such adverse EDX consequences. A few of these will now be discussed.

First, all the lower extremity NCSs that are most sensitive for detecting axon loss polyneuropathies are unelicitable in a certain percentage of normal elderly patients. Thus, medial plantar responses may be unelicitable "normally" in patients over the age of 45 years, whereas the sural and superficial peroneal SNAPs, as well as the H response, are unelicitable in a substantial portion of presumably healthy persons over the age of 60 years (Table 18–9). The absence of these NCS responses in young to middle-aged patients, particularly when they are unelicitable bilaterally, is almost pathognomonic of a polyneuropathy when found in the appropriate clinical situation. However, the absence of these same responses in patients over the age of 60 years is of uncertain significance.[10, 28] In fact, in the elderly, the presence of these

TABLE 18–9	Combination of NCS Changes Encountered in the Lower Extremity with Some Frequency								
			Amplitude		Latency (msec)		Conduction Velocity (m/s)		F waves (mS)
			uV/mV		Peak/Distal				
Nerve Stimulated	Stimulation Site	Recording Site	R	L	R	L	R	L	
Sural (S)	Distal. post. leg	Lat. ankle	NR*		—				
SPS (S)	Dist. ant. leg	Ant. ankle	NR*		—				
Peroneal (M)	Ankle	Ext. dig. brevis	NR*		—				
	Pop. fossa		NR*		—		—		
Tibial (M)	Ankle	Abd. hallucis	8.5		3.5				
	Pop. fossa		5.0		12.9		40		51
Tibial	Pop. fossa	Gastroc./soleus							
(1) H response			NR*		—				
(2) M response			8.4		5.9				

*Abnormal response.
The significance of these findings depends on the NEE, the age of the patient, and the upper extremity SNAPs. The unelicitable peroneal (EDB) NCS responses may be of no clinical significance, if NEE abnormalities are restricted to the EDB muscle. The unelicitable sural NCS response, SPS NCS response, and H response are reliable evidence of a polyneuropathy in patients of <60 years; in elderly patients they are of uncertain significance. However, if the upper extremity SNAPs (median; ulnar; radial) are low in amplitude or unelicitable, a polyneuropathy can still be diagnosed.
S, sensory; M, motor; NEE, needle electrode examination; SPS, superficial peroneal sensory; EDB, extensor digitorum brevis; NCS, nerve conduction studies; SNAPs, sensory nerve action potentials.

lower extremity NCS responses is more helpful than their absence; if they are present, and particularly if their amplitudes are substantial, then most types of polyneuropathy can be excluded.

If upper extremity (e.g., median, ulnar, and radial) SNAPs are abnormally low or unelicitable in patients with unelicitable lower extremity SNAPs, then the diagnosis of a polyneuropathy becomes likely, even in patients of advanced age. Unfortunately, however, the upper extremity SNAPs are not affected by axon loss polyneuropathies until the latter are relatively severe. Consequently, whenever the lower extremity NCS responses are unelicitable while those in the upper extremity are normal, the possibility of a moderate axon loss polyneuropathy cannot be excluded. Moreover, in many patients the value of the upper extremity SNAPs in the assessment of polyneuropathies is severely compromised by the presence of bilateral CTS and UN-ES.

Second, not only does polyneuropathy occur more commonly in elderly patients, but so does the disorder most often confused with it, both clinically and electrically: bilateral, chronic L5 and S1 radiculopathies. The latter, as with polyneuropathies, can cause bilateral distal lower extremity symptoms, such as pain and paresthesias. Moreover, in the population over the age of 60 years, these focal lesions appear to be almost as common as polyneuropathies. Finally, it is in patients in exactly the same age group in whom neuroimaging abnormalities of the lumbar spine are so common, so their value for separating bilateral L5 and S1 radiculopathies from axon loss polyneuropathies is seriously compromised as well. Consequently, whenever elderly patients are encountered who have bilaterally unelicitable lower extremity SNAPs, bilateral unelicitable H responses, and fibrillation potentials and MUAP drop out limited to the distal lower extremity muscles bilaterally, the findings are always open to multiple interpretations:

1. They are all due to an axon loss polyneuropathy;
2. The sural and superficial peroneal sensory SNAP abnormalities are due to age alone, whereas the NEE findings are due to a lumbar intraspinal canal lesion, and
3. The absent H responses are due to advanced age, a polyneuropathy, or bilateral S1 root compromise.

No definite conclusions can be drawn based on the EDX findings. However, the NEE sometimes is very helpful for distinguishing chronic bilateral L5, S1 lumbar intraspinal canal lesions from axon loss polyneuropathies. This occurs whenever active denervation is present, not only in the more distal lower extremity muscles bilaterally (which it may be with either entity), but also in the more proximal L5- and S1-innervated muscles, such as the hamstrings and glutei. Often the diagnosis rests on these muscles when patients such as this are encountered.

A number of other confounding PNS disorders—again, far more commonly found in the elderly—can seriously compromise the diagnosis of polyneuropathy in the EDX laboratory. These include such problems as limb edema or gross obesity (with either, low amplitude or unelicitable NCS responses in the lower extremities may be due to technical factors, rather than to actual nerve pathology); remote poliomyelitis; remote lumbar laminectomies; and prior hip, leg, and foot operations or injuries. With all these situations, low amplitude or unelicitable CMAPs may be found during motor NCS recorded from distal lower extremity muscles, along with chronic neurogenic MUAP changes and MUAP drop out on NEE of the same muscles; moreover, with many, the lower extremity SNAPs may be unelicitable as well. Although the presence of some of these disorders can be overcome, at least partially, by performing extensive comparative assessments of both lower extremities, in many instances the re-

sults are still inconclusive, and in all cases the necessary time expenditure is excessive. It is an unfortunate but undeniable fact that, while diagnosing an axon loss polyneuropathy in young to middle-aged patients often is a relatively simple task, attempting to diagnose the same disorder in elderly patients can be a very frustrating, time-consuming endeavor.

Diabetic Polyneuropathy

Diabetes mellitus is the most common serious metabolic disorder that has significant PNS complications, and the most common PNS complication of it, by far, is a polyneuropathy. The incidence of diabetic polyneuropathy is related to the duration of the disease; approximately 50% of patients have PNS abnormalities after 25 years.[6, 30] The earliest and most prominent symptom is burning pain in the feet, shading proximally. Usually, sensory changes are rather significant in the more distal lower extremities before substantial lower extremity weakness appears. Both the sensory and motor abnormalities are in a distal-to-proximal gradient.

On EDX examination, the earliest finding most often is an unelicitable H response which, unfortunately, is of significance only in those patients under the age of 60 years. Other early findings include low amplitude or unelicitable sural and superficial peroneal sensory NCS responses, and a modest number of fibrillation potentials restricted to the intrinsic foot muscles. The peroneal and tibial motor CVs, recording intrinsic foot muscles, usually are borderline to slightly slow; on stimulation of the peroneal nerve distal to the fibular head, the CV becomes slightly slower, rather than faster. The motor CV slowing is due to metabolic factors, rather than to either mild demyelination or to loss of the fastest conducting fibers resulting from axon degeneration.[1, 30] In general, excluding the mild motor CV slowing that often is present, and the marked propensity for bilateral superimposed CTS and often UN-ES as well, diabetic polyneuropathy generally manifests as a typical axon loss polyneuropathy of various degrees of severity. Noteworthy is that it almost never presents as a "pure" sensory polyneuropathy, even though the symptoms are restricted to the sensory sphere in many patients. When the process is advanced, the peroneal and tibial motor NCS responses, recording from the TA and GTN, respectively, become unelicitable or very low in amplitude, the upper extremity SNAPs become unelicitable, and the motor CMAPs, recording intrinsic hand muscles, become low in amplitude or unelicitable. At this point, not only are severe neurogenic changes present in the more distal limb muscles, shading proximally, but also nonspecific MUAP changes often are seen in the proximal lower and upper extremity muscles; the latter can be mistaken for evidence of a non-necrotizing myopathy.[30]

The EDX assessment of diabetic polyneuropathy entails all the confounding factors associated with polyneuropathies in general, most particularly including advanced patient age. Moreover, an additional technical factor is commonly encountered, namely, patients being grossly obese. Also, when assessing the upper extremities for evidence of a polyneuropathy, the electromyographer often is stymied by the seemingly ever-present superimposed CTS and UN-ES. In a relatively small percentage of patients with diabetic polyneuropathy, the clinical and EDX presentations are complicated by the superimposition of a diabetic polyradiculopathy. Most often these affect the L2–L4 roots, but occasionally they extend more caudally, to involve the L5 or even the S1 roots.[30]

Conclusions

The EDX examination is valuable in the assessment of multiple mononeuropathies and, particularly, polyneuropathies. Similar to focal peripheral nerve lesions, the underlying pathophysiology with these various disorders is either axon loss, resulting in conduction failure, or demyelination, causing conduction slowing or block.

REFERENCES

1. Albers JW: Evaluation of the patient with suspected peripheral neuropathy. In Pease WS, Johnson EW (eds): Practical Electromyography, 3rd ed. Baltimore, Williams & Wilkins, 1997, pp 311–338.
2. Albers JW: Clinical neurophysiology of generalized polyneuropathy. J Clin Neurophysiol 10:149–166, 1993.
3. Bouche P, Leger JM, Travers MA: Peripheral neuropathy in systemic vasculitis: Clinical and electrophysiologic study of 22 patients. Neurology 36:1598–1602, 1986.
4. Bromberg MB, Albers JW: Patterns of sensory nerve conduction abnormalities in demyelinating and axonal peripheral nerve disorders. Muscle Nerve 16:262–266, 1993.
5. Brown WF: Electrophysiologic features of Guillain-Barré syndrome. In Syllabus 1991 AAEM Course B: Demyelinating polyneuropathies and the electrophysiology of conduction block. Rochester, MN, American Association of Electrodiagnostic Medicine, 1991, pp 21–23.
6. Campbell WW: Essentials of Electrodiagnostic Medicine. Baltimore, Williams & Wilkins, 1999.
7. Cornblath DR, Sumner AJ, Daube J: Conduction block in clinical practice. Muscle Nerve 14:869–871, 1991.
8. Cwik VA, Wilbourn AJ, Rorick M: Acute brachial neuropathy: Detailed EMG findings in a large series. Muscle Nerve 13:859, 1990.
9. Donofrio PD, Albers JW: AAEM Minimonograph #34: Polyneuropathy: Classification by nerve conduction studies and electromyography. Muscle Nerve 13:889–903, 1990.
10. Dumitru D: Electrodiagnostic Medicine. Philadelphia, Hanley & Belfus, 1995.
11. Ferrante MA, Wilbourn AJ: The characteristic electrodiagnostic features of Kennedy's disease. Muscle Nerve 20:323–329, 1997.
12. Gilliatt RW: Experimental peripheral neuropathy. In Scientific Basis of Medicine; Annual Review. London, Athlone Press (University of London), 1969, pp 202–219.
13. Gilliatt RW: Applied electrophysiology in nerve and muscle disease. Proc R Soc Med 59:989–993, 1966.
14. Kaeser HE, Lambert EH: Nerve function studies in experimental neuritis. Electroencephalogr Clin Neurophysiol (Suppl)22:29–35, 1962.
15. Lambert EH: Electromyography and electrical stimulation of peripheral nerves and muscle. In Mayo Clinic Staff: Clinical Examinations in Neurology. Philadelphia, WB Saunders, 1956, pp 287–317.
16. Lewis RA, Sumner AJ, Brown MJ, Asbury AK: Multifocal demyelinating neuropathy with persistent conduction block. Neurology 32:958–964, 1982.
17. Marinacci AA: Clinical Electromyography. Los Angeles, San Lucas Press, 1955, p 143.
18. McDonald WI: The effects of experimental demyelination on conduction in peripheral nerves; A histological and electrophysiological study. II. Electrophysiological observations. Brain 86:501–524, 1963.
19. McLeod JG, Prineas JW, Walsh JC: The relationship of conduction velocity to pathology in peripheral nerves. In Desmedt JE (ed): New Developments in Electromyography and Clinical Neurophysiology. Basel, S Karger, 1973, pp 248–258.
20. Mitsumoto H, Wilbourn AJ: Causes and diagnosis of sensory neuropathies: A review. J Clin Neurophysiol 11:553–567, 1994.
21. Morgan-Hughes JA: Changes in motor nerve conduction velocity in diphtheritic polyneuritis. Rev Patol Nerv Ment 86:253–260, 1965.
22. Murray NMF, Wade DT: The sural sensory nerve action potential in Guillain-Barré syndrome. Muscle Nerve 3:444, 1980.
23. Parry GJ: Motor neuropathy with multifocal conduction block. In Dyck PJ, Thomas PK (eds): Peripheral Neuropathy, 3rd ed. Philadelphia, WB Saunders, 1993, pp 1518–1524.
24. Parry GJG: Mononeuropathy multiplex (AAEE Case Report #11). Muscle Nerve 8:493–498, 1985.
25. Preston DC, Shapiro BE: Electromyography and Neuromuscular Disorders: Clinical-Electrophysiologic Correlations. Boston, Butterworth-Heinemann, 1998.
26. Stevens JC, Sun S, Beard CM, et al: Carpal tunnel syndrome in Rochester, Minnesota, 1961–1980. Neurology 38:134–138, 1988.
27. Wilbourn AJ, Shields RW: Generalized polyneuropathies and other nonsurgical peripheral nervous system disorders. In Omer GE, Spinner M, VanBeek AL (eds): Management of Peripheral Nerve Problems, 2nd ed. Philadelphia, WB Saunders, 1998, pp 648–660.
28. Wilbourn AJ, Ferrante MA: Clinical electromyography. In Joynt RJ, Griggs RC (eds): Clinical Neurology. Philadelphia, Lippincott-Raven, 1997, pp 1–76.
29. Wilbourn AJ: Nerve injuries caused by injections and tourniquets. In Syllabus 1996 AAEM Plenary Session: Physical Trauma to Peripheral Nerves. Rochester, MN, American Association of Electrodiagnostic Medicine, 1996, pp 15–33.
30. Wilbourn AJ: Diabetic neuropathies. In Brown WF, Bolton CF (eds): Clinical Electromyography, 2nd ed. Boston, Butterworth-Heinemann, 1993, pp 477–515.
31. Wilbourn AJ, Levin KH: Ischemic neuropathy. In Brown WF, Bolton CF (eds): Clinical Electromyography, 2nd ed. Boston, Butterworth-Heinemann, 1993, pp 369–390.
32. Wilbourn AJ: Brachial plexus disorders. In Dyck PJ, Thomas PK (eds): Peripheral Neuropathy, 2nd ed. Philadelphia, WB Saunders, 1993, pp 911–950.
33. Wilbourn AJ: Differentiating acquired from familial segmental demyelinating neuropathies by EMG. EEG Clin Neurophysiol 43:616, 1977.

Chapter 19

Motor Neuron Disorders

Erik P. Pioro, MD, DPhil, FRCPC

INTRODUCTION
COMMON ELECTRODIAGNOSTIC
  FEATURES
Nerve Conduction Studies
Needle Electrode Examination
SPECIFIC MOTOR NEURON DISORDERS
Amyotrophic Lateral Sclerosis
Variants of Amyotrophic Lateral Sclerosis
X-Linked Recessive Bulbospinal
  Neuronopathy (Kennedy's Disease)
Poliomyelitis and Postpolio Syndrome
Hereditary Spinal Muscular Atrophy

DIFFERENTIAL DIAGNOSIS
Secondary Motor Neuron Disorders
Non-Motor Neuron Disorders
AN APPROACH TO ELECTRODIAGNOSIS
  OF SUSPECTED MOTOR NEURON
  DISORDERS
Focal or Segmental Weakness
Generalized or Multifocal Weakness
Decision Tree of Electrodiagnosis

INTRODUCTION

Disorders in which anterior horn cells (AHCs), or motor neurons, undergo degeneration are diverse and numerous but relatively uncommon. An abbreviated list of the motor neuron disorders with lower motor neuron dysfunction is presented in Table 19–1, although more complete lists have been compiled.[24] Specific diagnosis in the patient with focal, multifocal, or generalized weakness can be delayed because of clinical similarities seen among many of the motor neuron disorders, as well as other neuromuscular disorders. Motor neuron disorders can be characterized clinically by the extent to which they cause dysfunction of the upper motor neuron (UMN) and lower motor neuron (LMN) (Table 19–2). Amyotrophic lateral sclerosis (ALS), the prototypical motor neuron disorder, results in both UMN and LMN dysfunction.

After a thorough history and neurological examination, much information on the degree of LMN dysfunction can be derived from the electrodiagnostic examination. It can greatly assist the clinician in achieving an accurate and timely diagnosis as well as objectively monitoring disease progression. Nerve conduction studies (NCS) also provide information on sensory nerve fiber integrity distal to the dorsal root ganglion, which is normal in most motor neuron disorders. Use of more specialized techniques, including single fiber electromyography (SFEMG), macro electromyography (EMG), and transcranial magnetic stimulation (TMS), is rarely necessary to obtain a diagnosis of motor neuron degeneration and is usually reserved for research purposes. Except for TMS, all other electrodiagnostic, techniques directly assess only the motor unit, comprised of the AHC (in the spinal cord and brain stem motor nuclei), motor axon, neuromuscular junction, and muscle fibers.

Anterior horn cell degeneration produces fairly consistent pathophysiological changes on EMG with characteristic sparing of sensory nerve fibers, the so-called intraspinal canal pattern, an electrodiagnostic picture also shared by most spinal nerve root lesions (Table 19–3). The following discussion begins with electrodiagnostic features common to most motor neuron disorders, and then turns to clinical and electromyographic presentations specific to each condition. Of the motor neuron disorders listed in Table 19–1, those with prominent LMN involvement and of particular diagnostic

and clinical interest will then be emphasized, including ALS and its variants, X-linked recessive bulbospinal neuronopathy (Kennedy's disease), poliomyelitis and postpolio syndrome (PPS), and hereditary spinal muscular atrophy (SMA). Next, the electrodiagnostic identification of conditions mimicking motor neuron disorders is discussed in the context of weakness patterns or specific distinguishing features. Finally, a problem-oriented electrodiagnostic approach to the patient with focal (or segmental), and generalized (or multifocal) weakness follows.

COMMON ELECTRODIAGNOSTIC FEATURES

Nerve Conduction Studies

Motor NCS assess the capacity of large myelinated motor axons to transmit electrical impulses from the AHCs to the innervated muscle fibers across the neuromuscular junction. As degeneration of AHCs occurs in motor neuron disorders, their motor axons degenerate and the target muscle fibers become denervated and eventually atrophy. In the setting of motor neuron disorders, reduction of the compound muscle

TABLE 19–1	Motor Neuron Disorders with Lower Motor Neuron Dysfunction

Motor neuron diseases
  Amyotrophic lateral sclerosis (ALS)
  Monomelic amyotrophy
  Progressive bulbar palsy
  Progressive muscular atrophy
  X-linked recessive bulbospinal neuronopathy (Kennedy's disease)
Poliomyelitis and postpolio syndrome
Hereditary spinal muscular atrophy (SMA)
  SMA I (Werdnig-Hoffman disease)
  SMA II (intermediate form)
  SMA III (Kugelberg-Welander disease)
  SMA IV (adult-onset form)
Other
  Hexosaminidase deficiency
  Motor neuronopathy and malignancy
  Postirradiation motor neuron syndrome
  Toxins and heavy metals

TABLE 19-2	Clinical Features of Upper and Lower Motor Neuron Dysfunction in Motor Neuron Disorders

Upper Motor Neuron	Lower Motor Neuron
Weakness	Weakness
Loss of dexterity	Muscle atrophy
Spasticity	Hypotonia
Hyperreflexia	Hyporeflexia
Pathological reflexes	Fasciculations
(e.g., extensor plantar)	Cramps

action potential (CMAP) amplitude is the result of motor axon loss and muscle atrophy. Early in the course of disease, when axon loss is minimal or sufficiently compensated by collateral sprouting and reinnervation of muscle fibers, the CMAP amplitude may remain in the normal range. With disease progression and continued loss of motor neurons and axons, the CMAP amplitude decreases significantly, particularly when muscle atrophy is pronounced. If the degeneration is sufficiently slow to allow for reinnervation and motor unit enlargement, the CMAP amplitude may be relatively preserved despite significant muscle fiber denervation and reinnervation. This occurs in static or slowly progressing motor neuron degenerations, such as late progression of poliomyelitis or SMA. Generally, conduction velocities and distal latencies are unaffected except in cases of nearly total axon loss, when loss of all the fastest conducting fibers results in mild or moderate velocity slowing. In these circumstances, the CMAP amplitudes are usually dramatically reduced.[19, 20]

Routine sensory NCS reveal the integrity of large myelinated sensory axons distal to the dorsal root ganglion. Whether the sensory nerve action potential (SNAP) is recorded orthodromically or antidromically, no abnormalities of amplitude or peak latency are detected in the majority of motor neuron disorders. The presence of SNAP amplitude loss in the setting of otherwise typical AHC dysfunction (as discussed later) should raise the possibility of X-linked

recessive bulbospinal neuronopathy (Kennedy's disease) (see Table 19–3), as well as neuromuscular disorders such as polyradiculoneuropathies, including chronic inflammatory demyelinating polyneuropathy.

Table 19–4 lists the sensory and motor nerves we, in the Cleveland Clinic EMG laboratory, routinely study the arm and leg on the side most affected in individuals with suspected motor neuron disorders. F wave responses and H reflexes are also obtained. This provides a general survey of C6–8 dermatomes and C8–T1 myotomes in the upper extremities and L5–S1 dermatomes and myotomes in the lower extremities.

Needle Electrode Examination

Assessing motor unit integrity by needle electrode examination (NEE) is vital to the electrodiagnostic evaluation of motor neuron disorders. The progressive loss of AHCs and their motor units, along with motor axon collateral sprouting and muscle fiber reinnervation, results in a characteristic "neurogenic" pattern on NEE (see Chap. 12).

If denervation is actively occurring, resting muscle shows spontaneous activity in the form of fibrillation potentials. In addition, hyperexcitability of motor units, which is relatively common in AHC disorders, produces fasciculations that can occur in multiple muscles for extended periods. Fasciculation potentials usually occur in less denervated muscles because they are generated by intact motor units. Therefore, there tends to be an inverse relationship between the relative abundance of fibrillation potentials and fasciculation potentials. Except for ALS, fasciculation potentials are not an

TABLE 19-3	Electrodiagnostic Findings Common to Most Motor Neuron (Intraspinal Canal) Disorders

Diagnostic Study	Finding
Nerve Conduction Studies	
Compound motor action potential (CMAP) amplitude	Normal (early) or decreased
Motor conduction velocity and distal latency	Normal[a1]
Sensory nerve action potential (SNAP) amplitude	Normal[a2]
Needle Electrode Examination	
Fibrillations and positive insertional waves	Present[a3]
Fasciculations	Present[a4]
Motor unit potential	
Number	Reduced
Firing rate	Rapid[b]
Duration ± amplitude	Increased
Polyphasia	Increased

[a]Usually.
[b]Upper motor neuron pathology in amyotrophic lateral sclerosis (ALS) often restricts the firing rate.
[1]Advanced cases with loss of fastest conducting fibers may result in some velocity slowing and distal latency prolongation.
[2]X-linked recessive spinobulbar neuronopathy (Kennedy's disease) has characteristically diminished SNAP amplitudes.
[3]Very slowly advancing or chronic motor neuron disorders may not have evidence of ongoing denervation.
[4]Most frequently detected in ALS.

TABLE 19-4	Peripheral Nerves Routinely Examined by Nerve Conduction Studies

	Recording Site	Measurement	Site of Neuron Assessed[a]
Upper Extremity			
Sensory nerve			
Median	Index finger	Amp, PL	C6, C7 DRG[b]
Ulnar	Fifth finger	Amp, PL	C8 DRG
Radial	Thumb	Amp, PL	C6, C7 DRG
Motor nerve			
Median	Abductor pollicis brevis m.	Amp, DL, CV	C8, **T1** spinal cord
Ulnar	Abductor digiti minimi m.	Amp, DL, CV, F wave	**C8**, T1 spinal cord
Lower Extremity			
Sensory nerve			
Sural	Lateral malleolus	Amp, PL	S1 DRG
H response (H reflex)	Soleus m.	H response amplitude, DL	S1 spinal cord
Motor nerve			
Peroneal	Extensor digitorum brevis m.	Amp, DL, CV	**L5**, S1 spinal cord
Posterior tibial	Abductor hallucis m.	Amp, DL, CV, F wave	**S1**, S2 spinal cord
M response (H reflex)	Soleus m.	Amp, DL	**S1** spinal cord

Amp, amplitude; CV, conduction velocity; DL, distal latency; DRG, dorsal root ganglion; m, muscle; PL, peak latency.
[a]Spinal cord segments in boldface indicate usual major innervation for the muscles specified.

essential component of any of the AHC disorders. Other relatively common forms of spontaneous activity in motor neuron disorders, particularly if chronic and slowly evolving (e.g., SMA), are complex repetitive discharges and myotonic discharges. These have no specific diagnostic value in motor neuron disorders, because they can be seen in other chronic neuromuscular conditions (e.g., radiculopathies, myopathies).[37]

In contracting muscle that has undergone partial denervation within a relatively recent period of time (e.g., less than 1 or 2 months), the remaining motor unit potentials fire at an abnormally rapid rate (neurogenic recruitment pattern) but have a relatively normal morphology (duration and amplitude) because significant motor unit remodelling has not yet occurred. On the other hand, chronically denervated muscle in which motor axon sprouting and motor unit enlargement has extensively occurred, reveals neurogenic recruitment of motor unit potentials with increased duration, amplitude, and polyphasia. A neurogenic recruitment pattern may be masked in cases with prominent UMN pathology, such as UMN-predominant ALS. The amplitudes of motor unit potentials are generally increased only in very chronic AHC disorders, such as remote poliomyelitis or SMA, when extensive motor unit remodelling has occurred. Muscle with ongoing reinnervation frequently demonstrates motor unit potentials with moment-to-moment configuration variation indicative of unstable neurotransmission across immature, freshly reinnervated neuromuscular junctions.

In disorders of motor neurons, these characteristic changes are present in a widespread distribution and are not confined to a single nerve root or peripheral nerve territory. However, early in the course of disease they may be more localized. Therefore, definite diagnosis by EMG may not be possible in the early stage of a widespread disorder of the AHCs. It is the challenge of the electrodiagnostician to weigh the NCS and NEE findings with the clinical features, reach a diagnosis that is true to the data collected but also reflects the general neuromuscular context in which less than widespread findings may be seen, and at the same time assess for mimickers of motor neuron disease that may be treatable. As discussed next in the section on ALS, certain electrodiagnostic criteria have been proposed,[2, 10] which should be fulfilled before a definite diagnosis of ALS can be made by NCS and NEE.

Table 19-5 lists the muscles routinely studied in our EMG laboratory with NEE in at least three regions (lumbosacral, thoracic, cervical, and cranial) of the patient with a suspected motor neuron disorder. Although the total number of muscles studied varies from laboratory to laboratory, and from patient to patient, we have found this selection of muscles to be useful because of the variety of nerve root levels and peripheral nerve territories that must be assessed. Other muscles may be added as directed by specific symptoms or previous findings. As indicated for NCS, when possible it is important to study clinically involved muscles and thereby enhance the diagnostic yield.

SPECIFIC MOTOR NEURON DISORDERS

Amyotrophic Lateral Sclerosis

Clinical Features

ALS, also known as Lou Gehrig's disease (United States) or motor neuron disease (United Kingdom), is a progressive neurodegenerative disorder of primarily motor neurons of the brain and spinal cord, resulting in widespread weakness, atrophy, and spasticity of skeletal muscles. Its onset peaks in

TABLE 19-5	Needle Electrode Examination of Muscles in Four Regions Routinely Studied in Patients with Suspected Motor Neuron Disorders

Muscle	Nerve	Innervation (Cord Segments)[a]
Lumbosacral		
Extensor digitorum brevis	Peroneal	**L5**, S1
Abductor hallucis	Tibial	**S1**, S2
Tibialis anterior	Peroneal	L4, **L5**
Flexor digitorum longus	Tibial	**L5**, S1
Medial gastrocnemius	Tibial	**S1**, S2
Vastus lateralis	Femoral	**L2, L3, L4**
Gluteus medius or maximus	Superior or inferior gluteal	**L5, S1**
Paraspinal	Anterior root	L2–S1 (usually)
Thoracic		
Paraspinal	Anterior root	T4–11 (usually)
Cervical		
Abductor pollicis brevis	Median	**T1**
First dorsal interosseous	Ulnar	**C8**, T1
Extensor indicis proprius	Radial	C7, **C8**
Flexor pollicis longus	Median	**C8, T1**
Pronator teres	Median	**C6, C7**
Biceps	Musculocutaneous	**C5, C6**
Triceps	Radial	**C6, C7**
Deltoid	Axillary	**C5, C6**
Paraspinal	Anterior root	C5–8 (usually)
Cranial		
Orbicularis oris	Facial	Cranial nerve VII
Masseter	Trigeminal	Cranial nerve V
Mentalis	Facial	Cranial nerve VII
Tongue	Hypoglossal	Cranial nerve XII

[a]Spinal cord segments in boldface indicate usual major innervation for the muscles specified.

the seventh decade of life although it can present as early as the third decade and as late as the ninth decade. Men are about twice as likely as women to develop ALS. The diagnosis of ALS requires the presence of both UMN and LMN dysfunction (see Table 19-2) at multiple levels of the neuraxis. The onset of ALS is usually focal or segmental, although symptoms can seemingly begin in a widespread or generalized fashion. Nonetheless, symptoms and signs eventually become generalized as multiple muscles of the extremities, trunk, head, and neck become involved to varying degrees. Dysarthria and dysphagia in patients with ALS result from bulbar muscle dysfunction and are frequent causes of morbidity. Death usually results from respiratory failure or intercurrent illness in the debilitated patient. Despite active research and recent promising breakthroughs, no cure or effective treatment is available yet for this devastating disease.

The most significant recent discovery has been the identification of mutations in the Cu/Zn-superoxide dismutase (SOD1) gene in 10% to 20% of familial ALS patients.[30] Although only 5% to 10% of all patients with ALS have a familial form, those with an SOD1 mutation have a nearly identical clinical and pathological phenotype to those with nonfamilial (sporadic) ALS. Therefore, by understanding how SOD1 mutations cause ALS, we may also understand the mechanisms causing sporadic ALS.

Because there is no definitive test to confirm the diagnosis of ALS, either sporadic or familial, a combination of clinical and laboratory evaluations is required to arrive at the correct diagnosis. The EMG plays a major role in assisting the clinician to achieve this goal, by identifying look-alike conditions that may be treatable, characterizing the extent and activity of motor neuron degeneration present at a particular time, assisting in prognostication, and potentially monitoring drug efficacy.

Electromyographic Findings

The routine electrodiagnostic procedures of NCS and NEE identify LMN degeneration in ALS. Evidence of UMN pathology can also be obtained, although only indirectly. ALS is the most common motor neuron disorder studied in the majority of EMG laboratories. In addition to the general features of the EMG examination presented earlier (see "Common Electrodiagnostic Features"), other findings particular to ALS are presented here. The presence or absence of various abnormalities on NCS and NEE is usually dependent on the stage of disease, as indicated in Table 19–6.

NCS reveal preserved SNAP amplitudes in the vast majority of instances, unless there is a coexisting neuropathy (e.g., carpal tunnel syndrome, diabetic polyneuropathy). However, infrequently individuals with ALS have been described with reduced SNAP responses in the absence of an underlying polyneuropathy; such sensory abnormalities appear to parallel the motor decline.[13] Of note, healthy individuals over the age of 60 years may normally have unelicitable sensory responses in their feet and absent H reflexes bilaterally. This factor is important when studying the patient with ALS, who is often older than 60 years. The H reflex is usually elicitable in ALS, in part because the sensory component remains intact, but also because the associated UMN disorder tends to increase the H reflex amplitude.[1] In fact, finding a relatively high H response amplitude (in our laboratory, above 2 or 3 μV), in the presence of a reduced soleus CMAP response, should raise the suspicion of UMN pathology. CMAP amplitudes can be normal, usually in early disease, or severely diminished and unelicitable, in very weak and usually atrophic muscles. Motor nerve conduction velocities and distal latencies are unaffected except when the fastest conducting fibers are lost, which usually happens only when the CMAP amplitudes are reduced more than 75%.[19, 20] When

possible, motor responses should always be recorded from clinically weak muscles. This is most easily carried out with the more distal muscles. In ALS, progressive motor axon loss results in CMAPs of reduced amplitude. However, conduction block along motor nerve trunks has been identified in patients with suspected ALS by finding intact CMAPs recorded from weak muscles. In one study, distal latency prolongation, multifocal motor conduction slowing, and conduction blocks occurred in up to 20% of patients (6 of 31) with clinically classic ALS who had no other features of multifocal motor neuropathy with conduction block.[39]

Patients with advanced disease and low amplitude CMAPs have a poor prognosis.[26] The pattern of generalized reduction of CMAP amplitudes and normal SNAPs has been termed the "generalized low motor-normal sensory" (GLMNS) pattern.[35] Although uncommon, this pattern is noteworthy because it is almost invariably associated with a poor outcome. Most patients with this pattern have progressive ALS or some other widespread intraspinal canal lesion, but some have a potentially treatable disorder such as Guillain-Barré syndrome, Lambert-Eaton myasthenic syndrome, or distal myopathy.

Another pattern seen exclusively in some patients with ALS is the so-called split-hand, characterized by decreased CMAP amplitudes in the lateral half of the hand (median motor to the abductor pollicis brevis and ulnar motor to the first dorsal interosseous), compared with less involvement in the medial half (ulnar motor to abductor digiti minimi)[38] (Fig. 19–1). This phenomenon is not seen in other forms of motor neuron disorders. The cause is not clear, but may be related, in part, to the contiguous nature of spread of motor neuron damage in ALS, and to the somatotopic orientation and selective vulnerability of motor neurons destined to the abductor pollicis brevis and first dorsal interosseous.

NEE is essential to secure the diagnosis of ALS. During the examination of muscle at rest, ongoing motor axon loss is manifested by fibrillation potentials. Fasciculation potentials, another form of spontaneous activity, are so frequently seen in ALS that their absence should cause the electromyographer to seriously consider other etiologies.[19] The etiology of fasciculation potentials in ALS is debatable; they may arise from the AHC, nerve trunk, or distal nerve terminal (for review, see Mitsumoto and associates[23]). During the examination of voluntary motor unit activation, a picture of chronic and active motor axon loss is identified. This is characterized by neurogenic motor unit potential changes, reduced recruitment of motor unit firing, and moment-to-moment motor unit configuration variability. The classic recruitment pattern is sometimes obscured by a superimposed central disorder of motor unit control caused by UMN involvement.

Diagnostic Criteria

Arriving at a diagnosis of definite ALS is not difficult in the patient with generalized or advanced findings. However, patients with limited findings in the early stages of disease or with only focal abnormalities often undergo extensive investigations before the correct diagnosis can be reached. This is because there is no definitive test for accurate diagnosis except in the 10% to 20% of familial ALS individuals with an SOD1 mutation. Table 19–7 lists three common electrodiagnostic patterns encountered in the EMG laboratory when evaluating patients with suspected ALS, only the first being consistent with a diagnosis of widespread AHC degeneration. Even with the help of EMG, therefore, diagnosis of early or restricted disease is difficult. Because the electrodiagnostic presentation of ALS is of an intraspinal canal lesion, it may be electrically indistinguishable from other motor neuron disorders (e.g., remote poliomyelitis and PPS), motor root disorders (e.g., polyradiculopathies),

TABLE 19–6	Electrodiagnostic Changes at Different Stages of Amyotrophic Lateral Sclerosis		

	Stage		
Study	Early	Mid	Advanced
Motor NCS			
CMAP amplitude	N – ↓	↓ – ↓↓	↓↓ – ↓↓↓
Conduction velocity	N	N	N – ↓
Distal latency	N	N – ↑	N – ↑
Sensory NCS			
SNAP amplitude	N	N	N – ↓
Conduction velocity	N	N	N – ↓
H reflex	N – ↑	N – ↑	N – ↓↓
NEE			
Fasciculations	N – +	N – + +	N – +
Insertional positive sharp waves	N – + +	+ – + + +	N – +
Fibrillations	N – +	+ – + + +	N – + + +
Motor unit potentials			
Recruitment	N – ↓	↓ – ↓↓	↓↓ – ↓↓↓
Duration	N – ↑	↑ – ↑ ↑	↓ – ↑ ↑ ↑
Amplitude	N	↑ – ↑ ↑	↓ – ↑ ↑ ↑
Complexity	N	+ – + +	+ – + +

N, normal/none; ↓ / ↑, decreased/increased: one arrow = mild, two arrows = moderate, three arrows = marked; +, mild; + +, moderate; + + +, marked; CMAP, compound motor action potential; NEE, needle electrode examination; NCS, nerve conduction study; SNAP, sensory nerve action potential.

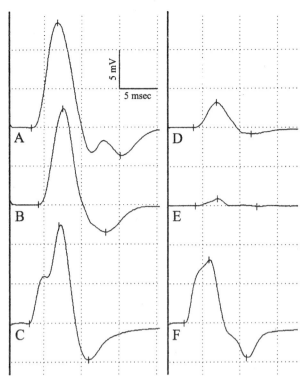

FIGURE 19–1. Compound muscle action potential (CMAP) amplitudes recorded from intrinsic hand muscles (stimulating median or ulnar nerve at the wrist) including the abductor pollicis brevis *(A, D)*, first dorsal interosseus *(B, E)*, and abductor digiti minimi *(C, F)* of a healthy individual *(A–C)* and of a paient with amyotrophic lateral sclerosis (ALS) *(D–F)*. Compared to the healthy age-matched individual, the CMAPs of the patient are generally lower. Of note in the patient, the amplitudes from the abductor pollicis brevis *(D)* and first dorsal interosseus *(E)* muscles are disproportionately lower than the amplitude from the abductor digiti minimi muscle *(F)*. This difference reflects the split-hand pattern of greater denervation of muscles in the lateral half of the hand that is occasionally seen in ALS.

TABLE 19–8	Lambert Criteria for the Electromyographic Diagnosis of Amyotrophic Lateral Sclerosis

1. Fibrillation and fasciculation potentials in muscles of the lower and the upper extremities, or in the extremities and head
2. Reduced number and increased amplitude and duration of motor unit potentials
3. Normal excitability and conduction velocity of sensory nerve fibers even in severely affected extremities
4. Normal excitability of remaining motor fibers, and motor fiber conduction velocity within the normal range in nerves of relatively unaffected muscles and not less than 70% of the average normal value according to age in nerves of more severely affected muscles

From references 2, 19.

In order to secure an earlier diagnosis of ALS, particularly for timely enrollment in research-treatment protocols, the World Federation of Neurology (WFN) proposed electrodiagnostic guidelines at a meeting in El Escorial, Spain, in 1990.[10] Since their original introduction in 1994, the El Escorial electrodiagnostic criteria for ALS have come under considerable criticism, in part because of lack of clarity and precision.[36] Therefore, revised criteria have been proposed by the WFN (El Escorial Revisited: Revised Criteria for the Diagnosis of Amyotrophic Lateral Sclerosis, Airlie House, Warrenton, Virginia, April 2–4, 1998) in an effort to provide the electromyographer with a useful framework for more accurate and timely electrodiagnosis of ALS. Briefly, these criteria rely on conventional EMG studies with NCS and NEE to (1) confirm LMN dysfunction in clinically affected regions, (2) detect electrophysiological evidence of LMN dysfunction in clinically uninvolved regions, and (3) exclude other pathophysiological processes.[41] Tables 19–9 and 19–10 list the features required for the diagnosis and exclusion of ALS, respectively.

Variants of Amyotrophic Lateral Sclerosis

The symptoms and signs of ALS usually begin focally with subsequent generalization but can also remain localized to a single limb segment (focal motor neuron disease). In a similar way, the presentation of ALS may be limited to dysfunction of only corticospinal and corticobulbar tracts (UMN-predominant ALS), bulbar muscles (bulbar-onset ALS), or spinal cord AHCs (LMN-predominant ALS). Although the majority of these cases eventually evolves into typical generalized ALS, with varying degrees of combined UMN and LMN involvement, a minority (usually < 5%) remain restricted as they began. These are presented next, emphasizing the

or even some neuromuscular junction transmission disorders (e.g., Lambert-Eaton syndrome) and myopathies (e.g., inclusion body myositis). Therefore, efforts have been made to establish electrodiagnostic criteria to assist in this diagnostic process.

The earliest such criteria for diagnosis had emerged from the extensive experience of Lambert[19] and were widely used until recently (Table 19–8). However, the demanding nature of the Lambert criteria excluded from diagnosis all but the most clinically advanced or rapidly progressive ALS cases.[2]

TABLE 19–7	Three Common Patterns of Electrodiagnostic Findings in Amyotrophic Lateral Sclerosis: Muscles Demonstrating Motor Axon Loss[a] and Fasciculations			
Bulbar	Upper Extremity	Mid-Thoracic PSP	Lower Extremity	Electrodiagnostic Pattern Consistent with MND
If symptoms present	Most[b]	Yes	Most[b]	Yes
No	No	Variable	Multiple bilaterally	No[c]
No	C8–T1 ± C7 levels bilaterally	Variable	No	No[c]

MND, motor neuron disease; PSP, paraspinal.
[a]Fibrillation potentials, chronic neurogenic motor unit potentials.
[b]Bilaterally in either upper or lower extremity.
[c]Even if paraspinal fibrillation potentials are present.
Adapted from Wilbourn AJ: Electrodiagnostic evaluation of the suspected ALS patient. In Belsh JM, Schiffman PL (eds): Amyotrophic Lateral Sclerosis: Diagnosis and Management for the Clinician. Armonk, NY, Futura, 1996.

TABLE 19–9	Revised El Escorial World Federation of Neurology Criteria[a] for Electrodiagnosis of Amyotrophic Lateral Sclerosis

Features on Nerve Conduction Study
1. Nerve conduction studies should generally be normal or near normal
2. Sensory nerve conduction studies can be abnormal in the presence of entrapment syndromes and coexisting peripheral nerve disease
3. Lower extremity sensory nerve responses can be difficult to elicit in the elderly

Features on Needle Electrode Examination
1. Combination of active denervation findings and chronic denervation findings is required but the relative proportion may vary from muscle to muscle
2. Signs of active denervation are fibrillation potentials and positive sharp waves
3. Signs of chronic denervation consist of large MUPs, reduced interference pattern with firing rates higher than 10 Hz (unless there is a significant UMN component), and unstable MUPs
4. Presence of fasciculation potentials, especially when polyphasic and of long duration, supports the diagnosis; their absence raises diagnostic doubts but does not preclude the diagnosis of ALS

Topography of Active and Chronic Denervation and Reinnervation
1. In brain stem region: changes in at least one muscle (e.g., tongue, facial, or jaw muscles)
2. In thoracic spinal cord region: changes in either paraspinal muscles at or below the T6 level or in abdominal muscles
3. In cervical and lumbosacral spinal cord: changes in at least two muscles innervated by different roots and peripheral nerves

MUPs, motor unit potentials; large MUPs, long duration and large amplitude; UMN, upper motor neuron.
From World Federation of Neurology Research Group on Motor Neuron Diseases: El Escorial revisited: Revised criteria for the diagnosis of amyotrophic lateral sclerosis. www.wfnals.org/Articles/elescorial1998.htm.

usefulness of the electrodiagnostic evaluation in their recognition.

Monomelic Amyotrophy

At least 20 synonyms have been attached to this form of LMN disease, first described by Hirayama in 1959,[16] including monomelic amyotrophy, focal motor neuron disease, and benign focal amyotrophy. It is also termed juvenile segmental muscular atrophy because of its frequent onset between 20 and 35 years of age.[8, 31] It is more than twice as frequent in males as in females and preferentially affects hand, forearm, and arm muscles (75%). Most cases remain restricted to

TABLE 19–10	Revised El Escorial World Federation of Neurology Criteria for ALS Electrodiagnosis: Exclusion Criteria

1. Evidence of motor conduction block
2. Motor conduction velocities < 70%, and distal latencies > 30%, of the lower and upper limit of normal values, respectively
3. Abnormal sensory nerve conduction studies, except because of coexisting entrapment syndromes, peripheral neuropathies, or advanced age affecting lower extremity responses
4. F wave or H wave latencies > 30% above established normal values
5. Decrements > 20% on repetitive stimulation
6. Somatosensory evoked response latency > 20% above established normal values
7. Full interference pattern in a clinically weak muscle
8. Significant abnormalities in autonomic function or electronystagmography

From World Federation of Neurology Research Group on Motor Neuron Diseases: El Escorial revisited: Revised criteria for the diagnosis of amyotrophic lateral sclerosis. www.wfnals.org/Articles/elescorial1998.htm.

one (50%) or both limbs (~45%) but a minority become generalized. After an initial rapid progression for 2 or 3 years after onset, it has a relatively stable course. Magnetic resonance imaging of the spinal cord can be normal but often reveals atrophy, corresponding to the clinically affected level.

In addition to neuroimaging, electrodiagnostic studies are especially useful in identifying conditions that resemble monomelic amyotrophy, including cervical radiculopathy (degenerative disc disease, tumor), cervical syringomyelia, multifocal motor neuropathy with conduction block, and ALS. The EMG findings are very similar to those seen in other motor neuron disorders except that the C5 to T1 myotomes are preferentially involved in most cases. Although sensory NCS are usually normal, reductions in amplitudes can be seen on side-to-side comparison in up to one-third of patients. Decreased CMAP amplitudes, when present, reflect the degree of axon loss and muscle atrophy. Evidence of conduction block is not found. NEE reveals prominent chronic neurogenic changes although active denervation (fibrillation potentials) is seen in varying amounts depending on the stage of disease. It is important to study the contralateral limb, even when asymptomatic, because a minority of cases with unilateral findings clinically have bilateral involvement by EMG.[8]

Primary Lateral Sclerosis

Rarely, patients presenting with a UMN-onset motor neuron disorder never develop any LMN features and are considered to have primarily lateral sclerosis (PLS). Symptoms usually begin insidiously with gait difficulties caused by leg stiffness and spasticity, but hand clumsiness or bulbar symptoms may occasionally be the earliest complaints. Clinical deterioration is usually slow and gradual, with most patients surviving for several years or even decades. With a family history of similarly affected individuals, other diagnoses such as hereditary spastic paraplegia should be considered.

The diagnosis of PLS is one of exclusion, being made only after extensive investigations have been performed. One of the most important tests is EMG, especially NEE, to exclude the presence of LMN changes. The NEE most commonly shows a nonspecific pattern of slow motor unit activation characteristic of a central nervous system disorder of motor unit control. Occasionally, some patients with PLS have sparse fibrillation potentials in isolated distal lower extremity muscles, such as the intrinsics of the feet. Fasciculation potentials are not seen. Other causes of UMN pathology should be searched for with the appropriate blood, cerebrospinal fluid (CSF), and neuroimaging studies. These include spinal cord compression, especially at the cervical and foramen magnum levels (e.g., tumors, Chiari malformation, spinal stenosis), demyelinating disease (e.g., multiple sclerosis), and retroviral infections (e.g., HIV-1, HTLV-I).

Progressive Bulbar Palsy

Patients with motor neuron dysfunction beginning in the muscles of speech and swallowing have pathology in either the corticobulbar tract (UMN) or the brain stem nuclei (LMN), or both. The vast majority of these patients, who are more often female than male (1.5:1), eventually develops more generalized findings of ALS. Up to 40% of patients who eventually develop ALS, have a bulbar onset to their symptoms. It is, therefore, best to think of progressive bulbar palsy (PBP) as an early symptom of ALS, although rare cases of true PBP exist. If there has been no involvement of more widespread regions after a substantially long period of time, and investigations have excluded other etiologies, PBP is likely.

Electrodiagnosis plays a pivotal role in the diagnosis of

PBP by determining whether LMN pathology also exists in nonbulbar muscles. The NEE is particularly useful because it can detect fibrillation potentials and neurogenic motor unit potentials in extremity or paraspinal muscles, even when clinical evidence of widespread denervation is still lacking. Of patients presenting to our institution with bulbar dysfunction who on follow-up developed ALS, only 20% had neurogenic changes outside of their bulbar muscles on initial NEE examination.[15] However, most of the 80% of patients whose findings were initially restricted to bulbar muscles subsequently developed LMN signs in a generalized distribution consistent with a widespread disorder of the AHCs. In practice, therefore, the passage of time is required to ultimately distinguish between bulbar-onset ALS and PBP.

Progressive Muscular Atrophy

Progressive muscular atrophy (PMA) is the opposite of PLS in that it is a pure LMN syndrome, with no evidence of UMN pathology. Similar to PLS, however, it has a slow clinical progression with patients surviving for many years. All but about 8% of patients who first develop LMN symptoms and signs go on to develop ALS. On the other hand, only about 20% of patients who eventually develop ALS present with a pure LMN syndrome. If the family history is positive, one should consider the possibility of a hereditary adult-onset spinal muscular atrophy (SMA type IV). Because some look-alike conditions are potentially treatable, specific blood tests and electrodiagnosis are very useful to make the correct diagnosis. These include multifocal motor neuropathy with conduction block (MMNCB), Lambert-Eaton myasthenic syndrome, hexosaminidase A deficiency, and paraneoplastic neuronopathies. Therefore, like ALS and its other variants, PMA is a diagnosis of exclusion.

EMG confirms the widespread and predominantly chronic nature of the AHC degeneration in PMA, with neurogenic motor unit potential changes and reduced recruitment of motor unit firing. Complex repetitive discharges are occasionally encountered, consistent with the chronicity of the disease. Fibrillation potentials are also prominent, indicating active motor axon loss. In the presence of conduction block or other demyelinating features on NCS, disorders such as MMNCB, chronic inflammatory demyelinating polyradiculoneuropathy (CIDP), and paraneoplastic polyneuropathy should be considered.

X-Linked Recessive Bulbospinal Neuronopathy (Kennedy's Disease)

Clinical Features

A recessive form of LMN degeneration occurs in males who possess a trinucleotide repeat (CAG) expansion in the androgen receptor gene on the X chromosome.[5] This produces a rare and unique syndrome that resembles LMN-predominant ALS or PMA, but also includes endocrine abnormalities and loss of SNAPs on NCS without significant sensory symptoms or signs. Initially termed "spinal and bulbar muscular atrophy" by Kennedy in 1968,[18] involvement of extramotor structures prompted the name X-linked recessive bulbospinal neuronopathy. Initial symptoms, which usually begin by the fourth or fifth decade, are variable but usually include prominent muscle cramps, fasciculations, and slowly progressive limb-girdle weakness. Other early findings often include hand tremor and gynecomastia, and less commonly diabetes mellitus, infertility, and impotence. Dysphagia and dysarthria, which usually develop later, can be absent in approximately 50% of patients. However, neurological examination often reveals facial weakness with characteristic peri-

oral fasciculations. In contrast to ALS, there is no clinical or electrophysiological evidence of UMN dysfunction.[34]

Included in the differential diagnosis of Kennedy's disease are LMN-predominant ALS, PMA, adult-onset SMA (type IV), hereditary sensorimotor neuropathy (HMSN type II) limb-girdle dystrophy, and facioscapulohumeral dystrophy.

Electromyographic Findings

Although a thorough history and physical examination should alert the experienced neurologist to the diagnosis of X-linked recessive bulbospinal neuronopathy, electrodiagnostic testing is very helpful in confirming the clinical suspicion and in excluding other look-alike conditions. Almost invariably, one finds diffusely low amplitude or absent SNAPs on NCS (including the H reflex), and chronic neurogenic motor unit potential changes on NEE.[12] Motor NCS are usually normal, although one-third of patients have reduced CMAP amplitudes. On NEE, motor units show characteristic chronic neurogenic features. Fibrillation potentials and insertional positive sharp waves may be sparse in extremity muscles, whereas fasciculation and cramp potentials are common. Characteristically, perioral facial muscles contain grouped repetitive motor unit potential discharges, which are often accentuated by voluntary pursing of the lips. Although the electrodiagnostic findings are essentially pathognomonic for Kennedy's disease, the diagnosis should be verified by genetic testing.

CASE STUDY 1

A 44-year-old man began experiencing progressive distal upper extremity weakness and muscle atrophy 4 years earlier. In retrospect, he first experienced frequent thigh cramps at 10 years of age and was noted to have prominent breasts at 12 years of age. The cramping persisted and spread to involve widespread muscle regions, including the neck and jaw. Progressive proximal leg weakness began approximately 8 years ago, and frequent falls necessitated the recent use of a cane. In the past 4 years, his voice has become hoarse and he has developed numbness and tingling in the hands. Examination revealed active fasciculations in facial muscles, especially periorally, and in all limbs. There was a noticeable facial diplegia, especially periorally, and the tongue showed fasciculations and mild atrophy. Moderate atrophy was noted in proximal muscles and in intrinsic hand muscles. Strength was minimally decreased in the legs, except for moderate weakness of hip girdle muscles; arm muscles, especially distally, were severely weak. Stretch reflexes were absent throughout except for a brisk jaw jerk, and plantar responses were flexor. Gait was slow and waddling and pinprick sensation was diminished in the legs to mid-thighs, in a graded stocking-glove distribution. There was prominent gynecomastia. EMG revealed markedly diminished to absent sensory responses in the legs and hands. Motor nerve conductions were essentially normal. The NEE revealed moderate chronic neurogenic changes in all extremity muscles examined and active motor axon loss in most of them. Grouped repetitive discharges were persistent in the lower facial muscles, especially the orbicularis oris and mentalis. Genetic analysis for suspected Kennedy's disease revealed expansion of the trinucleotide (CAG) repeat region of the androgen receptor gene, consistent with this diagnosis. The proband has two similarly affected maternal male cousins.

Poliomyelitis and Postpolio Syndrome

Clinical Features

Worldwide epidemics of the poliovirus, a neurotropic RNA picornavirus (genus *Enterovirus*), were common until the mid-1950s when the injectable Salk vaccine was introduced. As a result, in the past 40 years there have been only rare cases of acute poliomyelitis in North America and other developed countries, although it is still a significant health problem in many developing regions. Present-day sporadic cases in the United States have been described after polio vaccine administration or infection by other related viruses. A small proportion of individuals in North America, mainly children, exposed to the virus before the mid-1950s developed a flu-like illness and subsequently experienced fever, headache, and neck stiffness, consistent with an aseptic meningitis. Approximately half of these patients developed a paralytic disease as the neurotropic poliovirus infected motor neurons of the brain stem and spinal cord. This resulted in focal and asymmetrical weakness of the limb, trunk, and bulbar muscles in various combinations, including muscles of deglutition and respiration. Associated acute dysautonomia sometimes caused life-threatening blood pressure instability and cardiac arrhythmias.

After acute poliomyelitis, patients are left with varying degrees of residual weakness and muscle atrophy, termed remote poliomyelitis. Such patients often present to neuromuscular specialists and electromyographers 20 or more years later with increasing weakness. The diagnostic challenge is to determine whether the symptoms can be attributed to progression of the remote poliomyelitis, as has been found in some patients, and termed "postpolio syndrome" (PPS),[17] or whether they are caused by some other process. This differentiation relies heavily, if not exclusively, on clinical judgment because no significant electrodiagnostic differences exist between the remote poliomyelitis patients with new weakness and those without. In part, this is because the severe electrophysiological residuals left by poliomyelitis significantly compromise the electromyographer's ability to detect further deterioration or to diagnose other neuromuscular disorders (e.g., ALS, myopathy, radiculopathy) in such patients. However, some authors have reported differences in single fiber EMG (SFEMG) characteristics between the two groups, as discussed next.

Three risk factors have been identified for the development of PPS: onset of poliomyelitis after 10 years of age, severe initial paralysis, and substantial motor recovery. Clinical features of the PPS usually include fatigue, myalgia, arthralgia, cold intolerance, depression, and sleep apnea syndrome. In addition, new weakness can occur in about 75% and new atrophy in about 50% of patients with PPS. Both changes usually occur in muscles that had been previously affected and improved significantly, but can also occur in muscles that had no previous clinical involvement.

Electromyographic Findings

Electrodiagnostic changes in patients with remote poliomyelitis and PPS are nearly identical to those previously described for other AHC disorders except that chronic neurogenic changes predominate and evidence of ongoing motor axon loss is minimal or nonexistent. On NCS, SNAPs are normal and CMAPs are diminished over markedly atrophic muscles. Low amplitude or absent CMAPs in remote poliomyelitis reflect severe denervation and few, if any, surviving motor units. However, CMAP amplitudes may not be as low as expected when recorded from weak muscles, owing to extensive collateral sprouting and motor unit remodelling, both of which require surviving motor units, absence of ongoing AHC degeneration, and an extended recovery pe-

riod. This explains why conditions such as ALS, where AHC loss is progressive and sometimes rapid, are associated with low amplitude CMAPs in weak muscles at mid and late stages of disease.

The NEE of muscles in the patient with remote poliomyelitis reveals two characteristic changes: (1) reduced recruitment patterns of motor units with marked neurogenic changes, referred to as giant (>10 mV) motor unit potentials, approaching 20 to 30 mV in amplitude, and (2) widespread distribution of such abnormalities not only in limbs affected during the acute attack but also in reputedly unaffected regions. Of note, no features exist on any single NEE examination that differentiate between remote poliomelitis that is clinically stable and that which is associated with progressive weakness (i.e., PPS). Giant motor unit potentials occur in both instances and can occasionally be unstable with variation in morphology. Modest numbers of small fibrillation potentials (usually less than 50 μV) can be found in isolated regions of atrophic muscle. Muscles that have undergone severe atrophy from total denervation are replaced with fibrofatty tissue, causing a combination of decreased insertional activity and increased mechanical resistance. Although tiny fibrillation potentials may sometimes be detected in such muscles, activated motor unit potentials are absent. SFEMG and macro EMG (a modification SFEMG that records from all muscle fibers innervated by the same motor unit) have been used to study the electrophysiology of PPS.[29, 32] Although SFEMG has reportedly demonstrated slightly increased innervation ratios and more abnormal jitter and blocking in PPS compared to asymptomatic remote poliomyelitis,[29] the differences are insufficient to be considered significant. Careful longitudinal macro EMG studies of leg muscles in 18 patients who underwent muscle biopsy and repeated strength testing over 4 years demonstrated that PPS causes ongoing denervation but only limited or insufficient reinnervation.[32]

Hereditary Spinal Muscular Atrophy

Clinical Features

Hereditary (SMA) can have either a childhood or adult onset, resulting in progressive LMN degeneration in the spinal cord and occasionally lower brain stem, but not in upper brain stem (e.g., extraocular or facial nuclei). Childhood SMA is one of the most common hereditary neuromuscular disorders. It can be classified into three types depending on age of onset and course of the disease, as follows: Werdnig-Hoffman disease (SMA I), intermediate (SMA II), and Kugelberg-Welander disease (SMA III). Over 98% of cases are autosomal recessive and occur at an average frequency of 1:13,000 births.[6] This classification is somewhat artificial because of the clinical overlap and common chromosomal region (5q11.2-q14) linked to all three types. Details of the exact genetic defect, which involves at least two genes, *SMN* (spinal motor neuron) and *NAIP* (neuronal apoptosis inhibitor protein), may explain the clinical variability.[7] Adult-onset SMA (type IV), which constitutes less than 10% of all SMA, has a prevalence of approximately 3:1,000,000. It has several forms but primarily affects motor neurons in the spinal cord. Although autosomal recessive in two-thirds of cases, it is likely not genetically related to childhood SMA.[28]

SMA I is identified between birth and 6 months, although it probably begins in utero. It is the most severe form, preventing any degree of normal motor development in the affected hypotonic infant. Death from respiratory compromise usually occurs before 2 years of age. SMA II resembles a milder form of SMA I and is usually noted between 6 and 18 months, as normal motor milestones are not being achieved (e.g., unassisted standing or walking). With aggressive medical attention to musculoskeletal deformities (scoliosis and contractures) and pulmonary care, survival into the fourth or fifth decade is possible. SMA III almost always begins after 18 months of age and usually between 5 and 15 years. Gait difficulty owing to pelvic girdle weakness is usually the first sign of disease in children who had previously attained normal motor milestones with independent ambulation. The disease may appear rather static for years and may not necessitate use of a wheelchair before 30 or 40 years of age.[6] In SMA IV, weakness begins in the fourth decade of life or later, usually in the proximal legs, quadriceps femoris muscles. Less often, other forms of adult-onset SMA affect muscles distally (distal SMA) or in a generalized fashion. Its course is relatively benign, rarely causing musculoskeletal deformities (scoliosis) and infrequently confirming the patients to a wheelchair 20 or 30 years after onset.[28]

Electromyographic Findings

NCS and NEE findings in all forms of SMA are similar to those previously described for other motor neuron disorders. Sensory NCS are usually normal. Some specific differences among the SMAs are related, at least in part, to the patient's age, muscle size, and duration of disease.

CMAPs can be normal or reduced in amplitude, depending on when the patient is studied in relation to disease onset and whether significant reinnervation of muscle fibers has occurred. For example, infants with SMA I are often studied when the disease is advanced and NCS reveal very low CMAP amplitudes. Individuals with SMA II, III, and especially IV may have relatively normal CMAPs because of collateral motor axon sprouting. CMAP amplitudes in these later SMAs also tend to be preserved because routinely recorded muscles are distal and most severely affected muscles are proximal. Motor conduction velocities are preserved although mild to moderate slowing may be seen when motor amplitudes have significantly diminished, likely owing to loss of the fastest conducting fibers.

The NEE reveals typical neurogenic abnormalities of insertional spontaneous activity (fasciculations, insertional positive sharp waves, fibrillation potentials), and features of chronic motor axon loss (neurogenic firing pattern and neurogenic motor unit changes). Fibrillation potentials may be prominent in patients with SMA III and IV, even though progression of the disease is slow. Of note, facial muscles are spared in all SMAs, but tongue muscles show neurogenic changes in SMA I and II. Tongue fasciculations are frequent in children with SMA I (up to 50%) and especially SMA II (70%). On the other hand, extremity muscles have numerous fasciculations in patients with SMA IV.

The increasing availability of testing for the *SMN* and *NAIP* gene abnormalities in the various forms of SMA (I to IV) will reduce the importance of routine electrodiagnostic studies and muscle biopsy in securing the correct diagnosis. Although presently these latter tests are considered essential for the workup of SMA in most centers, future genetic testing will likely make them redundant.

DIFFERENTIAL DIAGNOSIS

Secondary Motor Neuron Disorders

Findings on NCS and NEE in all the following motor neuron disorders are similar to those previously discussed, except for the specific exceptions discussed with each condition.

Hexosaminidase Deficiency

Hexosaminidase deficiency is an autosomal recessive GM2 gangliosidosis presenting in youth or adulthood with an SMA- or PMA-like clinical picture characterized by slowly progressive weakness, muscle atrophy, cramps, and fasciculations.[11] It is caused by a partial deficiency of β-hexosaminidase (Hex), a lysosomal enzyme, resulting from 1 of over 50 identified point mutations of either (1) the alpha subunit coding gene on chromosome 15, resulting in Hex-A deficiency, or (2) the beta subunit coding gene on chromosome 5, resulting in Hex-B deficiency. Although the neuromuscular deficit is prominent, other features that indicate a more widespread process include cerebellar dysfunction (with atrophy on neuroimaging), extrapyramidal signs, progressive encephalopathy, and psychosis. The occasional presence of UMN features may raise the suspicion of ALS, although the generally slow course, predilection for the Ashkenazi Jewish population, and the aforementioned clinical features, are more consistent with Hex-A or Hex-B deficiency. Definitive diagnosis is obtained by demonstrating reduced hexosaminidase A or B activity in serum, leukocytes, or cultured fibroblasts.

NCS reveal reduced SNAP amplitudes. Predominantly chronic neurogenic changes are seen on NEE, although active denervation and reinnervation are also found. Of note, complex repetitive discharges can be frequent and widespread. Although complex repetitive discharges represent nonspecific features of chronic motor unit loss, when combined with abnormal SNAPs, the question of hexosaminidase deficiency should be raised.

Motor Neuronopathy and Malignancy

The association of motor neuron disorders and malignancy has been debated in the literature. This is particularly true for cases of "typical" ALS, occurring in the setting of lung cancer or lymphoma, in which a causal relationship is difficult to prove (for review, see Levin[21]). However, a few patients with an LMN-predominant form of motor neuron disease have been found to have lymphoma, including Hodgkin's disease and other malignant lymphomas. Such a progressive subacute motor neuronopathy affects primarily the legs in an asymmetrical and patchy fashion. Sensory symptoms are minimal or absent and findings are usually limited to the LMN. Rarely, patients with lymphoma have combined LMN and UMN signs, making differentiation from ALS difficult unless the malignancy is readily diagnosed.[42] Findings of paraproteinemia, elevated CSF protein, and CSF oligoclonal bands suggest an immunological process and should prompt a bone survey and bone marrow examination for definitive diagnosis. Approximately half the patients with an LMN form of motor neuron disease and lymphoma have experienced either a relatively stable course or some improvement in motor symptoms.[42]

The electrodiagnostic findings are consistent with an intraspinal canal lesion causing chronic and active motor axon loss primarily in the lower extremities. SNAP and CMAP amplitudes are usually normal. The NEE findings of fasciculations, fibrillation potentials, and chronic neurogenic motor unit potential changes are often patchy.

Postirradiation Lower Motor Neuron Syndrome

Radiation of the spinal cord, particularly in the lumbar region, has been associated with a delayed LMN syndrome.[14, 33] Motor dysfunction usually develops after a latency period of 3 months to 23 years when more than 4000 rads have been administered. The local radiotherapy has usually been for malignant testicular tumors and occasionally lymphoma or other tumors. Most patients with this syndrome develop an asymmetrical weakness in the lower extremities with muscle atrophy, cramping, fasciculations, and areflexia. The clinical features usually stabilize after progressing slowly over 1 or 2 years but may worsen for several more years in rare patients (KH Levin, personal communication). Sensation and bladder and bowel functions are spared. The likely pathogenesis is radiation-induced vasculopathy caused by endothelial damage, resulting in demyelination and degeneration of nerve roots and AHCs. Cases with a very delayed onset suggest that additive effects of cellular aging contribute to the lethal effect of direct irradiation on motor neurons and other neural structures.[14]

Electrodiagnostic findings of spared sensory nerve responses and motor axon loss are consistent with a predominantly intraspinal canal process. Increased F wave latencies point to damage of the AHCs or proximal nerve root segments.[33] Active and chronic denervation and reinnervation are noted on NEE. Myokymia and other grouped repetitive discharges are characteristic findings in radiation-induced damage to nerve roots and adjacent plexus (see Chap. 12).

Toxins and Heavy Metals

Significant exposure to heavy metals such as inorganic lead and organic mercury has been rarely associated with a motor neuron disorder (for review see Mitsumoto and associates[25]). Although prolonged exposure to either can result in combined UMN and LMN dysfunction and make differentiation from ALS difficult, distinguishing features include a history of toxin exposure, slow clinical progression, gastrointestinal complaints (anorexia, abdominal pain, constipation) and a blue leadline on the gums. Changes on EMG are essentially indistinguishable from typical motor neuron disease.

Non-Motor Neuron Disorders

Benign Fasciculation Syndrome

The presence of fasciculations is common, probably occurring in up to 70% of healthy individuals. In the absence of clinical muscle weakness and motor axon loss on EMG, widespread fasciculations are inconsequential and benign. They may persist in one region of the body for several days or weeks before disappearing or shifting to another location. Individuals who recognize these symptoms often have some amount of medical knowledge, and worry that they are developing ALS. It is important to reassure them that fasciculations are generally not a precursor to motor neuron disease in the presence of a normal neurological examination and normal EMG.[4] Absence of any progression over time is particularly reassuring, because a small percentage of patients have fasciculations as the initial and sole manifestation of ALS.[9]

Cervical Spondyloradiculomyelopathy

Combined cervical motor nerve root and spinal cord impingement, especially when present bilaterally and at several levels, can be difficult to differentiate clinically and electrodiagnostically from ALS. This is particularly true if sensory nerve root involvement is minimal or absent and only motor symptoms occur. Consequently, LMN changes occur at the level(s) of the nerve root and spinal cord compression while UMN findings occur caudally. Because such lesions can mimic ALS, the WFN Revised Criteria for the Diagnosis of ALS limit the level of certainty of diagnosis to only possible ALS when UMN signs are *below* LMN signs.[41] Electrodiagnostic features in this disorder can resemble early changes in an evolving AHC disorder: normal sensory NCS, low amplitude

CMAPs, and an NEE picture of active and chronic neurogenic changes in the cervical myotomes, often with fasciculation potentials. In such cases, careful neuroimaging of the cervical spine, preferably by magnetic resonance imaging, is essential to reveal the compressive lesion.

Multifocal Motor Neuropathy

Multifocal motor neuropathy (MMN) with or without conduction block occurs predominantly in males in the fifth decade and can superficially resemble LMN-predominant motor neuron disease.[27] This rare motor neuropathy causes a slowly progressive asymmetrical weakness, predominantly in distal upper extremities with variable degrees of muscle atrophy and hyporeflexia. Minimal atrophy in the setting of marked weakness, especially if it conforms to a single nerve trunk territory, is an important clue to the diagnosis. Fasciculations can occur. Stretch reflexes are occasionally brisk, although definite UMN signs are not found. Sensory symptoms may be absent intially but eventually occur in up to 40% of patients. Very high titer of anti-monosialoganglioside (GM1) antibodies is seen in the sera of about 25% of patients and strongly supports the diagnosis but the intermediate levels more commonly detected (~45%) can also be found in some patients with ALS.

Electrodiagnosis plays a major role in the diagnosis of MMN. NCS are particularly useful in differentiating it from motor neuron disease, especially when motor conduction block is present. Sensory responses are characteristically normal despite the presence of sensory symptoms. Recording from a weak muscle, a CMAP amplitude reduction of at least 40% between two sites across a long nerve trunk segment not prone to compression is consistent with conduction block. Stimulation along nerves studied in the upper and lower extremity, as listed in Table 19–4, should include proximal sites in the axilla and supraclavicular region. If nerve conductions are normal to a weak muscle, the conduction block may be proximal to the most proximal stimulation site. The NEE can indirectly support the presence of conduction block by having a reduced recruitment pattern of motor unit firing in the absence of significant chronic or active motor axon loss features. Fasciculation potentials are often seen.

Less common electrodiagnostic presentations of MMN include diffuse low amplitude CMAPs, or mild demyelinating features such as conduction velocity slowing. When marked axon loss has occurred, fibrillation potentials and denervation-reinnervation motor unit changes appear.

CASE STUDY 2

A 43-year-old right-handed male presents with a 10-year history of left foot drop and weakness of the fifth digit on the left hand. An EMG done elsewhere revealed "nerve damage" and details are not available. These initial symptoms have remained relatively stable for the past 7 years. Three years ago, he developed weakness of left elbow flexors and right hand grip. He also noticed atrophy of intrinsic hand muscles on the left and thenar eminence on the right. For several years, he has experienced cramping with exertion in the left hand. More recently, he has noticed fasciculations in the left forearm and both legs. He denies right leg weakness, dysphagia, dysarthria, sensory symptoms, or sphincter dysfunction. Review of systems is otherwise negative and there is no significant past medical and medication history. He is a nonsmoker and has remained very active hunting and fishing. He works in a sporting goods store where, for the past 18 years, he has used his teeth to bite closed sinkers made of lead. There is no family history of neuromuscular problems.

The neurological examination showed atrophy that was severe in the left anterior foreleg muscles, moderate in the left hand intrinsics and right thenar eminence. Scattered fasciculations were noted in the left deltoid, triceps, first dorsal interosseous, and vastus lateralis muscles. Upper extremity weakness on the left was very severe in intrinsic hand muscles, moderately severe in elbow flexors, and mild to moderate in other muscles; on the right, upper extremity weakness was only very mild. Lower extremity weakness on the left was very severe in ankle dorsiflexors, severe in toe extensors and flexors, and only mild elsewhere; on the right, the equivalent distal muscles were moderately weak and normal elsewhere. Shoulder girdle muscles had normal strength bilaterally. Coordination and cerebellar testing were normal. Stretch reflexes were present in the upper extremities, except at the left triceps; they were slightly hyperactive in the legs with a crossed adductor to the right, although the left ankle jerk was absent. The plantar responses were mute and no other pathological reflexes were seen. A prominent left foot drop was noted on gait testing. Sensation to all modalities was normal with no Romberg sign.

Electrodiagnostic testing revealed normal sensory responses but significant motor axon loss and multifocal conduction blocks, especially along the right peroneal, left median, and bilateral ulnar nerve trunks. M wave configurations were temporally dispersed along these nerves at proximal sites of stimulation. The NEE showed features of active and chronic motor axon loss in some muscles, and supported conduction block in others.

Myopathy

Some myopathies may be difficult to differentiate from LMN-predominant ALS, at least in the early stages. Inclusion body myositis (IBM) generally begins insidiously in the seventh decade, causing distal (as well as proximal) weakness with muscle atrophy. Initial symptoms may be rather focal, especially in thumb and finger flexors, resembling a distal and segmental onset of motor neuron disease. Further resemblance with a motor neuron disorder is seen when neck and bulbar muscle weakness occur, which can be an early feature in IBM. Polymyositis can present in youth and early adult life with generalized muscle weakness, especially in proximal lower extremities. Muscle atrophy is not prominent in poliomyositis, at least initially, and bulbar involvement is rare. Fasciculations and cramps generally do not occur in either type of myopathy and UMN signs are not found.

Motor NCS are frequently normal in myopathies, particularly in the upper extremities. However, CMAP amplitudes may be significantly reduced in advanced cases or with more distal involvement, as is often seen in IBM. On NEE, IBM characteristically shows fibrillation potentials (indicative of myonecrosis) and motor unit potentials of short duration, low amplitude, and prominent polyphasia. In some patients with IBM, there is a striking subpopulation of neurogenic motor unit potential changes, sometimes obscuring the identification of myopathic motor unit potential changes, and further confusing the distinction from motor neuron disease. Muscle biopsy is the definitive test to distinguish IBM from motor neuron disease. Motor unit morphology in severe myopathies may occasionally be difficult to distinguish from early reinnervation in advanced neurogenic conditions.

Neuromuscular Transmission Disorders
Myasthenia Gravis
Patients with myasthenia gravis, particularly older males with generalized or bulbar symptoms, can be mistakenly thought to have LMN-predominant ALS. In the absence of ocular involvement, the dysarthria and dysphagia of bulbar myasthenia gravis can closely resemble bulbar-onset ALS. In myasthenia gravis, antibody-mediated damage to the postsynaptic acetylcholine receptors results in increasing weakness with muscle use (fatiguability) and improvement in muscle strength after rest. These features are also occasionally seen in early ALS, probably because neurotransmission is inconsistent across the newly formed neuromuscular junctions in collateral sprouts of the remaining motor units. Some patients with early ALS may even have a falsely positive Tensilon test, further delaying the correct diagnosis. Electrodiagnostic studies can reveal similar abnormalities in neuromuscular transmission in myasthenia gravis and up to 50% of patients with ALS, particularly those with rapidly progressing disease.[3] This is true for both slow repetitive stimulation of peripheral nerves at 2 to 3 Hz, which demonstrates a decremental CMAP response, and SFEMG, which demonstrates unstable motor unit potentials with increased jitter and blocking. NEE in myasthenia gravis reveals motor unit potentials with normal or myopathic configurations and moment-to-moment configuration variability. Fibrillation potentials are uncommon.

Lambert-Eaton Myasthenic Syndrome
A rare cause of generalized weakness, mostly of limb-girdle muscles, is the Lambert-Eaton myasthenic syndrome (LEMS). Although it may be confused with ALS, the prominent proximal muscle weakness and hyporeflexia or areflexia are more suggestive of a myopathy. In contrast to myasthenia gravis, LEMS is a presynaptic abnormality in which antibodies interfere with voltage-gated calcium channels and prevent acetylcholine release. NCS demonstrate normal SNAPs but globally diminished CMAP amplitudes, a pattern that can also be seen in advanced ALS. However, muscle bulk is usually preserved in LEMS (as it is in myasthenia gravis). The diagnosis of LEMS is confirmed by demonstrating marked facilitation of the CMAP, usually greater than 100% increment, after exercise or fast repetitive stimulation (50 Hz). A characteristic progressive increment in CMAP amplitude occurs during the rapid train of stimuli in cases of LEMS but not ALS.

Syringomyelia

Usually a developmental disorder with slowly progressive upper extremity sensorimotor symptoms, syringomyelia is often associated with craniovertebral anomalies, particularly Chiari I malformation. Syringomyelia is a cavity (or syrinx) within the central spinal cord, usually cervical, which gradually expands and compresses decussating spinothalamic tract fibers, ventral horn motor neurons, and eventually the corticospinal tracts. The prominent sensory loss helps differentiate syringomyelia from motor neuron disorders. Other distinguishing clinical features of syringomyelia include a relatively young age at onset (often in the third decade), long duration of disease (spanning decades), and frequent neck pain. The muscle weakness and atrophy, which are usually prominent in intrinsic hand muscles innervated by C8 and T1 spinal segments, can develop rapidly and be most prominent.[22]

Sensory NCS are normal because, as in most radiculopathies, the sensory nerve lesion is proximal to the dorsal root ganglion. CMAPs may be normal but frequently show diminished amplitudes when recording over intrinsic hand muscles. The NEE reveals a pattern of very chronic motor axon loss with minimal fibrillation potentials. These NEE changes are often most prominent in the C8 and T1 segments. Although such changes can be found bilaterally at these levels, the lack of widespread muscle involvement is a distinguishing feature from motor neuron disorders. Neuroimaging of the cervical spinal cord, preferably by magnetic resonance imaging, reveals the intramedullary syrinx.

AN APPROACH TO ELECTRODIAGNOSIS OF SUSPECTED MOTOR NEURON DISORDERS

A problem-oriented approach to the patient with suspected motor neuron weakness is useful in planning the electrodiagnostic examination and providing a list of possible causes, including non-motor neuron–related ones. This approach is presented in the context of focal/segmental and multifocal/generalized weakness, beginning with routine NCS and ending with NEE examination of appropriate muscles.

Focal or Segmental Weakness

Almost any motor neuron disorder can present with focal or segmental weakness in the early stages of disease, although more generalized involvement usually follows. This is particularly true of ALS, which often begins focally and spreads to involve many regions. Therefore, it is important to determine whether more widespread electrophysiological changes are present, even though obvious symptoms and signs may be lacking. The absence of more widespread abnormalities, especially on follow-up studies, would suggest a non-motor neuron disorder etiology and warrant a search for other causes (e.g., radiculopathy). Specific examples of pathologies presenting as focal or segmental weakness are listed in Table 19–11.

Nerve Conduction Studies

The approach in our laboratory to evaluating a focally affected muscle begins with screening NCS of the involved extremity (as indicated in Table 19–4). Additional NCS to evaluate a specific weak muscle can be added to this basic study. For example, electrodiagnostic investigation of a foot drop would be comprised of the routine lower extremity NCS, with additional studies of the peroneal motor response to tibialis anterior and the sensory response along the superficial peroneal nerve, followed by comparison studies on the opposite side. Stimulation of motor nerves proximally (if possible) is essential to exclude a proximal conduction block causing focal muscle weakness in cases where distal responses are normal. Examples of frequently affected muscles, presenting symptoms, and nerves stimulated for CMAP recording are presented in Table 19–12.

Needle Electrode Examination

The focally weak muscles and their contralateral counterparts are the most important muscles to be tested by NEE. However, as with the NCS, other presumably asymptomatic muscles are also examined in the affected limb to determine whether pathological changes are present, indicating a more widespread process. In our EMG laboratory, we routinely study specific limb muscles innervated by different nerves from various spinal levels (see Table 19–5) and add muscles in the distribution of clinical symptoms. With results of the NCS in hand, the electromyographer can formulate a general approach to the NEE, based on likely causes of the focal weakness, but as the NEE is carried out sometimes additional, unexpected findings are made.

The NEE is more sensitive than the NCS in detecting early changes of denervation in clinically "normal" muscles. The CMAP amplitude does not decrease until a substantial number of motor axons and the muscle fibers they innervate have degenerated,[40] usually sufficient to result in muscle

TABLE 19–11	Problem-Oriented Approach to Motor Neuron Disease and Its Differential Diagnosis

Distribution of Weakness	Symptom	Pathology
Focal/Segmental	Foot drop	*Motor neuron disease (e.g., ALS)*
		Lumbosacral radiculopathy
		Sacral plexopathy
		Peroneal mononeuropathy
		Focal myopathy (e.g., IBM)
	Grip weakness (± atrophy)	*Motor neuron disease (e.g., ALS)*
		Syringomyelia
		Cervical radiculopathy
		Brachial plexopathy
		Median mononeuropathy (e.g., carpal tunnel syndrome)
		Ulnar mononeuropathy
		Myopathy, distal (e.g., IBM)
		Brain lesion (e.g., stroke, demyelination)
	Dysarthria ± dysphagia	*Motor neuron disease (e.g., bulbar-onset ALS)*
		Syringobulbia
		Myasthenia gravis
		Myopathy (e.g., IBM)
		Dystrophy (e.g., oculopharyngeal)
		Brain stem lesion (e.g., stroke, tumor, demyelinating disease)
		Pseudobulbar palsy
Multifocal/ Generalized	Without UMN signs	*Motor neuron disease (e.g., PMA; Kennedy's disease)*
		Poliomyelitis and postpolio syndrome
		Chronic inflammatory demyelinating polyneuropathy
		Guillain-Barré syndrome
		Hereditary neuropathy with predisposition to pressure palsies
		Lambert-Eaton myasthenic syndrome
		Multifocal motor neuropathy with conduction block
		Myasthenia gravis
		Myopathy
	With UMN signs	*Motor neuron disease (e.g., ALS)*
		Structural lesion of spinal cord (e.g., cervical spondylosis with myeloradiculopathy, tumor, vascular malformation, syringomyelia)
		Multiple sclerosis
		Vasculitis (e.g., polyarteritis nodosa)
		Hereditary spastic paraparesis
		HTLV-1 associated myelopathy

ALS, amyotrophic lateral sclerosis; HTLV-1, human T cell leukemia virus type 1; IBM, inclusion body myositis; PMA, progressive muscular atrophy; UMN, upper motor neuron.
Conditions primarily affecting lower motor neurons are italicized.

weakness and atrophy. In contrast, the NEE can detect evidence of ongoing denervation and reinnervation after a small number of motor axons have degenerated and have not yet resulted in muscle weakness. However, fibrillation potentials are most often found in muscles manifesting some degree of clinical weakness. This was demonstrated by Lambert, who observed fibrillation potentials in 54% and 97% of muscles with three-quarters and less than one-half strength,

TABLE 19–12	Focal Weakness and the Responsible Muscles and Nerves Examined by Motor Nerve Conduction Studies

Symptom	Nerve Stimulated	Additional Muscle Recorded
Upper Extremity		
Inability to abduct arm	Axillary	Deltoid
Inability to flex elbow	Musculocutaneous	Biceps brachii
Wrist drop	Radial	Extensor digitorum communis
Weak hand grip	Ulnar	First dorsal interosseous
Weak hand grip	Median	Abductor pollicis brevis
Lower Extremity		
Foot drop	Peroneal (deep branch)	Tibialis anterior
Knee buckling	Femoral	Quadriceps femoris
Inability to stand on toes	Posterior tibial	Gastrocnemius-soleus complex

respectively; only 25% of muscles with normal strength had fibrillation potentials.[19]

LMN involvement of bulbar muscles can be difficult to confirm electrodiagnostically in early-onset disease. Abnormalities of motor unit potential firing and morphology can be difficult to interpret unless they are clearly neurogenic, mainly because electromyographers have less experience with the normal motor unit morphology in these frequently examined muscles. Visualization of fibrillation potentials in bulbar musculature, especially the tongue, is challenging because of the difficulty in establishing relaxation in the muscles. Failure to identify clear neurogenic changes in bulbar muscles may also result from the fact that symptomatology at the time may be owing to UMN involvement. About 20% of patients with symptoms limited to bulbar muscles have evidence of more widespread denervation at initial electrodiagnostic evaluation.[37]

Generalized or Multifocal Weakness

By the time an individual with a motor neuron disorder comes to the attention of a neurologist, there is frequently evidence of multifocal or generalized weakness. An initial focal deficit may have been present, which has progressed to involve multiple or generalized sites, although some patients may develop symptoms in multiple muscles almost simultaneously. Several treatable neuromuscular conditions can mimic motor neuron disorders by causing global weakness (e.g., myasthenia gravis, Lambert-Eaton myasthenic syndrome, Guillain-Barré syndrome, and multifocal motor conduction block (see Table 19–11).

Nerve Conduction Studies

Patients presenting with either multifocal or generalized weakness should undergo screening NCS of one arm and leg as listed in Table 19–4. Additional NCS are added depending on the distribution of the multifocal weakness. Significant abnormalities of sensory or motor nerve latency, conduction velocity, or temporal dispersion of M waves suggest a non-motor neuron cause for the multifocal or generalized weakness, such as the acquired inflammatory demyelinating polyneuropathies. Sensory nerve involvement has been described only very rarely in motor neuron disorders.[13] One

exception is X-linked recessive bulbospinal neuronopathy (Kennedy's disease), which typically demonstrates diminished or absent SNAP amplitudes[12] (see earlier discussion).

Finding diminished CMAP amplitudes in most or all muscles in the setting of generalized or multifocal weakness should prompt a search for the Lambert-Eaton myasthenic syndrome, a presynaptic disorder of acetylcholine release that may resemble an LMN-predominant motor neuron disorder (see earlier discussion).

Needle Electrode Examination

In patients with generalized or multifocal weakness, we routinely examine select limb and paraspinal muscles, which are innervated by different nerves from various spinal levels (see Table 19–5). This is particularly important in patients with suspected motor neuron disorders, where diagnosis depends on the presence of widespread motor axon loss. We perform NEE of muscles from at least three spinal regions: cervical (upper extremity and related paraspinal muscles), lumbar (lower extremity and related paraspinal muscles), and thoracic (paraspinal muscles). The importance of finding axon loss (especially ongoing) in thoracic paraspinal muscles is that compressive radiculopathies at these levels are distinctly rare. In addition, NEE of facial and tongue muscles may be informative when there is clinical bulbar dysfunction.

The NEE can be limited to a single side if diffuse motor axon loss is present, although muscles in the contralateral limb may need to be examined if the abnormality is more restricted. More rapidly progressing motor neuron degeneration displays a predominance of acute motor axon loss with fibrillations, whereas more slowly evolving motor neuron disorders (e.g., adult forms of spinal muscular atrophy) have

TABLE 19–13	Electrodiagnostic Workup of Patients Presenting with Focal Weakness

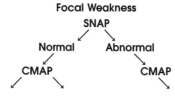

Focal Weakness
SNAP
Normal / Abnormal
Normal → CMAP ← Abnormal → CMAP ←
Normal / Abnormal / Normal–Abnormal

Normal

Mild/Early ISC Lesion or Focal MND
NEE: active and chronic neurogenic changes

Focal UMN Lesion (e.g., lacunar infarct)
NEE: normal except for nonspecific abnormality of MUP activation

Abnormal

ISC Lesion
Radiculopathy
Myelopathy with VR exit zone lesion (e.g., MS)
Focal MND
Syringomyelia
NEE: active and chronic neurogenic changes

Focal Myopathy (e.g., IBM)
NEE: short duration and low amplitude MUPs ± fibrillations

Normal–Abnormal

Nerve Trunk Lesion Distal to DRG
Plexopathy
Mononeuropathy
HNPP
NEE: active and chronic neurogenic changes

CMAP, compound muscle action potential; DRG, dorsal root ganglion; HNPP, hereditary neuropathy with predisposition to pressure palsies; IBM, inclusion body myositis; ISC, intraspinal canal; MND, motor neuron disorder; MS, multiple sclerosis; MUPs, motor unit potentials; NEE, needle electrode examination; PNS, peripheral nervous system; SNAP, sensory nerve action potential; UMN, upper motor neuron; VR, ventral root.

TABLE 19–14	Electrodiagnostic Workup of Patients Presenting with Generalized Weakness

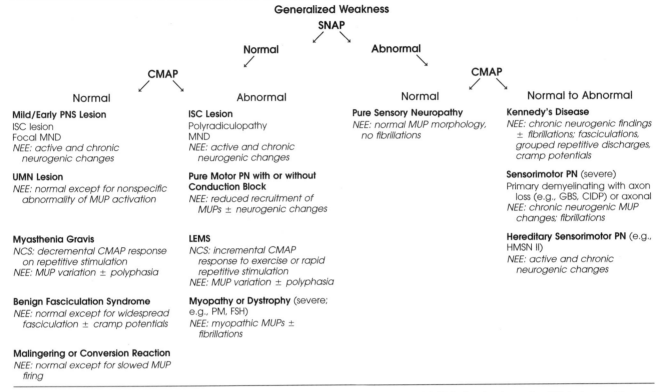

Generalized Weakness

SNAP

Normal / \ **Abnormal**

CMAP (under Normal SNAP)

Normal — **Abnormal**

CMAP (under Abnormal SNAP)

Normal — **Normal to Abnormal**

Mild/Early PNS Lesion
ISC lesion
Focal MND
NEE: active and chronic neurogenic changes

UMN Lesion
NEE: normal except for nonspecific abnormality of MUP activation

Myasthenia Gravis
NCS: decremental CMAP response on repetitive stimulation
NEE: MUP variation ± polyphasia

Benign Fasciculation Syndrome
NEE: normal except for widespread fasciculation ± cramp potentials

Malingering or Conversion Reaction
NEE: normal except for slowed MUP firing

ISC Lesion
Polyradiculopathy
MND
NEE: active and chronic neurogenic changes

Pure Motor PN with or without Conduction Block
NEE: reduced recruitment of MUPs ± neurogenic changes

LEMS
NCS: incremental CMAP response to exercise or rapid repetitive stimulation
NEE: MUP variation ± polyphasia

Myopathy or Dystrophy (severe; e.g., PM, FSH)
NEE: myopathic MUPs ± fibrillations

Pure Sensory Neuropathy
NEE: normal MUP morphology, no fibrillations

Kennedy's Disease
NEE: chronic neurogenic findings ± fibrillations; fasciculations, grouped repetitive discharges, cramp potentials

Sensorimotor PN (severe)
Primary demyelinating with axon loss (e.g., GBS, CIDP) or axonal
NEE: chronic neurogenic MUP changes; fibrillations

Hereditary Sensorimotor PN (e.g., HMSN II)
NEE: active and chronic neurogenic changes

CIDP, chronic inflammatory demyelinating polyneuropathy; CMAP, compound muscle action potential; DRG, dorsal root ganglion; FSH, facioscapulohumeral dystrophy; GBS, Guillain-Barré syndrome; HMSN, hereditary motor sensory neuropathy; ISC, intraspinal canal; LEMS, Lambert-Eaton myasthenic syndrome; MND, motor neuron disorder; MUP, motor unit potential; NCS, nerve conduction studies; NEE, needle electrode examination; PM, polymyositis; PN, polyneuropathy; SNAP, sensory nerve action potential.

a predominance of the chronic changes including prolonged motor unit potentials. The presence of unstable motor units (moment-to-moment configuration variation or block) indicates early reinnervation, which can be seen in a variety of motor neuron disorders.

Decision Tree of Electrodiagnosis

A logical approach to the electrodiagnostic workup of patients with suspected motor neuron disorders begins by noting whether the symptoms are predominantly focal (Table 19–13) or generalized (Table 19–14). Focal regions of greater weakness can be superimposed on a background of global weakness, particularly in more advanced stages of disease.

Beginning with NCS, and specifically sensory responses, one can already differentiate between most processes originating in the intraspinal canal versus more distally. Motor NCS further refine the differential diagnosis which is subsequently explored by the NEE (see Tables 19–13 and 19–14).

REFERENCES

1. Angel RW, Hofmann W: The H reflex in normal, spastic, and rigid subjects. Arch Neurol 8:591–596, 1963.
2. Behnia M, Kelley JJ: Role of electromyography in amyotrophic lateral sclerosis. Muscle Nerve 14:1236–1241, 1991.
3. Bernstein LP, Antel JP: Motor neuron disease: Decremental responses to repetitive nerve stimulation. Neurology 31:202–204, 1981.
4. Blexrud MD, Windebank AJ, Daube JR: Long-term follow-up of 121 patients with benign fasciculations. Ann Neurol 34:622–625, 1993.
5. Brook BP, Fischbeck KH: Spinal and bulbar muscular atrophy: A trinucleotide-repeat expansion neurodegenerative disease. Trends Neurosci 18:459–461, 1995.
6. Byers RK, Banker BQ: Infantile muscular atrophy. Arch Neurol 5:140–164, 1961.
7. Crawford TO: From enigmatic to problematic: The new molecular genetics of childhood spinal muscular atrophy. Neurology 46:335–340, 1996.
8. Donofrio PD: AAEM case report 28: Monomelic amyotrophy. Muscle Nerve 17(10):1129–1134, 1994.
9. Eisen A, Stewart H: Not-so-benign fasciculation. Ann Neurol 35:375–376, 1994.
10. El Escorial World Federation of Neurology: Criteria for the diagnosis of amyotrophic lateral sclerosis. J Neurol Sci 124(suppl):96–107, 1994.
11. Federico A, Palmeri S, Malandrini A, et al: The clinical aspects of adult hexosaminidase deficiencies. Dev Neurosci 13(4–5):280–287, 1991.
12. Ferrante M, Wilbourn AJ: The characteristic electrodiagnostic features of Kennedy's disease. Muscle Nerve 20:323–329, 1997.
13. Gregory R, Mills K, Donaghy M: Progressive sensory nerve dysfunction in amyotrophic lateral sclerosis: A prospective clinical and neuropsychological study. J Neurol 240:309–314, 1993.
14. Grunewald RA, Chroni E, Panayiotopoulos CP, Enevoldson TP: Late onset radiation-induced motor neuron syndrome. J Neurol Neurosurg Psychiatr 55(8):741–742, 1992.
15. Heitzman D, Wilbourn AJ, Mitsumoto H: A retrospective study examining the clinical and electrodiagnostic features of patients with bulbar-onset amyotrophic lateral sclerosis [abstract]. Neurology 45(suppl 4):A447, 1995.
16. Hirayama K, Toyokura Y, Tsubaki T: Juvenile muscular atrophy of unilateral upper extremity: A new clinical entity: Psychiatric Neurol Jpn 61:2190–2197, 1959.
17. Jubelt B, Drucker J: Post-polio syndrome: an update. Semin Neurol 13:283–290, 1993.

18. Kennedy WR, Alter M, Sung JH: Progressive proximal spinal and bulbar muscular atrophy of late onset: a sex-linked recessive trait. Neurology 18:671–680, 1968.
19. Lambert EH: Electromyography in amyotrophic lateral sclerosis. In Norris FH, Kurland LT (eds): Motor Neuron Diseases: Research on Amyotrophic Lateral Sclerosis and Related Disorders. New York, Grune & Stratton, 1969, pp 135–153.
20. Lambert EH, Mulder DW: Electromyographic studies in amyotrophic lateral sclerosis. Staff Meet Mayo Clin 32:441–446, 1957.
21. Levin K: Paraneoplastic neuromuscular syndromes. Neurol Clin 15(3):597–614, 1997.
22. Mariani C, Cislaghi MG, Barbieri S: The natural history and results of surgery in 50 cases of syringomyelia. J Neurol 238:433–438, 1991.
23. Mitsumoto H, Chad DC, Pioro EP: Amyotrophic Lateral Sclerosis. Philadelphia, Oxford University Press, 1998, pp 65–86.
24. Mitsumoto H, Chad DC, Pioro EP: Amyotrophic Lateral Sclerosis. Philadelphia, Oxford University Press, 1998, pp 3–17.
25. Mitsumoto H, Chad DC, Pioro EP: Amyotrophic Lateral Sclerosis. Philadelphia, Oxford University Press, 1998, pp 270–284.
26. Mitsumoto H, Schwartzman MJ, Levin K, et al: Electromyographic (EMG) changes and disease progression in ALS [abstract]. Neurology 40(suppl 1):318, 1990.
27. Parry GJ: Motor neuropathy with multifocal conduction block. Semin Neurol 13:266–275, 1993.
28. Pearn JH, Hudgson P, Walton JN: A clinical and genetic study of spinal muscular atrophy of adult onset: The autosomal recessive form as a discrete disease entity. Brain 101:591–606, 1978.
29. Rodriquez AA, Agre JC, Franke TM: Electromyographic and neuromuscular variables in unstable postpolio subjects, stable postpolio subjects, and control subjects. Arch Phys Med Rehab 78:986–991, 1997.
30. Siddique T, Deng H-X: Genetics of amyotrophic lateral sclerosis. Hum Molec Genet 5:1465–1470, 1996.
31. Sobue I, Saito N, Iida M, Ando K: Juvenile type of distal and segmental muscular atrophy of upper extremities. Ann Neurol 3:429–432, 1978.
32. Stålberg E, Grimby G: Dynamic electromyography and muscle biopsy changes in a 4-year follow-up: Study of patients with a history of polio. Muscle Nerve 18:699–707, 1995.
33. Tallaksen CME, Jetne V, Fossa S: Postradiation lower motor neuron syndrome: A case report and brief literature review. Acta Oncologica 36(3):345–347, 1997.
34. Weber M, Eisen A: Assessment of upper and lower motor neurons in Kennedy's disease: Implications for corticomotoneuronal PSTH studies. Muscle Nerve 22:299–306, 1999.
35. Wilbourn AJ: Generalized low motor-normal sensory conduction responses: The etiology in 55 patients [abstract]. Muscle Nerve 7:564, 1984.
36. Wilbourn AJ: Clinical neurophysiology in the diagnosis of amyotrophic lateral sclerosis: The Lambert and the El Escorial criteria. J Neurol Sci 160(suppl 1):S25–S29, 1998.
37. Wilbourn AJ, Belsh JM, Schiffman PL (eds): Amyotrophic Lateral Sclerosis: Diagnosis and Management for the Clinician. Armonk, NY, Futura, 1996, pp 163–202.
38. Wilbourn AJ, Sweeney PJ: Dissociated wasting of the medial and lateral hand muscles with motor neuron disease [abstract]. Can J Neurol Sci 21 (suppl 2):S9, 1994.
39. Wirguin I, Breener T, Argov Z, Steiner I: Multifocal motor nerve conduction abnormalities in amyotrophic lateral sclerosis. J Neurol Sci 112:199–203, 1992.
40. Wohlfart GL: Collateral regeneration from residual nerve fibers in ALS. Neurology 7:124–134, 1957.
41. World Federation of Neurology Research Group on Motor Neuron Diseases: El Escorial revisited: Revised criteria for the diagnosis of amyotrophic lateral sclerosis. www.wfnals.org/Articles/elescorial1998.htm. 10-26-1998.
42. Younger DS, Rowland LP, Latov N, et al: Lymphoma, motor neuron diseases, and amyotrophic lateral sclerosis. Ann Neurol 29(1):78–86, 1991.

Chapter 20

Disorders of Neuromuscular Junction Transmission

Kerry H. Levin, MD

ANATOMY OF THE NEUROMUSCULAR
   JUNCTION
NEUROPHYSIOLOGY OF
   NEUROMUSCULAR JUNCTION
   TRANSMISSION
MYASTHENIA GRAVIS
Pathophysiology
Repetitive Stimulation
Needle Electrode Examination
Clinical Features
Diagnosis
Management
Myasthenic and Cholinergic Crisis
Drug Interference with Neuromuscular
   Junction Transmission
Neonatal Myasthenia Gravis
LAMBERT-EATON MYASTHENIC SYNDROME
Pathophysiology
Repetitive Stimulation
Needle Electrode Examination
Clinical Features

Diagnosis
Management
BOTULISM
Pathophysiology
Repetitive Stimulation and Needle
   Electrode Examination
Clinical Features
Diagnosis
Management
CONGENITAL MYASTHENIA
Familial Infantile Myasthenia
Congenital Acetylcholinesterase
   Deficiency
Classic Slow Channel Syndrome
Congenital Acetylcholine Receptor
   Deficiency
OTHER DISORDERS OF NEUROMUSCULAR
   TRANSMISSION
Organophosphate Poisoning
Magnesium Intoxication
Tick Bite Paralysis

ANATOMY OF THE NEUROMUSCULAR JUNCTION

The neuromuscular junction (NMJ) includes the peripheral motor nerve terminal, the synapse linking the terminal and the muscle membrane, and the muscle end plate. The presynaptic region of the nerve terminal includes the apparatus for synthesis of acetylcholine (ACh), areas for the storage of ACh vesicles, and active zones for the immediate release of ACh vesicles. Zones for the release of ACh are located at the presynaptic nerve membrane, closely associated with voltage-gated calcium channels (VGCCs). The postsynaptic portion of the NMJ, or muscle end plate, is characterized by a specialized highly infolded muscle membrane holding the nicotinic ACh receptor (AChR) molecules. Each AChR is a complex of multiple glycoprotein subunits configured to form an ion channel. New AChRs replace old ones about every 8 to 11 days. AChRs are concentrated at the peaks of the junctional folds, directly opposite the active zones of ACh release. Acetylcholinesterase is localized in the troughs of the junctional folds. Figure 20–1A shows an electron photomicrograph of a normal NMJ.

NEUROPHYSIOLOGY OF NEUROMUSCULAR JUNCTION TRANSMISSION

The signal for muscle fiber contraction passes across the NMJ in the form of the neurotransmitter ACh. ACh is synthesized in the endoplasmic reticulum of the nerve terminal, and is packaged into vesicles, or quanta, each containing about 10,000 molecules. Vesicles of ACh attach to active zones, where they are released into the synapse in response to influx of calcium through closely connected VGCCs. The probability and amount of ACh release depend upon many factors, including the uptake of choline and acetate by the nerve terminal membrane, synthesis of ACh, storage of ACh, mobilization of ACh into the immediately releasable active zones, and the concentration of calcium in the microenvironment of the active zones.[35]

There are three independent mechanisms of ACh release.[45] First, there is spontaneous release of single vesicles or quanta of ACh that occurs continuously and independently of nerve terminal depolarization. This process of continuous slow release of ACh is required for the integrity of the muscle end plate, for without it (as in cases of severe botulism) the muscle end plate degenerates and the muscle fiber becomes denervated. The second mechanism of ACh release is a nonquantal leakage of ACh from sites other than the active release zones. The third mechanism is triggered by a descending nerve fiber action potential that depolarizes the nerve terminal. In response to a nerve terminal depolarization, VGCCs open, causing an influx of calcium. Calcium binds with and activates the intracellular protein calmodulin, which in turn initiates a series of events that triggers the release of 60 to 100 quanta of ACh. The term quantal content refers to the number of vesicles of ACh released in response to a single nerve terminal depolarization. At slow rates of repetitive stimulation (2 to 5 Hz), there is depletion of the immediately releasable ACh pool, and little calcium accumulates over the first four stimuli. As slow repetitive stimulation continues, calcium accumulation and the immediate releasable pool of ACh increase. With rapid rates of

251

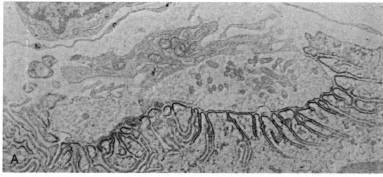

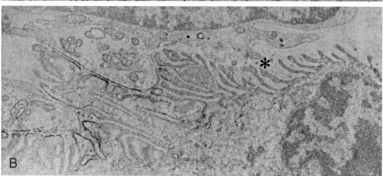

FIGURE 20–1. *(A)* Electron photomicrograph (×11,000 magnification) of a normal appearing neuromuscular junction, showing a tightly apposed junction between the nerve terminal and the postsynaptic surface that contains highly infolded end plate membrane folds. The specimen was treated to highlight acetylcholine receptors, visualized as dark outlining of the peaks of the membrane folds, but not the troughs. *(B)* Electron photomicrograph (×15,400 magnification) showing the pathological features of myasthenia gravis. In the region of the asterisk there is widening of the synaptic cleft between the nerve terminal membrane and the postsynaptic membrane, simplification of the postsynaptic end plate folds, and absence of the dark outlining of the membrane peaks corresponding to loss of acetylcholine receptors. (From Engel A, Lindstrom JM, Lambert EH, et al: Ultrastructural localization of the acetylcholine receptor in myasthenia gravis and in its experimental autoimmune model. Neurology 27: 307–315, 1977, with permission.)

repetitive stimulation (>20 Hz), calcium influx and the quantal content increase for as long as 1 minute, a phenomenon known as postactivation facilitation.

Upon reaching the postsynaptic membrane, ACh molecules bind to AChRs. When both active binding sites are occupied by ACh molecules, a conformational change occurs in the receptor polypeptide subunits, opening the central ion channel, allowing sodium influx, and producing a tiny depolarization of the muscle membrane. The change in end plate potential (EPP) of the muscle fiber membrane resulting from the release of a single vesicle of ACh is about 1 mV, and is called a miniature end plate potential (MEPP). The EPP resulting from the summation of MEPPs generated from the release of quanta from a single nerve terminal depolarization is about 40 mV, or 25 mV above the threshold

EPP required for the generation of a propagated muscle fiber action potential.[13, 36] This overshoot constitutes the safety factor of NMJ transmission.

In normal individuals, a number of physiological factors can reduce the safety margin of NMJ transmission. At the slow rates of 2 to 5 Hz repetitive nerve stimulation, the safety margin can be reduced to 5 to 10 mV, due to the depletion of the immediately releasable pool of ACh vesicles at the nerve terminal. The greatest decline in ACh release occurs between the first and second nerve terminal depolarization in a train; subsequent decline progressively lessens over the next three to five stimuli, until a steady amount of release occurs thereafter. At normal NMJs, up to 10 seconds are required for repletion of the immediately releasable pool of ACh after a volley of nerve terminal stimuli. In spite of the

EFFECTS OF PAIRED STIMULI

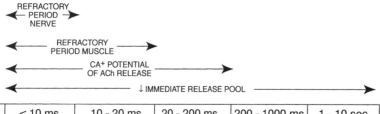

	< 10 ms	10 - 20 ms	20 - 200 ms	200 - 1000 ms	1 - 10 sec
NL	2 <1	2 < 1 (by less)	66 % 1 = 2 33 % 2 > 1 1 by < 10 %	1 = 2	1 = 2
MG	2 << 1	2 < 1 (by less) 2 > 1 (in some)	VARIABLE	2 <<< 1	2 < 1
LEMS	2 >> 1 or 2 < 1	2 >> 1	2 > 1	2 <<< 1	

FIGURE 20–2. Two sequential nerve trunk stimuli are delivered, and the time interval between the two is progressively lengthened from 10 msec (the equivalent of a repetitive stimulation rate of 100 Hz) to 10 seconds (the equivalent of a repetitive stimulation rate of 0.1 Hz). The numbers 1 and 2 signify the CMAP amplitudes resulting from the first and second stimuli, respectively. (Derived from data published by Lambert EH, Rooke ED, Eaton LM, et al: Myasthenic syndrome occasionally associated with bronchial neoplasm: Neurophysiologic studies. In Viets HR [ed]: Myasthenia Gravis. Springfield, IL, Charles C Thomas, 1961, pp 364–410.)

reduction of immediately releasable ACh, at this frequency of nerve terminal depolarization each stimulus produces a threshold depolarization of the muscle membrane, and results in a propagated muscle fiber action potential. In normal individuals, after stimulation at rapid rates above 20 Hz, or after brief forceful exercise, ACh release is enhanced by greater calcium influx into the nerve terminal, and postactivation facilitation of NMJ transmission is seen for as long as 1 minute, but thereafter the EPP falls for up to 4 minutes (postactivation exhaustion). With continued rapid stimulation ACh release falls owing to (1) progressive depletion of the immediately releasable pool of ACh, and (2) progressive refractoriness of action potential propagation along nerve and muscle fibers, resulting from the increased frequency of action potential propagation. This results in loss of the safety factor and progressive reduction of the amplitude of compound muscle action potentials (CMAPs) recorded during a train of nerve stimuli.

In pathological conditions, the safety margin is lost during routine operation of the NMJ, resulting in weakness. The failure to reach threshold postsynaptic EPPs results in the failure to generate propagated muscle fiber action potentials. In myasthenia gravis there is loss of AChRs, in Lambert-Eaton syndrome and botulism there is deficiency of ACh release, and in other conditions such as drug toxicity and immaturity of freshly formed NMJs there is dysfunction at both the presynaptic and postsynaptic sites. In all these situations, the safety factor of NMJ transmission can be lost and CMAP amplitudes progressively decline during trains of nerve stimuli. The effects on CMAP amplitude resulting from increasing frequencies of nerve trunk stimulation in normal and pathological situations are reviewed in Figure 20–2.[22]

MYASTHENIA GRAVIS

Pathophysiology

Autoimmune myasthenia gravis (MG), the most common postsynaptic defect of NMJ transmission, is the result of antibody attack leveled against the AChR. The resulting pathology is characterized by loss of AChRs, simplification and scarring of the postsynaptic junctional folds, and widening of the NMJ synaptic cleft (see Fig. 20–1B).

MG is a disorder of muscle fatigability that results from the loss of the safety margin of NMJ transmission during a voluntary contraction. In MG, ACh is released in normal amounts from the nerve terminal. When the normal decline in the release of ACh over the course of a train of slow repetitive stimulation is combined with the reduced numbers of AChRs at the myasthenic end plate, a point is reached during slow repetitive stimulation when insufficient numbers of AChR ion channels are opened to produce a threshold EPP. At this point the safety factor of NMJ transmission is lost. In the absence of a propagated muscle fiber action potential, the muscle fiber will not contract, and muscle power diminishes.

In the patient with MG, AChR antibody damage to AChRs is not evenly distributed over all muscle fibers and all muscle bellies. Some NMJs have lost so many AChRs that not even the first release of ACh in a train of stimuli will produce a threshold EPP, whereas transmission across other NMJs is nearly normal. When measuring the effect in the whole muscle belly with a recording electrode over the muscle's motor point, an averaging of the effect of AChR loss in all muscle fibers takes place, producing a picture of progressive

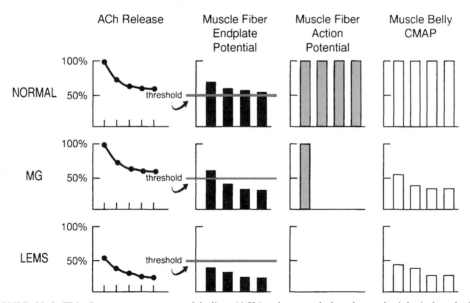

FIGURE 20–3. This figure compares acetylcholine (ACh) release and the electrophysiological end plate response in the normal individual with two pathological states: myasthenia gravis (MG) and Lambert-Eaton syndrome (LEMS). The columns entitled ACh release, muscle fiber end plate potential, and muscle fiber action potential pertain to events at a single NMJ, while the column entitled muscle belly compound muscle action potential (CMAP) pertains to responses obtained from a whole muscle belly. At the muscle fiber end plate, if sufficient ion channels open and sufficient sodium influx occurs, the resulting end plate potential will be above the threshold required to generate a propagated muscle fiber action potential. The response recorded from a whole muscle belly (CMAP) after nerve trunk stimulation is a summation of all the individual muscle fiber action potentials generated after nerve trunk stimulation, and represents an averaging of the responses from all NMJs in the muscle belly, some of which have severe impairment of transmission while others have little or no transmission defect. (Adapted from Desmedt JE: The neuromuscular disorder in myasthenia gravis. II. Presynaptic cholinergic metabolism, myasthenia-like syndromes, and a hypothesis. In Desmedt JE [ed]: New Developments in Electromyography and Clinical Neurophysiology, vol 1. Basel, S. Karger, 1973, pp 305–342.)

decline of muscle belly contraction during a train of stimuli, mirroring the natural decline in ACh release during a train of nerve terminal depolarizations[8] (Fig. 20–3).

Repetitive Stimulation

Clinical electrodiagnostic techniques have been developed to measure the degree to which patients suspected of having MG demonstrate loss of the safety factor of NMJ transmission. In the normal individual, slow repetitive stimulation (RS) at 2 to 5 Hz, with recording electrodes applied over the muscle belly innervated by the stimulated nerve trunk, results in the recording of a train of identical CMAPs, each representing the summation in the volume conductor of the propagated muscle fiber action potentials activated by the nerve trunk stimulus. In MG, during a train of stimuli, a classic decrementing response is seen: there is progressive loss of the CMAP amplitude with each successive stimulus in the train, reflecting the result of progressive reduction of releasable ACh in the setting of a reduced number of available AChRs. The greatest drop in CMAP amplitude occurs between the first and second stimuli in the train, with lesser amounts of decrement with subsequent stimuli. This pattern of decrement mirrors the natural decline in presynaptic ACh release, and is the classical decremental response seen in true defects of NMJ transmission (Fig. 20–4).

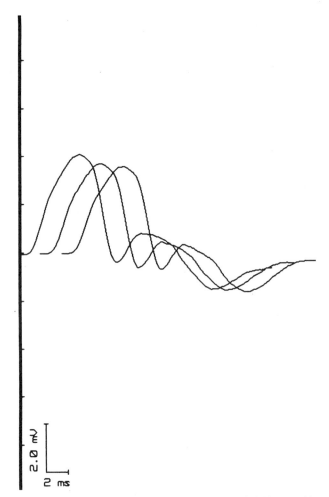

FIGURE 20–4. Classical decremental response with 2-Hz repetitive stimulation, showing maximal decline in amplitude between the first and second stimuli in the train.

In the EMG laboratory, the usual procedure for RS involves delivery of a train of three to five stimuli at 2 Hz with the recorded muscle at rest, followed by repeat trains immediately after forceful exercise of the muscle, and at 30 seconds, 1 minute, 2 minutes, 3 minutes, and 4 minutes thereafter. In patients with no decremental response during stimulation at rest, exercise is given continuously for 1 to 2 minutes. In patients with decrement at rest, exercise is given for 15 to 20 seconds to avoid exhaustion of the NMJs. Decremental responses during RS are calculated by comparing either the amplitude or area of the first CMAP in the train to that of the last CMAP in the train. The percentage of decrement in a train is calculated by subtracting the amplitude of the last response in a train from the first, then dividing that value by the amplitude of the first response. The same calculation can be performed using the area of each response. A maximal decremental response greater than 10% traditionally defines a defect of NMJ transmission, a value chosen largely to take into account the small technical error that can be introduced during the RS procedure, resulting from patient movement and slippage of the stimulating electrodes. In normal individuals there is never a decremental response at slow rates of RS under ideal recording conditions. The classic finding in a generalized postsynaptic defect of NMJ transmission of moderate severity is a three-stage pattern: (1) decremental response in CMAP amplitudes during a train of stimuli delivered at rest, (2) repair of the decremental response following a 15- to 20-second period of forceful exercise of the recorded muscle, and (3) return and worsening of the decremental response within 2 to 4 minutes after the exercise period (Fig. 20–5). In patients without a decremental response in rested muscle, a decremental response can sometimes be provoked by exhausting the recorded muscle belly with 1 to 2 minutes of forceful isometric exercise.

In almost all cases of MG, the CMAP amplitude at rest is normal. Brief forceful exercise may increase the CMAP amplitude by up to 15% to 20% for up to 30 seconds, a phenomenon known as pseudofacilitation. This is thought to result from synchronization of muscle fiber action potentials in the muscle belly, narrowing and increasing the amplitude of the recorded CMAP, but not significantly changing the CMAP area. In very severe postsynaptic defects of NMJ transmission the safety margin may be lost at rest, resulting in a reduced resting CMAP amplitude. Exercise or rapid RS may produce a marked incremental response in amplitude of 100% or more. Although this phenomenon is most commonly seen in presynaptic disorders such as Lambert-Eaton syndrome, it can also occur in severe postsynaptic disorders.[21, 38]

Several factors can adversely affect the reliability of RS studies. These studies are most reliably carried out on distal nerve trunks, as they are better tolerated by the patient and there tends to be less technical artifact. Technical difficulties include displacement of the stimulating or recording electrodes, submaximal stimulation, and poor muscle relaxation, all of which are more likely to occur at proximal sites of nerve trunk stimulation. Technical difficulties are likely to be the source of baseline instability, CMAP configuration changes, and nonphysiological decremental or incremental changes (Fig. 20–6). Cooling of muscle will decrease the decremental response, and can lead to a false negative RS study, especially when testing distal muscles (Fig. 20–7). Several factors contribute to this phenomenon. Cooling can prolong the duration of nerve terminal depolarization and the open time of the AChRs, leading to increased ACh release and more complete depolarization of the end plate, thus improving the NMJ safety margin.[21] Cooling inactivates

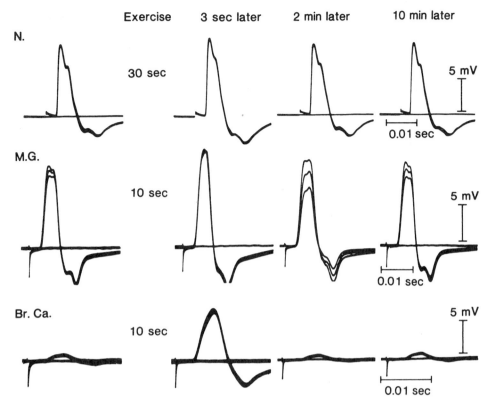

FIGURE 20–5. Effect of exercise on the action potential of the hypothenar muscles evoked by maximal stimulation of the ulnar nerve at the wrist. The response of the rested muscle *(far left)* is compared with responses 3 seconds, 2 minutes, and 10 minutes after the end of a maximal voluntary contraction of the muscle. Each record consists of three superimposed action potentials evoked at a rate of three per second. N., responses of a normal subject; M.G., responses of a patient with generalized myasthenia gravis. In the rested muscle, a progressive decline of amplitude occurs during stimulation at a rate of three per second. Three seconds after exercise, this defect is repaired and there is some increase in the amplitude of response (post-tetanic facilitation). Two minutes after exercise, the defect is more marked than it was initially. At 10 minutes, the response is returning to its original level. Br. Ca., patient with the myasthenic syndrome associated with a small-cell bronchogenic carcinoma. The slight progressive decline in amplitude of response during stimulation at a rate of three per second that occurred in the rested muscle is not evident in the reproduced record. There is marked post-tetanic facilitation 3 seconds after exercise, but a depression of the response is seen 2 minutes after exercise. (From Lambert EH, Rooke ED, Eaton LM, et al: Myasthenic syndrome occasionally associated with bronchial neoplasm: Neurophysiologic studies. In Viets HR [ed]: Myasthenia Gravis. Springfield, IL, Charles C Thomas, 1961, pp 364–410.)

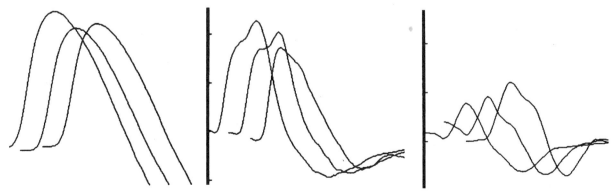

FIGURE 20–6. Three examples of technical artefact introduced during repetitive stimulation studies. In the first panel, there is baseline drift. In the second panel, there is baseline drift and stimulating electrode slide off the nerve trunk, producing configuration change in the CMAP. In the third panel, there is lack of stable placement of the stimulating electrode, leading to intermittent volume conduction from activation of nearby muscle, as well as CMAP configuration change.

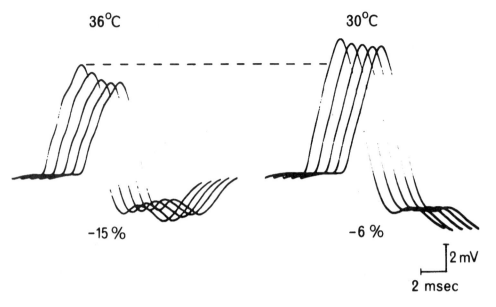

FIGURE 20–7. Effect of temperature on decremental responses during 2 Hz repetitive stimulation. Cooling produces increased CMAP amplitude and duration, and repairs the decremental response, making the test less sensitive. (From Denys EH: Minimonograph #14: The influence of temperature in clinical neurophysiology. Muscle Nerve 14:795–811, 1991, with permission.)

the temperature-dependent acetylcholinesterase enzyme system, leading to an effect similar to that produced by cholinesterase inhibitors such as edrophonium chloride and pyridostigmine, both of which can diminish or normalize the decremental response. Distal limb temperature should be maintained above 32°C, and cholinesterase inhibitors should be discontinued for at least 24 hours prior to the study.

The observation of decremental responses in two separate nerve distributions is recommended for the diagnosis of a defect of NMJ transmission. It is appropriate to begin with RS in the distal arm, such as the ulnar nerve at the wrist (recording abductor digiti minimi) or the median nerve (recording abductor pollicis brevis), because of their technical reliability and acceptability. If these studies are normal, RS at proximal sites is necessary, keeping in mind that symptomatically weak muscles are usually the most rewarding. For example, RS should be performed on the peroneal nerve (recording tibialis anterior) in the patient with prominent leg weakness, and on the facial nerve (recording nasalis) in the patient with ocular-bulbar weakness. A systematic approach to nerve conduction studies and RS in the diagnosis of MG is found in Table 20–1.

Decremental responses may be seen in conditions other than primary defects of NMJ transmission (Table 20–2). Synaptic transmission failure often occurs at newly formed or immature NMJs. Decremental responses to RS can be seen in muscles undergoing active reinnervation, and can be prominent in progressive denervating-reinnervating disorders such as amyotrophic lateral sclerosis. In myotonic dystrophy and myotonia congenita, decremental responses are seen at RS rates greater than 3 Hz, increasing with higher rates of stimulation. In contrast to the decremental response seen in MG, decremental responses in the myotonias may develop only after prolonged or high rates of RS, but may progress over longer periods of stimulation than is seen in MG[42] (Fig. 20–8). With continued RS beyond 1 minute, the CMAP amplitude eventually returns to the prestimulation level.[1] After a brief forceful exercise, the CMAP amplitude can markedly decline. In all forms of periodic paralysis, the CMAP amplitude is reduced during episodes of weakness. During RS of a weak muscle at rates of 5 to 10 Hz for up to 10 minutes, the CMAP amplitude gradually increases, approaching a normal amplitude toward the end of the simulation period.[4] Between episodes, a characteristic pattern is seen after 3 minutes of forceful contraction, characterized by greater than 50% decline in CMAP amplitude in response to a stimulus delivered each minute over 30 min-

TABLE 20–1	Nerve Conduction Studies in the Evaluation of Myasthenia Gravis

Basic Studies
One motor nerve conduction study in the upper extremity and lower extremity, including amplitude, distal latency, and conduction velocity
One sensory nerve conduction study in the upper extremity and lower extremity, including amplitude and distal latency
Motor nerve conduction study, including amplitude and distal latency, on each nerve trunk chosen for repetitive stimulation

Repetitive Stimulation Studies
Performed on at least one nerve trunk recording over a distal muscle, such as the hypothenar (ulnar nerve) or thenar (median) eminence
Performed on at least one nerve trunk recording over a proximal muscle, such as the trapezius (spinal accessory nerve), deltoid (axillary nerve), or biceps (musculocutaneous nerve)
Performed on other nerve trunks recording over muscles in the distribution of the patient's weakness, such as the nasalis (facial nerve), tibialis anterior (peroneal nerve), or quadriceps (femoral nerve)

TABLE 20–2	Disorders Causing Decremental Responses on Repetitive Stimulation

Primary Defects of Neuromuscular Junction (NMJ) Transmission
   Myasthenia gravis
   Lambert-Eaton myasthenic syndrome
   Botulism
   Congenital myasthenic syndromes
   Specific NMJ toxins
Neurogenic Disorders Undergoing Reinnervation
   Motor neuron diseases with active reinnervation
   Reinnervating axon loss mononeuropathies and radiculopathies
   Reinnervating axon loss polyneuropathies
Myopathic Disorders
   Hereditary myotonias
   Periodic paralysis
   Myophosphorylase deficiency
   Other muscle disorders subject to physiological contracture

FIGURE 20–8. Twenty Hz repetitive stimulation in a patient with autosomal dominant myotonia congenita (Thomsen's disease). (From Streib EW: AAEE Minimonograph #27: Differential diagnosis of myotonic syndromes. Muscle Nerve 10:603–615, 1987, with permission.)

utes.[27] In McArdle's disease (myophosphorylase deficiency), prolonged RS at rates greater than 15 Hz for more than 40 seconds can produce progressive loss of CMAP amplitude associated with a prolonged painful physiological contracture of the muscle.[11, 28]

Needle Electrode Examination

The needle electrode examination (NEE) is a mandatory component of the electrodiagnostic workup for a defect of NMJ transmission, and serves several purposes. In the patient with abnormal RS studies, the presence of motor unit action potential variation confirms a defect of NMJ transmission by an independent method. In the patient with equivocal or normal RS studies, the NEE provides the opportunity to examine both proximal and distal muscles that cannot be assessed by RS, and is superior to RS in its ability to identify limited defects of NMJ transmission. In the patient with normal RS studies, the NEE is also performed to search for other neuromuscular causes of weakness.

Evidence of increased insertional and spontaneous activity in MG is uncommon. Fibrillation potentials in MG result from scarring and complete loss of the end plate, and can be seen in severely affected muscles. They are usually confined to proximal muscles such as the paraspinals. Fasciculation potentials are seen in patients being treated with cholinesterase inhibitors such as pyridostigmine. Clinical and NEE evidence of fasciculation is especially prominent in cases of overdosage with cholinesterase inhibitors, and is a classical feature of poisoning with organophosphates such as malathion and parathion, compounds that have cholinesterase inhibitory function.

In MG, motor unit action potential changes are the most common pathological feature of the NEE. In mild MG no clear abnormalities may be seen. In moderately affected muscles, motor unit action potentials may be reduced in duration and amplitude and may have an increased number of turns or phases. In MG of moderate or greater severity, a pathognomonic feature is motor unit action potential variation: the tendency for motor unit action potential configuration to change from one firing to the next (Fig. 20–9). This phenomenon reflects the momentary failure of transmission across one or more affected NMJs, and results in the loss of the component of the motor unit action potential derived from the blocked muscle fiber action potential(s). The particular blocked muscle fiber action potentials change from one activation of the motor unit action potential to the next, giving the characteristic ever-changing configuration of the motor unit action potential. Although motor unit action potential variation is a classic finding in all disorders of NMJ transmission, it is more likely to be seen in patients with freshly reinnervating neurogenic disorders. In the process of reinnervation, newly created NMJs are immature and frequently demonstrate synaptic failure, leading to motor unit action potential variation during NEE.

Single fiber EMG is a powerful technique that provides additional sensitivity in the identification of mild defects of NMJ transmission.[41] The technique is described in detail in Chapter 13. The relative sensitivity of various tests in the diagnosis of MG is reviewed in Table 20–3.

TABLE 20–3	Relative Value of Diagnostic Modalities in Myasthenia Gravis

Single Fiber Electromyogram
  Most sensitive test
  Abnormal in up to 85% of all MG patients in extensor digitorum
    communis
  Abnormal in up to 99% of all MG patients when frontalis and
    orbicularis oculi muscles are also studied[39]
Acetylcholine Receptor Antibody Detection
  Most specific test
  Present in up to 80% of pure ocular MG patients
  Present in over 90% of generalized MG patients[23]
Repetitive Stimulation Tests
  Neither specific nor highly sensitive
  Abnormal in about 50% of patients with mild generalized MG
Tensilon Test
  Easiest test to perform, but useful only in the presence of objective
    signs of weakness and not entirely specific for MG

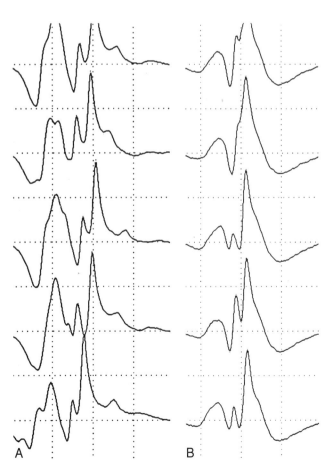

FIGURE 20–9. Five consecutive firings of two different motor unit action potentials, demonstrating moment-to-moment variation in waveform configuration.

CASE STUDY 1

Symptoms but No Signs

A 29-year-old woman has had a 1-year course of end-of-the-day fatigue. Over the past 2 months she has noticed occasional left ptosis and a tendency to lose saliva from the corner of her mouth after a long meal and in bed at night. AChR antibody binding and modulating assays gave borderline positive results. Routine nerve conduction studies and RS studies performed on the median and spinal accessory nerves were normal. RS studies on the facial nerve produced a decrement of 9% 3 minutes after a 1-minute period of facial exercise. The NEE showed mild motor unit instability in the frontalis muscle. The single fiber EMG study of the frontalis and orbicularis oculi muscles showed mean jitter values of 30 μsec and 50 μsec, respectively.

In this patient's case, routine studies suggested the possibility of a defect of NMJ transmission but were insufficient for reliable diagnosis. The single fiber EMG study of the orbicularis oculi muscles was diagnostic of a defect, although the study of the frontalis was not.

Clinical Features

MG is a disorder of weakness and fatigability of skeletal muscle, and is the most commonly occurring defect of NMJ transmission, with a prevalence of 1 per 10,000 to 20,000 population. A diagnostic cornerstone of the disease is the tendency for symptoms to be most prominent with physical activity and exercise, and least noticeable after rest.[32] Cranial muscles are often involved first, producing ptosis and diplopia due to weakness of levator palpebrae and extraocular muscles, respectively. Weakness of mastication and swallowing often develops, and the natural facial expression may change, resulting in a loss of facial tone and a horizontal smile. Subsequently, weakness may develop in extremity muscles, and finally in muscles of respiration. Some patients may present with extremity fatigability and weakness, or with fragmentary symptoms such as dysphagia or dysarthria, but without classic ocular features. Some patients have a pure ocular form that never generalizes.

Early in the disease process symptoms tend to vary remarkably from hour to hour and day to day, depending on the patient's activity level. Patients often have fewest symptoms upon arising from rest, and maximal symptoms late in the day or after activity. When symptoms are mild, no objective evidence of the disorder will be found on clinical examination. Fluctuating symptoms of fatigue, blurred vision, and trouble swallowing without objective findings can lead the physician to the erroneous diagnosis of a somatization disorder or depression.

Diagnosis

Aside from the patient's history, support for the diagnosis can come from the neurological examination. In advanced cases, there is evidence of bifacial weakness, ptosis, dysarthria, disconjugate gaze, and extremity weakness. In subtle cases, there may be slight asymmetry of eyelid position, less than a full smile, and mild weakness of orbicularis oculi muscles (tested by opening eyelids closed forcefully by the patient), frontalis muscles (tested by depressing eyebrows forcefully elevated by the patient), or neck flexor muscles. Strong support for the diagnosis can come from observing fatigability of muscle strength. Sustained upgaze may produce progressive ptosis or disconjugate gaze. During sus-

tained upgaze, a classic finding is fleeting eyelid contractions, known as Cogan's eyelid twitches. Fatigability can also be elicited in extremity muscles. Myasthenic weakness found on the examination can be distinguished from other causes of muscle weakness by intravenous administration of edrophonium chloride. Two milligrams are injected as a test dose. In the absence of a bradycardic response or clear improvement, 8 mg is injected about 1 minute after the test dose. Clear change in the pattern of weakness should be apparent within 1 to 2 minutes. One milligram of injectable atropine should be available in case of symptomatic bradycardia from cholinergic excess.

Identification of acetylcholine receptor antibody in the patient's serum is a sensitive and highly specific method of diagnosis. By means of a series of immunoassays, antibody can be detected in the sera of 70% to 80% of patients with pure ocular MG, and in 90% to 95% of those with generalized MG (see Table 20–3). Antibody detection in the absence of MG has been encountered rarely in patients exposed to the α-bungarotoxin found in cobra venom, and as the result of certain intravenous NMJ blocking agents such as suxamethonium chloride or dimethyltubocurarine iodide. False positive antibody tests have also been reported rarely in patients with amyotrophic lateral sclerosis, pernicious anemia, and in patients taking the drug D-penicillamine.

MG may occur in the setting of other autoimmune disorders, most commonly immune thyroid disease (associated with antithyroglobulin antibodies) and pernicious anemia (associated with antiparietal cell antibody). Less common associations include rheumatoid arthritis, systemic lupus erythematosus, diabetes, and pemphigus. The serological workup for MG should include thyroid function studies and vitamin B_{12} concentration.

When the presentation of myasthenic symptoms occurs rapidly over weeks, the differential diagnosis is relatively narrow, and includes brain stem disorders that are malignant (lymphoma, glioma, metastasis) or inflammatory (sarcoidosis, Lyme disease, chronic meningeal diseases); botulism; and the Miller Fisher variant of acute inflammatory demyelinating polyradiculoneuropathy (Guillain-Barré syndrome).

When the onset of myasthenic symptoms is more slowly progressive, and includes a prominence of ocular-bulbar manifestations, the differential diagnostic considerations include Graves' disease with extraocular muscle involvement, oculopharyngeal dystrophy, myotonic dystrophy, and mitochondrial myopathy. When the symptoms are primarily in the extremity muscles, disorders to be considered include Lambert-Eaton myasthenic syndrome, polymyositis, chronic inflammatory demyelinating polyradiculoneuropathy (CIDP), and weakness associated with metabolic derangements, such as hypercalcemia, hypermagnesemia, hypophosphatemia, cortisol deficiency, hypothyroidism, or hyperthyroidism. The differential diagnosis of MG is summarized in Table 20–4.

Management

After the diagnosis of MG has been established, structural abnormalities of the thymus gland must be sought. In patients under 50 to 60 years of age, thymectomy is an appropriate consideration whether or not there is evidence of thymic enlargement on computerized tomography (CT) scans of the chest. Seventy-five percent of patients undergoing thymectomy have evidence of thymic hyperplasia. About 15% of patients have evidence of thymoma, a locally invasive malignancy for which surgery is indicated irrespective of the age of the patient. Many scientific reports have confirmed that patients are more likely to achieve remission or reduce their requirement for immunosuppressant medication after thymectomy, compared to those not having surgery.

TABLE 20–4	Differential Diagnosis of Myasthenia Gravis

Chronic/Progressive Presentation
  Brain stem mass lesions
  Inflammatory/infectious diseases of the cranial nerves
  Guillain-Barré syndrome (Miller Fisher variant)
  Graves' disease
  Hypothyroidism/hyperthyroidism
  Oculopharyngeal dystrophy
  Myotonic dystrophy
  Mitochondrial myopathy (progressive external ophthalmoplegia)
  Lambert-Eaton myasthenic syndrome
  Polymyositis
Acute Presentation
  Botulism
  Organophosphate poisoning
  Tick bite paralysis
  Magnesium intoxication
  Pharmacological neuromuscular blockade

For patients with mild, stable symptoms, an oral acetylcholinesterase inhibitor such as pyridostigmine may be sufficient treatment. The drug is usually started at a dose of 30 to 60 mg every 6 hours, and can be increased to up to 120 mg every 3 to 4 hours, but higher doses usually produce intolerable side effects of gastrointestinal upset and diarrhea. Pyridostigmine increases the concentration of ACh in the neuromuscular synapse by inhibiting the action of acetylcholinesterase, but it is ineffective in combating the autoimmune process that leads to destruction of the NMJ. It should be replaced by more potent therapies if symptoms are progressive or life threatening.

For patients with progressive symptoms, corticosteroids have been the cornerstone of modern treatment. Treatment regimens vary widely and should be individualized to the patient's needs. In the stable patient a slowly incrementing dosage of prednisone, starting at 10 or 20 mg every other day, minimizes the risk of acute myasthenic worsening, which can occur in as many as 50% of marginally compensated patients during the first several weeks of therapy.[10] In patients with rapidly progressive or severe symptoms, high-dose corticosteroids should be instituted along with plasmapheresis. During plasmapheresis, AChR antibody is removed along with 2 to 3 L of the patient's plasma, which is replaced by saline and albumin. Clinical improvement from plasmapheresis is often apparent after 1 week of every-other-day treatment, while the benefits of corticosteroids alone are usually delayed by at least 2 weeks. Plasmapheresis should not be used as the sole immune-modulating therapy in marginally compensated patients, as return or progression of the symptoms is not uncommon within several weeks of the last treatment, likely due to rebound accumulation of antibody in the circulation. The risk of relapse is minimized by continuing with plasmapheresis until corticosteroid therapy takes effect. In patients requiring chronic immunosuppression, azathioprine can be added to reduce the long-term requirement of corticosteroids. Cyclosporine has also been used to minimize the corticosteroid requirement.

Myasthenic and Cholinergic Crisis

Rapidly progressive myasthenic weakness is a potentially life-threatening event, and is commonly described as myasthenic crisis. The patient with the new onset of the illness may present in crisis, although MG is usually slowly progressive. Patients with stable myasthenic symptoms may suddenly decompensate for a number of reasons: intercurrent infection, drug interaction, pregnancy, the postpartum state, and idiopathic progression of the illness.

Many drugs adversely affect NMJ transmission, but the safety factor of NMJ transmission protects normal individuals in most cases. In the patient with MG or other disorders of NMJ transmission, drugs may have a profound effect. Overdosage with acetylcholinesterase inhibitors is especially common in patients whose myasthenic weakness has markedly improved on immunosuppressant therapy. This paradoxical worsening weakness in the improving patient results from an excess concentration of ACh in the synapse, producing a state of constant depolarization of the motor end plate, failure of end plate repolarization, and eventual neuromuscular blockade and paralysis. Clinically, cholinergic crisis should be suspected in the presence of autonomic symptoms such as excessive abdominal cramping, diarrhea, increased bronchial secretions, sweating, salivation, bradycardia, and hypotension. Patients suspected of cholinergic crisis should be admitted to the hospital and withdrawn from acetylcholinesterase inhibitors under close supervision.

Drug Interference with Neuromuscular Junction Transmission

Numerous drugs can interfere with NMJ transmission.[43] Some act by producing reversible depolarizing blockade, such as overdosage of therapeutic cholinesterase inhibitors, and succinylcholine. Some produce irreversible neuromuscular blockade, such as curare. Some drugs have been shown to induce MG in nonmyasthenic individuals with rheumatoid disorders, including D-penicillamine, pyritinol, and captopril.[33] Other drugs unmask or aggravate a preexisting myasthenic state, such as antiarrhythmics, aminoglycosides, and others.

Some agents exert effects at the nerve terminal. Aminoglycoside antibiotics such as neomycin, gentamycin, and tobramycin are thought to interfere with calcium influx at presynaptic VGCCs.[35] Iodinated contrast agents may produce similar effects due to the sodium citrate and sodium edetate used as sequestering agents in the compounds.

Other drugs have combined presynaptic and postsynaptic, or undetermined mechanisms of NMJ transmission inhibition. Drugs in this category include antiarrhythmic and anesthetic agents containing procaine, lidocaine, and quinine. Drugs that have been occasionally reported to increase myasthenic weakness include calcium channel blocking agents, β-adrenergic blocking agents (propranolol), lithium, phenothiazines, erythromycin, ciprofloxacin, tetracycline, and phenytoin. A listing of agents that interfere with NMJ transmission can be found in Table 20–5.

Neonatal Myasthenia Gravis

This condition occurs in up to 20% of the offspring of mothers with autoimmune MG. The likelihood of developing neonatal MG does not correlate with the severity of disease in the mother, but does correlate to some extent with the maternal concentration of AChR antibody. In the newborn, AChRs appear to be of the fetal type, differing in one of the five AChR subunits from adult AChRs. Mothers with a higher fetal-AChR-antibody to adult-AChR-antibody ratio seem most likely to transfer the illness.[19] It is not clear whether newborn AChR antibody is the result of transplacental transfer from the mother. Patients are symptomatic from birth, but usually show spontaneous resolution within 3 weeks, rarely as long as 7 weeks. Findings include facial weakness, ptosis, dysphagia, weak cry, and respiratory insufficiency. Rarely, ventilator assistance is required. The electrodiagnostic features of neonatal MG are identical to MG.

TABLE 20–5	Agents that Interfere with Neuromuscular Junction Transmission

Metabolic States
  Hypermagnesemia
  Hypocalcemia
Neuromuscular Blockers
  Curare
  Succinylcholine
Cholinesterase Inhibitors
  Pyridostigmine (high dose)
  Organophosphate compounds
Antibiotics
  Aminoglycosides
  Polymixins
  Clindamycin
  Lincomycin
  Tetracycline
  Erythromycin
  Ciprofloxacin
  Carbapenems (Imepenem)[33]

Antiarrhythmics
  Procaine-based drugs
  Quinine-based drugs
  Lidocaine
  β-adrenergic blockers
  Cibenzoline[47]
Antihypertensives
  *Captopril[33]
  β-adrenergic blocking agents
    (propranolol)
  Calcium channel blockers
Miscellaneous
  Lithium
  Iodinated contrast agents
  Phenytoin[33]
  *D-penicillamine
  *Pyritinol[33]

*Single or rare cases.

LAMBERT-EATON MYASTHENIC SYNDROME

The Lambert-Eaton myasthenic syndrome (LEMS) was first reported as a myasthenic disorder associated with bronchial carcinoma in 1953. Edward Lambert and colleagues first differentiated LEMS from MG in 1957, and subsequently discovered that it is caused by reduced release of Ach from the presynaptic nerve terminal membrane.[12]

Pathophysiology

LEMS is the most common presynaptic disorder of NMJ transmission and is one of the major paraneoplastic neurological disorders, being associated with small-cell carcinoma of the lung in over 50% of cases. On freeze-fracture electron microscopy, the main pathology can be seen as a disorganization of the normally linear arrays of active zone particles constituting the sites of immediate release of Ach vesicles. These pathological changes can be produced by exposing normal presynaptic NMJs to the serum of patients with LEMS (Fig. 20–10). These findings, as well as the ability to produce LEMS-like clinical features in mice receiving immunoglobulin from LEMS patients, demonstrated the immune-mediated nature of this illness.[18]

The target to which LEMS IgG is directed appears to be the presynaptic membrane VGCCs. LEMS IgG crosslinks active zone particles on the nerve terminal membrane, the likely site of the VGCCs responsible for facilitating ACh release. Antibody directed against the P/Q subtype of calcium channel has been detected in the vast majority of LEMS patients both with and without small-cell lung cancer.[24] P/Q calcium channels are found on both the presynaptic membranes of NMJs and the membranes of small-cell carcinoma cells. The link between small-cell lung carcinoma and LEMS is thought to involve antigenic determinants shared between this neuroectodermal tumor and the presynaptic nerve terminal membrane. Small-cell lung cancer cells express VGCCs on their surface, and calcium influx through these channels is greatly reduced when exposed to LEMS IgG.

The electrophysiological features of LEMS result from hampered release of ACh vesicles. In the resting state of the nerve terminal, the nonquantal release of ACh is reduced, measured by a reduced rate of ACh vesicle release and a reduced number of MEPPs recorded at the muscle end plate. The size of individual vesicles and the size of individual MEPPs remain normal. By contrast, in MG the number of MEPPs is normal whereas the size of MEPPs is reduced. In LEMS, with a single nerve terminal depolarization, the quantal content of ACh is reduced. This results in the generation of a smaller EPP at the muscle end plate. The release of ACh progressively diminishes in a train of depolarizations of the nerve terminal, at which point the safety factor of NMJ transmission fails and propagated muscle fiber action potentials are not produced because threshold EPPs are not generated. In this way, the electrophysiological appearance of a presynaptic disorder such as LEMS closely resembles that of MG (see Fig. 20–3).

Repetitive Stimulation

In the EMG laboratory, the classic finding in LEMS is a reduced amplitude CMAP recorded from a muscle belly at

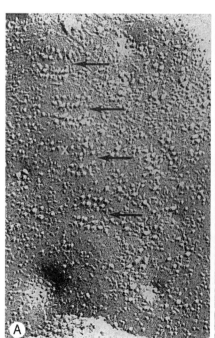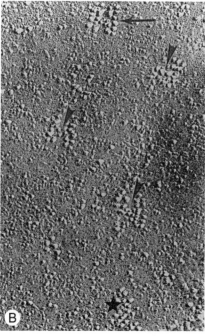

FIGURE 20–10. *(A)* Freeze-fracture electron photomicrograph of the presynaptic nerve terminal membrane, showing active zones identified by arrows, consisting of double parallel arrays of large membrane particles. *(B)* Disorganization of active zones after treatment with IgG from patients with LEMS. The arrow indicates one normal array, the arrowheads abnormal active zones, and the asterisk a particle cluster. (From Fukuoka T, Engel AG, Lang B, et al: Lambert-Eaton myasthenic syndrome: I. Early morphological effects of IgG on the presynaptic membrane active zones. Ann Neurol 22:193–199, 1987, with permission.)

rest in response to a single nerve trunk stimulus, followed by a markedly increased amplitude response when the second nerve trunk stimulus is preceded by 10 to 20 seconds of forceful voluntary contraction of the muscle (see Fig. 20–5). The electrodiagnosis of a presynaptic defect of NMJ transmission such as LEMS is suggested by a doubling of the CMAP amplitude (100% facilitation), and is certain with 200% facilitation or more. In some patients the facilitation can exceed 1000%. When patients are unable to deliver maximal voluntary contraction, a diagnostic incremental response can be elicited with tetanic repetitive electrical stimulation at 30 to 50 Hz (Fig. 20–11), but this is uncomfortable and should be undertaken only if other diagnostic maneuvers fail. Diagnostic incremental responses may not be seen in all muscles; in some cases proximal muscle responses must be recorded to confirm the diagnosis.

With slow repetitive electrical stimulation (2 to 5 Hz), decremental responses like those in MG are typical, but may not be obvious when the baseline CMAP amplitude is below 3 mV. In a study of 49 patients reported by Tim and associates, an abnormal decremental response was seen in hand muscles in 98% of cases, and the resting CMAP amplitude was reduced in the abductor digiti minimi muscle in 94%.[46] In one patient, the resting CMAP was normal in all muscles examined. Postexercise facilitation of 100% or more was seen in at least one muscle in 88%. In patients with early disease and in those whose disease is partially treated or under the influence of cholinesterase inhibitors, an incomplete electrodiagnostic pattern can be seen, with elements suggesting a postsynaptic defect or a mixed picture with features of both presynaptic and postsynaptic dysfunction. In such cases, one must rely on clinical information from the history and examination, and the presence or absence of AChR antibody to make a final diagnosis. During the course of the illness, the electrodiagnostic pattern may change in a presynaptic or postsynaptic direction.[38]

Needle Electrode Examination

The NEE in LEMS is similar to that seen in MG. Motor unit action potential configuration variation is a prominent feature in cases of moderate severity. There may be mild morphological changes in the motor unit action potential, characterized by short duration, low amplitude, and polyphasia. Fibrillation is even less commonly seen than in MG, but can occur in severe cases.

Clinical Features

LEMS is characterized primarily by progressive generalized weakness, most prominent in the proximal leg muscles. Fati-

TABLE 20–6	Clinical Manifestations of Lambert-Eaton Myasthenic Syndrome

Common
Proximal > distal extremity weakness
Diminished or absent deep tendon reflexes
Transient improvement with exercise of the muscle group
Cholinergic autonomic failure
    Dry eyes, dry mouth
    Constipation
    Erectile dysfunction

Uncommon
Craniobulbar dysfunction
    Diplopia, ptosis, dysphagia
Respiratory insufficiency

gability is less commonly described than in MG. In the later stages of disease craniobulbar involvement may occur, producing diplopia, ptosis, and dysphagia. Respiratory failure is an uncommon feature, although rarely it may be the presenting symptom of LEMS. Symptoms of autonomic cholinergic failure, such as mouth dryness, eye dryness, constipation, and erectile dysfunction, may be presenting symptoms in up to 6% of patients.[30]

The neurological examination often reveals a waddling gait. Weakness is most prominent in proximal muscles of the arms and legs, and a characteristic improvement in strength occurs after a few seconds of full voluntary muscle contraction, corresponding to the facilitation of ACh release due to calcium influx at the presynaptic site. Muscle stretch reflexes are reduced or absent, and may improve after brief, forceful exercise of the appropriate muscle group. Ptosis may be seen in some patients. The clinical manifestations of LEMS are summarized in Table 20–6.

Over 50% of patients with LEMS harbor a small-cell lung carcinoma. Other malignancies have been associated with LEMS, including adenocarcinoma of the lung, mixed parotid tumor, systemic mastocytosis, lymphoma, and renal cell carcinoma. In many cases the neurological symptoms antedate the diagnosis of the malignancy by as much as 2 years, often because the responsible tumor is initially extremely small. In some cases the malignancy is found only at postmortem examination. Rarely, paraneoplastic LEMS may be accompanied by progressive cerebellar ataxia, another paraneoplastic neurological disorder.[6]

Close to 50% of patients have no identifiable malignancy, and are thought to have LEMS on an autoimmune basis. These patients tend to be young women, in contrast to those with paraneoplastic LEMS. Patients with autoimmune LEMS,

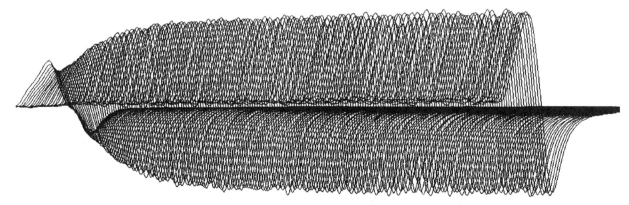

FIGURE 20–11. Incremental response to 50 Hz repetitive stimulation in a patient with LEMS. From Russ Bodner, M.D., with permission.

like those with MG, have an increased incidence of other autoimmune disorders, especially thyroid disease and pernicious anemia, but also vitiligo, celiac disease, type I diabetes, rheumatoid arthritis, psoriasis, asthma, ulcerative colitis, scleroderma, and multiple sclerosis.

Diagnosis

The generalized weakness of LEMS is usually insidiously progressive, and is often mistaken for more common conditions such as myopathy, MG, and CIDP. The combination of progressive generalized weakness, areflexia, and dry eyes and mouth of several weeks' to months' duration should raise the suspicion of LEMS. Electrodiagnostic studies to confirm a presynaptic defect of NMJ transmission and to rule out other neuromuscular disorders is the most direct and sensitive approach to diagnosis.

The majority of patients with LEMS harbor antibodies to P/Q VGCCs, the likely cause of calcium channel dysfunction and reduced release of ACh. This antibody can be measured in the serum, and is present in virtually all patients with paraneoplastic LEMS, and about 90% of patients with autoimmune LEMS.[24] It is measurable in small concentration in a few patients with small-cell cancer without LEMS and in patients with autoimmune disorders such as systemic lupus erythematosus.

CASE STUDY 2

Myasthenia Gravis or Lambert-Eaton Syndrome?

A 48-year-old man with a 3-year history of Crohn's disease developed left ptosis, diplopia, and generalized weakness over a 6-month period. On neurological examination there was fluctuating left ptosis, incomplete adduction of the right eye, and weakness of neck flexor muscles and proximal arm and leg muscles. Deep tendon reflexes were unelicitable in the arms, but present at the knees.

Routine nerve conduction studies showed resting distal CMAP amplitudes of 3.7 mV, 5.9 mV, and 2.9 mV along the peroneal (tibialis anterior), median (thenar), and ulnar (abductor digiti minimi) nerve trunks, respectively, with amplitudes of 3.9 mV, 12.4 mV, and 7.0 mV, respectively, after 15 seconds of brief forceful exercise of the muscle. Prominent decremental responses were seen with RS studies along multiple nerve trunks. Figure 20–12 summarizes the RS studies along the median nerve. The NEE showed muscle unit action potential changes characterized by polyphasia and instability.

This patient's clinical presentation suggested myasthenia gravis, and there were prominent decremental responses on RS studies. However, resting CMAP amplitudes were reduced and there was significant facilitation after exercise, although no incremental response exceeded 140%. Subsequent serological testing showed marked elevation of the P/Q VGCC antibody concentration, with absence of AchR antibodies, confirming the diagnosis of LEMS.

Management

After the diagnosis of LEMS has been confirmed, a search for malignancy is indicated, including a CT scan of the chest, blood screen, and comprehensive physical examination. In the absence of an abnormality on CT scan in a patient with a history of smoking, repeat scans every 6 months are indicated, especially in the face of progressive disease.[25] In patients with diagnosed malignancy, antineoplastic therapy is the most effective treatment.[5] In many patients with paraneoplastic LEMS, the weakness is not disabling; however, most die from their malignancy within 2 years of the diagnosis of LEMS.

Medical treatment of LEMS can be initiated with pyridostigmine in the same doses used for MG, in an attempt to improve the safety margin of NMJ transmission by increasing the ACh concentration in the synapse. 3,4-Diaminopyridine prolongs the period of nerve terminal depolarization by inhibiting the function of presynaptic potassium channels, resulting in improved ACh release. Patients can experience striking improvement in motor function with dosages of 5 to 25 mg, 3 to 4 times per day.[26, 37] Minor side effects include perioral and acral paresthesias, epigastric distress, and insomnia. Seizures have been documented in a few patients and can be treated either with dosage reduction or with the addition of an anticonvulsant.

In patients with aggressive symptoms of LEMS, immunosuppressant treatment is appropriate. Corticosteroids such as prednisone are a reasonable first step, to which azathioprine can be added for additional immunosuppressant effect. Plasmapheresis, especially when administered long term in conjunction with corticosteroids, can be beneficial in severe cases, although its onset of action may be delayed compared with its effect in MG. The benefit of high-dose intravenous gamma globulin (IVIG) has been suggested in several anecdotal reports.[2, 37]

Pharmacological agents that adversely affect neuromuscular transmission in MG have similar effects in LEMS. Table 20–5 lists a number of potentially hazardous agents.

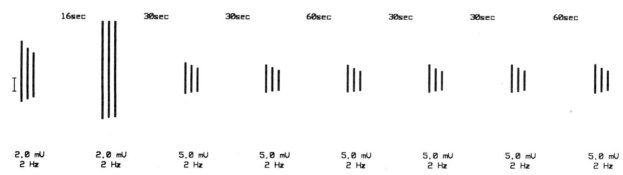

FIGURE 20–12. A 48-year-old man with a 6-month history of progression of generalized weakness and left ptosis. The bars indicate the amplitudes of median CMAPs during trains of 2 Hz repetitive stimulation before and after 16 seconds of forceful voluntary exercise of the thenar eminence. The initial amplitude is borderline low, and the initial train of stimuli produces a decremental response of 27%. After brief forceful exercise, there is an incremental response in amplitude of 144% and repair of the decremental response. Vertical bar represents the amplitude indicated under each train of stimuli.

BOTULISM

Botulism is a life-threatening presynaptic disorder of NMJ transmission produced by neurotoxins elaborated by the bacterium *Clostridium botulinum*. The disease occurs in three situations: (1) by ingestion of toxin in food, (2) by exposure to toxin produced in a traumatic wound infected by *C. botulinum*, and (3) by exposure to toxin produced by *C. botulinum* colonization of the gastrointestinal tract of infants.

Pathophysiology

Botulinum toxin is an extremely potent blocker of ACh release from the presynaptic nerve terminal membrane. There may be more than one mechanism by which botulinum toxin inhibits ACh release. The botulinum toxin molecule acts as a protease specifically inactivating synaptobrevin, a protein on the synaptic vesicle membrane important in the attachment of the vesicle to the nerve terminal membrane. The failure of ACh vesicle fusion with the nerve terminal membrane leads to failure of ACh release.[20]

As few as ten molecules of botulinum toxin can irreversibly stop ACh release. The result is complete failure of NMJ transmission across the NMJ, followed by degeneration of the motor end plate and denervation of the muscle fiber. Full recovery from botulism is delayed because new nerve sprouts are required to reinnervate denervated muscle fibers, creating new NMJs.

The electrophysiological manifestations of botulism are those of a presynaptic defect of NMJ transmission, and are similar to those of LEMS, except that the reduction in ACh release is even more profound. Little or no ACh may be released in response to a nerve terminal depolarization, and the first stimulus of a train may not produce sufficient ACh release to result in a threshold EPP. In the face of severe botulinum toxin inactivation of the NMJ, calcium influx into the nerve terminal does not significantly improve ACh release.

Repetitive Stimulation and Needle Electrode Examination

In the EMG laboratory, the RS findings in botulism depend on the amount of toxin exposure. In severe poisoning, the resting CMAP elicited by a single nerve trunk stimulus is extremely small or unrecordable. Because such patients are too weak to deliver adequate voluntary forceful exercise, rapid RS is performed, but without resulting significant increment in the CMAP amplitude. In situations where the exposure to toxin is less profound, most often in the setting of endogenous toxin production in the course of infantile *C. botulinum* colonization, a pattern with rapid RS very similar to LEMS is seen[7] (Fig. 20–13). The resting CMAP is low in amplitude, and rapid RS yields a striking incremental response. A unique electrophysiological feature in infantile botulism is the tendency for prolongation of the post-tetanic facilitation effect. The post-tetanic increase in CMAP amplitude can persist for as long as 20 minutes in some cases, compared with the 30 to 60 seconds seen in LEMS.[17]

The NEE will also vary depending upon the amount of toxin exposure. In severe cases, little or no voluntary motor unit potential activation is seen. Motor unit action potentials that do appear are low in amplitude and short in duration, reflecting the severe dropout of muscle fiber action potentials. In less severe exposure, more voluntary activation of small motor unit action potentials is seen, and motor unit action potential variation is apparent.

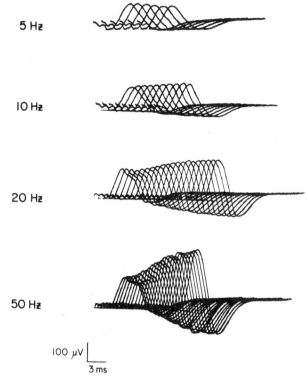

FIGURE 20–13. Repetitive stimulation studies performed on the median nerve, recording abductor pollicis brevis, in a patient with infantile botulism. At 5 Hz stimulation an 8% decrement is seen, at 20 Hz, a 38% increment is seen, and at 50 Hz, a 94% increment is seen. (From Cornblath WR, Sladky JT, Sumner AJ: Clinical electrophysiology of infantile botulism. Muscle Nerve 6:448–452, 1983, with permission.)

CASE STUDY 3

A Case of Severe Botulism

A 6-month-old boy was seen by his pediatrician for a 3-day history of poor oral intake, increased drooling, and reduced motor activity. The patient was afebrile, and there were no systemic signs of infection. Over the next 2 days, the child became quadriplegic and required ventilatory assistance. A neurological consultant raised the question of botulism. Nerve conduction studies showed normal sensory responses, but no motor responses could be obtained. Rapid RS studies failed to elicit motor responses. The NEE identified no abnormal insertional or spontaneous activity, and there were no motor unit action potentials under voluntary control.

This EMG picture defines a severe disorder of the motor unit, but does not specify the site of the lesion; peripheral motor nerves or NMJs may have been affected primarily, and the study was performed too early after onset of symptoms to reveal evidence of fibrillation potentials. The patient's serum was subsequently found to transmit botulism to rabbits, and a clear diagnosis was made.

Clinical Features

In those who ingest contaminated food, nausea and vomiting may occur within 12 to 36 hours. Neurological symptoms may be delayed up to 3 days. Cranial nerves are affected first: diplopia occurs as a result of paralysis of extraocular muscles, blurred vision occurs due to fixed dilation of the pupils. Bulbar involvement results in dysphonia, dysarthria, and dysphagia. Later, weakness of extremity and respiratory muscles occurs. In addition to the effects at nicotinic synapses, there is failure of ACh transmission across muscarinic cholinergic synapses, producing autonomic dysfunction in the form of paralysis of pupillary function, loss of salivation (sometimes producing severe drying of the pharyngeal mucosa), ileus, constipation, urinary retention, and orthostatic hypotension.

Diagnosis

The diagnosis of botulism rests on clinical suspicion and confirmatory laboratory tests. Toxin can be identified in the blood, stool, gastric contents, or food by means of a bioassay in mice. Culture of the organism anaerobically is difficult. With electrodiagnostic studies as noted earlier, a single nerve trunk stimulus at rest will produce a very low amplitude or absent CMAP. Low amplitude CMAPs in the absence of incremental responses to rapid RS or forceful exercise is a nonspecific electrophysiological finding. The differential diagnosis includes severe myopathy, acute demyelinating peripheral polyneuropathy (Guillain-Barré syndrome), polyradiculopathy, and advanced motor neuron disease. Early on, botulism can be confused with disorders such as streptococcal pharyngitis (owing to severe drying and redness of the oral mucosa) and intestinal obstruction producing ileus. The differential diagnosis also includes anticholinergic intoxication from exogenous sources, such as atropine, belladonna, jimsonweed, and mushroom poisoning.

Management

Antitoxin therapy has been recommended in adult cases, and should be administered as soon as possible. Delayed use may still be beneficial, owing to the finding of toxin in the blood of some patients as long as 30 days postintoxication. All preparations of antitoxin are of equine origin, and up to 20% of patients have untoward reactions, so that hypersensitivity testing and desensitization may be necessary before use.

The use of penicillin to eradicate *C. botulinum* in the bowel of colonized patients is of uncertain value. Even in the case of wound infection by *C. botulinum*, use of antibiotics does not prevent intoxication.[40]

Treatment with drugs that increase ACh release may yield some clinical benefit. Guanidine hydrochloride can improve general strength and respiratory function in as many as 50% of patients, but its use is severely limited by the potential complications of bone marrow suppression and renal toxicity. 3,4-Diaminopyridine has been used successfully in the treatment of LEMS, and could potentially have a positive effect on the presynaptic release of ACh in patients with botulism.

CONGENITAL MYASTHENIA

A heterogeneous group of disorders is described as congenital myasthenia. They all share in common the absence of detectable AChR antibody and the presence of the defect from birth, although the first clinical manifestations may appear in childhood, adolescence, or adulthood. Each disorder represents a different genetic defect affecting some molecular component of NMJ transmission. New specific defects continue to be described in studied families. A brief description of four major entities follows.[21] These disorders are summarized in Table 20–7.

Familial Infantile Myasthenia

Familial infantile myasthenia is an autosomal recessive disorder characterized by intermittent hypotonia, fatigable generalized weakness, ptosis, dysphagia, and respiratory insufficiency. Symptoms are exacerbated by infection or excitement.[15] It can present in infancy and childhood. In older childhood, exacerbations lessen, and the clinical picture becomes similar to MG. At times the disorder can be difficult to distinguish from antibody-negative MG, since it responds to acetylcholinesterase inhibitors. The cause is presynaptic, stemming from impaired reuptake of the ACh precursor choline by the nerve terminal, or some other step in ACh synthesis or release.

With a single nerve trunk stimulus, the resting CMAP amplitude is normal, and repetitive CMAPs are not seen. In weak muscles, a decremental response is seen with RS that repairs with brief forceful exercise. In normal muscle and in older individuals, the decremental response requires prolonged RS at 10 Hz for 5 to 10 minutes.[16]

Congenital Acetylcholinesterase Deficiency

Congenital acetylcholinesterase deficiency is an autosomal recessive disorder that often presents in the neonatal period, but may be delayed to later infancy or childhood.[14, 15] Poor cry and feeding, hypotonia, and severe generalized weakness may occur, sometimes requiring ventilatory support. The pupillary light response is sluggish. Older children and adults may have fatigable ptosis, extraocular muscle weakness, dysarthria, and dysphagia. In general, there is reduced muscle bulk and deep tendon reflexes may be hypoactive. There is a characteristic tendency to develop scoliosis and lumbar lordosis after brief standing. The disorder occurs as the result of deficiency in acetylcholinesterase in the troughs of the postsynaptic junctional folds, leading to excessive concentrations of synaptic ACh. Additionally, nerve terminals are reduced in size, the synaptic cleft is widened, and junctional folds are simplified. These features likely contribute to the finding of reduced quantal content and MEPP size. This disorder does not respond to acetylcholinesterase inhibitors.

RS studies show prominent decremental responses at slow rates of stimulation, with little repair after exercise. A classical electrodiagnostic feature in this condition is the tendency to record multiple CMAPs after a single electrical stimulus to the nerve trunk (Fig. 20–14). The repetitive CMAPs trail the initial potential by 3 to 6 msec, and are much lower in amplitude. With slow RS, decrement is more marked in the trailing potentials. Deficiency of acetylcholinesterase leads to excessive synaptic ACh and continuous occupation of the AChR binding sites, resulting in prolonged duration of the muscle end plate EPP. When a suprathreshold EPP persists beyond the muscle fiber refractory period produced by the previous propagated action potential, additional action potentials are generated.[21] This unique electrophysiological phenomenon also occurs in patients with excess of exogenous acetylcholinesterase inhibitors (pyridostigmine, malathion, parathion) and in the congenital myasthenic syndrome associated with a prolonged open time of the AChR, a condition also known as classic slow channel syndrome.

The identification of repetitive CMAPs may be hampered by several factors. Because trailing responses decrement quickly, stimulus rates as low as 0.5 to 1 Hz can abolish the repetitive potentials. In young infants and in those severely affected, the repetitive CMAPs may be lost as a result of the additional effects of reduced quantal content and small MEPP size.

TABLE 20-7	Major Forms of Congenital Myasthenia Gravis			
	Familial Infantile Myasthenia	Congenital Acetylcholinesterase Deficiency	Slow Channel Syndrome	Acetylcholine Receptor Deficiency
Presentation	Autosomal recessive Infancy-childhood	Autosomal recessive Neonatal-childhood	Autosomal dominant Childhood-adulthood	Variable inheritance Infancy-adulthood
Findings	Hypotonia Oculobulbar, generalized Exacerbations	Hypotonia, oculobulbar Generalized weakness Postural scoliosis Poor pupillary response	Oculobulbar, generalized Extensor weakness	Identical to MG
Cause	Presynaptic	Deficiency of AchE	Prolonged AChR channel open time	Various
EMG	No repetitive CMAPs Decrement in early life, in later life only after prolonged RS	Repetitive CMAPs Decremental responses	Repetitive CMAPs Decremental responses	No repetitive CMAPs MG pattern of RS
Response to anticholinesterase	Positive	Negative	Negative clinically Positive on repetitive CMAPs	Positive

AchE, acetylcholinesterase; AChR, acetylcholine receptor; CMAP, compound muscle action potential; EMG, electromyogram; MG, myasthenia gravis; RS, repetitive stimulation.

Classic Slow Channel Syndrome

The slow channel syndrome is an autosomal dominant disorder with onset from childhood to adulthood. The clinical manifestations may be identical to MG, with fluctuating ptosis, extraocular muscle and extremity weakness, that can be asymmetrical. Many patients have characteristic weakness of neck, wrist, and finger extensor muscles.[14, 15] Patients do not respond to acetylcholinesterase inhibitors, but there is a single report of improvement with the calcium channel antagonist flunarizine.[31]

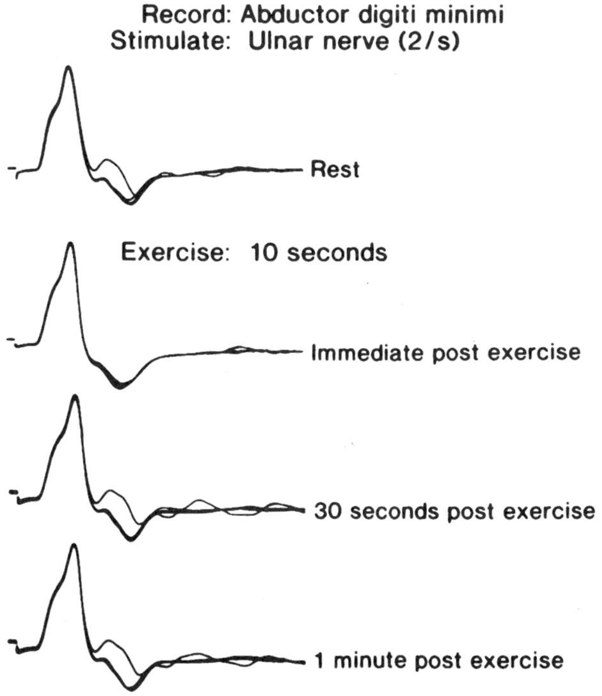

The pathophysiology of this disorder is characterized by prolongation of the open time of AChR channels, leading to prolonged duration of the muscle EPP. In addition to excess influx of sodium through the channel, the excess entry of calcium leads to degeneration of the end plate and the synapse, producing a secondary reduction of AChRs.

RS studies show prominent decrement at slow rates of stimulation, with little repair after forceful exercise. Repetitive CMAPs occur after single nerve trunk stimuli, similar to the pattern in the congenital acetylcholinesterase deficiency syndrome, except that they may be recorded in asymptomatic muscles, and in asymptomatic relatives. Although acetylcholinesterase inhibitors do not decrease the decremental response of the main CMAP to RS, they tend to increase the amplitude and number of repetitive CMAPS, in contrast to the absence of any effect of these inhibitors in the congenital acetylcholinesterase deficiency syndrome.[21] The NEE shows small motor unit action potentials, which vary in configuration.

Record: Abductor digiti minimi
Stimulate: Ulnar nerve (2/s)

Rest

Exercise: 10 seconds

Immediate post exercise

30 seconds post exercise

1 minute post exercise

FIGURE 20–14. Repetitive stimulation studies in a patient with myasthenia gravis who took Mestinon shortly before the study. Note the repetitive CMAP that follows the main M wave by several milliseconds. (From Harper CM Jr: Neuromuscular transmission disorders in childhood. In Jones HR Jr, Bolton CF, Harper CM Jr [eds]: Pediatric Clinical Electromyography. Philadelphia, Lippincott-Raven, 1996, pp 353–385, with permission.)

Congenital Acetylcholine Receptor Deficiency

Several different congenital disorders result in AChR deficiency.[14, 15] Both autosomal recessive and sporadic cases have been identified, ranging in onset from infancy to adulthood. In young patients hypotonia and respiratory insufficiency can be seen, whereas in older patients there may be ocular, bulbar, and proximal extremity weakness. The clinical picture may be identical to MG. Patients tend to respond to acetylcholinesterase, inhibitors, but not to immunosuppressants.

The electrodiagnostic findings in these disorders mimic those in MG. There are no repetitive CMAPs to single nerve trunk stimuli. The NEE shows small motor unit action potentials that vary in configuration.

OTHER DISORDERS OF NEUROMUSCULAR TRANSMISSION

Organophosphate Poisoning

Exposure to organophosphates occurs primarily through the use of insecticides such as malathion and parathion.

They act by irreversibly inhibiting acetylcholinesterase, producing weakness, muscle cramps and fasciculation, autonomic cholinergic hyperactivity, disorientation, and seizures. In mild cases, the electrodiagnostic findings can resemble pyridostigmine intoxication and congenital acetylcholinesterase deficiency, with repetitive CMAPs after single nerve trunk stimuli. The treatment of this condition includes pralidoxine and atropine.

Magnesium Intoxication

Magnesium acts as a calcium antagonist, competing for uptake at the VGCCs. In states of magnesium excess such as renal failure or after magnesium sulfate treatment for eclampsia during pregnancy, calcium influx drops, ACh release is inhibited, and generalized weakness can develop. This effect can be reversed with calcium infusion.

Tick Bite Paralysis

The toxin produced by the pregnant female tick of several species has been shown to produce progressive generalized weakness over several days, sometimes affecting swallowing and respiratory muscles. In North America, the tick *Dermacentor andersoni* has been implicated in a number of cases. Deep tendon reflexes are lost and sensation is spared, giving rise to a clinical picture that resembles Guillain-Barré syndrome. CMAPs following nerve trunk stimulation are low in amplitude and have been reported to have slowed conduction velocities.[9, 44] Only one report suggests an abnormality of RS, in the form of decrement to fast RS.[29] Thus, the disorder has not been clearly demonstrated to be a defect of NMJ transmission. However, the rapid recovery of CMAP amplitude and clinical strength that occurs after tick removal (sometimes within 24 hours) leads to the speculation that some easily reversible blockade of the NMJ, or other nerve or muscle membrane channel, may be the cause.

REFERENCES

1. Aminoff MJ, Layzer RB, Satya-Murti S, Faden AF: The declining electrical response of muscle to repetitive nerve stimulation in myotonia. Neurology 27:812–816, 1977.
2. Bird SJ: Clinical and electrophysiologic improvement in the Lambert-Eaton syndrome with intravenous immunoglobulin therapy. Neurology 42:1422–1423, 1992.
3. Rivner MR, Swift TR: Electrical testing in disorders of neuromuscular transmission. In Brown WF, Bolton CF (eds): Clinical Electromyography, 2nd ed. Boston, Butterworth-Heinemann, 1993, pp 625–652.
4. Campa JF, Sanders DB: Familial hypokalemic periodic paralyses. Arch Neurol 31:110–115, 1974.
5. Chalk CH, Murray NMF, Newsom-Davis J, et al: Response of the Lambert-Eaton myasthenic syndrome to treatment of associated small-cell lung carcinoma. Neurology 40:1552–1556, 1990.
6. Clouston PD, Saper CB, Arbizu T, et al: Paraneoplastic cerebellar degeneration. III. Cerebellar degeneration, cancer and the Lambert-Eaton syndrome. Neurology 42:1944–1950, 1992.
7. Cornblath DR, Sladky JT, Sumner AJ: Clinical electrophysiology of infantile botulism. Muscle Nerve 6:448–452, 1983.
8. Desmedt JE: The neuromuscular disorder in myasthenia gravis. II. Presynaptic cholinergic metabolism, myasthenia-like syndromes, and a hypothesis. In Desmedt JE (ed): New Developments in Electromyography and Clinical Neurophysiology, vol 1. Basel, S. Karger, 1973, pp 305–342.
9. Donat JR, Donat JF: Tick paralysis with persistent weakness and electromyographic abnormalities. Arch Neurol 38:59–61, 1981.
10. Drachman DB: Myasthenia gravis. N Engl J Med 330:1797–1810, 1994.
11. Dyken ML, Smith DM, Peake RL: An electromyographic diagnos-

tic screening test in McArdle's disease and case report. Neurology 17:45–50, 1967.
12. Eaton LM, Lambert EH: Electromyography and electric stimulation of nerves in diseases of motor unit: Observations on myasthenic syndrome associated with malignant tumors. JAMA 163:1117–1124, 1957.
13. Engel AG: The neuromuscular junction. In Engel AG, Banker BQ (eds): Myology. New York, McGraw-Hill, 1986, pp 209–254.
14. Engel AG: Congenital disorders of neuromuscular transmission. Semin Neurol 10:12–26, 1990.
15. Engel AG: Myasthenic syndromes. In Engel AG, Franzini-Armstrong C (eds): Myology, 2nd ed. New York, McGraw-Hill, 1994, pp 1806–1835.
16. Engel AG, Lambert EH: Congenital myasthenic syndromes. Electroencephalogr Clin Neurophysiol Suppl 39:91–102, 1987.
17. Fakadej V, Gutmann L: Prolongation of post-tetanic facilitation in infant botulism. Muscle Nerve 5:727–729, 1982.
18. Fukuoka T, Engel AG, Lang B, et al: Lambert-Eaton myasthenic syndrome: I. Early morphological effects of IgG on the presynaptic membrane active zones. Ann Neurol 22:193–199, 1987.
19. Gardnerova M, Eymard B, Morel E, et al: The fetal/adult acetylcholine receptor antibody ratio in mothers with myasthenia gravis as a marker for transfer of the disease to the newborn. Neurology 48:50–54, 1997.
20. Hall ZW, Sanes JR: Synaptic structure and development: The neuromuscular junction. Neuron 10:99–122, 1993.
21. Harper CM Jr: Neuromuscular transmission disorders in childhood. In Jones HR Jr, Bolton CF, Harper CM Jr (eds): Pediatric Clinical Electromyography. Philadelphia, Lippincott-Raven, 1996, pp 353–385.
22. Lambert EH, Rooke ED, Eaton LM, et al: Myasthenic syndrome occasionally associated with bronchial neoplasm: Neurophysiologic studies. In Viets HR (ed): Myasthenia Gravis. Springfield, IL, Charles C Thomas, 1961, pp 364–410.
23. Lennon VA: Serologic profile of myasthenia gravis and distinction from the Lambert-Eaton myasthenic syndrome. Neurology 48(suppl 5):S23–S27, 1997.
24. Lennon VA, Kryzer TJ, Griesmann GE, et al: Calcium-channel antibodies in the Lambert-Eaton syndrome and other paraneoplastic syndromes. N Engl J Med 332:1467–1474, 1995.
25. Levin KH: Paraneoplastic neuromuscular syndromes. Neurol Clin 15:597–613, 1997.
26. McEvoy KM: Diagnosis and treatment of Lambert-Eaton myasthenic syndrome. Neurol Clin 12:387–399, 1994.
27. McManis PG, Lambert EH, Daube JR: The exercise test in periodic paralysis. Muscle Nerve 9:704–710, 1986.
28. Miller RG, Maxfield M: McArdle's disease: Treatment and pathophysiology. Electroencephalogr Clin Neurophysiol 52:S131–S132, 1981.
29. Morris H: Tick paralysis: Electrophysiologic measurements. Southern Med J 70:121–122, 1977.
30. O'Neill JH, Murray NM, Newsom-Davis J: The Lambert-Eaton myasthenic syndrome. A review of 50 cases. Brain 111:577–596, 1988.
31. Oosterhuis JGH, Newsom-Davis J, Wokke JHJ, et al: The slow channel syndrome. Two new cases. Brain 110:1061–1079, 1987.
32. Osserman KE, Genkins G: Studies in myasthenia gravis: Review of twenty-year experience in over 1200 patients. Mt Sinai J Med 38:497–537, 1971.
33. Penn A: Drug-induced autoimmune myasthenia gravis. In: Myasthenia Gravis and Related Diseases. Ann NY Acad Sci 841:433–449, 1998.
34. Pickett JB. Neuromuscular transmission. In Sumner AJ (ed): The Physiology of Peripheral Nerve Disease. Philadelphia, WB Saunders, 1980, pp 238–264.
35. Rivner MR, Swift TR: Electrical testing in disorders of neuromuscular transmission. In Brown WF, Bolton CF (eds): Clinical Electromyography, 2nd ed. Boston, Butterworth-Heinemann, 1993, pp 625–651.
36. Rutecki PA: Neuronal excitability: Voltage-dependent currents and synaptic transmission. J Clin Neurophysiol 9:195–211, 1992.
37. Sanders DB: 3,4-Diaminopyridine (DAP) in the treatment of Lambert-Eaton myasthenic syndrome (LEMS). In: Myasthenia Gravis and Related Diseases. Ann NY Acad Sci 841:811–816, 1998.

38. Sanders DB, Stalberg E: The overlap between myasthenia gravis and Lambert-Eaton myasthenic syndrome. Ann NY Acad Sci 505:864–865, 1987.

39. Sanders DB, Stalberg EV: AAEM Minimonograph #25: Single-fiber electromyography. Muscle Nerve 19:1069–1083, 1996.

40. Schaffner W: *Clostridium botulinum* (botulism). In Mandell RG, Douglas RG Jr, Bennett JE (eds): Principles and Practice of Infectious Diseases, 3rd ed. New York, Churchill Livingstone, 1990, pp 1847–1850.

41. Stalberg E, Trontelj JV: Single Fiber Electromyography, 2nd ed. New York, Raven, 1994.

42. Streib EW: AAEE Minimonograph #27: Differential diagnosis of myotonic syndromes. Muscle Nerve 10:603–615, 1987.

43. Swift TR: Disorders of neuromuscular transmission other than myasthenia gravis. Muscle Nerve 4:334–353, 1981.

44. Swift TR, Ignacio OJ: Tick paralysis: Electrophysiologic studies. Neurology 25:1130–1133, 1975.

45. Thesleff S, Molgo J: A new type of transmitter release at the neuromuscular junction. Neuroscience 9:1–8, 1983.

46. Tim RW, Massey JM, Sanders DB: Lambert-Eaton myasthenic syndrome (LEMS): Clinical and electrodiagnostic features and response to therapy in 57 patients. In: Myasthenia Gravis and Related Diseases. Ann NY Acad Sci 841:823–826, 1998.

47. Wakutani Y, Matsushima E, Son A, et al: Myasthenia-like syndrome due to adverse effects of cibenzoline in a patient with chronic renal failure. Muscle Nerve 21:416–417, 1998.

Chapter 21

Myopathies

Mark A. Ferrante, MD • Asa J. Wilbourn, MD

INTRODUCTION
PERTINENT ANATOMY AND PHYSIOLOGY
NERVE CONDUCTION STUDIES
LATE RESPONSES
REPETITIVE STIMULATION STUDIES
NEEDLE ELECTRODE EXAMINATION
Insertion Phase
Rest Phase
Activation Phase
AN APPROACH TO DIAGNOSIS
THE ELECTRODIAGNOSTIC
    EXAMINATION—LIMITATIONS

THE ELECTRODIAGNOSTIC
    EXAMINATION—BENEFITS
ELECTRODIAGNOSTIC FINDINGS IN
    SELECTED MYOPATHIES
Polymyositis
Inclusion Body Myopathy
Duchenne Muscular Dystrophy
Myotonic Dystrophy
Limb Girdle Dystrophy
AIDS Myopathy
Acute Illness/ICU Myopathy

INTRODUCTION

Of the various portions of the peripheral neuromuscular system (PNMS)—anterior horn cells (AHCs), axons, neuromuscular junctions (NMJs), and muscles—the early electromyographers had the least interest in disorders affecting muscle fibers. This is ironic, considering that all the information they were acquiring about the other components of the PNMS was obtained by recording the electrical activity of muscle, either with needle or surface electrodes. Nonetheless, it was only after Kugelberg's two papers on the needle electrode examination (NEE) findings with myopathies appeared, in 1947 and 1949, that electromyographers developed a sustained interest in using the electrodiagnostic (EDX) examination to assess patients with suspected muscle disorders.[12, 13]

The term myopathy encompasses a large, heterogeneous group of primary muscle fiber disorders. The EDX manifestations of myopathies can range from normal to grossly abnormal. Moreover, even when abnormal, the EDX examination does not supply an etiological diagnosis. Nonetheless, it can exclude other disorders with which myopathies can be confused, and sometimes it provides information that is not otherwise available, for example, the presence of electrical myotonia in the absence of clinical myotonia.[26]

PERTINENT ANATOMY AND PHYSIOLOGY

A motor unit, defined as a single AHC and all the muscle fibers that it innervates by means of its axons, terminal nerve branches, and intervening NMJs, is the smallest unit of muscle contraction over which the central nervous system has control. The average number of muscle fibers in a motor unit (i.e., the average number of muscle fibers innervated by a single axon) determines the *innervation ratio* of the muscle. This value varies greatly from one muscle to another and is inversely related to the degree of dexterity required

of the particular muscle; that is, the number is quite small for intrinsic hand muscles (± 200/1), whereas it is much higher for large leg muscles (± 1600/1). The muscle fibers belonging to a single motor unit, under normal circumstances, seldom are contiguous.[28] Rather, they are scattered over a circular-to-oval territory, termed the *motor unit territory,* whose cross-sectional area is large enough to contain the muscle fibers of approximately 20 to 30 other motor units[3] (Fig. 21–1).

Whenever a lower motor neuron is activated, the action potential that is generated propagates centrifugally along the motor axon, ultimately causing a muscle fiber action potential (MFAP) to be generated in each of the muscle fibers it innervates. The summation of these MFAPs is termed the motor unit action potential (MUAP). With myopathies, individual muscle fibers are functionally lost, resulting in a reduction of the number of muscle fibers within each motor unit. Thus, the motor unit becomes *internally fragmented* or *disintegrated.* As discussed later in this chapter, this motor

Concepts Regarding Motor Unit

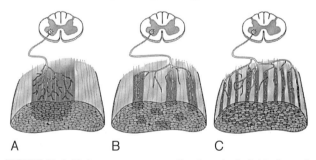

A   B   C

FIGURE 21–1. Various concepts regarding how the individual muscle fibers of a motor unit are distributed within a muscle. *A,* Early theory: all muscle fibers of the same motor unit are grouped together. *B,* Later theory: muscle fibers of same motor unit are grouped in small "subunits," consisting of 8 to 30 fibers. *C,* Current theory: muscle fibers of same motor unit are scattered throughout the muscle, so that two fibers seldom are contiguous. (From Wilbourn AJ, Ferrante MA: Clinical electromyography. In Joynt RJ, Griggs RC (eds.): Clinical Neurology, vol 1. Philadelphia, Lippincott-Raven, 1997, pp 1–76; used with permission.)

The opinions and assertions contained herein are the private views of the authors and are not to be construed as the official policy or position of the U.S. Government, the Department of Defense, or the Department of the Air Force.

unit change is quite different from the motor unit disruption caused by neurogenic disorders that result in all the muscle fibers of the motor unit simultaneously becoming inoperative ("all or none" principle), owing either to conduction failure or to conduction block.[27, 28]

The EDX examination typically consists of motor and sensory nerve conduction studies (NCS), one or more late responses, and the NEE. Of these, only the NEE plays a consistent, meaningful role in diagnosing myopathies; all other components of the EDX examination, including the NCS, are of limited value. Nonetheless, they are important in patient assessment, principally because they frequently are abnormal with other disorders that may be mistaken clinically for myopathies. How each of these is affected with myopathies is discussed next.

NERVE CONDUCTION STUDIES

The NCS, both motor and sensory, usually are normal with myopathies. Occasionally, however, they are altered. Thus, the sensory NCS may be abnormal in the presence of a coexisting peripheral nervous system (PNS) disorder or if the patient is elderly (i.e., unelicitable lower extremity sensory responses bilaterally). Nonetheless, because the sensory NCS do not assess skeletal muscle tissue, they are never directly affected by myopathies. In contrast, the compound muscle action potentials (CMAPs) elicited during the motor NCS may be abnormally low in amplitude with a myopathy if the latter causes a significant inactivation or loss of individual muscle fibers in the recorded muscle. Nonetheless, low-amplitude CMAPs seldom are seen with myopathic disorders, primarily because the routine motor NCS are recorded from distal upper and lower extremity muscles, whereas myopathies have a predilection for the proximal musculature.[26] Some noteworthy exceptions include (1) the uncommon generalized myopathies that affect muscles diffusely, without a striking proximal-to-distal gradient (e.g., acute illness myopathy[14]); (2) generalized myopathies that are sufficiently severe and chronic that they have extended distally enough along the limbs to involve the intrinsic hand and foot muscles (e.g., far-advanced polymyositis); and (3) myopathic processes that preferentially affect the distal muscles (myotonic dystrophy, Welander's distal myopathy).[26, 27] If motor NCS are performed using more proximal muscles as recording sites, the CMAPs with many myopathies may be abnormally low in amplitude—they almost invariably are if the recorded muscles are clinically weak. Thus, in patients with bilateral "myopathic" foot drop, the CMAP recorded from either tibialis anterior (TA) muscle while stimulating the ipsilateral common peroneal nerve characteristically is very reduced in amplitude. Similarly, in patients with myopathic weakness of the proximal upper extremity and shoulder girdle, the musculocutaneous and axillary CMAPs obtained by recording from biceps and deltoid muscles, respectively, usually are of low amplitude.

LATE RESPONSES

H reflexes can be obtained consistently only by stimulating the tibial nerve in the popliteal fossa, while recording from the gastrocnemius and soleus muscles. When this technique is used, the H reflexes generally are unaffected by myopathies. However, in the uncommon instances in which these muscles are severely involved, the H reflex typically is unelicitable, and the direct M wave amplitude from the soleus is reduced. F waves usually are elicited from distal muscles;

consequently, they seldom are affected by myopathies. If the distally recorded muscles undergo substantial myopathic change, F waves may be unelicitable, or the number elicited for a given number of stimuli may be reduced. These changes characteristically do not occur until the direct M waves recorded from the affected muscles are severely reduced.[26–28]

Because of their low yield, H reflex and F wave tests usually are not included in myopathy assessments.

REPETITIVE STIMULATION STUDIES

To detect presynaptic and postsynaptic defects in neuromuscular transmission, repetitive stimulation studies are performed at both slow rates (e.g., 2 Hz) and fast rates (e.g., 40 Hz), respectively. Repetitive stimulation studies typically are normal with myopathies, unless they manifest changes in the excitability of the muscle membranes in response to activation, as with some myotonic disorders. Incrementing responses are seldom seen with myopathies.

The principal reason to perform repetitive stimulation studies when patients are referred for the evaluation of a presumed myopathy is to exclude myasthenia gravis, because many patients with it are referred to the laboratory with the presumptive diagnosis of a generalized myopathy.[25, 26]

NEEDLE ELECTRODE EXAMINATION

Of the various components of the EDX examination, the NEE is by far the most important in the diagnosis of myopathies. In fact, myopathies are one of the few generalized neuromuscular disorders in which customarily the only abnormal component of the entire EDX study is the NEE. (Other such disorders include radiculopathies and early AHC disease.)

The NEE can be divided into three phases: insertion, rest, and activation. These have been discussed in detail elsewhere (see Chap. 12). Virtually any or all phases of the NEE can be abnormal with myopathies. Moreover, the mechanical resistance to needle movement may be altered substantially. The changes each of these phases manifest as a result of myopathies are reviewed next.

Insertion Phase

The brief burst of electrical activity observed during this period, termed *insertional activity,* is caused by needle electrode–induced mechanical depolarization of muscle fibers. It signifies that the needle electrode has entered viable muscle tissue, and is defined as normal, decreased, or increased. Normally, these bursts persist less than 50 msec after needle advancement ceases. Because this activity is the product of muscle fibers, the quantity generated is reduced by any condition that (1) decreases the amount of muscle tissue present (e.g., fibrous or fatty tissue replacement), (2) causes muscle fibers to be less compact (e.g., intramuscular edema), or (3) impairs muscle fiber depolarization (e.g., periodic paralysis during a bout of weakness). Occasionally the electrical insertional activity seen with myopathies is described as "increased." Most often this consists of insertional positive sharp waves that are not accompanied by actual spontaneously firing potentials. These unsustained (by definition) potentials usually are either very minimal myotonic discharges or the forerunner of fibrillation potentials.

The degree of mechanical resistance encountered during

TABLE 21–1	Types of Spontaneous Activity Potentially Observed with Various Myopathies	

Fibrillation Potentials	Myotonic Potentials	Complex Repetitive Discharges
Inflammatory myopathies (i.e., PM, DM, IBM)	Myotonia congenita	Inflammatory myopathies (PM, DM, IBM)
Muscular dystrophy (e.g., Duchenne, myotonic)	Myotonic dystrophy	Muscular dystrophy (Duchenne, limb-girdle)
Infectious myopathies (e.g., trichinosis, AIDS)	Paramyotonia congenita	Acid maltase deficiency
Toxic myopathies (e.g., alcohol, chloroquine, zidovudine, aminocaproic acid)	Proximal myotonic myopathy	Proximal myotonic myopathy
Acute illness/ICU myopathy	Myotonia fluctuans	Debrancher myopathy
Proximal myotonic myopathy	Hyperkalemic periodic paralysis	Myxedema
Muscle trauma (e.g., injection, ischemia)	Polymyositis[a]/dermatomyositis[a]	Schwartz-Jampel syndrome
Rhabdomyolysis	Debrancher myopathy[a]	
Acid maltase deficiency	Acid maltase deficiency[a]	
Debrancher myopathy	Myotubular myopathy[a]	
Myotubular myopathy	Certain drugs (e.g., chloroquine, diazocholesterol, clofibrate)[a]	
Late-onset rod myopathy	Hypothyroid myopathy[a]	
Nemaline myopathy		
Muscle carnitine deficiency		
Sarcoid myopathy		
Hyperkalemic periodic paralysis		

AIDS, acquired immunodeficiency syndrome; DM, dermatomyositis; IBM, inclusion body myopathy; ICU, intensive care unit; PM, polymyositis.
[a]Electrical myotonia without accompanying clinical myotonia.
Modified from Daube JR: AAEM Minimonograph #11: Needle examination in clinical electromyography. Muscle Nerve 14:685–700, 1991; and Wilbourn A: The electrodiagnostic examination with myopathies. J Clin Neurophysiol 10(2):132–148, 1993. Additional material from Ricker K, Koch MC, Lehmann-Horn F, et al: Proximal myotonic myopathy. Arch Neurol 52:25–32, 1995.

needle electrode advancement also is evaluated. This is unaltered with most myopathies. Decreased mechanical resistance, however, is noted during the initial insertion, as the needle penetrates the skin, in patients with subcutaneous edema and in those with corticosteroid excess, either of exogenous (e.g., prednisone treatment) or endogenous (e.g., Cushing syndrome) origin. Moreover, it is observed when the needle advances within the muscle itself whenever intramuscular edema exists. Conversely, increased mechanical resistance is detected on the initial needle advancement in patients with thickened subcutaneous tissues (e.g., scleroderma), and when the needle electrode is within what formerly was muscle substance but now has been replaced by fibrous tissue. Fibrotic change in muscle may result from (1) severe chronic denervation (e.g., remote poliomyelitis), (2) very chronic muscle disorders that have caused muscle fiber destruction (e.g., end-stage polymyositis), (3) traumatic myopathies (e.g., repeated intramuscular injections of certain substances, such as pentazocine), and (4) ischemia-induced muscle necrosis (e.g., compartment syndromes).[25, 27]

Rest Phase

During this period of the NEE, composed of the time interval between needle electrode advancements, when the needle electrode is maintained motionless within the completely relaxed muscle, any electrical activity observed (other than miniature end plate potentials and end plate spikes) is abnormal, and is referred to as *spontaneous activity*. Of the many types of spontaneous activity that occur, only three are encountered with myopathies in any significant amount: (1) fibrillation potentials, (2) myotonic discharges, and (3) complex repetitive discharges (CRDs)[27] (Table 21–1). None of these is specific for myopathies, as all are seen with neurogenic disorders as well.

Fibrillation potentials are spontaneous repetitive action potentials generated by single muscle fibers. They are, overall, the most common type of spontaneous activity seen with myopathies, just as they are the most common type seen with neuropathies. There is a certain irony in this fact, given that the early electromyographers considered that fibrillation potentials were essentially pathognomonic of a PNS disorder

and, therefore, that they never occurred with myopathies. It was not until 1950 that Lambert and colleagues and Guy and associates independently reported that fibrillation potentials frequently were found with dermatomyositis.[8, 15] Since then, it has become obvious that these potentials are seen with a variety of myopathic processes, including polymyositis, inclusion body myopathy, acute illness/intensive care unit (ICU) myopathy, trichinosis, and the more rapidly progressing muscular dystrophies. One common thread linking these myopathies is that muscle necrosis is characteristic of all of them. (It is pertinent to note that not all of them show inflammation on muscle biopsy.) For this reason, an appropriate shorthand term for muscle disorders in which fibrillation potentials occur is "necrotizing myopathy."[26, 27]

Why spontaneous action potentials so closely associated with muscle fiber denervation are found so frequently with primary muscle fiber disease is unclear, although several theories have been proposed to explain it[25, 26] (Table 21–2). Whatever the mechanism, it must be common to the great number of myopathies in which fibrillation potentials occur. At present, the most widely accepted view is that "myopathic" fibrillation potentials are generated by a muscle fiber that has undergone segmental necrosis, specifically, that portion of the muscle fiber that no longer contains the end

TABLE 21–2	Several Theories Proposed to Explain Fibrillation Potentials That Occur with Myopathies

Degenerating muscle fibers
Increased membrane irritability
Terminal nerve fiber abnormalities
Low intracellular potassium
Regenerating muscle fibers
Local membrane defect
Segmental muscle fiber necrosis
Longitudinal fiber splitting with obliterated bridge
De novo formation of myotubules

From Wilbourn AJ: The EMG examination with myopathies. In Syllabus 1987 AAEE Course A: Myopathies, Floppy Infant, and Electrodiagnostic Studies in Children. Rochester, MN, American Association of Electromyography and Electrodiagnosis, 1987, pp 7–20.

plate.[1] The various theories presume that fibrillation potentials develop in denervated, or at least "noninnervated," muscle segments. Thus, changes suggestive of reinnervation in the terminal innervation pattern have been found, using both single-fiber electromyography and histochemical techniques, in polymyositis.[8, 9] Consequently, the concept of the early electromyographers, that fibrillation potentials are always a sign of denervation, may ultimately prove to be correct. Nonetheless, to refer to disorders in which the primary defect is in the muscle fiber as "denervating diseases" can only cause confusion among clinicians. For this reason, fibrillation potentials, when seen with myopathies, should be labeled as such, and not *denervation potentials*.

Similar to the fibrillation potentials manifested by neurogenic disorders, those that occur with myopathies have two distinct forms—positive sharp wave and brief biphasic spike—depending principally on whether the tip of the needle recording electrode has actually injured the muscle fiber or has simply approached near it.

In general, the fibrillation potentials seen with myopathies are identical in appearance to those encountered with denervating neurogenic disorders. Thus, "myopathic" and "neurogenic" fibrillation potentials cannot be distinguished from one another by either their form or their size. The one notable difference between the two concerns their firing pattern. Rather frequently, "myopathic" fibrillation potentials fire at very slow rates.[6, 25, 26] Such "slow ticking" fibrillation potentials are, in our opinion, the only EDX feature essentially pathognomonic for myopathies, because they are almost never seen with neurogenic disorders.

The size of fibrillation potentials is mainly determined by two factors: (1) the diameter of the discharging muscle fiber, and (2) the relationship of the discharging fiber to the recording area of the needle electrode, particularly the distance between them. With some myopathies, the muscle fiber diameter is very reduced and, consequently, the fibrillation potentials generated are extremely small. Such "tiny" fibrillation potentials often are seen in adults with very chronic myopathies and, particularly, in boys with Duchenne muscular dystrophy. For the detection of these minute electrical potentials, two conditions must be met: (1) the electromyogram (EMG) machine sensitivity must be high, with 50 μV per division preferred, and no greater than 100 μV per division used; and (2) the muscle being sampled must be completely relaxed, so that an absolutely quiet field is present. In the absence of these conditions, fibrillation potentials are easily overlooked, as they are so readily obscured by both baseline noise and voluntary MUAPs.[25, 26]

The distribution of fibrillation potentials varies with the particular myopathy and with its severity. Most often they are found in the paraspinal muscles, limb girdle muscles, proximal limb muscles, and some of the mid-limb muscles (particularly the brachioradialis and TA). Less often, they are generalized in distribution, although often more prominent in the more proximal muscles. Sometimes they are very patchy in distribution, being present in significant numbers in one proximal muscle while undetectable in another one. Because of this fact, whenever fibrillation potentials are sought in a patient with myopathy, several muscles must be sampled, particularly including the paraspinal muscles, before any conclusions are drawn regarding their presence or absence. Otherwise, they can be missed, thereby resulting in the particular muscle disorder present being assigned to an incorrect etiological category. They also may be patchy in appearance within individual muscles, so at times an extensive search (i.e., several needle insertions in the same muscle) is required before they are detected.[25, 26]

With myopathies, the density of fibrillation potentials (i.e., the amount encountered in a particular muscle) usually is a reliable indicator of the level of activity of the necrotizing muscle disease producing them. Thus, if they are abundant, the process is quite active; conversely, if they are sparse, the process is relatively inactive (e.g., regressing, only slowly progressive, or far advanced). (MUAP changes, in contrast, appear to reflect the overall severity of the myopathic process, rather than the degree of its activity at the time of the EDX examination.) This fact has practical importance whenever patients with treatable myopathies, such as polymyositis, are developing progressive weakness in spite of what is presumed to be adequate therapy.[25] These points are discussed later in the chapter.

Widespread fibrillation potentials alone (i.e., not accompanied by any substantial MUAP changes) occasionally are encountered in patients who are experiencing progressive weakness. Often, these patients clinically are diagnostic dilemmas. Because fibrillation potentials per se are nonspecific changes, their presence does not provide a definite diagnosis. Nonetheless, in these instances, they are much more likely to be caused by a myopathic, rather than a neurogenic, process, because recent-onset myopathies can generate fibrillation potentials without significantly altering the MUAPs. In contrast, fibrillation potentials usually are detected with chronic progressive neurogenic disorders only when or after chronic neurogenic MUAP changes appear.[26]

Myotonic discharges are repetitive action potentials generated by single muscle fibers. Usually they are produced by mechanical stimulation of muscle fibers, including movement of the needle recording electrode. Myotonic discharges are the second most prevalent type of insertional or spontaneous activity associated with myopathies. Historically, they were the first type of spontaneous activity linked to abnormalities of muscle fibers, being described in 1939. They have two different forms, consisting of trains of (1) positive sharp waves, and (2) negative brief sharp spikes. The positive sharp wave form is the type customarily associated with clinical myotonia; these potentials vary, both in firing frequency and amplitude, as they repetitively discharge. Such alterations produce corresponding changes in the sound they generate, resulting in the distinctive "diving airplane" sound associated with them. The negative brief sharp spike type follows brief contraction of a rested muscle, and is referred to as "prolonged afterdischarge." Although less appreciated than the positive sharp wave type, it is the electrical phenomenon directly responsible for impaired muscle relaxation (i.e., the major symptom of clinical myotonia). It is suppressed by sustained or repeated muscle activation, a process referred to as the "warm-up phenomenon"; consequently, it is best observed following brief, vigorous contraction of a rested muscle.[5, 26]

Although all myotonic discharges result from a muscle membrane disorder, the muscle diseases in which they are seen are quite heterogeneous in terms of their underlying pathogenesis. The myotonia associated with myotonia congenita is caused by a primary disturbance of chloride channels, whereas that associated with myotonic dystrophy, paramyotonia congenita, and hyperkalemic periodic paralysis results from sodium channel dysfunction.[1, 7] The EDX attributes of myotonic discharges (i.e., their quantity, distribution, and duration) differ among the various myotonic disorders. With myotonia congenita, they are intense and generalized; with myotonic dystrophy, they are less pronounced and have a distal prominence; and with many cases of adult acid maltase deficiency, they are sparse, of brief duration, and restricted to the paraspinal muscles.[27]

Two points regarding myotonia merit comment. First, clinical myotonia and electrical myotonia are not synonymous terms and, for this reason, they should not be used interchangeably. Only the predominantly negative spike type of myotonic discharge is constantly linked to clinical myotonia; the positive sharp wave type occurs with several myopathies which, on clinical examination, have no demonstrable impairment of muscle relaxation, including hyperkalemic periodic paralysis, adult acid maltase deficiency, and polymyositis. Nonetheless, whenever only electrical myotonia occurs, it generally tends to be limited in both quantity and distribution. Second, electrical myotonia is not restricted to just myopathies. Occasionally, it is found in muscles that have undergone severe, chronic denervation. However, its presence in these circumstances almost never causes diagnostic confusion, because it is usually relatively sparse and always accompanied by prominent chronic neurogenic MUAP changes.[26]

Complex repetitive discharges are generated by a group of adjacent muscle fibers, with a single muscle fiber, termed the "principal pacemaker," ephaptically activating the group and then being ephaptically reactivated by one of the previously activated muscle fibers ("co-pacemaker").[24] They are the third most common type of spontaneous activity observed with myopathies.

Similar in morphology to both fibrillation potentials and myotonic discharges, CRDs are nonspecific in nature, in that they can occur with both neurogenic and myopathic disorders. They differ from the other two, however, in that they are only very rarely of assistance in diagnosing any particular neuromuscular disorder. Thus, characteristically, the diagnosis neither depends on their presence, nor is it altered in any way by their absence. Consequently, when they occur with myopathies, they are usually superfluous; their only positive contribution is that they indicate that the disorder is chronic, as they rarely are seen with acute processes. Also, they occasionally are found in one muscle, the iliacus, in normal persons. Unfortunately, that muscle often is among the proximal muscles assessed in patients with suspected myopathies. Consequently, if CRDs are detected only in it, they should be considered a normal variant, of no significance.[25–27]

Activation Phase

During this period of the NEE, the patient activates MUAPs on request, using different degrees of force, so that they can be assessed in regard to their firing pattern, external configuration (duration, amplitude), internal configuration (number of turns or phases), and stability on repeated firings. As noted, the early electromyographers recorded the electrical activity of muscle not to assess the muscle itself but, rather, to determine the status of either the motor nerve innervating it, or the neuromuscular junctions linking it with the terminal motor nerve fibers. Kugelberg was the first electromyographer to delineate the MUAP alterations seen with primary muscle disease. Based on the NEE findings he found in patients with "muscular dystrophy" (a term used at that time for almost any muscle disorder), he described almost all of the MUAP changes now considered typical of those seen with myopathies. These include MUAPs of decreased duration; MUAPs of decreased amplitude; an increased proportion of polyphasic MUAPs, often with their overall duration shorter than normal; MUAPs firing out of proportion to effort; and, when muscles are affected by end-stage disease, MUAPs firing in severely decreased numbers at a rapid rate.[12, 13] For several decades, all of these MUAP changes were attributed solely to the loss of functioning muscle fibers within individual motor units. More recently, however, principally as a result of computer simulation, other causes for them have been appreciated. Thus, regenerated muscle fibers may play a major role, as may variations in conduction velocity (CV) along the individual muscle fibers,

owing to differences in their diameter.[17, 18, 22] These MUAP alterations are briefly considered here.

An increased proportion of polyphasic MUAPs (i.e., MUAPs having more than four phases), probably is the earliest recognizable change with myopathies, although at least conceivably, MUAPs with multiple turns (i.e., serrated MUAPs), but not polyphasic, could be an even earlier change. These MUAP alterations probably are caused by increased temporal dispersion (desynchronization) of the MFAPs generated in the relatively few muscle fibers near the needle recording surface. This increased temporal dispersion could result from variations in muscle fiber diameter, longitudinal muscle fiber splitting, or the spread of end plate regions in the unaffected muscle fibers. Moreover, a dropout of muscle fibers from each motor unit with the loss of a few near the recording electrode tip could also play a role.

Decreased MUAP duration is considered one of the most reliable findings with myopathies. It probably results from functional loss of a substantial number of muscle fibers within the motor unit. This affects the MUAP duration because all the muscle fibers of the motor unit contribute about equally to the onset and terminal portions of the MUAPs. Even though segments of a few of the muscle fibers are near the tip of the recording electrode, the MFAPs are initiated so relatively distant from those segments that, overall, these muscle fibers individually contribute no more than any of the other muscle fibers do to the MUAP duration. Because far more muscle fibers of the motor unit are at a distance from the recording electrode than near it, the distant muscle fibers are responsible for most of the beginning and ending components of the MUAP.

Decreased MUAP amplitude is considered less reliable evidence of a myopathy, because it is so dependent on needle electrode factors. This MUAP change probably results from one or more alterations in the relatively few muscle fibers in the motor unit that are near the recording surface of the electrode; it is these fibers that are principally responsible for the spike component of the MUAP. Other potential causes include the smaller diameter of muscle fibers, producing smaller MFAPs; dispersion of muscle fibers in the motor unit with intervening muscle fibrosis; and variations in muscle fiber CVs, leading to desynchronization of the spike component of the summated MUAP.

Early recruitment of MUAPs is considered a very reliable indicator of substantial disruption of the motor unit. However, this characteristically is seen only with severe myopathies, because it indicates that each individual motor unit has been depleted of a significant number of functioning muscle fibers. As a result, far more motor units than normal must fire to generate a given degree of force. Because marked muscle fiber dropout is required to produce early recruitment, it is *never* the initial EDX change found with myopathies; whenever it is evident, major alterations in MUAP appearance should be obvious as well.

MUAPs firing in severely decreased numbers, at faster than their basal firing rate (reduced recruitment), are seen only with "end-stage" muscle disease. With very severe disruption of the motor unit, all the muscle fibers of individual motor units may be lost on a random basis. Consequently, a dropout of entire MUAPs is seen on activation, mimicking the MUAP firing pattern observed with severe neurogenic disorders. Under these circumstances, the few MUAPs that can be activated usually are very abnormal in appearance, being of decreased duration and amplitude, and thereby often having the appearance of single fibrillation potentials.

Increased MUAP duration, which characteristically is linked to chronic neurogenic disorders, occasionally is seen with some chronic myopathies, especially those in which substantial muscle fiber destruction and regeneration occur with subsequent desynchronization and delay of some components of the MUAP (e.g., polymyositis, inclusion body myopathy, Duchenne muscular dystrophy). These external configurational changes in the MUAPs result from their undergoing considerable remodeling.[1, 4, 7, 17, 18, 23, 25, 26]

The term "myopathic" MUAP is firmly entrenched in the literature. Nonetheless, its use is discouraged,[6] and when it is used, it should always be bracketed with quotation marks because none of the MUAP changes seen with myopathies is pathognomonic of a muscle fiber disorder. Instead, all such alterations merely are indicative of various disruptive processes occurring within the individual motor units. The early electromyographers referred to disturbances of this nature as "disintegration of the motor unit." Even though this term seldom is used today, it is a most apt descriptive phrase, because it emphasizes the differences between disorders that characteristically affect entire motor units, such as most PNS lesions, and those that affect fractions of the motor unit, as myopathies characteristically do. Motor unit disintegration, which results in the generation of "myopathic" MUAPs, can be produced by a number of disorders of the terminal nerve fibers, the NMJs, and the muscle fibers themselves. The only requirement is that a variable number of the muscle fibers composing the motor unit not fire, or not fire at the appropriate time, whenever the motor unit is activated. There may be several causes for this, including: (1) early or partial reinnervation following denervation; in this case, the regenerating axons are narrow and therefore conducting at slower rates than normal, and only some of the NMJs are mature enough to transfer impulses; (2) conduction blocks or axon degeneration affecting some of the terminal motor nerve fibers, such as is seen with early Guillain-Barré syndrome and with botulism; (3) neuromuscular transmission abnormalities, particularly those in which transmission is blocked along many of the NMJs, such as frequently occurs with myasthenic syndrome and some of the muscle-relaxing drugs; and (4) many primary muscle diseases in which individual muscle fibers are inactivated, destroyed, or have regenerated (Fig. 21–2).

All of these etiologies can produce "disintegrated MUAPs," which are very similar in appearance. Those caused by early reinnervation and defects in neuromuscular transmission may show instability on repetitive firing, but this can be difficult to determine unless individual MUAPs are isolated. The needle recording electrode simply registers the fact that some of the muscle fibers of the motor unit are not firing, not why they are failing to do so. For this reason, it

can be almost impossible to distinguish, in the ICU, a very early Guillain-Barré syndrome from acute illness/ICU myopathy, based on the NEE findings. In these situations, unfortunately, the NCS may be of little help as well, because the sensory nerve action potentials (SNAPs) may be normal with either (although later, with Guillain-Barré syndrome they usually would be affected), while the CMAPs may be uniformly low in amplitude with either as well, owing to either very distal conduction blocks or to muscle fiber inactivation. In any case, the motor latencies and CVs would be normal. Similarly, it is difficult to diagnose a superimposed or coexisting myopathy in the presence of a known defect in neuromuscular transmission (e.g., to diagnose "myasthenic myopathy"). With both the aforementioned situations, the "myopathic" MUAPs seen on NEE may not be caused by intrinsic abnormalities in the muscle fibers but, rather, by normal muscle fibers failing to generate MFAPs because the impulses that would cause them to do so are blocked, either along the terminal nerve fibers or at the NMJs;[25, 26] the latter point was first recognized by Kugelberg in 1947.[12]

The MUAP alterations seen on NEE with myopathies indicate the severity of the myopathic process, just as the density of fibrillation potentials seen with necrotizing myopathies reflects how active the process is at that time. With treatable necrotizing myopathies (e.g., polymyositis), various combinations of fibrillation potentials and MUAP changes can be observed, depending on when the NEE is performed in relationship to the progression-and-regression cycle of the myopathy (Fig. 21–3). Thus, with recent-onset necrotizing myopathies, fibrillation potentials predominate; with established cases, both fibrillation potentials and MUAP changes are prominent; while with inactive disease, MUAP changes persist far longer than do the fibrillation potentials.[15, 25, 26]

AN APPROACH TO DIAGNOSIS

Most patients referred to the EDX laboratory for myopathy assessment are thought to have a generalized disease. Consequently, as when all patients with suspected generalized disorders are evaluated, an upper extremity and a lower extremity should be studied; that is, the assessment should be "vertical," rather than "horizontal" (both upper or both lower extremities). Unless circumstances dictate otherwise, the limbs assessed should be ipsilateral to one another. As the NEE progresses, if it does appear that a myopathy is

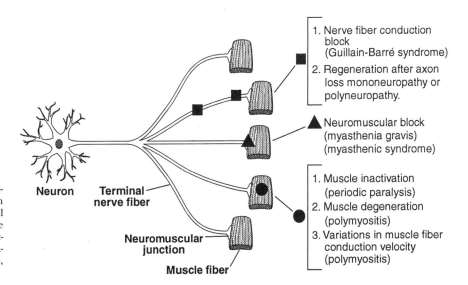

FIGURE 21–2. The various causes of "myopathic" motor unit potentials being seen on the needle electrode examination. All of these produce "disintegration of the motor unit." (From Wilbourn A: The electrodiagnostic examination with myopathies. J Clin Neurophysiol 10:132–148, 1993; used with permission.)

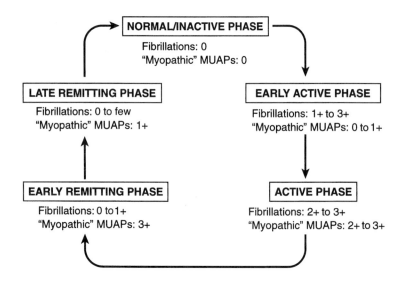

FIGURE 21–3. The cycle of changes seen on needle electrode examination during the various stages of a reversible necrotizing myopathy, such as polymyositis. (From Wilbourn A: The electrodiagnostic examination with myopathies. J Clin Neurophysiol 10:132–148, 1993; used with permission.)

present, the assessment should not be extended to the contralateral limbs. Otherwise, muscles will not be available for biopsy.

Both NCS and NEE should be performed on every patient referred to the EDX laboratory for a myopathy assessment. As previously noted, the NCS usually are normal. Nonetheless, one of the functions of the EDX examination in evaluating patients with suspected myopathies is to exclude other disorders of the PNMS. The NCS are quite helpful for doing so, especially in regard to those affecting peripheral nerve fibers. They are of particular benefit when NEE abnormalities are present that are equally consistent with both a myopathic and a neurogenic process; when caused by the latter, most often they result from early reinnervation or from conduction blocks along the terminal nerve fibers.

Concerning NCS, at least one motor and one sensory study should be performed on each of the two limbs assessed. Additional NCS sometimes are indicated, particularly motor NCS when clinically weak or atrophic muscles can serve as recording sites. In this regard, one of the most helpful is the peroneal motor NCS, recording TA, when foot drop is present.

Generally, little is gained by performing late responses during myopathy assessments. However, if the gastrocnemius muscles appear atrophic, the amplitude of the M wave elicited during the H reflex test can serve as an indicator of the number of muscle fibers that have degenerated.

The NEE is the critical component of the EDX examination in the diagnosis of myopathies. Not only is it typically the only portion of the EDX examination likely to show abnormalities, but also the abnormalities seen may be very patchy in their distribution from one muscle to another, and within a given muscle. Consequently, more time must be expended performing the NEE on patients with suspected myopathies than on those with suspected neurogenic disorders. This increased time requirement applies to all aspects of the NEE, as abnormalities may be observed, either singularly or in various combinations, during all of its phases: insertion, rest, and MUAP activation.

Several features of the NEE performed on patients with suspected myopathies are not standardized, including (1) the optimal number of muscles to be sampled, (2) the number of muscles to be studied in each limb, and (3) the specific muscles to be surveyed. Nonetheless, when it is appreciated that the paraspinal muscles, certain limb girdle muscles, and some proximal limb muscles in both the upper and lower extremities are likely to show abnormalities, it is

apparent that, at minimum, several muscles per limb must be studied for the NEE to be considered adequate. In our laboratory the standard NEE survey in these instances consists of sampling approximately 12 muscles (Table 21–3). Regarding the paraspinal muscles, we prefer to assess the mid-lumbar region, because we have found that (1) patient discomfort is often less than when other paraspinal regions are evaluated, (2) obtaining a quiet field is relatively easy, and (3) the area usually is cephalad to any lumbar laminectomy scars. Two mid-limb muscles, the brachioradialis and TA, so commonly show abnormalities with myopathies— frequently more often than some proximal muscles—that they always are sampled. Two of the muscles in our 12-muscle "myopathy assessment" are sampled for reasons other than to diagnose a myopathy. The first dorsal interosseous is included to ensure that an upper extremity muscle innervated by other than the C5–7 segments is sampled, and at least one distal limb muscle is sampled. The medial gastrocnemius is included to prevent the lower extremity NEE from being restricted solely to L2– through L5–innervated muscles. Importantly, whenever a myopathy that affects muscles in other than a proximal distribution is suspected, the NEE survey must include those muscles that are clinically weak, such as the pronator teres and flexor pollicis longus in inclusion body myopathy, distal muscles in myotonic dystrophy and distal myopathies, the sternocleidomastoid in neck flexor weakness, and cervical paraspinals in neck extensor weakness.

An extensive NEE is mandatory with myopathies, particu-

TABLE 21–3	One Set of Muscles That Can Be Sampled on Needle Electrode Examination Whenever the Clinical Question Is a Generalized Myopathy Having Proximal Emphasis, Such As Polymyositis

Upper Extremity	Lower Extremity	Paraspinals
First dorsal interosseous	Tibialis anterior	Mid-lumbar
Brachioradialis	Medial gastrocnemius	Low cervical
Biceps	Vastus lateralis/ rectus femoris	
Triceps	Iliacus	
Deltoid	Gluteus maximus	
Infraspinatus		

TABLE 21-4	Changes on Nerve Conduction Studies with Necrotizing Myopathy (Polymyositis)[a]

Nerve Stimulated	Stimulation Site	Recording Site	Amplitude (μV/mV)	Latency (msec) Peak	Latency (msec) Distal	Conduction Velocity (msec)
Median (S)	Wrist	Digit 2	25	3.0		
Sural	Distal calf	Lateral malleolus	15	3.8		
Median (M)	Wrist	Thenar	9.0		3.0	
	Elbow		8.2		8.3	56
Peroneal (M)	Ankle	Extensor digitorum brevis	3.9		4.9	
	Popliteal fossa		2.8		15.0	44
Peroneal (M)	Below fibular neck	Tibialis anterior	2.1[b]		3.0	
	Popliteal fossa		1.9[b]		5.5	48

[a]The fact that the peroneal motor responses, recording tibialis anterior, are low in amplitude indicates that substantial muscle fiber loss has occurred in the anterior (and possibly lateral) compartment muscles of the leg, and explains the patient's bilateral foot drop. (The latter caused the peroneal motor nerve conduction studies, recording tibialis anterior, to be performed.)
[b]Abnormal response.
M = motor, S = sensory.

larly if the initial muscles surveyed yield negative results, because so many muscle disorders tend to affect individual muscles, including proximal ones, in a very unpredictable, haphazard manner. This includes myopathies that generally are reported to produce clinical findings in a proximal, symmetrical fashion, such as polymyositis (Tables 21-4 and 21-5). Probably the major error made in myopathy assessment in the EDX laboratory is inadequate muscle sampling. Such statements as " . . . it is impossible to sample . . . more than a few (muscles) if due regard is paid to the patient's comfort,"[2] or "it is rarely necessary to examine more than four muscles"[23] reflect a viewpoint that virtually guarantees a high (and, to many electromyographers, an unacceptable) number of false-negative EDX examinations with myopathies. Such restricted muscle sampling is responsible for very misleading data appearing in the literature concerning the reported frequency with which certain NEE abnormalities are seen with specific muscle disorders. The most obvious misrepresentation concerns the published incidence of fibrillation potentials occurring with various myopathies, especially polymyositis. This undoubtedly is substantially understated in the literature, primarily because the series that reported low incidences usually were composed of patients who had only one or, at most, two muscles sampled on NEE (e.g., just the biceps, deltoid, or vastus lateralis). No conclusions can justifiably be drawn from such meager muscle sampling, particularly in regard to negative results.[25, 26]

THE ELECTRODIAGNOSTIC EXAMINATION—LIMITATIONS

Several limitations are encountered whenever the EDX examination is used to assess patients with possible myopathies. The first, and the one which overshadows all others, is that none of the changes that occur with myopathies is specific for disorders of the muscle fiber. Rather, the diagnosis rests principally on detecting certain nonspecific changes in various "patterns and combinations."[6] This limitation is not unique to muscle disorders. Nonetheless, the fact that none of the EDX abnormalities manifested by myopathies is diagnostic of such disorders makes the expectations for electrodiagnosis of myopathy different from those for PNS disorders. For this reason, the level of diagnostic uncertainty must always be somewhat higher when attempting to diagnose myopathic disorders compared to neurogenic ones.

Second, the diagnosis rests solely on the NEE. In this regard, myopathies are similar to single radiculopathies, but quite different from most other disorders of the PNMS, where the NCS can also be of positive diagnostic benefit. Rendering the diagnosis even more difficult is that it often completely depends on detecting MUAP changes on the NEE. Unfortunately, judging whether the appearance of MUAPs has been altered sufficiently so that they can be considered "myopathic" is one of the more difficult tasks confronting EDX consultants. In fact, the MUAPs in many

TABLE 21-5	Needle Electrode Examination Findings Seen with Nerve Conduction Studies Reported in Table 21-4[a]

Muscle (left)	Spontaneous Activity (FIBs)	Motor Unit Action Potentials Recruitment	Duration	Amplitude	Polyphasic
1st dorsal interosseous	0	Normal	Normal	Normal	None
Brachioradialis	2+	Early	Most ↓	Most ↓	Most + +
Biceps	2+	Early; reduced	All ↓	All ↓	All + + +
Triceps	1+	Normal	Some ↓	Normal	Some + +
Deltoid	0-1+	Normal	Few ↓	Normal	Few + +
Infraspinatus	1+	Normal	Normal	Normal	Many + +
Tibialis anterior	2+	Early; reduced	All ↓	All ↓	All + + +
Medial gastrocnemius	0-1+	Normal	Normal	Normal	Few + +
Vastus lateralis	1+	Normal	Normal	Normal	Many + +
Iliacus	3+	Early; reduced	All ↓	All ↓	All + + +
Gluteus maximus	2+	Early	Most ↓	Normal	All + + +
L4 paraspinals	3+	—	—	—	—

[a]Note the variation in degree of both fibrillation potentials and motor unit action potential (MUAP) changes from one muscle to another, even among those at the same limb level. In some muscles (e.g., biceps, tibialis anterior, and iliacus) the muscle fiber loss has been so severe that, on maximal effort, only a relatively few highly "myopathic" MUAPs can be activated. The severe MUAP changes present in the tibialis anterior muscle were predictable, based on the low amplitude peroneal motor NCS responses recorded from that muscle.

muscles of patients with suspected myopathies are not simply either "normal" or "myopathic." Rather, they lie in a continuum, at various points between these two possibilities. As with other situations, the extremes are readily recognized, whereas the area between the two poles often causes major difficulties in interpretation. If the myopathy is of a type that produces abnormal insertional activity or spontaneous activity (e.g., widespread myotonic discharges or fibrillation potentials) then this zone of uncertainty is narrowed considerably. However, if the myopathy manifests only MUAP changes, then in a certain percentage of patients the NEE findings, although consistent with a myopathy, simply will not be abnormal enough for definite diagnosis. Those performing EDX examinations should not hesitate to admit this, as it is an undeniable fact.

Third, the EDX findings with myopathies are not only very nonspecific, but they also are very variable from one myopathy to another, ranging anywhere from no detectable physiological abnormality (e.g., metabolic myopathies and some congenital myopathies) to substantial abnormalities manifested during all phases of the NEE (e.g., many of the myotonic and necrotizing myopathies). Presumably, those myopathies that cause no structural alteration of the MUAPs tend to affect the contractile elements rather than the muscle membrane (and therefore the electrical properties) of muscle fibers.

Fourth, muscle fibers, similar to nerve fibers, are very restricted in their electrical responses to disease. Consequently, almost all the EDX changes, and combination of changes, that occur with them can be seen with more than one muscle disorder. Thus, the EDX consultant's ability to differentiate one myopathy from another, based on the EDX findings, is limited; for instance, it has been appreciated for many years that the NEE findings with polymyositis may be indistinguishable from those with Duchenne muscular dystrophy. More recent experience indicates that EDX changes essentially identical to those found with these two myopathies may also be seen with inclusion body myopathy at times, and with the myopathy associated with human immunodeficiency virus (HIV) infection. The recognition of specific myopathies by EDX examination is complicated further by the fact that not only many different myopathies share the same EDX appearance, but the same myopathy, at different stages in its evolution or resolution, may have markedly different EDX appearances. The best example of this is the varying presentations of polymyositis, already discussed. Also, some myopathies, such as periodic paralysis, may manifest both their clinical and electrophysiological abnormalities on an intermittent basis only. As a result, the EDX examination with them can vary substantially over a period of just a few hours or days. From the foregoing, it is obvious that a specific myopathy, just as myopathies in general, cannot be diagnosed with certainty by EDX examination.

A fifth limitation is that the value of the EDX examination for recognizing changes consistent with a primary muscle fiber abnormality can be impaired, or even completely negated, in the presence of many other disorders of the motor unit, including such diverse conditions as the residuals of poliomyelitis, diabetic polyneuropathy, and neuromuscular transmission disorders. Some of these confounding conditions distort the MUAP configurations so drastically that any superimposed "myopathic" changes cannot be discerned, whereas others produce "myopathic" MUAP changes that are essentially identical to those seen with myopathies.

Still another limitation, and one that appears to particularly frustrate physicians with little knowledge of the EDX examination, is that the degree of EDX changes seen with a particular muscle disease often parallels the amount of other laboratory changes (e.g., muscle enzymes, muscle

biopsy findings) seen with it. Consequently, it all too frequently happens that the muscle disorder that produces very little, if any, muscle biopsy changes happens to be exactly the same one that produces very little, if any, EDX abnormalities.[1, 6, 11, 16, 26, 27]

THE ELECTRODIAGNOSTIC EXAMINATION—BENEFITS

Despite all its limitations, the EDX examination remains one of the linchpins in the assessment of patients with suspected myopathies because it has substantial diagnostic value, for several reasons. First, it often can definitely differentiate other disorders of the PNMS from myopathies, even when their clinical presentations are very similar. Thus, polymyositis usually is readily separated from amyotrophic lateral sclerosis by EDX examination, even though clinically both may present with proximal limb weakness and dysphagia. Consequently, at times the EDX examination can convert what clinically appears to be a progressive, untreatable disorder to one that responds very satisfactorily to medical therapy.

Second, the particular changes, or combination of changes, seen on NEE with a myopathy—especially the presence or absence of certain types of spontaneous activity, along with the presence and severity of MUAP alterations—narrow the possible causes of a myopathy, because they allow it to be assigned to one of several broad categories of muscle disorders. As a result, certain potential etiologies for the "myopathic" EDX changes seen are excluded, whereas others not only remain possibilities but, because of the elimination process, become more likely candidates[11, 25] (Table 21–6). Thus, a myopathy that manifests fibrillation potentials on NEE is very likely to be a toxic or inflammatory myopathy, or a rapidly progressive muscular dystrophy; conversely, it is extremely unlikely to be a steroid myopathy.

Third, the NEE component of the EDX examination permits widespread sampling of the skeletal musculature. This feature is in marked contrast to muscle biopsy, in which sampling is restricted to one or, at most, two muscles. This is of particular value with those muscle disorders in which the degree of involvement varies substantially from one muscle to another, even among the proximal muscles.

Fourth, because of the widespread sampling ability of the NEE, the EDX examination findings can be used, at least in those patients whose muscle abnormalities appear to be bilaterally symmetrical, to select the most appropriate muscle for biopsy. The goal is to biopsy a muscle that is moderately involved, rather than one that is either unaffected, or one with end-stage disease. Typically, to avoid confusing microscopic changes caused by the needle recording electrode itself, the homologous muscle in the contralateral limb is biopsied, rather than the one assessed on NEE.

Fifth, abnormalities of diagnostic value may be seen on the NEE, which otherwise would be undetectable. Thus, myotonic discharges may be present, even though clinical myotonia is not evident. This electrophysiological and clinical discordance is found with very few disorders, so the differential diagnosis list is immediately markedly narrowed (see Table 21–1).[25–27]

ELECTRODIAGNOSTIC FINDINGS IN SELECTED MYOPATHIES

Polymyositis

This is the most common type of myopathy seen in adults in our EDX laboratories. The NEE abnormalities frequently

TABLE 21-6 An Approach to Identifying the Likely Cause of a Myopathy, Based on the Needle Electrode Examination Findings

Normal	"Myopathic" MUAPs Only	Fibrillation Potentials Only	"Myopathic" MUAPs and Fibrillation Potentials	Myotonic Discharges Alone	Myotonic Discharges and "Myopathic" MUAPs (+/− Fibrillation Potentials)
Congenital myopathies (some)	Congenital myopathies (some)	Inflammatory myopathies (early; mild)	Inflammatory myopathies	Myotonia congenita	Myotonic dystrophy
Central core[a]	Myotubular[a]	Acute rhabdomyolysis	Myositis (poly/dermato)[a]	Thomsen's disease	Proximal myotonic myopathy
	Multicore[a]	Toxic myopathies (early; mild)	Inclusion body myopathy[a]	Becker's disease	Colchicine myopathy
Metabolic myopathies	Endocrine myopathies (some)		HIV myopathy	Hypothyroidism (some)	Paramyotonia congenita
Lipid storage diseases	Steroid[a]		Muscular dystrophies (some; rapidly progressive)	Adult acid maltase deficiency	Myotubular myopathy
Glycogen storage diseases	Muscular dystrophies (some): slowly progressive		Duchenne[a]		Acid maltase deficiency
Mitochondrial myopathies	Toxic myopathies (few)		Becker[a]		Hyperkalemic periodic paralysis
Adenine nucleotide disorders	Treated inflammatory myopathies		Congenital myopathies (some)		Polymyositis
Endocrine myopathies (some)	Polymyositis[a]		Centronuclear[a]		Welander's distal myopathy
Steroid[a]			Acute illness/ICU myopathy		
Periodic paralysis (between attacks)			Myotubular myopathy		
Fiber-type disproportion			Infectious myopathy		
			Trichinosis		
			Toxoplasmosis		
			AIDS		
			Toxic myopathies (most)		
			Zidovudine		
			Alcohol		
			Chochicine		

AIDS, acquired immunodeficiency syndrome; HIV, human immunodeficiency virus; ICU, intensive care unit.

[a]Common myopathy in category.

Material from Katirji B: Electromyography in Clinical Practice: A Case Study Approach. St. Louis, Mosby, 1998, pp 219–232; Dumitru D: Electrodiagnostic Medicine. Philadelphia, Hanley and Belfus, 1995; Wilbourn A: The EMG examination with myopathies. In Syllabus: 1987 AAEE Course A: Myopathies, Floppy Infant, and Electrodiagnosis Studies in Children. Rochester, MN, American Association of Electromyography and Electrodiagnosis, 1987, pp 7–20.

are restricted to the paraspinal, limb girdle, and proximal limb muscles, along with the brachioradialis and TA muscle. In established, untreated cases, both fibrillation potentials and "myopathic" MUAP changes usually are prominent, although each often varies considerably from one muscle to another in an unpredictable fashion (see Tables 21–4 and 21–5). If the condition is chronic, the fibrillation potentials may be very small. In some patients, other types of abnormal insertional or spontaneous activity are seen, either CRDs or, less commonly, myotonic discharges, but these are never as prominent as the fibrillation potentials. In this regard, it is pertinent to note that some investigators consider CRDs to be necessary for diagnosis, just as are fibrillation potentials and "myopathic" MUAP changes. We disagree with this view, as in our experience CRDs are found in only a small percentage of patients, and invariably only in those with chronic disease.

Patients with polymyositis tend to be referred to the EDX laboratory for two reasons: (1) when they are initially being evaluated and the diagnosis has not been established; or (2) months after the diagnosis has been made, when they are on steroid therapy and yet they are either unresponsive to it, or have become even more symptomatic. Lack of response suggests three explanations: (1) correct diagnosis but poor response to treatment; (2) correct diagnosis and response to treatment, but supervening steroid myopathy; and (3) treatment failure because of incorrect diagnosis. Often the question is easily answered by the NEE. If fibrillation potentials remain widespread and fairly abundant, the disease is still quite active. Conversely, if no or only a very few fibrillation potentials are observed, then the primary myopathy is extremely unlikely to be very active, and the current muscle weakness probably is secondary to steroid myopathy.

Finally, progression of the disease, especially into new muscles such as finger flexors, suggests the condition is inclusion body myositis and not polymyositis. One scenario, seen with some frequency and yet rather difficult to interpret, is the presence of just a modest number of fibrillation potentials restricted to the paraspinal muscles. It is uncertain whether this represents inadequately treated disease. Nonetheless, if an earlier EDX examination had revealed widespread, abundant fibrillation potentials, then the process is at least much improved. It is noteworthy that once effective treatment begins, fibrillation potentials tend to disappear much earlier than the MUAP changes. As a result, a NEE can be performed on a patient with treated polymyositis at a time when fibrillation potentials essentially are absent, even though MUAP changes remain prominent; these findings may be mistaken for a non-necrotizing myopathy.[15, 27, 28]

Inclusion Body Myopathy

This probably is the second most common type of myopathy we encounter, and the leading cause of onset of myopathy after 50 years of age. Without question, many patients considered with instances of "treatment-resistant" polymyositis in the past actually had this disorder. Often, patients with inclusion body myopathy have NEE findings identical to those seen with polymyositis; in them it is virtually impossible to distinguish one disorder from another by EDX examination. In other patients, however, the findings are atypical in the type of MUAP alterations present, or in their distribution, or both. Thus, limb muscles that rarely are substantially involved in polymyositis (e.g., pronator teres, flexor pollicis longus) may be significantly affected, even when more proximal muscles in the same limb show less severe abnormalities. Also, in some patients, chronic neurogenic MUAP changes are intermixed with the "myopathic" MUAP changes, either within contiguous proximal muscles, or, sometimes, even in different portions of the same muscle.[10, 20, 27, 28]

Duchenne Muscular Dystrophy

The NEE findings with this disorder usually are identical to those seen with polymyositis, although the involvement of

| TABLE 21–7 | Nerve Conduction Changes with Acute Illness/ICU Myopathy[a] |

Nerve Stimulated (Right)	Stimulation Site	Recording Site	Amplitude (μV/mV)	Latency (msec) Peak	Latency (msec) Distal	Conduction Velocity (msec)
Median (S)	Wrist	Digit 2	28	3.1		
Ulnar (S)	Wrist	Digit 5	21	2.8		
Radial (S)	Distal forearm	Thumb base	25	2.4		
Sural	Distal calf	Lateral malleolus	NR[b] (edema)		—	
Super peroneal sensory	Ankle	Distal leg	NR[b] (edema)		—	
Median (M)	Wrist	Thenar	0.9[b]		3.2	—
	Elbow		NR		—	
Ulnar (M)	Wrist	Hypothenar	2.0[b]		3.0	49
	Elbow		1.2		8.8	
Musculocutaneous (M)	Axilla	Biceps	0.5[b]		2.7	
Radial (M)	Elbow	Proximal extensor forearm	0.3[b]		2.3	
Peroneal (M)	Ankle	Extensor digitorum brevis	NR[b]		—	—
	Popliteal fossa		NR[b]		—	—
Tibial (M)	Ankle	Abd. hallucis	NR[b]		—	—
	Popliteal fossa		NR[b]		—	—
Peroneal (M)	Below fibular neck	Tibialis anterior	1.5[b]		3.2	47
	Popliteal fossa		1.0		6.0	
M wave of H reflex	Popliteal fossa	Gastrocnemius	0.5[b]		3.5	
Femoral	Inguinal ligament	Rectus femoris	1.3[b]		4.0	

NR, No response.
[a]Note that all the motor nerve conduction study (NCS) responses are very low in amplitude or unelicitable; in contrast, the motor nerve conduction velocities that can be calculated are either normal or just slightly slow. The sensory NCS responses are normal in the upper extremity, while unelicitable in the lower extremities; the latter may be due to either limb edema or advanced age (patient age 56).
[b]Abnormal response.
S = sensory; M = motor.

TABLE 21-8	Needle Electrode Examination Findings Seen with Nerve Conduction Studies Reported in Table 21-7[a]

Muscle (right)	Spontaneous Activity (FIBs)	Motor Unit Action Potentials			
		Recruitment	Duration	Amplitude	Polyphasic
First dorsal interosseous	2+	Early; reduced	Most ↓	All ↓	All + + +
Abductor pollicus brevis	3+	No MUAPs	—	—	—
Pronator teres	2+	Early	Most ↓	All ↓	All + + +
Extensor Indicis proprius	2+	Early	Most ↓	All ↓	All + + +
Brachioradialis	3+	No MUAPs	—	—	—
Biceps	2+	No MUAPs	—	—	—
Triceps	2+	Early; reduced	All ↓	All ↓	All + + +
Deltoid	3+	Early; reduced	All ↓	All ↓	All + + +
Infraspinatus	2+	Early; reduced	All ↓	All ↓	All + + +
Tibialis anterior	3+	Early; reduced	All ↓	All ↓	All + + +
Medial gastrocnemius	2+	Early	Most ↓	Most ↓	All + + +
Extensor digitorum brevis	2+	No MUAPs	—	—	—
Abductor hallucis	3+	No MUAPs	—	—	—
Vastus lateralis	3+	Early; reduced	All ↓	All ↓	All + + +
Gluteus maximus	3+	No MUAPs	—	—	—
L4 paraspinals	3+	—	—	—	—

FIBs, fibrillation potentials.

[a]Note that abundant fibrillation potentials and prominent "myopathic" motor unit action potentials (MUAPs) are generalized in distribution; that is, they lack a proximodistal gradient. In many muscles, only a few MUAPs fire on maximal effort. When taken in conjunction with the nerve conduction study results reported in Table 21-7, these changes are suggestive of severe disintegration of the motor unit, but it is unclear whether the disorder is one of terminal nerve fibers, neuromuscular junctions, or muscle fibers. Fast repetitive stimulation studies (40/sec) would exclude most presynaptic neuromuscular junction problems.

the proximal muscles tends to be more uniform. A non-necrotizing myopathy can be erroneously diagnosed if fibrillation potentials are not detected on the NEE, an oversight that can readily occur because the fibrillation potentials are very low in amplitude and therefore may be easily obscured by baseline noise or voluntary muscle activation. This is particularly likely to happen if the EDX examination is performed on an unsedated affected child.[25, 27, 28]

Myotonic Dystrophy

The myotonic discharges seen on NEE in these patients tend to be relatively widespread in distribution, but of rather brief duration—the prolonged "dive airplane" sound considered so characteristic of myotonic disorders seldom is encountered. "Myopathic" MUAPs usually are more prominent in some of the more distal muscles (e.g., the forearm and leg muscles) compared to the arm and thigh muscles. Because of the limb distribution of the muscle abnormalities, the typical NEE "myopathy assessment" (see Table 21–3), with its heavy emphasis on proximal muscles, is not adequate; several distal limb muscles must be assessed as well, in both the upper and lower extremities.[27, 28]

Limb Girdle Dystrophy

This particular heterogeneous group of muscle disorders at times can be one of the most frustrating to evaluate in the EDX laboratory, because frequently muscles are either not affected at all, or they are "end stage" (i.e., totally degenerated). A common scenario in these cases is for the clinically normal muscles to appear normal on NEE, whereas the markedly wasted muscles, being completely atrophic, show no insertional activity, no fibrillation potentials, and no voluntary MUAPs. As a result, in spite of the obvious clinical findings, there is nothing seen on the NEE that is particularly suggestive of a myopathy.[27]

AIDS Myopathy

Two separate types of necrotizing myopathy occur with acquired immunodeficiency syndrome (AIDS), one a direct manifestation of the disease and the other caused by a drug used to treat it (zidovudine). These apparently can be difficult to separate from one another at times. Unfortunately, the EDX examination is of no help in distinguishing them because the NEE findings are very similar with both. These consist of fibrillation potentials and suggestive "myopathic" MUAP changes present in the axial, limb girdle, and more proximal limb muscles.[16, 21, 27]

Acute Illness/ICU Myopathy

This is almost invariably the disorder found whenever patients in the ICU are assessed because they cannot be weaned from the respirator. A characteristic combination of EDX changes is seen, including generalized low-amplitude or unelicitable CMAPs; normal SNAPs (although allowance must be made for bilaterally absent lower extremity SNAPs in the elderly, as well as low-amplitude or unelicitable SNAPs if limb edema is present); widespread fibrillation potentials; and either no voluntary MUAPs or, if some can be activated, MUAPs that are highly "myopathic" in configuration[14, 19] (Tables 21–7 and 21–8). The findings are those of severe disintegration of the motor unit, and the differential diagnosis includes early Guillain-Barré syndrome, myasthenic syndrome, or some other NMJ disorder, and a severe myopathy. In an attempt to exclude an NMJ abnormality, fast repetitive stimulation studies are performed. These are necessary because the more benign test for post-tetanic facilitation, the Lambert-Eaton test (i.e., single supramaximal stimulus to motor nerve, followed by 10 seconds of exercise of the recorded muscle, and then another supramaximal stimulus), is unreliable in these patients, as they typically are too weak to exercise effectively.

REFERENCES

1. Aminoff MJ: Electromyography in Clinical Practice, 3rd ed. New York, Churchill Livingstone, 1998.
2. Barwick DD: Clinical electromyography. In Walton J (ed.): Disorders of Voluntary Muscle, 4th ed. Edinburgh, Churchill Livingstone, 1981, p 953.
3. Brandstater ME, Lambert EH: Motor unit anatomy. In Desmedt

JE (ed.): New Developments in Electromyography and Clinical Neurophysiology, vol. 1. Basil, Karger, 1973, pp 14–22.

4. Buchthal F, Guld C, Rosenfalck P: Volume conduction of the spike of the motor unit potential investigated with a new type of multielectrode. Acta Physiol Scandinav 38:331–354, 1957.

5. Daube JR: AAEM Minimonograph #11: Needle examination in clinical electromyography. Muscle Nerve 14:685–700, 1991.

6. Daube JR: Electrodiagnosis of muscle disorders. In Engel AG, Franzini-Armstrong C (eds.): Myology, 2nd ed. New York, McGraw-Hill, 1994, pp 764–794.

7. Dumitru D: Electrodiagnostic Medicine. Philadelphia, Hanley and Belfus, 1995.

8. Guy E, Lefebre V, Lerique J, Scherrer J: Les signes electromyographique des dermatomyosites. Rev Neurol (Paris) 83:278–279, 1950.

9. Henriksson KG, Stalberg E: The terminal innervation pattern in polymyositis: A histochemical and SF EMG study. Muscle Nerve 1:3–13, 1978.

10. Joy JL, Oh SJ, Baysal AI: Electrophysiological spectrum of inclusion body myopathy. Muscle Nerve 13:949–951, 1990.

11. Katirji B: Electromyography in Clinical Practice: A Case Study Approach. St. Louis, Mosby, 1998, pp 219–232.

12. Kugelberg E: Electromyogram of muscular disorders. J Neurol Neurosurg Psychiatry 10:122–133, 1947.

13. Kugelberg E: Electromyogram in muscular dystrophies. J Neurol Neurosurg Psychiatry 12:129–136, 1949.

14. Lacomis D, Giuliani MJ, Cott AV, Kramer DJ: Acute myopathy of intensive care: Clinical electromyographic and pathological aspects. Ann Neurol 40:645–654, 1996.

15. Lambert EH, Beckett S, Chen W, Eaton LM: Unipolar electromyograms of patients with dermatomyositis. Fed Proc 9:73, 1950.

16. Lange DJ: AAEM Minimonograph #41: Neuromuscular diseases associated with HIV-1 infection. Muscle Nerve 17:16–30, 1994.

17. Nandedkar SD, Sanders DS: Simulation of myopathic motor unit action potential. Muscle Nerve 12:197–202, 1989.

18. Nandedkar SD, Sanders DS, Stalberg E: Simulation of concentric needle EMG motor unit action potentials. Muscle Nerve 11:151–159, 1988.

19. Ruff RL: Why do ICU patients become paralyzed? Ann Neurol 43:154–155, 1998.

20. Shields RW, Wilbourn AJ, Levin KH, et al: Inclusion body myopathy: The EMG features. Neurology 39(suppl 1):233, 1989.

21. Simpson DM, Citak KA, Godfrey E, et al: Myopathies associated with human immunodeficiency virus and zidovudine: Can their effects be distinguished? Neurology 43:971–976, 1993.

22. Stalberg E: Invited review: Electrodiagnostic assessment and monitoring of motor unit changes in disease. Muscle Nerve 14:293–303, 1991.

23. Swash M, Swartz SM: Neuromuscular Disorders, 2nd ed. Berlin, Springer-Verlag, 1988, p 16.

24. Trontelj J, Stalberg EV: Bizarre repetitive discharges recorded with single fiber EMG. J Neurol Neurosurg Psychiatry 46:310–316, 1983.

25. Wilbourn A: The EMG examination with myopathies. In Syllabus: 1987 AAEE Course A: Myopathies, Floppy Infant, and Electrodiagnosis Studies in Children. Rochester, MN: American Association of Electromyography and Electrodiagnosis, 1987, pp 7–20.

26. Wilbourn A: The electrodiagnostic examination with myopathies. J Clin Neurophysiol 10(2):132–148, 1993.

27. Wilbourn A: The electrodiagnosis examination with myopathies. In Syllabus: 1996 AAN Course 245: Clinical EMG. 48th Annual Meeting of the American Academy of Neurology, San Francisco, 1996, pp 85–108.

28. Wilbourn AJ, Ferrante MA: Clinical electromyography. In Joynt RJ, Griggs RC (eds.): Clinical Neurology, vol. 1. Philadelphia, Lippincott-Raven, 1997, pp 1–76.

IV EVALUATION OF CENTRAL INFLUENCES ON PERIPHERAL FUNCTION

Chapter 22

Electrodiagnosis in Movement Disorders*

Mark Hallett, MD

INTRODUCTION
EVALUATION OF INVOLUNTARY
   MOVEMENTS
Myoclonus, Startle, and Asterixis
Tic
Chorea, Dyskinesia, and Ballism
Dystonia and Athetosis
EVALUATION OF TREMOR
Parkinsonian Tremor at Rest
Exaggerated Physiological Tremor
Essential Tremor
Cerebellar Tremor
Other Tremors

Task-specific Tremors
Hysterical Tremor
EVALUATION OF DISORDERS OF
   VOLUNTARY MOVEMENT
Slowness
Clumsiness
EVALUATION OF TONE
Spasticity
Rigidity
Stiffman Syndrome
BOTULINUM TOXIN TREATMENT OF
   INVOLUNTARY MUSCLE SPASMS

INTRODUCTION

Movement disorders is an area of growing attention in neurological practice, in part because of the increasing therapeutic options. In disorders that look superficially similar, it is critical to make the right diagnosis because therapies might differ. Clinical neurophysiological methods can be helpful in extending the physical examination. One reason is that small differences in timing, easily measured with simple techniques, can be impossible to tell by eye. What is the burst duration of the electromyogram (EMG) underlying an involuntary movement? What is the frequency of tremor and how does it change with an intervention? What is the latency of a muscle jerk after a stimulus? This chapter is organized by type of problem that a neurophysiologist might face.

EVALUATION OF INVOLUNTARY MOVEMENTS

Involuntary movements can be characterized by the body part affected, the frequency of the movements, the triggers for the movements, and the duration of each movement. It is the latter feature that is best studied with EMG methods. Because mechanical events take so long compared with the electrical events that control them, observations of the mechanical events by themselves can be very difficult to sort out. Additionally, EMG can determine the relationship between the activity in antagonist muscle pairs, another difficult observation to make only clinically.

EMG data can be measured with surface, needle, or wire electrodes.[19, 23] The advantages of surface electrodes are that they are not painful and they record from a relatively large volume of muscle. The advantage of needle electrodes is that they are more selective, sometimes a necessity when recording from small or deep muscles. Traditional needle electrodes are stiff, and it is best to use them when recording from muscles during movements that are close to isometric. Pairs of fine wire electrodes have the advantage of selectivity similar to that of needle electrodes, and are flexible, permitting free movement with only minimal pain. In any case, it is important to avoid movement artifact, which can contaminate the EMG signal. Wire movement should be limited. Low-frequency content of the EMG signal can be restricted with filtering. Impedance of surface electrodes should be reduced.

There are three EMG patterns that may underlie involuntary movements.[17, 19, 21, 23] One pattern, which can be called "tonic," resembles slow voluntary movements and is characterized by continuous or almost continuous EMG activity lasting for the duration of the movement, from 200 to 1000 msec or longer. Activity can be solely in the agonist muscle, or there can be some co-contraction of the antagonist muscle with the agonist. Another pattern, which can be called "ballistic," resembles voluntary ballistic movements with a triphasic pattern; there is a burst of activity in the agonist muscle lasting 50 to 100 msec, a burst of activity in the antagonist muscle lasting 50 to 100 msec, and then return of activity in the agonist, often in the form of another burst. The third pattern, which can be called "reflex," resembles the burst occurring in many reflexes, including H reflexes and stretch reflexes. The EMG burst duration is 10 to 40 msec, and EMG activity in the antagonist muscle is virtually always synchronous.

*All material in this chapter is in the public domain, with the exception of any borrowed figures or tables.

Myoclonus, Startle, and Asterixis

Myoclonus is characterized by quick muscle jerks, either irregular or rhythmic.[17, 19, 23] There are many types of myoclonus and no common etiological, physiological, or therapeutic features bind them together. Myoclonus can be focal, involving only a few adjacent muscles; generalized, involving many or most of the muscles in the body; or multifocal, involving many muscles but in different jerks. Myoclonus can be spontaneous; activated or accentuated by voluntary movement (action myoclonus); and activated or accentuated by sensory stimulation (reflex myoclonus). Rhythmic (segmental) myoclonus has the appearance of a rest tremor but is typically unaffected by action, stimulation, or even sleep. In this disorder, a segment of the spinal cord (spinal myoclonus) or brain stem (palatal myoclonus) produces persistent, rhythmic, repetitive discharges usually unaffected by sleep. A number of contiguous muscles produce synchronous contractions at a rate of 1 to 3 Hz. Because of the slow speed of the movements, palatal myoclonus is now preferentially called palatal tremor (see later discussion).

By defining epileptic myoclonus as myoclonus that is a fragment of epilepsy, it is possible to divide irregular myoclonus into epileptic and nonepileptic myoclonus.[17, 19, 23] The physiological characteristics of epileptic myoclonus are as follows: (1) EMG burst length of 10 to 50 msec, (2) synchronous antagonist activity, and (3) an electroencephalographic (EEG) correlate (the technique of EMG–EEG correlation is described later). The EMG shows a reflex pattern. Nonepileptic myoclonus shows (1) EMG burst lengths of 50 to 300 msec, (2) synchronous or asynchronous antagonist activity, and (3) no EEG correlate. The EMG patterns are either ballistic or tonic.

Examples of epileptic myoclonus are cortical reflex myoclonus, reticular reflex myoclonus, and primary generalized epileptic myoclonus; these are discussed later. Examples of nonepileptic myoclonus include dystonic myoclonus; essential myoclonus, such as ballistic movement overflow myoclonus; exaggerated startle; physiological phenomena, such as hypnic jerks; and periodic movements of sleep. Frequent myoclonus may have the appearance of tremor. In the case of action myoclonus, this may be confusing clinically, but EMG analysis is definitive.

For sorting out the different types of myoclonus, it can be useful to look for EEG events at the time of a movement.[19] Events in the ongoing EEG can be correlated with EMG events, but it is more informative to average the EEG with respect to the EMG.[23] Just as sensory-evoked cerebral potentials are time-locked to the stimulus, these movement-related EEG potentials must be time-locked to a phase of the EMG, such as its onset. A great deal of attention is devoted to that part of the potential preceding movement onset because it may relate to generation of the movement; the part of the potential after movement onset includes feedback from the movement itself. The movement potential can be analyzed for the presence of consistent positive and negative waves, and the topography and time relationship of these to the movement can be determined.

Stimulation may produce involuntary movements, such as reflex myoclonus, and evoke responses in the EEG. The waves in the evoked response that precede the provoked movement can be analyzed for their relationship to the movement. If the timing and topography of an event in the movement potential prior to a spontaneous involuntary movement are similar to the timing and topography of an event in the evoked response prior to the provoked movement, a similarity of the physiological mechanism can be suggested.

With reflex myoclonus, a late response may appear in a relaxed muscle after stretch, mixed nerve stimulation, or cutaneous nerve stimulation. This response, which would not normally be present, may also be seen in muscles outside of the region of the nerve stimulated or even throughout the body. This additional response, sometimes called a C reflex, is a myoclonic movement produced by the stimulation.[17, 36] Such responses are manifestations of hyperexcitability of the nervous system and typically reflect exaggerations of a normal reflex.

The use of these techniques can distinguish the three types of epileptic myoclonus described earlier.[17, 19, 36] Cortical reflex myoclonus is a fragment of focal or partial epilepsy. Each myoclonic jerk involves only a few adjacent muscles, but larger jerks with involvement of more muscles can be seen. The disorder is commonly multifocal and accentuated by action and sensory stimulation. The genesis of cortical reflex myoclonus is thought to be hyperexcitability of sensorimotor cortex, with each jerk representing the discharge of a small region activated by a paroxymal depolarization shift (PDS). The EEG recognizes the discharge as a focal negative event preceding spontaneous and reflexly induced myoclonic jerks. The event with reflex jerks is a giant P1–N2 component of the somatosensory evoked potential.

Reticular reflex myoclonus is a fragment of a type of generalized epilepsy. These jerks are usually generalized with proximal more than distal and flexor more than extensor predominance. Voluntary action and sensory stimulation increase the jerking. The genesis of the myoclonus is thought to be hyperexcitability of a portion of the caudal brain stem reticular formation. A spike can be seen in the EEG often associated with the myoclonic jerk; but since it follows the first EMG manifestation and is not time-locked to the jerk, it does not seem responsible for the jerk. The first activated muscles are those innervated by the 11th cranial nerve; this strongly suggests the brain stem origin. The somatosensory evoked potential is not enlarged, but there can be a C reflex.

Primary generalized epileptic myoclonus is a fragment of primary generalized epilepsy. The most common clinical manifestation is a small, focal jerk that often involves only the fingers and has sometimes been called minipolymyoclonus. Generalized body jerks can also be seen. This type of myoclonus is thought to arise from the firing of a hyperexcitable cortex driven synchronously by ascending subcortical impulses. The EEG correlate is a slow, bilateral, frontocentrally predominant negativity similar to the wave of a primary generalized paroxysm. In this circumstance, there is neither an enlarged somatosensory evoked response nor a C reflex.

The startle reflex is a rapid, generalized motor response to a sudden, surprise stimulus.[7, 19, 27, 29] The most extensively studied human startle response is that which occurs to loud noises. It is an oligosynaptic reflex mediated in the brain stem. The startle response is distinctive on EMG testing with surface electrodes. The pattern is bilaterally symmetrical with an invariable blink; other craniocervical muscles almost always are activated, but recruitment in the limbs is variable. The onset latency of EMG activity is 30 to 40 msec in orbicularis oculi, 55 to 85 msec in masseter and sternocleidomastoid, 85 to 100 msec in biceps brachii, 100 to 125 msec in hamstrings and quadriceps, and 130 to 140 msec in tibialis anterior. There is synchronous activation of antagonist muscles with an EMG burst duration of 50 to 400 msec. Habituation generally occurs after four or five stimuli. Increased startle responses are recognized by being excessive or being evoked by stimuli that would not be effective in most people. This is most easily identified by loss of habituation. Increased startle reflexes are characteristic of a variety of disorders, including hereditary hyperekplexia.

Asterixis is a brief lapse in tonic innervation.[19, 34] It appears as an involuntary jerk superimposed on a postural or intentional movement. Careful observation often reveals that the jerk is in the direction of gravity, but this can be difficult as the lapse is frequently followed by a quick compensatory antigravity movement to restore limb position. The involuntary movement is usually irregular, but when asterixis comes rapidly there may be the appearance of tremor. EMG analysis shows characteristic synchronous pauses in antagonist muscles. Asterixis is also called negative myoclonus.[34] When there is an EEG correlate, the physiology is likely similar to epileptic myoclonus as previously described.

Tic

Tics are quick, involuntary, repetitive movements that occur at irregular intervals.[19] The unique feature of a tic is that it is not completely involuntary. Most patients describe a psychic tension that builds up inside them and can be relieved by the tic movement. Hence, the tics can be voluntarily suppressed for some period of time at the expense of increasing psychic tension; patients "let the tic happen" (or perhaps even "make the tic") to relieve the tension. Tic movements, which can be simple or complex, look like quick voluntary movements both clinically and electromyographically. EMG bursts vary from 50 to 200 msec in duration and may have a ballistic or tonic pattern.

Chorea, Dyskinesia, and Ballism

The most appropriate adjective to describe chorea is "random."[19] Random muscles throughout the body are affected at random times and make movements of random duration. Movements can be brief, such as myoclonus, or long, such as dystonia. Usually they are totally beyond voluntary control, but in some mild cases the movements can be temporarily suppressed. EMG patterns are reflex, ballistic, and tonic. Dyskinesia describes choreic movements seen in selected circumstances, such as a late consequence of neuroleptic drugs or with levodopa toxicity. Ballism describes wild, large-amplitude choreic movements; these usually involve one side of the body and are then called hemiballismus.

Dystonia and Athetosis

The involuntary movements of dystonia and athetosis are similar, and the use of one term rather than the other seems more a matter of situation and semantics than physiology.[19] The movements are typically slow but can be quick and may be "sustained for a second or longer at the height of the involuntary contraction." Dystonia is often used to describe proximal "twisting" movements; athetosis is often used to describe distal "flowing" movements. Dystonic and athetotic movements are frequently characterized by co-contraction of antagonist muscles.[8, 18] Although normal voluntary movement is often characterized by reciprocal inhibition, there may be some co-contraction. The co-contraction of dystonia and athetosis is excessive, with the appearance of increased tension at the joint. Some dystonic and athetotic movements are fully involuntary, arising at rest independent of will. Other movements arise as excessive, unwanted concomitants to voluntary movements. This phenomenon is called overflow, with the implication that the motor control command is sent to too many muscles with too much intensity. EMG studies can document these phenomena. The shortest EMG bursts seen even with dystonic myoclonus are in the range of 100 to 300 msec.

CASE STUDY 1

A Patient with an Involuntary Movement

In approaching a patient with an involuntary movement, the first consideration is, of course, the usual neurological history and physical examination (Fig. 22–1). If the movement is rhythmic, other than segmental myoclonus, it would generally be considered to be tremor, and this topic is reviewed next. The involuntary movements considered in this section have generally been irregular. Some voluntariness suggests tic, and unusual features and psychopathology suggest a psychogenic movement disorder. If completely involuntary, a neurophysiological assessment may be helpful. EMG pattern evaluation is the first step. Subsequent EEG and reflex testing might have additional value in some circumstances as described above.

A 44-year-old woman presented with jerking of the right hand. The movement had been bothering her for 6 months and seemed to be getting worse. Jerks were present at rest but seemed worse when she tried to use her hand for actions. Otherwise, she had no complaint, but her family was a little concerned that her memory might be a problem. On physical examination, the only abnormality was the right hand. Routine mental status testing was normal. There were occasional jerks of the fingers or the wrist in a multifocal fashion. The jerks worsed with postural or kinetic action. It was not clear, but it appeared that there might be reflex jerks with quick stretches of the fingers. Routine investigations were normal, including magnetic resonance imaging (MRI) brain scan.

EMG assessment showed EMG discharges of 30 to 40 msec duration in individual muscles (Fig. 22–2). When muscles were activated in the same jerk, they were synchronous. Electrical stimulation of the right index finger produced a consistent C reflex at about 50 msec latency and a second, less consistent, response at about 125 msec (Fig. 22–3). EEG studies showed that there was a brief positive-negative event over the left central area prior to a muscle jerk (Fig. 22–4), and there was a giant somatosensory potential of 15 μV with right median nerve stimulation. These findings are indicative of cortical reflex myoclonus.

The patient and her family were very keen on knowing the diagnosis. As the physiology had showed a cortical abnormality, a brain biopsy was thought useful. The biopsy revealed significant Alzheimer changes. Over the next few months the patient's memory became clearly worse, revealing the more classical findings of Alzheimer's disease. Cortical myoclonus is a finding in Alzheimer's disease, although more typically it appears late rather than as the presenting feature.[37]

EVALUATION OF TREMOR

Physiology differs in different forms of human tremor.[12, 13] Tremors may come from mechanical oscillations, mechanical-reflex oscillations (EMG activity is entrained with a mechanical oscillation), normal central oscillators, and pathological central oscillators.[20] Any physical object has mechanical properties that obey the laws of physics, and this certainly will include a joint with its associated muscles. Any perturbation of a mechanical system will cause it to oscillate at its resonant frequency. Muscles are connected to the central nervous system via reflex loops. These loops operate over a period of time and may oscillate at frequencies inversely proportional to the loop delay. In certain circumstances, when the frequencies of the mechanical and reflex oscillations are similar, the two frequencies will be mutually entrained to a single frequency. Such a mutually entrained

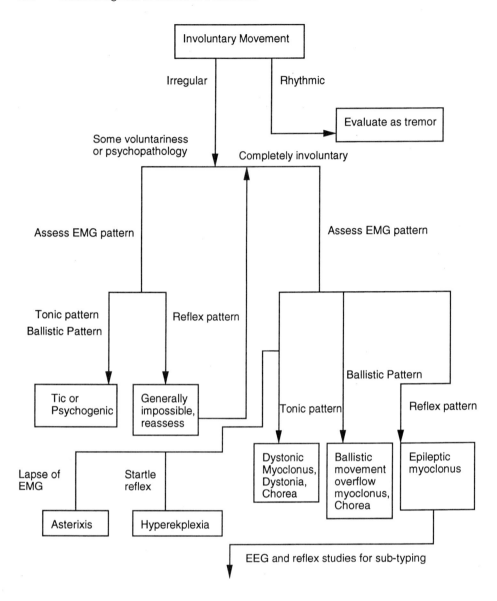

FIGURE 22–1. Flow diagram for evaluating an involuntary movement.

oscillation may behave just like a mechanical system, and is referred to as mechanical-reflex. Central oscillators differ, at least theoretically, from central nodes of reflexes, by their independence of peripheral input. Central oscillators might arise from the properties of individual neurons, from the properties of neuronal networks, or a combination of the two.

Tremor can be examined with accelerometry. The output of an accelerometer is a time-varying analog signal proportional to instantaneous acceleration.[23] Accelerometers in common use measure movement only in one axis so that they must be oriented carefully in the direction of movement. Also available are triaxial accelerometers that measure movement in three orthogonal axes and permit the accurate three-dimensional description of movement.

An excellent method for determining the physiology consists of combined accelerometry and EMG using spectral analysis and study of the tremor with and without weighting of the body part.[12, 13, 20] Since mechanical or mechanical-reflex tremors reduce frequency with increasing inertia, it is possible to separate such tremors using weighting from those coming from a central oscillator. In a pure mechanical tremor, such as a physiological tremor, there will be a tremor peak in the acceleration trace, but the EMG will be relatively the same over different frequencies. With weight, the tremor

peak will shift down and the EMG will remain nondescript. In a mechanical-reflex tremor, seen when a physiological tremor is enhanced, the EMG now participates at a similar frequency as the accelerometry record. With weight in this circumstance, the peaks for both the accelerometer and the EMG shift down similarly. These two situations can be compared with the result when there is a central oscillator. For example, some normal subjects have a contribution to their physiological tremor from an 8- to 12-Hz central oscillator.[12, 13] In this circumstance without weighting, the EMG will show an 8- to 12-Hz peak and the accelerometer record may be the same or different, depending on the mechanical characteristics. With weight, the accelerometer record will show two peaks, one shifting downward, the mechanical peak, and the other staying at 8 to 12 Hz. A component of tremor with a constant frequency with weighting presumably comes from a central oscillator.

Tremor can be divided into those occurring at rest and those occurring with action. "Rest" is only relative; some slight tonic postural maintenance is often required. Action tremors must be subdivided into those seen only with postural maintenance (postural or static tremor) and those requiring goal-directed movement (intentional or kinetic tremor). A third division of action tremors is those seen only with specific types of kinetic movement, such as handwriting.

INTENTION MYOCLONUS

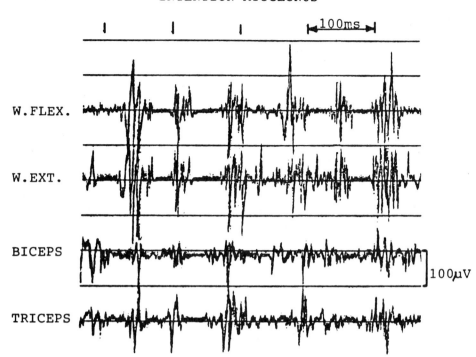

FIGURE 22–2. EMG recording of myoclonus in a patient with cortical reflex myoclonus. The right arm was extended and the surface EMG was recorded from four muscles in the arm. Note the synchronous activity in the different muscles.

"C-Reflex"
to cutaneous stimulation

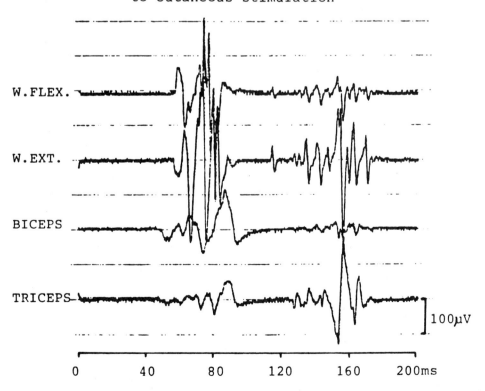

FIGURE 22–3. C reflex in a patient with cortical reflex myoclonus (same patient as in Fig. 22–2). The right index finger is stimulated electrically at three times sensory threshold at time 0. Note that the C reflex can be repetitive.

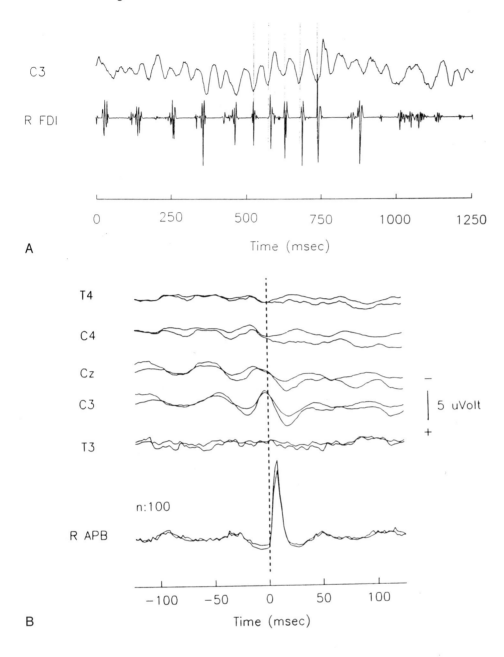

FIGURE 22–4. EEG recordings in a patient with cortical reflex myoclonus. Part *A* shows a single sweep with simultaneous recording from C3 and right first dorsal interosseous. Often the synchrony of the cortical positive-negative transient and the onset of the myoclonus is clear. Part *B* shows the results of backaveraging from spontaneous jerks of the right abductor pollices brevis. A focal positive-negative transient occurs just before the onset of the EMG, corresponding to the myoclonic jerk. Recordings were made by C. Toro.

Parkinsonian Tremor at Rest

The most frequent tremor at rest is that seen with Parkinson's disease or other parkinsonian states, such as that produced by neuroleptics.[12, 13, 16, 19] The tremor is usually seen in the context of other basal ganglia symptoms, but on rare occasions can be the sole clinical finding. It is present at rest and disappears with action, but it may resume with a static posture, particularly late in the disease. The frequency is 3 to 7 Hz. EMG studies show antagonist muscles to be active alternately. The tremor frequency is not altered by weighting and, hence, has its origin in a central oscillator, but its location is not known.

Exaggerated Physiological Tremor

Physiological tremor is a normal postural action tremor. In certain circumstances, such as anxiety, fatigue, thyrotoxicosis, and excessive use of caffeine, the tremor can be increased in magnitude and may be symptomatic. The frequency usually is in a range from 5 to 12 Hz, varying in part because of

the weight of the tremulous body part. The EMG in mild cases may look just like a normal interference pattern without well-defined bursting. In more severe cases, bursting may appear; it is usually synchronous in antagonist muscles. The accelerometric and EMG spectral peaks will be the same and shift together with weighting. There may also be a component from an 8- to 12-Hz central oscillator also seen in this condition.[12, 13, 16, 19]

Essential Tremor

Essential tremor is a common neurological disorder that often runs as an autosomal dominant trait in families.[19] It may appear in childhood or late life and runs a slowly progressive course. Typically it is a postural tremor; in some patients there is some increase in tremor with intention (kinetic movement) and in others the tremor occurs primarily with goal-directed movement (intentional essential tremor). Rarely, it appears to persist with rest. In most circumstances it is seen as the sole neurological abnormality; there is no known pathology and the pathophysiology is not clear. The frequency ranges from 4 to 12 Hz. EMG studies commonly show synchronous activity in antagonist muscles, but alternating activity is also possible. Sometimes it is clinically difficult to separate exaggerated physiological tremor and essential tremor. Using accelerometry and EMG, there is a constant frequency with weighting, which is clearly different from that seen with exaggerated physiological tremor.[12, 13] This observation indicates that essential tremor originates from a generator in the central nervous system. The cerebellum and cerebellar circuits seem to be involved,[25] and one commonly considered candidate is the inferior olivary nucleus.[22]

Cerebellar Tremor

Tremor with cerebellar lesions can be postural as well as the more well-known kinetic tremor. Postural cerebellar tremor has been separated into two groups, designated mild and severe.[15, 16, 19] The more characteristic is severe postural cerebellar tremor, which has commonly been called rubral tremor, unfortunately an inaccurate term. Typically it has a frequency of 2.5 to 4 Hz, affects proximal muscles more than distal muscles, waxes and wanes, and has a tendency to increase progressively in amplitude with prolonged posture. It persists or worsens with goal-directed movement and is associated with dysmetria. EMG studies show bursts of activity lasting 125 to 250 msec and alternation of activity in antagonist muscles. The responsible lesion appears to be in the superior cerebellar peduncle. The most common etiology in general neurological practice is multiple sclerosis, but there may be other causes, such as strokes or tumors. One type of post-traumatic tremor seems to be caused by a lesion here.

Mild postural cerebellar tremor is less well defined. The group includes tremors that are transient and more rapid (up to 10 Hz), and have distal predominance.

Kinetic tremor without postural tremor is usually ascribed to cerebellar dysfunction, and is often called cerebellar intention tremor. It certainly can also be seen together with a postural tremor. The lesions can be in the cerebellum or cerebellar pathways. Kinetic tremor is characterized by rhythmic oscillations about the target of movement; EMG studies show alternating activity in antagonist muscles. It should be differentiated from sequential irregular, inaccurate movements toward the target, which have been named serial dysmetria.[15, 16, 19]

Other Tremors

Wilson's disease can present with tremor as its sole manifestation, although other movement disorders and psychiatric disturbances are frequently present as well. The tremor is an action tremor present with posture and kinetic movement. Physiological analysis shows alternating activity in antagonist muscles at 3 to 5 Hz.[15]

Postural tremor may be seen in the setting of congenital or acquired peripheral neuropathies.[19] The pathophysiology is obscure. The tremor seen with hereditary sensorimotor neuropathy type 1 may simply be essential tremor that is co-inherited. Physiologically, the frequency is in the range of 6 to 8 Hz, and EMG studies show a mixture of synchronous and alternating activity in antagonist muscles. It is likely that in many circumstances, slowing of nerve conduction in the peripheral loop will give rise to delays that create instability.[2] In other circumstances, studies suggest a coexisting central oscillator.[33]

Palatal tremor, also known as palatal myoclonus, is characterized by rhythmic movements of the palate at approximately 3 Hz. There are two separate disorders, essential palatal tremor (EPT), which manifests as an ear click, and symptomatic palatal tremor (SPT), which is associated with cerebellar disturbances.[11, 19] The palatal movements are caused by activation of the tensor veli palatini muscle in EPT and of the levator veli palatini muscle in SPT. In SPT, the palatal movements may be accompanied by synchronous movements of adjacent muscles such as the external ocular muscles, tongue, larynx, face, neck, diaphragm, or even limb muscles. SPT is associated with hypertrophy of the inferior olive, and many authorities consider it to arise there, but the generator for EPT likely differs.

Task-specific Tremors

Primary writing tremor is task specific and appears only with handwriting and a few additional skilled tasks.[19] It is not seen with all skilled tasks, and is not produced by posture or goal-directed movement in general. The condition is probably underdiagnosed, as it is frequently confused with essential tremor and tremulous writer's cramp. Although the originally described patient had tremor with synchronous activity, subsequent experience suggests that the tremor is more commonly alternating in antagonist muscles with a frequency of 5 to 6 Hz.[1]

Orthostatic tremor is a tremor of the legs only when standing; it is not present when making voluntary movements of the legs when lying down nor is it present with walking.[6, 19, 30] The tremor in leg muscles is about 16 Hz, not influenced by peripheral feedback and synchronous between homologous leg muscles. The site of the central generator is unknown.

Hysterical Tremor

Tremor can be a conversion symptom.[19] Such tremors can take many forms, but the most common are action tremors with alternating activity in antagonist muscles. Often they violate rules of clinical behavior and look "funny," but the distinction can be difficult. Hysterical tremors vary in amplitude more than expected. Accelerometry in most tremors shows a narrow peak frequency that does not change over short time periods; in hysterical tremors, the peak may be broad and change frequently.

CASE STUDY 2

A Patient with Tremor

In the approach to the patient, the first consideration, of course, is a careful neurological history and physical examination. In particular, the situation in which the tremor is most prominent should be determined: rest, posture, kinetic movement, or task-specific situation (Fig. 22–5). The next step would be accelerometry and EMG with weighting of the limb, and this would be most useful for differentiating the postural tremors.

A 50-year-old man presented with tremor of both hands that had been present for the past 5 years, but was clearly getting worse. It was now interfering with fine motor tasks, but was not present at rest. His health was generally good. Note was made of the fact that he drank 4 cups of black coffee each morning. None of his relatives had obvious tremor. His father died at age 85 and may have had some shaking in his final years, but any abnormality was attributed to old age. On physical examination, the patient had a postural action tremor without other neurological sign. A standard laboratory screen did not reveal any abnormality.

EMG evaluation of the right index finger tremor showed 8 Hz bursts in finger flexor and extensor muscles that were usually, but not always, synchronous (Fig. 22–6).

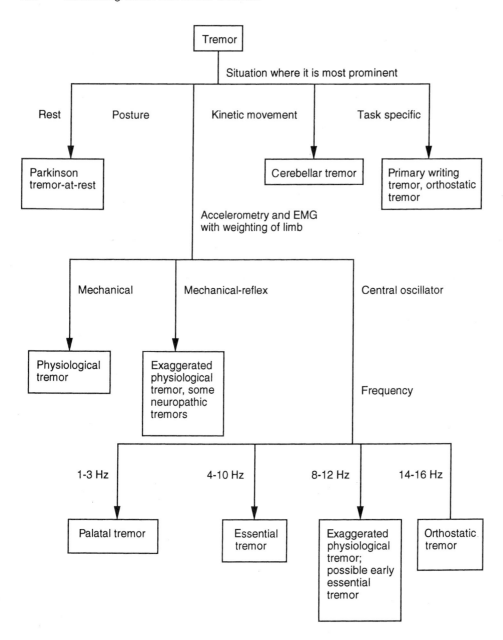

FIGURE 22–5. Flow diagram for evaluating tremor.

Combined accelerometry and EMG study of the tremor of the right wrist showed a peak at 8 Hz for both recordings (Fig. 22–7). With a 500-g weight, both recordings continued to show a dominant peak at 8 Hz. (With the weight, there also is a small mechanical component that shifts to a lower frequency.) The findings are consistent with essential tremor.

EVALUATION OF DISORDERS OF VOLUNTARY MOVEMENT

Not uncommonly patients complain that their movements are not normal; the movements have become clumsy, slow, stiff, or small. In general, patients feel as though they are losing control. Typically the physical examination is sufficient to define the problem, but there are characteristic neurophysiological features of the different disorders.

As with involuntary movements, the first approach with voluntary movement is to evaluate the EMG with defined voluntary movements.[4, 19] Normal patterns vary with the

speed of movement. A slow, smooth movement is characterized chiefly by continuous activity in the agonist. A movement made as rapidly as possible, a so-called ballistic movement, has a triphasic pattern with a burst of activity in the agonist lasting 50 to 100 msec, a burst of activity in the antagonist lasting 50 to 100 msec, and return of activity in

FIGURE 22–6. EMG recording of a right index finger, flexion-extension tremor in a patient with essential tremor. Recordings are from the first lumbrical and extensor indicis using wire electrodes. Note the synchronous activity in the muscles.

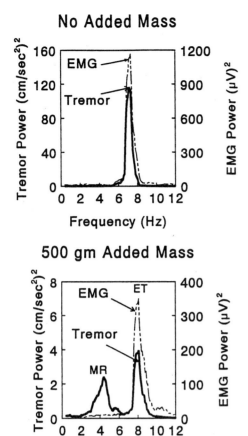

No Added Mass

500 gm Added Mass

FIGURE 22–7. Frequency analysis of accelerometer records and EMG recordings in a patient with essential tremor. The recordings are made of wrist tremor, and EMG is from the wrist extensor muscles. Top shows the results without weighting. Bottom shows the results with a 500-g weight on the dorsum of the wrist. ET is the essential tremor central oscillator component, and MR is the mechanical-reflex component. (From Elble RJ: The pathophysiology of tremor. In Watts RL, Koller WC (eds): Movement Disorders: Neurologic Principles and Practice. New York, McGraw Hill, 1977.)

the agonist often in the form of a burst. The mechanical features of the movement itself can also be evaluated. Measurements can be made of the movement time (time from initiation to completion of a movement) and accuracy. If the movement is triggered by a stimulus, then the reaction time (time from stimulus to initiation of movement) can also be measured.

Slowness

With corticospinal tract lesions, there is generally prominent weakness. Assessment of the EMG pattern shows that there is prolongation of the first agonist or the first antagonist EMG bursts, or both. In addition, there is a reduced interference pattern on maximal effort in routine electromyography. It appears that the lack of power is compensated by prolongation of the activation time of the muscle.

With parkinsonian bradykinesia, there is abnormal patterning, with multiple bursts having the appearance of repetitive cycles of the triphasic pattern to complete the movement. The patients do not put sufficient EMG activity into each burst to generate sufficient force. However, different from the patients with corticospinal tract lesions, they com-

pensate by making more bursts. Interestingly, these repetitive cycles of bursts look like their resting tremor. Study of reaction time and movement time is helpful for quantification of Parkinson's disease, because slowness is such a central aspect of the disease.

Clumsiness

When the principal complaint is that of clumsiness, cerebellar dysfunction should be the first thought. The clinical appearance is classic with dysmetria (more commonly hypermetria) and dysdiadochokinesia of the limbs as well as saccadic dysmetria, nystagmus, dysarthria, and gait instability. The EMG pattern with attempted rapid movement shows prolongation of the first agonist or antagonist EMG bursts, or both, but without a reduced interference pattern as seen with corticospinal tract disease. The prolongations can be marked, and there is a good correlation of the acceleration time of the movement with the duration of the first agonist burst. Unwanted prolongation of acceleration time should predispose to hypermetria. The antagonist burst can be delayed as well. Patients with cerebellar disorders do better when they move slowly. Hence, they often slow down to help compensate their deficit. They can, however, move quickly.

It is important to realize that dystonia and athetosis are characterized by abnormal voluntary movement as well as involuntary movements. Indeed, often the involuntary movements are largely brought out by attempted voluntary efforts. With attempted rapid movement, the EMG pattern shows excessive activity, including co-contraction activity, in the antagonist. Excessive activity also overflows into muscles not needed for the action. EMG burst length can be prolonged. In athetosis particularly, there are a variety of abnormal patterns of antagonist activity that appear to block the movement from occurring.

It is sometimes difficult to make the diagnosis of dystonia. The movements of the patient may be unusual, and psychogenic disease is often considered. Several reflex studies and other neurophysiologic tests are abnormal in dystonic patients.[18] Two reflex studies might be employed on an individual basis to help in the differential diagnosis: the blink reflex and reciprocal inhibition.

In normal subjects, if a second blink reflex is produced at an interval of less than 3 seconds after a first blink reflex, the R2 component of the second blink reflex is reduced in amplitude compared with that of the first blink reflex.[3, 19, 23] The amount of inhibition is proportional to the interval between the two stimuli for production of the blink reflexes. The curve relating the amplitude of the second R2 to the interval between the stimuli is called the blink reflex recovery cycle. Normal values can be determined, and patients with blepharospasm show less inhibition than normal. Abnormalities of blink reflex recovery have also been demonstrated in generalized dystonia, spasmodic torticollis, and spasmodic dysphonia. In the last two conditions, abnormalities can be found even without clinical involvement of the eyelids. The abnormality is not specific, however, as it is also found in Parkinson's disease.

Reciprocal inhibition is the spinal process of inhibition of a motoneuron pool when the antagonist motoneuron pool is activated.[19, 23, 31, 32] This can be studied by assessing the influence on an H reflex of stimulation of a nerve with afferents from muscles antagonist to the muscle where the H reflex is produced. There are several normal periods of inhibition, depending on the interval between the stimulus to the antagonist nerve and that eliciting the H reflex. The period of inhibition best understood is that occurring when the two nerves are stimulated close to the same time. This inhibition is mediated by the Ia inhibitory interneuron. In

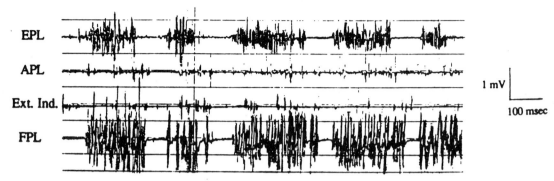

FIGURE 22–8. EMG recordings from a patient with focal hand dystonia. Recordings are from four muscles in the right arm during motor performance. (Modified and retouched from Cohen LG, Hallett M: Hand cramps: Clinical features and electromyographic patterns in a focal dystonia. Neurology 38:1005, 1988.)

the arm, reciprocal inhibition has been studied looking at the effects of radial nerve stimulation upon the H of the flexor carpi radialis (FCR). Via various pathways, and therefore at various time intervals after the radial nerve stimulus, the radial afferent traffic can inhibit the motoneuron pool of the FCR. The first period of inhibition is caused by disynaptic Ia inhibition, the second period of inhibition is likely presynaptic inhibition, and little is known about the third period of inhibition. Reciprocal inhibition is reduced in patients with dystonia, including those with generalized dystonia, writer's cramp, spasmodic torticollis, and blepharospasm. It should be noted that reciprocal inhibition can be abnormal even in asymptomatic arms, as in the situation of blepharospasm. Reciprocal inhibition studies can be used as a sensitive method for detecting abnormality in patients with dystonia.[32] However, the method is not specific, as abnormal results are also seen in patients with Parkinson's disease.[26]

CASE STUDY 3

A Patient Who Is Clumsy

A 29-year-old professional pianist complained that the thumb of his right hand was not working properly. It seemed to stick on the key and could not be lifted easily. The problem seemed to worsen over the past 3 months despite additional practice time. He had no other difficulty with his right hand for any other task. History and physical examination were fully normal except for performance of the right hand when playing the piano. The movements of the right thumb were as he described.

Analysis of the EMG patterns of the right extensor pollices longus and flexor pollices longus muscles using wire electrodes while playing the piano showed prolonged EMG bursts and synchronous contractions (Fig. 22–8). Reciprocal inhibition studies in the right arm showed reduced inhibition in all three periods (Fig. 22–9).

The patient has a task-specific focal dystonia of the right hand.

EVALUATION OF TONE

Tone is defined as the response to passive stretch. Study of EMG responses to controlled stretches can provide physiological insights and permit quantitative analysis. Controlled stretches can be delivered by devices containing torque motors. The stretch can be produced by altering the force, or torque, of the motor or by altering the position of the shaft

of the motor. The perturbation can be a single step or more complex, such as a sinusoid. The mechanical response of the limb can also be measured: the positional change if the motor alters force, or the force change if the motor alters position. Such mechanical measurements can directly mimic and quantify the clinical impression.[19, 23]

An extensive literature has developed about the short and long latency EMG responses to controlled stretches and the changes that occur in spasticity and rigidity. The short latency reflex is the monosynaptic reflex. Reflexes occurring at a longer latency than this are designated long latency. If a pathway ascends to the brain stem or higher, it could be called "long loop." When a relaxed muscle is stretched, in general only a short latency reflex is produced. When a muscle is stretched while it is active, one or more distinct long latency reflexes are produced following the short latency reflex and prior to the time needed to produce a voluntary response to the stretch. These reflexes are recognized as separate because of brief time gaps between them, giving rise to the appearance of distinct "humps" on a rectified EMG trace. Each component reflex, short or long in latency, has about the same duration, approximately 20 to 40 msec. They appear to be true reflexes in that their appearance and magnitude depend primarily on the amount of background force that the muscle was exerting at the time of the stretch and the mechanical parameters of the stretch;

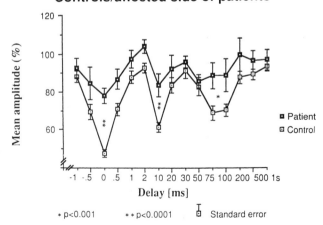

Controls/affected side of patients

FIGURE 22–9. Reciprocal inhibition study of the affected hands of patients with focal hand dystonia compared with a group of normal control subjects. (From Panizza ME, Hallett M, Nilsson J: Reciprocal inhibition in patients with hand cramps. Neurology 39:85, 1989.)

they do not vary much with whatever the subject might want to do after experiencing the muscle stretch. By contrast, the voluntary response that occurs after a reaction time from the stretch stimulus is strongly dependent on the will of the subject.

Spasticity

Studies of stretch reflexes in spasticity show that the short latency reflex is enhanced and the magnitude of enhancement is correlated with the clinical impression of increased tone, as might have been anticipated. A prominent feature of the enhancement is a lowering of the threshold stretch that produces a response. Long latency reflexes are frequently absent or reduced, but they may be present or even enhanced. In some patients, they do appear to contribute to increased tone; the clinical correlation is not yet well understood.

The principal abnormality can also be analyzed by the tendon jerk or H reflex. To obtain a meaningful measure of the response, the amplitude of the maximal reflex must be compared with the amplitude of the EMG in maximum voluntary effort or the amplitude of the EMG produced by supramaximal stimulation of the nerve to that muscle (H/M ratio).[19, 23] Unfortunately, there is a large interindividual variability that makes the measurement less useful than it might be. The H/M ratio is enhanced in spasticity, but not in rigidity or in dystonia. Furthermore, in spasticity, H reflexes may appear in muscles in which they are not ordinarily seen, such as the small hand muscles.

Another clinically useful test is vibratory inhibition of the H reflex.[5] In normal subjects, the amplitude of the H reflex is markedly inhibited by vibration of the muscle. Vibration is a strong stimulus for Ia afferents, although other afferents are activated as well. The inhibition has been thought to be caused largely by presynaptic inhibition from polysynaptic effects of the homonymous Ia afferents, but postactivation depression may also be responsible for some of the effect. H reflexes are tested with and without vibration. Vibration is applied to the tendon of the muscle being studied. The effect can be seen over a wide range of vibrations, but typically a frequency of about 100 Hz is used with an excursion of about 1 mm. The effect can be studied over the range of H reflex amplitudes. The percent inhibition is considered a measure of intensity of presynaptic inhibition. Vibratory inhibition is often dramatically reduced in spasticity. It is also reduced in the stiffman syndrome and dystonia. On the other hand, it is normal in parkinsonian rigidity.

It is important to recognize that some of the impression of increased tone in spasticity in the chronic situations results from changes in the muscle itself. The extreme of this change is a contracture.

Rigidity

Studies of stretch reflexes in rigidity show that the long latency reflexes are enhanced and that, at least in some studies, the magnitude of the enhancement is correlated with the clinical impression of increased tone. In rigidity, long latency reflexes appear even at rest, when normally no long latency reflexes are seen. The physiology of these long latency reflexes is not clear, although enhancement of group II–mediated reflexes may play a role. Thus investigation of stretch reflexes in patients with increased tone would serve to implicate short or long latency mechanisms (roughly correlating with spasticity and rigidity) and provide a means of quantification.

The enhancement of long latency reflexes can also be brought out by electrical stimulation of a mixed nerve. Such stimulation while the limb is at rest produces only an M wave and F response in the muscles innervated by that nerve. If a mixed nerve is stimulated while the muscles are active, however, additional responses are produced.[19, 23] With mixed nerve stimulation, there is a short latency response that seems analogous to the H reflex (HR) and one or more long latency responses. One of these long latency responses, called LLRII by Deuschl and associates,[10] may have a transcortical pathway similar to some of the long latency reflexes to stretch. A long latency response, the LLRI, intermediate in latency between the HR and LLRII, is enhanced in about half of patients with Parkinson's disease.

Not all studies find a significant relationship between reflex responsiveness and increased tone. It seems clear that the failure of the Parkinson patient to relax at rest gives rise to some of the clinical phenomenon. Additionally, just as with spasticity, there can be changes in the muscle itself with longstanding disease.

Stiffman Syndrome

Stiffman (stiffperson) syndrome is a disorder characterized by continuous muscle contraction giving rise to severe stiffness, most severe axially. Patients have difficulty moving, and walk and make voluntary movements slowly. Sensory stimulation often induces a worsening of the spasms.[28] The disorder is often associated with circulating antibodies to glutamic acid decarboxylase (GAD), and the resultant deficiency of gamma-aminobutyric acid (GABA) may play a role in its pathophysiology.

The EMG demonstrates continuous activity of motor units in a normal pattern although it cannot be silenced even with contraction of the antagonist muscle.[35] Deficiency of vibratory inhibition of the H reflex is a prominent reflex abnormality, a feature shared with spasticity.

CASE STUDY 4

A Patient with Increased Tone

A neurological consultation was requested from a rheumatologist for a 68-year-old woman who was complaining of stiffness. Stiffness had been developing over the past year, and there had been some concomitant difficulty in moving. She had a long history of back pain and other joint pains, and these pains were also getting worse. The rheumatologist, however, could not explain the extent of the symptoms on the basis of joint disease and wondered if there might be a neurological cause.

Neurological examination revealed increase in tone without cogwheeling. The increase in tone did not seem velocity dependent, but the tendon reflexes were brisk. Plantar responses were mute. Movements were slow, but the patient complained of pain. In addition, the patient was not able to exert maximal force.

Clinical neurophysiological assessment revealed absent EMG activity at rest. Studies of the H reflex in the FCR showed a normal H/M ratio and normal vibratory inhibition of the H reflex (Fig. 22–10). Long latency reflexes in the abductor pollices brevis were increased in magnitude after stimulation of the median nerve at the wrist (Fig. 22–11). These findings were compatible with increase in tone caused by rigidity.

The patient was thought likely to have Parkinson's disease. A trial of levodopa was successful, resulting in significant improvement in the patient's stiffness and slowness.

recruitment curves M and H reflex

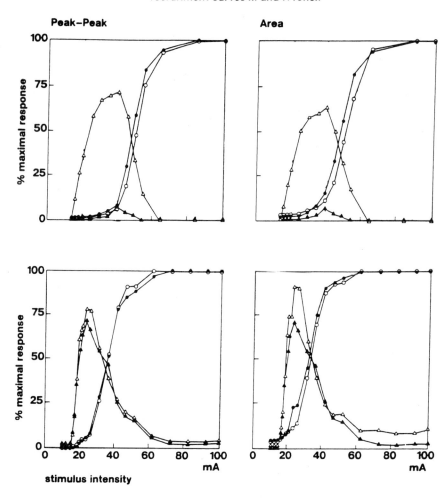

FIGURE 22–10. Recruitment curves for the M wave *(triangles)* and H *(circles)* with *(filled symbols)* and without *(open symbols)* vibration. Left side shows peak to peak amplitude and right side shows area. Top is from a normal subject and bottom is from a patient with spasticity. Patients with Parkinson's disease show normal behavior. (Modified from Bour LJ, Ongerboer de Visser BW, Koelman HTM, et al: Soleus H-reflex tests in spasticity and dystonia: A computerized analysis. J Electromyogr Kinesiol 1:9, 1991.)

BOTULINUM TOXIN TREATMENT OF INVOLUNTARY MUSCLE SPASMS

Botulinum toxin is now a first line treatment for involuntary muscle spasms such as focal dystonias, hemifacial spasm, and spasticity.[24] Minute amounts of the toxin are injected directly into the muscle, where it interferes with acetylcholine release and weakens the muscle. With botulinum toxin type A, the effect lasts about 3 months.

EMG techniques might be utilized both to help identify the muscles for therapy and to guide the delivery of toxin into the appropriate muscles. If the muscles with the most severe spasms are not obvious from clinical inspection, then multichannel EMG recordings might be undertaken. In some circumstances, it is possible to use surface electrodes or needle electrodes. In other circumstances, such as in the forearm, it is necessary to use fine wire electrodes. The fine wires allow the patient to move freely and to perform the task that provokes the dystonia while at the same time permitting isolation of individual muscles. Needle electrodes would be too painful and surface electrodes are often not sufficiently selective.

The importance of EMG guidance in performing injections is not yet fully accepted. Botulinum toxin does diffuse in the tissues for at least several centimeters. Some authorities suggest simply making the injection based on a knowledge of the anatomy. In some studies, however, all injections have been made only with EMG guidance. Certainly, with

deep muscles or small, closely packed muscles, it would seem that EMG guidance would help assure that the medication was delivered to the correct target and that nearby muscles were not weakened unnecessarily. In one study that made a comparison, for the treatment of spasmodic torticollis, EMG guidance has been demonstrated to be valuable.[9] In the treatment of spasticity, one group has suggested that magnitude of EMG activity should be used as a criterion for whether to inject the muscle.[14]

There are several techniques for EMG guidance. They all employ a hollow needle coated with Teflon except for the tip and hub. The hub is connected to the active input of an amplifier, and the reference electrode can be a surface electrode. Once the site is identified, the toxin is injected through the same needle. There are three methods for verifying the correct location of the needle:

1. Asking the patient to activate the muscle voluntarily and selectively and looking for the EMG interference pattern. This seems straightforward, but presents some difficulties as it is not always easy to activate a muscle selectively. Moreover, patients with dystonia seem to have a particular difficulty with selective activation.
2. Passive movement of the joint on which the muscle acts. The passive movement stretches and relaxes the muscle, which should cause the needle to move back and forth. This is an excellent technique for superficial muscles and gives very precise localization. (The syringe must be disconnected from the needle for this maneu-

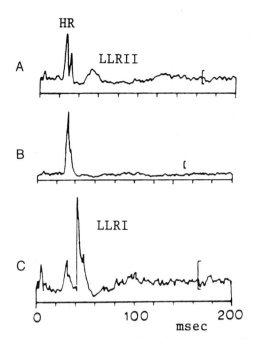

FIGURE 22–11. Averaged rectified EMG responses in APB after stimulation of the median nerve at the wrist. There is a background contraction of the APB and the stimulation is subthreshold for activation of an M wave. Part *A* is from a normal subject, part *B* from a patient with spasticity, and part *C* from a patient with Parkinson's disease. HR is the H reflex, LLRI and LLRII are long latency reflexes. Note the absence of LLRs in the spastic patient and the enlarged LLRI in the patient with Parkinson's disease. Voltage calibration bar is 100 μV. (Modified from Deuschl G, Lücking CH: Physiology and clinical applications of hand muscle reflexes. Electroencephalogr Clin Neurophysiol Suppl 41:84, 1996.)

ver as the weight of the syringe makes the movements difficult to assess.)

3. Stimulating through the needle. With low currents this causes a small localized muscle twitch that should identify where the tip of the needle is located.

Knowledge of anatomy that is sufficient for routine electromyography is not always enough for injections. For example, if the problem affects only the middle finger, it is not adequate to find the flexor digitorum superficialis (FDS); the injection should be into the fascicle of that muscle specific for the third finger. An example of a difficult anatomical point is the location of the second finger fasicle of FDS. This fasicle runs diagonally across the arm below the third finger fascicle; distal to proximal, it runs lateral to medial. In the superficial layer of muscles moving lateral to medial, the third finger fascicle of the FDS lies next to the flexor pollicis longis (FPL). The novice, encountering this anatomy, might well wonder how the index finger is flexed!

The toxin has its effect only at the end plate, and in quantitative studies it has been demonstrated that the toxin is more effective if delivered at the end plate. It would be useful, therefore, to identify and inject into the end plate zone, but given the large size of the end plate zone and the diffusion of the toxin, most authorities have been content with injections into the mid-belly of the muscle. Criteria for the end plate would be end plate spikes and motor unit potentials, beginning with a negative phase.

ACKNOWLEDGMENT

Many portions of the text of this chapter are similar to portions of a chapter in *Electrodiagnosis in Clinical Neurology*, edited by M. J. Aminoff.[19]

REFERENCES

1. Bain PG, Findley LJ, Britton TC, et al: Primary writing tremor. Brain 118:1461, 1995.
2. Bain PG, Britton TC, Jenkins IH, et al: Tremor associated with benign IgM paraproteinaemic neuropathy. Brain 119:789, 1996.
3. Berardelli A, Rothwell JC, Day BL, et al: Pathophysiology of blepharospasm and oromandibular dystonia. Brain 108:593, 1985.
4. Berardelli A, Hallett M, Rothwell JC, et al: Single-joint rapid arm movements in normal subjects and in patients with motor disorders [review article]. Brain 119:661, 1996.
5. Bour LJ, Ongerboer de Visser BW, Koelman HTM, et al: Soleus H-reflex tests in spasticity and dystonia: A computerized analysis. J Electromyogr Kinesiol 1:9, 1991.
6. Britton TC, Thompson PD, van der Kamp W, et al: Primary orthostatic tremor: Further observations in six cases. J Neurol 239:209, 1992.
7. Brown P, Rothwell JC, Thompson PD, et al: The hyperekplexias and their relationship to the normal startle reflex. Brain 114:1903, 1991.
8. Cohen LG, Hallett M: Hand cramps: Clinical features and electromyographic patterns in a focal dystonia. Neurology 38:1005, 1988.
9. Comella CL, Buchman AS, Tanner CM, et al: Botulinum toxin injection for spasmodic torticollis: Increased magnitude of benefit with electromyographic assistance. Neurology 42:878, 1992.
10. Deuschl G, Lücking CH: Physiology and clinical applications of hand muscle reflexes. Electroencephalogr Clin Neurophysiol Suppl 41:84, 1990.
11. Deuschl G, Toro C, Valls-Solé J, et al: Symptomatic and essential palatal tremor. 1. Clinical, physiological, and MRI analysis. Brain 117:775, 1994.
12. Elble RJ: The pathophysiology of tremor. In Watts RL, Koller WC (eds): Movement Disorders. Neurologic Principles and Practice. New York, McGraw-Hill, 1997, p 405.
13. Elble RJ, Koller WC: Tremor. Baltimore, Johns Hopkins University Press, 1990.
14. Finsterer J, Fuchs I, Mamoli B: Automatic EMG-guided botulinum toxin treatment of spasticity. Clin Neuropharm 20:195, 1997.
15. Hallett M: Differential diagnosis of tremor. In Vinken PJ, Bruyn GW, Klawans HL (eds): Extrapyramidal Disorders. Handbook of Clinical Neurology, vol 5(49). Amsterdam, Elsevier Science, 1987, p 583.
16. Hallett M: Classification and treatment of tremor. JAMA 266:1115, 1991.
17. Hallett M: Myoclonus and myoclonic syndromes. In Engel JJ, Pedley TA (eds): Epilepsy: A Comprehensive Textbook, vol 3. Philadelphia, Lippincott-Raven, 1997, p 2717.
18. Hallett M: Physiology of dystonia. In Fahn S, Marsden CD, DeLong M (eds): Advances in Neurology. Philadelphia, Lippincott-Raven, 1998, p 11.
19. Hallett M: Electrophysiologic evaluation of movement disorders. In Aminoff MJ (ed): Electrodiagnosis in Clinical Neurology, 4th ed. New York, Churchill Livingstone, 1999, p 365.
20. Hallett M: Overview of human tremor physiology. Mov Disord 13(Suppl 3):43, 1998.
21. Hallett M, Marsden CD, Fahn S: Myoclonus. In Vinken PJ, Bruyn GW, Klawans HL (eds): Extrapyramidal Disorders. Handbook of Clinical Neurology, vol 5(49). Amsterdam, Elsevier Science, 1987, p 609.
22. Hallett M, Dubinsky RM: Glucose metabolism in the brain of patients with essential tremor. J Neurol Sci 114:45, 1993.
23. Hallett M, Berardelli A, Delwaide P, et al: Central EMG and tests of motor control. Report of an IFCN Committee. Electroencephalogr Clin Neurophysiol 90:404, 1994.
24. Jankovic J, Hallett M, eds: Therapy with Botulinum Toxin. New York, Marcel Dekker, 1994.
25. Jenkins IH, Bain PG, Colebatch JG, et al: A positron emission tomography study of essential tremor: Evidence for overactivity of cerebellar connections. Ann Neurol 34:82, 1993.
26. Lelli S, Panizza M, Hallett M: Spinal cord inhibitory mechanisms in Parkinson's disease. Neurology 41:553, 1991.
27. Matsumoto J, Fuhr P, Nigro M, et al: Physiological abnormalities in hereditary hyperekplexia. Ann Neurol 32:41, 1992.

28. Matsumoto J, Caviness JN, McEvoy KM: The acoustic startle reflex in stiff-man syndrome. Neurology 44:1952, 1994.
29. Matsumoto J, Hallett M: Startle syndromes. In Marsden CD, Fahn S (eds): Movement Disorders 3. Oxford, Butterworth-Heinemann, 1994, p 418.
30. McManis PG, Sharbrough FW: Orthostatic tremor: Clinical and electrophysiologic characteristics. Muscle Nerve 16:1254, 1993.
31. Nakashima K, Rothwell JC, Day BL, et al: Reciprocal inhibition in writer's and other occupational cramps and hemiparesis due to stroke. Brain 112:681, 1989.
32. Panizza ME, Hallett M, Nilsson J: Reciprocal inhibition in patients with hand cramps. Neurology 39:85, 1989.
33. Pedersen SF, Pullman SL, Latov N, et al: Physiological tremor analysis of patients with anti-myelin-associated glycoprotein associated neuropathy and tremor. Muscle Nerve 20:38, 1997.
34. Shibasaki H: Pathophysiology of negative myoclonus and asterixis. In Fahn S, Hallett M, Lüders HO, et al (eds): Negative Motor Phenomena (Advances in Neurology; vol 67), Philadelphia, Lippincott-Raven, 1995, p 199.
35. Toro C, Jacobowitz DM, Hallett M: Stiffman syndrome. Semin Neurol 14:154, 1994.
36. Toro C, Hallett M: Pathophysiology of myoclonic disorders. In Watts RL, Koller WC (eds): Movement Disorders. Neurologic Principles and Practice. New York, McGraw-Hill, 1997, p 551.
37. Wilkins DE, Hallett M, Berardelli A, et al: Physiologic analysis of the myoclonus of Alzheimer's disease. Neurology 34:898, 1984.

Chapter 23

Silent Period Studies and Long Latency Reflexes

A. Arturo Leis, MD

INTRODUCTION
MIXED NERVE SILENT PERIOD
Procedure
Physiological Mechanisms
Clinical Applications
CUTANEOUS SILENT PERIOD
Procedure
Physiological Mechanisms
Clinical Applications

LONG LATENCY RESPONSES
Long Latency Responses in Normal
  Subjects
Long Latency Responses in Myoclonus
Pathways
Techniques
Normal Subjects
Myoclonus
Other Techniques
Clinical Utility

INTRODUCTION

This chapter provides a summary of the electromyographic silent periods and other long latency reflexes recorded in voluntarily contracting muscles following electrical stimulation to peripheral nerves. The term *silent period* refers to a transient, relative or absolute suppression in the electromyogram (EMG) activity of a voluntarily contracting muscle following stimulation to a mixed nerve (*mixed nerve silent period*) or to a cutaneous nerve (*cutaneous silent period*). Stimulation to a nerve carrying mixed afferents also typically produces an excitatory response at a latency similar to the H reflex (or F wave), and a second excitation, the *long latency reflex (LLR)*, with an onset about 20 to 25 msec later. In addition, during a weak voluntary contraction, stimulation to cutaneous afferents can elicit a sequence of low-amplitude excitatory and inhibitory responses. Because different techniques may produce long latency responses with similar latency, it is sometimes assumed that these responses are mediated along the same pathways. However, it cannot be excluded that responses with different underlying mechanisms may have the same latency. Although silent period and LLR studies are used uncommonly in clinical practice, they complement routine neurophysiological studies and, in some instances, may be more sensitive in demonstrating abnormality than routine studies.

I will first review certain physiological properties of axons that contribute to our understanding of silent periods. It should be recognized that axons within a mixed peripheral nerve that are directly activated by an electrical stimulus carry two volleys of impulses, one that ascends to the spinal cord and the other that descends toward the muscle. The ascending volley consists of orthodromic sensory or antidromic motor impulses, while the descending volley contains orthodromic motor or antidromic sensory impulses. In addition, axons within a mixed or cutaneous nerve that are activated by an electrical stimulus differ markedly in their diameter, ability to conduct impulses, conduction velocity, and thresholds for activation.[9] The recorded responses depend largely on the fiber types activated by the stimulus; for example, a low-intensity stimulus will generate responses mediated primarily by the lower threshold, larger diameter fibers, whereas a high-intensity stimulus to the same nerve will produce responses that are also mediated by higher threshold, smaller diameter fibers. It should also be noted that the term *collision*, when used with reference to nerve conduction studies, refers to the interaction of two action potentials propagated toward each other from opposite directions on the same nerve fiber so that the *refractory periods* of the two potentials prevent propagation past each other.[6] The refractory membrane behind an advancing action potential prevents passage of action potentials propagated toward each other. Collision of antidromic with orthodromic motor impulses accounts for a portion of the mixed nerve silent period.

MIXED NERVE SILENT PERIOD

Despite continued effort, EMG activity from a voluntarily contracting muscle undergoes a transient suppression following an electrical stimulus to the mixed nerve innervating that muscle (homonymous nerve) or to a nearby mixed nerve (nonhomonymous nerve). This period of electrical inactivity is designated the mixed nerve silent period (MNSP). For many years, it was generally accepted that the MNSP was largely dependent on orthodromic motor impulses in the descending action potential volley and the mechanical changes created by the superimposed direct muscle twitch. Matthews[43] in 1931 showed that a mechanical twitch of an isolated frog muscle modified the discharges from the muscle spindles, and postulated that a "spindle pause" could be responsible for the MNSP in humans. In 1951, Merton[46] concluded that the MNSP resulted from the descending motor volley and that the ascending volley was not even partially responsible for the silent period. Higgins and Lieberman[25] (1968) also attributed the MNSP to a pause in spindle discharges and considered the burst of EMG activity that terminates the MNSP to result from spindle excitation during stretch of the muscle in the falling phase of the twitch. Shahani and Young,[58] in 1973, also contended that much of the second half of the MNSP was related to changes produced in spindle discharges by the superimposed twitch, but also stressed the importance of hitherto ignored cutaneous mechanisms in the genesis of the MNSP. Indeed, subsequent investigators argued that much of the MNSP was de-

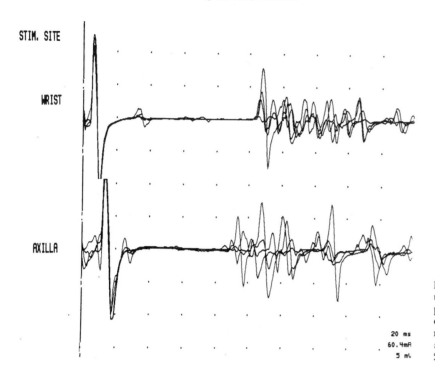

STIM. SITE

WRIST

AXILLA

20 ms
60.4mA
5 mV

FIGURE 23–1. Compound muscle action potential, F wave, and silent period in the abductor pollicis brevis muscle after stimulation of the median nerve at the wrist (*top*, four superimposed responses). Stimulation at the axilla (*bottom tracings*) produced a silent period that ended 15 to 20 msec earlier.

pendent on orthodromic sensory impulses in the ascending volley. Also in 1973, McLellan[45] found no correlation between the degree of muscle shortening during the twitch contraction and the duration of the MNSP. He also observed that the MNSP end point produced by stimulation of a mixed nerve at a proximal site was consistently shorter than that produced by distal stimulation (Fig. 23–1). This indicates that the impulses destined to produce this portion of the MNSP, like the impulses that generate the F wave, first travel away from the recording electrodes toward the spinal cord. If the descending volley contributed to the latter portion of the MNSP, then proximal stimulation, which generates an M wave and muscle twitch of increased latency, should have produced an MNSP that ended later. McLellan concluded that the MNSP was determined by the ascending volley via an inhibitory spinal reflex, with no indication that the descending volley or changes in spindle activity contributed to the MNSP.[45] More recently, Leis and colleagues[37] (1991) used selective nerve blocks to reassess the role of the descending and ascending action potential volleys in the genesis of the MNSP. They confirmed McLellan's observations in homonymous and nonhomonymous nerves (Table 23–1) and provided evidence that the MNSP could be elicited in the absence of a descending volley. They also concluded that the MNSP resulted from the effects of the ascending volley, without apparent contribution from the descending volley.

Procedure

The technique for recording the MNSP is described using the combination of median nerve and abductor pollicis brevis (APB) muscle. However, other muscle and mixed nerve combinations can be used to study the silent period. The MNSP is usually evoked by electrical stimulation of the median nerve at the wrist or elbow while the subject maintains constant isometric or isotonic abduction of the thumb. The MNSP is recorded with the active surface electrode (G1) placed over the belly of the APB muscle, and the reference electrode (G2) over the tendon. Single shocks are delivered to the median nerve to elicit the MNSP. Stimuli are square

waves 0.1 to 0.4 msec in duration with a stimulus intensity 25% or greater above supramaximal for the direct M response recorded from the APB muscle. Stimuli of lower intensity, however, may also generate a MNSP.[37] In addition, single shocks delivered to a nonhomonymous nerve such as

TABLE 23–1	Silent Periods Recorded in the Abductor Pollicis Brevis Muscle Under Isotonic Conditions			
Site of Stimulation	Subject	Individual Mean SP End Point (msec)	Individual Range SP End Point (msec)	Group Mean (msec)
Median nerve				
Wrist	1	116	112–114	
	2	104	98–108	114
	3	120	114–126	
	4	116	108–120	
Elbow	1	99	94–108	
	2	86	84–90	96
	3	100	96–106	
	4	98	90–104	
Arm	1	88	80–94	
	2	*	*	88
	3	84	80–88	
	4	92	90–96	
Ulnar nerve				
Wrist	1	110	98–114	
	2	102	97–106	108
	3	112	108–116	
	4	106	102–110	
Elbow	1	94	86–98	
	2	84	82–88	87
	3	86	82–90	
	4	86	80–90	

SP, Silent period.
*, Data missing.
From Leis AA, Ross MA, Emori T, et al: The silent period produced by electrical stimulation of mixed peripheral nerves. Muscle Nerve 14:1202–1208, 1991, with permission.

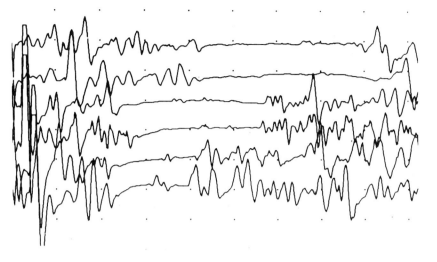

FIGURE 23–2. Silent periods recorded in the abductor pollicis brevis muscle after stimulation of the ulnar innervated fifth digit *(top two tracings)*, ulnar nerve stimulation at the wrist *(middle two tracings)*, and ulnar nerve stimulation at the elbow *(bottom two tracings)*. Note that the silent period produced by proximal stimulation started and ended earlier than the silent period elicited by more distal stimulation. Sweep: 20 msec/division; gain: 1 mV/division. (From Leis AA, Ross MA, Emori T, et al: The silent period produced by electrical stimulation of mixed peripheral nerves. Muscle Nerve 14:1202–1208, 1991.)

the ulnar nerve at the wrist or elbow may also elicit a silent period from the contracting APB muscle (Fig. 23–2).

Physiological Mechanisms

The silent period following mixed nerve stimulation appears to be comprised of at least three periods of EMG suppression. The first period is known to result from collision of antidromic with orthodromic motor impulses.[46, 58] The appearance of the F wave or H reflex signals the end of this first suppression. The next portion of the MNSP corresponds to the segment from the end of the F wave or H reflex to the beginning of LLR (also designated the V2 wave,[63, 68] cortical C response,[11] second excitatory response E2,[19, 21] or second reflex response R2[17]). This segment of the SP occurs too soon to be attributable to cutaneous effects,[58] and is most likely dependent on the ability of the antidromic motor volley to maximally activate Renshaw cell inhibition.[54] The last segment of the MNSP may be attributable to cutaneous afferent impulses which, in isolation, produce a complete silent period between 70 and 120 msec after digital stimulation[37, 38] (see later discussion of cutaneous silent period). With more proximal stimulation, these same cutaneous fibers are presumably activated within the mixed nerve. Figure 23–3 shows the MNSP in the APB muscle after median nerve stimulation in a patient with polyneuropathy. In this

patient, the neuropathy resulted in marked temporal dispersion of the MNSP into the three distinct periods of EMG suppression. In normal subjects, such a distinct separation of the MNSP is not usually seen (compare with Fig. 23–1).

Clinical Applications

Several studies have described the MNSP in patients with a variety of diseases affecting the motor systems.[1, 15, 17, 18, 26, 35, 36, 44] Because the precise physiological mechanisms responsible for the various components of the MNSP are only now being elucidated, one should be cautious in interpreting reports that attempt to explain complex motor phenomena such as posture, control of movement, and pathological movements (e.g., spasticity, rigidity, tremor, dystonia) on the basis of studies of silent periods. Several investigators have reported that the MNSP is normal in Parkinson's disease patients with prominent rigidity,[1, 26] but prolonged in Parkinson's disease patients with predominant tremor.[44] Following levodopa treatment, there was significant shortening of the silent period on the side of the tremor but not on the contralateral side, suggesting that the response to levodopa may be linked with reduced inhibition of the motor neuron by peripheral stimulation.[44] The MNSP has also been investigated in patients with Huntington's disease and found to be shortened,[17, 55] implying that abnormalities in basal ganglia

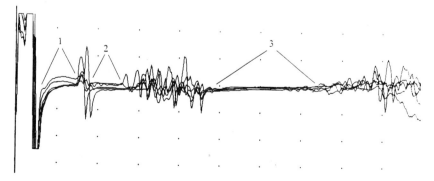

FIGURE 23–3. Mixed nerve silent period recorded in the abductor pollicis brevis muscle after median nerve stimulation at the wrist in a 31-year-old woman with polyneuropathy. Marked temporal dispersion resulted in the three distinct periods of EMG suppression (numbers 1 to 3) that make up the mixed nerve silent period. Two potentials can be recorded during this silent period: a response analogous to the F wave, which separates the first two periods of EMG suppression, and the long latency response, which occurs between the second and third periods of EMG suppression. (From Leis AA: Conduction abnormalities detected by silent period testing. Electroenceph Clin Neurophysiol 93:444–449, 1994.)

can influence the spinal segmental mechanisms that produce the silent period. In dystonias, silent periods have been reported to be variably absent,[3, 47] supporting the notion that central inhibitory mechanisms are decreased. It has been suggested that silent period measurements may differentiate between groups of patients with dystonia and groups with psychogenic dystonia.[18] In spasticity, a condition marked by deranged processing of muscle spindle and other afferent input to the spinal cord, both normal and prolonged MNSPs have been reported.[15, 35] However, as noted earlier, the MNSP cannot be used as a measure of spindle function, because the MNSP can be evoked without apparent contribution from the descending volley.[37]

Currently, I use the MNSP technique primarily to determine if collision is occurring between antidromic and orthodromic motor impulses. The presence of the first component of the MNSP, due to collision, provides quick physiological documentation that orthodromic impulses are carried in the same fibers that are being electrically stimulated. Conversely, the first component of the MNSP is absent when collision does not occur, as in stimulation to a nonhomonymous nerve (compare the MNSP in the APB following median and ulnar nerve stimulation in Figures 23–1 and 23–2, respectively). Analysis of the MNSP can therefore identify orthodromic impulses traveling in anomalous nerves (Case Study 1) or anastomotic branches. For example, in the Martin-Gruber anastomosis, the communicating branch may infrequently supply thenar hand muscles. The presence of the first portion of the MNSP recorded in thenar muscles, following stimulation to the ulnar nerve at the wrist, indicates that orthodromic impulses are carried in communicating fibers.

CASE STUDY 1

History. A 16-year-old boy awoke from left axillary surgery (cyst removal) with weakness and numbness in the left hand and weakness of forearm flexors. The neurosurgeon admitted to "cutting the median nerve."

Examination. There was loss of sensory and motor functions mediated by the left median nerve except for preserved thumb flexion and abduction.

EMG and Nerve Conduction Studies. The left median motor and sensory responses were absent recorded from thenar muscles and second and third digits, respectively. Other nerve conduction studies were normal. EMG revealed no voluntarily recruited motor unit potentials (MUPs) in forearm muscles innervated by the left median nerve or in the left opponens pollicis muscle. A normal pattern of MUPs was seen in the left APB and flexor pollicis brevis.

Silent Period Studies. The pattern of voluntarily recruited MUPs from the left APB was not affected by left median nerve stimulation at the wrist (i.e., no evidence of collision). Muscle activity was suppressed by stimulation to the left ulnar nerve at the wrist (Fig. 23–4), providing quick physiological documentation that orthodromic impulses to APB were carried in the ulnar fibers that were being electrically stimulated.

CUTANEOUS SILENT PERIOD

EMG activity from a voluntarily contracting muscle may undergo a transient suppression following electrical stimulation of a cutaneous nerve. This period of inactivity is designated the cutaneous silent period (CSP). Most investigators agree that the afferent impulses that generate the CSP in upper and lower extremity muscles are carried primarily by

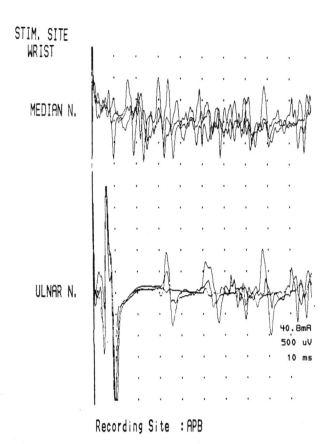

FIGURE 23–4. Stimulation to the left median nerve at the wrist during voluntary contraction of the abductor pollicis brevis muscle *(top tracings)*. The stimuli did not alter the EMG. Stimulation to ulnar nerve produced antidromic motor impulses that collided with orthodromic impulses to produce the first component of the mixed nerve silent period *(bottom tracings)*.

the smaller, slower conducting delta fibers.[8, 27, 38, 59, 67] Insight on the fiber-type contribution to the CSP was provided by the systematic analysis of silent periods in patients with pure sensory neuropathy who had absent sensory nerve action potentials (SNAPs) or somatosensory evoked potential (SEPs).[8, 38] In such subjects, CSPs, often indistinguishable from those of healthy subjects, have been recorded in voluntarily contracting muscles following stimulation to cutaneous nerves. Since SNAPs and SEPs evoked by conventional nerve conduction techniques primarily reflect conduction through large myelinated fibers, such findings argue against large fiber contribution to the CSP and implicate the smaller, slower conducting fibers.

The fibers that originate in the fingertips to produce the CSP are presumably activated during more proximal stimulation of the mixed nerve to generate the third portion of the MNSP. Indeed, impulses traveling at a rate of 10 to 15 m/sec, in the range of delta fibers, have been independently reported to produce the latter portion of the MNSP.[45] Accordingly, there is evidence that conduction abnormalities that selectively delay or abolish the CSP also selectively delay or abolish the latter portion of the MNSP (Case Study 2).

CASE STUDY 2

History. A 31-year-old woman complained of painful dysesthesias, ataxia, and mild weakness involving all limbs. Symptoms began at age 8 with slight dragging of the right foot. Symptoms related to dysfunction of sensory or motor systems slowly progressed so that by age 18 the patient was wheelchair bound. Magnetic resonance imaging (MRI) scan of brain at age 28 was normal. Sural nerve biopsy showed marked loss of large myelinated fibers. The small myelinated fibers were also affected, but to a lesser extent. Biceps brachii muscle biopsy was normal.

Examination. There was profound loss of vibration and joint position sense in all joints. Areflexia was present in all limbs and muscle tone was hypotonic. Motor examination revealed 4/5 weakness diffusely.

Nerve Conduction Studies. There were absent SNAPs in all limbs with normal motor nerve studies. SEPs following stimulation to median and tibial nerves failed to evoke peripheral, spinal, or cortical potentials.

Silent Period Studies. Stimulation to all digits produced CSPs in the voluntarily contracting APB muscle, despite absent SNAPs and SEPs. Stimulation to the median nerve elicited the MNSP. The CSP and latter portion of the MNSP were markedly but comparably delayed (Fig. 23–5).

An important variant of the CSP is the masseter silent period, also called the masseter inhibitory reflex or exteroceptive suppression, which is evoked by electrical stimulation of the mental nerve of the chin during maximal clenching of the jaw. This inhibitory reflex consists of an early and late phase of electrical silence (S1 and S2), that interrupts the voluntary EMG activity in ipsilateral and contralateral masseter and temporalis muscles (Fig. 23–6). Ongerboer De Visser and colleagues[49] in 1989 studied the effects of various brain stem lesions on the masseter silent period and found that some lesions selectively affected one component of the reflex, proving that S1 and S2 were relayed by independent circuits. Their results suggested that the afferent fibers for S1, which reach the pons via the trigeminal sensory root, enter the ipsilateral trigeminal spinal tract and terminate at the level of the mid-pons; impulses are then relayed by inhibitory interneurons to the ipsilateral and contralateral trigeminal motor nuclei. The afferent fibers for S2 follow a similar path, but descend to the pontomedullary junction; at this level impulses are conducted along bilateral interneuronal pathways, which probably ascend through the lateral

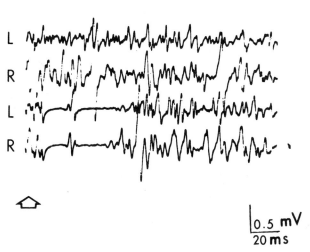

FIGURE 23–6. Voluntary contraction of the masseter muscles *(top two traces)*. EMG was recorded simultaneously from left (L) and right (R) sides with surface electrodes. A single shock applied to the left mental nerve of the chin at the beginning of the sweep *(open arrow)* elicited the masseter silent period *(bottom two traces)*. Note the early and late phases of electrical silence that interrupt the voluntary EMG activity in ipsilateral and contralateral masseter muscles.

reticular formation, before connecting with trigeminal motor nuclei. The S1 response (10 to 15 msec latency) is probably mediated by A beta fibers through an oligo- or disynaptic circuit.[49] The afferents for S2 are probably mediated by similar, but independent fibers to those of S1. S2 impulses are relayed through a chain of interneurons before reaching the inhibitory interneuron.[49]

Procedure

The CSP is evoked by electrical stimulation of a cutaneous nerve while the subject maintains voluntary muscle contraction. Several cutaneous nerve and muscle combinations can be used to study the CSP. For purposes of demonstration,

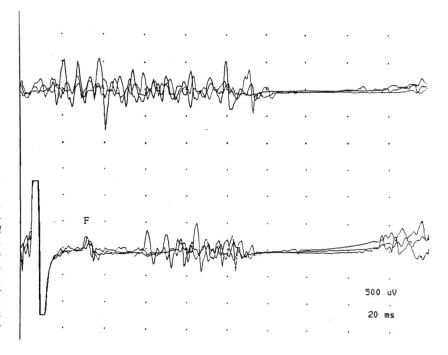

FIGURE 23–5. Cutaneous silent period in the abductor pollicis brevis muscle following stimulation to the second digit *(top superimposed tracings)*. Note that the cutaneous silent period is markedly delayed in onset. Stimulation to the median nerve at the wrist produced a comparable marked delay in the latter portion of the mixed nerve silent period *(bottom tracings)*. F, enhanced F wave. (From Leis AA: Conduction abnormalities detected by silent period testing. Electroenceph Clin Neurophysiol 93:444–449, 1994.)

the CSP is easily recorded with the active surface electrode placed over the belly of the APB muscle and the reference electrode over the tendon. Single shocks are then delivered to the digital nerves of the fingers while the subject maintains thumb abduction. The CSP can be elicited irrespective of the digit that is stimulated; stimulation to the ulnar innervated fifth digit will elicit a CSP in the APB muscle similar to the one evoked by second digit stimulation (Fig. 23–7). The CSP latency onset and end point in thenar muscles typically occurs at about 70 msec and 120 msec, respectively[38, 67] following stimulation to the digits. However, the CSP onset latency is reduced and total duration increases with increased stimulus strength.[59] Stimuli are square waves 0.1 or 0.2 msec in duration with filters set from 20 Hz to 5 KHz. Stimulus intensity is maintained near the pain threshold, although lower stimulus intensities may elicit a shorter CSP. The sweep is 20 msec/cm. A relative or absolute decrease in the voluntary EMG activity identifies the CSP.

To elicit the masseter silent period, electrical stimulation is delivered to the mental nerve of the chin during maximal clenching of the jaw. EMG activity is recorded simultaneously from both masseter or temporalis muscles by needle or surface electrodes.[49] An early (S1) and late phase (S2) of electrical silence interrupt the voluntary EMG activity in ipsilateral and contralateral masseter and temporalis muscles. One-sided stimulation evokes symmetric bilateral responses (see Fig. 23–6).

Physiological Mechanisms

Shefner and Logigian[59] reported that activation of sensory axons with conduction velocities in the range of A

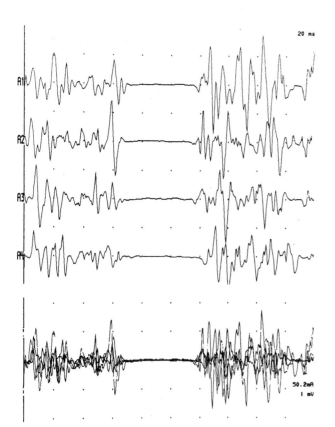

FIGURE 23–7. Cutaneous silent periods recorded from the abductor pollicis brevis muscle after stimulation of the distal index finger *(top tracing, A1)*, middle finger *(A2 tracing)*, ring finger *(A3 tracing)*, and little finger *(A4 tracing)*. Bottom tracings are superimposed trials. Note the reproducible and complete silent periods irrespective of the digit stimulated.

delta fibers was necessary to evoke the CSP. Other investigators[27, 67] also analyzed the CSP and, likewise, concluded that it was generated by sensory afferents with conduction velocities of 12 to 13 m/sec. The long latency of the CSP therefore depends on a long afferent conduction time rather than a long central delay, and suggests an inhibitory spinal reflex.[8, 27, 38, 67] However, the precise physiological mechanism by which inhibition occurs at a spinal level remains to be fully elucidated. Eccles[16] used intracellular recordings to show that inhibition from cutaneous afferent volleys was associated with hyperpolarization of the motoneuron membrane. This effect was regarded as typical of an inhibitory postsynaptic potential (IPSP). However, those observations were recorded in the lower limbs of animals and even Eccles acknowledged that the stimuli used were "too weak to excite group II or III afferent fibers." Uncini and colleagues[67] postulated that motor neuron inhibition from nociceptive stimulation could be mediated by Renshaw cells directly activated by high threshold cutaneous afferents. They cited experimental evidence by Piercey and Goldfarb[50] that Renshaw cells may be excited, without prior discharge of motoneurons, by stimulation of high threshold afferent fibers to produce motoneuron inhibition. Moreover, they noted that Renshaw cells may discharge repetitively for up to 50 msec after a single excitation,[52] may have a several millimeter rostral–caudal extension of their axons,[53] and may show a degree of convergence of excitation from different peripheral afferent segments.[52] Such features resemble some characteristics exhibited by the silent period, and implicate Renshaw cells in the genesis of the CSP. However, Ryall and Piercey[52] also showed that the Renshaw cell excitation evoked by stimulation of afferent fibers was quickly followed by profound depression of excitability. This inhibition of Renshaw cell activity lasted up to 1 second, and was attributed to an inhibitory phenomenon rather than just postactivation depression of the cell. This feature of Renshaw cells argues against a role in the CSP, as in humans the CSP does not habituate with trains of stimuli from 0.5 to 5 Hz or with paired shock interstimulus intervals of only 100 msec.[67] Leis and colleagues[39, 40] assessed spinal motoneuron excitability during the CSP and found that motoneurons remain excitable to an antidromic motor volley at the same time that the corticospinal volley is profoundly inhibited (Fig. 23–8). They suggested that the CSP may be a consequence of inhibition of corticospinal fibers or inhibition of spinal interneurons that relay the corticomotoneuronal command to the motoneuron pool. Spinal interneurons are known to receive convergent input from multisensorial afferents and also from descending pathways,[29, 42, 62] and cutaneous input to interneurons receiving corticospinal convergence may be quite common.[7, 28] Digital stimulation may therefore produce the silent period by inhibiting common interneurons that relay the corticomotoneuronal command to the spinal motor neurons. Inghilleri and colleagues,[27] however, provided recent evidence that cutaneous activation may postsynaptically change the level of spinal motor neuron excitability. Additional studies are, therefore, warranted to determine the precise physiological mechanism by which inhibition occurs.

Clinical Applications

The masseter silent period has been reported to be characteristically absent in patients with tetanus.[64, 65] Depression of the masseter silent period has also been reported in recovering alcoholics, supporting the hypothesis that chronic alcoholism may interfere with inhibitory mechanisms in the central nervous system.[69] Schoenen and associates[57] (1987) found that the mean duration of the second phase (S2) of the temporalis silent period was significantly decreased or

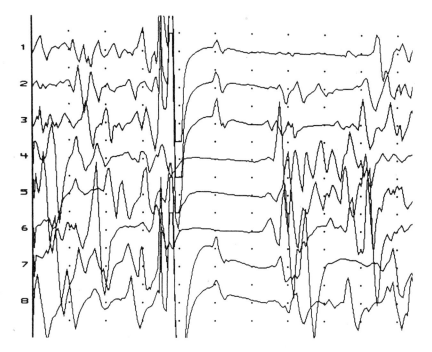

FIGURE 23–8. F wave responses (F) recorded from the voluntarily contracting abductor pollicis brevis muscle following supramaximal stimulation to the median nerve at the wrist *(tracings 1 to 3)*. Stimulation to the fifth digit elicits the cutaneous silent period *(tracings 4 to 6)*. F wave responses are easily elicited in the midst of the cutaneous silent period *(tracings 7 and 8)*. The wrist stimulus was delayed so that the F wave would appear during the cutaneous silent period. Sweep: 20 msec/division, gain 500 µV/division. (From Leis AA, Stetkarova I, Beric A, Stokic DS: Spiral motor neuron excitability during the cutaneous silent period. Muscle Nerve 18:1464–1470, 1995.)

abolished in chronic tension-type headache sufferers compared to migraine patients or healthy controls. A number of investigators have confirmed these findings, and the temporalis and masseter silent periods have proven an important method for investigating pathophysiological mechanisms of headache (for a review, see references 22, 56, and 70). Analysis of the masseter silent period can also be of particular value in documenting various brain stem lesions.[49] Several other studies have described changes in the temporalis or masseter silent periods in patients with a variety of disorders, including temporomandibular joint dysfunction,[4, 5] trigeminal sensory neuropathy,[2] and movement disorders.[12] However, one must caution against premature extrapolation from alterations in the silent period in various pathological situations to explanations of even more complex motor phenomena. A major problem when evaluating temporalis and masseter silent period studies has been the lack of an objective and precise measure for the onset and end point of the silent period. To overcome this problem, a fully automated system for the evaluation of temporalis and masseter silent periods has recently been described.[10]

In upper and lower limbs, the CSP technique can be used to assess conduction in the smaller, slower conducting A delta fibers[8, 27, 38, 59, 67] following peripheral nerve injury (Fig. 23–9). Conduction in these fibers is not generally assessed by routine nerve studies. The CSP can also be used as a "poor man's SEP" to monitor afferent sensory impulses passing through the various roots to reach the spinal cord, and to assess the integrity of the intraspinal pathways that mediate this inhibitory response (Fig. 23–10). Conditions that interrupt this pathway, including radiculopathy, root avulsion, or myelopathy, may be associated with absence or delay of the CSP (Case Study 3). However, such lesions may not alter the CSP, because profound axonal loss of the smaller diameter A delta fibers may be required before the CSP is delayed or abolished.[41] In some cases of suspected Erb's palsy (avulsion of C5 and C6 roots of the brachial plexus), the CSP may provide the only evidence of preserved conduction through the upper trunk (Fig. 23–11).

CASE STUDY 3

History. A 51-year-old woman was referred for evaluation of right carpal tunnel syndrome. She had a carpal tunnel release 5 years ago but complained of persistent numbness and pain involving the hand.

Examination. There was normal function in right upper limb muscles, with no atrophy. Sensory examination did not reveal definite abnormalities. Deep tendon reflexes were diffusely hypoactive at all sites.

Nerve Conduction and Silent Period Studies. Routine conduction studies in the right median and ulnar nerves were normal, including palmar studies. Stimulation to the little, middle, and index fingertips, and distal thumb during voluntary contraction of the APB produced the responses shown in Figure 23–12; the CSP was normal following stimulation to fingers in the C8 and C7 dermatomes, but abolished following C6 (thumb) stimulation. These findings suggested the possibility of a C6 radiculopathy.

EMG Needle Examination. There were rare fibrillation potentials, complex repetitive discharges or excessive polyphasic MUPs in the biceps, extensor carpi radialis, deltoid, and cervical paraspinal muscles, supporting the diagnosis of a C6 radiculopathy.

LONG LATENCY RESPONSES

The long latency reflex (LLR) is muscle activity induced by an electrical stimulus to a nerve, which occurs after the F wave or H reflex and before the end of the silent period. These responses were first noted by Dawson in 1947 in the study of myoclonus.[13] The responses can be obtained under certain circumstances in normal individuals, and are enhanced in certain types of myoclonus and states of cortical hyperexcitability.

Long Latency Responses in Normal Subjects

In order to elicit long loop reflexes in normal individuals, some muscle contraction is usually necessary.[11, 13] Following stimulation of a mixed nerve in a normal individual at rest, a direct (M) response is recorded. Depending on stimulation parameters, an F wave due to antidromic activation of motor axons or an H reflex may be elicited. In a normal subject at rest, there is usually no muscle electrical activity after the F wave.[32] Following electrical stimulation to a mixed nerve during voluntary activation of the muscle, the background EMG activity is suppressed because of the silent period (MNSP). In the median nerve, the MNSP lasts for approximately 100 msec after the delivery of the stimulus. Voluntary activity resumes after the silent period. Two potentials can be recorded during this silent period (see Fig. 23–3); a response analogous to the F wave and the LLR. The "enhanced" F wave has been termed the voluntary (V1) response[68] or the spinally mediated (S) response.[11] In the hand, the LLR is fairly constant in each subject, but varies between 38 and 51 msec,[11] 40 and 58 msec,[32] or 48 and 60 msec[68] for different subjects. The LLR is not present at rest in most normal individuals, but is facilitated by activation of the muscle under investigation. The LLR is more pronounced during isotonic contractions (e.g., moving the distal phalanx of the thumb smoothly against the little finger while stimulating the median nerve at the wrist and recording from the APB) than in isometric contraction (e.g., pressing the thumb against a constant resistance).[11] The response is

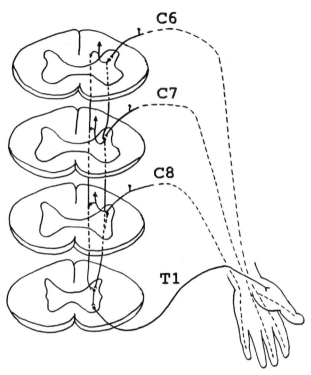

FIGURE 23–10. Diagram of the proposed pathways mediating the cutaneous silent period following digital nerve stimulation. Afferent impulses reach the spinal gray matter in the dorsal horn and give rise to propriospinal projections that descend in grey matter or dorsal columns to suppress activity in motor nuclei of intrinsic hand muscles.

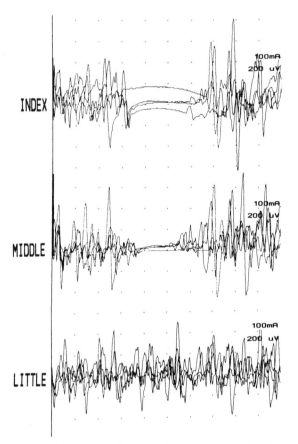

FIGURE 23–9. Cutaneous silent period (CSP) responses in a 19-year-old man with penetrating injury to the left arm 3 days prior to evaluation. Examination revealed complete loss of function in muscles supplied by the ulnar nerve. EMG and nerve conduction studies showed a complete conduction block at the mid-arm level and absent F waves in the ulnar nerve. There were no voluntarily recruited motor unit potentials in ulnar innervated muscles. The CSP in the abductor pollicis brevis (APB) could not be elicited following fifth digit stimulation, confirming a conduction block in the smaller, slower conducting A delta fibers of the ulnar nerve.

not commonly obtained in the leg, but when present, the latency is between 70 and 80 msec.[32]

During a weak tonic contraction of intrinsic hand muscles, stimulation to cutaneous afferents can also elicit a sequence of low-amplitude excitatory and inhibitory responses. Numerous terms have been used to describe these LLRs, including E1, I1, E2[19, 21, 30] (where E = excitatory, I = inhibitory), V1 and V2[51, 63, 68] (first and second volitional responses following the M wave), S and C[11] (S = spinal, C = cortical), R1 and R2[17] (R = reflex), and I1, E1, I2[8] (where E1 = E2 of other investigators). The variety of names for these responses reflects both the different mechanisms by which they can be elicited and the controversy regarding their origin and pathways. Moreover, it should be noted that long latency responses may not be demonstrated in all experiments.

This review primarily emphasizes the higher amplitude LLRs occurring during movement in normal individuals as well as at rest in myoclonic syndromes. LLRs that occur at rest in myoclonic syndromes are also frequently termed cortical (C) reflexes. Some authors restrict the term LLR to potentials elicited in normal individuals, whereas others use the term C reflex to signify the response elicited by limb or body stimulation in patients with cortical reflex myoclonus.[23, 60, 66] To avoid confusion, only the term LLR is used here. Whereas LLRs can be obtained in normal individuals by activation procedures, LLRs have received the greatest attention in the evaluation of myoclonus.

Long Latency Responses in Myoclonus

Myoclonic jerks can be elicited by somatosensory stimulation to a peripheral nerve, and are associated with an electroencephalogram (EEG) transient that precedes, but is time-locked to the myoclonic movement.[24] If a mixed nerve is stimulated in individuals with certain types of myoclonus, in conjunction with simultaneous recording of muscle EMG activity and cortical EEG activity, a high-amplitude cortical potential known as a giant SEP can be recorded. The giant SEP is followed in fixed temporal relationship by a muscle myoclonic jerk,[24, 66] which occurs at the same time as the expected LLR. The latency of these stimulus-evoked LLRs is about twice that of the latency of the corresponding cerebral evoked response.

LLRs (or C reflexes) are not elicited in every type of myoclonus. Myoclonus characterized by giant SEPs, LLRs at rest, and the presence of cortical spikes preceding spontaneous myoclonic jerks at a fixed latency[24, 31, 32] has been termed cortical reflex myoclonus[23] or pyramidal myoclonus.[24] LLRs at rest are not seen in myoclonic jerking which is either not accompanied by a cortical discharge, or if associated with a cortical discharge, does not have a fixed latency between the timing of the cortical potentials and the EMG bursts.[24] LLRs are also absent in spinal or segmental myoclonus, produced in the spinal cord or the brain stem. This type of myoclonus tends to be rhythmic and is not accompanied by abnormal potentials on the EEG.[24]

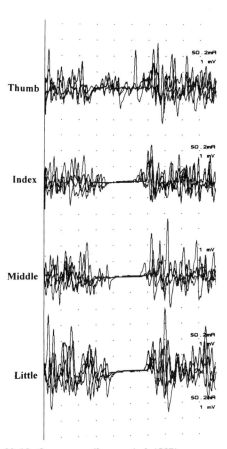

FIGURE 23–12. Cutaneous silent period (CSP) responses in a 51-year-old woman referred for right carpal tunnel syndrome. Stimulation to the distal thumb during voluntary contraction of the abductor pollicis brevis muscle (APB) failed to elicit the CSP. This finding suggested a possible C6 radiculopathy, which was confirmed by EMG needle examination.

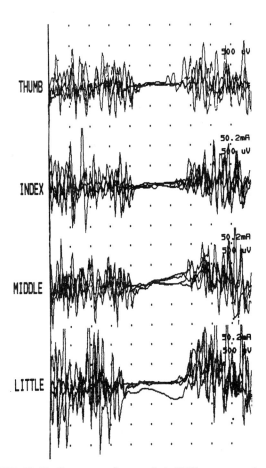

FIGURE 23–11. Cutaneous silent period (CSP) responses in a 34-year-old man involved in a car accident 4 days prior to evaluation. Examination revealed complete loss of function in all muscles innervated by the C5 and C6 roots. EMG showed no voluntarily recruited motor unit potentials in C5–6 innervated muscles. A diagnosis of avulsion of the C5 and C6 roots (upper trunk of brachial plexus) was made by the surgeons. However, stimulation to the thumb and index finger elicited a CSP in the abductor pollicis brevis (APB), providing quick physiological evidence of preserved conduction through the upper trunk.

Pathways

The pathways involved in the LLR are controversial. The LLR may be a reflex, as it occurs with a latency much shorter than that of a voluntary response and falls within the mixed nerve silent period. The LLR may represent a long loop reflex mediated via the sensory motor cortex[11] or a spinal reflex that requires cortical facilitation associated with voluntary activity.[51] As the latency to the muscle discharge is about twice that of the cortical response, there is time for the impulses to travel first cephalad to the cortex and then caudally before producing the jerk.[24] The proposed ascending pathway includes the peripheral nerve, posterior columns, medial lemniscus, and contralateral ventral posterior nucleus of the thalamus and sensorimotor cortex. The descending pathway is presumed to be the corticospinal tract, anterior horn cell, and peripheral nerve to the muscle.[61]

Halliday[24] suggested that one could determine spinal transit time by determining the difference in arrival of the LLRs in the arm compared to the leg, and subtracting the peripheral distance. Conversely, Upton and colleagues[68] found that the LLR (which they termed V2) occurred about 25 msec after the modified F wave (V1), whether tested in the arm or leg. They argued that the intervals between V1 and V2 should have been longer in the leg if the pathway included conduction through the spinal cord to the brain. They doubted that the second response was via a long loop reflex involving the brain, and proposed that the LLR was mediated via a spinal mechanism that required a certain amount of descending facilitation to become demonstrable. Irrespective of the pathway, it appears that excitability of the corticospinal tract or cortex, either during voluntary activity or associated with pyramidal tract myoclonus, is required to consistently evoke LLR responses.

Techniques

The technique for stimulation of the median nerve and recording EMG activity from thenar eminence is described here. Other upper extremity and lower extremity muscles can be evaluated using similar methods.

Normal Subjects

To record LLRs in normal individuals, surface electrodes are placed on the thenar muscles of the stimulated side. A single supramaximal stimulus is delivered either to the median nerve at the wrist or to a sensory digital nerve.[11, 23] The LLR is elicited either during an isotonic contraction (moving the distal phalanx of the thumb smoothly against the middle phalanx of the fifth finger, with the movement lasting 1 to 2 seconds)[11, 14] or during an isometric contraction (pushing the distal phalanx of the thumb against the middle phalanx of the fifth finger). The stimulus can be delivered either manually, when the movement is half completed, or by a switch attached to the moving thumb that is activated by movement.[14] Recording parameters should include filters appropriate for motor studies. The gain should be 500 μV/cm, and the sweep duration should be 100 to 150 msec.[35]

Myoclonus

Simultaneous recording of an EEG and EMG is the most basic, yet most important, test in the evaluation of myoclonus. The EEG can be recorded by placing electrodes over the scalp in accordance with the international 10–20 system; the central scalp region should be covered by at least three electrodes, including C3, Cz, and C4. More recently, backward averaging of EEGs time-locked to myoclonic EMG discharges (jerk-locked back averaging) has been found effective for detecting EEG correlates of myoclonus (for a review, see reference 60). Since Dawson,[13] SEPs have been widely used for the study of stimulus-sensitive myoclonus. The conventional method for recording SEPs can be used. The reflex myoclonus itself is recorded in the surface EMG as the LLR following the stimulus. EMG recording is as described for normal individuals, although LLRs occur at rest in subjects with certain myoclonic syndromes. Electrical shocks to the median nerve at the wrist are most commonly used with the EMG response best recorded from thenar muscles of the stimulated hand. However, in the evaluation of myoclonus, it is important to find, by clinical observation, the most appropriate muscles from which the myoclonic jerk can be recorded. The EEG and EMG are averaged using the stimulus as a trigger.

In cortical reflex myoclonus or pyramidal myoclonus,[23, 24] a markedly enhanced LLR is recorded from thenar muscles at a latency of 31 to 58 msec after stimulation to the median nerve at the wrist. The SEP to the same stimulus shows a characteristic waveform, its main feature being extremely large P25 and N33 peaks.[60] The SEP is considered to be "giant" when the component corresponding to the P25 of the normal SEP is larger than 8.6 μV or the corresponding N33 is larger than 8.4 μV.[61] However, the peak-to-peak amplitude of P25 and N33 may be as large as 34 and 63 μV, respectively.[60, 61] The initial negative peak N20 is not enhanced at all.[60] The LLR is recognized in thenar muscles of the stimulated side at a time interval of 11 to 24 msec after the P25 peak of the giant SEP. This EMG burst is time-locked to the cortical event. In some patients, an LLR can also be recorded from the thenar muscles of the opposite (nonstimulated) hand, with an onset latency of approximately 10 msec longer than that of the LLR of the stimulated side.[60, 61]

Other Techniques

Any region of the body known to be sensitive to cutaneous stimulation can be stimulated, and cortical and EMG events can be recorded.[48] Stimulation of tendon reflexes by tapping the tendon can be used as the stimulus.[23, 48] Computer data collection can be triggered not only at the time of delivery of a stimulus, but also by the myoclonic jerks themselves, by giant cortical potentials, or by external stimuli such as light flashes. Event-related potentials before and after the trigger can be recorded.[23] Analysis determines the temporal relationship of the recorded potentials.

Clinical Utility

The LLR is not generally elicited in normal individuals at rest; however, potentials breaking through the silent period

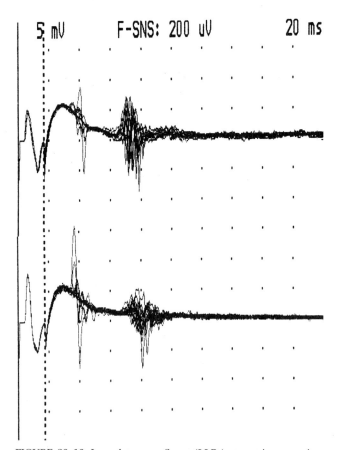

FIGURE 23–13. Long latency reflexes (LLRs) at rest in an anxious individual. LLRs at rest repeatedly occurred at a latency of about 60 msec in this anxious but otherwise neurologically normal subject. Electrical stimulation was delivered to the median nerve at the left wrist (*top tracings superimposed*) and right wrist (*bottom tracings*), recording over the respective abductor pollicis brevis muscle (APB). Three muscle potentials were recorded at rest: the direct muscle response (M wave, left of dotted line, gain 5 mV/division), followed by F waves and LLR responses (gain 200 μV/division).

at the time of the LLR may be seen in individuals who are very anxious (Fig. 23–13). Whether these reflect cortical hyperexcitability or muscles that are not fully relaxed is speculative. Typically, LLRs at rest and giant SEPs are observed almost exclusively in patients with pyramidal myoclonus,[24, 61] which is also called cortical reflex myoclonus.[23] These syndromes include progressive myoclonic epilepsy,[33] cherry-red spot myoclonus syndrome (lipidosis),[61] Lafora body disease,[61] neuronal ceroid lipofuscinosis,[61] olivopontocerebellar atrophy with myoclonus,[48] and postanoxic myoclonus (Lance-Adams syndrome).[34] Spontaneous myoclonic jerks in patients with these myoclonic conditions may represent enhanced LLRs in response to unnoticed stimuli, although the possibility of spontaneous discharge in the sensory motor cortex causing spontaneous myoclonus cannot be excluded. Giant SEPs and enhanced LLRs are not observed in patients with epilepsy with myoclonus, essential myoclonus, palatal myoclonus, or the opsoclonus polymyoclonia syndrome, suggesting that these conditions are not associated with cortical hypersensitivity.[60]

The LLR during movement may be absent or delayed on the paretic side in patients with stroke.[11] It should be mentioned, however, that the amount of muscle contraction is also reduced on the paretic side, and LLRs are not expected in subjects at rest. Thus, the absence or prolongation of LLRs in stroke patients does not necessarily imply a cortical origin for the LLR, as suppression of the LLR could reflect diminished cortical facilitation. Changes in LLRs evoked by a variety of methodologies have also been reported in other disorders of the central nervous system, including multiple sclerosis, Parkinson's disease, Huntington's chorea, dystonia, cerebellar disorders, and essential tremor.[20] A detailed analysis of these techniques is beyond the scope of this review, although interested readers are referred to the review by Fuhr.[20]

REFERENCES

1. Angel RW, Hofmann WW, Eppler W: Silent period in patients with Parkinsonian rigidity. Neurology 16:529–532, 1966.
2. Auger RG, McMavis PG: Trigeminal sensory neuropathy associated with decreased oral sensation and impairment of the masseter inhibitory reflex. Neurology 40:759–763, 1990.
3. Berardelli A, Day BL, Marsden CD, Rothwell JC: Pathophysiology of blepharospasm and oromandibular dystonia. Brain 108:593–608, 1985.
4. Bessette RW, Bishop B, Mohl ND: Duration of masseter silent periods in patients with TMJ syndrome. J Appl Physiol 30:864–869, 1971.
5. Bessette RW, Shatkin SS: Predicting by electromyography the results of non-surgical treatment of temporomandibular joint syndrome. Plast Reconstr Surg 64:232–238, 1979.
6. Borg J: Refractory period of single motor nerve fibers in man. J Neurol Neurosurg Psychiatry 47:344–348, 1984.
7. Brink E, Jankowska E, Skoog B: Convergence onto interneurones subserving primary afferent depolarization of group I afferents. J Neurophysiol 51:432–449, 1984.
8. Caccia MR, McComas AJ, Upton ARM, Blogg T: Cutaneous reflexes in small muscles of the hand. J Neurol Neurosurg Psychiatry 36:960–977, 1973.
9. Collins WF, Nulsen FE, Randt CT: Relation of peripheral nerve fiber size and sensation in man. Arch Neurol 3:381–385, 1960.
10. Connemann BJ, Urban PP, Luttkopf V, Hopf HC: A fully automated system for the evaluation of masseter silent periods. Electroencephalogr Clin Neurophysiol 105:53–57, 1997.
11. Conrad B, Aschoff JC: Effects of voluntary isometric and isotonic activity on late transcortical reflex components in normal subjects and hemiparetic patients. Electroencephalogr Clin Neurophysiol 42:107–116, 1977.
12. Cruccu G, Pauletti R, Agostino R, et al: Masseter inhibitory reflex in movement disorders. Huntington's chorea, Parkinson's disease, dystonia, and unilateral masticatory spasm. Electroencephalogr Clin Neurophysiol 81:24–30, 1991.
13. Dawson GD: Investigations on a patient subject to myoclonic seizures after sensory stimulation. J Neurol Neurosurg Psychiatry 10:141–162, 1947.
14. Diener H, Dichgans J, Bacher M, Guschlbauer B: Characteristic alterations of long-loop "reflexes" in patients with Friedreich's disease and late atrophy of the cerebellar anterior lobe. J Neurol Neurosurg Psychiatry 47:679–685, 1984.
15. Dietrichson P: The silent period in spastic, rigid, and normal subjects during isotonic and isometric muscle contractions. Acta Neurol Scand 47:183–193, 1971.
16. Eccles JC: The Neurophysiological Basis of Mind: The Principles of Neurophysiology. Oxford, Clarendon Press, 1953, pp 150–192.
17. Eisen A, Bohlega S, Bloch M, Hayden M: Silent periods, long-latency reflexes and cortical MEPs in Huntington's disease and at-risk relatives. Electroencephalogr Clin Neurophysiol 74:444–449, 1989.
18. Ford B, Fahn S, Pullman S: Peripherally induced EMG silent periods: Normal physiology and disorders of motor control. In Fahn S, Hallett M, Luders HO (eds): Negative Motor Phenomena. Advances in Neurology, vol 67. Philadelphia, Lippincott-Raven, 1995, pp 321–328.
19. Fuhr P: Cutaneous reflexes in Parkinson's disease. Muscle Nerve 15:733–739, 1992.
20. Fuhr P: Cutaneous and other long-latency reflexes. In: 1994 AAEM Course E: Electrodiagnostic Techniques and Their Clinical Utility. Rochester, MN, American Association of Electrodiagnostic Medicine, 1994, pp 23–33.
21. Fuhr P, Friedli WG: Electrocutaneous reflexes in upper limbs—reliability and normal values in adults. Eur Neurol 27:231–238, 1987.
22. Gobel H, Dworschak M, Wallasch TM: Exteroceptive suppression of temporalis muscle activity: Perspectives in headache and pain research. Cephalalgia 13:15–19, 1993.
23. Hallett M, Chadwick D, Marsden CD: Cortical reflex myoclonus. Neurology 29:1107–1125, 1979.
24. Halliday AM: The electrophysiological study of myoclonus in man. Brain 90:241–283, 1967.
25. Higgins DC, Lieberman JS: The muscle silent period and spindle function in man. Electroencephalogr Clin Neurophysiol 25:238–243, 1968.
26. Higgins DC, Haidri NH, Wilbourn AJ: Muscle silent period in Parkinson's disease. J Neurol Neurosurg Psychiatry 34:508–511, 1971.
27. Inghilleri M, Cruccu G, Argenta M, et al: Silent period in upper limb muscles after noxious cutaneous stimulation in man. Electroencephalogr Clin Neurophysiol 105:109–115, 1997.
28. Jankowska E, Johannisson T, Lipski J: Common interneurones in reflex pathways from group Ia and Ib afferents of ankle extensors in the cat. J Physiol 310:381–402, 1981.
29. Jankowska E, Lundberg A: Interneurones in the spinal cord. TINS 4:230–233, 1981.
30. Jenner JR, Stephens JA: Cutaneous reflex responses and their central nervous pathways studied in man. J Physiol 333:405–419, 1982.
31. Kakigi R, Shibasaki H: Generator mechanisms of giant somatosensory evoked potentials in cortical reflex myoclonus. Brain 110:1359–1373, 1987.
32. Kelly JJ, Sharbrough FW, Daube JR: A clinical and electrophysiological evaluation of myoclonus. Neurology 31:581–589, 1981.
33. Kugelberg E, Widen L: Epilepsia partialis continua. Electroenceph Clin Neurophysiol 6:503–506, 1954.
34. Lance JW, Adams RA: The syndrome of intention or action myoclonus as a sequel to hypoxic encephalopathy. Brain 86:111–133, 1963.
35. Laxer K, Eisen A: Silent period measurement in the differentiation of central myelination and axonal degeneration. Neurology 25:740–744, 1975.
36. Leis AA: Conduction abnormalities detected by silent period testing. Electroencephalogr Clin Neurophysiol 93:444–449, 1994.
37. Leis AA, Ross MA, Emori T, et al: The silent period produced by electrical stimulation of mixed peripheral nerves. Muscle Nerve 14:1202–1208, 1991.
38. Leis AA, Kofler M, Ross MA: The silent period in pure sensory neuronopathy. Muscle Nerve 15:1345–1348, 1992.
39. Leis AA, Stetkarova I, Beric A, Stokic DS: Spinal motor neuron

excitability during the cutaneous silent period. Muscle Nerve 18:1464–1470, 1995.

40. Leis AA, Stetkarova I, Beric A, Stokic DS: The relative sensitivity of H-reflexes and F-waves to similar changes in motoneuronal excitability. Muscle Nerve 19:1342–1344, 1996.

41. Leis AA, Stokic DS: The cutaneous silent period in cervical radiculopathy. Muscle Nerve 20:1076, 1997.

42. Lundberg A: Multisensorial control of spinal reflex pathways. In Granit R, Pomeiano D (eds): Reflex Control of Posture and Movement. Progress in Brain Research. Amsterdam, Elsevier, 1979, pp 11–28.

43. Matthews BHC: The response of a muscle spindle during active contraction of a muscle. J Physiol 72:153–174, 1931.

44. McLellan DL: Levodopa in Parkinsonism: Reduction in the electromyographic silent period and its relationship with tremor. J Neurol Neurosurg Psychiatry 35:373–378, 1972.

45. McLellan DL: The electromyographic silent period produced by supramaximal electrical stimulation in normal man. J Neurol Neurosurg Psychiatry 36:334–341, 1973.

46. Merton PA: The silent period in a muscle of the human hand. J Physiol 114:183–198, 1951.

47. Nakashima K, Thompson PD, Rothwell JC, et al: An exteroceptive reflex in the sternocleidomastoid muscle produced by electrical stimulation of the supraorbital nerve in normal subjects and patients with spasmodic torticollis. Neurology 39:1354–1358, 1989.

48. Obeso JA, Rothwell JC, Marsden CD: The spectrum of cortical myoclonus. Brain 108:193–224, 1985.

49. Ongerboer De Visser BW, Cruccu G, et al: Effects of brainstem lesions on the masseter inhibitory reflex. Brain 113:781–792, 1989.

50. Piercey MF, Goldfarb J: Discharge patterns of Renshaw cells evoked by volleys in ipsilateral cutaneous and high-threshold muscle afferents and their relationship to reflexes recorded in ventral roots. J Neurophysiol 37:294–302, 1974.

51. Rossinni PM, Paradiso C, Zarola F, et al: Bit-mapped somatosensory evoked potentials and muscular reflex responses in man: Comparative analysis in different experimental protocols. Electroenceph Clin Neurophysiol 77:266–276, 1990.

52. Ryall R, Piercey MF: Excitation and inhibition of Renshaw cells by impulses in peripheral afferent nerve fibers. J Neurophysiol 34:242–251, 1971.

53. Ryall R, Piercey MF, Polosa C. Intersegmental and intrasegmental distribution of mutual inhibition of Renshaw cells. J Neurophysiol 34:700–707, 1971.

54. Ryall RW, Piercey MF, Polosa C, Goldfarb J: Excitation of Ren-

shaw cells in relation to orthodromic and antidromic excitation of motorneurons. J Neurophysiol 35:137–148, 1972.

55. Sandyk R, Bamford CR, Iacono RP, et al: The electromyographic silent period is reduced in individuals at risk for Huntington's disease. Int J Neurosci 40:109–110, 1988.

56. Schoenen J: Exteroceptive suppression of temporalis muscle activity: Methodological and physiological aspects. Cephalalgia 13:3–10, 1993.

57. Schoenen J, Jamart B, Gerard P, et al: Exteroceptive suppression of temporalis muscle activity in chronic headache. Neurology 37:1834–1836, 1987.

58. Shahani BT, Young RR: Studies of the normal human silent period. In Desmedt JE (ed): New Developments in Electromyography and Clinical Neurophysiology, vol 3. Basel, Karger, 1973, pp 589–602.

59. Shefner JM, Logigian EL: Relationship between stimulus strength and the cutaneous silent period. Muscle Nerve 16:278–282, 1993.

60. Shibasaki H: Electrophysiologic studies of myoclonus. AAEE Minimonograph #30. Muscle Nerve 11:899–907, 1988.

61. Shibasaki H, Yamashita Y, Neshige R et al: Pathogenesis of giant somatosensory evoked potentials in progressive myoclonic epilepsy. Brain 108:225–240, 1985.

62. Shomberg ED: Spinal sensorimotor systems and their supraspinal control. Neurosci Res 7:265–340, 1990.

63. Stanley EF: Reflexes evoked in human thenar muscles during voluntary activity and their conduction pathways. J Neurol Neurosurg Psychiatry 41:1016–1023, 1978.

64. Steinegger TH, Wiederkehr M, Ludin HP, Roth F: Elektromyogramm als diagnostische hilfe beim tetanus. J Suisse Med 126:379–385, 1996.

65. Struppler A, Struppler E, Adams D: Local tetanus in man: Its clinical and neurophysiological characteristics. Arch Neurol 8:162–178, 1963.

66. Sutton GG, Mayer RF: Focal reflex myoclonus. J Neurol Neurosurg Psychiatry 37:207–217, 1974.

67. Uncini A, Kujirai T, Gluck B, Pullman S: Silent period induced by cutaneous stimulation. Electroencephalogr Clin Neurophysiol 81:344–352, 1991.

68. Upton ARM, McComas AJ, Sica REP: Potentiation of 'late' responses evoked in muscles during effort. J Neurol Neurosurg Psychiatry 34:699–711, 1971.

69. Vernon CE, Hirsch JA, Bishop B, et al: Depression of an inhibitory reflex, the masseteric silent period, in recovering alcoholics. Alcohol Clin Exp Res 19:527–532, 1995.

70. Wallasch TM, Gobel H: Exteroceptive suppression of temporalis muscle activity: Findings in headache. Cephalalgia 13:11–14, 1993.

Chapter 24

Autonomic Nervous System Testing

Robert W. Shields, Jr., MD

FUNCTIONAL ANATOMY OF THE
    AUTONOMIC NERVOUS SYSTEM
The Baroreceptor Reflex
AUTONOMIC REFLEX TESTS

Cardiovascular Autonomic Tests
Sudomotor Function Tests
Pupillometry
Pharmacological Testing

The autonomic nervous system (ANS) is a rather complex anatomical and physiological system that is largely responsible for the regulation of visceral functions and the maintenance of homeostasis of the internal environment. The ANS accomplishes these functions via its interaction with the endocrine system, as well as through various autonomic reflexes. The ANS is involved in a wide array of disorders, including neurodegenerative disorders such as Parkinson's disease, multisystem atrophy, and pure autonomic failure.[6] In addition, there is a diverse group of peripheral autonomic disorders, including those related to hereditary, inflammatory, infectious, metabolic, nutritional, paraneoplastic, and toxic mechanisms.[76] The role of ANS testing is to identify the presence of an autonomic disorder, investigate the various components of the autonomic nervous system that may be involved, and provide an estimate of the severity of involvement.[58, 59, 73, 74] ANS testing is valuable not only in the assessment of patients with generalized autonomic failure, but also in the evaluation of patients with more restricted or milder dysautonomias.

Over the past decade there has been increasing interest among clinical neurophysiologists in the assessment of the ANS. This has been largely due to the development of highly sensitive and specific autonomic tests that are safe, noninvasive, and clinically relevant.[59, 74] This chapter reviews many of the autonomic nervous system tests that are currently in wide clinical application. Before describing the specific tests, a review of the functional anatomy of the autonomic nervous system is provided.

FUNCTIONAL ANATOMY OF THE AUTONOMIC NERVOUS SYSTEM

The ANS has been traditionally divided into the sympathetic nervous system (SNS) and the parasympathetic nervous system (PNS) (see Johnson and Spalding,[37] Loewy,[56] Shields,[106] and Williams and Warwick[123] for reviews). A third major division has been proposed, the enteric ANS, which is composed primarily of neurons located in the wall of the gastrointestinal tract, which function relatively independent of the SNS and PNS to regulate motility and fluid and electrolyte homeostasis.[124]

The anatomical features of the efferent components of the sympathetic and parasympathetic divisions of the ANS are illustrated in Figure 24–1. Both divisions have similar anatomical arrangements with cell bodies of origin located in the brain stem and spinal cord. In both divisions, efferent axons leave the central nervous system via cranial nerves or ventral roots to synapse upon specialized ganglia, where second order neurons give rise to axons that directly inner-

vate smooth muscle, cardiac muscle, or various glands. Thus, both divisions have pre- and postganglionic fibers.

In the SNS, sometimes referred to as the thoracolumbar system, the cell bodies of origin are found in the intermediolateral (IML) cell column of the spinal cord, which is located from the T1 to L2 segments. The preganglionic sympathetic fibers that exit in the ventral root are small myelinated fibers, the white rami communicantes, which typically synapse on second order neurons in the paravertebral ganglia, paired sympathetic ganglia lying adjacent to the spine and extending from the cervical to sacral segments (Fig. 24–2). Preganglionic fibers may pass up or down many segments before synapsing in their ganglia. The second order, or postganglionic fibers, originating from these paravertebral ganglia, leave the ganglia as thin unmyelinated fibers, the gray rami communicantes. These fibers, which may travel in various segmental spinal nerves or in specialized autonomic nerves, innervate nearly all the organs of the body. Some preganglionic fibers pass through the paravertebral ganglia chain and synapse on special sympathetic ganglia in the abdomen, such as the celiac, superior mesenteric, and inferior mesenteric ganglia. Second order postganglionic fibers from these specialized ganglia innervate the glands of the abdomen and pelvis. Some preganglionic fibers traveling in the splanchnic nerves may directly innervate the adrenal medulla. In a sense, the adrenal medulla may be regarded as a specialized sympathetic ganglion, which is capable of secreting sympathetic neurotransmitters directly into the blood in response to the preganglionic stimulation.

The PNS has its cell bodies of origin in specialized nuclei of cranial nerves III, VII, IX, and X, as well as from cells in the IML cell column of the sacral segments, S2, S3, and S4. Thus, the PNS has been referred to as the craniosacral system. Unlike the SNS, the PNS ganglia are located in close proximity to their target organs. Thus, the PNS preganglionic fibers are rather long, traveling to various ganglia close to the target organ, and their postganglionic fibers are rather short.

The ANS was traditionally thought of as a pure efferent or motor system; however, autonomic afferent fibers are also an integral part of the ANS. Although many sensory afferent fibers may evoke an autonomic reflex, autonomic afferent fibers are those that relay special sensory information from various viscera to the ANS centers.[85] Axons of these afferent autonomic fibers may travel in somatic peripheral nerves or in specialized autonomic nerves along with the autonomic efferent or motor fibers.

The neurotransmitter for all ANS preganglionic fibers, both SNS and PNS, is acetylcholine. In addition, all PNS postganglionic fibers and the SNS postganglionic fibers innervating sweat glands (sudomotor fibers) also utilize acetyl-

AUTONOMIC NERVOUS SYSTEM

PARASYMPATHETIC DIVISION   **SYMPATHETIC DIVISION**

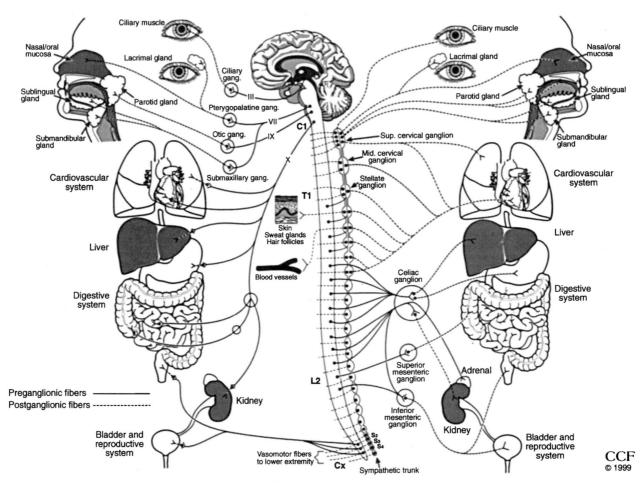

FIGURE 24–1. The efferent systems of the sympathetic and parasympathetic divisions of the autonomic nervous system. (Copyright © Cleveland Clinic Foundation, 1999.)

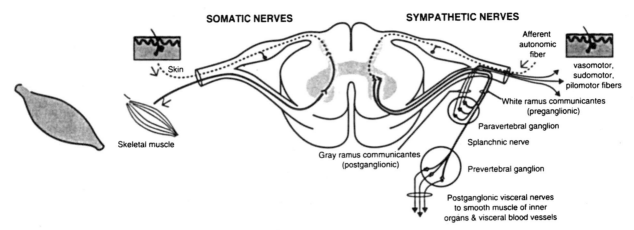

FIGURE 24–2. The sympathetic outflow from the spinal cord and the course of the sympathetic fibers. (Modified with permission from Pick J: The Autonomic Nervous System: Morphological, Comparative, Clinical and Surgical Aspects. Philadelphia, Lippincott, 1970, pp 3–21.)

choline as their neurotransmitter. The remainder of the SNS postganglionic fibers utilize norepinephrine.

There are numerous complex anatomical relationships between the various supraspinal centers that serve to integrate ANS function.[57] These centers reside in the brain stem, mesencephalon, and cortical areas, and include specific structures such as the nucleus tractus solitarius, nucleus ambiguus, dorsal motor nucleus of vagus, dorsal raphe nucleus, medullary reticular formation nuclei, locus ceruleus, hypothalamus, limbic system, and primary sensory and motor cortex. The hypothalamus, which receives a wide array of sensory afferents from the other higher centers, has been regarded as the principal higher center for the integration of ANS function. The hypothalamus exerts its effect via interaction with the pituitary and endocrine system as well as via its descending pathways to the midbrain, which give rise to the reticulospinal pathways. These fibers synapse upon interneurons in the spinal cord that influence the IML cells. The nucleus tractus solitarius also appears to have an important integrating function of descending information from higher centers as well as from spinal afferents.[57]

Many of the important functions of the ANS are listed in Table 24–1. Most target organs receive both sympathetic and parasympathetic innervation, but there are many important exceptions including sweat glands, piloerector muscles, and most small blood vessels or arterioles, all of which receive only sympathetic innervation. When dual innervation occurs, the SNS and PNS often have antagonistic functions. However, this is not uniformly true, particularly in those organs such as the lacrimal, parotid, and submandibular glands, which are stimulated to secret by both divisions of the ANS. In these glands, however, the PNS typically produces a more potent secretory effect.

The SNS is anatomically organized via its numerous ascending and descending preganglionic fibers along the paravertebral chain to permit a rather diffuse sympathetic discharge. Furthermore, the direct innervation of the adrenal medulla with SNS preganglionic fibers may stimulate it to secret epinephrine and norepinephrine into the blood, thus producing a diffuse and far-reaching effect. A diffuse sympa-

thetic discharge results in pupillary dilatation, increased heart rate and contractility, bronchodilatation, vasoconstriction (particularly in the mesenteric circulation), and vasodilatation of skeletal muscle arterioles. This type of diffuse sympathetic discharge is encountered in the "fight or flight" reaction.[1] The PNS is anatomically organized to produce more focal and discrete functions.

The Baroreceptor Reflex

Both the SNS and PNS accomplish their regulatory functions by virtue of various autonomic reflexes. Possibly the most thoroughly studied of the autonomic reflexes is the baroreceptor reflex, which is responsible for maintenance of homeostasis of arterial blood pressure.[48] This reflex is also important in clinical autonomic function testing, because many of the cardiovascular tests, especially the Valsalva maneuver[50, 82] and the standing and tilt tests,[9, 122] assess its integrity.

A brief review of the baroreceptor reflex will illustrate the basic components of an autonomic reflex. The principal components of the baroreceptor reflex are illustrated in Figure 24–3. The afferent limb of the reflex begins with specialized mechanoreceptors, located in the aortic arch and carotid sinuses, which are capable of responding to changes in arterial blood pressure. Responses from these receptors are carried via cranial nerves IX and X to the medulla, where they synapse upon neurons in the nucleus tractus solitarius (NTS).[92] The NTS receives a wide range of afferent information in addition to the arterial baroreceptor inputs, and considerable processing of this information occurs at this level.[38, 77] Activation of the arterial baroreceptors by increased systemic blood pressure results in an increased discharge of NTS fibers innervating the caudal ventrolateral medulla, which in turn results in gamma aminobutyric adrenergic (GABAergic) inhibition of the rostral ventral lateral quadrant of the medulla,[4] the so-called vasomotor center, and specifically the adrenergic neurons in this center referred to as the C1 area.[93] The C1 area sends descending fibers to directly stimulate the cells in the IML column, resulting in increased sympathetic tone to blood vessels and the heart. In addition, activation of arterial baroreceptors also results in NTS stimulation of cardiovagal neurons in the nucleus ambiguus, which in turn causes slowing of heart rate.[4] Thus, an increase in systemic blood pressure results in inhibition of the C1 area with reduced sympathetic tone and a subsequent fall in blood pressure and slowing of heart rate. If there is a sudden drop in systemic pressure, the C1 area becomes disinhibited and results in increased sympathetic tone to blood vessels and the heart, resulting in increased blood pressure and heart rate. Thus, the arterial baroreceptor reflex functions as a negative feedback reflex arc, designed to maintain homeostasis of blood pressure.

AUTONOMIC REFLEX TESTS

A wide variety of autonomic reflex tests have been devised to assess different components of the ANS.[59, 74] Although in some respects these tests appear to be deceptively straightforward methods that are simple to perform and interpret, they are in fact quite complex and may be influenced by a wide range of variables as well as other interacting autonomic reflexes.[59, 68] Thus, the proper interpretation of these tests requires a comprehensive understanding of the many variables that may influence each specific test.[68] Unfortunately, many of the confounding variables cannot be easily controlled during the testing procedure. However, there are

TABLE 24–1	Functions of the Autonomic Nervous System	

Organ	SNS	PNS
Eye		
Pupil	Dilatation	Construction
Ciliary muscle	Relax (far vision)	Constrict (near vision)
Lacrimal gland	Slight secretion	Secretion
Parotid gland	Slight secretion	Secretion
Submandibular gland	Slight secretion	Secretion
Heart	Increased rate	Slowed rate
	Positive inotropism	Negative inotropism
Lungs	Bronchodilation	Bronchoconstriction
Gastrointestinal tract	Decreased motility	Increased motility
Kidney	Decreased output	None
Bladder	Relax detrusor	Contract detrusor
	Contract sphincter	Relax sphincter
Penis	Ejaculation	Erection
Sweat glands	Secretion	None
Piloerector muscles	Contraction	None
Blood vessels		
Arterioles	Contraction	None
Muscle		
Arterioles	Constriction or dilatation	None
Metabolism	Glycogenolysis	None

PNS, parasympathetic nervous system; SNS, sensory nervous system.
(Reproduced from Shields RW: Functional anatomy of the autonomic nervous system. J Clin Neurophysiol 10:2–13, 1993, with permission.)

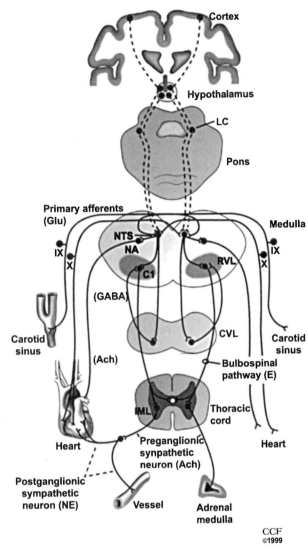

FIGURE 24–3. Proposed pathways and neurotransmitters of the baroreceptor reflex. NTS, nucleus tractus solitarius; NA, nucleus ambiguus; CVL, caudal ventrolateral medulla; RVL, rostral ventrolateral medulla; C1, ventrolateral epinephrine cell area; IML, intermediolateral cell column; IX, glossopharyngeal nerve; X, vagus nerve; LC, locus ceruleus; RFN, retrofascial nucleus; GABA, gamma-aminobutyric acid; Glu, glutamic acid; E, epinephrine; Ach, acetylcholine; NE, norepinephrine. (Copyright © Cleveland Clinic Foundation, 1999.)

guidelines that may help reduce variability of the autonomic reflex test results.[59, 68] It is very important to consider medications, food and diet, emotional stress and anxiety, and intravascular volume status in all subjects undergoing ANS testing.[68] It is generally recommended that patients should avoid food, coffee, tea, and nicotine for at least 3 hours prior to testing. Medications, especially anticholinergic drugs, including antidepressants, antihistamines, and other medications that may influence or alter sympathetic and parasympathetic function, should be withheld for at least 48 hours prior to the testing, if possible. Patients who are in the midst of acute illness, or who are emotionally upset or in pain, should not be tested. Furthermore, autonomic reflex testing should be conducted in a calm and quiet environment to reduce stress and anxiety, which can affect the results of the tests.

The proper administration and interpretation of ANS tests also require strict adherence to standardized methods of testing and the proper use of normal values. The develop-

ment of normal values for any individual test typically requires the study of a large number of age-stratified normal subjects. One set of normal values for all age groups is inappropriate, as in most cases, the various autonomic nervous system tests display changes in response to advancing age.[29, 59, 65, 67] If normal values are to be used from the literature, it is important that great care be observed in preparing patients for autonomic testing, utilizing the same guidelines as those used by the investigators who collected the normal values. Failure to conform to these guidelines could clearly result in greater variability in responses and lead to misinterpretation of the test results and erroneous conclusions.

Most autonomic laboratories utilize a battery of tests designed to assess different components of the ANS. The battery of five bedside cardiovascular reflex tests reviewed by Ewing and Clarke in 1982 has served as the nucleus of cardiovascular tests used by many autonomic laboratories today.[23] This battery includes the heart rate response to deep breathing, the Valsalva ratio, diastolic blood pressure response to sustained handgrip, the heart rate response to standing (30:15 ratio), and systolic blood pressure response to standing. Most modern autonomic laboratories incorporate sudomotor as well as cardiovascular tests into their battery of autonomic tests.[59, 68, 74, 89] In general, tests should be both sensitive and specific, with regard to the detection of an ANS abnormality. In addition, autonomic tests should be reliable and reproducible, and most importantly, clinically and physiologically relevant.[58] Lastly, autonomic screening batteries should utilize safe, noninvasive methods. The tests that are reviewed in this chapter are those that have gained wide acceptance in the clinical assessment of ANS disorders.

Cardiovascular Autonomic Tests

Heart Rate Response to Deep Breathing

The variation of heart rate with deep breathing (HRdb) has proven to be a very useful and reliable test of cardiovagal function.[59, 74, 89] In fact, it has been regarded as the most sensitive test of cardiovagal function.[59] Although several autonomic reflexes contribute to the HRdb reflex, it is clear that the efferent pathway of the reflex is mediated via the vagus nerve.[121] The reflex results in an increase in heart rate during inspiration and a decrease in heart rate with expiration[59] (Fig. 24–4). This reflex, which is responsible for normal sinus arrhythmia, is maximized during deep respirations and in this context is utilized for ANS testing.

There are numerous factors that affect the HRdb, including the rate and depth of breathing, position of the patient, duration of deep breathing, and heightened sympathetic activity.[59, 68] The optimal or maximum heart rate variability with deep breathing is obtained with the patient lying flat, deep breathing at a tidal volume of greater than 1.2 liters, and breathing at a rate of five to six breaths per minute.[59, 84, 121] Prolonged deep breathing can induce hypocapnia, which lessens the HRdb.[68] Furthermore, patients who are anxious or have increased sympathetic tone can also have a reduction in their HRdb.

A variety of methods have been used to quantitate HRdb. The most widely used method is the HR range, which is obtained by subtracting the maximum heart rate at inspiration from the minimum heart rate at expiration.[59] Typically, at least six cycles are recorded and the mean of the heart rate range is reported. Some protocols measure eight cycles and report the maximum mean HR range for five consecutive cycles.[59] Some authors prefer to measure the heart rate range of the first respiration, which is often the largest.[7] HRdb can also be assessed via the expiratory to inspiratory ratio (E:I ratio), which is measured by dividing the minimum

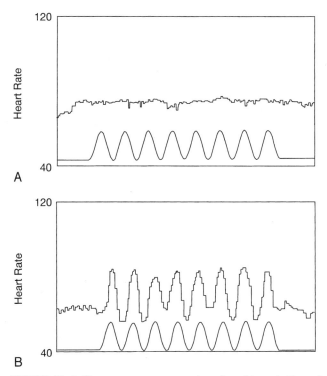

FIGURE 24–4. Heart rate response to deep breathing. *A.* Normal control. *B.* Patient with autonomic neuropathy resulting from amyloidosis. Respiratory rate is six breaths per minute.

heart rate expressed as the maximum RR interval during expiration by the maximum heart rate expressed as the minimum RR interval during inspiration.[59, 107] The E:I ratio can be measured as a mean of several cycles or the value of a single deep respiration.[107] Another method utilizes the mean circular resultant, which assesses the synchronization between heart rate variation and respiration.[59] The HR range has proven to be the most popular method, as it is less affected by the mean heart rate and irregularity of respiratory breathing rate than the E:I ratio.[59] Furthermore, it has some advantages over the mean circular resultant in that the latter method requires several minutes of deep breathing, which may induce hypocapnia and subsequently a reduction in the HRdb.[59]

Many autonomic laboratories utilize computers to collect the heart rate and automatically analyze and calculate the HR range. However, this test can be performed at the bedside using an electrocardiogram (ECG) or electromyogram (EMG) unit and making measurements directly off the ECG or EMG printout.[23] Patients must be trained to breath deeply at a regular rate of five to six breaths per minute. This can be accomplished by direct instruction from the individual administering the test, or by using a visual cue such as a sine wave on a computer screen.

Ewing and Clarke reported non–age stratified normal values for HRdb utilizing a protocol of six consecutive cycles of deep breathing at a rate of six breaths per minute.[23] The mean heart rate range was considered normal for values greater than 15 beats per minute (bpm), borderline for values of 11 to 14 bpm, and abnormal for values less than 10 bpm.[23] However, the HRdb is markedly affected by the age of the patient. With advancing age, the HRdb is significantly reduced.[59, 65, 67] Age-stratified control values provide a more precise measure of normal limits and increase the sensitivity of the test. A very popular protocol utilizes eight cycles of deep breathing at a rate of six breaths per minute.[59] The

largest five consecutive cycles are used to determine the mean heart rate range. Normal values for this protocol are as follows: age 10 to 40 years, ≥18 bpm; age 41 to 50 years, ≥16 bpm; age 51 to 60 years, ≥12 bpm; and age 61 and over, ≥8 bpm.[59] Age-stratified normal values (95th percentile) for single breath HRdb expressed as E:I ratio are as follows: 16 to 20 years, >1.23; 21 to 25 years, >1.20; 26 to 30 years, >1.18; 31 to 35 years, >1.16; 36 to 40 years, >1.14; 41 to 45 years, >1.12; 46 to 50 years, >1.11; 51 to 55 years, >1.09; 56 to 60 years, >1.08; 61 to 65 years, >1.07; 66 to 70 years, >1.06; 71 to 75 years, >1.06; and 76 to 80 years >1.05.[107]

The HRdb has proven to be a very sensitive and reliable method for the early detection of cardiovagal dysfunction in a wide variety of autonomic disorders, including diabetic neuropathy,[24, 84, 107, 114] uremic neuropathy,[120] small-fiber neuropathies,[112] familial neuropathies,[8] idiopathic autonomic neuropathy,[113] and in patients with pure autonomic failure,[91] multisystem atrophy,[11] and other central neurodegenerative disorders.[98]

Valsalva Maneuver

The Valsalva maneuver (VM) occurs when intrathoracic and intra-abdominal pressure is increased during a straining maneuver.[5, 82] Intrathoracic pressures of approximately 40 mm Hg for 15 seconds appear ideal to produce an adequate stimulus to assess the cardiovascular responses to the VM.[5, 59] The VM and the cardiovascular responses that follow can be divided into four phases[5, 82] (Fig. 24–5). Phase I occurs at the onset of the maneuver and is accompanied by an abrupt elevation of arterial pressure reflecting increased mechanical pressure on the aorta and great vessels. Accompanying this brief increase in pressure is slowing of the heart rate mediated by increased parasympathetic efferent activity. This phase of the Valsalva maneuver lasts approximately 4 seconds. Phase II may be divided into an early and late phase.[99] During early phase II, blood pressure and cardiac output are reduced, primarily owing to reduced venous return to the heart, secondary to the increased intrathoracic pressure. During this phase, there is a reduction in left ventricular stroke volume and cardiac output. The tachycardia that occurs in early phase II is primarily a result of cardiovagal inhibition. The reduction in blood pressure during early phase II triggers the arterial baroreceptor reflex, which causes peripheral vasoconstriction and compensatory tachycardia. Thus, in late phase II, approximately 4 to 5 seconds into phase II, compensatory vasoconstriction causes a leveling off or recovery of arterial blood pressure and there is a sympathetically mediated tachycardia.[99] Phase III marks the sudden cessation of the maneuver with release of straining. This results in a rather brief decrease in arterial blood pressure lasting 1 to 2 seconds, owing to the reduction in pressure on the aorta and great vessels. This is followed by phase IV, which is marked by an overshoot of arterial blood pressure beyond baseline, typically lasting less than 10 seconds and apparently caused by increased venous return to the heart and increased cardiac output in the setting of persistent vasoconstriction. The blood pressure overshoot triggers a reflex bradycardia, again mediated by the arterial baroreceptor reflex via the vagus nerve and a subsequent reflex-mediated decrease in arteriolar vasoconstriction. Blood pressure and heart rate usually return to baseline after approximately 90 seconds.

Although many facets of the VM can be measured, the cardiovagal-induced heart rate changes during the VM are readily quantitated without invasive testing or special equipment.[23, 24] ECG monitoring during the VM permits the assessment of the Valsalva ratio (VR): the ratio of the longest RR interval after the maneuver (reflecting the bradycardia in

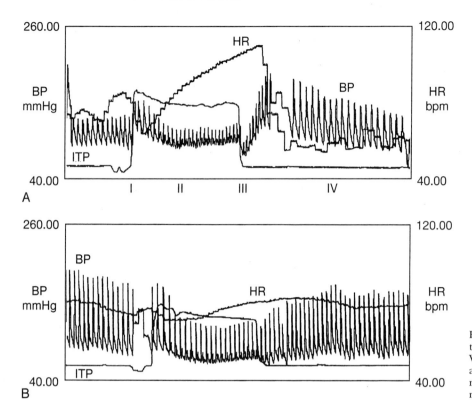

FIGURE 24–5. Valsalva maneuver. Duration of straining is 15 seconds. *A.* Normal Valsalva maneuver with phases I, II, III, and IV noted. *B.* Abnormal Valsalva maneuver in a patient with autonomic neuropathy resulting from amyloidosis.

phase IV) to the shortest RR interval during the maneuver (the maximum tachycardia during late phase II).[23, 50]

There are many variables that may influence the VM, including position of the subject, duration of the strain, the level of intrathoracic pressure, and the phase of respiration prior to the maneuver.[59] Maximum responses are obtained with the patient in a supine position performing a maneuver of 10 to 15 seconds in duration, and with intrathoracic pressures of 40 to 50 mm Hg.[59] In addition, the cardiovascular responses to VM are greater when the maneuver is performed following a normal tidal inspiration, rather than after a maximum inspiration.

The advent of noninvasive continuous beat-to-beat assessment of blood pressure by a photoplethysmographic technique utilizing the Finapres has enabled assessment of the sympathetically mediated blood pressure responses to the VM.[5, 34, 35] In fact, assessment of the heart rate responses to VM without monitoring of the blood pressure responses may lead to misinterpretation of the VR as an index of cardiovagal function.[59] For example, if the VR is abnormal because of the absence of significant bradycardia in phase IV, it may be because of failure of the sympathetically mediated blood pressure overshoot (the stimulus to the vagally mediated reflex bradycardia). Thus, an abnormal VR cannot be interpreted as an isolated cardiovagal abnormality without monitoring beat-to-beat blood pressure during the maneuver. In addition, the VR may still fall in the normal range in a patient with normal cardiovagal function who has isolated sympathetic vasomotor deficits.[119] In such a setting, the VM may generate a prominent tachycardia in phase II despite the absence or diminution of the reflex bradycardia in phase IV. In some laboratories, the blood pressure response in late phase II is used as an index of adrenergic function; that is, the stabilization or recovery of blood pressure in late phase II is an index of the sympathetic component of the baroreceptor reflex.[59, 99] In fact, measuring beat-to-beat blood pressure changes during the VM adds sensitivity to the test in the detection of mild adrenergic dysfunction[99] (Fig. 24–6).

Responses to VM may be abnormal owing to factors other than autonomic dysfunction. The VM may be abnormal in patients with congestive heart failure or increased pulmonary blood volume, and in patients with reductions in intravascular volume. Although VM is a relatively safe test, it should not be performed in patients with proliferative retinopathy because of the risk of intraocular hemorrhage. In addition, caution must be used during the VM in elderly patients and in patients with cardiac disease, as the VM may provoke syncope, angina, or cardiac arrhythmia.

The influence of age on the VR is somewhat controversial. Clearly, the VR is less influenced by age than HRdb.[59] Although some authors have not reported significant variations with age,[24] others have noted age and gender differences.[59, 65, 67]

Non–age-stratified normal values reported by Ewing and Clarke for VR are as follows: normal, > 1.21; borderline, 1.11 to 1.20; and abnormal, < 1.20.[24] Other laboratories regard a VR > 1.50 as normal.[50] Age-stratified normal values reported for VR by Low are as follows: 10 to 40 years, > 1.50; 41 to 60 years, > 1.45; and 61 and over, > 1.35.[59] When monitoring beat-to-beat blood pressure during the VM, a drop of blood pressure ≥ 20 mm Hg in early phase II with an absent late phase II (failure of recovery of blood pressure) or an absent phase IV (failure of blood pressure overshoot) is regarded as a definite sign of adrenergic failure.[59]

The VM has proven to be a useful test of cardiovagal function, and with beat-to-beat measurements of blood pressure, it can also be used as an index of adrenergic function.[90, 99] The VM is typically abnormal in most patients with pure autonomic failure and multisystem atrophy,[11] and has been applied as a valuable test in diabetic autonomic neuropathy.[7, 84]

Heart Rate Response to Standing

The heart rate response to standing has been used as a reliable cardiovagal bedside test for many years.[21, 24] The normal response begins with a tachycardia during and

---

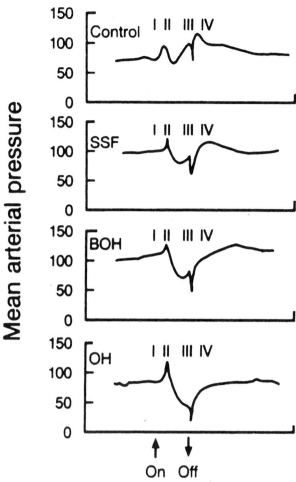

FIGURE 24–6. Mean arterial pressure tracings of a control subject and patients with sympathetic sudomotor failure (SSF), borderline orthostatic hypotension (BOH), and orthostatic hypotension (OH), respectively. (Reproduced from Sandroni T, Benarroch EE, Low PA: Pharmacologic dissection of components of the Valsalva maneuver in adrenergic failure. J Appl Physiol 71:1563–67, 1991, with permission.)

able to have the subject lie flat for at least 5 to 10 minutes before the maneuver.[122]

Ewing and Clarke reported non–age-stratified normal values for the 30:15 ratio.[23] Ratios greater than 1.04 are considered normal, ratios between 1.01 and 1.03 are borderline, and ratios less than 1.00 are abnormal. Age-stratified normal values are as follows: 10 to 29 years, > 1.17; 30 to 49 years, > 1.09; and 50 to 65 years, > 1.03.[121] Wieling has proposed assessing the Δ HR max, the maximum heart rate change from baseline during the first 15 seconds after standing, as an index of cardiovagal function.[122] The reflex bradycardia is assessed by determining the ratio of the maximum heart rate over the minimum heart rate during the first 30 seconds after standing. Normal values are as follows:

10 to 14 years, Δ HR max < 20, HR max/HR min < 1.20
15 to 19 years, Δ HR max < 19, HR max/HR min < 1.18
20 to 24 years, Δ HR max < 19, HR max/HR min < 1.17
25 to 29 years, Δ HR max < 18, HR max/HR min < 1.15
30 to 34 years, Δ HR max < 17, HR max/HR min < 1.13
35 to 39 years, Δ HR max < 16, HR max/HR min < 1.11
40 to 44 years, Δ HR max < 16, HR max/HR min < 1.09
45 to 49 years, Δ HR max < 15, HR max/HR min < 1.08
50 to 54 years, Δ HR max < 14, HR max/HR min < 1.06
55 to 59 years, Δ HR max < 13, HR max/HR min < 1.04
60 to 64 years, Δ HR max < 13, HR max/HR min < 1.02
65 to 69 years, Δ HR max < 12, HR max/HR min < 1.01
70 to 74 years, Δ HR max < 12, HR max/HR min < 1.00
75 to 80 years, Δ HR max < 11.

Blood Pressure Response to Standing and Tilt

Measurement of blood pressure following standing or tilt primarily assesses sympathetic reflexes. The blood pressure

shortly after standing.[9, 22, 122] This abrupt increase in heart rate peaks at about 3 seconds after standing and is caused by inhibition of cardiovagal tone induced by the exercise reflex[9, 122] (Fig. 24–7). The second more gradual increase in heart rate begins at 5 seconds and peaks around 12 seconds.[9, 122] This is caused by increased sympathetic tone as well as further inhibition of cardiovagal tone as a result of a transient drop in blood pressure,[111] and is mediated via the arterial baroreceptor reflex.[122] The delayed bradycardia, occurring approximately 20 seconds after standing, is also mediated via the baroreceptor reflex and is triggered by a blood pressure overshoot resulting in cardiovagal disinhibition.[122] Thus, although the heart rate response to standing is a cardiovagal test, it requires an intact sympathetic nervous system response to standing[122] (see Fig. 24–7).

The heart rate response to standing may be assessed by dividing the longest RR interval at or around the 30th beat, by the shortest RR interval at or around the 15th beat; that is, the 30:15 ratio.[21, 22] The physiology of the heart rate response to standing is somewhat more complex than that of the heart rate response to deep breathing, and the variables that may influence this test have been less carefully delineated and studied.[59] Clearly, the response is attenuated with advancing age. In addition, the period of supine rest before the maneuver may affect the result. It is usually desir-

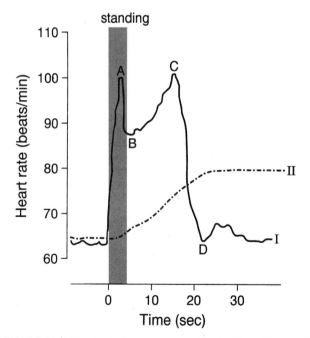

FIGURE 24–7. Heart rate changes provoked by standing. *I.* A normal control illustrating the initial peak tachycardia (A), the primary bradycardia (B), the secondary peak tachycardia (C), and the secondary or reflex bradycardia (D). *II.* Heart rate responses to standing in a patient with diabetic autonomic neuropathy illustrating the absence of the tachycardia at A and C, along with an absent reflex bradycardia at D. (Reproduced with modifications from Wieling W, Borst C, van Lieshout JJ, et al: Assessment of methods to estimate impairment of vagal and sympathetic innervation of the heart in diabetic autonomic neuropathy. Neth J Med 28:383–392, 1985, with permission.)

response to standing can be measured with the patient lying supine and then immediately upon standing. The difference in systolic blood pressure from lying to standing is noted. Although the criteria for an abnormal response remain somewhat controversial, it is generally agreed that a drop in systolic blood pressure upon standing of greater than 30 mm Hg is usually abnormal.[23, 24] More recent guidelines suggest a fall of 20 mm Hg systolic pressure or 10 mm Hg diastolic pressure within 3 minutes is abnormal.[12] Similar recordings can be measured using a tilt table. The tilt test is more sensitive than the standing test as it reduces the effect of the exercise reflex and the contraction of leg muscles that occurs with standing, which can increase peripheral resistance and obscure drops in blood pressure. Maximum responses to standing and tilt are achieved when the patient has been lying supine for 20 minutes.[122] In many laboratories, the convention is to tilt the table to a level of 80 degrees rapidly, typically within several seconds. Blood pressure is measured with a sphygmomanometer cuff at 1-minute intervals or continuously beat to beat, using a Finapres. It appears that similar results can be achieved with testing at 60, 80, or 90 degrees of tilt.[45] During upright tilt, normal subjects will undergo a mild reduction in systolic, diastolic, and mean arterial pressure, which recovers within 1 minute.[35] In patients with adrenergic failure, blood pressure will continue to drop during the tilt and, typically, there will be an absence of reflex tachycardia resulting from an abnormality of the arterial baroreceptor reflex. The optimal duration for the tilt is somewhat controversial. There is evidence that 5 minutes' duration is adequate to detect orthostatic hypotension[45] but longer durations are needed for assessment of postural orthostatic tachycardia and neurally mediated syncope.[59]

For the assessment of syncope caused by vasovagal or vasodepressor responses, patients are typically tilted to 60 to 80 degrees and remain tilted for durations of up to 20 to 60 minutes.[45] During this time, patients subject to vasovagal or vasodepressor responses may have a positive response (hypotension, bradycardia, or both). In some laboratories, subjects who do not demonstrate a vasovagal or vasodepressor response to the tilt are given an infusion of isoproterenol, typically at a dose of 1 to 5 μg/min, and then retilted.[2] Using isoproterenol during head-up tilting will clearly increase the sensitivity of detecting vasovagal or vasodepressor responses.[2] However, up to 10% of normal subjects may experience syncope during headup tilting[41] and a greater percentage may experience positive responses under the influence of isoproterenol.[39] This reduction in specificity has led some authors to question the value of the tilt test with isoproterenol in assessing patients with syncope.[39]

Blood Pressure Response to Sustained Hand Grip

Another cardiovascular test of sympathetic function is the blood pressure response to sustained hand grip. This test is performed with the aid of a dynamometer. The blood pressure is measured with the patient at rest, and then after 5 minutes of sustained hand grip at 30% of maximum contraction.[20] During the muscular contraction, heart rate and blood pressure typically increase owing to the exercise reflex.[15] Although there is some evidence suggesting that the pressor response to exercise may be caused by local muscle afferent reflex activity, there is clear evidence that the early heart rate increase during sustained hand grip is, in fact, due to vagal withdrawal, whereas the tachycardia that follows is a result of sympathetic activation.[46]

This reflex has not been studied as carefully as the Valsalva maneuver and the HRdb and thus, confounding variables and influences are not well known. It appears that the greatest sensitivity of this test occurs after 4 to 5 minutes of the maneuver.[20, 46] Thus, it is essential that subjects be encouraged to complete the test despite the fatigue and discomfort that usually occurs after a few minutes of sustained hand grip. The response tends to be uniform with increasing age,[29] but there are significant differences between men and women, with women showing a smaller rise in diastolic blood pressure than men.[46] In addition, subjects with lower grip strength have a reduced blood pressure response to hand grip.[46] The response to sustained hand grip has been studied in diabetic neuropathy[20] and has been found to have comparable sensitivity and specificity in the detection of adrenergic failure as the tilt test.[46]

Non–age-stratified normal values indicate that an increase in diastolic blood pressure from baseline to the conclusion of the hand grip maneuver of greater than 16 mm Hg is normal, whereas an increment of 11 to 15 mm Hg is considered borderline, and an increment of 10 mm Hg or less is considered definitely abnormal.[20, 23]

Other Cardiovascular Tests

In addition to the cardiovascular reflex studies previously described, there are numerous other tests that have been applied to the study of cardiovascular autonomic function. The mental arithmetic test, performed by asking a subject to subtract 7 from 100 as quickly as possible, has been utilized as a test of adrenergic function.[69] The mental stress of performing arithmetic typically results in emotional stress and a pressor response, due primarily to an increase in cardiac output with no change or a fall in peripheral sympathetic nerve activity and a fall in total peripheral resistance of approximately 30%.[59] This results in an increase in blood pressure and heart rate.[69] This test, however, is less reliable than the tests discussed earlier, as the stimulus intensity is quite variable from subject to subject and adaption can occur.[69] An absent response may occasionally be encountered in normal control subjects.[59] The cold face test or diving response utilizes a cold stimulus to the first divisions of the trigeminal nerve bilaterally, typically resulting in reflex bradycardia.[44] It is a test of trigeminal-vagal-cardiac and trigeminal-sympathetic-vasomotor function. This test has been applied to a wide range of disorders,[44] but it has not been studied systematically and confounding factors have not been defined.

The cold pressor test is performed by the subject submerging a hand in ice cold water while blood pressure is assessed. The efferent limb of the reflex is sympathetic and a normal response is an elevation of blood pressure.[100] The test lacks sensitivity and it is difficult for many patients to tolerate the cold stimulus long enough for the blood pressure response to be noted.[59] Although the cold pressor test has been reported to be valuable in the assessment of baroreceptor failure[94] and diabetic autonomic neuropathy,[100] it appears to lack reproducibility.[25]

Skin blood flow may be measured by a laser Doppler flowmeter or by plethysmography in order to record the vasoconstrictor response to an autonomic maneuver or stimulus.[60] Measurements are usually made on the toe or finger pads following a variety of stimuli, including Valsalva maneuver, inspiratory gasping, cold stimulus, and so on. These reflex tests have been used primarily for the assessment of autonomic neuropathies; however, technical factors in recording responses in the feet and the amount of variability in the responses relating to various environmental factors make the test technically difficult and less reliable.[58]

Spectral analysis of heart rate has also been applied as a test of autonomic function. At rest, in normal subjects, spectral analysis of heart rate discloses power at two frequencies, a low-frequency component (frequencies less than 0.15 Hz) and a higher frequency component at approximately 0.25

Hz.[88] The high-frequency component of the spectrum appears to reflect heart rate variability with respiration and, thus, has been regarded as an index of cardiovagal function.[101] The lower frequency component appears to be the product of both sympathetic and parasympathetic activity and in part is reflective of baroreceptor feedback activity and circulating catecholamines.[101] In the supine position, the spectral analysis of high frequency, low frequency, and total power are good predictors of parasympathetic function as measured by HRdb, Valsalva ratio, and the 30:15 ratio.[28] After tilt, there is a shift of power from higher to lower frequencies, presumably reflecting increased sympathetic influence on heart rate. Although this has been used as an index of sympathetic function, it is only a modest predictor of blood pressure fall during tilt.[28] Spectral analysis has been successfully applied to the study of diabetic autonomic neuropathy.[28, 55, 83] However, the precise role of spectral analysis on autonomic function testing has yet to be clearly defined,[53] especially with regard to the early detection of sympathetic dysfunction.[101]

CASE STUDY

A 44-year-old man presented with a 3-month history of paresthesia in the feet in conjunction with weight loss, abdominal cramps, diarrhea alternating with constipation, and erectile impotence. Immediately prior to evaluation, he complained of light-headedness and fainting when standing. There were no symptoms of dryness of the eyes or mouth, and the patient did not notice any alteration in sweating or bladder function. There was no generalized weakness or alteration in gait or station.

The neurological examination revealed a normal mental status and cranial nerve function. The motor examination was notable for mild weakness of toe extensors and toe flexors bilaterally. Tendon reflexes were unelicitable in the lower extremities, but were present and symmetrical in the upper extremities. Plantar signs were flexor bilaterally. The sensory examination disclosed absent vibratory sense at the toes and reduced pinprick to the level of the ankles bilaterally. Gait and station were normal, and there were no cerebellar signs.

Electrodiagnostic studies disclosed evidence of a generalized sensorimotor polyneuropathy, axon loss in type, and graded moderate in degree electrically. Superimposed upon these findings were changes consistent with a nonnecrotizing myopathy, predominantly involving the proximal lower extremity muscles. General laboratory studies disclosed a normocytic normochromic anemia with hemoglobin 11.0 g/dl, and hematocrit 31.2%. Chemistry profiles of the blood were unremarkable. Other tests, such as creatine kinase, vitamin B[12], folic acid, thyroid-stimulating hormone and hemoglobin A1C were normal. Serum protein immunoelectrophoresis with immunofixation disclosed evidence of a monoclonal gammopathy, lambda light chain type. The skeletal bone survey was unremarkable. A bone marrow biopsy revealed normocellular marrow with focal amyloid deposition. An echocardiogram was unremarkable. A rectal mucosal biopsy was positive for amyloidosis.

A battery of cardiovascular autonomic tests was performed. Heart rate response to deep breathing was markedly abnormal with essentially no variability of heart rate (see Fig. 24–4). The Valsalva maneuver was also abnormal, disclosing failure of recovery of blood pressure in late phase II, coupled with an absent blood pressure overshoot in phase IV and a subsequent abnormal Valsalva ratio (see Fig. 24–5). Blood pressure response to sustained hand

grip was also abnormal with essentially no increase in diastolic blood pressure at the conclusion of 5 minutes of hand grip at 30% maximum force. A tilt test disclosed a baseline blood pressure of 99/67 mm Hg and heart rate of 95 bpm. The patient tolerated 30 degrees of tilt for 2 minutes, during which time he complained of light-headedness. The tilt test was terminated owing to severe light-headedness, at which time blood pressure was 65/39 mm Hg and heart rate, 93 bpm. Recovery was rapid and uneventful. The tilt test disclosed progressive orthostatic hypotension without compensatory tachycardia. Norepinephrine and epinephrine levels supine were 186 pg/ml and 13 pg/ml, respectively, both within normal limits. At the conclusion of the tilt, there was no increase in these levels. The pattern of findings on autonomic testing was consistent with a generalized dysautonomia with features of both parasympathetic and sympathetic failure.

The patient's evaluation was consistent with primary systemic amyloidosis with generalized dysautonomia. He was treated with a high-salt diet, thigh-high pressure stockings, and subsequently Florinef 0.1 mg bid. Ultimately, midodrine 2.5 mg 3× daily, was added to his regimen, with some improvement in his orthostasis. In addition, a course of chemotherapy, including melphalan and prednisone was initiated. Approximately 6 months later, the patient experienced profound syncope, followed by a cardiopulmonary arrest and expired. Autopsy disclosed changes consistent with systemic amyloidosis, polyclonal AL type, with involvement of the gastrointestinal tract, liver, spleen, heart, lungs, adrenals, testes, thyroid, and peripheral nerves.

Sudomotor Function Tests
Thermoregulatory Sweat Test

The thermoregulatory sweat test (TST) is a semiquantitative test of sudomotor function that provides information regarding the pattern of sweating and the integrity of both central and peripheral sympathetic sudomotor pathways.[26] The test is performed by subjecting the patient to whole body heating, often utilizing a thermal cabinet that can maintain an air temperature of 45 to 50°C and a relative humidity of 35% to 40%. Typically, patients are heated for approximately 30 to 40 minutes, until their body temperatures increase by an average of 1.5°C up to a target of 38°C, or an increase of 1°C from baseline.[26] Prior to the heating, the patient is dusted with a powder or a solution is applied that will react with sweat, causing a change in color. Various indicators have been utilized, including iodinated corn starch, mixtures of alizarin red, corn starch and sodium carbonate, and starch and iodine in solution.[26] When the sweating is complete, the pattern of sweating provoked by the thermostimulation is recorded. Various patterns of normal and abnormal sweating may be recognized[26, 27, 64] (Fig. 24–8). This test has been applied to the diagnosis of a wide range of central and peripheral autonomic disorders (Figs. 24–9 and 24–10). Its principle advantage is that it can disclose abnormalities of sweating in regions of the skin that are not easily tested by other techniques. Its disadvantages are that it cannot differentiate between pre- and postganglionic lesions, or peripheral and central lesions. Furthermore, it is a relatively messy test and exposes the patient to some discomfort. As with the cardiovascular autonomic reflex tests, patients should be properly prepared prior to the test. Dehydration should be avoided and drugs that may reduce thermoregulatory sweating, such as anticholinergics, including tricyclic antidepressant drugs, should be discontinued at least 48 hours before the test is performed.[26]

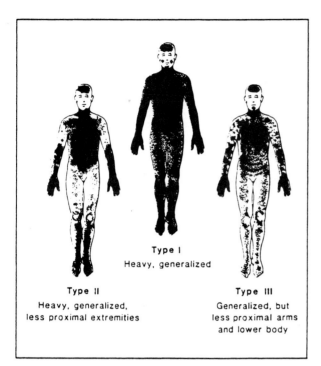

FIGURE 24–8. Normal thermoregulatory sweat test patterns. Dark areas represent areas of sweating. (Reproduced from Fealey RD, Low PA, Thomas JE: Thermoregulatory sweating abnormalities in diabetes mellitus. Mayo Clin Proc 64:617–628, 1989, with permission.)

Quantitative Sudomotor Axon Reflex Test

The quantitative sudomotor axon reflex test (QSART) was introduced by Low and colleagues as a quantitative measure of postganglionic sympathetic sudomotor function.[61] The test involves iontophoresis of acetylcholine onto the skin to stimulate sympathetic C fibers antidromically. As the impulse reaches a branching point, the action potential then reverses, traveling orthodromically along another fiber to stimulate a nearby sweat gland[61, 66] (Fig. 24–11). A special multicompartmental sweat cell is utilized for the iontophoresis and the measurement of the sweat response over time (Fig. 24–12). Thus, the QSART is a dynamic sweat test that measures the integrity of the distal postganglionic sympathetic sudomotor axons and sweat glands.[59]

Extensive data are available on QSART that document different normal values at various standard sites.[59] Men tend to have higher sweat production than women, and there is a very modest reduction in sweat production in the foot with advancing age.[59] Various abnormal patterns have been described, including general reductions in sweat output, an absence of sweat output, a persistence or "hung up" response, and an excessive response[58] (Fig. 24–13). QSART has been applied to a wide range of autonomic disorders, including diabetic and other polyneuropathies,[61, 63, 64] primary systemic anhidrosis,[62] multiple system atrophy and pure autonomic failure,[11] and Parkinson's disease and other central nervous system degenerative disorders.[98] It has been particularly valuable in the diagnosis of distal small-fiber neuropathy.[112] Persistent and excessive responses often associated with short latencies may occur in painful polyneuropathies and reflex sympathetic dystrophy.[58] In conjunction with the TST, the QSART may be helpful in localizing the site of a lesion as being pre- or postganglionic.[59] Absent TST and QSART in a region indicate a postganglionic lesion, whereas a preserved QSART in the setting of an absent TST indicates

a preganglionic lesion. The major disadvantage of QSART is that it is a time-consuming method that requires special equipment and thus has not been widely used in many autonomic laboratories.

Silastic Imprint Method

The silastic imprint method involves stimulating the sweat gland receptors directly via iontophoresis of pilocarpine.[43] A silastic material is spread over the stimulated surface and after several minutes, it hardens and may be removed. The imprint on the silastic material made by the sweat droplets can be seen under a dissecting microscope and counted with a grid[40] (Fig. 24–14). More sophisticated analysis can be

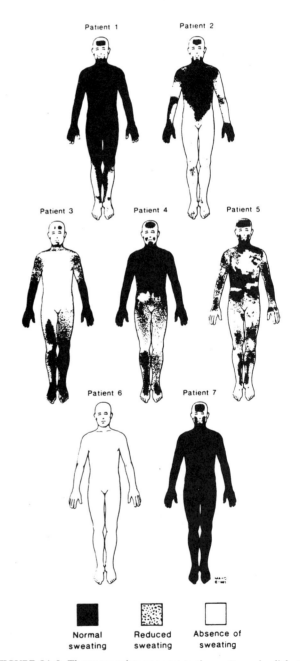

FIGURE 24–9. Thermoregulatory sweat testing patterns in diabetic neuropathy. (Reproduced from Low PA, Fealey RD: Sudomotor neuropathy. In Dyck PJ, Thomas PK, Asbury AK, et al (eds): Diabetic Neuropathy. Philadelphia, WB Saunders, 1987, p 140, with permission.)

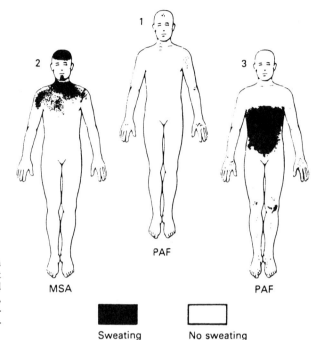

FIGURE 24–10. Thermoregulatory sweat test patterns in multisystem atrophy (MSA) and pure autonomic failure (PAF). Dark areas represent areas of sweating. (Reproduced from Low PA, Fealey RD: Structure and function of pre- and post-ganglionic neurons in pure autonomic failure, In Bannister R (ed): Autonomic Failure, A Textbook of Clinical Disorders of the Autonomic Nervous System, 2nd ed. Oxford, Oxford University Press, 1988, p. 549, with permission.)

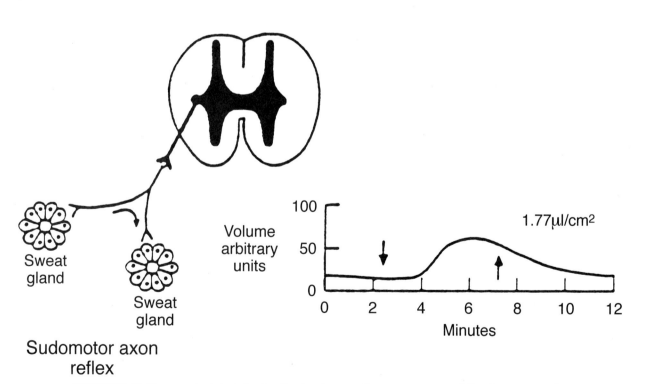

FIGURE 24–11. The neural substrate for the direct and axon reflex sweat responses *(left)* and a representative normal axon reflex sweat response *(right)*. (Reproduced from Low PA, Opfer-Gehrking TL, Kihara M: In vivo studies on receptor pharmacology of the human eccrine sweat gland. Clin Auton Res 2:29–34, 1992, with permission.)

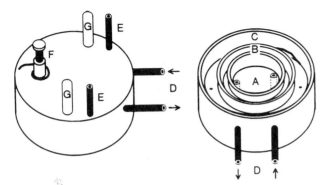

FIGURE 24–12. Multicompartment sweat cell for QSART. Acetylcholine is loaded into the cell via (E) and iontophoresed via the anode (F) in the 4-mm wide circular stimulus moat (C). Nitrogen gas passes through the 1-cm diameter recording compartment (A). A 1.5-mm wide air gap (B) separates the simulating and recording compartments. The nitrogen gas evaporates the sweat from (A) and the sweat output is measured dynamically via a sudorometer. (Reproduced from Low PA, Fealey RD: Sudomotor neuropathy. In Dyck PJ, Thomas PK, Asbury AK, et al (eds): Diabetic Neuropathy. Philadelphia, WB Saunders, 1987, p 142, with permission.)

performed by digitizing the image and using computerized imaging techniques to measure the number of droplets and the volume of sweating.[42] This method, which was developed and pioneered by Kennedy and colleagues, has been systematically applied to patients with diabetic polyneuropathy.[40, 42] Normal values were obtained for standard sites on the dorsum of the hand and foot. Results from clinical studies in type I diabetic patients indicate that a sudomotor abnormality may be detected in the forearm in 24% of patients and in the foot in 56% of patients.[42] In patients with mild neuropathy, sweat droplet size may be increased, possibly due to multi-innervation receptor alterations or other mechanisms.[47] Abnormal silastic imprint studies appear to correlate with abnormal thermal detection thresholds on quantitative sensory testing, indicating that small-fiber function may be distinctly and separately involved in the course of diabetic polyneuropathy.[43] The silastic imprint method is a safe and relatively simple technique that can represent a sensitive indicator of sweat gland function. It is not effective in the assessment of central and preganglionic disorders. This technique also requires special equipment and has not been widely used in many autonomic laboratories. A variety of other imprint methods have also been described.[43]

Sympathetic Skin Response

The sympathetic skin response (SSR) is a method that measures the sweating-induced change in the skin potential provoked by electrical stimulation or other activation maneuvers.[3, 30, 102] The SSR has also been referred to as the galvanic skin response (GSR) or peripheral autonomic skin potential (PASP).[30, 102] The precise pathways of the reflex have not been well-defined in humans, but it apparently is a reflex mediated by higher centers; thus, the SSR may be useful in detecting sudomotor abnormalities from central degenerative disorders as well as peripheral sympathetic disorders.[30] The origin of the response appears to reflect a rather complex interaction between potentials generated from sweat glands and epidermis.[102] Local factors such as callous formation, trophic changes of the skin, and degree of hydration of the epidermis may influence the SSR.[102] It has been recently demonstrated that the SSR is influenced by skin temperature.[13] SSR amplitude is linearly correlated with temperature, whereas latency is inversely correlated.[13]

The response can easily be measured using standard elec-

tromyographic equipment[30, 49, 104] (Fig. 24–15). Standard 1-cm circular disc electrodes used for nerve conduction studies can be applied to the hand or foot. The recording electrode is typically placed on the palm or sole with the reference electrode placed on the dorsum of the hand or foot respectively. Gain is usually set at 100 to 500 μV per division with a sweep speed of 500 msec to 1 second per division. Recommended filter settings range from 0.1 to 0.5 Hz for the low-frequency filter settings and from 32 to 3200 Hz for the high-frequency filter settings.[13, 16, 18, 30, 104, 105, 117] Larger amplitudes are usually obtained with lower low-frequency filter settings of 0.1 to 0.2 Hz. Activation maneuvers include inspiratory gasp, a loud noise, a cough, or a noxious stimulus.[30] It is usually most convenient to use an electrical shock from the nerve conduction stimulator. The responses vary considerably with regard to waveform and amplitude and tend to habituate with repeated stimulations.[30, 102] Utilizing stimuli at irregular intervals of greater than 30 seconds will lessen habituation and provide more consistent responses.[33] Responses from the palm are larger in amplitude and shorter in latency than those recorded from the sole.[30] Amplitudes typically range between 300 μV and 1200 μV with latencies of 1.3 to 1.5 seconds for the palm and 1.9 to 2.09 for the sole.[16, 30] Most laboratories will measure at least four responses and report the mean or the largest response.

Recently, techniques have been described utilizing magnetic stimulation of the brain or neck to evoke an SSR.[54, 96, 97] Experience with these techniques indicates that the SSR is more reproducible with regard to amplitude and latency than electrically evoked SSRs.[75] The magnetically evoked SSRs may provide more selective assessment of the efferent sympathetic sudomotor pathways and are less dependent upon integrity of afferent sensory pathways.[54, 96, 117] The clini-

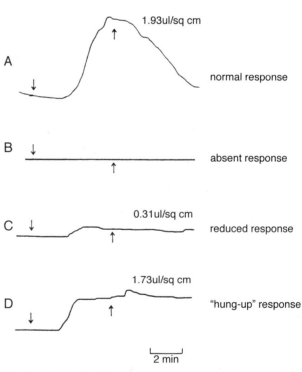

FIGURE 24–13. QSART responses. (A) normal, (B) absent, (C) reduced, and (D) persistent sweat activity. Arrows indicate onset and cessation of stimulus. (Reproduced from Low PA, Fealey RD: Sudomotor neuropathy. In Dyck PJ, Thomas PK, Asbury AK, et al (eds): Diabetic Neuropathy. Philadelphia, WB Saunders, 1987, p 140, with permission.)

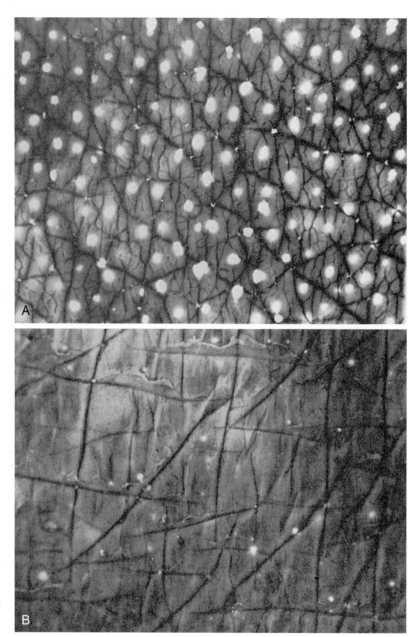

FIGURE 24–14. Silastic imprint sweating studies in (A), a normal control, and (B), a patient with diabetic autonomic neuropathy. Note the remarkable reduction in sweat droplet numbers in the diabetic neuropathy patient. (Reproduced from Kennedy WR, Sakuta M, Sutherland D, et al: Quantitation of the sweating deficiency in diabetes mellitus. Ann Neurol 15:482–488, 1984, with permission.)

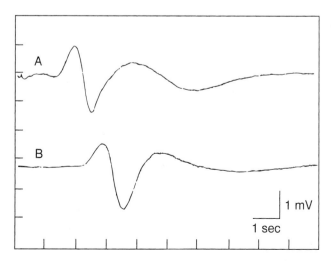

FIGURE 24–15. Sympathetic skin response tracings in a normal control subject. Upper trace recording from palm; lower trace recording from sole.

cal value of magnetically evoked SSRs remains to be precisely defined.

There is some controversy as to what actually constitutes an abnormal sympathetic skin response. Although some authors have used amplitude and latency to define a normal range,[19, 30, 51] most consider the response abnormal only when it is absent.[30] However, it is essential to consider the possibility that an absent SSR may be caused by an abnormality of sensory afferent function in patients with polyneuropathy in whom the noxious stimulus is delivered via a nerve conduction study.[75, 93, 102] Confirming an absent SSR utilizing other modes of stimulation is very important.[102] It is also important to recognize that an absent SSR may occur as a result of aging. The SSR is expected to be present at the palm and sole in all normal subjects below 60 years of age, but in those over 60 years of age, the SSR may be absent at the sole in 50% and at the palm in 27%.[16]

Because of its simplicity and the fact that the SSR can be performed on most electromyographic units, the SSR has been a popular method for assessment of sudomotor function. It has been applied to a wide range of peripheral and central disorders. The SSR appears to be abnormal in patients with axonal polyneuropathies[51, 104, 118] and tends to correlate with clinical features of dysautonomia.[105] However, SSRs do not appear to be helpful in the diagnosis of small-fiber neuropathies.[19] The SSR has been studied extensively in diabetic polyneuropathy and is reported to be absent in nearly 80% of patients.[81, 110] The SSR has been shown to correlate well with the results of QSART in patients with advanced diabetic polyneuropathy.[72] However, other reports have indicated a lack of correlation between the SSR and other tests of sudomotor function[102, 103] and symptoms of autonomic dysfunction.[70] Clearly, the SSR is a much less sensitive test of sudomotor function than QSART.

The SSR has been reported to be abnormal in a wide range of other polyneuropathies, including those related to chronic renal failure,[120] alcohol,[80] Guillain-Barré syndrome,[115] familial amyloid polyneuropathy,[78] and Charcot-Marie-Tooth disease (HSMN type I).[109]

The SSR has also been found to be abnormal in a variety of central nervous system disorders, including Parkinson's disease,[36] multisystem atrophy and progressive autonomic failure,[91] amyotrophic lateral sclerosis,[14, 86] nonfamilial olivopontocerebellar atrophy and striatonigral degeneration,[125] and multiple sclerosis.[18, 31]

The SSR has been reported to be reduced or absent in the majority of patients with reflex sympathetic dystrophy (RSD).[17, 95] However, the value of the SSR in the diagnosis of RSD has yet to be precisely established.[102] The SSR has also been used to study disorders of sweating, including idiopathic generalized anhidrosis[79] and palmar hyperhidrosis.[52]

Pupillometry

The assessment of pupillary function via infrared pupilometers has provided valuable information regarding the autonomic function of the pupil. Measurement of the dark-adapted pupil size conveniently expressed as the pupil diameter percentage (derived from dividing the horizontal pupil diameter by the horizontal iris diameter) can be used as an index of sympathetic innervation of the dilator pupillae.[108] This test has been applied to patients with diabetic autonomic neuropathy and has disclosed a significant percentage of patients with small pupil diameters compared to age-matched controls[108] (Fig. 24–16). Pupil diameter is influenced greatly by age and it is essential that age-stratified normal values are obtained for all laboratories performing this test. Abnormalities of pupil diameter tend to correlate with clinical signs of polyneuropathy as well as alterations in

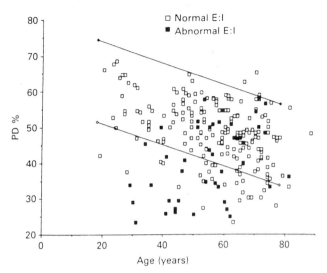

FIGURE 24–16. Pupil diameter percentages (PD%) in darkness in 243 diabetic patients with normal and abnormal expiratory to inspiratory ratios during heart rate response to deep breathing tests. The PD% was abnormally small in 21% of diabetic patients and abnormally large in 1% when compared to normal age-related control values defined by the lines. There was a significant correlation between small PD% and abnormal beat-to-beat heart rate variation measured by expiratory to inspiratory ratio. (Reproduced from Smith SA: Pupillary function in autonomic failure. In Bannister R (ed): Autonomic Failure, A Textbook of Clinical Disorders of the Autonomic Nervous System, 2nd ed. Oxford, Oxford University Press, 1988, pp 393–412, with permission.)

heart rate variability.[107] Other measures of pupillary autonomic function include the light reflex latency and velocity.[108] Pupillometry can quantitate pupil diameter at baseline and after a flash of light stimulating the pupillary light reflex. In patients with autonomic neuropathy, the light reflex has a delayed latency and time course with reduced maximum velocities of constriction and dilatation[108] (Fig. 24–17). In particular, the velocity of redilatation in the second half of the redilatation phase of the light reflex has proven to be an important index of sympathetic function.[108] The pupil cycle time has been used as an index of parasympathetic innervation of the pupil.[71, 116] This test is performed by focusing a narrow beam of light on the pupil margin using a slit lamp. Stimulation of the retina results in pupillary constriction, which in turn blocks further stimulation of the retina. The pupil then dilates to the point when the light again strikes the retina and the cycle repeats. The period of these oscillations, the pupil cycle time, is prolonged in diabetics with autonomic neuropathy.[71, 116] Pupillary function tests are very reproducible and compare favorably with cardiovascular reflex tests such as Valsalva ratio and HRdb.[108] Because of the need for special equipment and other technical factors relating to their utility, pupillometry has not been widely used in autonomic laboratories.

Pharmacological Testing

The measurement of various plasma catecholamines and their metabolites has been utilized as a means to better characterize and localize the site of a sympathetic disorder.[87] Plasma norepinephrine levels are derived in part from a spillover from sympathetic postganglionic nerve terminals, and the supine value of norepinephrine has been used as an index of sympathetic activity. In fact, plasma norepinephrine levels have been used to help distinguish between a pre- and postganglionic sympathetic disorder. In preganglionic disorders, the resting supine norepinephrine level is typically

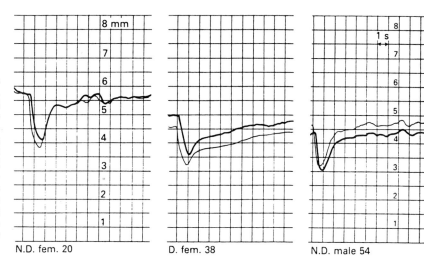

FIGURE 24–17. Pupilograph tracings from diabetic (D) and nondiabetic (ND) subjects. The bolderline indicates the right pupil and the responses from the two pupils are separated in the time axis (1 vertical bar = 1 second) for convenience. Measurements were made in darkness interrupted by a 1-second light flash. Note the redilatation lag and absent hippus in the diabetic subject. (Reproduced from Smith SA: Pupillary function in autonomic failure. In Bannister R (ed): Autonomic Failure, A Textbook of Clinical Disorders of the Autonomic Nervous System, 2nd ed. Oxford, Oxford University Press, 1988, pp 393–412, with permission.)

N.D. fem. 20   D. fem. 38   N.D. male 54

normal, and after standing, there is no increment due to failure of activation of the sympathetic ganglion. In contrast, in a postganglionic disorder, the supine levels are reduced, particularly if the lesion is widespread, and there is no increment with standing. In many laboratories, plasma norepinephrine levels are incorporated in the tilt-testing procedure. Additional methods have been described to increase the sensitivity of catecholamine analysis. This entails measuring urinary and plasma metabolites, including vanillylmandelic acid and plasma dihydroxyphenylglycol, as well as norepinephrine.[10, 32]

REFERENCES

1. Abrahams VC, Hilton SM, Malcolm JL: Sensory connections to the hypothalamus and midbrain and their role in the reflex activation of the defense reaction. J Physiol (Lond) 164:1–16, 1962.
2. Almquist A, Goldenberg IF, Milstein S, et al: Provocation of bradycardia and hypotension by isoproterenol and upright posture in patients with unexplained syncope. N Engl J Med 320:346–351, 1989.
3. Arunodaya GR, Taly AB: Sympathetic skin response: a decade later. J Neurol Sci 129:81–89, 1995.
4. Benarroch EE: The central autonomic network. In Low PA (ed): Clinical Autonomic Disorders, 2nd ed. Philadelphia, Lippincott-Raven, 1997, p 17.
5. Benarroch EE, Opfer-Gehrking TL, Low PA: Use of the photoplethysmographic technique to analyze the Valsalva maneuver in normal man. Muscle Nerve 14:1165–1172, 1991.
6. Benarroch EE, Chang FF: Central autonomic disorders. J Clin Neurophysiol 10:39–50, 1993.
7. Bennett T, Farquhar IK, Hosking DJ, et al: Assessment of methods for estimating autonomic nervous control of the heart in patients with diabetes mellitus. Diabetes 27:1167–1174, 1978.
8. Bird TD, Reenan AM, Pfeifer M: Autonomic nervous system function in genetic neuromuscular disorders. Hereditary motor-sensory neuropathy and myotonic dystrophy. Arch Neurol 41:43–46, 1984.
9. Borst C, Wieling W, Van Brederode JF, et al: Mechanisms of initial heart rate response to postural change. Am J Physiol 243:H676–H681, 1982.
10. Christensen NJ, Dejgaard A, Hilsted J: Plasma dihydroxyphenylglycol (DHPG) as an index of diabetic autonomic neuropathy. Clin Physiol 8:577–580, 1988.
11. Cohen J, Low P, Fcaley R, et al: Somatic and autonomic function in progressive autonomic failure and multiple system atrophy. Ann Neurol 22:692–699, 1987.
12. The Consensus Committee of the American Autonomic Society and the American Academy of Neurology: Consensus statement on the definition of orthostatic hypotension, pure autonomic failure and multiple system atrophy. Neurology 46:1470, 1996.
13. Deltombe T, Hanson P, Jamart J, et al: The influence of skin temperature on latency and amplitude of the sympathetic skin response in normal subjects. Muscle Nerve 21:34–39, 1998.
14. Dettmers C, Fatepour D, Faust H, et al: Sympathetic skin response abnormalities in amyotrophic lateral sclerosis. Muscle Nerve 16:930–934, 1993.
15. Donald KW, Lind AR, McNicol GW, et al: Cardiovascular responses to sustained (static) contractions. Cir Res 20:15, 1967.
16. Drory VE, Korczyn AD: Sympathetic skin response: Age effect. Neurology 43:1818–1820, 1993.
17. Drory VE, Korczyn AD: The sympathetic skin response in reflex sympathetic dystrophy. J Neurol Sci 128:92–95, 1995.
18. Elie B, Louboutin JP: Sympathetic skin response (SSR) is abnormal in multiple sclerosis. Muscle Nerve 18: 185–189, 1995.
19. Evans BA, Lusky D, Knezevic W: The peripheral autonomic surface potential in suspected small fiber peripheral neuropathy [abstract]. Muscle Nerve 11:982, 1988.
20. Ewing DJ, Irving JB, Kerr F, et al: Cardiovascular responses to sustained handgrip in normal subjects and in patients with diabetes mellitus: A test of autonomic function. Clin Sci Mol Med 46:295–306, 1974.
21. Ewing DJ, Campbell IW, Murray A, et al: Immediate heart rate response to standing: Simple test of autonomic neuropathy in diabetes. Br Med J 1:145–147, 1978.
22. Ewing DJ, Hume L, Campbell IW, et al: Autonomic mechanisms in the initial heart rate response to standing. J Appl Physiol 49:809–814, 1980.
23. Ewing DJ, Clarke BF: Diagnosis and management of diabetic autonomic neuropathy. Br Med J 285:916–918, 1982.
24. Ewing DJ, Martyn CN, Young RJ, et al: The value of cardiovascular autonomic function tests: 10 years experience in diabetes. Diabetes Care 8:491–498, 1985.
25. Fasano ML, Sand T, Brubakk AO, et al: Reproducibility of the cold pressor test: Studies in normal subjects. Clin Auton Res 6:249–253, 1996.
26. Fealey RD: Thermoregulatory sweat test. In Low PA (ed): Clinical Autonomic Disorders, 2nd ed. Philadelphia, Lippincott-Raven, 1997, p 245.
27. Fealey RD, Low PA, Thomas JE: Thermoregulatory sweating abnormalities in diabetes mellitus. Mayo Clin Proc 64:617–628, 1989.
28. Freeman R, Saul JP, Roberts MS, et al: Spectral analysis of heart rate in diabetic autonomic neuropathy. A comparison with standard tests of autonomic function. Arch Neurol 48:185–190, 1991.
29. Goldstraw PW, Warren DJ: The effect of age on the cardiovascular responses to isometric exercise: A test of autonomic function. Gerontology 31:54–58, 1985.
30. Gutrecht JA: Sympathetic skin response. J Clin Neurophysiol 11:519–524, 1994.
31. Gutrecht JA, Suarez GA, Denny BE: Sympathetic skin response in multiple sclerosis. J Neurol Sci 118:88–91, 1993.
32. Hoeldtke RD, Cilmi K, Reichard G, et al: Assessment of norepinephrine secretion and production. J Lab Clin Med 101:772–782, 1983.

33. Hoeldtke RD, Davis KM, Hshieh PB, et al: Autonomic surface potential analysis: Assessment of reproducibility and sensitivity. Muscle Nerve 15:926–931, 1992.

34. Imholz BP, van Montfrans GA, Settels JJ, et al: Continuous noninvasive blood pressure monitoring: Reliability of Finapres device during Valsalva maneuver. Cardiovasc Res 22:390–397, 1988.

35. Imholz BP, Settels JJ, van der Meiracker AH, et al: Noninvasive continuous finger blood pressure measurement during orthostatic stress compared to intra-arterial pressure. Cardiovas Res 24:214–221, 1990.

36. Johns DR, Gress DR, Shahani BT, et al: Electrophysiologic evaluation of autonomic function in Parkinson's disease. Muscle Nerve 9:657, 1986.

37. Johnson RH, Spalding JMK: Disorders of the Autonomic Nervous System. Oxford, Blackwell Scientific Publications, 1974.

38. Jordan D, Spyer KM: Brainstem integration of cardiovascular and pulmonary afferent activity. Prog Brain Res 67:295–314, 1986.

39. Kapoor WN, Brant N: Evaluation of syncope by upright tilt testing with isoproterenol. A nonspecific test. Ann Intern Med 116:358–363, 1992.

40. Kennedy WR, Sakuta M, Sutherland D, et al: Quantitation of the sweating deficiency in diabetes mellitus. Ann Neurol 15:482–488, 1984.

41. Kenny RA, Ingram A, Bayliss J, et al: Head-up tilt: a useful test for investigating unexplained syncope. Lancet 1:1352–1355, 1986.

42. Kennedy WR, Navarro X: Sympathetic sudomotor function in diabetic neuropathy. Arch Neurol 46:1182–1186, 1989.

43. Kennedy WR, Navarro X: Evaluation of sudomotor function by sweat imprint methods. In Low PA (ed): Clinical Autonomic Disorders. Boston, Little, Brown, 1993, p 253.

44. Khurana RK, Watabiki S, Hebel JR, et al: Cold face test in the assessment of trigeminal-brainstem-vagal function in humans. Ann Neurol 7:144–149, 1980.

45. Khurana RK, Nicholas EM: Head-up-tilt table test: How far and how long? Clin Auton Res 6:335–341, 1996.

46. Khurana RK, Setty A: The value of the isometric hand-grip test—studies in various autonomic disorders. Clin Auton Res 6:211–218, 1996.

47. Kihara M, Opfer-Gehrking TL, Low PA: Comparison of directly stimulated with axon reflex-mediated sudomotor responses in human subjects and in patients with diabetes. Muscle Nerve 16:655–660, 1993.

48. Kirchheim HR: Systemic arterial baroreceptor reflexes. Physiol Rev 56:100–176, 1976.

49. Knezevic W, Bajada S: Peripheral autonomic surface potential. A quantitative technique for recording sympathetic conduction in man. J Neurol Sci 67:239–251, 1985.

50. Levin AB: A simple test of cardiac function based upon the heart rate changes induced by the Valsalva maneuver. Am J Cardiol 18:90–99, 1966.

51. Levy DM, Reid G, Rowley DA, et al: Quantitative measures of sympathetic skin response in diabetes: Relation to sudomotor and neurological function. J Neurol Neurosurg Psychiatry 55:902–908, 1992.

52. Lin TK, Chee EC, Chen HJ, et al: Abnormal sympathetic skin response in patients with palmar hyperhidrosis. Muscle Nerve 18:917–919, 1995.

53. Linden D, Diehl RR: Comparison of standard autonomic tests and power spectral analysis in normal adults. Muscle Nerve 19:556–562, 1996.

54. Linden D, Weng Y, Glocker FX, et al: Sympathetic skin responses evoked by magnetic stimulation of the neck: Normative data. Muscle Nerve 19:1487–1489, 1996.

55. Lishner M, Akselrod S, Avi VM, et al: Spectral analysis of heart rate fluctuations. A noninvasive sensitive method for the early diagnosis of autonomic neuropathy in diabetes mellitus. J Auton Nerv Syst 19:119–125, 1987.

56. Loewy AD: Anatomy of the autonomic nervous system: An overview. In Loewy AD, Spyer KM (eds): Central Regulation of Autonomic Functions. New York, Oxford University Press, 1990, pp 4–16.

57. Loewy AD: Central autonomic pathways. In Loewy AD, Spyer KM (eds): Central Regulation of Autonomic Functions. New York, Oxford University Press, 1990, pp 88–103.

58. Low PA: Autonomic nervous system function. J Clin Neurophysiol 10:14–27. 1993.

59. Low PA: Laboratory evaluation of autonomic function. In Low PA (ed.): Clinical Autonomic Disorders, 2nd ed. Philadelphia, Lippincott-Raven, 1997, p 179.

60. Low PA, Neumann C, Dyck PJ, et al: Evaluation of skin vasomotor reflexes by using laser Doppler velocimetry. Mayo Clin Proc 58:583–592, 1983.

61. Low PA, Caskey PE, Tuck RR, et al: Quantitative sudomotor axon reflex test in normal and neuropathic subjects. Ann Neurol 14:573–580, 1983.

62. Low PA, Fealey RD, Sheps SG, et al: Chronic idiopathic anhidrosis. Ann Neurol 18:344–348, 1985.

63. Low PA, Zimmerman BR, Dyck PJ: Comparison of distal sympathetic with vagal function in diabetic neuropathy. Muscle Nerve 9:592–596, 1986.

64. Low PA, Fealey RD: Sudomotor neuropathy. In Dyck PJ, Thomas PK, Asbury AK, et al (eds): Diabetic Neuropathy. Philadelphia, WB Saunders, 1987, p 140.

65. Low PA, Opfer-Gehrking TL, Proper CJ, et al: The effect of aging on cardiac autonomic and postganglionic sudomotor function. Muscle Nerve 13:152–157, 1990.

66. Low PA, Opfer-Gehrking TL, Kihara M: In vivo studies on receptor pharmacology of the human eccrine sweat gland. Clin Auton Res 2:29–34, 1992.

67. Low PA, Denq JC, Opfer-Gehrking TL, et al: Effect of age and gender on sudomotor and cardiovagal function and blood pressure response to tilt in normal subjects. Muscle Nerve 20:1561–1568, 1997.

68. Low PA, Pfeifer MA: Standardization of autonomic function. In Low PA (ed): Clinical Autonomic Disorders, 2nd ed. Philadelphia, Lippincott-Raven, 1997, p 287.

69. Ludbrook J, Vincent A, Walsh JA: Effects of mental arithmetic on arterial pressure and hand blood flow. Clin Exp Pharmacol Physiol 2(suppl):67–70, 1975.

70. Martyn CN, Reid W: Sympathetic skin response [letter]. J Neurol Neurosurg Psychiatry 48:490, 1985.

71. Martyn CN, Ewing DJ: Pupil cycle time: A simple way of measuring an autonomic reflex. J Neurol Neurosurg Psychiatry 49:771–774, 1986.

72. Maselli RA, Jaspan JB, Soliven BC, et al: Comparison of sympathetic skin response with quantitative sudomotor axon reflex test in diabetic neuropathy. Muscle Nerve 12:420–423, 1989.

73. Mathias CJ: Autonomic disorders and their recognition. N Eng J Med 336:721–724, 1997.

74. Mathias CJ, Bannister R: Investigation of autonomic disorders. In Bannister R, Mathias CJ (eds): Autonomic Failure: A Textbook of Clinical Disorders of the Autonomic Nervous System, 3rd ed. Oxford, Oxford University Press, 1992, p 225–290.

75. Matsunaga K, Uozumi T, Tsuji S, et al: Sympathetic skin responses evoked by magnetic stimulation of the neck. J Neurol Sci 128:188–194, 1995.

76. McLeod JG: Autonomic dysfunction in peripheral nerve disease. J Clin Neurophysiol 10:51–60, 1993.

77. Mifflin SW, Spyer KM, Withington-Wray DJ: Baroreceptor inputs to the nucleus tractus solitarius in the cat: Modulation by the hypothalamus. J Physiol (Lond) 399:369–387, 1988.

78. Montagna P, Salvi F, Liquori R: Sympathetic skin response in familial amyloid polyneuropathy [letter]. Muscle Nerve 11:183–184, 1988.

79. Murakami K, Sobue G, Iwase S, et al: Skin sympathetic nerve activity in acquired idiopathic generalized anhidrosis. Neurology 43:1137–1140, 1993.

80. Navarro X, Miralles R, Espadaler JM, et al: Comparison of sympathetic skin sudomotor and skin responses in alcoholic neuropathy. Muscle Nerve 16:404–407, 1993.

81. Niakan E, Harati Y: Sympathetic skin response in diabetic peripheral neuropathy. Muscle Nerve 11:261–264, 1988.

82. Nishimura RA, Tajik AJ: The Valsalva maneuver and response revisited. Mayo Clin Proc 61:211–217. 1986.

83. Pagani M, Malfatto G, Pierini S, et al: Spectral analysis of heart rate variability in the assessment of autonomic diabetic neuropathy. J Autonom Nerv Syst 23:143–153, 1988.

84. Pfeifer MA, Cook D, Brodsky J, et al: Quantitative evaluation of cardiac parasympathetic activity in normal and diabetic man. Diabetes 31:339–345, 1982.

85. Pick J: The Autonomic Nervous System. Morphological Comparative, Clinical and Surgical Aspects. Philadelphia, Lippincott, 1970, pp 3–21.
86. Pisano F, Miscio G, Mazzuero G, et al: Decreased heart rate variability in amyotrophic lateral sclerosis. Muscle Nerve 18:1225–1231, 1995.
87. Polinsky RF, Kopin IJ, Ebert WH, et al: Pharmacologic dissection of different orthostatic lypotension syndromes. Neurology 31:1–7, 1981.
88. Pomeranz B, Macaulay RJ, Caudill MA, et al: Assessment of autonomic function in humans by heart rate spectral analysis. Am J Physiol 248:H151–H153, 1985.
89. Ravits JM: Autonomic nervous system testing. Muscle Nerve 20:919–937, 1997.
90. Ravits JM, Baker M, Hallet M, et al: A comparative study of electrophysiologic tests of autonomic function in patients with primary autonomic failure. Muscle Nerve 9:657, 1986.
91. Ravits J, Hallett M, Nilsson J, et al: Electrophysiological tests of autonomic function in patients with idiopathic autonomic failure syndromes. Muscle Nerve 19:758–763, 1996.
92. Reis DJ, Granata AR, Joh TH, et al: Brain stem catecholamine mechanisms in tonic and reflex control of blood pressure. Hypertension 6 (suppl II): 7–15, 1984.
93. Reis DJ, Morrison S, Ruggiero DA: The C1 area of the brain stem in tonic and reflex control of blood pressure. Hypertension 11(suppl I): 8–13, 1988.
94. Robertson D, Hollister AS, Biaggioni I, et al: The diagnosis and treatment of baroreflex failure. N Engl J Med 329:1449–1455, 1993.
95. Rommel O, Tegenthoff M, Pern U, et al: Sympathetic skin response in patients with reflex sympathetic dystrophy. Clin Auton Res 5:205–210, 1995.
96. Rossini PM, Massa R, Sancesario G, et al: Sudomotor skin responses to brain stimulation do not depend on nerve sensory fiber functionality. Electroencephalogr Clin Neurophysiol 89:447–451, 1993.
97. Rossini PM, Opsomer RJ, Boccasena P: Sudomotor skin responses following nerve and brain stimulation. Electroencephalogr Clin Neurophysiol 89:442–446, 1993.
98. Sandroni P, Ahlskog JE, Fealey RD, et al: Autonomic involvement in extrapyramidal and cerebellar disorders. Clin Auton Res 1:147–155, 1991.
99. Sandroni P, Benarroch EE, Low PA: Pharmacologic dissection of components of the Valsalva maneuver in adrenergic failure. J Appl Physiol 71:1563–1567, 1991.
100. Sayinalp S, Sözen T, Özdogan M: Cold pressor test in diabetic autonomic neuropathy. Diabetes Res Clin Prac 26:21–28, 1994.
101. Schondorf R: New investigations of autonomic nervous system function. J Clin Neurophysiol 10:28–38, 1993.
102. Schondorf R: Skin potentials: normal and abnormal. In Low PA (ed.): Clinical Autonomic Disorders, 2nd ed. Philadelphia, Lippincott-Raven, 1997, p 221.
103. Schondorf R, Gendron D: Evaluation of sudomotor function in patients with peripheral neuropathy. Neurology 40:386, 1990.
104. Shahani BT, Halperin JJ, Boulu P, et al: Sympathetic skin response—a method of assessing unmyelinated axon dysfunction in peripheral neuropathies. J Neurol Neurosurg Psychiatry 47:536–542, 1984.
105. Shahani BT, Day TJ, Cros D, et al: RR interval variation and the sympathetic skin response in the assessment of autonomic function in peripheral neuropathy. Arch Neurol 47:659–664, 1990.
106. Shields RW: Functional anatomy of the autonomic nervous system. J Clin Neurophysiol 10:2–13, 1993.
107. Smith SA: Reduced sinus arrhythmia in diabetic autonomic neuropathy: Diagnostic value of an age-related normal range. Br Med J 285:1599–1601, 1982.
108. Smith SA: Pupillary function: Tests and disorders. In Bannister R, Mathias CJ (eds): Autonomic Failure. A Textbook of Clinical Disorders of the Autonomic Nervous system, 3rd ed. Oxford, Oxford University Press, 1992, pp 421–441.
109. Solders G, Andersson T, Persson A: Central conduction and autonomic nervous function in HMSN I. Muscle Nerve 14:1074–1079, 1991.
110. Soliven B, Maselli R, Jaspan J, et al: Sympathetic skin response in diabetic neuropathy. Muscle Nerve 10:711–716, 1987.
111. Sprangers RLH, Wesseling KH, Iruholz ALT, et al: Initial blood pressure fall on stand up and exercise explained by changes in total peripheral resistance. J Appl Physiol 70:523–530, 1991.
112. Stewart JD, Low PA, Fealey RD: Distal small fiber neuropathy; results of tests of sweating and autonomic cardiovascular reflexes. Muscle Nerve 15:661–665, 1992.
113. Suarez GA, Fealey RD, Camilleri M, et al: Idiopathic autonomic neuropathy: Clinical, neurophysiologic, and follow-up studies on 27 patients. Neurology 44:1675–1682, 1994.
114. Sundkvist G, Almer LO, Lilja B: Respiratory influence on heart rate in diabetes mellitus. Br Med J 1:924–925, 1979.
115. Taly AB, Arunodaya GR, Rao S: Sympathetic skin response in Guillain-Barré syndrome. Clin Auton Res 5:215–219, 1995.
116. Thompson HS: The pupil cycle time. J Clin Neuro-ophthalmol 7:38–39, 1987.
117. Uncini A, Pullman SL, Lovelace RE, et al: The sympathetic skin response: Normal values, elucidation of afferent components and application limits. J Neurol Sci 87:299–306, 1988.
118. Van den Bergh P, Kelly JJ: The evoked electrodermal response in peripheral neuropathies. Muscle Nerve 9:656–657, 1986.
119. van Lieshout JJ, Wieling W, Wesseling KH, et al: Pitfalls in the assessment of cardiovascular reflexes in patients with sympathetic failure but intact vagal control. Clin Sci 76:523–528, 1989.
120. Wang SJ, Liao KK, Liou HH, et al: Sympathetic skin response and R-R interval variation in chronic uremic patients. Muscle Nerve 17:411–418, 1994.
121. Wheeler T, Watkins PJ: Cardiac denervation in diabetes. Br Med J 4:584–586, 1973.
122. Wieling W: Standing orthostatic stress and autonomic function. In Bannister R, Mathias CJ (eds): Autonomic Failure. A Textbook of Clinical Disorders of the Autonomic Nervous System, 3rd ed. Oxford, Oxford University Press, 1992, pp 291–311.
123. Williams PL, Warwick R: Gray's Anatomy. Philadelphia, WB Saunders, 1980, pp 1121–1138.
124. Wood JD: Physiology of the enteric nervous system. In Johnson LR (ed): Physiology of the Gastrointestinal Tract, vol 1, 2nd ed. New York, Raven Press, 1987, pp 67–109.
125. Yokota T, Hayashi M, Tanabe H, et al: Sympathetic skin response in patients with cerebellar degeneration. Arch Neurol 50:422–427, 1993.

Chapter 25

Emerging Applications in Neuromagnetic Stimulation

Paul J. Maccabee, MD • Vahe E. Amassian, MB • Ulf Ziemann, MD
Eric Wassermann, MD • Vedran Deletis, MD, PhD

INTRODUCTION
BIOPHYSICAL BACKGROUND
PHYSIOLOGY OF MOTOR CORTEX
   EXCITATION
Electrical Shock
Magnetic Stimulation
MOTOR RESPONSES TO CORTICAL AND
   PERIPHERAL STIMULATION
Cortical Response and Background
   Facilitation
Cortical Response Variability
Factors Affecting the Peripheral
   Response
Interpretation of CMCT
CORTICAL THRESHOLD

CORTICAL SILENT PERIOD
Physiology
Procedural Issues
Clinical Correlation
Drug Studies
PAIRED-PULSE STIMULATION
TEMPORAL I WAVE FACILITATION
SPATIAL FACILITATION
PLASTICITY
Lesion Studies (Pathological, Iatrogenic)
Motor Training
Cross-modal Plasticity in the Blind
SAFETY CONCERNS: sTMS VERSUS rTMS
Seizure Induction in Normals
Physiological, Hormonal, and
   Immunological Effects
Direct Exclusions
THERAPEUTIC REPETITIVE STIMULATION
COMMENT

INTRODUCTION

Following the introduction of single pulse magnetic stimulators by Barker and colleagues in the early 1980s, there has been a significant outpouring of clinical and fundamental research to evaluate the peripheral and central nervous system in health and disease.[17, 135] These applications range from the detection of physiological responses using single pulse or paired-pulse stimuli to the modification and treatment of neurological and psychiatric disease using repetitive stimuli. This chapter summarizes some recent advances now being implemented in the clinical research laboratory.

BIOPHYSICAL BACKGROUND

A magnetic coil (MC) is typically planar (i.e., flat) with either a round or figure-of-eight design. A very brief current in the coil windings (approximately 100 μsec or less) generates a changing magnetic field, which induces a brief electrical field into an adjacent wire circuit or volume conductor (Faraday's law). If the conductivity within a volume conductor is sufficient, then an electrical current flows that, if large enough, depolarizes and discharges neural membrane. The high-resistance skull and vertebral bone are transparent to the magnetic field and the induced current flows internally, minimizing pain. By contrast, pain afferents in the skin and deeper tissue are more readily excited when using conventional electrical stimulation because the current must be driven from the surface to the interior. An exception is neuromagnetic stimulation over the distal wrist which can, at 100% output, be quite painful, as is anodal (scalp) or cathodal (spine) electrical stimulation.

Possible mechanisms involved in neuromagnetic stimulation have been elucidated chiefly using theoretical techniques (i.e., computer simulation based on Maxwell's equations), and by studies of mammalian peripheral nerve in vitro. Briefly, the following principles are proposed to account for many observations made when stimulating the human:

1. The maximal induced electrical field is oriented in a plane that is always predominantly parallel to the inner surface of the volume conductor nearest to or in contact with the MC.[48, 181, 205] The design or orientation of the MC does *not* influence the predominant plane of the induced current, and radial currents are either small, or, in a perfect sphere, nonexistent.[48]
2. The induced electrical field is maximal at the inner surface of the volume conductor in contact with the coil and falls off significantly with increasing distance.
3. Three-dimensional focusing at a desired depth ("targeting") is not possible within the volume conductor.[89]
4. The greater the size of the volume conductor, the greater the magnitude of the induced current.[228] This influences studies of animals with small cranial vaults as compared to humans.
5. The induced current generally follows the trajectory of the coil windings, but the direction of current flow is opposite to that in the windings (e.g., a round MC held flat to the volume conductor surface as in Fig. 25–1, *top*). By contrast, the magnetic field is maximal at the center of the MC. Edge (orthogonal) stimulation (by holding the MC perpendicular to the surface of the volume conductor) induces two current loops, which are joined under the contacting edge of the MC (Fig. 25–1, *middle*); this orientation is not clinically

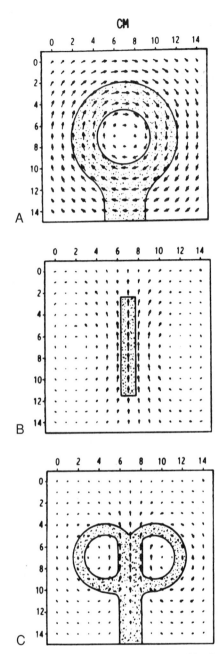

FIGURE 25–1. Current arrow diagrams computed in the plane tangential to the bottom surface of a cylindrical tank filled with isotonic saline. Electric fields induced by: a tangentially orientated round magnetic coil (MC) (*top*); an orthogonally orientated, round MC (*middle*); and a tangentially orientated figure-of-eight MC (*bottom*). Distance from inner surface of tank to recording electrodes was 0.5 cm. MC stimulator output was 20% of maximum. Magnitude of electrical field is proportional to arrow length. In all three plots, maximal values are normalized. (Reprinted from Maccabee PJ, Amassian EE, Eberle LP, et al: Measurement of the electric field induced into inhomogeneous volume conductors by magnetic coils. Application to human spinal neurogeometry. Electroencephalogr Clin Neurophysiol 81: 224–237, 1991, with permission.)

useful as insufficient flux is conveyed to the volume conductor compared to a flat orientation. The figure-of-eight coil (two adjacent coils) held flat to the volume conductor induces two current loops with induced current superimposed under the junction of coils[213] (see Fig. 25–1); this provides a precise direc-

tional probe used to map the motor and sensory homunculus.

6. In the artificial paradigm of a straight nerve within a homogeneous media volume conductor, excitation occurs at the negative-going first spatial derivative of the induced electrical field parallel to the nerve[68, 123, 147, 171, 180] (Fig. 25–2). Unexpectedly, significant excitation was reported when the junction of a figure-of-eight MC was oriented perpendicular to a straight nerve.[187] It was proposed that excitation was achieved by the significantly lesser electrical field component initially perpendicular to the MC junction and parallel to the nerve fibers.[108]

7. As first shown by Maccabee and colleagues,[125] low threshold excitation of nerve roots where they exit the cervical, thoracic, and lumbar spine occurs by channeling of the induced current at their respective neuroforamina, which are oriented medial-lateral (Fig. 25–3). This is best accomplished by directing the induced current (generated by a round or figure-of-eight MC) perpendicular or at a small angle to the long axis of the vertebral spine.

8. Low threshold excitation also takes place at a bend in a nerve (and also at a cut nerve ending), if the induced current is directed along the nerve towards the bend (or nerve end)[123] (Fig. 25–4). Probable examples of bend excitation in the human nervous system include the proximal cauda equina (lumbosacral roots), distal cauda equina (sacral roots),[125] motor cortex (see later discussion), and visual cortex.[9]

9. Excitation of an axon at a low threshold point optimally occurs near the peak of the induced electrical field and not at the spatial derivative site.[123] Moreover, when adjusting MC output and moving the junction toward or away from a low threshold point, responses are elicited at nearly identical latency, implying a common site of excitation (Fig. 25–5). Thus, there may be no latency shift of elicited compound muscle action potentials (CMAPs) or motor units when moving the MC along the spinal axis.[28, 40, 70, 122, 214]

10. A monophasic pulse confers distinct advantages and disadvantages compared to a polyphasic pulse. For example, different elements in human motor cortex (to upper limb) are optimally excited by a posterior-to-anterior (P-A) induced current[92, 184] compared with lateral-to-medial (L-M) induced current.[7, 229] The root exit zone of the human conus medullaris requires a caudal-to-rostral induced current[125] (Fig. 25–6). By contrast, the induced polyphasic pulse is not as direction sensitive because either the first or the opposite polarity second phase may be effective.[46] In addition, the polyphasic pulse yields a larger amplitude response when the first phase (quarter cycle I) is in the hyperpolarizing direction and the second phase (quarter cycles II and III) is in the depolarizing direction (see Fig. 25–6). This reflects the greater duration of the depolarizing half cycle compared to other waveforms in which the depolarizing phase is one quarter cycle.[126] Moreover, in a healthy human at normal temperature, both directions of a polyphasic pulse give larger amplitude responses compared to a monophasic pulse. This is seen over distal median nerve in the forearm, at spinal neuroforamina, and also over motor cortex, and implies two separate but closely adjacent excitation sites. Finally, polyphasic pulses may yield the "pull-down" phenomenon, in which the depolarization elicited by the first phase is attenuated ("pulled down") by the second hyperpolarizing phase.[126] Pull-down is readily seen in vitro at low temperature and

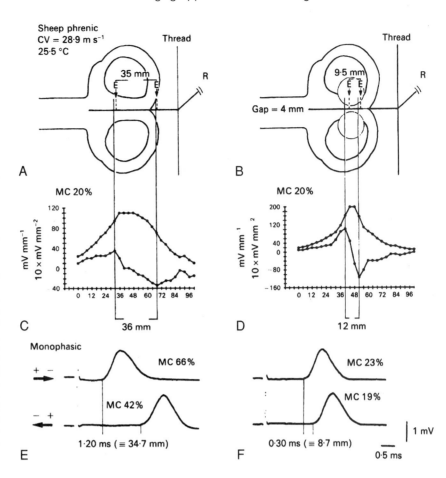

FIGURE 25–2. MC stimulation of sheep phrenic nerve immersed in homogeneous and inhomogeneous media volume conductors. *A.* Accurate dimensional relationship of the figure-of-eight MC to the nerve trajectory within the trough. E indicates sites of excitation by monopolar MC pulses in both directions, obtained by matching latencies of MC-induced responses with those elicited by direct electrical stimulation. *B.* Accurate dimensional relationship of nerve trajectory in trough and lucite cylinders astride the nerve. *C* and *D.* The measured electrical field (mV mm^{-1}) and its first spatial derivative (mV mm^{-2}; note ten times increase in amplification), corresponding to experimental setups illustrated at left and right, respectively. *E* and *F.* Nerve responses elicited by monophasic current pulses. The electrical fields were measured and the nerve responses were elicited in the same Ringer solution, at ambient room temperature. MC immediately beneath trough. (Reprinted from Maccabee PJ, Amassian VE, Eberle LP, Cracco RO: Magnetic coil stimulation of straight and bent amphibian and mammalian peripheral nerve in vitro: Locus of excitation. J Physiol 460:201–219, 1993, with permission.)

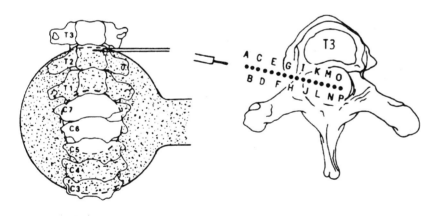

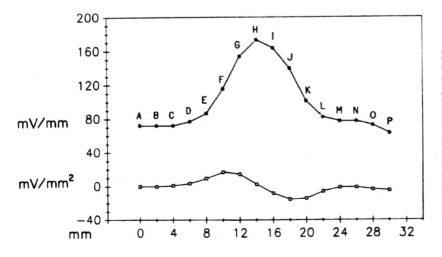

FIGURE 25–3. The transneuroforaminal electric field recorded proximal, within, and distal to the intervertebral foramen between T2–T3. A 4-mm linear coaxial cable probe was used. Each data point designated by an alphabetical letter, obtained at 2-mm intervals, corresponds to the midpoint of the probe. The lower trace indicates the first spatial derivative (mV/mm^2) of the electrical field. The MC was orientated symmetrical-tangentially. (Reprinted from Maccabee PJ, Amassian VE, Eberle LP, et al: Measurement of the electric field induced into inhomogeneous volume conductors by magnetic coils. Application to human spinal neurogeometry. Electroencephalogr Clin Neurophysiol 81:224–237, 1991, with permission.)

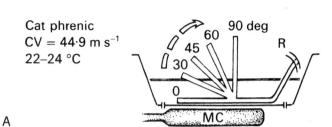

A

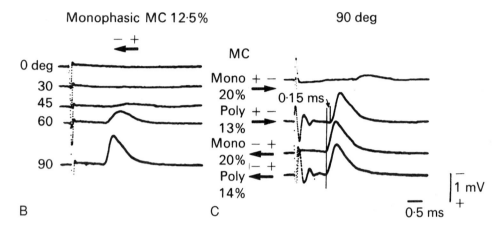

B

C

FIGURE 25–4. MC stimulation of straight and bent nerve (cat phrenic). Experimental set-up as in Figure 25–7. *A.* Various bend angles used. *B.* Nerve responses are elicited by monophasic current, which is outward at the bend. *C.* Nerve responses are elicited by monophasic and polyphasic current, both polarities, with a 90-degree bend in the nerve. (Reprinted from Maccabee PJ, Amassian VE, Eberle LP, Cracco RQ: Magnetic coil stimulation of straight and bent amphibian and mammalian peripheral nerve in vitro: Locus of excitation. J Physiol 460:201–219, 1993, with permission.)

FIGURE 25–5. Effect on response latency of MC stimulation at low threshold sites created by straddling the nerve (cat phrenic) with nonconducting solid lucite cylinders (*A*) and by a 90-degree bend (*B*). With the nerve in place, the MC junction is located approximately beneath the cylinders or the bend, the divergence away from the handle being referred to 0. Maximal amplitude responses and those at 50% (*left*) and 60% (*right*) of maximum were obtained at position 0. The MC was then moved 3 cm to either side and the output intensity adjusted to obtain a response amplitude equal to the submaximal response obtained at position 0. (Reprinted from Maccabee PJ, Amassian VE, Eberle LP, Cracco RQ: Magnetic coil stimulation of straight and bent amphibian and mammalian peripheral nerve in vitro: Locus of excitation. J Physiol 460:201–219, 1993, with permission.)

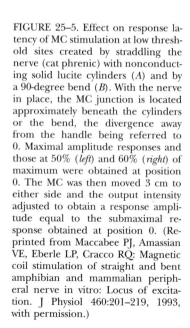

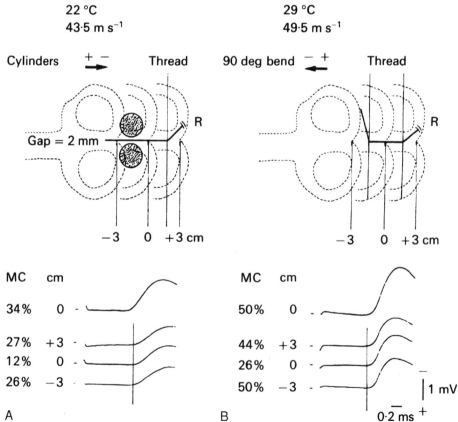

A

B

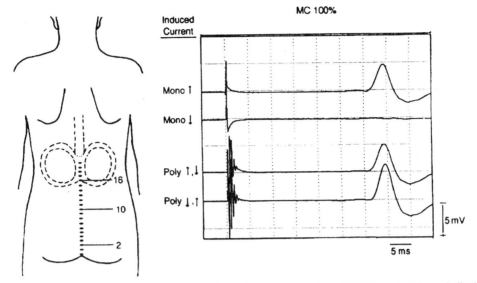

Subject 2, Normal male
Stim: Proximal Cauda Equina
Distal Divergence position 16 cm
Rec: Lt. FHB

FIGURE 25–6. Proximal cauda equina compound muscle action potentials (CMAPs) in left flexor hallucis brevis elicited by the midsize twin MC held vertically at the optimal level. *Top trace*: monophasic current is induced in a caudal-to-cranial direction. *Second trace from top*: reversal of monophasic current in this typical, thin, normal subject elicited no response, despite the same 100% MC output. *Third trace from top*: a slightly larger amplitude CMAP is elicited by a polyphasic pulse (phase 1 directed cranially, that is, depolarizing at proximal cauda equina; phase 2 directed caudally, that is, hyperpolarizing at proximal cauda equina). *Bottom trace*: a substantially larger amplitude response is elicited by the reverse polyphasic pulse yielding a hyperpolarizing-depolarizing sequence at proximal cauda equina. (Reprinted from Maccabee PJ, Lipitz ME, Desudchit T, et al: A new method using neuromagnetic stimulation to measure conduction time within the cauda equina. Electroencephalogr Clin Neurophysiol 101:153–166, 1996, with permission.)

is thought to indicate delay in action potential generation, permitting the second hyperpolarizing phase to reduce excitation.

PHYSIOLOGY OF MOTOR CORTEX EXCITATION

Electrical Shock

The elements defining the response of the motor cortex to a single *electrical* shock consist of direct (D) and indirect (I) waves.[2, 90, 106, 159] When recording from the ipsilateral medullary pyramid or crossed lateral corticospinal tract in the monkey, multiple descending waves are recorded. The first of these, the D wave, occurs too soon to have been transmitted through an intervening synapse. The lowest threshold D discharge is elicited by a focal anodic compared to a cathodic stimulus. Anodic current applied to the cortical surface was thought to enter the dendrites of pyramidal neurons in the outer cortical layers and more vertically toward the inner layers to emerge at the initial segment;[113] more likely the anodal current enters nearer the cell body region where core resistance is lower and exits a few nodes below the initial segment.[6] This notion is supported by the trivial variability in D wave amplitude to successive anodal stimuli, implying an origin in cortical white matter immediately beneath the motor cortex and isolated from fluctuations in membrane potential near the cell body. By contrast, D waves elicited by cathodal stimuli show great amplitude variability consistent with an origin nearer the initial segment. Following the D wave, there may be several I waves with a period a

little greater than 1 msec. The multiple I waves originate *within* cortex and are thought to reflect transsynaptic activation of pyramidal tract and corticospinal tract neurons in a vertical chain of cortical interneurons. This is suggested by experiments in the monkey in which stimulation through a microelectrode in superficial cortex selectively elicits later I waves; these are selectively depressed by cooling the pial surface. Asphyxia or cortical ablation that suppresses I waves leaves D waves relatively unaffected.[2] A D wave followed by multiple I waves are necessary to raise the *resting* membrane potential of the anterior horn cell to threshold, and increasing stimulus intensity activates more motor columns with an increase in motor-evoked response amplitude.[59, 161] (In the human, background voluntary activity most likely permits a D wave to elicit a motor response.)

Descending waves to a transient electrical shock are also easily recorded from the dorsal epidural space of patients undergoing neurosurgical procedures. With upper thoracic cord recording, the anode-cathode pair placed coronally preferentially elicits a D wave (Fig. 25–7, *top*) compared with P-A anode-cathode orientation, which also easily elicits I waves (see Fig. 25–7, *bottom*). The recruitment of additional I waves and a shorter D wave occurs with typically increasing stimulus intensity[18, 21, 30] (Fig. 25–8).

Magnetic Stimulation

Initially, electrical stimuli and vertex-tangential magnetic stimuli were shown to act differently on human motor cortex, with magnetic stimuli resulting only in transsynaptic excitation of corticospinal neurons.[58, 59] However, Amassian and colleagues[3] demonstrated that a lateral-sagittally ori-

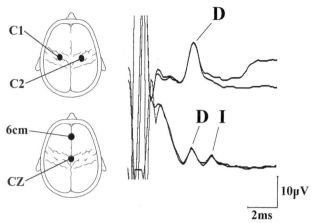

FIGURE 25–7. Upper thoracic epidural recordings of D and I waves in a 14-year-old female during surgery for low cervical intramedullary tumor. The upper trace was obtained after transcranial electrical stimulation over C1 (anode) and C2 (cathode) using 140-mA stimulus intensity and stimulus duration of 0.5 msec. The lower trace was obtained after anodic stimulation at Cz and cathodal stimulation 6-cm anteriorly, using the same stimulus duration but at 200 mA. Note appearance of the D and I wave with this electrode arrangement. Negative up.

ented round MC could directly excite human corticospinal neurons by eliciting similar latency responses to that elicited by focal anodic stimuli. In the monkey, the lateral-sagittal round MC preferentially excited corticospinal fibers directly to elicit D waves, whereas a round MC held flat, centered at the vertex (vertex-tangential orientation) preferentially elicited I waves, probably by excitation of corticocortical afferents.[6]

Modeling peripheral nerve in a skull-shaped volume further implied direct nerve excitation with L-M induced current flow to upper limbs and with medial-to-lateral (M-L) flow to lower limbs.[7] Figure 25–9 illustrates the optimal current direction in the coronal plane for arm and leg responses. Werhahn and associates[229] established that D wave latency responses in a hand muscle occurs *only* with L-M induced current flow. These hypotheses are strongly sup-

ported by recordings from the cervical epidural space in conscious humans.[66, 104, 144] Typically, when targeting upper limbs, induced L-M current elicits D waves followed by I waves, whereas P-A induced current elicits I waves alone or a small D wave followed by larger I waves at higher stimulation intensity. When targeting lower limbs, I waves were apparently elicited at lowest threshold by all directions of induced current.[145] However, it is likely that M-L induced current is optimal for corticospinal lower limb excitation and, therefore, for obtaining the earliest muscle response during voluntary contraction.

In the clinical laboratory, lower threshold responses are traditionally activated with a round MC held in the vertex-tangential orientation. This presumably elicits P-A current flow beneath the windings over the hand area regions of motor cortex to elicit I waves. When using the figure-of-eight MC, the *lowest* threshold responses to upper limbs are elicited with the junction rotated approximately 45 degrees from the midline, thereby inducing P-A current perpendicular to the central sulcus.[22, 142, 189] Nevertheless, for analytical purposes it may be better to orient the MC junction in a strictly P-A direction so as to avoid, as much as possible, evoking D waves.

MOTOR RESPONSES TO CORTICAL AND PERIPHERAL STIMULATION

Cortical Response and Background Facilitation

The methodology of motor cortical stimulation for central motor conduction time (CMCT) measurement has been extensively reviewed elsewhere.[179, 185] Perhaps the most important variable is the influence of preexisting voluntary contraction, which could occur at the motor neuron, or at the motor cortex level, or at both. Approximately 5% to 10% of maximal voluntary contraction is considered sufficient by some and will greatly reduce threshold, increase the amplitude, and also, significantly, shorten response latency.[59, 92, 131, 135, 140, 183] Cortical latencies may be shortened by as much as 4 msec in active as compared to relaxed muscles. It is likely that this latency shift in part reflects the ambient threshold

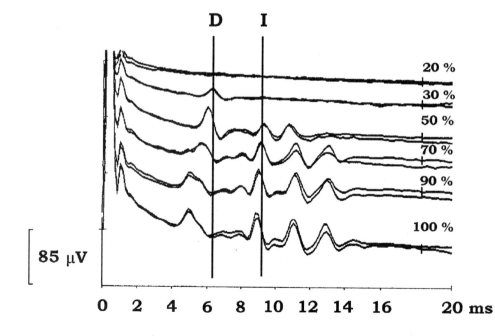

FIGURE 25–8. Epidural recordings or D and I waves in a 14-year-old male during surgery for idiopathic scoliosis. Percutaneous electrical stimulation was over Cz (anode) and Fz (cathode); recordings were by catheter electrode placed in the epidural space over middle thoracic spinal cord (100% corresponds to 750 V). Note that the D wave latency decreases with the increase in stimulation intensity. Negative up. (Reprinted from Deletis V: Intraoperative monitoring of the functional integrity of the motor pathways. In Devinsky O, Beric A, Dogali M (eds): Advances in Neurology: Electrical and Magnetic Stimulation of the Brain. New York, Raven Press, 1993, pp 201–214, with permission.)

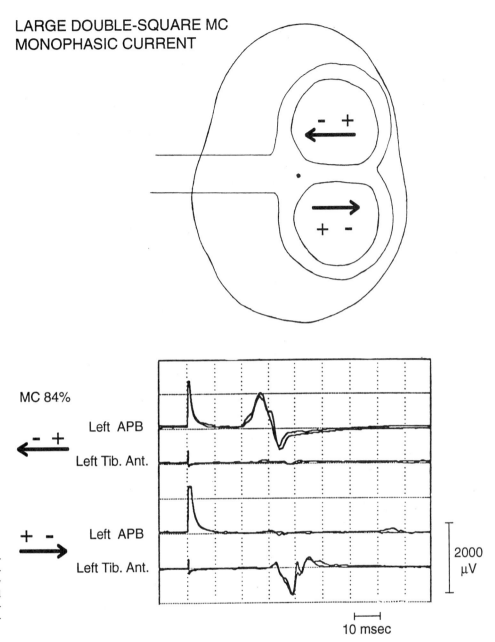

FIGURE 25–9. Effect of lateral-to-medial versus medial-to-lateral induced monophasic current flow in the coronal plane. 63-year-old male physician at rest. APB, abductor pollicis brevis; tib. ant., tibialis anterior.

of the motor neuron. If the motor neuron is relatively hyperpolarized, then an initial set of excitatory postsynaptic potentials (EPSPs) will be inadequate to achieve threshold. Repetitive EPSPs mediated by multiple I waves will, through temporal summation, achieve motoneuron threshold more readily. Excitation at the motor cortex can further be enhanced by increasing the cortical stimulus intensity, thereby eliciting larger corticospinal D and I waves and an increased number of I waves. Proof that background facilitation decreases threshold at the motoneuron is provided by stimulating corticospinal fibers at the foramen magnum.[215, 216]

Direct (D wave) activation by appropriate stimulation during voluntary contraction results in *minimal* central delay and *should be employed whenever possible.* Background voluntary contraction ensures that any corticospinal discharge can dependably discharge some alpha motoneurons. Two kinds of information may be obtained, depending on whether the corticospinal neurons are directly excited or trans-synaptically excited. Direct excitation of corticospinal neurons during voluntary activity provides information about conduction velocity that may be useful in demonstrating corticospinal pathophysiology in diseases including multiple sclerosis (MS), amyotrophic lateral sclerosis (ALS), and stroke. Alternatively, transsynaptic stimulation provides information concerning the effects of corticocortical afferents on motor cortical interneurons as well as corticospinal neurons. Thus, stimulation orientations that yield direct versus transsynaptic excitation give different but complementary information. Indirect (I wave) activation without background facilitation may also be valuable when assessing the corticocortical afferent input. Detecting thresholds during relaxation and comparing them during active contraction could in principle give useful information about the degree of background facilitation of the motoneuron at rest. Normal values in the relaxed state may only be obtainable when assessing patients who cannot, for psychiatric or neurological reasons, provide preexisting background facilitation. It also appears useful to calculate the amplitude ratio between the motor-evoked potential (MEP) and the maximal distal M response elicited by conventional electrical stimulation.[217]

In the preceding discussion, it was assumed that the magnetic stimulator induces a substantially monophasic electrical field and that the direction of the induced current may significantly affect the responses. Notably, some commercial stimulators and all repetitive magnetic stimulators induce substantially polyphasic fields. When using a polyphasic pulse with a round MC at the vertex, excitation in the depolarizing direction occurs at a given site in one hemisphere by the first phase (quarter cycle I), then 100 to 150 μsec later depolarization occurs in the other hemisphere by the second phase (quarter cycles II and III). The slight delay in initiation of excitation by the second phase of a polyphasic pulse is not detectable in the clinical laboratory. The same limitations apply to a figure-of-eight MC held directly over one motor cortex.

Cortical Response Variability

The amplitude and latency of consecutive MEPs elicited by the same stimulus vary over time, although latency is much more stable during either relaxation or contraction.[3, 107, 146] By contrast, amplitude varies much more in relaxed than contracting muscles. In addition, the amplitudes of MEPs from two adjacent or other muscles in the same subject also do not covary.[107] Presumably, there is a constantly changing fluctuation of resting motor threshold which is unique in each muscle. Such variation is intrinsic to the human motor system and initially was thought not to be appreciably modified by attempts to rigidly control the degree of background facilitation or by any choice of pulse configuration, pulse delivery rate, coil orientation, or coil design. However, MEP size may change in predictable directions with tasks related to the type and execution of movement and mental imagery.[1, 56, 76, 114, 194] Moreover, amplitude variability may be reduced by (1) stimulating at optimal compared to suboptimal scalp positions;[24] (2) employing paired-pulse stimulation at the scalp, which is most effective in relaxed muscle;[146] and (3) performing "triple stimulation technique" at the scalp, distal wrist nerve, and Erb's point.[129] Using triple stimulation, Magistris and colleagues[129] concluded that a major source of MEP variability arises from desynchronization of action potentials within the corticospinal tract or spinal cord, which causes phase cancellation of consecutive MEPs. Furthermore, desynchronization mainly accounts for the reduced MEP amplitude compared to that of the distally elicited M response.

Factors Affecting the Peripheral Response

Much of the early clinical emphasis using neuromagnetic stimuli explored peripheral motor conduction time (PMCT) and CMCT, which usually requires that stimuli be applied to motor cortex and to cervical or lumbar spine, or both (Fig. 25–10). CMCT is calculated by subtracting peripheral motor conduction time latency (i.e., PMCT) from absolute cortical latency.[17] However, neither CMCT nor PMCT accurately measure the segments they purportedly represent. The PMCT latency may be elicited using either electrical cathodal or neuromagnetic stimuli. In either case, excitation of cervical motor roots occurs at or distal to the neuroforamina.[74, 139, 196] These studies estimate the excitation site to vary from 3 to 4 cm distal to the spinal cord to as far as 7.7 cm distal to the intervertebral foramen (i.e., 7.0 to 8.4 cm proximal to Erb's point). By contrast, conventional round or large figure-of-eight MC stimuli applied over the distal lumbar or sacral cauda equina excites nerve rootlets even further from the spinal cord, at a distance of 11 to 16 cm or more from the root exit zone of the conus medullaris (depending on

the nerve root). Moreover, at the cervical spine, the site of neuromagnetic excitation shifts distally toward the proximal brachial plexus if the cathodal stimulus intensity is significantly increased, or if the MC shifts distally, and *especially* if the coil diameter increases.[162] This in part reflects the relatively superficial location of the brachial plexus. Such large shifts do not occur, however, at the lumbosacral spine, probably because of the greater depth of the proximal nerve roots from the skin surface, which *requires* excitation at low threshold sites near or at neuroforamina.

Therefore, CMCT calculated for both upper and lower limbs typically includes a segment representing proximal nerve root conduction time and synaptic delay at the motoneuron. Depending on the precise location of neuropathology, this could significantly impact and confound the interpretation of results when attempting to separate latencies into categories of PMCT versus CMCT. Concerning leg responses, it is now possible to stimulate proximal cauda equina at the lumbosacral root exit zone with a large figure-of-eight MC, which improves the estimate of peripheral motor conduction time and therefore of CMCT.[125] The cauda equina stimulation technique also has a significant advantage in that it provides a direct measure of conduction time in proximal nerve roots (obtained by subtracting distal from proximal cauda equina latency, Fig. 25–11). Thus, in acute Guillain-Barré syndrome, a cauda equina conduction time of 10 msec (more than 200% above the upper limits of normal and corresponding to a motor conduction velocity of 15 to 20 m/sec) may be seen when conduction time in more distal segments is normal.[124] Another approach employs neuromagnetic stimulation at the foramen magnum, making it possible to analyze CMCT and detect upper versus lower corticospinal pathology.[216]

PMCT may also be estimated by calculating the F wave latency from the anterior horn to the target muscle (usually defined as $1/2 \times$ (F latency + M latency -1)). However, F wave estimates are problematic for several reasons: (1) there is confusion as to which fiber population conveys the F response, (2) F responses may be absent in acute proximal demyelinating neuropathies, (3) F responses are not easily recorded in proximal limb muscles, and (4) F responses are difficult to elicit when the M response amplitude is low. Despite such limitations, the F response method may be preferable to electrical or neuromagnetic stimuli delivered to the *cervical* spine in the presence of vertebral instability, vertebral fusions, painful radiculopathy, spinal cord compression, or a significant spinal stenosis.

Unfortunately, neuromagnetic stimulation of the proximal nerve roots and brachial plexus does not consistently reach supramaximal levels, so measurement of response amplitudes and detection of conduction block are not reliable.[55, 125] However, despite wide variations in amplitude, CMAP onset latency remain stable.

Interpretation of CMCT

CMCT obtained by subtracting PCMT from absolute cortical latency includes not only corticospinal conduction time, but also monosynaptic delay at the alpha motoneuron; with relaxation, additional delay affects temporal summation. CMCT can be calculated to muscles of upper and lower limbs as well as to muscles innervated by cranial nerves. In the clinical setting, cortical responses to three or more stimuli are usually superimposed. Corresponding CMCT latencies depend on the methods used to obtain peripheral latencies at the cervical or lumbosacral spine (high-voltage electrical stimulation, neuromagnetic stimulation, one-way F response estimate). In normal subjects, absolute motor cortical latencies to hand muscles usually range from 19 to

Magnetic Stimulation of Motor System

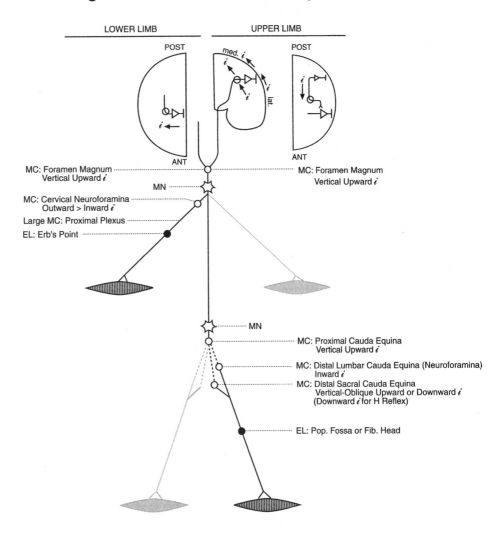

FIGURE 25–10. Illustration of proposed low threshold sites and optimal induced current flow for neuromagnetic excitation of the human motor system (see text).

MC = Magnetic exitation site in CNS($\ominus$) and PNS($\bigcirc$)
EL = Electric exitation site in PNS($\bullet$)
i = Optimal direction of induced monophasic current
MN = (Motor Neuron); med.(medial); lat.(lateral)

21 msec; to tibialis anterior, the range is usually 26 to 30 msec. The results from different laboratories are tabulated in Rothwell and associates.[185] Corresponding CMCTs in normal subjects range from 5 to 8 msec when recording from hand muscles and 13 to 15 msec when recording from tibialis anterior.

It cannot be overemphasized that CMCT obtained with a *conventional* round MC is overestimated when stimulating over the spine to estimate PMCT by either the segment (1) from the cervical root exit zone to cervical neuroforamina, or (2) from the lumbosacral root exit zone to distal cauda equina. By contrast, using a large figure-of-eight MC over conus medullaris directly excites lumbosacral roots at their exit zone, resulting in an accurate calculation of CMCT to lower limbs.

CMCT has been estimated in a great number of disease states, including stroke, MS, ALS, and other neurodegenerative disorders, and a detailed summary is beyond the scope of this chapter.[34, 91, 179, 198] One emerging and potentially significant application, however, is to determine CMCT in patients with suspected cervical myelopathy whose magnetic resonance imaging (MRI) scans reveal ventral ridging. The data indicate that slowing consistent with demyelination may be an important factor in selecting patients for proposed decompressive surgery.[63, 100, 127, 204] Lower limb recordings may be more diagnostically sensitive than upper limb recordings.[204]

CORTICAL THRESHOLD

Despite attempts at standardizing the procedure for measuring the cortical motor threshold, various approaches are employed and new methods continue to be introduced.[42, 143, 178] As recommended by an international committee, the

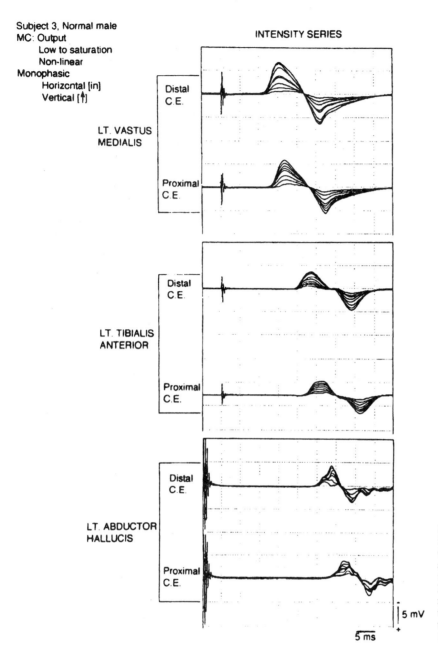

Subject 3, Normal male
MC: Output
   Low to saturation
   Non-linear
Monophasic
   Horizontal [in]
   Vertical [↑]

INTENSITY SERIES

LT. VASTUS MEDIALIS
Distal C.E.
Proximal C.E.

LT. TIBIALIS ANTERIOR
Distal C.E.
Proximal C.E.

LT. ABDUCTOR HALLUCIS
Distal C.E.
Proximal C.E.

5 mV
5 ms

FIGURE 25–11. Response to stimulus intensity increase. Distal and proximal cauda equina CMAPs are elicited in left vastus medialis, tibialis anterior, and abductor hallucis by a large twin MC. (Reprinted from Maccabee PJ, Lipitz ME, Desudchit T, et al: A new method using neuromagnetic stimulation to measure conduction time within the cauda equina. Electroencephalogr Clin Neurophysiol 101:153–166, 1996, with permission.)

absolute threshold can be found in a relaxed target muscle by increasing stimulus intensity in 5% steps until reproducible, approximately 100-µV MEPs are elicited in nearly 50% of 10 to 20 consecutive stimuli.[178] However, a carefully performed study of repeatability estimates revealed a greater than 10% range of absolute threshold variation (at the 5% probability level) in healthy relaxed subjects who were retested 1 to 3 months later.[143] Another report suggested that cortical threshold variation may be substantially reduced by stimulating at the "hot spot" (i.e., the locus of greatest amplitude and shortest latency) identified by cortical mapping.[42] Whatever the approach, less test-retest variability might occur if a background of strong voluntary contraction were present.

The term "cortical threshold" is also misleading in the physiological sense.[143] Although the muscle responses depend on impulses initiated at the motor cortex, the descending volleys traverse the corticospinal axon, undergo temporal and probable spatial summation at the anterior horn, and

continue along the peripheral nerve, to the neuromuscular junction and target muscle. Obviously, a change at any of these levels could reduce the recorded response threshold, amplitude, and latency and alter the magnitude of variation of these parameters. Moreover, the direction of induced current (e.g., L-M versus P-A for hand muscles) and possibly pulse waveform (monophasic versus polyphasic) could significantly affect the evoked response.

The following clinical correlations have been observed. In normal right-handed subjects, the threshold for activation of muscles of the right arm was found to be lower than the threshold for activation of the left arm,[211] and the lower transcranial magnetic stimulation threshold for one hand is associated with greater dexterity.[212] However, more recent studies have not reproduced differences of hemisphere asymmetry in the same subject.[42, 143] In ALS, a low cortical threshold with large amplitude MEPs may be seen early, but with disease progression cortical threshold rises and the MEP amplitude becomes smaller or unelicitable.[34, 69, 198, 217] Eisen and colleagues[69] observed an inverse exponential correlation between cortical threshold and MEP/M wave ratio in ALS. Cortical motor threshold may be increased in MS,[34, 170] cerebellar lesions,[64, 65] and migraine (both ictal and interictal[20, 128]). Following stroke the MEP may be absent.[234]

In addition, cortical motor threshold is increased following administration of sodium channel blockers (phenytoin, carbemazepine, lamotrigine, and losigamone) but is not affected by drugs that affect gamma aminobutyric acid (GABA)ergic systems (ethanol, lorazepam, baclofen, gabapentin, vigabatrin).[38, 132, 238]

CORTICAL SILENT PERIOD

Physiology

During sustained voluntary activation of a limb muscle, a silent period (SP) may be present in the tonic electromyogram (EMG) following an electrical shock delivered to the peripheral innervating nerve,[134, 201] following electrical or magnetic stimulation of motor axons at the cervicomedullary junction,[96] and following electrical shock or magnetic stimulation over the contralateral motor cortex.[5, 31, 33, 78, 93, 96, 131, 166, 218] The SP follows the M response if elicited peripherally, or the MEP if elicited centrally, and is terminated by a return to preexisting baseline EMG activity. The cortical SP is optimally elicited in contralateral hand muscles at intensities ranging from 1.1 to 1.5 times the relaxed motor threshold and typically lasts approximately 100 to 150 msec, but is shorter in more proximal muscles[166] (Fig. 25–12). SPs are also recorded in cranial musculature.[230] The SP to magnetic cortical stimulation is longer than that elicited by electrical cortical shock and is shortest when elicited at the cervicomedullary junction.[96]

It is generally proposed that the first 50 msec of the magnetic cortical SP reflects both cortical and spinal inhibitory mechanisms, whereas the latter 100 msec or more exclusively reflects cortical physiology.[85, 96] Spinal mechanisms implicated in the first 50 msec include Renshaw inhibition following synchronized discharge of spinal motor neurons,[59, 85] and activation of inhibitory Ia interneurons by descending input from motor cortex.[53, 85] Prolonged spinal motor neuron afterhyperpolarization is less likely.[53, 59, 85, 141] Reflex mechanisms mediated by Ib afferent volley from Golgi tendon organs probably do not contribute, based on reports of patients with severe large-fiber sensory neuropathy whose SPs were preserved.[51, 78, 208] Nevertheless, both H reflexes and

F responses are usually inhibited within the first 50 msec,[33, 78, 235] although in one study, H reflexes were facilitated in the first 50 msec,[177] and in another, H reflexes were inhibited well beyond 50 msec.[210] The notion that cortical mechanisms are primarily responsible for the latter two-thirds of the magnetic cortical SP derive from studies showing that a second magnetic stimulus given during this time elicits either an inhibited MEP or no MEP,[33, 96, 177, 209, 219] whereas a single *strong* electrical shock or a pair of electrical shocks (given 1.8 msec apart) at this time, or even earlier, elicit a preserved MEP.[33, 60, 96] The pair of electrical shock stimuli presumably elicit D waves.[85] These results are interpreted to indicate that electrical shock stimuli excite corticospinal axons in white matter just beneath motor cortex, thereby short-circuiting the cortical inhibition arising from the initial conditioning magnetic stimulus. Moreover, the SP elicited by electric shock is shorter than that elicited by a magnetic pulse, consistent with the absence of contributions from motor cortical inhibitory mechanisms.[85]

However, if the test magnetic stimulus intensity is *significantly* increased, then not only can a second MEP be elicited during the SP, but it may be facilitated.[210] Possibly, other motor pathways not susceptible to inhibition are accessed at higher stimulation intensity.

There is also strong evidence that the cortical MEP and SP are physiologically dissociated, reflecting separate mechanisms.[85] The SP has a lower threshold and may occur without a preceding MEP.[33, 43, 57, 208, 224] Motor map studies also reveal separate topographical loci for MEPs versus SPs,[224] but this was not reproduced in other studies.[57, 233] Also, the SP can be easily elicited in diseases affecting peripheral motor nerves, especially ALS.[208] Furthermore, *ipsilateral* SPs without preceding MEPs are elicited in voluntary activated muscles, but require greater stimulation intensity.[224]

Procedural Issues

When eliciting the SP, stimulation parameters strongly influence reproducibility, as follows: (1) Stimulus intensity can significantly affect SP duration, and it is recommended that stimulation be performed in terms of multiples of relaxed motor threshold, using two intensities of stimulation.[166] Typically, the SP duration increases linearly with increasing stimulation intensity although the preceding MEP saturates;[33] but see Inghilleri and associates.[96] (2) An increase in background isometric contraction decreases SP duration.[33, 96, 232] It may be useful to standardize and perform SP measurements at active motor threshold in subject and patient populations.

RELAXED   CONTRACTED

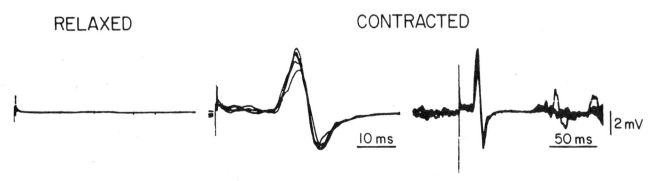

FIGURE 25–12. Effect of voluntary contraction on digit flexor CMAP elicited by MC stimulation of contralateral motor cortex and ensuing silent period. Right records sweep slower to show duration of silent period and rebound. Five superimposed sweeps in each record. (Reprinted from Amassian VE, Maccabee PJ, Cracco RQ, Cracco JB: Basic mechanisms of magnetic coil excitation of nervous system in humans and monkeys: Applications in focal stimulation of different cortical areas in humans. In Chokroverty S (ed): *Magnetic Stimulation in Clinical Neurophysiology.* Stoneham, MA, Butterworth, 1990, pp 73–112, with permission.)

(3) Five or more superpositions are usually sufficient to obtain a well-defined response. Measurement intervals vary with different laboratories. SP onset may be defined as either the onset or the return to baseline of the MEP. SP termination may be defined as either the end of absolute EMG silence or when the EMG returns to its prestimulus baseline level.[166]

Clinical Correlation

The following clinical correlations have been observed:
1. The SP is absent or shortened in motor cortex ischemia, possibly indicating damage to cortical inhibitory neurons.[197] Paradoxically, the SP is reportedly prolonged following stroke, more prominently in patients with upper motor neuron signs.[26, 88, 217]
2. The SP is shortened in lesions of primary sensorimotor cortex but prolonged with lesions elsewhere, including the cerebellum, premotor cortex, parietal and temporal lobes, internal capsule, and thalamus.[65, 220, 221]
3. In some hemiparetic patients with prolonged SPs studied by Classen and associates,[44] a "motor arrest" phenomenon was observed following a magnetic stimulus, characterized by "a complete inability to initiate voluntary muscle activity for several seconds." Furthermore, some patients manifested "motor neglect," suggesting that the severe motor dysfunction reflects hyperactivity of cortical inhibitory interneurons (from damage to one or more motor cortical afferents) rather than direct lesioning of descending motor tracts.
4. Significant shortening of the SP has been reported in ALS by some investigators but contradicted by others.[62, 167] Apparently, the SP may also significantly shorten with advancing age in normals.[167] In either case, such findings could signify damage to cortical inhibitory mechanisms.
5. The SP is shortened in Parkinson's disease and neuroleptic drug-induced parkinsonism but is prolonged in Parkinson's disease patients after the administration of L-dopa.[32, 165, 173] L-dopa given to normal subjects also prolongs the SP, but to a lesser degree. Similarly, in normal subjects given the dopamine receptor agonist pergolide, the SP is prolonged.[239]
6. The SP is prolonged in Huntington's chorea[164] but shortened during motor tasks in writer's cramp.[75, 133, 202]

Drug Studies

A coherent perspective has not emerged relating the proposed main actions of some drugs in normal subjects to their effects on SP duration. For example, one study showed that although the peripheral SP remained stable, the cortical SP was shortened by intravenous diazepam.[97] The shortening was reversed by giving flumazenil, a benzodiazepine antagonist, suggesting an effect on GABA receptors. However, additional reports indicate that the SP is prolonged by lorazepam,[237] gabapentin,[238] and ethanol,[236] whereas other drugs with GABAergic activity (vigabatrin, baclofen) do not affect SP duration.[238] Also, carbamazepine reportedly prolongs the SP,[199, 238] but phenytoin, also a sodium channel blocker, has no effect.[38] Such contradictory experimental evidence may reflect differences in drug dosing. Moreover, measurement of the SP in subjects taking sodium channel blockers is further complicated by elevations in motor threshold.

PAIRED-PULSE STIMULATION

Studies of motor responses to paired-pulse stimuli induced by the same MC, given over a range of interstimulus intervals (ISIs), may be performed with either a rapid rate stimulator or two separate stimulators connected via a "bistim" module.[47, 184, 219] Unlike the rapid rate stimulator, the "bistim" approach enables each pulse intensity to be independently varied.[47] In 1993 this paradigm was significantly advanced by Kujirai and colleagues[112] also using separate stimulators for each pulse, but systematically setting the conditioning stimulus to subthreshold intensity, followed by a test stimulus at suprathreshold intensity. Typically, the subthreshold conditioning stimulus was approximately 80% of resting motor threshold or 95% of active motor threshold. Using this weak pulse–strong pulse paradigm, in muscles at rest, the test MEP is inhibited at short ISIs of 1 to 5 msec (referred to later as intracortical inhibition, or ICI) and is facilitated at slightly longer ISIs of 7 to 20 msec (referred to later as intracortical facilitation, or ICF); both are illustrated in Figure 25–13 (normal subject curve). Rothwell[186] and Ziemann and colleagues[240] provide arguments to support the notion that the excitatory and inhibitory activity are each produced in separate neural circuits and that both reflect activity within cortical rather than subcortical circuits, as follows: (1) the conditioning threshold stimulus required for inhibition is lower than that of facilitation, so that the facilitation is not a rebound of the preceding inhibition; (2) the inhibitory effect does not depend on the direction of induced current, whereas the facilitatory effect is optimal with P-A as compared with L-M induced current; (3) inhibition and excitation summate in an approximately linear fashion; (4) consistent with a cortical origin, the low intensity of the conditioning stimulus probably does not evoke any descending activity in corticospinal axons, as shown by the lack of effect on the H reflex or on the evoked response to simultaneous anodal cortical shock. In addition, raising the intensity of the conditioning stimulus produces facilitation instead of increase of the inhibition.

The two pulse paradigm eliciting an inhibitory-facilitatory sequence has proved a valuable tool for investigating neurological disorders.[19, 29, 35, 77, 87, 172, 173, 242, 243] The inhibitory phase may be diminished in various epilepsies (juvenile myoclonic, cortical myoclonus, epilepsia partialis continua), Parkinson's disease, focal dystonia, Tourette's syndrome, and ALS (see Fig. 25–13) (patient curve). ICF appears to be increased in stiff-person syndrome.[190]

A further and potentially extremely useful application of this paradigm relates to studies of the effects of various drugs on ICI and ICF, pioneered by Ziemann and associates.[238, 248] In normal subjects, the later transitional phase of ICI is increased and the full phase of ICF is decreased by single doses of GABAergic drugs (lorazepam, diazepam, baclofen, gabapentin, ethanol), and antiglutamergic drugs (memantine, riluzole, gabapentin).[120, 238] Possibly, the data on ICI reflects difficulty in increasing an already strong inhibitory control effect, or unmasking a longer latency inhibitory effect, which was previously obscured by facilitation manifest after 5 msec, or both. Alternatively, atropine (which acts as a GABA antagonist as well as an anticholinergic), and zolmitriptan (a serotonin receptor agonist) both reduce ICI but do not affect ICF.[119, 231] Vigabatrin, a GABA potentiator, appears to decrease ICF but does not affect ICI.[238] It is expected that other modulators of neurotransmission in the central nervous system will be shown to affect ICI and ICF.

Conversely, drugs whose primary mechanism of action is targeted at inhibiting voltage-gated sodium channels (carbamazepine, phenytoin, lamotragine) affect threshold but specifically do not affect ICI or ICF. The direct dopamine receptor agonist bromocriptine increases ICI and motor threshold; the dopamine receptor blocker haloperidol both reduces ICI and increases ICF.[241] In the near future, testing for ICI and ICF in a specific patient may possibly predict

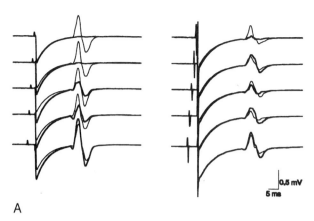
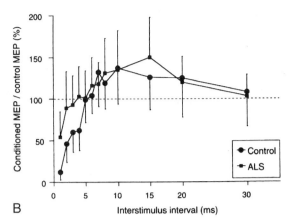

A

B

FIGURE 25–13. *A.* Surface EMG recordings from the abductor digiti minimi in one healthy control subject (*left*) and one representative with amytrophic lateral sclerosis (ALS) patient (*right*). All traces are averages of 10 single trials. For each pair of traces, the thin curve denotes the control-motor-evoked potential (MEP; response to the suprathreshold magnetic test stimulus alone), whereas the thick curve indicates the conditioned MEP (response to paired stimulation). The interstimulus interval between conditioning and test stimulus was varied in 1-msec steps from 1 (*top traces*) to 5 msec (*bottom traces*). Note the reduced inhibitory effect of the conditioning stimulus on the control MEP in the ALS patient. *B.* Mean intracortical excitability curves for the groups of healthy control subjects (*left*) and ALS patients (*right*). Vertical bars are standard deviations. Interstimulus intervals are given on the abscissa, the size of the conditioned mean MEPs as a percentage of the unconditioned mean on the ordinate. Horizontal dashed line (100%) indicates the size of the unconditioned MEP, lower and higher values indicate intracortical inhibition and facilitation, respectively. Note the reduced intracortical inhibition in the ALS group. (Reprinted from Ziemann U, Winter M, Reimers CD, et al: Impaired motor cortex inhibition in patients with amyotrophic lateral sclerosis. Evidence from paired transcranial magnetic stimulation. Neurology 49:1292–1298, 1997, with permission.)

the usefulness of specific medications in epileptic and basal ganglia disorders.

TEMPORAL I WAVE FACILITATION

When using pairs of identical near-threshold stimuli (nearly 0.9 to 1.1 times resting motor threshold), three short latency facilitation peaks appear at ISIs of approximately 1.0 to 1.5, 2.5 to 3.0, and 4.5 or greater msec.[206, 244] At these intervals, the size of the responses to each pair of stimuli was greater than the algebraic sum of the responses to each stimulus alone. Given the timing of the facilitation, the authors suggested that the response reflected activation of circuits responsible for generating I waves.

The three short latency facilitatory peaks are significantly reduced by drugs that enhance GABA effects in the neocortex (lorazepam, vigabatrin, phenobarbitol, ethanol) whereas the GABA-B receptor agonist baclofen, antiglutamate drugs (gabapentin, memantine), and sodium channel blockers (carbemazepine, lamotragine) unexpectedly had no effect.[247] Despite the reduction in facilitation with GABAergic drugs, the optimal ISIs for the three peaks were not affected by these drugs. It was suggested that I wave production is primarily controlled by GABA-related neuronal circuits. However, inhibitory postsynaptic potentials (IPSPs) typically have a much longer duration than the I wave periodicity. Furthermore, the earliest inhibition with the weak pulse–strong pulse paradigm was unaffected by GABA potentiators. It is more likely that ongoing inhibitory neuronal activity is potentiated by the drugs resulting in diminution of excitatory interneuronal effects, which are intrinsically periodic.

Amassian and associates[11] further extended the notion of I wave facilitation by giving multiple near-threshold stimuli to create tuning curves (Fig. 25–14). Three or four consecutive weak, near-threshold stimuli gave a sharp peak at a specific ISI for each subject, provided that the period be-

tween stimuli matched I wave periodicity. In three normal subjects, optimal ISIs were 1.6 to 1.7 msec, and 1.3 msec in another subject. Possibly, these ISIs corresponded to interneuron chain activity at approximately 600 Hz, supporting the hypothesis of a clock-like time quantizer in motor cortex.[2]

When giving near-threshold induced paired pulses at short interval, such as 1 msec, it is likely that the first conditioning stimulus excites cortico-cortical afferents at bends, but the second test stimulus, when smaller than the first, fires the motor cortical neurons at a *different* site, probably at the initial segment usually of interneurons.[14]

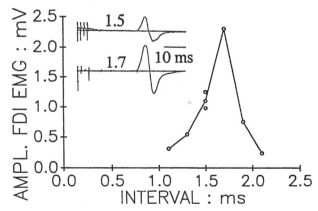

FIGURE 25–14. In the same subject, 10 summed responses are plotted versus interstimulus intervals (ISIs); *inset,* summed responses at 1.5 and 1.7 msec ISIs. At 1.5 msec, averages taken throughout the series to control for drift in motor cortical excitability. (Reprinted from Amassian VE, Rothwell JC, Redding M, et al: A high frequency resonant circuit in human motor cortex. Soc Neuroscience Abstracts 22:658, 1996, with permission.)

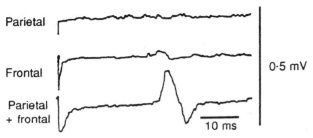

Parietal

Frontal

Parietal + frontal

0·5 mV

10 ms

FIGURE 25–15. Spatial facilitation of first dorsal interosseous responses to magnetic stimulation of first parietal then frontal areas at 0.5 msec ISI; five responses summed on each trace. (Reprinted from Amassian VE, Cracco RQ, Maccabee PJ, et al: spatial facilitation of human motor responses by near-threshold magnetic stimulation of parietal and frontal areas. J Physiol 504.P, 115P, 1997, with permission.)

SPATIAL FACILITATION

It is also possible to demonstrate convergence of cortico-cortical inputs from premotor and parietal areas onto motor cortex (Fig. 25–15). This is performed by delivering stimuli via two MCs located caudal and rostral to the optimal scalp focus for the target muscle under study (e.g., first dorsal interosseous). After identifying the target muscle focus with clearly suprathreshold stimuli, the MCs are moved apart until a near threshold or no response is elicited by each stimulus.[12]

Facilitation is marked at ISIs of 0.5 to 1 or 2 msec. An ISI of 0.5 msec is too long for summation of either the electrical fields or nodal electrotonus and too short for repetitive activation of the same cortico-cortical fiber. This interareal facilitation is observed during both relaxation and voluntary contraction. During voluntary contraction, when one of the inputs does not elicit a detectable motor response (i.e., probably no corticospinal discharge occurs), any facilitation *must* include a cortical site. The functional implication is that commanding a voluntary movement from anterior frontal regions interacts with parietal information systems as to the state of the body when movement is commanded. Similar studies have recently revealed spatial inhibition by parietal elicited responses to premotor or supplementary motor area stimulation.[15]

PLASTICITY

Lesion Studies (Pathological, Iatrogenic)

Traditionally, cortical maps obtained by direct electrical stimulation on the cortical surface were regarded as fairly static representations in the adult mammal.[160] However, the concept was disproved with the demonstration in the rat and monkey of changes in both the sensory and motor homunculus following peripheral nerve interruption and reconnection.[67, 136, 137, 163, 193] This was further extended in the human by mapping the motor cortex of adult humans with long-term (approximately 2-year) histories of traumatic cervical quadriplegia (Fig. 25–16);[115] and subsequently in amputees.[49, 84, 157] In one experiment by Levy and associates,[115] the figure-of-eight MC at 100% output intensity was moved in 2-cm steps over a scalp grid with the subject's muscles relaxed. The reference was the midpoint of the MC junction. These maps revealed motor responses to be elicited from a much wider cortical area compared to normal in spared muscles (biceps, deltoid) immediately proximal to the traumatic level. It was inferred that the motor system reorganized such that the immediate proximal muscles now elicited responses from cortical regions normally eliciting digit movement. Despite the long-standing history of quadriplegia, mechanisms postulated to account for the reorganization included the loss of tonic inhibitory input from below the lesions (an immediate effect), loss of presynaptic inhibition or denervation supersensitivity to proximal muscle anterior horn cells (a 2- to 4- day effect), and collateral sprouting of axons (a long-term effect).

It was then shown that digit representation of the motor homunculus to neuromagnetic stimuli changed in response to acute limb ischemia, performed by inflating a blood pressure cuff to just above systolic for approximately 40 minutes; the lack of such change with focal anodic stimulation implied that acute as well as delayed facilitation of motor outputs to muscles immediately above the deafferentation level were predominately mediated at a cortical level.[23, 25] Possibly, these

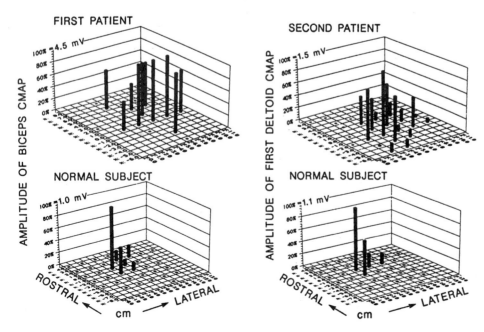

FIGURE 25–16. Amplitudes of biceps (*left*) and deltoid (*right*) CMAPs of two patients and two normal subjects plotted as a function of the X, Y coordinates of the MC on the scalp. CMAP amplitude is represented by height of bar. Largest responses correspond to 100%. (Reprinted from Levy WJ, Amassian VE, Traad M, Cadwell J: Focal magnetic coil stimulation reveals motor cortical system reorganized in humans after traumatic quadriplegia. Brain Res 510:130–134, 1990, with permission.)

acute changes in cortical motor representations indicate disinhibition or unmasking of preexisting connections. Significant modification of cortical motor maps were subsequently observed in subacute cervical spinal cord injury (i.e., within 6 to 17 days postinjury),[203] delayed thoracic spinal cord injury,[207] and Bell's palsy.[176]

Mapping during background voluntary contraction reveals increased areas of representation in both proximal and distal muscles, and it was suggested that quantitative changes in preexisting connections accounted in part for the change in maps.[117, 174] In an elegant series of experiments, Ridding and Rothwell[175] constructed stimulus-response curves by eliciting responses at a single cortical site, over a range of intensities. Recordings at rest from spared forearm flexors in a subject with a low right arm amputation showed increased map area and stimulus-response curve slope compared to the intact side. Notably, such differences disappeared during background voluntary contraction suggesting that preexisting connections were activated. However, it is unclear whether the extreme changes observed in long-standing quadriplegia involve more than preexisting mechanisms experienced during voluntary contraction.[117]

In a remarkable longitudinal series of observations, Mano and colleagues[130] evaluated four patients with severe traumatic cervical root avulsion resulting in complete upper limb paralysis. Surgical anastomosis of an intercostal nerve into the musculocutaneous nerve was performed to restore function in the biceps muscles. Four to six months after surgery, motor units from the biceps were recorded only during deep respiration and in response to cortical magnetic stimulation. In other words, the patients could not recruit motor units independent of respiration and there was no movement at the elbow. Gradually, 1 to 2 years after the operation, during which time the patients received intensive feedback therapy, voluntary elbow movements were observed. Notably, biceps muscle cortical maps during and beyond this period shifted from the normal intercostal nerve region to a more lateral position, representing the arm area of motor cortex. The association between the return of voluntary control with cortical map lateralization to the arm region is striking and suggests an evolving concordance between function and cortical location.

More recent studies in amputees (not earlier than 7 months) revealed significantly reduced motor threshold to transcranial magnetic stimulation (TMS), but not to electrical stimulation, and reduced ICI on the amputated compared to the normal side.[39] Prior work also demonstrated reduced motor threshold to neuromagnetic stimuli in amputation.[49, 84] These findings suggest that a decrease in GABAergic inhibition at the cortical level is involved in long-term plasticity in the human. Similarly, acute forearm ischemic nerve block following low frequency (0.1 Hz) repetitive TMS of the deafferented motor cortex also showed a decrease in ICI, implying decreased GABAergic inhibition as well, in short-term plasticity.[246] Furthermore, ischemic nerve block alone produced a moderate increase in MEP size, which was much enhanced following low frequency repetitive TMS. Such changes in ICI and MEP size are abolished after single oral doses of lorazepam (a benzodiazepine agonist) or lamotrigine (a sodium and calcium channel blocker) and ICI was suppressed but MEP size not affected after dextromethorphan (an NMDA receptor blocker).[245] These pharmacological data more firmly establish that GABAergic inhibition is directly involved in both short-term and long-term plasticity as defined by magnetic stimulation studies.

Motor Training

The acquisition of motor skills with repetitive practice requires reorganization within the central nervous system.

This has been documented in animal studies, and in humans using neuroimaging.[16, 81, 101, 103, 105, 138, 168, 195, 200]

TMS has also been used to study rapid plastic change in the human motor system with the acquisition of new fine motor skills. Pascual-Leone and colleagues[153] mapped the cortical motor areas targeting the contralateral long finger flexor and extensor muscles in subjects learning a one-handed, five-finger exercise on the piano. Over 5 days, with daily 2-hour manual practice sessions, the cortical maps enlarged and activation threshold decreased[153] (Fig. 25–17). Notably, (1) changes in motor maps were limited only to the cortical representation of the exercised hand, (2) similar changes were not observed in control subjects undergoing daily TMS mapping who did not perform practice exercises on the piano, and (3) compared to control subjects, mental practice alone resulted in a significant enlargement of motor maps and improved performance of the five-fingered exercise. However, these changes were less than that of the physical practice group. The authors suggest that, as has previously been shown in animal studies, rapid cortical modulation may be affected via unmasking of latent intracortical connections and long-term potentiation.[98, 99] Another study suggests that rapid plasticity of human cortical movement representation may be induced by motor practice over 15 to 30 minutes and in some subjects, as briefly as 5 to 10 minutes.[45]

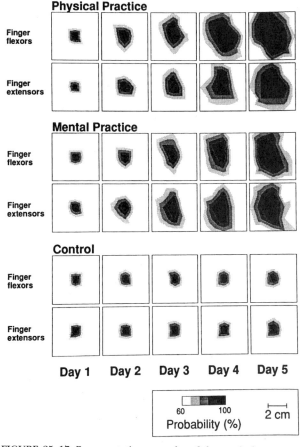

FIGURE 25–17. Representative examples of the cortical motor output maps of the long finger flexor and extensor muscles from days 1-5 in a subject performing one-handed five-finger exercises on a piano. (Reprinted from Pascual-Leone A, Dang N, Cohen LG, et al: Modulation of muscle responses evoked by transcranial magnetic stimulation during the acquisition of new fine motor skills. J Neurophysiol 74:1037–1045, 1995, with permission.)

Alternatively, the motor cortical maps to the tibialis anterior muscles showed a reduced area in subjects who had immobilization of the ankle joint (secondary to distal lower limb fracture without peripheral nerve lesion).[118] The area reduction was correlated to the duration of the immobilization and could be quickly reversed by voluntary muscle contraction.

Cross-modal Plasticity in the Blind

Deafferentiation may also be studied in blind subjects. Initially, the cortical motor representation of the reading finger of proficient Braille readers (recording from the first dorsal interosseous) was found to be significantly larger than that of the nonreading hand or in either hand of blind control subjects.[149] This effect was more robust following an intense day of reading compared to days off in which reading was not routinely performed.[154] Hypotheses were then advanced that deafferentation in one sensory modality modifies or expands cortical motor map representations in other sensory modalities.[110, 111, 169] Neuroimaging studies further revealed activation of the occipital cortex of early blind subjects by tactile discrimination tasks such as Braille reading.[188]

To further specify whether such cross-modal plasticity was *functionally* relevant, Cohen and colleagues[50] used high frequency repetitive TMS (rTMS) (10 Hz, 3 seconds' duration) to disrupt function of visual cortical regions in early blind subjects as they attempted to identify Braille and embossed Roman letters (Fig. 25–18). Earlier studies had shown that single or multiple pulse neuromagnetic stimuli to striate or extrastriate cortex reversibly suppressed visual perception.[4, 8, 71, 72, 121] In blind subjects, transient stimulation of the occipital visual cortex induced errors in both Braille and embossed Roman letter reading as well as distorted tactile perception (subjects reported "missing dots," "phantom dots," and "dots don't make sense"). By contrast, identical studies in normal sighted subjects had no effect on tactile performance. The authors concluded that "blindness from an early age can cause the visual cortex to be recruited to a role in somatosensory processing," which may account in part for their "superior tactile perceptual abilities."

SAFETY CONCERNS: sTMS VERSUS rTMS

Seizure Induction in Normals

Single pulse transcranial magnetic stimulation (sTMS) may be defined as nonrhythmic stimulation delivered at low frequencies to a single scalp locus (usually at rates not higher than 1 Hz).[37] Following the introduction of commercial magnetic stimulators in the mid-1980s, seizures to sTMS have occurred extremely rarely, and always in the context of preexisting epilepsy or stroke.[94, 95, 226] By contrast, rTMS, strictly defined at rates exceeding 1 Hz, has resulted in at least seven seizures worldwide as of 1996.[225, 226] At the NINDS (National Institute of Neurological Disease and Stroke), seizures were initiated by high frequency rTMS, consisting of one or more burst trains of single stimuli given as rapidly as 15 to 25 Hz.[226] However, seizures have been produced at frequencies as low as 3 Hz.[226]

Practical guidelines arising from the pioneering NINDS safety studies include the following:

1. Lowest threshold MEPs are elicited in hand muscles and remain relatively uniform at stimulation trains of low frequency and intensity.[150, 223]
2. Persistent motor activity following high frequency rTMS of the motor cortex may indicate after-discharges.[150]

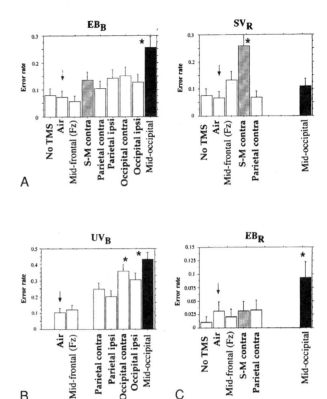

FIGURE 25–18. Error rates (mean ± s.e.) during rTMS at different scalp locations of subjects performing tactile identification of Braille letters and embossed Roman letters in four studied groups: EB$_B$, early blind subjects reading Braille; EB$_R$, early blind subjects reading Roman; UV$_B$, a different subset of early blind subjects reading Roman, at higher stimulation intensity: and SV$_R$, sighted volunteers reading Roman. Black bars indicate error rates induced by stimulation of the mid-occipital position, and gray bars the error rates induced by stimulation of the contralateral sensorimotor cortex. In both groups of early blind subjects, stimulation of the mid-occipital position induced more errors in reading Braille and Roman letters than stimulation of any other position, whereas in the sighted volunteers stimulation of the contralateral primary sensorimotor region induced more errors than stimulation of any other position. Asterisks indicate scalp positions where significantly more errors occurred than control (air, marked with arrows). S-M, sensorimotor cortex; contra, contralateral; ipsi, ipsilateral. Asterisk, < .001. (Reprinted from Cohen LG, Celnik P, Pascual-Leone A, et al: Functional relevance of cross-modal plasticity in blind humans. Nature 389:180–183, 1997, with permission.)

3. Intracortical spread of excitation and after-discharge precede seizures elicited by rTMS and may be identified by the appearance of EMG activity spreading out from the target muscle(s) to other muscles contralateral to the stimulus.[150] High frequency rTMS may result in "temporal summation of EPSPs without compensatory IPSPs," leading to a net excitation of neighboring pyramidal cells with horizontal spread of excitation.[150, 152]
4. Pascual-Leone and associates[150, 152] defined the minimal number of pulses required for the spread of cortical excitability as a function of rTMS intensity (% resting motor threshold) and frequency. These parameters were more recently abridged to emphasize that short inter-train intervals (1 second or less) are not safe for trains given at 20 Hz for 1.6 seconds, using stimulus intensities greater than 100% of relaxed motor threshold.[36] The otherwise identical parameters are considered to be safe at inter-train intervals of 5 seconds or more.

Physiological, Hormonal, and Immunological Effects

A detailed summary of other safety-related issues including TMS in epilepsy is found in Wasserman and colleagues.[226] Briefly, sTMS and rTMS do not significantly change the neurological examination, electroencephalogram, electrocardiogram, or serum ACTH, LH, or FSH levels.[27, 41, 95, 109, 116, 150] Single TMS does not alter cognition or memory. However, right prefrontal rTMS (5 Hz) increases TSH, which parallels a subjective decrease in sadness,[79] and one subject showed an increased serum prolactin following mid-frontal rTMS (accompanied by "acute dysphoria.")[226] Subjects occasionally complain of a mild, transient headache. Hearing loss is possible and ear plugs are recommended.[52] Prolonged change in cerebral blood flow is not seen in sTMS, but increased cerebral blood flow to rTMS has been documented in both the anterior and posterior cerebral circulations.[86, 191, 192] During rTMS, scalp burns may arise from surface electrode heating.[182] Following sTMS there may be transient (under 48 hr) elevations or depression of T lymphocyte subsets according to whether the left or right posterior hemisphere is stimulated.[10]

Direct Exclusions

Subjects or patients with metallic objects or implants in the scalp, brain, neck, vertebral spine, sacrum, or elsewhere in the body are generally excluded from near or direct exposure to magnetic stimuli. These objects include aneurysm clips, cochlear implants, cardiac pacemakers, bullets or their remnants, devices that monitor intracranial pressure, bone plates, sheet metal, foreign bodies in the eye, recently implanted joint prostheses, stents, sacral screws, vertebral spinal rods and wires, or any other electronic, structural, or mechanical implant. Evidence of elevated intracranial pressure and the presence of one or more skull defects are additional exclusions. Systematic studies have also been performed in normal young children and neonates (sTMS) as well as in adults with a history of seizures or epilepsy (sTMS and rTMS); however, these groups do not routinely undergo TMS in most laboratories. An enhanced photoconvulsive response may also be a contraindication. In addition, rTMS in the very young should be restricted as it could theoretically provoke changes in developmental plasticity. In pregnancy, the standard position is that TMS should not be performed unless there are overriding clinical considerations (i.e., diagnosis, or treatment of depression) where the benefits outweigh any perceived risks.

THERAPEUTIC REPETITIVE STIMULATION

Repetitive TMS at relatively high frequencies (10 to 20 Hz) delivered intermittently in short trains over several minutes may have lasting effects on mood. George and colleagues[79] and Pascual-Leone and associates[156] found that rTMS of the prefrontal cortex of healthy subjects affected their mood as measured by self-rating scales. In both studies, right lateral prefrontal stimulation produced subclinical but statistically significant changes toward happiness, whereas left-sided stimulation appeared to have the opposite effect. Other studies suggest an opposite lateralization for the production of therapeutic mood changes in patients with depression. Sham-controlled trials of several days of treatment suggest potentially therapeutic effects of rTMS in severely depressed patients.[80, 155] However, only one of the studies tested right-versus left-sided stimulation, and this study

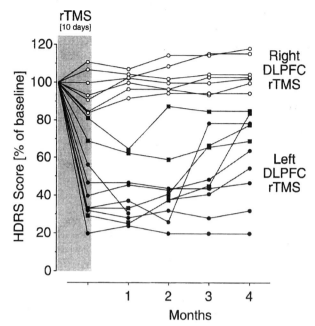

FIGURE 25–19. Effects of repetitive transcranial magnetic stimulation (rTMS) to the right or left dorsolateral prefrontal cortex (DLPFC) on score in the 16-item Hamilton Depression Rating Scale (HDRS) in patients with medication-resistant depression. Patients receiving left DLPFC rTMS are indicated by solid symbols. HDRS score was obtained and is plotted at the end of the 10 days of stimulation and monthly thereafter. The difference between the two study groups was marked. Right DLPFC rTMS had no effect, whereas left DLPFC rTMS resulted in marked improvement in HDRS scores and a sustained effect for 4 months in some patients. (Reprinted from Pascual-Leone A, Rubio B, Pallardó F, Catalá MD: Beneficial effect of rapid-rate transcranial magnetic stimulation of the left dorsolateral prefrontal cortex in drug-resistant depression. J Clin Neurophysiol 15(4):333–343, 1998, with permission.)

showed the more dramatic improvement with the left side[155] (Fig. 25–19). Encouraging preliminary data have also come from a small controlled treatment trial in obsessive compulsive disorder, which found improvements in mood (as in normal subjects) and a transient but significant reduction in obsessions after a single session of right prefrontal rTMS treatment at 20 Hz).[83]

Important practical questions presently under investigation include the following: (1) Is antidepressant action significantly improved by raising stimulation intensity from subthreshold to suprathreshold? (2) Is low frequency TMS also therapeutic (e.g., over the right prefrontal cortex)? (3) Is rTMS also therapeutic in mania or in severely depressed schizophrenics? (4) Is rTMS therapeutic in the presence of appropriate anticonvulsant medications? (5) Furthermore, are seizures *purposefully* induced by rTMS as efficacious as electroconvulsive therapy in the treatment of depression? (rTMS as currently performed is not as efficacious.)

COMMENT

As demonstrated by Amassian and colleagues in 1989,[4] suprathreshold sTMS applied to occipital cortex reversibly suppresses visual perception; the effect is prolonged using consecutive subthreshold stimuli given at rTMS frequencies.[8] Since then, transient neurological defects using sTMS or rTMS have been described in numerous other experiments, including, for example, color vision,[121] vocalization,[54] parietal extinction,[151] left temporal lobe visual naming,[227] speech arrest,[73, 102, 148] and verbal recall.[82] The extensive and expanding experience in visual function is summarized in two recent reviews.[13, 222] At this writing, permanent adverse consequences have not been described, even in those few normal subjects experiencing seizures, or in patients after as many as 50,000 cumulative rTMS stimuli for the treatment of depression.[155] The central dogma that these stimuli have transient and reversible effects will no doubt result in an enormous number of similar studies in the coming decades. Nevertheless, researchers should continue to assess their subjects for any long-term, or delayed effects that may occur.

The authors are indebted to Leonardo G. Cohen and Alvaro Pascual-Leone for providing figures and other materials. We also thank Mary Lombardo for secretarial assistance.

REFERENCES

1. Abbruzzese G, Trompetto C, Schieppati M: The excitability of the human motor cortex increases during execution and mental imagination of sequential but not repetitive finger movements. Exp Brain Res 111:465–472, 1996.
2. Amassian VE, Stewart M, Quirk GJ, Rosenthal JL: Physiologic basis of motor effects of a transient stimulus to cerebral cortex. Neurosurgery 20:74–93, 1987.
3. Amassian VE, Cracco RO, Maccabee PJ: Focal stimulation of human cerebral cortex with the magnetic coil: A comparison with electrical stimulation. Electroencephalogr Clin Neurophysiol 74:401–416, 1989.
4. Amassian VE, Cracco RQ, Maccabee PJ, et al: Suppression of visual perception by magnetic coil stimulation of human occipital cortex. Electroencephalogr Clin Neurophysiol 74:458–462, 1989.
5. Amassian VE, Maccabee PJ, Cracco RQ, Cracco JB: Basic mechanisms of magnetic coil excitation of nervous system in humans and monkeys: Applications in focal stimulation of different cortical areas in humans. In Chokroverty S (ed): Magnetic Stimulation in Clinical Neurophysiology. Stoneham, MA, Butterworth, (1990), pp 73–112.
6. Amassian VE, Quirk GJ, Stewart M: A comparison of corticospinal activation by magnetic coil and electrical stimulation of monkey motor cortex. Electroencephalogr Clin Neurophysiol 77:390–401, 1990.
7. Amassian VE, Eberle LP, Maccabee PJ, Cracco RQ: Modelling magnetic coil excitation in human cerebral cortex with a peripheral nerve immersed in a brain-shaped volume conductor: The significance of fiber bending in excitation. Electroencephalogr Clin Neurophysiol 85:291–301, 1992.
8. Amassian VE, Maccabee PJ, Cracco RQ, et al: Measurement of information processing delays in human visual cortex with repetitive magnetic coil stimulation. Brain Res 605:317–321, 1993.
9. Amassian VE, Maccabee PJ, Cracco RQ, et al: The polarity of the induced electric field influences magnetic coil inhibition of human visual cortex: Implications for the site of excitation. Electroencephalogr Clin Neurophysiol 93:21–26, 1994.
10. Amassian VE, Cracco RQ, Maccabee PJ, et al: Some positive effects of transcranial magnetic stimulation. Adv Neurol 67:79–106, 1995.
11. Amassian VE, Rothwell JC, Redding M, et al: A high frequency resonant circuit in human motor cortex. Soc Neurosci Abstracts 22:658, 1996.
12. Amassian VE, Cracco RQ, Maccabee PJ, et al: Spatial facilitation of human motor responses by near-threshold magnetic stimulation of parietal and frontal areas. J Physiol 504.P, 115P, 1997.
13. Amassian VE, Cracco RQ, Maccabee PJ, et al: Transcranial magnetic stimulation in study of the visual pathway. J Clin Neurophysiol 15:288–304, 1998.
14. Amassian VE, Rothwell JC, Cracco RQ, et al: What is excited by near-threshold twin magnetic stimuli over human cerebral cortex? J Physiol 506.P, 122P, 1998.
15. Amassian VE, Maccabee PJ, Cracco RQ, Rothwell JC: Spatial inhibition of human motor responses to magnetic stimulation of parietal lobe by premotor stimulation. J Physiol 515.P, 156P, 1999.
16. Asanuma H, Keller A: Neuronal mechanisms of motor learning in mammals. Neuroreport 2:217–224, 1991.
17. Barker AT, Freeston IL, Jalinous R, Jarratt JA: Magnetic stimulation of the human brain and peripheral nervous system: An introduction and the results of an initial clinical evaluation. Neurosurgery 20:100–109, 1987.
18. Berardelli A, Inghilleri M, Cruccu G, Manfredi M: Descending volley after electrical and magnetic transcranial stimulation in man. Neurosci Lett 112:54–58, 1990.
19. Berardelli A, Rona S, Inghilleri M, Manfredi M: Cortical inhibition in Parkinson's disease. A study with paired magnetic stimulation. Brain 119:71–77, 1996.
20. Betucci D, Cantallo R, Gianelli M, et al: Menstrual migraine without aura: Cortical excitability to magnetic stimulation. Headache 32:345–347, 1992.
21. Boyd SG, Rothwell JC, Cowan JMA, et al: A method of monitoring function in cortical pathways during scoliosis surgery with a note on motor conduction velocities. J Neurol Neurosurg Psychiatry 49:251–257, 1986.
22. Brasil-Neto JP, Cohen LG, Panizza M, et al: Optimal focal transcranial magnetic activation of the human motor cortex: Effects of coil orientation, shape of the induced current pulse, and stimulus intensity. J Clin Neurophysiol 9:132–136, 1992.
23. Brasil-Neto JP, Cohen LG, Pascual-Leone A, et al: Rapid reversible modulation of human motor outputs after transient deafferentation of the forearm: A study with transcranial magnetic stimulation. Neurology 42:1302–1306, 1992.
24. Brasil-Neto JP, McShane LM, Fuhr P, et al: Topographic mapping of the human motor cortex with magnetic stimulation: Factors affecting accuracy and reproducibility. Electroencephalogr Clin Neurophysiol 85:9–16, 1992.
25. Brasil-Neto JP, Valls-Solé J, Pascual-Leone A, et al: Rapid modulation of human cortical motor outputs following ischemic nerve block. Brain 116:511–525, 1993.
26. Braune HJ, Fritz C: Transcranial magnetic stimulation-evoked inhibition of voluntary muscle activity is impaired in patients with ischemic hemispheric lesion. Stroke 26(4):550–553, 1995.
27. Bridgers SL: The safety of transcranial magnetic stimulation reconsidered: Evidence regarding cognitive and other cerebral effects. In Levy WJ (ed): Magnetic Motor Stimulation: Basic Principles and Clinical Experience. Amsterdam, Elsevier Science, 1991, pp 170–179.
28. Britton TC, Meyer BU, Herdman J, Benecke R: Clinical use of the magnetic stimulator in the investigation of peripheral conduction time. Muscle Nerve 13:396–406, 1990.
29. Brown P, Ridding MC, Werhahn KJ, et al: Abnormalities of the balance between inhibition and excitation in the motor cortex of patients with cortical myoclonus. Brain 119:309–317, 1996.
30. Burke D, Hicks RG, Stephen PH: Corticospinal volleys evoked by anodal and cathodal stimulation to the human motor cortex. J Physiol 425:283–299, 1990.
31. Calancie B, Nordin M, Wallin U, Hagbarth K-E: Motor-unit responses in human wrist flexor and extensor muscles to transcranial cortical stimuli. J Neurophysiol 58:1168–1185, 1987.
32. Cantello R, Gianelli M, Bettucci D, et al: Parkinson's disease rigidity: Magnetic motor evoked potentials in a small hand muscle. Neurology 41:1449–1456, 1991.
33. Cantello R, Gianelli M, Civardi C, Mutani R: Magnetic brain stimulation: The silent period after the motor evoked potential. Neurology 42:1951–1959, 1992.
34. Caramia MD, Cicinelli P, Paradiso C, et al: 'Excitability' changes of muscular responses to magnetic brain stimulation in patients with central motor disorders. Electroencephalogr Clin Neurophysiol 81:243–250, 1991.
35. Caramia MD, Gigli G, Iani C, et al: Distinguishing forms of generalized epilepsy using magnetic brain stimulation. Electroencephalogr Clin Neurophysiol 98:14–19, 1996.
36. Chen RF, Gerloff CF, Classen JF, et al: Safety of different intertrain intervals for repetitive transcranial magnetic stimulation and recommendations for safe ranges of stimulation parameters. Electroencephalogr Clin Neurophysiol 105:415–421, 1997.
37. Chen R, Classen J, Gerloff C, et al: Depression of motor cortex excitability by low frequency transcranial magnetic stimulation. Neurology 48:1398–1403, 1997.

38. Chen R, Samii A, Caños M, et al: Effects of phenytoin on cortical excitability in humans. Neurology 49:881–883, 1997.

39. Chen R, Corwell B, Yaseen Z, et al: Mechanisms of cortical reorganization in lower-limb amputees. J Neurosci 18(9):3443–3450, 1998.

40. Chokroverty S, Spire JP, DiLullo J, et al: Magnetic stimulation of the human peripheral nervous system. In Chokroverty S (ed): Magnetic Stimulation in Clinical Neurophysiology. Boston Butterworth, 1990, pp 249–273.

41. Chokroverty S, et al: Magnetic brain stimulation: Safety studies. Electroencephalogr Clin Neurophysiol 97:36–42, 1995.

42. Cicinelli P, Traversa R, Bassi A, et al: Interhemispheric differences of hand muscle representation in human motor cortex. Muscle Nerve 20:535–542, 1997.

43. Classen J, Benecke R: Inhibitory phenomena in individual motor units induced by transcranial magnetic stimulation. Electroencephalogr Clin Neurophysiol 97:264–274, 1995.

44. Classen J, Schnitzler A, Binkofski F, et al: The motor syndrome associated with exaggerated inhibition within the primary motor cortex of patients with hemiparetic stroke. Brain 120:605–619, 1997.

45. Classen J, Liepert J, Wise SP, et al: Rapid plasticity of human cortical movement representation induced by practice. J Neurophysiol 79:1117–1123, 1998.

46. Claus D, Murray NMF, Spitzer A, Flugel D: The influence of stimulus type on the magnetic excitation of nerve structures. Electroencephalogr Clin Neurophysiol 75:342–349, 1990.

47. Claus D, Weis M, Jahnke U, et al: Corticospinal conduction studied with magnetic double stimulation in the intact human. J Neurol Sci 111:180–188, 1992.

48. Cohen D, Cuffin BN: Developing a more focal magnetic simulator. Part 1: Some basic principles. J Clin Neurophyiol 8:102–111, 1991.

49. Cohen LG, Bandinelli S, Findlay TW, Hallett M: Motor reorganization after upper limb amputation in man. Brain 114:615–627, 1991.

50. Cohen LG, Celnik P, Pascual-Leone Al, et al: Functional relevance of cross-modal plasticity in blind humans. Nature 389:180–183, 1997.

51. Cole JD, Merton WL, Barrett G, et al: Evoked potentials in a subject with a large-fiber sensory neuropathy below the neck. Can J Physiol Pharmacol 73(2):234–245, 1995.

52. Counter SA, Borg E, Lofqvist L, Brismar T: Hearing loss from the acoustic artifact of the coil used in extracranial magnetic stimulation. Neurology 40:1159–1162, 1959.

53. Cowan JMA, Day BL, Marsden CD, Rothwell JC: The effect of percutaneous motor cortex stimulation on H reflexes in muscles of the arm and leg in intact man. J Physiol 377:333–347, 1986.

54. Cracco RQ, Amassian VE, Maccabee PJ, Cracco JB: Flow of symbolic visual information from retina to vocalization. In Kimura J, Shibasaki H (eds): Recent Advances in Clinical Neurophysiology. Amsterdam, Elsevier Science, 1996, pp 962–969.

55. Cros D, Gominak S, Shahani B, et al: Comparison of electric and magnetic coil stimulation in the supraclavicular region. Muscle Nerve 15:587–590, 1992.

56. Datta AK, Harrison LM, Stephens JA: Task dependent changes in the size of response to magnetic brain stimulation in human first dorsal interosseous muscle. J Physiol (Lond) 418:13–23, 1989.

57. Davey NJ, Romaiguère P, Maskill DW, Ellaway PH: Suppression of voluntary motor activity revealed using transcranial magnetic stimulation of the motor cortex in man. J Physiol 477.2:223–253, 1994.

58. Day BL, Thompson PD, Dick JP, et al: Different sites of action of electric and magnetic stimulation of the human brain. Neurosci Lett 75:101–106, 1987.

59. Day BL, Rothwell JC, Thompson PD, et al: Motor cortex stimulation in intact man. 2. Multiple descending volleys. Brain 110:1191–1209, 1987.

60. Day BL, Rothwell JC, Thompson PD, et al: Delay in the execution of voluntary movement by electrical or magnetic brain stimulation in intact man. Brain 112:649–663, 1989.

61. Deletis V: Intraoperative monitoring of the functional integrity of the motor pathways. In Devinsky O, Beric A, and Dogali M (eds): Advances in Neurology: Electrical and Magnetic Stimulation of the Brain. New York, Raven Press, 1993, pp 201–214.

62. Desiato MT, Caramia MD: Towards a neurophysiological marker of amyotrophic lateral sclerosis as revealed by changes in cortical excitability. Electroencephalogr Clin Neurophysiol 105:1–7, 1997.

63. Di Lazzaro V, Restuccia D, Colosimo C, Tonali P: The contribution of magnetic stimulation of the motor cortex to the diagnosis of cervical spondylotic myelopathy. Correlation of central motor conduction to distal and proximal upper limb muscles with clinical and MRI findings. Electroencephalogr Clin Neurophys 85:311–320, 1992.

64. Di Lazzaro V, Restuccia D, Molinari M, et al: Excitability of the motor cortex to magnetic stimulation in patients with cerebellar lesions. J Neurol Neurosurg Psychiatry 57:108–110, 1994.

65. Di Lazzaro V, Restuccia D, Nardone R, et al: Motor cortex changes in a patient with hemicerebellectomy. Electroencephalogr Clin Neurophysiol 97:259–263, 1995.

66. Di Lazzaro V, Oliviero A, Profice P, et al: Comparison of descending volleys evoked by transcranial magnetic and electric stimulation in conscious humans. Electroencephalogr Clin Neurophys 109:397–401, 1998.

67. Donoghue JP, Suner S, Sanes JN: Dynamic organization of primary motor cortex output to target muscles in adult rats. II. Rapid reorganization following motor nerve lesions. Exp Brain Res 79:492–503, 1990.

68. Durand D, Ferguson AS, Dalbasti T: Induced electric fields by magnetic stimulation in non-homogeneous conducting media. Proc IEEE Eng Med Biol Soc 11:1252–1253, 1989.

69. Eisen A, Pant B, Stewart H: Cortical excitability in amyotrophic lateral sclerosis: A clue to pathogenesis. Can J Neurol Sci 20:11–16, 1993.

70. Epstein CM, Fernandez-Beer E, Weissman JD, Matsuura S: Cervical magnetic stimulation: The role of the neural foramen. Neurology 41:677–680, 1991.

71. Epstein CM, Zangaladze A: Magnetic coil suppression of extrafoveal visual perception using disappearance targets. J Clin Neurophysiol 13:242–246, 1996.

72. Epstein CM, Verson R, Zangaladze A: Magnetic coil suppression of visual perception at an extracalcarine site. J Clin Neurophysiol 13:247–252, 1996.

73. Epstein CM, Lah JK, Meador K, et al: Optimum stimulus parameters for lateralized suppression of speech with magnetic brain stimulation. Neurology 47:1590–1593, 1996.

74. Evans BA, Daube JR, Litchy WJ: A comparison of magnetic and electrical stimulation of spinal nerves. Muscle Nerve 13:414–420, 1990.

75. Filipovic SR, Ljubisavljevic M, Svetel M, et al: Impairment of cortical inhibition in writer's cramp as revealed by changes in electromyographic silent period after transcranial magnetic stimulation. Neurosci Lett 222:167–170, 1997.

76. Flament D, Goldsmith P, Buckley CJ, Lemon RN: Task dependence of responses in first dorsal interosseous muscle to magnetic brain stimulation in man. Physiol (Lond) 464:361–378, 1993.

77. Fong JKY, Werhahn KJ, Rothwell JC, et al: Motor cortex excitability in focal and generalized epilepsy [abstract]. J Physiol (Lond) 459:468P, 1993.

78. Fuhr P, Agostino R, Hallet M: Spinal motor neuron excitability during the silent period after cortical stimulation. Electroencephalogr Clin Neurophysiol 81:257–262, 1991.

79. George MS, Wasserman EM, Williams WA, et al: Changes in mood and hormone levels after rapid-rate transcranial magnetic stimulation (rTMS) of the prefrontal cortex. J Neuropsychiatry Clin Neurosci 8:172–180, 1996.

80. George MS, Wasserman EM, Kimbrell TA, et al: Daily left prefrontal repetitive transcranial magnetic stimulation improves mood in depression: A placebo-controlled crossover trial. Am J Psychiatry 154:1752–1776, 1997.

81. Grafton ST, Mazziotta JC, Presty S, et al: Functional anatomy of human procedural learning determined with regional cerebral blood flow and PET. Neurosci 12:2542–2548, 1992.

82. Grafman J, Pascual-Leone A, Alway D, et al: Induction of a recall deficit by rapid-rate transcranial magnetic stimulation. Neuroreport 5:1157–1160, 1994.

83. Greenberg BD, George MS, Martin JD, et al: Effect of prefrontal repetitive transcranial magnetic stimulation in obsessive-compulsive disorder: A preliminary study. Am J Psychiatry 154:867–869, 1997.

84. Hall EJ, Flament D, Fraser C, Lemon RN: Non-invasive brain stimulation reveals reorganized cortical outputs in amputees. Neurosci Lett 116:379–386, 1990.

85. Hallett M: Transcranial magnetic stimulation negative effects. In Fahn S, Hallett M, Lüders O, Marsden CD (eds): Negative Motor Phenomena. Advances in Neurology. Philadelphia, Lippincott Raven, 1995, pp 107–113.

86. Hamano T, et al: Lack of prolonged cerebral blood flow change after transcranial magnetic stimulation. Electroencephalogr Clin Neurophysiol 89:207–210, 1993.

87. Hanajima R, Ugawa Y, Terao Y, et al: Ipsilateral cortico-cortical inhibition of the motor cortex in various neurological disorders. J Neurol Sci 140:109–116, 1996.

88. Haug BA, Schonle PW, Knobloch C, Kohne M: Silent period measurement revives as a valuable diagnostic tool with transcranial magnetic stimulation. Electroencephalogr Clin Neurophysiol 85:158–160, 1992.

89. Heller L, van Hulsteyn DB: Brain stimulation using electromagnetic sources: Theoretical aspects. Biophys J 63:129–138, 1992.

90. Hern JEC, Landgren S, Phillips CG, Porter R: Selective excitation of corticofugal neurones by surface-anodal stimulation of the baboon's motor cortex. J Physiol (Lond) 161:73–90, 1962.

91. Hess CW, Mills KR, Murray NM, Schriefer TN: Magnetic brain stimulation: Central motor conduction studies in multiple sclerosis. Ann Neurol 22:744–752, 1987.

92. Hess CW, Mills KR, Murray NM: Responses in small hand muscles from magnetic stimulation of the human brain. J Physiol (Lond) 388:397–419, 1987.

93. Holmgren H, Larrson LE, Pedersen S: Late muscular responses to transcranial cortical stimulation in man. Electroencephalogr Clin Neurophysiol 75:161–172, 1990.

94. Hömberg V, Netz J: Generalized seizures induced by transcranial magnetic stimulation of motor cortex [letter] Lancet 2:1223, 1989.

95. Hufnagel A, Elger CE, Klingmuller D, et al: Activation of epileptic foci by transcranial magnetic stimulation: Effects on secretion of prolactin and luetinizing hormone. J Neurol 237:242–246, 1990.

96. Inghilleri M, Berardelli A, Marchetti P, Manfredi M: Silent period evoked by transcranial stimulation of the human motor cortex and cervicomedullary junction. J Physiol (Lond) 161:112–125, 1993.

97. Inghilleri M, Berardelli A, Marchetti P, Manfredi M: Effects of diazepam, baclofen and thiopental on the silent period evoked by transcranial magnetic stimulation in humans. Exp Brain Res 109: 467–472, 1996.

98. Iriki A, Pavlides C, Keller A, Asanuma H: Long-term potentiation of motor cortex. Science (Wash DC) 245:1385–1387, 1989.

99. Jacobs KM, Donoghue JP: Reshaping the cortical motor map by unmasking latent intracortical connections. Science (Wash DC) 251:944–947, 1991.

100. Jaskolsky DJ, Jarratt JA, Jakubowsky J: Clinical evaluation of magnetic stimulation in cervical spondylosis. Br J Neurosurg 3:541–548, 1989.

101. Jenkins WM, Merzenich MM, Recanzone G: Neocortical representational dynamics in adult primates: Implications for neuropsychology. Neuropsychologia 28:573–584, 1990.

102. Jennum P, Friberg L, Fuglsang-Frederiksen A, Dam M: Speech localization using repetitive transcranial magnetic stimulation. Neurology 44:269–273, 1994.

103. Kaas JH: Plasticity of sensory and motor maps in adult mammals. Annu Rev Neurosci 14:137–167, 1991.

104. Kaneko K, Kawai S, Fuchigami Y, et al: The effect of current direction induced by transcranial magnetic stimulation on the corticospinal excitability in human brain. Electroencephalogr Clin Neurophysiol 101:478–482, 1996.

105. Karni A, Meyer G, Jezzari P, et al: Functional MRI evidence for adult motor cortex plasticity during motor skill learning. Nature 377:155–158, 1995.

106. Kernell D, Wu CP: Responses of the pyramidal tract to stimulation of the baboon's motor cortex. J Physiol 191:653–672, 1967.

107. Kiers L, Cros D, Chiappa KH, Fang J: Variability of motor potentials evoked by transcranial magnetic stimulation. Electroencephalogr Clin Neurophysiol 89:415–423, 1993.

108. Kobayashi M, Ueno S, Kurokawa T: Importance of soft tissue inhomogeneity in magnetic peripheral nerve stimulation. Electroencephalogr Clin Neurophysiol 105:406–413, 1997.

109. Krain L, et al: Consequences of cortical magnetoelectric stimulation. In Chokroverty S (ed): Magnetic Stimulation in Clinical Neurophysiology. Boston, Butterworth, 1990, pp 157–163.

110. Kujala T, et al: Visual cortex activation in blind subjects during sound discrimination. Neurosci Lett 183:143–146, 1995.

111. Kujala T, Alho K, Huotilainen M, et al: Electrophysiological evidence for cross-modal plasticity in humans with early- and late-onset blindness. Psychophysiology 34:213–216, 1997.

112. Kujirai T, Caramia MD, Rothwell JC, et al: Corticocortical inhibition in human motor cortex. J Physiol (Lond) 471:501–519, 1993.

113. Landgren S, Phillips CG, Porter R: Cortical fields of origin of the monosynaptic pyramidal pathways to some alpha motoneurones of the baboon's hand and forearm. J Physiol 161:112–115, 1962.

114. Lemon RN, Johansson RS, Westling G: Corticospinal control during reach, grasp, and precision lift in man. J Neurosci 15:6145–6156, 1995.

115. Levy WJ, Amassian VE, Traad M, Cadwell J: Focal magnetic coil stimulation reveals motor cortical system reorganized in humans after traumatic quadriplegia. Brain Res 510:130–134, 1990.

116. Levy WJ, et al: Safety studies of electrical and magnetic stimulation for the production of motor evoked potentials. In Chrokroverty S (ed): Magnetic Stimulation in Clinical Neurophysiology. Boston, Butterworth, 1990, pp 165–172.

117. Levy WJ, Amassian VE, Schmid UD, Jungreis C: Mapping of motor cortex gyral sites non-invasively by transcranial magnetic stimulation in normal subjects and patients. In Levy WJ, Cracco RQ, Barker AT, Rothwell J (eds): Magnetic Motor Stimulation: Basic Principles and Clinical Experience (EEG Suppl 43). Amsterdam Elsevier Science, 1991, pp 51–75.

118. Liepert J, Tegenthoff M, Malin JP: Changes of cortical motor area size during immobilization. Electroencephalogr Clin Neurophysiol 97:382–386, 1995.

119. Liepert J, Schardt S, Balogh A, Weiller C: Atropine suppresses intracortical inhibition. Electroencephalogr Clin Neurophysiol 107(3):82P, 1998.

120. Liepert J, Schwenkreis P, Witscher K, et al: Modulation of intracortical facilitation and intracortical inhibition during 7 days of riluzole ingestion. Electroencephalogr Clin Neurophysiol 107(3):83P, 1998.

121. Maccabee PJ, Amassian VE, Cracco RQ, et al: Magnetic coil stimulation of human visual cortex: Studies of perception. Electroencephalogr Clin Neurophysiol 43:111–120, 1991.

122. Maccabee PJ, Amassian EE, Eberle LP, et al: Measurement of the electric field induced into inhomogeneous volume conductors by magnetic coils. Application to human spinal neurogeometry. Electroencephalogr Clin Neurophysiol 81:224–237, 1991.

123. Maccabee PJ, Amassian VE, Eberle LP, Cracco RQ: Magnetic coil stimulation of straight and bent amphibian and mammalian peripheral nerve in vitro: Locus of excitation. J Physiol 460:201–219, 1993.

124. Maccabee PJ, Lipitz ME, Desudchit T, et al: Detection of proximal demyelinating neuropathy in cauda equina by neuromagnetic stimulation [abstract]. Neurology 45(suppl 4):A170, 1995.

125. Maccabee PJ, Lipitz ME, Desudchit T, et al: A new method using neuromagnetic stimulation to measure conduction time within the cauda equina. Electroencephalogr Clin Neurophysiol 101:153–166, 1996.

126. Maccabee PJ, Nagarajan SS, Amassian VE, et al: Influence of pulse sequence, polarity and amplitude on magnetic stimulation of human and porcine peripheral nerve. J Physiol 513.2:571–585, 1998.

127. Maertens de Noordhout A, Remacle JM, Pepin JL, et al: Magnetic stimulation of the motor cortex in cervical spondylosis. Neurology 41:75–80, 1991.

128. Maertens de Noordhout A, Pepin J, Schoenen J, Delwaide PJ: Percutaneous magnetic stimulation of the motor cortex in migraine. Electroencephalogr Clin Neurophysiol 85:110–115, 1992.

129. Magistris MR, Rosler HM, Truffert A, Myers JP: Transcranial stimulation excites virtually all motor neurons supplying the target muscle. Brain 121:437–450, 1998.

130. Mano Y, Nakamuro T, Tamura R, et al: Central motor reorgani-

zation after anastomosis of the musculocutaneous and intercostal nerves following cervical root avulsion [see comments]. Ann Neurol 38:15–20, 1995.

131. Marsden CD, Merton PA, Morton HB: Direct electrical stimulation of corticospinal pathways through the intact scalp in human subjects. In Desmidt JE (ed): Motor Control Mechanisms in Health and Disease. New York, Raven Press, 1983, pp 387–391.

132. Mavroudakis N, Caroyer JM, Brunko E, Zegers de Beyl D: Effects of diphenylhydantoin on motor potentials evoked with magnetic stimulation. Electroencephalogr Clin Neurophysiol 93:428–433, 1994.

133. Mavroudakis N, Caroyer JM, Brunko E, Zegers de Beyl D: Abnormal motor evoked responses to transcranial magnetic stimulation in focal dystonia. Neurology 45:1671–1677, 1995.

134. Merton PA: The silent period in a muscle of the human hand. J Physiol 114:183–198, 1951.

135. Merton PA, Hill DK, Marsden CD, Morton HB: Scope of a technique for electrical stimulation of human brain, spinal cord and muscle. Lancet ii:597–598, 1982.

136. Merzenich MM, Kaas MH, Wall JT, et al: Progression of change following median nerve section in the cortical representation of the hand in areas 3b and 1 in adult owl and squirrel monkeys. Neuroscience 10:639–665, 1983.

137. Merzenich MM, Nelson RJ, Stryker MP, et al: Somatosensory cortical map changes following digit amputation in adult monkeys. J Comp Neurol 224:591–605, 1984.

138. Merzenich MM, Recanzone GH, Jenkins WM, Grajsh KA: Adaptive mechanisms in cortical networks underlying cortical contributions to learning and nondeclarative memory. Cold Spring Symp Quant Biol 55:873–887, 1990.

139. Mills KR, Murray NMF: Electrical stimulation over the human vertebral column: Which neural elements are excited? Electroencephalogr Clin Neurophysiol 63:582–589, 1986.

140. Mills KR, Murray NMF, Hess CW: Magnetic and electrical transcranial brain stimulation: Physiological mechanisms and clinical applications. Neurosurgery 20:164–168, 1987.

141. Mills KR, Boniface SJ, Schubert M: Origin of the secondary increase in firing probability of human motor neurons following transcranial magnetic stimulation. Brain 114:2451–2463, 1991.

142. Mills KR, Boniface SJ, Schubert M: Magnetic brain stimulation with a double coil: The importance of coil orientation. Electroencephalogr Clin Neurophysiol 85:17–21, 1992.

143. Mills KR, Nithi KA: Corticomotor threshold to magnetic stimulation: Normal values and repeatability. Muscle Nerve 20:570–576, 1997.

144. Nakamura H, Kitagawa H, Kawaguchi Y, Tsuji H: Direct and indirect activation of human corticospinal neurones by transcranial magnetic and electric stimulation. Neurosci Lett 210:45–48, 1996.

145. Nakamura H: D- and I-waves in humans. Electroencephalogr Clin Neurophysiol 107(3):76P, 1998.

146. Nielsen JF: Improvement of amplitude variability of motor evoked potentials in multiple sclerosis patients and healthy subjects. Electroencephalogr Clin Neurophysiol 101:404–411, 1996.

147. Nilsson J, Panizza M, Roth BJ, et al: Determining the site of stimulation during magnetic stimulation of a peripheral nerve. Electroencephalogr Clin Neurophysiol 85:253–264, 1992.

148. Pascual-Leone A, Gates JR, Dhuna A: Induction of speech arrest and counting errors with rapid-rate transcranial magnetic stimulation. Neurology 41:697–702, 1991.

149. Pascual-Leone A, Torres F: Plasticity of the sensorimotor cortex representation of the reading finger in Braille readers. Brain 116:39–52, 1993.

150. Pascual-Leone A, Houser CM, Reeves K, et al: Safety of rapid-rate transcranial magnetic stimulation in normal volunteers. Electroencephalogr Clin Neurophysiol 89:120–130, 1993.

151. Pascual-Leone A, Gomez-Tortosa E, Grafman J, et al: Induction of visual extinction by rapid-rate transcranial magnetic stimulation of parietal lobe. Neurology 44:494–498, 1994.

152. Pascual-Leone A, Valls-Sole J, Wassermann EM, Hallett M: Responses to rapid-rate transcranial stimulation of the human motor cortex. Brain 117:847–858, 1994.

153. Pascual-Leone A, Dang N, Cohen LG, et al: Modulation of muscle responses evoked by transcranial magnetic stimulation during the acquisition of new fine motor skills. J Neurophys 74:1037–1045, 1995.

154. Pascual-Leone A, Wassermann EM, Sadato N, Hallett M: The role of reading activity on the modulation of motor cortical outputs to the reading hand of Braille readers. Ann Neurol 34:33–37, 1995.

155. Pascual-Leone A, Rubio B, Pallardō F, Catalá MD: Beneficial effect of rapid-rate transcranial magnetic stimulation of the left dorsolateral prefrontal cortex in drug-resistant depression. Lancet 348:233–238, 1996.

156. Pascual-Leone A, Catalá MD, Pascual-Leone A: Lateralized effect of rapid-rate transcranial magnetic stimulation of the prefrontal cortex on mood. Neurology 46:499–502, 1996.

157. Pascual-Leone A, Peris M, Tormos JM, et al: Reorganization of human cortical motor output maps following traumatic forearm amputation. Neuroreport 7:2068, 1996.

158. Pascual-Leone A, Tormos JM, Keenan J, et al: Study and modulation of human cortical excitability with transcranial magnetic stimulation. J Clin Neurophysiol 15(4):333–343, 1998.

159. Patton HD, Amassian VE: Single and multiple unit analysis of cortical stage of pyramidal tract activation. J Neurophysiol 17:345–363, 1954.

160. Penfield W, Boldrey E: Somatic motor and sensory representation in the cerebral cortex of man as studied by electrical stimulation. Brain 60:389–443, 1937.

161. Phillips CG, Porter RR: Corticospinal Neurones London, Academic Press, 1977.

162. Plassman BL, Gandevia SC: High-voltage stimulation over the human spinal cord: Sources of latency variation. J Neurol Neurosurg Psychiatry 52:213–217, 1989.

163. Pons TP, Garraghty PE, Ommaya AK, et al: Massive cortical reorganization after sensory deafferentation in adult macaques. Science 252:1857–1860, 1991.

164. Priori A, Berardelli A, Inghilleri M, et al: Electromyographic silent period after transcranial brain stimulation in Huntington's disease. Mov Dis 9(2):178–182, 1994.

165. Priori A, Berardelli A, Inghilleri M, et al: Motor cortical inhibition and the dopaminergic system. Pharmacological changes in the silent period after transcranial brain stimulation in normal subjects, patients with Parkinson's disease and drug induced parkinsonism. Brain 117:317–323, 1994.

166. Priori A: Clinical applications of silent period measurements. In Maugeri S (ed): Advances in Occupational Medicine and Rehabilitation. Pavia, Fondazione, 1996, pp 91–97.

167. Prout AJ, Eisen AA: The cortical silent period and amyotrophic lateral sclerosis. Muscle Nerve 17:217–223, 1994.

168. Raichle ME, Fiez JA, Videen TO, et al: Practice-related changes in human brain functional anatomy during nonmotor learning. Cerebral Cortex 4:8–26, 1994.

169. Rauschecker JP: Compensatory plasticity and sensory substitution in the cerebral cortex. Trends Neurosci 18:36–43, 1995.

170. Ravnborg M, Blinkenberg M, Dahl K: Standardization of facilitation of compound muscle action potentials evoked by magnetic stimulation of the cortex. Results in healthy volunteers and in patients with multiple sclerosis. Electroencephalogr Clin Neurophysiol 81:195–201, 1991.

171. Reilly JP: Peripheral nerve stimulation by induced electric currents: exposure to time-varying magnetic fields. Med Biol Eng Comp 27:101–110, 1989.

172. Ridding MC, Sheean G, Rothwell JC, et al: Changes in the balance between motor cortical excitation and inhibition in focal, task specific dystonia. J Neurol Neurosurg Psychiatry 59:493–498, 1995.

173. Ridding MC, Inzelberg R, Rothwell JC: Changes in excitability of motor cortical circuitry in patients with Parkinson's disease. Ann Neurol 37:181–188, 1995.

174. Ridding MC Rothwell JC: Reorganization in human motor cortex. Can J Physiol Pharmacol 73:218–222, 1995.

175. Ridding MC, Rothwell JC: Stimulus response curves as a method of measuring motor cortical excitability in man. Electroencephalogr Clin Neurophysiol 105:340–343, 1997.

176. Rijntjes M, Tegenthoff M, Liepert J, et al: Cortical reorganization in patients with facial palsy. Ann Neurol 41:621–630, 1997.

177. Roick H, von Giesen HJ, Benecke R: On the origin of the postexcitatory inhibition seen after transcranial magnetic brain

stimulation in awake human subjects. Exp Brain Res 94:489–498, 1993.

178. Rossini PM, Barker AT, Berardelli A, et al: Non-invasive electrical and magnetic stimulation of the brain, spinal cord and roots: Basic principles and procedures for routine application. Report of an IFCN committee. Electroencephalogr Clin Neurophysiol 91:79–92, 1994.

179. Rossini PM, Rossi S: Clinical applications of motor evoked potentials. Electroencephalogr Clin Neurophysiol 106:180–194, 1998.

180. Roth BJ, Basser P: Model of the stimulation of a nerve fiber by electromagnetic induction. IEEE Trans Biomed Eng 37:588–597, 1990.

181. Roth BJ, Cohen LG, Hallett M, et al: A theoretical calculation of the electrical field induced by magnetic stimulation of a peripheral nerve. Muscle Nerve 13:734–741, 1990.

182. Roth BJ, Pascual-Leone A, Cohen LG, Hallett M: The heating of metal electrodes during rapid-rate magnetic stimulation: A possible safety hazard. Electroencephalogr Clin Neurophysiol 85:116–123, 1992.

183. Rothwell JC, Thompson PD, Day BL, et al: Motor cortex stimulation in intact man. 1. General characteristics of EMG responses in different muscles. Brain 110:1173–1190, 1987.

184. Rothwell JC, Ferbert A, Caramia MD, et al: Intracortical inhibitory circuits studied in the human. Neurology 41(suppl 1):192, 1991.

185. Rothwell JC, Thompson PD, Day BL, et al: Stimulation of the human motor cortex through the scalp. Exp Physiol 76:159–200, 1991.

186. Rothwell JC: The use of paired pulse stimulation to investigate the intrinsic circuitry of human motor cortex. In Nilsson J, Panizza M, Grandori F (eds): Advances in Magnetic Stimulation. Mathematical Modelling and Clinical Applications. Pavia, PI-ME Press, 1996, pp 99–104.

187. Ruohonen J, Panizza M, Nilsson J, et al: Transverse-field activation mechanism in magnetic stimulation of peripheral nerves. Electroencephalogr Clin Neurophysiol 101:167–174, 1996.

188. Sadato N, Pascual-Leone A, Grafman J, et al: Activation of the primary visual cortex by Braille reading in blind subjects Nature 380:526–528, 1996.

189. Sakai K, Ugawa Y, Terao Y, et al: Preferential activation of different I waves by transcranial magnetic stimulation with a figure-of-eight shaped coil. Exp Brain Res 113:24–32, 1997.

190. Sandbrink F, Syed NA, Hallet M, Bloeter MK: Increased motor cortex excitability in stiff person syndrome–a paired pulse TMS study. Electroencephalogr Clin Neurophysiol 107(3):90P, 1998.

191. Sander D, Meyer B-U, Röricht S, et al: Effect of hemisphere-selective repetitive transcranial magnetic brain stimulation on middle cerebral artery blood flow velocity. Electroencephalogr Clin Neurophysiol 97:43–48, 1995.

192. Sander D, Meyer B-U, Röricht S, et al: Increase in posterior cerebral artery blood flow during threshold repetitive magnetic stimulation of the human visual cortex: Hints for neuronal activation without cortical phosphenes. Electroencephalogr Clin Neurophysiol 99:473–478, 1996.

193. Sanes JN, Suner S, Donoghue JP: Dynamic organization of primary motor cortex output to target muscles in adult rats. I. Long-term patterns of reorganization following motor or mixed peripheral nerve lesions. Exp Brain Res 79:479–491, 1990.

194. Schieppati M, Trompetto C, Abbruzzese G: Selective facilitation of responses to cortical stimulation of proximal and distal arm muscles by precision tasks in man. J Physiol (Lond) 491:551–562, 1996.

195. Schlaug G, Knorr U, Seitz RJ: Inter-subject variability of cerebral activations in acquiring a motor skill: A study with positron emission tomography. Exp Brain Res 98:523–534, 1994.

196. Schmid UD, Walker G, Schmid-Sigron J, Hess CW: Transcutaneous magnetic and electric stimulation over the cervical spine: Excitation of plexus roots rather than spinal roots. In Levy WJ, Cracco RQ, Barker AT, Rothwell J (eds): Magnetic Motor Stimulation: Basic Principles and Clinical Experience (EEG Suppl. 43). Amsterdam Elsevier Science 1991, pp 369–384.

197. Schnitzler A, Benecke R: The silent period after transcranial brain stimulation is of exclusive cortical origin: Evidence from isolated cortical ischemic lesions in man. Neurosci Lett 180(1):41–45, 1994.

198. Schriefer TN, Hess CW, Mills KR, Murray NM: Central motor conduction studies in motor neuron disease using magnetic brain stimulation. Electroencephalogr Clin Neurophysiol 74:431–437, 1989.

199. Schulze-Bonhage A, Knott H, Ferbet A: Effects of carbemazepine on cortical and inhibitory phenomenon: A study with paired transcranial magnetic stimulation, Electroencephalogr Clin Neurophysiol 99:267–273, 1996.

200. Seitz RJ, Roland E, Bohm C, et al: Motor learning in man: A positron emission tomographic study. Neuroreport 1:57–60, 1990.

201. Shahani BT, Young RR: Studies of the normal human silent period. In Desmedt JE (ed): New Developed EMG Clinical Neurophysiology. Basel, Karger, 1973, pp 589–602.

202. Siebner HR, Tormos JM, Ceballos-Baumann AO, et al: Low-frequency repetitive transcranial magnetic stimulation of the motor cortex in writer's cramp. Neurology 52:529–537, 1999.

203. Streletz LJ, Belevich JK, Jones SM, et al: Transcranial magnetic stimulation: Cortical motor maps in acute spinal cord injury. Brain Topogr 7:245–250, 1995.

204. Tavy DLJ, Wagner GL, Keunen RWM, et al: Transcranial magnetic stimulation in patients with cervical spondylotic myelopathy: Clinical and radiological correlation. Muscle Nerve 17:235–241, 1994.

205. Tofts PS: The distriction of induced currents in magnetic stimulation of nervous system. Physics Med Biol 35:1119–1128, 1990.

206. Tokimura H, Ridding MC, Tokimura Y, et al: Short latency facilitation between pairs of threshold magnetic stimuli applied to human motor cortex. Electroencephalogr Clin Neurophysiol 101:263–272, 1996.

207. Topka H, Cohen LG, Cole RA, Hallett M: Reorganization of corticospinal pathways following spinal cord injury. Neurology 41:1276–1283, 1991.

208. Triggs WJ, Macdonell RAL, Cros D, et al: Motor inhibition and excitation are independent effects of magnetic cortical stimulation. Ann Neurol 32:345–351, 1992.

209. Triggs WJ, Cros D, Macdonell RAL, et al: Cortical and spinal motor excitability during the transcranial magnetic stimulation silent period in humans. Brain Res 628:39–48, 1993.

210. Triggs WJ, Kiers L, Cros D, et al: Facilitation of magnetic motor evoked potentials during the cortical stimulation silent period. Neurology 43:2615–2620, 1993.

211. Triggs WJ, Calvanio R, Macdonell RAL, et al: Physiological motor asymmetry in human handedness: Evidence from transcranial magnetic stimulation Brain Res 636:270–276, 1994.

212. Triggs WJ, Calvanio R, Levine M: Transcranial magnetic stimulation reveals a hemispheric asymmetry correlate of intermanual differences in motor performance. Neuropsychologia 35:1355–1363, 1997.

213. Ueno S, Tashiro Y, Harada K: Localized stimulation of neural tissues by means of a paired configuration of time-varying magnetic fields. J Appl Physiol 64:5862–5864, 1988.

214. Ugawa Y, Rothwell JC, Day BL, et al: Magnetic stimulation over the spinal enlargements. J Neurol Neurosurg Psychiatry 51:1025–1032, 1989.

215. Ugawa Y, Rothwell JC, Day BL, et al: Percutaneous electrical stimulation of corticospinal pathways at the level of the pyramidal decussation in humans. Ann Neurol 29:418–427, 1991.

216. Ugawa Y, Uesaka Y, Terao Y, et al: Magnetic stimulation of corticospinal pathways at the foramen magnum level in humans. Ann Neurol 36:618–624, 1994.

217. Uozumi T, Tsuji S, Murai Y: Motor potentials evoked by magnetic stimulation of the motor cortex in normal subjects and patients with motor disorders. Electroencephalogr Clin Neurophysiol 81:251–156, 1991.

218. Uozumi T, Ito Y, Tsuji S, Murai Y: Inhibitory period following motor potentials evoked by magnetic cortical stimulation. Electroencephalogr Clin Neurophysiol 85:273–279, 1992.

219. Valls-Sole J, Pascual-Leone A, Wassermann EM, Hallett M: Human motor evoked responses to paired transcranial magnetic stimuli. Electroencephalogr Clin Neurophysiol 85:355–364, 1992.

220. Van der Linden C, Bruggeman R, Goldman WH: Alterations of motor evoked potentials by thalamotomy. Electromyogr Clin Neurophysiol 33(6):329–334, 1993.

221. Von Giesen HJ, Roick H, Benecke R: Inhibitory actions of the motor cortex following unilateral brain lesions as studied by magnetic brain stimulation. Exp Brain Res 99:84–96, 1994.
222. Walsh V, Cowey A: Magnetic stimulation studies of visual cognition. Trends Cognitive Sci 2:103–110, 1998.
223. Wassermann EM, McShane LM, Hallett M, Cohen LG: Noninvasive mapping of muscle representations in human motor cortex. Electroencephalogr Clin Neurophysiol 85:1–8, 1992.
224. Wassermann EM, Pascual-Leone A, Valls-Solé J, et al: The topography of the inhibitory and excitatory effects of transcranial magnetic stimulation in a hand muscle. Electroencephalogr Clin Neurophysiol 89:424–433, 1993.
225. Wassermann EM, Cohen LG, Flitman SS, et al: Seizures in healthy people with repeated 'safe' trains of transcranial magnetic stimuli [letter]. Lancet 347:825, 1996.
226. Wasserman EM: Risk and safety of repetitive transcranial magnetic stimulation: Report and suggested guidelines from the International Workshop on the safety of repetitive transcranial magnetic stimulation, June 5–7, 1996. Electroencephalogr Clin Neurophysiol 108:1–16, 1998.
227. Wassermann EM, Blaxton TA, Hoffman EA, et al: Repetitive transcranial magnetic stimulation of the dominant hemisphere can disrupt visual naming in temporal lobe epilepsy patients. Neuropsychologia 37(5):537–544, 1999.
228. Weissman JO, Epstein CM, Davey KR: Magnetic brain stimulation and brain size: Relevance to animal studies. Electroencephalogr Clin Neurophysiol 85:215–219, 1992.
229. Werhahn KJ, Fong JK, Meyer BU, et al: The effect of magnetic coil orientation on the latency of surface EMG and single motor unit responses in the first dorsal interosseous muscle. Electroencephalogr Clin Neurophysiol 93:138–146, 1994.
230. Werhahn KJ, Classen J, Benecke R: The silent period induced by transcranial magnetic stimulation in muscle supplied by cranial nerves: Normal data and changes in patients. J Neurol Neurosurg Psychiatry 59(6):586–596, 1995.
231. Werhahn KJ, Förderreuther S, Straube A: Effects of the serotonin 1B/1D receptor agonist zolmitriptan on motor cortical excitability in healthy humans. Neurology 51:896–898, 1998.
232. Wilson SA, Lockwood RJ, Thickbroom GW, Mastaglia FL: The muscle silent period following transcranial magnetic cortical stimulation. J Neurol Sci 114:216–222, 1993.
233. Wilson SA, Thickbroom GW, Mastaglia FL: Topography of excitatory and inhibitory muscle responses evoked by transcranial magnetic stimulation in the human motor cortex. Neurosci Lett 154:52–56, 1993.
234. Xing J, Katayama Y, Yamamoto T, et al: Quantitative evaluation of hemiparesis with corticomyographic motor evoked potential recorded by transcranial magnetic stimulation. J Neurotrauma 7:57–64, 1990.

235. Ziemann U, Netz J, Szelenyi A, Hömberg V: Spinal and supraspinal mechanisms contribute to the silent period in the contracting soleus muscle after transcranial magnetic stimulation of human motor cortex. Neurosci Lett 156:167–171, 1993.
236. Ziemann U, Lönnecker S, Paulus W: Inhibition of the human motor cortex by ethanol. A transcranial magnetic stimulation study. Brain 118(6):1437–1446, 1995.
237. Ziemann U, Lönnecker S, Steinhoff BJ, Paulus W: The effect of lorazepam on the motor cortical excitability in man. Exp Brain Res 109:127–135, 1996.
238. Ziemann U, Lönnecker S, Steinhoff BJ, Paulus W: Effects of antiepileptic drugs on motor cortex excitability in humans: A transcranial magnetic stimulation study. Ann Neurol 40:367–378, 1996.
239. Ziemann U, Bruns D, Paulus W: Enhancement of human motor cortex inhibition by the dopamine receptor agonist pergolide: Evidence from transcranial magnetic stimulation. Neurosci Lett 208:187–190, 1996.
240. Ziemann U, Rothwell JD, Ridding MC: Interaction between intracortical inhibition and facilitation in human motor cortex. J Physiol (Lond) 496:873–881, 1996.
241. Ziemann U, Tergau F, Bruns D, et al: Changes in human motor cortex excitability induced by dopaminergic and antidopaminergic drugs. Electroencephalogr Clin Neurophysiol 105:430–437, 1997.
242. Ziemann U, Paulus W, Rothenberger A: Decreased motor inhibition in Tourette disorder: Evidence from transcranial magnetic stimulation. Am J Psychiatry 154:1277–1284, 1997.
243. Ziemann U, Winter M, Reimers CD, et al: Impaired motor cortex inhibition in patients with amyotrophic lateral sclerosis. Evidence from paired transcranial magnetic stimulation. Neurology 49:1292–1298, 1997.
244. Ziemann U, Tergau F, Wasserman EM, et al: Demonstration of facilitatory I wave interaction in the human motor cortex by paired transcranial magnetic stimulation. J Physiol 511:181–190, 1998.
245. Ziemann U, Hallett M, Cohen LG: Mechanisms of deafferentation-induced plasticity in human motor cortex. J Neurosci 18(17):7000–7007, 1998.
246. Ziemann U, Corwell B, Cohen LG: Modulation of plasticity in human motor cortex after forearm ischemic nerve block. J Neurosci 18(3):1115–1123, 1998.
247. Ziemann U, Tergau F, Wischer S, et al: Pharmacological control of facilitatory I-wave interaction in the human motor cortex. A paired transcranial magnetic stimulation study. Electroencephalogr Clin Neurophysiol 109:321–330, 1998.
248. Ziemann U, Steinhoff BJ, Tergau F, Paulus W: Transcranial magnetic stimulation: Its current role in epilepsy research. Epilepsy Res 30:11–30, 1998.

V   ELECTROENCEPHALOGRAPHY

Chapter 26

Basic Cellular and Synaptic Mechanisms Underlying the Electroencephalogram

Thomas H. Swanson, MD

INTRODUCTION
GENERATION OF CORTICAL POTENTIALS
Anatomy
Electrophysiology of Potential
    Generation
Ion Channels Underlying Potential
    Generation
Attenuation of Extracellular Potentials
Glial Cells and Cerebral Electrical
    Potentials
RHYTHM GENERATION
Barbiturate-induced Sleep Spindles in Cats
Reticular Nucleus of the Thalamus

SUBCORTICAL MODULATION OF RHYTHMS
Arousal Response
SPECIFIC FREQUENCY BANDS OF THE
    ELECTROENCEPHALOGRAM
Delta
Theta
Alpha
Beta
PATHOLOGICAL RHYTHMS
Epileptiform Bursting
Pathological Slow Waves
CONCLUSION

INTRODUCTION

To draw clinically rational diagnostic conclusions from the electroencephalogram (EEG) requires a fundamental understanding of how and where the scalp-recorded EEG potentials are generated. This chapter is an attempt to distill and summarize a wealth of data concerning the brain regions and neurons responsible for the generation of the EEG potentials, the cellular mechanisms involved in this potential generation, the mechanisms of production of the various rhythms recorded with typical EEG amplifiers, and the various factors that modify these basic rhythms. Our understanding of these mechanisms is incomplete owing to the relative paucity of information concerning synaptic circuits, where and how synapses are formed, what controls their formation and function, and how they are remodeled. The physiological basis of the EEG has been the subject of many excellent, well-written chapters and reviews, which are highly recommended to the interested reader (see refs. 1, 36, 46, and 50).

GENERATION OF CORTICAL POTENTIALS

Although the electrical origin of the EEG was first described in 1929 by Hans Berger,[5] only recently have we begun to understand the cellular and network properties of neurons that give rise to the various brain oscillations which summate to form the EEG. The basic mechanisms that generate the electricity recorded by the surface EEG have been well studied. However, substantially less is known about the basic mechanisms responsible for generating rhythmic activity from the asynchronous, seemingly random potentials that comprise the background EEG. The current thinking on how the electricity is generated is reviewed first.

Anatomy

The anatomical arrangement of the cortex plays an important role in the ultimate configuration of scalp-recorded potentials. The cortex contains densely packed neurons in discrete lamina, arranged in parallel with extensive, apically oriented dendritic arborizations that interdigitate with adjacent pyramidal cells. This configuration is ideally suited to generate large potentials of sufficient magnitude to be recorded on the scalp (Fig. 26–1).

Activity from deeper nuclear groups such as the thalamus cannot be detected with scalp electrodes, owing in part to the distance from the recording electrode[9] and in part to the architecture of the thalamic groups, which tends to give rise to closed potential fields,[1, 9] Because of the anatomy of the cortex and the inability of the scalp electrodes to detect deep nuclear activity, it is felt that cortical neurons most likely generate the potentials recorded on the EEG.

Electrophysiology of Potential Generation

Simultaneous intracellular neuronal and scalp EEG recording show that cortical pyramidal activity in layers II, III, and IV corresponds with EEG wave generation.[9] It is thought that the action potential time course of several milliseconds is too short, and that the amount of membrane depolarized by the action potential too small, to generate the scalp EEG potential. Postsynaptic potentials (PSPs) are of longer duration, and involve more membrane space, and are thus likely responsible for the scalp potential. These observations and others gave rise to the currently accepted hypothesis that synchronous excitatory postsynaptic potentials (EPSPs) and inhibitory postsynaptic potentials (IPSPs) of cortical pyramidal cells generate the scalp-recorded potentials.[11, 12, 17, 20, 21, 36] It should be noted, however, that most intracellular micro-

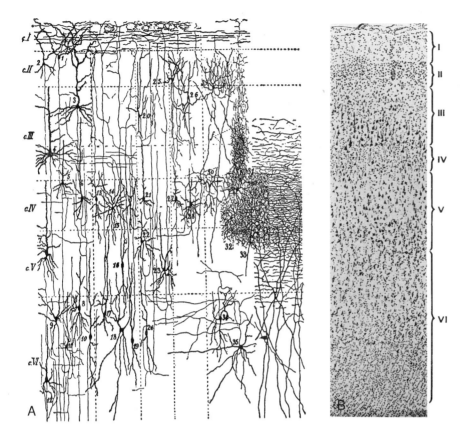

FIGURE 26–1. *A* is a drawing of the human cerebral cortex demonstrating the various cell types which interconnect. Layers II, III, and IV contain the neurons that likely generate the EEG rhythms recorded from scalp electrodes. Note the horizontal orientation of the axons and apically oriented dendrites. *B* is a Nissil stain of the cortex, which demonstrates the closely packed cell bodies of layers II and III. It is this packing of cells that produces the large summed extracellular potentials of the EEG. (Reprinted from Coppenhaber WM, Kelly DE, Wood RL (eds): Bailey's Textbook of Histology, 17th ed. Baltimore, Williams & Wilkins, 1978, with permission.)

electrode studies have not examined the role of nonpyramidal cells because they are technically more difficult to study; thus, their contributions to the network properties of this area are unknown. Of particular interest are glial cells, which are being increasingly studied, and appear to influence neuronal activity in several ways (see later discussion).

Ion Channels Underlying Potential Generation

Much of the study of individual neurons has been at the ion channel level. Neurons of the cortex appear as a homogenous group of cells, and share many similar membrane properties and voltage-dependent ion channel conductances. There are two ion conductances of particular interest in the cortex that participate in cortical "pacing." The first is the "rebound" calcium conductance, which has also been described in thalamic cells[31, 47] (see later discussion); the second is an inactivating sodium conductance, which renders pyramidal cells capable of repetitive firing.[50] These conductances enhance the ability of cortical pyramidal cells to respond to rhythmic subcortical input. It is tempting to model the cortex as a homogenous layer, and this type of thinking is helpful in conceptualizing how the EEG is generated. It should be kept in mind, however, that there is a great deal of diversity in both the molecular makeup of channels and in their single-channel gating characteristics.[24] In addition, ligand-gated ion channels are present in many cortical neurons. Because many different neurotransmitter substances are known to be present in the cortex,[32] it is unlikely that all cortical neurons share a common receptor and voltage-dependent ion channel composition. It is, therefore, likely that our understanding of the EEG will be greatly enhanced as we delineate the unique ion channel "phenotypes" or various neuronal groups.

How do cortical PSPs produce a signal recordable on the scalp? The cortex receives major synaptic input from the thalamus. Afferent excitatory EPSPs on the apically oriented pyramidal cell dendrites cause an inward current to flow at the dendrites, creating a current sink, which leaves a net negative charge in the extracellular space. When a corresponding source deeper near the cell body produces an outward current flow, a net extracellular positivity develops, creating an extracellular dipole. Figure 26–2 diagrams the field potentials generated by excitatory and inhibitory input onto cortical pyramidal cells.

Attenuation of Extracellular Potentials

The net effect of synchronous thalamocortical synaptic input is the creation of an apical dendritic dipole layer. This dipole layer is propagated to the surface, and detected by a scalp electrode. These scalp-recorded signals are attenuated and modified by the extracellular fluid, cerebrospinal fluid, meninges, bone, muscle, and scalp. Factors that govern the amplitude of the scalp-recorded potentials include not only the amplitude of the neuronal potentials, but the area of cortex involved, the synchronization of the potentials, and the orientation of the dipoles produced.[9] An attenuation factor of as much as 5000:1 from cortex to scalp for small, localized potentials may exist. On the other hand, for coherent activity over a wide area, the signal attenuation factor may be as low as 2:1.[9] This means that the amplitude of scalp-recorded activity is not related directly to its distance from the generator, even on a referential recording. A more distant, synchronous transient may appear higher in amplitude than a closer, less synchronous discharge. Areas of thin skull or thickened tissue will also alter the amplitude. Focal and transient alterations of cerebrospinal fluid ion concentrations by drugs or neuronal activity can also theoretically alter PSP amplitude, in turn altering scalp-recorded potentials.

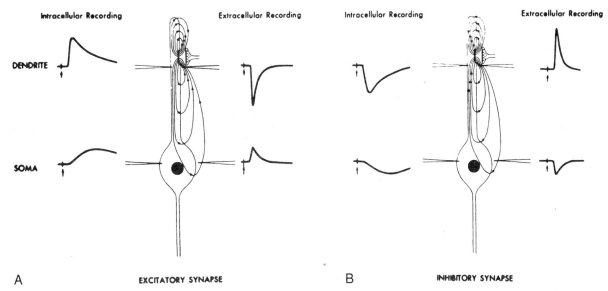

Intracellular Recording   **Extracellular Recording**   **Intracellular Recording**   **Extracellular Recording**

DENDRITE

SOMA

A   **EXCITATORY SYNAPSE**   B   **INHIBITORY SYNAPSE**

FIGURE 26–2. Diagram of fields generated by excitatory (A) and inhibitory (B) synaptic currents in central nervous system neurons. Intracellularly in A (left), a dendritic excitatory postsynaptic potential (EPSP) generates a rapid depolarization at the synaptic site and a slower and smaller depolarization at the soma. Extracellularly in A (right), the same EPSP generates a sink (negative) near the synapse and a simultaneous source (positive) at a distance along the core conductor. Intracellularly in B (left), a dendritic inhibitory postsynaptic potential (IPSP) generates a hyperpolarization at dendritic and somatic levels with a characteristic time course change. Extracellularly in B (right), the dendritic IPSP generates a source (positive) near the synapse and a simultaneous sink (negative) along the core conductor. The lines of current flow give a general idea of the direction and distribution of the intracellular and extracellular longitudinal currents. The number of current lines per area represents the current density at different points in the path. Note that the extracellular potential is larger at the site of current convergence in the vicinity of the synapse. In general, the extracellular potentials are much smaller and faster than their intracellular counterparts. All potentials are shown with positive up. (Reprinted from Pedley T, Traub R: Physiologic basis for EEG. In Daly DD, Pedley TA (eds): Current Practice of Clinical Electroencephalography, 2nd ed. New York, Raven Press, 1990, p 110, with permission.)

The magnitude of the potential produced by a dipole layer decreases as a function of distance from its source. In addition to physical modification of these potentials, if the dipole is not oriented exactly vertical to the cortex, for example in a sulcus, it may create a potential with a longitudinally oriented field, which could lower the amplitude of the recorded potential. The potential at a given distance from a dipole at a certain angle can be approximated by the formula:

$$P = \pm(e/4\pi)\Omega$$

where P is the recorded potential, Ω is the solid angle subtended at P by the dipole layer, and e is the potential across the dipole layer. It can be seen from the equation that a scalp electrode maximally detects activity that is directly perpendicular; that is, with a solid angle of 180 degrees (Fig. 26–3). Because the cortex is convoluted, EEG electrodes will detect varying amounts of a potential depending on the angle of the generated dipole (cortical surface) to the electrode. If a dipole layer activates only a small portion of the cortex, no matter how it is oriented, the scalp electrode may not record a potential. It has been estimated that, in order to be detected by scalp EEG electrodes, at least 6 square centimeters of cortex must be activated.[1, 9]

Glial Cells and Cerebral Electrical Potentials

Although cortical potentials are the source of the EEG, glial cells have an important role in the spread of extracellular potentials, and thus in the generation of the EEG. Glia differ from neurons in many important aspects. They do not form synaptic connections, fire action potentials, or generate graded potentials such as EPSPs or IPSPs. A unique Na^+-K^+ adenosine triphosphate (ATP)ase exists in the glial membrane, activated by extracellular K^+, which transports K^+ inward and Na^+ outward. It is thought that the increase in extracellular K^+ observed during neuronal activity is buffered by such a mechanism.[4] In addition to ionic influences on glia, neurotransmitters, and metabolites released during neuronal activity can affect glial K^+, Na^+, and Cl^- handling.[4] These changes in ion uptake can alter local regions of the glial membrane. Lastly, neurotransmitter receptors are present on astrocytes,[6, 30, 35, 55, 56] suggesting that these cells play a more complex role in brain signal processing than was once thought.

The ability to take up ions gives glial cells the unique capability to respond to changes in the extracellular ion milieu. For example, with pathological increase of neuronal firing, such as during seizures, the K^+ concentration in the extracellular space increases dramatically (Fig. 26–4). This excess K^+ is taken up by the glia, altering the polarity of the glial cell surface, leading to local patches of glial membrane that are in a relative state of depolarization. Because such local membrane polarities spread less readily than in neurons (which have complex mechanisms for spreading graded potentials), glia can exist in a polarized state much longer than can neurons. The high degree of glial interconnectivity may then lead to large polarized areas of the cortex, which could be responsible for propagating local currents, and recruiting local bundles of neurons to fire. The net result would be a stronger, more synchronized output.

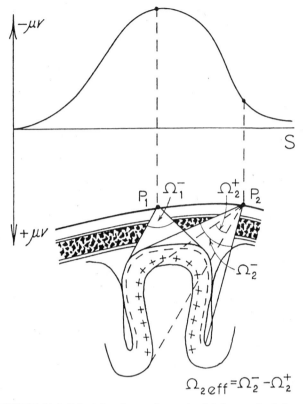

FIGURE 26–3. Principles of recording potentials from a dipole layer. A potential, P, when recorded in a monopolar fashion from a distant reference point, is proportional to the solid angle Ω subtended by a dipole layer at the position of the recording electrode. The magnitude of the potential P at any point around the dipole layer is determined by the formula $P = \pm (e/4\pi)\Omega$, where e is the potential across the dipole layer. This figure illustrates a dipole layer occupying the crown of a gyrus and its two sides, each forming one of the walls of the two adjacent sulci. At P_1, the potential depends only on the solid angle $-\Omega_1$, since at this point the recording electrode only detects a portion of the negative side of the dipole layer. At P_2, two solid angles have to be taken into account, $-\Omega_2$ and $+\Omega_2$, since the electrode at position P_2 detects portions of both the negative and positive sides of the folded dipole layer. The potential P_2 is therefore proportional to the effective solid angle $\Omega_{2\,\mathrm{eff}}$, which equals the difference between $-\Omega_2$ and $+\Omega_2$. (Reprinted from Gloor P: Neuronal generators and problems of localization in electroencephalography: Application of volume conductor theory to electroencephalography. J Clin Neurophysiol 1985, 2:327–354, with permission.)

RHYTHM GENERATION

The pyramidal cell layer generates potentials that are recorded on the scalp. But how are these potentials modulated? In particular, how are rhythms generated? Examples of clinically relevant rhythms are regularly occurring patterns such as periodic complexes, the alpha rhythm, theta rhythms, sleep spindles, hypnogogic hypersynchrony, pathological delta rhythms, drug-induced spindle-like activity, and ictal bursts. The answers to most of these questions are unknown. However, data exist concerning certain types of rhythms.

Barbiturate-induced Sleep Spindles in Cats

Our understanding of rhythm generation assumes that regardless of the frequency of the discharge, underlying

mechanisms are similar. It is difficult to study basic physiology in awake, freely moving animals. Accordingly, the asleep EEG has been most widely studied until recently. One useful preparation for studying sleep has proven to be the barbiturate-anesthetized cat. Sleep spindles have been particularly well scrutinized owing to their striking rhythmicity, and much is known about the underlying circuitry of spindle generation as compared to other brain rhythms. It is tempting to generalize these mechanisms to other types of rhythmical activity arising from different brain regions; however, this should be done cautiously as such mechanisms may be dissimilar.

Spindles, as other rhythms, are probably generated in the thalamus and projected to the cortex. Summarized here is the experimental work supporting this hypothesis. Isolated feline cortical slabs produce no spontaneous potentials.[8] If the thalamus is lesioned on one side, the ipsilateral cortically recorded spindles disappear.[25] Furthermore, removing brain stem influences by midbrain transection in cats does not abolish spontaneous cortical activity.[7] Therefore, the "pacemaker" of spindles resides somewhere in the thalamus. Subsequent studies have shown that the nonspecific thalamic nuclei directly drive the cortex, and other thalamic cells in turn drive the nonspecific thalamic nuclei cells.

The mechanisms of spindle generation have been extensively studied and reviewed in detail elsewhere.[48, 49] Recent mathematical models[18] incorporating anatomical and electrophysiological data on thalamic cells involved in spindling support the early notions of Andersen and Andersson[2] and Steriade[48]; spindling appears to be a network-generated oscillation of thalamocortical-thalamic reticular (RE) nucleus-thalamus neurons.

Reticular Nucleus of the Thalamus

The RE is a thin layer of gamma aminobutyric acid (GABA)ergic neurons that covers, the rostral, lateral, and ventral surfaces of the thalamus.[23] The RE receives input from thalamus, cortex, rostral brain stem, and basal forebrain, and projects mainly to the dorsal thalamus, with minor output to the rostral brain stem. Figure 26–5 is a schematic diagram of these interconnections. The dorsal thalamic nuclei project to the cortex, and both structures resonate within the spindle frequency and reinforce the oscillation.

The pacemaker driving sleep spindles is thought to reside in the RE cells. This is likely owing to a unique property of these cells, not to the ion channel characteristics of the thalamocortical cells described earlier. Evidence that the RE cells are the spindle pacemakers derives from the observation that anterior thalamic nuclei, which also project to cortex but are not innervated by RE cells, do not spindle. Nor can spindling be recorded in cingulate cortex, to which anterior thalamic cells project. Other pacemaker cells do not drive RE nucleus cells, as removing their input fails to abolish spindling. Thus, the RE cells themselves generate a spontaneous rhythm, and drive other thalamic cells to produce spindles.

The RE cells generate slowly growing and decaying depolarizations and fire in bursts at about 7 to 14 Hz. RE neurons contact thalamocortical neurons through axo-dendritic and dendrito-dendritic GABAergic synapses, and possibly gap junctions.[34] Owing to the peculiar interconnectivity within the RE nucleus, large groups of cells form these inhibitory contacts on thalamocortical cells, which may be one of the key features allowing resonance. Bursting RE cells generate simultaneous bursts of IPSPs in the thalamocortical follower cells. Such phasic inhibitory inputs cause the thalamocortical cells to burst through a process called postinhibitory re-

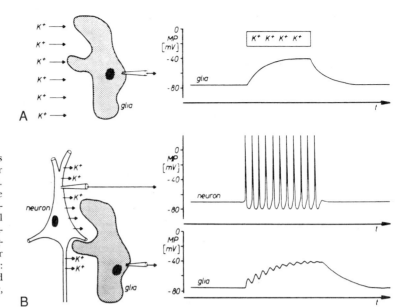

FIGURE 26–4. The membrane potential (MP) changes of glial cells induced by an increase in the extracellular K⁺ concentration (arrows in the schematic drawings). A. K⁺ is applied extracellularly to the glial cell. B. The K⁺ concentration is increased as a result of an activation of a neighboring neuron. (Drawings after original tracings from Kuffler et al, 1966, reprinted from Speckman EJ, Elger CE: Introduction to the neurophysiological basis of the EEG and DC potentials. In Niedermyer E, Lopes da Silva F (eds): Electroencephalography: Basic Principles, Clinical Applications and Related Fields, 2nd ed. Baltimore, Urban & Schwarzenberg, 1987, pp 1–13, with permission.)

bound excitation; they are hyperpolarized, and then burst as they become more depolarized.

At least two mechanisms underlie the rebound phenomenon of thalamocortical cells. Intracellular recording from thalamocortical cells shows that an inward rectifying current, which is inactive at the resting membrane potential, becomes activated by hyperpolarization to −75 mV. This inward rectifier triggers a calcium-dependent slow spike, the low-threshold spike (LTS), which resembles the T-type calcium current in that it is blocked by ethosuximide.[10] This LTS, in turn, activates a burst of fast, Na⁺-channel dependent, action potentials in the thalamocortical cells. The thalamocortical cells also possess a voltage-dependent, very slowly inactivating persistent Na⁺ current. Such a current would be a constant driving force towards depolarization, particularly from relatively hyperpolarized states. Finally, a Ca²⁺-dependent K⁺ current has been identified in these neurons, which would prolong the long-lasting IPSP and affect the duration of hyperpolarization and thus the frequency at which these cells oscillate.[41]

The thalamocortical cells appear to directly drive cortical pyramidal cells. The frequency of the spindle thus depends

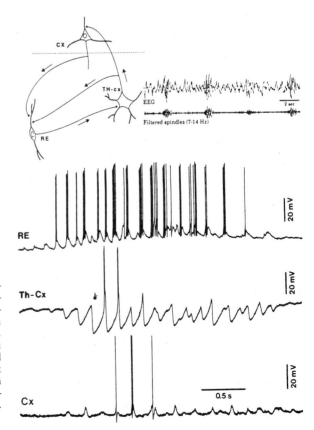

FIGURE 26–5. Spindle oscillations in reticular thalamic (RE), thalamocortical (Th-Cx, ventrolateral nucleus), and cortical (Cx, motor area) neurons. Top, circuit of the three critical populations of neurons that participate in spindle formation. Two rhythms (7 to 14 and 0.1 to 0.2 Hz) can be recorded in the cortex during spindle oscillations (see text for details). Bottom, intracellular recordings in cats under barbiturate anesthesia from the three types of neurons depicted in the top sketch. Each of the cell types is driven in complex ways by other cells in the network (see text for details). (Reprinted from Steriade, M: Cellular substrates of brain rhythms. In Niedermeyer E, Lopes da Silva F (eds): Electroencephalography: Basic Principles, Clinical Applications and Related Fields, 3rd ed. Baltimore, Urban & Schwarzenberg, 1993, pp 27–62, with permission.)

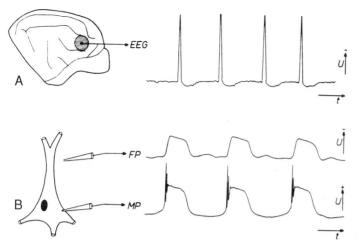

FIGURE 26–6. EEG (*A*) and membrane potential (MP) changes of a pyramidal tract neuron and extracellular field potential (FP) recorded in the vicinity of the impaled neuron (*B*) during focal interictal activity elicited by focal application of penicillin to the cortical surface (hatched area in *A*). Drawings of original tracings from experiments in the rat. The sweep speed in *B* is five times that in *A*. The recording sites are shown in the schematic drawings. (Reprinted from Speckman EJ, Elger CE: Introduction to the neurophysiological basis of the EEG and DC potentials. In Niedermyer E, Lopes da Silva F (eds): Electroencephalography: Basic Principles, Clinical Applications and Related Fields, 2nd ed. Baltimore, Urban & Schwarzenberg, 1987, pp 1–13, with permission.)

directly on the duration of the hyperpolarization induced by the RE cells. For example, long duration (150 to 200 msec) hyperpolarizations generate 7-Hz rhythms (Fig. 26–6).

SUBCORTICAL MODULATION OF RHYTHMS

Normal rhythmic activity on the EEG is often desynchronized during transitions to different levels of arousal. During sleep, the thalamocortical cells drive the cortex and are, in turn, driven by the pacemaker cells of the RE. Mechanisms of altering this sequence of neuronal activation must exist which allow the cortex to respond to various stimuli regardless of the state of consciousness. Some of these mechanisms have been well studied and have shed light on how rhythms are modulated. One such system is the so-called arousal response.

Arousal Response

Thalamic and cortical rhythmic activity is altered during transitions between different states of arousal. The reticular activating system plays an important role in these transitional states, functioning to alert the cortex and thalamus with brain stem–processed information. Many modulatory substances play a role in the regulation of thalamic activity. In particular, certain cholinergic nuclei appear to play a crucial role in thalamic regulation and arousal.

Cholinergic input to the thalamus is thought to convert the brain from a state of oscillation and sleep to desynchronized arousal, either rapid eye movement (REM) sleep or alertness. The pedunculopontine tegmental nucleus (PPT) and the lateral-dorsal tegmental nucleus (LDT) lie between the midpontine and collicular regions, and contain neurons that send important cholinergic input to the thalamus. Both nicotinic and muscarinic receptors play a role in thalamic neuron responses to acetylcholine.

Cholinergic inputs inhibit spindling by a variety of mechanisms, one such being hyperpolarization of RE cells through increase of a K^+ conductance. This K^+ conductance increase results in inhibition of the depolarization required to generate spindles. The net result is that they collectively act to desynchronize the RE network, leading to activation of the cortex and arousal (see ref. 50, section 4.1, for a more complete discussion of cholinergic influences on thalamic function).

The locus ceruleus and raphe nuclei lie in close proximity to the PDT and LDT, and project directly to the cortex.

Noradrenaline from the locus ceruleus and serotonin from the raphe nucleus depolarize RE cells, but also induce single spike "arousal-like" activity. These monoamine nuclei act in concert with the cholinergic nuclei to activate the cortex during wakefulness. The exact details of how these various neuromodulators interact to produce arousal remain unclear.

SPECIFIC FREQUENCY BANDS OF THE ELECTROENCEPHALOGRAM

Delta

In deeper stages of sleep, spindles progressively disappear and slow waves of between 0.1 and 4 Hz predominate. The mechanisms for production of rhythmical slow waves may be different from those that produce spindles as slow waves continue in athalamic preparations, whereas spindles are abolished. Recently, however, it was shown that two oscillations can occur at different membrane potentials in the same thalamocortical neuron. At a resting membrane potential of about −60 mV, thalamic cells display cortically elicited spindles, while at a more negative −79 mV, they produce delta oscillations. Thus, as sleep progresses, thalamic cells may become more hyperpolarized.[48] Cortical input onto these cells could then potentiate a delta oscillation and synchronize previously uncoupled delta-oscillating thalamic cells.

It has become clear that the mechanism of slow wave production differs depending on the frequency of the wave. Delta waves in the 1- to 4-Hz range are felt to result from different mechanisms than do slower 0.1- to 0.8-Hz waves. Waves of 1- to 4-Hz frequency are thought to be produced by dipoles set up between cortical pyramidal cells in layers II and III, and layer V.[37] EEG-recorded delta waves are probably summated afterhyperpolarizations (AHPs) mediated by a variety of K^+ currents in these pyramidal cells.[43–45] Oscillations in the delta frequency are thought to arise intrinsically from thalamocortical cells dependent on two types of ionic currents, the transient Ca^{2+} current, which underlies the LTS, and a hyperpolarization activated cation current, a noninactivated inward rectifier current carried by Na^+ and K^+.[33] These oscillations are blocked by barbiturates and spindles.[48] The slower frequency oscillations of 0.1 to 0.8 Hz are believed to arise from intrinsic properties shared by neocortical, RE, and thalamocortical cells,[48] generated by cortical cells and transmitted to RE cells. The predominant frequency of 0.3 Hz in experimental cat studies could be detected simultaneously in cortical neurons, RE cells, and thalamocortical cells synchronous with the scalp-recorded EEG.[48] In summary, although there is some information regarding slow wave production, it is conflicting, and the precise mechanisms remain obscure.

Theta

Theta rhythms often represent a slowing of the alpha frequency, such as occurs during drowsiness. However, there is also a hippocampal-paced theta rhythm that was first described in rabbits during sensory or brain stem/RE-stimulated arousal, particularly exploratory behavior.[19] Mostly studied in rodents, this rhythm is less prominent in other species, and thought not to exist in humans, although isolated studies report a theta rhythm in the hippocampus of epileptic patients.[3, 22] In animals, two independent theta rhythms have been described: one in the CA1 pyramidal cells, and the other emanating from the dentate granule cells.[36] The relevance of these hippocampal theta rhythms to scalp-recorded theta rhythms in humans is unclear.

Alpha

First described by Berger,[5] the cellular basis of origin of the alpha rhythm is still poorly understood. Andersen and Andersson devoted an entire volume to this subject, and put forth the classic theory that the thalamus is the primary pacemaker of the alpha rhythm, using barbiturate-induced feline spindles as a model.[2] However, the significance and origin of these two wave types is quite different, both in topographical distribution and in spectra. In addition, sleep spindles occur during unconsciousness, associated with decreased thalamic synaptic transmission, whereas the alpha rhythm is most prominent during relaxed wakefulness. Recent work suggests that the role of the midline thalamic nuclei in the production of the alpha rhythm may not be as important as once thought.[48] No human data, other than scalp EEG recordings, have been published regarding the alpha rhythm, but studies in monkey and dog have elucidated some mechanisms of alpha wave generation.

In awake dogs, alpha rhythms of similar peak frequency, bandwidth, and reactivity can be recorded from occipital cortex, the pulvinar, and the lateral geniculate nuclei.[27] This activity seems to be generated in small cortical islands in the basal dendrites and soma of pyramidal neurons in layers IV and V, with spread to adjacent cortical areas.[28] The intracortical coherence of this activity is much greater than corticothalamic coherence.[27] In summary, it appears that the occipital cortex can generate alpha rhythms, but that these rhythms are modulated by the visual thalamus.

Another hypothesis regarding the generation of the alpha rhythm is that it represents thalamic-processed retinal noise. According to this notion, random noise generated by the retina is processed through the thalamus, and inhibitory oscillations are imposed before the rhythm is "projected" to the cortex.[26, 29]

Beta

Waves in the beta frequency band can be recorded from a variety of preparations during arousal and vigilance associated with increased alertness. However, little is known about the cellular sites of origin or network properties of this rhythm aside from the fact that cortical neurons can generate 40-Hz frequency rhythms. Since arousal from sleep is characterized by transformation of the EEG from synchro-nized slow activity to low-voltage fast, the mechanisms of such beta waves produced upon arousal have been studied in an effort to understand other faster rhythms.

PATHOLOGICAL RHYTHMS

Epileptiform Bursting

Sharp waves, spikes, and electrographic seizures arise from abnormally functioning cells. They are hallmarks of epilepsy, although the mechanism of their generation is far from understood. Animal models of epilepsy have shown that the surface interictal spike produced by local application of penicillin is associated with abnormal bursts in neurons near the penicillin focus.[13, 14, 40] Such neurons undergo spontaneous, paroxysmal depolarizing shifts, which are sustained EPSPs with bursts of action potentials riding on the crest of the depolarization.[38, 40] These EPSPs are followed by a period of hyperpolarization called an afterhyperpolarization, associated with a slow wave on the scalp-recorded EEG.[39] Figure 26–6 demonstrates the correlation between cortical potentials and underlying cellular events during the penicillin-induced paroxysmal depolarizing shift.

The paroxysmal depolarizing shift may or may not explain spike and slow wave complexes observed on scalp recordings. The complex connections and modulatory influences of neurotransmitters and modulators in the cortex likely render this simple model inadequate, but it is a useful foundation on which to build further experiments. One type of experimental design that is proving fruitful in understanding the complexities of cortical connections and ionic currents is mathematical modeling of neurons guided by in vitro slice electrophysiological observations (for review, see refs. 36 and 58).

The hippocampus is particularly prone to epileptic discharges. This area has been studied extensively, at least in animals, and contains two populations of neurons that intrinsically burst, the CA2 and CA3 pyramidal neurons. In acute rat models of epilepsy, these neuron bursts have scalp EEG correlates. However, in humans, extracellular unit recordings of epileptogenic cortex show that cellular bursting can only be detected on the scalp about 50% of the time.[59] It is therefore likely that other properties of the epileptic brain modify normal rhythms, participating in the production of abnormal discharges. A number of changes in epileptic hippocampus have been identified that could contribute to decreased inhibition: reorganization of dentate granule cell axons, loss of inhibitory interneurons as well as principle cells, and alterations of neurotransmitter receptor numbers and distribution, all of which likely contribute to epileptogenicity (for reviews, see refs. 15 and 51). Direct recordings from cells in acute brain slices taken from epileptic patients have failed to produce convincing evidence of a cellular defect that can explain seizure initiation or propagation (for review, see ref. 51).

Pathological Slow Waves

Two types of pathological slow waves are encountered clinically: theta and delta. Pathological theta waves can be grouped into three types. The first type is a slowing of the underlying alpha rhythm, often secondary to hypoxia, hypoglycemia, cerebrovascular disease, dementia, and metabolic encephalopathy.[42] Another type is intermittent, rhythmic, often frontally predominant theta, likely produced by a mechanism similar to other "projected" rhythms (see subsequent text). Lastly, focal theta is thought to represent a mild form of focal delta, discussed later.

Delta waves are hallmarks of mass lesions such as brain tumors, abscesses, and infarctions. Because such lesions are electrically silent, the mechanism for delta wave production likely involves disruption of underlying rhythms. This has been postulated to occur by local disruption of blood supply, energy substrate, and the microenvironment. Loss of synaptic inputs into cortical neurons may also play a part. The precise mechanism of production of these pathological slow waves is unclear, but certain observations have been made that indicate which parts of the brain are involved. In particular, it has been shown that white matter disruption produces higher amplitude, polymorphic slowing without paroxysmal activity, whereas gray matter involvement tends to produce paroxysmal synchronous activity.[16] Mechanisms that generate these rhythms remain unstudied.

Intermittent rhythmic delta activity is associated with a variety of insults and is very commonly encountered in the hospital clinical practice. Dysfunction of the dorsal medial nucleus of the thalamus has been implicated in production of these rhythms.[36]

CONCLUSION

Much of what is known about the physiological basis of the EEG is derived from animal study. Though some recent advancements have been made in our understanding of rhythm generation, a great deal is unknown. Until routine recording from human neurons in situ is accomplished, it will be difficult to untangle the mysteries of EEG generation. Conversely, using the EEG to understand how electrical activity is generated in the brain is also hampered by a lack of correlation with cellular events, and by the technical constraints of EEG recording. One such constraint is the inability to routinely record direct current (DC) potentials due to slow potential artifact.

The EEG is typically recorded with a low-pass filter of 100 Hz, and a high-pass filter with a time constant of 1 second or less. If recorded with an infinite time constant, by using a DC amplifier, one can record very slow waves intermixed with higher frequency waves. Such recordings have demonstrated more accurately how cellular potentials produce the scalp-recorded EEG. For example, one difference between the two types of recording devices is that EEG only records fast-changing events. Sustained DC shifts are represented as flat-line EEG. However, sustained depolarization in a given brain region will activate a large number of voltage-sensitive ion channels with profound effects on brain electrical activity. These effects will not be "seen" with conventional EEG.

Lastly, it is clear that the ongoing work incorporating observations about individual neuronal ion channels into mathematical models of neuron ensemble function will provide a tremendous amount of information regarding rhythmical behavior of groups of cells (see refs. 52, 53, 54, 57, and 58).

REFERENCES

1. Abraham K, Marsan CA: Patterns of cortical discharges and their relation to routine scalp electroencephalography. Electroencephalogr Clin Neurophysiol 10:447–461, 1958.
2. Andersen P, Andersson SA: Physiologic Basis of the Alpha Rhythm. New York, Appleton-Century-Crofts, 1968.
3. Arnolds DEAT, Lopes de Silva FH, Aitink JW, et al: The spectral properties of hippocampal EEG related to behaviour in man. Electroencephalogr Clin Neurophysiol 50:324–328, 1980.
4. Ballanyi K: Modulation of glial potassium, sodium, and chloride activities by the extracellular milieu. In Kettenmann H, Ransom BR (eds): Neuroglia. New York, Oxford University Press, 1995, pp 289–298.
5. Berger H: Über das Elektrenkephalogramm des Menschen. Arch Psych Nervenkrankheiten 87:527–570, 1929.
6. Blankenfeld G, Enkvist K, Kettenmann H: Gamma-aminobutyric acid and glutamate receptors. In Kettenmann H, Ransom BR (eds): Neuroglia. New York, Oxford University Press, 1995, pp 335–345.
7. Bremer F: Cerveau "isolé" et physiologie du sommeil. C R Soc Biol Paris 118:1235–1241, 1935.
8. Burns BD: Some properties of isolated cerebral cortex. J Physiol 112:156–175, 1951.
9. Cooper R, Winter AL, Crow HJ, Walter WG: Comparison of subcortical, cortical and scalp activity using chronically indwelling electrodes in man. Electroencephalogr Clin Neurophysiol 18:218–228, 1965.
10. Coulter DA, Huguenard JR, Prince DA: Characterization of ethosuximide reduction of low-threshold calcium current in thalamic neurons. Ann Neurol 25:582–593, 1989.
11. Creutzfeld O, Watanabe S, Lux HD: Relations between EEG phenomena and potentials of single cortical cells. I. Evoked responses after thalamic and epicortical stimulation. Electroencephalogr Clin Neurophysiol 20:1–18, 1966.
12. Creutzfeld O, Watanabe S, Lux HD: Relations between EEG phenomena and potentials of single cortical cells. II. Spontaneous and convulsoid activity. Electroencephalogr Clin Neurophysiol 20:19–37, 1966.
13. Dichter M, Spencer WA: Penicillin-induced interictal discharges from the cat hippocampus: I. Characteristics and topographical features. J Neurophysiol 32:649–662, 1969.
14. Dichter M, Spencer WA: Penecillin-induced interictal discharges from the cat hippocampus: II. Mechanisms underlying origin and restriction. J Neurophysiol 32:663–687, 1969.
15. Glass M, Dragunow M: Neurochemical and morphological changes associated with human epilepsy, Brain Res Rev 21:29–41, 1995.
16. Gloor P, Ball G, Schaul N: Brain lesions that produce delta waves in the EEG. Neurology 27:326–333, 1977.
17. Goldensohn ES, Purpura DP: Intracellular potentials of cortical neurons during focal epileptogenic discharges, Science 193:840–842, 1983.
18. Golomb D, Wang X-J, Rinzel J: Synchronization properties of spindle oscillations in a thalamic reticular nucleus model. J Neurophysiol 72:1109–1126, 1994.
19. Green JD, Arduina A: Hippocampal electrical activity in arousal. J Neurophysiol 17:533–557, 1954.
20. Humphrey DR: Re-analysis of the antidromic cortical response. II. On the contribution of cell discharge and PSPs to evoked potentials. Electroencephalogr Clin Neurophysiol 25:421–442, 1968.
21. Humphrey DR: Re-analysis of the antidromic cortical response. I. Potentials evoked by stimulation of the isolated pyramidal tract. Electroencephalogr Clin Neurophysiol 24:116–119, 1968.
22. Isokawa-Akesson M, Wilson CL, Babb TL: Diversity in periodic pattern of firing in human hippocampal neurons. Exp Neurol 98:137–151, 1987.
23. Jones EG: The Thalamus. New York, Plenum Press, 1985.
24. Jones SW, Swanson TH: Basic cellular neurophysiology. In Wyllie E (ed): The Treatment of Epilepsy: Principles and Practice. Philadelphia, Lea & Febiger, 1996.
25. Kristiansen K, Courtois G: Rhythmic electrical activity from isolated cerebral cortex. Electroencephalogr Clin Neurophysiol 1:265–272, 1949.
26. Levick WR, Williams WO: Maintained activity of lateral geniculate neurones in darkness. J Physiol 170:582–597, 1964.
27. Lopes de Silva FH, van Lierop THMT, Schrijer CFM, Storm van Leewen W: Organization of thalamic and cortical alpha rhythm: Spectra and coherences, Electroencephalogr Clin Neurophysiol 35:627–639, 1973.
28. Lopes de Silva FH, Storm van Leewen W: The cortical source of the alpha rhythm, Neurosci Lett 6:237–241, 1977.
29. Mastronarde DN: Correlated firing of cat retinal ganglion cells. II. Responses of X- and Y-cells to single quantal events. J Neurophysiol 49:325–349, 1983.
30. McCarthy KD, Enkvist K, Shao Y: Astroglial adrenergic receptors. In Kettenmann H, Ransom BR (eds): Neuroglia, New York, Oxford University Press, 1995, pp 354–366.
31. McCormick DA, Connors BW, Lighthall JW, Prince DA: Compar-

ative electrophysiology of pyramidal and sparsely spiny stellate neurons of the neocortex. J Neurophysiol 54:782–806, 1985.

32. McCormick DA, Williamson A: Convergence and divergence of neurotransmitter action in human cerebral cortex. Proc Natl Acad Sci USA 86:8098–8102, 1989.

33. McCormick DA, Pape H-C: Properties of a hyperpolarization-activated cation current and its role in rhythmic oscillation in thalamic relay neurones. J Physiol 431:291–318, 1990.

34. Nagy JI, Yamamoto T, Shiosaka S, et al: Immunohistochemical localization of gap junction protein in rat CNS: A preliminary account. In Hertzberg EL, Johnson RG (eds): Gap Junctions. New York, Alan R. Liss, 1988, pp 375–389.

35. Pearce B, Wilkin GP: Eicosanoid, purine, and hormone receptors. In Kettenmann H, Ransom BR (eds): Neuroglia. New York, Oxford University Press, 1995, pp 377–384.

36. Pedley TA, Traub RD: Physiological basis of the EEG. In Daly DD, Pedley TA (eds): Current Practice of Clinical Electroencephalography, 2nd ed. New York, Raven Press, 1990, pp 107–137.

37. Petsche H, Pockberger H, Rappelsberger P: On the search for the sources of the electroencephalogram. Neuroscience 11:1–27, 1990.

38. Prince DA: The depolarization shift in "epileptic" neurons. Exp Neurol 21:467–485, 1968.

39. Prince DA: Inhibition in epileptic neurons. Exp Neurol 21:307–321, 1968.

40. Prince DA: Neurophysiology of epilepsy, Annu Rev Neurosci 1:395–415, 1978.

41. Roy JP, Clercg M, Steriade M, Deschenes M: Electrophysiology of neurons of the lateral thalamic nuclei in cat: Mechanisms of long-lasting hyperpolarizations. J Neurophysiol 51:1220–1235, 1984.

42. Saunders MG, Westmoreland BF: The EEG in evaluation of disorders affecting the brain diffusely. In Klass DW, Daly DD (eds): Current Practice of Clinical Electroencephalography. New York, Raven Press, 1979, pp 343–379.

43. Schwindt PC, Spain WJ, Foehring RC et al: Multiple potassium conductances and their functions in neurons from cat sensorimotor cortex in vitro. J Neurophysiol 59:424–449, 1988.

44. Schwindt PC, Spain WJ, Foehring RC, et al: Slow conductances in neurons from cat sensorimotor cortex in vitro and their role in slow excitability changes. J Neurophysiol 59:450–467, 1988.

45. Schwindt PC, Spain WJ, Crill WE: Long-lasting reduction of excitability by a sodium-dependent potassium current in cat neocortical neurons. J Neurophysiol 61:233–244, 1989.

46. Speckmann E-J, Elger CE: Introduction to the neurophysiological basis of the EEG and DC potentials. In Niedermeyer E, Lopes da Silva F (eds): Electroencephalography: Basic Principles, Clinical Applications and Related Fields, 2nd ed. Baltimore, Urban and Schwarzenberg, 1987, pp 1–13.

47. Stafstrom CE, Schwindt PC, Crill WE: Repetitive firing in layer V neurons form cat neocortex in vitro. J Neurophysiol 52:264–277, 1984.

48. Steriade M: Cellular substrates of brain rhythms. In Niedermeyer E, Lopes da Silva F (eds): Electroencephalography: Basic Principles, Clinical Applications, and Related Fields, 3rd ed. Baltimore, Williams & Wilkins, 1993, pp 27–62.

49. Steriade M, Llinas RR: The functional states of the thalamus and the associated neuronal interplay. Physiol Rev 68:649–742, 1988.

50. Steriade M, Gloor P, Llinas RR, et al: Basic mechanisms of cerebral rhythmic activities. Electroencephalogr Clin Neurophysiol 76:481–508, 1990.

51. Swanson TH: The pathophysiology of human mesial temporal lobe epilepsy. J Clin Neurophysiol 12:, 1995.

52. Traub RD, Wong RKS: Synaptic mechanisms underlying interictal spike initiation in a hippocampal network. Neurology 33:257–266, 1983.

53. Traub RD, Diongledine R: Model of synchronized epileptiform bursts induced by high potassium in CA3 region of rat hippocampal slice. Role of spontaneous EPSPs in initiation. J Neurophysiol 64:1009–1018, 1990.

54. Traub RD, Miles R, Jefferys JGR: Synaptic and intrinsic conductances shape picrotoxin-induced synchronized after-discharges in the guinea-pig hippocampal slice. J Physiol 461:525–547, 1993.

55. Walz W: Acetycholine and serotonin receptor activation. In Kettenmann H, Ransom BR (eds): Neuroglia. New York, Oxford University Press, 1995, pp 346–353.

56. Wilkin G, Marriott DR, Pearce B: Peptide receptors and astrocytes. In Kettenmann H, Ransom BR (eds): Neuroglia, New York, Oxford University Press, 1995, pp 367–376, 1995.

57. Wong RKS, Traub RD: Synchronized burst discharge in disinhibited hippocampal slice. I. Initiation in CA2-CA3 region. J Neurophysiol 49:442–458, 1983.

58. Wong RKS, Traub RD, Miles R: Cellular basis of neuronal synchrony in epilepsy. In Delgado-Escueta AV Jr, Ward AA, Woodbury DM, Porter RJ (eds): Basic Mechanisms of the Epilepsies: Molecular and Cellular Approaches. New York, Raven Press, 1986, pp 583–592.

59. Wyler AR, Ojemann GA, Ward AA: Neurons in human epileptogenic cortex: Correlation between unit and EEG activity. Ann Neurol 11:301–308, 1982.

Chapter 27

Electrode Montages and Localization of Potentials in Clinical Electroencephalography

Hajo M. Hamer, MD • Hans O. Lüders, MD, PhD

INTRODUCTION
PEN DEFLECTIONS AND POLARITY
ELECTRODE MONTAGES
RULES OF LOCALIZATION
Localization in Bipolar Montages
Localization in Referential Montages
Contour Map
Bipolar Versus Referential Montages
Special Rules of Localization
Localization and Distribution of EEG
Rhythms

POLARITY OF POTENTIAL GENERATORS
RECOGNITION OF ARTIFACTS AND
IDENTIFICATION OF MONTAGES
DIPOLES
Benign Focal Epileptiform Discharges of
Childhood
Eye Movements
Electrocardiogram
MULTIPLE GENERATORS

INTRODUCTION

Electroencephalograms (EEGs) consist of recordings of brain waves from different amplifiers using various combinations of electrodes. Montages specify which electrode is connected to which amplifier in any given EEG recording. Within these recordings, localization of potentials plays a critical role in clinical interpretation of EEG tracings. The methodology used to localize a potential will vary according to the montage used in the recording. Accurate localization of interictal and ictal epileptiform discharges helps in defining the epileptogenic zone. The cortical area that generates the maximum potential of a slow wave focus may point to the localization of a structural lesion. In addition, the voltage distribution of an EEG transient may assist in identification of artifacts.

Most reviews of techniques in EEG localization have stressed the significance of phase reversals in bipolar montages and the importance of amplitudes in referential montages.[3, 8, 9, 10, 13, 16, 22] Usually the reader is introduced to a known hypothetical distribution of voltages. It is important, however, to remember that the problem encountered when interpreting EEGs is just the opposite. EEG tracings show a certain type of pen deflections, and it is our task to deduce the distribution of the potential field and the polarity of the generator that produced these pen deflections. Therefore, EEG interpretation and localization are analyzed from this point of view in this discussion.

PEN DEFLECTIONS AND POLARITY

In clinical electroencephalography, differential amplifiers are used which have a dual input, identified here as input 1 and input 2. The output voltage is defined as the difference "input 1 − input 2" multiplied by the gain of the amplifier. It is safe to assume that owing to volume conduction effects, both inputs always are active, at least to a certain degree. This applies to so-called bipolar and referential montages

(see later discussion). However, in some cases one of the two inputs may be so much less active than the other that it can be considered a "quiet reference electrode" for practical purposes.

The direction of the pen deflection depends on the polarity of the potential difference between the two inputs. For a single EEG channel, there are always two possibilities. By definition, an upward pen deflection is caused by either (1) a relative negativity at input 1 or (2) a relative positivity at input 2. A downward deflection is caused by either (1) a relative positivity at input 1 or (2) a relative negativity at input 2.[16] All clinical EEG machines operate according to this convention by international agreement[12] (Fig. 27–1).

Assuming that the potential was produced by a single generator, we can choose for any given pen deflection between the two previously listed possibilities. Either the polarity of the generator, or which of the two inputs is the more active one, can be deduced. Knowledge of either one of the two options will permit determination of the other one from the deflections in the EEG tracing. For example, if an upward deflection was generated by a negative potential, the activity was generated mainly in input 1. Conversely, if an upward deflection was generated by a positive potential, then the activity was generated mainly in input 2. It is important to emphasize that the direction and extent of a given deflection will be a function of *relative* positivities or negativities at input 1 or 2. Assuming still a single generator, a small upward deflection could be generated by a small negative potential in input 1 with a relative quiet input 2 or a large negative potential affecting both input 1 and 2 but with a slightly higher amplitude at input 1.

The situation becomes significantly more complex when dipoles or multiple generators, or both, are involved. In the setting of dipoles, an upward deflection may be produced by recording the negative pole at input 1 and the positive pole at input 2. From the analysis of the EEG alone, it then becomes impossible to define exactly which of the poles has a greater influence on the pen deflection. This point is discussed in more detail at the end of the chapter. In this

358

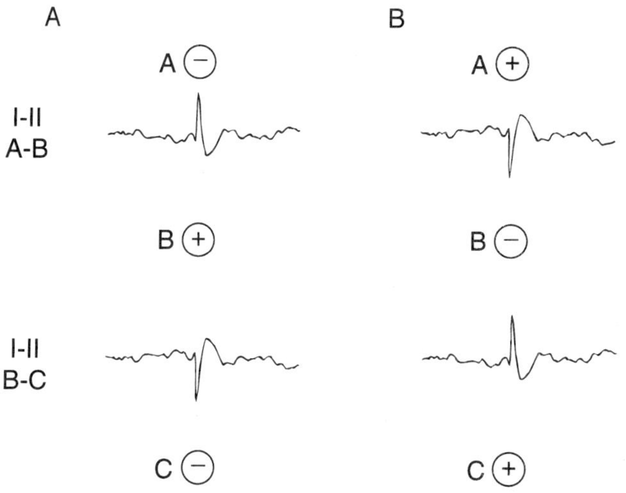

A

I-II
A-B

A ⊖

B ⊕

B

A ⊕

B ⊖

I-II
B-C

C ⊖

C ⊕

FIGURE 27–1. In channel 1, the electrode A is connected to input 1 and electrode B to input 2. In channel 2, electrode B is connected to input 1 and electrode C to input 2. *A.* In channel 1, the upward deflection shows that electrode A is relatively more negative than electrode B or electrode B is relative more positive than electrode A. Channel 2 shows a downward deflection, which represents a relative positivity at electrode B compared to electrode C or a relative negativity at electrode C compared to electrode B. In conclusion, electrode B is relatively more positive than electrodes A and C or electrodes A and C are more negative than electrode B. Both possibilities principally exist. *B.* The reversed situation is shown compared to Figure 27–1*A.* Channel 1 displays a downward deflection and channel 2 an upward deflection. According to the polarity convention, this represents a relative negativity at electrode B compared to A and C or relative positivities at A and C compared to B. (With permission of Lüders HO, Noachter S: Atlas und Klassifikation der Elektroenzephalographie. Wehr, Ciba-Geigy Verlag, 1994.)

initial discussion, we always assume that the EEG deflections are produced by a single generator.

In clinical electroencephalography, it is extremely important to be able to deduce quickly the *relative polarity* of a unipolar generator. Therefore, it is useful to have some visual clues that may assist in defining the polarity of the potential at a given electrode when we see a certain pen deflection. This is shown in Figure 27–2. If the name of the electrode connected to input 1 (Fp$_1$ in Fig. 27–2*A* and F$_3$ in Fig. 27–2*B*) is written above the deflection and the name of the electrode connected to input 2 (F$_3$ in Fig. 27–2*A* and C$_3$ in Fig. 27–2*B*) is written below the deflection, then the pointed part of the

deflection identifies the "relative negativity" whereas the concave portion identifies the "relative positivity." In Figure 27–2*A*, Fp$_1$ is "relative positive" and F$_3$ is "relative negative," compared to Figure 27–2*B* in which F$_3$ is "relative negative" and C$_3$ is "relative positive." This simple procedure may later assist in fast identification and localization of the polarity of a generator. Similar rules help in defining the polarity in bipolar and referential montages.

ELECTRODE MONTAGES

The placement of electrodes on the scalp defines the spatial sampling. Electrode density must be generous enough to provide adequate information from all areas of the scalp. However, electrodes should not be so closely spaced as to overwhelm with redundant data. The International 10–20 system[1, 6] is generally accepted for standardized electrode placement. The nomenclature of each electrode site follows a logical alphabetical abbreviation that identifies the lobe or area of the brain to which it refers: F stands for frontal, Fp for frontopolar, T for temporal, C for central, P for parietal, O for occipital, and A for auricular (ear electrode). The following subscripts in the electrode annotation refer to the specific location of the electrode site within one brain hemisphere. All even numbers identify electrode positions on the right side of the head, whereas all the odd numbers refer to the left hemisphere. The subscript "z" (for zero) points to electrode sites in the midline of the skull. For example F$_z$ identifies the mid-frontal electrode, T$_8$ refers to the right temporal electrode, and O$_1$, to the left occipital electrode. In order to locate the exact electrode position, this system uses four anatomical landmarks from which measurements can be made. The nasion is the indentation between the forehead and the nose, the inion is a ridge or knob that can be felt at the midline of the occipital skull, and the preauricular points are defined as the indentations just above the cartilage (tragus) that covers the external ear openings. The electrode locations and interelectrode distances are then defined as 10% or 20% of these anatomical distances (Fig. 27–3).

In routine EEG, a montage is a combination of channels arranged on an EEG page, which usually displays simultaneously electrode derivations from the entire scalp. Even with logically organized series of channels, montages distort spatial information by converting complex three-dimensional EEG activity to arrays of channels that provide two-dimensional and spatially discontinuous data. However, a given

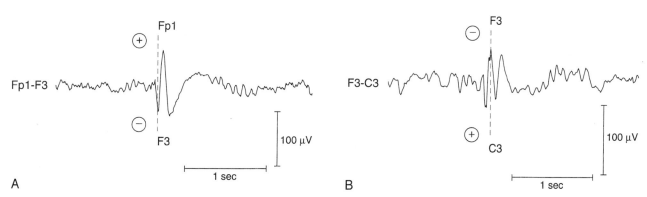

FIGURE 27–2. *A.* This EEG channel shows a downward deflection caused by a sharp wave which has a negativity at electrode F$_3$ compared to a relative less negative (or more positive) electrode Fp$_1$. *B.* This EEG channel shows an upward deflection caused by the same sharp wave shown in Figure 27–2*A* which has a negativity at electrode F$_3$ compared to a relative less negative (or more positive) electrode C$_3$.

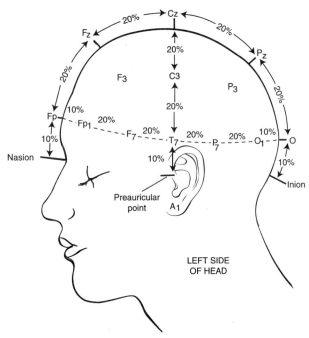

FIGURE 27–3. In the International 10–20 system, the distance between electrodes is defined by placing marks at either 10% or 20% of the total distance between skull landmarks.

EEG montage tends to preserve spatial relationships of electrodes in one direction better than in others. Therefore, montages can be classified as longitudinal montages if they better display potential characteristics in the sagittal direction, and as transverse if they better preserve spatial relations in the coronal direction (Fig. 27–4). According to the recommendations of the American Electroencephalographic Society,[2] no less than eight channels should be used per montage and the inclusion of all 21 electrodes according to the 10–20 system in every montage is encouraged. The pattern of electrode connections should be made as simple as possible to facilitate comprehension. The electrode connections for each channel should be clearly indicated at the beginning of each montage. Both bipolar and referential montages should be used.

In bipolar montages, adjacent channels have one electrode in common, but no single electrode is common to all channels (Fig. 27–4). Bipolar montages link serial pairs of electrodes by connecting the electrode at input 2 of one channel to input 1 of the subsequent channel. A set of channels connected in this way is called a bipolar chain. Derivations should be arranged so that bipolar chains connect electrodes in the longitudinal anterior-posterior (Fig. 27–4A) or transverse (left to right) direction (Fig. 27–4B). Electrode connections should preferentially run in straight, unbroken lines and the interelectrode distances should be kept equal. Bipolar recordings avoid common references which potentially may be contaminated with artifacts or potentials of less clinical interest (see later discussion).

In referential montages, different electrodes are connected to input 1 of each amplifier with a reference electrode connected to input 2 of each amplifier. All electrodes can be referred to one single electrode (e.g., C_z), one common electrode on each side of the head (Fig. 27–4C), or the electrically combined activity from two or more electrodes (e.g., linked ear reference: $A_1 + A_2$). Ideally, electrodes that are uninvolved in the electrical field of the target potentials are chosen as references. However, completely inactive electrodes often do not exist (see the detailed discussion later).

In an attempt to avoid potentially active reference electrodes, the common average of all electrodes applied to the head may be calculated and used as reference.[5] The disadvantages of this reference are twofold: Depending on the number of the electrodes included, the amplitude of the potential under study will change. In addition, high-voltage transient (e.g., eye blinks) may contaminate the common average, leading to a distortion of the actual amplitude and polarity of the transients. Therefore, they may be difficult to recognize. The Laplacian montage represents a montage that compares voltages of each electrode to a "local reference" derived from potentials at the adjacent electrodes.[17] For example, the simplest value of the "local reference" for the electrode F_3 is the average of the voltages recorded at the neighboring electrodes Fp_1, F_z, C_3, and F_7. These discontinuous measurements from closely spaced electrodes approximate the continuous Laplacian gradient of the electrical field. Thus, the output of any channel is proportional to the intensity of the local current. Local peaks are localized to the electrode of maximal amplitude and widely distributed activity is deemphasized. These source derivations eliminate concerns for an active reference. However, excessive interelectrode distances and asymmetries among the neighboring electrodes, as is the case for frontopolar, temporal, and occipital electrodes, may lead to field distortions with falsely localized voltage peaks and inaccurate phase reversals.[19]

Visual pattern recognition of EEG potentials is an important factor in the clinical practice of EEG interpretation. However, as illustrated later, a given potential generator causes completely different deflections when recorded with different montages. Therefore, it is advisable to work with a standardized set of montages that facilitate the accurate and fast interpretation of EEG recordings. The American Electroencephalographic Society[2] has proposed several "standard montages" for use in clinical practice. During an EEG recording, at least one of a longitudinal referential (Fig. 27–4C), longitudinal bipolar (Fig. 27–4A), and transverse bipolar montage (Fig. 27–4B) should be used.

Montages can be arranged in an unpaired, paired-grouped, or paired-channel format. Unpaired longitudinal or transverse montages group channels in anatomically neighboring chains. An unpaired bipolar longitudinal montage may display the left temporal chain over the left parasagittal chain and the midline derivations F_z-C_z, and C_z-P_z followed by the right parasagittal chain over the right temporal chain. Figure 27–4B shows a transverse bipolar montage in an unpaired format. The unpaired format minimizes spatial distortion and best reveals the spatial distribution of the potentials. However, it does not easily allow comparison of homologous left and right regions. The paired-group format of montages compares corresponding left and right areas (Fig. 27–4A). This format has a fair emphasis on symmetry between homologous left and right regions without producing excessive spatial distortion. Paired-group arrangements apply only to longitudinal montages, as functional or anatomical symmetry does not exist in the coronal direction. It is the convention in North America to place left-sided channels or groups over right-sided ones whereas in Europe, the reverse convention, right over left, is frequently used. Paired-channel montages focus strongly on comparison of homologous left- versus right-sided channels but may blur the spatial distribution of potentials (Fig. 27–4C). According to the American Electroencephalographic Society,[2] longitudinal bipolar montages are recommended in an unpaired or paired-group format as "phase reversals" then occur in adjacent channels (see later discussion). Transverse montages should be displayed in an unpaired fashion (Fig. 27–4B). Referential montages may be arranged in an unpaired, paired-group, or paired-channel fashion.

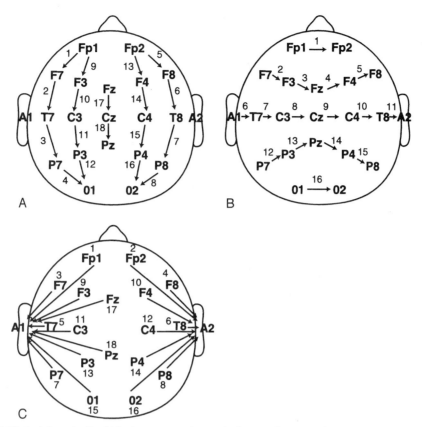

FIGURE 27–4. *A.* Longitudinal bipolar montage in a paired-group format. Left temporal chain (channels 1 to 4) over right temporal chain (channels 5 to 8) and left parasagittal chain (channels 9 to 12) over right parasagittal chain (channels 12 to 16). Channels 17 and 18 connect midline electrodes. This longitudinal montage displays best potential characteristics in the sagittal direction. The paired-group format of this montage compares corresponding left and right areas (left versus right temporal and left versus right parasagittal regions). This format has a fair emphasis on symmetry between homologous regions without producing excessive spatial distortion. *B.* Transverse bipolar montage in an unpaired format. Channels 2 to 5 form the frontal chain from left to right side. Channels 6 to 11 build the centrotemporal chain, including the ear electrodes A_1 and A_2, and channels 12 to 15 focus on the parietal regions. Transverse montages concentrate on potential distributions in the coronal direction. The unpaired format minimizes spatial distortion but does not easily allow comparison of homologous left and right regions. *C.* Referential montage in channel-paired format. Left-sided electrodes and midline electrodes are connected to ipsilateral ear electrode A_1 as reference and right-sided electrodes are linked to right ear electrode A_2. The channel-paired format focuses on comparison of homologous left versus right channels but tends to distort the spatial distribution of potentials.

RULES OF LOCALIZATION

As previously mentioned, we assume at the beginning of this discussion that the potentials detected by surface electrodes are caused by single unipolar generators. In reality, all biological generators consist of dipoles.[20] However, they are usually oriented perpendicularly with respect to the cortical surface so that scalp electrodes tend to detect only one pole.[7] If the scalp EEG records both poles of a dipole, one is usually clearly predominant, as explained later (see Fig. 27–32). Hence, interpretation of EEG findings assuming a unipolar generator still approximates the correct solution.

The following discussion outlines a set of rules that should help in the systematic identification and localization of potentials when analyzing EEG tracings.

Localization in Bipolar Montages

Each chain of a bipolar montage should first be analyzed independently. Assuming a single generator, the interpretation of a bipolar chain of electrodes will differ whether or

not there is a "phase reversal." Phase reversals are defined as pen deflections of two channels within a chain pointing in opposite directions (Fig. 27–5*A,B*). There are, by definition, two types of phase reversals: phase reversals in which the deflections between channels converge (so-called negative phase reversals; Fig. 27–5*A*) and those in which the deflections between channels diverge (so-called positive phase reversals; Fig. 27–5*B*). Notice that the concave portions of the two adjacent channels meet in positive phase reversals. The pointed parts of the two adjacent channels converge in negative phase reversal. This is just an extension of the visual clues explained earlier for the deflections in a single channel. The meaning of a negative phase reversal for the localization of the generator can easily be explained from the discussion regarding the pen deflections in a single channel. A negative phase reversal is usually produced by a negativity at the electrode that is common to both channels. This electrode is called "the electrode at which the phase reversal occurs" (electrode F_3 in Fig. 27–5*A*). On the other hand, a negative phase reversal can also be produced by a positivity at the two other electrodes that are not common to both channels (electrodes Fp_1 and C_3 in Fig. 27–5*A*). In

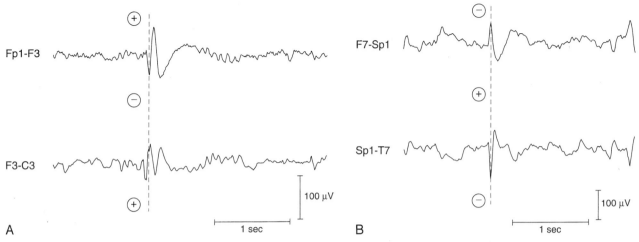

FIGURE 27–5. *A.* Left frontal sharp wave. Same recordings as shown in Figure 27–2*A* and *B.* Example of negative phase reversal, which points to a maximum negativity (or minimum positivity) at electrode F_3. *B.* Left temporal sharp wave. Example of positive phase reversal, which represents a maximum positivity (or minimum negativity) at electrode Sp_1.

other words, *a negative phase reversal points either to a maximum negativity or to a minimum positivity at the electrode of the phase reversal* (Table 27–1). Notice that Figure 27–5*A* is a composite bipolar montage which includes the two channels that were analyzed separately in Figure 27–2*A* and *B.* The interpretation of a positive phase reversal in two channels of a bipolar chain is exactly the opposite. This finding can be explained assuming either a maximum positivity or minimum negativity at the electrode of phase reversal (Fig. 27–5*B;* Table 27–1). The appearance of a phase reversal with one or more channels in between that do not show a pen deflection is called wide phase reversal (Fig. 27–6). It can be explained by recalling that EEG machines use differential amplifiers. The absence of a deflection between the two inputs of a differential amplifier does not imply absence of activity. It indicates that the two inputs of the amplifier record exactly the same potential. Therefore, a wide (negative) phase reversal can be interpreted by assuming that the maximum negativity (or minimum positivity) is equally located at the electrodes that are within the wide phase reversal. Figure 27–6*A* illustrates such a case, in which the electrodes F_8 and FT_{10} are maximum and isoelectric. This produces a phase reversal across the isopotential channel 2 (F_8–FT_{10}) and between channel 1 (FP_2–F_8) and channel 3 (FT_{10}–T_8). This situation is the reflection of a maximum negative potential that has a relatively extensive distribution in the anterior temporal region. A referential montage confirms this assumption, showing simi-

larly large deflections at electrodes F_8 and FT_{10} (Fig. 27–6*B*). Additional closely spaced electrodes (10–10 system[2]) show frequently in such cases that the true maximum is located between the electrodes that originally recorded the maximal discharges.[15] In some cases, a maximal isopotential field may extend even further, involving more than two adjacent electrodes, and produce "flat" channels over a large area. In these cases, it is necessary to look for phase reversals across two or more channels of a bipolar chain.

Figure 27–7 illustrates the case in which there is no phase reversal in a bipolar chain of electrodes. Two possibilities must be considered; either all the deflections point upward or they all point downward. In Figure 27–7*A*, all deflections point upward. Notice also how a bipolar chain can be rewritten placing the name of the common electrode between the corresponding two adjacent channels (Fig. 27–7*A*). Applying the rules of polarity explained earlier, we can deduce that the first electrode of the chain has to be the most negative and the last electrode in the chain has to be the least negative (or the most positive). Assuming that the deflections are produced by a single unipolar generator, there must be a maximal negativity at the first electrode of the chain or a maximal positivity at the last electrode of the chain. *Therefore, the two voltage extremes (maximum and minimum) are located at the two ends of a bipolar chain when there is no phase reversal.* Which of both possible hypotheses is true, depends on the polarity of the generator. If a negative generator is involved, the maximum lies at the first electrode (F_3 in Fig. 27–7*A*). Vice versa, the last electrode will be maximum (O_1 in Fig. 27–7*A*) if the generator is of positive polarity. The interpretation of a chain in which all deflections point downward (Fig. 27–7*B*) follows the same rules but deduces reversed polarities (see Table 27–1). There is either a maximum negativity at the end of the chain (O_2 in Fig. 27–7*B*) or a maximum positivity at the beginning of the chain (Fp_2 in Fig. 27–7*B*), depending on the polarity of the unipolar generator.

Figure 27–8*A* shows a phase reversal in a bipolar chain of electrodes that consists of more than two channels. The phase reversal can be interpreted in the same way as in a setting of two channels: the electrode at which the phase reversal occurs is the relatively most negative electrode. It represents the maximum negativity if the polarity of the generator is negative, and it is minimum when the generator is positive. Assuming a positive generator, the question re-

TABLE 27–1	Potential Localization and Distribution in Surface Electroencephalography[a]

Montage	Phase Reversal	Conclusion
Bipolar chain	Yes	Maximum or minimum is located at the electrode of the phase reversal
Bipolar chain	No	Maximum or minimum is located at the end of the chain
Referential recording	Yes	Referential electrode is neither maximum nor minimum
Referential recording	No	Referential electrode is either maximum or minimum

[a]Rules apply to a single unipolar generator.

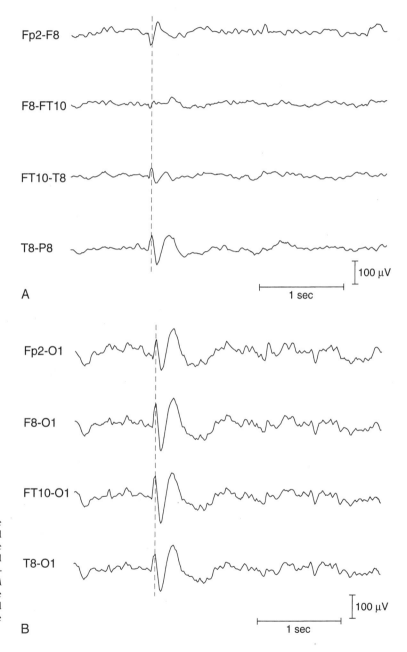

FIGURE 27–6. *A.* Right temporal sharp wave with wide phase reversal in a bipolar montage. Electrodes F_8 and FT_{10} are similarly negatively charged. Therefore, there is no significant potential difference between F_8 and FT_{10} and, consequently, no deflection in channel F_8–FT_{10}. *B.* The same sharp wave shown in Figure 27–6A is given here in a referential montage. Note that the sharp wave has similarly large negativities at F_8 and FT_{10} as was already predicted in the bipolar montage (Fig. 27–6A).

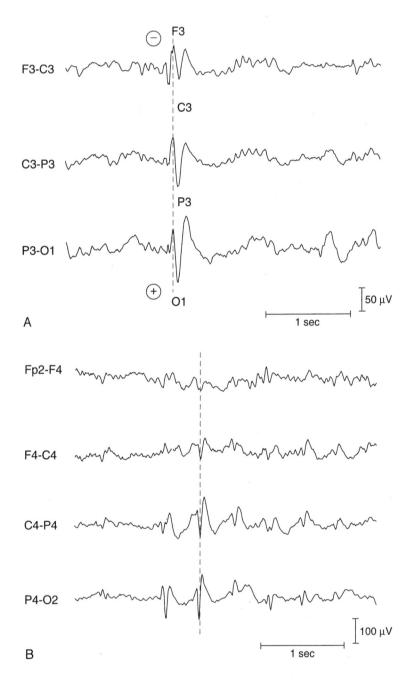

A

B

FIGURE 27–7. *A.* Left frontal sharp wave. Example of bipolar chain without phase reversal (all deflections pointing upward). Same potential as in Figure 27–2*B* with additional channels C_3–P_3 and P_3–O_1. Notice also how a bipolar chain can be rewritten, placing the name of the common electrode between two adjacent channels. In this montage, the pointed part of the deflection identifies the relatively more negative electrode whereas the concave portion identifies the more positive electrode. *B.* Right occipital sharp wave. Example of bipolar chain without a phase reversal (all deflections pointing downward). Maximum negativity (or minimum positivity) at O_2.

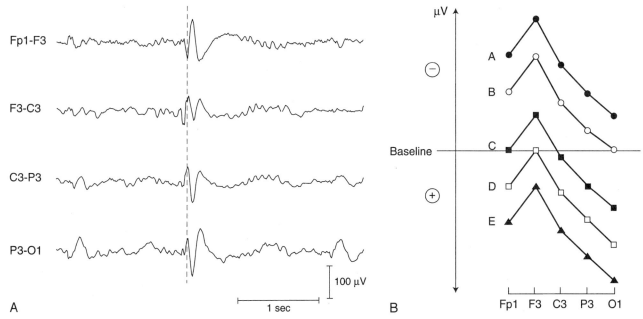

FIGURE 27–8. *A*. Bipolar longitudinal chain composed from Figures 27–2*A* and 27–7*A*. The sharp wave shows a negative phase reversal at electrode F_3. *B*. Voltage/electrode position diagram for sharp wave shown in Figure 27–8*A*. The polarity of the potentials changes according to the position of the baseline. In graph A, all electrodes are negatively charged and in graph E, positively. In graph C, F_3 is negative and the remainder, positive (dipole).

mains where the maximum positivity is located. This can be deduced by analyzing separately both "half chains" above and below the phase reversal. The bipolar chain shown in Figure 27–8*A* is actually a composite chain in which the channel analyzed in Figure 27–2*A* completed the bipolar chain shown in Figure 27–7*A*. If the rules explained earlier are applied separately to both half chains (one of the half chains consists only of channel Fp_1–F_3), it can be concluded that the maximum positivity lies at either end of the chain (at the electrode Fp_1 or at O_1 shown in Figure 27–8*A*). However, it still has to be defined which of the two ends shows the actual maximum positivity in the complete chain (Fp_1–F_3–C_3–P_3–O_1). Rules to determine the maximum in this situation are discussed later.

Further analysis of Figure 27–8*A* can be done by plotting the different electrodes on the x axis and their voltages on the y axis. The resulting graphs have been labeled "voltage/electrode position graphs" in this report. Positioning of the baseline is arbitrary as voltages derived from the EEG are not absolute but relative, differential values. However, if we assume a single unipolar generator, all electrodes must be either positive (plotted below the baseline) or negative (plotted above the baseline; Fig. 27–8*B*). Assuming first that the phase reversal was produced by a maximum negativity at electrode F_3, we can arbitrarily plot the voltage of F_3 at any point within the negative voltage area (example shown by graph C in Fig. 27–8*B*). To define the voltages of the other two electrodes adjacent to the point of phase reversal, the amplitude of the deflections (in channel Fp_1–F_3 and F_3–C_3 in this example) are measured. By comparison with the calibration signal, it can be determined how much more positive the adjacent electrodes are as compared to the maximal negative electrode (F_3 in Fig. 27–8*B*). This allows us to define the voltages of electrodes Fp_1 and C_3. The voltages of P_3 and O_1 can be deduced analogously by the deflections in channel C_3–P_3 and P_3–O_1. An analysis of graph C in Figure 27–8*B* shows, however, that some electrodes have positive voltages and others recorded negative voltages, which violates the initial assumption that the deflections shown in

Figure 27–8*A* were produced by a single unipolar generator. An increase in negativity at electrode F_3 will result in upward shifting of the entire graph as shown in diagrams A and B (Fig. 27–8*B*). In graph A, all electrodes are active and relatively negative. In graph B, all electrodes are negative except O_1, which remains neutral. Graph B represents the maximum possible shift toward positivity, assuming still a single negative generator. If the electrode at the phase reversal does not represent the maximum negativity but rather a minimum positivity, the deflections are produced by a positive unipolar generator and the entire graph has to be placed below the baseline. This is shown in graphs D and E (Fig. 27–8*B*). Any graph displaced in the positive direction below graph D would explain the situation. It can easily be determined from this voltage/electrode position graph in Figure 27–8*B* that the maximum positivity occurs at electrode O_1 rather than at Fp_1. Notice that the position of the entire graph changes with respect to the baseline but not the location of its relative minimum and maximum, assuming generators of different polarities. Notice also that phase reversals are represented by a change in direction of the voltage/electrode position graph. A negative phase reversal is a relative maximum pointing upward (at F_3 in Fig. 27–8*B*), and a positive phase reversal is represented by a relative minimum pointing down. When there is more than one phase reversal in a bipolar chain of electrodes, there will be the same number of changes in direction in the voltage/electrode position graph. A typical example is given in Figure 27–9. The bipolar chain of electrodes (Fig. 27–9*A*) shows three phase reversals (two negative phase reversals at F_7 and T_7 and one positive at Sp_1). These deflections are explained best by a negativity involving F_7 and T_7 and affecting Sp_1 only to a lesser extent. This renders Sp_1 relatively more positive, as compared to F_7 and T_7, and leads to a positive phase reversal at Sp_1. Figure 27–9*B* shows how such triple phase reversals occur in temporal bipolar chains, including the sphenoidal electrode linked between F_7 and T_7. A referential montage with additional closely spaced electrodes confirms this interpretation and reveals the maximum negativity at

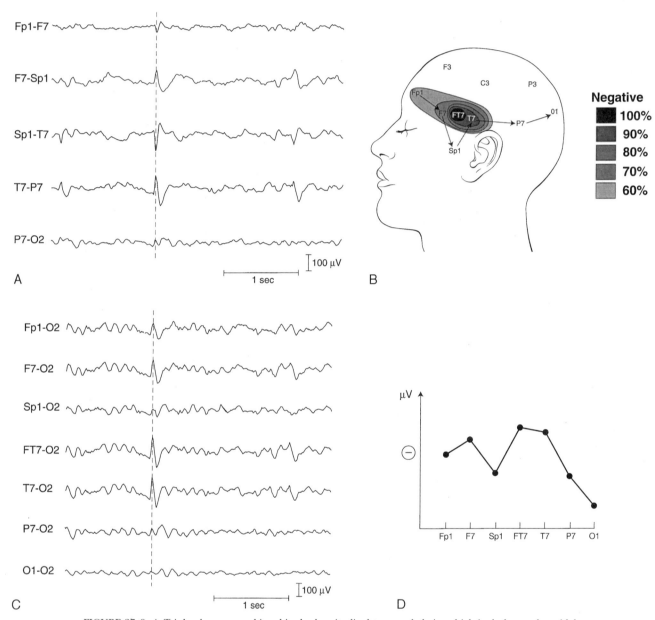

FIGURE 27–9. *A.* Triple phase reversal in a bipolar longitudinal temporal chain, which includes a sphenoidal electrode (Sp_1). Two negative phase reversals are seen at F_7 and T_7 and one positive phase reversal at Sp_1. *B.* The isopotential map of this right temporal sharp wave explains the triple phase reversal in Figure 27–9A. The montage includes Sp_1, which was located at the periphery of the sharp wave's negative field. F_7 and T_7, however, were similarly involved so that both electrodes were relatively more negative than Sp_1. *C.* Reference montage with additional closely spaced electrodes using O_2 as reference. The maximum negativity is located at FT_7 (slightly more active than T_7). Electrode F_7 also shows a large negativity whereas Sp_1 is relatively inactive. *D.* The corresponding voltage/electrode position graph shows the triple phase reversal with two negative and one positive peak. A negative generator is assumed.

electrode FT_7 (Fig. 27–9C). The voltage/electrode position diagram in Figure 27–9D gives a graphic representation of this triple phase reversal and shows the expected three changes of direction (relative maximum negativity at F_7 and T_7 and relative minimum negativity at Sp_1). The other extreme (electrode O_1 in Fig. 27–9) is the relatively least affected electrodes. Notice that Fp_1 is relatively more negative than Sp_1, P_7, and O_1.

Localization in Referential Montages

In referential montages, different electrodes are connected to input 1 of each amplifier with a single common reference electrode connected to input 2 of each amplifier. All the electrodes in a referential montage connected to the same reference electrode (input 2) should be analyzed together.

We first analyze an EEG sample in which all channels of a referential montage produce deflections in the same direction (Fig. 27–10). From the previous discussion of how to analyze a single channel (see Fig. 27–2), it can be concluded that in this sample all the electrodes are relatively more negative than the reference electrode as there are only upward deflections (Fig. 27–10). All are relatively more positive compared to the reference electrode when there are only downward deflections. Electrodes that are isoelectric to the reference will not show any deflection. It may be useful to represent the deflections shown in Figure 27–10 in a voltage/electrode position graph in order to illustrate the maximum and distribution of the potential (Fig. 27–11). If there are only upward deflections, the reference electrode will always have the lowest value and represent either the minimum or the maximum, depending on the polarity of the generator (see Table 27–1). A single unipolar generator positions the entire graph either above (negative generator: graph A in Fig. 27–11) or below the baseline (positive generator: graph B in Fig. 27–11). The reference electrode represents a positive maximum, as in graph B or a negative minimum as shown in graph A (Fig. 27–11). *Assuming a negative generator, the electrode with the largest upward deflection represents the negative maximum in a referential montage* (electrode FC_1 in Figs. 27–10 and 27–11).

If there is a phase reversal in a referential montage (Fig. 27–12), it can easily be deduced that some electrodes will be

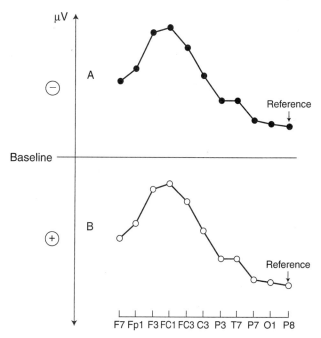

FIGURE 27–11. Voltage/electrode position graph of sharp wave shown in Figure 27–10. Graph A reflects a negative generator and graph B, a positive generator. All electrodes are more negative compared to reference electrode P_8.

more positive and others more negative than the reference electrode. In other words, *the reference electrode is neither maximum nor minimum,* independently of the polarity of the generator (see Table 27–1). It is again useful to view this situation in a voltage/electrode position graph (Fig. 27–13). The graph is identical to the graph shown in Figure 27–11 except for the different reference electrode. The new reference represents an electrode that recorded a potential that was neither maximum nor minimum. Notice that the referential montage in Figures 27–10 and 27–12 included the same electrodes shown in the bipolar montage in Figure 27–8, except that three additional electrodes (FC_1, FC_3, AF_3) were added in this montage. In both cases (Figs. 27–10 and 27–

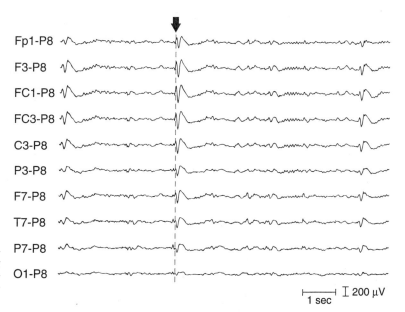

FIGURE 27–10. Example of referential montage displaying a sharp wave without a phase reversal (only upward deflections). Therefore, the reference electrode P_8 represents a minimum negativity (or maximum positivity) compared to the other electrodes.

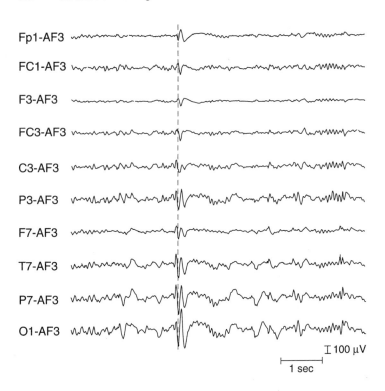

FIGURE 27–12. Example of phase reversal in a referential montage. The same sharp wave as given in Figure 27–10 is here shown in a referential montage using AF_3 as reference. Electrodes FC_1, F_3, and FC_3 show upward deflections and are, therefore, more negative than reference AF_3. The remaining electrodes reveal downward deflections. These electrodes are less negative (or more positive) than the reference electrode.

12), we were able to deduce the same voltage/electrode position graph. Again, the electrode with the largest upward deflection represents the negative maximum in a referential montage if the potential was produced by a negative generator. Electrode F_3 is maximum in Figure 27–8 and electrode FC_1 is maximum in Figures 27–11 and 27–13, giving a more exact potential distribution by adding closely spaced electrodes.

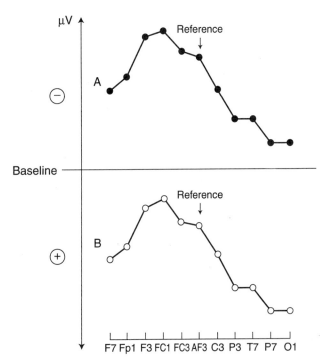

FIGURE 27–13. Voltage/electrode position graph of the sharp wave shown in Figure 27–10. Graph A represents a negative and graph B, a positive generator. The reference electrode AF_3 records a potential that is neither maximum nor minimum.

These examples show that the use of different referential electrodes leads to the same voltage/electrode position graph and, therefore, to the same conclusions. The choice of the reference electrode does not make any difference as long as the results are appropriately interpreted. There are, however, other practical considerations in the selection of a reference electrode. In practical electroencephalography, electrodes not only record brain activity but also are affected by extracerebral sources of potentials, which may interfere with the EEG recordings. Typical examples are electrocardiogram (ECG) and electromyogram (EMG) potentials. These extraneous potentials tend to affect certain electrodes in a relatively selective manner. For example, EMG potentials are preferentially recorded at frontopolar (Fp_1, Fp_2), inferior frontal (F_7, F_8), and temporoparietal (T_7, T_8, P_7, P_8) electrodes. The ECG signal affects primarily the ear electrodes (A_1, A_2). If such extracerebral potentials are of high amplitude and coincide with the brain-generated potentials, difficulties may occur in assessing the EEG (Fig. 27–14A). However, the use of a different electrode as reference may solve this problem to a great extent (Fig. 27–14B) and allows one to establish a voltage/electrode position graph. A bipolar montage may also show less artifact interference (Fig. 27–14C) as there is no common electrode that potentially can contaminate all channels with artifacts. Interference between "target potentials" and others may not only be limited to the extracerebral artifacts as discussed earlier. Other brain potentials of less clinical interest may also make the localization of target potentials difficult. Typical examples are background activity (Fig. 27–14B) and vertex sharp transients in sleep. Vertex sharp transients affect preferentially the midline electrodes and may interfere with the recording of abnormal potentials in a reference montage using midline electrodes as reference. In such cases, it is useful to look for other referential electrodes that are not affected by such high-amplitude "physiological" transients.

In summary, the choice of the reference electrode does not change the potential distribution, and an analysis of the voltage/electrode position graph leads to the same conclusions, independent of the reference used. However, from a

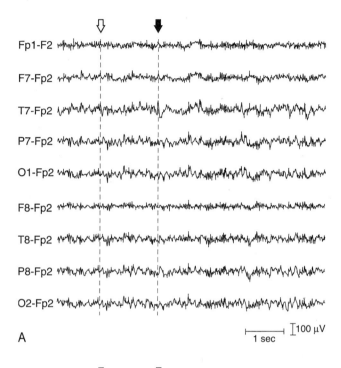

A

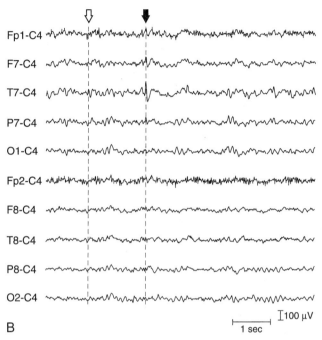

B

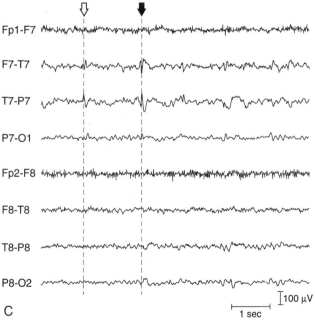

C

FIGURE 27–14. *A.* Referential montage using Fp₂ as reference electrode. The reference electrode is contaminated by EMG artifacts. One sharp wave is hardly seen (*solid arrow*) and a second sharp wave is not identifiable (*open arrow*). *B.* Referential montage using artifact free electrode C₄ as reference. One sharp wave is clearly seen (*solid arrow*) and the second sharp wave (*open arrow*) is still blurred by background activity. *C.* Bipolar longitudinal montage permits identification of both sharp waves (*solid and open arrow*) without difficulties.

practical point of view, *selection of a reference that is free of artifacts and inactive is advisable and facilitates accurate localization.* Inactive references are those electrodes that are not or only minimally involved in the target potential. In addition, an inactive and quiet reference has the advantage that all deflections point in the same direction, including the largest deflection at the electrode with the potential's maximum (see Fig. 27–10). The polarity of the generator is easily defined when we have a relatively inactive and quiet reference. In such a setting, an upward deflection represents a negative generator and a downward deflection a positive generator. However, it is important to emphasize again that accurate interpretation is independent of the choice of reference electrodes as long as artifacts do not obscure analysis.

Contour Map

A useful graphic representation of potential distributions can be obtained by contour maps. Figure 27–15A shows the contour map of the left frontal sharp wave illustrated in Figure 27–10. The easiest way to establish such a map is to plot first the corresponding voltage/electrode position graph as shown in Figure 27–15B. The graph should be placed completely above (graph A in Fig. 27–15B) or below the baseline (graph B in Fig. 27–15B), depending on the polarity of the generator. The electrode that shows the relative maximum of either negativity (FC_1 in graph A of Fig. 27–15B) or positivity (O_1 in graph B of Fig. 27–15B) are set at the 100% value of the potential field. The potential field can then be scaled in 10% ranges which are transferred as "isopotential lines" to the contour map of the head. Each of these isopotential lines includes all electrodes that recorded at least 90% or 80%, and so on, of the charge of the maximal electrode (Fig. 27–15A,C). For each of the two possible assumptions (negative or positive generator), two different maps can be generated. Both maps represent mirror images of the same potential field focusing on the relatively negative (Fig. 27–15A) or positive side of the field (Fig. 27–15C). Thus, the maximum negativity in Figure 27–15A becomes minimum positivity in Figure 27–15C. Strategies to decide which of both solutions are a more accurate reflection of the actual generator, are discussed later.

Contour maps may vary slightly according to the reference electrode used. This can be explained by considering that the amplitude of the maximal recorded potential that de-

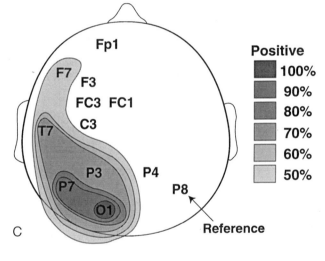

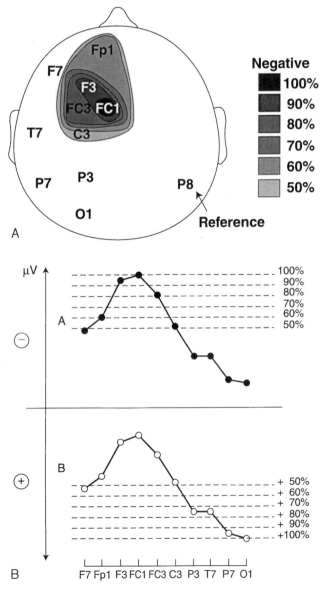

FIGURE 27–15. *A.* Left frontal sharp wave shown in Figures 27–7 and 27–10. Isopotential map was drawn assuming a negative generator. Additional electrodes according to the International 10–10 system[1, 2] were included (FC_1, FC_3). Maximum negativity occurs at FC_1 (100%). *B.* Voltage/electrode position graph of sharp wave shown in Figures 27–7, 27–10, and 27–15A. (A, negative generator; B, positive generator.) *C.* Isopotential map of sharp wave shown in Figure 27–15A assuming a positive generator. "Mirror image" of Figure 27–15A.

fines the 10% scales of the map is the difference between the potentials recorded at the maximal electrode and the reference electrode. The amplitude, therefore, varies according to the absolute potential recorded at the reference site. This again emphasizes the importance of selection of a quiet and inactive reference.

Bipolar Versus Referential Montages

As previously mentioned, the correct analysis of a potential field will yield the same results independently of the use of bipolar or referential montages. Different montages include the same information but present it in different ways. Any specific montage does not add more or leave out any data that others contain as long as the same electrodes are included. However, there are practical considerations why one or another montage may be preferred in certain situations. This again relates to the difficulty of interpreting certain potentials of interest when they are intermixed with artifacts or other brain potentials. Bipolar montages generally have advantages when potentials of relatively low amplitude and limited spatial distribution have to be displayed, especially in the presence of widely distributed, high-amplitude background activity. Bipolar montages tend to cancel out these usually widely distributed potentials due to small interelectrode distances and allow visualization of the intermixed low-amplitude potentials, which are less widely distributed (Fig. 27–16; first arrow). On the other hand, referential montages display best widely distributed potentials with a small voltage gradient, especially when large interelectrode distances are chosen (Fig. 27–16; second arrow).

Special Rules of Localization

In bipolar montages, the interpretation of the EEG begins with the analysis of one channel followed by consideration of one "chain" of electrodes at a time. A chain of electrodes was defined as a series of electrodes in which one electrode is always common to adjacent channels but there is no single electrode common to all channels. In general, however, we are interested in the voltage distribution of a generator potential across all the electrodes used in that particular montage. A typical example is shown in Figure 27–17, which shows four chains of electrodes in a standardized longitudinal bipolar montage (from above: left and right temporal chain, left and right parasagittal chain). The left parasagittal chain was already introduced in Figures 27–8 and 27–10. Figures 27–18 and 27–19 illustrate the corresponding voltage/electrode position graphs and the isopotential maps separately for each of the four "chains" of electrodes. It is possible to link both left-sided chains in order to achieve a more complete representation of the potential field. The combination of two chains can be done when there is a common electrode represented in both bipolar chains. In Figure 27–17, the electrodes Fp_1 and O_1 are common to both left-sided chains. We can choose, therefore, either one of the two electrodes to "join" both chains and construct the common voltage/electrode position graph shown in Figure 27–20 and the common isopotential graph shown in Figure 27–21. This is possible since the recorded voltage at electrode O_1 has to be the same, independently of whether the left temporal or parasagittal chain is used to calculate its voltage. In two bipolar chains of electrodes that share the same electrodes at both ends, the sum of upward and downward deflections will always be identical (Fig. 27–22).

The same analysis is possible for both right-sided chains of Figure 27–17. There is, however, no common electrode to left- and right-sided chains, which theoretically makes it impossible to link left- and right-sided chains. For practical purposes, however, approximations can be done so that one voltage distribution can be established including all left- and right-sided electrodes. The assumption is made that electrodes more distant from the generator are relatively inactive and record approximately equal potentials, especially if they are in close proximity. In the example given in Figure 27–17, electrode O_1 and O_2 are both relatively distant from the left frontal generator and at the same time close to each other. Therefore, we assume that O_1 and O_2 record equal potentials and are then able to construct a common voltage/electrode position graph and isopotential map linking the chains at O_1 and O_2 (Fig. 27–23A,B).

All electrodes of a referential montage connected to the same reference electrode should be analyzed simultaneously, as previously mentioned. In addition, it is possible to link two chains of different referential electrodes when one channel connects one electrode of each chain (channel Fp_1–TP_{10} in Figs. 27–24, 27–25, and 27–26). Thus, a relationship between right- and left-sided electrodes of the referential montage in Figure 27–24 can be established (Fig. 27–26). Another possibility is to assume that the two electrodes revealing the smallest potentials in both chains are relatively indifferent

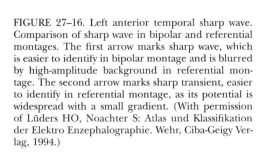

FIGURE 27–16. Left anterior temporal sharp wave. Comparison of sharp wave in bipolar and referential montages. The first arrow marks sharp wave, which is easier to identify in bipolar montage and is blurred by high-amplitude background in referential montage. The second arrow marks sharp transient, easier to identify in referential montage, as its potential is widespread with a small gradient. (With permission of Lüders HO, Noachter S: Atlas und Klassifikation der Elektro Enzephalographie. Wehr, Ciba-Geigy Verlag, 1994.)

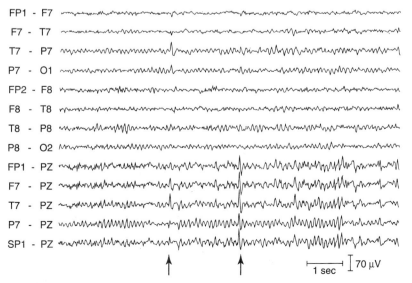

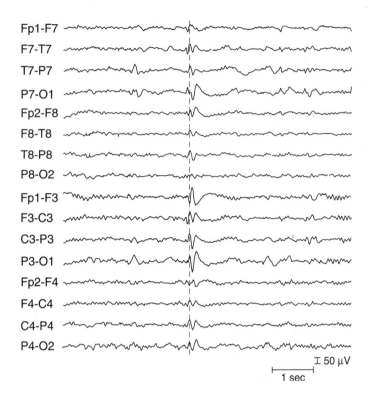

Fp1-F7
F7-T7
T7-P7
P7-O1
Fp2-F8
F8-T8
T8-P8
P8-O2
Fp1-F3
F3-C3
C3-P3
P3-O1
Fp2-F4
F4-C4
C4-P4
P4-O2

Ⅰ 50 µV

1 sec

FIGURE 27–17. Left frontal sharp wave. Bipolar longitudinal montage including 16 channels in four chains. Same sharp wave as shown in Figures 27–8*A* and 27–10.

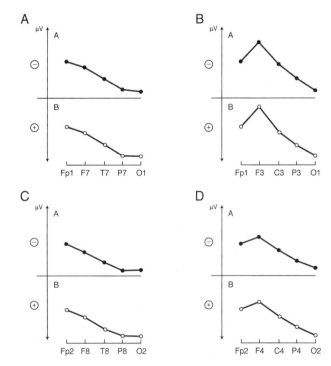

FIGURE 27–18. Voltage/electrode position graph of the four bipolar chains shown in Figure 27–17. *A.* Left temporal chain. *B.* Left parasagittal chain. *C.* Right temporal chain. *D.* Right parasagittal chain. (A, negative generator; B, positive generator.)

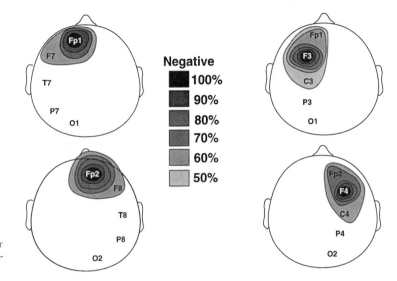

FIGURE 27–19. Separate isopotential maps of the four bipolar chains shown in Figures 27–17 and 27–18. Negative generator is assumed.

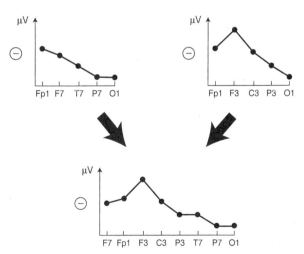

FIGURE 27–20. Voltage/electrode position graph of both left-sided bipolar chains shown in Figure 27–17. Linkage of both chains is possible at common electrodes (Fp₁ or O₁) as they have identical charges in both chains.

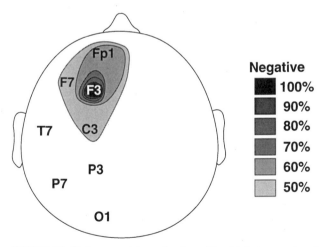

FIGURE 27–21. Isopotential graph of combined left temporal and parasagittal chain assuming a negative generator.

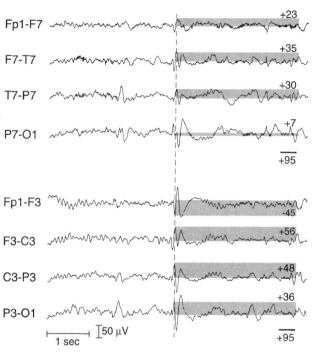

Fp1-F7   +23

F7-T7   +35

T7-P7   +30

P7-O1   +7

                           +95

Fp1-F3   -45

F3-C3   +56

C3-P3   +48

P3-O1   +36

⊢——————⊣ ⌈50 μV
  1 sec   +95

FIGURE 27–22. Left frontal sharp wave in two bipolar longitudinal chains sharing common electrodes at both ends. Left: potentials' differences per channel (in μV). Note that the sum is equal for both chains.

and have, therefore, equal potentials in the setting of a quiet reference. Notice that Figures 27–17 to 27–23 and Figures 27–24 to 27–26 are obtained from identical EEG data and come to the same conclusion independently of the montages used (Fig. 27–23 versus Fig. 27–26). Only the level of the entire voltage/electrode position graph with respect to the absolute voltages of the y axis is arbitrary and can certainly vary. If slight differences occur, they may be due to the fact that some of the assumptions are approximations that were used to permit linking of different chains. However, for all practical purposes, the final distribution maps are sufficiently precise to give a realistic reflection of the maximum and distribution of the potential.

Localization and Distribution of EEG Rhythms

All the rules that apply for localizing isolated transients can also be used in the evaluation of the voltage distribution of EEG rhythms. This is illustrated for an EEG seizure pattern in Figure 27–27. Figure 27–27A and B show the same EEG seizure pattern recorded with two different references. Notice that the interpretation is easier when a relatively indifferent reference (here as C_z) is used, as compared to a reference electrode that is partially involved in the seizure pattern and contaminates all channels with the pattern. However, the resulting isopotential maps, derived from both referential montages should be identical (Fig. 27–27C).

POLARITY OF POTENTIAL GENERATORS

As previously discussed, the rules of localization always allow for two solutions, one mirroring the other, depending on the assumption of the polarity or localization of the

generator. There are, however, several clues that may help to select the correct solution.

The usual polarity of a potential may be well known. For example, the main peak of vertex sharp transients or epileptiform discharges tends to be negative; POSTS (*Positive Occipital Sharp Transients in Sleep*) tend to be positive.

In other instances, the expected maximum of a potential may be well known. For example, POSTS tend to be maximum in occipital leads and sleep spindles in the frontocentral region. Patients with automotor seizures tend to have maximal epileptiform discharges in anterior temporal leads, whereas patients with clonic seizures frequently show spikes or sharp waves in central leads. Moreover, the EEG of patients with a structural lesion and seizures tends to reveal epileptiform discharges in the immediate neighborhood of the structural lesion.

The distribution map itself may also be helpful in localizing the maximum potential. The potential field of an ideal point generator decreases as an inverse square function of the distance away from it. Therefore, potential differences are large between pairs of successive electrodes near the generator and are small when the electrodes are located far from it. In practical terms, this means that channels of bipolar montages showing large deflections are closer to the generator, as compared to channels revealing a relatively small interelectrode difference. However, the EEG records potentials from surface generators of the cortex, which may significantly differ from ideal point generators[21] and lead at times to a relatively shallow fall-off of the potential field. Closely spaced electrodes especially may be at risk of recording only small voltage differences even when close to the generator. In clinical practice, however, a shallow fall-off close to the generator still is unusual and typically represents a minimum of the field. Moreover, voltage maps of true generators show a logical, orderly distribution (Fig. 27–15A and Fig. 27–23B, negative generator) in contrast to the maps of falsely poled or localized generators, which reveal unnatural distributions (Fig. 27–15C, and Fig. 27–23B, positive generator). A combination of all these different clues will, in most cases, permit deduction of the correct localization and polarity of a generator.

RECOGNITION OF ARTIFACTS AND IDENTIFICATION OF MONTAGES

In clinical electroencephalography, it is of great importance to differentiate artifacts from brain potentials. In addition montages may sometimes not be specifically labeled or may be mislabeled. Voltage distributions can assist in solving some of these problems.

We have previously shown that the deflections in Figure 27–17 result in a "logical voltage distribution map" (Fig. 27–23B, negative generator) given the correct montage. "Nonlogical, unnatural" maps will result if the same deflections are interpreted under the false assumption of a different montage (Fig. 27–28). The drawing of a voltage distribution map may be sometimes time consuming, but it is a reliable means to demonstrate whether or not the assumed montage is correct.

Similar principles can also be used to recognize artifacts. Artifacts are generally divided into two groups. One group consists of physiological artifacts such as ECG, eye, or tongue movements. The second group comprises electrical artifacts such as sudden changes of electrode impedances ("electrode pops"), recording of 60-Hz line current, or telephone ringing. In contrast to electrical artifacts, physiological artifacts follow logical voltage distributions at specific locations and

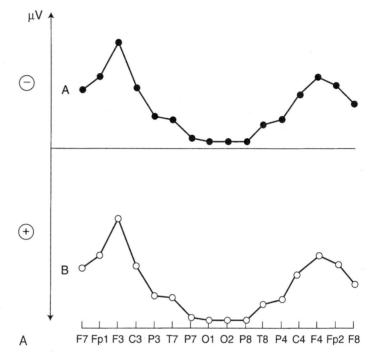

F7 Fp1 F3 C3 P3 T7 P7 O1 O2 P8 T8 P4 C4 F4 Fp2 F8

FIGURE 27–23. *A.* Voltage/electrode position graph of left frontal sharp wave including all four bipolar chains shown in Figure 27–17. Right- and left-sided chains are linked assuming relatively equal potentials in O_1 and O_2 as they are distant electrodes from the left frontal generator and at the same time close to each other. (A, negative generator; B, positive generator.) *B.* Corresponding isopotential maps assuming negative generator on the left and positive generator on the right side. Both isopotential maps are mirror images of each other.

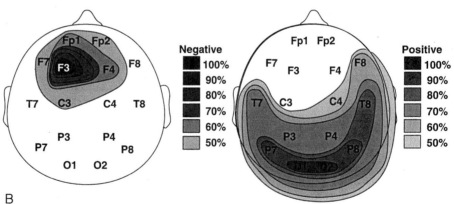

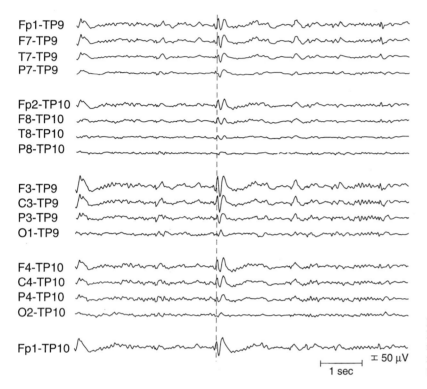

FIGURE 27–24. Left frontal sharp wave (same as in Figure 27–17) shown in referential montage using ipsilateral inferior temporoparietal electrodes as references. In additional last channel, both referential chains are linked together.

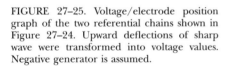

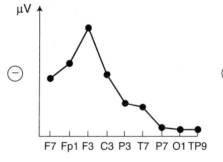

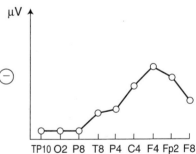

FIGURE 27–25. Voltage/electrode position graph of the two referential chains shown in Figure 27–24. Upward deflections of sharp wave were transformed into voltage values. Negative generator is assumed.

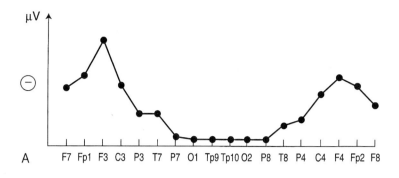

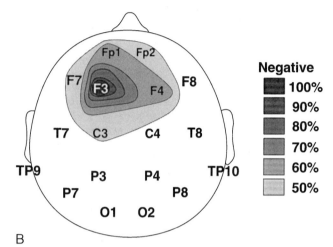

FIGURE 27–26. *A.* Voltage/electrode position graph of left frontal sharp wave displayed in Figures 27–17 and 27–24. Both referential montages of Figure 27–24 are combined as the relationship between both chains can be deduced by the deflection in the last channel (Fp$_1$–TP$_{10}$) of Figure 27–24. *B.* Corresponding isopotential map.

are produced by extracerebral biological generators such as heart, muscle, eye, or tongue (glossokinetic artifact). Blinking, for example, represents a typical physiological artifact in EEG recordings. Its positive maximum typically is at Fp$_1$ and Fp$_2$ and it consistently shows an amplitude decrease of about 50% in the successive channels (Fig. 27–29). In longitudinal bipolar montages, artifacts caused by blinking or other eye movements never show higher amplitudes in channels 10 and 14 as compared to channels 9 and 13 (Fig. 27–29) because the electrodes Fp$_1$ and Fp$_2$ are closer to the generator (eye bulb) as compared to F$_3$ and F$_4$. This observation can be helpful in differentiating between eye movement artifacts and intermittent slowing which frequently is of equal or even higher amplitude in channels 10 and 14 as compared to channels 9 and 13. This is illustrated in Figure 27–30, which shows frontal intermittent rhythmic delta activity (FIRDA) as an EEG abnormality and isolated vertical eye movements in the later part of the tracing.

In contrast to physiological artifacts, interferences caused by electrical artifacts do not follow the rules of biological generators. For example, current induced by movement of the electrodes or their wires produces random potentials that result in illogically distributed voltage maps (Fig. 27–31). An additional feature characteristic of many electrical artifacts such as electrode pops, pulse artifacts, or contamination with 60-Hz line is its presence limited to one electrode only. This usually contrasts with brain potentials, which reveal a peak value in one or more electrodes and show a fall-off in adjacent leads.

DIPOLES

All of the previously discussed rules are valid when we assume that a potential is produced by a single pole genera-

tor. This is certainly a simplification as biological generators are dipoles, as mentioned earlier. Nevertheless, this approximation is useful in clinical practice and leads to reliable results in most cases. In certain instances, however, careful analysis permits detection of both poles. Therefore, the analysis of any distribution map should not exclude the possibility of a bipolar distribution. Typical examples of bipolar distributions are benign focal epileptiform discharges of childhood (BFEDC), sharp waves after epilepsy surgery, eye movements, and the ECG (Figs. 27–32, 27–34, and 27–39).

Benign Focal Epileptiform Discharges of Childhood

The typical distribution of BFEDC is caused by a bipolar generator in the rolandic region (Fig. 27–32*C*) and is an important EEG feature of the syndrome in addition to stereotyped typical spike morphology, preferential location, activation by sleep, and relatively low epileptogenicity.[11] The negative pole of BFEDC is most frequently located in the centrotemporal region and the positive pole in the frontal area. This can best be appreciated in a referential montage with a noncephalic or contralateral mid-temporal electrode as reference (Fig. 27–32*B*). Although the maximums of both poles can be determined, the position of the baseline cannot be defined exactly. This is visualized most easily in a voltage/electrode position graph (Fig. 27–32*D*). The baseline can be shifted up and down without changing the maximums or the relationship of the electrodes to each other (Fig. 27–32*D*). It is important to remember that the position of the baseline can change the polarity of the generator, but it does not alter the potential distribution derived by differential recordings.

Eye Movements

Movements of the eyes cause bipolar potential fields because the eye itself is charged with a steady state bipolar

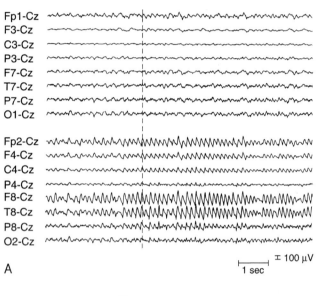

A

B

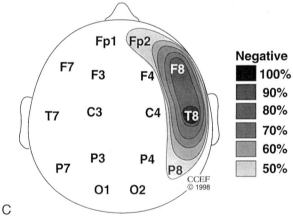

C

FIGURE 27–27. *A.* Right frontotemporal seizure pattern is shown in referential montage with C_z as reference. No phase reversal of rhythmic activity is present, indicating that the reference is either maximum or minimum. Negative maximum at T_8 ($> F_8$). *B.* Same EEG recording as in Figure 27–27*A* but with Fp_2 as reference that is partially involved. Fp_2 is neither maximum nor minimum, resulting in a phase reversal. T_8 still shows maximal upward deflections and represents a negative maximum. *C.* Corresponding isopotential map.

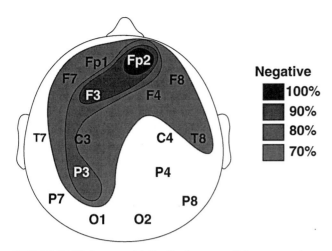

FIGURE 27–28. Isopotential distribution map of sharp wave shown in Figure 27–17 assuming *falsely* that the acquiring montage was the *channel-paired* bipolar longitudinal montage shown below. Assuming this bipolar montage, the deflections of each channel would be interpreted as follows: Falsely assumed montage:

Fp_1–F_7	Fp_1 more negative than F_7
Fp_2–F_8	Fp_2 more negative than F_8
F_7–T_7	F_7 more negative than T_7
F_8–T_8	F_8 isoelectric to T_8
T_7–P_7	T_7 more negative than P_7
T_8–P_8	T_8 more negative than P_8
P_7–O_1	P_7 more negative than O_1
P_8–O_2	P_8 isoelectric to O_2
Fp_1–F_3	Fp_1 more positive than F_3
Fp_2–F_4	Fp_2 more negative than F_4
F_3–C_3	F_3 more negative than C_3
F_4–C_4	F_4 more negative than C_4
C_3–P_3	C_3 more positive than P_3
C_4–P_4	C_4 more negative than P_4
P_3–O_1	P_3 more negative than O_1
P_4–O_2	P_4 more negative than O_2

potential. The eyeball has a positive pole oriented anteriorly toward the cornea and a negative pole pointing posteriorly toward the retina.[18] Differential AC (alternating current) amplifiers which are used in EEG machines do not detect steady state DC (direct current) potentials. However, when the eyeball rotates about its horizontal or vertical axis, it creates a large AC potential field, detectable by surface electrodes close to the eye. Blinking is associated with a conjugate upward deviation of the eyes (Bell's phenomenon) during eye closure, which induces a positive potential (downward deflection in bipolar montage) in frontopolar electrodes with a characteristic fall-off (see Fig. 27–29; see earlier discussion). Moreover, the eyelid acts as a resistor and its movement over the eyeball changes voltage outside the eye, which contributes an additional positivity at the frontal electrodes during blinking.[22] Eye opening induces downward eye movements so that blinking artifacts end with a symmetrical bifrontopolar negativity (see Fig. 27–29). Isolated vertical eye movements are shown in Figure 27–33. Blinking typically induces symmetrical artifacts which may, however, become asymmetrical in the setting of enucleation of one eye, severe damage to one eye (loss of DC potential), or external ophthalmoplegia (Fig. 27–34).

Lateral conjugate eye movements induce bilateral deflections which are out of phase since the positive pole of one eyeball (cornea) will approach the lateral frontal electrodes on one side while the negative pole (retina) of the other eyeball will come closer to lateral frontal electrodes of the other side. Thus, a left-sided eye movement induces a maximal positivity in electrode F_7 and a synchronous maximal negativity in F_8 (Fig. 27–35). Frontopolar electrodes (Fp_1, Fp_2) are quiet in true lateral eye movements since horizontal rotation of the eyeballs does not change the potential field in the vertical direction. Therefore, lateral eye movements induce asymmetrical deflections in longitudinal bipolar montages. Channels Fp_1–F_7 and Fp_2–F_8 show large deflections representing large potential differences between the maximal electrodes F_7 and F_8 and the quiet electrodes Fp_1 and

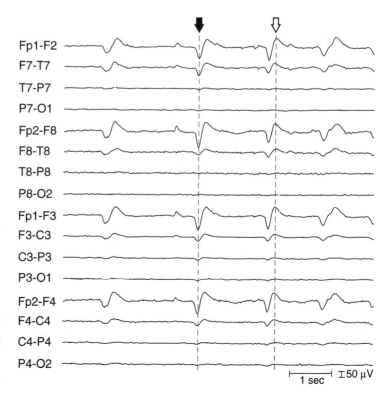

FIGURE 27–29. Blinking artifacts in a bipolar longitudinal montage. Notice a symmetrical positivity at the beginning of the artifact, which is maximal at Fp_1 and Fp_2 *(solid arrow)*. Amplitude decrease of about 50% occurs in successively more posterior channels. Artifact ends with a frontopolar negativity *(open arrow)*.

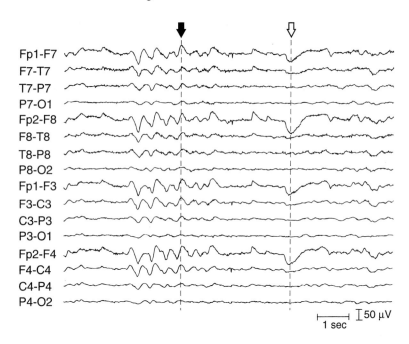

FIGURE 27–30. Frontal intermittent rhythmic delta activity (FIRDA) in the first part of the EEG *(solid arrow)* and isolated vertical eye movement in the later part of the tracing *(open arrow)*. Compare the different voltage distributions.

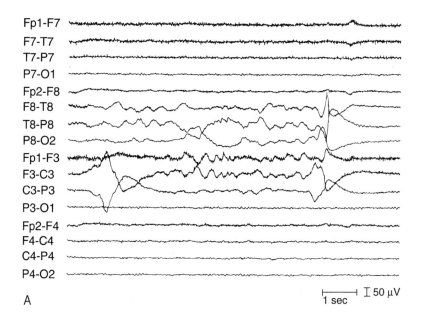

A

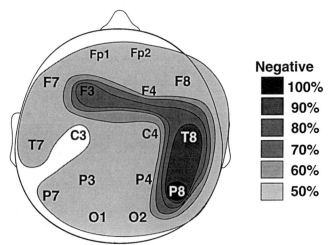

FIGURE 27–31. *A.* Movement artifacts at electrodes T₈, P₈, F₃, and C₃ shown in a bipolar montage. *B.* Corresponding isopotential map revealing illogical potential distribution.

B

Negative
100%
90%
80%
70%
60%
50%

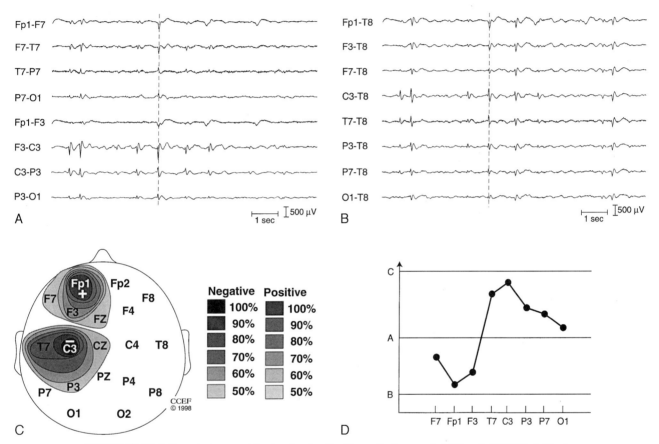

FIGURE 27–32. Left centrotemporal benign focal epileptiform discharges of childhood (BFEDC) *A.* Bipolar montage. *B.* Referential montage showing negative pole of dipole maximum at C_3 and T_7 and positive pole maximum at Fp_1 and F_3. *C.* The generator is localized approximately in the rolandic area (probably in the sylvian fissure[11]). Electrodes T_7 and C_3 record negativity, Fp_1 and F_3 positivity, and T_8 is relatively quiet. *D.* Corresponding voltage/electrode position graph displaying the dipole (baseline A assumes bipolar generator). The graph could also be explained by a pure negative (B) or positive generator (C) if the baseline is arbitrarily changed. However, the relatively acute potential changes at *both* poles make unipolar generator unlikely.

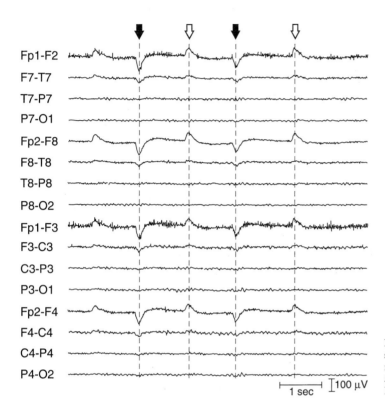

FIGURE 27–33. Upward eye movements *(solid arrow)* elicit a symmetrical frontopolar positivity and downward eye movements *(open arrow)* induce a negativity. A fall-off of at least 50% occurs in successively more posterior channels. Notice EMG artifact in Fp1.

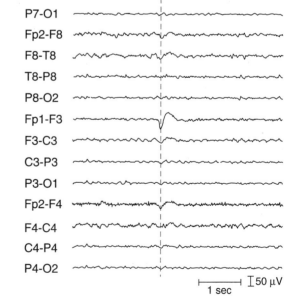

FIGURE 27–34. Asymmetrical blinking artifact after enucleation of the right eye. Note blinking artifact in Fp1, F7, and F3, but only small deflections on the right side, which most probably is caused by volume conduction (spread) of the left eye's potential.

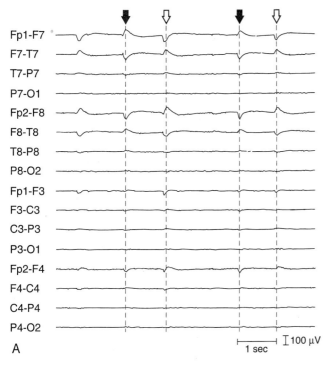

A

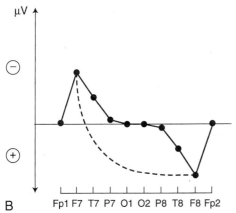

B

FIGURE 27–35. *A.* Bipolar montage: eye movements to the left *(solid arrow)* and to the right *(open arrow).* Eye movements shown here are not ideally horizontal but have a slight oblique component so that Fp₁ and Fp₂ detect small potentials. *B.* Voltage/electrode position graph of horizontal eye movement to the right. Potential distribution characteristic for dipole with two steep fall-offs. Exponential fall-off close to negative pole at electrode F_7 and exponential fall-off also close to positive pole at electrode F_8. Eye movements to the left show same distribution but reversed polarities. Assuming a unipolar generator, exponential fall-off would follow the dotted line in the voltage/electrode position graph.

Fp_2. The channels F_7–T_7 and F_8–T_8 reveal smaller deflections compared to Fp_1–F_7 and Fp_2–F_8. This reflects smaller amplitude differences between the maximal electrodes F_7 and F_8 and T_7 and T_8, which are not quiet but still record fall-off potentials of the eye movement (see Fig. 27–37C).

Muscle activity of extraocular muscles generally cannot be recorded by scalp electrodes, because these muscles lie deep in the orbits. The exception is the lateral rectus muscle. At the beginning of lateral eye movements, a low-amplitude sharp artifact of brief duration ("lateral rectus spike") frequently is recorded in the ipsilateral frontal electrodes F_7 or F_8 (Fig. 27–36). This artifact represents isolated motor unit potentials in the lateral rectus muscle.

The deflections caused by oblique eye movements in longitudinal bipolar montages can be explained by arithmetic summation of the effects of its composing lateral and vertical components on frontotemporal electrodes (Fig. 27–37A,B,C). Both components can be separately displayed in different montages. Transverse montages focus on the lateral component whereas parasagittal bipolar chains show the vertical component of oblique eye movements (Fig. 27–37B).

Notice also that the distribution maps of eye movements appear less "logical" if a unipolar generator is assumed. Voltage drops exponentially as an inverse square function of the distance from a unipolar generator (Fig. 27–35B). This rule is violated when we represent a dipole as a unipole. From a practical point of view, this observation can be used to differentiate between unipolar and bipolar distributions (Fig. 27–38). Unfortunately, this is not always possible, especially when electrodes are relatively close to one of the poles of the bipolar generator and relatively distant to the other pole, as is frequently the case in routine electroencephalography. A noncephalic reference may be at times useful in these cases.

Electrocardiogram

In EEG tracings, identification of the ECG signal (QRS complex) usually causes few problems owing to its regularity and morphology. Its electrical field extends to the base of the skull and is detected best by inferior electrodes such as ear electrodes. In persons with short necks, however, the ECG signal can also be easily recorded by parasagittal electrodes. In recordings for suspected electrocerebral silence, the EEG is amplified at maximal sensitivities, which frequently allows the ECG to intrude in some or all channels, obscuring at times low-voltage brain activity. The vector of the ECG projecting on the scalp can change over time as respiration causes the electrical axis of the heart to vary slightly in orientation.[4] However, the ECG signal usually appears as a dipole on the scalp. Its negative maximum is found in the right frontotemporal region and positivity is found over the left posterior temporoparietal area (Fig. 27–39).

MULTIPLE GENERATORS

Intracortical recordings could show that in reality, the potential fields that are detected by routine electroencephalography are composed by multiple layers of bipolar generators of different strengths. They mostly are located in the gyral and sulcal cortex.[21] The recorded voltages of such overlapping dipoles depend on the orientation ("solid angle") and distance of the recording electrode with respect to the vector and location of the potential field.[14] Even with highly sophisticated mathematical calculations, an exact definition of the generators producing the EEG at a given time cannot be made. Therefore, the "visual approximation rules" discussed in this chapter constitute only a crude approximation, assuming usually an "equivalent" unipole and in certain circumstances, a dipole. However, these rules still lead to useful and reliable interpretations of the clinical EEG in most cases. Progressive digitization and computerization of the EEG may provide mathematically more eleborated methods to define with more precision the generators of specific EEG signals. Integration with functional and anatomical neuroimaging techniques should eventually allow us to limit the mathematically possible solutions and increase their predictive value. There is no doubt, however, that these highly elaborated techniques will continue to greatly benefit

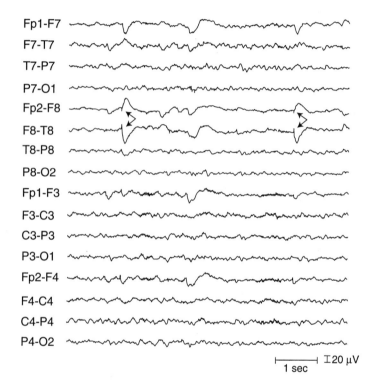

FIGURE 27–36. "Lateral rectus spike" *(arrows)* in lateral eye movement to the right during rapid eye movement (REM) sleep. Notice the absence of other muscle artifacts.

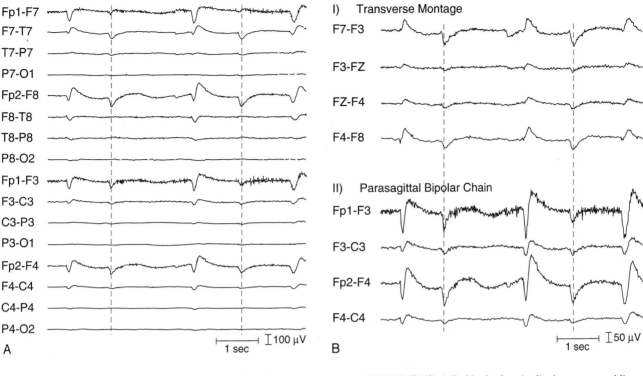

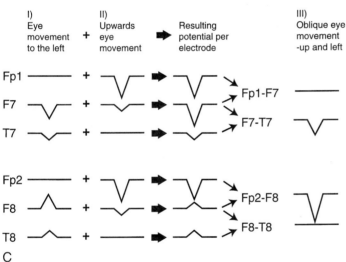

FIGURE 27–37. *A.* In bipolar longitudinal montages, oblique eye movements show asymmetrical deflections. Here, eye movements up and to the left are recorded (dashed lines). Notice the downward deflection in channels F_7–T_7 and Fp_2–F_8 and the relative lack of deflections in channels Fp_1–F_7 and F_8–T_8 (for explanation see legend for part *C*). *B.* Same oblique eye movements as shown in *A.* Transverse montage (I) shows lateral component (to the left) of oblique movement: positivity at F_7 and negativity at F_8. Parasagittal bipolar chains (II) reveal vertical component (upward): positivity at Fp_1 and Fp_2. *C.* The asymmetrical deflections in temporal bipolar chains caused by oblique eye movements (up and to the left) are shown on the right (III; compare to *A*). The deflections can be explained by the effects of the lateral (I) and vertical (II) components on frontotemporal electrodes. Deflections in I and II are recorded versus an inactive reference. Notice that the deflections in an EEG channel represent the arithmetic difference of the potentials recorded at the sites of the corresponding electrodes.

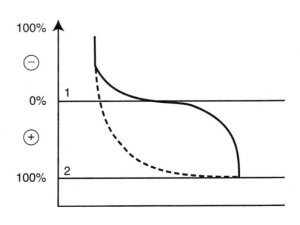

FIGURE 27–38. Potential field of a dipole showing two exponential fall-offs (solid line). There is a steep fall-off near the negative pole and a gradual decline close to the baseline (position 1). When the field comes closer to the dipole's positive pole, there is again an exponential increase. If we assume that the baseline is at position 2, the field polarity would indicate a unipolar negativity. However, the expected fall-off for a unipolar distribution (dotted line) differs significantly from the one corresponding to a dipole (solid line).

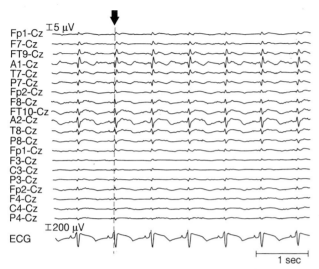

FIGURE 27–39. Recording of electrocerebral silence in a referential montage using the midline electrode C$_z$ as reference. Notice that the ECG signal projects as a dipole on the scalp. Its negative maximum is found in the right frontotemporal region (A$_2$ > FT$_{10}$, T$_8$) and the positivity over the left posterior temporoparietal area (A$_1$ > T$_7$, P$_7$).

from the practical experience of electroencephalographers who have analyzed the EEG using fundamental rules, as previously described, for more than 60 years.

REFERENCES

1. American Electroencephalographic Society: Guidelines for standard electrode position nomenclature. J Clin Neurophysiol 8:200–202, 1991.
2. American Electroencephalographic Society: Guidelines in electroencephalography, evoked potentials, and polysomnography. J Clin Neurophysiol 11(1):30–36, 111–113, 1994.
3. Binne CD, MacGillivray BB, Osselton JW: Derivations and montages. In Remond A (ed): Handbook of Electroencephalography and Clinical Neurophysiology, vol 3C. Amsterdam, Elsevier, 1974, pp 22–57.
4. Brittenham DM. Artifacts. Activities not arising from the brain. In Daly DD, Pedley TA (eds): Current Practice in Clinical Electroencephalography, 2nd ed. New York, Raven Press, 1990, pp 85–105.
5. Goldman D: The clinical use of the "average" electrode in monopolar recording. Electroencephalogr Clin Neurophysiol 2:211–214, 1950.
6. Jasper H: Report of committee on methods of clinical exam in EEG. Electroencephalogr Clin Neurophysiol 10:370–375, 1958.
7. Jayakar P, Duchowny M, Resnick TJ, Avarez LA: Localization of seizure foci: Pitfalls and caveats. J Clin Neurophysiol 8:414–431, 1991.
8. Knott JR: Electrode montages revisited: How to tell up from down. Am J EEG Technol 9:33–45, 1969.
9. Knott JR: Further thoughts on polarity, montages and localization. J Clin Neurophysiol 2:63–75, 1985.
10. Lesser RP, Lüders H, Dinner DS, et al: An introduction to the basic concepts of polarity and localization. J Clin Neurophysiol 2:46–61, 1985.
11. Lüders H, Lesser RP, Dinner DS, Morris HH: Benign focal epilepsy of childhood. In Lüders H, Lesser RP (eds): Epilepsy. Electroclinical Syndromes. London, Springer, 1987, pp 303–346.
12. Lüders HO, Noachter S: Atlas und Klassifikation der Elektroenzephalographie. Wehr, Ciba-Geigy Verlag, 1994.
13. Maulsby RL: Basics for visual analysis. Polarity convention, principles of localization and electrical fields. In Klass DW, Daly DD (eds): Current Practice of Clinical Electroencephalography. New York, Raven Press, 1979, pp 27–36.
14. Morris HH, Lüders H: Electrodes. Electroencephalogr Clin Neurophysiol 37:S3–S26, 1985.
15. Morris HH, Lüders H, Lesser RP, et al: The value of closely spaced scalp electrodes in the localization of epileptiform foci: A study of 26 patients with complex partial seizures. Electroencephal Clin Neurophysiol 63:107–111, 1986.
16. Niedermeyer E: The EEG signal: Polarity and field determination. In Niedermeyer E, Lopes da Silva F (eds): Electroencephalography: Basic Principles, Clinical Applications, and Related Fields. Baltimore, Urban & Schwarzenberg, 1987, pp 79–83.
17. Pernier J, Perrin F, Bertrand O: Scalp current density fields: Concepts and properties. Electroencephalogr Clin Neurophysiol 69:385–389, 1988.
18. Peters JF: Surface electrical fields generated by eye movements. Am J EEG Technol 7:27–40, 1967.
19. Sharbrough FW: Electrical fields and recording techniques. In Daly DD, Pedley TA (eds): Current Practice of Clinical Electroencephalography, 2nd ed. New York, Raven Press, 1990, pp 29–49.
20. Speckmann EJ, Elger CE: Introduction to the neurological basis of the EEG and DC potentials. In Niedermeyer E, Lopes da Silva F (eds): Electroencephalography: Basic Principles, Clinical Applications, and Related Fields. Baltimore, Urban & Schwarzenberg, 1987, pp 1–14.
21. Speckmann EJ, Elger CE, Altrup U: Neurophysiologic basis of the EEG. In Wyllie E (ed): The Treatment of Epilepsy: Principles and Practice. Philadelphia, Williams & Wilkins, 1997, pp 204–217.
22. Tyner FS, Knott JR, Mayer WB: Fundamentals of EEG Technology, vol 1: Basic Concepts and Methods. New York, Raven Press, 1983, pp 146–159, 288–294.

Chapter 28

Development of the Normal Electroencephalogram

Richard A. Hrachovy, MD

INTRODUCTION
ELECTROENCEPHALOGRAPHY OF THE
  NEONATE
GENERAL APPROACH TO VISUAL
  ANALYSIS OF THE NEWBORN
  ELECTROENCEPHALOGRAM
ELECTROENCEPHALOGRAPHIC FEATURES
  OF PREMATURITY
General Developmental Features of
  Analysis
Specific Developmental Landmarks of
  Prematurity (Graphoelements)
ELECTROENCEPHALOGRAPHIC FEATURES
  OF THE TERM INFANT
Awake
Quiet Sleep
Active Sleep

PATTERNS OF UNCERTAIN SIGNIFICANCE
  IN THE TERM INFANT
ELECTROENCEPHALOGRAPHIC CHANGES
  FROM TERM TO THREE MONTHS
Awake
Sleep
NORMAL ELECTROENCEPHALOGRAM
  FROM THREE MONTHS TO
  ADOLESCENCE
Features of the Normal Awake
  Electroencephalogram
Features of the Normal Sleep
  Electroencephalogram
Changes Seen with Arousal

INTRODUCTION

Of the various challenges that the clinical electroencephalographer faces, none may be more daunting, yet rewarding, than the interpretation of pediatric, particularly neonatal, electroencephalograms (EEGs). It is not possible to convey the fund of knowledge required to adequately interpret such recordings in an entire volume, much less a brief chapter such as this. To develop such interpretative skills requires that an individual acquire at least 2 years of formal training in pediatric electroencephalography, with exposure to many thousands of recordings in order to establish the appropriate mental templates necessary to properly interpret neonatal and pediatric recordings.

In this chapter, I review the major normal developmental EEG patterns from prematurity through adolescence. Because of the limited space allowed for presentation of such a broad topic, I restrict my discussion to what I consider the most pertinent information concerning each waveform or pattern. I illustrate most of the important patterns or waveforms encountered in the interpretation of pediatric EEGs. For readers seeking a more detailed discussion of neonatal and pediatric electroencephalography, several chapters and atlases are recommended (see refs. 10, 17, 25, 26, 30, 53, 58, 69, 70, 84, 98, 109).

ELECTROENCEPHALOGRAPHY OF THE NEONATE

The information concerning normal developmental electroencephalographic features of the neonate presented in this chapter is based on a collective experience of over 70 years of studying neonates at our institution, as well as a review of the work of other workers in the field.[10, 14–27, 30, 32, 68–70, 75–77, 90, 97–101, 104–106] Obviously, documenting normality in prematures can be difficult. In past years we considered premature infants to be "normal" if their metabolic condition was normal, they had no abnormal neurological signs or symptoms, and they had uneventful clinical courses prior to discharge from the hospital. More recently, information provided by ultrasound, computerized tomography and magnetic resonance imaging have improved our ability to separate normal from abnormal infants.

GENERAL APPROACH TO VISUAL ANALYSIS OF THE NEWBORN ELECTROENCEPHALOGRAM

Prior to interpreting the EEG in older infants, children, and adults, the electroencephalographer must know two essential pieces of clinical information: the age and clinical state of the patient. However, when analyzing the EEG of the newborn, the state of the patient (i.e., awake, lethargic, comatose, and so on) is the only clinical information that the electroencephalographer should know before interpreting the record. The age of the patient should not be known prior to interpreting the newborn's EEG because estimating the conceptional age (CA) of the patient on the basis of certain developmental features or landmarks present in the EEG is one of the most crucial aspects of neonatal EEG interpretation. If the EEG does not show developmental features that allow the determination of CA to be made, this is evidence of significant brain dysfunction. Also, if a mixture of developmental features are present (dyschronism), features specific to the earliest developmental period may indicate when a brain insult occurred.[53]

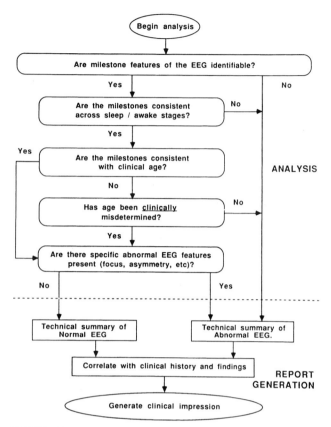

FIGURE 28–1. Flow diagram showing the process used to analyze neonatal EEGs. (From Hrachovy RA, Mizrahi EM, Kellaway P: Electroencephalography of the newborn. In Daly DO, Pedley TA [eds]: Current Practice of Clinical Electroencephalography, 2nd ed. New York, Raven Press, 1990, pp 201–242, with permission.)

The electroencephalographer experienced in interpreting premature EEGs should be able to estimate CA to within ± 2 weeks between 27 and 33 weeks' CA and to within ± 1 week between 34 and 37 weeks CA. Figure 28–1 summarizes the process used to analyze neonatal EEGs.

ELECTROENCEPHALOGRAPHIC FEATURES OF PREMATURITY

General Developmental Features of Analysis

Continuity

When brain electrical activity first appears, there are long periods of quiescence or flattening during which no brain electrical activity can be identified. This pattern has been called tracé discontinu, and before 30 weeks' CA is present in both the waking and sleep states (Fig. 28–2). The average duration of the discontinuous periods seen throughout prematurity is shown in Figure 28–3. The maximal periods of discontinuity are seen during deep non–rapid eye movement (NREM) sleep and may be more prolonged than these shown for any given age; however, the upper limit of "normal" is difficult to determine.[2, 12, 49]

With increasing age, the periods of inactivity shorten, and at approximately 30 weeks' CA, EEG activity becomes continuous during rapid eye movement (REM) sleep. At 34 weeks' CA, the EEG becomes continuous in the waking state. Continuity appears in slow wave (NREM) sleep at about 37 to 38 weeks (Fig. 28–4). From that time until about 46 and 48 weeks' CA, NREM sleep continues to show intermittent periods of attenuation, but not complete flattening, a pattern referred to as tracé alternant (Fig. 28–5, see later discussion).

Bilateral Synchrony

Before 30 weeks' CA, the EEG activity in homologous regions of the hemispheres is asynchronous. In general, the greater the distance from the midline, the greater the degree of asynchrony. With increasing maturity, synchrony increases. Degree of synchrony is also affected by state (Fig. 28–6). Asynchrony is most prominent and persists to the latest age in NREM sleep and initially decreases and is least prominent in REM sleep.

Specific Developmental Landmarks of Prematurity (Graphoelements)

Beta Delta Complexes

Beta delta complexes have been referred to as brushes, delta brushes, ripples of prematurity, spindle delta bursts,

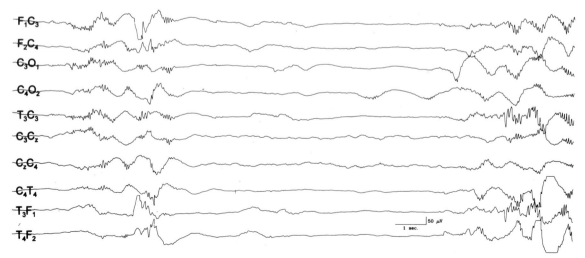

FIGURE 28–2. Tracé discontinu pattern in an infant with a conceptional age (CA) of 30 weeks.

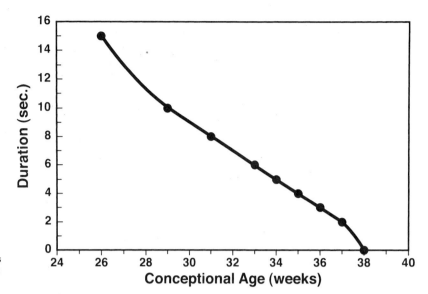

FIGURE 28–3. Average duration of discontinuous periods in non–rapid eye movement (NREM) sleep in the EEG of the premature infant.

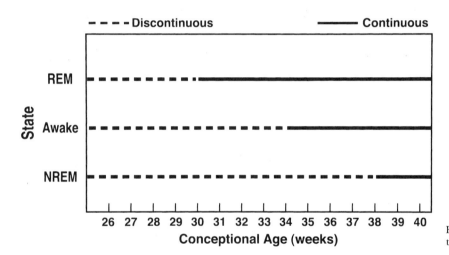

FIGURE 28–4. Development of continuity in the EEG of the premature infant.

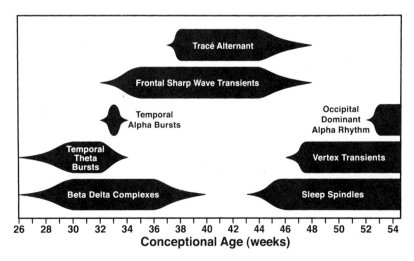

FIGURE 28–5. Major developmental landmarks from prematurity to 3 months' post-term.

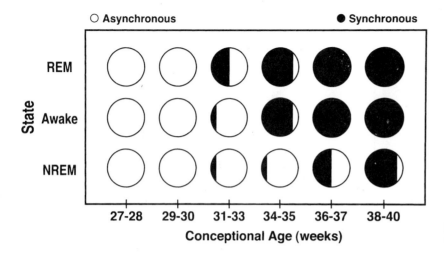

FIGURE 28–6. Development of bilateral synchrony in the EEG of the premature infant.

and rapid bursts. They first appear at approximately 26 to 27 weeks' CA (see Fig. 28–5). They consist of random, 0.5- to 1.5-Hz, 50- to 250-μV waves upon which are superimposed bursts of rhythmic 8- to 12-Hz or, more commonly, 18- to 22-Hz activity (Fig. 28–7). The voltage of these faster frequencies waxes and wanes and rarely exceeds 75 μV. Both the alpha and beta forms may be seen in the same infant. They usually occur asynchronously between homologous regions and may show consistent voltage differences between hemispheres.

When beta delta complexes first appear, they are predominant in the central regions. Between the ages of 29 and 33 weeks' CA, they increase in voltage and are most prominent in REM sleep. During this time the spacial distribution of the complexes also changes, extending to the occipital and temporal regions. Between 33 and 38 weeks' CA, beta delta complexes become most prominent in NREM sleep. Occasional remnants of these complexes may be seen during NREM sleep in term infants.

Temporal Theta and Alpha Bursts

Temporal theta bursts first appear at approximately 26 weeks' CA and are maximally expressed between the ages of 30 and 32 weeks' CA (see Fig. 28–5). These bursts consist of rhythmical 4- to 6-Hz waves, often of sharp configuration. Voltages range from 20 to 200 μV and bursts rarely last more than 2 seconds (Fig. 28–8). At 33 weeks' CA the frequency of some bursts increases to alpha frequency (Fig. 28–9), and at this age, the alpha frequency bursts are more prominent. The theta and alpha bursts decline rapidly beyond 33 weeks' CA.[55]

Frontal Sharp Wave Transients

Random frontal diphasic sharp waves first appear at about 34 weeks' CA and attain maximal expression at 35 weeks' CA. They remain until 44 weeks' CA, then decline rapidly, and are rarely seen beyond 6 weeks post-term (see Fig. 28–5). Frontal sharp wave transients are bilaterally synchronous and symmetrical, although some shifting asymmetry between hemispheres is common. The initial surface negative component may last up to 200 msec, and the following surface positive phase is more prolonged (Fig. 28–10). The amplitude of the transients may range from 50 to 150 μV. In some individuals, the waves become very sharp or have a spike-like appearance.

Frontal sharp wave transients may occur in isolation or in brief serials, and may be intermixed with bifrontal delta

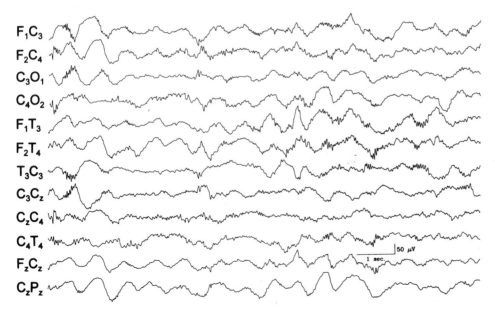

FIGURE 28–7. Beta delta complexes in an infant with a CA of 34 weeks.

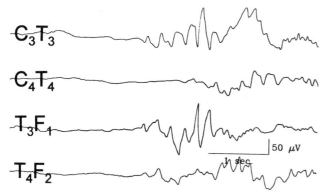

FIGURE 28–8. Temporal theta bursts in an infant with a CA of 32 weeks.

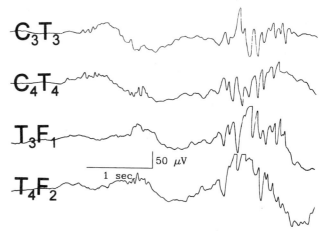

FIGURE 28–9. Temporal alpha bursts in an infant with a CA of 33 weeks.

activity (Fig. 28–11). The transients are most prominent during transitional sleep.

Sleep-Wake Differentiation

Although differences between states may be seen at 30 to 32 weeks' CA, clear distinction between the awake and sleep EEG occurs at about 36 weeks. At this age the background activity in the awake EEG shifts from predominantly beta-delta complex activity to a polyfrequency record similar to that seen in the term infants (see later discussion). The NREM sleep pattern, tracé alternant (see later discussion) also appears between 36 and 37 weeks' CA (see Fig. 28–5).

Reactivity

EEG changes to stimuli do not clearly occur until approximately 33 to 34 weeks' CA and are clearly present at 36 to 37 weeks' CA. The response consists of a generalized attenuation (flattening) of activity in all regions, particularly to loud sounds (Fig. 28–12). This flattening may last 8 to 10 seconds. Such responses to stimuli should not be misinterpreted as evidence of immaturity in the term infant, or be confused with abnormal attenuation episodes that occur in abnormal conditions (e.g., metabolic derangements, hypoxia, and so on).

ELECTROENCEPHALOGRAPHIC FEATURES OF THE TERM INFANT

Awake

The background activity of the awake state in the term infant is relatively low in voltage (25 to 50 μV) and consists chiefly of random and semirhythmic 4- to 8-Hz activity mixed with slower and faster frequencies (polyfrequency record) (Fig. 28–13). Although brief runs of semirhythmic activity may be seen, sustained or prolonged runs of rhythmic activity do not occur in the normal term infant's awake EEG.

Quiet Sleep

Two types of quiet sleep patterns (NREM) may be distinguished in the term infant:
1. Continuous, 0.5- to 1.5-Hz, 50- to 100-μV activity in all regions, but maximum in amplitude posteriorly with superimposed low voltage theta and fast activity (Fig. 28–14), and

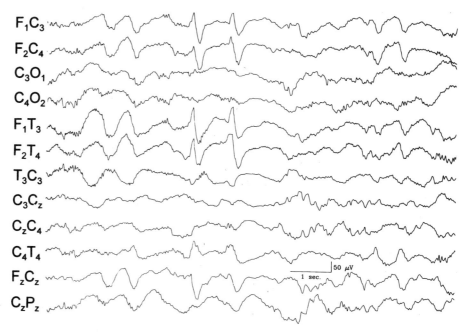

FIGURE 28–10. Frontal sharp wave transients in an infant with a CA of 35 weeks.

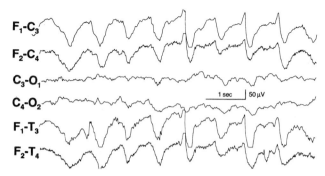

FIGURE 28–11. Frontal sharp wave transients mixed with rhythmic bifrontal delta activity in a term infant.

2. Periodic runs of such activity, interspersed with episodes of generalized voltage attenuation. These interburst episodes are not "flat" and usually have a duration of from 3 to 15 seconds (Fig. 28–15). This pattern has been called tracé alternant, distinguishing it from tracé discontinu (see earlier discussion). Stimulation during NREM sleep results in a generalized voltage attenuation as described earlier in the discussion of reactivity.

As previously mentioned, frontal sharp waves transients, which first appear around 34 to 35 weeks' CA, are seen in the term infant usually during transitional sleep. They decline rapidly after 44 weeks' CA and are usually not seen after 6 weeks' post-term (see Fig. 28–5).

Active Sleep

The background activity during active sleep is essentially indistinguishable from that seen in the awake state (Fig. 28–16). In fact, without the use of other polygraphic data (e.g., electrooculogram and electromyogram) it is often not possible to differentiate between the awake and REM states on the basis of the EEG findings alone.

PATTERNS OF UNCERTAIN SIGNIFICANCE IN THE TERM INFANT

The patterns and waveforms described next share certain common features: (1) They may appear in the EEGs of term infants; (2) they may occur as isolated phenomena, in combination with each other or in association with other clearly defined EEG abnormalities; (3) they may occur in infants who appear normal, or in infants who have experienced CNS insults; and (4) they appear to have no prognostic significance.

Low Voltage, Differentiated Background Activity

In most instances in which the background EEG activity is consistently less than 20 μV, there is also a reduction or disappearance of the polyrhythmic activities normally present. Such an EEG is often referred to as "depressed and undifferentiated" and is usually associated with a variety of other abnormalities, including multifocal sharp waves, electrographic seizure discharges, absence of sleep cycling, and so on. Patients whose EEGs exhibit this pattern have altered levels of responsiveness. This pattern is seen with a variety of cerebral insults, and if seen within the first 24 hours after birth, usually signifies a poor prognosis. Occasionally, one will encounter a term infant's EEG in which the waking and REM background activity is consistently lower in voltage than expected for a term infant (< 20 μV); however, the polyfrequency character of the record is preserved. The voltage of the background activity during quiet sleep in such patients may also be somewhat lower in voltage than expected, but may, at times, be normal.

Infants demonstrating such an EEG pattern are usually alert and active. If no other abnormalities are present (e.g., multifocal sharp waves), such low voltage records probably should be interpreted as normal.

Excessive Rhythmical Theta or Alpha Frequency Activity

As mentioned earlier, the EEGs of normal term infants show much random and semirhythmic 4- to 8-Hz activity.

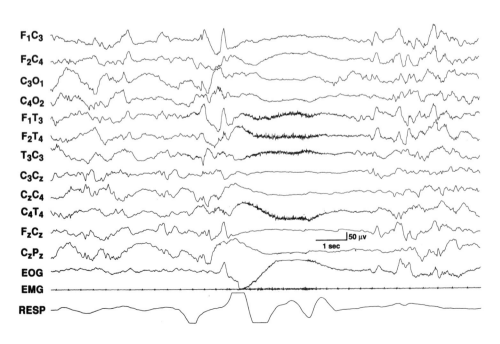

FIGURE 28–12. Arousal from NREM sleep in a term infant. Generalized attenuation of the background activity occurs in all regions with superimposed myogenic artifact in the temporal derivations.

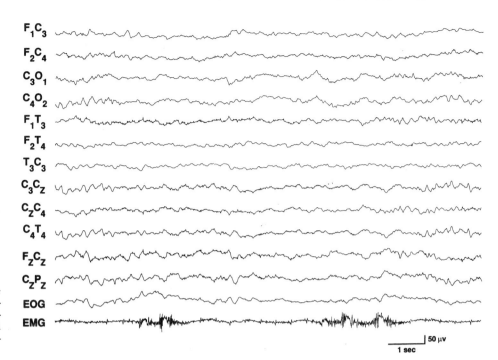

FIGURE 28–13. Normal awake pattern in a term infant. Background activity consists of a mixture of alpha, theta, beta, and delta activity (polyfrequency record).

Sustained rhythmic generalized or regional theta or alpha frequency activity in the awake or sleep tracings, or both, is clearly an abnormal finding and may be associated with various disorders, including chromosomal abnormalities and inborn errors of metabolism (Fig. 28–17). However, various degrees of generalized or regional rhythmic theta or alpha activity may be seen in the normal term infant's EEG. At present, there are no specific criteria for determining when the amount of rhythmic theta or alpha activity is excessive. Currently, we consider such rhythmic activity as being abnormal if it occurs continuously as the predominant activity present, or if it occurs in recurrent runs, each run lasting greater than 8 to 10 seconds. In the latter instance, the amplitude of the rhythmic activity is greater than the ongoing background activity.[53]

Midline Frontal and Central Rhythmic Theta/Alpha Bursts

These bursts consist of rhythmic 50- to 200-µV, 5- to 9-Hz activity that occurs in the midline frontal or central regions (Fig. 28–18). The duration of such bursts is usually less than 1.5 seconds. The bursts usually occur in transitional and quiet sleep. Hayakawa and associates[51] reported these bursts in 55% of neurologically normal infants at 38 to 42 weeks' CA. However, these bursts may also be seen in infants who have experienced various central nervous system (CNS) insults. In such instances, this rhythmic midline theta and alpha frequency activity is frequently mixed with well-defined sharp waves or spikes and the duration of the bursts in such infants may be more prolonged (Fig. 28–19).

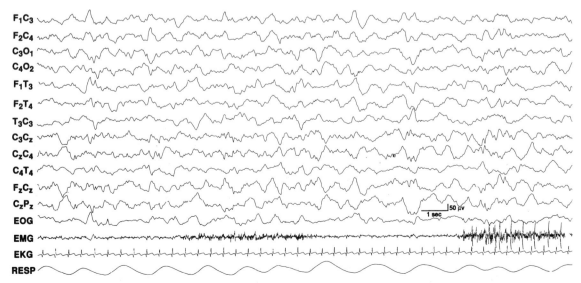

FIGURE 28–14. Continuous slow wave activity in NREM sleep in a term infant.

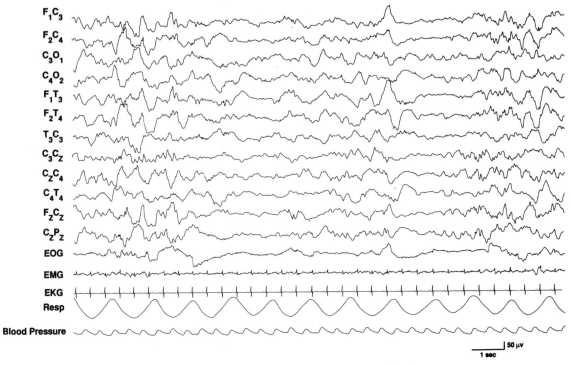

FIGURE 28–15. Tracé alternant pattern in a term infant.

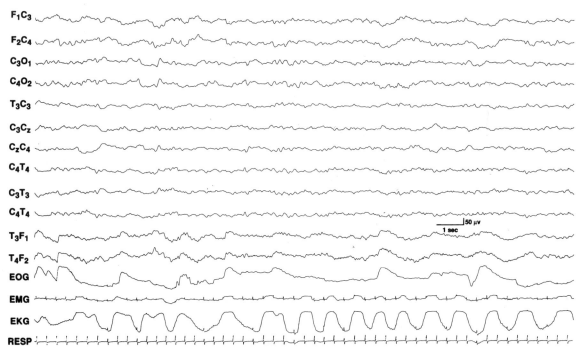

FIGURE 28–16. Rapid eye movement (REM) sleep in a term infant.

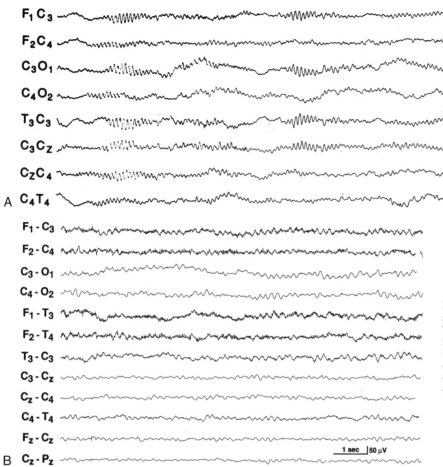

FIGURE 28–17. Excessive rhythmic alpha (*A*) and theta (*B*) activity. *A*, Term infant with a chromosomal abnormality. *B*, Term infant with congenital heart disease. The significance of excessive rhythmic alpha and theta frequency in the term infant has yet to be determined.

Rhythmic Bifrontal Delta Activity

In some infants, from about 37 weeks' CA until about 6 weeks post-term, brief serials of rhythmic 50- to 250-μV, 1.5- to 2-Hz monomorphic waves may appear synchronously in the frontal leads bilaterally (Fig. 28–20). This activity usually is seen in transitional sleep when it may be intermixed with frontal sharp wave transients. However, it may also appear in the waking state. Such bifrontal delta activity is probably a benign variant in the newborn.

Rhythmic Occipital Theta Bursts

A relatively uncommon finding in the term infant's EEG is the presence of brief bursts of rhythmic 50- to 100-μV, 4-

to 6-Hz activity in the occipital region (Fig. 28–21). This activity usually occurs asynchronously on the two sides and, in some instances, may occur unilaterally. It may be seen in the waking or sleep states. Such activity may occur as an isolated feature in an otherwise normal EEG, but it may also occur in association with abnormal waveforms. The significance of such rhythmic occipital activity requires further study.

Transient Unilateral Attenuation During Sleep

This unusual phenomenon occurs within minutes of entering NREM sleep. It begins and ends abruptly and lasts

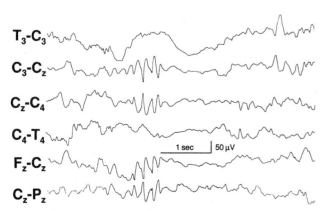

FIGURE 28–18. Midline central rhythmic theta burst in a term infant. Note the brief duration (< 1.5 seconds) of this activity. Such activity is probably a normal variant in the term infant.

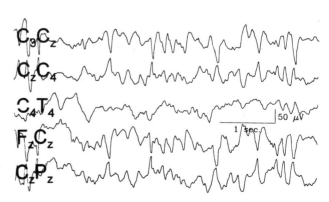

FIGURE 28–19. Midline central rhythmic theta activity mixed with sharp waves in term infant who experienced hypoxia at the time of delivery. Note the longer duration of this rhythmic midline central activity compared to that in Figure 28–18. Such activity does not occur in the normal term infant.

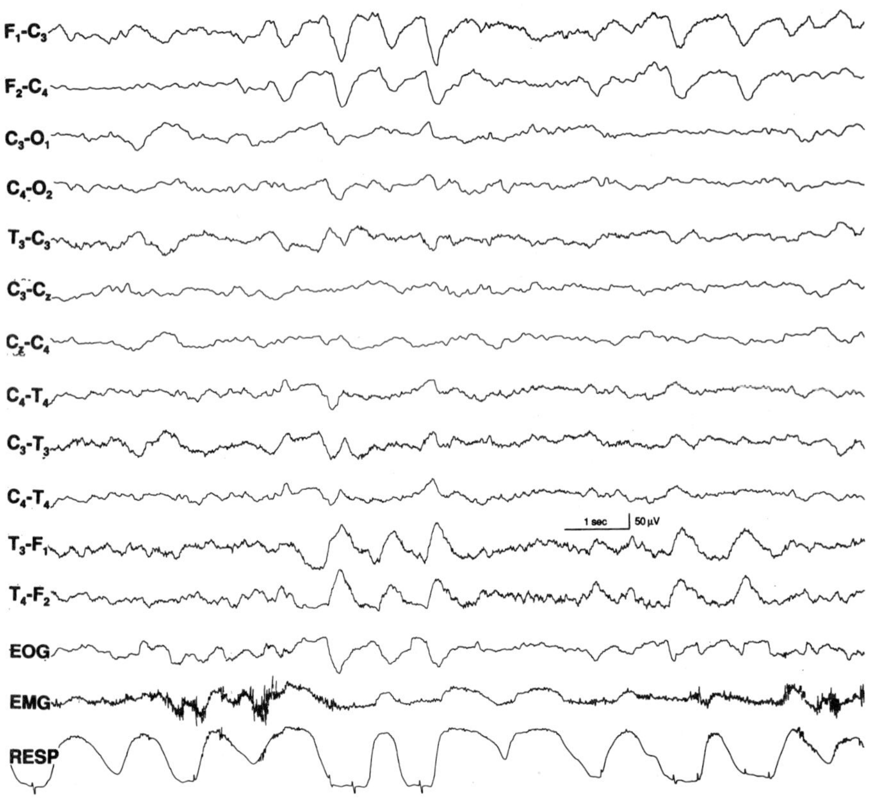

FIGURE 28–20. Rhythmic bifrontal delta activity in a term infant. Such activity may be mixed with frontal sharp wave transients (see Figure 28–11). Such rhythmic bifrontal delta activity is probably a benign variant in the newborn.

around 60 seconds (Fig. 28–22). It usually occurs only once during a recording session. The EEG background activity is usually normal. This pattern must be distinguished from the transient attenuation pattern that occurs with arousal from NREM sleep discussed earlier. Also, it must not be confused with episodes of unilateral or generalized attenuation that occur in association with a variety of CNS insults. The latter attenuation episodes occur repeatedly and may be present in all states.

Challamel and colleagues[9] and O'Brien and associates[80] reported that transient unilateral attenuation during sleep occurred in 3.17% and 4% of newborns, respectively. The pathophysiological mechanism underlying this interesting phenomenon and its significance have yet to be determined.

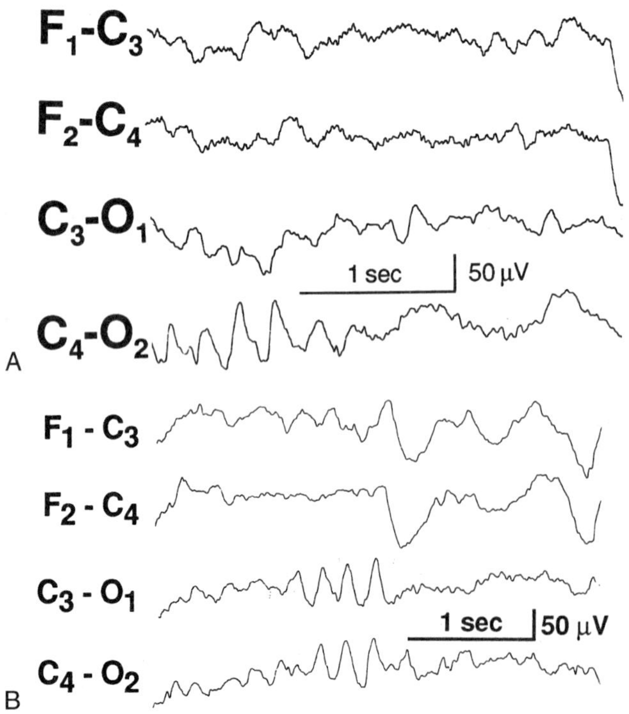

Focal Sharp Waves

One of the most difficult tasks in interpreting neonatal EEGs is differentiating "normal" from "abnormal" sharp waves. Numerous descriptions of "normal" sharp waves are found in the literature;[31, 48, 50, 63, 76, 88, 92, 93, 96] however, no stringent criteria have been established to distinguish normal from abnormal.

Temporal Sharp Waves

Table 28–1 summarizes the major features that can be used to differentiate between normal and abnormal temporal sharp waves. Figure 28–23 illustrates some normal and abnormal temporal sharp waves. Temporal sharp waves meeting all of the criteria of "normality" occur in at least 20% of normal healthy term infants during sleep. These sharp waves decrease rapidly during the first month of life and disappear after 6 weeks post-term. Unfortunately, many sharp waves share features of both groups and fall in the "gray" zone. In such instances, it is probably best to be very conservative in one's interpretation of the significance of such waves, especially because such sharp waves by themselves have no prognostic significance.

Central Rapid Spikes

These relatively low amplitude ($< 50 \mu V$), short duration (< 75 msec) spikes (Fig. 28–24) are uncommon and are

FIGURE 28–21. Rhythmic occipital theta bursts in an asymptomatic term infant. Such activity may occur unilaterally (A) or bilaterally (B). The significance of such rhythmic occipital activity is not known.

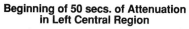

**Beginning of 50 secs. of Attenuation
in Left Central Region**

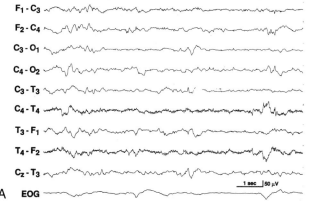

A   EOG

Middle of Attenuation

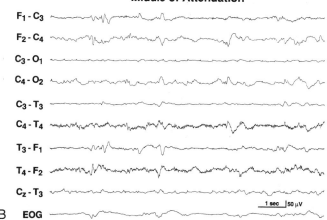

B   EOG

End of Attenuation

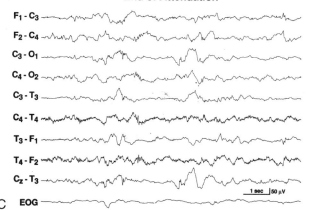

C   EOG

FIGURE 28–22. Transient unilateral attenuation during sleep in a term infant. *A.* Beginning of the attenuation. *B.* Middle of the attenuation. *C.* End of the attenuation. This is a relatively uncommon finding seen during NREM sleep in the EEGs of term infants. The pathophysiological mechanism underlying this phenomenon and its significance are not known.

usually encountered in premature infants. However, they may be seen in NREM sleep until about 1 week post-term.

ELECTROENCEPHALOGRAPHIC CHANGES FROM TERM TO THREE MONTHS

Awake

There is no perceptible change in the awake pattern during the first 2 months post-term. Approximately 3 months post-term, poorly sustained, rhythmic 3- to 4-Hz activity may appear in the occipital leads bilaterally (see Fig. 28–5). The rhythmic activity may be partially or completely abolished by

eye opening and disappears with sleep onset. This rhythmic activity represents the appearance of the occipital alpha rhythm (see later discussion).

Sleep

Frontal sharp wave transients decrease in number and voltage during the first 5 to 6 weeks after birth and are no longer evident after 7 to 8 weeks post-term. The tracé alternant pattern of quiet sleep evolves to a continuous pattern over a period ranging from 2 to 6 weeks after term (see Fig. 28–5).

Rudimentary sleep spindles may appear in the central regions before 44 weeks' CA; however, well-defined central

TABLE 28–1	Features Used to Distinguish Normal from Abnormal Temporal Sharp Waves in the Newborn	
Feature	**Normal**	**Abnormal**
Amplitude	< 75 μV	Usually > 75 μV
Duration	< 100 msec	Usually > 150 msec
Manner of occurrence	Random, bilateral, asynchronous, symmetrical or asymmetrical	Variable, but frequently occur in serials or runs, sharp waves occurring in only one temporal region
Frequency	Intervals > 1 minute	Occur every few seconds, or at least several times per minute
Morphology	Mono- or diphasic	Variable morphology, frequently polyphasic, and may be followed by a high voltage slow wave
Polarity	Initial component surface negative	Either surface negative or positive
State	NREM sleep	Awake and NREM sleep, diminished or absent in REM sleep

NREM, non-rapid eye movement; REM, rapid eye movement.

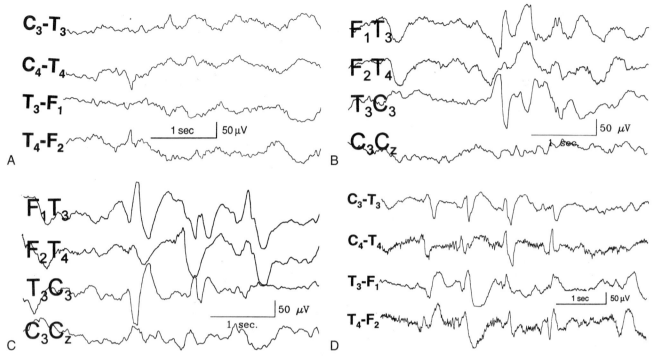

FIGURE 28–23. Examples of temporal sharp waves from a group of term infants. *A,* normal temporal sharp wave; *B, C,* and *D,* abnormal temporal sharp waves. See Table 28–1 for a list of features used to distinguish normal from abnormal temporal sharp waves in the newborn's EEG.

14-Hz spindles usually appear at 5 to 6 weeks post-term and should be present in all term infants by 8 weeks.[58, 59] Vertex transients also appear between 6 and 8 weeks after term. Sleep spindles and vertex transients are discussed in more detail in the following section.

NORMAL ELECTROENCEPHALOGRAM FROM THREE MONTHS TO ADOLESCENCE

Unlike the rapid and marked changes that occur in the EEG of the premature and term infant, the developmental

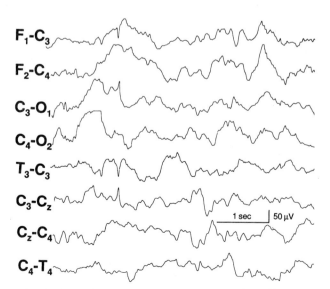

FIGURE 28–24. Central rapid spike in a term infant. The significance of such spikes is not known.

changes during the remainder of infancy and childhood occur at a much slower rate. During this time period certain patterns (e.g., hypnagogic hypersynchrony) appear and disappear; however, most patterns and waveforms that develop during childhood persist throughout the remainder of the individual's life.

Features of the Normal Awake Electroencephalogram

Alpha Rhythm

Identification of the alpha rhythm is the starting point of visual analysis in any patient more than 3 months of age. Rarely, apparently normal children show no alpha activity, and other individuals may show only brief runs of alpha activity during hyperventilation or only transiently on eye closure.

As previously mentioned, when the alpha rhythm first appears at approximately 3 months of age, its frequency is 3 to 4 Hz (see Fig. 28–5). It may be poorly sustained and initially lack reactivity to eye opening. However, shortly thereafter in most individuals the alpha rhythm attenuates with eye opening. Also, it initially slows and then disappears as the child becomes drowsy. These two features can help differentiate the alpha rhythm from other occipital rhythms that may be present in the EEG of the infant or child.

Throughout the first years of life, the frequency of the alpha rhythm rapidly increases so that by the age of 1 year its frequency has reached 6 Hz in 70% of normal children. The lower limit of the adult range (8 Hz) is reached in 80% of normal individuals by age 3 years. Thereafter, there is a much slower increase in frequency, with the usual adult frequency of 9 Hz being reached in 65% of individuals by age 9 years[28, 57, 84] (Fig. 28–25).

Slowing of the alpha rhythm is one of the earliest and most important changes that occurs with diffuse brain dys-

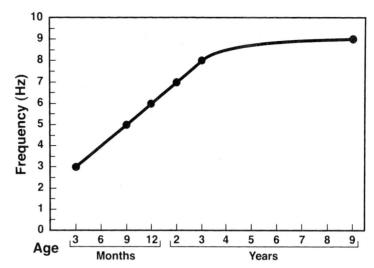

FIGURE 28–25. Normal development of the occipital dominant alpha rhythm between the ages of 3 months and 9 years.

function. However, certain drugs (e.g., phenytoin and carbamazepine) may slow the alpha rhythm without other changes in the EEG, even when the drugs are within the therapeutic range.[36]

Although a significant number of adults show low voltage alpha rhythm voltages (< 15 μV) at the scalp,[72] children rarely show alpha rhythms with voltages less than 30 μV.[84] The average voltage of the alpha rhythm in children between 3 and 15 years of age ranges from 50 to 60 μV. Occipital dominant alpha rhythm of 100 μV or more may be seen in about 10% of children.[84] The voltage of the alpha rhythm slowly deceases with age, probably owing to increasing thickness of the skull and overlying tissues. However, in some normal adults the voltage of the occipital dominant rhythm may be similar to that seen in childhood.

The maximal area of expression of the alpha rhythm is the occipital region in 95% of children.[58] However, the alpha activity may be equally distributed in the central and temporal regions in a small number of children.

The alpha rhythm may show an asymmetry with the lower voltage side being not less than 80% of the higher voltage side in 95% of normal children. None show a ratio of greater than 2:1. In almost all individuals, the right side is the higher voltage side.[28, 84] This asymmetry is not related to handedness, but is probably related to skull thickness with the left side of the skull usually being the thicker side.[66] In the clinical setting, asymmetries of the alpha rhythm (ratio less than 2:1) should be considered as insignificant. However, the presence of a significant asymmetry (ratio greater than 2:1) does not necessarily imply focal brain damage. In children and infants, in particular, subgaleal swelling caused by trauma or IV infiltration may produce marked attenuation of the underlying EEG activity recorded by scalp electrodes.

Between the ages of 6 months and 3 years, the alpha rhythm is almost monorhythmic (i.e., there is good voltage and frequency regulation) (Fig. 28–26). However, from 3 years to mid-adolescence there is rather poor regulation of the voltage of the alpha rhythm, and during this time the alpha rhythm becomes mixed with the so-called posterior slow waves of youth, which increase the complexity of the waveforms (see subsequent text).

Occipital Alpha Rhythm Variants

Occasionally in older children, the alpha rhythm may show waves with a frequency exactly half of the alpha rhythm. This produces a pattern that has been referred to as the "slow" alpha variant. Also, the individual alpha waves may bifurcate,

resulting in a harmonic rhythm twice the basic frequency. This latter pattern has been called the "fast" alpha variant.[45] Both the "slow" and "fast" alpha variants block with eye opening.

Posterior Slow Waves of Youth

One of the more difficult tasks one faces in the interpretation of EEGs of children is differentiating between normal and abnormal slow activity in the posterior leads. There has been much confusion concerning this area of interpretation, and an excellent review of this topic is provided by Kellaway.[58]

Posterior slow waves of youth are maximally expressed in the central-occipital derivations. They consist of slow fused waves (1.5 to 3 Hz) intermixed with the occipital alpha rhythm (the alpha rhythm is usually superimposed on the slower fused wave)[1, 28, 84] (Fig. 28–27). The voltage of this activity is not more than 120% of the alpha rhythm. As with the alpha rhythm, these waves block with eye opening and with onset of drowsiness, and may be asymmetrical, with the right side usually being the higher voltage side. The asymmetry between sides should not exceed a ratio of greater than 2:1. These waves are accentuated by anxiety and hyperventilation.[8, 11, 34, 56, 64, 65, 78, 107]

Posterior slow waves of youth rarely occur before 2 years

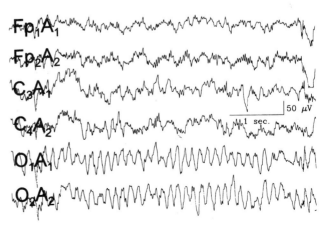

FIGURE 28–26. Occipital dominant alpha rhythm in an asymptomatic 18-month-old boy. Note the monorhythmic character of the alpha rhythm.

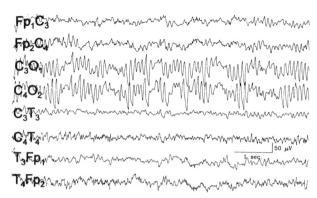

FIGURE 28–27. Posterior slow waves of youth in an asymptomatic 9-year-old girl.

of age. The frequency of this activity increases in the first decade and reaches a maximum between ages 8 and 14 years. Thereafter, this activity declines and is rarely seen after age 21 years. Interestingly, this occipital slow activity occurs more often in females. The degree of expression of posterior slow waves of youth often decreases as the EEG recording continues. The reason for this decrement is not clear, but may be related to a diminished anxiety level of the child as he or she becomes more familiar with the testing procedure.[8, 11, 34, 56, 64, 65, 78, 107]

In Table 28–2 the features of posterior slow waves of youth are compared with abnormal occipital slow activity. Some examples of abnormal occipital slow activity are shown in Figure 28–28.

Although high voltage rhythmic slow waves (see Table 28–2) are an abnormal feature of the child's EEG, brief runs (< 3 seconds) of rhythmic, 2.5- to 4.5-Hz, monorhythmic waves with a voltage less than 100 μV are reported to occur in about 25% of normal children between the ages of 1 and 15 years.[29, 84]

Posterior Slow Transients Associated with Eye Movements

In some children less than 10 years of age, most commonly between 2 and 3 years of age, monophasic or diphasic transients appear in the occipital leads bilaterally in association with eye blinks or movements[108] (Fig. 28–29). The latency of these transients varies from 100 to 500 msec following the eye blink or movements. They have a duration of 200 to 400 msec and their voltage may reach 200 μV. The surface negative component of these transients has a sharp configuration.

Lambda Waves

Lambda waves are sharp biphasic or triphasic waves that appear in the occipital region bilaterally and occur in association with saccadic eye movements (e.g., scanning a geometric design) (Fig. 28–30). There has been some debate about the polarity of the most prominent phase of these waves. Most studies from human scalp recordings describe them as predominantly surface positive at the occipital electrodes.[3, 33, 47, 58, 86, 89, 95, 102] Others report the most prominent phase is negative.[37, 67, 87] In children either the positive or negative phase may be more prominent. Lambda waves are commonly seen in children between the ages of 2 and 15 years, occasionally younger. The amplitude of these waves in children may reach 70 μV with a duration of 160 to 250 msecs. They may be asymmetrical and may appear in only one occipital region. Lambda waves can be differentiated from pathological occipital spikes by having the patient close his or her eyes, at which time lambda waves disappear.

Mu Rhythm

The mu rhythm is an alpha frequency rhythm (most commonly 8 to 10 Hz) that occurs in the central or centroparietal regions. It has also been called the wicket rhythm, comb rhythm, or rhythm en arceau. It frequently occurs asynchronously on the two sides, and in some individuals may appear in only one central region. The surface negative components of the mu rhythm may have a very sharp or spike configuration (Fig. 28–31). The voltage of this rhythm

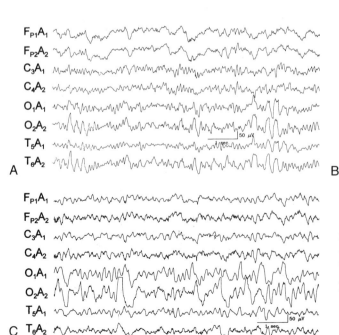

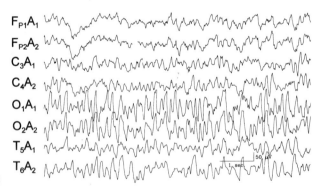

FIGURE 28–28. Examples of abnormal occipital slow wave activity. *A.* A 9-year-old girl with a history of basilar migraine headaches. *B.* A 4-year-old boy with a history of seizures of unknown type. *C.* An 18-month-old boy who experienced a febrile convulsion 24 hours prior to the EEG recording.

TABLE 28–2	Features Differentiating Posterior Slow Waves of Youth from Abnormal Occipital Slow Activity	
Feature	**Posterior Slow Waves of Youth**	**Abnormal Occipital Slowing**
Morphology	Fused waveform, superimposed alpha rhythm	Frequently mono- or polymorphic, disrupts alpha rhythm
Manner of occurrence	Random, scattered throughout each EEG epoch	May occur randomly, in serials, rhythmically or continuously
Voltage	< 1.5 times the voltage of the alpha rhythm	Variable, but frequently > 1.5 times the voltage of the alpha rhythm
Persistence	Blocks with eye opening and onset of drowsiness	May or may not block with eye opening and onset of drowsiness
Synchrony	May be synchronous or asynchronous	May be synchronous or asynchronous
Symmetry	May be symmetrical or asymmetrical (< 50% between sides), never focal	May be symmetrical, asymmetrical, or focal

is similar to the occipital alpha rhythm; however, the mu rhythm does not block with eye opening. Instead, it attenuates or is blocked by contralateral movement or thought of movement of the contralateral extremity.

The mu rhythm is uncommon before age 4 years with about 5% of children showing this pattern.[28] However, in infants as young as 1 year of age, an 8- to 10-Hz rhythm may occur in the central leads which, although not having the typical morphology of the mu rhythm seen in older patients, may be related[81] (Fig. 28–32). After age 4 years, the incidence of the mu rhythm slowly increases, and during mid-adolescence reaches the adult incidence of 17% to 19%.[28, 38, 72, 84]

Beta Activity

Two forms of beta activity may be seen in the normal child's EEG: an 18- to 25-Hz band and a less common 14- to 16-Hz band. In almost all children, the voltage of this activity in the waking state does not exceed 25 μV.[28, 84] Beta activity in the 18- to 25-Hz band is most evident in the fronto-central region, where it is intermixed with theta and alpha frequency waves (Fig. 28–33). The frequency of beta activity should be the same in both hemispheres; however, a voltage asymmetry with the lower voltage side being not less than 65% of the higher voltage side may be seen in normal individuals and may be related to skull thickness. Although a significant asymmetry of beta activity usually suggests focal pathology on the lower voltage side, it must be remembered that significant asymmetries of beta activity can occur in the absence of focal brain pathology. For example, beta activity may be enhanced at the site of a skull defect, resulting in a significant asymmetry between hemispheres. Also, scalp swelling

may attenuate the voltage of beta activity, a finding frequently encountered in infants.

Beta activity in the 14- to 16-Hz band may be predominant in either fronto-central or occipital regions and is frequently enhanced by hyperventilation.

Fronto-central Theta Activity

As the child's age increases, the amount of theta activity in the anterior regions slowly decreases.[41] The criteria used

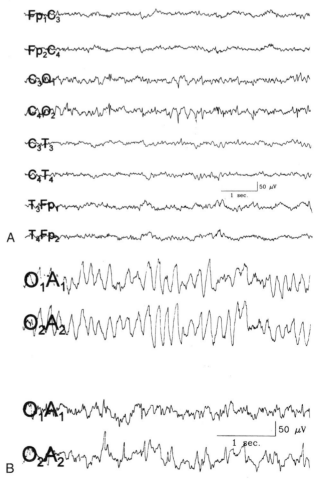

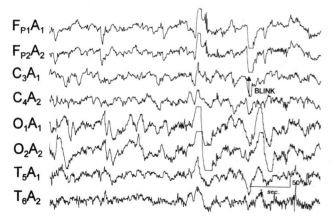

FIGURE 28–29. Posterior slow transients associated with eye movements in a 2-year-old asymptomatic girl.

FIGURE 28–30. Lambda waves. *A.* A 14-year-old boy with behavioral problems. The patient's eyes are open and he is looking at ceiling tiles. *B.* An 11-month-old asymptomatic boy. Top sample—The patient's eyes are closed. Bottom sample—The patient's eyes are open and he is looking around the laboratory.

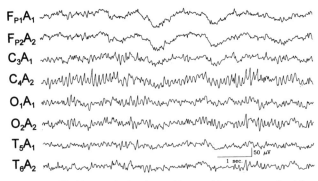

FIGURE 28–31. Mu rhythm in an 11-year-old asymptomatic girl.

to determine what amount of anterior slow activity is abnormal or excessive for a given age are not easily described. Considerable exposure to children's EEGs is required to develop the mental templates necessary to make this determination. Frequency is not the only factor that must be considered. Voltage, degree of persistence, and manner of occurrence of the waveforms (random versus rhythmic) must also be assessed.

Figure 28–34 provides representative examples of "normal" anterior slow activity for different age ranges. It is important to remember that the background activity in the anterior leads may be slow for age in the presence of a normal frequency occipital alpha rhythm and vice versa. Slowing of the alpha rhythm in the presence of normal frequency anterior activity suggests brain dysfunction.

One form of theta activity in the fronto-central leads is easily recognizable as being abnormal. It consists of paroxysmal, monomorphic, rhythmic, 4- to 7-Hz activity with an amplitude greater than 100 μV (Fig. 28–35). Such activity is not seen in normal control populations.[29, 84]

Hyperventilation

The hyperventilation response obtained in children is more pronounced than that usually seen in adults.[40] Two factors that affect the degree of response to deep breathing are low blood glucose levels and reduced blood CO_2.[5, 6, 46, 82]

Children younger than 5 years of age show very little change in the EEG during hyperventilation. The most pronounced response occurs between the ages of 8 and 12 years.[7, 35, 40, 84] The usual initial change seen in the EEG in children is the appearance of rhythmic, 2.5- to 4-Hz activity in the posterior leads, which shortly thereafter becomes

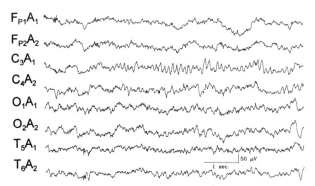

FIGURE 28–32. Central alpha frequency rhythm in a 15-month-old asymptomatic boy. This central rhythm may be related to the mu rhythm seen in older patients (see Fig. 28–31).

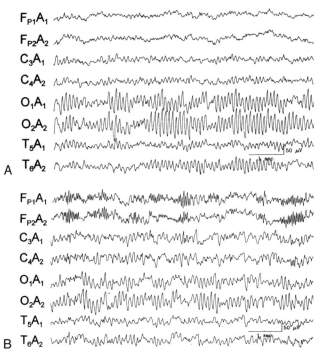

FIGURE 28–33. *A.* Normal 18- to 25-Hz activity in an asymptomatic 6-year-old boy. *B.* Abnormally high voltage beta activity in an 8-year-old boy receiving phenobarbital.

higher in voltage and generalized (Fig. 28–36). The generalized slowing may be continuous or paroxysmal in character. There is some persistence of the rhythmic slow activity for variable durations of time (seconds to 1 to 2 minutes) following cessation of hyperventilation. Although an excessively prolonged response to hyperventilation raises suspicions of abnormality, the only EEG changes that can be unequivocally classified as abnormal are the induction of epileptiform waveforms (either generalized spike and wave or rarely, focal spikes) or the induction of focal or lateralized slowing.

Features of the Normal Sleep Electroencephalogram

In this chapter, discussion is limited to patterns and waveforms seen during drowsiness and NREM sleep in infancy and childhood.

Hypnagogic Hypersynchrony

At approximately the time the occipital alpha rhythm first appears, at around 3 months of age, the onset of drowsiness is characterized by the disappearance of the occipital alpha rhythm and the appearance of relatively continuous rhythmic generalized monomorphic 3- to 4-Hz activity (hypnagogic hypersynchrony). In older children the frequency of this activity increases to 4 to 5 Hz. This activity is usually central or centro-frontal dominant; however, in some young infants similar rhythmic activity may appear in the temporal regions during drowsiness. Hypnagogic hypersynchrony may appear when there is still an occipital alpha rhythm present. However, in such instances, the frequency of the alpha rhythm will usually have slowed 1 to 2 Hz at the time the drowsy pattern appears. In Kellaway's series,[58] 30% of 3-month-old infants showed hypnagogic hypersynchrony and the pattern could be seen in children up to 12 to 13 years of age. However, only 10% of children showed this pattern

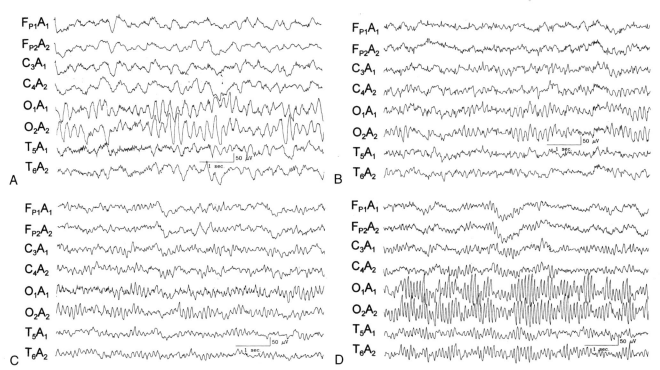

FIGURE 28–34. Examples of normal fronto-central theta activity at various ages. *A*, 6 months; *B*, 16 months; *C*, 3 years; *D*, 9 years. Also note increasing frequency of the occipital dominant rhythm with increasing age.

after the age of 11 years. These findings are similar to those reported by others.[29, 42] Examples of hypnagogic hypersynchrony are shown in Figure 28–37.

In addition to the more continuous hypnagogic pattern just described, paroxysmal slow activity may also be seen during drowsiness.[4, 29, 42, 59] This "paroxysmal" hypnagogic hypersynchrony pattern consists of brief bursts of 100- to 300-μV monomorphic rhythmic waves occurring at a frequency of 2.5 to 4.5 Hz. This activity is maximally expressed in the central or centro-frontal region. Intermixed with or superimposed on the paroxysmal slow waves may be poorly defined low voltage spikes or sharp components. The expres-

sion of the spike or sharp components is variable (Fig. 28–38). Paroxysmal hypnagogic hypersynchrony may be seen between the ages of 1 and 15 years, but occurs most commonly between 4 and 9 years.[29, 58]

These patterns serve as constant sources of misinterpretation for the electroencephalographer not experienced in reading pediatric EEGs. The most common error made is to call paroxysmal hypnagogic hypersynchrony an epileptic pattern (i.e., generalized spike and wave activity).

A thorough discussion of EEG changes associated with the drowsy state can be found in the book by Santamaria and Chiappa.[91]

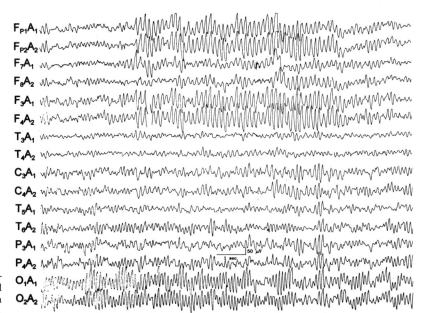

FIGURE 28–35. Paroxysmal high voltage fronto-central theta frequency activity in an 8-year-old boy with a history of behavioral disorder. Such activity is not seen in normal control populations.

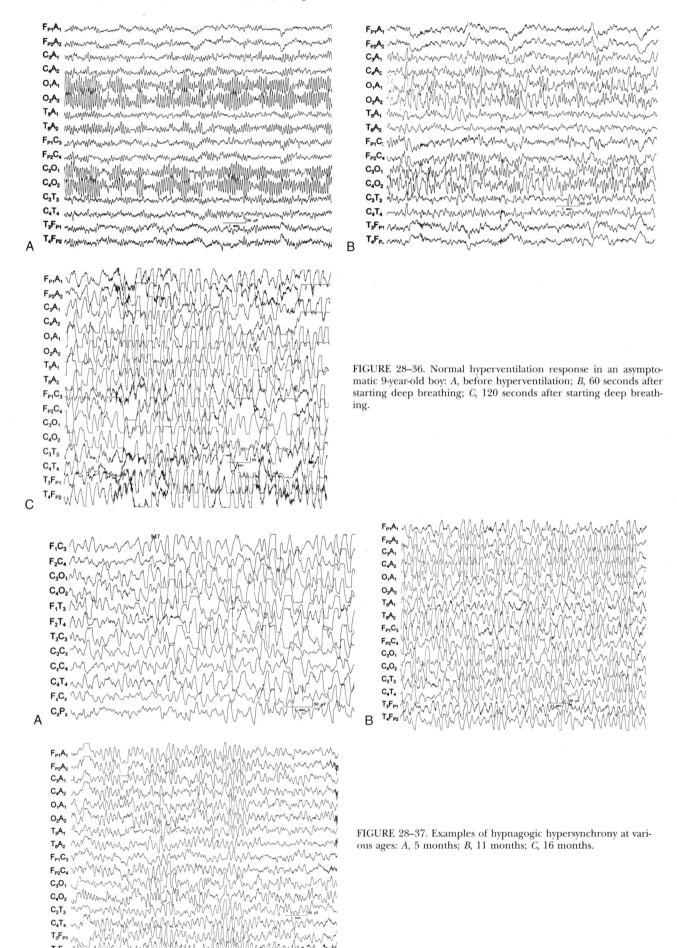

FIGURE 28–36. Normal hyperventilation response in an asymptomatic 9-year-old boy: *A,* before hyperventilation; *B,* 60 seconds after starting deep breathing; *C,* 120 seconds after starting deep breathing.

FIGURE 28–37. Examples of hypnagogic hypersynchrony at various ages: *A,* 5 months; *B,* 11 months; *C,* 16 months.

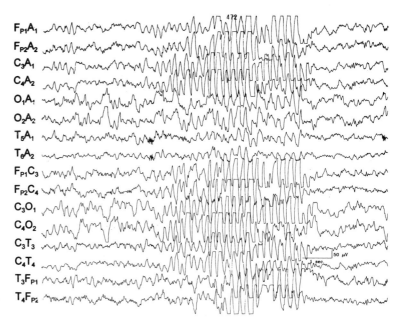

FIGURE 28–38. Paroxysmal hypnagogic hypersynchrony in an 11-year-old asymptomatic girl.

Vertex Transients

Vertex transients appear in stage I sleep and persist into deeper stages of NREM sleep. They are diphasic sharp waves with an initial negative phase followed by a positive deflection. They may be followed by a surface negative slow wave or a spindle in deeper stages of NREM sleep. Vertex transients are maximally expressed at the central or vertex regions, thus the source of their name (Fig. 28–39).

As mentioned earlier, vertex transients appear by 6 to 8 weeks post-term and are bilaterally symmetrical and synchronous at this time. It is not uncommon to see some variable asymmetry between sides in children; however, a constant asymmetry with the lower voltage side being less than 80% of the higher voltage side is considered significant, the lower voltage side being abnormal.

Vertex transients may occur in isolation or in runs lasting many seconds (Fig. 28–40). In children they frequently are higher in voltage (100 to 200 μV) and spike-like in appearance and must be distinguished from abnormal sharp waves or spike discharges arising from the central region.

Sleep Spindles

As previously mentioned, sleep spindles first appear between 6 and 8 weeks post-term. These spindles have a frequency of approximately 14 Hz and are maximally expressed in the central and vertex regions. They are bilaterally symmetrical, but asynchronous in the young infant. Sleep spindle synchrony increases during the first year of life, and between 1 and 2 years of age they become bilaterally synchronous. The sleep spindles of infants show some differences from those of older children or adults. Infant's spindles are longer in duration (up to 6 to 8 seconds), lack the waxing and waning quality of spindles in older individuals, and frequently have a sharp negative component (Fig. 28–41).

Two types of sleep spindles are seen in older normal children. The most common type is the 14-Hz spindle, which is maximal in its expression in the central region (Fig. 28–42). These spindles usually have a duration of 2 to 4 seconds, but occasionally may be more prolonged. They first appear in stage II NREM sleep, but persist into stages III and IV. On occasion, they may be seen during REM sleep (i.e., spindle intrusion). These 14-Hz central spindles persist into adult life. A less common sleep spindle seen in 5% of normal children up to 12 years of age has a frequency of 10 to 12 Hz, and is maximally expressed in the frontal regions. The duration of these "frontal" spindles is usually less than 3 seconds[58] (Fig. 28–43).

Positive Occipital Sharp Transients of Sleep

Positive occipital sharp transients of sleep (POSTS) are surface positive sharp waves that appear in the occipital leads during sleep. The initial positive waves may be followed by a

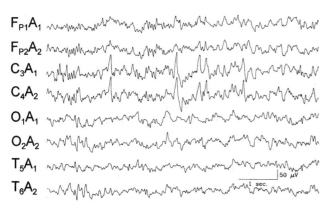

FIGURE 28–39. Vertex transients in a 10-year-old asymptomatic girl.

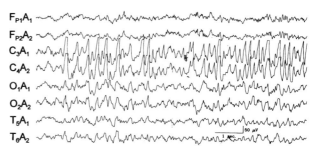

FIGURE 28–40. Run of vertex transients in an 8-year-old asymptomatic boy.

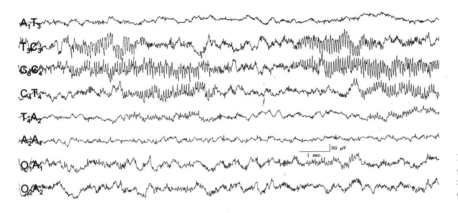

FIGURE 28–41. Sleep spindles in 6-month-old boy with a history of questionable seizures. Note the asynchrony and spike-like quality of these spindles.

lower voltage, longer duration, negative component with the entire wave having a check mark appearance (Fig. 28–44). POSTS are bilaterally synchronous, but voltage asymmetries between sides with the lower voltage side being greater than 40% of higher voltage side are common.[103] They may occur in isolation, but frequently occur in runs.

POSTS usually appear around 3 to 4 years of age, but may occur in some infants less than 1 year of age. They do not become maximally expressed until mid-adolescence.

In children, POSTS may be quite sharp in appearance. This feature, coupled with the fact that they frequently are asymmetrical, may lead to the erroneous interpretation of POSTS as focal epileptic discharges.

Beta Activity of Sleep

Beyond 6 months of age, the amount of beta activity increases with sleep onset. The frequency of this activity is usually 18 to 25 Hz and is usually maximally seen in the centro-occipital regions. This activity reaches a maximal voltage in stages I and II NREM sleep, then decreases in deeper NREM sleep stages. Between the ages of 12 and 18 months (Fig. 28–45), the voltage of this activity may reach 50 to 60 μV. The amplitude of this activity then progressively diminishes, and by age 6 to 7 years, the amplitude of such activity is rarely more than 10 to 15 μV.

In children, high voltage 16- to 25-Hz activity mixed with 12- to 14-Hz sleep spindle activity may occur in a continuous fashion. This pattern has been referred to as "continuous spindling and fast activity"[58] or "extreme spindles,"[44] if the sleep spindles are the most dominant feature. Such patterns have been associated with mental retardation or cerebral

palsy, or both. However, it must be remembered that high voltage fast activity may be seen during sleep in the EEGs of children receiving various medications. In such children, the resulting mixture of beta and sleep spindle activity may produce a pattern very similar to the "continuous spindling and fast activity" or "extreme spindle" patterns.

Occipital Slow Transients

In children, high voltage slow transients occur bilaterally in the occipital leads in NREM sleep.[59] These waves may appear as monophasic (cone-shaped) or diphasic waves (Fig. 28–46). During light sleep these waves occur sporadically, but as NREM sleep deepens, they become much more frequent.

Delta Activity in Deep Sleep

In children, the amplitude of the delta frequency activity seen in stages III and IV NREM sleep is much greater than that seen in adults. The voltage of the slow activity may reach up to 500 μV in deep sleep in some normal children (Fig. 28–47).

Fourteen- and Six-per-Second Positive Bursts

The 14- and 6-per-second positive burst pattern has a long and interesting history in the field of EEG. Initially the pattern was considered abnormal and was associated with a variety of clinical complaints, including so-called neurovegetative symptoms (e.g., dizziness, headaches, and so forth), behavioral disorders, and possible seizures.[39, 52, 54, 60, 61, 62, 74, 79, 85] Some authors considered the pattern to be evidence of hypothalamic epilepsy.[43, 61, 62] However, 14- and 6-per-second positive bursts have also been reported in normal control populations.[13, 71, 73, 83, 94] Currently the pattern is considered by most to be a benign normal variant.

In the routine clinical setting, the 14- and 6-per-second positive burst pattern, also called the 14- and 6-per-second positive spike pattern, or "ctenoids," is seen predominantly during drowsiness and light sleep. However, overnight recordings reveal that the pattern may be seen in deeper stages of NREM sleep and, in some children, during REM sleep as well. The pattern consists of brief bursts of waves exhibiting a spike-like positive component. The pattern is similar to a sleep spindle, except for the prominent spike positive phase. Two basic frequencies are seen. The "fast" form consists of waves in the 12- to 16-Hz range, most commonly 14 Hz, and a "slow form" consists of waves in the 6- to 8-Hz range, most commonly 6 Hz (Fig. 28–48). On occasion the bursts of positive spikes may be followed by generalized high voltage slow waves, which may be easily misinterpreted as generalized spike and wave activity to those unaware of this variant. The

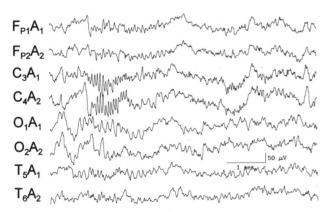

FIGURE 28–42. Central spindles of 14 Hz in an asymptomatic 3-year-old boy.

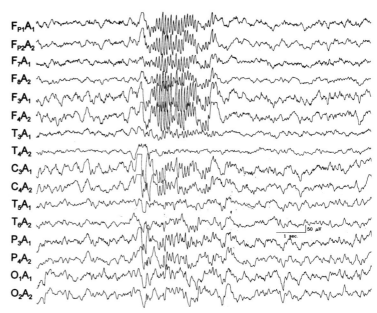

FIGURE 28–43. Frontal spindles of 10 to 12 Hz in an asymptomatic 5-year-old girl.

waveform is maximally expressed in the occipital or posterior temporal electrodes and best identified when these electrodes are referred to the ipsilateral or contralateral ear.

The duration of the bursts is usually less than one second. In a given patient, one may see either the "14" Hz burst, the "6" Hz burst, or both. The bursts usually occur independently on the two sides, and in some patients, may be seen only unilaterally.

The 14- and 6-per-second positive burst pattern first appears around 3 years of age. The maximal expression of this pattern occurs between the ages of 12 and 15 years, during which time approximately one half of control populations show this pattern. Thereafter, the incidence decreases, although the pattern may still be seen in a small percentage of adults.

Changes Seen with Arousal

K Complexes

K complexes are high voltage diphasic slow waves that occur during sleep in response to sudden sensory stimuli

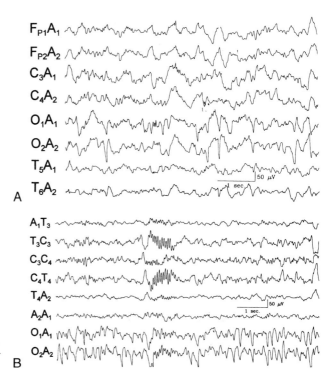

FIGURE 28–44. Positive occipital sharp wave transients (POSTS). *A.* Isolated POSTS in a 6-year-old boy with a history of behavioral problems. *B.* Runs of POSTS in an asymptomatic 14-year-old girl.

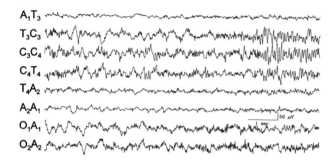

FIGURE 28–45. Normal beta activity during NREM sleep in an asymptomatic 15-month-old girl.

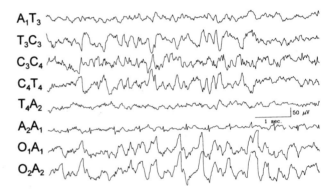

FIGURE 28–46. Occipital slow transients occurring in NREM sleep in a 4-year-old boy with a history of behavioral problems.

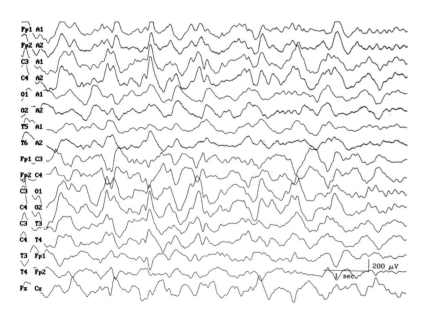

FIGURE 28–47. Very high voltage slow wave activity in NREM sleep in an asymptomatic 10-year-old boy. Note the voltage calibration. Such high voltage activity is not uncommonly seen during deep sleep in normal children.

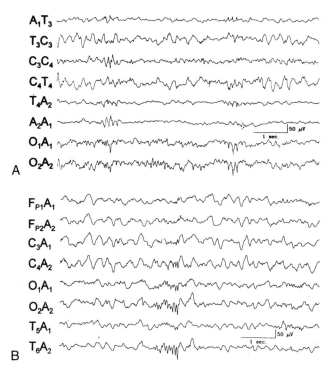

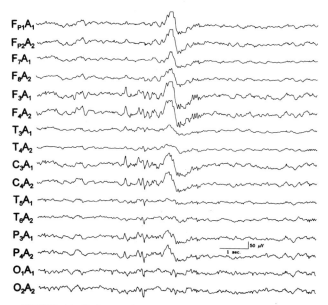

FIGURE 28–49. K complex in an asymptomatic 12-year-old girl.

A

B

FIGURE 28–48. Fourteen- and 6-per-second positive bursts. *A.* Six-hertz variety in an asymptomatic 12-year-old boy. *B.* Fourteen-hertz variety in a 7-year-old girl with a history of headaches.

(usually auditory) or, apparently spontaneously. Rudimentary K complexes first appear between the ages of 8 and 12 weeks post-term.[58] They consist of high voltage diphasic (negative-positive) slow waves of sharp configuration, often followed by a brief run of sleep spindle activity. The complex shows a wide distribution at the scalp, but is maximally expressed in the centro-frontal region (Fig. 28–49). If present, the sleep spindle activity typically follows the diphasic slow wave, but on occasion may precede the slow wave. The diphasic slow wave may also occur in the absence of sleep spindle activity.

As arousal continues following a stimulus, fronto-central dominant faster frequencies follow the K complex. At

around 7 to 8 months of age, the frequency of this activity is approximately 4 to 5 Hz, and with increasing age, this activity increases in frequency and eventually reaches 8 to 10 Hz, the same rate seen in adults (Fig. 28–50).

Post-arousal Slowing

Beginning around 3 months of age, rhythmic high voltage centro-frontal dominant 3- to 4-Hz waves, similar in appearance to that seen during drowsiness, occurs in the EEG during arousal (Fig. 28–51).

In older children, the faster component of arousal following the K complex is followed by rhythmic, high voltage, monorhythmic 2.5- to 3.5-Hz activity in the frontal leads (Fig. 28–52). As the child continues to arouse, the rhythmic slow activity increases in frequency and shifts its maximal expression to the occipital leads. As arousal continues, the rhythmic slow activity disappears from the frontal regions, and subsequently, from the occipital leads where it is replaced by the occipital alpha rhythm. Spontaneous arousal may occur in the absence of the K complex, previously described.

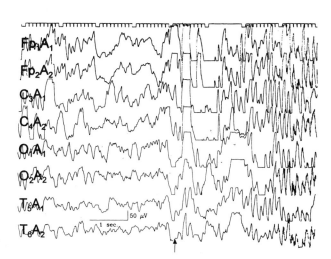

FIGURE 28–50. Arousal in an asymptomatic 6-year-old boy. Note the 7- to 8-Hz activity occurring immediately following the K complex (*arrow*).

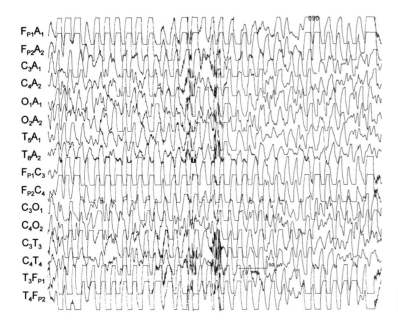

FIGURE 28–51. Postarousal slowing in an asymptomatic 4-month-old boy.

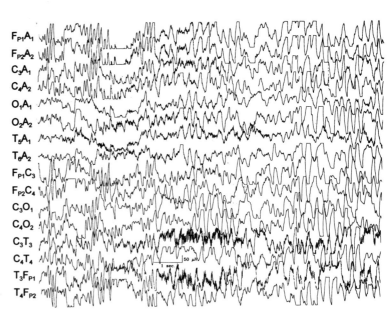

FIGURE 28–52. Monorhythmic slowing following the faster component of arousal in an asymptomatic 4-year-old girl.

REFERENCES

1. Aird RB, Gastaut Y: Occipital and posterior electroencephalographic rhythms. Electroencephalogr Clin Neurophysiol 11:637–656, 1959.
2. Anderson CM, Torres F, Faoro A: The EEG of the early premature. Electroencephalogr Clin Neurophysiol 60:95–105, 1985.
3. Barlow JS, Cigánek L: Lambda responses in relation to visual evoked responses in man. Electroencephalogr Clin Neurophysiol 26:183–192, 1969.
4. Brandt S, Brandt H: The electroencephalographic patterns in young healthy children from 0 to 5 years of age. Acta Psychiatr Scand 30:77–89, 1955.
5. Brazier MAB, Finesinger JE, Schwab RS: Characteristics of the normal electroencephalogram. II. The effect of varying blood sugar levels on the occipital cortical potentials in adults during quiet breathing. J Clin Invest 23:313–317, 1944.
6. Brazier MAB, Finesinger JE, Schwab RS: Characteristics of the normal electroencephalogram. II. The effect of varying blood sugar levels on the occipital cortical potentials in adults during hyperventilation. J Clin Invest 23:319–323, 1944.
7. Brill NQ, Seidemann H: The electroencephalogram of normal children: Effect of hyperventilation. Am J Psychiatry 98:250–256, 1941.
8. Carels G: Les ondes lentes postérieures de l'électroencéphalogramme du'un jeune adulte et leur variation quantitative dans le temps. Acta Neurol Belg 59:409–413, 1959.
9. Challamel MJ, Isnard H, Brunon AM, et al: Asymétrie EEG transitoire a l'entrée dans le sommeil calme chez le nouveau-né: étude sur 75 observations. Rev Electroencephalogr Neurophysiol Clin 14:17–23, 1984.
10. Clancy RR, Chung HJ, Temple JP: Neonatal electroencephalography. In Sperling MR, Clancy RR (eds): Atlas of Electroencephalography, vol I. Amsterdam, Elsevier, 1993.
11. Cohn R, Nardini JE: The correlation of bilateral occipital slow activity in the human EEG with certain disorders of behavior. Am J Psychiatry 115:44–54, 1958.
12. Connell JA, Oozeer R, Dubowitz V: Continuous 4-channel EEG monitoring: A guide to interpretation, with normal values, in preterm infants. Neuropediatrics 18:138–145, 1987.
13. Demerdash A, Eeg-Olofsson O, Petersen I: The incidence of 14 and 6 per second positive spikes in a population of normal children. Develop Med Child Neurol 10:309–316, 1968.
14. Dreyfus-Brisac C: Activité électrique cérébrale du foetus et du tres jeune pre-maturé. IVᵉ Congr. Int. Electroencephalogr. Neurophysiol. Clin. Bruxelles. Acta Med Belg Ed 163–171, 1957.
15. Dreyfus-Brisac C: Electroencephalography in infants. In Linneweh F (ed): Die physiologische Entwicklung des Kindes. Berlin, Springer, 1959, pp 29–40.
16. Dreyfus-Brisac C: The electroencephalogram of the premature infant. World Neurol 3:5–15, 1962.
17. Dreyfus-Brisac C: The electroencephalogram of the premature infant and full-term newborn: Normal and abnormal development of waking and sleeping patterns. In Kellaway P, Petersén I (eds): Neurological and Electroencephalographic Correlative Studies in Infancy. New York, Grune & Stratton, 1964, pp 186–206.
18. Dreyfus-Brisac C: Sleep ontogenesis in early human prematurity from 24 to 27 weeks of conceptional age. Dev Psychobiol 1:162–169, 1968.
19. Dreyfus-Brisac C: Ontogenesis of sleep in human prematures after 32 weeks of conceptional age. Dev Psychobiol 3:91–121, 1970.
20. Dreyfus-Brisac C: Ontogenesis of brain bioelectrical activity and sleep organization in neonates and infant. In Faulkner F, Tanner JM (eds): Human Growth, Vol III: Neurobiology and Nutrition. New York, Plenum Press, 1978, pp 157–182.
21. Dreyfus-Brisac C, Blanc C: Electroencéphalogramme et maturation cerebrale. Encephale 45:205–241, 1956.
22. Dreyfus-Brisac C, Fischgold H, Samson-Dollfus D, et al: Veille, sommeil, réactivité sensorielle chez le prématuré, le nouveau-né et le nourrisson. Electroencephalogr Clin Neurophysiol S6:417–440, 1957.
23. Dreyfus-Brisac C, Monod N, Salama P, et al: L'EEG dans les six premiers mois de la vie, aprés réanimation prolongée et état de mal néonatal. Recherches d'éléments de pronostic. Vth International Congress on Electroencephalography and Clinical Neurophysiology. Excerpta Med Int Congr Ser 37:228–229, 1961.
24. Dreyfus-Brisac C, Flescher J, Plassart E: L'électroencéphalogramme. Critère d'áge conceptionnel du nouveau-né á terme et du prématuré. Biol Neonat 4:154–173, 1962.
25. Dreyfus-Brisac C, Monod N: Sleeping behavior in abnormal newborn infants. Neuropadiatrie 1:354–366, 1970.
26. Dreyfus-Brisac C, Curzi-Dascalova L: The EEG during the first year of life. In Remond A (ed): The Evolution of the EEG from Birth to Adulthood, Handbook of EEG and Clinical Neurophysiology, vol 6B. Amsterdam, Elsevier, 1975, pp 24–30.
27. Dreyfus-Brisac C, Monod N: The electroencephalogram of full-term newborns and premature infants. In Remond A (ed): The Evolution of the EEG from Birth to Adulthood, Handbook of EEG and Clinical Neurophysiology, vol 6B. Amsterdam, Elsevier, 1975, pp 6–23.
28. Eeg-Olofsson O: The development of the electroencephalogram in normal adolescents from the age of 16 through 21 years. Neuropaediatrie 3:11–45, 1971.
29. Eeg-Olofsson O, Petersén I, Selldén U: The development of the EEG in normal children from the age of 1 to 15 years: Paroxysmal activity. Neuropaediatrie 4:375–404, 1971.
30. Ellingson RJ: EEGs of premature and full-term newborns. In Klass DW, Daly DD (eds): Current Practice of Clinical Electroencephalography. New York, Raven Press, 1979, pp 149–177.
31. Engel R: Abnormal Electroencephalograms in the Neonatal Period. Springfield, IL, Charles C. Thomas, 1975.
32. Engel R, Butler BV: Appraisal of conceptual age of newborn infants by electroencephalographic methods. J Pediatr 63:386–393, 1963.
33. Evans CC: Spontaneous excitation of the visual cortex and association areas—lambda waves. Electroencephalogr Clin Neurophysiol 5:69–74, 1953.
34. Faure J, Guérin A: Au sujet de l'électroencéphalogramme des enfants caractériels. Rev Neurol (Paris) 99:209–219, 1958.
35. Fiedlerová D: Der Einfluss der Hyperventilation von 3 Minuten auf das EEG-Bild bei gesunden Kindern im Alter von 7 bis 11 Jahren. Sb Lek 4:417–422, 1967.
36. Frost JD Jr, Glaze DG, Hrachovy RA, et al: EEG and neuropsychological changes associated with carbamazepine therapy in children with partial seizures. J Clin Neurophysiol 5:336–337, 1988.
37. Gastaut Y: Un signe électroencéphalographique peu connu: Les pointes occipitales survenant pendant l'ouverture des yeux. Rev Neurol (Paris) 84:640–643, 1951.
38. Gastaut H, Lee MC, Laboureur P: Comparative EEG and psychometric data for 825 French naval pilots and 511 control subjects of the same age. Aerospace Med 31:547–552, 1960.
39. Gibbs FA: Clinical correlates of 14 and 6 per second positive spikes (1865 cases). Electroencephalogr Clin Neurophysiol 8:149, 1956.
40. Gibbs FA, Gibbs EL, Lennox WG: Electroencephalographic response to overventilation and its relation to age. J Pediatr 23:497–505, 1943.
41. Gibbs FA, Knott JR: Growth of the electrical activity of the cortex. Electroencephalogr Clin Neurophysiol 1:223–229, 1949.
42. Gibbs FA, Gibbs EL: Atlas of Electroencephalography: Normal Controls, vol l. Cambridge, MA, Addison-Wesley. 1950.
43. Gibbs EL, Gibbs FA: Electroencephalographic evidence of thalamic and hypothalamic epilepsy. Neurology 1:136–144, 1951.
44. Gibbs EL, Gibbs FA: Extreme spindles: Correlation of electroencephalographic sleep pattern with mental retardation. Science 138:1106–1107, 1962.
45. Goodwin JE: The significance of alpha variants in the EEG, and their relationship to an epileptiform syndrome. Am J Psychiatry 104:369–379, 1947.
46. Gotoh F, Meyer JS, Takagi Y: Cerebral effects of hyperventilation in man. Arch Neurol 12:410–423, 1965.
47. Green J: Some observations on lambda waves and peripheral stimulation. Electroencephalogr Clin Neurophysiol 9:691–704, 1957.
48. Hagne I: Development of the EEG in normal infants during the first year of life. Acta Paediatr Scand S232:1–53, 1972.
49. Hahn JS, Monyer H, Tharp BR: Interburst interval measurements in the EEGs of premature infants with normal neurological outcome. Electroencephalogr Clin Neurophysiol 73:410–418, 1989.

50. Harris R, Tizard JPM: The electroencephalogram in neonatal convulsions. J Pediatr 57:502–520, 1960.

51. Hayakawa F. Watanabe K, Hakamada S, et al: FZ theta/alpha bursts: A transient EEG pattern in healthy newborns. Electroencephalogr Clin Neurophysiol 67:27–31, 1987.

52. Henry CE: Positive spike discharges in the EEG and behavior abnormality. In Glaser GH (ed): EEG and Behavior. New York and London, Basic Books, 1963, pp 315–344.

53. Hrachovy RA, Mizrahi EM, Kellaway P: Electroencephalography of the newborn. In Daly DD, Pedley TA (eds): Current Practice of Clinical Electroencephalography, 2nd ed. New York, Raven Press, 1990, pp 201–242.

54. Hughes JR, Gianturco D, Stein W: Electro-clinical correlations in the positive spike phenomenon. Electroencephalogr Clin Neurophysiol 13:599–605, 1961.

55. Hughes JR Jr, Fino JJ, Hart LA: Premature temporal theta (PTθ). Electroencephalogr Clin Neurophysiol 67:7–15, 1987.

56. Igert C, Lairy GC: Intérêt pronostique de l'EEG au cours de l'évolution des schizophrénes. Electroencephalogr Clin Neurophysiol 14:183–190, 1962.

57. Kellaway P: Introduction to plasticity and sensitive period. In Kellaway P, Noebels JL (eds): Problems and Concepts in Developmental Neurophysiology. Baltimore, Johns Hopkins University Press, 1989, pp 3–28.

58. Kellaway P: An orderly approach to visual analysis: Characteristics of the normal EEG of adults and children. In Daly DD, Pedley TA (eds): Current Practice of Clinical Electroencephalography, 2nd ed. New York, Raven Press, 1990, pp 139–199.

59. Kellaway P, Fox BJ: Electroencephalographic diagnosis of cerebral pathology in infants during sleep. I. Rationale, technique, and the characteristics of normal sleep in infants. J Pediatr 41:262–287, 1952.

60. Kellaway P, Moore FJ, Kagawa N: The "14 and 6 per second positive spike" pattern of Gibbs and Gibbs. Electroencephalogr Clin Neurophysiol 9:165–166, 1957.

61. Kellaway P, Crawley JW, Kagawa N: A specific electroencephalographic correlate of convulsive equivalent disorders in children. J Pediatr 55:582–592, 1959.

62. Kellaway P, Crawley JW, Kagawa N: Paroxysmal pain and autonomic disturbances of cerebral origin: A specific electro-clinical syndrome. Epilepsia 1:466–483, 1959/60.

63. Kellaway P, Crawley JW: A Primer of Electroencephalography of Infants, Sections I & II: Methodology and Criteria of Normality. Houston, Baylor University College of Medicine, 1964.

64. Lairy GC: E.E.G. et neuropsychiatrie infantile. Psychiatr Enfant 3:525–608, 1961.

65. Lairy GC: L'EEG comme moyen d'investigation des modalité individuelles d'adaptation aux situations de stress. Electroencephalogr Clin Neurophysiol S25:282–298, 1967.

66. Leisser P, Lindholm L-E, Petersén I: Alpha amplitude dependence on skull thickness as measured by ultrasound technique. Electroencephalogr Clin Neurophysiol 29:392–399, 1970.

67. Lesévre N: Étude de réponses moyennes recueillies sur la région postérieure du scalp chez l'homme au cours de l'exploration visuelle ("complexe lambda"). Psychol Franc 12:26–36, 1967.

68. Lombroso CT: Quantified electrographic scales on 10 pre-term healthy newborns followed up to 40–43 weeks of conceptional age by serial polygraphic recordings. Electroencephalogr Clin Neurophysiol 46:460–474, 1979.

69. Lombroso CT: Neonatal electroencephalography. In Niedermeyer E, Lopes da Silva F (eds): Electroencephalography: Basic Principles, Clinical Applications and Related Fields. Baltimore, Urban & Schwarzenberg, 1982, pp 599–637.

70. Lombroso CT: Neonatal polygraphy in full-term and premature infants: A review of normal and abnormal findings. J Clin Neurophysiol 2:105–155, 1985.

71. Lombroso CT, Schwartz IH, Clark DM, et al: Ctenoids in healthy youths: controlled study of 14- and 6-per-second positive spiking. Neurology 16:1152–1158, 1966.

72. Maulsby RL, Kellaway P, Graham M, et al: The Normative Electroencephalographic Data Reference Library. Final Report, Contract NAS 9-1200, National Aeronautics and Space Administration, 1968.

73. Metcalf DR: Controlled studies of the incidence and significance of 6 and 14 per sec positive spiking. Electroencephalogr Clin Neurophysiol 15:161, 1963.

74. Milstein V, Small JG: Psychological correlates of 14 and 6 positive spikes, 6/sec spike-wave, and small sharp spike transients. Clin Electroenceph 2:206–212, 1971.

75. Monod D, Dreyfus-Brisac C, Ducas P, et al: L'EEG du nouveaué à terme. Étude comparative chez le nouveau-né en présentation céphalique et en presentation de siége. Rev Neurol 102:375–379, 1960.

76. Monod D, Pajot N: Le sommeil du nouveau-né et du prématuré. I. Analyse des études polygraphiques (mouvements oculaires, respiration et EEG chez le nouveau-né á terme). Biol Neonat 8:281–307, 1965.

77. Monod D, Pajot N, Guidasci S: The neonatal EEG: Statistical studies and prognostic value in full-term and pre-term babies. Electroencephalogr Clin Neurophysiol 32:529–544, 1972.

78. Netchine S, Lairy GC: The EEG and psychology of the child. In Lairy GC (ed): Handbook of Electroencephalography and Clinical Neurophysiology, Vol. 6: The Normal EEG Throughout Life. Part B: The Evolution of the EEG from Birth to Adulthood. Amsterdam, Elsevier Scientific, 1975, pp 69–104.

79. Niedermeyer E, Knott JR: The incidence of 14 and 6/sec positive spikes in psychiatric material. Electroencephalogr Clin Neurophysiol' 14:285–286, 1962.

80. O'Brien MJ, Lems YL, Prechtl HFR: Transient flattenings in the EEG of newborns—a benign variation. Electroencephalogr Clin Neurophysiol 67:16–26, 1987.

81. Pampiglione G: Development of rhythmic EEG activities in infancy (waking state). Rev Electroencephalogr Neurophysiol Clin 7:327–334, 1977.

82. Patel VM, Maulsby RL: How hyperventilation alters the EEG: A review of controversial viewpoints emphasizing neurophysiological mechanisms. J Clin Neurophysiol 4:101–120, 1987.

83. Petersén I, Akesson HO: EEG studies of siblings of children showing 14 and 6 per second positive spikes. Acta Genet (Basel) 18:163–169, 1968.

84. Petersén I, Eeg-Olofsson O: The development of the electroencephalogram in normal children from the age of 1 through 15 years—non-paroxysmal activity. Neuropaediatrie 2:247–304, 1971.

85. Poser CM, Ziegler DK: Clinical significance of 14 and 6 per second positive spike complexes. Neurology 8:903–912, 1958.

86. Rémond A, Lesévre N: Remarques sur les conditions d'apparition et l'importance statistique des ondes lambda chez les individus normaux. Rev Neurol (Paris) 94:160–161, 1956.

87. Rémond A, Lesévre N, Torres, F: Étude chronotopographique de l'activité occipitale moyenne recueilli sur le scalp chez l'homme en relation avec le déplacement du regard (complexe lambda). Rev Neurol (Paris) 113:193–226, 1965.

88. Rose AL, Lombroso CT: Neonatal seizure states. Pediatrics 45:404–425, 1970.

89. Roth M, Green J: The lambda wave as a normal physiological phenomenon in the human electroencephalogram. Nature 172:864–866, 1953.

90. Sainte-Anne-Dargassies S, Berthault F, Dreyfus-Brisac C, et al: La convulsion du tout jeune nourrisson; aspects électroencéphalographiques du probleme. Presse Med 46:965–966, 1953.

91. Santamaria J, Chiappa KE: The EEG of Drowsiness. New York, Demos Publications, 1987.

92. Scher MS, Bova JM, Dokianakis SG, et al: Physiological significance of sharp waves transients on EEG recording of healthy pre-term and full-term neonates. Electroencephalogr Clin Neurophysiol 90:179–185, 1994.

93. Schulte FJ: Neonatal EEG and brain maturation: Facts and fallacies [letter to the editor]. Dev Med Child Neurol 12:396–399, 1970.

94. Schwartz IH, Lombroso CT: 14 & 6/second positive spiking (ctenoids) in the electroencephalograms of primary school pupils. J Pediat 72:678–682, 1968.

95. Scott DF, Groetheysen UC, Bickford RG: Lambda responses in the human electroencephalogram. Neurology 17:770–778, 1967.

96. Statz A, Dumermuth G, Mieth D, Duc G: Transient EEG patterns during sleep in healthy newborns. Neuropediatrics 13:115–122, 1982.

97. Stockard-Pope JE, Werner SS, Bickford RC: Atlas of Neonatal Electroencephalography, 2nd ed. New York, Raven Press. 1992.

98. Tharp BR: Neonatal and pediatric electroencephalography. In Aminoff M (ed): Electrodiagnosis in Clinical Neurology. Edinburgh, Churchill-Livingstone, 1980, pp 67–117.

99. Tharp BR, Cukier F, Monod N: The prognostic value of the electroencephalogram in premature infants. Electroencephalogr Clin Neurophysiol 51:219–236, 1981.

100. Torres F, Blaw ME: Longitudinal EEG-clinical correlations in children from birth to 4 years of age. Pediatrics 41:945–954, 1968.

101. Torres F. Anderson C: The normal EEG of the human newborn. J Clin Neurophysiol 2:89–103, 1985.

102. Tsai H-J, Liu S-Y: Lambda waves of human subjects of different age levels. Acta Psychol (Sin) 4:343–352, 1965. Translated from Chinese by Barlow JS: Contemporary Brain Research in China. New York, Consultants Bureau, 1971, pp 50–61.

103. Vignaendra V, Matthews RL, Chatrian GE: Positive occipital sharp transients of sleep: Relationships to nocturnal sleep cycle in man. Electroencephalogr Clin Neurophysiol 37:239–246, 1974.

104. Watanabe K, Iwase K: Spindle-like fast rhythms in the EEGs of low birth weight infants. Dev Med Child Neurol 14:373–381, 1972.

105. Watanabe K, Iwase K, Hara K: Development of slow-wave sleep in low birth weight infants. Dev Med Child Neurol 16:23–31, 1974.

106. Weerd AW de: Atlas of EEG in the First Month of Life. Amsterdam, Elsevier, 1995.

107. Werre PF: The Relationships Between Electroencephalographic and Psychological Data in Normal Adults. Leiden, Universitaire Presse Leiden, 1957.

108. Westmoreland BF, Sharbrough FW: Posterior slow wave transients associated with eye blinks in children. Am J EEG Technol 15:14–19, 1975.

109. Westmoreland BF, Stockard JE: The EEG in infants and children: Normal patterns. Am J EEG Technol 17:187–218, 1977.

Chapter 29

Normal Electroencephalogram and Benign Variants

Nancy Foldvary, DO

NORMAL RHYTHMS
Alpha Rhythm and Alpha Variants
Beta Rhythm
Mu Rhythm
Theta Rhythm
Posterior Slow Waves of Youth
Lambda Waves
Low Voltage Electroencephalogram
SLEEP RHYTHMS
Sleep Architecture
Non–Rapid Eye Movement Sleep
Rapid Eye Movement Sleep
Arousal Patterns
ACTIVATION METHODS
Hyperventilation
Intermittent Photic Stimulation

BENIGN VARIANTS
Rhythmic Midtemporal Discharges
Subclinical Rhythmic Electrographic
   (Theta) Discharge in Adults
Midline Theta Rhythm
Frontal Arousal Rhythm
Fourteen- and Six-Hertz Positive Bursts
Small Sharp Spikes
Six-Hertz Spike and Wave
Wicket Spikes
PHYSIOLOGICAL ARTIFACTS
Ocular Artifacts
Cardiac Artifacts
Myogenic Artifact
Glossokinetic Artifact
Miscellaneous Physiological Artifacts

NORMAL RHYTHMS

Alpha Rhythm and Alpha Variants

First described by Hans Berger in 1929, the alpha rhythm is comprised of 8- to 13-Hz waves appearing over the posterior head regions during relaxed wakefulness with the eyes closed[1, 12] (Fig. 29–1). The lower limit of the frequency band is usually reached by 8 years of age. In most normal adults, the frequency lies between 9 and 11 Hz. The alpha frequency decreases slightly with advancing age. However, in healthy individuals in the seventh and eighth decades, the mean is maintained at or above 9 Hz.[13] A posterior dominant rhythm of less than 8 Hz during wakefulness in an adult is abnormal. The frequency of the alpha rhythm is similar over the two hemispheres. An interhemispheric asymmetry of 1 Hz or greater is abnormal.

On referential montages, the distribution of the alpha rhythm is usually maximal at the occipital electrodes (O_1, O_2). However, the amplitude may be highest in the parietal or posterior temporal regions and occasionally is more widespread. Using the P_4 to O_2 derivation, the voltage of the alpha rhythm ranges from 15 to 45 μV in adults. Higher voltages are observed in younger individuals. The voltage

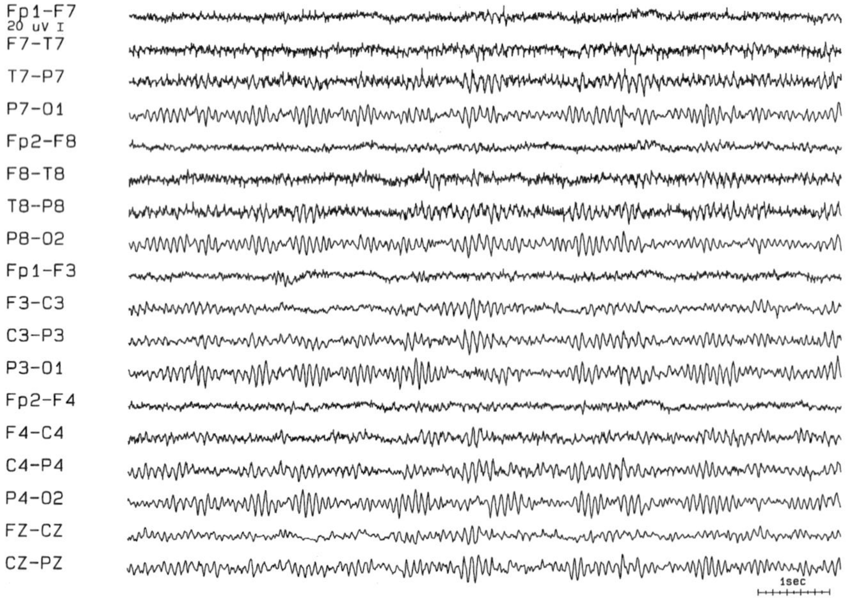

FIGURE 29–1. Alpha rhythm. An 11-Hz alpha rhythm in a healthy 41-year-old male.

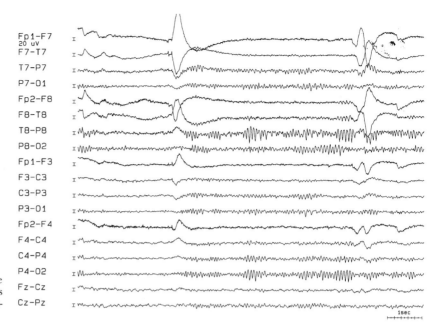

FIGURE 29–2. Alpha rhythm asymmetry. The amplitude of alpha rhythm on the left is less than 50% of that on the right, suggesting dysfunction involving the left hemisphere.

decreases with age, reflecting changes in bone density and increased electrical impedance of intervening tissue. Mild voltage asymmetry is common. The right hemisphere typically shows higher amplitudes. Asymmetries are considered significant when the amplitude over one hemisphere is less than 50% of that of the other hemisphere (Fig. 29–2). To be considered abnormal, voltage asymmetries should be present on both referential and bipolar montages.

Reactivity to sensory stimuli and mental activation is one of the features that distinguishes the normal alpha rhythm from abnormal activity in the alpha frequency range. The alpha rhythm is best observed during relaxed wakefulness with the eyes closed. Attenuation or blocking of the alpha rhythm occurs with eye opening, concentrated mental effort, heightened levels of alertness, and auditory or tactile stimulation. Immediately after eye closure, lower amplitude alpha activity of higher frequency may appear for a few seconds.

This phenomenon is known as alpha squeak or the squeak effect (Fig. 29–3). Unilateral failure of the alpha rhythm to attenuate with eye opening is known as Bancaud's phenomenon. It suggests dysfunction in the nonreactive hemisphere. During drowsiness, the alpha rhythm begins to wax and wane until it disappears with the onset of sleep. When a drowsy subject is alerted, the alpha rhythm may appear with eye opening, only to disappear again with eye closure when sleep ensues. This is known as paradoxical alpha rhythm.

Activity having features of the alpha rhythm and frequencies of half or twice that of the dominant alpha rhythm are known as alpha variants. Slow alpha variant has a frequency one-half that of the alpha rhythm, usually 4- to 5-Hz, with a notched appearance (Fig. 29–4). Fast alpha variant is twice that of the dominant alpha rhythm, usually ranging from 16- to 20-Hz (Fig. 29–5). Both patterns attenuate with eye opening, mental activation, and other sensory stimuli. Alpha vari-

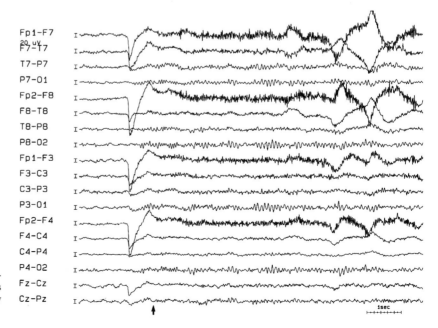

FIGURE 29–3. Alpha squeak. The alpha frequency immediately after eye closure (arrow) is lower in amplitude and higher in frequency than the dominant alpha frequency.

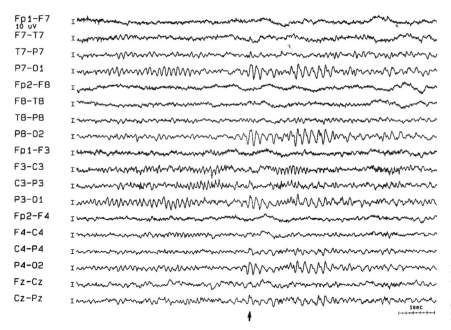

FIGURE 29–4. Slow alpha variant. Notched 4- to 4.5-Hz activity in the occipital regions (*arrow*) is approximately one-half the frequency of the alpha rhythm seen in the first half of the tracing.

ants typically appear sporadically in a record intermixed with the dominant alpha frequency and are considered normal.

Beta Rhythm

Beta activity includes frequencies greater than 13 Hz and is divided into 14- to 16-Hz, 18- to 25-Hz, and 35- to 40-Hz bands. The 18- to 25-Hz band is most common, whereas frequencies of 35 Hz and higher are rarely encountered in the surface electroencephalogram (EEG). Beta activity is typically most prominent in the fronto-central regions during relaxed wakefulness and drowsiness (Fig. 29–6). The voltage of beta activity is less than 20 μV in 98% and less than 10 μV in 70% of normal individuals.[9] Voltages above 25 μV are considered abnormal. Benzodiazepines, barbiturates, neuroleptics, antidepressants, antihistamines, stimulants, and chloral hydrate produce higher amplitude and more widespread beta activity, which preferentially affects the 18- to 25-Hz band. Beta activity is enhanced in the region of a skull defect and attenuated over focal cerebral lesions and intra- and extracranial fluid collections. Cognitive tasks such as arithmetic, light sleep, and rapid eye movement (REM) sleep produce higher amplitude beta activity, whereas the amplitude is usually decreased during deep sleep.

Mu Rhythm

The mu rhythm is a precentral rhythm of 7 to 12 Hz having a comb-like or archiform morphology (Fig. 29–7). Mu activity is usually unaffected by eye opening or closure but attenuates with the thought of movement or tactile stimulation of the extremities, more commonly of the contralateral limbs. Fatigue and problem solving also attenuate the mu rhythm, whereas immobility during relaxed wakefulness

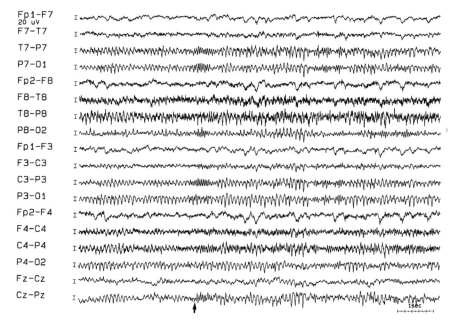

FIGURE 29–5. Fast alpha variant. A 20-Hz rhythm in the posterior regions bilaterally (*arrow*) is intermixed with 10- to 11-Hz alpha activity.

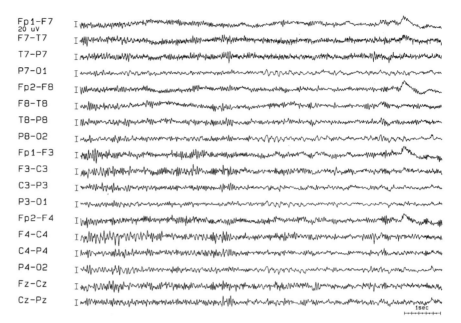

FIGURE 29–6. Beta rhythm. Fronto-central beta during drowsiness in a normal 32-year-old female.

enhances it. Previously referred to as rhythme rolandique en arceau, wicket rhythm, or comb rhythm, mu is observed in approximately 20% of normal young adults and is less common in children and older individuals.[9] It is slightly more common in females than males. The mu rhythm may be asymmetrical and asynchronous and is particularly prominent in the area overlying a skull defect. High amplitude mu, alpha, and beta activity usually confined to a single electrode overlying a burr hole or along the edge of a craniotomy, is referred to as breach rhythm[3] (Fig. 29–8). In the absence of slowing or epileptiform activity, breach rhythms have no clinical significance other than suggesting the presence of a skull defect.

Theta Rhythm

The theta rhythm, consisting of activity of 4- to less than 8-Hz waves, is frequently seen in the waking EEG of normal adolescents and adults. Theta activity is both age and state dependent, predominating in the first months of life. Most normal young adults will have low voltage 6- to 7-Hz theta activity in the fronto-central regions during wakefulness. Theta activity becomes more sustained and higher in amplitude during drowsiness. Hypnagogic hypersynchrony refers to high amplitude, monomorphic and rhythmic theta which characterizes the drowsy state in infancy and childhood. This pattern is sometimes seen in adolescents and young adults (Fig. 29–9). Some normal individuals over the age of 40 years show intermittent theta activity in the temporal regions during wakefulness. Temporal theta activity of the elderly is usually more pronounced on the left side and may be enhanced by hyperventilation (Fig. 29–10). There are no established criteria for determining how much temporal theta activity is acceptable. Abnormal findings are temporal theta activity in subjects under 40 years of age and continuous theta or delta rhythms during wakefulness in adults of any age.

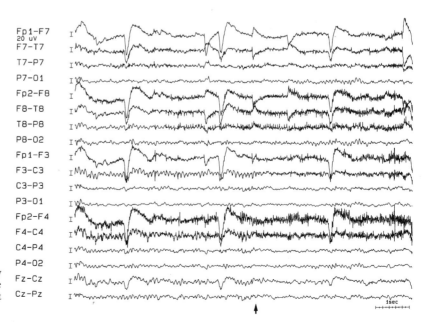

FIGURE 29–7. Mu rhythm. Arciform 9-Hz activity in the left central region attenuates when the subject is asked to make a fist with the right hand (*arrow*).

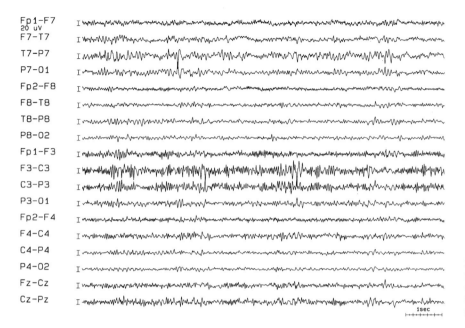

FIGURE 29–8. Breach rhythm. High amplitude beta and mu activity is seen in the left central region in a subject with a history of a left subdural hematoma following surgical evacuation.

Posterior Slow Waves of Youth

Posterior slow waves of youth are delta waves with superimposed alpha frequencies in the posterior head regions that attenuate with eye opening and drowsiness (Fig. 29–11). Posterior slow waves may be asymmetrical and asynchronous, but asymmetries should not exceed 50%. They are most common in children and adolescents but can be seen in approximately 15% of young adults.[9]

Lambda Waves

Lambda waves are bi- or triphasic sharp transients having an initial small positive phase followed by a prominent negative component that appears bilaterally over the occipital regions (Fig. 29–12). Lambda waves are elicited by looking at a patterned design in a well-illuminated room. They are more common in children and adolescents than adults.

Lambda waves may be asymmetrical leading to their misinterpretation as occipital spikes. Eye closure, reducing the illumination of the room, or having the subject stare at a blank card will attenuate lambda waves but not affect the appearance of epileptiform activity.

Low Voltage Electroencephalogram

The low voltage EEG consists of cerebral activity having amplitudes no greater than 20 μV over all head regions. Typically an admixture of beta, theta, and to a lesser degree delta waves, with or without alpha activity, is observed. Low voltage records are considered normal unless focal or diffuse slowing of higher amplitude, asymmetries, or paroxysmal activity are present. The incidence of low voltage EEGs in normal adult subjects ranges from 4% to 9%.[12] This pattern is exceedingly rare in subjects under the age of 20 years. Low voltage records have been correlated with vertebrobasi-

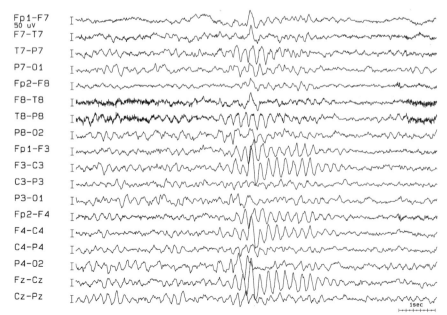

FIGURE 29–9. Hypnagogic hypersynchrony in a 19-year-old normal subject.

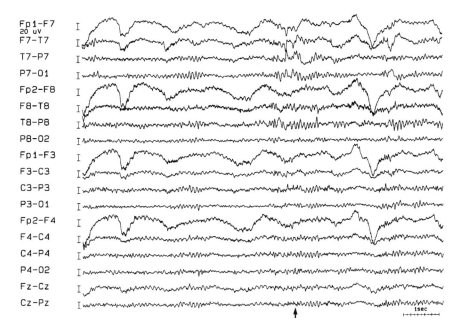

FIGURE 29–10. Temporal theta activity of the elderly. Intermittent theta rhythm in the left temporal region (*arrow*) is seen during wakefulness in this 53-year-old woman.

lar artery insufficiency, extrapyramidal syndromes, psychiatric disturbances, alcoholism, and head injury, but are now considered normal.

SLEEP RHYTHMS

Sleep Architecture

Sleep is comprised of non–rapid eye movement (NREM) and REM stages with NREM further subdivided into stages I through IV. The normal sleep cycle begins in stage I, or drowsiness, which is typically brief, comprising 5% to 10% of sleep time. Stage I is followed by stage II, which is followed by stages III and IV, collectively known as slow wave sleep (SWS) or deep sleep. After a brief return to stage II, REM sleep ensues. Most adults have five to seven sleep cycles per night, each lasting approximately 90 minutes, with the first being the shortest. SWS predominates in the first third of the night while REM sleep increases during the last few hours of sleep. Stage II comprises 30% to 50%, stages III and IV 20% to 40%, and the REM stage 20% to 25% of sleep time in normal adults. Sleep architecture is altered with advancing age. In the elderly, the percentage of SWS is reduced, while stage I is increased and stage II and REM sleep percentages remain unchanged. Elderly individuals have an increased number of arousals and awakenings. Neonates, on the other hand, spend nearly 50% of sleep time in stage REM.

Non–Rapid Eye Movement Sleep

Stage I is a transitional state characterized by slow eye movements, gradual waning of the waking background

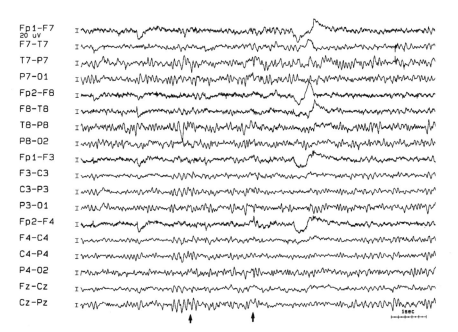

FIGURE 29–11. Posterior slow waves (*arrows*) in a 16-year-old male.

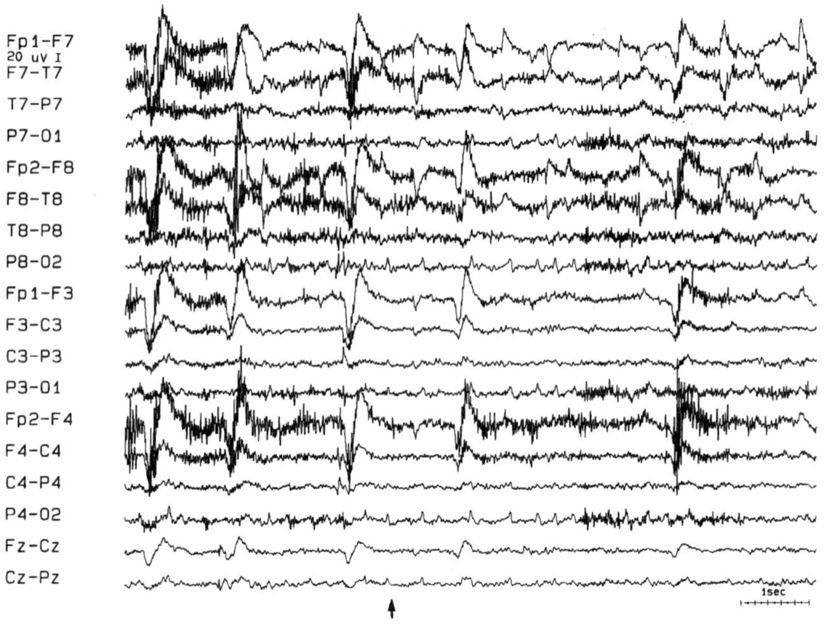

Fp1-F7
20 uV I
F7-T7

T7-P7

P7-O1

Fp2-F8

F8-T8

T8-P8

P8-O2

Fp1-F3

F3-C3

C3-P3

P3-O1

Fp2-F4

F4-C4

C4-P4

P4-O2

Fz-Cz

Cz-Pz

1sec

FIGURE 29–12. Lambda waves. A run of lambda waves is seen in the occipital regions bilaterally (*arrow*) as this subject reads.

rhythm, fronto-central theta activity, vertex waves, and positive occipital sharp transients (POSTs) (Fig. 29–13). The predominant background of the drowsy state is low to medium amplitude delta and theta activity intermixed with faster frequencies. Slow pendular eye movements having a frequency of less than 0.5 Hz are observed. There is usually a reduction in muscle artifact. Owing to its sporadic and sometimes abrupt appearance, drowsiness is sometimes confused with abnormal paroxysmal activity.

Vertex waves are diphasic, sharp transients with negative polarity ("surface negative"), seen in the central regions (Fig. 29–14). First appearing in drowsiness, vertex waves may persist into stage II and SWS. In adults, vertex activity is bilaterally synchronous and symmetrical. Persistent asymmetries are abnormal, usually suggesting dysfunction on the side of lower amplitude.

Positive occipital sharp transients are surface positive triangular waves occasionally followed by a lower voltage negative

phase seen in light sleep (Fig. 29–15). POSTs usually appear in runs of bilaterally synchronous 4- to 5-Hz waves; however, mild asymmetries are common. They first appear in drowsiness but are seen in deep stages of NREM sleep as well.

Stage II sleep is characterized by sleep spindles and K complexes on a background of mixed frequencies with relatively little delta activity (Fig. 29–16). Sleep spindles, or sigma activity, are rhythmic sinusoidal waves of 10 to 14 Hz over the central regions. Their duration in adults ranges from 0.5 to 2 seconds. They are asynchronous at the time of their first appearance at 6 to 8 weeks post-term until the age of 2 years. K complexes are diphasic waves having an initial negative sharp component followed by a slow positive phase, which often ends with a sleep spindle. Their duration is 0.5 second or greater. Appearing symmetrically over the central regions, K complexes may occur singly or in trains, spontaneously, or following an auditory stimulus.

Stage III sleep is scored when 20% to 50% of the back-

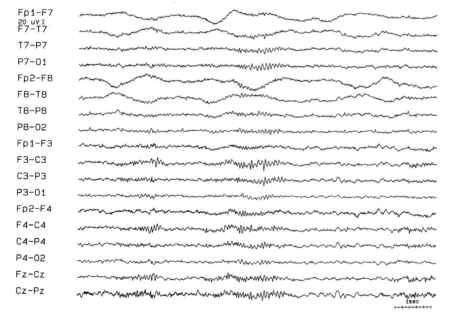

Fp1-F7
20 uV I
F7-T7

T7-P7

P7-O1

Fp2-F8

F8-T8

T8-P8

P8-O2

Fp1-F3

F3-C3

C3-P3

P3-O1

Fp2-F4

F4-C4

C4-P4

P4-O2

Fz-Cz

Cz-Pz

1sec

FIGURE 29–13. Stage I sleep showing slow pendular eye movements and a background of theta and alpha frequencies.

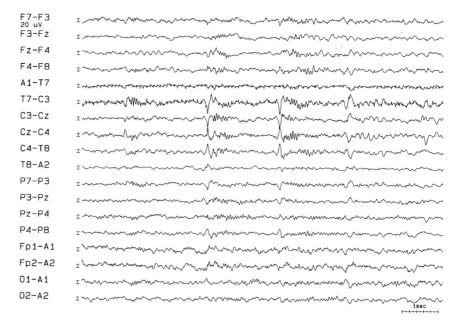

FIGURE 29–14. Vertex waves. A series of vertex waves during the transition from stage 1 to stage 2 sleep.

ground is comprised of delta activity of less than 2 Hz and greater than 75 μV (Fig. 29–17). Spindles may be present, but are less frequent and lower in frequency than in stage II sleep. Stage IV sleep is scored when at least 50% of the background is comprised of delta activity of less than 2 Hz and amplitude greater than 75 μV. In general, stages III and IV need not be differentiated in the routine EEG.

Rapid Eye Movement Sleep

Rapid eye movement sleep normally occurs 60 to 90 minutes after sleep onset, and, consequently, is usually only seen in the EEG laboratory in conditions of reduced REM latency. The latency to REM sleep is shorter in severely sleep-deprived individuals, neonates, patients with narcolepsy, and subjects withdrawing from alcohol and REM suppressant medications. The REM stage consists of low voltage, mixed frequencies with admixed alpha activity usually 1- to 2-Hz

slower than the waking alpha rhythm, bursts of rapid eye movements, and suppression of electromyographic (EMG) activity (Fig. 29–18). Runs of 2- to 6-Hz notched waves, referred to as saw-tooth waves, are observed in the fronto-central regions, usually in association with bursts of eye movements.

Arousal Patterns

Arousals from sleep are commonly seen in routine EEGs. Arousals from stage I consist of low voltage fast activity or a reversion to the waking background rhythm. Once stage II is reached, an arousal begins with a K complex followed by a brief train of alpha-range frequencies with or without intermixed high voltage delta transients and an increase in EMG activity. Arousals from SWS are generally not observed in normal adults. Their presence should raise the suspicion of a NREM parasomnia. Because bursts of alpha activity are

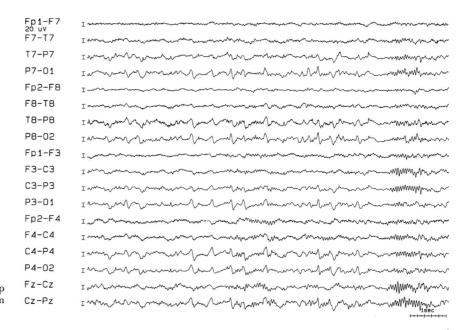

FIGURE 29–15. Positive occipital sharp transients seen during the transition from stage 1 to stage 2 sleep.

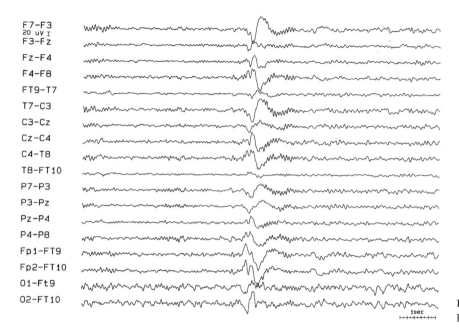

FIGURE 29–16. Stage II sleep with a K complex followed by sleep spindles.

seen normally in REM sleep, a simultaneous increase in EMG activity is required for arousals in this stage.[15]

ACTIVATION METHODS

Hyperventilation

The technique of hyperventilation consists of deep breathing at a rate of 20 per minute for 2 to 5 minutes to elicit abnormalities on EEG. Hyperventilation causes a drop in plasma carbon dioxide and cerebral vasoconstriction. The magnitude of the EEG response depends on the adequacy of the technique, age of the subject, serum glucose level, body position, body habitus, and pulmonary status. Vigorous air exchange will accentuate the effect. More prominent and persistent responses are observed in younger, taller, and thinner individuals and in persons with normal pulmonary function. Lower serum glucose levels tend to enhance the response, whereas hyperglycemia may inhibit it. The effect is often accentuated when performed in the erect position.

A typical response consists of generalized semirhythmic or rhythmic polymorphic delta or theta activity, 100 to 300 μV in amplitude, beginning usually within 1 minute of hyperventilation (Fig. 29–19). Delta activity tends to be maximal in the frontal regions in adults and in the occipital regions in children. The effect usually dissipates within 60 seconds and sometimes immediately upon cessation of hyperventilation. The technique is effective in eliciting epileptiform discharges in individuals with idiopathic generalized epilepsy, particularly in childhood absence epilepsy. Hyperventilation rarely produces epileptiform discharges in patients with focal epilepsy. The response may be exaggerated or suppressed in a region overlying a focal cerebral lesion. A hyperventilation response is considered abnormal only when unambiguous

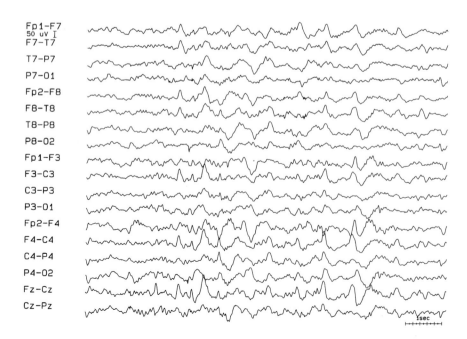

FIGURE 29–17. Slow wave sleep.

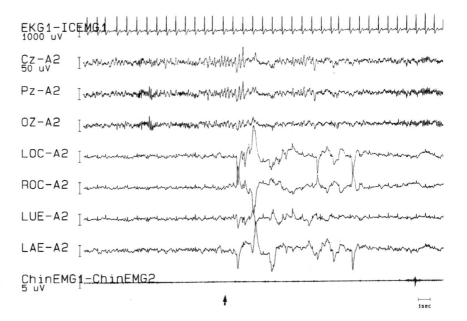

FIGURE 29–18. REM sleep. A 30-second epoch of REM sleep with rapid eye movements, saw-tooth waves (*arrow*), and electromyographic suppression.

epileptiform activity or persistent focal or lateralizing changes are observed. Prominent or persistent generalized slowing should be interpreted with caution, as this can be caused by hypoglycemia. The technique should be avoided in patients with recent stroke, subarachnoid hemorrhage, or cardiopulmonary disease in whom cerebral vasoconstriction or sudden acid base fluctuations may be detrimental.

Intermittent Photic Stimulation

Intermittent photic stimulation (IPS) utilizes a strobe light with 1.5 million candlepower, and flash duration of 10 μsec to present flashing lights at variable rates during an EEG. The light is placed 20 to 30 cm from the subject's eyes and presented during eye closure in a room with low illumination. The patterns produced by IPS are described here.

Rhythmic activity in the occipital regions that is time-locked to or a harmonic of the flash rate is referred to as

photic driving (Fig. 29–20). Photic driving is a normal response. It tends to occur at flash rates nearest to a subject's alpha frequency, although rates ranging from 5 to 30 Hz can elicit the response. Very high amplitude photic driving at frequencies of 0.5 to 3 Hz has been described in children with neuronal ceroid lipofusinosis. Cortical lesions may cause unilateral suppression of the response. Driving may be accentuated ipsilateral to an epileptogenic focus. Mild asymmetries and absence of the response may be seen in normal individuals. Unilateral absence of photic driving is abnormal, suggesting dysfunction in the suppressed hemisphere. During IPS, light flashes can produce a photoelectric artifact in which the light and electrodes react to produce spike-like transients time-locked to the flash rate (Fig. 29–21). This phenomenon is usually observed in frontal electrodes and can be eliminated by shading the electrode from the light.

The photomyoclonic or photomyogenic response (PMR) is another normal response to IPS. Muscle potentials typically

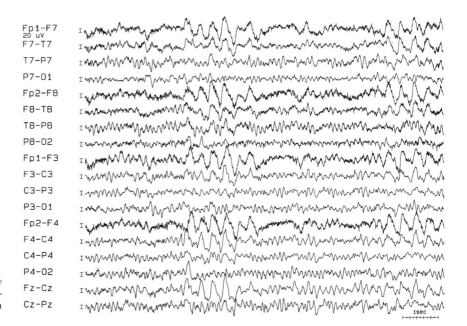

FIGURE 29–19. Generalized bifrontally predominant rhythmic delta activity 40 seconds after the onset of hyperventilation in a 38-year-old subject.

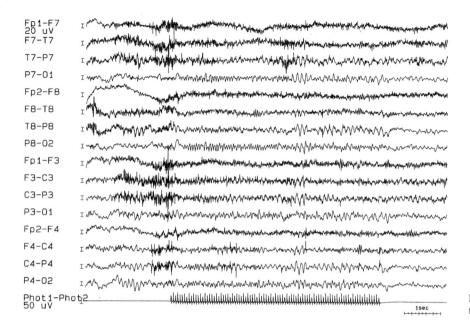

FIGURE 29–20. Photic driving at 15-Hz flash frequency.

occur 50 to 60 msec after each flash over the anterior head regions, as a result of contraction of the facial muscles. The potentials increase in amplitude with increasing flash rates and are most prominent at flash frequencies of 12 to 18 Hz (Fig. 29–22). Eyelid fluttering or contractions of head and neck musculature may be observed. The response terminates abruptly with the end of stimulation. The photomyoclonic response is observed most often in anxious subjects, patients with parkinsonian syndromes and blepharospasm, and in subjects experiencing alcohol or drug withdrawal.

The photoconvulsive response (PCR) or photoparoxysmal response is characterized by generalized or posteriorly dominant spike or polyspike and wave complexes in response to IPS (Fig. 29–23). Activity that is time-locked to the stimulus and limited to the occipital regions is considered normal. Generalized epileptiform activity and discharges that are independent of the stimulus or outlast the termination of the stimulus, are abnormal. Photic stimulation may provoke

myoclonic and tonic-clonic seizures, particularly if the stimulus is maintained once generalized epileptiform activity is produced. The response is most commonly elicited with flash rates of 15 to 20 Hz in patients with idiopathic generalized epilepsy, particularly in juvenile myoclonic epilepsy. It is also seen in patients with toxic or metabolic encephalopathy and following drug withdrawal. A self-sustained response is highly suggestive of a genetic predisposition for epilepsy. However, photoconvulsive responses are observed in 2% of normal subjects.[16]

BENIGN VARIANTS

Rhythmic Midtemporal Discharges

This pattern was originally named "psychomotor variant" for its resemblance to the ictal pattern of temporal lobe

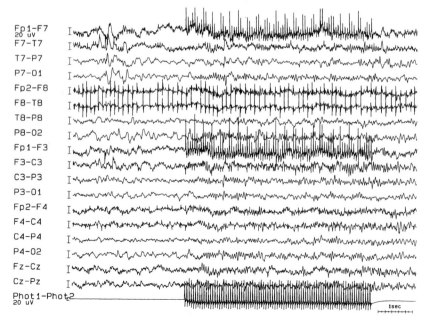

FIGURE 29–21. Photoelectric artifact. Repetitive spike-like transients time-locked to the flash are seen at FP₁.

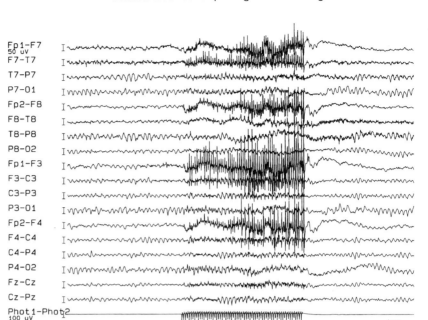

FIGURE 29–22. Photomyogenic artifact.

seizures.[6] Bursts or trains of rhythmic 4- to 7-Hz theta waves, having a flat-topped or notched appearance often with superimposed faster frequencies, are seen in the midtemporal regions during drowsiness or stage 2 sleep and rarely during wakefulness (Fig. 29–24). Bursts may be uni- or bilateral, usually shifting from one temporal region to the other. The typical duration ranges from 1 to 10 seconds, although bursts as long as 1 minute have been described. The amplitude tends to gradually increase in the first portion of the burst and gradually decrease toward the end, sometimes leading to its misinterpretation as a seizure. Rhythmic midtemporal discharges (RMTDs) are differentiated from seizures by the lack of evolution of frequency and morphology, and absence of behavioral changes. The pattern is observed in 0.3% to 2% of normal adolescents and adults.[10, 19]

Subclinical Rhythmic Electrographic (Theta) Discharge in Adults

This is a rare pattern seen primarily in individuals over the age of 50 years. Rhythmic, sharply contoured 5- to 7-Hz theta waves 40 to 100 μV in amplitude appear in a widespread, bilaterally synchronous manner typically maximal in the parietal and posterior temporal regions or over the vertex[20] (Fig. 29–25). The average duration is 40 to 80 seconds; however, subclinical rhythmic electrographic discharge in adults (SREDA) may last for several minutes. SREDA may begin and end abruptly or gradually. It can be differentiated from a seizure by the absence of associated symptoms and behavioral changes. The pattern is usually observed during wakefulness but has also been described during sleep.

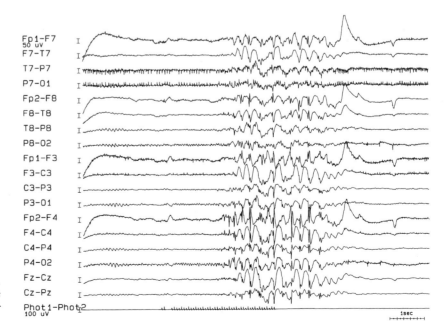

FIGURE 29–23. Photoparoxysmal response. Self-sustained generalized spike and wave complexes in the 25-year-old female with juvenile myoclonic epilepsy.

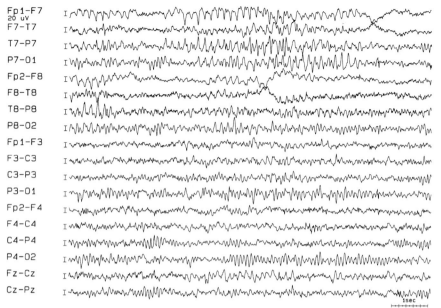

FIGURE 29–24. Rhythmic midtemporal discharges (RMTD). Runs of rhythmic notched theta are seen in the midtemporal regions bilaterally left greater than right.

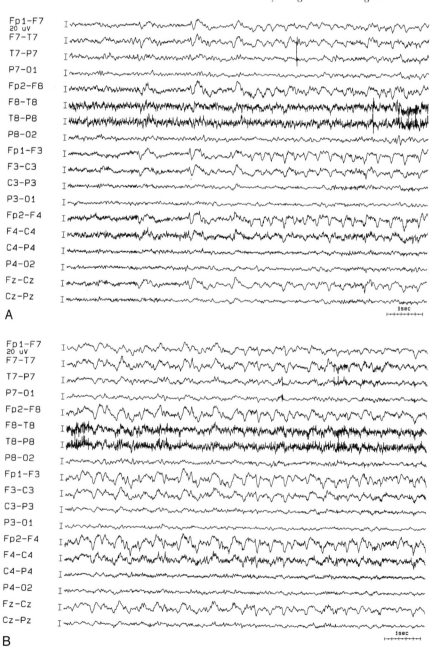

FIGURE 29–25. Subclinical rhythmic electrographic discharge in adults (SREDA). Serial 10-second tracings in a 66-year-old female. The pattern recurred multiple times in the absence of clinical signs lasting from 20 to 60 seconds.

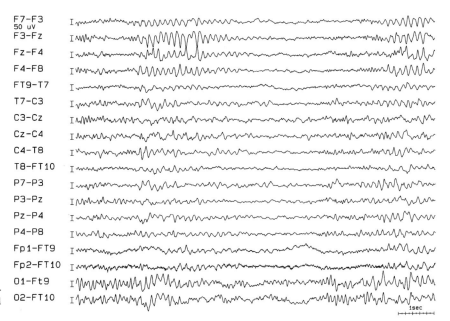

FIGURE 29–26. Midline theta rhythm. Arciform 6-Hz theta rhythm is largely confined to Fz.

SREDA tends to occur more than once during a single recording, particularly during or shortly after hyperventilation, and in subsequent EEGs of the same individual.

Midline Theta Rhythm

Rhythmic sinusoidal, archiform, or sharply contoured theta of 4 to 7 Hz focally over midline anterior electrodes (maximal at Fz or Cz) is seen during wakefulness, drowsiness, or after arousal from sleep[21] (Fig. 29–26). The pattern may spread to the parasagittal regions but is not seen over the temporal lobes. Midline theta rhythm tends to wax and wane, typically lasting for several seconds. It is variably reactive to eye opening and movement of the extremities. The pattern was originally described in patients with temporal lobe epilepsy and was believed to represent a projection of mesiobasal temporal epileptiform activity to the cingulate gyrus.[2]

This pattern was found in 0.01% of EEGs performed in one laboratory over 20 years in subjects with a variety of symptoms and is considered normal.[21]

Frontal Arousal Rhythm

The frontal arousal rhythm (FAR) is characterized by bursts or trains of monophasic 7- to 10-Hz waves several seconds in duration appearing focally at the F_3 and F_4 electrodes[22] (Fig. 29–27). The amplitude of FAR tends to wax and wane producing a spindle-like morphology. The pattern may evolve from beta activity of progressively increasing amplitude resulting in a superimposed notch on the ascending and descending limbs of the individual waves. This pattern was first described in children with cerebral dysfunction during arousal from stage II sleep, but is now considered normal. It was found in eight of several thousand EEGs performed in one laboratory over several years.[22]

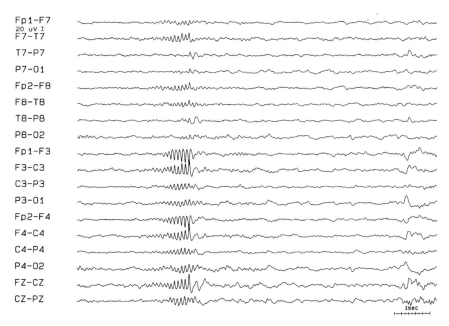

FIGURE 29–27. Frontal arousal rhythm.

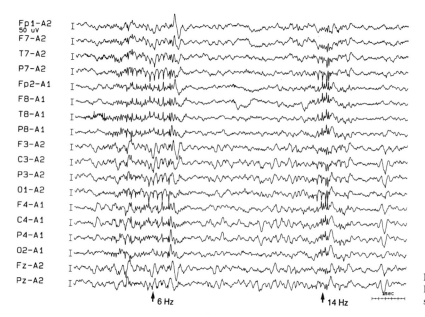

FIGURE 29–28. Fourteen- and 6-Hz positive bursts. Bursts of predominately 6 Hz and 14 Hz positive spikes are seen in this normal 20-year-old subject.

Fourteen- and Six-Hertz Positive Bursts

Also known as 14- and 6-Hz positive spikes and ctenoids, from the Greek *ktenos* meaning "comb," these bursts consist of rhythmic arciform waves 0.5 to 1 second in duration seen in wakefulness and light sleep[5, 11] (Fig. 29–28). The two frequencies may be intermixed or one or the other may predominate. Fourteen-hertz activity is more common in adults, and 6 Hz more common in children. This pattern is seen best in derivations with long interelectrode distances and is maximal in amplitude over the posterior temporal regions. Bursts are usually unilateral or bilateral and independent, but may be bilaterally synchronous. The pattern is more common in adolescents than in adults or young children. Early studies reported an association with a variety of neurological, psychiatric, and medical disorders.[5, 11] The reported incidence of 14- and 6-Hz positive spikes varies widely, ranging from 2% to 58% of healthy individuals.[5, 7, 11]

Small Sharp Spikes

Also referred to as benign epileptiform transients of sleep (BETS) and benign sporadic sleep spikes (BSSS), small sharp spikes (SSS) were originally named for their low amplitude (less than 50 μV) and short duration (less than 50 msec). However, transients higher in amplitude and longer in duration have been described.[4, 23] SSS have a characteristic morphology of a monophasic or diphasic spike with an abrupt ascending and a steep descending limb, occasionally followed by a lower amplitude slow wave (Fig. 29–29). Best seen in the anterior to midtemporal regions in derivations with long interelectrode distances, isolated spikes occur sporadically in adults during drowsiness and stage II sleep. They can be unilateral or bilateral and typically shift from one hemisphere to the other. Small sharp spikes are not associated with background slowing, tend to occur singly and not in trains, and diminish in deep sleep. These features are

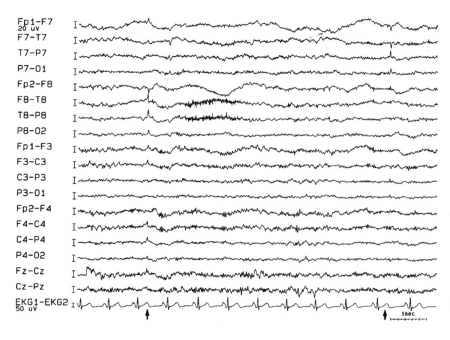

FIGURE 29–29. Small sharp spikes are seen in the right and left temporal regions (*arrows*).

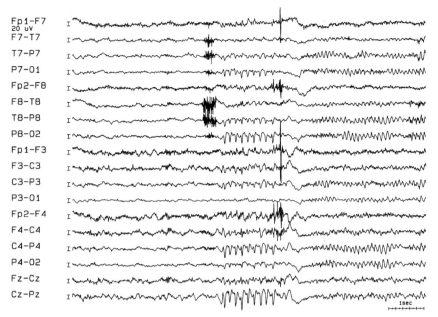

FIGURE 29–30. Six-hertz spike and wave. Rhythmic 5-Hz wave preceded by low amplitude spikes occur before an arousal in this 28-year-old female.

useful in their differentiation from epileptic discharges. They are observed in the EEGs of 20% to 25% of healthy subjects.[19, 23]

Six-Hertz Spike and Wave

Six-hertz spike and wave is characterized by brief bursts of 5- to 7-Hz generalized spike and wave discharges seen primarily in adolescents and young adults during wakefulness or drowsiness[18] (Fig. 29–30). Also known as phantom spike and wave, the spike is characteristically low in amplitude in comparison to the slow wave, frequently under 25 μV. Two types of 6-Hz spike and wave based on differences in voltage, distribution, gender, and state have been described.[8] FOLD was described in *f*emales in the *o*ccipital regions, is *l*ow in amplitude, and present during *d*rowsiness. WHAM, on the other hand, occurs during *w*akefulness, is *h*igher in ampli-

tude, more *a*nterior in distribution, and seen primarily in *m*ales. WHAM was once believed to be more common in patients with epilepsy, whereas FOLD occurred predominantly in individuals with psychiatric illness, headaches, and other neurological and medical disorders.[8] Both patterns are now considered normal variants and are found in 0.2% to 4.5% of healthy subjects.[8]

Wicket Spikes

Wicket spikes are monophasic, arciform 6- to 11-Hz transients that occur singly or in trains in the anterior to mid-temporal regions during wakefulness or sleep[14] (Fig. 29–31). Longer trains tend to occur in wakefulness and drowsiness, while isolated spikes or short trains are more common in sleep. They are typically bilateral and independent, usually predominating on one side. Wicket spikes are seen in 0.9%

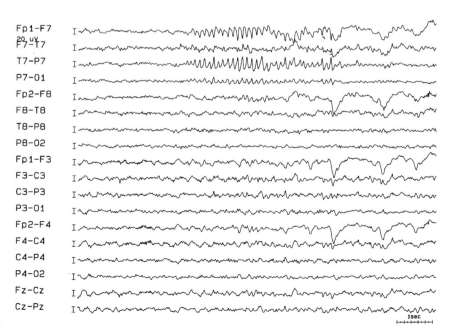

FIGURE 29–31. Wicket spikes.

to 2.9% of normal adults.[14] They are differentiated from temporal lobe spikes by the absence of aftercoming slow waves and background slowing, and by their characteristic morphology.

PHYSIOLOGICAL ARTIFACTS

Ocular Artifacts

Ocular artifacts arise from the direct current (DC) potential of the eyeball or myogenic potentials of the orbital and periorbital muscles. The eyeball acts as a dipole with the positive pole anteriorly toward the cornea and the negative pole toward the retina.[17] With movement of the globe, an alternating current (AC) potential is recorded by nearby electrodes. Eye movements are distinguished by their frontal distribution, bilateral symmetry, and stereotypical morphology. Vertical eye movements are highest in amplitude at FP_1 and FP_2 or at electrodes placed directly above or below the eye. Horizontal eye movements are recorded best at F_7 and F_8 or by additional electrodes placed at the outer canthi. One can determine the direction of gaze by analyzing the distribution and polarity of ocular artifacts at these electrodes (Fig. 29–32). Whereas some eye movements are purely vertical or horizontal, most are oblique, with both vertical and horizontal components. Eye closure or blinking results in upward vertical eye movements that produce downward (positive) deflections at FP_1 and FP_2 as the positively charged cornea rotates upward. Downward vertical eye movements produce upward (negative) deflections maximal at the frontopolar electrodes as the negative end of the dipole rotates upward. The amplitude of eye movement deflections diminishes rapidly and progressively, from an anterior to posterior direction, such that no significant deflections are observed in occipital regions. Horizontal conjugate eye movements produce deflections at F_7 and F_8 that are opposite in polarity with the positive end of the dipole recorded by electrodes on the side of the direction of gaze. Horizontal eye movements may be preceded by lateral rectus muscle potentials (lateral rectus spikes) seen best at F_7 and F_8 with left and right gaze respectively. Artifacts caused by nystagmus and repetitive blinking can be differentiated from abnormal cerebral activity through the use of infraorbital electrodes. Orbital disease causing restriction of globe mobility, enucleation, unilateral retinal destruction, unilateral skull defect near the globe, and asymmetrical electrode placement can cause asymmetries of eye movements. This may lead to their misinterpretation as abnormal cerebral activity.

Cardiac Artifacts

Because the potential field of the heart extends toward the skull base, electrocardiographic (ECG) artifact is detected in ear and other basal electrodes frequently contaminating referential recordings (Fig. 29–33). The field may extend into the parasagittal regions in subjects with short necks. ECG artifacts may have positive polarity in the left posterior temporal region since the main vector producing the R wave is positive, directed posteriorly toward the left. The amplitude of ECG artifact changes with the respiratory cycle, sometimes disappearing altogether. ECG artifact is variable in morphology, sometimes resembling spikes or sharp waves. Pulse artifact is produced when electrodes are placed directly over small scalp arteries and the systolic pulse wave causes an alteration in electrode impedance (Fig. 29–34). Repetitive slow transients occurring 200 to 300 msec after the QRS complex are observed that can be eliminated by repositioning the electrode.

Myogenic Artifact

Artifact from the frontalis and temporalis muscles is a common source of EEG contamination. It is especially prominent in referential recordings using the ear or mastoid as reference. Muscle potentials are repetitive rapid spikes usually shorter in duration than epileptic discharges. They can be reduced by having the subject relax the jaw or reposition the head. Large body movements produce erratic, high am-

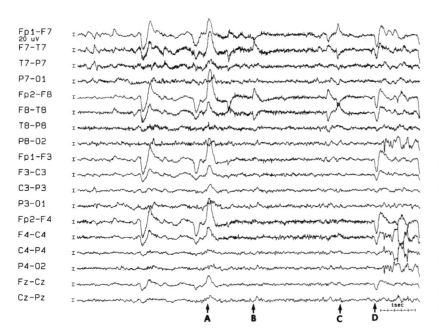

FIGURE 29–32. Series of eye movements. Downgaze (A) produces a negative deflection maximal in the frontopolar electrodes as the relative negative retina rotates upward. Horizontal gaze to the right (B) results in a surface positive transient at F_8 preceded by a lateral rectus spike caused by activation of the right lateral rectus muscle. A negative deflection is seen at F_7 as the cornea rotates toward the right and the retina toward the left. Horizontal gaze to the left (C) produces a transient of opposite polarity to that seen in B. In D, the eyes look up producing symmetrical positive deflections in the frontopolar regions.

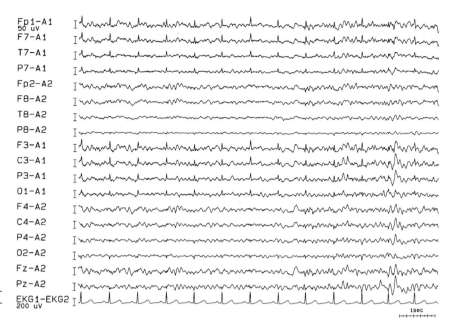

FIGURE 29–33. Electrocardiographic artifact. The left ear reference (A₁) is contaminated.

plitude, and arrhythmic waveforms and can usually be readily identified as artifactual.

Glossokinetic Artifact

Movements of the tongue and oropharyngeal structures produce artifact that may resemble intermittent generalized or temporal slow activity. Like the orbit, the tongue has a DC potential, the tip negative with respect to the base. Bilaterally symmetrical, repetitive slow waves seen maximally in the temporal regions are produced with tongue movement (Fig. 29–35). Glossokinetic artifact can be confirmed by asking the subject to repeat words beginning with the letter "l" or recording from an electrode on the cheek.

Miscellaneous Physiological Artifacts

Respiration can produce rhythmic, usually low amplitude slow waves, or sharp activity that is synchronous with body movement. This is usually maximal in electrodes on which the subject is lying and can be eliminated by repositioning the head.

Perspiration alters the impedance of the skin producing slow waves usually greater than 2 seconds in duration and appearing in one or multiple channels. Cooling the head or lowering the temperature in the room will eliminate this type of artifact.

Dental fillings with dissimilar metals can produce transients resembling spikes or sharp waves when the metal pieces come into contact. This usually occurs when a subject

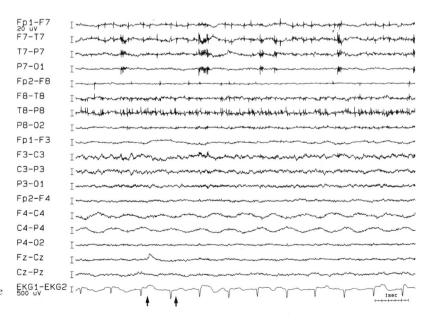

FIGURE 29–34. Pulse artifact at C₄ (arrows). Note its consistent relationship to the QRS complex.

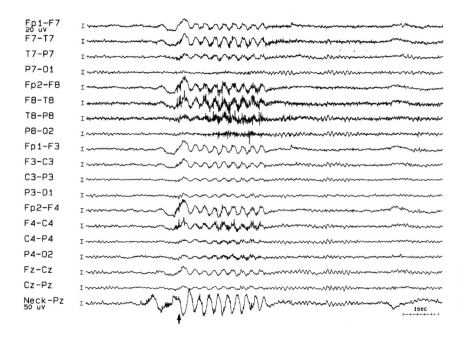

FIGURE 29–35. Glossokinetic artifact. Rhythmic activity is caused by movement of the tongue as the subject speaks. It is confirmed by placement of an additional electrode in the submandibular region (neck).

is speaking or chewing and can be eliminated by partially opening the mouth.

REFERENCES

1. Berger H: On the electroencephalogram of man. Second report. J Psychiatry Neurol 40:160–179, 1930.
2. Ciganek L: Theta-discharges in the middle-line—EEG symptom of temporal lobe epilepsy. Electroencephalogr Clin Neurophysiol 13:669–673, 1961.
3. Cobb WA, Guiloff RJ, Cast J: Breach rhythm: The EEG related to skull defects. Electroencephalogr Clin Neurophysiol 47:251–271, 1973.
4. Gibbs FA, Gibbs EL (eds): Atlas of Electroencephalography, vol 1. Reading, MA: Addison-Wesley, 1951.
5. Gibbs FA, Gibbs EL: Fourteen- and six per second positive spikes. Electroencephalogr Clin Neurophysiol 15:553–558, 1962.
6. Gibbs FA, Rich CL, Gibbs EL: Psychomotor variant type of seizure discharge. Neurology 13:991–998, 1963.
7. Gibbs EL, Gibbs FA: Clinical significance of 14 and 6 per second positive spikes in the electroencephalograms of patients over 29 years of age. Clin Electroencephalogr 4:140–144, 1973.
8. Hughes JR: Two forms of the 6/sec spike and wave complex. Electroencephalogr Clin Neurophysiol 48:535–550, 1980.
9. Kellaway P: An orderly approach to visual analysis: Characteristics of the normal EEG of adults and children. In Daly DD, Pedley TA (eds): Current Practice of Clinical Electroencephalography, 2nd ed. New York, Raven Press, 1990, pp 139–199.
10. Lipman IJ, Hughes JR: Rhythmic mid-temporal discharges. An electro-clinical study. Electroencephalogr Clin Neurophysiol 27:43–47, 1969.
11. Lombroso CT, Schwartrz IH, Clark DM, et al: Ctenoids in healthy youths: Controlled study of 14- and 6-per-second positive spiking. Neurology 16:1152–1158, 1966.
12. Markand ON: Alpha rhythms. J Clin Neurophysiol 7:163–189, 1990.
13. Obrist WD: Problems of aging. In Remond A (ed): Handbook of Electroencephalography and Clinical Neurophysiology, vol 6A. Amsterdam, Elsevier, 1976, pp 275–292.
14. Reiher J, Lebel M: Wicket spikes: Clinical correlates of a previously undescribed EEG pattern. Can J Neurol Sci 4:39–47, 1977.
15. Sleep Disorders Atlas Task Force of the American Sleep Disorders Association: EEG arousals: scoring rules and examples. Sleep 15(2):174–183, 1992.
16. Takahashi T: Activation methods. In Neidermeyer E, Lopes da Silva F (eds): Electroencephalography: Basic Principles, Clinical Applications, and Related Fields, 3rd ed. Baltimore, Williams & Wilkins, 1993, pp 241–262.
17. Tyner FS, Knott JR, Mayer WB Jr: Artifacts. In Tyner FS, Knott JR, Mayer WB Jr (eds): Fundamentals of EEG Technology, vol 1: Basic Concepts and Methods, New York, Raven Press, 1983, pp 280–311.
18. Walter WG, Hill JDN, Park G (eds): Electroencephalography. New York, Macmillan, 1950.
19. Westmoreland BF: Benign EEG variants and patterns of uncertain clinical significance. In Daly DD, Pedley TA (eds): Current Practice of Clinical Electroencephalography, 2nd ed. New York, Raven Press, 1990, pp 243–252.
20. Westmoreland BF, Klass DW: A distinctive rhythmic EEG discharge of adults. Electroencephalogr Clin Neurophysiol 51:186–191, 1981.
21. Westmoreland BF, Klass DW: Midline theta rhythm. Arch Neur 43:139–141, 1986.
22. White JC, Tharp BR: An arousal pattern in children with organic cerebral dysfunction. Electroencephalogr Clin Neurophysiol 37:265–268, 1973.
23. White JC, Langston JW, Pedley TA: Benign epileptiform transients of sleep. Clarification of the small sharp spike controversy. Neurology 27:1061–1068, 1977.

Chapter 30

Long-Term Electroencephalographic/Video Monitoring

Harold H. Morris, MD • George Klem, MD, R EEG T

INTRODUCTION
DESIGN OF THE LONG-TERM MONITORING
   UNIT
STAFFING OF THE MONITORING UNIT
ELECTROENCEPHALOGRAM MONITORING
   EQUIPMENT
DATA ACQUISITION SYSTEMS
Data Storage
Video Recording
Video Cassette Recorders

ELECTRODES
Depth Electrodes
Subdural Electrode Grid Arrays
ARTIFACT RECOGNITION
ACTIVATING PROCEDURES
MANAGEMENT OF THE PATIENT IN THE
   MONITORING UNIT

INTRODUCTION

Epilepsy is one of the most common neurological diagnoses and carries a host of implications for the patient. Not only does the patient with epilepsy have to take medication on a chronic basis, but often the patient's driving, working, and recreational activities are restricted to some extent. For women in the childbearing years, the diagnosis and treatment of epilepsy has implications affecting both birth control and pregnancy.

The physician taking the patient's history must consider a variety of other intermittent and paroxysmal neurological conditions, including nonepileptic seizures (pseudoseizures), syncope, cardiac rhythm disturbances, transient ischemic attacks, migraine, narcolepsy, and rapid eye movement (REM) behavior disorder before making a diagnosis of epilepsy. Frequently a detailed history from both the patient and an observer is sufficient to make a diagnosis of epilepsy with reasonable certainty, but in a significant number of patients events are unwitnessed or the history is unclear. Routine electroencephalography is helpful in making a diagnosis if definite epileptiform activity is recorded, but a routine 20- or 30-minute electroencephalogram (EEG) is relatively insensitive in detecting these discharges. All neurologists know a normal routine EEG does not "rule out" epilepsy. Ambulatory, computerized, EEG monitoring with 16 or more channels is helpful in diagnosis because it permits a greater sampling time; however, frequently the "spells" must be observed along with a high-quality, attended EEG recording for an accurate diagnosis to be made.

In patients with a firm diagnosis of epilepsy but poor response to treatment, recording of the clinical events along with the EEG will permit accurate classification of seizure type(s) and EEG characterization of both interictal epileptiform activity and ictal onset. Such information may be helpful in selecting the optimal antiepileptic drug and is required in the evaluation for surgical suitability.

DESIGN OF THE LONG-TERM MONITORING UNIT

Epilepsy monitoring units developed because of the needs expressed in the preceding paragraphs. The purpose of an epilepsy monitoring unit is to characterize events that may be epileptic or nonepileptic while simultaneously recording the EEG and electrocardiogram (ECG) associated with these events. An epilepsy monitoring unit may be a dedicated or nondedicated facility. A dedicated monitoring unit has both a trained EEG technologist and nursing staff with in-depth knowledge and experience in the recognition of different seizure types as well as in the interpretation of interictal and ictal EEG patterns. They are trained in first aid for seizures and parenteral and oral administration of medications. A nondedicated monitoring facility may be a room or group of rooms equipped with video cameras and EEG machines located on a medical or neurological hospital floor; the floor nurses may or may not be especially knowledgeable about epilepsy and a trained EEG technologist may or may not be present on a continuous basis. A third type of monitoring consists of portable EEG equipment that may be placed in any type of nursing unit, such as an intensive care unit, or in a nonhospital setting.

The advantage of a dedicated unit is that the personnel become "expert" in the technical and clinical management of patients with epilepsy. The advantage of a nondedicated unit is that there is more freedom for personnel changes, and the hospital rooms (an expensive and valuable commodity in the United States) may be used for other purposes: A nondedicated monitoring unit is less expensive to staff but has disadvantages in that the staff are not as familiar with epilepsy and the management of patients with seizures. The objectives of monitoring are to record both EEG interictal abnormalities and EEG/video ictal events in a safe, cost-effective environment.

STAFFING OF THE MONITORING UNIT

The physician director of an epilepsy monitoring unit should have expertise in both clinical epilepsy and in clinical neurophysiology. Postneurology residency training of at least 1 but preferably 2 years at an active epilepsy program with training in both noninvasive and invasive long-term monitoring is mandatory.

The staffing of the monitoring unit includes nurses familiar with different seizure types and an intimate knowledge of the anticonvulsants used in treatment, along with an EEG technologist. The technologist should be able to recognize all clinical seizure types, ictal and interictal epileptiform patterns, and nonepileptiform activity recorded from either scalp or invasive electrodes. These special requirements are in addition to the usual performance expectations of an EEG technologist. A monitoring technician may be part of the team, especially during off hours when there is usually less need for the more experienced (and expensive) EEG technologist. The monitoring technician may be a nursing care assistant or other individual whose role is primarily to recognize clinical seizures or electrographic seizures and ensure that the EEG and video tapes are saved for review by the medical staff. All the staff should be trained to interact with patients during ictal events for assessment of cognitive function, to assist with first aid, provide routine care of the electrodes, and perform other "trouble shooting" duties. Specific guidelines for staffing and equipment for long-term monitoring have been published by the American Electroencephalographic Society.

ELECTROENCEPHALOGRAM MONITORING EQUIPMENT

Approximately 10 to 15 years ago, much of the EEG monitoring was performed using paper writing EEG machines. Because seizures occur unpredictably, the EEG machines had to run continuously. On those occasions when recording of large numbers of channels was required (e.g., 64 channels for an 8×8 subdural grid array), multiple machines had to be simultaneously used for a single patient. This method of collecting data required personnel to be present to keep the paper in the equipment and ink in the pens. It required the epileptologist to manually search for interictal sharp waves and analyze ictal events. This technology was wasteful of both paper and personnel. Because of the reduced cost, increased convenience, and ease and advantages of manipulating digital data, most monitoring units now use digital EEG equipment.

EEG amplifiers should meet or exceed the technical requirements for clinical electroencephalography. For routine noninvasive monitoring, 32-channel equipment is recommended; for invasive records 96- or 128-channel equipment may be required. The ability to record frequencies below 1 Hz and above 70 Hz is desirable when recording from subdural and depth electrodes.

DATA ACQUISITION SYSTEMS

Basic equipment includes a minimum 12-bit computer with sufficient memory to store a minimum of 32 separate channels of continuous data obtained at a sampling rate of at least 200 points per second. In addition to recording EEG data, recording of the ECG should be routine; the recording of other physiological variables may be useful depending on the clinical circumstances. Postacquisition analysis is facilitated by the ability to reformat, digital filter as needed, change the time display, and adjust the gain. Software for the automatic detection of seizures and interictal spikes is ideal. Software designed for topographical mapping, Laplacian and other source derivations, and fast Fourier transform may be desired. Periodic samples of the EEG collected should be available for review by the electroencephalographer. The majority of EEG data collected is of no clinical

help and does not need to be permanently stored. At the end of the recording period, adequate samples of background EEG, interictal spikes or sharp waves, and all seizures should be permanently stored.

Data Storage

Several data storage media are available at the present time. These include DAT (digital analog tapes), magneto optical discs, and compact disc options. All of the storage media have advantages and disadvantages and the user will have to decide which option is best suited to his or her particular use. Decisions should be based on management of the data, cost, ease of retrieval and archiving, and the amount of data to be stored.

Video Recording

One of the main purposes of video monitoring in long-term monitoring units is to record the EEG and behavioral changes that occur during seizures or other clinical events. Cameras and lighting sources should meet or exceed those described in the Minimal Guidelines of the American Electroencephalographic Society. It is important to carefully analyze the clinical events along with the simultaneously recorded EEG. The information obtained during the behavioral events is increased by the interactions of the unit staff with the patients during events. Interview protocols allow systematic evaluation of awareness, speech, memory, and motor functions. The protocols should be age specific and appropriate for the type of activity being recorded.

Video Cassette Recorders

A variety of video cassette recorders (VCRs) are currently being used in long-term monitoring, including formats using ¾-inch tapes, ½-inch VHS, and ½-inch S VHS systems. The VHS systems use tapes that may record 6 to 8 hours of data, as opposed to the ¾-inch tapes, which record for only 45 to 60 minutes. The ¾-inch systems are commercial equipment, are more durable, and give better video resolution than the ½-inch tapes. Both the ¾-inch tapes and the ¾-inch VCRs are significantly more expensive than are the ½-inch tapes and recorders. Half-inch VHS systems may be "stacked," allowing recordings of more than 8 hours. Digital recording of video images offer even higher definition, less degradation, and the fast retrieval necessary for detailed analysis of seizures. The cost of digital video is decreasing to the point that it is frequently being requested as part of new monitoring systems.

ELECTRODES

Scalp electrodes are attached to the patient using the International 10–20 electrode placement system. Generally these are disc electrodes attached to a clean scalp using the collodion technique of application. Needle electrodes, and various cap electrodes do not meet the needs of long-term monitoring. Most electrodes have a hole at the top of the electrode to allow for re-jelling. The electrolyte used for jelling the electrodes should not contain harsh salts, which may cause skin irritation when used for long periods of time. Mild abrasion of the scalp should be performed to reduce and equalize interelectrode impedances. Reduction of electrode artifact starts with proper attachment and electrode care. The electrode wires should be wrapped in gauze from just below the scalp back to the electrode board. This technique prevents electrostatic movement artifact and the risk of electrode wires being inadvertently pulled during the recording. Additional electrodes chosen from the designated 10–10 electrode placement system should be generously used to sample the region(s) of interest depending on the clinical seizure type(s), the magnetic resonance imaging (MRI) findings, and information from earlier EEGs. Sphenoidal electrodes should be considered in all cases in which a mesial temporal lobe epileptogenic focus is a possibility. Sphenoidal electrodes are inserted by a physician after premedicating the patient. Accurate localization of ictal and interictal activity by scalp recordings is critical in deciding if invasive electrodes should be used, and where they should be placed. Application of EEG electrodes for more than 5 days may produce scalp irritation or temporary patchy hair loss at electrode sites.

Depth Electrodes

Most patients undergoing presurgical evaluation do not require invasive electrode recordings. Depth electrodes are most commonly used in patients with temporal lobe epilepsy when the site of ictal onset is either ambiguous or is in disagreement with neuroimaging or other studies. These special electrodes often have six to eight platinum (MRI compatible) contacts, usually placed 2 to 3 millimeters apart. Each contact is approximately 0.5 cm in length. Usually six depth electrodes are stereotactically placed in the left and right amygdala, and the anterior and posterior hippocampus. Scalp electrodes may be used in association with the depth electrodes to ensure that the activity recorded has a similar distribution to previously recorded seizures. Depth electrodes may be used to record from other regions, provided the physician has a good sense of which areas may be the source of the patient's seizures. Random placement of the electrodes in various brain regions is not recommended.

Data from depth electrodes should be recorded referentially. In our laboratory an epidural "peg" electrode is placed near the vertex at the time of depth electrode implantation. The "peg" electrode provides a relatively inactive reference and eliminates most of the EMG artifact. Reformatting is uncomplicated with digital recording equipment, allowing for the interchange of other electrodes with the common reference and also permitting the construction of bipolar montages. The epidural "peg" electrode reference also eliminates the need for high-frequency filtering and enables one to be confident of the nature of the high frequencies frequently seen at seizure onset when recording with depth electrodes.

Subdural Electrode Grid Arrays

Subdural electrode grid arrays are used to delineate a suspected cortical epileptogenic focus or multiple foci, and to evaluate extensive cortical areas, but are most often used to map primary motor, sensory, or eloquent cortex. The electrodes and wires may be stainless steel, but platinum is preferable because it is MRI compatible. Several arrays may be simultaneously inserted to record from widely spaced cortical regions. The advantage of subdural electrodes in the long-term monitoring unit is that a good sample of interictal activity can be recorded along with the patients typical seizures; neither of these objectives can be accomplished with acute electrocorticography in the operating room. Electrical cortical stimulation is performed in the patient's room in a relaxed environment and without the stresses present during stimulation in the operating room. Stimulation sessions may be repeated at different times to confirm the initial results, avoiding mistakes or misinterpretations which may occur with the relatively brief stimulation session performed in the surgical setting. Evoked potential recordings from the subdural electrodes may also be helpful in localizing primary sensorimotor or visual cortex.

ARTIFACT RECOGNITION

An important part of the long-term monitoring technologist's job is the recognition of the wide variety of physiological artifacts invariably recorded during long-term monitoring. This situation is in direct contrast to the role of the "routine" EEG technologist, whose patients are not carrying out normal daily activities during the recording process. Artifacts are common in the monitoring situation owing to the freedom of activity allowed each patient. Common artifacts include vertical and horizontal eye movements, frequently seen as patients watch television, read, or play video games. There may be an increased amount of EMG artifact related to eating, talking, brushing of teeth, and so on. Some activities may be eliminated to reduce EMG artifact (e.g., gum chewing), but it will not be possible to reduce many of the artifacts produced during a normal waking period. Movement artifacts may be reduced by proper electrode application and the "wrapping" of electrode wires. Electrode artifacts or "electrode pops" should be corrected when they interfere with the ability to recognize underlying EEG activity. Certain artifacts, particularly those that mimic rhythmic activity, created by rubbing or scratching the head may be misinterpreted as ictal discharges and must be properly identified. Occasionally artifact recognition can be accomplished by reviewing video tapes during the time of the artifact. Identification of other artifacts may be aided by using additional electrodes.

ACTIVATING PROCEDURES

Hyperventilation and photic stimulation are routinely carried out during long-term monitoring as they may precipitate a seizure. Other activating procedures include sleep deprivation and physical exercise. Sleep deprivation may require a total absence of sleep for more than 24 hours or very limited periods of sleep. Discontinuation of medications is discussed in the next section. In patients in whom nonepileptic seizures are a consideration, various methods of suggestion may be employed to precipitate an event.

MANAGEMENT OF THE PATIENT IN THE MONITORING UNIT

Before the patient is admitted to the epilepsy monitoring unit, the physician or nurse clinician should educate the patient about the unit environment and procedures. If appropriate, the hospital's smoking and visitation policies should be discussed beforehand. Nicotine patches should be provided if the smoking patient is not allowed to smoke in the hospital. If sphenoidal or invasive electrodes are to be used these topics should also be discussed with the patient. The patient should bring books or other activities to pass the time in the unit. The patient needs to understand that his or her medications may be reduced in dosage or discontinued to provoke seizures. The patient needs to know that usually an exact date of discharge cannot be given, as the staff cannot accurately predict when the patient will have a recordable seizure.

As soon as the patient is admitted, planning should begin regarding the type and dosage of anti-epileptic medication to be used at discharge. The goal is to accomplish this as rapidly and as safely as possible. If a patient was receiving carbamazepine or primidone on admission, he or she will have tolerance to those medications and will be able to take large doses with rapidly increasing blood levels at discharge. However, if a new medication is prescribed, it may not be possible to obtain reasonable blood levels in only 1 or 2 days.

In addition to caring for the patient, the physician in charge of the monitoring unit is under pressure from the hospital to accomplish his or her job in as few days as possible in order to reduce costs. In the adult epilepsy monitoring unit at the Cleveland Clinic the average time in the hospital is just under 5 days. The time in the monitoring unit will vary according to the question being asked. For example, if the question is whether or not a patient has epilepsy, the recording of interictal epileptiform activity and one seizure may be sufficient. If the question is whether or not a patient is a surgical candidate, the recording of more than one seizure is optimal, although one may suffice if other studies such as the MRI scan, positron emission tomography (PET) scan, intracarotid sodium amobarbital study, and magnetic resonance spectroscopy provide concordant information. Lastly, if the question is what percentage of seizures are of left temporal compared to right temporal origin, multiple seizures need to be recorded. Gone are the days when patients were monitored for a fixed time period.

Reduction or discontinuation of the patient's anticonvulsant medication is frequently carried out to decrease the length of hospitalization and increase the number of seizures recorded. However, it is best to avoid clusters of seizures and secondarily generalized seizures. If a patient has daily seizures as an outpatient no reduction of medication is necessary. If a patient has only one or two seizures a month, discontinuation of the anticonvulsant should be considered. A history of status epilepticus should make the physician cautious about abrupt discontinuation of medications but it is not a contraindication to it. Abrupt discontinuation of high dose benzodiazepines should not be done, whereas abrupt discontinuation of phenobarbital may be done cautiously because of the drug's long half life. Discontinuation or reduction of anticonvulsants does carry some risk, and this should be discussed with the patient beforehand, as a generalized tonic-clonic seizure can be induced in an individual with no prior history, or only remote history, of such events. Medications are reduced in the majority of monitoring units, and most epileptologists are of the opinion that while the seizure frequency and intensity may increase, new seizure types and new epileptogenic foci do not occur. Opinion on this issue is not, however, uniform. The monitoring unit team should be immediately available to intervene and administer medication if the seizure frequency is too high (e.g., more than two to four complex partial seizures per day) or if severity increases (generalized tonic-clonic seizures). Small oral doses of the patient's routine anticonvulsant, or 1 to 2 mg of intravenous lorazepam may be used to control the patient's seizures.

If a patient is suspected of having temporal lobe epilepsy, sphenoidal electrodes should be inserted on admission. In our unit the patient is premedicated with an oral benzodiazepine, a local anesthetic cream is applied to the skin region where the electrode will be placed, and parenteral analgesia is given with small doses of intravenous fentanyl. This method of sphenoidal insertion significantly increases patient comfort during a painful procedure.

Depending on the circumstances, an ictal single photon emission computed tomography (SPECT) scan to show focal increased cerebral blood flow may be helpful for seizure localization. Obtaining an ictal scan can be logistically difficult as the isotope must be premixed and available in the monitoring unit for instant injection and requires a trained, prepared nurse/technologist team. Ideally the isotope should be injected within 30 seconds of seizure onset. The patient will need to be scanned approximately 2 hours' post-injection.

Long-term monitoring technologists and nurses commonly establish great rapport with the patients and their family and friends. Details of patients' personal lives and anxieties about their disorder may be voiced to the monitoring staff. The staff are frequently asked to explain various procedures and the reason for specific tests. The staff should be well informed about all of the steps involved in the evaluation of patients being considered for epilepsy surgery. Frequently patients ask to "see their seizures." This is permitted at the Cleveland Clinic, but the physicians decide which seizure(s) the patient will see, and a nurse or technologist sits with the patient and family during viewing so that questions can be answered. It is good practice to ask family and friends if the seizures being recorded are similar to those the patient experiences in his or her normal environment.

Complications of EEG/video monitoring include biting of the tongue (with seizures), vertebral compression fractures (following generalized tonic-clonic seizures), other fracture/dislocations (e.g., humerus), status epilepticus, phlebitis, and postictal psychosis (especially following a cluster of seizures). Most complications can be prevented or reduced in severity if the physicians and nurses taking care of the patient respond quickly to an increased frequency or severity of seizures.

A careful explanation of the results of the monitoring should be given to the patient and family. A detailed report summarizing the essential elements of the history, examination, neuroimaging, and the monitoring should be expediently prepared. The report should include EEG examples of the abnormalities.

REFERENCES

American Electroencephalographic Society Guidelines Committees: Guideline twelve: Guidelines for long-term monitoring for epilepsy. J Clin Neurophysiology 11:88–110, 1994.

Binnie CD, Mizrahi EM: The epilepsy monitoring unit. In Engel J, Pedley TA (eds): Epilepsy: A Comprehensive Textbook, vol 1. Philadelphia, Lippincott-Raven, 1997, pp 1011–1019.

Gotman J, Ives J: Computer-assisted data collection and analysis. In Wyllie E (ed): The Treatment of Epilepsy: Principles and Practice, 2nd ed. Baltimore, Williams & Wilkins, 1996, pp 280–291.

Kaplan PW, Lesser RP: Noninvasive EEG. In Wyllie E (ed): The Treatment of Epilepsy: Principles and Practice, 2nd ed. Baltimore, Williams & Wilkins, 1996, pp 976–987.

Klem, GH: EEG technology. In Lüders HO (ed): Epilepsy Surgery. New York, Raven Press, 1992, pp 265–273.

Legatt AD, Ebersole JS: Options for long-term monitoring. In Engel J, Pedley TA (eds): Epilepsy: A Comprehensive Textbook, vol 1. Philadelphia, Lippincott-Raven, 1997, pp 1001–1010.

Mizrahi EM: Electroencephalographic-video monitoring in neonates, infants, and children. J Child Neurol 9(suppl 1):S46–S56, 1994.

Morris HH, Kanner A, Lüders H, et al: Can sharp waves localized at the sphenoidal electrode accurately identify a mesiotemporal epileptogenic focus? Epilepsia 30(5):532–539, 1989.

Morris HH, Klem G, Gilmore-Pollak W: Hair loss after prolonged EEG/video monitoring. Neurology 42:1401–1402, 1992.

Sanders PT, Cysyk BJ, Bare MA: Safety in long-term EEG/video monitoring. J Neurosci Nursing 28:305–313, 1996.

So EL, Fisch BJ: Drug withdrawal and other activating techniques.

In Engel J, Pedley TA (eds): Epilepsy: A Comprehensive Textbook, vol I. Philadelphia, Lippincott-Raven, 1997, pp 1021–1027.

Spurling MR: Clinical challenges in invasive monitoring in epilepsy surgery. Epilepsia 38(suppl 4):S6–S12, 1997.

Todorov A, Lesser R, Uematsu S, et al: Distribution in time of seizures during presurgical EEG monitoring. Neurology 44:1060–1064, 1994.

Wyllie E: Intracranial EEG and localization studies. In Wyllie E (ed): The Treatment of Epilepsy: Principles and Practice, 2nd ed. Baltimore, Williams & Wilkins, 1996, pp 988–999.

Chapter 31

Generalized Disturbances of Brain Function: Encephalopathies, Coma, Degenerative Diseases, and Brain Death

Eric B. Geller, MD

USE OF ELECTROENCEPHALOGRAPHY IN
  ALTERED MENTAL STATUS
ELECTROENCEPHALOGRAPHIC PATTERNS
  IN CEREBRAL DYSFUNCTION
Generalized Slowing
Excessive Beta Activity
Focal Slowing
Rhythmic Patterns in Coma
  (Alpha/Theta, Spindle)
Burst-suppression
Triphasic Waves
Periodic Complexes
Epileptiform Patterns
Normal Patterns
Electrocerebral Inactivity
ELECTROENCEPHALOGRAPHIC PATTERNS
  IN SPECIFIC CONDITIONS
Metabolic Encephalopathies

Hepatic Encephalopathies
Renal Disease
Toxic Encephalopathies
Anoxic Encephalopathy
Idiopathic Recurring Stupor
Nonconvulsive Status Epilepticus
Syncope
EVALUATION OF BRAIN DEATH USING
  ELECTROENCEPHALOGRAPHY
NEURODEGENERATIVE DISEASES
Alzheimer's Disease
Movement Disorders
Prion Diseases
Subacute Sclerosing Panencephalitis
Acquired Immune Deficiency Syndrome
Pseudodementia
CONCLUSIONS

USE OF ELECTROENCEPHALOGRAPHY IN ALTERED MENTAL STATUS

Electroencephalography is a commonly used test in the evaluation of patients with altered mental status, ranging from mild memory difficulties to coma. Because there is a continuum of severity in such patients, the divisions made in this chapter are rather artificial. Although many patients will have nonspecific electroencephalogram (EEG) findings, several recognizable patterns have high diagnostic or prognostic utility. Neuroimaging has to some extent supplanted the use of electroencephalography in the evaluation of brain disorders; however, electroencephalography still remains an excellent test of brain function which can be performed portably on critically ill patients.

Mental status can be affected in several general ways. *Diffuse conditions*, such as toxic or metabolic abnormalities, affect the entire brain, and typically show generalized EEG abnormalities. *Supratentorial lesions* can cause focal dysfunction (e.g., memory and speech problems from a left temporal lobe tumor) or may cause global dysfunction through intracranial pressure effects such as herniation. Such lesions may have focal EEG findings, which are discussed at greater length in Chapter 32. *Infratentorial (brain stem) lesions* affect consciousness via suppression of the reticular activating system, and may show only diffuse EEG changes. Thus, the EEG must be interpreted in the framework of potential abnormalities of supratentorial or infratentorial structures.

The EEG shows a variety of patterns in cerebral dysfunction. Some are quite nonspecific, while others are diagnostic. In this chapter, I first describe the EEG patterns and their clinical correlations; I then discuss the use of electroencephalography in commonly encountered conditions. Although electroencephalography may be useful in assessing sick neonates and young infants, the patterns are quite different from those in adults and older children, and are beyond the scope of this chapter.

An EEG classification scheme should reflect the severity of the EEG abnormality as well as its degree of specificity. At the Cleveland Clinic, we use a four-level EEG classification scheme, to which I refer in this chapter, as follows:

Normal

Abnormal I: Mild, nonspecific abnormalities (e.g., background slowing)

Abnormal II: Abnormalities that are more severe or more specific (e.g., continuous slowing)

Abnormal III: Abnormalities that reflect a severe abnormality, have high localizing value, or are highly correlated with certain conditions (e.g., a temporal lobe sharp wave, alpha coma)

ELECTROENCEPHALOGRAPHIC PATTERNS IN CEREBRAL DYSFUNCTION

Generalized Slowing

Generalized slowing is the most common finding in diffuse brain dysfunction, and is quite nonspecific; that is, numerous

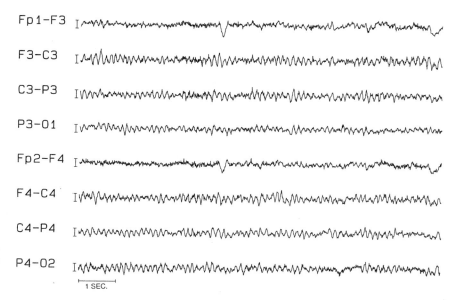

FIGURE 31–1. Background slowing of 7 to 8 Hz in a 66-year-old man with excessive drowsiness on antiepileptic medication. Scales: 20 μV, 1 second.

conditions can cause similar findings. The degree of slowing has a good correlation with the severity of the patient's clinical state. Slowing can be divided into the following categories, in order of increasing severity.

Background Slowing. In background slowing (BS), the frequency of the posterior dominant rhythm is slower than normal (Fig. 31–1). The normal of course is somewhat age-related, with at least 8 Hz being seen in older children and adults. The mean alpha rhythm frequency has been reported as 10.2 ± 0.9 Hz,[47] and 8 Hz is taken as the lower limit of normal in adults in our laboratory, although some consider this as borderline slowing. BS is graded as abnormal I if it is 6 to 8 Hz and abnormal II if less than 6 Hz. Care should be taken to rule out a slow alpha variant (harmonic), which is normal.

Intermittent Slowing. Intermittent slowing (IS) refers to excessive irregular slowing (delta or theta frequencies), appearing intermittently in less than 80% of the record (Fig. 31–2). Generalized IS is classified as abnormal I. IS may attenuate with arousal and increase with drowsiness.

Intermittent Rhythmic Slowing. Also known as intermittent rhythmic delta activity (IRDA), intermittent rhythmic slowing (IRS) describes slow waves that come in bursts with a rhythmic appearance (Fig. 31–3). In adults this is usually seen maximum in the frontal regions (FIRDA), whereas in children it may be expressed more occipitally (OIRDA). IRS is classified as abnormal I. It may sometimes be confused with eye movements. Sharp transients (including muscle and electrocardiogram [ECG] artifacts) intermixed with IRS can be mistaken for 3 per second spike wave complexes. IRS has been associated with numerous etiologies, including toxic, metabolic, focal brain lesions, and increased intracranial pressure.[56] IRS is more likely to occur in an active; fluctuating condition, associated with widespread cerebral dysfunction; it is less likely to occur in a chronic, stable encephalopathy.[57]

Continuous Slowing. In continuous slowing (CS), slowing occupies more than 80% of the record, is unreactive to stimuli, and is excessive for the patient's age (Fig. 31–4). We classify this according to the associated background rhythms: abnormal I with preserved alpha background, abnormal II

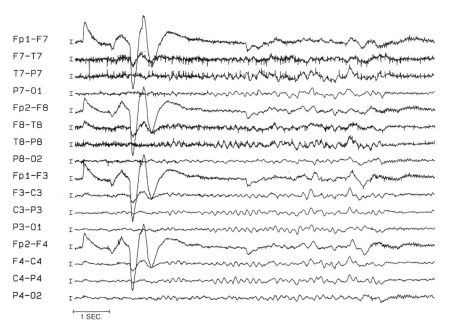

FIGURE 31–2. Generalized intermittent delta slowing and 6-Hz background slowing in a 44-year-old man with cirrhosis. Scales: 20 μV, 1 second.

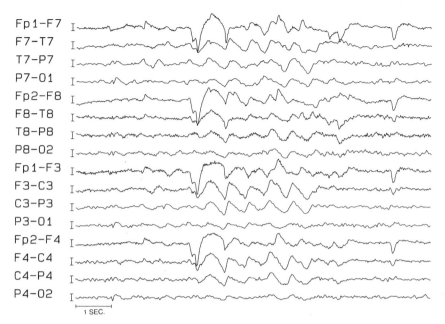

FIGURE 31–3. Intermittent rhythmic slowing (IRS), generalized maximum bifrontally, in addition to intermittent left temporal slowing. This 40-year-old woman had fluctuating mental status after a left middle cerebral artery aneurysmal hemorrhage. There is an eye blink artifact at the onset of the IRS burst; note the difference in distribution between the two. Scales: 50 μV, 1 second.

with BS, and abnormal III with no preservation of a reactive posterior background. Slowing may be caused by drowsiness or lethargy, so activating the patient as much as possible is necessary to rule out state-related changes. Activation with sensory stimulation (eye opening, sound, pain) is important and should always be assessed in patients with severe encephalopathies or coma. Reactivity consists of changes in the dominant EEG frequencies and amplitude, either increasing or decreasing. Even without a clinical response, a reactive EEG suggests a better prognosis than an unreactive EEG. In serial EEGs in a comatose patient, appearance of reactivity may be the first sign of improvement.

Excessive Beta Activity

Diffuse beta activity may be seen in intoxication with sedatives such as benzodiazepines and barbiturates (Fig. 31–5). Excessive beta activity is of higher than normal amplitude, distributed more diffusely than the normal waking frontal beta, and is frequently of slower frequency than the normal 15 to 20 Hz.[3, 40] Care must be taken to assess excessive beta activity in the awake EEG, as beta amplitude often increases in normal stage I sleep.

Focal Slowing

Slowing seen only in one region of the brain is indicative of focal hemispheric dysfunction. In the setting of a comatose patient it is suggestive of a *supratentorial lesion* as the cause of coma. Focal EEG abnormalities are covered in more detail in Chapter 32.

Rhythmic Patterns in Coma (Alpha/Theta, Spindle)

Comatose patients may have EEGs with a predominance of faster rhythms (theta, alpha, and spindle-like frequencies).

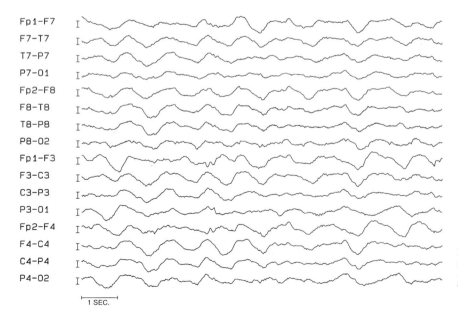

FIGURE 31–4. Continuous generalized delta slowing in a 20-year-old man 2 days after prolonged cardiopulmonary resuscitation. The patient did not recover. Scales: 20 μV, 1 second.

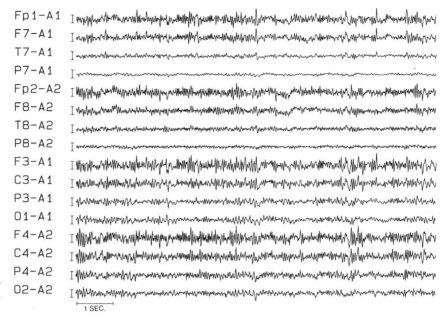

FIGURE 31–5. Excessive beta activity in a 70-year-old man on primidone. Scales: 50 μV, 1 second.

Although such patterns may resemble those seen in the normal waking or sleep state, they are abnormally distributed and unreactive to stimulation.[40]

Alpha or theta frequencies, or both, tend to be diffuse, sometimes with an anterior maximum (Fig. 31–6). Alpha coma has been described in postanoxic encephalopathy and associated with intrinsic brain stem lesions, such as infarct or hemorrhage.[40] Although it has been associated with grave prognosis in this setting, some survivors have been reported.[26, 31, 78] Young and associates[78] found that alpha, theta, and mixtures of alpha and theta can be seen. Using serial recordings, they found that these patterns are transient. The different frequencies did not differ in etiology or prognosis. Alpha and spindle coma patterns may be seen in drug-induced coma, with potentially excellent recovery.[36]

Spindle coma has been described in patients with lesions between the pontomesencephalic junction and the thalamus (Fig. 31–7). Like alpha coma, spindle coma has a variable prognosis depending on the etiology.[9, 30]

Burst-suppression

Burst-suppression consists of bursts of high amplitude intermixed slow and sharp waves, alternating with periods of depressed background or inactivity[63] (Fig. 31–8). Burst-suppression may be seen after severe anoxic brain injury or in the presence of general anesthesia or hypothermia. Steriade and colleagues[63] produced experimental evidence that burst-suppression represents complete disconnection of the thalamocortical circuits, which generate the EEG. The clinical significance of burst-suppression is discussed later in this chapter.

Triphasic Waves

Periodically repeating waves were first described by Foley and associates[22] in association with hepatic encephalopathy. Termed "triphasic waves" by Bickford and Butt,[4] this pattern has a prominent surface positivity which is preceded and

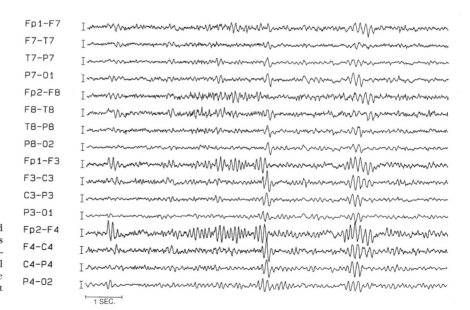

FIGURE 31–6. Alpha coma in a 34-year-old woman after cardiac arrest and 20 minutes of resuscitation. The patient was unresponsive, and CT scan showed diffuse cerebral edema. Note the anterior predominance of the 10-Hz rhythmic activity. She did not survive. Scales: 10 μV, 1 second.

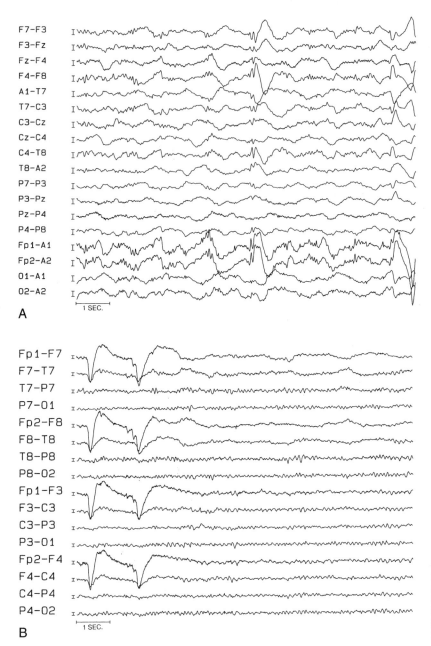

FIGURE 31–7. *A.* Spindle coma in carbamazepine overdose. Note the relatively normal appearance of sleep patterns, including a K complex. Spindles are seen best in the frontopolar derivations. Scales: 20 μV, 1 second. *B.* Normal awake EEG in the same patient, 2 days later. Scales: 20 μV, 1 second.

followed by smaller amplitude negative waves (Fig. 31–9). Triphasic Waves (TW) usually have an anterior predominance, but may be posterior or diffuse.[64] They occur in bursts or runs which are bilaterally synchronous, repeating at about 2 to 3 Hz. Associated background activity is usually abnormal, with delta and theta slowing.[64] Although TW are generalized, the wave sometimes appears to have a time lag going from anterior to posterior or vice versa.[64]

The specificity of TW for hepatic encephalopathy has been debated. TW have been reported in a wide variety of conditions other than hepatic encephalopathy, including renal disease, anoxia, hyperosmolarity, levodopa-induced encephalopathy, temporal arteritis, and postictal stupor after generalized seizures.[33, 46, 49, 69] The periodic sharp wave complexes seen in Creutzfeldt-Jakob disease often have triphasic morphology.[37] Sundaram and Blume[64] studied 63 patients with TW. They found that patients with TW from metabolic causes (anoxia, hepatic, renal) were more likely to be in coma, whereas patients with nonmetabolic causes (predominantly dementia) were awake but confused.

In conclusion, the appearance of TW in a patient with undiagnosed coma should suggest metabolic encephalopathy, especially from hepatic disease, but it is not pathognomonic. Triphasic-appearing waves in a dementing patient may suggest Creutzfeldt-Jakob disease.

Periodic Complexes

Periodic lateralized epileptiform discharges (PLEDs) were described in the 1950s, but were first well characterized by Chatrian and colleagues.[12] Morphologically, PLEDs may consist of spikes, multiple spikes, slow waves, or a complex mixture of spikes and slow waves[12] (Fig. 31–10). Voltage is medium to high. Repetition rate is often about 1 per second. Between the PLEDs, the EEG pattern is characterized by slowing, often of low amplitude. Usually abnormal rhythms are seen in the contralateral hemisphere as well. In a series of 33 patients, PLEDs had a high correlation (29/33) with the presence of seizures during the illness.[12] Seizures were

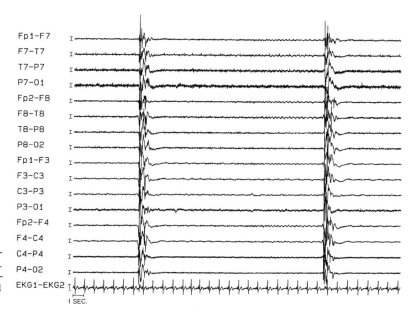

FIGURE 31–8. Burst-suppression with associated myoclonus, occurring 24 hours after cardiac arrest. There is muscle artifact from the myoclonus superimposed on the epileptiform bursts. Scales: 20 μV, 1 second (30-second page).

very often focal, and often resistant to medication.[12] PLEDs usually evolve, appearing some time after the onset of illness, then decreasing in amplitude and repetition rate and eventually disappearing. They usually outlast the clinical seizures.

In adults, PLEDs are usually associated with a comatose state in an acute illness. Etiologies vary but have in common an *acute destructive lesion of the hemisphere* ipsilateral to the PLEDs, such as stroke, hemorrhage, or viral encephalitis.[12, 16] Walsh and Brenner[74] found a mortality rate of 41% in 39 patients with PLEDs, and 30% of the survivors had poor functional outcome. Outcome was particularly poor (10/13) in those with stroke. PLEDs have been reported in children as young as 2 months, and may be associated with chronic conditions and severe diffuse brain dysfunction, but with less impaired consciousness than in adults.[50]

PLEDs may be seen bilaterally and independently (BIPLEDs) (Fig. 31–11). De la Paz and Brenner[16] compared patients with BIPLEDs to those with PLEDs. BIPLEDs were more commonly associated with anoxic encephalopathy and central nervous system (CNS) infection, whereas PLEDs were

seen more frequently in acute stroke. Chronic epilepsy was equally common in both groups (22%). Focal abnormalities on neurological examination, focal abnormalities on neuroimaging, and focal seizures were more common with PLEDs. Coma and increased mortality were seen more often in the BIPLEDs group, suggesting the presence of BIPLEDs is a poor prognostic sign.

Although PLEDs are usually considered an interictal phenomenon, Handforth and colleagues[27] have argued that PLEDs may represent a form of focal status epilepticus. They reported a patient who had left temporal lobe PLEDs after focal status epilepticus, associated with intense hypermetabolism in this region on fluorodeoxy glucose–positron emission tomography (FDG-PET). When the PLEDs resolved, the left temporal lobe became hypometabolic. They did note that the PLEDs did not improve with aggressive antiepileptic treatment.

Periodic patterns may also be bilaterally synchronous or generalized (Fig. 31–12). Generalized periodic sharp wave complexes (PSWC) are the classic finding in Creutzfeldt-

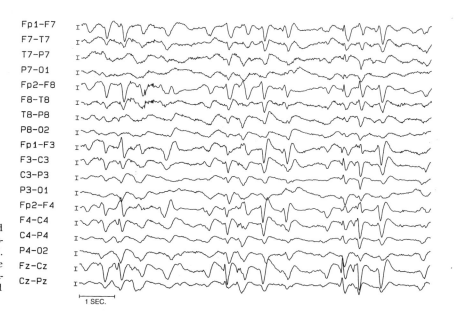

FIGURE 31–9. Triphasic waves in a 63-year-old woman with cirrhosis, awaiting liver transplantation, who recently became obtunded. Note the apparent "time lag" of the positive peak with posterior derivations delayed relative to anterior derivations. Scales: 20 μV, 1 second.

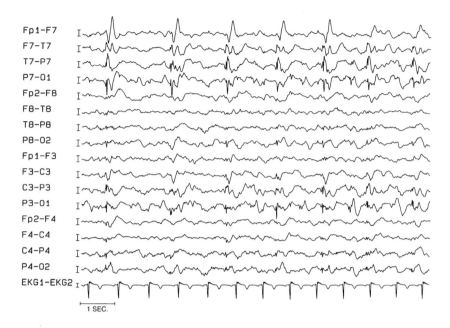

FIGURE 31–10. Left hemisphere periodic lateralized epileptiform discharges (PLEDs). Note the complex morphology of the discharges, and the diffuse slowing. This 71-year-old diabetic woman was found unconscious, and had three generalized tonic-clonic seizures on hospital admission, without recovery of consciousness. Other portions of the record showed status epilepticus with repeated EEG and clinical seizures. No etiology was found, and the patient was discharged to a rehabilitation facility after a prolonged hospital stay. Scales: 20 µV, 1 second; 60-Hz notch filter used.

Jakob disease and subacute sclerosing panencephalitis. They may be mono-, di-, or triphasic, with the sharp component lasting 90 to 200 msec with an amplitude of 50 to 200 µV.[37] PSWC differ in repetition rate between the two conditions, with Creutzfeldt-Jakob disease having about one per second (ranging 8 to 20 per 10 seconds, according to Levy and associates[37]), and subacute sclerosing panencephalitis having about one per 4 to 10 seconds.[40] Myoclonic jerks are more likely to occur synchronously with PSWC in subacute sclerosing panencephalitis than in Creutzfeldt-Jakob disease.[37, 40] Generalized periodic patterns may also be seen in coma owing to severe brain injury such as anoxia, and suggest a poor prognosis.[35] Distinguishing PSWC from triphasic waves may sometimes be difficult, as noted earlier.[64]

Epileptiform Patterns

The presence of epileptiform activity may help to diagnose the cause of altered mental status as a postical state or as

status epilepticus (see later discussion). In patients not known to have seizure disorders, the EEG may suggest the need for antiepileptic drug therapy.

Normal Patterns

A normal EEG may be seen in patients with psychogenic unresponsiveness (Fig. 31–13). The EEG will also be normal in patients with "locked-in syndrome," in which there is a lesion in the pontine tegmentum causing quadriparesis. Such lesions lie below the level of the reticular activating system, allowing normal hemispheric function and sleep-wake cycling.

Electrocerebral Inactivity

Electrocerebral inactivity (ECI; electrocerebral silence, iso-electric EEG, "flatline") is defined as the absence of any

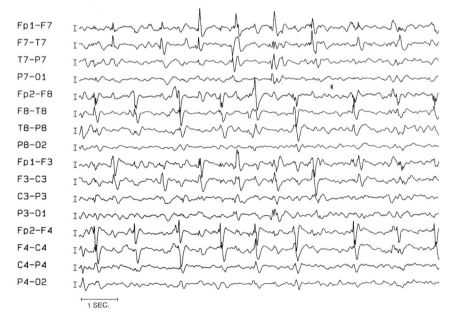

FIGURE 31–11. Bilateral independent periodic lateralized epileptiform discharges (BIPLEDs). This 69-year-old woman suffered cerebral anoxia as a result of severe respiratory distress. Scales: 50 µV, 1 second.

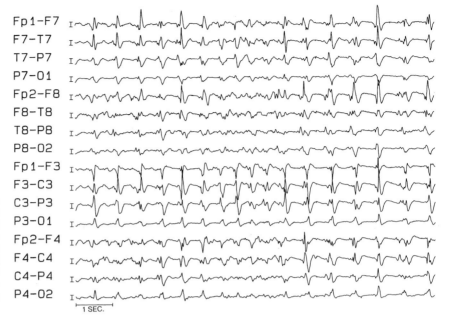

FIGURE 31–12. Periodic pattern, bilaterally synchronous but higher in amplitude on the left. This 67-year-old woman underwent emergent surgery for an ascending aortic aneurysm and was comatose postoperatively when this EEG was performed. CT showed multiple embolic infarcts as well as global ischemic injury. She was discharged to a hospice in a persistent vegetative state. Scales: 50 μV, 1 second.

brain waves greater than 2 μV in amplitude (Fig. 31–14). Considered the hallmark of brain death, ECI may also be seen in severe intoxications, general anesthesia, and hypothermia. Because of the diagnostic significance of ECI, proper EEG recording technique is critical. ECI and the technical requirements for recording it are discussed later in this chapter.

ELECTROENCEPHALOGRAPHIC PATTERNS IN SPECIFIC CONDITIONS

Metabolic Encephalopathies

Any metabolic abnormality that can affect mentation may affect the EEG. Abnormalities of sodium or glucose, for example, will cause nonspecific slowing.[72] Hypoglycemia may

enhance the rhythmic slowing seen during hyperventilation, a common phenomenon that should prompt the technologist to give some juice or crackers to the patient and repeat hyperventilation. Hypothermia may lead to slowing, and in severe cases may cause burst-suppression or even ECI, all of which is potentially reversible.[72]

Hepatic Encephalopathies

Portal systemic encephalopathy is associated with chronic liver disease (usually cirrhosis), often with portal hypertension and shunting, and is often *reversible* with appropriate treatment.[38] Onset may be insidious with attentional disturbances, depressed consciousness, or delirium. The pathophysiology remains unclear, and a variety of factors have been implicated, including elevated ammonia levels, toxin hypersensitivity, and endogenous benzodiazepines (endozepines).[38] *Ful-*

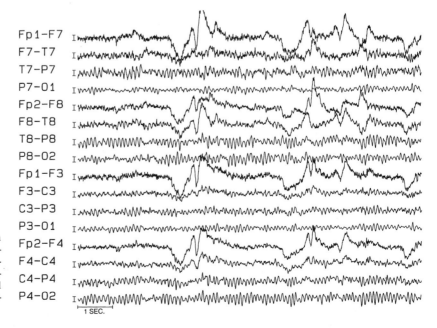

FIGURE 31–13. Normal EEG in a patient with juvenile myoclonic epilepsy who remained clinically unresponsive after a generalized tonic-clonic seizure. Note normal alpha rhythm posteriorly, eye blinks, and frontal muscle artifact, all indicating an awake state. Scales: 20 μV, 1 second.

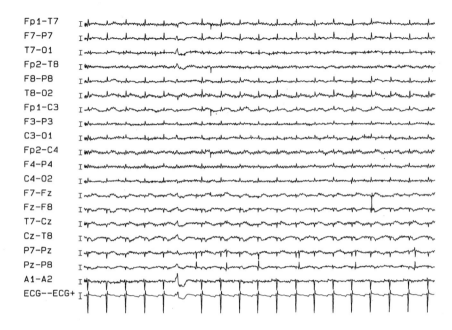

Fp1–T7
F7–P7
T7–O1
Fp2–T8
F8–P8
T8–O2
Fp1–C3
F3–P3
C3–O1
Fp2–C4
F4–P4
C4–O2
F7–Fz
Fz–F8
T7–Cz
Cz–T8
P7–Pz
Pz–P8
A1–A2
ECG––ECG+

FIGURE 31–14. Electrocerebral inactivity in a brain-dead patient. The montage used has > 10 cm interelectrode distances. Scale is 2 μV/ mm, 10-second page. A 60-Hz notch filter causes electrical artifact to appear as low amplitude beta activity. Note the widespread ECG artifact, along with a rhythmic pulse artifact more prominent in the right hemisphere channels. What appears to be a phase-reversing left temporal sharp wave is actually ECG artifact of a premature ventricular contraction, illustrating the need for a separate ECG channel.

minant hepatic failure is an irreversible condition with rapidly progressive encephalopathy and coma, usually fatal without liver transplantation.

In mild hepatic encephalopathy, the EEG shows background slowing and progressive increase in generalized slowing (see Fig. 31–2). Triphasic waves are typically seen in the moderate stage when patients are stuporous (see Fig. 31–9). In frank coma, there is high amplitude delta slowing with decrease or absence of triphasic waves.[40] As discussed previously, triphasic waves have been traditionally attributed to hepatic encephalopathy but may be seen in a variety of other conditions. Fourteen- and 6-Hz positive spikes (ctenoids) are usually seen as a benign variant in healthy children, but may also be seen in metabolic encephalopathies, especially Reye's syndrome. Drury[17] described this pattern in 7 of 111 children with encephalopathy and suggested it may represent the persistence of a normal pattern rather than an abnormality.

Electroencephalography may be useful in assessing the severity of hepatic encephalopathy, and may help document effectiveness of therapy. For example, Bansky and associates[2] found that in patients who improved transiently after flumazenil injection, the mean background frequency increased significantly from 4.2 to 5.2 Hz. The presence of triphasic waves in hepatic disease has been associated with unfavorable prognosis.[40]

Renal Disease

Uremic encephalopathy is characterized by inattention and disordered sleep, restless legs, tremors, and fasciculations.[42] Asterixis is classic. Progressive disease may lead to psychotic behavior and may cause stroke-like events. Seizures, usually generalized tonic-clonic, may also occur. EEG slowing parallels the deterioration of the mental state, and abnormalities such as triphasic waves, spike waves, photomyoclonic and photoparoxysmal responses, and increased slowing with hyperventilation may be seen.[40]

Dialysis dysequilibrium syndrome occurs as a result of brain swelling from a lag in osmolar shifts between brain and blood during dialysis.[42] Neurological findings include restlessness, headache, nausea, confusion, and seizures. The EEG may show high amplitude IRDA with or without background slowing.

Dialysis dementia is a chronic disorder that causes speech impairment, seizures (generalized tonic-clonic, focal, or multifocal), and motor abnormalities (myoclonus, apraxia, or immobility).[29, 40, 42] Hughes and Schreeder[29] studied 26 patients with dialysis dementia compared to 51 patients on dialysis without dementia. They found that the presence of bisynchronous spike wave complexes was highly sensitive and specific for clinical dementia, and in some cases predicted the appearance of dementia by months. The authors noted difficulty sometimes in distinguishing the spike wave complexes from IRDA or triphasic waves, features that were not diagnostic. Clinical and EEG findings of dialysis dementia appear to correlate with serum aluminum concentrations.[55]

Toxic Encephalopathies

Drug intoxication is a common cause of coma in the emergency setting. The EEG parallels the degree of mental alteration, depending on the severity of the intoxication.[3, 40] Background slowing progresses into more generalized theta, then delta frequencies. Sedative intoxications may show excessive beta activity (see Fig. 31–5). In the most severe cases (including general anesthesia), burst-suppression and ECI may be produced. Alpha, theta, and spindle patterns may be seen (see Fig. 31–7). Determination of coma etiology cannot be made by EEG as the same patterns may be seen in drug intoxication and anoxia, and from other causes. Thus, in patients with severe EEG changes every effort must be made to find the cause of the coma, as drug intoxication has a relatively good prognosis no matter how severely abnormal the EEG appears (see Fig. 31–7).

Drugs that may lower the seizure threshold and cause EEG epileptiform activity include lithium, tricyclic antidepressants, neuroleptics, clozapine, aminophylline, isoniazid, and high or intrathecal doses of penicillin.[3]

Anoxic Encephalopathy

A variety of EEG patterns can be seen after anoxia. As previously described, there may be generalized slowing, monorhythmic patterns such as alpha or theta coma, or both, periodic patterns (triphasic waves, PLEDs or BIPLEDs), burst-suppression, and ECI. Numerous studies have attempted to use EEG patterns to predict clinical outcome, in the hope of better allocating resources to those patients with the best chance of survival. Although poor prognosis has been associated with all these patterns, survivors do occur.[26, 78] Generalized periodic patterns represent a severe disruption of normal EEG generation and indicate involvement of cortical and subcortical structures;[63] as such, they have an especially poor prognosis.[35] Burst-suppression seen alone or in association with other patterns such as alpha coma is highly predictive of death (34 of 35 patients in the study of Young and colleagues[78]).

The timing of EEG studies may be important in the significance of the findings. EEGs performed immediately after resuscitation may still show ECI.[40] Alpha coma detected and resolved in the first 24 hours after coma may have a better prognosis than that detected later.[31]

Generalized myoclonic status epilepticus may be seen in postanoxic coma, associated with burst-suppression (Fig. 31–15). In one study, all of 40 patients with these findings died.[75]

Idiopathic Recurring Stupor

Tinuper and associates[67] described a patient who had spontaneous episodes of stupor lasting 8 to 10 hours, without clear toxic or metabolic cause and without neurological deficits. EEGs during the episodes showed monomorphic diffuse 14-Hz activity. The episodes were rapidly reversed with flumazenil, a benzodiazepine antagonist. In three patients with this condition, termed "idiopathic recurring stupor," an endogenous substance with affinity for the benzodiazepine receptor on the gamma aminobutyric acid (GABA) channel ("endozepine") was found.[54, 68] Endozepines were found in serum and cerebrospinal fluid only in the patients and not in controls. In the patients, endozepines were markedly elevated during episodes compared to interictal periods, suggesting that this was the cause of the condition. The same investigators reported similar findings of endozepine-related stupor responsive to flumazenil in two children.[61] Although idiopathic recurring stupor remains a rarely reported condition, the identification of a biochemical correlate of a clinical and EEG abnormality is intriguing.

Nonconvulsive Status Epilepticus

Nonconvulsive status epilepticus (NCSE) is a condition in which continued seizures lasting more than 30 or 60 minutes are responsible for altered mental status.[32] Although considered rare, NCSE is likely underdiagnosed, particularly in patients without a history of epilepsy. Both partial and primary generalized seizure types can cause NCSE; electroencephalography is helpful in distinguishing these types.[32, 70] The clinical features can vary widely, from inattention and confusion to coma, and can include altered affect, subtle myoclonus of face or limbs, eye deviations or nystagmus, and automatisms.[21, 32, 39] Simple partial status epilepticus may cause persistent aura symptoms or focal motor seizures (epilepsia partialis continua); this discussion focuses on NCSE that presents as acute confusion or coma.

Prolonged complex partial seizures may present as NCSE, and are more commonly seen in patients with a prior history of epilepsy.[32] The EEG may have varied patterns, from isolated spikes to continuous lateralized discharges or ictal patterns[32, 70] (Fig. 31–16).

Generalized absence status epilepticus may occur in patients with a history of generalized absence epilepsy, but has also been described as occurring de novo in adults[32, 66, 70] (Fig. 31–17). It is controversial whether absence status in adults represents the late manifestation of a primary generalized (genetic) epilepsy.[32] De novo absence status seems to have a female predominance, middle-age or elderly onset, and is often associated with a history of alcohol or drug abuse.[66] It may be triggered by drug withdrawal, especially from benzodiazepines, but may also be associated with other drug use or metabolic abnormalities. There is a relatively low risk of recurrence (6 of 41 reported cases). Thomas and colleagues suggested that de novo absence status be classified as a situation-related seizure syndrome.[66]

Prolonged confusion after a generalized or partial seizure may represent NCSE, but be mistaken for postictal confusion.[21] Such cases raise the difficult question of how to treat a patient in coma of unwitnessed onset, who has periodic generalized or lateralized epileptiform activity on EEG. Treiman[70] noted that periodic discharges can be seen in the late stages of experimental and clinical status epilepticus, hypermetabolism can be associated, and they may resolve

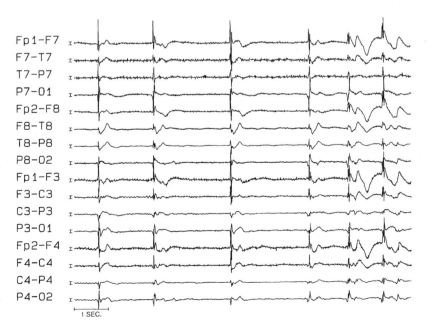

FIGURE 31–15. Myoclonic status epilepticus associated with burst-suppression, in a comatose patient after asystole. Scales: 20 μV, 1 second.

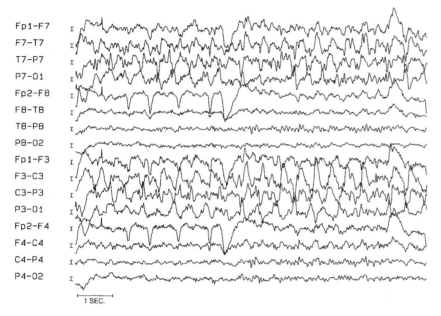

FIGURE 31–16. Focal nonconvulsive status epilepticus in a 60-year-old man who had acute worsening of aphasia a day after a left hemisphere stroke. CT scan showed no acute changes, so EEG was performed. During the EEG, the patient was alert and responsive to commands, but had continuous oral and hand automatisms and was unable to speak. He improved within minutes of lorazepam injection. Note the left hemisphere ictal pattern of high amplitude periodic sharp waves with intermixed fast activities, and preservation of the right-sided alpha rhythm. Scales: 20 μV, 1 second.

with antiepileptic treatment. Because of these findings, Treiman[70] argued that periodic discharges may represent a form of NCSE, and may require treatment (Fig. 31–18). The aggressiveness of such treatment is controversial. Lowenstein and Aminoff[39] examined 47 comatose patients in whom NCSE was suspected. They identified three groups: (1) clinical and EEG seizures present (n = 33); (2) clinical activity (e.g., clonic twitching), without EEG seizures (n = 9); and (3) no clinical seizures but with EEG seizures (n = 5). There was no difference among the groups in etiology, response to treatment, or outcome. They did not find a correlation between the duration of NCSE and the EEG patterns, suggesting that the progression of EEG changes in status epilepticus is not predictable.

In conclusion, NCSE may present with subtle clinical findings and requires a high degree of suspicion. Electroencephalography is critical and necessary in making the diagnosis. It may also suggest whether NCSE is of a focal or generalized nature, allowing appropriate treatment choices.

Syncope

The EEG changes seen in syncope reflect the effects of cerebral hypoperfusion, regardless of the mechanism (vasovagal, orthostatic, cardiac arrhythmia, and so on). There is a progression of generalized slowing, from initial background slowing to theta and then delta frequencies.[7, 14] With persistent hypoperfusion, severe generalized amplitude attenuation may be seen (Fig. 31–19). When perfusion returns, the EEG recovers in the reverse order of its disappearance. These observations highlight the importance of clinical assessment of the patient during an EEG if episodes of generalized slowing occur, as well as the need for simultaneous recording of the ECG.

EVALUATION OF BRAIN DEATH USING ELECTROENCEPHALOGRAPHY

The concept of brain death has important legal, ethical, and philosophical implications. The need to determine brain

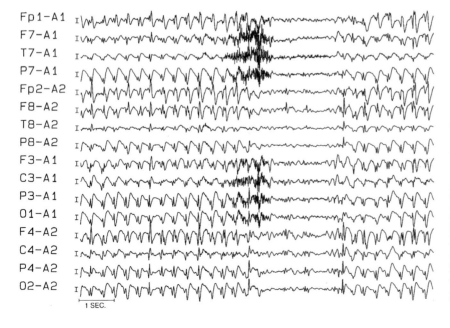

FIGURE 31–17. Generalized absence status epilepticus in a 12-year-old boy. Several months prior he had been evaluated for episodic eye blinking and head twitching, which were felt to be benign tics. Several days before this EEG, he began having prolonged periods of confusion and disorientation. The first occurred during a written examination, which he did not fill out at all. He was found wandering in the street after school. The next day he woke up normally and had no memory of the prior day's events. On the day of evaluation, he again had confusion which fluctuated over hours. Mini-mental state examination score was 17/30. Valproate was begun, and the patient recovered and has remained seizure-free. Note the persistent generalized 3-Hz spike wave complexes. During the brief break in the pattern the EEG is nearly normal. Scales: 100 μV, 1 second.

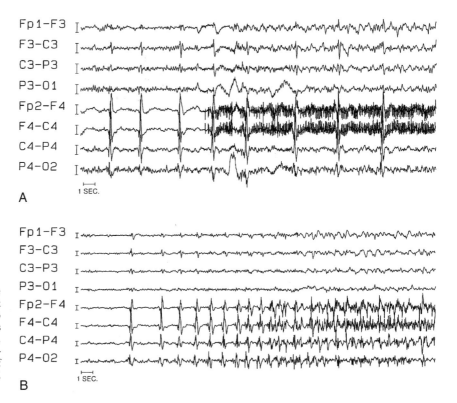

FIGURE 31–18. PLEDs and nonconvulsive status epilepticus in a patient with a right hemisphere glioblastoma multiforme who stopped his antiepileptic medication. Status epilepticus resolved with phenytoin loading. *A.* Right hemisphere PLEDs, with some volume conduction to the left. *B.* Evolution of PLEDs into an ictal pattern. Scales: 50 μV, 1 second.

death arose from the ability to prolong life with advanced technology, the desire to minimize the use of limited resources in patients who would not benefit, and the needs of organ transplantation programs to harvest organs as soon as possible. The determination of brain death has been defined as: "An individual who has sustained either (1) irreversible cessation of circulatory and respiratory functions or (2) irreversible cessation of all functions of the entire brain, includ-

ing the brainstem, is dead. A determination of death must be made in accordance with accepted medical standards."[65]

Electroencephalography has commonly been used as supporting evidence of brain death. The necessary demonstration of electrocerebral inactivity (ECI) can be technically challenging, as the intensive care unit is often a hostile electrical environment and there may be numerous artifacts. The American Electroencephalographic Society has formu-

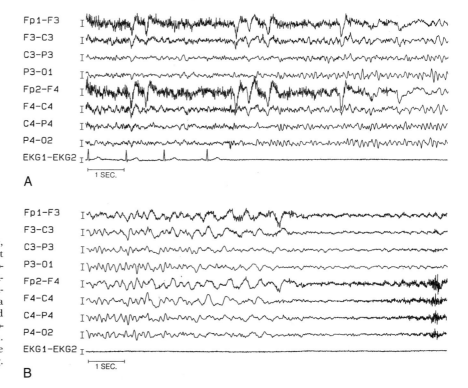

FIGURE 31–19. EEG changes in syncope, in a 44-year-old woman who underwent video-EEG monitoring for evaluation of episodes of loss of consciousness and shaking. *A.* Note asystolic arrest on ECG channel, rapidly followed by increasing theta slowing. *B.* In the subsequent 10-second epoch, higher amplitude delta slowing appears, followed by generalized attenuation. At the end of the page, there is muscle artifact, reflecting myoclonic jerking. Scales: 50 μV, 1 second.

lated 10 guidelines for the minimal technical standards to record EEGs in suspected cerebral death.[1] Full discussion of the rationale for these recommendations can be found in the Guidelines.

1. *Use a minimum of eight scalp electrodes.* The full 10–20 placement should be used if possible. This is to avoid mistaking focal attenuation for brain death.

2. *Ensure interelectrode impedances of between 100 and 10,000 ohms.* Excessively low impedances reduce EEG potentials, and excessively high impedances and impedance mismatch may increase artifact.

3. *Test the integrity of the entire recording system.* Testing can be performed by tapping each electrode in turn, and identifying the electrode artifact. This is to verify that the machine is actually recording scalp activity and that the montage is correct.

4. *Ensure interelectrode distances of at least 10 cm.* Longer distances may bring out low amplitude potentials which might cancel out with shorter distances. Such montages should be used for at least a portion of the recording. The montage used at Cleveland Clinic is illustrated in Figure 31–14.

5. *Ensure sensitivity of at least 2 μV/mm for at least 30 minutes, with appropriate calibrations.* This is *the most important parameter,* as the recording must rule out brain activity above 2 μV. The recording must be long enough to exclude intermittent low voltage activity.

6. *Use appropriate filter settings.* Low frequency filters should not be set above 1 Hz, to avoid eliminating very slow potentials. High frequency filters should not be set below 30 Hz. A 60-Hz notch filter may be used, particularly as there may be a great deal of artifact in ECI recordings.

7. *Use additional monitoring techniques when necessary.* Common artifacts can be monitored with ECG, electromyogram (EMG), and respiratory channels. The technologist should make every effort to eliminate or identify sources of artifact (intravenous drips, other machines, electric beds, and so on).

8. *Evaluate EEG reactivity to intense somatosensory, auditory, or visual stimuli.* There should be none.

9. *Have only a qualified technologist make recordings.* ECI recordings are technically difficult to perform, and require skilled, experienced personnel.

10. *Perform a repeat EEG if there is doubt about ECI.* There may be doubt about the technical quality of the recording, or there may be the possibility of confounding factors such as hypothermia or sedative drug levels.

The use of digitally-recorded EEG may affect the need for the technologist to observe all the preceding recommendations. Specifically, sensitivity adjustment, digital filtering, and montage changes can be performed by the EEG reader as postprocessing. Checking of electrode integrity and placement, patient activation, and recording time need to be carried out no matter how the EEG is recorded.

The finding of ECI must be placed in the clinical context. *ECI is not synonymous with brain death,* as it may occur in sedative drug intoxication or hypothermia, which are potentially reversible. Several studies have documented the presence of residual EEG activity in patients with clinical brain death and absence of cerebral perfusion on radiological studies.[20, 25, 43] Coma may be mimicked by severe deefferentation and deafferentation, such as in Guillain-Barré syndrome; the EEG may be normal in such patients, challenging the clinical impression of brain death.[18]

The concept of brain death as the cessation of function in the entire brain must be distinguished from "brain-stem death," in which all functions of the brain stem are absent, but the cortex may still function. The EEG may show residual activity, including the appearance of sleep-like activity.[25, 34]

Brain death criteria have been proposed for children, with modifications from the adult protocols.[65] EEG modifications include decreasing interelectrode distance in proportion to head size in small children. Determination of brain death in neonates remains controversial. The EEG may show discontinuous records in normal neonates, and delayed development may mimic severe dysfunction.[73]

Because of its shortcomings, electroencephalography may not be the best test to demonstrate brain death. Other tests such as radionuclide perfusion studies, magnetic resonance angiography, and transcranial Doppler have been proposed.[20, 43] Compared to such techniques, however, electroencephalography remains a low-risk, easily performed procedure in critically ill patients who may not tolerate travel outside the intensive care unit.

NEURODEGENERATIVE DISEASES

The causes of subacute and chronic cognitive changes are too numerous to cover exhaustively in a brief review. The remainder of this chapter discusses the more common causes of dementia and their EEG findings, as well as rarer conditions in which the EEG is diagnostically helpful.

Any consideration of EEG changes in dementia must begin with those changes to be expected in normal aging.[71] Alpha rhythm frequency may decline progressively over the age of 60, but should not be less than 8 Hz. Background amplitude may also decrease. Rhythmic focal intermittent theta slowing in the temporal regions, usually left, is more common in middle age and in the elderly, and may be nonspecific. Small amounts of intermittent theta and delta slowing have been seen in normal elderly, and are of uncertain significance. Care must be taken to distinguish state-related changes from encephalopathy.

Alzheimer's Disease

Alzheimer's disease (AD) is the most common dementing illness in the elderly. There have been a large number of studies of electroencephalography in AD. In general, AD patients have increased slow activity, including background slowing, and generalized continuous or intermittent slowing, although mildly impaired patients or those studied early on may have normal EEGs.[6] Rare findings include epileptiform discharges, triphasic waves, or generalized periodic patterns.[6]

Several longitudinal studies have followed EEG changes over time in AD patients. These studies, using both visual and quantitative analysis in 13 to 43 patients over 1 to 3 years, documented progressive decrease in alpha power and increases in theta and delta power.[13, 28, 59, 60] However up to 50% had no change over 1 year.[59] Nondemented controls also had some increases in delta power but did not have changes in alpha or theta.[13] It may be difficult to compare some of these results precisely, as not all studies used the same derivations for quantitative analysis. Rodriguez and associates[53] claimed that quantitative analysis could predict clinical course; right-sided delta predicted loss of activities of daily living and death, whereas right-sided theta predicted onset of incontinence, and left-sided changes were not predictive. Such unexpected findings probably reflect the fact that statistical significance will appear if enough variables are analyzed.

Some studies have attempted to define the utility of electroencephalography as a screening test in making the diagnosis of AD. Robinson and associates[52] compared 86 AD patients to 17 with mixed AD and multi-infarct dementia. Comparing normal and abnormal EEGs in AD versus controls, an abnormal EEG had a sensitivity of 0.872, specificity of 0.633, positive predictive value of 0.806 and negative predictive value of 0.738. For more abnormal EEG grades, specificity improved to 0.959 and the negative predictive value to 0.825. These authors suggested that if the EEG is normal in a clinically demented patient, the diagnosis of AD should be questioned, and if there is more severe abnormality on EEG, AD is more likely. Ettlin and associates[19] compared electroencephalography to computed tomography and clinical features in dementia diagnosis. They found computed tomography and electroencephalography had similarly high sensitivity in AD (12 of 15 patients), whereas clinical features were much better at detecting multi-infarct dementia, and all tests were poor at distinguishing those with mixed AD and multi-infarct dementia.

Forstl and colleagues[23] compared cognitive testing, quantitative electroencephalography, and magnetic resonance imaging (MRI) volumetry in predicting changes over time in 55 Alzheimer's patients versus 66 healthy elderly controls. They found over a 2-year period that cognitive performance predicted not only future performance but also brain atrophy on imaging, whereas electroencephalography did not. They suggested that electroencephalography may be more important for initial diagnosis, but is not useful for prediction of progression.

In conclusion, the EEG findings in AD are generally nonspecific but may help support the diagnosis. The quest for quantitative EEG variables to improve specificity has not yielded dramatically better results than visual analysis,[6] probably reflecting the relatively nonspecific nature of the EEG findings. The role of electroencephalography in diagnosing AD will likely decline as biochemical marker testing becomes more widely available and accurate.

Movement Disorders

Degenerative diseases characterized by movement disorders may be associated with dementia. These diseases are often classified as "subcortical dementias." The EEG in such conditions is usually nonspecific. Parkinson's disease may have background slowing and an increase in generalized slow frequencies, which is worse in demented than cognitively normal patients.[6] In Huntington's disease, EEG amplitude may be very low voltage (<10 μV).[6] Brenner[6] points out the importance of considering rhythmic head movement artifact when assessing the background rhythms of patients with movement disorders.

Prion Diseases

Prions are proteinaceous infectious agents that have been implicated as the causative agent of certain rapidly progressive dementia syndromes. These conditions include Creutzfeldt-Jakob disease (CJD), Gerstmann-Straussler-Scheinker (GSS), kuru, fatal familial insomnia, and recently bovine spongiform encephalopathy (BSE). Prion diseases share the pathological picture of neuronal loss, astrocytic proliferation and gliosis, and vacuolization leading to a "spongiform" appearance of the cortex. They may occur in sporadic, inherited, or transmissible forms. Clinical features include a nonspecific prodome followed by rapidly progressive cognitive decline, cerebellar ataxia, visual disturbances, movement disorders, long tract signs, and myoclonus, which is often stimulus-sensitive.

CJD is the most common prion disease, with an annual incidence of 0.5 to 2.0 cases per million. Because neuroimaging and cerebrospinal fluid examination are often nondiagnostic, the EEG remains an important diagnostic test (Fig. 31–20).

Bortone and colleagues[5] followed serial EEGs in 15 CJD patients, of whom 12 were pathologically confirmed. All patients showed PSWCs, usually generalized but sometimes asymmetrical. In 14 patients, PSWC appeared in the "full stage" of the disease (dementia present among other symptoms), and in one, PSWC did not appear until the terminal stage (coma or vegetative state). In the earliest records, nonspecific delta slowing was seen. PSWC tended to be brought out by auditory or intermittent photic stimulation.

Steinhoff and associates[62] studied the accuracy of electroencephalography in CJD comparing 15 pathologically confirmed cases of CJD to 14 patients with rapid dementias thought not to be CJD. Using strict criteria for periodic sharp wave complexes and blinded EEG readers, they found a sensitivity of 67%, a specificity of 86%, and a high interrater reliability. However, some patients were evaluated on EEG samples rather than the original record, possibly leading to a lower sensitivity value.

Levy and associates[37] reviewed 36 cases of CJD (21 pathologically proven) and extensively reviewed the literature. PSWC were seen in 28 cases, first appearing anywhere from 3 to 32 weeks after illness onset. In four cases, PSWC were not present in the initial EEG but developed later. PSWC patterns ranged from infrequent runs lasting 5 to 10 seconds to nearly continuous PSWC for long periods. PSWC amplitude ranged from 50 to 200 μV, with waxing and waning. The morphology could be mono-, bi-, or triphasic, with the duration of the sharp component varying from 90 to 200 msec. PSWC frequency was 14 per 10 seconds, ranging from 8 to 20 per 10 seconds, and was never perfectly regular. PSWC had an inconsistent relation to myoclonic jerks. In intervals without PSWC, alpha background activity was present with generalized theta or delta slowing. Background activity diminished and slowing increased as the disease progressed. In the later, comatose stage, PSWC diminished.

Brown and colleagues[10] described a familial form of CJD associated with a mutation at codon 178 of the prion protein. These patients were different from those with the usual mutation at codon 200 in that they had an earlier age of onset with insidious progressive memory loss and a longer duration of illness. Only one of 44 patients had PSWC on electroencephalography, whereas myoclonus was common, suggesting a dissociation between the two phenomena.

Recently a new variant of CJD in 10 patients has been described in the United Kingdom.[76] A unique neuropathological feature was the finding of abundant PrP plaques surrounded by spongiform change, something not seen in sporadic CJD. The clinical histories were atypical for CJD with a young age of onset and relatively long duration of illness (up to 2 years). The presenting symptoms were predominantly behavior changes and patients were often seen first by psychiatrists, although all eventually developed dementia and most had myoclonus.[76, 79] None of the cases had the typical periodic complexes of CJD, although EEGs were abnormal in most of the patients.[79] It is possible that this CJD variant may represent a human form of bovine spongiform encephalopathy, which has been epidemic in cattle in the United Kingdom.[11, 58]

In summary, CJD is associated with characteristic PSWC in most patients at some point in the illness, along with progressive loss of background activity and increased generalized slowing. Absence of PSWC in clinically advanced patients should cause reconsideration of the diagnosis, although long duration variants or other prion diseases may not show PSWC.

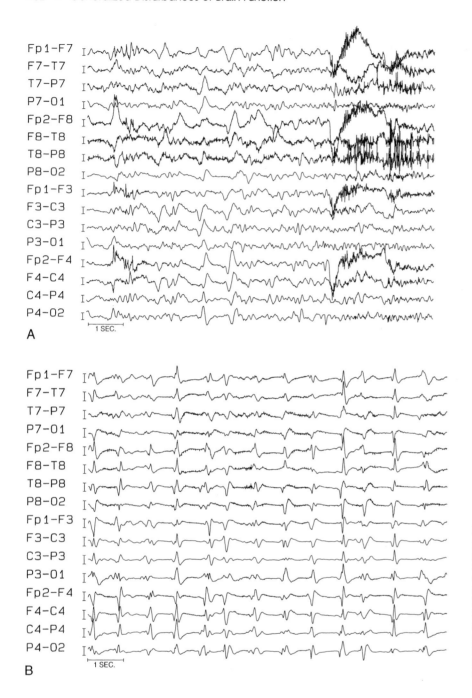

FIGURE 31–20. This 75-year-old woman presented with a 3-month history of progressive dementia, gait ataxia, diplopia, and recent onset of startle-sensitive myoclonus. The patient rapidly worsened and died within a week of admission. *A.* Initial EEG on admission, showing generalized irregular continuous slowing, with occasional sharply contoured waves. Scales: 50 μV, 1 second. *B.* Three days later, generalized periodic sharp wave complexes are present, diagnostic of Creutzfeldt-Jakob disease. Scales: 100 μV, 1 second.

Subacute Sclerosing Panencephalitis

Approximately one per million people infected with measles virus develops subacute sclerosing panencephalitis (SSPE). Incidence has dropped with the use of measles vaccination. SSPE begins an average of 7 years after the primary disease,[15] with a subacutely progressive dementia. Myoclonus, seizures, hyperkinesia, and ataxia often develop, with progression to coma or vegetative state and death. Chorioretinitis is common.

The EEG in SSPE characteristically shows stereotyped periodic complexes that are bilaterally synchronous and usually symmetrical (Fig. 31–21). They may be bi-, tri-, or polyphasic, with amplitudes of 200 to 500 μV. PSWC in SSPE differ from those in CJD in that they occur with a lower frequency, repeating every 4 to 10 seconds,[40] and myoclonic jerks correlate with the PSWC.[41] Background activity becomes progres-

sively slower and of lower amplitude. Other EEG features in SSPE include electrodecremental periods, FIRDA, and bisynchronous or rarely focal epileptiform activity.[41] In a serial study of 42 EEGs in five patients, Wulff[77] found that the interval between PSWCs shortened with progression of the disease. The author suggested that other studies had not found a similar progression because recordings were not done early enough to demonstrate the phenomenon. Wulff[77] also noted that the only EEGs without PSWC in his patients were at the onset of disease (one) or in the terminal stage (six).

Acquired Immune Deficiency Syndrome

Human immunodeficiency virus (HIV) infection has been associated with AIDS dementia complex (ADC), a syndrome

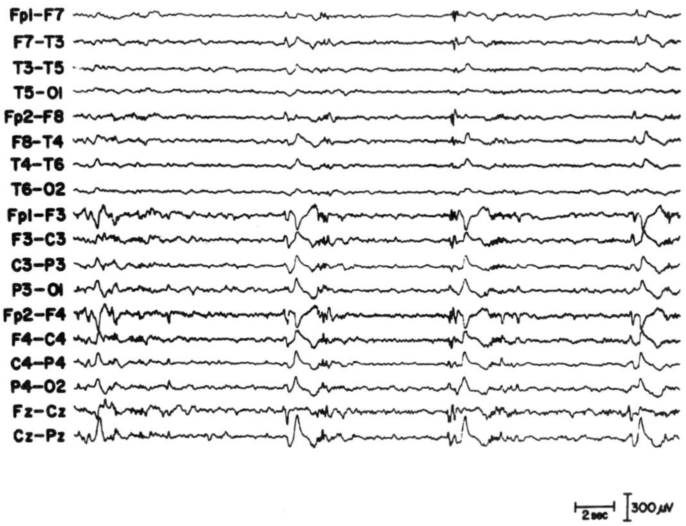

FIGURE 31–21. EEG in a patient with subacute sclerosing panencephalitis (SSPE). Note high amplitude, periodic complexes comprised of sharp and slow activity, which is generalized but maximum in the frontocentral regions and bilaterally synchronous. These complexes are superimposed on a diffusely slow background with absence of normal rhythms. Scales: 300 μV, 2 second.

of progressive cognitive decline with subcortical features. ADC may be the presenting illness in an otherwise healthy HIV-positive patient,[45] and may eventually affect 15% to 20% of all patients with AIDS.[51] Clinically patients present with memory loss, psychomotor slowing, apathy, long tract signs, and ataxia.[51] Neuroimaging reveals cortical atrophy and ventricular dilatation, and cerebrospinal fluid protein may be elevated.[44] The EEG may be abnormal in patients with ADC, showing mild generalized and background slowing.[24, 44, 45] Because of these nonspecific EEG changes, EEG abnormalities have not become a diagnostic criteria for ADC.[51]

Focal EEG abnormalities in HIV-positive patients should prompt investigation for focal CNS processes such as toxoplasmosis or CNS lymphoma.[24]

Nuwer and colleagues[48] found that there are no EEG changes associated with HIV serostatus in asymptomatic homosexual men, whereas patients with cognitive impairment have diffuse slowing. In addition, patients who were left-handed or had a history of learning disability were overrepresented in the abnormal EEG group. These results suggest that an abnormal EEG should not be attributed to asymptomatic HIV infection, and other causes (including long-standing abnormalities) should be sought.

Pseudodementia

Depression may present with prominent cognitive dysfunction, and demented patients frequently have concurrent depression, making accurate diagnosis difficult. Electroencephalography may be helpful in distinguishing depression from dementia. Brenner and colleagues[8] reported on a cohort of 33 patients with mixed symptoms of depression and dementia, followed longitudinally to determine diagnosis. Patients with depression only or depression with pseudodementia had normal EEGs or only mild slowing. The majority of the patients with dementia had abnormal EEGs, with about one-third showing moderate or severe abnormalities (intermittent or generalized continuous slowing).

CONCLUSIONS

Although many EEG patterns seen in encephalopathies and degenerative diseases are nonspecific, the EEG can be very useful in a number of situations, as follows:
1. To document seizure activity in a comatose patient

2. To suggest the cause of an undiagnosed coma as being due to toxic, metabolic, or focal abnormalities
3. To suggest prognosis after anoxic brain injury
4. To evaluate brain death
5. To diagnose psychogenic coma or the locked-in syndrome as the cause of unresponsiveness
6. To diagnose CJD and SSPE
7. To diagnose organic disease in a patient with altered mental status

REFERENCES

1. American Electroencephalographic Society: Guideline three: Minimum technical standards for EEG recording in suspected cerebral death. J Clin Neurophysiol 11:10–13, 1994.
2. Bansky G, Meier PJ, RiedererE, et al: Effects of benzodiazepine receptor antagonist flumazenil in hepatic encephalopathy in humans. Gastroenterology 97:744–750, 1989.
3. Bauer G: EEG, drug effects, and central nervous system poisoning. In Niedermeyer E, Lopes da Silva F (eds): Electroencephalography: Basic Principles, Clinical Applications, and Related Fields, 3rd ed. Baltimore, Williams & Wilkins, 1993, pp 631–642.
4. Bickford RG, Butt HR: Hepatic coma: The electroencephalographic pattern. J Clin Invest 34:790–799, 1955.
5. Bortone E, Bettone L, Giorgi C, et al: Reliability of EEG in the diagnosis of Creutzfeldt-Jakob disease. Electroencephalogr Clin Neurophysiol 90:323–330, 1994.
6. Brenner RP: EEG and dementia. In Niedermeyer E, Lopes da Silva F (eds.): Electroencephalography: Basic Principles, Clinical Applications, and Related Fields, 3rd ed. Baltimore, Williams & Wilkins, 1993, pp 339–349.
7. Brenner RP: Electroencephalography in syncope. J Clin Neurophysiol 14:1987–209, 1997.
8. Brenner RP, Reynolds CF, Ulrich RF: EEG findings in depressive pseudodementia and dementia with secondary depression. Electroencephalogr Clin Neurophysiol 72:298–304, 1989.
9. Britt CW: Nontraumatic "spindle coma": Clinical, EEG, and prognostic features. Neurology 31:393–397, 1981.
10. Brown P, Goldfarb LG, Kovanen J, et al: Phenotypic characteristics of familial Creutzfeldt-Jakob disease associated with the codon 178Asn PRNP mutation. Ann Neurol 31:282–285, 1992.
11. Bruce ME, Will RG, Ironside JW, et al: Transmissions to mice indicate that 'new variant' CJD is caused by the BSE agent. Nature 389:498–501, 1997.
12. Chatrian GE, Shaw CM, Leffman H: The significance of periodic lateralized epileptiform discharges in EEG: An electrographic, clinical and pathological study. Electroencephalogr Clin Neurophysiol 17:177–193, 1964.
13. Coben LA, Danziger W, Storandt M: A longitudinal EEG study of mild senile dementia of Alzheimer type: Changes at 1 year and 2.5 years. Electroencephalogr Clin Neurophysiol 61:101–112, 1985.

14. Daly DD: Epilepsy and syncope. In Daly DD, Pedley TA (eds): Current Practice of Clinical Electroencephalography, 2nd ed. New York: Raven Press, 1990, pp 274–334.

15. Davis LE: Nervous system complications of systemic viral infections. In Aminoff MJ (ed): Neurology and General Medicine, 2nd ed. New York, Churchill Livingstone, 1995, pp 731–756.

16. De la Paz D, Brenner RP: Bilateral independent periodic lateralized epileptiform discharges. Clinical significance. Arch Neurol 38:713–715, 1981.

17. Drury I: 14- and 6-Hz positive bursts in diverse encephalopathies of childhood. Electroencephalogr Clin Neurophysiol 72:479–485, 1989.

18. Drury I, Westmoreland BF, Sharbrough FW: Fulminant demyelinating polyradiculoneuropathy resembling brain death. Electroencephalogr Clin Neurophysiol 67:42–43, 1987.

19. Ettlin TM, Staehelin HB, Kischka U, et al: Computed tomography, electroencephalography, and clinical features in the differential diagnosis of senile dementia. Arch Neurol 46:1217–1220, 1989.

20. Fackler JC, Rogers MC: Is brain death really cessation of all intracranial function? J Pediatrics 110:84–86, 1987.

21. Fagan KJ, Lee SI: Prolonged confusion following convulsions due to generalized nonconvulsive status epilepticus. Neurology 40:1689–1694, 1990.

22. Foley JM, Watson CW, Adams R: Significance of the electroencephalographic changes in hepatic coma. Trans Am Neurol Assoc 75:161–163, 1950.

23. Forstl H, Sattel H, Besthorn C, et al: Longitudinal cognitive, electroencephalographic and morphological brain changes in ageing and Alzheimer's disease. Br J Psychiatry 168:280–286, 1996.

24. Gabuzda DH, Levy SR, Chiappa KH: Electroencephalography in AIDS and AIDS-related complex. Clin Electroencephalogr 19:1–6, 1988.

25. Grigg MM, Kelly MA, Celesia GG, et al: Electroencephalographic activity after brain death. Arch Neurol 44:948–954, 1987.

26. Grindal AB, Suter C, Martinez AJ: Alpha-pattern coma: 24 cases with 9 survivors. Ann Neurol 1:371–377, 1977.

27. Handforth A, Cheng JT, Mandelker MA, et al: Markedly increased mesiotemporal lobe metabolism in a case with PLEDs: Further evidence that PLEDs are a manifestation of partial status epilepticus. Epilepsia 35:876–881, 1994.

28. Hooijer C, Jonker C, Posthuma J, et al: Reliability, validity and follow-up of the EEG in senile dementia: Sequelae of sequential measurement. Electroencephalogr Clin Neurophysiol 76:400–412, 1990.

29. Hughes JR, Schreeder MT: EEG in dialysis encephalopathy. Neurology 30:1148–1154, 1980.

30. Hulihan JF Jr, Syna DR: Electroencephalographic sleep patterns in postanoxic stupor and coma. Neurology 44:758–760, 1994.

31. Iragui VJ, McCutchen CB: Physiologic and prognostic significance of "alpha coma." J Neurol Neurosurg Psychiatr 46:632–638, 1983.

32. Kaplan PW: Nonconvulsive status epilepticus. Semin Neurology 16:33–40, 1996.

33. Karnaze DS, Bickford RG: Triphasic waves: a reassessment of their significance. Electroencephalogr Clin Neurophysiol 57:193–198, 1984.

34. Kaukinen S, Makela K, Hakkinen VK, et al: Significance of electrical brain activity in brain-stem death. Int Care Med 21:76–78, 1995.

35. Kuroiwa Y, Celesia GG: Clinical significance of periodic EEG patterns. Arch Neurol 37:15–20, 1980.

36. Kuroiwa Y, Furukawa T, Inaki K: Recovery from drug-induced alpha coma. Neurology 31:1359–1361, 1981.

37. Levy SR, Chiappa KH, Burke CJ, et al: Early evolution and incidence of electroencephalographic abnormalities in Creutzfeldt-Jakob disease. J Clin Neurophysiol 3:1–21, 1986.

38. Lockwood AH: Hepatic encephalopathy and other neurological disorders associated with gastrointestinal disease. In Aminoff MJ (ed): Neurology and General Medicine, 2nd ed. New York, Churchill Livingstone, 1995, pp 247–266.

39. Lowenstein DH, Aminoff MJ: Clinical and EEG features of status epilepticus in comatose patients. Neurology 42:100–104, 1992.

40. Markand ON: Electroencephalography in diffuse encephalopathies. J Clin Neurophysiol 1:357–407, 1984.

41. Markand ON, Panszi JG: The electroencephalogram in subacute sclerosing panencephalitis. Arch Neurol 32:719–726, 1975.

42. Moe SM, Sprague SM: Uremic encephalopathy. Clin Nephrol 42:251–256, 1994.

43. Nau R, Prange HW, Klingelhofer J, et al: Results of four technical investigations in fifty clincally brain dead patients. Int Care Med 18:82–88, 1992.

44. Navia BA, Jordan BD, Price RW: The AIDS dementia complex: I. Clinical features. Ann Neurol 19:517–524, 1986.

45. Navia BA, Price RW: The acquired immunodeficiency syndrome dementia complex as the presenting or sole manifestation of human immunodeficiency virus infection. Arch Neurol 44:65–69, 1987.

46. Neufeld MY: Periodic triphasic waves in levodopa-induced encephalopathy. Neurology 42:444–446, 1992.

47. Niedermeyer E: The normal EEG of the waking adult. In Niedermeyer E, Lopes da Silva F (eds): Electroencephalography: Basic Principles, Clinical Applications, and Related Fields, 3rd ed. Baltimore, Williams & Wilkins, 1993, pp 131–152.

48. Nuwer MR, Miller EN, Visscher BR, et al: Asymptomatic HIV infection does not cause EEG abnormalities: results from the Multicenter AIDS Cohort Study (MACS). Neurology 42:1214–1219, 1992.

49. Ogunyemi A: Triphasic waves during post-ictal stupor. Can J Neurol Sci 23:208–212, 1996.

50. PeBenito R, Cracco JB: Periodic lateralized epileptiform discharges in infants and children. Ann Neurol 6:47–50, 1979.

51. Power C, Johnson RT: HIV-1 associated dementia: clinical features and pathogenesis. Can J Neurol Sci 22:92–100, 1995.

52. Robinson DJ, Merskey H, Blume WT, et al: Electroencephalography as an aid in the exclusion of Alzheimer's disease. Arch Neurol 51:280–284, 1994.

53. Rodriguez G, Nobili F, Arrigo A, et al: Prognostic significance of quantitative electroencephalography in Alzheimer patients. Preliminary observations. Electroencephalogr Clin Neurophysiol 99:123–128, 1996.

54. Rothstein JD, Guidotti A, Tinuper P, et al: Endogenous benzodiazepine receptor ligands in idiopathic recurring stupor. Lancet 340:1002–1004, 1992.

55. Rovelli E, Luciani L, Pagani C, et al: Correlation between serum aluminum concentration and signs of encephalopathy in a large population of patients dialyzed with aluminum-free fluids. Clin Nephrol 29:294–298, 1988.

56. Schaul N, Lueders H, Sachdev K: Generalized, bilaterally synchronous bursts of slow waves in the EEG. Arch Neurol 38:690–692, 1981.

57. Sharbrough FW: Nonspecific abnormal EEG patterns. In Niedermeyer E, Lopes da Silva F (eds): Electroencephalography: Basic Principles, Clinical Applications, and Related Fields, 3rd ed. Baltimore, Williams & Wilkins, 1993, pp 197–215.

58. Smith PG, Cousens SN: Is the new variant of Creutzfeldt-Jakob disease from mad cows? Science 273:748, 1996.

59. Soininen H, Partanen J, Laulumaa V, et al: Longitudinal EEG spectral analysis in early stage of Alzheimer's disease. Electroencephalogr Clin Neurophysiol 72:290–297, 1989.

60. Soininen H, Partanen J, Laulumaa V, et al: Serial EEG in Alzheimer's disease: 3 year follow-up and clinical outcome. Electroencephalogr Clin Neurophysiol 79:342–348, 1991.

61. Soriani S, Carrozzi M, De Carlo L, et al: Endozepine stupor in children. Cephalalgia 17:658–661, 1997.

62. Steinhoff BJ, Racker S, Herrendorf G, et al: Accuracy and reliability of periodic sharp wave complexes in Creutzfeldt-Jakob disease. Arch Neurol 53:162–166, 1996.

63. Steriade M, Amzica F, Contreras D: Cortical and thalamic cellular correlates of electroencephalographic burst-suppression. Electroencephalogr Clin Neurophysiol 90:1–16, 1994.

64. Sundaram MBM, Blume WT: Triphasic waves: clinical correlates and morphology. Can J Neurol Sci 14:136–140, 1987.

65. Task Force on Brain Death in Children: Guidelines for the determination of brain death in children. Pediatrics 80:298–300, 1987.

66. Thomas P, Beaumanoir A, Genton P, et al: 'De novo' absence status of late onset: Report of 11 cases. Neurology 42:104–110, 1992.

67. Tinuper P, Montagna P, Cortelli P, et al: Idiopathic recurring stupor: A case with possible involvement of the gamma-aminobutyric acid (GABA) ergic system. Ann Neurol 31:503–506, 1992.

68. Tinuper P, Montangna P, Plazzi G, et al: Idiopathic recurring stupor. Neurology 44:621–625, 1994.
69. Tomer Y, Neufeld MY, Shoenfeld Y: Coma with triphasic wave pattern in EEG as a complication of temporal arteritis. Neurology 42:439–440, 1992.
70. Treiman DM: Electroclinical features of status epilepticus. J Clin Neurophysiol 12:343–362, 1995.
71. Van Sweden B, Wauquier A, Niedermeyer E: Normal aging and transient cognitive disorders in the elderly. In Niedermeyer E, Lopes da Silva F (eds): Electroencephalography: Basic Principles, Clinical Applications, and Related Fields, 3rd ed. Baltimore, Williams & Wilkins, 1993, pp 329–338.
72. Vas GA, Cracco JB: Diffuse encephalopathies. In Daly DD, Pedley TA (eds): Current Practice of Clinical Electroencephalography, 2nd ed. New York, Raven Press, 1990, pp 372–399.
73. Volpe JJ: Brain death determination in the newborn. Pediatrics 80:293–297, 1987.
74. Walsh JM, Brenner RP: Periodic lateralized epileptiform discharges—long-term outcome in adults. Epilepsia 28:533–536, 1987.
75. Wijdicks EFM, Parisi JE, Sharbrough FW: Prognostic value of myoclonus status in comatose survivors of cardiac arrest. Ann Neurol 35:239–243, 1994.
76. Will RG, Ironside JW, Zeidler M, et al: A new variant of Creutzfeldt-Jakob disease in the UK. Lancet 347:921–925, 1996.
77. Wulff CH: Subacute sclerosing panencephalitis: Serial electroencephalographic studies. J Neurol Neurosurg Psychiatr 45:418–421, 1982.
78. Young GB, Blume WT, Campbell VM, et al: Alpha, theta and alpha-theta coma: A clinical outcome study utilizing serial recordings. Electroencephalogr Clin Neurophysiol 91:93–99, 1994.
79. Zeidler M, Stewart GE, Barraclough CR, et al: New variant Creutzfeldt-Jakob disease: Neurological features and diagnostic tests. Lancet 350:903–907, 1997.

Chapter 32

Focal Nonepileptic Disturbances of Brain Function

Selim R. Benbadis, MD

INTRODUCTION
SLOW ACTIVITY
AMPLITUDE ASYMMETRY
PERIODIC PATTERNS AND PERIODIC
    LATERALIZED EPILEPTIFORM DISCHARGES

OTHER LESS COMMON FOCAL PATTERNS
FOCAL DYSFUNCTION WITHOUT
    STRUCTURAL LESION
CONCLUSIONS

INTRODUCTION

Before the advent of neuroimaging techniques such as computed tomography (CT) and magnetic resonance imaging (MRI), when pneumoencephalography and angiography were the neuroimaging techniques of choice, electroencephalography was the best noninvasive tool to search for focal lesions. In the past few decades, with the recent and ongoing progress in imaging (structural and functional), the role of electroencephalography outside of epilepsy has markedly declined. Nevertheless, electroencephalograms (EEGs) are routinely performed in various clinical situations, so that the neurophysiologist must be familiar with the EEG findings even in conditions where it is of relatively limited value. This chapter describes focal nonepileptic abnormalities, and also highlights the limitations of electroencephalography in the setting of nonepileptiform focal abnormalities.

Like most neurophysiological tests, electroencephalography is a test of cerebral *function*, and as such is for the most part nonspecific as to etiology. Although earlier investigators have attempted to identify the reliability of electroencephalography in differentiating types of lesions,[14, 29] this has clearly become a senseless and futile exercise in the modern era. The exercise of describing EEG abnormalities by pathology (stroke, abscess, tumor, and so on), which was common in old electroencephalography texts,[34] is not followed here. Instead, the different patterns of focal (nonepileptic) disturbances of brain function and their clinical significance are reviewed.

SLOW ACTIVITY

Abnormal slow activity is by far the most common EEG manifestation of focal brain dysfunction. The abnormality that correlates best with the presence of a structural lesion is polymorphic or arrhythmic (as opposed to monomorphic or rhythmic) delta (0 to 3 Hz) slowing.[22] This is all the more reliable when it is continuous, unreactive (lack of change with state such as wake or sleep, or with external stimuli),[19] of high amplitude, polymorphic, and unilateral (Figs. 32–1 and 32–2). The likelihood of a structural lesion (i.e., specificity) diminishes when the slow activity lacks these characteristics and is, for example, intermittent, in the theta rather than the delta range, and of low amplitude (Figs. 32–3 and 32–4), as this type of slowing may be normal (e.g., temporal slowing of the elderly, see Chap. 29).

Continuous focal slow activity is the only nonepileptiform abnormality that can unequivocally be interpreted as abnormal as an isolated finding. Other focal abnormalities, which will be discussed later in the chapter, are quite frequent and of such low specificity that they almost never constitute an abnormality *in themselves*. To be interpreted as abnormal, these usually require the coexistence of a more definite abnormality, such as slowing or epileptiform discharges.

Focal slowing is completely nonspecific as to etiology, and in the era of imaging the EEG has no role in diagnosing the *nature* of a lesion. Focal slowing is the most common abnormality associated with focal lesions *of any type*, including (but not limited to) neoplastic lesions,[33, 36] vascular lesions,[36] subdural collections,[51] traumatic lesions,[36, 38] and infectious processes.[37, 50] It may even occasionally be seen in more subtle structural abnormalities such as mesiotemporal sclerosis[24] or focal malformations of cortical development.[44]

The physiological basis for focal polymorphic delta activity caused by focal cortical lesions is not fully understood. It is probably the result of abnormalities in the underlying white matter rather than in the cortex itself.[27] When present, focal slow activity correlates highly with the side of the lesion, but it is not reliable for lobar localization.[46]

A recent addition to the interpretation of slow EEG activity has been computerized frequency analysis. Correlating CT or MRI changes with EEG changes has been found by some to increase the sensitivity of diagnosis,[7] although this has been refuted by others.[40, 45] In the days of imaging this seems to be a relatively minor advance, and it is not likely that computerized frequency analysis will play a major role in the evaluation of focal nonepileptic dysfunction. It may be more useful in the evaluation of focal seizures.

In addition to the appearance of slow activity, focal dysfunction can also manifest by slowing of the frequencies of normal rhythms on one side. Though this could be called frequency asymmetry, it is best to reserve the term asymmetry for asymmetries in amplitude (see later discussion). In general, a difference in frequency of 1 Hz or more is considered significant. In the vast majority of cases, this regional slowing of normal rhythms is associated with polymorphic delta (see Fig. 32–1). This slowing is dependent on location, so that, for example, a posterior lesion may be expected to cause slowing of the alpha rhythm, whereas a central lesion would cause slowing of the mu rhythm.[15] The localization of slow potentials follows the same rules as that of epileptiform discharges. Thus, for example, "phase reversals" are useful to localize slow potentials and do not imply epileptogenicity.

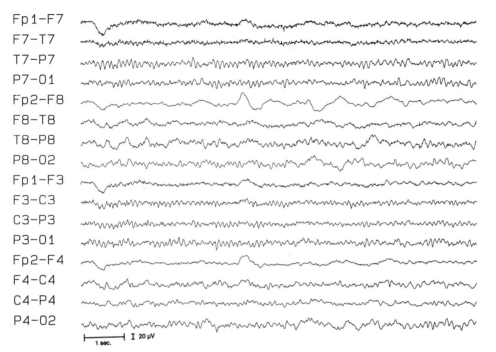

FIGURE 32–1. A 65-year-old woman with a large infarct in the right middle cerebral artery distribution. There is prominent polymorphic delta over the right hemisphere. This was continuous throughout the record, and nonreactive. This type of slowing is almost invariably associated with a structural lesion. Although this patient had a stroke, similar findings could be seen with any other lesion (e.g., tumor, abscess). Note that the amplitude of the alpha rhythm is reduced on the abnormal side, but this would be insufficient in itself to be considered abnormal.

AMPLITUDE ASYMMETRY

In the following discussion, the term *asymmetry* is used to refer to an asymmetry of *amplitude*, and to normal rhythms. By contrast, a focal frequency asymmetry would be classified as focal slow.

Destructive lesions clearly attenuate normal rhythms, a phenomenon most easily seen during prolonged subdural recordings or electrocorticography. However, normal rhythms are never perfectly symmetrical in amplitude, so when to consider asymmetries significant is not always clear. A rule of thumb is that, with very few exceptions, significant focal asymmetries are associated with slowing (see Figs. 32–1 and 32–2). Asymmetry or suppression of normal rhythms is somewhat more likely to be seen in structural abnormalities that increase the distance or interfere with the conduction of the electrical signal between the cortex and the recording scalp electrodes. Examples include subdural collections (hematoma, empyema, and so on), epidural collections (hematoma, abscess, and so on), subgaleal collections, and calcifications such as those seen in Sturge-Weber syndrome. Occasionally, asymmetry is a more sensitive indicator of focal dysfunction than focal slowing. This may occur in Sturge-Weber syndrome, where EEG abnormalities may precede CT changes.[5] Asymmetry may also be more common than slowing in subdural hematomas.[41] However, caution must be

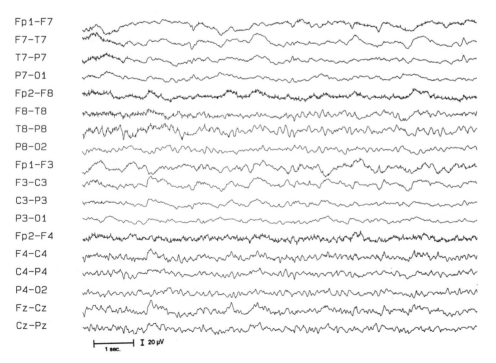

FIGURE 32–2. Another example of polymorphic delta activity, or continuous slowing. This 70-year-old patient presented with new-onset seizures, and had a large left hemisphere tumor shown pathologically to be a glioblastoma multiforme. In addition to polymorphic delta activity, there is also suppression of alpha activity on the left. Again, the same findings could be seen in other etiologies, and the EEG cannot reliably predict the nature of the lesion (compare with Fig. 32–1). There is also some diffuse slowing and background slowing, indicative of a diffuse encephalopathy.

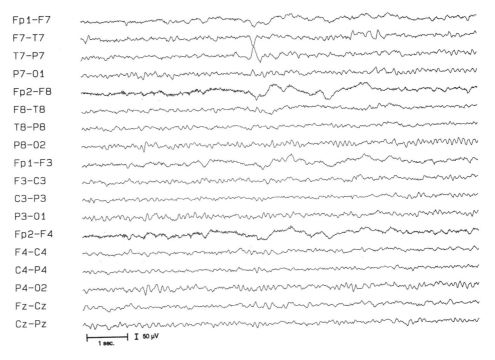

```
Fp1-F7
F7-T7
T7-P7
P7-O1
Fp2-F8
F8-T8
T8-P8
P8-O2
Fp1-F3
F3-C3
C3-P3
P3-O1
Fp2-F4
F4-C4
C4-P4
P4-O2
Fz-Cz
Cz-Pz
```

├─────┤ ⊥ 50 μV
1 sec.

FIGURE 32–3. A 55-year-old woman with headaches. This 10-second segment shows focal slowing over the left temporal region, which is intermittent. There is no slow activity in the first 4 seconds, but in the fifth and the eighth seconds focal slow waves are apparent. This type of intermittent slow activity has a much lower association with structural lesions than the polymorphic continuous delta activity seen in Figures 32–1 and 32–2. This patient's MRI was normal.

exercised before considering as abnormal isolated nonepileptiform focal findings other than slowing. In general, as with other types of focal EEG abnormalities such as slowing, asymmetry is nonspecific as to etiology.

The location of the dysfunction or lesion, rather than its cause, determines which rhythms are likely to be suppressed. Thus, a posterior process tends to affect the alpha rhythm, a central process, the mu rhythm, and a frontal process, the beta frequencies. In general, a reduction in amplitude of 50% is considered significant and may be indicative of focal dysfunction.[28] Because it is considered normal for the alpha rhythm to be of smaller voltage on the left, a smaller reduction on the right may be considered significant. The same can be said of asymmetries of most other normal rhythms, including sleep spindles, mu rhythm, and beta activity,[19, 42] and while asymmetry may suggest focal dysfunction, it is rarely sufficient to be considered abnormal without concomitant slowing.

Although asymmetry in amplitude is usually indicative of dysfunction on the side of depressed amplitude, one notable exception to this rule is the so-called breach rhythm.[16] This is caused by a skull defect, which attenuates the high frequency filter function of the intact skull. As a result, faster frequencies (alpha, spindles, beta) are of higher amplitude on the side of the defect (Figs. 32–5 and 32–6). With skull defects, the morphology of background rhythms is often sharply contoured. As a result, ruling out the epileptogenicity of these rhythms can be difficult (see Fig. 32–5 and Case Study 1), in which case erring with a nonepileptic diagnosis is preferable to an incorrect diagnosis of seizure disorder. Of interest is the fact that the breach rhythm may or may not disappear after bone replacement.[16] Because of a cancellation effect between fronto-polar (Fp_1/Fp_2) and frontal (F_3/F_4) electrodes, eye movements are often not increased on the side of a skull defect, and may indeed be of lesser amplitude on that side.[35]

Finally, it should be kept in mind that amplitude asymmetries are best interpreted on referential montages, as amplitude is highly dependent on interelectrode distances.

PERIODIC PATTERNS AND PERIODIC LATERALIZED EPILEPTIFORM DISCHARGES

Described in 1964,[10] periodic lateralized epileptiform discharges (PLEDs) are a special type of focal abnormality. As implied by their name, they are periodic, lateralized, and epileptiform. Periodicity is the most characteristic feature, and the one that sets PLEDs apart from other focal abnormalities. Periodicity refers to a relatively constant interval between two discharges, and varies between 0.5 and 3 seconds, most often around 1 second. The epileptiform morphology of the discharges is not invariable, as PLEDs are often closer to slow waves than to sharp waves in morphology (Fig. 32–7).

PLEDs are caused by acute destructive focal lesions, and are a transitory phenomenon: they tend to disappear in weeks, even if the causal lesion persists.[20, 47–49] With time, periodicity and duration increase,[47] and the record takes on a less specific focal slow appearance which is more likely to persist. By far the most common etiology is an acute cerebrovascular event, followed by focal infection, especially herpes simplex encephalitis.[48, 49] In a clinical context suggestive of viral encephalitis, the EEG can be of great value for diagnosis and can guide tissue biopsy.[11] Though most often associated with an acute destructive lesion, PLEDs, like other EEG findings, are not specific as to etiology and have been described in almost all types of structural lesions, including subdural hematoma[12] and chronic lesions, especially in the presence of a superimposed systemic disturbance.[13]

The pathophysiological substrate of PLEDs, like that of focal polymorphic delta, is thought to involve disconnection of the cortex from the diencephalic structures. There is some evidence that metabolic derangements, especially hyperglycemia and fever, increase the likelihood that a destructive lesion will generate PLEDs.[39]

In keeping with their epileptiform morphology, PLEDs have a high association with clinical seizures, and on average, about 80% of patients with PLEDs will have clinical seizures.[10, 43] The transition between PLEDs and a clear ictal

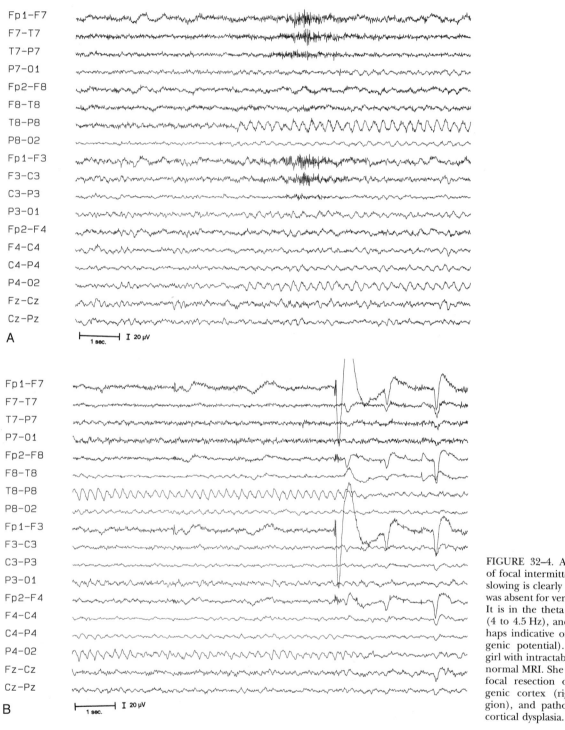

FIGURE 32–4. Another example of focal intermittent slowing. The slowing is clearly intermittent and was absent for very long segments. It is in the theta frequency band (4 to 4.5 Hz), and rhythmic (perhaps indicative of more epileptogenic potential). This 9-year-old girl with intractable seizures had a normal MRI. She eventually had a focal resection of her epileptogenic cortex (right parietal region), and pathology revealed a cortical dysplasia.

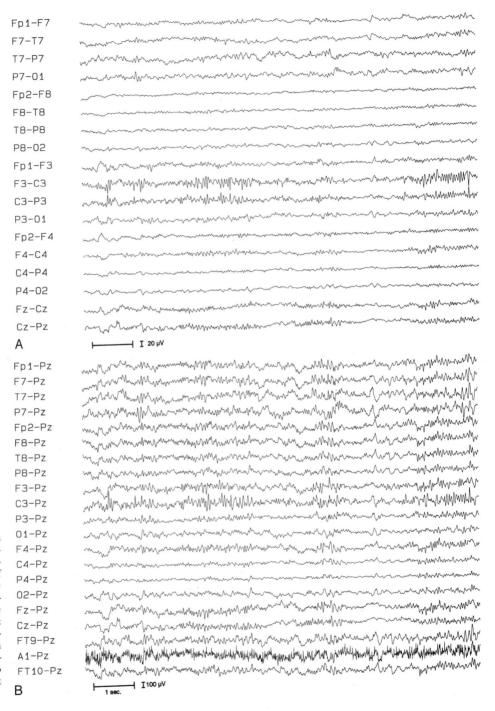

FIGURE 32–5. Breach rhythm. This 78-year-old woman was status post removal of a left frontal meningioma. The EEG shows significant asymmetry, especially of the fast (beta) activity in the fronto-central region. This is confirmed on the referential montage (*B*); compare amplitudes C3-Pz and C4-Pz. Note the sharp morphology of the increased beta rhythm, which should not be interpreted as spikes. There is also some delta slowing over the left hemisphere.

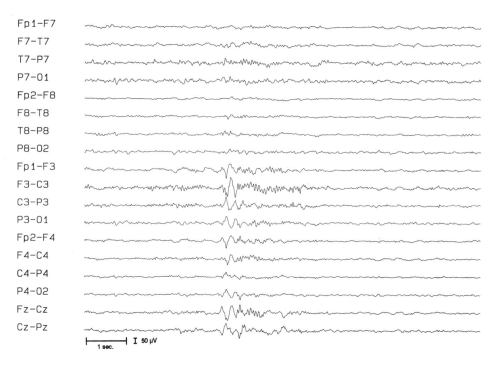

FIGURE 32–6. Breach rhythm. Same patient, with another example of significant asymmetry caused by a skull defect. In this sample, it is the sleep patterns that are markedly asymmetrical and predominant on the left side: vertex wave (fourth second) and sleep spindles.

seizure pattern is very gradual (Fig. 32–8), furthering the argument that PLEDs may indeed represent a subclinical ictal pattern. However, in clinical practice, PLEDs are usually managed as interictal discharges (spikes or sharp waves). They indicate a high risk of focal seizures, but are usually not treated with antiepileptic drugs unless there is *clinical* evidence for seizures. This position has been endorsed by several authors.[6] However, the question of treatment remains controversial, and some authors advocate treating all patients with PLEDs.[21] The controversy is comparable to that regarding treatment in the presence of a structural lesion with focal spikes.

Periodic patterns in Creutzfeldt-Jakob disease are usually generalized and bisynchronous, but occasionally, especially early in the course,[31] may be unilateral or markedly asymmetrical, and thus take on the appearance of PLEDs.[1, 23] Focal abnormalities are also more frequent in the Heidenhain's (occipital) variant of Creutzfeldt-Jakob disease.[23]

OTHER LESS COMMON FOCAL PATTERNS

An abnormal response to photic stimulation can be seen in focal lesions. It has long been known that normal photic driving may be reduced on the side of a lesion.[35] Posterior

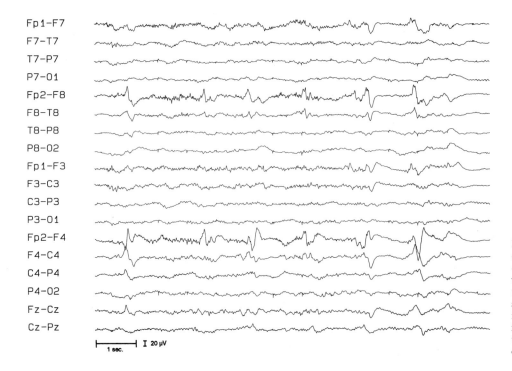

FIGURE 32–7. Right frontal periodic lateralized epileptiform discharges (PLEDs) in a 77-year-old woman with an acute stroke. Note the periodicity of the complexes, some of which are up to 400 msec in duration (slow waves). Periodicity is the most important characteristic that sets PLEDs apart from other focal abnormalities.

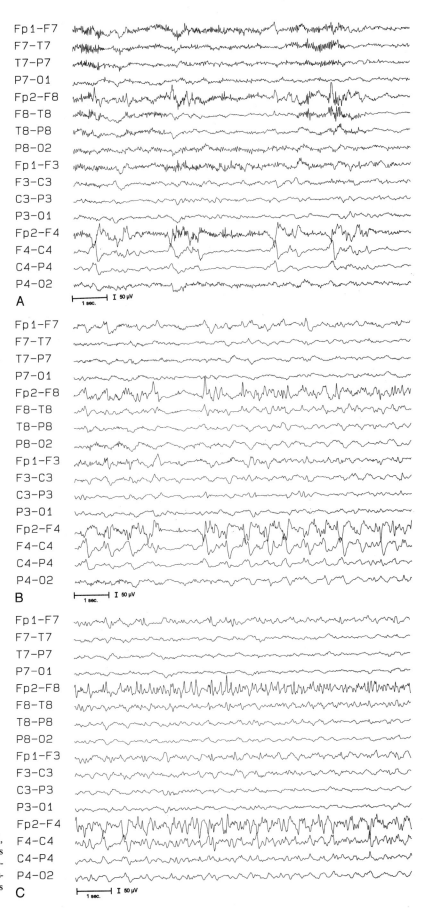

FIGURE 32–8. Same patient as in Figure 32–7. *A*, *B*, and *C* are consecutive epochs. The PLEDs gradually increase in frequency (decreased period) and evolve into a focal seizure. This illustrates the high risk of focal seizures when PLEDs are present.

destructive lesions are particularly likely to attenuate the driving response, but some reports have described an *enhanced* photic response on the side of dysfunction.[9] However, as the normal driving response can be quite asymmetrical (Fig. 32–9), such a finding should be accompanied by a more reliable abnormality, such as slowing, in order to be interpreted as abnormal.[18]

Bancaud's phenomenon refers to the unilateral loss of reactivity of a normal rhythm, and was initially described for the alpha rhythm.[2] It should be considered a pathological finding only when associated with other more definite abnormalities such as slowing.

FOCAL DYSFUNCTION WITHOUT STRUCTURAL LESION

There are a few situations in clinical neurology where the EEG may show clear evidence of focal dysfunction, but no structural abnormality is found. This is commonly the case with focal epilepsies (see Chap. 34), but may also occur with nonepileptiform abnormalities. In the presence of clear neurological symptoms, such as transient ischemic attack (TIA), migraine, and the postictal state, the likelihood that the EEG finding is significant increases, despite the absence of a structural lesion. Polymorphic delta activity in these cases may be indistinguishable from that caused by a structural lesion (Fig. 32–10), other than for its short lived nature (i.e., it disappears with time). The postictal state is the most common cause of *nonstructural* polymorphic delta activity,[26] but the activity disappears within minutes to hours of the ictal event.

Cerebrovascular disease represents a special situation, because focal alterations of blood flow cause alterations in clinical function and EEG activity before they cause structural damage. Thus, in this situation, the EEG could be said to be more "sensitive," or to have a better "temporal" (but not spatial) resolution than CT or MRI, with the possible exception of diffusion-weighted imaging. This has been well documented by experimental data.[4, 30] However, patients with

ongoing TIAs rarely undergo an EEG during the symptomatic period, so clinical data are scarce. As can be expected, focal slowing may be seen during a TIA.[8, 25] The close relationship between focal dysfunction (i.e., EEG slowing) and cerebral blood flow is further confirmed by data from continuous EEG monitoring during carotid endarterectomies.[3, 17]

Like those with TIAs, patients with migraine rarely undergo an EEG during the event, so data are scarce. As would be expected, the likelihood of focal EEG abnormalities in migraine increases with the presence of focal symptoms or signs, and is highest when lateralized sensory or motor symptoms are present.[32]

CASE STUDY 1

This 78-year-old woman was evaluated for dizzy spells. She was status post removal of a left frontal mass found to be a meningioma. The EEG tracing demonstrated in Fig. 32–11 shows a sharp transient at P7. If considered out of context, this could be interpreted as a sharp wave. However, the overwhelming majority of the tracing reveals significant asymmetry, especially of the fast (beta) activity in the fronto-central region, characteristic of a breach rhythm. The sharp transient is simply an isolated waveform of the breach rhythm. Figure 32–5 demonstrates another EEG tracing from the same patient.

CASE STUDY 2

This 49-year-old woman with a strong history of psychiatric disturbances and somatization disorder was admitted with a new onset of right-sided weakness of 6 hours' duration. An emergency CT scan was unremarkable, and it was suspected that symptoms were nonorganic. The EEG in Fig. 32–12 shows polymorphic delta slowing over the left hemisphere, maximum in the temporal chain. This supported an organic process, and an MRI the next day confirmed the presence of an infarct in the left middle cerebral artery territory.

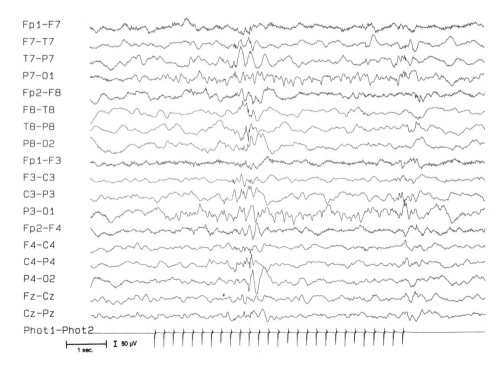

FIGURE 32–9. Asymmetrical photic driving. During photic stimulation at 5 Hz, clear driving is seen in the left occipital region, but not in the right occipital region. Although this suggests the possibility of an abnormality in the right occipital region, it is not sufficient to be interpreted as abnormal without associated abnormalities, such as slowing. This could also be seen with a skull defect in the left occipital region. This 67-year-old man was evaluated for confusion, which was most probably related to metabolic factors. MRI of the brain was normal.

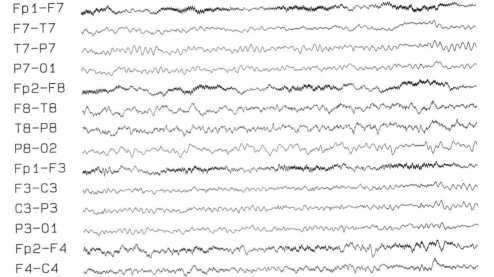

FIGURE 32–10. Postictal slowing. This EEG is that of a 35-year-old woman, and was recorded immediately following a seizure of right temporal origin. The continuous polymorphic delta seen here is undistinguishable from what could be caused by a structural lesion (similar to Figs. 32–1 and 32–2), except for the fact that it will disappear in minutes to hours. Also note slowing of alpha rhythm indicative of a diffuse encephalopathy.

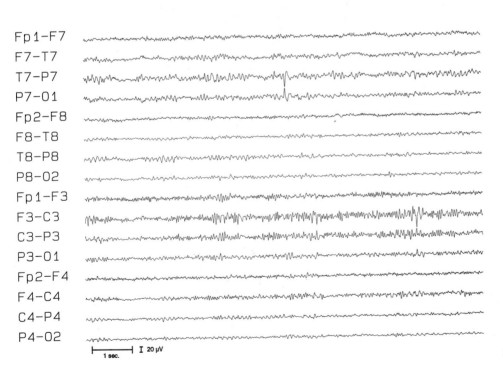

FIGURE 32–11. Case study 1.

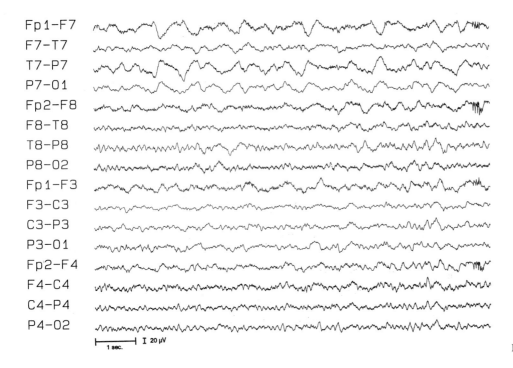

FIGURE 32–12. Case study 2.

CONCLUSIONS

In the era of neuroimaging, the role of EEG has changed significantly. Nonepileptic focal abnormalities are nonspecific in regard to etiology. Polymorphic delta activity and PLEDs are the only abnormalities that can be considered definitely abnormal when present in isolation. With rare exceptions, other abnormalities are not sufficient in themselves to be considered abnormal. Focal EEG abnormalities remain important in only a few conditions other than epilepsy, including focal (viral) encephalitis and Creutzfeldt-Jakob disease.

REFERENCES

1. Au WJ, Gabor AJ, Nazhiyath V, Markand ON: Periodic lateralized epileptiform complexes (PLEDS) in Creutzfeldt-Jakob disease. Neurology 30:611–617, 1980.
2. Bancaud J, Hécaen H, Lairy GC: Modifications de la réactivité EEG, troubles des fonctions symboliques et troubles confusionnels dans les lésions hémisphériques localisées. Electroencephalogr Clin Neurophysiol 7:179, 1955.
3. Blume WT, Ferguson GG, McNeil DK: Significance of EEG changes at carotid endarterectomy. Stroke 17:891–897, 1986.
4. Branston NM, Symon L, Crockard HA, Pasztor E: Relationship between the cortical evoked potential and local blood flow following acute middle cerebral artery occlusion in the baboon. Exp Neurol 45:195–208, 1974.
5. Brenner RP, Sharbrough FW: Electroencephalographic evaluation in Sturge-Weber syndrome. Neurology 26:629–632, 1976.
6. Brenner RP, Schaul N: Periodic EEG patterns: Classification, clinical correlation, and pathophysiology. J Clin Neurophysiol 7:249–267, 1990.
7. Brigell MG, Celesia GG, Salvi F, Clark-Bash R: Topographic mapping of electrophysiologic measures in patients with homonymous hemianopsia. Neurology 40:1566–1570, 1990.
8. Bruens JH, Gastaut H, Giove G: Electroencephalographic study of the signs of chronic vascular insufficiency of the Sylvian region in aged people. Electroencephalogr Clin Neurophysiol 12:283–295, 1960.
9. Cantor FK, Ilbag F: Facilitation of photic response by focal lesions. Electroencephalogr Clin Neurophysiol 34:77–79, 1973.
10. Chatrian GE, Shaw CM, Leffman F: The significance of periodic lateralized epileptiform discharges in EEG: An electrographic, clinical, and pathological study. Electroencephalogr Clin Neurophysiol 17:177–193, 1964.
11. Ch'ien LT, Boehm RM, Robinson H, et al: Characteristic early electroencephalographic changes in herpes simplex encephalitis. Arch Neurol 34:361–364, 1977.
12. Chu NS: Acute subdural hematoma and the periodic lateralized epileptiform discharges. Clin Electroencephalogr 10:145–150, 1979.
13. Chu NS: Periodic lateralized epileptiform discharges with preexisting focal brain lesions. Role of alcohol withdrawal and anoxic encephalopathy. Arch Neurol 37:551–554, 1980.
14. Churchill JA, Gonzalez S: The worth of serial electroencephalograms. Electroencephalogr Clin Neurophysiol 9:170, 1957.
15. Cobb W, Muller G: Parietal focal theta rhythms. Electroencephalogr Clin Neurophysiol 6:455, 1954.
16. Cobb WA, Guillof R, Cast J: Breach rhythm: the EEG related to skull defects. Electroencephalogr Clin Neurophysiol 47:251–271, 1979.
17. Collice M, Arena O, Fontana RA, et al: Role of EEG monitoring and cross-clamping duration in carotid endarterectomy. J Neurosurg 65:815–819, 1986.
18. Coull BM, Pedley TA: Intermittent photic stimulation. Clinical usefulness of nonconvulsive responses. Electroencephalogr Clin Neurophysiol 44:353–363, 1978.
19. Daly DD: The effect of sleep upon the electroencephalogram in patients with brain tumors. Electroencephalogr Clin Neurophysiol 25:521–529, 1968.
20. Erkulvrawatr S: Occurrence, evolution and prognosis of periodic lateralized epileptiform discharges in EEG. Clin Electroencephalogr 8:89–99, 1977.
21. Faught E: Current role of electroencephalography in cerebral ischemia. Stroke 24:609–613, 1993.
22. Fischer-Williams M, Last SL, Lyberi G, Northfield DWC: Clinico-EEG study in 128 gliomas and 50 intracranial tumours. Brain 78:42–58, 1962.
23. Furlan AJ, Henry CE, Sweeney PJ, Mitsumoto H: Focal EEG abnormalities in Heidenhain's variant of Jakob-Creutzfeldt disease. Arch Neurol 38:312–314, 1981.
24. Gambardella A, Gotman J, Cendes F, Andermann F: Focal intermittent delta activity in patients with mesiotemporal atrophy: A reliable marker of the epileptogenic focus. Epilepsia 36:122–129, 1995.
25. Gastaut H, Bruens JH, Roger J, Giove G: Etude éléctroencéphalographique des signes d'insuffisance circulatoire sylvienne chronique. Rev Neurol 100:59–65, 1959.
26. Gilmore PC, Brenner RP: Correlation of EEG, computerized

tomography, and clinical findings: A study of 100 patients with focal delta activity. Arch Neurol 38:371–372, 1981.

27. Gloor P, Ball G, Schaul N: Brain lesions that produce delta waves in the EEG. Neurology 27:326–333, 1977.
28. Green RL, Wilson WP: Asymmetries of beta activity in epilepsy, brain tumor and cerebrovascular disease. Electroencephalogr Clin Neurophysiol 13:75–78, 1961.
29. Grunnet GL, Goldensohn ES: Some differences in the EEG of brain tumors and cerebral vascular accidents. Electroencephalogr Clin Neurophysiol 23:493, 1967.
30. Heiss WD: Flow thresholds of functional and morphological damage of brain tissue. Stroke 14:329–331, 1983.
31. Heye N, Cervos-Navarro J: Focal involvement and lateralization in Creutzfeldt-Jakob disease: Correlation of clinical, electroencephalographic and neuropathological findings. Eur Neurol 32:289–292, 1992.
32. Hockaday JM, Whitty WM: Factors determining the electroencephalogram in migraine: A study of 560 patients, according to clinical type of migraine. Brain 92:769–788, 1969.
33. Joynt RJ, Cape CA, Knott JR: Significance of focal delta activity in adult electroencephalography. Arch Neurol 12:631–638, 1965.
34. Kiloh LG, Osselton JW: Clinical Electroencephalography. London, Butterworths, 1966.
35. Klem G, et al: Artifacts. In Pedley TA (ed): Current Practice of Clinical EEG. Philadelphia, Lippincott-Raven, 2000.
36. Kooi KA, Ecman HG, Thomas MH: Observations on the response to photic stimulation in organic cerebral dysfunction. Electroencephalogr Clin Neurophysiol 9:239–250, 1957.
37. Maytal J, Novak GP, Knobler SB, Schaul N: Neuroradiological manifestations of focal polymorphic delta activity in children. Arch Neurol 50:181–184, 1993.
38. Michel F, Gastaut JL, Bianchi L: Electroencephalographic cranial computerized tomographic correlations in brain abscess. Electroencephalogr Clin Neurophysiol 46:256–273, 1979.
39. Musil F: Development of focal changes in the EEG of those with cranio-cerebral injury. Electroencephalogr Clin Neurophysiol 26:229, 1969.
40. Neufeld MY, Vishnevskaya S, Treves TA, et al: Periodic lateralized epileptiform discharges (PLEDS) following stroke are associated with metabolic abnormalities. Electroencephalogr Clin Neurophysiol 102:295–298, 1997.
41. Oken BS, Chiappa KH, Salinsky MC: Computerized EEG frequency analysis: Sensitivity and specificity in patients with focal lesions. Neurology 39:1281–1287, 1989.
42. Petrovanu, I, Sandulescu G, Schwartzenberg T: The value of EEG in post-traumatic endocranial haematoma. Electroencephalogr Clin Neurophysiol 22:575, 1967.
43. Pfurtscheller G, Sager W, Wege W: Correlations between CT scan and sensorimotor EEG rhythms in patients with cerebrovascular disorders. Electroencephalogr Clin Neurophysiol 52:473–485, 1981.
44. Pohlmann-Eden B, Hoch DB, Cochius JI, Chiappa KH: Periodic lateralized epileptiform discharges: A critical review. J Clin Neurophysiol 13:519–530, 1996.
45. Raymond AA, Fish DR: EEG features of focal malformations of cortical development. J Clin Neurophysiol 13:495–506, 1996.
46. Salinsky MC, Oken BS, Kramer RE, Morehead L: A comparison of quantitative analysis and conventional EEG in patients with focal brain lesions. Electroencephalogr Clin Neurophysiol 83:358–366, 1992.
47. Schaul N, Green L, Peyster R, Gorman J: Structural determinants of electroencephalographic findings in acute hemispheric lesions. Ann Neurol 20:703–711, 1985.
48. Schwartz MS, Prior PF, Scott DF: The occurrence and evolution in the EEG of a lateralized periodic phenomenon. Brain 96:613–622, 1973.
49. Snodgrass SM, Tsuburaya K, Ajmone-Marsan C: Clinical significance of periodic lateralized epileptiform discharges: Relationship to status epileptics. J Clin Neurophysiol 6:159–172, 1989.
50. Upton A, Gumpert J: Electroencephalography in the early diagnosis of herpes simplex encephalitis. Rev Electroencephalogr Neurophysiol Clin 1:81–83, 1971.
51. Vignadnandra V, Ghee LT, Chawla J: EEG in brain abscess: Its value in localization compared to other diagnostic tests. Electroencephalogr Clin Neurophysiol 38:611–622, 1975.
52. Zenkov LR, Makarov VM: Differential diagnosis of brain tumors and chronic subdural hematomas according to echo and electroencephalographic findings. Zh Vopr Meirokhir 4:37–40, 1978.

Chapter 33

Pediatric Epilepsy Syndromes

Selim R. Benbadis, MD • *Elaine Wyllie, MD*

INTRODUCTION
IDIOPATHIC EPILEPSIES
Definition and General Features
Idiopathic Localization-related Epilepsy
Idiopathic Generalized Epilepsies

SYMPTOMATIC/CRYPTOGENIC EPILEPSIES
Definition and General Features
Symptomatic/Cryptogenic Generalized
  Epilepsies
Symptomatic/Cryptogenic Localization-
  related Epilepsies
LESS COMMON EPILEPTIC SYNDROMES
CONCLUSIONS: THE SYNDROMIC
  APPROACH TO EPILEPSY

INTRODUCTION

Epilepsy includes a group of heterogeneous conditions. In fact, it is more appropriate to speak of epilepsies. The terms *seizure* and *epilepsy* are not synonymous, and the distinction must be clear. A seizure is an abnormal behavior (with symptoms or signs) resulting from abnormal discharges of cortical neurons. It is an observable phenomenon that is finite in time. Epilepsy refers to a *chronic condition* characterized by recurrent seizures. A syndrome is a cluster of symptoms and signs that occurs together but, unlike a disease, does not have a single known etiology or pathology. Thus, an epileptic syndrome (or epilepsy), is a constellation of symptoms (including seizures), signs, and other findings, that tend to occur together, defining a phenotype. The diagnosis of an epileptic syndrome depends on a number of factors, including age of onset, etiology, family history, seizure frequency, imaging studies, precipitating factors, and electroencephalography (EEG). Why use the syndromic approach? It is well accepted that natural history, prognosis, and the choice of treatment are related to epileptic syndromes rather than to seizure type.[2, 8, 15, 29, 46, 56, 62, 63, 66]

With the recent advances in neuroimaging, the role of EEG has markedly declined in the evaluation of most neurological diseases. Even in epilepsy, imaging is playing an increasingly important role, especially in the evaluation for epilepsy surgery. However, because epilepsy is a disorder of electrical activity in the brain, EEG remains critical for the classification of seizure types[21] as well as the diagnosis of epileptic syndromes.[22, 28, 37, 70] This chapter describes the major pediatric epileptic syndromes, with emphasis on EEG findings.

IDIOPATHIC EPILEPSIES

Definition and General Features

The most recent version of the International Classification of Epileptic Syndromes and Epilepsies was published in

1989.[22] In addition to the fundamental dichotomy between partial and generalized seizures, it established an essential dichotomy between the idiopathic epilepsies on the one hand, and the symptomatic or cryptogenic epilepsies on the other. Idiopathic epilepsies are not truly of "unknown cause," but are genetically determined and have no apparent structural cause, with seizures as the only manifestation of the condition. Neurological examination is normal, imaging studies are normal, and EEG is normal other than for the epileptiform abnormalities. In some syndromes the genetic substrate has been proven and identified, while in most it remains elusive.

The most common types of idiopathic epilepsies are reviewed next, divided into the traditional localization-related and generalized categories.

Idiopathic Localization-related Epilepsy

Benign focal epilepsy of childhood is the only localization-related epilepsy that is idiopathic. Two varieties have been well described: centrotemporal and occipital.

Benign Childhood Epilepsy with Centrotemporal Spikes

Benign childhood epilepsy with centrotemporal spikes (BECTS) is by far the more common type of focal epilepsy in children. Age of onset is between 4 and 12 years, with a strong peak at ages 8 to 9. Seizures are simple partial with motor symptoms involving the face, and tend to occur during sleep or on awakening.[35, 50] Particularly common are seizures with symptoms caused by activation of the rolandic or "peri-sylvian" sensorimotor cortex (near the junction of the sylvian and rolandic fissures). Thus, characteristic ictal symptoms include guttural vocalizations, hypersalivation, drooling, sensations or movements of the mouth, dysphasia, and speech arrest.[50, 53] These symptoms are consistent with responses obtained on electrical stimulation of this region of cortex,[72] and the high incidence of facial and oropharyngeal symptoms are consistent with the large surface representa-

We thank George Klem and Barbara Wolgamuth for their help in gathering illustrations.

tion of these regions on the homonculus.[64, 72] Though these focal seizures are the most characteristic seizure types, they can be quite subtle and are easily missed. The most common mode of presentation is a nocturnal generalized tonic clonic seizure (during sleep),[50, 54] which is classified as a *secondary* generalized seizure due to spread of the ictal discharge.

As with all idiopathic epilepsy syndromes, neurological examination is normal. EEG findings are characteristic (Figs. 33–1 and 33–2), with stereotyped centrotemporal sharp waves. The "centrotemporal" label is related to the fact that the electrodes of the standard 10–20 system are above (central row) and below (temporal row) the region of maximal amplitude at the junction of the sylvian and rolandic fissures (see Fig. 33–2). These sharp waves (70 to 100 msec) have a characteristic morphology:[35, 54] diphasic transient with a

negative sharp peak followed by a positive rounded component (amplitude 50% of the negativity). They often occur in repetitive bursts, and can be bilateral and independent. These sharp waves are markedly activated by non-REM sleep,[35] and occasionally occur only in sleep[12] (Fig. 33–3).

Another characteristic feature of the sharp waves in BECTS is the presence of a "horizontal dipole." All spike generators have a positive and a negative pole, but typically only the negative one is identified with surface electrodes ("vertical" dipole).[41] These are generally oriented radially, perpendicular to the scalp, so that only one pole is recorded by surface electrodes. The frequent occurrence of a "horizontal" dipole in BECTS is related to the typical location of the epileptogenic zone situated in a convoluted region of cortex. The dipole of BECTS characteristically reveals a maxi-

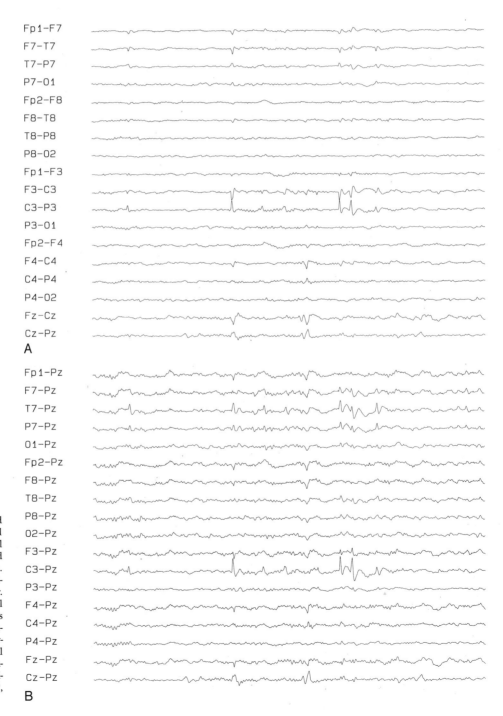

FIGURE 33–1. Benign childhood epilepsy with centrotemporal spikes (BECTS). An 8-year-old girl with new onset of generalized tonic-clonic seizures during sleep. She was taking Ritalin for attention deficit hyperactivity disorder. *A.* Bipolar montage. *B.* Referential montage. This segment shows BECTS with a typical centrotemporal location, and the characteristic dipole seen on a referential montage. Note a maximum negativity at C3>T7, and a low amplitude positivity in Fp1, F3, F7, Fp2, and F4.

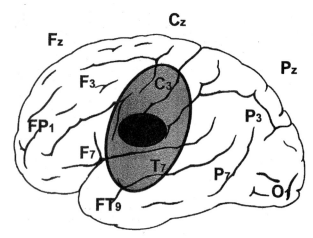

FIGURE 33–2. Location of the discharges in BECTS.

mum negativity in the centrotemporal region and a maximum positivity at the vertex or in the frontal region, which is best seen in referential montages (see Fig. 33–1*B*).

The sharp waves of BECTS are probably the only focal epileptiform discharges whose morphology is in itself associated with a specific syndrome.[30] These typical EEG findings, in the proper clinical setting, obviate the need for imaging. Therefore, typical BECTS is the single localization-related epilepsy in which neuroimaging may be deferred. However, EEG features may occasionally be less typical. Some less frequent variations of the EEG have been described, but their relationship with the "pure" syndrome is controversial. For example, some have described generalized spike wave complexes, but most investigators agree that this is a misinterpretation of high-amplitude repetitive BECTS discharges.[49, 54] If genuine generalized 3-Hz spike wave complexes are present, it most likely reflects the coexistence of a generalized and partial epilepsy[31] rather than an EEG variant.

It should be noted that the interictal sharp waves of BECTS often occur in asymptomatic children. In fact, only a minority of children with these discharges may have seizures,[54] and this is probably the main explanation for reports of focal spikes in children without epilepsy.[47] The natural history of BECTS is benign: Whether treated or not, this syndrome has an excellent prognosis, with spontaneous remission by ages 14 to 16.

In summary, the syndrome of BECTS is benign in several ways. Seizures are typically infrequent and nocturnal, and there is a very low incidence of status epilepticus. Children with BECTS are otherwise neurologically normal, and spontaneous remission is the rule. Most children with BECTS do not require treatment with antiepileptic drugs.[49] If treatment is elected because of greater seizure frequency or severity, then moderate doses of a single agent are usually sufficient. Options may include carbamazepine (CBZ), gabapentin (GBP), or valproic acid (VPA). Patients should be weaned from antiepileptic drugs by ages 14 to 16 in every case.

Childhood Epilepsy with Occipital Paroxysms

This type of focal epilepsy is less common than BECTS and less consistently benign. It shares all the characteristics of an idiopathic syndrome (normal examination, intelligence quotient [IQ], and neuroimaging studies). Age of onset is also 4 to 8 years. Seizures are rare and primarily nocturnal, and often involve visual symptoms. Sharp waves have a maximum occipital negativity, often occur in long bursts of spike wave complexes, and are markedly activated by eye closure.[32]

Idiopathic Generalized Epilepsies

General Features

These syndromes, formerly called "primary" generalized epilepsies, are the best known group of idiopathic epilepsies. They epitomize the meaning of the term *idiopathic*: genetic basis, normal neurological examination, and normal intelligence. EEG shows generalized epileptiform discharges and may show photosensitivity.[75] Seizure types include generalized tonic-clonic, absence, and myoclonic seizures. *Absence seizures* are clinically characterized by a brief (5- to 15-second)

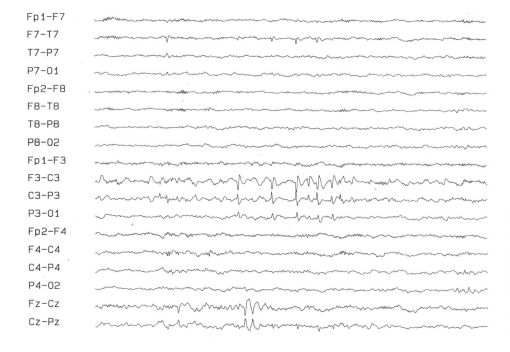

FIGURE 33–3. BECTS. An 8-year-old boy with a history of two seizures: one at age 2½, and one at age 4. Seizures consisted of inability to speak followed by whole body shaking and a Todd's paralysis on the right. This segment shows BECTS in repetitive runs during stage II sleep. Note normal vertex waves maximum in the midline channels (Fz-Cz, Cz-Pz).

impairment of consciousness, with few other symptoms. *Typical absence* seizures have an abrupt onset and termination and are associated with classic 3-Hz spike wave complexes (ictally and interictally) with no other EEG abnormality. *Atypical absence* seizures occur in the symptomatic generalized epilepsy of the Lennox-Gastaut type (see later discussion). *Myoclonic seizures* consist of sporadic jerks, usually appendicular and symmetrical, associated electrically with generalized epileptiform discharges, typically polyspikes. Patients often report them as an electric shock sensation. *Generalized tonic-clonic (GTC) seizures* are typically characterized by a tonic phase which is gradually interrupted by quiescence, then giving rise to the clonic phase with rhythmic jerking. During the clonic phase, the generalized spikes and spike wave complexes gradually decrease in frequency.

Within the group of idiopathic generalized epilepsies, distinct entities are distinguished, primarily based on the predominant seizure type(s) and the age of onset. Some syndromes are very well characterized, whereas others have less clear boundaries.[9, 11, 67] The major and well-defined types of idiopathic generalized epilepsies are reviewed here.

Childhood Absence Epilepsy

Also known as petit mal epilepsy, this syndrome typically presents between 4 and 10 years of age. The predominant seizure type is typical absences, which are brief (about 10 seconds), and occur at a high frequency (usually more than 10 daily) in untreated children. About 50% of patients also have GTC seizures,[14] and a minority also have myoclonic seizures. Typical absences are often accompanied by subtle motor phenomena, especially eye blinking or clonic twitches of the hands or face. Longer seizures typically involve mild oral or gestural automatisms.[65]

EEG shows generalized and symmetrical 3-Hz spike wave complexes, which typically have a maximum negativity frontally, almost always at F3 and F4[33] (Fig. 33–4). Since most spikes are followed by a slow wave, a run of at least 3 seconds should be required to classify a discharge as spike wave complexes. The frequency of the spike wave complexes may vary between 2.5 and 3.5 Hz, and is often faster at the onset (3.5 to 4.5 Hz).[33] Although typically symmetrical, minor asymmetries may be seen and should not lead to a diagnosis of partial seizures.[7]

Three-hertz spike wave complexes are seen in both the interictal and ictal periods. The border between the two is imprecise because the clinical impairment may be difficult to detect during short attacks. The use of a clicker to test responsiveness can be very useful to document very brief impairments in awareness. Clinical seizures and the 3-Hz spike wave complex discharges are often precipitated by hyperventilation. The 3-Hz frequency, and also the striking monomorphism of the spike wave complex discharges, help distinguish absence epilepsy from other syndromes. Because the correlation between the 3-Hz spike wave complexes and the clinical absence seizures is so strong in childhood absence epilepsy, the EEG may be useful in monitoring response to treatment, which is not the case in most other epilepsies.[4] These children do not require neuroimaging. In terms of management, a drug of choice for absence seizures is ethosuximide, but valproate is often preferred as children may also have generalized tonic-clonic seizures. Second-line drugs such as benzodiazepines, lamotrigine, and topiramate may be used (when first-line drugs fail or are not tolerated), although they are not yet approved for this indication in the United States.

Absence seizures in childhood absence epilepsy often remit around adolescence, but other seizure types, if present, may continue in adulthood. Complete remission (seizure freedom) is more likely in patients with only absence sei-

zures, than in those with GTC or myoclonic seizures.[14] Such patients probably do not have pure childhood absence epilepsy, but rather another type along the spectrum of idiopathic generalized epilepsies.[11, 67] Occasionally, typical absence seizures manifest de novo in adulthood.[71]

Juvenile Absence Epilepsy

This is a variant of generalized absence epilepsy, characterized by a later age of onset, peaking around age 12.[51] Other features include less frequent seizures and milder impairment of consciousness compared to those with the childhood form.[61] EEG findings are similar to childhood absence epilepsy, although the spike wave complexes may be faster and less monomorphic.[61] Management issues are similar to those for childhood absence epilepsy.

Juvenile Myoclonic Epilepsy

Juvenile myoclonic epilepsy (JME) is one of the best characterized syndromes. Its genetic basis has been documented, with an abnormality on the short arm of chromosome 6.[23] Onset is in adolescence, maximum between ages 12 and 18 years. The most common, and defining, seizure type is myoclonic. These tend to occur in the morning on awakening, and often in clusters that may lead to a GTC. Although myoclonic seizures occur earlier in life and more frequently, it is not until the first GTC that the patient comes to medical attention. Because myoclonic seizures are often not spontaneously reported by patients, this history must be specifically sought in every patient presenting with GTCs. Failure to identify the typical morning jerks often accounts for the striking delay in diagnosis of JME.[36, 44]

EEG characteristically reveals generalized, bilateral, and symmetrical polyspikes and generalized spike wave complexes at 4 to 6 Hz, usually maximal at the frontal electrodes (Fig. 33–5). It should be remembered that polyspikes are the electrical correlate of myoclonic seizures, and as such they are seen in other syndromes that include this seizure type. Other generalized epileptiform discharges may also be present (such as 3-Hz spike wave complexes or isolated generalized spikes). Clinical and EEG asymmetries may occasionally be present.[45] Photosensitivity (Fig. 33–6) is particularly frequent.[75]

As with other idiopathic epilepsies, neuroimaging is usually not necessary. The medication of choice for JME is VPA. In cases of failure (due to side effects or lack of efficacy), alternatives include benzodiazepines and newer antiepileptics such as lamotrigine and topiramate, although they are not approved for this use in the United States. Unlike other idiopathic epilepsies, remission rarely occurs in JME, even after many years of seizure freedom, so that lifelong treatment may be necessary.

Other Idiopathic Generalized Epilepsies

In addition to the preceding well-defined syndromes, many patients with idiopathic generalized epilepsy have features that do not meet criteria for a specific syndrome. Seizures typically begin in childhood or young adulthood, and neurological examination is normal. A positive family history is not uncommon. Seizures may predominate in the morning, including GTCs, myoclonic jerks, or absence seizures. The International League Against Epilepsy classification separates "epilepsy with grand-mal seizures on awakening" and "epilepsies with specific modes of precipitation," but these syndromes are less well defined than childhood absence epilepsy or JME.

EEGs (Fig. 33–7) typically show generalized epileptiform discharges, which may be of any type (spikes, sharp waves, spike wave complexes, and polyspikes). Photosensitivity is common.

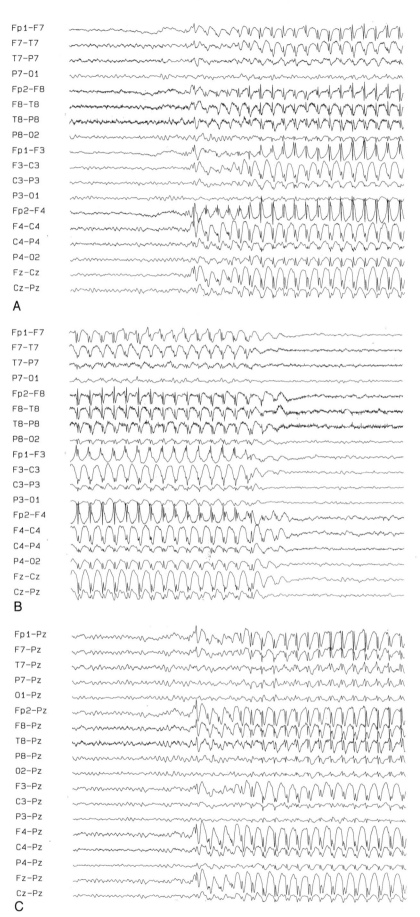

FIGURE 33–4. Childhood absence epilepsy. A 15-year-old boy with seizures since age 4, described as arrest of activity, eyelid flutter, and unresponsiveness for 5 to 15 seconds. They occur multiple times daily. *A* and *B.* Bipolar montage. *C.* Referential montage. This EEG shows typical monomorphic 3-Hz spike wave complexes. Note a frontal maximum, with phase reversal at F3 and F4 on the bipolar recording (*A* and *B*), and a maximum amplitude at these electrodes on the referential (*C*). This burst occurred during hyperventilation, lasted 12 seconds, and was associated with staring, unresponsiveness, cessation of activity, and postictal amnesia for a test word given during the event.

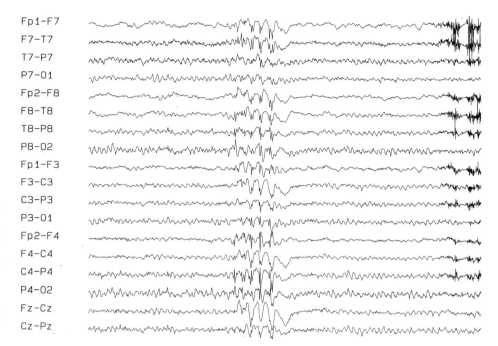

FIGURE 33–5. Juvenile myoclonic epilepsy. A 29-year-old woman with onset of seizures 15 years previously (age 14). She was evaluated for "lapses in consciousness" and arm jerks. EEG shows generalized polyspikes and fast (4 Hz) spike wave complexes. Note the normal background consistent with an idiopathic generalized epilepsy. Note again phase reversal at F3 and F4 (as in Fig. 33–4).

Patients whose idiopathic generalized epilepsies cannot be further classified share several characteristics with childhood absence epilepsy and JME, including unremarkable neuroimaging and a typically good response to treatment with VPA. Thus it is appropriate to view the idiopathic generalized epilepsies as a group.[9, 11, 40, 67]

SYMPTOMATIC/CRYPTOGENIC EPILEPSIES

Definition and General Features

These epilepsies are the result of a brain insult or lesion. If the damage is focal, it results in a localization-related epilepsy; if it is diffuse, it results in a generalized epilepsy.

The difference between symptomatic and cryptogenic is a subtle one: symptomatic epilepsies refer to disorders of known etiology, whereas in cryptogenic epilepsies, there appears to be an underlying etiology but it cannot be objectively documented. Categorization into one or the other group is largely dependent on the sensitivity of diagnostic and imaging techniques.

Symptomatic/Cryptogenic Generalized Epilepsies

General Features

These types of epilepsy are associated with diffuse brain dysfunction. Known etiologies include anoxic birth injury,

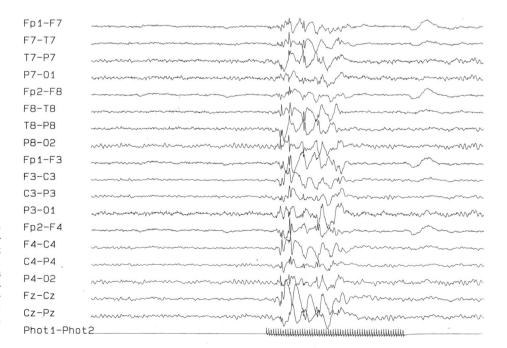

FIGURE 33–6. Photosensitivity. An 11-year-old boy with a single generalized tonic-clonic seizure that occurred after being up all night. Photoparoxysmal response. This may be seen in any of the idiopathic generalized epilepsies, or in patients without seizures. Note normal background just prior to the photoparoxysmal response.

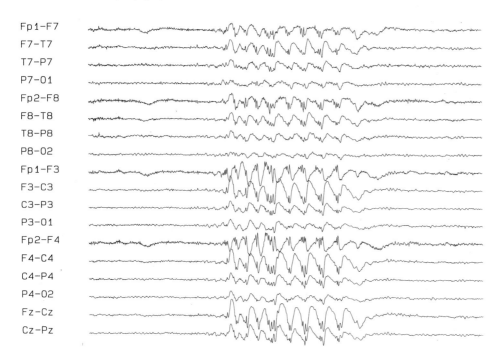

Fp1-F7
F7-T7
T7-P7
P7-O1
Fp2-F8
F8-T8
T8-P8
P8-O2
Fp1-F3
F3-C3
C3-P3
P3-O1
Fp2-F4
F4-C4
C4-P4
P4-O2
Fz-Cz
Cz-Pz

FIGURE 33–7. Idiopathic generalized epilepsy. A 29-year-old woman with a history of generalized tonic-clonic seizures, and episodes of staring and unresponsiveness since age 9. She was seizure free for 6 years at the time of this recording. The presence of 3-Hz spike wave complexes suggests absence seizures, and polyspikes may be in favor of juvenile myoclonic epilepsy, but only the clinical data can distinguish among the various syndromes of the group.

metabolic defect, or chromosomal defect. There is usually clinical evidence of diffuse brain dysfunction, either intellectual (developmental delay or mental retardation), or motor (developmental delay or cerebral palsy), or both. Similarly, the EEG usually shows evidence of diffuse brain dysfunction in addition to the epileptogenic discharges, in the form of slowing or amplitude abnormalities. The clinical and EEG manifestations are not specific as to etiology, but vary markedly with age, and are thus said to be age-dependent.

West Syndrome

West Syndrome is a phenotype of symptomatic or cryptogenic generalized epilepsy in the first year of life, characterized by infantile spasms, hypsarrhythmia, and developmental delay. It is best regarded as an age-specific response of the immature brain to a nonspecific focal or generalized insult. Age of onset peaks between 3 and 7 months of age, and there is a slight male predominance.[52] Clinically, there is usually developmental delay, plateau, or regression characterized by loss of early developmental milestones, and frequent seizures. The characteristic seizure type is an axial spasm in flexion, extension, or both, which occurs in clusters of 5 to 10 or more, especially in drowsiness or upon arousal. The individual spasms are brief (2 to 5 seconds) but can be distressing to the child and are often followed by a cry. Spasms are typically symmetrical, but various degrees of asymmetry may be seen.

Interictal EEG shows hypsarrhythmia (Fig. 33–8A). This is defined as continuous (during wakefulness) high amplitude (>200 Hz) generalized and polymorphic slowing with no organized background, associated with multifocal spikes. During non-REM sleep, the pattern becomes discontinuous and fragmented, resembling a burst-suppression or pseudoperiodic pattern, while it tends to disappear in REM sleep.[39] When some features are missing or atypical, the term "modified hypsarrhythmia" is sometimes used.[39] During spasms, there is typically a high amplitude sharp or slow wave, followed by diffuse attenuation (electrodecrement), often with superimposed low-amplitude paroxysmal fast activity (Fig. 33–8B).

As previously mentioned, West syndrome and hypsarrhythmia are self-limited phenotypes, and usually disappear within 12 months. The majority of patients evolve to other patterns of symptomatic generalized epilepsy, including the Lennox-Gastaut syndrome. Patients who were neurologically normal prior to the onset of spasms, and in whom no cause is found ("cryptogenic" type), have a somewhat better prognosis.

These children require a thorough evaluation for a possible underlying etiology (genetic, metabolic, or structural). Medications include ACTH, vigabatrin, VPA, and benzodiazepines. In addition to medical treatment, evidence has accumulated that in some infants, focal cortical lesions, most often cortical dysplasia, may manifest with typical infantile spasms and hypsarrhythmia.[19, 20, 57, 80] This has been ultimately documented in children with West syndrome becoming seizure-free after a focal brain resection.[20, 68, 73] Thus the traditional separation between partial and generalized seizures is difficult to apply in young children, with clinical and EEG findings differing from those seen in older patients.[24, 27, 76]

Lennox-Gastaut Syndrome

This syndrome with early childhood onset (ages 1 to 8 years) consists of intractable generalized epilepsy with multiple seizure types, mental retardation, and typical EEG findings dominated by generalized slow spike wave complexes. Like West syndrome, Lennox-Gastaut syndrome (LGS) can be considered an age-specific response of the immature brain to a variety of insults. Seizure types include atypical absences, tonic, atonic, myoclonic, and GTC seizures. Atypical absence seizures differ from typical ones in that the EEG shows slower and more irregular spike wave complexes, rather than the generalized 3-Hz spike wave complexes of childhood absence epilepsy. Atypical absences may also differ from typical ones clinically, with a less abrupt onset and termination and a longer duration. Finally, unlike typical absences, atypical ones are not usually precipitated by hyperventilation.[38] Atonic seizures consist of an abrupt and diffuse loss of tone, particularly of the axial musculature, associated on EEG with generalized epileptiform discharges (spikes, spike wave complexes) or an abrupt flattening of the cortical rhythms (electrodecrement). Tonic seizures are characterized by stiffening of the musculature, mostly axial but also appendicular, typically associated on EEG with low voltage paroxysmal fast activity (10 Hz). Like their EEG correlate,

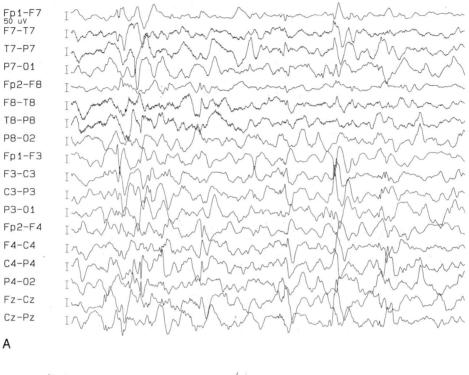

A

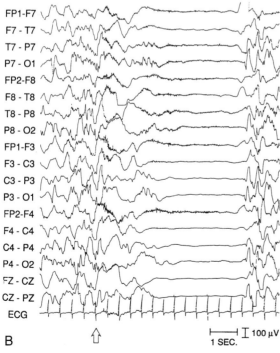

B

FIGURE 33–8. *A.* Hypsarrhythmia. A 1-month-old infant born with meconium aspiration and subsequent respiratory distress, evaluated for new episodes of body stiffening clustering over 5 minutes. Figure shows high amplitude slow waves, disorganized background, and multifocal spikes. On this segment, there are spikes in the first 3 seconds, in the following locations: left frontal (F3), left parietal (P7), and right parietal (P4). *B.* Infantile spasm. EEG during a spasm *(arrow).* Note the generalized high-amplitude slow transient followed by a generalized electrodecremental pattern for 3 seconds. The spasm involved tonic abduction and extension of both arms with flexion of the trunk and neck.

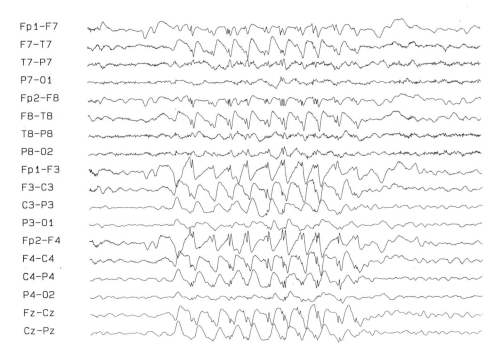

FIGURE 33–9. Lennox-Gastaut syndrome and symptomatic generalized epilepsies. A 6-year-old patient with seizures since age 4½, including convulsions, head drops, and staring spells. He also has behavioral problems and language delay. This segment shows 2.5-Hz generalized spike wave complexes. Mother identified a head bob during the burst.

tonic seizures are more frequent in non-REM sleep.[16] The atonic, myoclonic, tonic, and GTC seizures of LGS frequently result in unprotected falls (referred to as "drop attacks") with injury.

The classic EEG pattern of LGS consists of generalized slow spike and wave complexes, and was first noted by Gibbs, Gibbs, and Lennox.[34] This is an interictal pattern but may be seen ictally during atypical absences. In addition to being slower (<2.5 Hz), the slow spike wave complexes (Fig. 33–9) are more irregular (less monomorphic) than the 3-Hz spike wave complexes. During sleep, the EEG may show polyspikes and slow waves. Another typical feature of LGS is generalized paroxysmal fast (>10 Hz) activity during sleep (see Fig. 33–9). Other frequent but less specific interictal EEG findings

include background slowing, generalized slowing, and multifocal spikes (Fig. 33–10).

LGS, like West syndrome, requires a thorough search for an underlying etiology. The management of children with LGS is challenging. Although monotherapy is preferable, these patients often require multiple antiepileptic drugs, and seizures are often refractory. In selected cases, corpus callosotomy may be useful.

Other Symptomatic/Cryptogenic Epilepsies

Many patients with symptomatic generalized epilepsy do not meet all the criteria for LGS. Although the components of LGS are well described, its nosological boundaries are

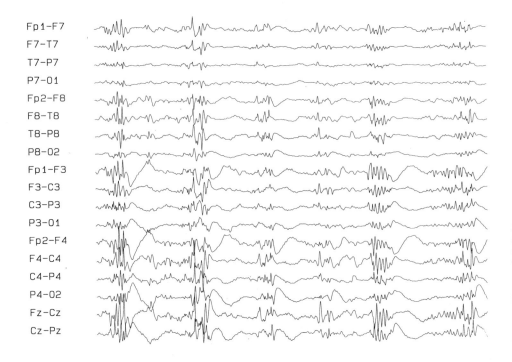

FIGURE 33–10. Lennox-Gastaut syndrome and symptomatic generalized epilepsies. A 16-year-old boy with mental retardation and multiple seizure types (generalized tonic-clonic and absence seizures). Patient had encephalitis at age 6 months. EEG shows bursts of generalized sharp waves and polyspikes.

rather controversial. Some authors[48] consider the clinical characteristics of the seizures (multiple generalized types including falls) and intractability sufficient for the diagnosis, and refer to this group as the myoclonic or myoclonic-astatic epilepsies. Other authors[13, 55, 59] emphasize that the EEG, with slow spike wave complexes, slow background, and multifocal spikes, is the primordial and sine qua non criterion. Although necessary, the generalized slow spike wave complexes alone do not seem to be sufficient for the diagnosis of LGS.[5, 60] One difficulty is that clinical seizures and EEG findings change with age and may no longer fulfill strict criteria for LGS. These cases may be best described as symptomatic or cryptogenic generalized epilepsy "not otherwise specified" or "of the LGS type," but they are often referred to as "mixed seizure disorder," a less specific term as most epilepsies have multiple seizure types.

Symptomatic/Cryptogenic Localization-related Epilepsies

General Features

These cases are characterized by seizures arising from a localized region of the brain. If the cause is found, they are said to be symptomatic. If imaging studies are normal, the cause remains elusive and they are said to be cryptogenic. As stated earlier, the boundary between the two is largely dependent on the sensitivity of diagnostic and imaging techniques, and etiologies such as low-grade tumors, hippocampal sclerosis, and subtle cortical dysplasias are being identified with more frequency as neuroimaging techniques advance.[10] Clinically, seizures may be *simple* partial or *complex* partial, with or without secondary generalization. Interictal EEG shows focal spikes or sharp waves, and ictal EEG shows a focal or regional discharge at onset.

Localization-related epilepsies can be divided into mesiotemporal and neocortical groups, based on distinctive electrical features, clinical semiology, and specific management. The most common localization-related epilepsy in adults is temporal lobe epilepsy, but this is less common than neocortical epilepsy in infants and young children.[25, 77, 81]

Mesiotemporal Lobe Epilepsy

Hippocampal sclerosis is the most common cause of temporal lobe epilepsy in adult candidates for epilepsy surgery, but it is relatively uncommon in pediatric candidates.[26, 79, 81] In this age group, etiologies are dominated by cortical dysplasias and low-grade neoplasms.[1, 26, 43] Seizures are typically complex partial with automatisms, often preceded by a simple partial phase with sensory symptoms (i.e., an aura). Auras commonly seen in temporal lobe epilepsy are epigastric (abdominal), psychic, and olfactory. Interictal EEG typically shows temporal sharp waves maximum at anterior temporal or sphenoidal electrodes (Fig. 33–11).

Neocortical Epilepsy

Seizures may be *simple* partial or *complex* partial, with symptoms depending on the area of cortex affected. Interictal EEG typically shows sharp waves or spikes outside the anterior temporal region. Ictal EEG often shows regional or widespread discharges. In general, localization of the epileptogenic zone by surface EEG (interictal and ictal) is less reliable for neocortical than for mesiotemporal epilepsy.[18, 69, 74] Epileptiform discharges can be absent on surface EEG when the focus is deep. A typical example of this is supplementary sensorimotor area epilepsy with the epileptogenic zone in the mesial frontal region.[58] In these cases, epileptiform discharges may be absent or may occur during sleep with a maximum at the vertex (Fig. 33–12) mimicking physiological transients.[58, 78] Frontal discharges can spread very rapidly and mimic "primary" generalized spike wave complexes. This is referred to as "secondary bilateral synchrony" (with interictal epileptiform discharges that appear to be generalized but are in reality "secondarily generalized" focal discharges). However, strict criteria should be required to attribute generalized discharges to secondary bilateral synchrony, such as coexistent focal or lateralized epileptiform discharges or a structural lesion.

Hemispheric Syndromes

These syndromes are localization-related and neocortical, but the epileptogenic zone is so widespread as to involve

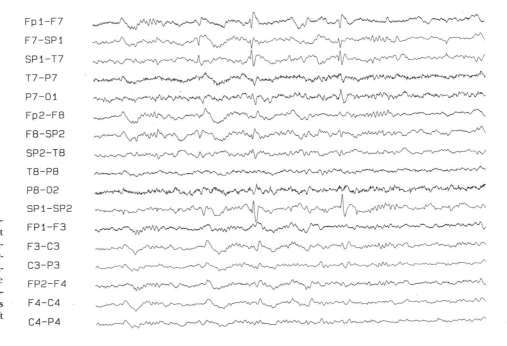

FIGURE 33–11. Temporal lobe epilepsy. A 16-year-old adolescent with a history of febrile convulsions in infancy, and complex partial seizures preceded by an epigastric aura. Magnetic resonance imaging (MRI) showed left hippocampal atrophy. This EEG shows sharp waves maximum at the left sphenoidal electrode.

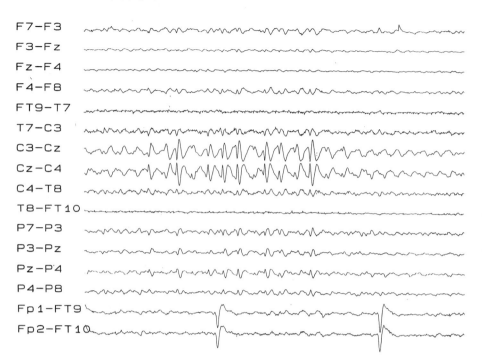

FIGURE 33–12. Supplementary motor area epilepsy. A 16-year-old boy with brief nocturnal body stiffening episodes with intact consciousness. MRI was normal. This EEG (transverse bipolar montage) shows repetitive sharp waves at the vertex. Vertex sharp waves can be difficult to differentiate from normal vertex waves of sleep, but in this case they have a definite epileptiform morphology and also occurred during wakefulness (see eye blinks on the bottom two channels).

nearly an entire hemisphere. In most cases structural abnormalities are identified on imaging, and EEG abnormalities (Fig. 33–13) include multifocal or widespread discharges lateralized to one hemisphere along with amplitude asymmetry, lateralized or focal slowing, and background slowing. Causes include perinatal infarction, Rasmussen's syndrome, Sturge-Weber syndrome, and hemimegalencephaly. In hemimegalencephaly, an extreme form of unilateral cortical dysgenesis, there may be obvious skull deformity or a more subtle asymmetry. Hemiparesis, hemianopia, and developmental delay often coexist. Rasmussen's syndrome is a progressive process that typically presents at between 14 months and 14 years of age with seizures, progressive hemiparesis, and unilateral cerebral atrophy on magnetic resonance im-

aging. Seizures are unilateral and intractable, frequently evolving into focal motor status epilepticus ("epilepsia partialis continua"). In Sturge-Weber syndrome (encephalotrigeminal angiomatosis), a sporadic disorder with leptomeningeal angiomatosis, seizures typically begin in the first year of life. There is a combination of facial angioma, cortical atrophy, gyriform calcifications, contralateral hemiparesis, and hemianopia. In addition to epileptiform discharges and slowing on the side of the angiomatosis, low-amplitude or absent background may be the most prominent abnormality.[17]

These hemispheric syndromes are often difficult to treat. Medications used against partial seizures are preferred. In the most refractory cases, functional hemispherectomy can be effective.[3]

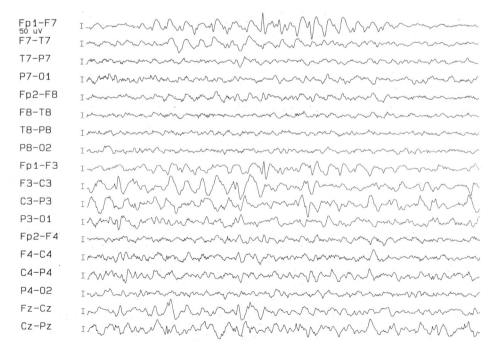

FIGURE 33–13. Hemispheric syndrome. A 5-year-old girl with epilepsia partialis continua and progressive hemiparesis secondary to Rasmussen's syndrome involving the left hemisphere. The diagnosis was documented by serial MRIs and eventually confirmed by histopathology after functional hemispherectomy. There is high-amplitude polymorphic continuous delta slowing in the left hemisphere, and sharp waves maximum in the left frontal or central region. The slowing was continuous, and sharp waves were seen frequently throughout the recording.

LESS COMMON EPILEPTIC SYNDROMES

Only those syndromes with noteworthy EEG findings are discussed in this section. *Landau-Kleffner syndrome,*[6] or acquired aphasia with epilepsy in childhood, and *epilepsy with continuous spikes and waves during slow sleep,*[6] previously known as "electrical status epilepticus of sleep," are epilepsies of childhood with focal seizures that remit after several years, while neuropsychological deficits persist. The hallmark is the presence of spike wave complexes during at least 85% of slow wave sleep, but this strict criterion seems somewhat arbitrary, and the boundaries between the two syndromes are unclear. Epilepsy with continuous spikes and waves during slow sleep is considered a syndrome "undetermined as to whether generalized or partial,"[22] but recent evidence suggests that it is probably focal with secondary bilateral synchrony.[42] Management is difficult. Seizures can usually be controlled with medications, but neuropsychological deficits are more difficult. Steroids may be tried. Some cases can be treated surgically, especially with multiple subpial transections.

CONCLUSIONS: THE SYNDROMIC APPROACH TO EPILEPSY

As this review of the major epileptic syndromes emphasizes, most patients with epilepsy have more than one type of seizures, and most types of seizures can occur in the context of various syndromes. This is because seizures are only symptoms. Although classification of seizures is important, the care of a patient requires a precise diagnosis of type of epilepsy (or syndrome). For example, JME, childhood absence epilepsy, benign focal epilepsy with centrotemporal spikes, and temporal lobe epilepsy may all present with generalized tonic-clonic seizures. Thus, a seizure diagnosis alone (or a diagnosis of "seizure disorder") would be uninformative. By incorporating other data including EEG, age of onset, etiology, family history, imaging studies, and precipitating factors, one obtains an epilepsy classification, which is a much more powerful tool for patient care than the seizure classification alone. The benefits of the syndromic approach have been emphasized and described in detail,[8, 82] and the International League Against Epilepsy explicitly states that the classification of syndromes, not of seizures, should be "used daily in communication between colleagues" and be the "subject of clinical trials and other investigations."[21]

REFERENCES

1. Adelson PD, Peacock WJ, Chugani HT, et al: Temporal and extended temporal resections for the treatment of intractable seizures in early childhood. Pediatr Neurosurg 18(4):169–178, 1992.
2. Aicardi J: Epileptic syndromes in childhood. Epilepsia 29(suppl 3):S1–S5, 1988.
3. Anderman F: Functional hemispherectomy: Clinical indications and outcome. In Wyllie E (ed): The Treatment of Epilepsy: Principles and Practice, 2nd ed. Baltimore, Williams & Wilkins, 1997, pp 1074–1080.
4. Appleton RE, Beirne M: Absence epilepsy in children: The role of EEG in monitoring response to treatment. Seizure 5:147–148, 1996.
5. Beaumanoir A: Les limites nosologiques du syndrome de Lennox-Gastaut. Rev Electroencephalogr Neurophysiol Clin 11:468–473, 1981.
6. Beaumanoir A, Bureau M, Deonna T, et al: Continuous spikes and waves during slow sleep. Electrical status epilepticus during slow sleep. In: Acquired Epileptic Aphasia and Related Conditions. London, John Libbey, 1995.
7. Benbadis SR: Observations on the misdiagnosis of generalized epilepsy as partial epilepsy: Causes and consequences. Seizure 8:140–145, 1999.
8. Benbadis SR, Lüders HO: Epileptic syndromes: An underutilized concept. Epilepsia 37:1029–1034, 1996.
9. Benbadis SR, Lüders HO: Generalized epilepsies [letter]. Neurology 46:1194–1195, 1996.
10. Bergin PS, Fish DR, Shorvon SD, et al: Magnetic resonance imaging in partial epilepsy: Additional abnormalities shown with the fluid attenuated inversion recovery (FLAIR) sequence. J Neurol Neurosurg Psychiatry 58:439–443, 1995.
11. Berkovic SF, Andermann F, Andermann E, et al: Concepts of absence epilepsies: Discrete syndromes or biological continuum. Neurology 37:993–1000, 1987.
12. Blom S, Heijbel J: Benign epilepsy of children with centrotemporal EEG foci: Discharge rate during sleep. Epilepsia 16:133–140, 1975.
13. Blume WT, David RB, Gomez MR: Generalized sharp and slow wave complexes: Associated clinical features and long term follow up. Brain 96:289–306, 1973.
14. Bouma PAD, Westendorp RG, van Dijk JG, et al: The outcome of absence epilepsy: A meta-analysis. Neurology 47:802–808, 1996.
15. Bourgeois BFD: Antiepileptic drugs in pediatric practice. Epilepsia 36(suppl 2):S34–S45, 1995.
16. Brenner RP, Atkinson R: Generalized paroxysmal fast activity: Electroencephalographic and clinical features. Ann Neurol 11:386–390, 1982.
17. Brenner RP, Sharbrough FW: Electroencephalographic evaluation in Sturge-Weber syndrome. Neurology 26:629–632, 1976.
18. Cascino GD, Hulihan JF, Sharbrough FW, et al: Parietal lobe lesional epilepsy: Electroclinical correlation and operative outcome. Epilepsia 34:522–527, 1993.
19. Chugani HT, Shewmon DA, Shields WD, et al: Surgery for intractable infantile spasms: Neuroimaging perspectives. Epilepsia 34:764–771, 1993.
20. Chugani HT, Shields WD, Shewmon DA, et al: Infantile spasms. I: PET identifies focal cortical dysgenesis in cryptogenic cases for surgical treatment. Ann Neurol 27:406–413, 1990.
21. Commission on classification and terminology of the International League Against Epilepsy: Proposal for revised clinical and electroencephalographic classification of epileptic seizures. Epilepsia 22:489–501, 1981.
22. Commission on classification and terminology of the International League Against Epilepsy: Proposal for revised classification of epilepsy and epileptic syndromes. Epilepsia 30(4):389–399, 1989.
23. Delgado-Escueta AV, Greenberg DA, Treiman L, et al: Mapping the gene for juvenile myoclonic epilepsy. Epilepsia 30(suppl 4):S8–S18, 1989.
24. Duchowny M: The syndrome of partial seizures in infancy. J Child Neurol 7(1):66–69, 1992.
25. Duchowny M: Epilepsy surgery in children. Curr Opin Neurol 8(2):112–116, 1995.
26. Duchowny M, Lewin B, Jayakar P, et al: Temporal lobectomy in early childhood. Epilepsia 33:298–303, 1992.
27. Duchowny MS, Resnick TJ, Alvarez LA, et al: Focal resection for malignant partial seizures in infancy. Neurology 40:980–984.32, 1990.
28. Engel J Jr: A practical guide for routine EEG studies in epilepsy. J Clin Neurophysiol 1:109–142, 1984.
29. Farrell K: Classifying epileptic syndromes: Problems and a neurobiologic solution. Neurology 43(suppl 5): S8–S11, 1993.
30. Frost JD, Hrachovy RA, Glaze DG: Spike morphology in childhood focal epilepsy: Relationship to syndromic classification. Epilepsia 33:531–536, 1992.
31. Gambardella A, Aguglia U, Guerrini R, et al: Sequential occurrence of benign partial epilepsy and childhood absence epilepsy in three patients. Brain Dev 18(3):212–215, 1996.
32. Gastaut H: A new type of epilepsy: Benign partial epilepsy of childhood with occipital spike-waves. Clin Electroencephalogr 13:13–22, 1982.
33. Gibbs FA, Gibbs EL: Atlas of Electroencephalography, 2nd ed, Vol 2. Cambridge, MA, Addison-Wesley, 1952.
34. Gibbs FA, Gibbs EL, Lennox WG: Influence of the blood sugar level on the wave and spike formation in petit mal epilepsy. Arch Neurol Psychiatry 41:1111–1116, 1939.

35. Gregory DL, Wong PK: Topographical analysis of the centrotemporal discharges in benign rolandic epilepsy of childhood. Epilepsia 25:705–711, 1984.

36. Grunewald RA, Panayiotopoulos CP: Juvenile myoclonic epilepsy: A review. Arch Neurol 50:594–598, 1993.

37. Holmes GL: Surgery for intractable seizures in infancy and childhood. Neurology 43(suppl 5):S28–S37, 1993.

38. Holmes GL, McKeever M, Adamson M: Absence seizures in children: Clinical and electroencephalographic features. Ann Neurol 21:268–273, 1987.

39. Hrachovy RA, Frost JD, Kellaway P: Hypsarrhythmia: Variations on the theme. Epilepsia 25:317–325, 1984.

40. Jantz D: Juvenile myoclonic epilepsy. Cleve Clin J Med 56(suppl 1):S23–S33, 1989.

41. Jayakar P, Duchowny M, Resnick TJ, et al: Localization of seizure foci: Pitfalls and caveats. J Clin Neurophysiol 8:414–431, 1991.

42. Kobayashi K, Nishibayashi N, Ohtsuka Y, et al: Epilepsy with electrical status epilepticus during slow sleep and secondary bilateral synchrony. Epilepsia 35:1097–1103, 1994.

43. Kuzniecki R: Magnetic resonance imaging in developmental disorders of the cerebral cortex. Epilepsia 35(suppl 6):S44–S56, 1994.

44. Lancman ME, Asconapé JJ, Brotherton T, et al: Juvenile myoclonic epilepsy: An underdiagnosed syndrome. J Epilepsy 8:215–218, 1995.

45. Lancman ME, Asconapé JJ, Penry JK: Clinical and EEG asymmetries in juvenile myoclonic epilepsy. Epilepsia 35:302–306, 1994.

46. Leppik IE: Epileptic syndromes: Genetic, diagnostic, and therapeutic aspects: Introduction remarks and symposium overview. Epilepsia 31(suppl 3):S1–S2, 1990.

47. Lerman P, Kivity S: Focal epileptic discharge in children not suffering from clinical epilepsy. Epilepsy Research-Supplement 6:99–103, 1992.

48. Livingston S, Eisner V, Pauli L: Minor motor epilepsy. Diagnosis, treatment and prognosis. Pediatrics 21:916–928, 1958.

49. Loiseau P: Idiopathic and benign partial epilepsies of childhood. In Wyllie E (ed): The Treatment of Epilepsy: Principles and Practice, 2nd ed. Baltimore, Williams & Wilkins, 1997, pp 442–450.

50. Loiseau P, Beaussart M: The seizures of benign childhood epilepsy with rolandic paroxysmal discharges. Epilepsia 14:381–389, 1973.

51. Loiseau P, Duché B, Pedespan JM: Absence epilepsies. Epilepsia 36:1182–1187, 1995.

52. Lombroso CT: A prospective study of infantile spasms: Clinical and therapeutic correlations. Epilepsia 24:135–158, 1983.

53. Lombroso CT: Sylvian seizures and mid-temporal spike foci in children. Arch Neurol 17:52–59, 1967.

54. Lüders H, Lesser RP, Dinner DS, et al: Benign focal epilepsy of childhood. In Lüders H, Lesser RP (eds): Epilepsy: Electroclinical Syndromes. London, Springer-Verlag, 1987, pp 303–346.

55. Markand ON: Slow spike-wave activity in EEG and associated clinical features: Often called "Lennox" or "Lennox-Gastaut" syndrome. Neurology 27:746–757, 1977.

56. Mattson RH: Efficacy and adverse effects of established and new antiepileptic drugs. Epilepsia 36(suppl 2):S13–S26, 1995.

57. Mimaki T, Ono J, Yabuuchi H: Temporal lobe astrocytoma with infantile spasms. Ann Neurol 14:695–696, 1983.

58. Morris HH, Dinner DS, Lüders H, et al: Supplementary motor seizures: Clinical and electroencephalographic findings. Neurology 38:1075–1082, 1988.

59. Niedermeyer E: The Lennox-Gastaut syndrome: A severe type of childhood epilepsy. Electroencephalogr Clin Neurophysiol 24:283, 1968.

60. Niedermeyer E, Walker AE, Burton C: The slow spike wave complex as a correlate of frontal and fronto-temporal post traumatic epilepsy. Eur Neurol 3:330–346, 1970.

61. Panayiotopoulos CP, Obeid T, Waheed G: Differentiation of typical absences in epileptic syndromes: A video EEG study of 224 seizures in 20 patients. Brain 112:1039–1056, 1989.

62. Pellock JM: Panel discussion. In McNamara JO (ed): Antiepileptic drugs: Optimal use and future prospects. Epilepsia 35(suppl 4):S58–S60, 1994.

63. Pellock JM: Antiepileptic drug therapy in the United States: A review of clinical studies and unmet needs. Neurology 45(suppl 2):S17–S24, 1995.

64. Penfield W, Jasper H: Epilepsy and the Functional Anatomy of the Brain. Boston, Little, Brown, 1954.

65. Penry JK, Porter RJ, Dreifuss FE: Simultaneous recording of absence seizures with videotape and electroencephalography: A study of 374 seizures in 48 patients. Brain 98:427–440, 1975.

66. Porter RJ: Recognizing and classifying epileptic seizures and epileptic syndromes. Neurol Clin 4(3):495–508, 1986.

67. Reutens DC, Berkovic SF: Idiopathic generalized epilepsy of adolescence: Are the syndromes clinically distinct? Neurology 45:1469–1476, 1995.

68. Ruggieri V, Caraballo R, Fejerman N: Intracranial tumor and West syndrome. Pediatr Neurol 5:327–329, 1989.

69. Salanova V, Morris HH, Van Ness PC, et al: Comparison of scalp electroencephalogram with subdural electrocorticogram recordings and functional mapping in frontal lobe epilepsy. Arch Neurol 50:294–299, 1993.

70. Spencer S: The relative contributions of MRI, SPECT, PET imaging in epilepsy. Epilepsia 35(suppl 6):S72–S89, 1994.

71. Thomas P, Beaumanoir A, Genton P, et al: 'De novo' absence status of late onset: Report of 11 cases. Neurology 42:104–110, 1992.

72. Uematsu S, Lesser R, Fisher RS, et al: Motor and sensory cortex in humans: Topography studied with chronic subdural stimulation. Neurosurgery 31:59–72, 1992.

73. Vivegano F, Bertini E, Boldrini R, et al: Hemimegalencephaly and intractable epilepsy: Benefits of hemispherectomy. Epilepsia 30:833–843, 1989.

74. Williamson PD, Spencer SS: Clinical and EEG features of complex partial seizures of extratemporal origin. Epilepsia 27(suppl 2):S46–S63, 1986.

75. Wolf D, Gooses R: Relationship of photosensitivity to epileptic syndromes. J Neurol Neurosurg Psychiatry 49:1386–1391, 1986.

76. Wyllie E: Developmental aspects of seizure semiology: Problems in identifying localized-onset seizures in infants and children. Epilepsia 36(12):1170–1172, 1995.

77. Wyllie E: Epilepsy surgery in infants. In Wyllie E (ed): The Treatment of Epilepsy: Principles and Practice, 2nd ed. Baltimore, Williams & Wilkins, 1997, pp 1087–1096.

78. Wyllie E, Bass NE: Supplementary sensorimotor area seizures in children and adolescents. Adv Neurol 70:301–308, 1996.

79. Wyllie E, Chee M, Granstrom ML, et al: Temporal lobe epilepsy in early childhood. Epilepsia 34:859–868, 1993.

80. Wyllie E, Comair Y, Kotagal P, et al: Epilepsy surgery in infants. Epilepsia 37:625–637, 1996.

81. Wyllie E, Comair YG, Kotagal P, et al: Seizure outcome after epilepsy surgery in children and adolescents. Ann Neurol 44:740–748, 1998.

82. Wyllie E, Lüders H: Classification of seizures. In Wyllie E (ed): The Treatment of Epilepsy: Principles and Practice, 2nd ed. Baltimore, Williams & Wilkins, 1997, pp 355–357.

Chapter 34

Focal Epilepsy and Surgical Evaluation

Nancy Foldvary, DO

INTRODUCTION
GENERAL CONCEPTS
ELECTROGRAPHIC FEATURES OF SIMPLE
 PARTIAL SEIZURES
ELECTROENCEPHALOGRAPHY IN
 TEMPORAL LOBE EPILEPSY
ELECTROENCEPHALOGRAPHY IN FRONTAL
 LOBE EPILEPSY

ELECTROENCEPHALOGRAPHY IN PARIETAL
 LOBE EPILEPSY
ELECTROENCEPHALOGRAPHY IN
 OCCIPITAL LOBE EPILEPSY
SURGICAL EVALUATION
ILLUSTRATIVE CASES

INTRODUCTION

The focal epilepsies are characterized by seizures in which the first clinical or electrographic manifestations, or both, indicate activation of a limited population of cortical neurons in part of one cerebral hemisphere.[12] The classification of epilepsies and epileptic syndromes proposed by the International League Against Epilepsy in 1989 subdivides the focal epilepsies into idiopathic and symptomatic forms.[12] The idiopathic focal epilepsies include the benign childhood epilepsies, which are addressed elsewhere. Chapter 33 discusses the electrographic features of the symptomatic focal epilepsies (Table 34–1).

GENERAL CONCEPTS

Electroencephalogram (EEG) abnormalities are divided into interictal and ictal epileptiform patterns. The term *epileptiform* describes waveforms or complexes distinguishable from background activity resembling those seen in patients with epilepsy. A spike is a transient having a pointed peak and a duration less than 80 msec (Fig. 34–1). A sharp wave is a pointed transient with a duration of 80 to less than 120 msec, usually with a steep ascending and a more prolonged descending slope. Spikes and sharp waves are usually of negative polarity because they are generated by the hypersynchronous depolarization of vertically oriented neurons. Surface positive epileptic discharges are more common in depth and electrocorticographic recordings; in patients with skull defects, head trauma, cortical dysplasia; and in cases where the generator produces a horizontally oriented dipole. In general, the frequency, repetition rate, morphologic characteristics, and state dependence of interictal epileptiform activity cannot be used to predict the etiology of the disorder. However, nearly continuous, repetitive sharp waves or spikes not associated with behavioral change have been found to be relatively specific for focal cortical dysplastic lesions.[21] Spikes and sharp waves convey an increased risk of epilepsy and must be differentiated from benign variants having epileptiform features.

The distribution of scalp-recorded epileptiform activity provides a reasonable estimate of the epileptogenic zone. The epileptogenic zone is defined as the region of cortex capable of generating seizures, the complete removal or disconnection of which is required to produce a seizure-free state.[36] However, spikes and sharp waves do not necessarily originate from cortical tissue directly beneath or even in the vicinity of the recording electrode(s). Their distribution on the scalp is dependent on the recording technique, the spatial characteristics of the generator, and the conductive properties of the surrounding tissue. Consequently, the distribution of interictal epileptic discharges and ictal patterns may fail to localize or even mislocalize the region or hemisphere of seizure origin.

Several factors influence the detection of interictal discharges on routine EEG. These include seizure type, location of the generator, patient age and state, recording technique, and seizure frequency.[13] Abnormal recordings are more common in children, particularly during non–rapid eye movement (NREM) or following sleep deprivation.[3, 50, 51] Prolonged sleep deprivation produces epileptiform activity in approximately 40% of epileptic subjects with normal awake EEGs.[45] As compared to wakefulness, spike rate is increased during NREM sleep and reduced during rapid eye movement (REM) sleep.[40, 56] In some cases, the extent of field is more restricted in REM as compared to wakefulness, while extension of the field and new spike foci are observed in slow wave sleep (stages 3 and 4 NREM).[56] Therefore, the localization of interictal discharges in REM sleep may be a better indicator of the epileptogenic zone than in NREM sleep or wakefulness. Spike rate increases markedly during the first 24 to 48 hours after a clinical seizure.[26, 66] Reduction in antiepileptic drug (AED) levels does not increase interictal epileptiform activity.[27] AED therapy does not reduce the yield of an EEG in patients with epilepsy.[66] Hyperventilation and photic stimulation rarely activate epileptiform activity in patients with focal epilepsy. Closely spaced scalp and semi-invasive electrodes improve the yield of spike detection over the standard 10–20 system.[42] Epileptic discharges are more common in patients with seizures of temporal lobe origin as compared to epilepsies arising from deep or midline regions. This is particularly true when sphenoidal or additional scalp electrodes are employed.[41]

Despite the use of activating methods and special electrode arrays, surface EEG has limitations.[10] The sensitivity of scalp recordings is determined by the depth, size, and orientation of the generator and the duration and synchronization of an epileptic discharge.[31] Cerebral activity is attenuated by the impedance characteristics of the intervening

TABLE 34–1	Symptomatic Epilepsies

Temporal lobe epilepsy
   Mesiobasal limbic
   Lateral temporal (neocortical)
Frontal lobe epilepsy
   Supplementary motor
   Cingulate
   Anterior frontopolar
   Orbitofrontal
   Dorsolateral
   Opercular
   Motor cortex
Parietal lobe epilepsy
Occipital lobe epilepsy

brain tissue, cerebrospinal fluid, meninges, skull, and scalp. Higher frequencies, which are commonly observed during the evolution of ictal patterns, are attenuated more than lower frequencies. There is no consistent or direct relationship between the amplitude of epileptiform discharges recorded simultaneously from intracranial and scalp electrodes. Spikes recorded from subdural or depth electrodes are frequently not seen on surface EEG.[1] Consequently, epileptic discharges, particularly those arising from deep regions, are apt to escape detection or produce widespread discharges of limited localizing value. The yield of an abnormal EEG in subjects with seizures increases from approximately 50% on the first tracing to over 90% by the fourth EEG.[55] Therefore, a normal interictal tracing does not exclude the diagnosis of epilepsy. Furthermore, epileptiform activity is observed in 2% of individuals without prior seizures.[76]

The transition from the interictal to the ictal state is variable and may be difficult to identify particularly when seizures are brief or occur during state changes. Clinical manifestations frequently precede the electrographic onset of scalp-recorded seizures. Repetitive interictal discharges which characterize some generalized seizures (typical absence seizures and 3-Hz spike wave complexes) are uncommon in focal epilepsy. Several ictal patterns of focal seizures have been described.[9, 57] The most common is rhythmic sinusoidal activity in the beta, alpha, or theta ranges or repetitive epileptiform discharges that evolve in frequency, field, or

morphology[23] (Fig. 34–2 A,B). A sudden generalized or lateralized suppression or attenuation of amplitude is observed in some focal seizures, particularly in those arising from deep structures, such as orbitofrontal or mesial temporal regions[30, 69] (Fig. 34–3). Sinusoidal activity was observed in 47% of seizures, repetitive epileptiform discharges in 39%, a combination of the two in 15%, and an attenuation of background activity in 11% of seizures in a series of patients with focal epilepsy.[9] Observed in over 90% of seizures was an initial increase or decrease in frequency, followed by a gradual decrease in frequency toward the end of the event.[9] In another series, the ictal onset consisted of rhythmic activity in the beta range or faster in 50% of seizures, theta range in 20%, and alpha or delta activity in 10% each, whereas no electrographic correlate was observed in 10% of seizures.[23] An abrupt cessation of interictal epileptiform activity was observed immediately before ictal onset in nearly 75% of seizures.[23] Ictal discharges limited to one or two electrodes are relatively uncommon in surface recordings. Regional or lateralized delta activity or attenuation postictally is a reliable predictor of the side of seizure origin.[29, 32, 70] Using lateralized rhythmic theta and alpha activity, postictal slowing and activity at seizure onset, 47% to 65% of extratemporal and 76% to 83% of temporal seizures were correctly lateralized in a series of patients with complex partial seizures (CPS).[70] In another series of focal epilepsy, ictal EEG correctly localized 40% of seizures, while the majority of the remainder were nonlateralized or uninterpretable.[47]

ELECTROGRAPHIC FEATURES OF SIMPLE PARTIAL SEIZURES

Simple partial seizures (SPS) are focal seizures characterized by motor, autonomic, somatosensory, special sensory, psychic, affective, or nonspecific symptoms during which consciousness is preserved. The term *aura* is defined as that portion of a seizure which occurs before loss of consciousness and for which memory is retained afterward. It is generally reserved for subjective sensations of epileptic origin reported by the patient in the absence of objective sings. The symptoms of SPS reflect the function of the cortex activated by the seizure, thereby providing clues to seizure localization.

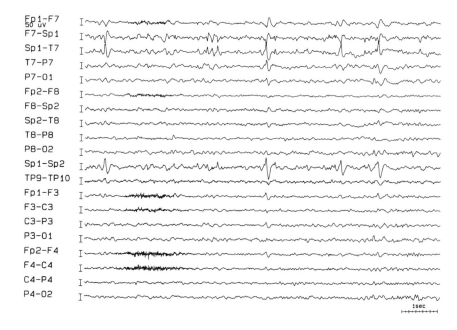

FIGURE 34–1. Sharp waves in the left mesial temporal region in a patient with epilepsy caused by left mesial temporal sclerosis.

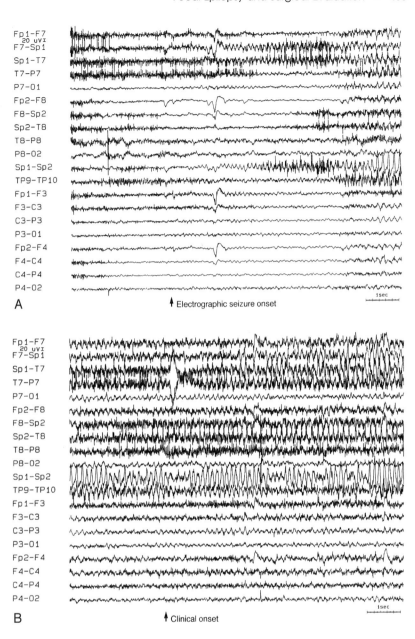

FIGURE 34–2. *A, B,* Rhythmic theta in the left mesial temporal region maximal at Sp1 in a patient with medically refractory seizures caused by mesial temporal sclerosis (see Case Study 1).

However, seizures originating in functionally silent cortex produce symptoms only when the ictal discharge spreads to symptomatic areas that may be at some distance from the generator. The electrographic manifestations of SPS depend on the recording technique and location of the epileptogenic zone. Electrographic changes were detected in only 21% of scalp recorded SPS in one series, more often in motor (33%) than nonmotor (15%) seizures.[14] The proximity of the recording electrode to the motor cortex is one proposed explanation for the higher yield of ictal changes during SPS with motor features. With the use of additional closely spaced scalp electrodes, electrographic correlates were detected in 61% of SPS.[6] In nearly half, the EEG change was maximal at an added electrode. These findings emphasize the limitations of surface EEG in patients with SPS. When consciousness is preserved, a normal EEG does not assure the nonepileptic nature of an event. Invasive recordings using depth or subdural electrodes are significantly better than surface EEG in this regard, detecting 41% to 90% of SPS.[15, 62, 63] In general, the onset of SPS detected by invasive electrodes reliably predicts the onset of complex

partial and secondary generalized tonic-clonic seizures.[15, 62, 63] Variability in the detection of EEG changes during invasively recorded SPS suggests that activation of very limited populations of neurons without significant distant propagation is sufficient to produce clinical symptoms.[5]

ELECTROENCEPHALOGRAPHY IN TEMPORAL LOBE EPILEPSY

Temporal lobe epilepsy (TLE) constitutes nearly two-thirds of the localization-related epilepsies that present during adolescence and adulthood. Seizures arising from the mesial temporal region (amygdala or hippocampus) are more common than those of lateral or neocortical temporal origin. Auras, including visceral sensations, fear, anxiety, olfactory disturbances, and psychic phenomena, are reported in mesial temporal lobe epilepsy (MTLE). Auditory auras and complex visual phenomena are more common in neocortical TLE. Complex partial seizures characterized by motionless

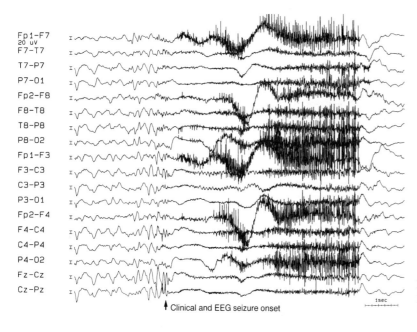

Fp1-F7
20 uv
F7-T7
T7-P7
P7-O1
Fp2-F8
F8-T8
T8-P8
P8-O2
Fp1-F3
F3-C3
C3-P3
P3-O1
Fp2-F4
F4-C4
C4-P4
P4-O2
Fz-Cz
Cz-Pz

↑ Clinical and EEG seizure onset

1sec

FIGURE 34–3. Abrupt generalized suppression is a patient with asymmetrical tonic seizures arising from the right supplementary sensory motor area.

staring and oral and manual automatisms are the most common seizure type observed in seizures arising from the mesial temporal region.[37] Approximately 50% of individuals have secondary generalized tonic-clonic seizures. In series comparing mesial and neocortical temporal seizure semiology, automatisms and dystonic posturing of the contralateral upper extremity in the initial part of the seizure were more common in MTLE, whereas early clonic activity of the contralateral upper extremity or facial grimacing suggested neocortical origin.[20, 25, 43]

During prolonged EEG monitoring, epileptic discharges localized to the sphenoidal or anterior temporal electrodes are observed in over 90% of patients with MTLE.[19, 73] Bitemporal independent spikes or sharp waves, maximal on the side of seizure origin, occur in 25% to 50% of cases.[19, 30, 73] Anterior temporal discharges are better defined using additionally placed electrodes designed to record from the basal or mesial temporal regions (Fig. 34–4 A,B). Surface electrodes placed 1 cm above a point one-third the distance between the external auditory meatus and the external canthus (T_1, T_2), FT_9, and FT_{10} in the 10–10 system, or sphenoidal electrodes are most commonly used.[58] Sphenoidal electrodes are placed beneath the zygomatic arch approximately 2.5 cm anterior to the incisura intertragica roughly 10 degrees superiorly from the horizontal plane and posteriorly from the coronal plane.[34] Sphenoidal electrodes are well tolerated, relatively artifact free, and have a low complication rate. Mesial temporal spikes have maximal negativity at anterior temporal or sphenoidal electrodes and a widespread positivity over the vertex.[8, 18, 19] Epileptic discharges are absent on serial recordings in approximately 10% of patients with TLE. Temporal intermittent rhythmic delta activity (TIRDA) consists of trains of 10 seconds or more of repetitive, rhythmic, saw-toothed or sinusoidal 1- to 4-Hz activity 50 to 100 μV in amplitude in the anterior temporal regions. It is observed in one-third of cases[48] (Fig. 34–5).

The distribution of epileptiform activity in patients with neocortical TLE varies with the location of the epileptogenic zone. Interictal spiking may be maximal at lateral or posterior temporal electrodes with minimal involvement of the sphenoidal and anterior temporal electrodes[16] (Fig. 34–6).

However, sphenoidal and anterior temporal interictal discharges have also been observed.[20, 43] Epileptiform activity must be differentiated from wicket spikes and benign epileptiform transients of sleep (BETS), normal variants seen in the temporal regions during light sleep.

Temporal lobe seizures are most often characterized by a gradual buildup of lateralized or localized rhythmic alpha or theta activity that may or may not be preceded by diffuse or lateralized suppression or arrhythmic activity[30, 49] (see Fig. 34–2). Lateralized rhythmic theta or alpha activity in the ipsilateral temporal region, appearing within the first 30 seconds of clinical or electrographic seizure onset, is observed in approximately 80% of patients with seizures of mesial temporal origin and correctly predicts an ipsilateral temporal onset in over 80% of patients with TLE confirmed by invasive recordings.[19, 49, 73] False lateralization with scalp electrodes has been observed in 3% to 13% of MTLE patients.[19, 73] Lateralized postictal slowing or background attenuation correctly predicts the side of seizure origin in 96% to 100% of cases.[70, 73] Focal ictal onset, characterized by rhythmic activity or repetitive spiking at one electrode with limited spread is observed in approximately 25% of subjects.[68] Depending on the series, diffuse suppression or nonlateralized rhythmic activity characterize one-third to three-quarters of temporal lobe seizures.[61, 68] Nonlateralized and falsely lateralized patterns are more common in patients with bitemporal independent epileptic discharges.[64]

It may be possible to differentiate mesial and neocortical temporal seizures based on ictal recordings.[17, 20] An initial, regular 5- to 9-Hz inferotemporal rhythm was found to be more specific for hippocampal-onset seizures. However, the presence of this pattern requires the synchronous recruitment of adjacent inferolateral temporal neocortex.[17] Seizures confined to the hippocampus by intracranial EEG produce no change on scalp recordings.[44] Owing to the vertical orientation of dipole sources, seizures of mesiobasal temporal origin may produce rhythmic activity of positive polarity at the vertex, and negative polarity at the sphenoidal electrode and electrodes located near the base of the skull.[17, 44] Neocortical seizures are more often associated with low-voltage activ-

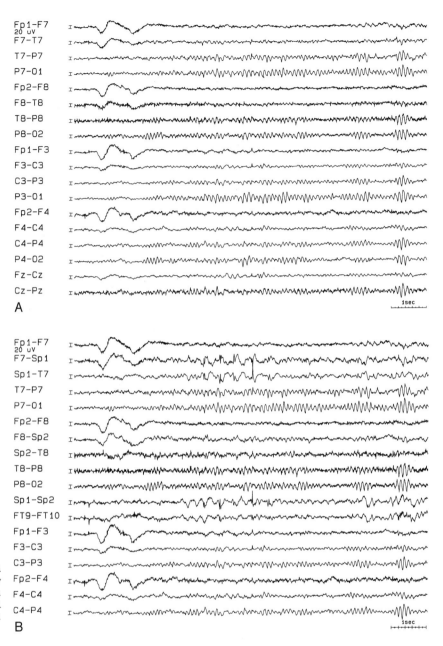

FIGURE 34–4. An interictal recording in a subject with left mesial temporal lobe epilepsy (MTLE) shown on A-P bipolar montages with (A) and without (B) sphenoidal electrodes. A series of Sp1 sharp waves escapes detection on the standard montage.

ity in the beta range or faster, rhythmic activity of less than 5 Hz, or nonlateralized patterns.[17, 20, 43]

ELECTROENCEPHALOGRAPHY IN FRONTAL LOBE EPILEPSY

The frontal lobes are the second most common site of origin of focal seizures. Ictal semiology is typically more helpful than EEG in seizure localization. Simple partial, complex partial, atonic, tonic, and tonic-clonic seizures are observed. Simple partial seizures are characterized by motor, sensory, autonomic, affective, and cognitive phenomena. Frontal lobe CPS tend to be brief with little or no postictal confusion, often occurring in clusters, many per day. Complex partial status epilepticus occurs in 40% of cases.[74] Complex motor automatisms, including bicycling and thrashing movements, sexual automatisms, and vocalizations may result in their misdiagnosis as psychogenic seizures. Supplementary motor area (SMA) seizures are characterized by sudden onset of bilateral, asymmetrical tonic posturing associated with vocalization and preservation of consciousness. Seizures similar to those seen in TLE, consisting of automatisms, autonomic, and affective symptoms, are observed in epilepsy of the cingulate gyrus and orbitofrontal cortex. Version, forced thinking, falls, and autonomic signs are observed in seizures originating from the frontopolar region. Activation of the dorsolateral frontal convexity produces tonic and clonic activity, version, and speech arrest. Opercular seizures are characterized by gustatory hallucinations, salivation, mastication, swallowing, speech arrest, and autonomic signs. Focal clonic activity, Todd's paralysis, and epilepsia partialis continua are observed in seizures involving the peri-rolandic area.

Several factors contribute to the frequent lack of EEG localization or mislocalization in frontal lobe epilepsy as compared to TLE. These include the inaccessibility of much of the frontal lobes to surface electrodes, the rapid spread of seizures within and outside the frontal lobe, secondary bilateral synchrony and bilateral epileptogenesis caused by

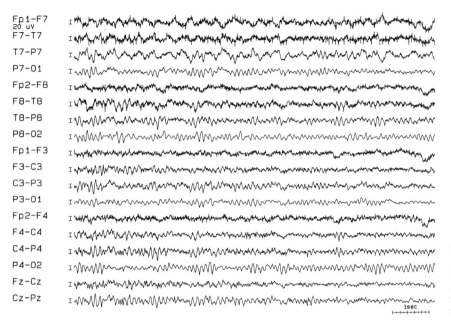

FIGURE 34–5. Temporal intermittent rhythmic delta activity (TIRDA). Intermittent rhythmic delta activity in the left temporal region seen in the interictal EEG of a subject with epigastric auras and partial seizures with oral and manual automatisms.

bifrontal injury, and variability in size of seizure onset zones.[46] The interictal EEG may be entirely normal or show generalized or lateralized slowing, focal, hemispheric, multiregional, or generalized spikes or polyspikes or low-voltage fast activity (Fig. 34–7). In one large series, interictal spiking was bilaterally synchronous in 37%, lobar in 32%, multilobar in 24%, focal in 12%, and hemispheric or bifrontal independent in 9% of cases.[46] Epileptiform activity appears exclusively outside the frontal lobes in 20% to 50% of cases and is restricted to the frontal lobes in only 25%.[35, 74] Spiking is absent in 20% to 70% of cases.[35, 74] Spike detection may be improved with supra- and infraorbital, sphenoidal, or additional closely spaced scalp electrodes in patients with suspected orbitofrontal or mesial frontal lobe epilepsy (Fig. 34–8). Midline epileptiform activity may be mistaken for vertex sharp transients that tend to appear spike-like in children or may escape detection if viewed on montages without midline electrodes. Secondary bilateral synchrony, the phe-

nomenon in which a unilateral epileptogenic focus near the midline produces seemingly bilaterally synchronous generalized epileptiform activity, is observed in up to 20% of cases[74] (Fig. 34–9). The presence of unilateral discharges that precede and initiate bilateral epileptiform activity, distinct morphology of focal and bilateral discharges, isolated unilateral discharges, and focal slowing differentiate SBS from the generalized epileptiform activity observed in the idiopathic generalized epilepsies.

Frontal lobe complex partial seizures are typically brief, beginning and ending abruptly, and are frequently characterized by excessive movement or tonic posturing, resulting in obscuration of the EEG. Approximately one-third of frontal lobe seizures are not accompanied by EEG changes.[35, 41, 74] Another one-third are characterized by nonlateralized slowing, rhythmic activity, or repetitive spiking[46] (Fig. 34–10 A, B). Focal ictal onsets are less common in frontal lobe seizures as compared to seizures of temporal lobe origin.[68] Lateral-

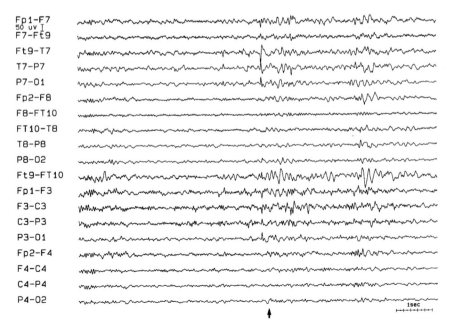

FIGURE 34–6. Interictal tracing from a patient with a left lateral temporal cavernous angioma and lateral temporal spikes (arrow).

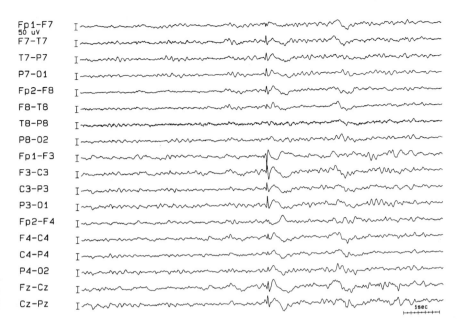

FIGURE 34–7. Left frontal spike in a subject with right arm clonic seizures resulting from a low-grade glioma.

ized or localized ictal patterns are observed in 33% to 50% of cases.[35, 46, 68] Occasionally, the ictal onset appears contralateral to the hemisphere of seizure origin.[41] Known as paradoxical lateralization, this phenomenon is observed when a generator is located within the interhemispheric fissure and produces an obliquely oriented dipole that projects to the opposite hemisphere.

ELECTROENCEPHALOGRAPHY IN PARIETAL LOBE EPILEPSY

Seizures arising from the parietal lobe constitute 5% to 6% of cases of focal epilepsy in medical and surgical series.[24, 39, 54] Simple partial seizures with somatosensory symptoms involving a contralateral extremity or face, occasionally experienced bilaterally, are described. These include paresthesias, numbness, pain, or thermal sensations of cold or burning.

Partial seizures with motor features, nonspecific cephalic sensations, visual hallucinations, vertigo, disturbances in body image, gustatory phenomena, ideomotor apraxia, nystagmus, and genital sensations have been reported.[53, 54, 67] Psychic auras, epigastric sensations, and ictal amaurosis suggesting activation of extraparietal regions may occur.[72] The symptoms of CPS of parietal lobe origin suggest different spread patterns. Seizures characterized by asymmetrical tonic posturing of the limbs, unilateral clonic activity, and contralateral version are observed with secondary activation of the frontal regions, whereas spread to the temporal lobes produces alteration of consciousness and automatisms.[22, 53, 54, 72]

The interictal EEG in parietal lobe epilepsy may be normal or show lateralized or generalized intermittent slow waves.[39] If present, interictal epileptiform activity is usually multiregional, suggesting involvement of areas distant from the epileptogenic zone[11, 24, 39, 53, 54, 72] (Fig. 34–11). Centro-parietal epileptiform activity as the sole interictal abnormality is ob-

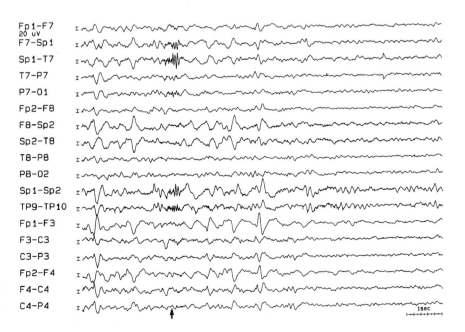

FIGURE 34–8. Polyspikes *(arrow)* at the left sphenoidal electrode in a patient with poorly controlled seizures caused by a left orbitofrontal cavernous angioma.

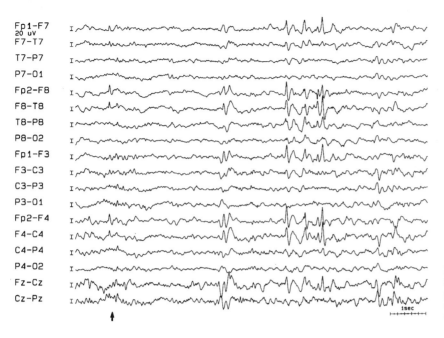

FIGURE 34–9. Secondary bilateral synchrony in a subject with a right mesial frontal low-grade astrocytoma. Bilateral synchronous sharp waves and isolated right frontal sharp waves *(arrow)* are observed.

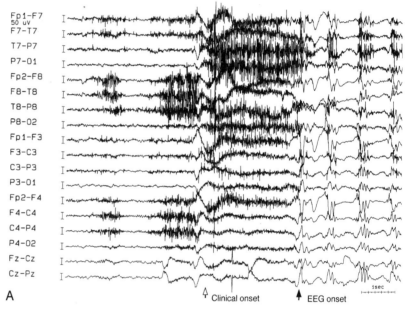

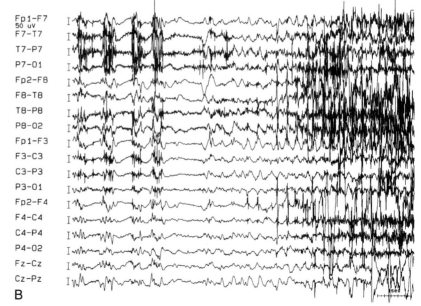

FIGURE 34–10. *A, B,* Serial 10-second tracings during a seizure of a patient with left frontal lobe epilepsy (see Case Study 2) characterized by abrupt onset of facial distortion, vocalization, repetitive leg movements.

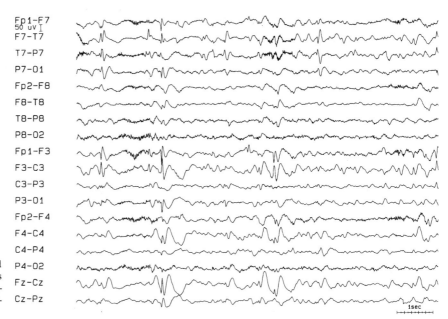

FIGURE 34–11. Left frontal, temporal, and bifrontal, left greater than right, sharp waves in a young girl with a left parietal oligodendroglioma and refractory seizures characterized by staring and behavioral arrest.

served in 5% to 20% of cases.[11, 53, 54, 72] Secondary bilateral synchrony occurs in one-third of patients.[53] Ictal patterns vary depending on the pathway of propagation, and may erroneously suggest a temporal or frontal lobe focus. SPS of parietal lobe origin frequently have no EEG correlate.[11, 14] CPS most commonly produce diffuse suppression or nonlateralized rhythmic activity, although lateralized ictal patterns have been reported[72] (Fig. 34–12). Localized seizure onsets were recorded in 6% of cases in a large surgical series.[53]

ELECTROENCEPHALOGRAPHY IN OCCIPITAL LOBE EPILEPSY

Occipital lobe epilepsy comprises 8% of cases of focal epilepsy.[24] SPS are characterized by elementary visual phenomena, sensations of ocular movement, nystagmus, eye flutter, forced eye blinking, ictal amaurosis, or versive head and eye movements. Elementary visual hallucinations are the most common manifestation, observed in 50% to 60% of cases.[52, 75] The clinical manifestations of occipital lobe CPS reflect different patterns of ictal propagation. Infrasylvian spread to the temporal lobe produces alteration of awareness and automatisms. Suprasylvian spread to the mesial frontal lobe produces asymmetrical tonic posturing, whereas propagation laterally results in focal motor or sensory seizures. Multiple seizure types are observed in one-third to one-half of patients.[52, 75]

The interictal EEG in occipital lobe epilepsy commonly reveals widespread epileptiform activity or activity maximal in the posterior temporal regions.[52, 75] Bitemporal independent, bilateral synchronous frontal complexes, biooccipital, and diffuse epileptiform activity is observed in 30% to 50% of cases.[52] Epileptiform activity is restricted to the occipital lobes in 8% to 18% of patients[52, 75] (Fig. 34–13). Ictal recordings show diffuse suppression or rhythmic activity that is usually generalized but may be lateralized or maximal over

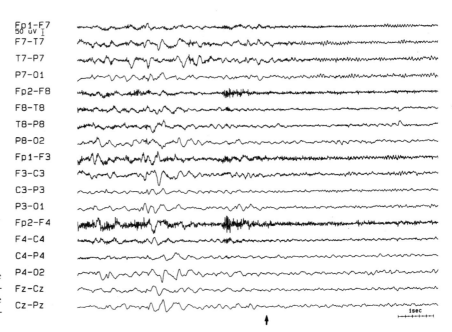

FIGURE 34–12. A typical seizure of the same subject as in Figure 34–11. Generalized suppression and low-voltage fast activity over the left hemisphere is shown. The *arrow* indicates clinical and EEG seizure onset.

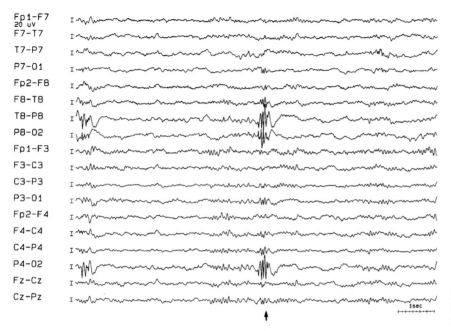

FIGURE 34–13. Right parieto-occipital poly-spikes *(arrow)* from a subject with left visual auras (see Case Study 3).

the temporo-occipital region[52, 75] (Fig. 34–14). Ictal onset restricted to the occipital lobe is seen in less than 20% of cases.[52] Invasive recordings show spread of seizure activity from the occipital region to the mesial temporal structures, the supplementary motor area, or dorsolateral frontal convexity before generalization.[75]

SURGICAL EVALUATION

The goal of surgical evaluation is to define the epileptogenic zone. The evaluation includes clinical history, ictal and interictal EEG, neuroimaging (magnetic resonance imaging [MRI], single photon emission computed tomography [SPECT], positron emission tomography [PET], computed tomography [CT], and magnetic resonance spectroscopy [MRS]), and neuropsychological assessment (neuropsychological battery, intracarotid sodium amytal test, psychiatric

evaluation). A recommendation for surgery is made when the evaluation successfully delineates the epileptogenic zone, and the resection of which carries a low risk of neurological and cognitive morbidity.

Long-term video EEG monitoring refers to the simultaneous recording of EEG and clinical behavior over several days for the evaluation of patients with suspected epilepsy and nonepileptic paroxysmal disorders. These include syncope, cardiac arrhythmias, transient ischemic attacks, narcolepsy, and psychiatric disorders.[2] The technique is also useful for quantifying seizure frequency and determining the response to therapeutic interventions. Long-term monitoring allows for the identification of interictal epileptiform and nonepileptiform abnormalities and the correlation of clinical behavior with EEG findings. During behavioral events, trained personnel interact with patients to determine the degree of responsiveness and language and motor function. These features may help to establish the epileptic nature of an

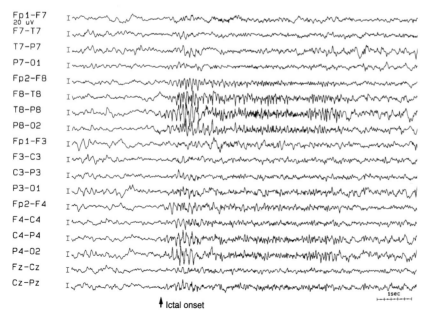

FIGURE 34–14. Ictal onset *(arrow)* in a seizure from the same subject as in Figure 34–13 with right occipital lobe epilepsy. Widespread sharply contoured alpha mixed with low-voltage fast activity appears over the right hemisphere, maximum right parieto-temporo-occipital.

event as well as assist in localization and lateralization of seizures. Modified electrode placements and semi-invasive techniques improve the localization of interictal and ictal activity. Withdrawal from anticonvulsant medication allows for the expedient recording of habitual seizures without affecting their clinical and electrographic manifestations.[38]

The extent of interictal epileptiform abnormalities tends to be larger than the area of cortex from which a clinical seizure originates.[36] Nevertheless, interictal epileptiform activity generally provides more reliable localizing information than ictal EEG. This is because a seizure has usually spread outside the zone of ictal onset to contiguous or even remote areas by the time an ictal pattern is detected by scalp electrodes. In a series of patients with temporal lobe epilepsy, the presence of a single interictal spike focus in the anteromesial temporal region accurately predicted temporal lobe onset whether the surface ictal EEG was focal, regional, or lateralized to the same hemisphere.[33] In cases with more than one temporal spike focus or in which the predominant focus was in the mid- to posterior temporal region, the interictal focus was less accurate in predicting the ictal onset zone.[33] Patients with unilateral temporal interictal spikes have a better outcome after temporal lobectomy than those with bitemporal independent discharges and extratemporal spike populations.[7]

The number of seizures needed to reliably determine seizure localization has not been standardized. Studies of the reproducibility and reliability of ictal patterns indicate that the first recorded seizure is representative of subsequent seizures in 68% of cases, and predictive of the ultimate clinical localization in 91% of patients with TLE.[59, 65] When the initial seizure is nonlocalized, particularly if it occurs in sleep, subsequent seizures are likely to remain nonlocalized.[65] Patients with unilateral interictal temporal epileptic discharges are more likely to have consistent ictal patterns than those with bitemporal independent epileptiform activity.[59] A minimum of four seizures was required to identify 100% of 18 patients with ictal patterns arising from both temporal regions independently.[59] However, in a series of medically refractory focal epilepsy not limited to the temporal lobe, 21% to 38% of scalp-recorded seizures were correctly localized and less than one-half correctly lateralized when depth recordings were used to define the epilepto-

genic zone.[61] Seizure outcome after anterior temporal lobectomy was found to be comparable whether or not video-EEG monitoring was performed in patients with unilateral anteromesial temporal interictal discharges concordant with neuroimaging and functional studies, emphasizing the importance of the interictal EEG.[28, 71]

Intracranial EEG is required to adequately delineate the epileptogenic zone when surface EEG and neuroimaging studies do not localize the lesion and epilepsy surgery is being considered. While more sensitive than surface EEG, intracranial EEG provides a limited view of cerebral activity as recordings are obtained only from areas where electrodes are placed. The successful use of intracranial EEG requires that prior noninvasive studies have provided enough information to assure adequate electrode placement. Invasive monitoring is performed with depth or subdural electrodes. Depth electrodes are bipolar or multiple-contact wires placed stereotactically into the brain. They are most commonly used for identifying the epileptogenic zone in patients with TLE and bilateral abnormalities on surface EEG or neuroimaging studies, or both (Fig. 34–15). Depth electrodes may be left in place for up to 4 weeks. The risks of depth electrode implantation include hemorrhage, infection, and infarction. Hemorrhage is the most common major complication, occurring in less than 2% of cases.[60] Subdural electrodes are embedded in thin Silastic plates arranged in various sizes of strips or grids. Electrodes are inserted through a burr hole or craniotomy into the subdural space over cortical regions of interest. This technique allows for the recording of multiple regions over one or both hemispheres and mapping of functional cortex. Complications, including infection of the central nervous system or bone flap and cerebral edema, occur in 1% to 4% of cases. The complication rate increases with the number of electrodes inserted.[4]

ILLUSTRATIVE CASES

CASE STUDY 1

A 31-year-old right-handed woman presented for evaluation of poorly controlled seizures. Her birth and develop-

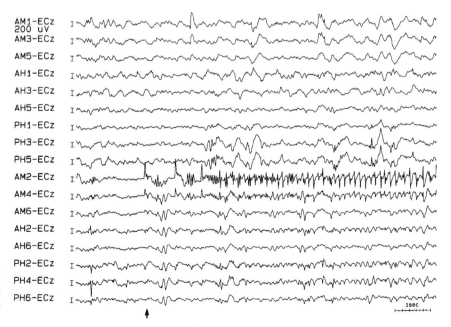

FIGURE 34–15. Bitemporal depth recording in a subject with bilateral mesial temporal sclerosis. Typical auras were preceded by rhythmic activity beginning in the right amygdala (AM2) *(arrow)*.

ment were normal. At the age of 6 months, she had a generalized tonic-clonic seizure with right Todd's paralysis during a febrile illness. During elementary school, she recalls episodes of an unusual odor akin to toothpaste or "the smell of a dentist's office" lasting for up to 30 seconds without alteration of consciousness. At the age of 21 years, she began having monthly tonic-clonic seizures not preceded by an aura. Later that year, a second seizure type developed characterized by the same odor she experienced in childhood. This was followed by a loss of awareness, oral and manual automatisms, and unintelligible speech occurring 10 to 15 times per month. Seizures remained poorly controlled despite adequate trials with phenytoin, carbamazepine, valproate, phenobarbital, primidone, and gabapentin in various combinations. General and neurological examinations were normal. Family history was negative for epilepsy and neurologic disorders.

The patient was admitted to the epilepsy monitoring unit where anticonvulsant medications were gradually tapered and one aura and one seizure were recorded. The seizure was characterized by an alteration of awareness, left hand automatisms, right arm dystonic posturing, evolving to head version to the right. The interictal EEG showed normal background waking and sleep rhythms. Intermittent slowing was observed in the left temporal region. Frequent sharp waves were observed having maximal amplitude at the left sphenoidal electrode (90%) and rarely maximal at the left anterior temporal (F9) region (5%) or right sphenoidal electrode (5%) (see Fig. 34–1). Rhythmic delta activity followed by repetitive spiking in the left temporal region maximal at the left sphenoidal electrode was observed during the aura. Rhythmic left temporal theta activity maximal at SP1 was observed during the seizure (see Fig. 34–2).

MRI of the brain demonstrated left hippocampal atrophy and increased signal in the left mesial temporal structures on fluid-attenuated inversion recovery sequence (FLAIR) (Fig. 34–16). Hypometabolism involving the left temporal lobe was observed on fluorodeoxyglucose-positron emission tomography (FDG-PET)[17] (Fig. 34–17). Intracarotid sodium amytal testing revealed bilateral speech (first verbal response occurred 10 seconds after both injections, but paraphasias persisted after the left injection). Memory retention was 50% after right injection and 63% after left injection (pretest 75%). On neuropsychological testing, cognitive abilities including memory were in the low average to average range with the exception of language, which was in the borderline range. Delayed free

recall of verbal and visual material was impaired to a similar degree. The patient underwent a left selective amygdalohippocampectomy and has been seizure free for over 2 years. Pathology revealed mesial temporal sclerosis.

Discussion. In this case, the history (febrile seizure with right Todd's paralysis followed by a latent period before onset of habitual seizures), seizure semiology (olfactory aura and seizures with automatisms and right arm dystonia), ictal and interictal EEG, and neuroimaging all supported the diagnosis of left mesial temporal lobe epilepsy. Based on deficient delayed recall of verbal material, the patient was considered not to be at risk for a decrement in memory should she proceed with resection of the left temporal lobe. She underwent a selective amygdalohippocampectomy, an alternative to anterior temporal lobectomy in patients with epilepsy arising from the mesial temporal structures. Memory was slightly improved postoperatively. When the surgical evaluation demonstrates abnormalities indicating unilateral mesial temporal dysfunction, particularly in the presence of a lesion on neuroimaging studies, the likelihood of a seizure-free outcome after resection of the anteromesial temporal lobe is approximately 65% to 80%.

CASE STUDY 2

A 36-year-old right-handed man presented to the epilepsy monitoring unit for evaluation of poorly controlled seizures. His birth and development were normal. At the age of 10 years, he was struck by an automobile and was comatose for 3 days. He sustained multiple superficial injuries to the face and head but no known intracranial hemorrhage. He subsequently recovered with only a mild right hemiparesis. At the age of 17 years, he developed seizures that proved to be refractory to all available AEDs. The most common seizure consisted of sudden facial distortion, rhythmic jerking of both lower extremities, and vocalization with loss of consciousness. These would occur several times per week and occasionally on a daily basis. Approximately once per month, he experienced right arm clonic seizures with alteration of consciousness lasting for 30 minutes or longer. Secondary generalized tonic-clonic seizures occurred once per week.

As part of the surgical evaluation, the patient was admitted to the epilepsy monitoring unit where medications were gradually tapered and five typical seizures were re-

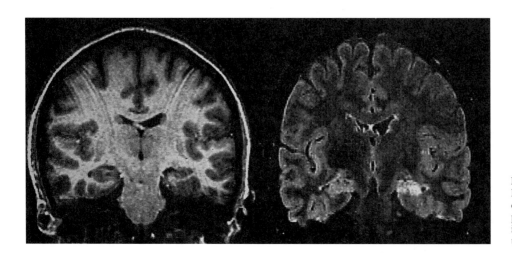

FIGURE 34–16. Case Study 1. MRI. T1-weighted and FLAIR sequences showing left hippocampal atrophy and increased signal in the left mesial temporal structures.

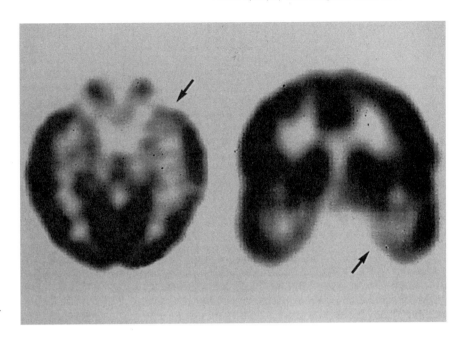

FIGURE 34–17. Case Study 1. Left temporal hypometabolism *(arrows)* on FDG-PET.

corded over a 3-day period. Seizures consisted of sudden tonic posturing of the upper extremities, vocalization, and facial distortion. One seizure evolved to head version to the right followed by generalized tonic-clonic activity. The interictal EEG showed frequent bursts of generalized slowing and generalized spike and sharp waves maximal in the frontal regions bilaterally but of higher amplitude on the left side (Fig. 34–18). During seizures, the EEG was nonlateralized or obscured by muscle and movement artifact (see Fig. 34–10). MRI of the brain revealed large areas of encephalomalacia involving both frontal lobes, left greater than right, and the left temporal lobe (Fig. 34–19). Neuropsychological testing revealed average intelligence, memory, and language function with impaired concentration and attention abilities. The intracarotid sodium amytal test demonstrated left hemisphere language dominance and bilateral memory representation, although better supported by the left hemisphere.

Owing to the extent of bilateral frontal injury found on MRI and the nonlateralizing surface EEG, invasive monitoring with subdural recording from the left frontal and temporal lobes was performed. Interictal spiking was seen maximally over the lateral aspect of the frontal lobe. Twenty-three seizures were recorded. Ictal onset involved multiple contacts over the frontal convexity anterior to the precentral gyrus with rapid spread to the temporal and mesial frontal regions. The patient underwent a left frontal lobectomy excluding primary and supplementary motor areas. Postoperatively, he had a mild right hemiparesis that resolved. Since surgery, he has had rare seizures only after missing a dose of his anticonvulsants.

Discussion. This patient represents a fairly typical case of frontal lobe epilepsy with widespread abnormalities on neuroimaging and EEG. Because the epileptogenic zone as defined by subdural grid evaluation was thought to be anterior to the motor cortex, resection of the frontal lobe

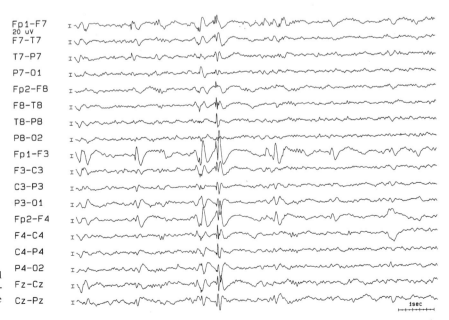

FIGURE 34–18. Generalized and bifrontal left greater than right sharp waves in a subject with left frontal lobe epilepsy (see Case Study 2).

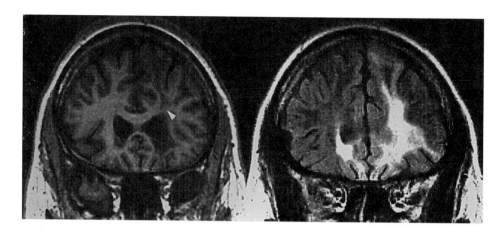

FIGURE 34–19. Case Study 2. MRI revealing bifrontal, left greater than right, and left temporal encephalomalacia secondary to head trauma.

sparing eloquent areas was recommended. The chance of a seizure-free outcome in this setting is significantly lower than in Case Study 1, estimated at 30% to 50% and largely dependent on the extent of resection of the epileptogenic zone.

CASE STUDY 3

A 23-year-old ambidextrous woman developed recurrent seizures at the age of 16 years. Her birth and development were normal. The first several seizures were tonic-clonic seizures in sleep. Within 1 year, she developed episodes of multicolored bright stars moving randomly in her left visual field lasting approximately 30 seconds followed by a throbbing headache. Within the past 2 years, seizures characterized by eye blinking, vocalization, and unresponsiveness without aura, sometimes evolving to generalized tonic-clonic activity, developed. At the time of evaluation, the frequency of the third seizure type was three to five per week and the patient had been involved in two motor vehicle accidents when these occurred while driving. Seizures remained poorly controlled despite trials of multiple AEDs, including phenobarbital, primidone, carbamazepine, phenytoin, valproate, felbatol, and gabapentin in various combinations. Family history was negative for epilepsy and neurological disorders. The patient has a 12th grade education level and currently is a homemaker and mother of three children. Physical and neurological examinations were normal.

Brain MRI was normal. Video-EEG monitoring revealed frequent multiregional sharp waves, spikes and polyspikes maximal in the right parieto-occipital (65%), left parieto-occipital (25%), and right temporal (10%) regions (see Fig. 34–13). Intermittent slowing was observed in the right parietal and temporal regions. Twelve seizures characterized by arousal from sleep, vocalization, unresponsiveness, and eye blinking, evolving to left head version and generalized tonic-clonic activity in two cases, were captured. Ten seizures were accompanied by fast activity over the right hemisphere (see Fig. 34–14). Two seizures consisted of generalized biposteriorly predominant rhythmic activity. Owing to the absence of a lesion of MRI, further investigations including invasive monitoring were recommended. The patient declined further evaluation.

Discussion. In this case, the history of left visual aura and distribution of polyspikes suggested right occipital lobe epilepsy. The MRI failed to identify a lesion. In this setting, invasive monitoring would be required to delineate the extent of the epileptogenic zone if surgery was

being considered. The risk of visual loss with occipital lobe resection must be weighed against the likelihood of producing a seizure-free state, which is generally less than 30% in cases of nonlesional extratemporal lobe epilepsy.

REFERENCES

1. Abraham K, Ajmone Marsan C: Patterns of cortical discharges and their relation to routine scalp electroencephalography. Elec Clin Neurophysiol 10:447–461, 1958.
2. AEEGS Guidelines for Long-term Monitoring for Epilepsy. J Clin Neurophysiol 11:88–110, 1994.
3. Ajmone Marsan C, Ziven LS: Factors related to the occurrence of typical paroxysmal abnormalities in the EEG records of epileptic patients. Epilepsia 11:361–381, 1970.
4. Arroyo S, Lesser RP, Awad IA, et al: Subdural and epidural grids and strips. In Engel J Jr (ed): Surgical Treatment of the Epilepsies, 2nd ed. New York, Raven Press, 1993, pp 377–386.
5. Babb TL, Wilson CL, Isokawa-Akesson M: Firing patterns of human limbic neurons during stereoencephalography (SEEG) and clinical temporal lobe seizures. Electroencephalogr Clin Neurophysiol 66:467–482, 1987.
6. Bare MA, Burnstine TH, Fisher RS, et al: Electroencephalographic changes during simple partial seizures. Epilepsia 35:715–720, 1994.
7. Barry E, Sussman NM, O'Connor MJ: Presurgical electroencephalographic patterns and outcome from anterior temporal lobectomy. Arch Neurol 49:21–27, 1992.
8. Baumgertner C, Lindinger G, Ebner A, et al: Propagation of interictal epileptic activity in temporal lobe epilepsy. Neurology 45:118–122, 1995.
9. Blume WT, Young GB, Lemieux JF: EEG morphology of partial epileptic seizures. Electroencephalogr Clin Neurophysiol 57:295–302, 1984.
10. Cascino GD, Herkes GK: Interpretation of extracranial EEG. In Wyllie E (ed): The Treatment of Epilepsy: Principles and Practice, 2nd ed. Baltimore, Williams & Wilkins, 1996, pp 265–279.
11. Cascino GD, Hulihan JF, Sharbrough FW: Parietal lobe lesional epilepsy: Electroclinical correlation and operative outcome. Epilepsia 34:522–527, 1993.
12. Commission on Classification and Terminology of the International League Against Epilepsy: Proposal for revised classification of epilepsies and epileptic syndromes. Epilepsia 30(4):389–399, 1989.
13. Daly DD: Epilepsy and syncope. In Daly DD, Pedley TA (eds): Current Practice of Clinical Electroencephalography, 2nd ed. New York, Raven Press, 1990, pp 269–334.
14. Devinsky O, Kelley K, Porter RJ, et al: Clinical and electrographic features of simple partial seizures. Neurology 38:1347–1352, 1988.
15. Devinsky O, Sato S, Kufta CV: Electroencephalographic studies of simple partial seizures with subdural electrode recordings. Neurology 39:527–533, 1989.
16. Duchowny M, Jayakar P, Resnick T, et al: Posterior temporal epilepsy: Electroclinical features. Ann Neurol 35:427–431, 1994.

17. Ebersole JS, Pacia SV: Localization of temporal lobe foci by ictal EEG patterns. Epilepsia 37:386–399, 1996.
18. Ebersole JS, Wade PB: Spike voltage topography identifies two types of frontotemporal epileptic foci. Neurology 41:1425–1433, 1991.
19. Ebner A, Hoppe M: Noninvasive electroencephalography and mesial temporal sclerosis. J Clin Neurophysiol 12:23–31, 1995.
20. Foldvary N, Lee N, Thwaites G: Clinical and electrographic manifestations of lesional neocortical temporal lobe epilepsy. Neurology 49:757–763, 1997.
21. Gambardella A, Palmini A, Andermenn F, et al: Usefulness of focal rhythmic discharges on scalp EEG of patients with focal cortical dysplasia and intractable epilepsy. Electroencephalogr Clin Neurophysiol 98:243–249, 1996.
22. Geier S, Bancaud J, Talairach J, et al: Ictal tonic postural changes and automatisms of the upper limb during epileptic parietal lobe discharges. Epilepsia 18:517–524, 1977.
23. Geiger LR, Harner RN: EEG patterns at the time of focal seizure onset. Arch Neurol 35:276–286, 1988.
24. Gibbs FA, Gibbs EL, Lennox WG: Epilepsy: A paroxysmal cerebral dysrhythmia. Brain 60:377–388, 1937.
25. Gil-Nagel A, Risinger MW: Ictal semiology in hippocampal versus extrahippocampal temporal lobe epilepsy. Brain 20:183–192, 1997.
26. Gotman J, Koffler DJ: Interictal spiking increases after seizures but does not decrease in medication. Electroencephalogr Clin Neurophysiol 72:7–15, 1989.
27. Gotman J, Marciani MG: Electroencephalographic spiking activity, drug levels, and seizure occurrence in epileptic patients. Ann Neurol 17:597–603, 1985.
28. Holmes MD, Dodrill CB, Ojemann LM, et al: Five-year outcome after epilepsy surgery in nonmonitored and monitored surgical candidates. Epilepsia 37:748–752, 1996.
29. Hufnagel A, Poersch M, Elger CE: The clinical and prognostic relevance of the postictal slow focus in the electrocorticogram. Electroencephalogr Clin Neurophysiol 94:12–18, 1995.
30. Jasper H, Pertuisset B, Flanigin H: EEG and cortical electrograms in patients with temporal lobe seizures. Arch Neurol 65:272–290, 1951.
31. Jayakar P, Duchowny M, Resnick TJ: Localization of seizure foci: Pitfalls and caveats. J Clin Neurophysiol 8:414–431, 1991.
32. Kaibara M, Blume WT: The postictal electroencephalogram. Electroencephalogr Clin Neurophysiol 70:99–104, 1988.
33. Kanner AM, Morris HM, Lüders H, et al: Usefulness of unilateral interictal sharp waves of temporal lobe origin in prolonged video-EEG monitoring studies. Epilepsia 34:884–889, 1993.
34. King DW, So EL, Marcus R, et al: Techniques and applications of sphenoidal recording. J Clin Neurophysiol 3:51–65, 1986.
35. Laskowitz DT, Sperling MR, French JA, et al: The syndrome of frontal lobe epilepsy. Neurology 45:780–786, 1995.
36. Lüders HO, Awad I: Conceptual considerations. In Lüders H (ed): Epilepsy Surgery. New York, Raven Press, 1991, pp 1063–1070.
37. Maldonado HM, Delgado-Escueta AV, Walsh GO: Complex partial seizures of hippocampal and amygdalar origin. Epilepsia 29:420–433, 1988.
38. Marciani MG, Gotman J: Effects of drug withdrawal on location of seizure onset. Epilepsia 27:423–431, 1986.
39. Mauguiere F, Courjon J: Somatosensory epilepsy: A review of 127 cases. Brain 101:307–332, 1978.
40. Montplaisir J, Laverdiere M, Saint-Hilaire J, et al: Nocturnal sleep recording in partial epilepsy: A study with depth electrodes. J Clin Neurophysiol 4:383–388, 1987.
41. Morris HH III, Dinner DS, Lüders H, et al: Supplementary motor seizures: Clinical and electroencephalographic findings. Neurology 38:1075–1082, 1988.
42. Morris HH III, Lüders H, Lesser RP, et al: The value of closely spaced scalp electrodes in the localization of epileptiform foci: A study of 26 patients with complex partial seizures. Electroencephalogr Clin Neurophysiol 63:107–111, 1986.
43. O'Brien TJ, Kilpatrick C, Murrie V, et al: Temporal lobe epilepsy caused by mesial temporal sclerosis and temporal neocortical lesions. A clinical and electroencephalographic study of 46 pathologically proven cases. Brain 119:2133–2141, 1996.
44. Pacia SV, Ebersole JS: Intracranial EEG substrates of scalp ictal patterns from temporal lobe foci. Epilepsia 38:642–654, 1997.
45. Pratt KL, Mattson RH, Weikers NJ, et al: EEG activation of epileptics following sleep deprivation: A prospective study of 114 cases. Electroencephalogr Clin Neurophysiol 24:11–15, 1968.
46. Quesney LF: Preoperative electroencephalographic investigation in frontal lobe epilepsy: Electroencephalographic and electrocorticographic recordings. Can J Neurol Sci 18:559–563, 1991.
47. Quesney LF, Gloor P: Localization of epileptic foci. In Gotman J, Ives JR, Gloor P (eds): Longterm Monitoring in Epilepsy. Amsterdam, Elsevier Science Publishers, 1985, p 165.
48. Reiher J, Beaudry M, Leduc CP: Temporal intermittent rhythmic delta activity (TIRDA) in the diagnosis of complex partial epilepsy: Sensitivity, specificity and predictive value. Can J Neurol Sci 16:398–401, 1989.
49. Risinger MW, Engel J Jr, Van Ness PC, et al: Ictal localization of temporal lobe seizures with scalp/sphenoidal recordings. Neurology 39:1288–1293, 1989.
50. Rossi GF, Colicchio G, Pola P: Interictal epileptic activity during sleep: A stereo-EEG study in patients with partial epilepsy. Electroencephalogr Clin Neurophysiol 58:97–106, 1984.
51. Rowan AJ, Veldhuisen RJ, Nagelkerke NJD: Comparative evaluation of sleep deprivation and sedated sleep EEGs as diagnostic aids in epilepsy. Electroencephalogr Clin Neurophysiol 54:357–364, 1982.
52. Salanova V, Adermann F, Olivier A, et al: Occipital lobe epilepsy: Electroclinical manifestations, electrocorticography, cortical stimulation and outcome in 42 patients treated between 1930 and 1991. Brain 115:1655–1680, 1992.
53. Salanova V, Andermann F, Rasmussen T, et al: Parietal lobe epilepsy: Clinical manifestations and outcome in 82 patients treated surgically between 1929 and 1988. Brain 118:607–627, 1995.
54. Salanova V, Andermann F, Rasmussen T, et al: Tumoural parietal lobe epilepsy: Clinical manifestations and outcome in 34 patients treated between 1934 and 1988. Brain 118:1289–1304, 1995.
55. Salinsky M, Kaner R, Dasheiff RM: Effectiveness of multiple EEGs in supporting the diagnosis of epilepsy: An operational curve. Epilepsia 28(4):331–334, 1987.
56. Sammaritano M, Gigli GM, Gotman J: Interictal spiking during wakefulness and sleep and the localization of foci in temporal lobe epilepsy. Neurology 41:290–297, 1991.
57. Sharbrough FW: Scalp-recorded ictal patterns in focal epilepsy. J Clin Neurophysiol 10(3):262–267, 1993.
58. Silverman D: The anterior temporal electrode and the ten-twenty system. Electroencephalogr Clin Neurophysiol 12:735–737, 1960.
59. Sirven JI, Liporace JD, French JA, et al: Seizures in temporal lobe epilepsy: I. Reliability of scalp/sphenoidal ictal recording. Neurology 48:1041–1046, 1997.
60. Spencer SS, So NK, Engel J Jr: Depth electrodes. In Engel J Jr (ed): Surgical Treatment of the Epilepsies, 2nd ed. New York, Raven Press, 1993, pp 359–376.
61. Spencer SS, Williamson PD, Bridgers SL, et al: Reliability and accuracy of localization by scalp ictal EEG. Neurology 35:1567–1575, 1985.
62. Sperling MR, Lieb JP, Engel J Jr, et al: Prognostic significance of independent auras in temporal lobe seizures. Epilepsia 30:322–331, 1989.
63. Sperling MR, O'Connor MJ: Auras and subclinical seizures: Characteristics and prognostic significance. Ann Neurol 28:320–328, 1990.
64. Steinhoff BJ, So NK, Lim S, et al: Ictal scalp EEG in temporal lobe epilepsy with unilateral versus bitemporal interictal epileptiform discharges. Neurology 45:889–896, 1995.
65. Sum JM, Morrell MJ: Predictive value of the first ictal recording in determining localization of the epileptogenic region by scalp/sphenoidal EEG. Epilepsia 36:1033–1040, 1995.
66. Sundarum M, Hogan T, Hiscock M, et al: Factors affecting interictal spike discharges in adults with epilepsy. Electroencephalogr Clin Neurophysiol 75:358–360, 1990.
67. Sveinbjornsdottir S, Duncan JS: Parietal and occipital lobe epilepsy: A review. Epilepsia 34(3):493–521, 1995.
68. Swartz BE, Walsh GO, Delgado-Escueta AV, et al: Surface ictal electroencephalographic pattern in frontal vs temporal lobe epilepsy. Can J Neurol Sci 18:649–662, 1991.
69. Tharp, BR: Orbital frontal seizures. An unique electroencephalographic and clinical syndrome. Epilepsia 3:627–642, 1972.

70. Walczak TS, Radtke RA, Lewis DV: Accuracy and interobserver reliability of scalp ictal EEG. Neurology 42:2279–2285, 1992.

71. Walczak TS, Radtke RA, McNamara JO, et al: Anterior temporal lobectomy for complex partial seizures: Evaluation, results, and long-term follow-up in 100 cases. Neurology 40:413–418, 1990.

72. Williamson PD, Boon PA, Thadani VM, et al: Parietal lobe epilepsy: Diagnostic considerations and results of surgery. Ann Neurol 31:193–201, 1992.

73. Williamson PD, French JA, Thadani VM: Characteristics of medial temporal lobe epilepsy: II. Interictal and ictal scalp electro-encephalography, neuropsychological testing, neuroimaging, surgical results, and pathology. Ann Neurol 34:781–787, 1993.

74. Williamson PD, Spencer DD, Spencer SS, et al: Complex partial seizures of frontal lobe origin. Ann Neurol 18:497–504, 1985.

75. Williamson PD, Thadani VM, Darcey TM, et al: Occipital lobe epilepsy: Clinical characteristics, seizure spread patterns, and results of surgery. Ann Neurol 31:3–13, 1992.

76. Ziven L, Ajmone-Marsan C: Incidence and prognostic significance of epileptiform activity in the EEG of nonepileptic subjects. Brain 91:751–778, 1969.

Chapter 35

Neonatal Seizures

Eli M. Mizrahi, MD

INTRODUCTION
CLINICAL CONSIDERATIONS
Incidence and Epidemiology
Characterization and Classification
Pathophysiology
Etiology

Therapy
Prognosis
ELECTROENCEPHALOGRAPHY OF
  NEONATAL SEIZURES
Interictal Features
Ictal Features

INTRODUCTION

Neonatal seizures represent an important problem: Seizures in this age group may be the first clinical signs of central nervous system (CNS) disturbance, they may be associated with significant morbidity and mortality, and they may be resistant to antiepileptic drugs (AEDs). Despite their importance, clinical seizures of the newborn may be difficult to recognize by direct observation. Thus, the electroencephalogram (EEG) is often utilized in the diagnosis and management of these seizures. However, there also may be difficulties in the clinical application of EEG in this age group. The findings of the neonatal EEG are age-dependent based upon gestational age (GA) or conceptional age (CA). There are EEG findings that are clearly normal and others that are clearly abnormal. There is also a wide range of findings, the precise clinical significance of which is not yet known.[14] In addition, the features of the ictal neonatal EEG have several unique characteristics. Thus, accurate interpretation of the neonatal EEG and the correlation of interictal and ictal EEG findings with clinical signs may be challenging.

There have been a number of detailed discussions of the many aspects of diagnosis and management of neonatal seizures elsewhere.[24] In addition, several reviews of neonatal electroencephalography have been published, detailing normal and abnormal EEG features, techniques of recording, methods of analysis and interpretation, and the strengths and limitations of the neonatal EEG.[5, 13, 20, 21, 31, 37, 38, 43] Thus, the objectives of this chapter are to provide a brief overview of the clinical aspects of neonatal seizures and then, in more detail, discuss the use of electroencephalography in their diagnosis and management.

The basic principle underlying the following discussion is that neonatal electroencephalography is most valuable when considered a correlative examination. When applied in the assessment of a neonate suspected of seizures, the EEG is best utilized when it is considered in relation to the infant's medical history; when preparations for recording are based on history and clinical findings; when the infant is directly observed and there is appropriate stimulation of the infant during the recording; when EEG findings are interpreted in relation to the infant's state of alertness and to relevant laboratory data; and when there is direct discussion between the clinical neurophysiologist and the physicians who have requested the EEG.[24]

Guidelines for the recording of neonatal EEG have been developed by the American Clinical Neurophysiology Society

(formerly the American Electroencephalographic Society).[1] Although some basic technical requirements for EEG recording in older children and adults apply to recording in newborns, there are some significant differences. Fewer electrode placements can provide appropriate coverage of frontal, temporal, central, occipital, and midline regions. Typically, the use of 12 channels of EEG is adequate, although additional channels are required for the recording of polygraphic data to aid in identification of sleep stages and clinical changes that may accompany seizures.

Respirations are monitored using either strain gauge or impedance techniques. Airway flow may be measured using a thermistor placed at the nares, mouth, or both. The electrocardiogram (ECG) is recorded using electrodes on the chest. The electro-oculogram (ECOG) is recorded from electrodes placed superior and medial to the inner canthus and inferior and lateral to the outer canthus of one eye. In addition, submental electromyography (EMG) is recorded with electrodes placed beneath the chin. Typically, a single montage is used throughout the recording, which must be of sufficient duration to include all stages of sleep and wakefulness. This usually requires approximately 1 hour but may take longer depending upon the infant and circumstances of recording.

The neurodiagnostic technologist's role is critical in the recording of neonatal EEG. Typically the technologist obtains and documents the medical history, identifies the reason for referral for electroencephalography, obtains descriptions of the character of the clinical events suspected of being seizures, notes current medications, assesses the infant's state of alertness, and maintains a log of clinical events during recording. After electrode application, the infant is left undisturbed until all cycles of sleep are recorded. The infant is then aroused (using tactile stimulation), and the awake EEG is recorded. The presence and character of paroxysmal clinical events and other abnormal behaviors are noted on the EEG and the log when they occur. Because some clinical seizure types may be provoked by stimulation or suppressed by restraint (see later discussion), these maneuvers are usually performed during recording and responses noted.

CLINICAL CONSIDERATIONS

Seizures may occur within the first month of life more than at any other time of life. This is due to inherent neurobiological factors that predispose the immature brain to generate and propagate seizure activity,[12, 25, 34, 36] to the rapid

This work was supported in part by the Peter Kellaway, PhD, Research Endowment of Baylor College of Medicine.

rate of brain growth and functional development in the perinatal period, and to external factors that may be the basis for the etiology of CNS dysfunction in newborns.

Incidence and Epidemiology

The reported incidence of neonatal seizures ranges between 1.8 and 5.1 per 1000 live births.[8, 10, 11, 18, 27, 30, 35] The incidence is thought to increase with decreasing birth weight and conceptional age and with increased severity of illness. Thus, the reported incidence of seizures in infants in neonatal intensive care units is substantially higher than in the general population.[3, 32, 33] When seizures do occur, most appear within the first week of life and with decreasing frequency with each successive week (Fig. 35–1).

Characterization and Classification

Seizures may be difficult to recognize in neonates because the clinical events may be poorly organized and fragmentary and may not have the distinct features that characterize seizures in older children or adults.[7] These features are, for the most part, a manifestation of the stage of development of the immature brain at the time of seizure onset. Thus, there have been a number of efforts over the years to characterize neonatal seizures and develop a distinct classification system.[7, 23, 24, 28, 39–42, 44]

Clinical seizures that are predominantly motor phenomena have been characterized as clonic, tonic, myoclonic, spasm, motor automatism,[23] and subtle seizures.[39] The clinical features of subtle seizures or motor automatism include random eye movements, oral-buccal-lingual movements, movements of progression such as swimming and pedaling, and paroxysmal complex purposeless movements. Clinical signs mediated by the autonomic nervous system (ANS) have also been considered seizures. These include changes in respiration, heart rate, and blood pressure; however, these

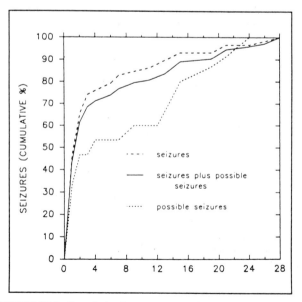

FIGURE 35–1. Cumulative frequency of neonatal seizures, expressed as a percentage of cases, according to day of onset. Frequency is given for confirmed seizures ("seizures"), those suspected but not confirmed ("possible seizures"), and both groups combined. (Reprinted with permission from Lanska MJ, Lanksa DJ, Baumann RJ, et al: A population-based study of neonatal seizures in Fayette County, Kentucky. Neurology 45:724–732, 1995.)

TABLE 35–1	Clinical Characteristics, Classification, and Presumed Pathophysiology of Neonatal Seizures
Classification	**Characterization**
Focal clonic	Repetitive, rhythmic contractions of muscle groups of the limbs, face, or trunk May be unifocal or multifocal May occur synchronously or asynchronously in muscle groups on one side of the body May occur simultaneously but asynchronously on both sides Cannot be suppressed by restraint *Pathophysiology: epileptic*
Focal tonic	Sustained posturing of single limbs Sustained asymmetrical posturing of the trunk Sustained eye deviation Cannot be provoked by stimulation or suppressed by restraint *Pathophysiology: epileptic*
Generalized tonic	Sustained symmetrical posturing of limbs, trunk, and neck May be flexor, extensor, or mixed extensor/flexor May be provoked or intensified by stimulation May be suppressed by restraint or repositioning *Presumed pathophysiology: nonepileptic*
Myoclonic	Random, single, rapid contractions of muscle groups of the limbs, face, or trunk Typically not repetitive or may recur at a slow rate May be generalized, focal, or fragmentary May be provoked by stimulation *Presumed pathophysiology: may be epileptic or nonepileptic*
Spasms	May be flexor, extensor, or mixed extensor/flexor May occur in clusters Cannot be provoked by stimulation or suppressed by restraint *Pathophysiology: epileptic*
Motor automatisms	
Ocular signs	Random and roving eye movements or nystagmus (distinct from tonic eye deviation) May be provoked or intensified by tactile stimulation *Presumed pathophysiology: nonepileptic*
Oral-buccal-lingual movements	Sucking, chewing, tongue protrusions May be provoked or intensified by stimulation *Presumed pathophysiology: nonepileptic*
Progression movements	Rowing or swimming movements Pedaling or bicycling movements of the legs May be provoked or intensified by stimulation May be suppressed by restraint or repositioning *Presumed pathophysiology: nonepileptic*
Complex purposeless movements	Sudden arousal with transient increased random activity of limbs May be provoked or intensified by stimulation *Presumed pathophysiology: nonepileptic*

Reprinted with permission from Mizrahi EM, Kellaway P: Diagnosis and Management of Neonatal Seizures. New York, Lippincott-Raven, 1998, p 21.

clinical events rarely occur in isolation. More often they occur in association with motor phenomena. One classification scheme, which has been recently updated[24] with characteristics of specific types, is presented in Table 35–1.

Pathophysiology

Some types of clinical seizures are consistently associated with electrical seizure activity and are clearly epileptic in origin. These include focal clonic, focal tonic, and some myoclonic seizures and spasms. Other seizure types have no consistent relationship to EEG seizure activity and most often occur in the absence of such electrical activity[15, 16, 23] (Table 35–2). These seizures demonstrate clinical features more characteristic of reflex physiology than of epileptic seizures:

TABLE 35–2	Classification of Neonatal Seizures Based on Electroclinical Findings

Clinical Seizures with Electrocortical Signatures
Focal clonic
  Unifocal
  Multifocal (alternating, migrating)
  Hemiconvulsive
  Axial
Focal tonic
  Asymmetrical truncal posturing
  Limb posturing
  Sustained eye deviation
Myoclonic
  Generalized
  Focal
Spasms
  Flexor
  Extensor
  Mixed extensor/flexor

Clinical Seizures without a Consistent Electrocortical Signature
(*Pathophysiology:* presumed nonepileptic)
Myoclonic
  Generalized
  Focal
  Fragmentary
Generalized tonic
  Flexor
  Extensor
  Mixed extensor/flexor
Motor automatisms
  Oral-buccal-lingual movements
  Ocular signs
  Progression movements
  Complex purposeless movements

Electrical Seizures without Clinical Seizure Activity

Reprinted with permission from Mizrahi EM, Kellaway P: Diagnosis and Management of Neonatal Seizures. New York, Lippincott-Raven, 1998, p 23.

TABLE 35–3	Most Frequently Encountered Etiologies of Neonatal Seizures[a]

Hypoxia-ischemia
Intracranial hemorrhage
  Intraventricular
  Intracerebral
  Subdural
  Subarachnoid
Infection (central nervous system)
  Meningitis
  Encephalitis
  Intrauterine
Infarction
Metabolic
  Hypoglycemia
  Hypocalcemia
  Hypomagnesemia

Chromosomal anomalies
Congenital abnormalities of the brain
Neurodegenerative disorders
Inborn errors of metabolism
Benign neonatal convulsions
Benign familial neonatal convulsions
Drug withdrawal or intoxication

[a]Listed in relative order of frequency. Not listed is "unknown" etiology, which is encountered in approximately 10% of cases.
Reprinted with permission from Mizrahi EM, Kellaway P: Diagnosis and Management of Neonatal Seizures. New York, Lippincott-Raven, 1998, p 49.

the clinical events can be provoked by stimulation; increasing the intensity of stimulation or the number of sites of stimulation increases the intensity of the provoked event. Spontaneous events can be suppressed by restraint or repositioning of the infants. These clinical observations suggest that these events are not epileptic in origin and have been referred to as "brain stem release phenomena."[15, 23]

Thus, terminology has been developed to address the problem of various neonatal seizure types generated by different pathophysiological mechanisms. The term *seizure* is used to designate paroxysmal, repetitive, and stereotypical abnormal behaviors in the neonate such as those listed in Tables 35–1 and 35–2. Their pathophysiology is designated by referring to clinical events as either "epileptic seizures" or "nonepileptic seizures."[16, 22–24]

Etiology

Whatever the pathophysiology of the clinical events in an individual infant, the onset of seizures in a neonate indicates the presence of significant CNS dysfunction and prompts an investigation for etiology. The list of potential etiologies of neonatal seizures is diverse and can be exhaustive, although they fall into several broad categories (Table 35–3). Although such factors are presented as single etiological entities, it should be noted that the causes of many neonatal seizures are multifactorial and that there is more than one coexisting condition at the time of seizure onset that may make management complex. Finally, despite every effort to uncover an etiology, for some infants, a cause of the seizures may not be identified.

Therapy

Once seizures are recognized and investigations into an etiology are either initiated or completed, seizures are typically treated with either etiology-specific therapy or AEDs. Decisions to treat with AEDs include considerations of seizure pathophysiology (nonepileptic seizures do not require AEDs), seizure frequency (infrequent epileptic seizures may not require AEDs, whereas those recurring frequently do), and seizure duration (brief seizures may not need AEDs, whereas prolonged seizures are vigorously treated). First- and second-line AEDs and dosing schedules are listed in Table 35–4.

Prognosis

The outcome of neonatal seizures has been the focus of a number of investigations, including those utilizing multivariant analyses. Issues such as features of interictal EEG from

TABLE 35–4	Dosages of First-Line, Second-Line AEDs in the Treatment of Neonatal Seizures[a]			
	Dose			
Drug	**Loading**	**Maintenance**	**Average Therapeutic Range**	**Apparent Half-life**
Diazepam	0.25 mg/kg IV (bolus), 0.5 mg/kg (rectal)	May be repeated one or two times		31–54 hours
Lorazepam	0.05 mg/kg (IV) over 2–5 minutes	May be repeated		31–54 hours
Phenobarbital	20 mg/kg IV (up to 40 mg/kg)	3–4 mg/kg in two doses	20–40 μg/L	100 hours after day 5–7
Phenytoin	20 mg/kg IV (over 30–45 minutes)	3–4 mg/kg in two to four doses	15–25 μg/L	100 hours (40–200)

[a]Listed alphabetically. Volpe JJ: Neonatal Seizures. In Neurology of the Newborn. Philadelphia, WB Saunders, 1995.

one or several recording periods, the ictal EEG, the neurological examination at the time of seizures, the character or duration of the seizures, etiology, neuroimaging findings, and conceptional age and weight have been considered.[2, 4, 6, 9, 26, 29] Although multiple rather than single factors appear to be most accurate in predicting long-term outcome, most of these variables relate to the degree of brain injury at the time of seizure occurrence. Thus, long-term outcome is primarily determined by seizure etiology, its associated severity, and its distribution of brain disturbance.

ELECTROENCEPHALOGRAPHY OF NEONATAL SEIZURES

Many electrographic features of the neonate are unique compared with those of older infants and children (see Chap. 28). Both interictal and ictal features of the EEG have relevance to the diagnosis and management of neonatal seizures.

Interictal Features

The characteristics of the interictal background activity provide basic information concerning the degree and distribution of CNS dysfunction that may be associated with neonatal seizures. In addition, focal features, such as voltage asymmetries and slow activity, may indicate regional areas of brain involvement. A more detailed discussion of these findings is contained in Chapter 28.

The relationship of these diffuse or focal abnormalities to seizure onset has not been fully explored. Laroia and colleagues[19] have suggested that the more abnormal the background activity of the EEG, the greater is the likelihood that the neonate may eventually develop seizures. In addition, it has often been assumed in neonates with seizures and focal interictal EEG features that the site of these focal features would be the site of seizure onset. However, electrographic onset may occur at the periphery or even at a distance from the location of these focal features.[24] This may be owing to the presence of a greater degree of epileptogenesis in neuronal tissue somewhat removed from the most severely injured cortex, or because some focal brain injury occurs in the presence of diffuse brain injury, with electrographic seizure onset related to the diffuse rather than focal process.

Another important issue in neonatal electroencephalography concerns the interpretation of the finding of interictal focal sharp wave transients. In older children and adults, a focal sharp wave or spike suggests the potential for an electrical seizure to arise from that region. However, in the neonate, interictal epileptiform discharges are rarely present to aid in diagnosis. Although isolated sharp waves do occur in the neonate, they are not interpreted as interictal epileptiform activity. Thus, they provide no useful information concerning the presence or absence of or the potential for electrical seizures in a given infant. On the other hand, focal sharp transients in the neonate may be considered abnormal, suggesting the presence of injury that is focal or even diffuse (when multifocal sharp waves are present). Such findings may also suggest the type of injury, depending upon the characteristics of the waveforms.[13]

The character of the background EEG may be helpful in determining the prognosis of a newborn with seizures. Generally, the greater the EEG abnormality, the more grave the prognosis. However, in making this correlation, it is critical to determine the timing between the EEG and the onset of the injury, as well as the rate of resolution of the EEG abnormalities over time. Although the initial EEG may be very abnormal, the accuracy of prognosis is based on the evolution (degree and rate of resolution) of the abnormality. On the other hand, a normal initial EEG (within 24 hours of birth) reliably suggests a good prognosis.

Ictal Features

Electrical seizure activity may be rare before the conceptional age of 34 weeks, and with increasing age electrical seizure activity becomes more frequent. However, the conceptional age at which the developing brain can consistently initiate and sustain electrical seizure activity has not been clearly defined. This observation is at odds with reports indicating a greater occurrence of seizures in premature infants. However, many of these studies may not have included EEG data, and the clinical observations may not have taken into account the pathophysiology of seizure types, nonepileptic seizures being more frequent in more premature infants. When electrographic seizure activity is present, it may be highly variable in appearance, well localized to relatively circumscribed regions of the brain, and, except for some discharges, associated with myoclonic jerks or spasms, exclusively focal.

Site of Onset

Electrical seizure activity in the neonate arises most often in the centrotemporal region of one hemisphere, less commonly in occipital and midline central regions (Fig. 35–2), and, rarely, in the frontal region. The midline central region is an important site of onset, as some recording montages utilized for routine clinical EEG recording in older children or adults may not provide such coverage (Fig. 35–3).

Focality

In a given infant, electrical seizure activity is usually stereotypically focal. The electrographic discharges repeatedly arise from the same brain region with each onset. Electrical seizure activity, however, can also be multifocal, with different seizures arising from different brain regions, either sequentially or simultaneously, but asynchronously. During sequential multifocal seizures, various sites are involved at different times with no overlapping periods when multiple sites are simultaneously generating seizure activity. During simultaneous multifocal seizures, electrical seizure activity occurs independently and asynchronously in different brain regions simultaneously, either in one hemisphere or, more typically, in both hemispheres (Fig. 35–4).

Frequency, Voltage, and Morphology

The frequency, voltage, and morphology of electrical seizure activity may vary greatly within the same seizure or from one seizure to the next in a given infant. The frequencies of activity can be in the alpha, theta, or beta ranges. The voltages of the activity may vary and may be related to frequency of the rhythmic activity. Voltages may range from extremely low, usually when faster frequencies are present, to very high, sometimes seen when slow frequencies are present. The morphology of the seizures may also range from monomorphically sharp to polymorphic, within a given seizure (Fig. 35–5).

Involvement of Specific Brain Regions

Once initiated, an individual electrical seizure discharge may be confined to a specific region or it may spread to involve other regions. The spread of electrical seizure activity

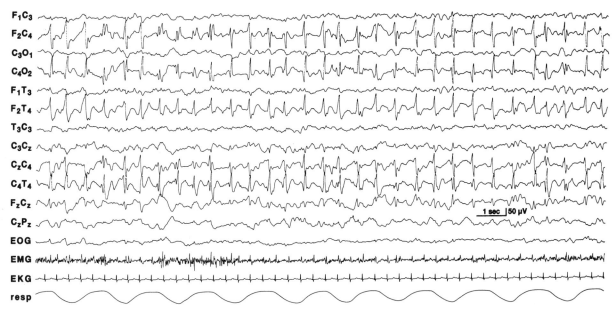

FIGURE 35–2. Repetitive spike and slow wave activity from the right centrotemporal region associated with focal clonic seizures involving the left arm, leg, and face in an 8-week-old, 34-week GA (42-week CA) female with a right frontal lobe intracerebral hemorrhage. (Reprinted with permission from Mizrahi EM, Kellaway P: Diagnosis and Management of Neonatal Seizures. New York, Lippincott-Raven, 1998, p 126.)

may have several appearances. There may be a gradual widening of the field of the focus, there may be abrupt changes from an initial small regional involvement to the entire hemisphere, or the spread may be by migration of the seizure discharge from one area to another, either in a jacksonian or more often a nonjacksonian fashion (Fig. 35–6).

Evolution of Appearance

The appearance of an electrical seizure may vary considerably. Some electrical seizure activity may begin abruptly and remain fairly constant throughout the seizure with similar frequencies, voltage, and morphology. More often, however, an electrical seizure may undergo an evolution in frequency, voltage, and morphology (Fig. 35–7). Electrical seizure activity has been described as having a beginning, a middle, and an end as regards its appearance.[37] This has been useful in differentiating focal rhythmic EEG activity in the neonate from seizures.

Special Ictal Patterns

In addition to the various features of electrographic seizures described earlier, there are also unique ictal patterns

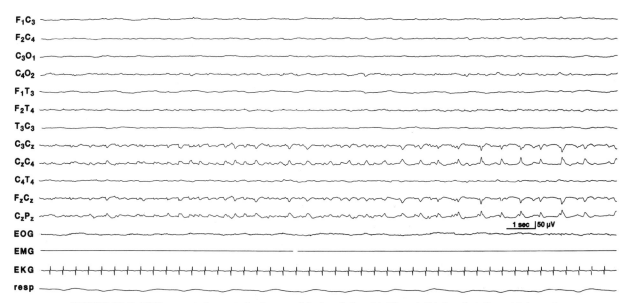

FIGURE 35–3. Midline central onset of seizure activity in a 1-day-old, 38-week GA female infant with hypoxic-ischemic encephalopathy. No clinical seizures were associated with this seizure discharge. The infant had been treated with phenobarbital prior to the recording. The background activity is depressed and undifferentiated. (Reprinted with permission from Mizrahi EM, Kellaway P: Diagnosis and Management of Neonatal Seizures. New York, Lippincott-Raven, 1998, p 119.)

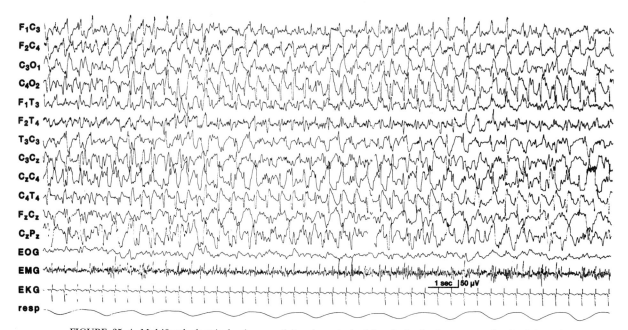

FIGURE 35–4. Multifocal electrical seizure activity, characterized by rhythmic sharp waves in the left and right central regions appearing independently and asynchronously, occurs in this 1-day-old, 39-week GA female infant with hypoglycemia and secondary hypoxic-ischemic encephalopathy. No clinical seizures accompanied these electrical seizures. The infant had been treated with phenobarbital prior to initiation of EEG recording. (Reprinted with permission from Mizrahi EM, Kellaway P: Diagnosis and Management of Neonatal Seizures. New York, Lippincott-Raven, 1998, p 120.)

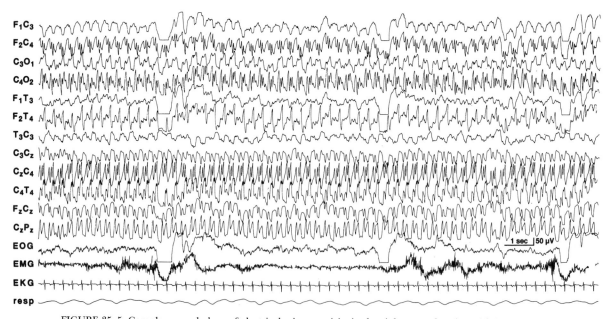

FIGURE 35–5. Complex morphology of electrical seizure activity in the right central region with independent spike wave activity in the right temporal region, associated with focal clonic seizure activity of the left arm and leg, in a 6-week-old, 37-week GA (43-week CA) infant with a right frontal lobe infarction. No antiepileptic drugs were administered prior to recording. (Reprinted with permission from Mizrahi EM, Kellaway P: Diagnosis and Management of Neonatal Seizures. New York, Lippincott-Raven, 1998, p 130.)

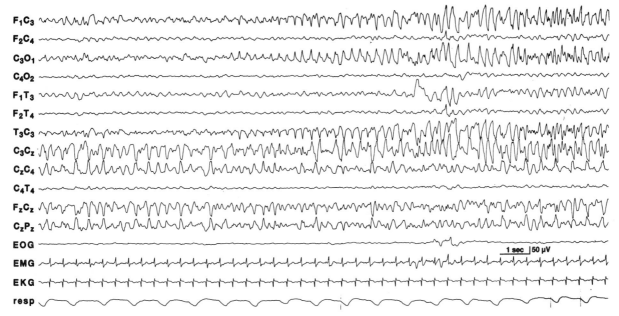

FIGURE 35–6. Electrical seizure activity spreading from the midline central region to the left central region in a 4-day-old term infant with hypoxic-ischemic encephalopathy. No clinical events were associated with the electrical seizure activity. Phenobarbital had been administered prior to EEG recording. (Reprinted with permission from Mizrahi EM, Kellaway P: Diagnosis and Management of Neonatal Seizures. New York, Lippincott-Raven, 1998, p 132.)

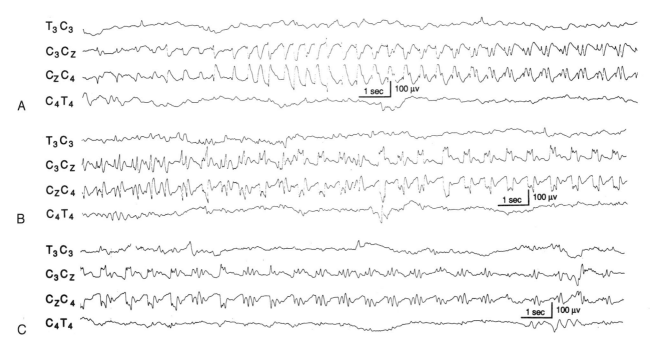

FIGURE 35–7. Electrical seizure discharges with changing frequency, amplitude, and morphology arising from the midline central region in a term infant with hypoglycemia. Selected samples are shown from a 16-channel recording from seizure onset (A), mid-seizure (B), and toward the end (C). (Reprinted with permission from Hrachovy RA, Mizrahi EM, Kellaway P: Electroencephalography of the newborn. In Daly D, Pedley TA (eds): Current Practice of Clinical Electroencephalography, 2nd ed. New York, Raven Press, 1990, p 231.)

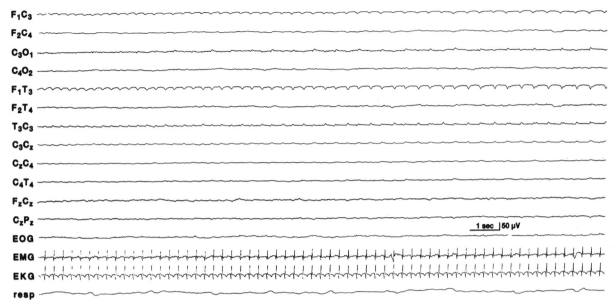

FIGURE 35–8. Seizure discharges of the depressed brain in the left temporal region, unassociated with clinical seizures in a 2-day-old, 38-week GA infant with hypoxic-ischemic encephalopathy. No antiepileptic drugs were administered prior to EEG recording. (Reprinted with permission from Mizrahi EM, Kellaway P: Diagnosis and Management of Neonatal Seizures. New York, Lippincott-Raven, 1998, p 134.)

that occur in the neonate. These typically occur in association with severe encephalopathies and can be categorized as seizure discharges of the depressed brain and alpha seizures.

Seizure discharges of the depressed brain are seen in neonates whose background EEG activity is depressed and undifferentiated (Fig. 35–8). The seizure discharges may be unifocal or multifocal. They are typically low in voltage, relatively long in duration, and highly localized. They show little tendency to spread to other brain regions from the site of origin or change in morphology, frequency, or waveform

morphology. This seizure pattern is one of a few electrical seizure patterns that are typically not accompanied by any clinical seizure activity (Table 35–5). Its presence suggests a poor prognosis.

The alpha seizure pattern is characterized by the sudden appearance of sustained, rhythmic 8- to 12-Hz, 20- to 70-µV activity in one temporal or central region[17, 45, 46] (Fig. 35–9). It can also evolve from rhythmic electrical activity that has more clear epileptiform features. In addition, alpha seizures may occur simultaneously with other electrical seizures oc-

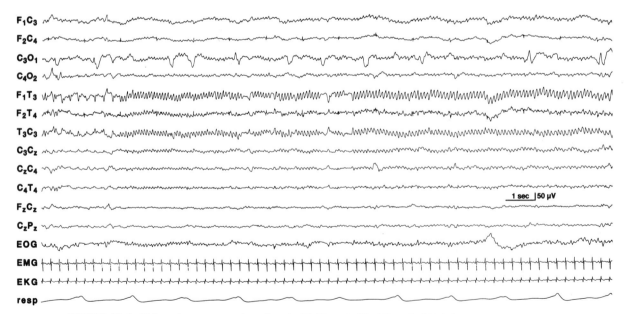

FIGURE 35–9. Alpha seizure pattern in a 4-week-old, 38-week GA (42-week CA) infant with pneumococcal meningitis. The alpha seizure pattern arises from the left temporal region. There are independent semiperiodic slow-wave transients in the left occipital region. No clinical seizures were associated with these electrical events. No antiepileptic drugs were administered prior to EEG recording. (Reprinted with permission from Mizrahi EM, Kellaway P: Diagnosis and Management of Neonatal Seizures. New York, Lippincott-Raven, 1998, p 135.)

TABLE 35-5	Electrical Seizure Activity that May Occur with No Accompanying Clinical Seizures

Seizure discharges of the depressed brain
Alpha seizure discharges
Decoupled electrical seizure activity
Electrical seizure activity in pharmacologically paralyzed neonates

curring at different sites with different characteristics. Alpha seizures typically indicate the presence of severe CNS dysfunction resulting from cerebral dysgenesis or severe hypoxic-ischemic injury. They are typically seen in neonates with encephalopathy and suggest a poor prognosis. In addition, they usually occur in the absence of any clinical seizures. The pattern must be differentiated from other rhythmic but nonseizure electrical activity.[14]

Electrical Seizure Activity and Medication Effect

Medications that are frequently given to neonates and may have an impact on the diagnosis and management of seizures include agents to paralyze infants requiring respiratory ventilation, and AEDs. With the administration of each class of drug, electrical seizures may be present in the absence of any clinical seizure. In pharmacologically paralyzed infants, movements are prevented and, thus, electrical seizures can be present with no clinical signs.

In an infant with an electroclinical seizure, the initial response to the administration of AEDs may be control of the clinical seizures with persistence of the electrical seizure. The phenomenon has been termed *decoupling* of the clinical from the electrical seizure[23] and must be considered in determining the response of the infant to AEDs. Seizure control must be defined in either clinical or electroencephalographic terms. Administration of AEDs and assessment of outcome by clinical observation may mean that electrical seizures may be occurring unaccompanied by clinical events. Alternatively, assessment of efficacy of AEDs with electroencephalography is also problematic as electrical seizure activity may be highly resistant to AEDs and, even at high doses and utilizing multiple drugs, may not be effectively extinguished. Vigorous AED therapy may lead to CNS or cardiovascular depression, further compromising an already ill neonate. Frequently, AED efficacy is determined by clinical assessment only. When electroencephalography is utilized, attempts are made to eliminate electrical seizure, but only with high therapeutic doses of first- and second-line AEDs.[24]

REFERENCES

1. American Electroencephalographic Society: Guidelines in EEG, 1–7 (revised 1985). J Clin Neurophysiol 3:131–168, 1986.
2. André M, Matisse N, Vert P, Debruille C: Neonatal seizures: Recent aspects. Neuropediatrics 19:201–207, 1988.
3. Bergman I, Painter MJ, Hirsch RP, et al: Outcome in neonates with convulsions treated in an intensive care unit. Ann Neurol 14:642–647, 1983.
4. Bye AME, Cunningham CA, Chee KY, et al: Outcome of neonates with electrographically identified seizures, or at risk of seizures. Pediatr Neurol 16:225–231, 1997.
5. Clancy RR, Chung HJ, Temple JP: Neonatal electroencephalography. In Sperling MR, Clancy RR (eds): Atlas of Electroencephalography. Amsterdam, Elsevier Science Publishers BV, 1993.
6. Clancy RR, Legido A: Postnatal epilepsy after EEG-confirmed neonatal seizures. Epilepsia 32:69–76, 1991.
7. Dreyfus-Brisac C, Monod N: Electroclinical studies of status epilepticus and convulsions in the newborn. In Kellaway P, Petersén I (eds): Neurological and Electroencephalographic Correlative Studies in Infancy. New York, Grune and Stratton, 1964, pp 250–272.
8. Ellenberg JH, Hirtz DG, Nelson KB: Age at onset of seizures in young children. Ann Neurol 15:127–134, 1984.
9. Ellison PH, Largent JA, Bahr JP: A scoring system to predict outcome following neonatal seizures. J Pediatr 99:455–459, 1981.
10. Eriksson M, Zetterström R: Neonatal convulsions. Incidence and causes in the Stockholm area. Acta Paediatr Scand 68:807–811, 1979.
11. Holden KR, Mellits ED, Freeman JM: Neonatal seizures. I. Correlation of prenatal and perinatal events with outcomes. Pediatrics 70:165–176, 1982.
12. Holmes GL: Epilepsy in the developing brain: Lessons from the laboratory and clinic. Epilepsia 38:12–30, 1997.
13. Hrachovy RA, Mizrahi EM, Kellaway P: Electroencephalography of the newborn. In Daly D, Pedley TA (eds): Current Practice of Clinical Electroencephalography, 2nd ed. New York, Raven Press, 1990, pp 201–242.
14. Hrachovy RA, O'Donnell DM: The diagnostic significance of rhythmic alpha and theta frequency activity in the EEG of the neonate. J Clin Neurophysiol 13(4):355, 1996.
15. Kellaway P, Hrachovy RA: Status epilepticus in newborns: A perspective on neonatal seizures. In Delgado-Escueta AV, Wasterlain CG, Treiman DM, Porter RJ (eds): Advances in Neurology, vol 34: Status Epilepticus. New York, Raven Press, 1983, pp 93–99.
16. Kellaway P, Mizrahi EM: Clinical, electroencephalographic, therapeutic, and pathophysiologic studies of neonatal seizures. In Wasterlain CG, Vert P (eds): Neonatal Seizures: Pathophysiology and Pharmacologic Management. New York, Raven Press, 1990, pp 1–13.
17. Knauss TA, Carlson CB: Neonatal paroxysmal monorhythmic alpha activity. Arch Neurol 35:104–107, 1978.
18. Lanska MJ, Lanska DJ, Baumann RJ, et al: A population-based study of neonatal seizures in Fayette County, Kentucky. Neurology 45:724–732, 1995.
19. Laroia N, Guillet R, Burchfiel J, McBride MC: EEG background as predictor of electrographic seizures in high-risk neonates. Epilepsia 39:545–551, 1998.
20. Lombroso CT: Neonatal electroencephalography. In Niedermeyer E, Lopes da Silva F (eds): Electroencephalography. Basic Principles, Clinical Applications and Related Fields. Baltimore, Urban & Schwarzenberg, 1982, pp 599–637.
21. Mizrahi EM: Electroencephalographic/polygraphic/video monitoring in childhood epilepsy. J Pediatr 105:1–9, 1984.
22. Mizrahi EM: Consensus and controversy in the clinical management of neonatal seizures. Clin Perinatol 16:485–500, 1989.
23. Mizrahi EM, Kellaway P: Characterization and classification of neonatal seizures. Neurology 37:1837–1844, 1987.
24. Mizrahi EM, Kellaway P: Diagnosis and Management of Neonatal Seizures. New York, Lippincott-Raven, 1998.
25. Moshé SL: Seizures in the developing brain. Neurology 43:S3–S7, 1993.
26. Ortibus EL, Sum JM, Hahn JS: Predictive value of EEG for outcome and epilepsy following neonatal seizures. Electroencephalogr Clin Neurophysiol 98:175–185, 1996.
27. Ronen GM, Penney S: The epidemiology of clinical neonatal seizures in Newfoundland, Canada: A five-year cohort. Ann Neurol 38:518–519, 1995.
28. Rose AL, Lombroso CT: A study of clinical, pathological, and electroencephalographic features in 137 full-term babies with a long-term follow-up. Pediatrics 45:404–425, 1970.
29. Rowe JC, Holmes GL, Hafford J, et al: Prognostic value of the electroencephalogram in term and preterm infants following neonatal seizures. Electroencephalogr Clin Neurophysiol 60:183–196, 1985.
30. Saliba RM, Annegers JF, Mizrahi EM: Incidence of clinical neonatal seizures. Epilepsia 37(suppl 5):13, 1996.
31. Scher MS: Pediatric electroencephalography and evoked potentials. In Swaiman KF (ed): Pediatric Neurology: Principles and Practice. St. Louis, CV Mosby, 1989, pp 67–103.
32. Scher MS, Aso K, Beggarly M, et al: Electrographic seizures in preterm and full-term neonates: Clinical correlates, associated brain lesions, and risk for neurologic sequelae. Pediatrics 91:128–134, 1993.
33. Scher MS, Hamid MY, Steppe DA, et al: Ictal and interictal

electrographic seizure durations in preterm and term neonates. Epilepsia 34:284–288, 1993.

34. Schwartzkroin PA: Plasticity and repair in the immature central nervous system. In Schwartzkroin PA, Moshé SL, Noebels JL, Swann JW (eds): Brain Development and Epilepsy. New York, Oxford University Press, 1995, pp 234–267.

35. Spellacy WN, Peterson PQ, Winegar A, et al: Neonatal seizures after cesarean delivery: Higher risk with labor. Am J Obstet Gynecol 157:377–379, 1987.

36. Swann JW: Synaptogenesis and epileptogenesis in developing neural networks. In Schwartzkroin PA, Moshé SL, Noebels JL, Swann JW (eds): Brain Development and Epilepsy. New York, Oxford University Press, 1995, pp 195–233.

37. Tharp BR: Neonatal and pediatric electroencephalography. In Aminoff M (ed): Electrodiagnosis in Clinical Neurology. Edinburgh, Churchill-Livingstone, 1980, pp 67–117.

38. Torres F, Anderson C: The normal EEG of the human newborn. J Clin Neurophysiol 2:89–103, 1985.

39. Volpe JJ: Neonatal seizures. N Engl J Med 289:413–416, 1973.

40. Volpe JJ: Neurology of the Newborn. Philadelphia, WB Saunders, 1981, pp 111–141.

41. Volpe JJ: Neonatal seizures: Current concepts and revised classification. Pediatrics 84:422–428, 1989.

42. Volpe JJ: Neonatal seizures. In: Neurology of the Newborn. Philadelphia, WB Saunders, 1995, pp 172–207.

43. Watanabe K: The neonatal electroencephalograph and sleep-cycle patterns. In Eyre JA (ed): The Neurophysiological Examination of the Newborn Infant. New York, Cambridge University Press, 1992, pp 11–47.

44. Watanabe K, Hara K, Miyazaki S, et al: Electroclinical studies of seizures in the newborn. Folia Psychiatr Neurol Jpn 31:383–392, 1977.

45. Watanabe K, Hara K, Miyazaki S, et al: Apneic seizures in the newborn. Am J Dis Child 136:980–984, 1982.

46. Willis J, Gould JB: Periodic alpha seizures with apnea in a newborn. Develop Med Child Neurol 22:214–222, 1980.

VIII EVOKED POTENTIALS AND INTRAOPERATIVE MONITORING

Chapter 36

Visual Evoked Potentials

Charles M. Epstein, MD

INTRODUCTION
FUNDAMENTALS OF VISUAL EVOKED
  POTENTIALS
Basic VEP Waveforms
Neuroanatomy of VEPs
Measurement of VEP Latency
Measurement of VEP Amplitude
Unusual Waveforms
STIMULUS PARAMETERS
Pattern Reversal
Pattern Onset
Check Size
Field Size
Luminance
Contrast
Color
Distance to Pattern
Reversal Frequency
RECORDING PARAMETERS
Passband
Sampling Rate
Recording Derivations
PATIENT CONSIDERATIONS
Preparation
Attention
Age
Gender

NEUROLOGICAL APPLICATIONS OF VISUAL
  EVOKED POTENTIALS
Optic Neuritis and Multiple Sclerosis
Other Diseases of the Optic Nerve
Leukodystrophies and Spinocerebellar
  Disorders
Tumors
Miscellaneous Central Nervous System
  Disorders
VEPs in Coma
VEPs in the Operating Room
OPHTHALMOLOGICAL APPLICATIONS OF
  VISUAL EVOKED POTENTIALS
Visual Acuity
Amblyopia
Glaucoma
Retinopathies
Hysteria and Malingering
PARTIAL FIELD STIMULATION
Background and Technique
Applications
FLASH VEPs
Background
Technique and Applications

INTRODUCTION

Visual evoked potentials (VEPs) produced by shifting patterns are exquisitely sensitive to disorders of the optic nerves and optic chiasm, which is the basis of their clinical utility. The major measurement used in VEP testing is the latency of a single scalp-positive waveform that in normal subjects occurs about 100 msec after the visual stimulus and is recorded over the mid-occiput. Often VEPs can document disease that is minimally or entirely asymptomatic. However, VEP abnormalities may have an extensive differential diagnosis, and the results may be affected by a large number of test and patient factors. These variables must be understood for VEPs to be applied with optimal accuracy and usefulness.

FUNDAMENTALS OF VISUAL EVOKED POTENTIALS

Basic VEP Waveforms

A typical clinical VEP is shown in Figure 36–1. The visual stimulus is a checkerboard pattern of black-and-white squares, which reverses one to three times per second: the black squares become white, and the white squares become

black. This pattern-reversal VEP (PVEP) usually has a negative-positive-negative configuration extending from 70 to 150 msec; the major positive peak occurs around 100 msec. The usual convention displays relative positivity at the occiput by a downward deflection. Peaks are labeled using the average

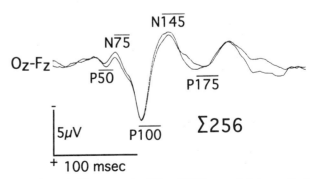

FIGURE 36–1. The normal midline PVEP, recorded in an O_z–F_z derivation from an average of 256 responses. The major positive and negative peaks are labeled according to their mean latency in normal subjects. By convention, in this figure the mean is indicated with a bar over the latency value.

latency values in normal subjects: N75, P100, and N145. Depending on test conditions, a smaller positive peak may be recorded between 50 and 60 msec. Most of the long delay from the visual stimulus to the positive peak of P100 occurs in the retina and cortex. The earliest signal recorded in the optic nerve appears around 45 msec,[105] which is not much before the onset of P50 in the occipital lobe. In the most typical PVEP abnormality, which is a latency prolongation, the entire complex of waveforms is delayed and shifted to the right.

Neuroanatomy of VEPs

In the optic nerve and cortex, the percentage of fibers devoted to the macula is magnified far out of proportion to its actual size (Fig. 36–2). Approximately one-third of the primary visual cortex is devoted to the 3 degrees of central

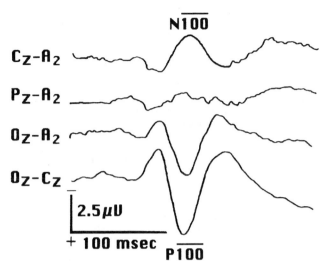

FIGURE 36–3. A normal PVEP recorded from a chain of sagittal derivations, showing the occipital P100, the frontocentral N100, and their summation in the O_z–C_z derivation. (Modified with permission of the publisher from Pedley T, Daly D: Current Practice of Clinical EEG, 2nd ed. New York, Raven Press, 1990.)

vision. The macular projection area occupies the posterior portion of the calcarine cortex, extending in the medial portion of the hemisphere approximately to the occipital pole. However, the visual areas show considerable individual variability.[154] Because the primary visual system is arranged to emphasize detection of edges and movement,[78] shifting patterns with multiple edges and contrasts are the most appropriate way to assess visual function, rather than bright flashes of unpatterned light.

Because signals from the temporal field of each eye cross in the optic chiasm, patterns appearing in only one hemifield go entirely to the opposite occipital lobe. With large stimulus fields (16 degrees across or more), pattern shift in one hemifield produces a net pattern VEP vector that points diagonally back across the opposite occipital lobe and is represented by the large arrow in Figure 36–2.[17, 89, 143] With pattern stimuli, the scalp VEP derives almost exclusively from the calcarine cortex;[111] the net vector reflects predominantly activity in macular and paramacular areas. When the full visual field is stimulated, vectors from the two hemispheres sum to produce a PVEP maximum at the occipital midline.

"Far-field" VEPs arising from subcortical structures have proven much more difficult to record than the corresponding short-latency potentials for auditory and somatosensory systems. A few studies using flash stimuli identified early oscillatory potentials that may arise subcortically.[45, 124] However, later depth electrode studies suggested that even these early oscillatory potentials may have a cortical origin.[47]

Electrodes placed anteriorly along the midline often show a negative N100 wave in the frontocentral region, which occurs at about the same time as the P100 (Fig. 36–3, top channel and Fig. 36–8). If a recording derivation interconnects the area of this N100 and the occiput (e.g., F_z–O_z), then the P100 and N100 will sum to give a larger and better-defined PVEP (Fig. 36–3, bottom channel).

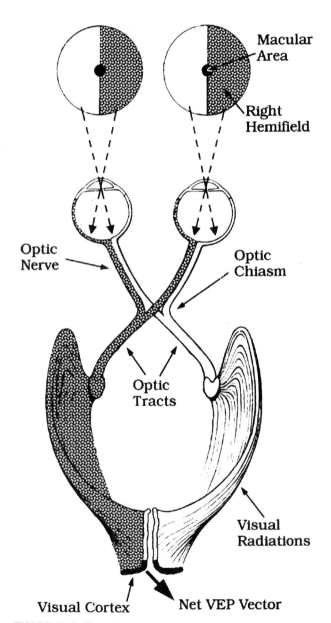

FIGURE 36–2. The primary visual pathways, showing the crossing of nasal fibers in the optic chiasm, the projections to the striate cortex, and the magnification of the macular area in the visual cortex (shown in black). The large arrow indicates the direction of the net VEP vector from hemifield stimulation, which slants obliquely back across the opposite hemisphere.

Measurement of VEP Latency

Latency at the center of P100 is the cardinal measurement in clinical VEPs. P100 may be narrow or broad, and symmetrical or asymmetrical in shape (Fig. 36–4). Although pathological broadening of P100 may be seen in some disease states, it usually accompanies latency increases and has not been used as an independent sign of disease. When the

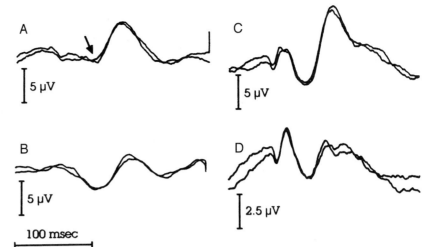

FIGURE 36–4. Normal waveform variability of the PVEP in four different subjects, all recorded from O_z–C_z. N75 may be entirely absent *(A)*, or the largest peak *(D)*. P50 is seen in *A*, *C*, and *D* but not in *B*. P100 is consistently present at about 100 msec. In the absence of N75, the latency of P100 is measured at the beginning of the upstroke to N145, as indicated by the arrow in *A*. (Modified with permission of the publisher from Pedley T, Daly D: Current Practice of Clinical EEG, 2nd ed. New York, Raven Press, 1990.)

VEP waveform is noisy or asymmetrical, the choice of "peak latency" may be ambiguous. One technique is to extrapolate the downslopes from N75 and N145, taking the P100 latency as the point of their intersection. Another is to estimate, by inspection, the center of the entire complex from N75 to N145. If N75 is absent (Fig. 36–4A), P100 should be considered the beginning of the upstroke to N145.

As discussed repeatedly throughout this chapter, the range of "normal" P100 latencies depends greatly on stimulus and patient factors. Normal individuals often show latency variations of several milliseconds on serial studies.[70, 118] However, intereye latency differences are less variable than are absolute latency measurements.[147, 155] Upper limits for interocular latency difference usually lie between 6 and 8 msec.

Measurement of VEP Amplitude

VEP amplitudes are more variable and less specific than latencies; thus, they should be used only with great caution as indicators of neurological disease. Different laboratories may measure amplitude from the peak of N75 to P100, from P100 to N145, as the sum of the two, or as whichever is best defined in a given subject. P100 amplitude distribution in normal individuals is nongaussian, so that attempts to calculate normal limits by parametric statistics such as mean and standard deviation can be misleading. The lower limit of normal for the amplitude of the N75–P100 component is zero (Fig. 36–4A).[147, 155] Additional channels may be used to compare amplitude in the left and right occipital regions; normal cooperative subjects may have left/right amplitude asymmetries of up to 2.5:1.

If average background noise is acceptably low (1 μV or less) a nonreproducible PVEP should be considered absent (see Fig. 36–7B). In cooperative subjects, this absence should be confirmed by additional recordings at P_z and the inion.

Unusual Waveforms

A VEP waveform is "aberrant" if it can be recorded reproducibly but a P100 peak cannot be identified. Many of these aberrant responses appear to contain too many peaks rather than too few, and the origin of each peak must be analyzed individually. Apparently bifid, or W, potentials may be recorded in normal subjects when the frontal N100 and occipital P100 components are asynchronous.[153, 155] This double peak will be seen only with the use of a frontal or central reference. In such cases, the true P100 latency can be identified easily in simultaneous ear-reference recording (Fig. 36–

5). Other W potentials disappear when the patient focuses on the upper edge of the pattern, rather than the center. Bifid potentials that fail to resolve with these simple maneuvers are rarely seen in normal individuals; many authors consider them pathological. Persistently bifid waveforms are best analyzed with hemifield stimulation[81] or with the techniques described later in this chapter.

Some supernumerary waveforms, such as those that follow P175 in the top tracing of Figure 36–5, appear to represent transient oscillations of the occipital cortex, the activity of which is entrained by each stimulus. Rarely, such oscillations can lead to the appearance of VEP waves earlier than P50.

STIMULUS PARAMETERS

Pattern Reversal

The most common stimulus for PVEP recording is a pattern of light and dark checks that is repeatedly reversed; that is, the light portions of the pattern abruptly become dark, and the dark portions light. The total luminance of the

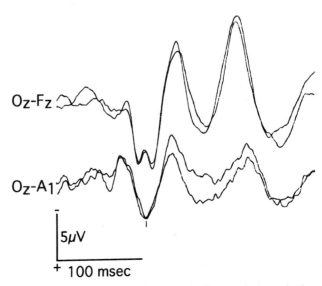

FIGURE 36–5. Bifid P100 with a vertex reference *(top)*, resolved to a single peak with a simultaneously recorded ear reference *(bottom)*. The W-shaped P100 is caused by an asynchronous N100 picked up by the vertex electrode. This subject also has a series of prominent oscillatory waves continuing after P175 in the upper tracing.

pattern does not change. Some laboratories use bars or stripes as patterns instead of checkerboards.

Video monitors are the most common and versatile devices for producing VEP patterns. However, video monitors draw patterns slowly enough to affect the features of evoked potentials. In the regular "interlace" mode, television generates only 30 new frames per second (25 in Europe), requiring over 30 msec to fully replace one picture with another. A faster "noninterlace" mode used in many PVEP systems requires about 15 msec. Either way, the exact timing of pattern reversal depends on what part of the screen one is looking at. This temporal smearing of video images affects the resulting PVEPs. The early, relatively narrow P50 is less likely to be seen with video pattern generators than with other techniques, and P100 latency is slightly prolonged.

Pattern reversal can also be performed with moving mirrors or with grids of light-emitting diodes (LEDs) which turn on and off in a few microseconds.[4, 52] LED stimulators are small, sturdy, portable, and reliable, making them suitable for mounting above a supine patient. However, the size of the elements is fixed, and the total pattern area is small.

Pattern Onset

A VEP pattern may be flashed on and off rather than reversed. It is possible to hold luminance constant with flashing patterns,[131] but pattern onset usually combines both pattern and luminance stimuli. Although used less widely than pattern reversal, pattern onset produces more robust responses.[2, 131, 139] Aminoff and Ochs[2] reported that this type of stimulus produced a higher yield of diagnostic abnormalities in patients with multiple sclerosis.

Check Size

Research on the contrast sensitivity function of the fovea suggests that checks with angular edge dimensions around 3 cycles per degree (cpd) represent an optimal stimulus to central vision. The same size checks give, on average, the largest P100 after pattern reversal. Other considerations, however, mitigate against use of a pattern this fine for routine clinical testing. Many patients referred to PVEP testing have less than 20/20 acuity because of presbyopia, refractive errors, or cataracts. Some patients arrive at the laboratory without their glasses or with the lingering effects of mydriatics that paralyze accommodation. With very small checks or gratings, latency of the PVEP is quite sensitive to blurring of the pattern by poor acuity;[151] to ophthalmological disorders such as central serous retinopathy (which can mimic optic neuritis);[84] to age;[152] and even to nonvisual disorders such as Parkinson's disease.[163] To the neurologist interested in searching for disease of the optic nerve or posterior visual pathways, latency prolongation produced by any of these factors is misleading and reduces the utility of the test. On the other hand, checks larger than 1 degree across approach the size of the entire fovea, with increased stimulation of parafoveal and peripheral portions of the retina. As a compromise, the American Clinical Neurophysiology Society guidelines suggest a check size close to 30 minutes of arc.[147] In practice, checks up to 50 minutes of arc have proved clinically useful for full-field stimulation and are preferred in many laboratories specifically to avoid the confounding factors noted previously. Checks between 50 minutes and 90 minutes of arc are recommended when PVEPs are used for testing visual fields. However, checks larger than about 1 degree of arc are more likely to produce bifid or otherwise aberrant waveforms. Check size in minutes of arc can be calculated from the formula $M = 3438*W/D$, where W is

the width of each check in millimeters and D is distance in mm from pattern to patient.

Field Size

Checkerboard patterns used for PVEPs have ranged in width from as small as 3.0[32] or 3.8[21] degrees up to 48 degrees of arc.[95] For full-field testing of central vision, a pattern 4 degrees square covers the macula and gives quite adequate results. Larger fields produce only a small increase in PVEP amplitude. Some research suggests that smaller fields increase sensitivity of the PVEP to optic nerve demyelination.[74] However, larger fields allow the patient's eye to fixate less rigorously during testing.

The size of the pattern significantly affects the direction of the PVEP net vector from hemifield stimulation. With large stimulus fields, the contribution of the peripheral retina directs the net vector obliquely through the opposite hemisphere (see Fig. 36–2), so that the hemifield VEP may appear to arise from the "wrong" side. With patterns only a few degrees across, the net vector tends to point straight back, so that this paradoxical lateralization does not occur. In the past, the effect of field size on the PVEP vector led to conflicting experimental results. Hemifield stimulation should be done only with large patterns.

Luminance

The brightness of light is measured in candelas; the luminance of a two-dimensional surface is expressed in candelas per square meter (cd/m^2) or in foot-lamberts (1.0 f-l = 3.426 cd/m^2). The contrast between two adjoining areas of a pattern is usually expressed as a percentage, where L_{max} is the brighter area and L_{min} the darker one:

$$\text{Contrast} = 100 \times \frac{L_{max} - L_{min}}{L_{max} + L_{min}}$$

In normal subjects, the average latency of P100 falls by about 12 msec per log increase in luminance.[28] That is, every time the brightness of the pattern is increased tenfold, latency decreases 12 msec. A luminance difference of 100 times can be expected to produce a P100 latency difference around 24 msec. The effect of luminance on latency is greater in patients with optic nerve disease, so that the use of lower luminance has been suggested as a method for increasing PVEP sensitivity.[28] Unfortunately, lower luminance may also increase sensitivity to other conditions such as retinal disease,[141] poor visual acuity, and age.[138]

The possibility of *accidental* luminance changes is critically important in the VEP laboratory, especially with TV pattern generators. Luminance of the TV screen can be altered by brightness and contrast controls, and even by aging of the picture tube. Laboratories collecting normal data sometimes neglect to measure the luminance of the screen before they start, or even to mark the positions of the controls. Later, visitors to the laboratory may randomly turn the adjustment knobs, altering the range of normal values and increasing the possibility of major errors in interpretation. Control settings should be clearly marked, and the screen luminance checked periodically with a simple light meter placed at a standard distance from the screen.

PVEP testing should be performed in a moderately well-illuminated room, where the technician can observe patient cooperation and fixation on the pattern. In such a photopic environment, vision is dominated by retinal cones, and uniform test conditions are relatively simple to attain. In a dark-adapted (scotopic) test situation, the acuity, color sensitivity, and integrating functions of the retina are different and, more importantly, are harder to control. Scotopic adaptation takes about 30 minutes, and may be altered unpredictably by the brightness of the pattern itself.

Contrast

The amplitude of the PVEP increases with increasing pattern contrast up to the range of 20% to 40%. Beyond this the effects of contrast saturate, and there is no further change. In most clinical laboratories, regardless of stimulator type, contrast is well above saturation level.

Color

Retinal cones sensitive to red light are heavily concentrated in the foveal area, and red desaturation is a sensitive clinical sign of paracentral scotomata. On this basis, red-and-black patterns have been suggested as a technique for increasing the sensitivity of the PVEP to central or paracentral lesions.[110] In general, different colors produce minor changes in PVEP amplitude and latency, but colored patterns have not been widely adopted for clinical testing.

Distance to Pattern

An eye-to-pattern distance between 0.7 and 1.5 meters is generally satisfactory. Longer distances may cause blurring from severe uncorrected myopia or may result in reduced attention in uncooperative patients. Shorter distances may cause difficulty from presbyopia.

Reversal Frequency

The practical range for PVEP testing is 0.5 to 4 reversals/second. The average latency of P100 rises by a few milliseconds with higher reversal frequency.[155] Conventional TV patterns begin to degrade around 4 per second, and American Clinical Neurophysiology Society guidelines recommend staying below this rate.[147] Above 4 per second, at the faster rates used in steady state recording, the net VEP vector from pattern reversal swings posteriorly and is seen over the ipsilateral rather than the contralateral hemisphere.[119] This effect is similar to that of reducing the stimulus field size, and implies that visual field defects are more likely to be identified at slower reversal rates.

RECORDING PARAMETERS

Passband

Common guidelines recommend a low filter (high-pass) setting of 1.0 Hz, and a high filter (low-pass) setting of 100 Hz.[147] However, the literature includes many studies done with the high filter set as low as 70 Hz.[81, 131] Waveform distortion may be noticeable at this point. Many laboratories use a high filter of 200 to 300 Hz for clinical testing. This corresponds to a smaller phase lag, but should have no significant effect on results—*provided* normative data are collected in the same way! 60-Hz filters should be avoided in VEP testing.

Sampling Rate

Analog data sampling should be fast enough to produce a new data point in each channel every 1 msec or less. For

the sensitivity setting, an optimal input range for the A/D converter is usually 50 to 100 μV. Sweep duration should allow for sampling to continue for at least 250 msec after each stimulus.

Typically 100 to 250 artifact-free samples are necessary to produce a reliable waveform. At least two averages should be taken in each test condition—more if responses are poorly reproduced. Ideally, well-reproduced responses should contain a P100 latency difference of no more than 2.5 msec, and an amplitude difference of no more than 15%.

Recording Derivations

All VEP montages must include a mid-occipital (MO) electrode. The popular "Queen Square" technique uses an MO position 5 cm above the inion[147] and is meant to occupy the average topographical center of P100. The 10–20 location O$_z$ lies about 1.4 cm below this. Neither MO nor O$_z$ can be considered a definitive location in all subjects, however, because the center of P100 may lie as high as P$_z$ or as low as the inion in normal individuals. An example of P100 centered at the inion is shown in Figure 36–6. Such atypical localization, and the additional recordings needed to clarify it, should be considered for all poorly defined waveforms.

A common reference for PVEPs is F$_z$ (or the Queen Square

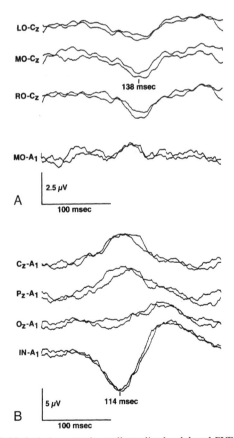

FIGURE 36–6. *A,* Apparently small-amplitude, delayed PVEP to full-field stimulation. However, the response from MO-A$_1$ is discordant, suggesting an atypical distribution of the electric field. *B,* The same subject recorded with an A$_1$ reference montage. P100 is located caudally at the level of the inion, and has normal latency for this laboratory. The transition zone between P100 and N100 is located near MO and O$_z$, producing small and aberrant waveforms at those sites. (Modified with permission of the publisher from Pedley T, Daly D: Current Practice of Clinical EEG, 2nd ed. New York, Raven Press, 1990.)

MF, which is 12 cm above the nasion). A vertex reference at C$_z$ produces similar results, and electromyographic (EMG) activity may be less prominent. Either site will record the anterior negativity, N100, that accompanies P100. This has the advantage of maximizing apparent P100 amplitude while minimizing noise (see Figure 36–3), but the disadvantage is that the contributions of N100 and P100 to an aberrant waveform may be impossible to separate. A large N100 can mask an ectopic, small, or absent P100; asynchrony of these two components may lead to an apparently bifid or ambiguous P100.[153] The easiest way to recognize and resolve such aberrant waveforms is to record routinely from mid-occiput to one or both ears in a second channel. The ears are minimally involved by the PVEP, and can be considered inactive for full-field recording. The ear-reference channel corroborates the validity of the midline recording and helps identify normal variants that otherwise might lead to serious misinterpretation[142] (Fig. 36–6). Frontal reference-to-ear is also used as a check channel, but its interpretation may be more convoluted.

Full-field testing generally includes additional electrodes situated lateral to the midline. The 10–20 locations O$_1$ and O$_2$ are too close together, so the American Clinical Neurophysiology Society recommends left occipital (LO) and right occipital (RO) positions 5 cm to each side of MO[147] (Fig. 36–6). These additional derivations help to identify P100 in difficult cases. They occasionally may be useful in screening for unsuspected field defects, which can then be better defined by subsequent hemifield recording. However, the sensitivity of full-field testing to field defects is limited,[11, 87, 129] and the range of normal variation between LO and RO is large. Even P100 amplitude asymmetries greater than 2.5:1 should not be considered abnormal in the absence of corroboration by hemifield studies.

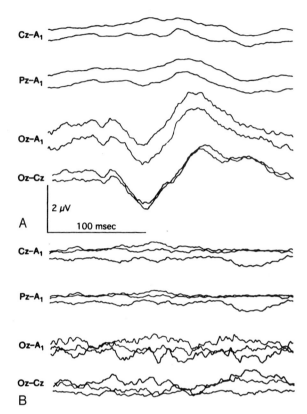

FIGURE 36–7. *A*, Normal monocular PVEP. *B*, Same subject, same eye, performing transcendental meditation. (Reproduced with permission of the publisher from Bumgartner J, Epstein C: Voluntary alteration of visual evoked potentials. Ann Neurol 12:475–478, 1982.)

PATIENT CONSIDERATIONS

Preparation

If at all possible, PVEP testing should not be performed within 24 hours of using mydriatics (drugs that dilate the pupils.) Patients who use glasses should try viewing the pattern with and without them, and be tested under the condition that produces the clearest image. Each eye is tested individually with the other occluded by a patch. Binocular testing is used only in patients who have difficulty cooperating otherwise.

Large pupillary asymmetries can alter the amount of light reaching the two retinae by a ratio of 10:1, and alter PVEP latencies by more than 10 msec. Miotics and cataracts can have similar effects. Unequal pupil size should be noted by the technician, and considered by the interpreter when evaluating small latency asymmetries.[147]

Attention

A PVEP will be recorded only if the subject is awake and concentrating on the stimulus. Amplitude and even latency can be affected if the subject's gaze is allowed to wander, or if there is excessive blinking and squinting.[171] Unfortunately, patient variables that are difficult to monitor may have an appreciable effect on PVEP amplitude and latency. The difference between relaxed and highly alert states can produce latency differences as large as 11 msec in one eye from the same subject.[43] Volunteers using maneuvers such as meditation and convergence with one eye covered (Figs. 36–7 and 36–8) have been able to abolish the PVEP or change its

latency by up to 15 msec without detection by an attentive technician.[25] Similar effects have been shown with different types of pattern stimuli in several other laboratories.[25, 107, 161] These problems may be reduced by larger stimulus fields, larger checks, and binocular testing.[161]

"Problem" patients often exhibit unusual difficulty in maintaining fixation on the pattern and repeatedly allow the eye to close unless it is held open by the technician. Some may be capable of more subtle manipulation, similar to the volunteers described above. In general, low amplitude, absent, or mildly delayed responses in patients with a possibility of factitious disease should be interpreted with caution.

Age

With checks of 30 minutes or larger, PVEP latencies appear to reach adult values by 20 weeks of age.[150] Latencies for all check sizes are at adult values by age 5[44, 150] and show no significant change from then until about age 40. In testing small children, cooperation may be the biggest problem. In later life, normative studies have given conflicting results concerning the nature and degree of latency changes in healthy middle-aged and elderly individuals. Some of these discrepant findings have been resolved by the observation that aging effects are dependent on both check size and luminance. Latency rises more rapidly with age using smaller checks[36, 152] and decreased luminance;[138] these effects are not explained by pupil size.[36, 152] They can be minimized by using checks of 30 minutes or larger and a luminance of 50 cd/m² or greater. When feasible, normative data should be calculated separately for older subjects.

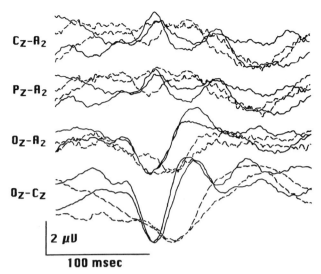

FIGURE 36–8. *Solid lines:* normal PVEP with the subject attending to the stimulus, which is a video pattern generator with 21-minute checks, 10-degree field. *Dashed lines:* same eye, same subject performing forced convergence with one eye covered. The apparent latency difference is 15 msec. The maneuver was inapparent to the technician. (Reproduced with permission of the publisher from Bumgartner J, Epstein C: Voluntary alteration of visual evoked potentials. Ann Neurol 12:475–478, 1982.)

Gender

Women consistently show average latencies a few milliseconds shorter than men.[42, 155] In some series, women average higher amplitude PVEPs.[42, 44] This difference is more apparent with checks smaller than 30 minutes.[51]

NEUROLOGICAL APPLICATIONS OF VISUAL EVOKED POTENTIALS

Disease of the anterior optic pathways may produce prolonged P100 latencies without detectable alteration in acuity,[68, 69] contrast sensitivity,[21, 109, 136] pupillary reactivity, color desaturation, fundoscopy, or perimetry.[24] Conversely, it is very rare for clinical examination to show any abnormality of the anterior visual pathways when the PVEP is normal.[24] If visual acuity is decreased but the PVEP is normal, a lesion of the optic nerve or chiasm is very unlikely to be the etiology. (The same high sensitivity of PVEPs is not, unfortunately, present for retrochiasmatic lesions.[77]) Figure 36–9 shows mild, moderate, and marked absolute latency increases for P100 using full-field monocular stimulation.

Optic Neuritis and Multiple Sclerosis

Screening for asymptomatic lesions in multiple sclerosis (MS) is the most common use of PVEPs. Approximately 90% of patients with a history of definite optic neuritis have abnormalities of the PVEP.[6, 28, 40, 68] During the acute attack, the number of patients with unobtainable PVEPs or prolonged P100 latencies approaches 100%. In the majority of patients, latency abnormalities persist for many years, long after recovery of normal acuity. Chiappa reviewed 26 clinical series totaling nearly 2000 patients with possible, probable, or definite MS.[40] Overall, PVEP abnormalities were present in 37%, 58%, and 85%, respectively. Among 774 patients without clinical or historical evidence of optic neuritis, 54%

had abnormal PVEPs, the majority of which were latency prolongations. Other series[54, 80, 104, 113, 117] document a similar yield. The classic PVEP pattern in MS is an asymmetrical, marked prolongation of P100 with relative preservation of amplitude (Fig. 36–9C). Such striking latency prolongations seldom occur in other disorders; however, MS can commonly produce mildly prolonged and absent PVEPs as well.

Various modifications have been tried to increase the yield of PVEP abnormalities in demyelinating disorders. These include decreased field size,[31, 74, 110] smaller checks,[113, 117] light-emitting diodes,[4] decreased luminance,[28] red/black patterns,[110] pattern onset stimulus,[2] hemifield stimulation,[113, 134] multiple orientations of gratings,[27] hyperventilation,[46] Fourier analysis,[34] and addition of pattern electroretinogram.[35] Hyperthermia has been applied as a provocative test; it usually decreases PVEP amplitudes in both normal subjects and MS patients but has no effect on latency.[97, 174]

Obviously it is not possible to use all these special techniques in every patient. Patient tolerance and the need for multiple additional control studies limit the number of variations that can be tried in a single individual.[146] A useful screen for patients of varying age, acuity, and cooperation might incorporate full-field testing with checks 50 minutes across and high luminance (at least 50 cd/m²). For these conditions, a series of 20 normal controls is adequate; the controls should include a healthy mix of different ages and sexes, reflecting the expected patient population. A higher sensitivity approach to cooperative patients with good acuity could involve a full-field pattern using low luminance (under 40 cd/m²),[28] small field size, and small checks. The latter approach would generate a higher yield of VEP abnormalities; it would also be expected to produce a higher percentage of false positives, and require a much larger control group compensated for age.

The advent of magnetic resonance imaging (MRI) has changed the indications for evoked potentials.[54, 59, 121] However, conventional MRI is relatively insensitive to lesions of the optic nerve. In one series, 6 of 11 MS suspects with negative MRIs had abnormal PVEPs, and of those, 4 of 6 had

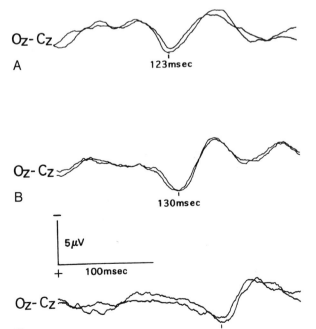

FIGURE 36–9. *A* to *C,* Mild, moderate, and marked midline latency prolongation in vertex-reference derivations.

cerebrospinal fluid (CSF) immunoglobulin abnormalities.[54] Especially in the chronic state, PVEPs remain at least as sensitive as MRI for detecting optic nerve lesions, and are more cost-effective.[103]

CASE STUDY 1

Lhermitte's Symptom

A 22-year-old woman presents with a sensation of electric tingling down her arms and legs whenever she coughs or bends her head forward. On questioning, she recalls that 2 years ago she had an episode of blurred vision in the right eye that lasted for several days, but did not see a doctor. Visual acuity is now 20/20 OU. MRI of the brain and cervical spinal cord shows a questionable area of gadolinium enhancement in the dorsal portion of the cord. Her midline PVEPs are shown in Figure 36–10.

Comment. For this laboratory, an absolute P100 latency of 137 msec on the right would be moderately prolonged; but compared to the normal latency of 108 msec on the left, that on the right is severely prolonged. This finding strongly corroborates the clinical suspicion of past optic neuritis on the right, despite normal visual acuity at the time of testing. It helps substantiate the presence of disseminated lesions in time and space required for a diagnosis of MS.

Other Diseases of the Optic Nerve

Anterior ischemic optic neuropathy is a common disorder of later life, often associated with arteriosclerosis. It may also occur with temporal arteritis and other forms of vasculitis, tertiary syphilis, or migraine. Thompson and associates[165] found that the PVEP was abnormal in all cases of ischemic optic neuropathy studied within 2 weeks of onset, and in 10 of 12 eyes studied more than 2 months later. Although the P100 often appeared bifid or delayed, they reported that detailed partial-field analysis revealed the true pathophysiology to be an alteration in PVEP topography resulting from macular field defects. In almost all cases of Leber's hereditary optic atrophy, PVEPs are absent or show mild latency increases; but only subtle changes are found in presumed asymptomatic carriers.[32]

Optic atrophy is occasionally found with peripheral neuropathies, and PVEP latency prolongations have been reported in Charcot-Marie-Tooth disease,[99, 159] giant axonal neuropathy,[94] and neuropathy associated with macroglobulinemia.[8]

Sarcoidosis has multiple central nervous system (CNS) and ocular manifestations including optic neuropathy. Streletz and colleagues[158] studied PVEPs in 50 patients with systemic sarcoidosis. Abnormalities of latency and amplitude were found in all 4 patients with clinical brain involvement, and in 7 of 29 patients who had no clinical evidence of ocular or neurological disease.

Increased intracranial pressure can produce papilledema and visual loss, so it is not surprising that P100 latency changes have been reported with hydrocephalus[49, 140] and pseudotumor cerebri,[156] although such abnormalities are not particularly frequent.

A variety of toxins may produce clinical or subclinical effects on the optic nerve. Latency prolongations have been described with quinine[58] and nitrous oxide[75] toxicity. PVEPs have been used to monitor therapy with ethambutol,[175] cisplatin,[92] and deferoxamine.[164] P100 delays are reported in chronic alcoholics;[38] PVEPs are uniformly absent or delayed in patients with tobacco-alcohol amblyopia.[85] In most of these conditions, improvement has occurred after discontinuation of therapy or abstinence.[50]

Leukodystrophies and Spinocerebellar Disorders

PVEPs are generally abnormal in disorders of central myelin. Many of these conditions share clinical features with MS, and abnormal evoked potentials should not be overinterpreted as proof of a specific disorder. Adrenoleukodystrophy, Pelizaeus-Merzbacher disease, and metachromatic leukodystrophy prolong PVEP latency[5, 30, 95, 96] and may produce abnormalities in presymptomatic affected children.[95] The site of the responsible lesion is not clear, given the predilection of these disorders for the cerebral hemispheres. Prolonged latencies have been found in varying proportions of patients with Friedreich's and other inherited ataxias,[13, 31, 115, 122] familial[13, 71, 122] and tropical[6] spastic paraparesis, subacute combined degeneration (vitamin B_{12} deficiency),[86] and vitamin E deficiency.[82, 101] Latency prolongation in this group of disor-

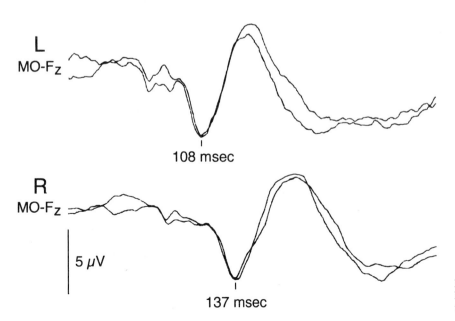

L
MO-F$_z$

108 msec

R
MO-F$_z$

5 μV

137 msec

FIGURE 36–10. PVEP of patient in Case 1. Right eye stimulation shows a marked relative delay of P100.

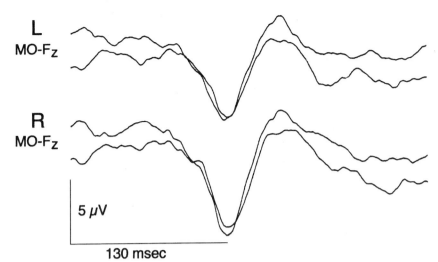

L
MO-F$_Z$

R
MO-F$_Z$

5 μV

130 msec

FIGURE 36–11. PVEP of patient in Case 2. Moderate, bilateral symmetrical prolongation of P100 is present.

ders is usually symmetrical. Abnormalities caused by vitamin E and B$_{12}$ deficiency may improve with treatment.[82, 86, 101]

CASE STUDY 2

Spastic Paraparesis

A 32-year-old man, an immigrant from Jamaica, presents with 3 years of gradually worsening gait, which he initially blamed on "arthritis." Neurological examination is positive for lower extremity hyperreflexia and positive Babinski signs. His VEP results are shown in Figure 36–11.

Comment. Prolonged VEP latencies indicate additional lesions in a patient whose clinical examination is consistent only with involvement of the spinal cord. However, this moderate, symmetrical prolongation of latencies is quite nonspecific. The wide differential diagnosis includes MS, familial spastic paraparesis, tropical spastic paraparesis, B$_{12}$ deficiency, CNS sarcoidosis, and late-onset adrenoleukodystrophy. Because responses are symmetrical, it is not possible to determine with certainty whether the lesions are anterior or posterior to the optic chiasm. However, the former is more likely.

Tumors

Extrinsic compression of the optic nerve or chiasm usually produces abnormal PVEPs.[60, 67, 76, 119] In general, such lesions are expressed by a decrease in amplitude, and by mild or moderate prolongations in latency. In patients with pituitary adenomas, PVEPs may improve the serial assessment of visual function.[76]

Miscellaneous Central Nervous System Disorders

PVEPs are delayed and "desynchronized" in cerebrotendinous xanthomatosis, and improve after treatment with chen-

odiol.[106] PVEP latencies were prolonged in a patient who suffered decompensation of phenylketonuria in adulthood.[98]

PVEP abnormalities are found in some CNS disorders whose major clinical and pathological features lie outside the sensory pathways. P100 latency was delayed in 3 of 8 patients with neurological involvement from Wilson's disease,[41] and in patients with idiopathic Parkinson's disease.[22] The pathophysiology of PVEP abnormalities in Parkinson's disease has been debated. Tartaglione and associates[163] reported that they occurred with sine wave gratings of 2 cpd, but not with 55-minute checks. In contrast, Bhaskar and colleagues[12] found that prolonged latencies occurred with both gratings and checks, but that improvement after L-dopa treatment occurred only with checks. The changes in Parkinson's disease have been correlated by some researchers with abnormalities of the electroretinogram (ERG), and ascribed to involvement of dopaminergic amacrine cells in the retina,[61] but others have found evidence for extraretinal involvement.[26] (As PVEPs have little value in the diagnosis of Parkinson's disease, one might choose to select test conditions that are less likely to be affected by it.)

High-amplitude PVEPs occur in familial myoclonic epilepsies[66] and other gray matter disorders. This finding proved useful in screening a large pedigree for mild or asymptomatic mitochondrial encephalomyopathy (Fig. 36–12).[132] VEP latency increases are reportedly common in photosensitive epilepsies.[100]

Migraine has been correlated with increased high-frequency photic following on electroencephalogram (EEG),[116, 145] with increased oscillatory wave amplitude of the PVEP 250 to 500 msec after a flash or pattern stimulus (see Fig. 36–5),[108] and with hemifield changes that match the visual aura.[160] However, the specificity and clinical usefulness of these changes have been disputed.[172, 173]

Both flash and pattern VEPs have been studied in Alzheimer's disease and other dementias. Changes in P100 latency are minimal[79] or nonexistent.[72] Compared to appropriate-aged controls, demented subjects have shown latency prolon-

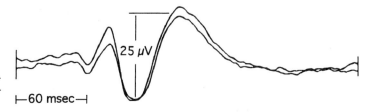

FIGURE 36–12. Increased amplitude of the PVEP in a minimally symptomatic carrier of mitochondrial encephalomyopathy.

25 μV

├─60 msec─┤

gation in the late components of the PVEP (N145 and after)[79] and in some components of the flash VEP (FVEP).[72] Similar changes have been described with anticholinergics in normal subjects.[7, 128] Group differences were shown in Alzheimer's patients using steady state stimulation and Fourier analysis of frequencies from 15 to 30 Hz.[166] These changes are not a consequence of diminished attention.[79] But because the affected components of the VEP are present less reliably than P100 and are more variable, the applicability of VEPs to individual cases of dementia is uncertain.

CASE STUDY 3

Disability Evaluation

A 38-year-old woman presents for evaluation complaining of bilateral visual loss. She has always had a "lazy eye" on the left, and has never seen normally with that eye. Now she describes gradually decreasing acuity on the *right* for 1 year, and is applying for Social Security disability benefits. An ophthalmologist reported amblyopia *ex anopsia* on the left, but no ocular pathology on the right, and no Marcus-Gunn pupil. A noncontrast computed tomographic (CT) scan of the brain was reported normal. The technologist who performed the PVEP said that during testing the patient had trouble keeping her right eye open and focused on the checkerboard pattern. Her right PVEP is shown in Figure 36–13. Check size is 45 minutes.

Comment. Disability from visual loss requires bilateral impairment, and the eye to examine in this case is clearly the right. With preexisting amblyopia on the left, an interocular latency comparison is not meaningful. However, normal latency of P100 on the right practically excludes a lesion in the optic nerve on anterior chiasm. Recording of a clearly definable P100 on the right, despite poor cooperation, also mitigates against severe acquired visual loss from any cause.

VEPs in Coma

VEPs have been used, along with somatosensory evoked potentials (SEPs) and brain stem auditory evoked potentials (BAEPs), to assess disability and attempt prognosis in prolonged coma, especially after closed head trauma. In most studies, the stimulus has been a stroboscopic flash, and the ERG has been monitored to verify retinal function. Different grading systems[62] make comparisons among reports difficult. Anderson and colleagues[3] found that severely abnormal VEPs accurately predicted an unfavorable outcome after closed head trauma. However, normal VEPs did not reliably predict a favorable result, and VEPs did not improve prognostic accuracy beyond that obtained using SEPs alone. Rappaport and associates[125, 126] reported that in head injury patients the FVEP had a weak correlation with disability ratings performed 1 year later. Pfurtscheller and colleagues[123] had difficulty making any correlation between VEPs and outcome. Gupta and associates[63] studied patients 6 to 24 months

after head injury, when they had recovered consciousness and were able to cooperate for PVEPs. They found abnormal PVEPs in one-third of all patients, and in one-half of those with severe residual cognitive impairment.

VEPs in the Operating Room

The VEP is used intraoperatively to monitor surgical procedures that might endanger the optic nerve and chiasm. As it is difficult to present true pattern reversal in the eyes of anesthetized patients, the typical stimulus is a flash from a strobe or from red LEDs mounted in small goggles, delivered through closed eyelids. Unfortunately, the middle-latency VEP is much more affected by level of consciousness, anesthesia,[39, 137, 144, 170] and hypothermia[130, 135] than are short-latency SEPs and BAEPs. As a result, operative VEPs are extremely labile, with a high incidence of false-positive alterations that detracts from their utility.[33, 127] Harding and colleagues[73] found that a combination of enflurane and nitrous oxide had minimal effect on VEPs, and that disappearance of a previously normal VEP for more than 4 minutes correlated significantly with postoperative visual loss.

OPHTHALMOLOGICAL APPLICATIONS OF VISUAL EVOKED POTENTIALS

Visual Acuity

The neurological applications of VEPs are due largely to their *lack* of correlation with visual acuity and ophthalmological disease. However, checkerboard patterns or gratings with sufficiently small elements can be used to estimate visual acuity. Reasonably accurate predictions can be made by plotting PVEP amplitude against spatial frequency for patterns of several different sizes.[112, 148] Sokol[148] used conventional PVEPs from checkerboard patterns to determine visual acuity in infancy. He estimated acuities of 20/150 at 2 months of age, improving to 20/20 by age 6 months. In adults, a well-defined response to 10 minute checks indicates essentially normal acuity, and a clear response to checks of 45 minutes or smaller suggests vision of 20/200 or better.[40]

More rapid estimates of PVEP amplitude can be obtained with steady state stimuli (i.e., rapid pattern reversal at 10 to 20 Hz) and signal analysis techniques such as the Fourier transform.[53, 162] Using a total test duration of 10 seconds, Norcia and Tyler[112] estimated an acuity of 4.5 cpd in the first month of life, reaching adult values of 20 cpd by 8 months.

Amblyopia

Using small checks or gratings, amblyopia *ex anopsia* in children is characterized by decreased amplitude and occasionally increased latency of PVEPs. Friendly and colleagues[56] used reversing checks 15 minutes on edge, or 2 cpd, and were able to identify the amblyopic eye correctly in 22 of 27 children. (They also misclassified one of four normal subjects as amblyopic.) Sokol and associates[149] found that PVEPs with 15 minute checks had a sensitivity of 96% when the acuity difference between eyes was three lines or more on a Snellen chart. However, sensitivity was only 50% with a two-line acuity difference, and 30% with one line.

Glaucoma

Midline PVEP latencies were abnormal in a little over one half of all eyes with glaucomatous field defects,[29, 167] but in

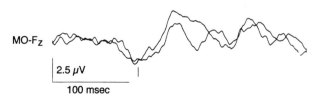

MO-F$_z$

2.5 µV

100 msec

FIGURE 36–13. PVEP of patient in Case 3. Normal latency P100 to right eye stimulation.

less than one-quarter of those with ocular hypertension.[167] Surprisingly, latency prolongation was most likely to be found with 48-minute checks, rather than with smaller checks of 12 minutes. Conventional ("transient") PVEPs were less sensitive than steady state PVEPs at 7.5 alternations per second. Increased PVEP latency was associated with visible cupping and pallor of the optic disc.

Retinopathies

Central serous retinopathy is an important differential diagnosis for optic neuritis. This benign retinopathy often produces symptoms of central scotoma and decreased acuity, and it commonly produces PVEP delays. Papakostopoulos and colleagues[120] reported that in six patients with central serous retinopathy PVEP latencies averaged 7 msec longer from the affected eye. Sherman and associates[141] found significantly increased latencies in 9 of 10 involved eyes. Compared to the normal eye, the mean delay was 10.6 msec with 56-minute checks and 21 msec with 14-minute checks. One subject showed a latency difference of 43 msec. Amplitudes were affected less than latencies, further mimicking PVEP changes seen in demyelinating disorders. Kraushar and Miller[84] reported three patients in whom ophthalmologists mistook central serous retinopathy for evidence of MS. The retinal disorder usually improves spontaneously—and, in contrast to the usual outcome in optic neuritis, PVEP abnormalities generally remit with it.

PVEP delays have been described with a variety of other acute and chronic retinal disorders.[20, 90] Because latency delays can be produced both by retinal diseases and by optic neuritis, complete ophthalmological examination is mandatory in patients with unexplained symptomatic visual loss, regardless of PVEP findings.

CASE STUDY 4

Acute Monocular Blindness

A 45-year-old woman presents for routine follow up in the epilepsy clinic, and mentions that she abruptly lost vision in the right eye 2 days ago. She describes flashing lights in that eye for several hours at the onset of symptoms. Her PVEP from the right eye is shown in Figure 36–14.

Comment. An absent PVEP is extremely nonspecific. The differential diagnosis includes inability to concentrate on the stimulus, malingering, and practically any severe ocular or retrobulbar disease. An ophthalmological examination is mandatory if symptoms are unilateral, and brain imaging may be indicated if they are bilateral. This patient had a retinal detachment.

Hysteria and Malingering

Hysterical visual loss or exaggeration of a mild impairment are frequent sources of referral to the VEP laboratory. Ensur-

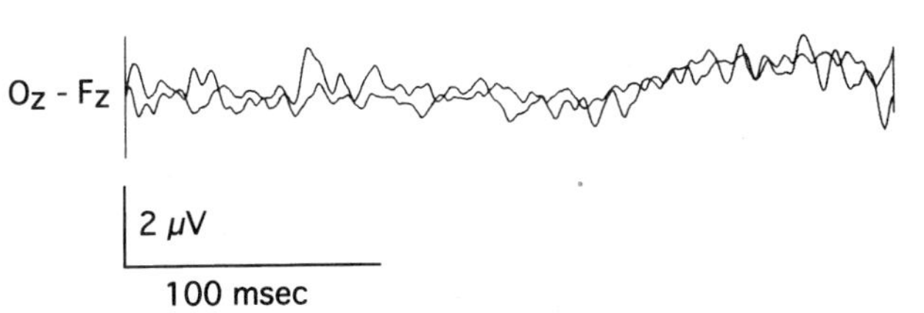

FIGURE 36–14. PVEP of patient in Case 4. Absent PVEP to right eye stimulation.

ing good cooperation is essential in these patients. A normal P100 latency makes lesions of the optic nerve or anterior chiasm very unlikely as the cause of subjective visual loss. Approximate benchmarks for visual acuity have been noted earlier. Long-latency PVEPs have also been used to document responses to stimuli that patients claim they are unable to see.[168] The technician and interpreter must be cautious: as previously noted, abolition of the PVEP and mild degrees of latency prolongation may be accomplished volitionally by some subjects despite apparently good cooperation.[25]

In addition, normal PVEPs do not rule out retrochiasmatic visual field defects, especially with full-field stimulation and recording at the occipital midline. Very rarely, patients with complete cortical blindness have recordable PVEPs when an island of striate cortex is spared by extensive parieto-occipital lesions.[19] In general, such cases are easily identified by clinical and neuroimaging evaluation.

PARTIAL FIELD STIMULATION

To this point, the discussion has centered on the VEP following full-field pattern stimulation, which appears to be a simple waveform. But its apparent simplicity is deceptive. The PVEP represents the sum of multiple components arising from both hemispheres and from both upper and lower quadrants. The true complexity of the PVEP can be appreciated—and clinically utilized—only when portions of the visual field are stimulated selectively. Hemifield PVEPs are considerably more difficult to analyze than full-field studies, with a smaller increment in benefit, so they are not used routinely in all laboratories. However, comprehending their features and origins may improve understanding of more straightforward techniques.

Background and Technique

Pattern reversal stimulation of one visual hemifield usually produces a net VEP vector that tends to point obliquely back through the opposite hemisphere (large arrow in Fig. 36–2). Thus, as illustrated in Figure 36–15, P100 is recorded maximally *ipsilateral* to the stimulus rather than contralaterally over the active occipital lobe.[18, 88, 143] To show this distribution properly, additional recording sites must be added 5- and 10-cm lateral to MO, in the form of LO, RO, left temporal (LT), and right temporal (RT), respectively.[147] Using these five posterior electrodes referenced to a frontal site, most normal subjects have an asymmetrical response across the temporal-occipital scalp. P100 is recorded at the ipsilateral occipital lead and a nearly simultaneous negative wave, N105, appears at the contralateral temporal area.[9, 14, 64] N105 is independent of the occipital P100 and more variable; it arises from projections of the peripheral visual field rather than the area of the macula.[15, 16, 18, 64] In about 50% of normal subjects the lateral N105 forms part of a full PNP complex, with surrounding P75 and P135 waves that roughly mirror the more familiar NPN waveform of the full-field response (Fig. 36–15, right field stimulation).[64]

Hemifield stimulation requires larger field sizes and larger checks than may be used for routine full-field testing. Small fields and checks emphasize contributions to the PVEP from the macular region, while minimizing input from the peripheral elements that produce the N105. Hemifield testing must be performed using a large stimulus field, subtending a visual angle of 16 degrees or more. Check sizes of at least 50 minutes are preferred. Ear-reference recording cannot be used, as it distorts the lateral waveforms present over LT and RT.

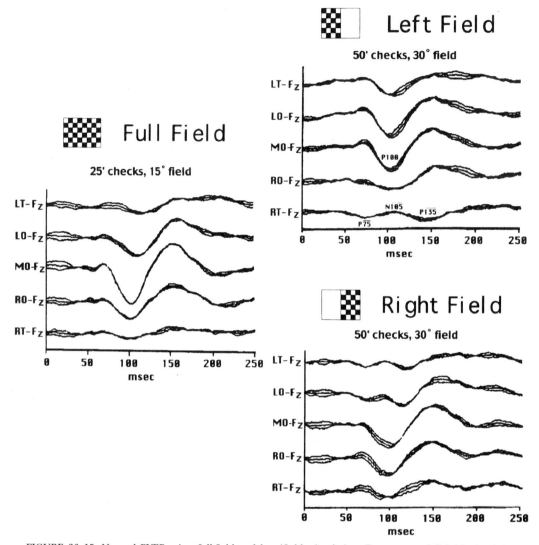

FIGURE 36–15. Normal PVEP using full-field and hemifield stimulation. Response to full-field stimulation *(left)* is maximal at the midoccipital electrode and symmetrical. Stimulation of the left hemifield *(upper right)* shows maximal positivity of P100 at the midline and at the ipsilateral occipital area (LO). A complementary asymmetry is seen on right hemifield stimulation *(lower right)*. The contralateral occipital (LO) and temporal (LT) electrodes register an almost synchronous negativity, N105, followed by a late positivity, P135. (Original figure courtesy of Dr. TA Pedley.)

Accurate fixation and a consistent level of attention by the subject are even more critical for partial-field recording than for full-field PVEPs. Partial-field studies produce the clearest results if visual fixation is maintained 1 to 2 degrees outside the edge of the reversing pattern, thereby preventing activation of the limited uncrossed fiber systems in the fovea.[91] If pattern reversal in the right and left half fields is done sequentially, activity from both hemispheres can be sampled during a single average.[133] This modification often improves the quality of hemifield recording.[147]

Half-field analysis has shown that some apparent PVEP latency prolongations are in fact caused by complete loss of the P100 from a central scotoma, with enhanced appearance of the remaining off-center P135.[15, 16] This may account for some of the P100 latency prolongations reported with macular disease[57] and partial central field defects.[165] Additionally, the presence of a prominent P75 component with the exaggerated P135 may lead to a W-shaped waveform on full-field testing; this pattern can then present as a bifid P100.[81]

Applications

Hemifield studies are more successful than full-field PVEPs for evaluating postchiasmatic visual field defects.[17, 65, 87, 119, 129, 157] A complete homonymous hemianopia often results in decreased amplitude or complete loss of the response from the affected field (Fig. 36–16, right hemifield stimulation, at lower right).[65, 87, 119] However, it is best to interpret amplitude asymmetries conservatively. In laboratories with limited experience in recording hemifield PVEPs, an abnormality should be confidently identified only when the hemifield response is absent.[65] In laboratories that accumulate their own series of normative studies, P100 amplitude asymmetries at LO and RO must still be quite large to represent a definite abnormality—generally greater than 3:1. Hemifield latency abnormalities were rare in the reports of Blumhardt and colleagues[17] and Haimovic and Pedley,[65] whose patients had destructive hemisphere lesions resulting from tumor, stroke, trauma, and surgery. Others groups, however, have reported

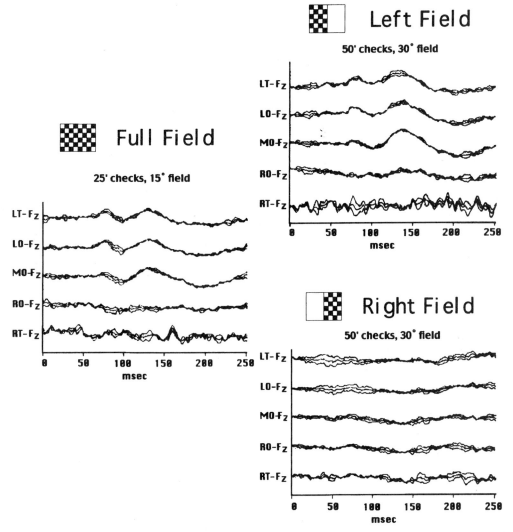

FIGURE 36–16. Abnormal PVEP to full-field and hemifield stimulation in a patient with left occipital hemorrhage. The P100 is paradoxically more prominent over the left hemisphere *(upper right)*, and absent with right hemifield stimulation *(lower right)*. (Original figure courtesy of Dr. TA Pedley.)

more frequent hemifield latency abnormalities in patients suspected of having MS[81, 113, 134] or compressive lesions of the optic chiasm.[23] These latency asymmetries were not often correlated with visual field defects on conventional perimetry.

Although most dense hemianopias can be identified by hemifield PVEPs, it has not been possible to duplicate the high sensitivity and precise localization routinely obtained with conventional perimetry and neuro-ophthalmologic examination.[93] PVEPs will not reliably detect changes involving much less than one hemifield, or identify anatomical features such as partial incongruity, von Willebrand's knee, and the loop of Meyer. Similarly, conventional perimetry is more sensitive to lesions that spare the macula.[141] Thus, the clinical utility of hemifield VEPs remains limited.

Unlike PVEPs from lateral hemifields, those from upper and lower fields are generally not symmetrical in normal individuals, and do not form mirror images above and below an MO electrode. Instead, the full-field PVEP reflects predominantly the lower visual field.[89, 102] Altitudinal PVEPs have not been widely used. The visual evoked spectral array (VESA)[10, 37] delivers stimuli simultaneously to all quadrants of the retina. VESA expedites attempts at VEP perimetry but

has not succeeded in matching standard ophthalmological techniques.

FLASH VEPs

Background

FVEPs have long been familiar as the photic "driving" response in EEG. Averaged FVEPs were studied for many years before evoked potentials became a common clinical procedure. Although FVEPs may resemble PVEPs (Fig. 36–17), the cortical response to flash stimuli is more widely distributed, complex, and variable than that resulting from pattern shift. There are multiple positive and negative peaks. Simple rules for interpretation have been difficult to develop despite extensive studies.[1, 83] The wide topographical distribution of FVEPs may be owing in part to the effects of activating additional cortical projection systems, including the retinotectal pathways. Axons from the superior colliculus project to the pulvinar, and from there to nonstriate areas of the occipital and parietal lobes.

The great variability of FVEPs limits their utility, because

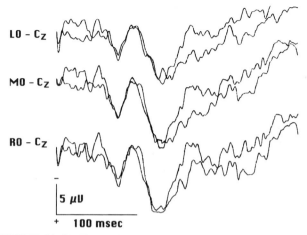

FIGURE 36–17. Flash VEP from a normal subject. Two different scalp-positive waveforms are present across the occipital lobes, neither showing the characteristics of P100.

it complicates the task of distinguishing normal responses from abnormal ones. A robust FVEP may occur even with cortical blindness.[55] Thus, the American Clinical Neurophysiology Society Guidelines have advised that only the complete absence of a flash response can be considered definitely abnormal.[147] Less dramatic alterations "must be interpreted with extreme caution." Although a few authors regard this recommendation as excessively conservative, the difficulty in interpreting FVEPs generally restricts their clinical use to situations in which pattern stimulation is not feasible, such as infancy. In young children, FVEP waveforms are simpler, and several laboratories have proposed maturational data.[169]

Technique and Applications

For monocular FVEP recording, the opposite eye must be fully occluded with a black patch. Some stroboscopic flash units generate a clearly audible "click" with each discharge; the patient must then wear sound-attenuation headphones to prevent the FVEP from being confounded with an auditory evoked potential. Common stimulus rates are 1 to 2 per second. The FVEP can easily be recorded through closed eyelids and through all but the densest ocular opacities. It is not affected by refractive errors and is robust in full-term neonates. These features make it useful for assessing function of the optic nerve and central visual pathways in patients with ocular scarring or hemorrhage. Preservation of a well-defined FVEP in these situations suggests that the visual pathways are at least partially intact, whereas its absence strongly suggests that no useful visual function can be salvaged. Intraoperative use of the FVEP is discussed earlier in the chapter.

FVEPs have been used with several computerized EEG systems, where their activation of large cortical areas is considered an advantage. Asymmetries of the FVEP may occur at the site of brain tumors and other structural lesions.[48] Increased amplitude of FVEPs occurred focally in some children with benign Rolandic epilepsy.[48] Appropriately localized abnormalities were described in almost 50% of patients with intractable focal seizures.[114] However, persistent problems with computerized EEG have prevented the extension of these observations in clinical practice.

REFERENCES

1. Allison T, Matsumiya Y, Goff G, Goff W: The scalp topography of human visual evoked potentials. Electroencephalogr Clin Neurophysiol 42:185–197, 1977.
2. Aminoff M, Ochs A: Pattern-onset visual evoked potentials in suspected multiple sclerosis. J Neurol Neurosurg Psychiatry 44:608–614, 1981.
3. Anderson D, Bundlie S, Rockswold G: Multimodality evoked potentials in closed head trauma. Arch Neurol 41:369–379, 1984.
4. Andersson T, Siden A: Comparison of visual evoked potentials elicited by light-emitting diodes and TV monitor stimulation in patients with multiple sclerosis and potentially related conditions. Electroencephalogr Clin Neurophysiol 92:473–479, 1994.
5. Apkarian P, Koetsveld-Baart JC, Barth PG: Visual evoked potential characteristics and early diagnosis of Pelizaeus-Merzbacher disease. Arch Neurol 50:981–985, 1993.
6. Asselman P, Chadwick D, Marsden C: Visual evoked responses in the diagnosis and management of patients suspected of multiple sclerosis. Brain 98:261–282, 1975.
7. Bajalan AA, Wright CE, van der Vliet VJ: Changes in the human visual evoked potential caused by the anticholinergic agent hyoscine hydrobromide: Comparison with results in Alzheimer's disease. J Neurol Neurosurg Psychiatry 49:175–182, 1986.
8. Barbieri S, Nobile-Orazio E, Baldini L, et al: Visual evoked potentials in patients with neuropathy and macroglobulinemia. Ann Neurol 22:663–666, 1987.
9. Barrett G, Blumhardt L, Halliday A, et al: A paradox in the lateralisation of the visual evoked response. Nature 261:253–255, 1976.
10. Baseler HA, Sutter EE, Klein SA, Carney T: The topography of visual evoked response properties across the visual field. Electroencephalogr Clin Neurophysiol 90:65–81, 1994.
11. Benbadis SR, Lancman ME, Wolgamuth BR, Cheek JC: Value of full-field visual evoked potentials for retrochiasmal lesions. J Clin Neurophysiol 13:507–510, 1996.
12. Bhaskar P, Vanchilingam S, Bhaskar E, et al: Effect of L-DOPA on visual evoked potential in patients with Parkinson's disease. Neurology 36:119–121, 1986.
13. Bird T, Crill W: Pattern-reversal visual evoked potentials in the hereditary ataxias and spinal degenerations. Ann Neurol 9:243–250, 1981.
14. Blumhardt L, Barrett G, Halliday A: The asymmetrical visual evoked potential to pattern-reversal in one half-field and its significance for the analysis of visual field defects. Brit J Ophthalmol 61:454–461, 1977.
15. Blumhardt L, Barrett G, Halliday A, Kriss A: The contralateral negativity of the half-field response and its association with central scotomata. Electroencephalogr Clin Neurophysiol 43:286P, 1977.
16. Blumhardt L, Barrett G, Halliday A, Kriss A: The effect of experimental "scotomata" on the ipsilateral and contralateral responses to pattern-reversal in one half-field. Electroencephalogr Clin Neurophysiol 45:376–392, 1978.
17. Blumhardt L, Barrett G, Kriss A, Halliday A: The pattern-evoked potential in lesions of the posterior visual pathways. Ann NY Acad Sci 388:264–289, 1982.
18. Blumhardt L, Halliday A: Hemisphere contributions to the composition of the pattern-evoked potential waveform. Exp Brain Res 36:53–69, 1979.
19. Bodis-Wollner I, Atkin A, Raab E, Wolkstein M: Visual association cortex and vision in man: Pattern-evoked potentials in a blind boy. Science 198:629–631, 1977.
20. Bodis-Wollner I, Feldman R: Old perimacular pathology causes VEP delays in man. Electroencephalogr Clin Neurophysiol 53:38P, 1982.
21. Bodis-Wollner I, Henley C, Mylin L, Thornton J: Visual evoked potentials and the visuogram in multiple sclerosis. Ann Neurol 5:40–47, 1979.
22. Bodis-Wollner I, Yahr M: Measurements of visual evoked potentials in Parkinson's disease. Brain 101:661–671, 1978.
23. Brecelj J: A VEP study of the visual pathway function in compressive lesions of the optic chiasm. Full-field versus half-field stimulation. Electroencephalogr Clin Neurophysiol 84:209–218, 1992.
24. Brooks E, Chiappa K: A comparison of clinical neuroophthalmological findings and pattern shift visual evoked potentials in multiple sclerosis. In Courjon J, Mauguiere F, Revol M (eds): Clinical Applications of Evoked Responses in Neurology. New York, Raven Press, 1982, pp 453–458.

25. Bumgartner J, Epstein C: Voluntary alteration of visual evoked potentials. Ann Neurol 12:475–478, 1982.

26. Calzetti S, Franchi A, Taratufolo G, Groppi E: Simultaneous VEP and PERG investigations in early Parkinson's disease. J Neurol Neurosurg Psychiatry 53:114–117, 1990.

27. Camisa J, Mylin L, Bodis-Wollner I: The effect of stimulus orientation on the visual evoked potential in multiple sclerosis. Ann Neurol 10:532–539, 1981.

28. Cant B, Hume A, Shaw N: Effects of luminance on the pattern visual evoked potential in multiple sclerosis. Electroencephalogr Clin Neurophysiol 45:496–504, 1978.

29. Cappin J, Nissim S: Pattern visual evoked response in the detection of field defects in glaucoma. Arch Ophthalmol 93:9–18, 1975.

30. Carlin L, Roach E, Riela A, et al: Juvenile metachromatic leukodystrophy: Evoked potentials and computed tomography. Ann Neurol 13:105–106, 1983.

31. Carroll W, Kriss A, Baraitser M, et al: The incidence and nature of visual pathway involvement in Friedreich's ataxia. Brain 103:413–434, 1980.

32. Carroll W, Mastaglia S: Leber's optic neuropathy. A clinical and visual evoked potential study of affected asymptomatic members of a six generation family. Brain 102:559–580, 1979.

33. Cedzich C, Schramm J, Fahlbusch R: Are flash-evoked potentials useful for intraoperative monitoring of visual pathway function? Neurosurgery 21:709–715, 1987.

34. Celesia G, Brigell M, Gunnink R, Dang H: Spatial frequency evoked visuograms in multiple sclerosis. Neurology 42:1067–1070, 1992.

35. Celesia G, Kaufman D, Cone S: Simultaneous recording of pattern electroretinography and visual evoked potentials in multiple sclerosis. Arch Neurol 43:1247–1252, 1986.

36. Celesia G, Kaufman D, Cone S: Effects of age and sex on pattern electroretinograms and visual evoked potentials. Electroencephalogr Clin Neurophysiol 68:161–171, 1987.

37. Celesia G, Meredith J, Pluff K: Perimetry, visual evoked potentials and visual evoked spectrum array in homonymous hemianopia. Electroencephalogr Clin Neurophysiol 56:16–30, 1983.

38. Chan Y, McLeod J, Tuck R, et al: Visual evoked responses in chronic alcoholics. J Neurol Neurosurg Psychiatry 49:945–950, 1986.

39. Chi O, McCoy C, Field C: Effects of fentanyl anesthesia on visual evoked potentials in humans. Anesthesiology 67:827–830, 1987.

40. Chiappa K: Evoked Potentials in Clinical Medicine, 3rd ed. Philadelphia, Lippincott-Raven, 1997.

41. Chu N-S: Sensory evoked potentials in Wilson's disease. Brain 109:491–507, 1986.

42. Chu N-S: Pattern-reversal visual evoked potentials: Latency changes with gender and age. Clin EEG 18:159–162, 1987.

43. Cohen S, Syndulko K, Tourtellotte W, et al: Volitional manipulation of visual evoked potential latency. Neurology 32:A209, 1982.

44. Cohn N, Kircher J, Emmerson R, Dustman R: Pattern reversal evoked potentials: Age, sex, and hemisphere asymmetry. Electroencephalogr Clin Neurophysiol 62:399–405, 1985.

45. Cracco R, Cracco J: Visual evoked potential in man: Early oscillatory potentials. Electroencephalogr Clin Neurophysiol 45:731–739, 1978.

46. Davies H, Carroll W, Mastaglia F: Effects of hyperventilation on pattern-reversal visual evoked potentials in patients with demyelination. J Neurol Neurosurg Psychiatry 49:1392–1396, 1986.

47. Ducati A, Fava E, Motti ED: Neuronal generators of the visual evoked potentials: Intracerebral recording in awake humans. Electroencephalogr Clin Neurophysiol 71:89–99, 1988.

48. Duffy F: Topographic display of evoked potentials—clinical applications of brain electrical activity mapping (BEAM). Ann NY Acad Sci 388:183–196, 1982.

49. Ehle A, Sklar F: Visual evoked potentials in infants with hydrocephalus. Neurology 29:1541–1544, 1979.

50. Emmerson R, Dustman R, Shearer D, Chamberlin H: EEG, visually evoked and event related potentials in young abstinent alcoholics. Alcohol 4:241–248, 1987.

51. Emmerson R, Dustman R, Shearer D, Steinhaus L: PREP amplitude of older women as a function of check size: Visual system responsivity. Electroencephalogr Clin Neurophysiol 69:88P, 1987.

52. Epstein C: True checkerboard pattern reversal with light-emitting diodes. Electroencephalogr Clin Neurophysiol 47:611–613, 1979.

53. Epstein C, Gammon A, Gemmill M, Till J: Visual evoked potential pattern generation, recording, and data analysis with a single microcomputer. Electroencephalogr Clin Neurophysiol 56:691–693, 1984.

54. Farlow M, Markand O, Edwards M, et al: Multiple sclerosis: Magnetic resonance imaging, evoked responses, and spinal fluid electrophoresis. Neurology 36:828–831, 1986.

55. Frank Y, Torres F: Visual evoked potentials in the evaluation of "cortical blindness" in children. Ann Neurol 6:126–129, 1979.

56. Friendly D, Weiss I, Barnet A, et al: Pattern-reversal visual-evoked potentials in the diagnosis of amblyopia in children. Am J Ophthalmol 102:329–339, 1986.

57. Fukui R, Kato M, Kuroiwa Y: Effect of central scotomata on pattern reversal visual evoked potentials in patients with maculopathy and healthy subjects. Electroencephalogr Clin Neurophysiol 63:317–326, 1986.

58. Gangitano J, Keltner J: Abnormalities of the pupil and visual-evoked potential in quinine amblyopia. Am J Ophthalmol 89:425–430, 1980.

59. Giesser B, Kurtzberg D, Vaughan GJ, et al: Trimodal evoked potentials compared with magnetic resonance imaging in the diagnosis of multiple sclerosis. Arch Neurol 44:281–284, 1987.

60. Gott P, Weiss M, Apuzzo M, Van Der Meulen J: Checkerboard visual evoked potentials in evaluation and management of pituitary tumors. Neurosurgery 5:553–558, 1979.

61. Gottlub I, Schneider E, Heider W, Skrandies W: Alteration of visual evoked potentials and electroretinograms in Parkinson's disease. Electroencephalogr Clin Neurophysiol 66:349–357, 1987.

62. Greenberg R, Becker D, Miller J, Mayer D: Evaluation of brain function in severe human head trauma with multimodality evoked potentials. Part 2: Localization of brain dysfunction with posttraumatic neurological conditions. J Neurosurg 47:163–177, 1977.

63. Gupta N, Verma N, Guidice M, Kooi K: Visual evoked response in head trauma: Pattern-shift stimulus. Neurology 36:578–581, 1986.

64. Haimovic I, Pedley T: Hemi-field pattern reversal visual evoked potentials. I. Normal subjects. Electroencephalogr Clin Neurophysiol 54:111–120, 1982.

65. Haimovic I, Pedley T: Hemi-field pattern reversal visual evoked potentials. II. Lesions of the chiasm and posterior visual pathways. Electroencephalogr Clin Neurophysiol 54:121–131, 1982.

66. Halliday A, Halliday E: Photosensitive epilepsy: The electroretinogram and visual evoked response. Arch Neurol 20:191–198, 1969.

67. Halliday A, Halliday E, Kriss A, et al: The pattern-evoked potentials in compression of the anterior visual pathways. Brain 99:357–374, 1976.

68. Halliday A, McDonald W, Mushin J: Delayed pattern-evoked responses in optic neuritis in relation to visual acuity. Trans Ophthalmol Soc UK 93:315–324, 1973.

69. Halliday A, McDonald W, Mushin J: The visual evoked response in the diagnosis of multiple sclerosis. Br Med J 4:661–664, 1973.

70. Hammond S, MacCallum S, Yiannikas C, et al: Variability on serial testing of pattern reversal visual evoked potential latencies from full-field, half-field, and foveal stimulation in control subjects. Electroencephalogr Clin Neurophysiol 66:401–408, 1987.

71. Happel L, Rothschild H, Garcia C: Visual evoked potentials in two forms of hereditary spastic paraplegia. Electroencephalogr Clin Neurophysiol 48:223–236, 1980.

72. Harding G, Wright C, Orwin A: Primary presenile dementia: The use of the visual evoked potential as a diagnostic indicator. Br J Psychiatry 147:532–539, 1985.

73. Harding GF, Bland JD, Smith VH: Visual evoked potential monitoring of optic nerve function during surgery. J Neurol Neurosurg Psychiatry 53:890–895, 1990.

74. Hennerici M, Wenzel D, Freund H-J: The comparison of small-size rectangle and checkerboard stimulation for the evaluation of delayed visual evoked responses in patients suspected of multiple sclerosis. Brain 100:119–136, 1977.

75. Heyer EJ, Simpson DM, Bodis-Wollner I, Diamond SP: Nitrous oxide: Clinical and electrophysiologic investigation of neurologic complications. Neurology 36:1618–1622, 1986.

76. Holder GE, Bullock PR: Visual evoked potentials in the assessment of patients with non-functioning chromophobe adenomas. J Neurol Neurosurg Psychiatry 52:31–37, 1989.

77. Hornabrook RS, Miller DH, Newton MR, et al: Frequent involvement of the optic radiation in patients with acute isolated optic neuritis. Neurology 42:77–79, 1992.

78. Hubel D, Wiesel T: Receptive fields and functional architecture of monkey striate cortex. J Physiol 195:215–243, 1968.

79. Huisman U, Posthuma J, Visser S, et al: The influence of attention on visual evoked potentials in normal adults and dementias. Clin Neurol Neurosurg 89:151–156, 1987.

80. Javidan M, McLean D, Warren K: Cerebral evoked potentials in multiple sclerosis. Can J Neurol Sci 13:240–244, 1986.

81. Jones D, Blume W: Aberrant waveforms to pattern reversal stimulation: Clinical significance and electrographic "solutions." Electroencephalogr Clin Neurophysiol 61:472–481, 1985.

82. Kaplan P, Rawal K, Erwin C, et al: Visual and somatosensory evoked potentials in vitamin E deficiency with cystic fibrosis. Electroencephalogr Clin Neurophysiol 71:226–272, 1988.

83. Kooi K: Visual Evoked Potentials in Central Disorders of the Visual System. Hagerstown, MD, Harper & Row, 1979.

84. Kraushar M, Miller E: Central serous choroidopathy misdiagnosed as a manifestation of multiple sclerosis. Ann Ophthalmol 4:215–218, 1982.

85. Kriss A, Carroll W, Blumhardt L, Halliday A: Pattern and flash-evoked potential changes in toxic (nutritional) optic neuropathy. In Courjon J, Mauguiere F, Revol M (eds): Clinical Applications of Evoked Responses in Neurology. New York, Raven Press, 1982, pp 11–19.

86. Krumholz A, Weiss H, Goldstein P, Harris K: Evoked responses in vitamin B12 deficiency. Ann Neurol 9:407–409, 1981.

87. Kuroiwa Y, Celesia G: Visual evoked potentials with hemifield pattern stimulation: Their use in the diagnosis of retrochiasmatic lesions. Arch Neurol 38:86–90, 1981.

88. Lehmann D, Darcey T, Skrandies W: Intracerebral and scalp fields evoked by hemiretinal checkerboard reversal, and modelling of their dipole generators. In Courjon J, Mauguiere F, Revol M (eds): Clinical Applications of Evoked Responses in Neurology. New York, Raven Press, 1982, pp 41–48.

89. Lehmann D, Meles H, Mir Z: Average multichannel EEG potential fields evoked from upper and lower hemi-retina: Latency differences. Electroencephalogr Clin Neurophysiol 43:725–731, 1977.

90. Lennerstrand G: Delayed visual evoked cortical potentials in retinal disease. Acta Ophthalmol 60:497–504, 1982.

91. Leventhal A, Ault S, Vitek D: The nasotemporal division in primate retina: The neural bases of macular sparing and splitting. Science 240:66–67, 1988.

92. Maiese K, Walker RW, Gargan R, Victor JD: Intra-arterial cisplatin–associated optic and otic toxicity. Arch Neurol 49:83–86, 1992.

93. Maitland C, Aminoff M, Kennard C, Hoyt W: Evoked potentials in the evaluation of visual field defects due to chiasmal or retrochiasmal lesions. Neurology 32:986–991, 1982.

94. Majnemer A, Rosenblatt B, Watters G, Andermann F: Giant axonal neuropathy: Central abnormalities demonstrated by evoked potentials. Ann Neurol 19:394–396, 1986.

95. Mamoli P, Graf M, Toifl K: EEG pattern-evoked potentials and nerve conduction velocity in a family with adrenoleukodystrophy. Electroencephalogr Clin Neurophysiol 47:411–419, 1979.

96. Markand O, DeMeyer W, Worth R, Warren C: Multimodality evoked responses in leukodystrophies. In Courjon J, Mauguiere F, Revol M (eds): Clinical Applications of Evoked Responses in Neurology. New York, Raven Press, 1982, pp 409–415.

97. Matthews W, Read D, Pountney E: Effect of raising body temperature on visual and somatosensory evoked potentials in patients with multiple sclerosis. J Neurol Neurosurg Psychiatry 42:250–255, 1978.

98. McCombe PA, McLaughlin DB, Chalk JB, et al: Spasticity and white matter abnormalities in adult phenylketonuria. J Neurol Neurosurg Psychiatry 55:359–361, 1992.

99. McLeod J, Low P, Morgan J: Charcot-Marie-Tooth disease with Leber optic atrophy. Neurology 28:179–184, 1978.

100. Mervaala E, Keranen T, Penttila M, et al: Pattern-reversal VEP and cortical SEP latency prolongations in epilepsy. Epilepsia 26:441–445, 1985.

101. Messenheimer J, Greenwood R, Tennison M, et al: Reversible visual evoked potential abnormalities in vitamin E deficiency. Ann Neurol 15:499–501, 1984.

102. Michael W, Halliday A: Differences between the occipital distribution of upper and lower field pattern-evoked responses in man. Brain Res 32:311–324, 1971.

103. Miller D, Newton M, Van der Poel J, et al: Magnetic resonance imaging of the optic nerve in optic neuritis. Neurology 38:175–179, 1988.

104. Miller J, Burke A, Bever C: Occurrence of oligoclonal bands in multiple sclerosis and other CNS diseases. Ann Neurol 13:53–58, 1983.

105. Moller A, Burgess J, Sekhar L: Recording compound action potentials from the optic nerve in man and monkeys. Electroencephalogr Clin Neurophysiol 67:549–555, 1987.

106. Mondelli M, Rossi A, Scarpini C, et al: Evoked potentials in cerebrotendinous xanthomatosis and effect induced by chenodeoxycholic acid. Arch Neurol 49:469–475, 1992.

107. Morgan R, Nugent B, Harrison J, O'Connor P: Voluntary alteration of pattern visual evoked responses. Ophthalmology 92:1356–1363, 1985.

108. Mortimer M, Good P, Marsters J, Addy D: Visual evoked responses in children with migraine: A diagnostic test. Lancet i:75–77, 1990.

109. Neima D, Regan D: Pattern visual evoked potentials and spatial vision in retrobulbar neuritis and multiple sclerosis. Arch Neurol 41:198–201, 1984.

110. Nilsson B: Visual evoked responses in multiple sclerosis: Comparison of two methods for pattern reversal. J Neurol Neurosurg Psychiatry 41:499–504, 1978.

111. Noachtar S, Hashimoto T, Luders H: Pattern visual evoked potentials recorded from human occipital cortex with chronic subdural electrodes. Electroencephalogr Clin Neurophysiol 88:435–446, 1993.

112. Norcia A, Tyler C: Spatial frequency sweep VEP: Visual acuity during the first year of life. Vision Res 25:1399–1408, 1985.

113. Novak G, Wiznitzer M, Kurtzberg D, et al: The utility of visual evoked potentials using hemifield stimulation and several check sizes in the evaluation of suspected multiple sclerosis. Electroencephalogr Clin Neurophysiol 71:1–9, 1988.

114. Nuwer M: Frequency analysis and topographic mapping of EEG and evoked potentials in epilepsy. Electroencephalogr Clin Neurophysiol 69:118–126, 1988.

115. Nuwer M, Perlman S, Packwood J, Pieter Kark R: Evoked potential abnormalities in the various inherited ataxias. Ann Neurol 13:20–27, 1983.

116. Nyrke T, Kangasniemi P, Lang AH: Difference of steady-state visual evoked potentials in classic and common migraine. Electroencephalogr Clin Neurophysiol 73:285–294, 1989.

117. Oishi M, Yamada T, Dickens Q, Kimura J: Visual evoked potentials by different check sizes in patients with multiple sclerosis. Neurology 35:1461–1465, 1985.

118. Oken BS, Chiappa KH, Gill E: Normal temporal variability of the P100. Electroencephalogr Clin Neurophysiol 68:153–156, 1987.

119. Onofrj M, Bodis-Wollner I, Mylin L: Visual evoked potential diagnosis of field defects in patients with chiasmatic and retrochiasmatic lesions. J Neurol Neurosurg Psychiatry 45:294–302, 1982.

120. Papakostopoulos D, Hart C, Cooper R, Natsikos V: Combined electrophysiological assessment of the visual system in central serous retinopathy. Electroencephalogr Clin Neurophysiol 59:77–80, 1984.

121. Paty D, Oger J, Kastrukoff L, et al: MRI in the diagnosis of MS: A prospective study with comparison of clinical evaluation,

evoked potentials, oligoclonal banding and CT. Neurology 38:180–185, 1988.

122. Pedersen L, Trojaborg W: Visual, auditory, and somatosensory pathway involvement in hereditary cerebellar ataxia, Friedreich's ataxia and familial spastic paraplegia. Electroencephalogr Clin Neurophysiol 52:283–297, 1981.

123. Pfurtscheller G, Schwartz G, Gravenstein N: Clinical relevance of long-latency SEPs and VEPs during coma and emergence from coma. Electroencephalogr Clin Neurophysiol 62:88–98, 1985.

124. Pratt H, Bleich N, Berliner E: Short latency visual evoked potentials in man. Electroencephalogr Clin Neurophysiol 54:55–62, 1982.

125. Rappaport M, Hall K, Hopkins K, et al: Evoked brain potentials and disability in brain-damaged patients. Arch Phys Med Rehabil 58:333–345, 1977.

126. Rappaport M, Hopkins H, Hall K, Belleza T: Evoked potentials and head injury. Clinical applications. Clin EEG 12:167–176, 1981.

127. Raudzens P: Intraoperative monitoring of evoked potentials. Ann NY Acad Sci 388:308–326, 1982.

128. Ray PG, Meador KJ, Loring DW, et al: Effects of scopolamine on visual evoked potentials in aging and dementia. Electroencephalogr Clin Neurophysiol 80:347–351, 1991.

129. Regan D, Milner B: Objective perimetry by evoked potential recording: Limitations. Electroencephalogr Clin Neurophysiol 44:393–397, 1978.

130. Reilly E, Kondo C, Brunberg J, Doty D: Visual evoked potentials during hypothermia and prolonged circulatory arrest. Electroencephalogr Clin Neurophysiol 45:100–106, 1978.

131. Riemslag F, Spekreijse H, Van Wessem T: Responses to paired onset stimuli: Implications for the delayed evoked potentials in multiple sclerosis. Electroencephalogr Clin Neurophysiol 62:155–166, 1985.

132. Rosing H, Hopkins L, Wallace D, et al: Maternally inherited mitochondrial myopathy and myoclonic epilepsy. Ann Neurol 17:228–237, 1985.

133. Rowe M: A sequential technique for half-field pattern visual evoked potential testing. Electroencephalogr Clin Neurophysiol 51:463–469, 1981.

134. Rowe M: The clinical utility of half-field pattern reversal visual evoked potential testing. Electroencephalogr Clin Neurophysiol 53:73–77, 1982.

135. Russ W, Kling D, Loesevitz A, Hempelmann G: Effect of hypothermia on visual evoked potentials (VEP) in humans. Anesthesiology 61:207–210, 1984.

136. Sanders E, Volkers A, Van der Poel J, Van Lith G: Visual function and pattern visual evoked response in optic neuritis. Br J Ophthalmol 71:602–608, 1987.

137. Sebel P, Flynn P, Ingram D: Effect of nitrous oxide on visual, auditory and somatosensory evoked potentials. Br J Anaesth 56:1403–1407, 1984.

138. Shaw N, Cant B: Age-dependent changes in the latency of the pattern visual evoked potential. Electroencephalogr Clin Neurophysiol 48:237–241, 1980.

139. Shearer D, Creel D, Dustman R: Efficacy of evoked potential stimulus parameters in the detection of visual system pathology. Am J EEG Technol 23:137–146, 1983.

140. Shearer D, Dustman R, Emmerson R: Hydrocephalus: Electrophysiological correlates. Am J EEG Technol 27:199–212, 1987.

141. Sherman J, Bass S, Noble K, et al: Visual evoked potential (VEP) delays in central serous choroidopathy. Invest Ophthalmol Vis Sci 27:214–221, 1986.

142. Shih PY, Aminoff MJ, Goodin DS, Mantle MM: Effect of reference point on visual evoked potentials: Clinical relevance. Electroencephalogr Clin Neurophysiol 71:319–322, 1988.

143. Sidman R, Smith D, Henke J, Kearfott B: Localization of neural generators in the visual evoked responses. Electroencephalogr Clin Neurophysiol 53:35P, 1982.

144. Silva I, Wang A, Symon L: The application of flash visual evoked potentials during operations on the anterior visual pathways. Neurol Res 7:11–16, 1985.

145. Simon R, Zimmerman A, Sanderson P, Tasman A: EEG markers of migraine in children and adults. Headache 23:21–25, 1983.

146. Skuse NF, Burke D: Sequence-dependent deterioration in the visual evoked potential in the absence of drowsiness. Electroencephalogr Clin Neurophysiol 84:20–25, 1992.

147. Society Guidelines: American EEG Society guidelines on evoked potentials. J Clin Neurophysiol 11:40–73, 1994.

148. Sokol S: Measurement of infant visual acuity from pattern reversal evoked potentials. Vis Res 18:33–39, 1978.

149. Sokol S, Hansen V, Moskowitz A, et al: Evoked potential and preferential looking estimates of visual acuity in pediatric patients. Ophthalmology 90:552–562, 1983.

150. Sokol S, Jones K: Implicit time of pattern evoked potentials in infants: An index of maturation of spatial vision. Vision Res 19:747–755, 1979.

151. Sokol S, Moskowitz A: Effects of retinal blur on the peak latency of the pattern evoked potential. Vision Res 21:1279–1286, 1981.

152. Sokol S, Moskowitz A, Towle V: Age-related changes in the latency of the visual evoked potential: Influence of check size. Electroencephalogr Clin Neurophysiol 51:559–562, 1981.

153. Spitz M, Emerson R, Pedley T: Dissociation of frontal N100 from occipital P100 in pattern reversal visual evoked potentials. Electroencephalogr Clin Neurophysiol 65:161–168, 1986.

154. Stensaas S, Eddington D, Dobelle W: The topography and variability of the primary visual cortex in man. J Neurosurg 40:747–755, 1974.

155. Stockard J, Hughes J, Sharbrough F: Visually evoked potentials to electronic pattern reversal: Latency variations with gender, age, and technical factors. Am J EEG Technol 19:171–204, 1979.

156. Stockard J, Iragui V: Clinically useful applications of evoked potentials in adult neurology. J Clin Neurophysiol 1:159–202, 1984.

157. Streletz L, Bae S, Roeshman R, et al: Visual evoked potentials in occipital lobe lesions. Arch Neurol 38:80–85, 1981.

158. Streletz L, Chambers R, Bae S, et al: Visual evoked potentials in sarcoidosis. Neurology 31:1545–1549, 1981.

159. Tackmann W, Radu E: Pattern shift visual evoked potentials in Charcot-Marie-Tooth disease, HMSN type I. J Neurol 224:71–74, 1980.

160. Tagliati M, Sabbadini M, Bernardi G, Silvestrini M: Multichannel visual evoked potentials in migraine. Electroencephalogr Clin Neurophysiol 96:1–5, 1995.

161. Tan C, Murray N, Sawyers D, Leonard T: Deliberate alteration of the visual evoked potential. J Neurol Neurosurg Psychiatry 47:518–523, 1984.

162. Tang Y, Norcia AM: Improved processing of the steady-state evoked potential. Electroencephalogr Clin Neurophysiol 88:323–334, 1993.

163. Tartaglione A, Pizio N, Bino G, et al: VEP changes in Parkinson's disease are stimulus dependent. J Neurol Neurosurg Psychiatry 47:305–307, 1984.

164. Taylor M, Keenan N, Gallant T, et al: Subclinical VEP abnormalities in patients on chronic deferoxamine therapy: Longitudinal studies. Electroencephalogr Clin Neurophysiol 68:81–87, 1986.

165. Thompson P, Mastaglia F, Carroll W: Anterior ischemic optic neuropathy. A correlative clinical and visual evoked potential study of 18 patients. J Neurol Neurosurg Psychiatry 49:128–135, 1986.

166. Tobimatsu S, Hamada T, Okayama M, et al: Temporal frequency deficit in patients with senile dementia of the Alzheimer type: A visual evoked potential study. Neurology 44:1260–1263, 1994.

167. Towle V, Moskowitz A, Sokol S, Schwartz B: The visual evoked potential in glaucoma and ocular hypertension: Effects of check size, field size, and stimulation rate. Invest Ophthalmol Vis Sci 24:175–183, 1983.

168. Towle V, Sutcliffe E, Sokol S: Diagnosing functional visual deficits with the P300 component of the visual evoked potential. Arch Ophthalmol 103:47–50, 1985.

169. Tsuneishi S, Casaer P, Fock JM, Hirano S: Establishment of normal values for flash visual evoked potentials (VEPs) in preterm infants: A longitudinal study with special reference to two components of the N1 wave. Electroencephalogr Clin Neurophysiol 96:291–299, 1995.

170. Uhl R, Squires K, Bruce D, Starr A: Effect of halothane anesthesia on the human cortical visual evoked response. J Anesthesiol 53:273–276, 1980.
171. Uren S, Stewart P, Crosby P: Subject cooperation and the visual evoked response. Invest Ophthalmol Vis Sci 18:648–652, 1979.
172. van Dijk JG, Dorresteijn M, Haan J, Ferrari MD: No confirmation of visual evoked potential diagnostic test for migraine. Lancet 337:517–518, 1991.
173. van Dijk JG, Dorresteijn M, Haan J, Ferrari MD: Visual evoked potentials and background EEG activity in migraine. Headache 31:392–395, 1991.
174. Wildberger H, Hofmann H, Siegfried J: Fluctuations of visual evoked potential amplitudes and of contrast sensitivity in Uhthoff's symptom. Doc Ophthalmol 65:357–365, 1987.
175. Yiannikas C, Walsh J: The use of visual evoked potentials in the detection of subclinical optic toxicity secondary to ethambutol. Neurology 32:A205, 1982.

Chapter 37

Auditory Evoked Potentials

Hans O. Lüders, MD, PhD • Kiyohito Terada, MD

INTRODUCTION
Electrocochleogram
Brain Stem Auditory Evoked Potentials
Middle Latency Auditory Evoked
    Potentials
Long Latency Auditory Evoked Potentials
TERMINOLOGY AND DEFINITIONS
TECHNIQUE
Stimulation
Electrode and Montage
Amplifier and Averager Settings
Subject
NORMAL WAVEFORM AND PEAK
    IDENTIFICATION
Waves V and IV
Wave I
Waves II and III
Waves VI and VII
GENERATORS OF BRAIN STEM AUDITORY
    EVOKED POTENTIALS
Wave I
Wave II
Wave III

Wave IV
Wave V
Waves VI and VII
ANALYSIS METHODS
NON-NEUROLOGICAL FACTORS
    AFFECTING BRAIN STEM AUDITORY
    EVOKED POTENTIALS
Age
Gender
Temperature
Parameters of Click
ABNORMALITIES OF BRAIN STEM
    AUDITORY EVOKED POTENTIALS IN
    NEUROLOGICAL DISEASES
Coma and Brain Death
Multiple Sclerosis
Other Demyelinating Disorders
Posterior Fossa Tumors
Cerebrovascular Brain Stem Lesions
Degenerative Diseases
Infectious and Inflammatory Disorders
Metabolic Disorders
Other Disorders

INTRODUCTION

In response to an abrupt auditory stimulus, several distinct neuronal responses can be recognized. These responses are classified into four categories based on their latencies.

Electrocochleogram

Electrocochleograms (ECoGs) are recorded by placing electrodes either transtympanically directly into the middle ear or extratympanically directly in contact with the tympanic membrane.[9, 45, 92, 110] The following potentials are recorded (Fig. 37–1):

Cochlear microphonics (CMs), which most probably correspond to receptor potentials generated in the hairy cells of the spiral organ.

Summation potentials (SPs), which have been interpreted as presynaptic potentials.

Action potentials (APs), which correspond to wave I of the brain stem auditory evoked potentials (BAEPs) and are an expression of the APs generated in the most distal portion of the cochlear nerve. The APs occur at least 1.2 msec after the click[10] (Fig. 37–2). All potentials occurring before 1.2 msec are either artifacts or CMs and SPs. The CMs and SPs can be isolated from the APs by alternating the polarity of the auditory stimulus; APs and SPs are always recorded as negative potentials in the active electrode of the ECoG, whereas the polarity of the CMs will change when the polarity of the stimulus

alternates (Fig. 37–3). ECoGs are valuable in diagnosing patients with various cochleovestibular disorders.[30]

Brain Stem Auditory Evoked Potentials

Normal BAEPs occur within the first 10 msec. They provide unique information about brain stem functions and have been used extensively in clinical practice. In this chapter we discuss primarily BAEPs.

GENERATORS OF ECoG

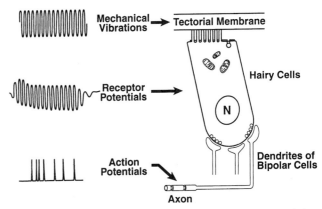

FIGURE 37–1. Diagram showing the probable generators of the human electrocochleogram (ECoG). (Modified from Spillmann T, Erdmann W, Leitner H: Clinical experiences with althesin sedation for ERA. Rev Laryngol Otol Rhinol (Bord) 96(3-4):192–198, 1975.)

ECOG AND BAEP

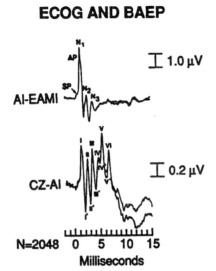

FIGURE 37–2. Electrocochleogram (ECoG) and brain stem auditory evoked potential (BAEP) recorded simultaneously. Notice that N1, N2, and N3 of the ECoG correspond to waves I (SP and N1), II, and III of the auditory evoked potential (AEP). Notice also the significantly higher amplitude of the ECoG. A1, left ear; EAM1, external auditory meatus (left); C$_z$, vertex. (With permission from Chatrian GE, Wirch AL, Lettich E, et al: Click-evoked human electrocochleogram. Noninvasive recording method, origin and physiologic significance. Am J EEG Technol 22:151–174, 1982.)

Middle Latency Auditory Evoked Potentials

These consist of auditory evoked potentials (AEPs) occurring between 10 and 50 msec. They are labeled as N0, P0, Na, Nb, and Pb.[78] N0 and P0 may overlap with myogenic potentials[83] (Fig. 37–4). The neurogenic potentials are probably a reflection of activation of both subcortical structures (thalamus) and auditory cortices (mainly the primary auditory cortex).[35]

Long Latency Auditory Evoked Potentials

These AEPs have latencies of more than 50 msec with widespread distribution over the scalp (N100, P200, N250, and P300). "P300" occurs only when a subject pays attention to and recognizes a special auditory signal. It was demonstrated that this potential arises from multiple cortical generator sources.[69] The subject is instructed to count the number of low-frequency tones embedded in a large sequence of

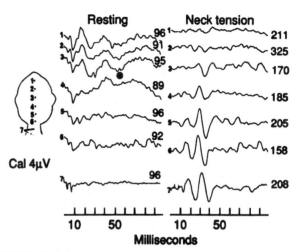

FIGURE 37–4. Myogenic and neurogenic middle latency auditory evoked potentials. Notice that at rest neurogenic potentials are recorded primarily from electrodes placed over the scalp, whereas with mock tension, myogenic components in electrodes placed over neck muscles are dominant. (With permission from Robinson K, Rudge P: Centrally generated auditory potentials. In Halliday AM (ed): Evoked Potentials in Clinical Testing, vol 10. London, Churchill Livingstone, 1982, pp 344–372.)

high-frequency tones. The "P300" consists of the difference in amplitude of the positive deflection occurring with a latency of approximately 300 msec when averages to the low- and high-frequency stimulus are subtracted. The amplitude of "P300" is inversely related to the probability of occurrence of the low-frequency signal with respect to the high-frequency signal. In elderly patients, "P300" is of lower amplitude and longer latency[101] (Fig. 37–5). Goodin and associates[32] and later Syndulko and colleagues[101] demonstrated that "P300" is significantly delayed in patients with dementia in comparison with age-matched controls (Fig. 37–6).

TERMINOLOGY AND DEFINITIONS

Decibel (dB = 1/10 Bel). This is defined as "20 log (P1/P2)," where P1 is the intensity of sound to be measured and P2 is the intensity of a reference sound. Therefore, if the sound pressure level of P1 is 10 times bigger than P2, P1 would be 20 dB. If P1 is 100 times bigger, it would be 40 dB.

ELECTROCOCHLEOGRAM

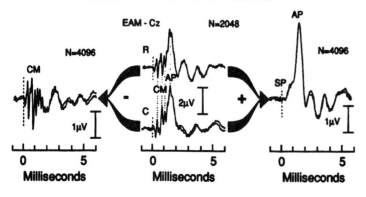

FIGURE 37–3. ECoG obtained with rarefaction (R) and condensation (C) clicks. The trace on the extreme left side shows the result of subtracting the ECoG obtained with condensation clicks from the ECoG obtained with rarefaction clicks. Notice that all components except the cochlear microphonics (CMs) tend to disappear. The graph on the extreme right side was obtained by adding the ECoG obtained by condensation and rarefaction clicks. In this case, the CMs disappear and all other components increase in amplitude. (With permission from Chatrian GE, Wirch AL, Lettich E, et al: Click-evoked human electrocochleogram. Noninvasive recording method, origin and physiologic significance. Am J EEG Technol 22:151–174, 1982.)

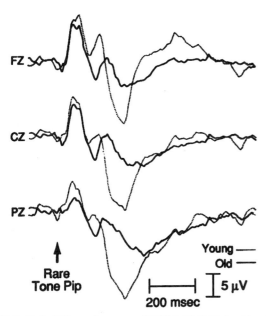

FIGURE 37–5. Effect of age on "P300." "P300" is of reduced amplitude and prolonged latency in older individuals. (With permission from Syndulko K, Hansch EC, Cohen SN, et al: Long-latency event-related potentials in normal aging and dementia. In Courjon J, Mauguière F, Revol M (eds): Advances in Neurology, vol 32. Clinical Applications of Evoked Potentials in Neurology. New York, Raven Press, 1982, p 281.)

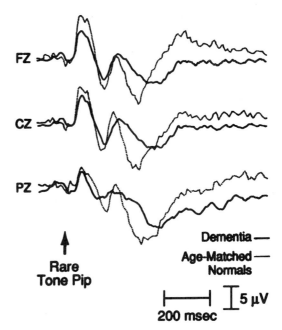

FIGURE 37–6. Effect of dementia on "P300." "P300" is of lower amplitude and of longer latency in demented patients. (With permission from Syndulko K, Hansch EC, Cohen SN, et al: Long-latency event-related potentials in normal aging and dementia. In Courjon J, Mauguière F, Revol M (eds): Advances in Neurology, vol 32. Clinical Applications of Evoked Potentials in Neurology. New York, Raven Press, 1982, p 284.)

SPL = sound pressure level. This is the standard physical reference for sound. It corresponds to the weakest sound heard by the most sensitive ear (20 micropascals or 0.0002 dyne/cm²). Sounds of a frequency of 100 Hz, which have the lowest threshold in humans, are used to define the SPL.

peSPL = peak equivalent SPL. Sound pressure meters usually require long duration signals for accurate measurements. The usual approach taken to measure a click is to use an oscilloscope and to match the amplitude of a sine wave (pure tone) with the peak amplitude of the click stimulus. The pure tone can then be measured with a sound level meter.

dBHL = hearing level. Zero dBHL corresponds to the average hearing threshold of a group of normal hearing young adults in an ideal listening environment (0 dBHL corresponds to approximately 30 dB peSPL).

dBSL = sensory level. This expresses the intensity of a sound as a function of the hearing threshold for an individual ear for any given subject.

TECHNIQUE

Stimulation

Clicks are generated by applying rectangular electrical pulses with a duration of 0.1 msec to a shielded earphone. Stimuli are presented monaurally with an intensity of 70 dBSL at around 11 Hz. In infants, it is reasonable to use 70 to 75 dBHL stimulation for the initial session to observe waveforms. To prevent the stimulus from activating the contralateral cochlea, through bone conductions, white noise at an intensity of 30 dB below the stimulation level is delivered to the contralateral ear, a technique known as "masking" (Fig. 37–7).

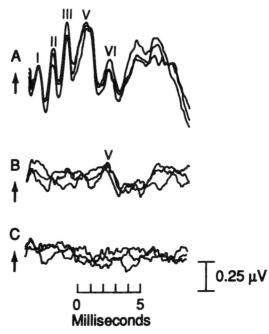

FIGURE 37–7. Cross-stimulation and masking. *A* and *B* show the response of the left and right ear to ipsilateral stimulation in an individual who is totally deaf on the right side. *A* shows a normal response and *B* shows a low-amplitude, long latency wave V. *C* shows the response of the right ear after applying a "masking" white noise to the left ear at the time of the stimulation of the right ear. Notice that now the low-amplitude, delayed wave V has disappeared, indicating that it was the result of cross-stimulation of the left ear by bone conduction. (With permission from Chiappa KH, Gladstone KJ, Young RR: Brainstem auditory evoked responses. Studies of waveform variation of 50 normal human subjects. Arch Neurol 36(2):81–87, 1979.)

Electrode and Montage

Shallow cup electrodes are affixed by using any conventional electroencephalogram (EEG) technique. The impedance should be kept less than 3 KΩ to reduce possible noises. Recordings are obtained by using the following montage: Channel 1—A_1–C_z; Channel 2—A_2–C_z (A_1 is the left earlobe electrode; A_2, the right earlobe electrode; and C_z, the vertex). The derivations should be connected in such a way that a negativity at the ear electrode produces an upward deflection.

Amplifier and Averager Settings

The bandpass filter is set to 10 to 3000 Hz (roll-off of less than -12 dB/octave). The notch filter may be used. The sensitivity is set to observe activities ranging from 0.5 to 5 µV. A dwell time of less than 0.1 msec should be used. An analysis time of 10 to 15 msec is recommended as all potentials generated in the brain stem occur within this time frame. To obtain clean recordings, 1000 to 4000 responses should be averaged. At least two waveforms should be obtained using the same stimulation conditions to check for reproducibility. Only responses in which peak latencies in repeated trials fall within 80 µsec should be considered acceptable reproducibility. To reduce artifacts, an "artifact rejection" program, which stops the averaging when the amplitude of the input signal exceeds a certain threshold amplitude, is useful.

Subject

The recording room should be quiet. Recordings are optimal when the patient is supine and relaxed. The aforementioned artifact rejection option will only eliminate relatively high-amplitude myogenic or other artifacts. Therefore, for maximal artifact reduction, BAEPs should be obtained with the subject asleep. Sleep can be induced by sedation with moderate doses of chloral hydrate (1000 to 1500 mg) or of valium (5 to 10 mg).

NORMAL WAVEFORM AND PEAK IDENTIFICATION

The normal BAEP consists of seven negative peaks, which are usually numbered using the Roman numerals I through VII[58] (Fig. 37–8). The first five peaks occur in all normal individuals at approximately 1-msec intervals starting with wave I, which has an average latency of 1.7 msec (Table

TABLE 37–1	Latency of Brain Stem Auditory Evoked Potential Peaks in Normals		
Wave	**Average Latency**	**S.D.**	**Average + 3 S.D.**
I	1.7 msec	0.15	2.2 msec
II	2.8 msec	0.17	3.3 msec
III	3.9 msec	0.19	4.5 msec
IV	5.1 msec	0.24	5.8 msec
V	5.7 msec	0.25	6.5 msec
VI	7.3 msec	0.29	8.2 msec
VII	9.6 msec	0.6	11.4 msec

S.D., standard deviation.
Modified from Chiappa KH, Gladstone KJ, Young RR: Brainstem auditory evoked responses. Studies of waveform variation of 50 normal human subjects. Arch Neurol 36(2):81–87, 1979.

37–1). Peaks VI and VII are less well defined, occurring respectively in only 58% and 43% of normal individuals.[14] Therefore, they have little diagnostic significance.

Waves V and IV

Wave V is usually the most prominent peak and always occurs after 5.0 msec. It is usually followed by a prolonged downward deflection on which waves VI and VII are superimposed. In approximately one-third of cases, waves IV and V are fused into a single wave (IV/V complex) whose latency tends to fall between waves IV and V. In other cases, wave IV is only a small notch in the ascending limb of wave V, or wave V is only a small notch in the descending limb of wave IV (Figs. 37–9 and 37–10). The following methods can be used to define waves IV and V:

Reduce the intensity of the click. All BAEP components tend to disappear except wave V, which tends to be the last to disappear.

Compare the waveform at Ai–C_z and Ac–C_z (Ai is the earlobe electrode ipsilateral to the stimulation; Ac, the earlobe electrode contralateral to the stimulation). It has been shown that wave V is electropositive at the vertex and distributes preferentially to the ipsilateral ear.[104] Therefore, wave V is usually better seen and more clearly differentiated from wave IV at Ac–C_z.

Wave I

After having identified wave V, we try to identify wave I. Wave I consistently occurs after 1 msec. Potentials occurring before 1 msec correspond either to artifacts or to cochlear microphonics. Wave I is produced as a negativity at the

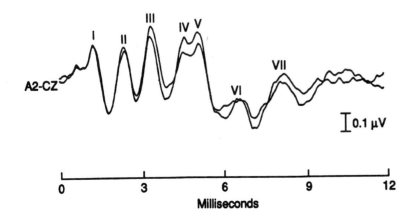

FIGURE 37–8. Normal auditory evoked potentials. A2, right ear; CZ, vertex. (With permission from Markand ON: Brainstem auditory evoked potentials. J Clin Neurophysiol 11(3):319–342, 1994.)

NORMAL VARIANTS OF PEAK IV/V

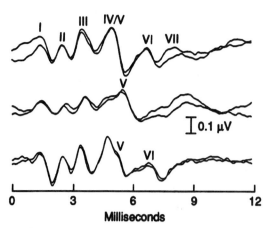

FIGURE 37–9. Variability of waves IV and V in different individuals showing a fused wave IV/V, a wave IV as a notch on the ascending phase of wave V, and a wave V as a notch on a descending limb of wave IV. See also Figure 37–10. (With permission from Markand ON: Brainstem auditory evoked potentials. J Clin Neurophysiol 11(3): 319–342, 1994.)

periaural area ipsilateral to the stimulus, and appears as a relatively low-amplitude positivity over the scalp. Therefore, it is best seen at Ai–C$_z$ and is usually very difficult to recognize at Ac–C$_z$. The polarity of wave I is always negative at Ai, independent of the polarity of the click (see Fig. 37–3).

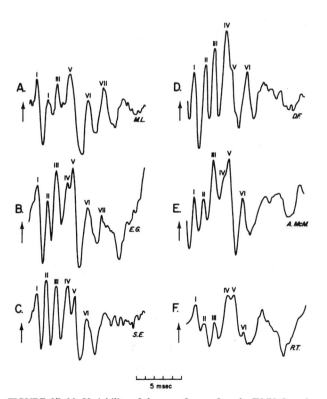

FIGURE 37–10. Variability of the waveform of peaks IV/V. See also Figure 37–9. (With permission from Chiappa KH: Brain stem auditory EPS: Methodology. In Chiappa KH (ed): Evoked Potentials in Clinical Medicine, 3rd ed. Philadelphia, Lippincott-Raven, 1997, pp 157–197.)

Waves II and III

Wave III is the most consistent component between waves I and V. Both waves II and III appear as widespread, relatively high-amplitude positivities over the scalp. Wave II extends preferentially into the ipsilateral ear, whereas wave III is seen preferentially in the contralateral ear. Therefore, wave II is best recorded at Ac–C$_z$, and wave III at Ai–C$_z$. A common normal variant is a fusion of waves II and III at Ac–C$_z$ due to a decrease of the latency of wave III and relative increase of the latency of wave II. Occasionally, a bilobed wave III may occur.[13] In these cases, the latency may be taken as the midpoint between the "two waves III."

Waves VI and VII

These components may be absent in normal subjects. Wave VI is often on the descending limb of wave V and is sometimes only an inflection. Wave VII is frequently discernible. These waves have average latencies of 7.3 and 9.6 msec, respectively (see Figs. 37–9 and 37–10 and Table 37–1).

GENERATORS OF BRAIN STEM AUDITORY EVOKED POTENTIALS

A detailed description of the general methods that can be applied to define the generators of evoked potentials recorded from the scalp has been published elsewhere.[57] The anatomical and neurophysiological background can be summarized as follows. The displacement of the tympanic membrane produced by the sound pressure wave is transmitted to the oval window through the inner ear ossicles. This produces movement of the perilymph (contained in the scala vestibuli and scala tympani, which are joined at the apex of the cochlea by the helicotrema) and secondarily of the endolymph, which is contained in the ductus cochlearis and also includes the basilar membrane, the spiral organ, and the tectorial membrane. The spiral organ contains hair cells that, when displaced by the movement of the tectorial membrane, produce auditory receptor potentials (Fig. 37–11). The receptor potentials lead to the release of neurotransmitters, which trigger APs in the dendrites of the afferent nerve fibers of the cochlear nerve. There are 20,000 to 30,000 hair cells distributed over a distance of approximately 31.5 mm in the 2.5 spirals of the cochlea. It has been established that high-frequency sounds activate the basal portion of the basilar membrane, whereas low-frequency sounds generate receptor potentials in the apical portion of the cochlea. The click, which is the most commonly used stimulus to produce BAEPs, contains mainly high-frequency components and, therefore, mainly activates the basal cochlear portion. The cell bodies of the cochlear nerve are located in the spiral organ, and the central branches terminate in the ventral and dorsal cochlei nuclei. The cochlear nuclei project second-order axons medially through the pons to end in either the ipsilateral or the contralateral posterior nucleus of the trapezoid body. The third-order axons ascend through the posterior pons and midbrain in the lateral lemniscus. Some of the fibers synapse at the nucleus of the lateral lemniscus. The lateral lemniscus ends in the inferior colliculus where some of the fibers synapse. Other fibers proceed directly to the medial geniculate body. Neurons in the medial geniculate body form the acoustic radiation of the internal capsule, synapsing in the Heschl gyrus of the primary auditory cortex. High-frequency tones, such as clicks, activate mainly the deep more mesial portion of the Heschl gyrus.

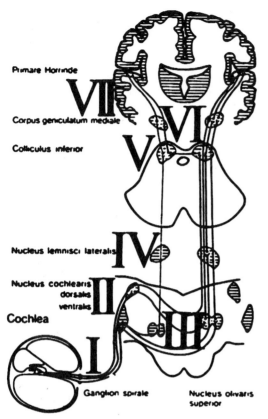

FIGURE 37–11. Diagrammatic representation of the auditory pathway and the most likely generators of the BAEPs. (Modified from Spillmann T, Erdmann W, Leitner H: Clinical experiences with althesin sedation for ERA. Rev Laryngol Otol Rhinol (Bord) 96(3-4):192–198, 1975.)

Wave I

Wave I corresponds to the compound AP recorded in the ECoG. This potential is generated in the most peripheral portion of the eighth cranial nerve.[37]

Wave II

Recent studies have shown that wave II consists of at least two different components:
APs in the intracranial portion of the eighth cranial nerve. This has been shown by direct intrasurgical recordings from the eighth cranial nerve[66] and by the observation that wave II may be preserved in brain-dead patients.[31]
Potentials generated in the intraparenchymal afferent auditory pathways in the ipsilateral rostral medulla. In animals the trapezoid body shows a prominent activity at the timing of wave II,[6] and lesions of the trapezoidal body may affect wave II.[1, 108]

Wave III

The origin of wave III is poorly defined but most probably is generated in the auditory pathways within the caudal pons ipsilaterally and contralaterally.[1, 6, 37, 55] Lesions at the mesencephalic level are usually associated with preservation of waves I through III.[59] Pontine lesions affect primarily the ipsilateral AEPs.

Wave IV

The generators of wave IV also remain uncertain. It has been suggested that this wave is generated in the superior olivary complex[67] or the inferior colliculus.[56]

Wave V

Wave V is most probably generated in the most rostral portion of the lateral lemnisci or in the inferior colliculi, or in both. Direct recordings in humans show a large potential with a latency similar to wave V at the collicular level following contralateral ear stimulation[37, 67] (Fig. 37–12). Clinical studies have shown, however, that in patients with unilateral mesencephalic lesions, wave V is consistently more attenuated or even absent[59] during ipsilateral ear stimulation.

Waves VI and VII

The origin of these inconsistent components is still poorly defined. Direct recordings from the Heschl gyrus reveal that the primary cortical response to click stimuli consists of a positivity with a latency of 12.7 to 19.4 msec.[54] It is, therefore, reasonable to assume that waves VI and VII, which occur with average latencies of 7.3 and 9.6 msec, are generated in subcortical structures such as the medial geniculate and the auditory radiations. In previous invasive studies, it was demonstrated that a positive peak was observed in the midbrain at the time of wave VI.[18, 37, 64, 65] Depth electrodes located in the medial geniculate body also showed a similar activity.[37, 107] In addition, invasive studies demonstrated a positive peak at a latency similar to wave VII in the midbrain.[35, 64, 65]

ANALYSIS METHODS

Identification of waves I, III, and V is essential for BAEP analysis. The other components cannot always be clearly

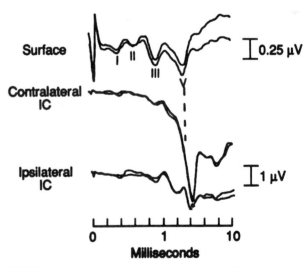

FIGURE 37–12. Simultaneous recordings of BAEPs with surface electrodes and electrodes placed close to the inferior colliculus (IC) ipsilateral and contralateral to the stimulus. Notice the large potential in the contralateral colliculus which has a latency similar to component V of the BAEP. (With permission from Hashimoto I, Ishiyama Y, Yoshimoto T, et al: Brain-stem auditory-evoked potentials recorded directly from human brain-stem and thalamus. Brain 104:84–859, 1981.)

TABLE 37-2	Interpeak Latencies of Brain Stem Auditory Evoked Potential Peaks[a]		
	Average	Average + 3 S.D.	AD–AS (Average + 3 S.D.)
I–III	2.1 msec	2.6 msec	0.37 msec
III–V	1.9 msec	2.4 msec	0.43 msec
I–V	4.0 msec	4.7 msec	0.43 msec

AD–AS, side-to-side difference; S.D., standard deviation.
[a]For females, I–V is 0.1 msec shorter; for older persons (>60 years), I–V is 0.1 msec longer; add 0.1 msec for each 20-Hz rate increase.
Modified from Chiappa KH, Gladstone KJ, Young RR: Brainstem auditory evoked responses. Studies of waveform variation of 50 normal human subjects. Arch Neurol 36(2):81–87, 1979.

identified even in normal subjects. BAEP analysis includes the interpeak latencies (IPLs) I–III, III–V, and I–V, and the amplitude ratio V/I. Absolute latencies of BAEP components are not reliable because they are affected by nonpathological factors. IPLs, however, are consistent and provide information on central conduction time. The I–III IPL is an index of the conduction time between the cochlea and the caudal pons, the III–V IPL reflects the conduction time from pons to midbrain, and the I–V IPL measures the conduction time between the cochlea and the midbrain. The upper limit of normal is usually set to 3.0 standard deviations (S.D.) beyond the mean (Table 37–2). Amplitudes are measured from each peak to the following peak of opposite polarity. Absolute amplitude values are not useful, and the ratio of V/I is commonly used. A value of less than 0.66 is generally considered abnormal. However, if the only abnormality is a low V/I ratio, the interpretation of the BAEP abnormality should be cautious.

NON-NEUROLOGICAL FACTORS AFFECTING BRAIN STEM AUDITORY EVOKED POTENTIALS

Age

Before 30 weeks' conceptual age (CA), BAEP waveforms are inconsistent. Only 50% of infants of less than 30 weeks' CA show reproducible BAEPs to 60 dBHL clicks.[85] The absence of BAEPs in these children is most probably related to desynchronization of the stimulus in the brain stem auditory pathways, because some of the children may show cortical AEPs even in the absence of BAEPs and the correlation of the absence of BAEPs in this age group with hearing pathology is poor. Between 30 and 40 weeks' CA, waves I, III, and V become progressively more clearly defined. At the same time, the latency of all the BAEP peaks decreases[50] (Fig. 37–13). The shortening of latency is most pronounced for wave V, producing a progressive shortening of I–V IPL from 6.05 msec at 30 to 32 weeks' CA to 4.92 msec at full term. Full-term infants consistently have well-developed waves I, III, and V but still do not attain fully mature BAEPs (Fig. 37–14). The following differences are noticed when comparing BAEPs of these infants with those of normal adults:

The latencies of all the components are prolonged[96] (Table 37–3).

Wave I may be double peaked and followed by a prominent downward deflection.

The downward deflection following wave III may be poorly defined.

Amplitude of wave V may be relatively low, resulting in a small V/I amplitude ratio.

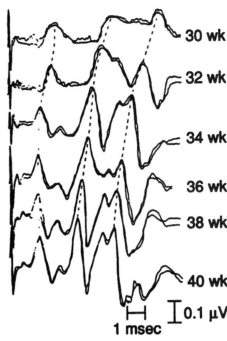

FIGURE 37–13. BAEPs in infants of 30 to 40 weeks' conceptual age. Notice how waves I, III, and V become progressively better defined and there is a significant decrease in latency of all components. (With permission from Krumholz A, Felix JK, Goldstein JP, et al: Maturation of the brain-stem auditory evoked potential in premature infants. Electroencephalogr Clin Neurophysiol 62:124–134, 1985.)

Scalp distribution may differ significantly from adults and waves III and V may be relatively small when using Ac–C$_z$.

Further maturation changes continue in the post-term period, until the middle or end of the second year of life.[40] Wave I has fully matured at ages 3 to 6 months. Waves III and V mature at a later age, attaining full maturity at the end of the second year. At that time, the I–III and I–V IPLs attain adult values[86] (Fig. 37–15). Click intensities of 30 dBHL produce clearly defined BAEPs in full-term newborns. At 3 months, click intensities of 20 dBHL produce BAEPs,

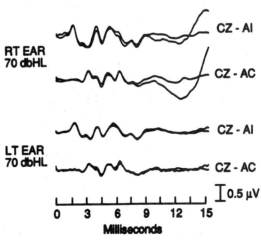

FIGURE 37–14. BAEPs obtained in a normal neonate. Notice the prominent downward deflection after wave I and the relatively low amplitude of wave V compared with wave I. (With permission from Markand ON, 11th Duke Symposium, 1992.)

TABLE 37–3	Latency of Brain Stem Auditory Evoked Potential Peaks in Normal Infants	
	Average	S.D.
Wave I	1.81 msec	0.22
Wave III	4.62 msec	0.29
Wave V	6.72 msec	0.32
I–V IPL	4.92 msec	0.26

IPL, interpeak latency; S.D., standard deviation.
From Stockard JE, Stockard JJ, Westmoreland BF, et al: Brainstem auditory-evoked responses: Normal variation as a function of stimulus and subject characteristics. Arch Neurol 36:823–831, 1979.

and at 12 months, the threshold decreases by another 5 dBHL.

With increasing age, there is a tendency for all the BAEP components to be delayed significantly. This is most probably the result of the presbycusis seen in most adults.[34] In addition, Rowe[84] described a selective prolongation of I–III IPL.

Gender

The BAEPs show significant differences between males and females.[102] Absolute latency of wave V and I–V IPL are consistently prolonged in males by 0.15 to 0.2 msec for 60 to 70 dBSL stimulation intensities. The amplitude of the BAEPs is significantly higher in females (up to 40% to 50%). These gender differences are seen in children only after 8 years of age. I–V IPL decreases by almost 1 msec at the time

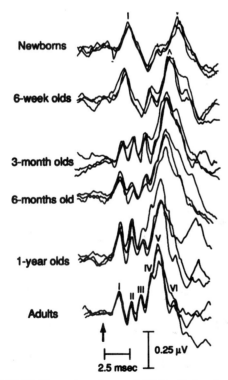

FIGURE 37–15. Maturation of normal BAEPs. Notice that wave I is almost fully matured at 3 to 6 months, whereas wave V does not reach maturity until 1 to 2 years. (With permission from Salamy A, McKean CM: Postnatal development of human brainstem potentials during the first year of life. Electroencephalogr Clin Neurophysiol 40:(4)418–426, 1976.)

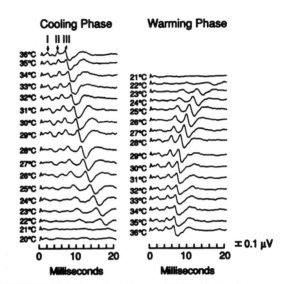

FIGURE 37–16. Effect of cooling on BAEPs. Notice the progressive increase of wave V, the latency of which approximately doubles at around 26°C. With further cooling, wave V also decreases progressively and disappears at 20 to 21°C. (With permission from Markand ON, Lee BI, Warren C, et al: Effects of hypothermia on brainstem auditory evoked potentials in humans. Ann Neurol 22:507–513, 1987.)

of the menstrual period. These differences between males and females are most probably related to differences of head size (corresponding to a thicker skull and longer auditory pathway in males) and possibly also to differences in core temperature, which tends to shorten IPLs (see later discussion).

Temperature

With decreasing central body temperature, the latencies of BAEPs are increased. The temperature-latency relationship is nonlinear. Latencies increase roughly 7% for each 1°C. The amplitude of the BAEPs first increases as the temperature decreases down to 27 to 28°C, and then amplitudes decrease linearly with temperature drop[60] and usually disappear at around 20°C (Fig. 37–16).

Parameters of Click

All BAEP peak latencies decrease at a rate of approximately 0.03 msec/dB as the click intensity increases (Fig. 37–17). Peak V has a latency of about 8.2 msec at 10 to 20 dBHL and then decreases to approximately 5.5 msec at 80 dBHL. All the other waves tend to decrease their latency by approximately the same value, resulting in relative constancy of the IPLs. However, wave I often decreases at a slightly faster rate, resulting in relatively more prolonged I–V IPL at higher stimulation intensities.[96] With decreasing click intensity all waves except I, III, and V tend to disappear. With even lower click intensities, only peak V, which may be seen with click intensities as low as 10 dBSL, may persist. In addition, it has been observed that the increase in latency of wave V is not linear, with a relative acute change in latency occurring between 40 and 60 dBSL clicks.[24] This is one of the reasons stimulation intensities of at least 70 dBSL are recommended for clinical studies. The changes in latency and I–V IPL described earlier are at least partially determined by the fact that BAEPs to clicks at different intensities will be generated by different frequencies with the higher

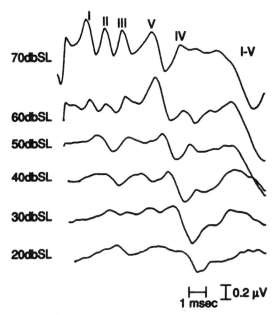

FIGURE 37–17. Effect of click intensity on BAEPs. Notice that all latencies increase with decreasing stimulus intensities and at the same time there is a progressive decrease of amplitude. I–V interpeak latency (IPL) remains constant and even becomes slightly shorter at lower stimulus intensities. (With permission from Markand ON, 11th Duke Symposium, 1992.)

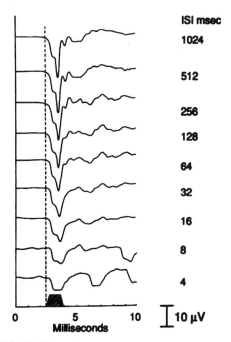

FIGURE 37–18. Effect of increasing stimulus frequency on the summation potential (SP) and action potential (AP) components (corresponding to wave I of the BAEP) of the ECoG. Notice that at stimulus frequencies of approximately 30 Hz (ISI of 32 msec) the SP and AP of the ECoG have only increased slightly in latency and their amplitude is still over 50% compared with an ISI of 1024 msec (≈ 1 Hz). (ISI, interstimulus interval.) (With permission from Eggermont JJ: Summating potentials in electrocochleography: Relation to hearing disorders. In Ruben RJ, Elberling C, Salomon G (eds): Electrocochleography. Baltimore, University Park Press, 1976, pp 67–87.)

frequency components playing a more significant part in the generation of clicks of higher intensity. It takes the traveling wave in the cochlea approximately 5 msec to travel from the basal cochlear region to the apex and, as previously mentioned, high-frequency sounds activate primarily the hair cells at the basal cochlear region. Wave I is predominantly produced by sound frequencies of 4 to 8 KHz and wave V by sounds of 1 to 4 KHz. This explains why wave I shows a bigger decrease in latency than wave V with higher stimulus intensities.

Increasing the click frequency results in slight prolongation of latency of all BAEP components, but the latency of wave I is minimally affected[14, 22] (Fig. 37–18). Increasing the rate from 10 to 80 Hz will increase the latency of peak I by only 0.14 msec, whereas peak III is increased by 0.23 msec and wave V by 0.39 msec. The net result is a slight increase in I–V interpeak latencies of approximately 0.1 msec for every 20-Hz increase in click frequency. Decrease in amplitude or loss of definition of individual waveforms occurs with stimulus rates over 30 Hz, but wave V remains least affected.[14] Stockard and Rossiter[98] reported a higher sensitivity of BAEPs obtained at higher click rates for detection of abnormalities in patients with multiple sclerosis. However, in patients with multiple sclerosis, Chiappa and colleagues[15] did not find any new abnormalities of BAEPs that had not been detected with BAEPs obtained using conventional click frequency rates.

Clicks are generated by applying rectangular electrical pulses of a duration of 100 µsec to a standard audiometric earphone. The sound pressure wave generated by this stimulus can be displayed on an oscilloscope and consists of an initial major wave followed by highly damped oscillations of alternating polarity which may last up to 2 msec or longer[62] (Fig. 37–19). Depending on the polarity of the click, they are divided into rarefaction clicks (R clicks), which result in an initial outward movement of the tympanic membrane, and condensation clicks (C clicks), which result in an initial inward movement of the tympanic membrane. In addition,

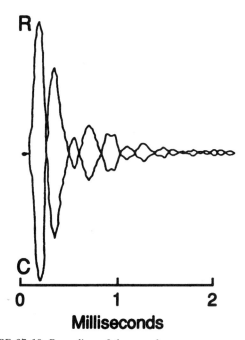

FIGURE 37–19. Recording of the sound pressure waves generated by a rarefaction (R) and condensation (C) click. Notice that the clicks consist of a large initial oscillation which is followed by highly dampened oscillations for 1 to 2 msec. Rarefaction and condensation clicks produce oscillations in opposite directions. (Modified from Mauer K: Akustisch evozierte Potentiale. In Lowtizsch von K, Mauer K, Hopf HC (eds): Evozierte Potentiale in der Klinischen Diagnostik, vol 1. Stuttgart, Georg Thieme Verlag, 1983, pp 177–195.)

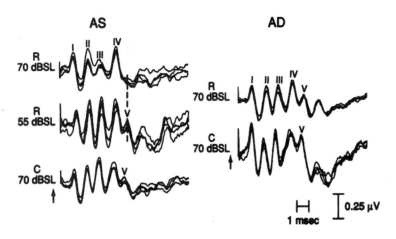

FIGURE 37–20. Effect of click polarity on BAEPs. Notice that when stimulating with a rarefaction click at 70 dBSL, no wave V is seen with AS stimulation but a clear wave V appears on AD stimulation. Wave V, however, emerges when reducing the stimulus intensity (AS). With condensation clicks wave V is seen with AS and AD stimulation even if AD stimulation elicits a more clearly defined wave V. (With permission from Emerson RG, Brooks EB, Parker SW, et al: Effects of click polarity on brainstem auditory evoked potentials in normal subjects and patients: Unexpected sensitivity of wave V. Ann NY Acad Sci 388:710–721, 1982.)

a click can be characterized by analyzing the frequency spectrum of the sound pressure wave recorded on an oscilloscope. It shows that most clicks consist of a wide spectrum of audio frequencies with the maximum energy between 2 and 5 KHz ("broad band" or "unfiltered" clicks). BAEPs elicited by condensation (C) clicks and rarefaction (R) clicks tend to show the following differences[26, 63] (Fig. 37–20):

   The latency of wave I is usually shorter and the amplitude of wave I is usually higher with R clicks.
   Wave IV is more prominent and more clearly distinctive from wave V with R clicks.
   Peak V tends to be smaller amplitude when R clicks are used.

The latency of peak V is usually not affected by the polarity of the click, resulting in lengthening of the I–V IPL when using R clicks. These differences tend to be relatively more accentuated in neonates and infants and also in older people, particularly when they have high-frequency hearing loss. There are reports that in some patients a BAEP abnormality, such as absence of wave V, occurred only with C clicks but not with R clicks.[26] It is not clear, however, whether these findings correspond to infrequent normal variants or an actual BAEP abnormality. Ideally, all patients should have BAEPs to alternating C and R clicks. The computer should be able to summate the evoked responses to each click modality separately, allowing independent analysis of the BAEPs to C and R clicks. In addition, this methodology permits summation of the two averages, which results in almost complete cancellation of the stimulus artifact (and also of the cochlear microphonics) whose polarity depends on the polarity of the stimulus. This methodology has advantages for those cases in which no clear wave I can be isolated from the electrical artifact, particularly when using high click intensities which result in relatively larger stimulus artifact. It has the disadvantage that some peaks may be obscured by partial peak cancellation when the two click polarities produce peaks of different latencies. It is generally recommended that rarefaction clicks be used when measuring a single click polarity. In addition, the routine protocol should call for additional use of the opposite click polarity when the BAEPs obtained with the first click polarity do not produce satisfactory results.

ABNORMALITIES OF BRAIN STEM AUDITORY EVOKED POTENTIALS IN NEUROLOGICAL DISEASES

Coma and Brain Death

BAEPs are usually normal in patients who are comatose as a result of a toxic or metabolic etiology and also in patients with a diffuse cortical process that spares the brain stem.[93] Patients with no clinical signs of brain stem dysfunction usually have normal BAEPs, whereas patients with clinical evidence of midbrain or lower pontine lesions will have abnormal BAEPs in almost 100% of the cases.[103] Several studies have investigated use of BAEPs as a prognostic tool in patients who are in a coma. In general, patients with bilaterally absent waves IV/V tend to have a very poor prognosis, but this is frequently already clear from careful analysis of the routine clinical information.[47, 90] In general, BAEP studies provide only additional prognostic information in a limited number of cases.

In brain-dead patients, BAEPs can give valuable information about brain stem function. Previous studies have clearly established that sedative drugs such as barbiturates or anesthetic overdoses, even to the point of producing isoelectric EEGs, will produce only minimal changes in the BAEPs.[87] In brain-dead patients, one would expect to find preservation of wave I but absence of all components generated within the brain stem, namely waves III and V (Fig. 37–21). Unfortunately, only approximately one-quarter of brain-dead patients show this pattern.[31] Slightly over three-quarters of the brain-dead patients show bilateral complete absence of all components including wave I. This is most probably because the vascular supply of the peripheral portion of the eighth cranial nerve is supplied via the intracranial circulation by the internal auditory artery. Brain death is consistently associated with significant brain edema, which frequently leads to occlusion of the internal auditory artery, a branch of either the anterior inferior cerebellar artery or the basilar artery. In the absence of a clearly identifiable wave I, it is not possible to determine whether the absence of BAEP components is the result of brain death or of other damage to the peripheral auditory system (examples include traumatic transverse fracture of the temporal bone with damage of the cochlea and preexisting deafness).

Multiple Sclerosis

In a review of the literature, Chiappa[12] found 1006 patients with varying classifications of multiple sclerosis (MS) who underwent BAEP studies; of these, 466 patients (46%) had abnormal BAEPs. The percentage of abnormal BAEPs increased with the degree of certainty of the clinical diagnosis: 30% in patients with possible MS, 41% in patients with probable MS, and 67% in patients with definite MS. BAEP studies in patients with MS tend to be either completely normal or markedly abnormal, suggesting that even small plaques that affect the auditory pathways are sufficient to produce marked conduction abnormalities. However, BAEPs may be completely normal in patients who present clear

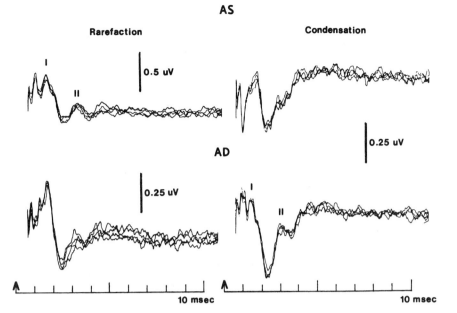

AS

Rarefaction   Condensation

0.5 uV

0.25 uV

AD

0.25 uV

10 msec   10 msec

FIGURE 37–21. Brain-dead patient. All waves after peak I and a small peak II are absent with rarefaction and condensation clicks. The preservation of peak II in brain-dead patients is one of the evidences suggesting that at least part of the peak is generated in the proximal auditory nerve. (With permission from Chiappa KH: Brainstem auditory evoked potentials methodology. In Chiappa KH (ed): Evoked Potentials in Clinical Medicine. New York, Raven Press, 1983, pp 105–202.)

clinical evidence of intrinsic brain stem lesions. Chiappa and associates[15] recorded normal BAEPs in almost half of the patients with MS and internuclear ophthalmoplegia.

BAEP abnormalities in MS patients can be categorized as follows. The most frequent abnormality is decreased amplitude or absence of waves IV and V, seen in more than 80% of MS patients with abnormal BAEPs (Fig. 37–22). Another frequently seen abnormality is increase of the I–V IPL (Fig. 37–23). III–V IPL prolongation is more common than I–III IPL prolongation, probably because the III–V IPL is a reflection of the longest intra-axial auditory pathway. Other less frequent abnormalities include absence of wave III (see Fig. 37–23) and, very rarely, no response at all.

It is important to note that in almost 50% of the cases, the abnormalities are only seen when stimulating unilaterally (see Figs. 37–22 and 37–23). Increasing the frequency of stimulation to 70 Hz does not increase the incidence of BAEP abnormalities in MS patients.[25]

Most patients with MS and abnormal BAEPs have subjec-

tively normal hearing. This suggests that the BAEP abnormalities are the result of temporal dispersion of the afferent auditory volley by demyelination and not the result of axon loss. MS patients tend to have normal click thresholds in routine audiological testing. However, abnormalities can be detected frequently on detailed auditory and vestibular testing, particularly on the interaural time discrimination test and on auditory localization testing.[38, 70]

BAEPs are particularly useful in the evaluation of patients in whom the diagnosis of MS is entertained but who have no signs or symptoms of brain stem involvement. In a series studied by Chiappa,[12] abnormal BAEPs were found in 21% of patients referred with the possible diagnosis of MS in whom no clear brain stem symptoms or signs were present. BAEPs can also be useful to document clinically silent lesions, adding support for a multicentric process.

BAEPs, and also the other modalities of evoked potentials, can be used to objectively detect progression of the disease process. This is particularly useful when monitoring disease

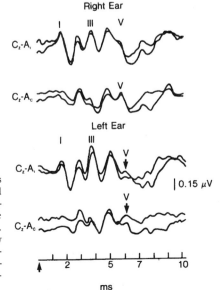

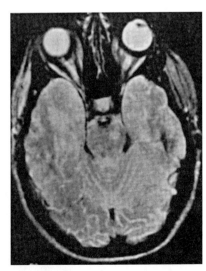

Right Ear

I   III   V

C₂-A₁

V

C₂-A_c

Left Ear

I   III

V

C₂-A₁

0.15 μV

V

C₂-A_c

2   5   7   10

ms

FIGURE 37–22. Patient with multiple sclerosis showing a decreased amplitude of peak V and some prolongation of III–V IPL when stimulating the left ear. The patient had a lesion in the lateral portion of the left midbrain on MRI. (With permission from Markand ON, Farlow MR, Stevens JC, et al: Brainstem auditory potential abnormalities with unilateral brainstem lesions demonstrated by magnetic resonance imaging. Arch Neurol 46:295–299, 1989.)

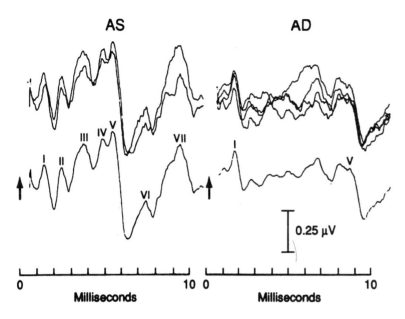

FIGURE 37–23. Patient with multiple sclerosis who on right ear stimulation shows poor definition of all components after peak I and a marked prolongation of I–V IPL. (With permission from Chiappa KH, Harrison JL, Brooks EB, et al: Brainstem auditory evoked responses in 200 patients with multiple sclerosis. Ann Neurol 7:135–143, 1980.)

change during experimental therapy, as BAEP deterioration can occur in the face of stable or improving clinical function.

Other Demyelinating Disorders

BAEP abnormalities were reported in patients with central pontine myelinolysis.[99, 109] At the time of maximum clinical deficit, marked prolongation of I–V IPL can be seen, with return to normal values upon recovery (Figs. 37–24 and 37–25).

Patients with leukodystrophies tend to have striking abnormalities, including absence of or marked abnormalities of all components after wave I. Ochs and associates[73] reported markedly abnormal BAEPs in seven patients with Pelizaeus-Merzbacher disease. Ten known carriers of the disease, however, had normal BAEPs. Similar findings have also been reported in patients with adrenoleukodystrophy and metachromatic leukodystrophy[4, 19, 39, 61] (Fig. 37–26), as well as in asymptomatic carriers. Patients with metachromatic leukodystrophy may also have abnormalities of wave I, as a result of peripheral nervous system demyelination.

Posterior Fossa Tumors

BAEPs were considered the most sensitive test to detect acoustic neuromas in the pre–magnetic resonance imaging

(MRI) era[3, 76] (Fig. 37–27). Hearing loss from cerebellopontine angle tumors can be distinguished by BAEPs from cochlear hearing loss (mostly Ménière's disease) and from labyrinthine disease because these disorders usually manifest normal BAEPs.[15, 23] However, recent studies have demonstrated that BAEPs are normal in 11% to 23% of patients with acoustic neuromas less than or equal to 1 cm.[8, 112]

About one-third of patients with acoustic neuroma have ipsilateral I–III or I–V IPL delay (Figs. 37–27 and 37–28), or both, due to the conduction abnormality produced by tumor compression from tumors of relatively small size. About one-third of patients have ipsilateral absence of all peaks following wave I (Fig. 37–29), probably due to compression from a tumor of sufficient size to produce a conduction block. The final one-third of patients have ipsilateral absence of all BAEP components including wave I, likely due to additional compression on the internal acoustic artery, leading to severe hearing problems. Contralateral BAEPs are usually normal except when the tumor is large and displaces the brain stem. The contralateral abnormality usually consists only of III–V IPL prolongation with normal earlier components.

Brain stem gliomas are almost invariably associated with BAEP abnormalities, including absence of or markedly decreased amplitude of all peaks after wave I (Fig. 37–30), and prolongation of I–III or III–V IPLs, or both.[4, 11, 19, 36, 44, 98] Other posterior fossa tumors such as metastatic tumors, me-

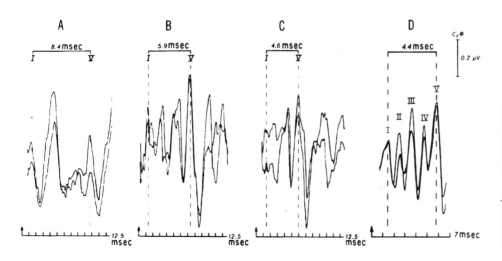

FIGURE 37–24. Patient with central portion myelinolysis who showed a marked prolongation of I–V IPL with return to normal values when the patient recovered clinically. See also Figure 37–25. (With permission from Stockard JJ, Rossiter VS, Wiederholt WC, et al: Brain stem auditory-evoked responses in suspected central pontine myelinolysis. Arch Neurol 33:726–728, 1976.)

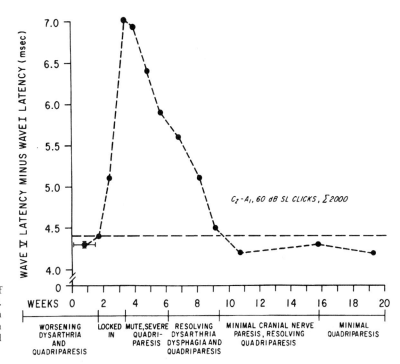

FIGURE 37–25. Clinical evolution and time course of I–V IPL in a patient with central pontine myelinolysis. Same patient as Figure 37–24. (With permission from Stockard JJ, Rossiter VS, Wiederholt WC, et al: Brain stem auditory-evoked responses in suspected central pontine myelinolysis. Arch Neurol 33:726–728, 1976.)

ningiomas, medulloblastomas, cerebellar astrocytoma, cerebellar hemangioblastomas, ependymomas, and pinealomas also frequently show abnormal BAEPs when there is significant brain stem involvement[33, 44, 74, 94, 98] (Fig. 37–31).

Cerebrovascular Brain Stem Lesions

Patients with transient ischemic attacks usually have normal BAEPs.[12, 49, 81] Patients with medullary infarcts below the

cochlear nuclei, such as Wallenberg's syndrome, also have normal BAEPs.[4, 74] Infarcts at higher brain stem levels frequently compromise the auditory pathways, leading to the expected BAEP abnormalities[36, 74, 98] (Fig. 37–32). However, patients with the locked-in syndrome may have normal BAEPs, as the auditory pathways are located more dorsally and laterally than the ventromedially located motor pathways in the pontine tegmentum.

Degenerative Diseases

Satya-Murti and colleagues[89] showed poorly defined BAEP waveforms in patients with Friedreich's ataxia, whereas Nuwer and associates[72] reported normal BAEPs in 20 patients. Abnormalities in patients with Friedreich's ataxia have been attributed to degeneration of the spiral ganglion, a

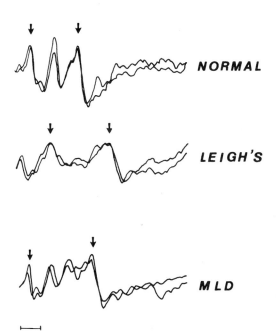

FIGURE 37–26. Abnormal BAEPs in patients with Leigh's disease and metachromatic leukodystrophy (MLD). (With permission from Hecox KE, Cone B, Blaw ME: Brainstem auditory evoked response in the diagnosis of pediatric neurologic disease. Neurology July: 832–840, 1981.)

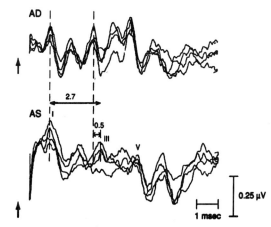

FIGURE 37–27. Abnormal I–III IPL in a patient with a normal computed tomographic scan and a left acoustic neuroma. (With permission from Chiappa KH: Brainstem auditory evoked potentials methodology. In Chiappa KH (ed): Evoked Potentials in Clinical Medicine. New York, Raven Press, 1983, pp 105–202.)

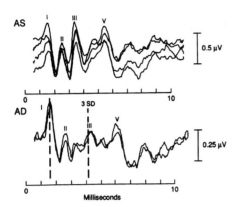

FIGURE 37–28. Unilateral BAEP abnormality (prolonged I–III and I–IV IPL) in a patient with inherited bilateral acoustic neuromas. (With permission from Chiappa KH: Brainstem auditory evoked potentials methodology. In Chiappa KH (ed): Evoked Potentials in Clinical Medicine. New York, Raven Press, 1983, pp 105–202.)

structure analogous to the dorsal root ganglion. Audiometric testing has also shown abnormalities of the eighth cranial nerve. Nuwer and colleagues[71] demonstrated abnormal BAEPs in patients with olivopontocerebellar degeneration. It was also reported that the latency of wave V is increased and the I–V IPL may be prolonged in patients with hereditary cerebellar ataxia.[77] Satya-Murti and Cacace[88] reported abnormal I–III latencies in four patients with Charcot-Marie-Tooth disease. BAEPs tend to be normal in patients with Parkinson's disease.[12, 105] However, patients with progressive supranuclear palsy may have BAEP abnormalities related to pathological changes in the brain stem.[12] It was reported that patients with facioscapulohumeral muscular dystrophy showed abnormal slowing of the later components of the BAEPs.[27] Normal BAEPs have been reported in Hallervorden-Spatz disease,[61] Huntington's disease,[79] and amyotrophic lateral sclerosis.[79]

Infectious and Inflammatory Disorders

Smith and colleagues[91] reported wave V latency prolongation in asymptomatic patients infected with the human im-

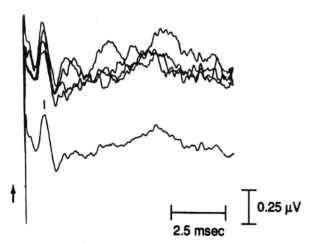

FIGURE 37–29. Absence of all peaks after a clearly defined peak I in a patient with an acoustic neuroma. (With permission from Chiappa KH: Brain stem auditory EPS: Methodology. In Chiappa KH (ed): Evoked Potentials in Clinical Medicine, 3rd ed. Philadelphia, Lippincott-Raven, 1997, pp 157–197.)

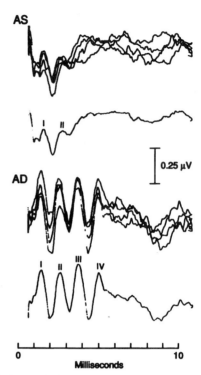

FIGURE 37–30. Markedly abnormal BAEP to left side stimulation with only preservation of peak I (and a small peak II) in a patient with a left brain stem glioma. (With permission from Chiappa KH: Physiologic localization using evoked responses: Pattern shift visual, brainstem auditory and short latency somatosensory. In Thompson RA, Green JR (eds): New Perspectives in Cerebral Localization. New York, Raven Press, 1982, pp 63–114.)

munodeficiency virus (HIV1). Chiappa[12] reported abnormally prolonged I–III IPL in patients with meningitis, attributable to inflammation of the subarachnoid space transversed by the acoustic nerve. Subacute sclerosing panencephalitis and Creutzfeldt-Jakob disease may be associated with BAEP abnormalities.[12, 61, 80] Behçet's disease can cause prolonged I–III or III–V IPL.[95] Chiappa[12] reported a patient with Miller Fisher variant of Guillain-Barré syndrome who had an abnormally prolonged I–III interpeak latency.

Metabolic Disorders

A patient with Gaucher's disease had abnormal BAEPs,[52] whereas patients with Batten's disease had normal BAEPs.[61]

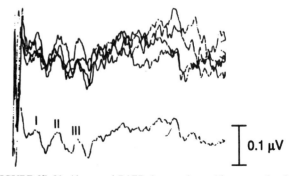

FIGURE 37–31. Abnormal BAEPs in a patient with a posterior fossa meningioma. Notice absence of all peaks (or markedly delayed peak V) after peak III. (With permission from Chiappa KH: Brainstem auditory evoked potentials methodology. In Chiappa KH (ed): Evoked Potentials in Clinical Medicine. New York, Raven Press, 1983, pp 105–202.)

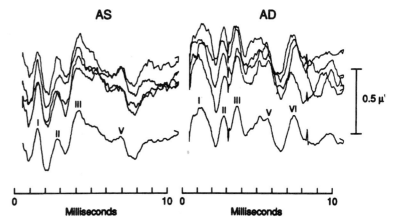

FIGURE 37–32. Abnormal BAEPs (poorly defined and prolonged wave V on left ear stimulation) in a patient with a left upper pontine-midbrain hemorrhage. Notice that the BAEP abnormality occurs when stimulating the ear ipsilateral to the hemorrhage. (With permission from Chiappa KH: Brainstem auditory evoked potentials methodology. In Chiappa KH (ed): Evoked Potentials in Clinical Medicine. New York, Raven Press, 1983, pp 105–202.)

Fujita and colleagues[29] reported abnormal BAEPs in patients with Wilson's disease. Acute alcoholism produces small but significant BAEP latency prolongations[2, 5, 16] attributable to lowered body temperature. Begleiter and colleagues[2] observed marked I–V IPL prolongations in chronic alcoholics who were abstinent for at least 3 weeks. They attributed this abnormality to demyelination or edema, or both. Patients with Wernicke-Korsakoff syndrome showed abnormal BAEPs.[7] Vitamin B_{12} deficiency may be associated with I–V IPL prolongations which are reversible with appropriate therapy.[28, 51] Patients with diabetes mellitus may have prolongations in I–III and I–V IPLs.[21] It was demonstrated that both hyperthyroidism and hypothyroidism produced abnormal BAEPs.[20] Renal disease produces little or no BAEP abnormality.[42, 82]

Other Disorders

Patients with cortical deafness caused by bitemporal infarcts may have completely normal BAEPs.[75] Patients with narcolepsy have normal BAEPs,[41] as do patients with obstructive sleep apnea, even at times when there is a significant fall of oxygen saturation.[68, 100] Patients with central sleep apnea may have abnormal BAEPs, as would be expected from the high frequency of brain stem lesions in this condition.[97, 100] Lee and colleagues[53] demonstrated abnormal BAEPs in patients with type II Arnold-Chiari malformation. Basilar migraine may be accompanied by wave V abnormalities during attacks.[111] Abnormal BAEPs were recorded in patients with cystic fibrosis.[106] Patients with Down's syndrome show significant reduction in the IV–V IPL and wave V amplitude.[46] In patients with Cockayne's syndrome, BAEPs can be normal in early life, turning abnormal with age.[43] Kaviani and Jafary[48] reported BAEP abnormalities in patients with Bell's palsy. In patients with blepharospasm, BAEPs showed prolonged wave III or V latency.[17]

REFERENCES

1. Achor LJ, Starr A: Auditory brain stem responses in the cat. I. Intracranial and extracranial recordings. Electroencephalogr Clin Neurophysiol 48:154–173, 1980.
2. Begleiter H, Porjesz B, Chou CL: Auditory brainstem potentials in chronic alcoholics. Science 211:1064–1066, 1981.
3. Bockenheimer S, Schmidt CL, Zollner C: Neuro-otological findings in patients with small acoustic neuromas. Arch Otorhinolaryngol 239:31–39, 1984.
4. Brown FR, Shimizu H, McDonald JM, et al: Auditory evoked brainstem response and high-performance liquid chromatography sulfatide assay as early indices of metachromatic leukodystrophy. Neurology 31:980–985, 1981.
5. Cadaveira F, Corominas M, Rodriguez Holduin S, et al: 1.

Reversibility of brain-stem evoked potential abnormalities in abstinent chronic alcoholics: One year follow-up. Electroencephalogr Clin Neurophysiol 90:450–455, 1994.
6. Caird D, Sontheimer D, Klinke R: Intra- and extracranially recorded auditory evoked potentials in the cat. I. Source location and binaural interaction. Electroencephalogr Clin Neurophysiol 61:50–60, 1985.
7. Chan YW, McLeod JG, Tuck RR, et al: Brain stem auditory evoked responses in chronic alcoholics. J Neurol Neurosurg Psychiatry 48:1107–1112, 1985.
8. Chandrasekhar SS, Brackmann DE, Devgan KK: Utility of auditory brainstem response audiometry in diagnosis of acoustic neuromas. Am J Otol 16:63–67, 1995.
9. Chatrian GE, Wirch AL, Edwards KH, et al: Cochlear summating potential to broadband clicks detected from the human external auditory meatus. A study of subjects with normal hearing for age. Ear Hear 6:130–138, 1985.
10. Chatrian GE, Wirch AL, Lettich E, et al: Click-evoked human electrocochleogram. Noninvasive recording method, origin and physiologic significance. Am J EEG Technol 22:151–174, 1982.
11. Chiappa KH: Physiologic localization using evoked responses: Pattern shift visual, brainstem auditory and short latency somatosensory. In Thompson RA, Green JR (eds): New Perspectives in Cerebral Localization. New York, Raven Press, 1982, pp 63–114.
12. Chiappa KH: Brainstem auditory evoked potentials methodology. In Chiappa KH (ed): Evoked Potentials in Clinical Medicine. New York, Raven Press, 1983, pp 105–202.
13. Chiappa KH: Brain stem auditory EPS: Methodology. In Chiappa KH (ed): Evoked Potentials in Clinical Medicine, 3rd ed. Philadelphia, Lippincott-Raven, 1997, pp 157–197.
14. Chiappa KH, Gladstone KJ, Young RR: Brainstem auditory evoked responses. Studies of waveform variation of 50 normal human subjects. Arch Neurol 36(2):81–87, 1979.
15. Chiappa KH, Harrison JL, Brooks EB, et al: Brainstem auditory evoked responses in 200 patients with multiple sclerosis. Ann Neurol 7:135–143, 1980.
16. Chu N-S, Squares KC, Starr A: Auditory brain stem potentials in chronic alcohol intoxication and alcohol withdrawal. Arch Neurol 35:596–602, 1978.
17. Creel DJ, Holds JB, Anderson RL: Auditory brain-stem responses in blepharospasm. Electroencephalogr Clin Neurophysiol 86:138–140, 1993.
18. Curio G, Oppel F: Intraparenchymatous ponto-mesencephalic field distribution of brain-stem auditory evoked potentials in man. Electroencephalogr Clin Neurophysiol 69:259–265, 1988.
19. Davis H, Aminoff MJ, Berg BO: Brain-stem auditory evoked potentials in children with brain-stem or cerebellar dysfunction. Arch Neurol 42:156–160, 1985.
20. DiLorenzo L, Foggia L, Panza N, et al: Auditory brainstem responses in thyroid diseases before and after therapy. Hormone Res 43:200–205, 1995.
21. Donald MW, Bird CE, Lawson JS, et al: Delayed auditory brainstem responses in diabetes mellitus. J Neurol Neurosurg Psychiatry 44:641–644, 1981.

22. Eggermont JJ: Summating potentials in electrocochleography: Relation to hearing disorders. In Ruben RJ, Elberling C, Salomon G (eds): Electrocochleography. Baltimore, University Park Press, 1976, pp 67–87.

23. Eggermont JJ, Don M, Brachmann DE: Electrocochleography and auditory brainstem electric responses in patients with pontine angle tumors. Ann Otol Rhinol Laryngol 89(6 Pt 2):1–19, 1980.

24. Elberling C: Transitions in cochlear action potentials recorded from the ear canal in man. Scand Audiol 2:151–159, 1973.

25. Elidan J, Sohmer H, Gafni M, et al: Contribution of changes in click rate and intensity on diagnosis of multiple sclerosis by brainstem auditory evoked potentials. Acta Neurol Scand 65:570–585, 1982.

26. Emerson RG, Brooks EB, Parker SW, et al: Effects of click polarity on brainstem auditory evoked potentials in normal subjects and patients: Unexpected sensitivity of wave V. Ann NY Acad Sci 388:710–721, 1982.

27. Fierro B, Daniele O, Alosio A, et al: Evoked potential study in facio-scapulo-humeral muscular dystrophy. Acta Neurol Scand 95:346–350, 1997.

28. Fine EJ, Hallett M: Neurophysiological study of subacute combined degeneration. J Neurol Sci 45:331–336, 1980.

29. Fujita M, Hosoki M, Miyazaki M: Brainstem auditory evoked responses in spinocerebellar degeneration and Wilson disease. Ann Neurol 9:42–47, 1981.

30. Ge NN, Shea JJ Jr, Orchik DJ: Cochlear microphonics in Meniere's disease. Am J Otol 18:58–66, 1997.

31. Goldie WD, Chiappa KH, Young RR, et al: Brainstem auditory and short-latency somatosensory evoked responses in brain death. Neurology 31:248–256, 1981.

32. Goodin DS, Squires CK, Starr A: Long latency event-related components of the auditory evoked potential in dementia. Brain 101:635–648, 1978.

33. Green JB, McLeod S: Short latency somatosensory evoked potentials in patients with neurological lesions. Arch Neurol 36:846–851, 1979.

34. Harkins SW, McEvoy TM, Scott ML: Effects of interstimulus interval on latency of the brainstem auditory evoked potential. Int J Neurosci 10:7–14, 1979.

35. Hashimoto I: Auditory evoked potentials from the human midbrain: Slow brain stem responses. Electroencephalogr Clin Neurophysiol 53:652–657, 1982.

36. Hashimoto I, Ishiyama Y, Tozuka G: Bilaterally recorded brain stem auditory evoked responses. Their asymmetric abnormalities and lesions of the brain stem. Arch Neurol 36:161–167, 1979.

37. Hashimoto I, Ishiyama Y, Yoshimoto T, et al: Brain-stem auditory-evoked potentials recorded directly from human brainstem and thalamus. Brain 104:84–85, 1981.

38. Hausler R, Levine RA: Brain stem evoked potentials are related to interaural time discrimination in patients with multiple sclerosis. Brain Res 191:589–594, 1980.

39. Hecox KE, Cone B, Blaw ME: Brainstem auditory evoked response in the diagnosis of pediatric neurologic disease. Neurology July 31:832–840, 1981.

40. Hecox K, Galambos R: Brain stem auditory evoked responses in human infants and adults. Arch Otolaryngol 99:30–33, 1974.

41. Hellekson C, Allen A, Greeley H, et al: Comparison of interwave latencies of brain stem auditory evoked responses in narcoleptics, primary insomniacs and normal controls. Electroencephalogr Clin Neurophysiol 47:742–744, 1979.

42. Hutchinson JC, Klodd DA: Electrophysiologic analysis of auditory, vestibular and brain stem function in chronic renal failure. Laryngoscope 92:833–843, 1982.

43. Iwasaki S, Kaga K: Chronological changes in auditory brainstem responses in Cockayne's syndrome. Int J Pediatr Otorhinolaryngol 30:211–221, 1994.

44. Jerger J, Neely JG, Jerger S: Speech, impedance and auditory brainstem response audiometry in brainstem tumors. Arch Otolaryngol 106:218–233, 1980.

45. Johansson RK, Haapaniemi JJ, Laurikainen EA: Transtympanic electrocochleography in evaluation of cochleovestibular disorders. Acta Oto-Laryngol Suppl 529:63–65, 1997.

46. Kakigi R, Kuroda Y: Brain-stem auditory evoked potentials in adults with Down's syndrome. Electroencephalogr Clin Neurophysiol 84:293–295, 1992.

47. Karnaze DS, Marshall LF, McCarthy CS, et al: Localizing and prognostic value of auditory evoked responses in coma after closed head injury. Neurology 32:299–302, 1982.

48. Kaviani M, Jafary AH: Auditory brain-stem response audiometry in patients with Bell's palsy. Clin Otolaryngol 20:135–138, 1995.

49. Kjaer M: Localizing brain stem lesions with brain stem auditory evoked potentials. Acta Neurol Scand 61:265–274, 1980.

50. Krumholz A, Felix JK, Goldstein JP, et al: Maturation of the brain-stem auditory evoked potential in premature infants. Electroencephalogr Clin Neurophysiol 62:124–134, 1985.

51. Krumholz A, Weiss HD, Goldstein PJ, et al: Evoked responses in vitamin B-12 deficiency. Ann Neurol 9:407–409, 1981.

52. Lacey DJ, Terplan K: Correlating auditory evoked and brainstem histologic abnormalities in infantile Gaucher's disease. Neurology 34:539–541, 1984.

53. Lee SI, Park TS, Dalmas AM: Brainstem auditory evoked potentials in Arnold-Chiari malformation. Electroencephalogr Clin Neurophysiol 61:20–21, 1985.

54. Lee YS, Lueders H, Dinner D, et al: Recording of auditory evoked potentials in humans using chronic subdural electrodes. Brain 107:115–131, 1984.

55. Legatt AD, Arezzo JC, Vaughn HG Jr: Short latency auditory evoked potentials in monkey. II. Intracranial generators. Electroencephalogr Clin Neurophysiol 64:53–73, 1986.

56. Lev A, Sohmer H: Sources of averaged neural responses recorded in animal and human subjects during cochlear audiometry (electro-cochleogram). Arch Klin Exp Ohr-Nas-u-Kehl Heilkd 201:79–90, 1972.

57. Lüders H, Lesser R, Dinner D, et al: Critical analysis of methods used to identify generator sources of evoked potentials (EP) peaks. In Lüders H (ed): Advanced Evoked Potentials. Boston, Martinus Nijoff, 1988, pp 29–63.

58. Markand ON: Brainstem auditory evoked potentials. J Clin Neurophysiol 11(3):319–342, 1994.

59. Markand ON, Farlow MR, Stevens JC, et al: Brainstem auditory potential abnormalities with unilateral brainstem lesions demonstrated by magnetic resonance imaging. Arch Neurol 46:295–299, 1989.

60. Markand ON, Lee BI, Warren C, et al: Effects of hypothermia on brainstem auditory evoked potentials in humans. Ann Neurol 22:507–513, 1987.

61. Markand ON, Ochs R, Worth RM, et al: Brainstem auditory evoked potentials in chronic degenerative central nervous system disorders. In Barber C (ed): Evoked Potentials. Baltimore, University Park Press, 1980, pp 367–375.

62. Maurer K: Akustisch evozierte Potentiale. In Lowitzsch von K, Maurer K, Hopf HC (eds): Evozierte Potentiale in der Klinischen Diagnostik, vol 1. Stuttgart, Georg Thieme Verlag, 1983, pp 177–195.

63. Maurer K, Schafer E, Leitner H: The effect of varying stimulus polarity (rarefaction vs. condensation) on early auditory evoked potentials. Electroencephalogr Clin Neurophysiol 50:332–334, 1980.

64. Møller AR, Jannetta PJ: Evoked potentials from the inferior colliculus in man. Electroencephalogr Clin Neurophysiol 201:79–90, 1982.

65. Møller AR, Jannetta PJ: Interpretation of brainstem auditory evoked potentials: Results from intracranial recordings in humans. Scand Audiol 12:125–133, 1983.

66. Møller AR, Jannetta PJ, Sekhar LN: Contributions from the auditory nerve to the brain-stem auditory evoked potentials (BAEPs): Results of intracranial recording in man. Electroencephalogr Clin Neurophysiol 71:198–211, 1988.

67. Møller AR, Jho HD, Yokota M, et al: Contribution from crossed and uncrossed brainstem structures to the brainstem auditory evoked potentials: A study in humans. Laryngoscope 105:596–605, 1995.

68. Mosko SS, Pierce S, Holowach J, et al: Normal brainstem auditory evoked potentials recorded in sleep apneics during waking and as a function of arterial oxygen saturation during sleep. Electroencephalogr Clin Neurophysiol 51:477–482, 1981.

69. Neshige R, Lüders H: Recording of event-related potentials (P300) from human cortex. J Clin Neurophysiol 9:294–298, 1992.

70. Noffsinger D, Olden W, Carhart R, et al: Auditory and vestibu-

lar aberrations of multiple sclerosis. Acta Otolaryngol Suppl (Stockh) 303:4–63, 1972.
71. Nuwer MR, Perlman SL, Packwood JW, et al: Evoked potential abnormalities in the inherited ataxias: Further results. Electroencephalogr Clin Neurophysiol 53:24P, 1982.
72. Nuwer MR, Perlman SL, Packwood JW, et al: Evoked potential in the various inherited ataxias. Ann Neurol 13:20–27, 1983.
73. Ochs R, Markand ON, DeMyer WE: Brainstem auditory evoked responses in leukodystrophies. Neurology 29:1089–1093, 1979.
74. Oh SJ, Kuba T, Soyer A, et al: Lateralization of brainstem lesions by brainstem auditory evoked potentials. Neurology 31:14–18, 1981.
75. Ozdamar O, Kraus N, Curry F: Auditory brain stem and middle latency responses in a patient with cortical deafness. Electroencephalogr Clin Neurophysiol 53:224–230, 1982.
76. Parker SW, Chiappa KH, Brooks EB: Brainstem auditory evoked responses in patients with acoustic neuromas and cerebello-pontine angle meningiomas. Neurology 30:413–414, 1980.
77. Pedersen L, Trojaborg W: Visual, auditory and somatosensory pathway involvement in hereditary cerebellar ataxia, Friedreich's ataxia and familial spastic paraplegia. Electroencephalogr Clin Neurophysiol 52:283–297, 1981.
78. Picton TW, Hillyard SA, Krausz HI, et al: Human auditory evoked potentials. I. Evaluation of components. Electroencephalogr Clin Neurophysiol 36:179–190, 1974.
79. Pierelli F, Pozzessere G, Bianco F, et al: Brainstem auditory evoked potentials in neurodegenerative diseases. In Morocutti C, Rizzo PA (eds): Evoked Potentials. Neurophysiological and Clinical Aspects. Amsterdam, Elsevier, 1985, pp 157–168.
80. Pollak L, Klein C, Giladi R, et al: Progressive deterioration of brainstem evoked potentials in Creutzfeldt-Jacob disease: Clinical and electroencephalographic correlation. Clin Electroencephalogr 27:95–99, 1996.
81. Ragazzoni A, Amantini A, Rossi L, et al: Brainstem auditory evoked potentials and vertebral-basilar reversible ischemic attacks. In Courjon J, Mauguiere F, Revol M (eds): Clinical Applications of Evoked Potentials in Neurology. New York, Raven Press, 1982, pp 187–194.
82. Rizzo PA, Pierelli F, Pozzessere G, et al: Pattern visual evoked potentials and brainstem auditory evoked responses in uremic patients. Acta Neurol Belg 82:72–79, 1982.
83. Robinson K, Rudge P: Centrally generated auditory potentials. In Halliday AM (ed): Evoked Potentials in Clinical Testing, vol 10. London, Churchill Livingstone, 1982, pp 344–372.
84. Rowe MJ: Normal variability of the brain-stem auditory evoked response in young and old adult subjects. Electroencephalogr Clin Neurophysiol 44:459–470, 1978.
85. Salamy A: Maturation of the auditory brainstem response from birth through early childhood. J Clin Neurophysiol 1:293–329, 1984.
86. Salamy A, McKean CM: Postnatal development of human brainstem potentials during the first year of life. Electroencephalogr Clin Neurophysiol 40(4):418–426, 1976.
87. Sanders RA, Duncan PG, McCullough DW: Clinical experience with brain stem audiometry performed under general anesthesia. J Otolaryngol 8:24–31, 1979.
88. Satya-Murti S, Cacace A: Brainstem auditory evoked potentials in disorders of the primary sensory ganglion. In Coujon J, Mauguiere F, Revol M (eds): Clinical Applications of Evoked Potentials in Neurology. New York, Raven Press, 1982, pp 219–225.
89. Satya-Murti S, Cacace A, Hanson P: Auditory dysfunction in Friedreich ataxia: Result of spiral ganglion degenerating. Neurology 30:1047–1053, 1980.
90. Seales DM, Rossiter VS, Weinstein ME: Brainstem auditory evoked responses in patients comatose as a result of blunt head trauma. J Trauma 19:347–352, 1979.
91. Smith T, Jakobsen J, Gaub J, et al: Clinical and electrophysiological studies of human immunodeficiency virus-seropositive men without AIDS. Ann Neurol 23:295–297, 1988.
92. Spillmann T, Erdmann W, Leitner H: Clinical experiences with althesin sedation for ERA. Rev Laryngol Otol Rhinol (Bord) 96(3-4):192–198, 1975.
93. Starr A, Achor LJ: Auditory brain stem responses in neurological disease. Arch Neurol 32:761–768, 1975.
94. Starr A, Hamilton AE: Correlation between confirmed sites of neurologic lesions and abnormalities of far-field auditory brainstem responses. Electroencephalogr Clin Neurophysiol 41:595–608, 1976.
95. Stigsby B, Bohlega S, Al-Kawi MZ, et al: Evoked potential findings in Behcet's disease. Brain-stem auditory, visual, and somatosensory evoked potentials in 44 patients. Electroencephalogr Clin Neurophysiol 92:273–281, 1994.
96. Stockard JE, Stockard JJ, Westmoreland BF, et al: Brainstem auditory-evoked responses: Normal variation as a function of stimulus and subject characteristics. Arch Neurol 36:823–831, 1979.
97. Stockard JJ: Brainstem auditory evoked potentials in adult and infant sleep apnea syndromes, including sudden infant death syndrome and near-miss for sudden infant death. Ann NY Acad Sci 388:443–465, 1982.
98. Stockard JJ, Rossiter VS: Clinical and pathologic correlates of brainstem auditory response abnormalities. Neurology 27:316–325, 1977.
99. Stockard JJ, Rossiter VS, Wiederholt WC, et al: Brain stem auditory-evoked responses in suspected central pontine myelinolysis. Arch Neurol 33:726–728, 1976.
100. Stockard JJ, Sharbrough FW, Staats BA, et al: Brain stem auditory evoked potentials (BAEPs) in sleep apnea. Electroencephalogr Clin Neurophysiol 50:167P, 1980.
101. Syndulko K, Hansch EC, Cohen SN, et al: Long-latency event-related potentials in normal aging and dementia. In Courjon J, Mauguière F, Revol M (eds): Advances in Neurology, vol 32. Clinical Applications of Evoked Potentials in Neurology. New York, Raven Press, 1982, pp 279–285.
102. Thivierge J, Cote R: Brain-stem auditory evoked responses (BAER): Normative study in children and adults. Electroencephalogr Clin Neurophysiol 68:479–484, 1987.
103. Tsubokawa T, Nishimoto H, Yamamoto T, et al: Assessment of brainstem damage by the auditory brainstem responses in acute severe head injury. J Neurol Neurosurg Psychiatry 43:1005–1011, 1980.
104. Tsuji S, Lüders H: Brainstem auditory evoked potentials recorded with noncephalic reference electrodes. Jpn J Electroencephalogr Electromyo 12:287–289, 1984.
105. Tsuji S, Muraoka S, Kuroiwa Y, et al: Auditory brainstem evoked response (ABER) of Parkinson-dementia complex and amyotrophic lateral sclerosis in Guam and Japan. Rinsho Shinkeigaku (Clin Neurol Tokyo) 21:37–41, 1981.
106. Vaisman N, Tabachnik E, Shahar E, et al: Impaired brainstem auditory evoked potentials in patients with cystic fibrosis. Dev Med Child Neurol 38:59–64, 1996.
107. Velasco M, Velasco F, Almanza X, et al: Subcortical correlates of the auditory brain stem potential in man: Bipolar EEG and multiple unit activity and electrical stimulation. Electroencephalogr Clin Neurophysiol 53:133–142, 1982.
108. Wada S-I, Starr A: Generation of auditory brain stem responses (ABRs) II. Effects of surgical section of the trapezoid body on the ABR in guinea pigs and cats. Electroencephalogr Clin Neurophysiol 56:340–351, 1983.
109. Wiederholt WC, Kobayashi RM, Stockard JJ, et al: Central pontine myelinolysis. A clinical reappraisal. Arch Neurol 34:220–223, 1977.
110. Wuyts FL, Van de Heyning PH, Van Spaendonck MP, et al: A review of electrocochleography: Instrumentation settings and meta-analysis of criteria for diagnosis of endolymphatic hydrops. Acta Otolaryngol Suppl (Stockh) 526:14–20, 1997.
111. Yamada T, Dickins S, Arensdorf K, et al: Basilar migraine: Polarity-dependent alternation of brainstem auditory evoked potential. Neurology 36:1256–1260, 1986.
112. Zappia JJ, O'Connor CA, Wiet RJ, et al: Rethinking the use of auditory brainstem response in acoustic neuroma screening. Laryngoscope 107:1388–1392, 1997.

Chapter 38

Somatosensory Evoked Potentials

Ronald G. Emerson, MD

SOURCES OF EVOKED POTENTIAL SIGNALS
Evoked Potential Signals Generated by
 White Matter Structures
Evoked Potential Signals Generated by
 Gray Matter Structures
NOMENCLATURE
MEDIAN NERVE SOMATOSENSORY
 EVOKED POTENTIALS
Brachial Plexus Volley
Cervical Potentials
P14 Complex
N18
Cortical Potentials

POSTERIOR TIBIAL NERVE
 SOMATOSENSORY EVOKED POTENTIALS
Spinal Potentials
Subcortical Components
Cortical Responses
RECORDING TECHNIQUES
Relaxation
Stimulation
Amplifier Parameters
Averager Parameters
Recording Montages
INTERPRETATIVE CRITERIA

Somatosensory evoked potentials (SEPs) are electrical signals generated by specific structures along the somatosensory pathways following stimulation of peripheral nerves. SEP waveforms are altered by disease processes that impair the function of these structures, or that produce conduction delays or blocks in the neural pathways connecting them.

SEP abnormalities result from a wide variety of pathological processes. They may be produced by focal lesions affecting the somatosensory pathways, including strokes, tumors, and spinal cord compression.[14, 17, 32, 80] Multiple sclerosis frequently produces SEP abnormalities,[13] as do other demyelinating diseases, including adrenoleukodystrophy and adrenomyeloneuropathy,[20, 41, 57, 109] metachromatic leukodystrophy,[104, 118] and Pelizaeus-Merzbacher disease.[73] Additionally, SEP abnormalities occur in some diseases that affect the nervous system diffusely, including human immunodeficiency virus (HIV) infection,[49, 79] vitamin E[8, 56] and B$_{12}$ deficiencies,[53] hereditary system degenerations,[85] amyotrophic lateral sclerosis,[42, 123] myotonic dystrophy,[44, 105] and diabetes.[37, 72]

SEPs are used diagnostically to demonstrate the existence, and often to suggest the locations, of neurological lesions. In this way, SEP testing can serve as an extension of the neurological examination. In addition to diagnostic testing, SEPs are increasingly being used to monitor neurological function during surgery, with the goal of detecting potentially reversible neurological insults.

SOURCES OF EVOKED POTENTIAL SIGNALS

Whereas electroencephalogram (EEG) signals are generated by the gray matter of the cerebral cortex, evoked potential (EP) signals are generated by both cortical and subcortical gray structures, as well as by white matter pathways. The anatomy of these generators largely determines the timing and topography of EP signals.

Evoked Potential Signals Generated by White Matter Structures

White matter fiber tracts generate two classes of EP signals—near-field potentials and far-field potentials. An electrode close to a nerve records a triphasic signal as a propagated volley approaches, reaches, and passes the electrode's location. An initial positivity reflects outward current flow from the nerve, ahead of the region of depolarization. A negativity, reflecting inward current flow, is recorded as the region of depolarization reaches the electrode. Finally, a positivity, reflecting outward current flow, is registered as the depolarized area moves past the point of recording (Fig. 38–1). Because this triphasic signal is the result of action potential propagation, its latency increases as the recording site is moved away from the point of stimulation. This type of signal, recorded from electrodes near a fiber tract and whose latency is position dependent, is termed a "near-field" potential.

White matter tracts also generate a distinct class of signals called "far-field" potentials. This term, borrowed from electrical engineering, was first used in neurophysiology by Jewett to describe brain stem auditory evoked potentials recorded at a distance from their neural generators.[50] In contrast to near-field potentials, the wave shapes and latencies of far-field potentials are not significantly affected by small changes in the location of the recording electrode. Whereas near-field signals may be detected in differential recordings between closely situated electrodes, "referential" recording technique, using widely spaced electrodes, is needed to detect far-field potentials.

Far-field potentials were classically attributed to the approaching positivity that precedes a propagated volley recorded beyond the end of a nerve fiber.[117] This type of far-field signal can be modeled theoretically in terms of an exploring electrode in a uniform volume conductor, and can be observed experimentally by recording past the crushed end of a nerve in a salt water bath.[68, 117]

Far-field potentials are also generated under other circumstances; "junctional" or "intercompartmental" far-field potentials can be produced at points where nerve trunks cross physical barriers. The mechanisms of generation of these signals are not fully understood. They appear, however, to be related to abrupt changes in current densities within the tissues surrounding nerves. These changes are thought to be produced when there is a sudden alteration in the geometry or conductive properties of these tissues.[58–62, 70, 81–83, 93] Far-field signals may also be produced at points where nerves

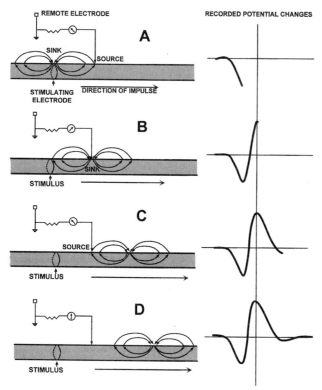

FIGURE 38–1. Generation of a near-field positive-negative-positive signal by a propagated volley. (From Brazier MAB: Electrical Activity of the Nervous System. Baltimore: Williams and Wilkins, 1997, p 68.)

bend or branch.[27] A model for far-field potential generation based on a moving quadrupole in a cylindrical volume conductor has recently been proposed. In this model, far-field signals are generated when there is an asymmetry of leading and trailing dipoles.[27]

Kimura has suggested that tissues on either side of a junction may serve as volume conductors connecting to a voltage source at the junction. A potential would be recorded between electrodes on either side of the junction, but not between electrodes on the same side.[58] Accordingly, there may be no inactive "reference" for junctional potentials, and assignment of polarity may be arbitrary.

Figure 38–2 illustrates the generation of both near-field and far-field signals following median nerve stimulation at the wrist.[122] Referential recordings from electrodes overlying the median nerve demonstrate a signal that increases in latency at electrodes placed progressively further from the wrist, along with three stationary, widely distributed, signals. The former reflects the near-field potential detected as the propagated volley passes under each electrode. The latter are far-field signals whose latencies correspond to the arrival of the afferent volley at the entry point of the median nerve into the brachioradialis muscle, the entry point of the median nerve into the deltoid muscle, and the junction of the arm and the body. These three far-field potentials appear similar in closely spaced electrodes. They therefore are not detected in the bipolar recording, which shows the near-field propagated volley in isolation.

Evoked Potential Signals Generated by Gray Matter Structures

As in the EEG, extracellular currents resulting from synchronous excitatory postsynaptic potentials (EPSPs) and in-

hibitory postsynaptic potentials (IPSPs) in cortical pyramidal cells and their dendrites also produce the cortical components of SEPs.[28, 63, 110] The cerebral cortex is an "open-field" system; the parallel orientation of cortical pyramidal cells allows summation of extracellular currents to produce EPs that are recordable at the scalp surface. Many subcortical nuclear structures are "closed-field" systems. Their internal geometries do not support generation of signals recordable beyond their limits.[3, 63, 67] Some subcortical gray matter structures, however, do contribute to EPs recordable at a distance. These include the central gray matter of cervical and lumbar spinal cords.[24, 33, 70, 97]

The surface topography of gray matter generated EP signals depends, in part, on the proximity of the recording electrode to the region of depolarization. Current gradients are steep in the vicinity of the region of depolarization, and the voltages of recorded signals vary considerably with changes in electrode position. In contrast, signals recorded far from the generator vary much less with changes in recording electrode position.[7] Gray matter generated signals may thus appear as near-field potentials when recorded from close to the generator, or as far-field signals when recorded from a distance. Whereas the terms *near-field* and *far-field* imply distinct mechanisms of generation for signals produced by white matter structures, for gray matter derived signals these same terms do not indicate different mechanisms of generation, but simply characterize the effects of recording from different locations.

NOMENCLATURE

By convention, EP waveforms are named by their polarity and their typical latency in normal individuals. This system is not adequate to uniquely identify all short latency SEP signals. The neural generators that contribute to an EP waveform recorded at a particular latency are selected, in part, by the recording electrode locations, including the location of the reference. These parameters are not taken into account by this naming system. Furthermore, within certain latency intervals, the number of discrete SEP components generated exceeds the number of unique integral latency values. It is not surprising that various authors have used different names to describe identical waves, and identical names to identify dissimilar signals. Confusion introduced by these ambiguities can be minimized by careful attention to the recording derivations used, and by examining the actual waveforms.

MEDIAN NERVE SOMATOSENSORY EVOKED POTENTIALS

Following stimulation of the median nerve at the wrist, a series of potentials is recorded from electrodes on the shoulder, neck, and head, that corresponds to the activation of specific structures along the afferent somatosensory pathways. The dorsal column–medial lemniscal system is largely responsible for the generation of the major components of the human median nerve SEP.[12] Major components of the median nerve SEPs are summarized in Table 38–1.

Brachial Plexus Volley

An electrode placed over Erb's point ipsilateral to the stimulated median nerve records a near-field potential as the afferent volley passes through the brachial plexus (Fig.

Median nerve stimulation at the wrist

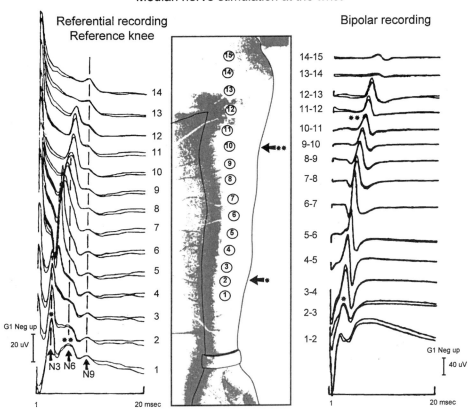

FIGURE 38–2. Generation of near-field and far-field signals following median nerve stimulation at the wrist. The right panel shows the propagated volley in a sequential bipolar recording from electrodes placed over the course of the median nerve in the arm and forearm. The left panel illustrates stationary far-field peaks N3, N6, and N9 recorded referentially from the same electrodes. These peaks correspond, respectively, to the times at which the afferent volley enters the brachioradialis muscle, enters the deltoid muscle, and crosses the boundary between the arm and the trunk. (From Yamada T, Machida M, Oishi M, et al: Stationary negative potentials near the source vs. positive far-field potentials at a distance. Electroencephalogr Clin Neurophysiol 60:512, 1985.)

38–3). This "Erb's point potential," or N9, has a positive-negative-positive morphology, corresponding to the volley approaching, reaching and then passing the recording electrode. The negative peak of N9 is generally used as a point of reference for measuring the latencies of the other median nerve SEP components. This permits conduction times central to Erb's point to be studied separately from the peripheral conduction time.

The afferent volley in the brachial plexus also produces a far-field potential, which is recorded between scalp electrodes and a noncephalic reference.[16, 51] This signal, the P9 (Fig. 38–3), is thought to represent an intercompartmental potential generated as the median nerve crosses the junction between the arm and the trunk.[122] The P9 precedes the arrival of the afferent volley at Erb's point slightly.[22, 121] Its exact latency can be influenced by changes in shoulder position.[26]

Cervical Potentials

The afferent volley may be recorded further proximally using electrodes on the lateral aspect the neck, ipsilateral to the stimulated median nerve. As shown in Figure 38–4, immediately following the Erb's point potential, electrodes along the anterior border of the sternocleidomastoid muscle record a propagated near-field signal. This signal, which is not recorded contralaterally, corresponds to activity in the brachial plexus or cervical roots. During the passage of this proximal plexus volley (PPV) through the neck, a far-field potential, the N10, is detected at constant latency in widely distributed scalp electrodes.[33] The exact point at which the passage of the PPV through the neck generates the far-field N10 is unknown, and there is considerable interindividual variability in the relation of the latency of the N10 to the PPV.

Posterior cervical electrodes record waveforms that are composites of the propagated volley in the cuneate tract along with postsynaptic activity in the dorsal gray matter of the cervical cord. It is often possible to distinguish two separate peaks in referential recordings. The latency of the first peak increases progressively at more rostral sites, while the latency of the second peak remains constant with changes of the recording electrode position (Fig. 38–5). The first peak corresponds to the afferent sensory volley

TABLE 38–1	Major Components of the Median Nerve Somatosensory Evoked Potential	
Waveform	**Recording Characteristics**	**Putative Generators**
N9	Traveling/Near-field	Brachial plexus
DCV	Traveling/Near-field	Cuneate tract
CERV N13/P13	Stationary/Near-field	Cervical cord, dorsal gray matter
P14 complex	Stationary/Far-field	Multiple generators, caudal medial lemniscus major contributor
N18	Stationary/Far-field	Multiple brain stem nuclei
N20	Stationary/Near-field	Primary somatosensory cortex

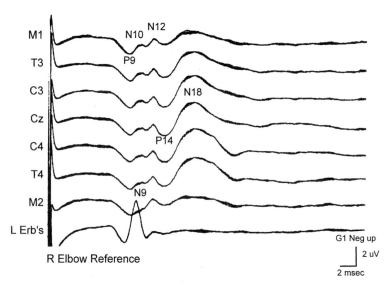

FIGURE 38–3. Median nerve SEP recorded in a normal subject, using an elbow reference. Amplifier passband was 30 to 3000 Hz (−6 dB). M1 and M2 correspond to electrode positions over the left and right mastoid processes, respectively. (From Emerson RG, Pedley TA: Somatosensory evoked potentials. In Daly DD, Pedley TA (eds): Current Practice of Clinical Encephalography, 2nd ed. New York, Raven Press, 1990, p 684.)

ascending the cuneate tract, and is designated DCV (dorsal column volley).[33]

The second, stationary peak, CERV N13, is attributed to postsynaptic activity in the dorsal gray matter of the cervical cord, produced in response to input from axon collaterals.[24, 70] The CERV N13 voltage is greatest in the posterior midline over the lower to middle cervical region. Referential recordings from a ring of electrodes around the neck, at the level of C5 posteriorly and the upper border of the thyroid cartilage anteriorly, show this stationary cervical potential to decrease in voltage in posterior lateral electrodes, become zero in lateral neck electrodes, and become positive (CERV P13) in anterior cervical electrodes[33] (Fig. 38–6). This relationship between the posterior cervical N13 and the anterior cervical P13 has also been observed in simultaneous recordings from posterior midline and intraesophageal electrodes.[24] Intramedullary recordings in cats[4, 15, 36] and monkeys[4, 6] confirm a polarity reversal between dorsal and ventral spinal horns, supporting the conclusion that CERV N13/P13 is a postsynaptic potential generated in the central gray of the cervical cord. Seyal and colleagues have demonstrated that at the upper cervical region, CERV N13 is selectively attenuated by finger movement, and have attributed this to an independent contribution from the cuneate nucleus.[98]

In referential scalp recording, N10 is followed by two small waves, P11 and N12. Based on extrapolated velocity measurements and intraoperative recordings, the P11 has been shown to correspond with the negative peak of the afferent volley at the C6 posterior rootlet.[22, 70] P11 demonstrates considerable interindividual variability, and is absent in some normal subjects.[102] The N12 peak on referential scalp recordings corresponds with the DCV peak at the first cervical vertebra[33] (see Fig. 38–5). The N12 may be a junctional potential generated at the interface of the spinal canal and the skull.

P14 Complex

Following N12, a prominent positivity, the P14 complex, is detected in referential scalp recordings. This signal com-

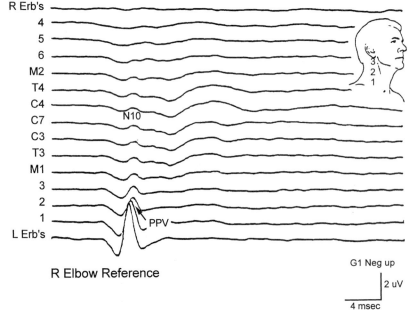

FIGURE 38–4. Normal median nerve SEP recorded referentially from scalp and lateral neck electrodes, illustrating the proximal plexus volley, and the N10 far-field signal. Amplifier passband is 1 Hz to 3000 Hz. M1 and M2 correspond to electrode positions over the left and right mastoid processes, respectively. (From Emerson RG, Pedley TA: Somatosensory evoked potentials. In Daly DD, Pedley TA (eds): Current Practice of Clinical Encephalography, 2nd ed. New York, Raven Press, 1990, p 685.)

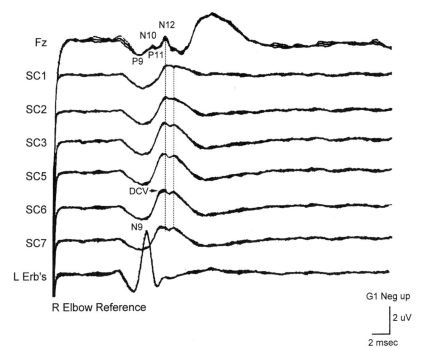

FIGURE 38–5. Normal median nerve SEP recorded referentially from a row of dorsal midline cervical electrodes (SC1 through SC7), F$_z$ and Erb's point. (From Emerson RG, Pedley TA: Somatosensory evoked potentials. In Daly DD, Pedley TA (eds): Current Practice of Clinical Encephalography, 2nd ed. New York, Raven Press, 1990, p 686.)

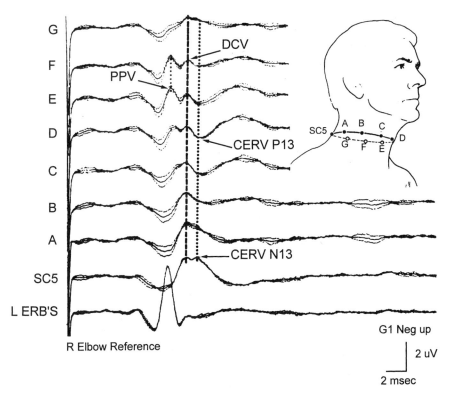

FIGURE 38–6. Normal left median nerve SEP recorded using a ring of electrodes around the neck, at the level of SC5 posteriorly and the superior border of the thyroid cartilage anteriorly. (From Emerson RG, Pedley TA: Somatosensory evoked potentials. In Daly DD, Pedley TA (eds): Current Practice of Clinical Encephalography, 2nd ed. New York, Raven Press, 1990, p 686.)

monly appears as a single positive deflection, but often one or more inflections or wavelets are superimposed upon it.[22, 31, 48, 91, 92, 100, 119, 121] Based on the frequency of occurrence of these wavelets as a function of latency following the P14 onset, Sonoo and colleagues suggested that the P14 complex consists of three components, denoted here as P14a, b, and c, which are differentially expressed among normal individuals (Fig. 38–7). Among 62 normal subjects, 22 subjects had one, 35 subjects had two, and 5 subjects had three distinct P14 components. Although its morphology varies among individuals, the P14 complex is highly reproducible and stable for individual subjects.[100]*

It is generally accepted that the P14 complex reflects activity in the medial lemniscus. This view is based on data derived from normal human subjects,[22–24] clinical observations,[74, 106, 119] and experimental animals.[2, 3] Tinazzi and associates concluded that the generator of P14 is located in the caudalmost portion of the medial lemniscus, prior to the decussation.[106] Sonoo and colleagues suggested that components P14a and P14b most likely originate in the caudal medial lemniscus.[100] Additionally, some authors have suggested a presynaptic contribution from the upper cervical cord.[54, 91, 92]

P14c may have a more rostral origin. Figure 38–8 demonstrates selective loss of P14c (and the cortical N20), with preservation of P14a and P14b in a patient with a mid pontine tumor. Although widely distributed, P14c is largest over the central region contralateral to the stimulated median nerve. Observing that it was lost in patients with lesions of the brain stem and thalamus, as well as with long-standing cortical lesions, Sonoo and colleagues suggested that the thalamocortical pathways were the likely generators of the P14c.[100] This interpretation is supported by intracranial recordings from monkeys showing the thalamocortical radiations to be the origin of the third of the three positive components, which together appear to correspond to the human P14 complex.[3]

In clinical testing, some ambiguity may be introduced by the existence of multiple components of the P14 complex, any one of which can be the most prominent component recorded in a particular patient. This ambiguity is diminished by the observation that when a single peak is present, it usually falls at a latency corresponding to either P14a or P14b; P14c is rarely the only peak present. When two peaks are present, they can represent any of the three P14 components, so that the initial component corresponds to either P14a or P14b.[100]

The foregoing discussion illustrates that even though postsynaptic cervical potentials and lemniscal far-field potentials are generated simultaneously, it is possible to record these signals separately by using appropriately placed electrodes. Despite this fact, it is common practice for clinical laboratories to employ a bipolar "neck to scalp" derivation (e.g., SC5–Fp$_z$) for the "cervical" channel. Bipolar recording combines these two signals into a composite waveform that obscures their individual identity. Figure 38–9 illustrates an SEP from a patient with a compressive cervical myelopathy showing a pronounced dissociation of the normally simultaneous CERV N13/P13 and far-field P14. These components are readily identified in referential (and bipolar SC5–Ant Cerv) channels. However, in the SC5–Fp$_z$ channel they appear as two sequential peaks that cannot be individually identified. In patients with cervical syringomyelia the only evoked potential abnormality may be selective loss of CERV

---

*In this discussion "P14a," "P14b," and "P14c" correspond respectively to Soono and colleagues' "P13" and "P14a" and "P14b".[100] Similarly, the "P14 complex" corresponds to the "P13/P14 complex" or "P13-P14 complex" of other authors.[24, 100, 119]

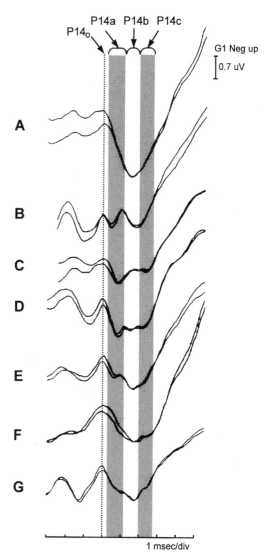

FIGURE 38–7. Tracings *A* through *G* are referential scalp recordings from seven normal individuals, illustrating the range of normal variability of the morphology of the P14 complex. Waveforms are aligned at point of onset of the P14 complex (P14o), and subcomponents P14a, P14b, and P14c are identified. (From Sonoo M, Genba-Shimizu K, Mannen T, et al: Detailed analysis of the latencies of median nerve somatosensory evoked potential components, 2: Analysis of subcomponents of P13/14 and N20 potentials. Electroencephalogr Clin Neurophysiol 104:298, 1997.)

N13/P13 (Fig. 38–10).[113] This finding would likely not be detected in an SC5-Fp$_z$ derivation.

N18

Following the P14 complex, referential scalp recordings demonstrate a widely distributed long duration signal, lasting approximately 20 msec, the N18. The N18 is symmetrical. With the minor exception of P14c, scalp to noncephalic recordings remain symmetrical until an added negativity, N20, generated by the primary somatosensory cortex, appears superimposed on N18 over the central parietal region opposite to the stimulated median nerve.[23] The N18, uncontaminated by superimposed cortical near field activity, is best recorded over the centroparietal scalp ipsilateral to the stimulated median nerve.[75]

The widespread distribution of N18 is characteristic of a

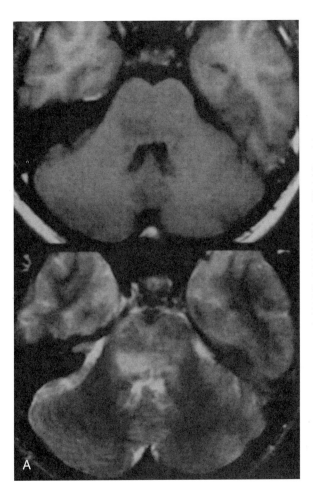

FIGURE 38–8. *A*, T1 *(upper)* and T2 *(lower)* weighed axial magnetic resonance images (MRIs) of a mid-pontine low-grade astrocytoma involving the region of the right medial lemniscus. *B,* The right median SEP is normal, demonstrating P14a, P14b, and P14c. The left median SEP shows preservation of P14a and P14b, but loss of P14c. (Modified from Sonoo M, Genba-Shimizu K, Mannen T, et al: Detailed analysis of the latencies of median nerve somatosensory evoked potential components, 2: Analysis of subcomponents of P13/14 and N20 potentials. Electroencephalogr Clin Neurophysiol 104:302, 1997.)

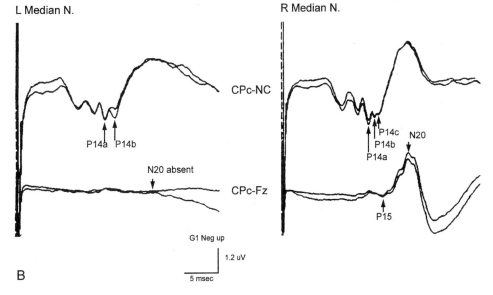

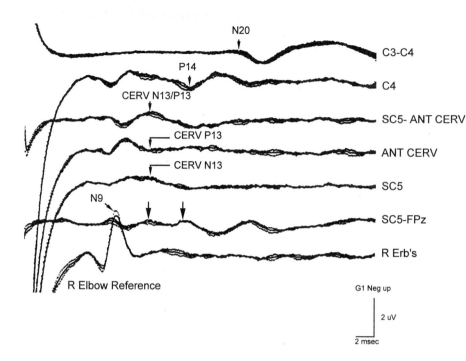

FIGURE 38–9. Median nerve SEP in a patient with a compressive cervical myelopathy. N13 and P14 are easily identified individually in referential neck (SC5–elbow) and scalp (C4–elbow) derivations, but cannot be distinguished in the bipolar SC5–Fp$_z$ derivation *(large arrows).* (From Emerson RG, Pedley TA: Somatosensory evoked potentials. In Daly DD, Pedley TA (eds): Current Practice of Clinical Encephalography, 2nd ed. New York, Raven Press, 1990, p 687.)

far-field potential of subcortical origin. Desmedt and Cheron[23] initially proposed the thalamus or caudal thalomocortical projections as possible sources for N18. Subsequently, it became clear that N18 remains present despite destructive lesions of the thalamus that caused the loss of cortical signals.[78] For example, Figure 38–11 demonstrates preservation of N18 despite loss of N20 in a patient with a bilateral thalamic astrocytoma. Postsynaptic activity has been recorded within the thalamus coinciding with the scalp-recorded N18. However, the signals generated were confined to the Vim nucleus and demonstrated large spatial gradients, characteristic of a closed field structure that does not generate a signal recordable beyond its borders.[114]

It is likely that N18 is generated by multiple brain stem

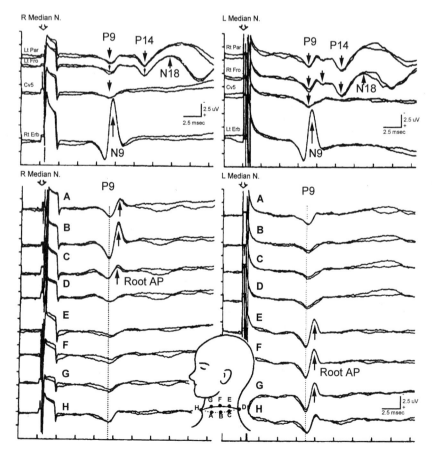

FIGURE 38–10. Median nerve SEP from a patient with cervical syringomyelia. All SEP components are normal, except for the absent stationary cervical N13/P13 potential. (From Urasaki E, Wada S, Kadoya C, et al: Absence of spinal N13–P13 and normal scalp far-field P14 in a patient with syringomyelia. Electroencephalogr Clin Neurophysiol 71:402, 1988.)

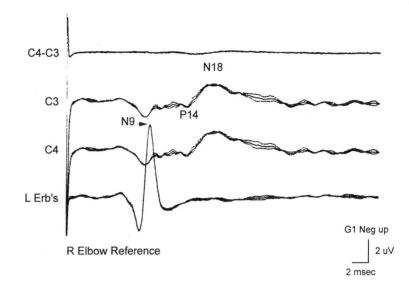

FIGURE 38–11. Median nerve SEP illustrating preservation of N18 and loss of N20 in a patient with a bilateral thalamic astrocytoma. (From Emerson RG, Pedley TA: Somatosensory evoked potentials. In Daly DD, Pedley TA (eds): Current Practice of Clinical Encephalography, 2nd ed. New York, Raven Press, 1990, p 691.)

structures. The view that several generators contribute to the N18 is supported by its complex morphology, with multiple small subpeaks, even when hemispherectomy has eliminated the possibility of near-field contamination.[75] Generation of the N18 by upper brain stem gray matter structures, particularly tectal and pretectal nuclei that receive collateral input from the medial lemniscus, has been proposed.[75, 112, 114, 115] Mauguiere and Desmedt observed that the tectum has an open-field structure compatible with the generation of signals that could be volume conducted to the scalp,[75] and loss of N18 amplitude has been demonstrated in patients with pontine and midbrain lesions[115] (Fig. 38–12). On the other hand, preservation of largely intact N18s has been reported in patients with tegmental pontine[90] and rostral medullary lesions, leading to the suggestion that major contributions to N18 may derive from the cuneate nuclei in the caudal medulla.[101]

Cortical Potentials

The earliest localized major component of the scalp-recorded median nerve SEP is the N20. It is recorded over the centroparietal region, contralateral to the stimulated median nerve. The local N20 is superimposed upon the widely distributed N18, and may be isolated from it either by numerical subtraction of referential recordings, or, more conveniently, by bipolar recording between contralateral and ipsilateral centroparietal scalp electrodes (Fig. 38–13). The separate identities of N18 and N20 were illustrated in two patients with small parietal lesions in whom isolated astereognosis and agraphesthesia were accompanied by loss of N20, but preservation of N18.[77]

The N20 represents the first cortical response to the afferent somatosensory volley.[1, 2, 23, 48, 75, 77] The manner in which activation of the primary somatosensory cortex generates the

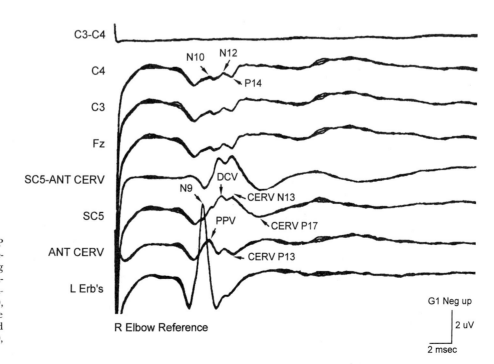

FIGURE 38–12. Median nerve SEP from a patient with a pontine arteriovenous malformation, illustrating loss of signals after P14. (From Emerson RG, Pedley TA: Somatosensory evoked potentials. In Daly DD, Pedley TA (eds): Current Practice of Clinical Encephalography, 2nd ed. New York, Raven Press, 1990, p 687.)

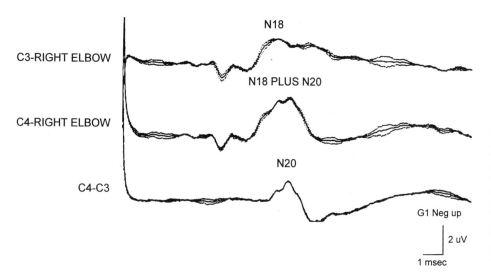

C3-RIGHT ELBOW

N18

C4-RIGHT ELBOW

N18 PLUS N20

C4-C3

N20

G1 Neg up

2 uV

1 msec

FIGURE 38–13. Derivation of the N20 by subtracting the underlying widespread N18 from a composite of N18 and N20. The lower trace was produced arithmetically by subtracting the top trace from the middle trace. (From Emerson RG, Pedley TA: Somatosensory evoked potentials. In Daly DD, Pedley TA (eds): Current Practice of Clinical Encephalography, 2nd ed. New York, Raven Press, 1990, p 690.)

N20 is explained using a model proposed by Lueders and co-workers based on recordings obtained from subdural electrodes in patients undergoing evaluation for the surgical treatment of intractable epilepsy.[69] Beginning about 18 msec after median nerve stimulation, simultaneous, phase opposite signals are recorded from electrodes straddling the central fissure. The negative signal is recorded from the posterior electrode (Fig. 38–14). The rising limb of N20 is thought to be generated as activation of area 3B produces a dipole within the posterior wall of the central sulcus, with the cortical surface positivity directed anteriorly and the corresponding negativity directed posteriorly (Fig. 38–15). The falling phase of N20, and the subsequent positivity, results from activation of adjacent cortical areas.[69, 89, 116] The phase reversal of N20 in cortical recordings is commonly used to identify functionally the central sulcus at surgery. Although direct cortical recordings sometimes demonstrate independent low-voltage cortical responses *ipsilateral* to the stimulated median nerve, it is unlikely that these signals make a significant contribution to scalp-recorded EPs.[84]

Similar to other EPs recorded from the body's surface, N20 is most likely a composite signal, with contributions from multiple generators. This view is supported by the frequent presence of small inflections superimposed on the N20 waveform. These inflections are most pronounced during wakefulness and tend to disappear during sleep. The latency of the voltage peak of the composite N20 signal is slightly longer (0.2 to 0.9 msec; mean, 0.6 msec) in stage II sleep than during wakefulness. We speculated that these inflections reflect contributions to N20 from multiple generators, probably within area 3B, activated by specific thalamocortical projections, and that some of these are down-modulated during sleep.[35, 39, 40, 103, 120] In bipolar scalp derivations, N20 is sometimes preceded by a low-voltage positivity, P15. P15 is also enhanced in waking recordings[34] (Fig. 38–16). P15 is recorded immediately following the far-field P14c, and may represent the approaching volley in subcortical thalamocortical fibers[100] (see Fig. 38–8B).

The parietal N20 is accompanied by an approximately simultaneous frontal positivity, P22. It was originally thought that N20 and P22 reflected opposite ends of a single dipole generator located in the posterior wall of the central sulcus.[1, 9] It is now clear that scalp-recorded N20 and P22 are generated by different structures. They have different onset and peak latencies[19, 21, 25, 94] and are influenced differently by variations in stimulation rate, anesthetic agents, and ischemia.[94] Furthermore, focal lesions in the internal capsule have been shown to cause selective loss of the parietal N20 with

preservation of the frontal P22, and vice versa (Figs. 38–17 and 38–18). Based on these observations, Mauguiere and Desmedt[76] suggested that N20 results from activation of sensory area 3B in response to input from VPLc, and P22 is generated separately by motor area 4 following input from VPLo. Over the central parietal region, N20 is followed by

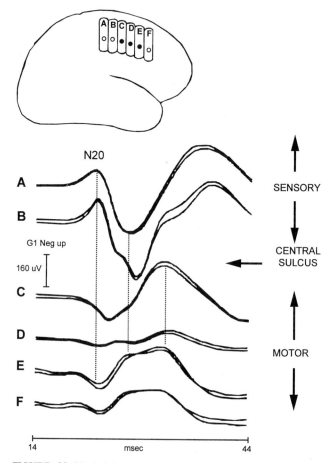

N20

A

B

G1 Neg up

160 uV

C

D

E

F

SENSORY

CENTRAL SULCUS

MOTOR

14   msec   44

FIGURE 38–14. Subdural electrode recording following median nerve stimulation, illustrating phase reversal of the N20 potential across the central fissure. The reference was a subdural electrode relatively distant from the central fissure. (Modified from Lueders H, Lesser RP, Hahn J, et al: Cortical somatosensory evoked potentials in response to hand stimulation. J Neurosurg 58:887, 1983.)

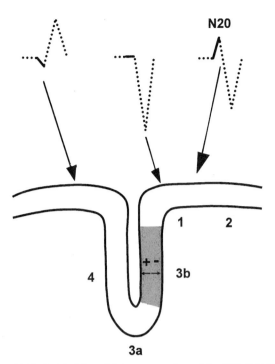

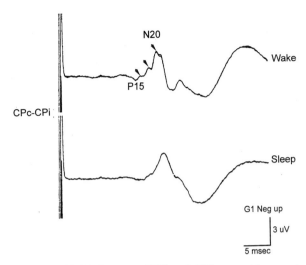

FIGURE 38–16. Median nerve P15 and N20 potentials recorded during wakefulness and during stage II sleep from the same normal subject. (From Emerson RG, Pedley TA: Somatosensory evoked potentials. In Daly DD, Pedley TA (eds): Current Practice of Clinical Encephalography, 2nd ed. New York, Raven Press, 1990, p 693.)

FIGURE 38–15. Model explaining generation of the N20. Solid lines correspond to potentials generated by depolarization of area 3B. Dotted lines correspond to signals generated subsequently by depolarization of adjacent cortical areas. Depolarization of area 3B produces the rising limb of N20, recorded from behind the central sulcus. A simultaneous positively is detected in cortical recordings from in front of the central sulcus. (Modified from Lueders H, Lesser RP, Hahn J, et al: Cortical somatosensory evoked potentials in response to hand stimulation. J Neurosurg 58:891, 1983.)

P27, which most likely is generated by area 1 on the crown of the postcentral gyrus.[76] The frontal P22 is followed by a negative signal, N30, which may be generated in premotor or supplementary motor areas.[76] The electrical fields of the frontal P22 and N30 extend to frontal scalp regions ipsilateral to the stimulated median nerve. However, recordings after hemispherectomy indicate that these potentials are generated only by the hemisphere contralateral to the stimulated median nerve.[75]

POSTERIOR TIBIAL NERVE SOMATOSENSORY EVOKED POTENTIALS

The components of the posterior tibial nerve (PTN) SEP are, in general, analogous to those of the median nerve SEP. Input responsible for generation of the PTN SEP travels through the dorsal columns, and possibly other pathways, including the dorsolateral funiculus.[11, 47, 52] The lumbar response to PTN stimulation resembles the cervical response to median nerve stimulation. Similarly, scalp-recorded responses consist of far-field subcortical and near-field cortical signals. Major components of the PTN SEPs are summarized in Table 38–2.

Spinal Potentials

The most prominent spinal component of the posterior tibial nerve SEP is the N22, analogous to the stationary CERV N13 of the median nerve SEP. N22 can be recorded over the lower thoracic and upper lumbar spine. Its latency is independent of electrode position, and its voltage is maximal over the lower thoracic spine, corresponding to the location of the lumbar enlargement[5] (Fig. 38–19). The dorsal N22 is accompanied by a simultaneous, phase opposite, ventral P22, analogous to the anterior cervical P13.[30, 97] In patients with tethering of the spinal cord, caudal displacement of the N22 may be seen.[30]

High quality recordings, particularly in children and thin adults, sometimes permit identification of a signal that precedes the peak of N22 and reflects the afferent propagated volley (PV). In contrast to the stationary N22, PV progressively increases in latency at more rostral electrode sites.

Subcortical Components

Referential scalp recordings following PTN stimulation demonstrate a widely distributed positive signal, P31, which is followed by a widely distributed, long duration, negative

TABLE 38–2	Major Components of the Posterior Tibial Nerve Somatosensory Evoked Potential		
	Waveform	Recording Characteristics	Putative Generators
	PV	Traveling/Near-field	Spinal roots, gracile tract
	N22/P22	Stationary/Near-field	Lumbar cord, dorsal gray matter
	P31	Stationary/Far-field	Multiple generators, caudal medial lemniscus major contributor
	N34	Stationary/Far-field	Multiple brain stem nuclei
	P38	Stationary/Near-field	Primary somatosensory cortex

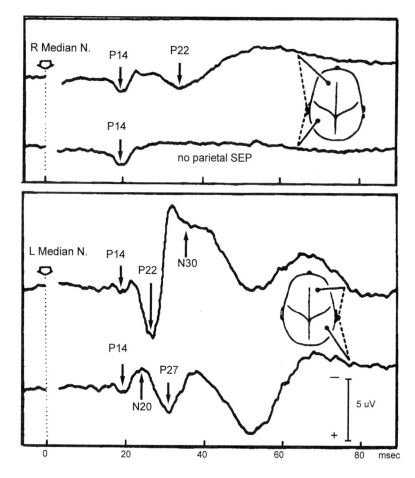

FIGURE 38–17. Median nerve SEPs from a patient who presented with severe right-sided motor and sensory deficits, and whose computed tomographic (CT) scan showed a small left internal capsule hematoma without cortical involvement. The right median SEP shows complete loss of the N20. The frontal P22, while present, is delayed and is of much lower voltage than the response obtained following stimulation of the normal left side. (From Mauguiere F, Desmedt JE: Focal capsular vascular lesions can selectively deafferent the prerolandic or pariental cortex: Somatosensory evoked potentials evidence. Ann Neurol 30:72, 1991.)

signal, N34 (Fig. 38–20). P31 and N34 correspond to the P14 and N18 components of the median nerve SEP. Similar to the multiple components of P14, P31 is preceded occasionally by another small positivity, P28. It is likely that P31 reflects activity with the caudal portion of the medial lemniscus, and that N34 originates in gray matter brain stem structures.[60, 97, 111]

The correspondence between posterior tibial and median nerve far-field signals is illustrated in Figure 38–21, which shows SEPs referentially recorded from the scalp and from a series of electrodes lying on the brain stem, within the fourth ventricle. Following both median and posterior tibial nerve stimulation, stationary positivities, corresponding in latency to P14 and P31 scalp-recorded signals, are recorded from electrodes rostral to the obex. Stationary negativities, corresponding in time to N18 and N34 scalp potentials, are recorded over the mid and upper pons.[111]

The correspondence between P14 and P31, and N18 and N34 is also illustrated by the effects of lesions. Figure 38–22B presents SEPs recorded before removal of a chondrosarcoma extending from C2 to C4, showing loss of far-field signals following both median and posterior tibial nerve stimulation, as well as loss or delay of the cortical responses. Figure 38–22C, obtained 7 months postoperatively, demonstrates normalization of both median and posterior tibial SEPs.[106] Similarly, Figure 38–23 demonstrates preservation of P14 and P31, with loss of cortical responses and partial preservation of N18 and N34, following an ischemic left medullary lesion.[106]

Cortical Responses

The cortical response to PTN stimulation is similar to the N20 of the median nerve SEP in the sense that it consists of localized scalp signals that are superimposed on underlying, widely distributed, far-field signals. The scalp distribution of the cortical response, however, is complex and includes localized near-field signals over both hemispheres. The bilateral distribution of the PTN SEP cortical response complicates the selection of a proper reference for recording the cortical response in isolation. As Fp$_z$ is nearly inactive for signals after N34, it is a useful reference for this purpose[96] (Fig. 38–24).

The most consistent component of the PTN SEP cortical response is the P38, a positive signal that typically is distributed asymmetrically about C$_z$, and is maximal in voltage over the hemisphere *ipsilateral* to the stimulated leg.[55, 96] This "paradoxical" lateralization of P38 can be explained by the location of the primary cortical receiving area for the leg and foot on the mesial aspect of the postcentral gyrus within the interhemispheric fissure. PTN stimulation causes depolarization of layer IV pyramidal cells in the primary sensory area and a corresponding passive current source at the cortical surface.[43] This cortical surface positivity projects toward the hemisphere ipsilateral to the stimulated leg, producing the P38 signal over the ipsilateral scalp.[18, 66, 96]

There is considerable interindividual variability in the P38 scalp topography, ranging from a positivity localized to the vertex to a positivity that is maximal over the ipsilateral scalp and *not* detectable at the vertex (Fig. 38–25A, B). This topographical variability of P38 likely reflects interindividual anatomical variability of the location of the primary somatosensory receiving area for leg and foot within the interhemispheric fissure.[87] Location of the foot area at the upper margin of the interhemispheric fissure would be expected to produce a vertically oriented P38, maximal in voltage at or close to the vertex. On the other hand, location of the

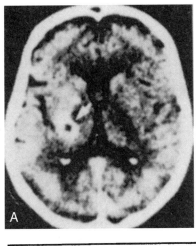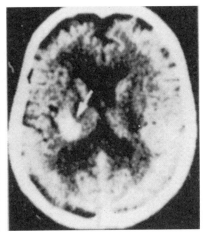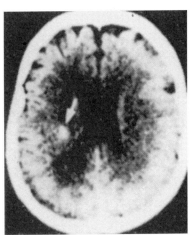

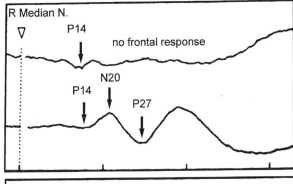

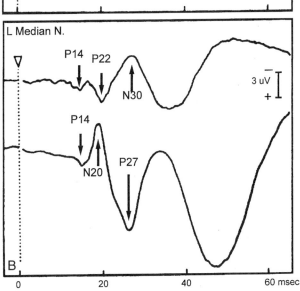

FIGURE 38–18. *A,* CT scan demonstrating a left capsular hematoma without cortical involvement in a patient who presented with sudden right hemiplegia and diminished but present touch, joint position, pain, and temperature sensation on the right. *B,* Corresponding median nerve SEPs showing an abnormal response to right-sided stimulation, with loss of the frontal P22. The parietal N20 is preserved, but of notably lower voltage than that recorded following left stimulation. The montage employed is identical to that illustrated in Figure 38–17. (From Mauguiere F, Desmedt JE: Focal capsular vascular lesions can selectively deafferent the prerolandic or parietal cortex: Somatosensory evoked potentials evidence. Ann Neurol 30:73, 1991.)

foot area deep within the fissure would produce a more horizontally oriented P38, maximal in voltage over the opposite (ipsilateral to the stimulated limb) centroparietal region.[18, 66, 96]

P38 is usually accompanied by a negative signal over the opposite hemisphere, maximal in voltage over the frontocentral scalp region[107] (Fig. 38–25*B*). It is tempting to view N38 as representing the "opposite end of the dipole" invoked to explain the P38.[18, 66, 96] This, however, is probably not accurate. Whereas N38 and P38 are approximately simultaneous, in many normal subjects P38 and N38 actually peak at slightly different times. Furthermore, P38 and N38 respond differently to changes in rates of stimulation, as well as to

passive and active foot motion during stimulation.[107, 108] These observations are inconsistent with P38 and N38 having a common single generator. They suggest that P38 and N38 result from nearly synchronous activation of several regions within the somatosensory cortex, with different orientations. Based on the greater sensitivity of P38 to increases in stimulus rate and to foot motion, Tinazzi and colleagues have suggested that the P38 cortical generator may receive both cutaneous and proprioceptive inputs, whereas the N37 cortical generator may be activated by cutaneous input alone.[107, 108] As with N20, P38 and N38 display state dependency. The latency of P38 increases by as much as 2 msec during stage II sleep.[99]

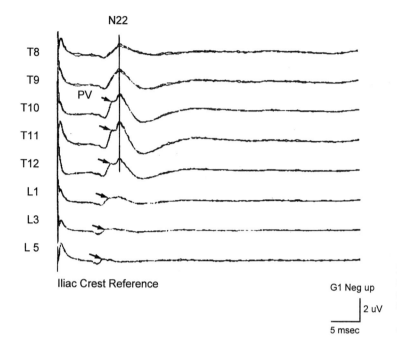

Iliac Crest Reference

G1 Neg up

⌐ 2 uV

5 msec

FIGURE 38–19. Normal posterior tibial SEP recorded referentially from a series of electrodes overlying the lumbar and lower thoracic spine. (From Emerson RG, Pedley TA: Somatosensory evoked potentials. In Daly DD, Pedley TA (eds): Current Practice of Clinical Encephalography, 2nd ed. New York, Raven Press, 1990, p 695.)

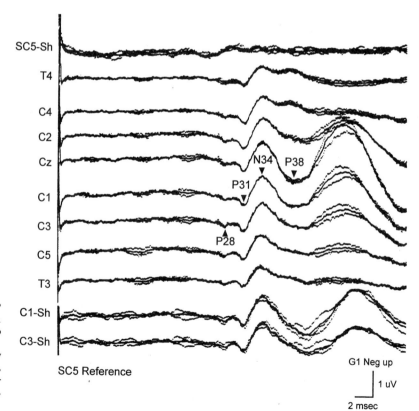

SC5 Reference

G1 Neg up

⌐ 1 uV

2 msec

FIGURE 38–20. Normal left posterior tibial SEP recorded referentially in the coronal plane through C_z. An SC5 reference is used except in the top and two bottom channels, where a shoulder (Sh) reference is employed. (From Emerson RG, Pedley TA: Somatosensory evoked potentials. In Daly DD, Pedley TA (eds): Current Practice of Clinical Encephalography, 2nd ed. New York, Raven Press, 1990, p 698.)

Bilateral Post. Tib. N.

R Median N.

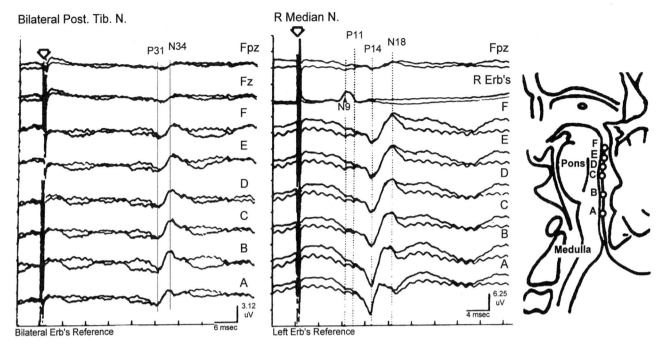

FIGURE 38–21. Simultaneous referential intracranial and scalp recordings from a patient undergoing surgery for a tentorial meningioma. Electrode locations are shown in the accompanying sketch. (From Urasaki E, Tokimura T, Yasukouchi H, et al: P30 and 33 of posterior tibial nerve SSEPs are analogous to P14 and N18 of median nerve SSEPs. Electroencephalogr Clin Neurophysiol 88:526, 1993.)

RECORDING TECHNIQUES

Relaxation

The degree of patient relaxation is a major determinant of the quality of SEP recordings. This is best illustrated in the operating room, where use of neuromuscular blocking agents makes it easy to record subcortical P31 and N34 components of the posterior tibial nerve SEP, signals that are often difficult to record in cooperative waking patients. Some laboratories perform routine EP recordings with the patient awake. Others prefer to use a mild sedative, for example, 20 mg diazepam administered orally in adults, to reduce muscle artifact. We favor the latter approach, as it enhances the quality of the recording, and facilitates evaluation of EP components that are likely to be unrecordable in the unsedated patient.

Stimulation

Median and posterior tibial nerves are usually stimulated transcutaneously using electrodes placed over the nerve, cathode proximal. Typically, square wave electrical pulses of 0.2-msec duration are delivered at rates between 3 and 5 Hz. Adjustment of the stimulus rate not to coincide with a multiple of the power mains frequency helps to eliminate power mains interference. Stimulation intensity is generally adjusted to produce a consistent but reasonably comfortable twitch.

Amplifier Parameters

A passband of 30 to 3000 Hz (−6 dB/octave) is most commonly employed.[45] A lower, low-frequency filter setting (e.g., 1 Hz) facilitates recording of certain long duration signals,[78] at the cost of introducing more low-frequency noise and increasing the number of trials that must be averaged.

More restrictive passbands may be useful for examining specific SEP components,[29, 71, 120] and may also be useful for intraoperative monitoring of cases in which electrical noise makes recording with the standard passband difficult.

Averager Parameters

For median nerve SEPs, an analysis time of 40 to 50 msec is appropriate. For PTN SEPs, 80 to 100 msec is usually employed. One to two thousand samples are usually averaged to produce a satisfactory recording. This number can vary considerably, depending on the characteristics of the noise accompanying the EP signal. Clinical EP recordings must be performed at least twice, and independent averages should be superimposed to demonstrate replicability.[45] Examination of independent superimposed averages is often required to determine whether a particular peak represents a true EP signal or an artifact.

Recording Montages

Median SEPs

A four-channel averager is generally necessary for standard laboratory recordings. The montage recommended by the American Clinical Neurophysiology Society[45] for median nerve SEPs is as follows:*

Channel 1	Erb's point–Noncephalic
Channel 2	SC5–Noncephalic
Channel 3	CPi–Noncephalic
Channel 4	CPc–CPi

---

*Erb's point opposite the stimulated median nerve generally serves as a satisfactory noncephalic reference. CPi and CPc refer to electrode locations midway between the standard C3 and P3, and C4 and P4 locations of the International 10–20 system, respectively ipsilateral and contralateral to the stimulated median nerve.[46]

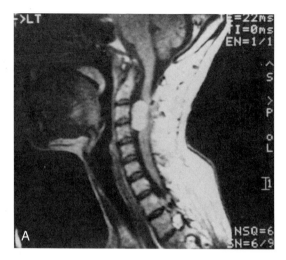

FIGURE 38–22. *A*, MRI showing the extent of a C2–C4 chondrosarcoma. *B*, Preoperatively, median SEPs show loss of P14 and all subsequent waveform to left stimulation, and attenuation of P14 to right stimulation. Posterior tibial SEPs demonstrate loss of subcortical components, and delay of the cortical P38.

Preoperative SEP

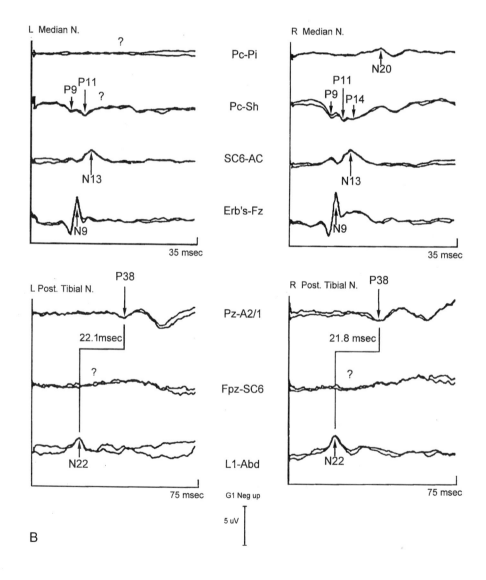

Postoperative SEP

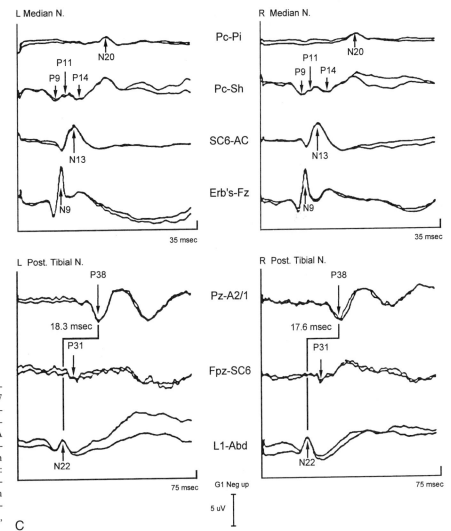

FIGURE 38–22 *Continued. C,* Normal median and posterior tibial SEPs recorded 7 months postoperatively. A bipolar derivation between SC6 and AC, an anterior cervical electrode, is used to record N13. A similar bipolar derivation using an abdominal electrode is used to record N22. (From Tinazzi M, Zanette G, Bonato C, et al: Neural generators of tibial nerve P30 somatosensory evoked potential studied in patients with a focal lesion of the cervicomedullary junction. Muscle Nerve 19:1540, 1996.)

Channel 1 records the afferent volley at Erb's point. Channel 2 records signals generated in the cervical spinal cord, the DCV and CERV N13. Channel 3 records subcortical far-field signals, most importantly P14 and N18. Channel 4 records N20 in isolation by employing a CPi reference to "subtract out" the underlying far-field signals. For clinical recording, the Erb's point electrode on the nonstimulated side serves as an effective noncephalic reference.

If additional channels are available, recording from an electrode at the superior margin of the thyroid cartilage may help to identify the CERV N13. A referential recording may demonstrate a synchronous, phase opposite, CERV P13. A bipolar SC5–"anterior cervical" recording may produce a higher amplitude signal by adding the out-of-phase CERV N13 and CERV P13.

The normal topographical variability of the cortical response to median nerve stimulation may occasionally make the N20 negativity difficult to identify.[64] In these cases, and in other situations in which anatomical abnormalities may distort the scalp topography of N20, the use of additional electrodes adjacent to the standard CP location may aid in N20 identification.

Some laboratories routinely substitute a bipolar SC5–Fp$_z$

derivation for channels 2 and 3. As discussed earlier, this practice should be discouraged. The signal recorded is a composite of the temporally overlapping stationary CERV N13 and the far-field P14 signals, making lesions that selectively affect either one difficult to detect. Nonetheless, in the patient who has not been sedated, or in whom muscle related artifact persists despite sedation, the use of an SC5–Fp$_z$ derivation may allow recording of a clinically useful signal of composite subcortical origin.

Posterior Tibial SEPs

A four-channel montage suitable for recording posterior tibial SEP is as follows:[45]

Channel 1	T12–Iliac crest
Channel 2	Fp$_z$–SC5
Channel 3	Cpi–Fp$_z$
Channel 4	CP$_z$*–Fp$_z$

Channel 1 registers the stationary lumbar potential, and may also record the afferent volley. Bipolar recording between closely spaced upper lumbar or lower thoracic elec-

*CP$_z$ is midway between C$_z$ and P$_z$.

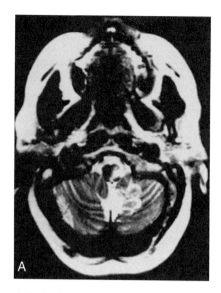

FIGURE 38–23. *A*, T2-weighted MRI images of an ischemia stroke in the left medulla. *B*, Right median and posterior tibial SEPs show preservation of P14 and P31 and loss of cortical responses and at least partial preservation of N18 and N34. Left-sided SEPs are normal. (From Tinazzi M, Zanette G, Bonato C, et al: Neural generators of tibial nerve P30 somatosensory evoked potential studied in patients with a focal lesion of the cervicomedullary junction. Muscle Nerve 19:1542, 1996.)

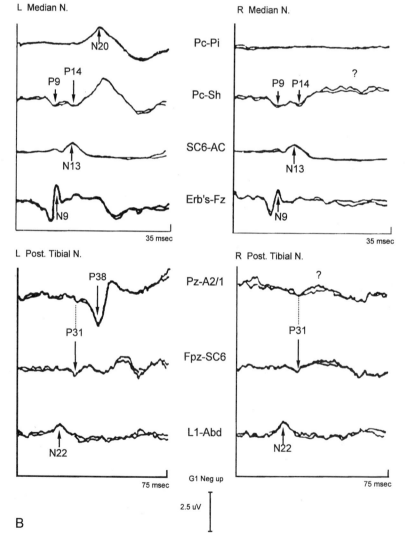

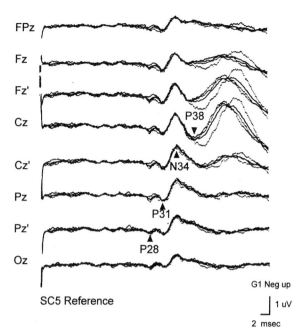

SC5 Reference

G1 Neg up

1 uV

2 msec

FIGURE 38–24. Normal posterior tibial nerve (PTN) SEP recorded referentially in the mid-sagittal plane. (From Emerson RG, Pedley TA: Somatosensory evoked potentials. In Daly DD, Pedley TA (eds): Current Practice of Clinical Encephalography, 2nd ed. New York, Raven Press, 1990, p 698.)

trodes is discouraged because in-phase cancellation attenuates the stationary lumbar potential.[65]

Channel 2 detects P31 and N34 far-field potentials. The choice of SC5 as a reference can be the source of some confusion. SC5 is used because it is less likely to be contaminated by artifact than alternative reference sites, such as the elbow or shoulder, and it is relatively inactive for PTN SEPs (see Fig. 38–20). The EPs recorded from the Fp$_z$–SC5 derivation following posterior tibial nerve stimulation consist al-

most entirely of far-field activity detected by the Fp$_z$ electrode.

Channels 3 and 4 are used to record the P38 primary cortical response. Two channels are recommended because in occasional subjects, P38 is recordable in only one of these two channels. The Fp$_z$ reference is used to "subtract out" the underlying far-field activity.

If additional channels are available, bipolar recording from the popliteal fossa is useful for detecting the peripheral afferent volley. This channel can be particularly helpful in the operating room for verifying the efficacy of stimulation in the pharmacologically paralyzed patient in whom SEPs have been lost. In standard laboratory recordings, adding one or more additional referential upper lumbar or lower thoracic channels may aid in identifying the N22. A contralateral central or frontal electrode referred to Fp$_z$ can be used to identify N37.

INTERPRETATIVE CRITERIA

Clinical interpretation of laboratory SEPs is primarily based on identification of the presence of the major component waveforms, and measurement of their absolute and interpeak latencies. Interpeak latency measures exceeding the mean for a normal control population by 2.5 or 3 standard deviations are usually considered abnormal. Relevant statistical considerations are discussed in the applicable guidelines of the American Clinical Neurophysiology Society.[45]

Interpeak latency measurements, made with respect to Erb's point for median SEP and the stationary lumbar potential for posterior tibial SEPs, are used to eliminate the effects of peripheral conduction times. For median SEPs, the Erb's point to N20 interpeak latency provides an overall assessment of conduction between the brachial plexus and the cortex. Separate measurement of Erb's point to P14, and P14 to N20 allow evaluation of conduction between the brachial plexus and the lower brain stem, and between the lower brain stem and cortex, respectively. For posterior tibial

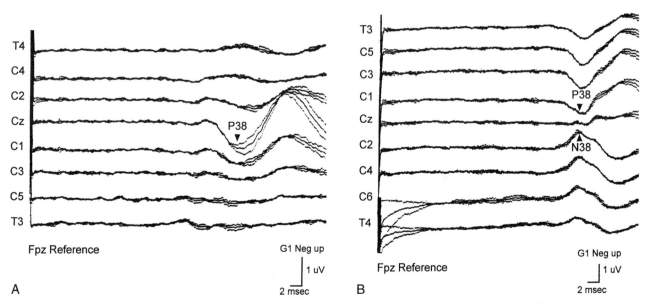

A Fpz Reference

G1 Neg up

1 uV

2 msec

B Fpz Reference

G1 Neg up

1 uV

2 msec

FIGURE 38–25. Left PTN SEP recorded in two different normal subjects, illustrating the range of normal variability of the scalp topography of the N38 and P38 cortical responses. (From Emerson RG, Pedley TA: Somatosensory evoked potentials. In Daly DD, Pedley TA (eds): Current Practice of Clinical Encephalography, 2nd ed. New York, Raven Press, 1990, p 699.)

SEPs, the lumbar to P38 interpeak latency provides a measurement of conduction between lumbar cord and cortex. N22 to P31 and P31 to P38 interpeak latencies provide separate measures of conduction between lumbar cord and lower brain stem, and lower brain stem and cortex.

Interpretation of median nerve SEPs also involves assessment of the presence of the CERV N13 and the far-field N18 signal. Although these signals are not generally used for calculation of interpeak latencies, they are still important features of the normal median nerve SEP. Similarly, N34 in a "required" feature of a normal PTN SEP.

The P14 to N20 and P31 to P38 interpeak latencies are sometimes referred to as "central conduction times."[48] Although this concept is useful, it is important to realize that these interpeak latencies measure more than simply conduction times in the central somatosensory pathways. They also measure the time required for generation of the peak N20 or P38 voltage following activation of somatosensory cortex.

An EP signal should not be considered absent when the noise level of a recording makes it difficult to detect. Although the P31 and N34 components of the PTN SEP are invariably present in normal subjects, and are easily recorded in anesthetized and pharmacologically paralyzed patients, they may be difficult or impossible to record in the waking or lightly sedated patient. A signal should not be judged abnormal because of poor reproducibility. This finding more likely reflects technical limitations of the recording. An absent signal will be reproducibly absent.

Standard interpretive criteria for the diagnostic laboratory rely largely on latency and interpeak latency measurements. Interpretation of intraoperative SEPs, on the other hand, relies much more heavily on assessment of waveform morphology and amplitude. Intraoperative injury commonly produces more prominent loss of SEP signal amplitude and deterioration of morphology than latency prolongation.[38, 86] Cortical SEP components can be substantially attenuated by general anesthetic agents; subcortical far-field signals are generally more appropriate for intraoperative monitoring of spinal cord and brain stem function.[10, 88, 95] Although interindividual variability of morphology and amplitude limits the utility of these features for assessing normality, they are appropriately used for intraoperative monitoring, where the patient serves as his or her own baseline.

REFERENCES

1. Allison T, Goff WR, Williamson PD, et al: On the neural origin of early components of the human somatosensory evoked potential. In Desmedt, JE (ed): Clinical Uses of Cerebral, Brainstem and Spinal Somatosensory Evoked Potentials. Basel, Karger, 1980, p 51.
2. Allison T, Hume AL: A comparative analysis of short-latency somatosensory evoked potentials in man, monkey, cat and rat. Exp Neurol 72:592, 1981.
3. Arezzo J, Legatt AD, Vaughn HG: Togography and intracranial sources of somatosensory evoked potentials in the monkey. I. Early components. Electroencephalogr Clin Neurophysiol 46:155, 1979.
4. Austin GM, McCouch GP: Presynaptic components of intermedullary cord potential. J Neurophysiol 18:441, 1955.
5. Barson AJ: The vertebral level of termination of the spinal cord during normal and abnormal development. J Anat 106:489, 1970.
6. Beall JE, Applebaum AE, Foremann RD, et al: Spinal cord potentials evoked by cutaneous afferents in the monkey. J Neurophysiol 40:199, 1977.
7. Brazier MAB: A study of the electrical field at the surface of the head. Electroencephalogr Clin Neurophysiol Suppl 2:38, 1949.
8. Brin MF, Pedley TA, Lovelace RE, et al: Electrophysiological features of abetalipoproteinemia: Functional consequences of vitamin E deficiency. Neurology 36:669, 1986.
9. Broughton RJ: Discussion. In Donchin E, DB Lindsley (eds): Average Evoked Potentials. Washington, DC, US Government Printing Office, 1969, p 79.
10. Browning JL, Heizer ML, Baskin DS: Variations in corticomotor and somatosensory evoked potentials: Effects of temperature, halothane anesthesia, and arterial partial pressure of CO_2. Anesth Analg 74(5):643, 1992.
11. Burke D, Gandevia SC: Muscle afferent contribution to the cerebral potentials of human subjects. In Cracco RQ, Bodis-Wollner I (eds): Evoked Potentials. New York, Alan R Liss, 1986, p 262.
12. Chiappa KH, Hill RA: Short latency-somatosensory evoked potentials: interpretation. In Evoked Potentials in Clinical Medicine, 3rd ed. New York, Lippincott-Raven, 1997, p 342.
13. Chiappa KH: Pattern shift visual, brainstem auditory and short latency somatosensory evoked potentials in multiple sclerosis. Neurology 30:110, 1980.
14. Chiappa KH, Choi SK, Young RR: Short-latency somatosensory evoked potentials following median nerve stimulation in patients with neurological lesions. In Desmedt JE (ed): Clinical Uses of Cerebral, Brainstem and Spinal Somatosensory Evoked Potentials. Basel, Karger, 1980, p 254.
15. Coombs JS, Curtis DR, Laudren S: Spinal cord potentials generated by impulses in muscle and cutaneous afferent fibers. J Neurophysiol 19:452, 1956.
16. Cracco RQ, Cracco JB: Somatosensory evoked potentials in man: Far-field potentials. Electroencephalogr Clin Neurophysiol 41:460, 1976.
17. Cracco RQ, Evans B: Spinal evoked potentials in the cat: Effects of asphyxia, strychnine, cord section and compression. Electroencephalogr Clin Neurophysiol 44:187, 1978.
18. Cruse R, Klem G, Lesser RP, et al: Paradoxical lateralization of the cortical potentials evoked by stimulus of the posterior tibial nerve. Arch Neurol 39:222, 1982.
19. Deiber MP, Giard MH, Mauguiere F: Separate generators with distinct orientations for N20 and P22 somatosensory evoked potentials to finger stimulation. Electroencephalogr Clin Neurophysiol 65:321, 1986.
20. DeMeirleir LJ, Taylor MJ, Logan WJ: Multimodal evoked potential studies in leukodystrophies of children. Can J Neurol Sci 15:26, 1988.
21. Desmedt JE, Bourguet M: Color imaging of parietal and frontal somatosensory potential fields evoked by stimulation of median or posterior tibial nerve in man. Electroencephalogr Clin Neurophysiol 62:1, 1985.
22. Desmedt JE, Cheron G: Central somatosensory conduction in man: Neural generators and interpeak latencies of the far-field components recorded from neck and right or left scalp and earlobes. Electroencephalogr Clin Neurophysiol 50:382, 1980.
23. Desmedt JE, Cheron G: Non-cephalic reference recording of early somatosensory potentials to finger stimulation in adult or aging normal man: Differentiation of widespread N18 and contralateral N20 from prerolandic P22 and N30 components. Electroencephalogr Clin Neurophysiol 52:553, 1981.
24. Desmedt JE, Cheron G: Prevertebral (esophageal) recording of subcortical somatosensory evoked potentials in man: The spinal P13 component and the dural nature of the spinal generators. Electroencephalogr Clin Neurophysiol 52:257, 1981.
25. Desmedt JE, Nguyen TH, Bourguet M: Bit-mapped color imaging of human evoked potentials with reference to the N20, P22, P27 and N30 somatosensory responses. Electroencephalogr Clin Neurophysiol 68:1, 1987.
26. Desmedt JE, Nguyen TH, Carmeliet J: Unexpected latency shift of the stationary P9 somatosensory evoked potential far field with changes in shoulder position. Electroencephalogr Clin Neurophysiol 56:628, 1983.
27. Dumitru D, King JC: Far-field potentials produced by quadrupole generators in cylindrical volume conductors. Electroencephalogr Clin Neurophysiol 88:421, 1993.
28. Eccles JC: Interpretation of action potentials evoked in the cerebral cortex. Electroencephalogr Clin Neurophysiol 3:449, 1951.
29. Eisen A, Roberts K, Low M, et al: Questions regarding the sequential neural generator theory of the somatosensory evoked potential raised by digital filtering. Electroencephalogr Clin Neurophysiol 59:388, 1984.

30. Emerson RG: The anatomic and physiologic bases of posterior tibial nerve somatosensory evoked potentials. In Gilmore R (ed): Neurologic Clinics of North America. Philadelphia, WB Saunders, 1988, p 735.

31. Emerson RG, Pedley TA: Generator sources of median somatosensory evoked potentials. J Clin Neurophysiol 2:203, 1984.

32. Emerson RG, Pedley TA: Effect of cervical spinal cord lesions on early components of the median nerve somatosensory evoked potential. Neurology 36:20, 1986.

33. Emerson RG, Seyal M, Pedley TA: Somatosensory evoked potentials following median nerve stimulation. I. The cervical components. Brain 107:169, 1984.

34. Emerson RG, Sgro JA, Pedley TA: Identification of a state-dependent Pre-N20 "near-field" positivity in the human median nerve SEP. Neurology 37 (suppl 1):359, 1987.

35. Emerson RG, Sgro JA, Pedley TA, et al: State-dependent changes in the N20 component of the median nerve somatosensory evoked potential. Neurology 38:64, 1988.

36. Fernandez de Molinda A, Gray JAB: Activity in the dorsal spinal grey matter after stimulation of cutaneous nerves. J Physiol 137:126, 1957.

37. Fierro B, Meli F, Brighina F, et al: Somatosensory and visual evoked potentials in young insulin-dependent diabetic patients. Electromyogr Clin Neurophysiol 36:481, 1996.

38. Forbes HJ, Allen PW, Waller CS, et al: Spinal cord monitoring in scoliosis surgery. J Bone Joint Surgery 73-B:487, 1991.

39. Gandolfo G, Arnaud C, Gottesmann C: Transmission processes in the ventrobasal complex of rat during the sleep-waking cycle. Brain Res Bull 5:553, 1980.

40. Gandolfo G, Gottesmann C: Transmission in the ventrobasal complex of thalamus during rapid sleep and wakefullness in the homolaterally neodecorticated rat. Acta Neurobiol Exp 42:443, 1982.

41. Garg BP, Markand ON, DeMyer WE, et al: Evoked response studies in patients with adrenoleukodystrophy and heterozygous relatives. Arch Neurol 40:356, 1983.

42. Georgesco M, Salerno A, Camu W: Somatosensory evoked potentials elicited by stimulation of lower-limb nerves in amyotrophic lateral sclerosis. Electroencephalogr Clin Neurophysiol 104:333, 1997.

43. Goff WR, Allison T, Vaughn HG: Functional neuroanatomy of event-related potentials. In Callaway E, Tueting P, Koslow SH (eds): Event Related Potentials in Man. New York, Adademic Press, 1978, p 1.

44. Gott PS, Karnaze DS: Short-latency somatosensory evoked potentials in myotonic dystrophy: Evidence for a conduction disturbance. Electroencephalogr Clin Neurophysiol 62:455, 1985.

45. Guidelines: American Electroencephalographic Society guideline nine: Guidelines on evoked potentials. J Clin Neurophysiol 11:40, 1994.

46. Guidelines: American Electrographic Society guidelines for standard electrode position nomenclature. J Clin Neurophysiol 8:200, 1991.

47. Halonen JP, Jones SJ, Edgar MA, et al: Multi-level epidural recordings of spinal SEPs during scoliosis surgery. Electroencephalogr Clin Neurophysiol Suppl 41:342, 1990.

48. Hume AL, Cant BR: Conduction time in central somatosensory pathways in man. Electroencephalogr Clin Neurophysiol 45:361, 1978.

49. Iragui VJ, Kalmijn J, Thal LJ, et al: Neurological dysfunction in asymptomatic HIV-1 infected men: Evidence from evoked potentials. Electroencephalogr Clin Neurophysiol 92:1, 1994.

50. Jewett DL, Williston JS: Auditory-evoked far fields averaged from the scalp of humans. Brain 94:681, 1971.

51. Jones SJ: Short latency potentials recorded from the neck and scalp in man. Electroencephalogr Clin Neurophysiol 43:853, 1977.

52. Jones SJ, Edgar MA, Ransford AO: Sensory nerve conduction in human spinal cord: Epidural recordings made during scoliosis surgery. J Neurol Neurosurg Psychiatry 45:466, 1982.

53. Jones SJ, Yu YL, Rudge P, et al: Central and peripheral SEP defects in neurologically symptomatic and asymptomatic subjects with low vitamin B12 levels. J Neurol Sci 82:55, 1987.

54. Kaji R, Sumner AJ: Vector short-latency somatosensory evoked potentials after median nerve stimulation. Muscle Nerve 13:1174, 1990.

55. Kakigi R, Shibasaki H: Scalp topography of the short latency somatosensory evoked potential following posterior tibial nerve stimulation in man. Electroencephalogr Clin Neurophysiol 56:430, 1983.

56. Kaplan PW, Rawal K, Erwin CW, et al: Visual and somatosensory evoked potentials in vitamin E deficiency with cystic fibrosis. Electroencephalogr Clin Neurophysiol 71:266, 1988.

57. Kaplan PW, Tusa RJ, Rignani J, et al: Somatosensory evoked potentials in adrenomyeopathy. Neurology 48:1662, 1997.

58. Kimura J, Ishida T, Suzuki S, et al: Far field recording of the junctional potential generated by median nerve volleys at the wrist. Neurology 36:1451, 1986.

59. Kimura J, Kimura A, Ishida T, et al: What determines the latency and amplitude of stationary peaks in far-field recordings. Ann Neurol 19:479, 1986.

60. Kimura J, Kimura A, Machida M, et al: Model for far-field recordings of SEP. In Cracco RQ, Bodis-Wollner I (eds): Evoked Potentials. New York, Alan R Liss, 1986.

61. Kimura J, Mitsudome A, Beck DO, et al: Field distribution of antidromically activated digital nerve potentials: Model for far field recording. Neurology 33:1164, 1983.

62. Kimura J, Mitsudome A, Yamada T, et al: Stationary peaks from a moving source in far-field recording. Electroencephalogr Clin Neurophysiol 58:351, 1984.

63. Klee M, Rall W: Computed potentials of cortically arranged populations of neurons. J Neurophysiol 40:647, 1977.

64. Legatt AD, Emerson RG, Labar DR, et al: Surface near-field mapping of the median nerve SEP N20 component. Neurology 37(suppl 1):366, 1987.

65. Legatt AD, Emerson RG, Pedley TA: Use of the stationary lumbar potential increases the diagnostic yield of posterior tibial nerve somatosensory evoked potentials. Electroencephalogr Clin Neurophysiol 64:72P, 1986.

66. Lesser RP, Lueders H, Dinner DS, et al: The source of 'paradoxical lateralization' of cortical evoked potentials to posterior tibial nerve stimulation. Neurology 37:82, 1987.

67. Lorente de No RA: Action potentials of motorneurons of the hypoglossus nucleus. J Cell Comp Physiol 29:207, 1947.

68. Lorente de No RA: A study of nerve physiology. Stud Rockefeller Inst 132:384, 1947.

69. Lueders H, Lesser RP, Hahn J, et al: Cortical somatosensory evoked potentials in response to hand stimulation. J Neurosurg 58:885, 1983.

70. Lueders H, Lesser R, Hahn J, et al: Subcortical sensory evoked potentials to median nerve stimulation. Brain 106:341, 1983.

71. Maccabee PJ, Pickhasov EL, Cracco RQ: Short latency somatosensory evoked potentials to median nerve stimulation: Effect of low frequency filter. Electroencephalogr Clin Neurophysiol 55:34, 1983.

72. Maetzu C, Villoslada C, Cruz Martinez A: Somatosensory evoked potentials and central motor pathway conduction after magnetic stimulation of the brain in diabetes. Electromyogr Clin Neurophysiol 35:443, 1995.

73. Markand ON, Garg BP, DeMyer WE, et al: Brain stem auditory, visual and somatosensory evoked potentials in leukodystrophies. Electroencephalogr Clin Neurophysiol 54:39, 1982.

74. Mauguiere F, Courjon J: The origins of short-latency somatosensory evoked potential in humans. Ann Neurol 9:607, 1981.

75. Mauguiere F, Desmedt JE: Bilateral somatosensory evoked potentials in four patients with long-standing surgical hemispherectomy. Ann Neurol 26:724, 1989.

76. Mauguiere F, Desmedt JE: Focal capsular vascular lesions can selectively deafferent the prerolandic of the parietal cortex: Somatosensory evoked potentials evidence. Ann Neurol 30:71, 1991.

77. Mauguiere F, Desmedt JE, Courjon J: Astereognosis and dissociated loss of frontal or parietal components of somatosensory evoked potentials in hemispheric lesions. Brain 106:271, 1983.

78. Mauguiere F, Desmedt JE, Courjon J: Neural generators of N18 and P14 far-field somatosensory evoked potentials studied in patients with lesions of thalamus or thalamo-cortical radiations. Electroencephalogr Clin Neurophysiol 56:283, 1983.

79. McAllister RH, Herns MV, Harrison MJ, et al: Neurological and neuropsychological performance in HIV seropositive men without symptoms. J Neurol Neurosurg Psychiatry 55:143, 1992.

80. Nagle KJ, Emerson RG, Adams DC, et al: Intraoperative motor

and somatosensory evoked potential monitoring: A review of 116 cases. Neurology 47:999, 1996.

81. Nakanishi T: Action potentials recorded by fluid electrodes. Electroencephalogr Clin Neurophysiol 53:343, 1982.

82. Nakanishi T: Origin of action potentials recorded by fluid electrodes. Electroencephalogr Clin Neurophysiol 55:114, 1983.

83. Nakanishi T, Tamaki M, Arasaki K, et al: Origins of the scalp-recorded far field potentials in man and cat. Electroencephalogr Clin Neurophysiol Suppl 36:336, 1982.

84. Noachtar S, Luders HO, Dinner DS, et al: Ipsilateral median somatosensory evoked potentials recorded from human somatosensory cortex. Electroencephalogr Clin Neurophysiol 104:189, 1997.

85. Nuwer RM, Perlman SL, Packwood JW, et al: Evoked potential abnormalities in the various inherited ataxias. Ann Neurol 13:20, 1983.

86. O'Brien MF, Lenke LG, Bridwell KH, et al: Evoked potentials monitoring of upper extremities during thoracic and lumbar spinal deformity surgery: A prospective study. J Spinal Disord 7:277, 1994.

87. Penfield W, Rasmussen T: The Cerebral Cortex of Man: A Clinical Study of Localization of Function. New York, MacMillan, 1950.

88. Perlik SJ, VanEgeren R, Fisher MA: Somatosensory evoked potential surgical monitoring. Observation during combined isoflurane-nitrous oxide anesthesia. Spine 17(3):273, 1992.

89. Peterson NN, Schroeder CE, Arezzo JC: Neural generators of early cortical somatosensory evoked potentials in the awake monkey. Electroencephalogr Clin Neurophysiol 96:248, 1995.

90. Raroque HG, Batjer H, White C, et al: Lower brain-stem origin of the median nerve N18 potential. Electroencephalogr Clin Neurophysiol 90:170, 1994.

91. Restuccia D, Di Lazarro V, Valeriani M, et al: Origin and distribution of P13 and P14 far-field potentials after median nerve stimulation. Scalp, nasopharyngeal and neck recording in healthy subjects and in patients with cervical and cervicomedullary lesions. Electroencephalogr Clin Neurophysiol 96:371, 1995.

92. Restuccia D, Di Lazarro V, Valeriani M, et al: Brain-stem somatosensory dysfunction in a case of long-standing left hemispherectomy with removal of the left thalamus: A nasopharyngeal and scalp study. Electroencephalogr Clin Neurophysiol 100:184, 1996.

93. Robinson BW, Bryan JS, Rosvold HE: Locating brain structures, extensions to the impedance method. Arch Neurol 13:477, 1965.

94. Rossini PM, Gigli GL, Marciani MG, et al: Non-invasive evaluation of input-output characteristics of sensorimotor cerebral areas in healthy humans. Electroencephalogr Clin Neurophysiol 68:88, 1987.

95. Sebel PS, Erwin CW, Neville WK: Effects of halothane and enflurane on far- and near-field somatosensory evoked potentials. Br J Anaesth 59:1492, 1987.

96. Seyal M, Emerson RG, Pedley TA: Spinal and early scalp-recorded components of the somatosensory evoked potential following stimulation of the posterior tibial nerve. Electroencephalogr Clin Neurophysiol 55:320, 1983.

97. Seyal M, Gabor AJ: The human posterior tibial somatosensory evoked potential: Synapse dependent and synapse independent spinal components. Electroencephalogr Clin Neurophysiol 62:323, 1985.

98. Seyal M, Ortstadt JL, Kraft LW, et al: Effects of movement on human spinal and subcortical potentials. Neurology 37:650, 1987.

99. Sgro JA, Emerson RG, Pedley TA: State dependent non-stationarity of the P38 cortical response following posterior tibial nerve stimulation. Electroencephalogr Clin Neurophysiol 69:77P, 1988.

100. Sonoo M, Genba-Shimizu K, Mannen T, et al: Detailed analysis of the latencies of median nerve somatosensory evoked potential components, 2: Analysis of subcomponents of P13/14 and N20 potentials. Electroencephalogr Clin Neurophysiol 104:296, 1997.

101. Sonoo M, Hagiwara H, Motoyoshi Y, et al: Preserved wide-spread N18 and progressive loss of P13/P14 of median nerve SEPs in a patient with unilateral medial medullary syndrome. Electroencephalogr Clin Neurophysiol 100:488, 1996.

102. Sonoo M, Kobayashi M, Genba-Shimizu K, et al: Detailed analysis of the latencies of median nerve somatosensory evoked potential components, 1: Selection of the best standard parameters and the establishment of normal values. Electroencephalogr Clin Neurophysiol 100:319, 1996.

103. Steriade M, Iosif G, Apostol V: Responsiveness of thalamic and cortical motor relays during arousal and various stages of sleep. J Neurophysiol 32:251, 1969.

104. Takakura H, Nakano C, Kasagi S, et al: Multimodality evoked potentials in progression of metachromatic leukodystrophy. Brain Dev 7:424, 1985.

105. Thompson DS, Woodward JB, Ringel SP, et al: Evoked potential abnormalities in myotonic dystrophy. Electroencephalogr Clin Neurophysiol 56:453, 1983.

106. Tinazzi M, Zanette G, Bonato C, et al: Neural generators of tibial nerve P30 somatosensory evoked potential studied in patients with a focal lesion of the cervicomedullary junction. Muscle Nerve 19:1538, 1996.

107. Tinazzi M, Zanette G, Fiaschi A, et al: Effects of stimulus rate on the cortical posterior tibial nerve SEPs: A topographic study. Electroencephalogr Clin Neurophysiol 100:210, 1996.

108. Tinazzi M, Zanette G, La Porta F, et al: Selective gating of lower limb cortical somatosensory evoked potentials (SEPs) during passive and active foot movements. Electroencephalogr Clin Neurophysiol 104:312, 1997.

109. Tobimatsu S, Fukui R, Kato M, et al: Multimodality evoked potentials in patients and carriers with adrenoleukodystrophy and adrenomyeloneuropathy. Electroencephalogr Clin Neurophysiol 62:18, 1985.

110. Towe AL: On the nature of the primary evoked response. Exp Neurol 15:113, 1966.

111. Urasaki E, Tokimura T, Yasukouchi H, et al: P30 and N33 of posterior tibial nerve SSEPs are analogous to P14 and N18 of median nerve SSEPs. Electroencephalogr Clin Neurophysiology 88:525, 1993.

112. Urasaki E, Uematsu S, Lesser RP: Short latency somatosensory evoked potentials around the human upper brain-stem. Electroencephalogr Clin Neurophysiol 88:92, 1993.

113. Urasaki E, Wada S, Kadoya C, et al: Absence of spinal N13-P13 and normal scalp far-field P14 in a patient with syringomyelia. Electroencephalogr Clin Neurophysiol 71:400, 1988.

114. Urasaki E, Wada S, Kadoya C, et al: Origin of the scalp far-field N18 of SSEPs in response to median nerve stimulation. Electroencephalogr Clin Neurophysiol 77:39, 1990.

115. Urasaki E, Wada S, Kadoya C, et al: Amplitude abnormalities in the scalp far-field N18 of SSEPs to median nerve stimulation in patients with midbrain-pontine lesion. Electroencephalogr Clin Neurophysiol 84:232, 1992.

116. Valeriani M, Restuccia D, Di Lazzaro V, et al: Giant central N20-P22 with normal area 3B N20-P20: An argument in favour of an area 3a generator of early median nerve cortical SEPs. Electroencephalogr Clin Neurophysiol 104:60, 1997.

117. Woodbury WJ: Potentials in a volume conductor. In Ruch TC, et al. (eds): Neurophysiology, 2nd ed. Philadelphia, WB Saunders, 1965, p 85.

118. Wulff CH, Trojaborg W: Adult metachromatic leukodystrophy: Neurophysiologic findings. Neurology 35:1776, 1985.

119. Yamada T, Ishida T, Kudo Y, et al: Clinical correlates of abnormal P14 in median SEPs. Neurology 36:765, 1986.

120. Yamada T, Kameyama S, Fuchigami Y, et al: Changes of short latency somatosensory evoked potential during sleep. Electroencephalogr Clin Neurophysiol 70:126, 1988.

121. Yamada T, Kimura J, Nitz DM: Short latency somatosensory potentials following median nerve stimulation in man. Electroencephalogr Clin Neurophysiol 48:367, 1980.

122. Yamada T, Machida M, Oishi M, et al: Stationary negative potentials near the source vs. positive far-field potentials at a distance. Electroencephalogr Clin Neurophysiol 60:509, 1985.

123. Zanette G, Tinazzi M, Polo A, et al: Motor neuron disease with pyramidal tract dysfunction involves the cortical generators of the early somatosensory evoked potentials to tibial nerve stimulation. Neurology 47:932, 1996.

Chapter 39

Intraoperative Monitoring

C. Michel Harper, MD

INTRODUCTION
INTRAOPERATIVE MONITORING
  TECHNIQUES
Auditory Evoked Potentials (AEPs)
Somatosensory Evoked Potentials
Motor Evoked Potentials
Electromyography
Compound Muscle and Nerve Action
  Potentials

APPLICATIONS OF INTRAOPERATIVE
  MONITORING
Cerebral Surgery
Spine Surgery
Cranial Nerve Surgery
Peripheral Nerve Surgery

INTRODUCTION

The purpose of intraoperative monitoring is to preserve function and prevent injury to the nervous system at a time when clinical examination is not possible. Systems currently in use produce reliable recordings that reflect the function of specific pathways in the nervous system. The electrophysiological potentials used for intraoperative monitoring are inherently variable and are affected by age, neurological disease, and a variety of technical factors (anesthetics, temperature, blood pressure, body position, electrical noise, and muscle artifact). These variables must be taken into account when determining the significance of any changes in the recordings that develop during surgery. The occurrence of a reversible change allows appropriate steps to be taken to prevent permanent damage to neural structures. Recognition of an irreversible change predicts the nature and severity of postoperative deficits and provides potential benefits for future patients when information about the mechanism of neurological damage is supplied.

Damage to neural structures during surgery can occur rapidly and may be irreversible. Methods that warn of impending injury before damage actually occurs are most desirable. If unrelated to technical problems, transient changes in monitored signals should not be considered to be "false-positives" because many, if not all, would result in loss of neurological function if the change failed to resolve. The rate of change in neural function will vary with the type and severity of damage. Some injuries occur abruptly whereas others develop gradually over minutes or hours. Abnormalities can occur at a variety of times in relation to the surgical procedure. Some are seen immediately after an injury; others, particularly those resulting from mild compression or stretch, may be delayed for several hours. Monitoring must therefore be carried out throughout the surgical procedure, even after a so-called critical period has passed.

As with most other techniques in medicine, surgical monitoring has its limitations. The most important of these is the rare occurrence of a postoperative deficit despite the absence of any intraoperative change in electrophysiological measurements. When technical causes are excluded, these so-called false-negatives occur when damage is confined to pathways that are not being monitored directly. Monitoring all modalities and all levels of the nervous system that are at risk during a particular surgical procedure can minimize this occurrence.

Many different electrophysiological techniques are available to monitor different levels of the nervous system (Table 39–1). The cerebral cortex can be assessed with somatosensory evoked potentials (SEPs) and electroencephalography (EEG). The nuclei of the basal ganglia and thalamus can be localized using depth electrode recordings of spontaneous or evoked activity during stereotactic surgery. Brain stem auditory evoked potentials (BAEPs), electromyography (EMG), compound muscle action potentials (CMAPs), and nerve action potentials (NAPs) are used to monitor the brain stem and cranial nerves. Attempts have been made to use visual evoked potentials (VEPs) during surgery in the region of the optic pathways, but signal variability caused by susceptibility to anesthetics and the need to use a flash stimulus has limited the effectiveness of VEP as an intraoperative monitoring tool. The spinal cord, including the anterior horn cells and spinal roots, can be monitored with motor evoked potentials (MEPs), SEP, EMG, H reflexes, and F waves. Peripheral nerves are monitored with EMG, CMAP, and NAP recordings.

A coordinated effort by a team made up of the clinical neurophysiologist, anesthesiologist, and surgeon is necessary to ensure development of an optimal monitoring protocol tailored to each patient. Methods are selected that minimize interference with the surgical procedure. Equipment should be capable of recording, storing, and displaying different types of recording modalities (e.g., evoked potentials and continuous EMG) simultaneously. Careful attention must be paid to electrical safety as patients are attached to multiple electrical instruments, grounded at multiple locations, and have intravenous and other catheters or electrodes in place that increase the risk of delivering current directly to the heart during surgery.

Despite some limitations, there is a growing body of data that supports the overall accuracy and utility of intraoperative monitoring. With the exception of VEPs and EEG, each of the major modalities currently used is described followed by a discussion of monitoring during different types of surgical procedures.

INTRAOPERATIVE MONITORING TECHNIQUES

Auditory Evoked Potentials (AEPs)

AEPs reflect activity in the cochlea, auditory nerve, cochlear nuclei in the medulla, and dorsolateral regions of the

TABLE 39–1	Intraoperative Monitoring Techniques	
Level	**Technique**	**Surgery**
Cerebral cortex	SEP, EEG, direct stimulation	Tumor, epilepsy
Basal ganglia, thalamus	SEP, VEP, direct stimulation	Pallidotomy, thalamotomy
Brain stem, cranial nerves	SEP, VEP, AEP, EMG, CMAP, NAP	Intra- or extra-axial tumors, AVM, MVD of cranial nerves, skull base, head and neck surgery
Spinal cord, nerve roots	SEP, MEP, EMG, CMAP, NAP, H reflex, F wave	Correction of spinal deformity, decompression or fusion for degenerative disease, tumors, congenital malformations, AVM, thoracoabdominal aneurysm repair
Plexus, peripheral nerve	SEP, EGM, CMAP, NAP	Primary or metastatic tumors, trauma, entrapment

AEP, auditory evoked potential; AVM, arteriovenous malformation; CMAP, compound muscle action potential; EMG, electromyography; MEP, motor evoked potential; MVD, microvascular decompression; NAP, nerve action potential; SEP, somatosensory evoked potential; VEP, visual evoked potential.

pons and midbrain.[14] AEP recordings are monitored primarily during surgical procedures that risk injury to the eighth cranial nerve. Brain stem injury may also produce changes in the BAEP, but only if the dorsolateral areas of the rostral medulla, pons, and midbrain are affected. Therefore, BAEP should be combined with SEP, EMG, and, if possible, MEP, when the brain stem is at risk.

Auditory Evoked Potential Techniques

The BAEP, electrocholeography (ECoG), and auditory NAP are the major techniques used to monitor auditory nerve function intraoperatively. Stimulation hardware may differ slightly, but stimulation parameters are similar for all three techniques. Several types of ear inserts are available to deliver the stimulus. The best of these incorporates an external ear canal recording electrode into the earphone, which improves the resolution of wave I of the BAEP. The external ear canal is sealed with bone wax to provide a watertight seal and reduce the introduction of ambient noise into the external ear canal.

Square wave clicks of 100 to 200 µsec duration provide the most reliable stimulus for intraoperative AEP recordings. Stimulus artifact is eliminated when a short polyvinyl tube is placed between the transducer and the external ear canal electrode. This introduces a fixed delay in the latency of all BAEP waves, but does not affect interpeak latencies or amplitude. Use of alternating click polarity also reduces stimulus artifact, but this cannot be used when recording the cochlear microphonics potential of the ECoG. Stimulus intensities of 60 to 80 dBSL are delivered to the monitored ear while white masking noise, 40 dB below that intensity, is applied to the contralateral ear. Optimal definition of the BAEP with rapid signal acquisition is accomplished with stimulation rates between 10 and 30 Hz. Stimulus rates that are even fractions of 60 Hz may result in synchronization of the stimulus with 60-cycle noise and should therefore be avoided.

Brain Stem Auditory Evoked Potentials. With minor modifications, standard BAEP recording techniques can be applied to surgical monitoring. Recording with surface or needle electrodes from the ipsilateral ear (A_1) or mastoid (M_1) referenced to the vertex (C_z) requires averaging of 500 to 1000 stimuli for clear definition of waves I through V. Wave I is generated in the most peripheral (lateral) portion of the

cochlear nerve within the auditory canal.[14, 105] With surface electrodes it can be recorded as a strong near-field negativity at and around the ipsilateral ear. Systems that utilize an electrode in the external ear canal frequently allow better definition of wave I because the electrode is closer to the generator source in the lateral portion of the auditory nerve (Fig. 39–1). Filter settings of 30 to 3000 Hz are optimal although restriction to 150 to 1500 Hz may be required to reduce 60-Hz and high-frequency artifact from muscle activity. If available, digital filtering may improve the signal-to-noise ratio without introducing latency shifts of the major peaks. When digital filters are used, it should be kept in mind that excessive filtering of any kind distorts BAEP morphology. An analysis time of 10 to 15 msec with a sampling interval of 20 to 100 µsec is needed for adequate definition of BAEP waveforms.

Electrocochleography. This technique requires placement of a needle electrode through the tympanic membrane into the soft tissue covering the bony promontory of the middle ear.[92] With a reference needle electrode in the skin of the pinna or over the mastoid, near-field potentials can be recorded from the cochlear membrane (CM potential) and the lateral segment of the auditory nerve (N_1 potential—analogous to wave I of the BAEP). Stimulus and recording parameters are similar to those described for the BAEP with the important exception that ECoG requires averaging of only 20 to 50 stimuli, thereby providing the surgeon with more rapid feedback.

Auditory Nerve Action Potential. The auditory NAP is recorded with an electrode placed onto the auditory nerve at or near the brain stem in the surgical field. A thin malleable wire with a small piece of cotton sutured onto a 2-mm segment of exposed tip or an electrode attached to the cerebellar retractor works equally well for recording the NAP.[80] A needle electrode in the subcutaneous tissue of the wound or a surface electrode on the scalp is used as a reference. Stimulus parameters are identical to those used for the BAEP and ECoG. The NAP is large (10 to 30 µV) compared to the BAEP and can be recorded with minimal or no averaging (Fig. 39–2). The amplitude of the NAP is variable unless spinal fluid is drained away from the nerve.

Comparison of Techniques

Each of the three major techniques of AEP monitoring has advantages and limitations. BAEP recordings are simple, noninvasive, can be used to monitor auditory function during the entire surgical procedure, and monitor the entire segment of the auditory system that is at risk during posterior fossa surgery. The BAEP is a sensitive indicator of damage to the auditory pathway, as preservation of wave V of the

BAEP: Acoustic Neuroma Surgery

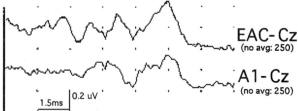

EAC-Cz (no avg: 250)

A1-Cz (no avg: 250)

1.5ms   0.2 uV

FIGURE 39–1. Brain stem auditory evoked potential (BAEP) recorded during surgery in a patient with an acoustic neuroma. Wave I of the BAEP is better defined with an external ear canal (EAC) electrode than with an electrode on the earlobe (A1). (Harper CM, Daube JR: Facial nerve EMG and other cranial nerve monitoring. J Clin Neurophysiol, 15:206–216, 1998.)

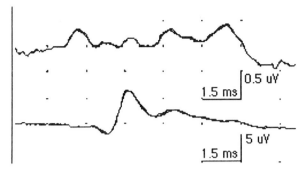

FIGURE 39–2. Brain stem auditory evoked potential (BAEP) and auditory nerve action potential (NAP) recorded simultaneously during surgery in a patient with an acoustic neuroma. The BAEP is recorded with an external ear canal electrode referenced to C_z (500 responses averaged at a stimulus rate of 10.1 Hz). The NAP is recorded with a wire electrode placed directly on to the auditory nerve as it enters the brain stem (20 responses averaged). (Harper CM, Daube JR: Facial nerve EMG and other cranial nerve monitoring. J Clin Neurophysiol 15:206–216, 1998.)

BAEP at the end of surgery is associated with preservation of useful hearing in the majority of cases.[18, 29, 34, 46, 81, 100, 101, 119] BAEP monitoring is sometimes limited by low-amplitude poorly defined responses on preoperative baseline studies in patients with acoustic tumors, a relatively long signal acquisition time, and relatively low specificity compared to other auditory evoked potentials. The latter is illustrated by reported cases of retained useful hearing despite the irreversible loss of wave V during surgery.[29, 92]

ECoG is an alternative to BAEP monitoring, particularly in the setting of acoustic neuromas, when waves I to V of the BAEP may be poorly defined.[92, 129, 130] High-amplitude reproducible CM and N_1 potentials are recorded within several seconds in the majority of patients. Some hearing is usually present when the N_1 potential is preserved at the end of surgery.[92] Technical problems and a relative lack of sensitivity compared to other techniques limit the ECoG. The ECoG electrode is difficult to place and maintain in proper position. Because the N_1 potential originates from the lateral portion of the auditory nerve, damage that is limited to the medial portion of the nerve or the brain stem will not be detected by ECoG monitoring. Thus, significant loss of hearing can occur despite the preservation of the N_1 at the end or surgery.[92] A compromise that avoids these limitations involves making BAEP recordings with an electrode placed deep within the external ear canal referenced to the vertex. This montage often enhances the definition of wave I of the BAEP, which is analogous to the N1 potential of the ECoG (see Fig. 39–1).

The auditory NAP recording is the most sensitive monitoring technique available. Changes in the auditory NAP have been correlated with specific operative procedures.[81, 88, 105, 129] Some changes have been reversible with specific actions taken by the surgeon. Even transient changes in the NAP may be associated with pathological evidence of eighth nerve injury. This technique provides a real-time monitor of auditory nerve physiology from the cochlea to the brain stem. Because of the limited space within the posterior fossa in patients with acoustic tumors, the use of direct auditory NAP monitoring is restricted to patients with small tumors or to those undergoing microvascular decompression (MVD) or sectioning of the fifth, seventh, or eighth cranial nerves. Some variation in the amplitude of the response occurs when the electrode is moved or momentarily covered with cerebrospinal fluid. This technique requires active participation of the surgeon to be successful.

None of the auditory monitoring systems available is ideal. At times, different techniques are used in combination to give the surgeon the best information regarding auditory nerve and brain stem function. The BAEP is monitored during exposure and manipulation of the proximal portions of the nerve in the cerebellopontine angle. During critical periods of the operation when the eighth nerve is exposed, the auditory NAP is monitored. The ECoG or BAEP can be monitored during wound closure. If the brain stem is at risk, then the BAEP provides the best method, preferably in combination with SEP, EMG, and, if possible, MEP monitoring.

Pathophysiological Correlation

Mechanisms that produce a change in the AEP include stretch, compression, ischemia, and transection of the auditory nerve or the brain stem. The latency of wave V may increase by 1 to 1.5 msec when the dura is opened due to cooling of the nerve (Fig. 39–3). This change occurs over several minutes, stabilizes, and is not associated with postoperative hearing loss. Mild traction during cerebellar retraction produces a stable increase in the latency of wave V with no significant change in the ECoG. Moderate traction or contusion of the nerve may produce a reproducible change in the latency and amplitude of wave V, N_1, and the auditory NAP.[81] The magnitude of change considered to be significant is still a matter of debate.[29, 81, 88, 92, 108] An increased latency of 1.0 msec or more or a 50% reduction in amplitude is usually considered significant enough to alert the surgeon (Fig. 39–3). The surgeon must interpret the changes in the context of the surgical circumstances at that moment and options for the remainder of the procedure. Changes in the position of the cerebellar retractor or in the surgical dissection may result in reversal or stabilization of the AEP abnormalities (Fig. 39–4). At other times, the BAEP is gradually lost despite efforts to modify the dissection. Wave I of the BAEP or the N_1 potential ECoG may be spared if the coch-

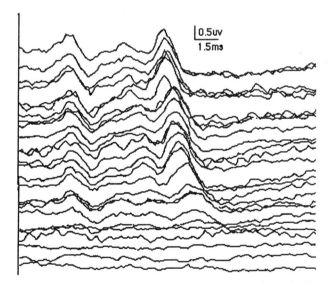

FIGURE 39–3. Brain stem auditory evoked potential monitoring during acoustic neuroma surgery (montage: external ear canal electrode referenced to C_z; stimulus rate: 10.1 Hz; number averaged: 500). After the sixth trace the latency of wave V increases by 1 msec. This is caused by cooling of the auditory nerve when it is exposed to air after the dura is opened and spinal fluid is drained. Later, during tumor dissection, there is a gradual drop in the amplitude (by 50%) and an increase in the latency (by > 1.5 msec) of waves II through V. Finally, over a 60-second interval all waves disappear suddenly and never return. (Copyright © Mayo Foundation.)

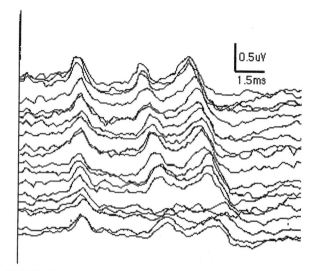

FIGURE 39–4. Brain stem auditory evoked potential monitoring during acoustic neuroma surgery (montage: external ear canal electrode referenced to C_z; stimulus rate: 10.1 Hz; number averaged: 500). There is a gradual increase in latency of wave V early during exposure and dissection of the tumor. Waves III through V then change more rapidly, but these changes reverse when the cerebellar retractor is adjusted and the approach to tumor dissection is altered. There was no loss of hearing after surgery. (Harper CM, Daube JR: Facial nerve EMG and other cranial nerve monitoring. J Clin Neurophysiol 15:206–216, 1998.)

lear blood supply is preserved and the lateral portion of the auditory nerve is not injured (Fig. 39–5). Avulsion of the nerve from the cochlea or interruption of the internal auditory artery results in rapid irreversible loss of all AEP recordings and hearing (see Fig. 39–3). Even though changes are not always reversible, correlation of AEP changes with surgical events helps to determine mechanisms of hearing loss. This information is used to modify surgical techniques with the hope of preventing the recurrence of similar mistakes and improving the rate of hearing preservation over time. Several controlled studies have shown that intraoperative AEP monitoring has contributed to the improved rate of hearing preservation in patients with acoustic neuromas,[46] microvascular decompression,[100] and vestibular neurectomy.[82]

Somatosensory Evoked Potentials

Somatosensory evoked potentials provide an accurate measure of function in peripheral and central pathways that mediate proprioceptive and other "large fiber" sensory functions.[15] This includes myelinated cutaneous, joint, and muscle afferents in the peripheral nervous system and posterior column, medial lemniscus, and thalamocortical pathways in the central nervous system. The SEP may also reflect activity in propriospinal and spinocerebellar pathways at the spinal level.[17, 98] The SEP monitors these sensory pathways directly, and may indirectly reflect changes in adjacent motor pathways.

Somatosensory Evoked Potential Techniques

The most reliable SEP recordings are produced by electrical stimulation of large mixed nerves of the limbs. Stimulation is applied to distal nerves (e.g., ulnar, median, and tibial) with surface electrodes or to proximal nerves (e.g., sciatic, cauda equina, brachial plexus, and nerve root) with

either needle electrodes or directly to the exposed nerve in the surgical field. Each nerve is stimulated unilaterally in a consecutive fashion so that those pathways carrying information from all potentially affected limbs are monitored. Bilateral simultaneous stimulation may miss a unilateral injury, and is therefore performed only when an adequate response cannot be obtained with unilateral stimulation. Stimulus duration and intensity is adjusted (0.2 to 0.5 msec, 5 to 90 mA) to produce maximal stimulation of sensory axons. The rate of stimulation is kept under 5 Hz to minimize rate-dependent attenuation of the SEP, which is accentuated by anesthetics. Rates under 2 Hz are sometimes required to record cerebral responses in children and adolescents, especially at deeper levels of anesthesia. Stimulus rates that are even fractions of 60 Hz are avoided to prevent averaging of 60-cycle interference into the recording. The number of stimuli required for averaging varies with the amount of background noise as well as with the size and reproducibility of the SEP. In the absence of a preoperative deficit and excessive artifact, 250 stimuli are usually adequate for recordings made from surface electrodes. The number of stimuli averaged should be kept to a minimum so that the surgeon receives feedback as rapidly as possible.

The type of recording electrode used depends on the location of recordings sites and the type of surgery. Scalp electrodes are standard and are similar to those used for SEP recordings in the outpatient setting. They are firmly attached with collodion and filled with conductive gel to ensure stability and low impedance during long surgical procedures. Esophageal or nasopharyngeal electrodes are used to record cervical cord potentials outside the surgical field in cervical spine surgery (Fig. 39–6). Needle electrodes inserted between spinous processes or over the laminae of

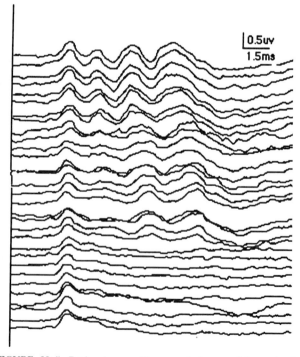

FIGURE 39–5. Brain stem auditory evoked potential monitoring during acoustic neuroma surgery (montage: external ear canal electrode referenced to C_z; stimulus rate: 10.1 Hz; number averaged: 500). There is a gradual increase in latency of waves II through V during exposure and dissection of the tumor. Eventually waves II through V disappear but wave I is preserved. This patient experienced diminished but preserved hearing after surgery. (Copyright © Mayo Foundation.)

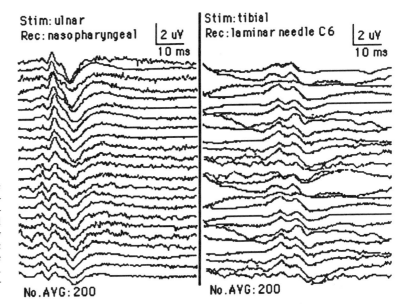

FIGURE 39–6. Somatosensory evoked potential (SEP) recordings during spine surgery. The cervical spine potential of the ulnar SEP recorded from a nasopharyngeal electrode during cervical spinal fusion is displayed on the left. The cervical potential of the tibial SEP recorded from a needle placed percutaneously onto the lamina of the C6 vertebra during thoracic spinal fusion is shown on the right. Surface and needle electrodes at F_z are used as a reference, respectively. (Stimulus rate: 1.9 Hz; filters: 30 to 3000 Hz.) (Copyright © Mayo Foundation.)

the spine can be used to record spinal cord activity[55] (see Fig. 39–6). Needle electrodes placed in the interspinous ligaments can be used within the operative field if posterior vertebral elements are left intact (Fig. 39–7). Small cotton-tipped electrodes or platinum electrodes are used to record directly from the surface of the spinal cord or cerebral cortex. Stereotactically placed microelectrodes are necessary when recording from thalamus or basal ganglia. For each of these active electrodes an appropriate reference must be chosen. Nearby electrodes reduce noise, but distant electrodes enhance signal amplitude. In general, active and reference electrodes should be of the same material to minimize impedance mismatch, which increases noise. Whenever possible, recordings should be made at multiple peripheral and central sites along the sensory pathways (Fig. 39–8). Adherence to this important principle will minimize the incidence of false-positive changes and make troubleshooting for technical errors more efficient.

Signal amplification and filter settings are similar to those used for diagnostic SEP recordings, although at times the

sensitivity must be reduced or the bandpass restricted because of the amount of noise in the surgical environment. Amplification of 5 to 10 μV/cm, sweep speeds of 2 to 10 msec/cm, low-frequency filters of 30 to 100 Hz, and high-frequency filters of 2000 to 3000 Hz are generally satisfactory. The equipment used for intraoperative SEP recordings must be versatile and easily tailored to the specific type of procedure being monitored. The ability to record other modalities

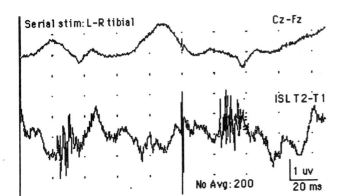

FIGURE 39–7. The left and right tibial SEP is recorded serially during the same sweep in a patient undergoing thoracic spinal fusion. The cortical potential is recorded from surface electrodes (C_z–F_z) while the cervical spinal cord potential is recorded from interspinous needle electrodes placed at the T2 and T1 interspaces in the surgical field. Stimulus delay between the left and right tibial nerves at the ankle is 100 msec. (Stimulus rate: 1.9 Hz; filters: 30 to 3000 Hz.) (Copyright © Mayo Foundation.)

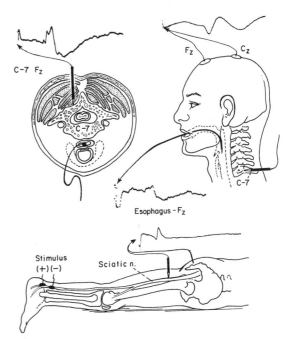

FIGURE 39–8. Multiple recording sites used during intraoperative SEP monitoring. The cortical potential is affected by anesthetics but not by muscle artifact. The spinal potential can be recorded with nasopharyngeal, esophageal, or laminar needle electrodes, and is refractory to the effects of anesthetics, but susceptible to muscle artifact. The peripheral "input" is monitored during the tibial SEP with a needle electrode over the sciatic nerve at the gluteal fold. (Harper CM, Daube JR: Surgical monitoring with evoked potentials: The Mayo experience. In Desmedt JD [ed]: Neuromonitoring in Surgery. Amsterdam, Elsevier, 1989, pp 257–303.)

(i.e., EMG, CMAP, NAP, BAEP, and MEP) concurrently with SEP is essential. Preamplifiers need to tolerate high current loads caused by cautery and other sources of electrical interference. Automatic cautery suppression, artifact rejection and software for digital filtering, trend analysis, data reproduction, and storage are desirable features.

Pathophysiological Correlation

Animal and human studies have shown that SEP changes occur when there is injury to "large fiber" sensory pathways at peripheral, spinal, and cortical levels.[17, 55, 75, 98, 124] In most instances, the SEP also changes when there is injury to adjacent motor pathways at the spinal and brain stem level.[6, 54, 91] Assuming that appropriate stimulation and recording can be accomplished, the major issue becomes the determination of what constitutes a significant change and how reliably this can be detected. The important steps in making this determination are summarized in Table 39–2. The primary disease often produces an SEP abnormality that can be recognized in baseline recordings. Depending on the abnormality, recording methods may have to be modified. Despite averaging, multiple sources of artifact result in unstable potentials that are different for each patient. It is therefore essential to determine the limits of SEP amplitude and latency variation with repeated samples during the early part of the surgery. Significance criteria can then be established that are beyond the baseline limits of variability. In patients with high-amplitude, well-defined potentials at peripheral, spinal, and cortical levels, a reproducible drop in amplitude of 50% or greater or an increase in latency of 2 msec or greater, or both, is considered significant. If baseline recordings under anesthesia show highly variable, low-amplitude cortical or spinal potentials, then all potentials recorded rostral to the area at risk would have to disappear for the change to be considered significant. Implicit in these judgments is the understanding that effects of a change in physiological variables, limb position, artifacts, and technical failures be identified and either corrected or accounted for before a final decision regarding the significance of a change in the SEP is made.

Early in the procedure, an effort is made to identify and eliminate all sources of noise, especially 60-cycle interference. Care must be taken to avoid ground loops. Any conductor in contact with the patient (including intravenous lines) or electrical equipment in the room can be a source of interference or a potential safety hazard. The recording system should suppress input during cautery and reject high-amplitude artifact. EMG activity from surrounding muscle can also produce unwanted artifact if neuromuscular activity is not blocked. Especially with light anesthesia, EMG activity may be so prominent as to obscure the SEP. A constant

controlled level of short-acting neuromuscular blocking agent or intermittent doses of benzodiazepines can be used to control muscle artifact in cases that require simultaneous monitoring of SEP and EMG.

Anesthesia reduces the amplitude and increases the latency of cortical SEP recordings. This is especially true in the presence of disease, and in children and adolescents.[45, 49] Because the reduction in amplitude is directly related to depth of anesthesia, the level should be kept as light as possible. Anesthetic effect varies with the agent used, being greatest with halogenated anesthetics, moderate with intravenous barbituates and nitrous oxide, and least with narcotics and benzodiazepines.[64, 83, 85, 95] Etomidate and ketamine have been shown to enhance the amplitude of the cortical SEP potential.[58, 85] These drugs can occasionally be used to record cortical potentials during surgery when responses are absent with standard anesthetics. Mean blood pressure below 70 mm Hg attenuates the amplitude of the cortical SEP. The effect of these variables on SEP measurements can be minimized by choosing an appropriate anesthetic regimen and by recording potentials from the spinal cord rostral to the area of surgery.

When surgery is performed in the sitting position and the dura is opened, subdural air can accumulate and serve as an insulator between the cortical surface and scalp electrodes used to record the SEP (Fig. 39–9). This can reduce the amplitude of the cortical SEP unilaterally or bilaterally. When this occurs, a normal potential is recorded with an electrode placed below the air level (i.e., over the parietal scalp just above the ear). SEP changes may also occur when there is loss of function in peripheral pathways secondary to limb compression or ischemia. Loss of the peripheral signal from stimulator failure or peripheral nerve injury is detected by monitoring the SEP at a peripheral recording site or by monitoring EMG activity in the limb at the same time as SEP monitoring.

Motor Evoked Potentials

SEPs have been the standard tool for intraoperative monitoring of the spinal cord for many years. When performed and interpreted properly, SEP recordings are highly accurate and sensitive, providing a direct measure of function in dorsal column sensory pathways and an indirect measure of function in motor pathways in the lateral columns of the spinal cord. Nevertheless, corticospinal pathways and anterior horn cells can be damaged without a change in the SEP recorded during surgery.[5, 31, 71] A variety of electrophysiological techniques are available to study the motor system more directly than SEP. The functional status of the lower motor neuron can be assessed with CMAP recordings, F waves, H reflexes, and EMG. These modalities are discussed later in this chapter. The MEP reflects activity in corticospinal tracts and other upper motor neuron pathways. Information regarding the integrity of lower motor neurons is also available when the MEP is recorded over peripheral nerve or muscle. Experience with these techniques in both animals and humans suggests that they are useful in predicting and preventing serious motor complications during surgery.[19, 26, 33, 61, 67, 72, 73, 89, 93, 94, 134]

Motor Evoked Potential Techniques

Transcranial Electrical Motor Evoked Potentials. Transcranial stimulation (either electrical or magnetic) is currently not approved by the Food and Drug Administration (FDA) and requires an investigative device exemption to be used in the United States. Direct transcranial electrical stimulation of the motor cortex can be accomplished by placing an

TABLE 39–2	**Process for Judging Significance of a Change in Somatosensory Evoked Potentials during Surgery**

1. Record the potential at peripheral, spinal, and cortical levels.
2. Take preoperative neurological deficits into account when planning intraoperative monitoring protocol.
3. Establish the level of somatosensory evoked potential variability in baseline recordings under anesthesia before critical parts of the surgery are undertaken.
4. Identify and eliminate sources of electrical noise and muscle artifact.
5. Consider the effect of anesthetics on the cortical potential of the SEP.
6. Consider subdural air if the patient is in sitting position.
7. Consider limb ischemia or peripheral nerve compression if peripheral potential is lost.

ULNAR AND TIBIAL SEP DURING FORAMEN MAGNUM MENGIOMA SURGERY

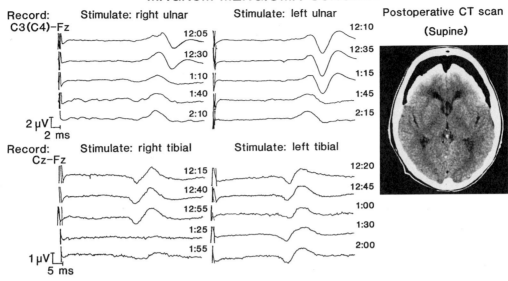

FIGURE 39–9. Left and right ulnar and tibial SEP recordings during surgery in the sitting position. After the dura is opened (approximately at time 1:00), there is a slight reduction in the amplitude of the right and left ulnar SEP cortical potentials (1:10) and then a large reduction in the amplitude of the tibial SEP cortical potentials (1:25). Cervical spine potentials recorded from a nasopharyngeal electrode did not change (not shown). The changes result from the accumulation of subdural air (demonstrated on the postoperative CT scan on the right). (Copyright © Mayo Foundation.)

anode over the motor strip and a cathode 5 cm anterior to this point ("bipolar method").[11, 56, 103] A 500 to 800 V stimulus of 50- to 100-μsec duration applied to the scalp activates the underlying motor cortex. The descending corticospinal tract potential can be recorded directly from the spinal cord or from the epidural space. Alternatively, the CMAP is recorded with surface electrodes over limb muscles. Muscle responses are enhanced by paired cortical stimulation with an interstimulus interval of 2 to 5 msec. Electrical stimulation using lower voltages can be obtained by using a circumferential ring cathode placed around the scalp with the anode over the motor strip ("unipolar method"),[125] or by stimulating the frontal cortex directly through a burr hole or when the cortex is exposed at craniotomy.[73]

Transcranial electrical MEP recordings have been performed during surgery by several investigators for the purpose of studying the physiology of the signal or the effects of anesthetics on the responses.[7, 8, 10, 11, 26, 51, 53, 57, 59, 99, 103, 106, 118, 121, 128, 132, 136] Using epidural or subdural spinal recordings, D waves and I waves of 10- to 15-μV amplitude are recorded with conduction velocities in the range of 55 to 75 m/sec.[10, 99] D waves result from direct activation of corticospinal neurons whereas I waves come from activation of collateral neurons within the cortex that secondarily stimulate corticospinal neurons.[7, 51] Transcranial electrical stimulation preferentially produces D waves, but I waves, which are needed to bring the lower motor neuron to firing threshold, are also elicited. Inhalation anesthetics and a variety of induction agents greatly reduce the number of I waves produced, leading to a reduction in the amplitude or total elimination of the transcranial electrical MEP.[51]

Transcranial Magnetic Motor Evoked Potentials. Transcranial magnetic stimulation can activate the motor cortex through the skull. A high-voltage discharge through a circular copper coil induces a 0.01-msec magnetic pulse of 2 to 3 Tesla that is not attenuated by the scalp or skull[7, 30, 50] The intensity of the field is greatest under the rim of the coil from which it extends in the shape of a doughnut for several

centimeters into the tissue. The electrical current induced by the magnetic field in tissue is poorly focused but in the awake patient is adequate enough to record an MEP over the spinal cord and muscles of the upper and lower extremities. Using the most effective tangential orientation, the magnetic stimulator tends to produce I waves instead of D waves in spinal recordings and lower amplitude muscle responses with longer latencies than electrical stimulation.[7, 122] Slight changes in the orientation of the stimulating coil produce extensive variability in the amplitude of the MEP recorded over the spinal cord and muscle. Although magnetic MEPs are less painful than electrical MEPs in the awake patient, the lower and more variable amplitude of magnetic MEPs makes them less reliable during surgery.[50, 109]

Motor Evoked Potentials Produced by Spinal Stimulation. Several investigators have reported their experience with MEPs produced by electrical stimulation of the spinal cord.[19, 66, 67, 77, 89, 94, 97, 117] The spinal cord can be stimulated with electrodes placed directly on the cord, in the epidural space, or within the interspinous ligaments of the cervical or thoracic vertebrae. We currently use a modification of these techniques in which the stimulating current is passed between an anode in the esophagus or nasopharynx and a needle cathode is placed onto the lamina of the cervical spine percutaneously.[21] This avoids the need to put stimulating electrodes in the surgical field where they often become dislodged or where there is potential for current to be shunted from the stimulating electrodes to spinal instrumentation. The latter can result in a loss of MEP secondary to insufficient stimulation.[107]

Following spinal stimulation, responses can be recorded over peripheral nerves or muscles in the upper and lower extremities. Peripheral nerve recordings are small (1 to 10 μV), require averaging, and must be recorded in the setting of complete neuromuscular blockade to avoid the potential for recording a volume-conducted muscle response. It is also theoretically possible that peripheral nerve recordings contain action potentials from antidromically activated sensory axons within the dorsal columns that are transmitted peripherally through the dorsal spinal root. Recordings over muscle are higher in amplitude (50 μV to 1mV) and avoid the potential problem of recording from sensory pathways. To avoid excessive patient movement secondary to muscle contraction, muscle action potentials are recorded in the setting of partial neuromuscular blockade. The degree of block is quantified and monitored by recording the ulnar or median CMAP following peripheral nerve stimulation. Reliable spinal MEPs can be recorded over limb muscles with a constant infusion of short-acting neuromuscular blocking drug adjusted to produce a peripheral CMAP that is 10% of the baseline response.[21, 66, 111]

Pathophysiological Correlation

As with SEP recordings, the MEP is affected by underlying disease of the motor system, anesthetics, and hypotension. The effect of disease and anesthetics on individual patients can be determined by performing baseline recordings intraoperatively before critical stages of surgery are reached. If possible the type and concentration of anesthetic agents should be kept relatively constant during MEP recordings made at baseline and at critical periods.

Anesthetic agents have significant effects on all types of MEP recordings. When MEP monitoring is contemplated, it is extremely important for the neurophysiologist, surgeon, and anesthesiologist to discuss the goals and risks of surgery in detail, so that the most appropriate anesthetic regimen can be chosen. The suppressive effects of anesthetics on MEP recordings occur at both the cortical and spinal level.[8, 21, 51, 74, 118, 120, 131] Halogenated anesthetics eliminate transcranial electrical and magnetic MEPs as well as spinal MEPs recorded over muscle in end-tidal concentrations as low as 0.2% to 0.5%, with 30% to 50% nitrous oxide having a similar effect.[26, 50, 51, 57, 94, 118, 128] Potentials recorded over the spinal cord are less susceptible to the effects of inhalation anesthetics.[77, 94, 97] Propofol and thiopental reduce all types of MEPs recorded over muscle by 70% to 100% with slow recovery over several hours after administration of a single dose or discontinuation of a constant infusion of the drug.[118] Etomidate and ketamine produce minor depression, and narcotics have no effect on MEP recordings.[132]

Using nitrous oxide–narcotic anesthesia, Edmonds and associates recorded transcranial magnetic MEPs over the anterior tibial muscle in neurologically normal patients undergoing scoliosis surgery.[26] CMAP amplitudes ranged from 20 to 500 μV but were highly variable, making interpretation of intraoperative changes difficult. Other investigators have had a similar experience.[50, 74, 109, 118] In our experience, transcranial magnetic MEPs are extremely sensitive to anesthetics, being eliminated by all inhalation and induction agents with the possible exception of etomidate.[74] We have been able to record magnetic MEPs successfully under narcotic anesthesia and partial neuromuscular blockade in patients undergoing thoracoabdominal aneurysm repair. Our preliminary experience suggests that early changes in the MEP recorded over the lower extremities following aortic clamping have a high correlation with postoperative paraplegia.[21]

Levy and colleagues described their experience with transcranial electrical MEP monitoring during a variety of neurosurgical procedures.[72, 73] A modification of the "bipolar method" was used, with the anode placed on the hard palate in some patients and direct cortical stimulation with anode and cathode placed through a burr hole in others. Responses were recorded from the spinal cord or the peripheral nerve. A good correlation between changes in the MEP and postoperative motor function was observed. In over 100 patients monitored, there were no cases of new motor deficits when the MEP was normal at the end of surgery. Zentner and

associates reported their experience with MEP monitoring in 50 neurosurgical operations on the spinal cord.[137] The "bipolar method" of transcranial electrical stimulation was used with CMAP recordings over the thenar and anterior tibial muscles. Reliable MEP recordings of 50 to 500 μV were possible by averaging 10 to 25 stimuli in the setting of fentanyl and low concentrations of nitrous oxide. Using a change in amplitude of 50% as a limit of significance, transient intraoperative changes in the MEP were noted in 20% of patients. None of these patients developed a new motor deficit after surgery. In addition, no new deficits were observed when the MEP remained unchanged during surgery. Of the four patients with new motor deficits postoperatively, the MEP disappeared in one and was reduced in amplitude by 60% to 85% in three. Kitagawa and co-workers reported a similar experience with transcranial electrical stimulation recording over the spinal cord in 20 patients undergoing cervical spine surgery.[61] The 3 patients with new postoperative deficits all had at least a 50% reduction in the averaged spinal MEP recorded from the epidural space below the level of surgery. Changes in latency of spinal cord or muscle responses have not been useful in predicting neurological outcome.

MEP recordings following spinal stimulation have been studied in animals and humans.[19, 66, 67, 89, 93, 94, 97] Owen and associates reported the findings of spinal MEP monitoring with peripheral nerve recording in 300 patients undergoing a variety of neurosurgical, orthopedic, and vascular procedures.[94] Ketamine or nitrous oxide–narcotic combination was used as anesthesia. Nitrous oxide concentrations greater than 60% or the use of halogenated anesthetics greatly reduced the amplitude and reliability of the MEP. Using a drop in amplitude of 60% or greater as a criteria for significant change, there were 18 patients who, early in their experience, had changes in the MEP recordings without developing postoperative deficits. These "false-positives" were attributed to technical difficulties that with experience were recognized and corrected. All 7 of the patients with new postoperative deficits had significant changes in the MEP amplitude intraoperatively. SEP recordings were "preserved" in all 7 patients with motor deficits. Two additional patients had new sensory deficits with changes in the SEP but not in the MEP recorded intraoperatively. Nagle and colleagues reported the findings of spinal MEP monitoring with muscle recordings in 116 cases of spine or spinal cord surgery. The MEP recordings correctly predicted motor outcome in all cases.[89] The method currently in use at our institution utilizes CMAP recordings over lower limb muscles following spinal cord stimulation with an esophageal and laminar needle electrode (Fig. 39–10). The recordings are made under controlled partial neuromuscular blockade using a narcotic-based anesthesia.

Electromyography

EMG records potentials generated by muscle. Although a number of different types of potentials are encountered during surgery, neurotonic discharges are the activity of major interest. Neurotonic discharges are high-frequency bursts of motor unit potentials recorded from a muscle when the nerve is mechanically or metabolically stimulated[20] (Fig. 39–11). Neurotonic discharges are distinguished from movement and electrical artifact as well as random motor unit potentials secondary to "light" levels of anesthesia by the pattern and rate of motor unit firing and the temporal relationship of the discharge to the inciting stimulus.

Electromyographic Techniques

A variety of electrodes can be used to record EMG activity during surgery. Surface and subcutaneous electrodes will

MEP During Scoliosis Surgery

Spinal stimulation: Paired ISI 3 ms, 80 mA, dur 1ms
ESOPHAGEAL-LEFT CERVICAL

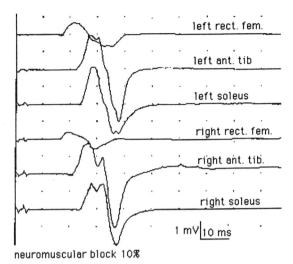

neuromuscular block 10%

FIGURE 39-10. Motor evoked potentials (MEPs) recorded from surface electrodes over lower extremity muscles following electrical stimulation of the cervical spinal cord. The cathode is an electrode in the esophagus and the anode is a needle electrode placed percutaneously on to the left lamina of the C6 vertebra. Paired electrical stimuli were administered with an interstimulus interval (ISI) of 3 msec (intensity 80 mA, duration 1 msec). Recordings were made from left and right rectus femoris (rect. fem.), anterior tibial (ant. tib.) and soleus muscles under a constant infusion of neuromuscular blocking agent adjusted to produce a 10% reduction in the amplitude of the baseline ulnar CMAP. (Copyright © Mayo Foundation.)

capture some neurotonic discharges, but cannot reliably record those from deep areas of the muscle. Standard concentric and monopolar EMG needle electrodes produce recordings of excellent quality, but they are bulky, can be easily dislodged, and may injure tissue if inadvertently moved during surgery. The most satisfactory EMG electrodes are malleable wires that can be inserted into the muscle with a small needle, which is then removed. These wires allow stable selective recordings from superficial or deep muscles. They are less bulky than needle electrodes, resulting in less movement artifact and less trauma to the muscle and surrounding tissues. Intraoperative EMG activity is recorded with the same sensitivities (200 to 500 μV), filter settings (30 to 20,000 Hz), and sweep speeds (10 msec/cm) that are used in standard needle electromyography. Recordings can be made from extraocular, facial, masseter, laryngeal, intercostal, abdominal, anal sphincter, and limb muscles. EMG recordings from multiple muscles are monitored simultaneously using a loudspeaker and an oscilloscope. EMG activity of interest can be photographed and recorded.

Neurotonic discharges can be recorded with a constant infusion of a short-acting neuromuscular blocking drug adjusted to produce a peripheral CMAP that is 25% of the baseline response.[70] Higher levels of inhalation anesthetics or intermittent doses of a narcotic or benzodiazepines can also be used to reduce background muscle activity that may interfere with the recording of neurotonic discharges.

Electrical interference can come from a variety of sources but cautery is of greatest concern. The latter can be suppressed to prevent damage to recording equipment, but EMG cannot be monitored when cautery is activated. Cavitron also interferes with EMG recordings. Movement artifacts are common and may be misinterpreted as brief neurotonic discharges by inexperienced personnel. They usually take the form of irregular, triangular waves and make a brief "popping" or "clicking" sound over the audio speaker (Fig. 39-12).

Pathophysiological Correlation

Neurotonic discharges have a variety of forms, but all consist of motor unit potentials that fire in a rapid and irregular manner (see Fig. 39-11). Each burst of motor unit potentials may last less than 50 msec or persist for many seconds. Long-duration bursts are more common after nerve stretching or irrigation of the nerve with saline. At times multiple neurotonic discharges firing asynchronously and independently are recorded from a single muscle.

Neurotonic discharges are often precipitated by mechanical stimulation of the axonal membrane of peripheral nerves. They are sensitive indicators of nerve irritation and occur in virtually all monitored patients. Sharp transection of a nerve may not produce neurotonic discharges, and damaged nerves are less likely to produce neurotonic discharges than are healthy nerves. In addition, irrigation of a nerve with saline frequently produces long trains of neurotonic discharges lasting 2 to 60 seconds. Therefore, the density and frequency of neurotonic discharges recorded during surgery correlate only roughly the severity of postoperative neurological deficit.[40] In addition to being a marker for

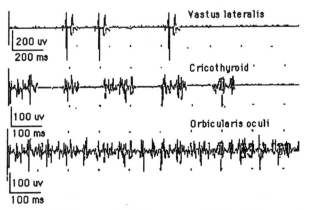

FIGURE 39-11. Neurotonic discharges recorded from intramuscular wire electrodes during surgery in three different patients during surgery. Examples are given of short, intermediate, and long duration bursts of motor unit potentials in response to inadvertent mechanical stimulation of the L4 root (vastus lateralis), vagus nerve (cricothyroid), and facial nerve (orbicularis oculi). (Copyright © Mayo Foundation.)

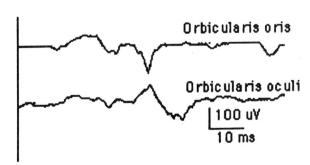

FIGURE 39-12. Continuous EMG monitoring with intramuscular wire electrodes in the orbicularis oris and oculi muscles during parotid surgery. Movement artifact is identified by the sharp triangular shape of the potential and the characteristic "popping" or "clicking" noise heard over the audio amplifier. (Copyright © Mayo Foundation.)

potential nerve injury, neurotonic discharges also help remind a surgeon of the location of peripheral nerves within the operative field.

Compound Muscle and Nerve Action Potentials

Compound Muscle and Nerve Action Potential Techniques

The integrity of motor axons within a peripheral nerve can be tested by stimulating the nerve in the surgical field while recording a CMAP from one or more distal muscles. CMAP recordings are made with intramuscular wire electrodes (as in EMG) or with surface electrodes (Fig. 39–13). Intramuscular electrodes are used when selective recordings are required from muscles that are located in close proximity to each other but are innervated by different nerves or fascicles. Because intramuscular electrodes record activity from only a small portion of the muscle, they are unsuitable for quantitative measurements. The latter require the use of surface electrodes, which record activity from the majority of muscle fibers. The resulting CMAP is a reproducible, biphasic potential with an amplitude and latency that are easy to measure and compare to CMAPs obtained when the nerve is stimulated at different locations or at different times during surgery. The optimal surface electrodes are 5-mm discs that are applied firmly to the skin with collodion. CMAP recordings are made with filter settings of 2 Hz to 20 KHz, sweep speeds of 1 to 10 msec/cm, and sensitivities of 5–100 μV/cm.

The NAP can also be recorded directly from peripheral nerves exposed at surgery (Fig. 39–14). This provides information when the nerve is purely sensory or when regeneration is underway, but has yet to reach the muscle. The NAP is more difficult to record than the CMAP because it is smaller in amplitude and is often recorded over short distances, leading to technical problems with excessive stimulus artifact. Although averaging helps distinguish the NAP from

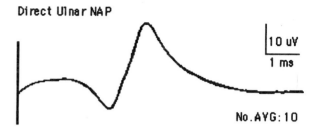

FIGURE 39–14. Nerve action potential (NAP) recorded directly from the ulnar nerve exposed at surgery. The recording and reference electrodes are hooked under the nerve with a fixed interelectrode distance at 3.5 cm. The stimulator is approximately 8 cm proximal to the recording electrode. Note the typical triphasic appearance and the relatively low amplitude of the NAP compared to the CMAP (see Fig. 39–13). (Copyright © Mayo Foundation.)

background noise, the amplitude of the response is more variable than that of CMAP recordings.

Recording the NAP outside the surgical field uses methods that are standard for routine clinical nerve conduction studies. When it is necessary to make NAP recordings within the surgical field, the nerve is isolated from surrounding tissue with small hook electrodes. The recording and reference electrodes are placed along the course of the nerve at a fixed distance apart to assure reproducible responses. NAP recordings are used to identify the presence of axonal continuity and to localize areas of focal conduction slowing or block. In this situation, stimulating and recording at several sites along the nerve provide the most useful information.

A number of different stimulators are used to activate peripheral nerve axons during surgery. Hand-held stimulators of various sizes and configurations that can be gassterilized are commercially available. Stimulators that are insulated to the very tip of the electrode have fewer problems with current shunting, but may also produce subthreshold stimuli if they are not applied to the surface of the nerve properly. Other stimulators have a hooked configuration that allows the nerve or fascicles within the nerve to be separated from surrounding tissue. This reduces artifact from the stimulus or surrounding muscles and allows the nerve elements of interest to be selectively stimulated. Bipolar stimulators have the cathode and anode attached to the same handle and within several centimeters of each other. This provides a localized stimulus that reduces the risk of current spread to adjacent nerves. The disadvantage of the bipolar stimulator is that activation may be inadequate if the nerve is distant or there is too much fluid in the operative field. Monopolar stimulators utilize a single hand-held cathode on the nerve with a separate anode placed some distance away, usually a needle in the edge of the surgical field or a distant surface electrode. Monopolar stimulation reduces the chance of inadequate stimulation, but increases the likelihood of current spread to other nerves and shock artifact. The size of the stimulating electrodes varies depending on the nerve that is stimulated. Small cranial nerves require stimulator tips as small as 1 mm; larger peripheral nerves may require 2- to 3-mm electrodes to provide an adequate stimulus. If a surgical forceps is modified for use as a bipolar stimulator, the surgeon can dissect tissue with the stimulator.

Stimulation is applied at rates of 1 to 5 Hz with a duration of 0.05 msec. Supramaximal responses are obtained from normal nerve with 1 to 5 mA of current (25 to 50 V if a constant voltage stimulator is used). Higher intensities are required if the nerve is damaged, if there is tissue between the nerve and the stimulating electrodes, or if there is excess fluid in the stimulating field. To minimize current spread to

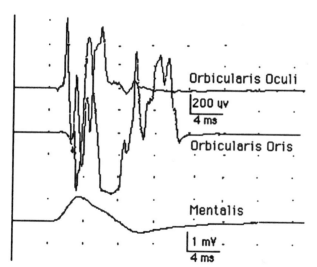

FIGURE 39–13. Compound muscle action potentials (CMAPs) recorded with intramuscular wire electrodes in the orbicularis oculi and oris muscles and from surface electrodes over the mentalis muscle in response to electrical stimulation of the facial nerve near the brain stem. Intramuscular electrodes record from fewer muscle fibers while surface electrodes represent activity in the entire muscle. The amplitude and area of the CMAP recorded with surface electrodes are easier to quantitate and are directly proportional to the number of functioning axons activated by the electrical stimulus. (Copyright © Mayo Foundation.)

nearby nerves, currents greater than 5 mA (50 V) should be avoided. An absent CMAP or NAP may indicate that there has been significant damage to the nerve, but technical problems must first be considered and eliminated. Absent or greatly attenuated responses can result from excessive neuromuscular blockade, use of a peripheral limb tourniquet, shorting of current, failure of stimulating electrodes to contact the nerve, faulty connections, or inappropriate instrument settings. In other situations a response may be obtained that is larger than expected or at times when one is not anticipated at all. This may occur if the stimulator has been placed on the wrong nerve, or if the stimulus is too strong, resulting in current spread to adjacent nerves.

Pathophysiological Correlation

The amplitude of the CMAP or NAP is proportional to the number of axons conducting the response. Therefore, when a goal of monitoring is to determine the number of intact axons, the amplitude or area of the response can be measured at any time and compared to values recorded earlier during surgery or to preoperative baseline measurements. A change in amplitude of 10% or more over a 1- to 2-cm segment of nerve indicates the presence of conduction block, whereas an increase in latency of 0.4 msec or more over a 1- to 2-cm segment signifies focal slowing of conduction (Fig. 39–15). The presence of a NAP across an injured segment of peripheral nerve indicates the presence of functioning residual or regenerating axons. If technical causes are excluded, the absence of both the CMAP and NAP indicates complete axon loss with no regeneration across the injured segment. The latter finding is usually an indication for resection of the injured segment with subsequent placement of a nerve graft or performance of a neurotization procedure.

F Wave and H Reflexes

F waves represent CMAPs that are recorded from one or a few motor units activated by an antidromic action potential in the motor axon transmitted from a distal site of stimulation to the anterior horn cell and back. F waves are most readily recorded from small foot or hand muscles following supramaximal stimulation of the tibial, peroneal, median,

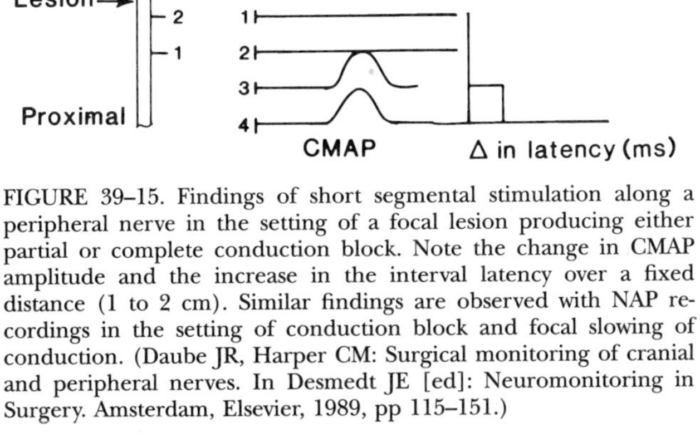

FIGURE 39–15. Findings of short segmental stimulation along a peripheral nerve in the setting of a focal lesion producing either partial or complete conduction block. Note the change in CMAP amplitude and the increase in the interval latency over a fixed distance (1 to 2 cm). Similar findings are observed with NAP recordings in the setting of conduction block and focal slowing of conduction. (Daube JR, Harper CM: Surgical monitoring of cranial and peripheral nerves. In Desmedt JE [ed]: Neuromonitoring in Surgery. Amsterdam, Elsevier, 1989, pp 115–151.)

and ulnar nerves. H reflexes represent CMAPs of anterior horn cells that are monosynaptically activated by large myelinated muscle afferents following a long duration submaximal stimulus to a mixed peripheral nerve. H reflexes are most readily recorded from the gastrocnemius and soleus muscles when the tibial nerve is stimulated, and at times can be recorded from the forearm flexor muscles with median nerve stimulation.

The presence of F waves or H reflexes indicates that at least a small proportion of anterior horn cells and motor axons are functioning in the distribution of the spinal segments stimulated. The main limitation in the use of these "late responses" intraoperatively is the exquisite sensitivity of both F waves and H reflexes to the majority of inhalation and intravenous anesthetics. Using narcotic anesthesia, we have monitored F waves and H reflexes along with MEPs and SEPs during thoracoabdominal aneurysm repair. We found that the H reflex in particular was as sensitive as the MEP in detecting ischemia of anterior horn cells within the lumbosacral segments of the spinal cord.[21]

APPLICATIONS OF INTRAOPERATIVE MONITORING

Cerebral Surgery

SEP monitoring is performed during a variety of surgical procedures directed at supratentorial structures. Direct recording of the SEP from the surface of the frontoparietal cortex is performed during cortical resections for epilepsy, neoplasms, or other structural lesions. When used in conjunction with direct electrical cortical stimulation, the location of the primary sensory and motor areas can be determined with a high degree of certainty.[2, 43] SEP recordings are an alternative method to EEG[16] as a monitor of cortical function during carotid endarterectomy and aneurysm surgery.[84, 114, 127] Methods for recording SEP as a monitor of cerebral function and the associated technical problems are similar to those described for their use in spine surgery. Direct cortical recordings are made using subdural electrode strips or grids (Fig. 39–16).

Direct recording of SEP from deep structures has been of help in placing stereotactic therapeutic lesions. Thalamotomy for Parkinson's disease and other movement disorders is facilitated by SEP recordings from depth electrodes within the thalamus in response to electrical or mechanical stimulation.[44] Responses from the ventroposterolateral thalamus (VPL) are extremely well localized and of high voltage (50 to 500 μV). Thalamic responses are specific to the nerve or dermatome stimulated and thereby help confirm the somatotopic organization of the VPL and corresponding areas of the ventralis intermedius. As the electrode leaves the thalamus and enters the medial lemniscus, the potential reverses polarity and becomes widespread in somatic distribution. This helps confirm the location of the inferior border of the thalamus. Microelectrode recordings of spontaneous and stimulus-evoked activity following limb movement, visual stimuli, and deep brain stimulation are useful in determining the location and size of the lesions made in stereotactic pallidotomy for Parkinson's disease.[4]

Spine Surgery

The spinal cord or nerve roots, or both, are at risk during a variety of surgical procedures performed on the spinal cord or surrounding structures. The risk varies with the underlying disease as well as the type and location of surgery.

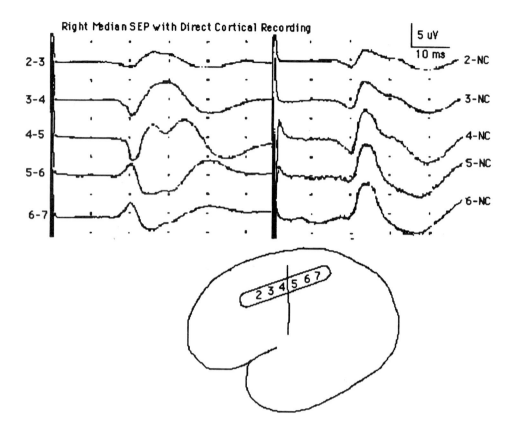

FIGURE 39–16. Intraoperative median SEP recorded from cortical electrodes for the purpose of localizing the primary sensory and motor cortex. The location of the central sulcus is confirmed by the phase reversal of the N20–P25 complex in the bipolar montage (*left panel*) and by the maximum amplitude of the N20–P25 potential in the referential montage (*right panel*). (Copyright © Mayo Foundation.)

Patients with intramedullary tumors, syringomyelia, spinal arteriovenous malformation, thoracoabdominal aneurysms, and any disorder associated with a baseline neurological deficit are at greatest risk. The frequency of neurological injury following scoliosis surgery, correction of congenital spinal deformities, and decompression with or without spinal fusion is low, but when damage occurs to the spinal cord, the resulting deficits are often severe, permanent, and devastating.[24, 31, 43, 71, 76, 91]

In the 1970s, the "wake up" test was developed in an attempt to reduce the risk of spinal cord injury in patients undergoing scoliosis surgery.[37, 126] This quickly became the standard to which other monitoring techniques were compared. Although it is helpful, the wake up test disrupts the surgical procedure, can be performed only intermittently, and is associated with some risk (e.g., extubation, pulmonary embolism). Furthermore, it is not applicable to patients undergoing surgical procedures in which there is no defined period of major risk, as in resection of spinal neoplasms.

In the 1980s, SEP monitoring was developed as an alternative to the wake up test.[1, 27, 28, 35, 78, 90] SEP recordings provided the means to monitor spinal cord function continuously without interfering with surgery or producing additional risk. Although no prospective controlled trial of SEP monitoring has ever been accomplished, a large body of data, including clinical experience in thousands of patients, supports the following conclusions regarding the utility and limitation of SEP monitoring during spinal surgery:[1, 23, 27, 28, 31, 35, 45, 47, 49, 71, 75, 78, 90]

- Technical problems related to SEP monitoring are well defined and can be readily identified and usually eliminated by experienced personnel.
- When technical problems are eliminated and the effect of anesthesia is accounted for, persistent changes in SEP recordings during surgery will lead to new sensory and at times motor deficits postoperatively.

- New neurological deficits are rare when intraoperative changes in SEP recordings resolve spontaneously or through action taken by the operative team and return to baseline at the end of surgery.
- Damage to upper or lower motor neuron pathways within the spinal cord can occur without concomitant changes in SEP recordings during surgery.

The experience with systems that monitor the motor pathways of the spinal cord directly (i.e., MEPs, EMG, F waves, H reflexes) has been more limited. However, the cumulative published experience involving over 1000 patients supports the following conclusions about motor system monitoring during spinal surgery:[19, 21, 26, 32, 46, 50, 61, 66–68, 72, 74, 75, 77, 89, 93, 97, 135]

- MEP recordings obtained with a variety of techniques are reproducible and reliable when special anesthetic regimens are utilized.
- The results of MEP recordings made during surgery correctly predict the presence or absence of new motor deficits after surgery.
- EMG monitoring of muscles innervated by spinal roots can help identify nerve roots and possibly prevent injury to roots or anterior horn cells, or both.
- MEPs, EMG, and SEPs provided complimentary information and when necessary should be used together in the same patient to comprehensively monitor motor and sensory function within the spinal cord.

The exact monitoring protocol differs for each patient depending on the underlying disease and neurological deficit, the type and location of surgery, and anesthesia requirements. After reviewing all pertinent data, the surgeon, anesthesiologist, and neurophysiologist should develop a monitoring plan tailored to the needs of the individual patient. Table 39–3 lists the most common monitoring protocols that are used at our institution for various types of spinal surgery. SEP recordings are monitored in the majority of

TABLE 39–3	Spinal Monitoring Protocols	
Level of Surgery	**Type of Surgery**	**Monitoring Protocol**
Cervical	Decompression (e.g., cervical spondylosis, Arnold-Chiari)	Left/right ulnar and tibial SEPs
	Tumor resection, syrinx repair, AVM, fusion with screw fixation	Left/right ulnar and tibial SEPs, EMG of upper limb muscles (spinal segments at risk)
Thoracic	Correction of spinal deformity, tumor resection, AVM	Left/right tibial SEPs, spinal MEP
	Thoracoabdominal aneurysm repair	Left/right tibial SEPs, spinal MEP, F waves, H reflexes
Lumbosacral	Lipomeningocele repair, tumor resection, AVM, fusion with screw fixation	Left/right tibial SEPs, EMG of lower limb muscles (spinal segments at risk)

AVM, arteriovenous malformation; EMG, electromyography; MEP, motor evoked potential; SEP, somatosensory evoked potential.

patients. MEP recordings from lower limb muscles following spinal cord stimulation are routinely monitored during thoracic spine surgery and thoracoabdominal aneurysm repair. H reflexes and F waves are monitored during thoracoabdominal aneurysm repair because they are sensitive indicators of lumbosacral spinal cord ischemia.[21] EMG monitoring is used when nerve roots or anterior horn cells within specific spinal segments are at risk.[20]

Nerve roots, spinal nerves, or elements of the plexus can be injured during spinal decompression or fusion especially when instrumentation is involved. Pedicle screw insertion may injure the nerve root directly or by narrowing the lateral recess or root foramen. Root injury can be prevented by placing a stimulating probe into the pedicle screw hole prior to screw insertion.[12, 33] The risk of root injury is low if the probe is stimulated with up to 20 mA or current at 0.1 msec duration to elicit an EMG response in muscles innervated by the root in question (Fig. 39–17). When anterior or posterior spinal fusion surgery, or both, is performed in the thoracolumbar area, the lumbar roots or plexus can be contused or stretched by surgical retractors, hooks, or other surgical

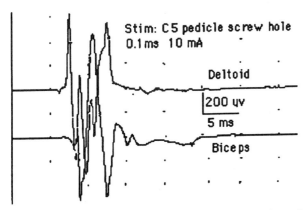

FIGURE 39–17. CMAP recordings in the biceps and deltoid muscles with intramuscular wire electrodes after stimulation of the C5 pedicle screw hole prior to screw placement. The low threshold for producing a CMAP (under 20 mA, 1 msec duration) suggested an increased risk of root injury by the pedicle screw. The screw hole was drilled at another location and the screw inserted after stimulation at 20 mA (1 msec duration) failed to elicit a CMAP. (Copyright © Mayo Foundation.)

instruments. EMG monitoring of muscles innervated by L2–4 segments can help detect mechanical trauma to these elements. Likewise L3–S1 roots are at risk for traction or ischemic injury, or both, when distraction is applied during scoliosis surgery. EMG monitoring can also help detect this type of injury. Of 184 patients who underwent thoracolumbar fusion for scoliosis at our institution between 1982 and 1984, we identified 4 who developed nerve root injury severe enough to be demonstrated on postoperative EMG studies.[45] Nerve root damage was often mistaken for spinal cord injury in the early postoperative period and added significant morbidity in some cases. There have been no further instances of nerve root injury in the 13 years since we began to perform EMG monitoring of L2–S1-innervated muscles routinely during scoliosis surgery.

The following cases represent examples of different types of spine procedures monitored as well as the use of multiple monitoring modalities in individual patients.

Case Examples

CASE STUDY 1

A 35-year-old woman with a slowly progressive myelopathy. Magnetic resonance imaging (MRI) of cervical spine revealed a large syrinx with an associated tumor in the cervical and thoracic regions. The surgical plan consisted of a multilevel laminectomy with decompression of the syrinx and biopsy of the tumor. Left and right tibial and ulnar SEPs, and EMG from biceps, triceps, abductor pollicis brevis, and first dorsal interosseous muscles were monitored continuously during surgery. Intraoperative baseline recordings revealed stable SEP recordings at the cervical spine (recorded from the esophageal electrode) and cortical levels. The amplitude of the ulnar and tibial SEPs fell after the dorsal myelotomy was made (Fig. 39–18). Neurotonic discharges were recorded in upper limb muscles when the tumor was biopsied and partially resected. The surgeon used the occurrence of neurotonic discharges to help determine the limits of tumor dissection. After surgery, the patient demonstrated a new proprioceptive deficit in the upper extremity but no new motor deficits were observed. This case shows that neurotonic discharges can be produced in extremity muscles when the anterior horn cells or intramedullary lower motor axons are stimulated mechanically. This phenomenon can be used to help guide the resection of intramedullary lesions.

CASE STUDY 2

A 12-year-old girl underwent resection of the T7 vertebra with a T2–T12 anterior and posterior fusion instrumentation for metastatic osteogenic sarcoma. One year earlier she had undergone a partial T7 corpectomy and anterior fusion without complication for metastatic osteogenic sarcoma. Despite radiation and chemotherapy, the tumor recurred at the T7 level and was associated with progressive kyphosis. Preoperative MRI scan showed narrowing of the central spinal canal at T6–T7 with mild to moderate spinal cord compression. There were no clinical manifestations of myelopathy before surgery. Monitoring consisted of left and right tibial SEPs recorded from a laminar needle at C6 referenced to F$_z$ and from C3–C4; and spinal MEPs recorded from surface electrodes over multiple lower limb muscles following electrical stimulation of the cervical spinal cord (cathode-esophageal electrode; anode-laminar needle at C6). Intraoperative base-

BIOPSY AND PARTIAL RESECTION OF CERVICAL INTRAMEDULLARY LESION

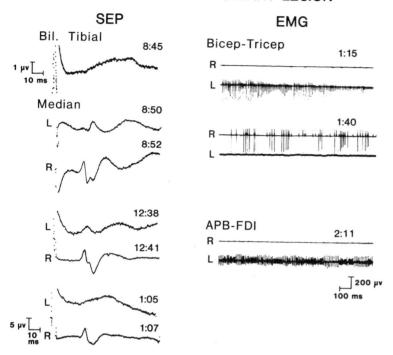

FIGURE 39–18. SEP and EMG recordings during cervical spine surgery. The cortical SEP following bilateral tibial nerve stimulation (Bil. Tibial) was absent on intraoperative baseline recordings. The median cortical SEP was lower on the left than the right at baseline and disappeared on the left after a dorsal myelotomy was performed. Neurotonic discharges were recorded in the biceps, triceps, abductor pollicis brevis (APB), and first dorsal interosseous (FDI) muscles during partial resection of the intramedullary tumor. Postoperatively the patient had a new proprioceptive deficit in the left upper extremity but no change in motor function. (Copyright © Mayo Foundation.)

line recordings showed well-defined SEPs and MEPs (Fig. 39–19).

During the posterior fusion, there was a loss of SEP and MEP responses over 40 to 60 minutes. The surgery was stopped and a "wake up" test confirmed weakness in both lower limbs. A postoperative MRI showed no change in the degree of spinal cord compression, but a new area of increased T2 signal in the adjacent thoracic spinal cord. The patient was initially paraplegic but improved dramatically within 24 hours at which time the anterior and posterior fusions were completed. No reliable SEP or MEP responses could be obtained during the second procedure. With physical rehabilitation she improved, but was left with moderate residual lower extremity weakness and sensory loss affecting all modalities with a level in the mid-thoracic region. This case shows a good correlation between intraoperative changes in MEP and SEP and the postoperative neurological status of the patient. The presumed mechanism of spinal cord injury in this patient was ischemia, as the postoperative MRI showed no further spinal cord compression than was evident on the preoperative scan.

CASE STUDY 3

A 69-year-old man underwent repair of a thoracoabdominal aneurysm. The aorta was cross-clamped without a shunt to the lower limbs. Monitoring consisted of left and right tibial SEPs recorded from a laminar needle at C6 referenced to F_z and from C_z–F_z; and spinal MEPs recorded from surface electrodes over multiple lower limb muscles following electrical stimulation of the cervical spinal cord (cathode-esophageal electrode; anode-laminar needle at C6). Intraoperative baseline recordings showed well-defined SEPs and MEPs (Fig. 39–20). (Only MEPs from left leg and right tibial SEP are displayed.) Approxi-

mately 25 minutes after the aorta was clamped, the MEP and SEP responses became diminished in amplitude and increased in latency. Over 15 minutes they disappeared but reappeared again within 10 minutes of unclamping. The tibial H reflex and CMAP following stimulation of the tibial nerve at the ankle also disappeared (not shown). This pattern is commonly observed in the setting of generalized transient lower body ischemia and is not typically associated with a postoperative neurological deficit. This patient's neurological examination after surgery was normal. If the SEP, MEP, and H reflex (but not the peripheral tibial CMAP) disappear within 15 minutes of clamping, the risk of postoperative paraplegia increases dramatically.

CASE STUDY 4

A 15-year-old boy underwent resection and repair of a lumbosacral lipomeningocele. There was no neurological deficit before surgery. Monitoring consisted of left and right tibial SEPs recorded from a laminar needle at C6 referenced to F_z and from C_z–F_z (not shown); and intramuscular wire electrodes in the tibialis anterior, abductor hallucis, and anal sphincter muscles bilaterally (Fig. 39–21). The surgeon used a hand-held electrical stimulator to identify functioning neural elements of the cauda equina. Neurotonic discharges were recorded when elements of the cauda equina were mechanically stimulated. No neurological deficit was noted after surgery.

Cranial Nerve Surgery

The modalities used to monitor different cranial nerves are outlined in Table 39–4. Location and the nerves at risk classify the types of surgery that are monitored.

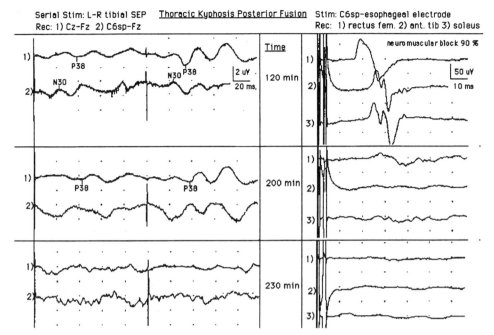

FIGURE 39–19. SEP and MEP monitoring during posterior fusion for kyphosis. The tibial SEP was monitored with serial stimulation of the left and right tibial nerve at the ankle (interstimulus interval:100 msec; stimulus rate: 1.9 Hz). The cortical potential was recorded with surface electrodes (C_z–F_z) and the cervical spine potential was recorded with a laminar needle at C6 referenced to a subcutaneous needle at F_z (number averaged 200). Motor evoked potentials were recorded in the setting of a constant infusion of neuromuscular blocking agent (peripheral ulnar CMAP maintained at 10% of the baseline). The cervical spinal cord was stimulated with an esophageal electrode as the cathode and a laminar needle at C6 as the anode (intensity: 100 mA; duration: 1 msec, paired stimulus with 3 msec interstimulus interval). The MEP was recorded with surface electrodes over the rectus femoris, anterior tibial, and soleus muscles bilaterally (no averaging; only the left side is shown). At the beginning of the posterior fusion (120 minutes after induction), both SEP and MEP recordings were well defined, although the left cortical SEP was lower in amplitude with a longer latency than the right cortical response. Over a 40- to 60-minute period, the MEP became unrecordable followed by disappearance of the SEP as well. (Copyright © Mayo Foundation.)

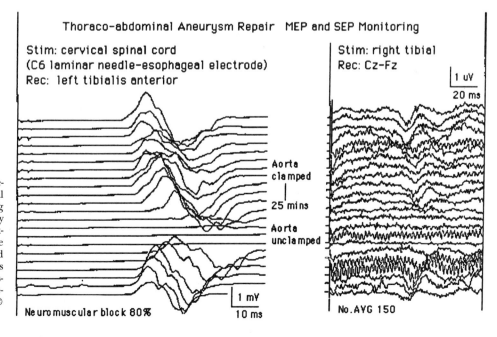

FIGURE 39–20. MEP and SEP recording during thoracoabdominal aneurysm repair demonstrating transient changes in both, likely related to generalized lower extremity ischemia (see text). The tibial SEP and MEP were recorded bilaterally but only the responses from the left tibialis anterior muscle and the right tibial cortical response are shown. (Copyright © Mayo Foundation.)

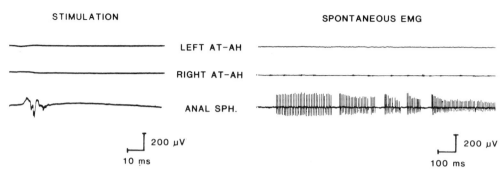

FIGURE 39–21. CMAP and EMG monitoring during resection and repair of a lipomeningocele. Recordings from intramuscular wire electrodes in the left and right anterior tibial (AT), abductor hallucis (AH), and anal sphincter (ANAL SPH.) muscles are displayed. Direct electrical stimulation with a hand-held stimulator was performed to help identify functional elements of the cauda equina (CMAP in the anal sphincter is displayed on the left). Mechanical stimulation of the roots produced neurotonic discharges in appropriate muscles (displayed in the anal sphincter on the right). (Daube JR, Harper CM: Surgical monitoring of cranial and peripheral nerves. In Desmedt JE [ed]: Neuromonitoring in Surgery. Amsterdam, Elsevier, 1989, pp 115–151.)

Middle Fossa Surgery

The occulomotor, trochlear, and abducens nerves are at risk during surgery in the region of the orbit, supraorbital ridge, cavernous sinus, or petrous portion of the temporal bone. Types of cases that may benefit from monitoring include tumors such as meningeomas, lymphomas, carcinomas, pituitary adenomas, and vascular lesions such as carotid or ophthalmic aneurysms. Especially in the case of neoplasms, the normal anatomy is distorted making it difficult to locate and identify the cranial nerves that control ocular motility. These nerves are monitored with EMG and CMAP monitoring. After the patient is anesthetized, intramuscular wire electrodes are put into the medial rectus muscle to monitor the occulomotor nerve, in the lateral rectus muscle to monitor the abducens nerve, and in the superior oblique muscle to monitor the trochlear nerve. An ophthalmologist can insert the wire electrodes into the extraoccular muscles if the neurophysiologist is unfamiliar with the technique.

Once inserted and connected to the amplifier, free-running EMG activity is monitored continuously for the occurrence of neurotonic discharges induced by mechanical irritation of the cranial nerves. In addition, selective direct stimulation of the nerves within the surgical field with a hand-held stimulator is used to distinguish the cranial nerves from each other and from non–nervous system tissue.

Posterior Fossa Surgery

The intracranial portions of cranial nerves VI through XII are at risk during a variety of surgical procedures involving

the posterior fossa. The most frequent procedure monitored is acoustic neuroma surgery, followed by microvascular decompression for hemifacial spasm or trigeminal neuralgia. Cranial nerve involvement is common with acoustic neuromas, especially in tumors that are over 2 cm in diameter.[41, 42, 52, 102] Hearing loss is the most frequent presenting symptom, but numbness and weakness of the face may also occur, especially with larger tumors.[41, 42] Electrophysiological signs of facial nerve damage (low facial CMAP, abnormal blink reflex, or abnormal EMG) are present in 75% of cases before surgery.[38–40] The extent of abnormality is proportional to the size of the tumor and is an excellent predictor of the extent of postoperative deficit.[39] In addition to cranial nerves V, VII, and VIII, the brain stem is also at risk during surgery for large acoustic neuromas.

Monitoring patients with acoustic neuroma requires the recording of multiple different modalities simultaneously (Fig. 39–22). EMG and CMAP recordings are used to monitor the trigeminal and facial nerves while auditory evoked potentials monitor cranial nerve VIII, and SEPs monitor brain stem function. The recording of neurotonic discharges in facial or trigeminal innervated muscles helps locate the nerves. Frequent neurotonic discharges during tumor dissection warn of potential mechanical trauma to the nerve. Before and after critical stages of surgery, the facial nerve is electrically stimulated proximal to the area of tumor dissection. The resulting CMAP recorded with surface electrodes over facial muscles provides a direct measure of the number of functioning facial nerve axons. This information is used to plan the remainder of surgery and to predict facial nerve function after surgery (Fig. 39–23). Patients with a preserved facial CMAP with proximal stimulation at the end of surgery have less severe postoperative deficits that recover more rapidly than those with an absent response.[38, 40]

The monitoring modalities chosen for each patient depend on the size of the tumor and the presence of any preoperative deficit. Preoperative BAEPs, facial nerve conduction studies, blink reflexes, and EMG define the amount of preoperative nerve damage and reliably predict the likelihood of further loss of function during surgery.[38, 39, 96] In small tumors, EMG from facial muscles, the facial CMAP, and AEPs are monitored. In larger tumors, EMG from muscles innervated by the facial and trigeminal nerves is monitored and SEPs are monitored when there are radiological or clinical signs of brain stem compression.

As surgical techniques have improved, there has been a

TABLE 39–4	Cranial Nerve Surgical Monitoring			
Cranial Nerve	EMG	CMAP	NAP	Evoked Potential
II				?VEP
III, IV, VI	+	+		
V	+	+	?	?
VII	+	+		
VIII			+	AEP
IX, X, XI, XII	+	+		

AEP, auditory evoked potential; CMAP, compound muscle action potential; EMG, electromyography; NAP, nerve action potential; VEP, visual evoked potential.
+, definite utility; ?, questionable utility.

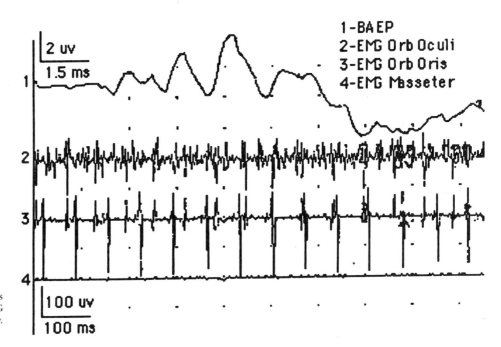

FIGURE 39–22. Simultaneous monitoring of BAEP and EMG during posterior fossa surgery. (Copyright © Mayo Foundation.)

greater emphasis placed on the preservation of facial nerve function during acoustic neuroma surgery. The outcome depends on tumor size and location and, to a lesser extent, on the specific surgical techniques utilized.[39, 60, 113] The majority of patients with large tumors (> 4 cm) and from 20% to 50% of patients with small to medium-sized tumors have complete facial paralysis postoperatively, even when the nerve is anatomically intact at the end of surgery.[39] If the nerve is left anatomically intact, recovery of function eventually occurs in up to two-thirds of patients.[38–40, 52] Using historical controls, several studies have reported improvement in anatomical preservation and function of the facial nerve during acoustic neuroma surgery attributable to intraoperative facial EMG monitoring.[25, 38, 69, 87, 123]

The monitoring of auditory nerve function during surgery has improved the preservation of hearing in patients with small acoustic neuromas.[46, 81, 100, 104, 129] When AEPs can be reliably recorded, monitoring should be attempted even in patients with relatively poor hearing preoperatively. Should the patient develop a contralateral acoustic neuroma or some other cause of hearing loss, then even poor hearing is better than total deafness. The BAEP is the most common AEP used during acoustic neuroma surgery. The BAEP monitors the full extent of the auditory pathway for the entire

surgical procedure and does not require recording electrodes in the surgical field. An external ear canal electrode can be used to enhance the resolution of wave I during surgery. In most patients with acoustic neuromas who have hearing preoperatively, waves I and V are readily identifiable using this method.

BAEP recordings are performed continuously and simultaneously with EMG monitoring once the dura is opened and the cerebellar retractor is put in place. The surgeon is alerted when there are reproducible changes in the amplitude or latency of waves I through V compared to baseline values. Gradual changes may be corrected by reducing the amount of cerebellar retraction or altering the approach to tumor dissection. Sudden loss of the BAEP indicates vascular injury or disruption of the nerve, which is irreversible. When the BAEP cannot be recorded reliably, the auditory NAP, which is typically much larger, can be recorded with an electrode placed on to the nerve directly within the surgical field. This technique is restricted to patients with small tumors during portions of the surgery where there is adequate exposure of the auditory nerve near the brain stem. The electrode can be easily dislodged or covered by cerebrospinal fluid, both of which will cause the amplitude of the NAP to vary considerably. The effect of BAEP monitoring on hearing preservation in patients with acoustic tumors has been studied using historical controls matched for preoperative hearing status and tumor size.[46] The rate of hearing preservation is significantly greater in monitored patients with tumors less than 2 cm in diameter even if wave I is the only component of the BAEP recorded at the beginning of surgery.

Cranial nerve monitoring is beneficial during surgery for tumors other than acoustic neuromas. Neoplasms of the cerebellopontine angle (e.g., meningiomas, epidermoid, metastases, etc.) are monitored using an identical protocol to that described for acoustic neuroma. Lesions in the region of the jugular foramen (e.g., glomus tumors), foramen magnum (e.g., meningioma), or clivus (e.g., chordoma) may risk injury to the lower cranial nerves. Often EMG monitoring is combined with AEP monitoring and SEP monitoring in individual cases. The spinal accessory nerve is monitored with intramuscular wire electrodes in either the sternocleidomastoid or trapezius muscle. The external and recurrent laryngeal nerves are monitored with electrodes in the crico-

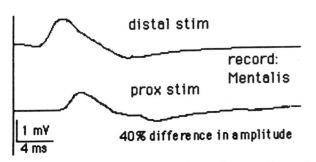

FIGURE 39–23. Facial CMAP recorded over the mentalis muscle during acoustic neuroma surgery. The difference in the amplitude/area of the CMAP obtained with direct stimulation proximal and distal to the area of tumor dissection is used to estimate the amount of injury that has occurred at various stages of the surgery. (Copyright © Mayo Foundation.)

thyroid and vocalis muscles, respectively. The hypoglossal nerve is monitored with electrodes placed in the tongue through the submental approach. The muscles are monitored for neurotonic discharges, while direct electrical stimulation can be used to identify specific nerves intraoperatively.

MVD is a commonly employed treatment for trigeminal neuralgia and hemifacial spasm and is sometimes used in chronic vertigo. During each of these procedures there is a slight risk of facial nerve injury and a 10% to 15% risk of hearing loss.[3] AEP and facial nerve monitoring are typically employed in all cases. Neurotonic discharges are a valuable indicator of mechanical trauma to the facial nerve. Changes in the amplitude and latency of wave V of the BAEP occur when the auditory nerve is stretched or mechanically injured. Electrical stimulation of the facial nerve is less useful because, unlike surgery for posterior fossa tumors, the cranial nerve anatomy is well preserved in trigeminal neuralgia and hemifacial spasm. Studies have shown that AEP monitoring reduces the risk of hearing loss in patients undergoing MVD surgery.[100]

In patients with hemifacial spasm the "lateral spread" reflex can be monitored during MVD. The "lateral spread" reflex represents ephaptic transmission between facial axons at the site of vascular compression.[88] It is recorded from a facial muscle innervated by one branch of the facial nerve when a different branch of the facial nerve is stimulated antidromically. Typically the response is recorded in the orbicularis oculi muscle when the mandibular branch of the facial nerve is stimulated or from the orbicularis oris (or mentalis) muscle when the zygomatic branch of the facial nerve is stimulated. The response is 100 to 200 μV with a latency of 7 to 10 msec. During surgery, the same wire electrodes are used for both EMG monitoring and recording of the "lateral spread" reflex. The "lateral spread" reflex persists under general anesthesia and can be monitored throughout surgery. The response typically disappears when the facial nerve is adequately decompressed[88] (Fig. 39–24). When hemifacial spasm is mild or intermittent before surgery, the "lateral spread" reflex may disappear when the dura is opened or with cerebrospinal fluid drainage. This presumably changes the relationship of the blood vessel causing the spasm and the facial nerve enough to interrupt ephaptic transmission. Absence of the "lateral spread" reflex at the end of surgery accurately predicts the absence of clinical spasms postoperatively.[36, 88, 110]

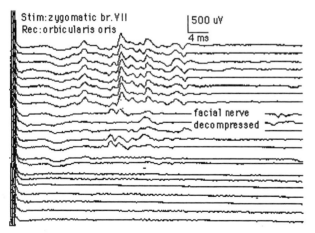

FIGURE 39–24. Recording of the "lateral spread" response in the oribicularis oris muscle in response to stimulation of the zygomatic branch of the facial nerve during microvascular decompression surgery for hemifacial spasm. The "lateral spread" response disappears after the arterial loop is moved away from the facial nerve. (Copyright © Mayo Foundation.)

Surgery is sometimes performed to section the glossopharyngeal and sensory portions of the vagus nerve in patients with intractable glossopharyngeal neuralgia.[115, 116] During these cases, it is may be difficult for the surgeon to tell where the rootlets that constitute the glossopharyngeal and sensory vagal nerves end and those that constitute the motor vagal nerve begin. These cases are monitored with wire electrodes in the cricothyroid or the vocalis muscles, both of which are innervated by the vagus nerve. With a small electrical stimulator, the surgeon is able to identify and avoid damage to the motor components of the vagus nerve.

Extracranial Surgery

The extracranial portions of the facial and laryngeal nerves are at risk during various types of head and neck surgery. Facial nerve monitoring is employed during surgery for parotid gland tumors. Intramuscular wire electrodes are placed in the frontalis, orbicularis oculi, orbicularis oris, and mentalis muscles so that each of the major branches of the facial nerve can be monitored. The wire electrodes are inserted near the midline of the face with the recording connections made on the side contralateral to the surgery in order to keep the recording system away from the surgical field. EMG activity is monitored continuously for neurotonic discharges, and intermittent direct electrical stimulation is performed to help identify branches of the facial nerve.

The superior and recurrent laryngeal nerves are monitored during complicated neck dissections such as those for recurrent thyroid cancer or during carotid endarterectomy.[20, 79] Intramuscular electrodes are placed into the cricothyroid and vocalis muscles. EMG activity is monitored continuously, and direct electrical stimulation is applied as outlined earlier for other cranial nerves.

Peripheral Nerve Surgery

SEP, EMG, CMAP, and NAP recordings are utilized during surgery for various entrapment neuropathies, repair of traumatic nerve injuries, and resection of tumors that affect peripheral nerves. The most proximal segments of the peripheral nervous system (roots and spinal nerves) are monitored during spinal surgery. Monitoring during surgery on more distal elements (i.e., brachial and lumbar plexus and individual nerves of the extremities) is performed to improve the localization of peripheral nerve lesions, determine the presence or absence of axonal regeneration, and to protect nerve fascicles from injury during surgery.

Preoperative nerve conduction studies (NCS) and needle EMG are used to localize and characterize the majority of lesions affecting the peripheral nervous system.[9, 86] When chronic lesions produce segmental demyelination, mechanical distortion of paranodal myelin, or impaired function of ion channels at the nodes of Ranvier, then conduction block and focal slowing of conduction velocity are observed on routine NCS. Two important criteria are required to demonstrate conduction block or focal slowing of conduction velocity. First, one must be able to stimulate the nerve in question proximal and distal to the lesion (preferably in short 1- to 2-cm segments through the area of the lesion). Second, the fascicles that contain the lesion have to be stimulated or recorded in isolation from fascicles belonging to other nerves in close proximity. These criteria cannot always be fulfilled when a lesion affects proximal or deep nerves. In this setting, performing CMAP or NAP recordings, or both, over short segments of exposed nerve at surgery is very helpful. When partial or complete axon loss is the major pathological substrate, then localization with preoperative NCS and EMG is much less precise. In this setting CMAP or

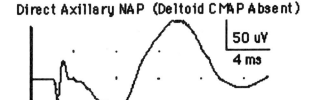

Direct Axillary NAP (Deltoid CMAP Absent)

50 uV
4 ms

No. AVG 10

FIGURE 39–25. NAP recording from the axillary nerve during surgical repair of a severe traumatic axillary neuropathy. Preoperative nerve conduction studies and needle EMG showed an absent axillary CMAP and fibrillation potentials with no motor unit potentials activated in the deltoid muscle. The branch to the teres minor muscle was spared. Intraoperative stimulation confirmed the absence of a deltoid CMAP, but a well-defined NAP was recorded from the axillary nerve distal to the teres minor branch. This indicated that axonal regeneration was taking place but had not yet reached the deltoid muscle. The patient was treated conservatively and recovered good function of the deltoid muscle. (Harper CM, Daube JR: Facial nerve EMG and other cranial nerve monitoring. J Clin Neurophysiol 15:206–216, 1998.)

NAP recordings, or both, performed during surgery can often help localize the main site of peripheral nerve pathology and can identify functioning nerve fascicles that are preserved or regenerating across the site of pathology (Fig. 39–25). Theoretically, a purely axonal lesion should not cause slowed segmental conduction. However, in most instances minor degrees of demyelination or axonal narrowing are associated with axonal lesions. Thus, slowed conduction along the short segment of nerve containing the lesion can frequently be demonstrated by recording CMAPs and NAPs. In the absence of conduction block, there is no change in CMAP amplitude between proximal and distal stimulation sites. In contrast, a segmental change in NAP amplitude can

frequently be demonstrated even in purely axonal lesions. To show this, a bipolar hooked nerve electrode is placed proximal to the lesion for NAP recording. This proximal placement eliminates movement artifact and CMAPs caused by contraction of adjacent muscles. The stimulator is moved proximally, in successive 1- to 2-cm segments. Lesion location is identified by a change in amplitude of the direct NAP, and by slowed conduction velocity within short nerve segments. This technique is useful even with complete axonal lesions that do not permit a CMAP to be recorded with either proximal or distal stimulation.

Entrapment Neuropathies

Localization by clinical signs and preoperative electrophysiological studies is adequate in the majority of cases of entrapment neuropathy.[9] Carpal tunnel release and most cases of ulnar transposition are not monitored. However, complicated cases of ulnar or median neuropathy in the forearm, or of radial, femoral, sciatic, tibial, and peroneal neuropathy are often monitored because of inherent difficulties in defining the number and location of lesions in these cases.

Ulnar transposition has traditionally been the most popular surgical procedure for ulnar neuropathy at the elbow. However, because transposition is not always successful and may increase morbidity, more conservative procedures such as simple cubital tunnel release have been advocated.[23] In some cases, precise localization of the lesion using preoperative and intraoperative electrophysiological studies may be very helpful in choosing the appropriate surgical procedure. A well-localized lesion in the region of the two heads of the flexor carpi ulnaris is treated with a simple cubital tunnel release, whereas a transposition is performed when there is a more diffuse lesion or one localized at or proximal to the medial epicondyle. Lesions at unusual sites, such as the distal aspect of the cubital tunnel, have also been localized with intraoperative NCS.[13] Figure 39–26 illustrates results of intra-

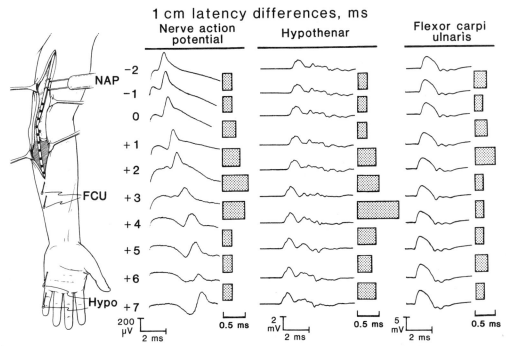

INTRAOPERATIVE ULNAR NERVE TESTING

1 cm latency differences, ms

Nerve action potential   Hypothenar   Flexor carpi ulnaris

FIGURE 39–26. Short segmental stimulation during surgery for ulnar neuropathy. The "0" marks the location of the medial epicondyle. The NAP and CMAP recordings both demonstrate focal slowing of conduction 1 to 3-cm distal to the medial epicondyle, which correlated with the origin of the cubital tunnel. (Copyright © Mayo Foundation.)

operative conduction studies performed during an ulnar nerve exploration. Preoperative NCS revealed localized slowing with increased CMAP dispersion approximately 3-cm distal to the medial epicondyle. CMAPs were recorded intraoperatively over the abductor digiti minimi and flexor carpi ulnaris muscles and the NAPs were recorded from the proximal ulnar nerve. There were changes in amplitude and latency of both CMAP and NAP over a 3-cm segment at the origin of the cubital tunnel. There were no areas of slowing proximal or distal to this point, so a cubital tunnel release was performed.

Repair of Traumatic Peripheral Nerve Injury

Intraoperative monitoring is particularly useful when multiple, deep, proximal nerves are injured and when multiple potential mechanisms of injury are involved (traction, contusion, ischemia, etc.). The primary purpose of monitoring in this setting is to localize the injured segment(s) and determine the presence or absence of axonal continuity across the injured area.

The utility of these techniques is best illustrated by examining their role in the surgical repair of traumatic injuries to the brachial plexus. The complexity of brachial plexus anatomy, multiplicity and severity of injury to its elements, and frequent occurrence of nerve root avulsion make lesions of this structure particularly difficult to evaluate and treat.[22, 63] The presence or absence of nerve root avulsion is one of the most important factors in determining prognosis and the need for surgical intervention in brachial plexus injuries. If root avulsion is present, then repair of postganglionic elements innervated by the avulsed root will be of no benefit. The clinical examination, NCS, needle EMG, and myelography are used preoperatively to assess the integrity of the cervical nerve roots.[22, 65] The combination of Horner's syndrome, denervation of paraspinal and other proximal muscles, preserved sensory nerve action potentials (SNAPs), and the presence of a meningocele on myelography in the setting of a paralyzed anesthetic limb strongly suggests the presence of multiple root avulsions. However, any one of these findings in isolation is much less specific. Examples of false-positive and false-negative myelograms have been well documented.[63] Because the posterior primary ramus of a given nerve root innervates paraspinal muscles at multiple levels, the distribution of fibrillation potentials may overestimate the number of roots involved. In addition, the presence of a postganglionic lesion with diminished SNAP may mask an associated lesion involving preganglionic segments. The predictive value of preoperative SEP in detecting root continuity as well as the presence of mixed pre- and postganglionic lesions has been disappointing.[133] These uncertainties can usually be resolved by performing SEP recordings during surgery.[112] In this setting, the exposed spinal nerve is stimulated directly by the surgeon while SEP recordings are made from the cervical spine or scalp, or both. If sensory fibers within the root are intact, then a well-defined SEP is recorded. The absence of a response confirms the presence of root avulsion at that level. Because this technique tests sensory function, it is theoretically possible that the ventral motor root could be avulsed with sparing of the dorsal root. If this occurred, the presence of an SEP would be misleading. Unfortunately, there is no way to directly assess the integrity of the ventral root. Therefore, when a well-defined SEP is present, it is probably in the patient's best interest to proceed with surgical repair of the postganglionic elements.

When root continuity is present, attention is turned to assessment and possible repair of postganglionic elements. Injured elements of the plexus are stimulated proximally, with attempts to record a CMAP from a distal muscle or an NAP from the nerve distal to the site of injury. When a CMAP is present, the surgeon can be confident that some motor axons are intact. In this setting, the lesion is left alone or at most a simple neurolysis is performed. When no CMAP is recorded, there may be complete disruption of axons at the site of injury or there may be regeneration across the lesion but insufficient time for the regenerating axons to reach a distal muscle. In this setting, the recording of an NAP across the injured segment suggests regeneration is taking place, whereas the absence of an NAP suggests severe and complete axon loss. The latter finding would suggest the need for lesion resection and subsequent grafting or a neurotization procedure.[62] Recording the NAP from individual nerve fascicles has been reported to help guide partial fascicular repair in the setting of incomplete lesions.[62] Technical difficulties related to the size of the electrodes, occurrence of shock artifact, and avoidance of current spread to adjacent fascicles have limited the widespread use of fascicular recordings.

Prevention of Injury During Peripheral Nerve Surgery

As with the cranial nerves, the peripheral nerves that innervate the trunk and extremity muscles are susceptible to mechanical or ischemic injury during a variety of surgical procedures. During these procedures, EMG monitoring for neurotonic discharges helps localize and warn of potential injury to the nerve, while direct electrical stimulation with a hand-held stimulator is used to identify viable nerves or fascicles.

Orthopedic procedures that involve disarticulation or extensive manipulation of the limbs are often monitored. EMG monitoring can help detect and prevent injury to the axillary and musculocutaneous nerves during shoulder surgery, and to the femoral, obturator, and sciatic nerves during high-risk hip surgery.[48]

EMG monitoring may also be useful during the resection of primary or metastatic peripheral nerve neoplasms. In this setting, the goal is to resect the tumor with as little damage to normal nerve fascicles as possible. During dissection, individual or groups of fascicles are stimulated mechanically or electrically while monitoring EMG activity in distal muscles. An attempt is made to preserve fascicles that produce a distal EMG response while those that do not are sacrificed.

REFERENCES

1. Allen A, Starr A, Nudleman K: Assessment of sensory function in the operating room utilizing cerebral evoked potentials: A study of fifty-six surgically anesthetized patients. Clin Neurosurg 28:457–481, 1981.
2. Allison T, Wood CC, Spencer DD, et al: Localization of sensorimotor cortex in surgery by SEP recording. Electroencephalogr Clin Neurophysiol 61:S74, 1985.
3. Auger RG, Piepgras DG, Laws ER, Jr: Hemifacial spasm. Results of microvascular decompression of the facial nerve in 54 patients. Mayo Clin Proc 61:640–644, 1986.
4. Baron MS, Vitek JL, Bakay RAE, et al: Treatment of advanced Parkinson's disease by posterior Gpi pallidotomy: 1-year results of a pilot study. Ann Neurol 40:355–366, 1996.
5. Ben-David B, Haller G, Taylor P: Anterior spinal fusion complicated by paraplegia: A case report of a false-negative somatosensory-evoked potential. Spine 12:536–539, 1987.
6. Bennett, M: Effects of compression and ischemia on spinal cord evoked potentials. Exp Neurol 80:508–519, 1983.
7. Berardelli A, Inghilleri M, Cruccu G, et al: Descending volley after electrical and magnetic transcranial stimulation in man. Neurosci Lett 112:54–58, 1990.
8. Bernard JM, Pereon Y, Fayet G, et al: Effects of isoflurane and desflurane on neurogenic motor and somatosensory-evoked potential monitoring for scoliosis surgery. Anesthesiology 85(5):1013–1019, 1996.

9. Brown WF, Ferguson GG, Jones MW, et al: The location of conduction abnormalities in human entrapment neuropathies. Can J Neurol Sci 3:111–122, 1976.
10. Burke D, Hicks R, Gandevia SC, et al: Direct comparison of corticospinal volleys in human subjects to transcranial magnetic and electrical stimulation. J Physiol 470:383–393, 1993.
11. Burke D, Hicks R, Stephen J: Corticospinal volleys evoked by anodal and cathodal stimulation of the human motor cortex. J Physiol 425:283–299, 1990.
12. Calancie B, Madsen P, Lebwohl N: Stimulus-evoked EMG monitoring during transpedicular lumbosacral spine instrumentation. Initial clinical results. Spine 19(24):2780–2786, 1994.
13. Campbell MD, Pridgeon RM, Sahni KS: Entrapment neuropathy of the ulnar nerve at its point of exit from the flexor carpi ulnaris muscle. Muscle Nerve 9:662, 1986.
14. Chiappa KH: Brain stem auditory evoked potentials: Interpretation. In Chiappa KH (ed): Evoked potentials in clinical medicine, 3rd ed. Philadelphia, Lippincott-Raven, 1997, pp. 199–251.
15. Chiappa KH: Short latency somatosensory evoked potentials: Interpretation. In Chiappa KH (ed): Evoked Potentials in Clinical Medicine, 3rd ed. Philadelphia, Lippincott-Raven, 1997, pp. 341–401.
16. Chiappa KH, Burke SR, Young RR: Results of electroencephalographic monitoring during 367 carotid endarterectomies. Use of a dedicated minicomputer. Stroke 10:381–388, 1979.
17. Cohen AR, Young W, Ransohoff J: Intraspinal localization of the somatosensory evoked potential. Neurosurgery 9:157–162, 1981.
18. Colletti V, Fiorino FG: Vulnerability of hearing function during acoustic neuroma surgery. Acta Oto-Laryngol 114:264–270, 1994.
19. Darden BV, Hatley MK, Owen JH: Neurogenic motor evoked potential monitoring in anterior cervical surgery. J Spinal Disord 9:485–493, 1996.
20. Daube JR, Harper CM: Surgical monitoring of cranial and peripheral nerves. In Desmedt JE (ed): Neuromonitoring in Surgery. Amsterdam, Elsevier, 1989, pp. 115–151.
21. Daube JR, Litchy WJ, Harper CM, et al: Monitoring spinal cord function during surgery with combined motor evoked potential and somatosensory evoked potential recordings. Neurology 43:A21, 1993.
22. Davis DH, Onofrio BM, MacCarty CS: Brachial plexus injuries. Mayo Clin Proc 53:799, 1978.
23. Dawson DM, Hallett M, Millender LH: Entrapment Neuropathies. Boston, Little, Brown, 1983, pp. 87–122.
24. Dawson EG, Sherman JE, Kanim LE, et al: Spinal cord monitoring. Results of the Scoliosis Research Society and the European Spinal Deformity Society survey. Spine 16(suppl)8:S361–S364, 1991.
25. Delgado TE, Buchheit WA, Rosenholtz HR, et al: Intraoperative monitoring of facial muscle evoked responses obtained by intracranial stimulation of the facial nerve. Neurosurg 418–420, 1979.
26. Edmonds H, Paloheimo M, Backman M, et al: Transcranial magnetic motor evoked potentials (tcMMEP) for functional monitoring of motor pathways during scoliosis surgery. Spine 14:683–686, 1988.
27. Engler G, Speilholz NJ, Bernhard WN, et al: Somatosensory evoked potentials during Harrington instrumentation for scoliosis. J Bone Joint Surg 60(4):528–532, 1978.
28. Forbes HJ, Allen PW, Waller CS, et al: Spinal cord monitoring in scoliosis surgery. Experience with 1168 cases. J Bone Joint Surg 73-B:487–491, 1991.
29. Friedman WA, Kaplan BJ, Gravenstein D, et al: Intraoperative brain stem auditory evoked potentials during posterior fossa microvascular decompression. J Neurosurg 62:552–557, 1985.
30. Fujiki M, Isono M, Hori S, et al: Corticospinal direct response to transcranial magnetic stimulation in humans. Electroencephalogr Clin Neurophysiol 101(1):48–57, 1996.
31. Ginsburg H, Shetter A, Raudzens P: Post-operative paraplegia with preserved intraoperative somatosensory evoked potentials. J Neurosurg 63:296–300, 1985.
32. Glassman SD, Dimar JR, Puno RM, et al: A prospective analysis of intraoperative electromyographic monitoring of pedicle screw placement with computed tomography scan confirmation. Spine 20:1375–1379, 1995.
33. Glassman SD, Zhang YP, Shields CB, et al: An evaluation of motor-evoked potentials for detection of neurologic injury with correction of an experimental scoliosis. Spine 20(16):1765–1775, 1995.
34. Grundy BL, Jannetta PJ, Lina A, et al: Intraoperative monitoring of brainstem auditory evoked potentials. J Neurosurg 57:674–677, 1981.
35. Grundy B, Nelson PB, Doyle E, et al: Intraoperative loss of somatosensory evoked potentials predicts loss of spinal cord function. Anesthesiology, 57(4):321–322, 1982.
36. Haines SJ, Torres F: Intraoperative monitoring of the facial nerve during decompressive surgery for hemifacial spasm. J Neurosurg 74:254–257, 1991.
37. Hall JE, Levine CR, Sudhir KG: Intraoperative awakening to monitor spinal cord function. Description of procedure and report of three cases. J Bone Joint Surg 60(4):533–536, 1978.
38. Harner SG, Daube JR, Beatty CW, Ebersold MJ: Intraoperative monitoring of the facial nerve. Laryngoscope 98:209–212, 1988.
39. Harner S, Daube JR, Ebersold M: Electrophysiologic monitoring of neural function during temporal bone surgery. Laryngoscope 96:65–69, 1986.
40. Harner SG, Daube JR, Ebersold MJ, Beatty CW: Improved preservation of facial nerve function with use of electrical monitoring during removal of acoustic neuromas. Mayo Clin Proc 69:92–102, 1987.
41. Harner S, Laws E: Posterior fossa approach for removal of acoustic neurinomas. Arch Otolaryngol 107:590–593, 1981.
42. Harner S, Laws, E: Clinical findings in patients with acoustic neurinomas. Mayo Clin Proc 58:721–728, 1983.
43. Harper CM, Daube JR: Surgical monitoring with evoked potentials: The Mayo Experience. In Desmedt JE (ed): Neuromonitoring in Surgery. Amsterdam, Elsevier, 1989, pp. 257–303.
44. Harper CM, Daube JR, Litchy WJ, et al: Monitoring thalamic SEP during surgery. Neurology 37(suppl 1):219, 1987.
45. Harper CM Jr, Daube JR, Litchy WJ, Klassen RA: Radiculopathy after spinal fusion for scoliosis. Muscle Nerve 11:386–391, 1988.
46. Harper CM, Harner SG, Slavit DH, et al: Effect of BAER monitoring on hearing preservation during acoustic neuroma resection. Neurology 42:1551–1553, 1992.
47. Harper CM, Nelson KR: Intraoperative electrophysiological monitoring in children. J Clin Neurophysiol 9:3423–3456, 1992.
48. Helfet DL, Anand N, Malkani AL, et al: Intraoperative monitoring of motor pathways during operative fixation of acute acetabular fractures. J Orthop Trauma 11(1):2–6, 1997.
49. Helmers SL, Hall JE: Intraoperative somatosensory evoked potential monitoring in pediatrics. J Pediatr Orthop 14(5):592–598, 1994.
50. Herdmann J, Lumenta CB, Huse KO: Magnetic stimulation for monitoring of motor pathways in spinal procedures. Spine 18(5):551–559, 1993.
51. Hicks RG, Woodforth IJ, Crawford MR, et al: Some effects of isoflurane on I waves of the motor evoked potential. Br J Anaesth 69:130–136, 1992.
52. House WF, Luetje, CM (eds): Acoustic Tumors. Baltimore, University Park Press, 1979.
53. Inghelleri M, Berardelli A, Cruccu G, Manfredi M: Corticospinal potentials after transcranial stimulation in humans. J Neurol Neurosurg Psychiatry 52:970–974, 1989.
54. Jones SJ, Carter L, et al: Experience of epidural spinal cord monitoring in 410 cases. In Schramm J, Jones S (eds). Spinal Cord Monitoring. Berlin, Springer, 1985, pp 207–227.
55. Jones SJ, Edgar MA, Ransford AO: Sensory nerve conduction in the human spinal cord: Epidural recordings made during scoliosis surgery. J Neurol Neurosurg Psychiatry 45:446–451, 1982.
56. Jones SJ, Harrison R, Koh KF: Motor evoked potential monitoring during spinal surgery: Responses of distal limb muscles to transcranial cortical stimulation with pulse trains. Electroencephalogr Clin Neurophysiol 100(5):375–383, 1996.
57. Kalkman C, Drummond J, Ribberink A: Low concentrations of isoflurane abolish motor evoked responses to transcranial electrical stimulation during nitrous oxide/opioid anesthesia in humans. Anesth Analg 73:410–415, 1991.
58. Kalkman CJ, van Rheineck-Leyssius AT, Hesselink EM, et al:

Effects of etomidate or midazolam on median nerve somatosensory evoked potentials. Anesthesiology 65(3A):A356, 1986.

59. Katayama Y, Tsubokawa T, Maejima S, et al: Corticospinal direct response in humans: Identification of the motor cortex during intracranial surgery under general anaesthesia. J Neurol Neurosurg Psychiatry 51:50–59, 1988.

60. Kirkpatrick PJ, Tierney P, Gleeson MJ: Acoustic tumor volume and the prediction of facial nerve functional outcome from intraoperative monitoring. Br J Neurosurg 7(6):657–664, 1993.

61. Kitagawa H, Itog T, Takano H, et al: Motor evoked potentials monitoring during upper cervical spine surgery. Spine 14:1078–1083, 1989.

62. Kline DG, DeJonge BR: Evoked potentials to evaluate peripheral nerve injuries. Surg Gynecol Obstet 127:1239, 1968.

63. Kline DG, Hackett ER, Happel LH: Surgery for lesions of the brachial plexus. Arch Neurol 43:170–181, 1986.

64. Lam AM, Sharar SR, Mayberg TS, et al: Isoflurane compared with nitrous oxide anaesthesia for intraoperative monitoring of somatosensory-evoked potentials. Can J Anaesth 41(4):295–300, 1994.

65. Landi A, Copeland SA, Wynn-Parry CD, et al: The role of somatosensory evoked potentials and nerve conduction studies in the surgical management of brachial plexus injuries. J Bone Joint Surg 62B:492, 1980.

66. Lang EW, Beutler AS, Chesnut RM, et al: Myogenic motor-evoked potential monitoring using partial neuromuscular blockade in surgery of the spine. Spine 21(14):1676–1686, 1996.

67. Lang EW, Chesnut RM, Beutler AS, et al: The utility of motor-evoked potential monitoring during intramedullary surgery. Anesth Analg 83(6):1337–1341, 1996.

68. Lee WY, Hou WY, Yang LH, et al: Intraoperative monitoring of motor function by magnetic motor evoked potentials. Neurosurgery 36(3):593–500, 1995.

69. Lenarz T, Ernst A: Intraoperative facial nerve monitoring in the surgery of cerebellopontine angle tumors: Improved preservation of nerve function. J Oto-Rhino-Laryngol 56:31–35, 1994.

70. Lennon RL, Hosking MP, Daube JR, Welna JO: Effect of partial neuromuscular blockade on intraoperative electromyography in patients undergoing resection of acoustic neuromas. Anesth Analg 75:729–733, 1992.

71. Lesser RP, Raudzens P, Luders H, et al: Postoperative neurological deficits may occur despite unchanged intraoperative somatosensory evoked potentials. Ann Neurol 19:22–25, 1986.

72. Levy W: Clinical experience with motor and cerebellar evoked potential monitoring. Neurosurgery 20:169–182, 1987.

73. Levy W, York D, McCaffrey M, Tanzer F: Motor evoked potentials from transcranial stimulation of the motor cortex in humans. Neurosurgery 15:214–227, 1984.

74. Litchy WJ, Harper CM, Daube JR: Electrically evoked motor potentials (MEP) during surgical monitoring. Electroencephalogr Clin Neurophysiol 87:S105, 1993.

75. Lorenzini NA, Schneider JH: Temporary loss of intraoperative motor-evoked potential and permanent loss of somatosensory-evoked potentials associated with a postoperative sensory deficit. J Neurosurg Anesthesiol 8:142–147, 1996.

76. MacEwen GD, Bunnell WP, Sriram K: Acute neurological complications in the treatment of scoliosis. A report of the Scoliosis Research Society. J Bone Joint Surg 57A:404–408, 1975.

77. Machida M, Weinstein S, Yamada T, Kimura J: Spinal cord monitoring. Spine 10:407–413, 1985.

78. Macon JB, Poletti CE: Conducted somatosensory evoked potentials during spinal surgery. Part 1: Control conduction velocity measurements. J Neurosurg 57:349–359, 1982.

79. Markand ON, Dilley RS, Moorthy SS, et al: Monitoring of somatosensory evoked responses during carotid endarterectomy. Arch Neurol 41:375–378, 1984.

80. Matthies C, Samii M: Direct brainstem recording of auditory evoked potentials during vestibular schwannoma resection: Nuclear BAEP recording. Technical note and preliminary results. J Neurosurg 86(6):1057–1062, 1997.

81. Matthies C, Samii M: Management of vestibular schwannomas (acoustic neuromas): The value of neurophysiology for intraoperative monitoring of auditory function in 200 cases. Neurosurgery 40(3):459–466, 1997.

82. McDaniel AB, Silverstein H, Norrell H: Retrolabyrinthine vestibular neurectomy with and without monitoring of eighth nerve potentials. Am J Otol Suppl 23–26, 1985.

83. McPherson RW, Mahla M, Johnson R, et al: Effects of enflurane, isoflurane, and nitrous oxide on somatosensory evoked potentials during fentanyl anesthesia. Anesthesiology 62:626–633, 1985.

84. McPherson RW, Niedermeyer EF, Otenasek RJ, Hanley DF: Correlation of transient neurological deficit and somatosensory evoked potentials after intracranial aneurysm surgery. J Neurosurg 59:146–149, 1983.

85. McPherson RW, Sell B, Traystman RJ, et al: Effects of thiopental, fentanyl, and etomidate on upper extremity somatosensory evoked potentials in humans. Anesthesiology 65:584–589, 1986.

86. Miller RG: The cubital tunnel syndrome: Diagnosis and precise localization. Ann Neurol 6:56, 1979.

87. Moller AR, Jannetta PJ: Preservation of facial function during removal of acoustic neuromas. Use of monopolar constant-voltage stimulation and EMG. J Neurosurg 61:757–760, 1984.

88. Moller AR, Jannetta PJ: Microvascular decompression in hemifacial spasm. Neurosurgery 16:612–618, 1985.

89. Nagle KJ, Emerson RG, Adams DC, et al: Intraoperative monitoring of motor evoked potentials: A review of 116 cases. Neurology 47(4):999–1004, 1996.

90. Nash CL Jr, Lorig RA, Schatzinger LA, et al: Spinal cord monitoring during operative treatment of the spine. Clin Orthoped 126:100–105, 1977.

91. Nuwer MR, Dawson EG, Carlson LG, et al: Somatosensory evoked potential spinal cord monitoring reduces neurologic deficits after scoliosis surgery: Results of a large multicenter survey. Electroencephalog Clin Neurophysiol 96:6–11, 1995.

92. Ojemann RG, Levine RA, Montgomery WM, et al: Use of intraoperative auditory evoked potentials to preserve hearing in unilateral acoustic neuroma removal. J Neurosurg 61:938–948, 1984.

93. Owen J, Jenny A, Naito M, et al: Effects of spinal cord lesioning on somatosensory and neurogenic-motor evoked potentials. Spine 14:673–682, 1989.

94. Owen JH, Bridwell K, Grubb R, et al: The clinical application of neurogenic motor evoked potentials to monitor spinal cord function during surgery. Spine 16:S384–390, 1991.

95. Pathak KS, Brown RH, Cascorbi HF, et al: Effects of fentanyl and morphine on intraoperative somatosensory cortical-evoked potentials. Anesth Analg 63:833–837, 1984.

96. Pavesi G, Macaluso GM, Tinchelli S, et al: Presurgical electrophysiological findings in acoustic nerve tumours. Electromyogr Clin Neurophysiol 32:119–123, 1992.

97. Phillips LH, Blanco JS, Sussman MD: Direct spinal stimulation for intraoperative monitoring during scoliosis surgery. Muscle Nerve 18:319–325, 1995.

98. Powers SK, Bolger CA, Edwards MSB: Spinal cord pathways mediating somatosensory evoked potentials. J Neurosurg 57:472–482, 1982.

99. Prestor B, Zgur T, Dolenc V: Epidural and subpial corticospinal potentials evoked by transcutaneous motor cortex stimulation during spinal cord surgery. In Rossini PM, Mauguiere F (eds): New Trends and Advanced Techniques in Clinical Neurophysiology, Amsterdam, Elsevier, 1990, pp 346–351.

100. Radtke RA, Erwin CW, Wilkins R, et al: Intraoperative brainstem auditory evoked potentials: Significant decrease in postoperative deficit. Neurology 37(suppl 1):219, 1987.

101. Raudzens PA, Shetter AG: Intraoperative monitoring of brain stem auditory evoked potentials. J Neurosurg 57:341–348, 1982.

102. Rhoton AL: Microsurgical removal of acoustic neuromas. Surg Neurol 6:211–219, 1976.

103. Rothwell JC, Thompson PD, Day BL, et al: Motor cortex stimulation in intact man. 1. General characteristics of EMG responses in different muscles. Brain 110:1173–1190, 1987.

104. Samii M, Matthies C: Management of 1000 vestibular schwannomas (acoustic neuromas): The facial nerve—preservation and restitution of function. Neurosurgery 40:684–694, 1997.

105. Sekiya T, Moller AR: Cochlear injuries caused by cerebellopontine angle manipulations. An electrophysiological and morphological study in dogs. J Neurosurg 67:244–249, 1987.

106. Schonle P, Isenberg C, Crozier T, et al: Changes of transcranially evoked motor responses in man by midazolam, a short acting benzodiazepine. Neurosci Lett 101:321–324, 1989.

107. Schwartz DM, Drummond DS, Ecker ML: Influence of rigid spinal instrumentation on the neurogenic motor evoked potential. J Spinal Disord 95(5):439–445, 1996.
108. Sekiya T, Moller AR: Cochlear injuries caused by cerebellopontine angle manipulations. An electrophysiological and morphological study in dogs. J Neurosurg 67:244–249, 1987.
109. Shields C, Paloheimo M, Backman M, et al: Intraoperative transcranial magnetic motor evoked potentials are difficult to obtain during lumbar disk and spinal tumor operations. Muscle Nerve 11:993, 1988.
110. Shin JC, Kim YC, Park CI, et al: Intraoperative monitoring of microvascular decompression in hemifacial spasm. Yonsei Med J 37(3):209–213, 1996.
111. Stinson LW, Murray MJ, Jones KA, et al: A computer-controlled, closed-loop infusion system for infusing muscle relaxants: Its use during motor-evoked potential monitoring. J Cardiothorac Vasc Anesth 8(1):40–44, 1994.
112. Sugioka H, Tsuyana N, Hara T, et al: Investigation of brachial plexus injuries by intraoperative cortical somatosensory evoked potentials. Arch Orthop Trauma Surg 99:143, 1982.
113. Sugita K, Kobayashi S, Matsuga N, et al: Microsurgery for acoustic neurinoma—lateral position and preservation of facial and cochlear nerves. Neurol Med Chir (Tokyo) 19:637–641, 1979.
114. Symon L, Wang AD, Costa e Silva IE, et al: Perioperative use of somatosensory evoked responses in aneurysm surgery. J Neurosurg 60:269–275, 1984.
115. Taha JM, Tew JM Jr: Long-term results of surgical treatment of idiopathic neuralgias of the glossopharyngeal and vagal nerves. J Neurosurg 36(5):926–930, 1995.
116. Taha JM, Tew JM Jr, Keith RW, et al: Intraoperative monitoring of the vagus nerve during intracranial glossopharyngeal and upper vagal rhizotomy: Technical note. Neurosurgery 35(4):775–777, 1994.
117. Tamaki T, Noquchi T, Takano H, et al: Spinal cord monitoring as a clinical utilization of spinal evoked potentials. Clin Orthop 184:58–64, 1984.
118. Taniguchi M, Nadstawek J, Langenbach U, et al: Effects of four intravenous anesthetic agents on motor evoked potentials elicited by magnetic transcranial stimulation. Neurosurgery 33(3):407–415, 1993.
119. Tator CH, Nedzelski JM: Preservation of hearing in patients undergoing excision of acoustic neuromas and other cerebellopontine angle tumors. J Neurosurg 63:168–174, 1985.
120. Taylor BA, Fennelly ME, Taylor A, et al: Temporal summation—the key to motor evoked potential spinal cord monitoring in humans. J Neurol Neurosurg Psychiatry 56(1):104–106, 1993.
121. Thompson P, Day B, Crockard H, et al: Intraoperative recording of motor tract potentials at the cervico-medullary junction following scalp electrical and magnetic stimulation of the motor cortex. J Neurol Neurosurg Psychiatry 54:618–623, 1991.
122. Thompson P, Rothwell J, Day B, et al: Mechanisms of electrical and magnetic stimulation of human motor cortex. In Chokreverty S (ed): Magnetic Stimulation in Clinical Neurophysiology. London, Butterworth, 1989, pp 121–143.
123. Torrens M, Maw R, Coakham H, et al: Facial and acoustic nerve preservation during excision of extracanalicular acoustic neuromas using the suboccipital approach. Br J Neurosurg 8:655–665, 1994.
124. Tsuyama N, Tsuzuki N, Kurokawa T, et al: Clinical application of spinal cord action potential measurement. Int Orthop 2:39–46, 1978.
125. Ubags LH, Kalkman CJ, Been HD, et al: The use of a circumferential cathode improves amplitude of intraoperative electrical transcranial myogenic motor evoked responses. Anesth Analg 82(5):1011–1014, 1996.
126. Vauzelle C, Stagnara P, Jouvinroux P: Functional monitoring of spinal cord activity during spinal surgery. Clin Orthop 93:173–178, 1973.
127. Wang AD, Cone J, Symon L, et al: Somatosensory evoked potential monitoring during the management of aneurysmal SAH. J Neurosurg 60:264–268, 1984.
128. Watt JW, Fraser MH, Soni BM, et al: Total i.v. anaesthesia for transcranial magnetic evoked potential spinal cord monitoring. Br J Anaesth 76(6):870–871, 1996.
129. Wazen JJ: Intraoperative monitoring of auditory function: Experimental observations and new applications. Laryngoscope 104:446–455, 1994.
130. Winzenburg SM, Margolis RH, Levine SC, et al: Tympanic and transtympanic electrocochleography in acoustic neuroma and vestibular nerve section surgery. Am J Otol 14:63–69, 1993.
131. Woodforth IJ, Hicks RG, Crawford MR, et al: Variability of motor-evoked potentials recorded during nitrous oxide anesthesia from the tibialis anterior muscle after transcranial electrical stimulation. Anesth Analg 82(4):744–749, 1996.
132. Yang LH, Lin SM, Lee WY, et al: Intraoperative transcranial electrical motor evoked potential monitoring during spinal surgery under intravenous ketamine or etomidate anaesthesia. Acta Neurochirurgica 127(3-4):191–198, 1994.
133. Yiannikas C, Shahani BT, Young RR: The investigation of traumatic lesions of the brachial plexus by electromyography and short latency somatosensory potentials evoked by stimulation of multiple peripheral nerves. J Neurol Neurosurg Psychiatry 46:1014, 1983.
134. Zappulla RA, Wang W: Neural origins of the motor evoked potential. Experimental approach. Adv Neurol 63:51–60, 1993.
135. Zentner J: Noninvasive motor evoked potential monitoring during neurosurgical operations on the spinal cord. Neurosurgery 24:709–712, 1989.
136. Zentner J, Kiss I, Ebner A: Influence of anesthetics—nitrous oxide in particular—on electromyographic response evoked by transcranial electrical stimulation of the cortex. Neurosurgery 24:253–256, 1989.
137. Zentner J, Thees C, Pechstein U, et al: Influence of nitrous oxide on motor-evoked potentials. Spine 22(9):1002–1006, 1997.

IX SLEEP DISORDERS

Chapter 40

Physiology of Sleep

Dudley S. Dinner, MD

INTRODUCTION
THE SLEEP CYCLE
NEURAL STRUCTURES INVOLVED IN SLEEP
  AND WAKING STATES
Activating System
Hypothalamus and Awaking
Brain Stem Center Involved in NREM
  Sleep
Thalamus and Sleep
Hypothalamus and Sleep
Neural Structures and REM Sleep

CHEMICALS INVOLVED IN SLEEP AND
  WAKEFULNESS
Catecholamines and Wakefulness
Acetylcholine and Wakefulness
Serotonin and NREM Sleep
Acetylcholine and REM Sleep

INTRODUCTION

Sleep refers to a behavioral state that differs from wakefulness by a readily reversible loss of reactivity to events in one's environment. It is this reversibility that differentiates sleep from other states of altered consciousness characterized by unresponsiveness such as coma or a state of anesthesia.

The first description of the use of the electroencephalogram (EEG) in humans was provided by Hans Berger.[4] He was the first to study the EEG patterns in relation to sleep. Soon thereafter, others began using the EEG in the electrophysiological evaluation of sleep. The first classification of sleep based on EEG patterns was developed several years later, dividing the stages into A through E.[32] In the 1950s, Nathaniel Kleitman at the University of Chicago made significant contributions to the electrophysiological study of sleep. He reported on regularly occurring periods of rapid eye movements and other associated phenomena occurring during sleep in a study of normal subjects.[1] A few years later, in a classic paper, Kleitman's group reported on the EEG during sleep and its relation to rapid eye movements, body movements, and dreaming and set out a new EEG classification of the stages of sleep.[11] According to this classification, sleep is divided into non–rapid eye movement (NREM) and rapid eye movement (REM) sleep. This classification has persisted to the present. The physiological parameters necessary for the definition of the stages of sleep are the EEG, the electro-oculogram (EOG), and the electromyogram (EMG).

THE SLEEP CYCLE

NREM sleep is subdivided into four stages, I through IV. Stages I and II are frequently referred to as light sleep and stages III and IV as deep sleep. The EEG during stage I sleep shows a loss of the posterior dominant alpha rhythm replaced by a slower theta rhythm. This rhythm is of low amplitude, is intermixed with low-amplitude delta activity, and is associated with slow eye movements predominantly in the horizontal direction (Fig. 40–1). In addition, the EEG may show positive occipital sharp transients (POSTs) and

high amplitude ($> 70\ \mu$V) vertex sharp transients. Stage II NREM sleep is characterized by the presence of theta and delta activity of low to medium voltage sleep spindles, and K complexes. Sleep spindles are sinusoidal waveforms at a frequency of 12 to 14 Hz occurring in the frontocentral region in bursts of at least 0.5 seconds. K complexes are biphasic high-amplitude transients distributed over the frontocentral region of at least 500 msec duration. The presence of sleep spindles or K complexes is essential for the definition of stage II NREM sleep (Fig. 40–2). Stage III NREM sleep is characterized by high-voltage ($> 75\ \mu$V) delta activity of 2 Hz or less which occupies at least 20% but not more than 50% of each epoch of the EEG being scored (usually 30 seconds in the interpretation of the polysomnogram). Stage IV NREM sleep is defined by the presence of the high-voltage delta activity occupying more than 50% of the EEG epoch. From a practical point, the differentiation of stages III and IV NREM sleep is not critical in the clinical EEG or sleep laboratory; thus, these stages frequently scored together (Fig. 40–3). There are no rapid or slow eye movements in stages II, III, or IV NREM sleep. The EMG activity generally decreases as one progresses through the various stages of NREM sleep, reflecting a general progressive decrease in muscle tone.

REM sleep is characterized by an EEG consisting of low-voltage irregular theta and delta activity (similar to stage I NREM sleep) and a dramatic decrease in the EMG activity, associated with bursts of rapid eye movements that recur in a periodic fashion (Fig. 40–4). In addition, in association with the bursts of rapid eye movements, the EEG frequently shows a burst of rhythmic delta or theta waves, which are referred to as saw-tooth waves.

The normal night's sleep is composed of constant cycling between NREM and REM sleep (Fig. 40–5). Normal young adults usually experience four to six cycles of sleep each night, beginning with a period of NREM sleep. Adults usually fall asleep without difficulty within 15 to 20 minutes, and a sleep latency of greater than 30 minutes is generally considered prolonged. Sleep latency is defined as the time to sleep onset from the time one goes to bed and turns the lights out. Stage I NREM sleep is entered for several minutes, then the individual enters stage II NREM sleep and subsequently stages III and IV NREM (deep) sleep. REM latency is a term

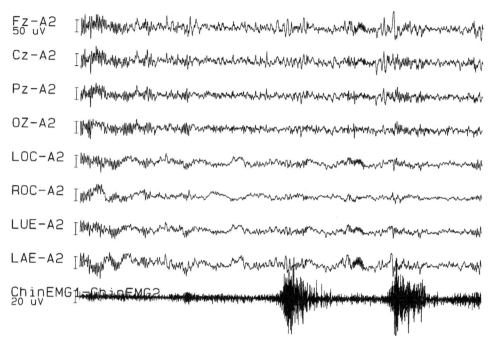

FIGURE 40–1. This is a 30-second epoch in which the EEG, EOG, and EMG demonstrate the features of stage I NREM sleep. FZ, CZ, PZ, and OZ are the mid-frontal, vertex, mid-parietal and mid-occipital electrode positions, respectively. A1 and A2 are the left and right ear electrode positions. LAE, left above eye; LOC, left outer canthus; LUE, left under eye; ROC, right outer canthus electrodes.

that refers to the interval between the onset of stage I NREM sleep and the first period of REM sleep. REM latency is approximately 90 minutes, with a range of 70 to 110 minutes. The interval to each subsequent REM period usually remains constant for a particular individual. The first sleep cycle is defined as that occurring from sleep onset to the end of the first REM period. The first REM period usually lasts only several minutes before the sleeper returns to NREM sleep. In a normal night's sleep, there is a predominance of stage III and IV NREM sleep during the first third of sleep, with REM sleep tending to predominate in the last third of sleep. Each REM period tends to increase in duration as the individual progresses through the sleep cycles.

In NREM sleep, the arousal threshold appears to increase with the depth of sleep. There is a parasympathetic predomi-

nance with a decrease in heart rate of approximately 10%. Blood pressure may also decrease. Respirations become slow and regular. Psychic activity is generally reduced, although hypnogogic hallucinations may occur during stage I NREM sleep.

In REM sleep, there is a dramatic decrease in muscle tone, which is punctuated by short bursts of rapid eye movements and other phasic movements including muscle twitches and jerks that may be observed clinically. Heart rate, systolic blood pressure, and respiratory rate increase, and respirations also may become irregular. Penile erections are characteristic of the REM state. Psychic activity in the form of vivid dreams is well recognized. In a study of dream recall, 80% of subjects awakened from REM sleep were able to recollect the vivid imagery of their dream content whereas only 7%

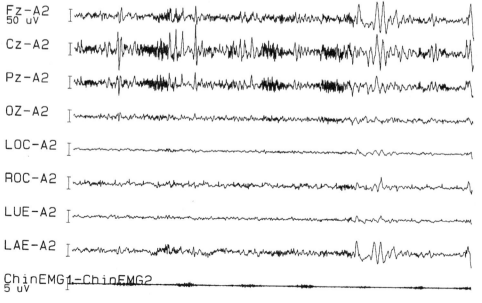

FIGURE 40–2. Thirty-second epoch demonstrating stage II NREM sleep.

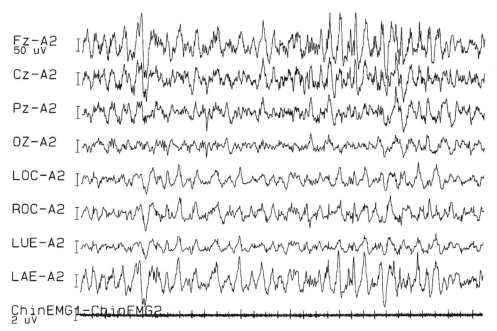

FIGURE 40–3. Thirty-second epoch demonstrating stage III/IV NREM sleep.

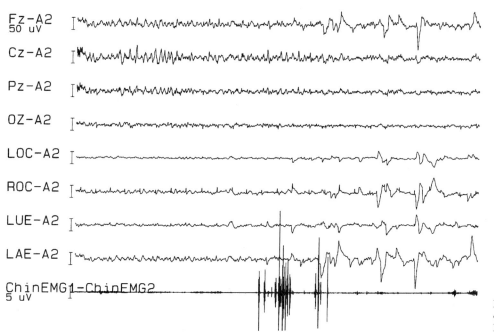

FIGURE 40–4. Thirty-second epoch demonstrating features of REM sleep.

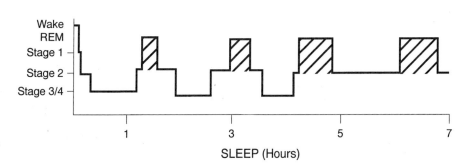

FIGURE 40–5. Sleep histogram of a young adult.

of subjects awakened from NREM sleep were able to recollect their dreams, and the recollection was more vague.[11]

The amount of time spent asleep in the 24-hour day and the proportion of time in the various stages vary with age (Fig. 40–6). The neonate sleeps for 15 to 16 hours each day with approximately 50% spent in NREM sleep and 50% in REM sleep. By the end of the first year, the total sleep time has decreased to 12 to 13 hours, with approximately 30% spent in REM sleep. By the age of 2 to 3 years, the total sleep time has decreased to 12 hours, with 25% spent in REM sleep. In young adults, the total sleep time is approximately 8 hours with 20% to 25% in REM sleep. The time in stages III and IV (deep) NREM sleep is between 30% and 40% of the total sleep time at the age of 2 to 3 years, and this percentage decreases progressively to the adult level of 20% to 25% in the teenage years.

In the young adult, the average time spent in each sleep stage as a percentage of the total sleep time is as follows: approximately 5% in stage I, 45% in stage II, 20% in stages III and IV NREM sleep, and 20% in REM sleep. There is no significant difference in the amount of sleep in males and females. However, some people appear to be long sleepers (more than 9½ hours per night) and some are short sleepers (less than 5½ hours per night). With advancing age, there may be an increase in the number of nocturnal awakenings, and the amount of light sleep may increase from 5% to as much as 15% in old age. A corresponding decrease in the amount of stage III and IV NREM sleep is also noted; however, the amount of REM sleep remains relatively constant.

The physiological role of sleep is as yet unknown. However, there appears to be a clear biological requirement for both NREM and REM sleep. If a person is selectively deprived of REM sleep by systematic awakening, a rebound of REM sleep occurs to levels higher than normal when undisturbed sleep is once again permitted. This increase in REM sleep occurs without affecting NREM sleep. Selective deprivation of NREM sleep produces a corresponding rebound of NREM during the recovery period. If subjects are deprived of all sleep, the recovery period is marked first by an excess of

NREM sleep followed by a rebound of REM sleep, which is delayed until the second or third recovery night. A primary function of NREM sleep may be to facilitate the conservation of metabolic energy and thermoregulation.[5, 43, 59] There appears to be little agreement, however, as to the function of REM sleep. Siegel has hypothesized that the periodic cessation of activity (monitored by cellular firing rates) in the noradrenergic neurons of the locus ceruleus and the serotonergic neurons in the raphe nuclei may be significant for the function of REM sleep.[51] He proposed that the periodic cessation of the cellular activity in the neurons in REM sleep prevents the desensitization of norepinephrine and possibly 5-hydroxytryptamine receptors, which are continuously activated during wakefulness.

In addition to the macroarchitecture of sleep in relation to the NREM and REM stages of sleep, one can further classify the different stages of NREM sleep into a microarchitecture based on the principles described by Terzano and co-workers. They described two patterns: cyclic alternating pattern (CAP) and non-CAP.[57] CAP is characterized by a polygraphic pattern consisting of a burst of activity that occurs with a periodicity of approximately 20 to 40 seconds, alternating with the underlying background of that particular stage of NREM sleep. Each cycle of CAP consists of two phases, A and B. Phase A is characterized by a short burst of activity lasting a few seconds and representing transient arousal; this is followed by phase B, characterized by recovery of the background theta and delta activity of that particular stage of NREM sleep, representing a return toward a deeper level of sleep. Phase A is also characterized by a burst of alpha frequency activity in stage I, which is superimposed on the underlying background; by a sequence of K complexes in stage II NREM sleep; and by a burst of high-amplitude delta activity, which stands out from the background delta of stage III/IV NREM sleep.

Polygraphically, non-CAP is characterized by the ongoing background activity of each stage of NREM sleep without the periodic occurrence of the burst of phase A activity. Each CAP cycle is accompanied by an increase in the heart and respiratory rate and muscle tone in phase A and a decrease of these physiological activities during phase B. Thus, it appears that phase A represents a state of greater arousal, which may facilitate epileptiform activity in patients with epilepsy, and that phase B produces an inhibitory influence.

NEURAL STRUCTURES INVOLVED IN SLEEP AND WAKING STATES

To understand the regulation of sleep, one must consider the regions concerned in the control of both sleep and wakefulness. Initially, the neurophysiological research aimed at understanding the regulation of sleep and wakefulness was based mainly on lesion and stimulation experiments. Over the past 25 years, however, there has been an increased focus on chemical neurotransmitters and corresponding neurons in the mechanisms of sleep and wakefulness.

Activating System

The concept of the ascending reticular activating system (ARAS) as responsible for wakefulness (Fig. 40–7) was based on a series of elegant animal experiments. Electrical stimulation of the reticular formation of the rostral pons and midbrain of anesthetized cats resulted in replacement of synchronized activity in the EEG with low-voltage fast activity.[40] Destruction of these brain stem areas resulted in a state of persistent hypersomnia with a sleep-like EEG pattern.[31]

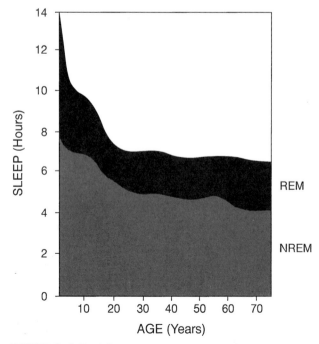

FIGURE 40–6. Total sleep time, NREM and REM sleep, as a function of age.

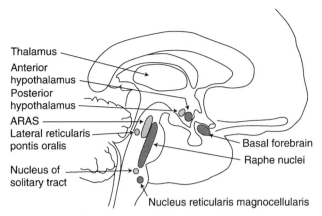

FIGURE 40–7. Location of the wake and sleep centers.

Thalamus
Anterior hypothalamus
Posterior hypothalamus
ARAS
Lateral reticularis pontis oralis
Nucleus of solitary tract
Basal forebrain
Raphe nuclei
Nucleus reticularis magnocellularis

Behavioral arousal in response to electrical stimulation was demonstrated in subsequent studies.[48] The reticular formation receives input from visceral, somatic, and special sensory systems, and has ascending projections into the forebrain along a dorsal pathway to the thalamic nuclei and a ventral pathway to and through the hypothalamus, subthalamus, and ventral thalamus up to the level of the basal forebrain. Single unit cellular recordings in the midbrain reticular formation have shown that the neurons projecting into the forebrain demonstrate a tonic high rate of discharge in association with cortical activation.[52] These cells decrease their firing rate before the onset of cortical slow wave activity.

Hypothalamus and Awaking

Initially experiments with midbrain transection in the cat, using the acute cerveau isolé preparation, showed resultant EEG sleep patterns.[6] However, in the chronic cerveau isolé preparation, there was recovery of sleep and waking cycles, suggesting the existence of areas rostral to this transection concerned with the sleep/wake cycle.[6] Bilateral lesions in the lateral hypothalamus extending to the caudal border of the mammillary bodies in monkeys produced a state of hypersomnolence.[42] Transection in the vicinity of the mammillary bodies in rats also resulted in a state of hypersomnolence.[41] With transection in the rostral half of the hypothalamus, the opposite effect was observed. Based on these studies, the presence of a waking center in the posterior hypothalamus was suggested (see Fig. 40–7). Subsequent chronic bilateral electrolytic lesions of the posterior hypothalamus in rats have supported this hypothesis.[35]

Brain Stem Center Involved in NREM Sleep

Subsequent to the identification and acceptance of a waking center in the brain stem reticular formation, it was thought that passive deactivation of this system might be responsible for the generation of sleep. However, subsequent transection studies of the brain stem indicated that sleep was, in fact, an active mechanism generated by its own specific neural structures. Pretrigeminal mid-pontine lesions in cats resulted in persistent insomnia associated with activated EEG activity.[2] Subsequently, subtotal destruction of the raphe system (see Fig. 40–7) from the upper medulla to the pontomesencephalic junction in cats was found to result in complete loss of sleep for 3 to 4 days.[28] Electrical stimulation studies in the region of the solitary tract and of the nucleus reticularis ventralis resulted in EEG synchronization.[3] These

studies also supported the existence of a sleep-inducing center in the brain stem. From single unit cellular recording studies, it has been observed that neurons in the region of the solitary tract nucleus become more active during slow wave sleep than during waking. Whether this increase in activity precedes sleep, which would suggest that the solitary tract nucleus is important in the generation of slow wave sleep, is not certain.[14] Fibers from the solitary tract nucleus have been demonstrated to project to the thalamus, subthalamus, hypothalamus, preoptic region, and basal forebrain.[44]

Thalamus and Sleep

Thalamic generators have been identified as the source of spindle activity in NREM sleep.[53] Studies have shown that the nucleus reticularis, which surrounds much of the thalamus, is necessary for the generation of the spindle activity in the thalamic nuclei as well as in the cortical projection regions. The reticular nuclei of the thalamus are under the control of the midbrain reticular system.

Hypothalamus and Sleep

As previously mentioned, the chronic cerveau isolè preparation demonstrated the recovery of the sleep/wake cycle in cat experiments, suggesting the presence not only of a wake center but also of a sleep center rostral to the lesion.[6] Transverse section of the rostral half of the hypothalamus resulted in persistent wakefulness.[41] Electrical stimulation studies of the basal forebrain have been shown to induce NREM sleep.[54, 55] Electrolytic and chemical lesions in this region have produced profound insomnia with permanent reduction in NREM sleep.[37, 56] Cells in the basal forebrain[56] and the preoptic region[29] have been shown to be extremely active in terms of their cellular recording studies in NREM sleep. These cells project to the cortex and also interact with midbrain regions.[37] They are thermosensitive, providing a basis for the facilitation of sleep by fever, and possibly play a role in normal body temperature oscillations in sleep regulation.

Neural Structures and REM Sleep

With experimental transection between the midbrain and pons, the brain stem manifestations of REM sleep (muscle atonia) can be recorded caudal to the transection. The forebrain of animals with transection at this level showed both synchronized EEG and desynchronized EEG states that alternated spontaneously.[58] The synchronized activity represented NREM sleep and the desynchronized state closely resembled wakefulness. Based on the absence of REM sleep rostral to such a transection and the presence of REM sleep manifestations caudal to the midbrain transection, it can be concluded that structures necessary for the production of REM sleep are caudal to the midbrain.

With transection between the medulla and the pons, the medulla cycles between an activated state and a quiescent state. The activated state is characterized by levels of higher muscle tone identical to those seen in active waking, while the quiescent state is characterized by lower levels of muscle tone, as is seen in NREM sleep. Muscle atonia is not seen after transection at the pontomedullary junction. Animals in which the medulla and spinal cord have been disconnected from rostral structures show spontaneous variations in levels of arousal and NREM sleep-like states, but do not show the medullary signs of REM sleep (muscle atonia). Thus, it appears that structures caudal to the pons are not sufficient to generate REM sleep.

Rostral to the pontomedullary junction, however, three

states can be distinguished. The first state is the EEG synchronized state without pontogeniculo-occipital (PGO) spikes, resembling NREM sleep. The second state is a desynchronized state without PGO spikes, resembling wakefulness. The third state is a desynchronized EEG state with PGO spikes. The PGO activity occurs in regular bursts in a manner very similar to that seen in intact animals during REM sleep. Lesions in the rostral pontine reticular formation, the nucleus reticularis pontis oralis, eliminated REM sleep.[36]

Further studies have shown that lateral rather than medial lesions in the reticularis pontis oralis eliminate REM sleep.[13, 15, 46, 47] Thus, it appears that the lateral region of the reticularis pontis oralis lateral to the locus ceruleus is the brain stem region most critical for the generation of REM sleep (see Fig. 40–7). Lesions in the dorsal pons can produce REM sleep without atonia.[17, 27] The lesions necessary to produce this are much smaller than those required to block the REM sleep state.

A second area in which stimulation can elicit atonia is in the medial medulla, the nucleus reticularis magnocellularis (see Fig. 40–7). Cellular unit recordings in the lateral pontine reticular formation[38, 45] and the medial medulla[45, 50] fire at a very high rate during REM sleep and have little or no activity during NREM sleep. The cells are referred to as REM "sleep-on" cells. These cells are generally inactive during wakefulness. The pontine REM sleep-on cells are distributed throughout the lateral pontine region.[45, 49] The REM sleep-on cells in the lateral pontine region appear to be cholinoceptive. The medullary REM sleep-on cells are located in an area receiving projections from the pontine REM sleep-on cells.[46]

CHEMICALS INVOLVED IN SLEEP AND WAKEFULNESS

Catecholamines and Wakefulness

The administration of the catecholamine precursor L-dihydroxyphenylalanine (L-dopa) stimulates prolonged arousal and cortical activation.[21] Amphetamines, which are known to produce an intense arousal, also act through catecholamines, the release of dopamine and norepinephrine. Cocaine also acts on catecholamine metabolism by blocking their inactivation through reuptake. Monoamine oxidase inhibitors prevent the catabolism of monoamines and also produce a significant prolonged arousal. Drugs that inhibit catechol-o-methyltransferase (responsible for the catabolism of catecholamine) will result in a prolonged and intense period of waking.[21] If catecholamine synthesis is inhibited, wakefulness is decreased.[20, 26] Dopamine-containing neurons are localized in the substantia nigra, the ventral tegmentum of the midbrain, and scattered through the posterior hypothalamus and subthalamus. Norepinephrine neurons are found in the pontine and medullary reticular formation. The largest concentration of norepinephrine neurons is in the locus ceruleus nucleus in the dorsolateral pontine tegmentum. These cells project in a diffuse fashion to the entire forebrain.[25] The dopamine neurons of the substantia nigra and ventral tegmental region project to the striatum and frontal cortex and appear to play an integral role in behavioral arousal. The norepinephrine neurons of the locus ceruleus and brain stem also project diffusely to the forebrain, including the cortex, and appear to play an integral role in the cortical activating system.

Cooling of the locus ceruleus has been found to produce sleep in a waking animal, and electrical stimulation has been found to produce arousal in an asleep animal.[9] Localized thermolytic lesions of the locus ceruleus did not produce significant loss of wakefulness or a change in cortical activation.[20, 24] Cellular unit recordings have shown that the firing rate of neurons in the locus ceruleus is most active during highly aroused states. Thus, it appears that the norepinephrine neurons of the locus ceruleus and perhaps of other pontomedullary regions have a modulatory role in the cortical activation of waking.

Acetylcholine and Wakefulness

Atropine, which blocks muscarinic cholinergic receptors, is well known to decrease vigilance. Physostigmine, which inhibits the enzyme acetylcholinesterase (responsible for the catabolism of acetylcholine [ACH]), enhances vigilance with prolonged cortical activation. Cholinergic agonists of either muscarinic or nicotinic type can prolong or elicit cortical activation and wakefulness. Two major groups of cholinergic neurons have been identified by immunohistochemical techniques. One group is composed of neurons located in the caudal mesencephalic-oral pontine reticular formation (laterodorsal and pedunculopontinetegmental nuclei), and project into the forebrain, the medial intralaminar thalamic nuclei, the lateral hypothalamus and basal forebrain, and up to the frontal cortex.[22, 23] The second group of cholinergic neurons is located within the basal forebrain. Release of ACH from the cortex has been found to be highest in association with spontaneous cortical activation during the waking state and REM sleep, and in response to electrical stimulation of the midbrain reticular formation or basal forebrain.[7, 8, 30]

Serotonin and NREM Sleep

Monoamine oxidase inhibitors such as pargyline and nialamide block serotonin catabolism and have been shown to enhance and to prolong slow wave sleep. Paracholorophenylalanine blocks the enzyme tryptophan hydroxylase in the synthesis of serotonin and has been shown to result in insomnia.[30] This effect can be reversed by the administration of the serotonin precursor 5-hydroxytryptamine, which bypasses tryptophan hydroxylase.[26] Histochemical studies demonstrated serotonin to be localized in the neurons of the raphe nuclei in the brain stem.[10] Lesions of the raphe serotonergic nuclei were shown to produce insomnia in the cat.[26] Recovery from raphe lesions has been found to occur in animals, particularly after lesions produced by the serotonin selective neurotoxins 5,6,-dihydroxytryptamine and 5,7-dihydroxytryptamine.[16, 20] Single cell-unit recordings have revealed that the serotonergic raphe nuclei decrease their firing rate with the onset of slow wave sleep and cease during REM sleep.[36] Thus, it has been concluded that serotonergic neurons may normally facilitate the onset of sleep but are not essential for the occurrence of slow wave sleep.

Acetylcholine and REM Sleep

The depletion of ACH with hemicholemium-3 has been demonstrated to decrease the amount of REM sleep as well as wakefulness, and also to eliminate the EEG desynchronization of these two states.[12] Atropine (a muscarinic ACH receptor blocker) abolished cortical desynchronization during REM sleep without eliminating muscle atonia or PGO activity.[18] Thus, ACH neurons appear to be involved in cortical activation in both the REM state as well as in wakefulness. Increasing the levels of ACH with acetylcholinesterase inhibitors such as physostigmine decreases the latency to and increases the duration of REM sleep episodes in normal and pontine cats.[3, 34] Direct injection of a cholinomimetic agent (carbachol) into the pontine tegmentum produced long-lasting muscle atonia and sometimes accompanying PGO waves and rapid eye movements in cats.[39] PGO waves displaced into the waking state following the administration of parachlorophenylalanine to cats were totally blocked by the administration of atropine.[19] These studies indicate that ACH may not be critical for the appearance of REM sleep, but may play an important role in components of the REM sleep state such as desynchrony of the EEG, production of PGO activity, and muscle atonia.

REFERENCES

1. Aserinsky E, Kleitman N: Regularly occurring periods of eye motility and concomitant phenomena during sleep. Science 118:273–274, 1953.
2. Batini C, Magni F, Palestini M, et al: Effects of complete pontine transection on the sleep-wakefulness rhythm. The midpontine pre-trigeminal preparation. Arch Ital Biol 97:1–12, 1959.
3. Belesky G, Henriksen S, McGarr K, et al: Effects of anticholinesterase (DFP) on the sleep-wakefulness cycle of the cat. Psychophysiology 5:243, 1968.
4. Berger H: Uber des EEG des menschen. Arch Psych Nervkrankh 87:527 1929.
5. Berger RJ: Bioenergetic functions of sleep and activity rhythms and their possible relevance to aging. Fed Proc 34:97–102, 1975.
6. Bremer F: Cerveau "isole" et physiologie du somneil. C R Soc Biol (Paris) 118:1235–1241, 1935.
7. Casamenti F, Diffenu G, Abbmondi AL, Pepeu G: Changes in cortical acetylcholine output induced by modulation of the nucleus basalis. Brain Res Bull 16:689, 1986.
8. Celesia GG, Jasper HH: Acetylcholine released from cerebral cortex in relation to state of activation. Neurology 16:1053, 1966.
9. Cespuglio R, Gomez ME, Faradji H, Jouvet M: Alterations in the sleep-wake cycle induced by cooling the locus coeruleus area. Electroencephalogr Clin Neurophysiol 54:570, 1982.
10. Dahlstrom A, Fuxe K: Evidence for the existence of monoamine containing neurons in the central nervous system. I: Demonstration of monoamines in the cell bodies of brain stem neurons. Acta Physio Scand 62(suppl 232):1, 1964.
11. Dement W, Kleitman N: Cyclic variations in EEG during sleep and their relation to eye movements, body motility and dreaming. Electroencephalogr Clin Neurophysiol 9:673–690, 1957.
12. Domino E, Yamamoto K, Dren A: Role of cholinergic mechanisms in states of wakefulness and sleep. Progr Brain Res 28:113–133, 1968.
13. Drucker-Colin R, Pedraza JGB: Kainic acid lesions of gigantocellular tegmental field (FTG) neurons do not abolish REM sleep. Brain Res 272:387–391, 1983.
14. Eguchi K, Satoh T: Characterization of the neurons in the region of solitary tract nucleus during sleep. Physiol Behav 24:99, 1980.
15. Friedman L, Jones BE: Computer graphics analysis of sleep-wakefulness state changes after pontine lesions. Brain Res Bull 13:53–68, 1984.
16. Froment J-L, Petitjean F, Bertrand N, et al: Effects de l'injection intracèrèbrale de 5,6-hydroxytryptamine sur les monoamines cèrèbrales et les ètats de sommeil du chat. Brain Res 67:405, 1974.
17. Henley K, Morrison AR: A re-evaluation of the effects of lesions of the pontine tegmentam and locus coeruleus on phenomena of paradoxical sleep in the cat. Acta Neurobiol Exp 34:215–232, 1974.
18. Henriksen S, Jacobs B, Dement W: Dependence of REM sleep PGO waves on cholinergic mechanisms. Brain Res 48:412–416, 1972.
19. Jacobs B, Henrikson S, Dement W: Neurochemical basis of the PGO Wave. Brain Res 48:406–411, 1972.
20. Jacobs BL, Jones BE: The role of central monoamine and acetylcholine systems in sleep-wakefulness states: Mediation or modulation. In Butcher LL (ed): Cholinergic-Monoaminergic Interactions in the Brain. New York, Academic Press, 1978, p 271.
21. Jones BE: The respective involvement of noradrenaline and its delaminated metabolites in waking and paradoxical sleep: A neuropharmacological model. Brain Res 39:121, 1972.
22. Jones BE, Beaudet A: Distribution of acetylcholine and catecholamine neurons in the cat brainstem: A choline acetlytransferase and tylosine hydroxylary immunohistochemical study. J Comp Neurol 261:15–32, 1987.
23. Jones BE, Cuello AC: Afferents to the basal forebrain cholinergic cell area from pontomesencephalic catecholamine, serotonin, and acetylcholine neurons. Neuroscience 31:37, 1989.
24. Jones BE, Harper ST, Halaris AE: Effects of locus coeruleus lesions upon cerebral monamine content, sleep-wakefulness states and the response to amphetamine in the cat. Brain Res 124:473, 1977.
25. Jones BE, Yang TZ: The efferent projections from the reticular formation and the locus coeruleus studied by anterograde and retrograde axonal transport in the rat. J Comp Neurol 242:56, 1985.
26. Jouvet M: The role of monoamines and acetylcholine-containing neurons in the regulation of the sleep-waking cycle. Ergeb Physiol 64:165, 1972.
27. Jouvet M, Delorme F: Locus coeruleus et sommeil paradoxal. C R Soc Biol 159:895–899, 1965.
28. Jouvet M, Renault J: Insomnie persistante apres lesions des moyaux du raphe chez let chat. C R Soc Biol (Paris) 160:1461–1465, 1966.
29. Kaitin KI: Preoptic area unit activity during sleep and wakefulness in the cat. Exp Neurol 83:347–357, 1984.
30. Kanai T, Szerb JC: Mesencephalic reticular activating system and cortical acetylcholine output. Nature 205:80, 1965.
31. Lindsley D, Bourden J, Magoun H: Effects upon the EEG of acute injury to the brain stem activating system. Electroencephalogr Clin Neurophysiol 1:475–486, 1949.
32. Loomis A, Harvey E, Hobert G: Cerebral states during sleep as studied by human brain potentials. J Exp Psychol 21:127–144, 1937.
33. Magnes J, Moruzzi G, Pompeiano O: Synchronization of the EEG produced by low frequency electrical stimulation of the region of the solitary tract. Arch Ital Biol 99:33–67, 1961.
34. Magherini P, Pompeiano O, Thoden U: Cholinergic mechanisms related to sleep. Arch Ital Biol 110:234–259, 1972.
35. McGinty D: Somnolence, recovery and hyposomnia following ventro-medial diencephalic lesions in the rat. Electroencephalogr Clin Neurophysiol 26:70–79, 1969.
36. McGinty D, Harper RM: Dorsal raphe neurons: Depression of firing during sleep in cats. Brain Res 101:569, 1976.
37. McGinty D, Sterman M: Sleep suppression after basal forebrain lesions in the cat. Science 160:1253–1255, 1968.
38. McGinty D, Szymusiak R: Discharge of pedunculopontine area neurons related to PGO waves. Sleep Res 17:9, 1988.
39. Milter M, Dement W: Cataplectic like behavior in cats after micro-injection of carbachol in pontine reticular formation. Brain Res 68:335–343, 1974.
40. Moruzzi G, Mogoun H: Brainstem reticular formation and activation of the EEG. Electroencephalogr Clin Neurophysiol 1:455–473, 1949.
41. Nauta W: Hypothalamic regulation of sleep in rats. Experimental study. J Neurophysiol 9:285–316, 1946.
42. Ranson SW: Somnolence caused by hypothalamic lesion in the monkey. Arch Neurol Psychiatry 41:1–23, 1939.
43. Rechtschaffen A, Bergman BM, Everson CA, et al: Sleep deprivation in the rat: X. Integration and discussion of the findings. Sleep 12:68–87, 1989.
44. Ricardo JA, Koh ET: Anatomical evidence of direct projections from the nucleus of the solitary tract to the hypothalamus, amygdala, and other forebrain structures in the rat. Brain Res 153:1, 1978.
45. Sakai K: Some anatomical and physiological properties of ponto-mesencephalic tegmental neurons with special reference to the PGO waves and postural atonia during paradoxical sleep in the cat. In Hobson JA, Brazier MA (eds): The Reticular Formation Revisited. New York; Raven Press, 1980, pp 427–447.
46. Sastre J, Sakai K, Jouvet M: Persistence of paradoxical sleep after destruction of the pontine gigantocellular tegmental field with kainic acid in the cat. C R Acad Sci (Paris) 289(13):959–964, 1979.
47. Sastre JP, Sakai K, Jouvet M: Are gigantocellular tegmental field neurons responsible for paradoxical sleep? Brain Res 229:147–161, 1981.
48. Segundo J, Arana R, French J: Behavioral arousal by stimulation of the brain in the monkey. J Neurosurg 12:601–613, 1995.

49. Shiromani PM, Armstrong DM, Bruce G, et al: Relation of pontine choline acetyltransferase immunoreactive neurons with cells which increase discharge during REM sleep. Brain Res Bull 18:447–455, 1987.
50. Siegel JM: Behavioral functions of the reticular formation. Brain Res Rev 1:69–105, 1979.
51. Siegel JM: Mechanisms of sleep control. Clin Neurophysiol 7(1):49–65, 1990.
52. Steriade M: Mechanisms underlying cortical activation: Neuronal organization and properties of the midbrain reticular core and intralaminar thalamic nuclei. In Pompeiano O, Ajmone-Marsan C (eds): Brain Mechanisms and Perceptual Awareness. New York, Raven Press, 1981, p 327.
53. Steriade M: Brain electrical activity and sensory processing during waking and sleep states. In Kryger MH, Roth T, Dement WD (eds): Principles and Practice of Sleep Medicine. Philadelphia, WB Saunders, 1989, pp 86–103.
54. Sterman M, Clemente C: Forebrain inhibitory mechanisms: Cortical synchronization induced by basal forebrain stimulation. Exp Neurol 6:91–102, 1962.
55. Sterman MB, Clemente CD: Forebrain inhibitory mechanisms: Sleep patterns induced by basal forebrain stimulation in the behaving cat. Exp Neurol 6:103–117, 1962.
56. Szymusiak R, McGinty DJ: Sleep-related neuronal discharge in the basal forebrain of cats. Brain Res 370:82–92, 1986.
57. Terzano MG, Mancia D, Salati MR, et al: The cyclic alternating pattern as a physiologic component of normal NREM sleep. Sleep 8(2):137–145, 1985.
58. Villablanca J: The electrocorticogram in the chronic cerveau isole cat. Electroencephalogr Clin Neurophysiol 19:576–586, 1965.
59. Zepelin H, Rechtschaffen A: Mammalian sleep, longevity and energy metabolism. Brain Behav Evol 10:425–470, 1974.

Chapter 41

Polysomnography and Multiple Sleep Latency Testing

Prakash Kotagal, MD

THE SLEEP LABORATORY
POLYSOMNOGRAPHY
Electrode Placement and Montage
Electro-oculogram
Chin Electromyogram
Anterior Tibialis Electromyogram
Respiration
Oximetry and Capnography
Electrocardiogram

Nocturnal Penile Tumescence
Esophageal pH
Videotape Monitoring
Recording Procedure
SCORING
Polysomnography During Parasomnias
Ambulatory Polysomnography
MULTIPLE SLEEP LATENCY TEST

Polysomnography (PSG) and the multiple sleep latency test (MSLT) are objective methods of evaluating nocturnal and diurnal sleep, respectively. Other methods such as ambulatory sleep recording, actigraphy, and body temperature monitoring have not been widely accepted into clinical practice. The polysomnogram provides information regarding sleep architecture, respiration, cardiac rhythms, body movements, and penile tumescence (as part of the nocturnal tumescence study). The MSLT has been used as an objective measure of daytime sleepiness and for the diagnosis of narcolepsy. The Association of Sleep Disorders Centers,[2] Rechtschaffen and Kales,[23] and the American EEG Society[1] have formulated useful guidelines in the selection, performance, and interpretation of these tests.

THE SLEEP LABORATORY

The sleep laboratory is usually a dedicated facility with one or more bedrooms. The patient's sleep is monitored from a separate room by the technologist using a polygraph. The patient's room must be conducive to sleep in terms of ambient temperature, environmental noise, decor, and lighting. It should resemble the patient's bedroom as much as possible. An intercom is used to monitor and communicate with the patient. Videotaping the night's sleep using an infrared video camera is helpful in looking at behavioral and respiratory events. Although a standard electroencephalogram (EEG) machine can record most sleep parameters, the polygraph has one or more direct current (DC) amplifier channels with wider pen excursions to allow recording of oxygen saturation, respiratory movements, airflow, and pH on the paper record. Most sleep laboratories in the United States are now changing from analog to digital EEG and sleep machines. It is essential that a qualified sleep technologist be present who is knowledgeable in EEG, polygraphic monitoring, and sleep scoring. After the studies are scored by the technologist, they must be interpreted by a qualified polysomnographer. The polysomnographer is responsible for overseeing the operation of the sleep laboratory and ensuring quality recordings.

POLYSOMNOGRAPHY

The patient should be told why the test is being done and what to expect during it. The patient should keep a log of his or her nocturnal sleep as well as naps, if any, for 1 to 2 weeks before the test and bring it to the laboratory. Ideally patients should be withdrawn from medications that alter sleep architecture such as hypnotics, antidepressants, and stimulants (medications such as tricyclic antidepressants can decrease REM sleep, whereas stimulant medications such as methylphenidate [Ritalin] and pemoline [Cylert] can influence daytime sleepiness). The patient arrives for the test 60 to 90 minutes before he or she is to sleep to allow sufficient time for application of electrodes and acclimatization to the surroundings. A bedtime and morning questionnaire should be completed by the patient. After electrode application, the patient may relax in a bed or chair, reading or watching television until he or she is ready to go to bed. Forcing a patient to go to sleep earlier than is usual for that individual, can produce spurious results, such as prolonged sleep latency. The PSG duration is usually 6 to 8 hours, and the patient should, if possible, be allowed to awaken on his or her own. This is important, especially when an MSLT is to follow the PSG. Cutting short the PSG may give rise to shortened sleep latency and even sleep-onset REM during the first MSLT trial. Table 41–1 shows the commonly used filter settings and recording gains in our laboratory.

Electrode Placement and Montage

EEG electrodes (nonpolarizable silver-silver chloride or gold electrodes) are used and are applied with collodion

TABLE 41–1	Filter Settings and Sensitivities		
Channel	LFF (Hz)	HFF (Hz)	Sens (µV/mm)
EEG	0.3	35	5
EOG	0.3	35	5
Leg EMG	5.0	70	5
Chin EMG	5.0	70	2
ECG	1.0	15	75
Airflow (AC amp)	0.1	15	50
Effort (AC amp)	0.1	15	50
O₂ saturation	DC amplifier		

AC, alternating current; DC, direct current; ECG, electrocardiogram; EEG, electroencephalogram; EMG, electromyogram; EOG, electro-oculogram; LFF, low-frequency filter; HFF, high-frequency filter; Sens, sensitivity.

using the International 10–20 system. At least two channels of EEG are necessary to score sleep stages: these should include one of the central or parasaggital electrodes connected to the opposite ear (i.e., C_3–A_2 or C_4–A_1) in order to record vertex waves, K complexes, and sleep spindles, as well as an occipital electrode (i.e., O_1–A_2) to record waking alpha activity and allow accurate identification of sleep onset. Placement of bilateral electrodes (C_3, C_4, O_1, O_2, and A_1, A_2) is helpful in case of electrode failure so that the recording can be continued without having to awaken the patient. Many laboratories use up to four channels of EEG, incorporating frontal leads as well. This technique may make delta activity more prominent and increase the time scored as stage III/IV sleep; therefore, control data must also be collected in the same way. A slow paper speed of 10 or 15 mm/sec is used to permit easier identification of alpha rhythms, sleep spindles, and K complexes. If the patient is suspected of having nocturnal seizures or parasomnias, a more extensive electrode placement is recommended. In many laboratories, simultaneous EEG/video monitoring is combined with a standard PSG when studying these patients.[1, 16] EEG sensitivity is usually 5 to 10 μV/mm with low-and high-frequency filter settings of 0.3 Hz and 70 Hz, respectively.

Electro-oculogram

The electro-oculogram (EOG) is essential to record the slow eye movements (0.25 to 0.5 Hz) occuring in drowsiness, as well as the rapid eye movements (faster than 1 Hz) seen in rapid eye movement (REM) sleep and wakefulness. EEG electrodes are applied with electrolyte paste and tape (collodion is best avoided because of possible corneal injury). If four channels are available, electrodes are placed over both outer canthi and above and below one of the eyes and referred to the ipsilateral ear. Alternatively, an electrode may be placed 1 cm above and lateral to the outer canthus of one eye referenced to the ear. The second electrode is placed 1 cm below and lateral to the outer canthus of the other eye also referenced to the ear. If only a single channel is available, the two electrodes may be combined in a single bipolar channel. The use of the ear reference introduces some EEG activity into the eye channels, but identification of eye movements is not difficult, as these movements will be out of phase whereas EEG activity will appear in phase. Filter and sensitivity settings are similar to those used for the EEG channels.

Chin Electromyogram

This is an essential parameter for the staging of sleep. Tonic electromyogram (EMG) decreases from the waking state through non–rapid eye movement (NREM) sleep and is abolished or at its lowest level in REM sleep. It is also helpful in detecting movement arousals, snoring, and bruxism. A pair of standard EEG electrodes 3 cm apart is placed submentally with collodion. A third electrode may be placed as backup. The EEG, EOG, and chin EMG wires can be drawn back and bundled conveniently in a "pony tail." The usual sensitivity setting is 2 μV/mm. The low-filter setting is 5 Hz and the high-filter setting is 70 Hz.

Anterior Tibialis Electromyogram

This EMG detects periodic leg movements. A pair of EEG electrodes 3 cm apart is placed over the anterior tibialis muscle belly, identified by asking the patient to dorsiflex the foot against resistance. A bipolar recording from each leg is standard; occasionally the two legs are combined on one

channel. Amplifier settings are similar to those for chin EMG.

Respiration

This includes measurement of ventilatory effort and airflow across the nostrils and mouth. Intercostal EMG electrodes similar to those used for chin EMG are placed over the lower rib cage and record diaphragmatic EMG activity. However, in obese individuals and those with emphysema, the signal may not be satisfactory. In addition to intercostal EMG, excursion of the chest wall and abdomen is measured using strain gauges, inductive plethysmography (Respitrace®), or impedance pneumography. Strain gauges are distensible tubes filled with mercury or packed graphite. The electrical resistance of the strain gauge is a function of the cross-sectional area of the tubing that varies with inspiration and expiration. Although convenient and inexpensive, they are prone to artifact from patient movement and position. Impedance pneumography is a simple and inexpensive method whereby a high-frequency electrical signal is applied to the chest wall between two electrodes. The electrical impedance of the chest wall varies with respiration and is recorded separately for the chest and abdomen. Inductive plethysmography uses a conductive wire sewn into an elastic band or mesh encircling the chest and abdomen. This is more expensive and technically demanding than other methods. Inductive currents in the loop of wire around the trunk change with respiration. This technique can also be used to provide quantitative measurement of total lung and tidal volumes. Two output channels measure chest wall and abdominal movements, while the third channel displays the sum of these two channels. During an obstructive apnea, owing to the paradoxical movement of the diaphragm, the sum channel becomes flat.

The most accurate method to assess ventilatory effort is the esophageal balloon or catheter, which measures the intrathoracic pleural pressure. The normal fluctuations are exaggerated with increased effort seen when airway obstruction occurs. The catheter or balloon is tolerated without much difficulty by most patients, although in some it produces sleep disruption. Although not indicated routinely for all patients, it is helpful in selected patients giving a history compatible with sleep apnea syndrome, having no measurable airflow reduction at the nose, yet showing arousals on EEG, body jerks, and oxygen desaturations regularly every 15 to 40 seconds (Fig. 41–1). These individuals are able to maintain normal airflow by working harder to overcome the increased airway resistance. This has been termed upper airway resistance syndrome by Guilleminault and colleagues.[9, 10] Figure 41–2A shows an example of upper airway resistance syndrome and Figure 41–2B demonstrates central and obstructive apneas detected by recording from the endoesophageal balloon.

Airflow measurement across both nostrils and the mouth is necessary in detection of apneas and hypopneas. Because airflow may change from one nostril to the other with change in body position, both nostrils should be monitored, summing both responses in one channel. Airflow across the mouth should be monitored on a separate channel. The change in temperature of expired air is detected by thermistors or thermocouples. In a thermocouple, a low-voltage electrical signal is generated between two dissimilar metals in contact with each other. Similarly, the electrical resistance of a thermistor (consisting of a small glass bead or wire) changes with the temperature of the expired air. The voltage drop is amplified and displayed on the polygraph. Other methods of airflow detection rely on measurement of expired CO_2 by using a capnograph, which samples expired air

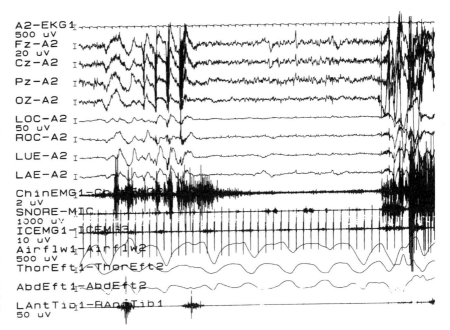

FIGURE 41–1. Repeated arousals and leg jerks occur despite maintained airflow. However, paradoxical movement of the thoracic and abdominal belts and subtle oxygen desaturations (not shown) raise the possibility of upper airway resistance syndrome.

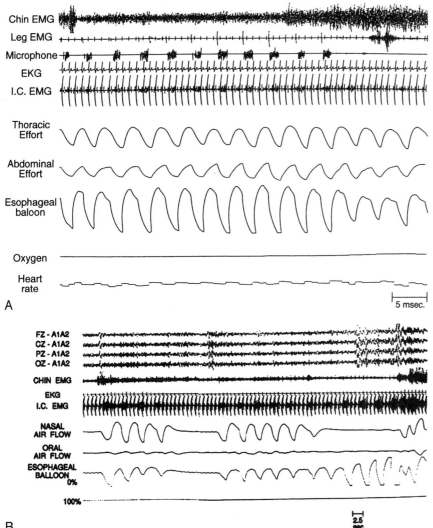

A

B

FIGURE 41–2. *A*, Endoesophageal balloon recordings performed in an adult with upper airway resistance syndrome illustrate a recurrent pattern of progressively negative intrathoracic pressure and snoring while airflow is maintained. Near the end of the epoch there is an arousal, the snoring stops abruptly, and the intrathoracic pressure returns to normal. Minimal deflections of oxygen saturation accompany these episodes. *B*, Endoesophageal balloon recordings from the same patient illustrate a central followed by an obstructive apnea.

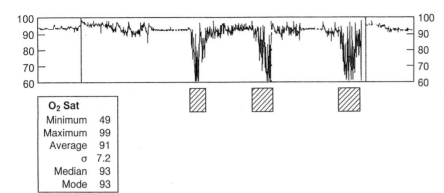

FIGURE 41–3. Oxygen desaturations are more profound during REM sleep and occur in a cyclical fashion in this patient with obstructive sleep apnea.

using a nasal cannula. The pneumotachograph measures resistance to the flow of air, using a flow-to-pressure transducer attached to a tight-fitting nasal-oral mask. The last two methods are more sensitive than thermistors in detecting airflow changes, but are also more expensive and cumbersome, and not used routinely. In addition, laryngeal or tracheal sound can be picked up by microphones taped to the neck and displayed on the paper record after rectification, integration, and amplification. Although useful in verifying snoring, they offer no advantage over airflow detection. More recently, a nasal cannula that incorporates a pressure transducer has been shown to detect episodes of decreased flow that were missed by nasal/oral thermistors.[21] These techniques may allow noninvasive detection of the upper airway resistance syndrome, during which airflow may be maintained while still producing arousals and subtle oxygen desaturations.

Oximetry and Capnography

Oxygen saturation measurements can be performed noninvasively with a pulse oximeter, which measures the transmission of light through a pulsating vascular bed. This is usually measured at the finger, toe, or earlobe. Oxygen desaturation occurs 7 to 15 seconds following an apneic episode, the time lag reflecting the circulation time. Review of oximetry data is very helpful in gauging the frequency and severity of the respiratory events as well as their relationship to REM sleep (Fig. 41–3) and body position (Fig. 41–4). Quantitative measurement of CO_2 concentration in the expired air can be performed using a capnograph, which provides a good measure of airflow as well as ventilation (concentration of CO_2 in the last one-fifth of expiration approximates the alveolar CO_2 tension). Newer devices allowing transcutaneous measurement of both oxygen saturation and CO_2 concentration are now available.

Electrocardiogram

The electrocardiogram (ECG) is an essential component of the sleep study. Not only does it reveal the typical pattern of bradycardia/tachycardia accompanying sleep apnea but it may detect cardiac arrhythmias such as frequent ventricular ectopic beats (see Fig. 41–12), ventricular tachycardia, ventricular fibrillation and asystole (Fig. 41–5). The technologist should be instructed to notify the physician immediately if serious arrhythmias such as ventricular tachycardia, ventricular fibrillation, or asystole of greater than 3 seconds duration are noted. The ECG is usually measured by placing one electrode over each subclavian fossa on the chest wall.

Nocturnal Penile Tumescence

Measurement of nocturnal penile tumescence (NPT) is helpful in distinguishing organic from psychogenic causes of impotence. In the latter, normal penile tumescence and rigidity are observed during REM sleep. Strain gauges are placed over the base and tip of the flaccid penis. The gauges should be 2 to 5 mm smaller than the circumference of the penis. Normally the base circumference will increase by 1.5 to 2 times the tip circumference. An increase of 12 to 15 mm in tip circumference indicates a full erection. Pulsations are present in the recording and helpful in excluding artifacts (Fig. 41–6). The duration of erections is also quantified (Table 41–2). A middle-aged man has about three maximum erections lasting 25 minutes each. Longitudinal rigidity is also measured, normally falling in the range of 1000 to

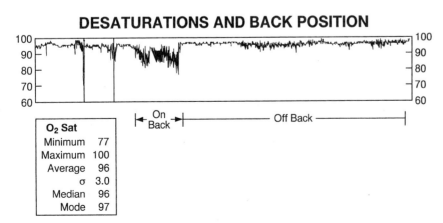

FIGURE 41–4. Oxygen desaturations are more frequent and severe while this patient is sleeping on his back, compared with lying off his back.

ASYSTOLE DURING OBSTRUCTIVE SLEEP APNEA

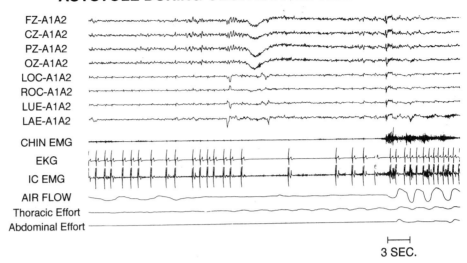

FIGURE 41–5. Several asystoles are noted during this polysomnogram of a patient with severe obstructive sleep apnea.

1500 g of force. An average minimum force of 500 g is needed to achieve penetration.[14, 15] A buckling pressure of more than 750 g indicates that the underlying mechanisms for achieving adequate erections are intact.[13] It may also be helpful to photograph the erect penis and ask if the patient would rate it as a full erection. Measurements of penile tumescence are performed on the second night of PSG to avoid the first night effect, which may decrease the amount of REM sleep.

Another recording device sometimes used for measurement of penile tumescence is the Rigiscan (not part of the polysomnogram study). It measures the circumferential rigidity but does not record sleep parameters such as sleep efficiency or amount of REM sleep. Thus, it cannot distinguish tumescence abnormalities secondary to sleep disruption, or differentiate organic from functional causes of impotence.

Esophageal pH

Reflux of acid into the esophagus may produce disturbed sleep or apnea, or both. This can be documented by introducing a pH probe into the distal sophagus, 5 cm above the lower esophageal sphincter. The probe is connected to a pH meter, which can interface with a DC amplifier on the polygraph. The severity of the reflux, temporal association with apneic events, and quality of sleep can be determined.

Videotape Monitoring

Videotaping of patients during PSG is highly recommended. It can be used to document the presence of unusual events such as parasomnias and seizures, intensity of snoring, and periodic leg movements. A time display should

appear on the videotape so that PSG/EEG events can be precisely correlated. Videotaping enables the technologist to watch several patients in different rooms, in which case an intercom system and call bell in each room are necessary.

Recording Procedure

After the polygraph has been calibrated, the biological or patient calibration is performed. The patient is asked to open and close the eyes to look for the alpha rhythm and its reactivity. He or she is asked to blink several times, inhale and exhale several times, and also hold his or her breath. While the patient grits the teeth and dorsiflexes both feet, the sensitivity of the channels should be adjusted to optimize the waveforms displayed. The lights are then turned out and the patient is asked to go to sleep. After concluding the study the next morning, the polygraph is again calibrated. The patient is asked to fill out another questionnaire regarding the night's sleep. During the night the technologist should periodically make a note of oxygen saturation and body position. Electronic body position monitors that interface with the polygraph are also available. If the airflow signal degrades, the technologist may adjust the channel gain or reapply the thermistor. The technologist should be on the lookout for severe desaturations, cardiac arrhythmias, and seizures that may require medical attention. Young children are reassured by having their favorite blanket or stuffed animal and having their parent sleep in the room on a cot. Continuous positive airway pressure (CPAP) titration may also be performed, by gradually increasing CPAP levels until apneas and desaturations are eliminated.

SCORING

The polysomnogram is first scored by the technologist and then reviewed by the polysomnographer, who needs to be

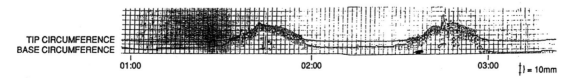

FIGURE 41–6. Strip chart recordings of nocturnal penile tumescence. The upper trace measures circumference from the tip of the penis while the lower trace records from the base of the penis. In this instance the penile tumescence is normal.

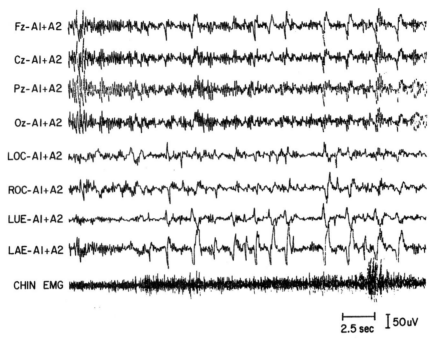

FIGURE 41–7. This awake EEG from the parietal and occipital regions shows an alpha rhythm (8 to 12 Hz). The eye leads show rapid eye blinks and the level of tonic EMG seen from the chin electrodes is relatively high.

equally knowledgeable in this area. The scoring is performed according to criteria established by the Association of Sleep Disorders Centers.[23] The record is scored one page (i.e., epoch) at a time. An epoch is an arbitrary time interval, usually 20 to 60 seconds. In most laboratories, a 30-second epoch is used for scoring. A sleep stage is scored when the activity characteristic of that stage occupies more than 50% of the epoch under study. The recognized sleep stages and scoring criteria follow.

Relaxed Wakefulness. The EEG channels show rhythmic alpha activity in the occipital areas that is attenuated by eye opening (Fig. 41–7). Elsewhere, the EEG is of mixed frequency and low in voltage. Eye movements are voluntary and consist of saccadic eye movements or blinks. The EMG shows high tonic activity, except in very relaxed individuals.

Stage I NREM Sleep. During the transition from wakefulness to stage I NREM sleep (or drowsiness), the occipital alpha activity disappears and the EEG is of mixed frequency and relatively low voltage (Fig. 41–8). In young children, rhythmic bursts of hypnagogic hypersynchrony may be seen (see Chapter 28). Slow rolling eye movements are observed and the EMG remains tonic, though at a lower level than during wakefulness. One may observe monophasic surface-negative vertex waves in the central regions; however, they do not have the characteristic high-voltage diphasic morphology of K complexes, which are the hallmark of stage II NREM sleep.

Stage II NREM Sleep. Stage II sleep is distinguished from stage I sleep by the appearance of K complexes, which are

high-voltage diphasic sharply contoured transients lasting more than 0.5 seconds over the frontocentral-parietal vertex and parasaggital central regions (Fig. 41–9). One also sees sleep spindles which are 12 to 14 Hz, of low to medium amplitude, often sinusoidal in shape, and last 0.5 to 1.5 seconds. Sleep spindles appear in the infant after 3 months of age. Sleep spindles, which are seen most commonly in stage II sleep, may intrude into REM sleep. One may also see delta activity, but this takes up less than 20% of the epoch. The EMG remains tonic; slow eye movements sometimes occur during the transition from stage I sleep.

Stage III and IV NREM Sleep. These stages are characterized by the appearance on the EEG of high-voltage delta (amplitude greater than 75 μv peak to peak) activity with a frequency of 2 Hz or less (Fig. 41–10). In stage III, this activity accounts for 20% to 50% of the epoch, whereas it predominates (more than 50% of the epoch) in stage IV sleep. The EMG remains tonically active.

Stage REM Sleep. REM sleep is notable for bursts of rapid eye movements, a low-amplitude mixed frequency pattern on the EEG (often with the appearance of "sawtooth waves" in the frontocentral region), and atonia. The chin EMG is at its lowest level during REM sleep and is often abolished, whereas twitches of phasic EMG persist (Fig. 41–11). More detailed scoring instructions can be found in the Rechtschaffen and Kales scoring manual.[23] Sleep-onset REM (SOREM) refers to the appearance of REM sleep within 15 minutes of sleep onset.

Movement Time. During gross postural adjustments in sleep, the EEG may be obscured. If this occupies more than 50% of the epoch, it is scored as movement time. If the epochs preceding and following it are wakefulness, then this epoch is also scored as awake.

Arousals. This activity refers to the appearance of the patient's waking alpha rhythm for more than 15 seconds. If it lasts longer than 30 seconds, it is called an awakening. A microarousal is scored when such EEG changes last at least 2 to 4 seconds and are accompanied by an increase in EMG activity in the leg or chin leads.

Other Sleep Staging Terminology. Recording time (RT) is

TABLE 41–2	Nocturnal Penile Tumescence	
NPT Parameter		Normal
Increase in tip circumference		>12 mm
Increase in base circumference		>15 mm
Longitudinal rigidity (buckling pressure)		>1000 g
Number of erection episodes (first night excluded)		>3

NPT, nocturnal penile tumescence.

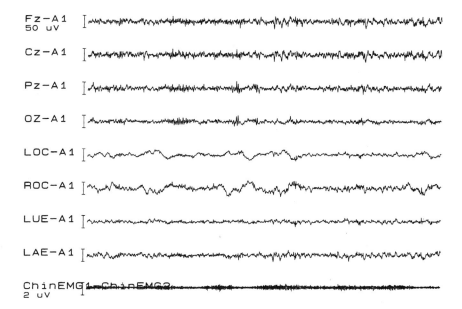

FIGURE 41–8. Stage I NREM sleep showing slow, rolling eye movements and lower EMG tone compared with wakefulness. Also seen in the middle of the recording are surface negative (upward deflections) vertex sharp transients. The alpha rhythm has been replaced by slower activity in the theta range.

from "lights out" at the start of the study to "lights on" at its conclusion. Total sleep time (TST) is the total time spent asleep in the various sleep stages. TST/RT expressed as a percentage is called sleep efficiency. Another measure of sleep efficiency is the wake after sleep onset time (WASO), which may be abnormal in some insomniacs who may have no trouble falling asleep, yet awaken several times during the night or awaken in the early morning hours and cannot return to sleep. Sleep latency is the time taken from "lights out" to the first epoch scored as stage I sleep. REM latency is the time taken from onset of this first stage I epoch to the first epoch scored as REM sleep, minus any intervening epochs scored as awake. Sleep stage percentages for stages I–IV NREM and REM sleep are obtained by dividing time spent in each stage by the TST, expressed as a percentage.

Apnea. This event is defined as cessation or greater than 90% reduction in airflow at the nose and mouth, for at least 10 seconds. Shorter interruptions in airflow for 4 to 9 seconds are called respiratory pauses. If the apnea occurs despite persisting effort noted on intercostal EMG or strain gauges, it is called an obstructive apnea (Fig. 41–12). A central apnea is caused by absence of any discernable effort (Fig. 41–13). In a mixed apnea, effort is absent for the first 6 seconds; thereafter, effort returns but is still insufficient to reestablish airflow—there is both a central and obstructive component (Fig. 41–14). "Physiological central apneas" follow sighs, body jerks during transitions from wakefulness to sleep, and during bursts of phasic EMG in REM sleep. In most laboratories, these are scored and included in the respiratory disturbance index (see the accompanying chart) but are not given clinical significance if they last less than 15 seconds. A partial reduction in airflow of less than 90% but greater than 50% is called a hypopnea when it is accompanied by arousal/awakening or a drop of more than 3% in oxygen saturation. The respiratory disturbance index (RDI) refers to the number of combined apneas and hypopneas divided by the hours of sleep. An RDI of greater than 5 is clinically significant in young and middle-aged adults.[3] However, up to 40% of healthy elderly subjects over the age of 60 will have an RDI above 5.[4, 7, 17] In this age group an

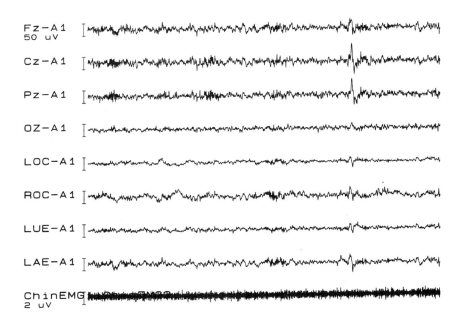

FIGURE 41–9. Stage II NREM sleep reveals sleep spindles and diphasic (negative-positive), high-voltage K complexes. No eye movements are seen although the eye leads show EEG activity from the reference A1 electrode. Tonic EMG is prominent.

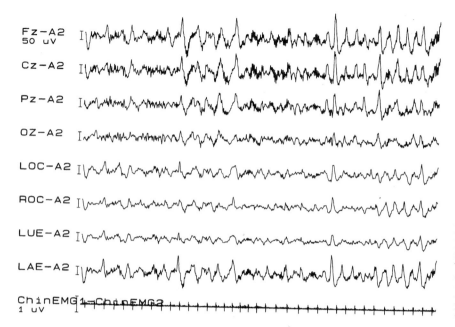

FIGURE 41–10. Stage IV NREM sleep shows prominent slowing less than 2 Hz and greater than 75 μV amplitude. The slow waves occupy more than 50% of the 30-second epoch during stage IV sleep and between 20% and 50% of the epoch in stage III sleep. In some laboratories, stages III and IV are scored together as stage III/IV.

RDI above 10 or 15 is more likely to be clinically significant. Only patients with RDIs greater than 20 have been found to have higher mortality during long-term follow-up.[12] The grading system used for quantifying the severity of apneas and hypopneas in our laboratory follows.

Severity	RDI	O₂Nadir
Mild	5–19	>85%
Moderate	20–49	75%–85%
Severe	>50	<75%

Periodic Movements of Sleep. These events are repetitive stereotyped movements of the legs, consisting of tonic extension of the great toe, dorsiflexion at the ankle, and flexion of the knee occurring during sleep. They were previously described as nocturnal myoclonus, which is a misnomer as the movements are not myoclonic. These movements are scored as periodic movements of sleep (PMS) if they meet

the following criteria.[8] The EMG activity must last 0.5 to 4 seconds, and at least five movements must occur in a cluster with an intermovement interval of 5 to 120 seconds; typically they occur every 20 to 40 seconds. The PMS may be accompanied by arousals or awakenings, which are also scored (Fig. 41–15A). The myoclonic index refers to the number of periodic leg movements per hour of sleep. The myoclonic arousal index (MAI) is the number of leg movements per hour that are associated with arousals. Periodic leg movements of sleep (PLMS) become increasingly common with advancing age; 30% of individuals over the age of 50 have PLMS. These movements are not clinically significant if they are unaccompanied by frequent arousals (Fig. 41–15B). If, however, the leg movements are accompanied by frequent arousals (MAI greater than 5), they are considered clinically significant and may contribute to daytime sleepiness. This clinical entity is called nocturnal myoclonus syndrome.[18] In

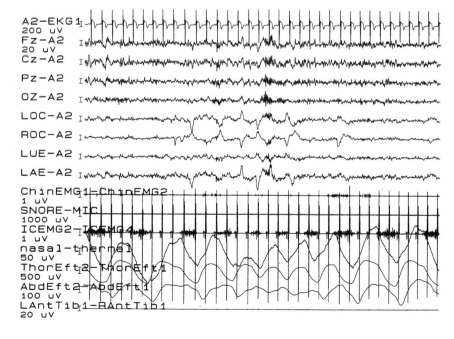

FIGURE 41–11. This 30-second epoch of REM sleep shows low-voltage fast-frequency activity on EEG channels. Rapid eye movements are seen on eye channels; meanwhile, chin EMG is at its lowest level during the study.

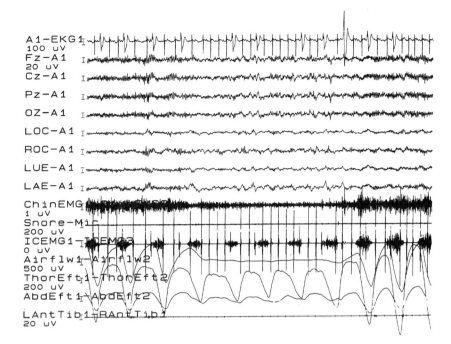

FIGURE 41–12. Absence of airflow (absent or less than 10% of baseline) in the presence of persistent effort characterizes obstructive sleep apnea. Effort can be noted in the thoracic and abdominal effort channels as well as in the intercostal EMG. At the end of the apnea an arousal on EEG is accompanied by increase in chin EMG. The ECG channel reveals frequent premature ventricular contractions (PVCs).

the restless legs syndrome (RLS), similar movements occur prior to sleep onset and are typically described as creepy, crawly, unpleasant sensations in the legs, resulting in an uncontrollable urge to move the legs about. This may result in difficulty falling asleep and prolongation of sleep latency. Some patients have both RLS and PMS.[18]

Alpha-Delta Sleep. This is an atypical sleep pattern, first described by Hauri,[11] which consists of a mixture of 5% to 20% delta waves (0.5 to 2 Hz, greater than 75 μV in amplitude) with a superimposed alpha-like rhythm of 7 to 10 Hz. Similar alpha activity has been described in other NREM stages and referred to as "alpha intrusion." It has been experimentally induced following sleep deprivation[20] in subjects who then complained of musculoskeletal and mood symptoms resembling those seen in fibromyalgia. It was postulated that "nonrestorative sleep" was responsible for the appearance of these symptoms in the fibromyalgia syndrome. However, not all patients with the fibromyalgia syndrome

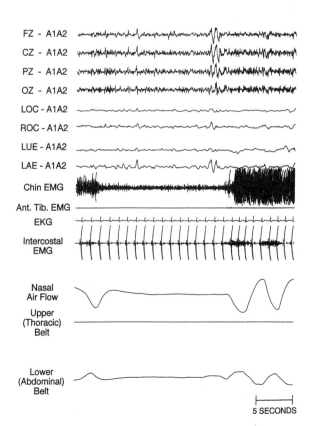

FIGURE 41–13. Central sleep apnea is indicated by absent airflow unaccompanied by respiratory effort. The apnea is terminated by an arousal.

5 SECONDS

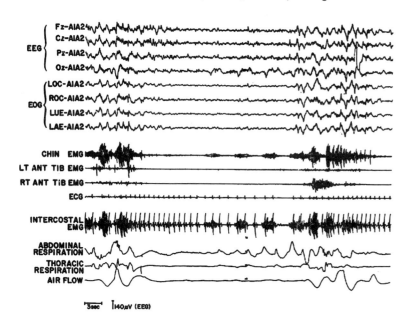

FIGURE 41–14. This mixed apnea shows absent effort in the first 4 or more seconds followed by resumption of effort. Airflow is not resumed for several more seconds despite increasing effort.

demonstrate this pattern. This pattern is also observed in patients with other psychiatric and medical conditions.[24, 25]

Polysomnography During Parasomnias

The PSG may provide helpful information about the occurrence of unusual behaviors during sleep, such as night terrors, sleepwalking, and REM sleep behavior disorder (RBD). NREM parasomnias such as night terrors and sleepwalking tend to occur out of the transition from slow wave sleep to REM sleep, typically within 1 to 2 hours of falling asleep. Figure 41–16 shows an example of repeated sudden stage IV arousals in a child with night terrors. On the other hand, REM parasomnias such as RBD and nightmares occur during REM sleep. In RBD, instead of atonia, there is persistence of or increase in tonic EMG as well as phasic EMG artifacts. It is vital to record the patient simultaneously on videotape for subsequent review. At times, patients having nightly episodes of parasomnias may not have any episodes when studied for a single night in the laboratory. In such cases, it is recommended that they have two or more consecutive nights of recording, as they are more likely to have clinical episodes on the second night as a result of better sleep quality. Episodes of RBD are directly correlated to the amount of REM sleep obtained by the patient. Because parasomnias by definition are nonepileptic events, it is essential to have not only adequate recording of parameters for a standard PSG but also a full set of EEG electrodes in the 10–20 distribution. Digital EEG machines permit a referentially recorded EEG to be subsequently reformatted into any montage preferred for viewing by the polysomnographer. Manual and automated detection of interictal spikes is also helpful. One should be mindful of the fact that lack of an observed ictal EEG pattern does not exclude the possibility of epileptic seizures; that is why a good quality videotape recording (even a home videotape) of the event is extremely helpful.[4]

Ambulatory Polysomnography

There has been recent interest in ambulatory techniques for diagnosis and follow-up of sleep disorders in the home setting. Obvious advantages include lower cost and improved sleep efficiency in a setting that is more comfortable for the patient. Disadvantages include loss of data from electrodes, artifacts, and malfunction of sensors that become dislodged or are improperly applied by the patient. Home oximetry during sleep had a sensitivity of 90% and specificity of 75% for screening sleep-related breathing disorders.[26] However, 46% patients judged "normal" by oximetry had obstructive sleep apnea/upper airway resistance syndrome, and 22% had other sleep disorders.[26] The American Sleep Disorders Association concluded that unattended portable monitoring is acceptable (1) in patients with severe obstructive sleep apnea who cannot be readily studied in the sleep laboratory and need urgent intervention, (2) in patients unable to be brought to the sleep laboratory, and (3) for follow-up after diagnosis has been established in the sleep laboratory.[2] It is not indicated in patients who (1) have isolated symptoms of excessive daytime sleepiness or loud snoring, (2) are mildly symptomatic patients, or (3) require home nasal CPAP titration.[2] Broughton has recently reviewed the pros and cons of home sleep studies.[5]

MULTIPLE SLEEP LATENCY TEST

The multiple sleep latency test (MSLT) provides an objective measure of patient sleepiness.[7] Its main usefulness is in the diagnosis of narcolepsy and other causes of daytime sleepiness. Sleep deprivation and delayed sleep phase (habitually going to bed late and waking up late) are common in our society. The patient should be asked to maintain a sleep log for 1 to 2 weeks prior to the test. Most laboratories require that a polysomnogram be performed the previous night to document the number of hours spent asleep and to uncover other reasons for daytime sleepiness, such as sleep apnea and periodic movements of sleep. Electrode application is similar to that for PSG, including EEG, EOG, and chin EMG. Although not essential, it is helpful to monitor ECG, respiratory sounds, airflow, and leg EMG. Most laboratories use the same paper speed as in PSG, 10 to 15mm/sec, while others use faster speeds of 30 mm/sec (control data must be collected accordingly).

The test is begun 1.5 to 3 hours after the patient awakens in the morning. It consists of a series of four to five naps every 2 hours over the day. In between the naps the subject is to remain awake and refrain from coffee and smoking

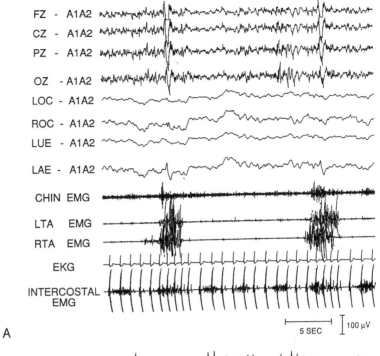

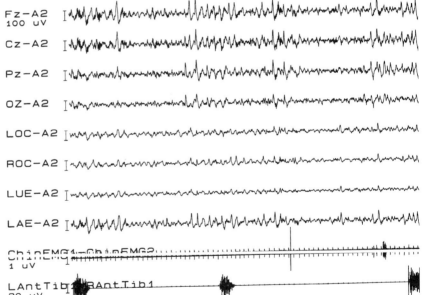

FIGURE 41–15. *A*, Repeated leg jerks every 20 to 40 seconds (periodic leg movements of sleep) are associated with microarousals. A myoclonic arousal index of greater than 5 per hour is clinically significant. *B*, Similar periodic leg movements are unaccompanied by signs of arousal. Periodic leg movements are increasingly common with advancing age and are not significant unless accompanied by repeated arousals.

(this may be problematic if the patient is a heavy nicotine or caffeine user). During each trial, the subject is told to try to fall asleep in a quiet, dark room in a comfortable bed. The study is continued for 15 minutes after the patient has reached stage I sleep; if the patient cannot fall asleep in 20 minutes, the trial is terminated. Figure 41–17 shows the onset of stage I sleep; Figure 41–18 shows the onset of SOREM occurring within 15 minutes of falling asleep. The average time taken to fall asleep in all of the naps is calculated and reported as the mean sleep latency (if there is no sleep on a trial, the sleep latency is assumed to be 20 minutes). A mean sleep latency of less than 5 minutes is considered pathological; a mean sleep latency between 5 and 10 minutes is in the "intermediate zone"; and a sleep latency

greater than 15 minutes is considered normal. The number of trials in which REM sleep occurs (SOREM periods) is noted. After extreme sleep deprivation or in the setting of sleep apnea, a patient may have one SOREM period; however, the presence of two or more SOREM periods is highly suggestive of a diagnosis of narcolepsy. One should bear in mind that individuals with a delayed sleep phase may show shortened sleep latency and the appearance of REM sleep in the first one or two naps. Ideally, such a person should be allowed to go to sleep and awaken at his or her habitual time and start the MSLT subsequently, though this is not always possible in all laboratories. If the patient has had only one SOREM period in the first four naps, it is advisable to do a fifth trial.

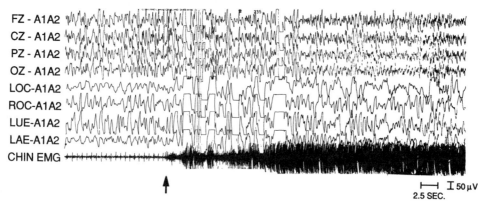

FIGURE 41–16. Repeated sudden arousals occur out of stage IV NREM sleep in this 5-year-old child with night terrors. This pattern provides supportive evidence of an NREM parasomnia such as night terrors or sleepwalking even when an actual episode is not captured during the night study. (From Bleasel A, Kotagal P: Paroxysmal nonepileptic disorders in children and adolescents. Semin Neurol 15:203–217, 1995.)

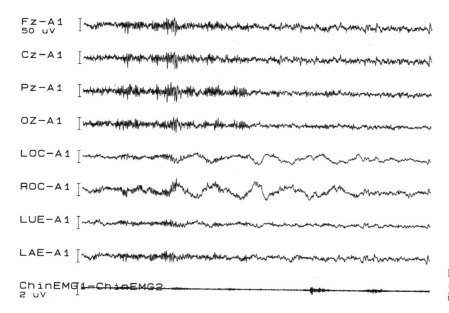

FIGURE 41–17. Multiple sleep latency test (MSLT) showing onset of stage I NREM sleep in the middle of this 30-second epoch.

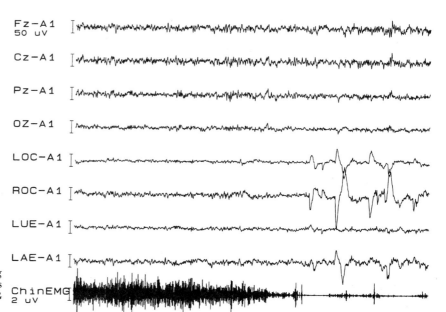

FIGURE 41–18. Onset of REM sleep during an MSLT occurs in the later portion of this epoch with the disappearance of chin EMG and rapid eye movements.

A modification of the MSLT is the maintenance of wakefulness test (MWT). The patient is asked to stay awake in a dark room in a comfortable chair.[19] This test is used to assess a patient's response to treatment, as it simulates a daily situation.

REFERENCES

1. American EEG Society: Guidelines for polygraphic assessment of sleep-related disorders (polysomnography). J Clin Neurophysiol 9:88–96, 1992.
2. American Sleep Disorders Association: Practice parameters for the use of portable recording in the assessment of obstructive sleep apnea. Sleep 17:372–377, 1994.
3. Berry DT, Webb WB, Block AJ: Sleep apnea syndrome: A critical review of the apnea index as a diagnostic criterion. Chest 86:529–531, 1984.
4. Bleasel A, Kotagal P: Paroxysmal nonepileptic disorders in children and adolescents. Sem Neurol 15:203–217, 1995.
5. Broughton R, Fleming J, Fleetham J: Home assessment of sleep disorders by portable monitoring. J Clin Neurophysiol 13:272–284, 1996.
6. Carskadon MA, Dement WC: Respiration during sleep in the aged human. J Gerontol 36:420–423, 1981.
7. Carskadon MA, Dement WC, Mitler MM, et al: Guidelines for the multiple sleep latency test (MSLT): A standard measure of sleepiness. Sleep 9:519–524, 1986.
8. Coleman RM, Pollak C, Weitzman ED: Periodic movements in sleep (nocturnal myoclonus). Relation to sleep-wake disorders. Ann Neurol 8:416–420, 1980.
9. Guilleminault C, Shimoni T, Stoohs R, Schnittger I: Echocardiographic studies in adults and children presenting with obstructive sleep apnea or heavy snoring. In Gaultier C, Escourrou P, Curzi-Descalova L (eds): Sleep and Cardiorespiratory Control. Montrouge, France: Colloques INSERM/John Libbey Eurotext, 1991, pp 95–103.
10. Guilleminault C, Stoohs R, Clerk A, et al: A cause of excessive daytime sleepiness. The upper airway resistance syndrome. Chest 104:781–787, 1993.
11. Hauri P, Hawkins DR: Alpha-delta sleep. Electroencephalogr Clin Neurophysiol 34:233–238, 1973.
12. He J, Kryger MH, Zorick FJ, et al: Mortality and apnea index in obstructive sleep apnea. Chest 94:9–14, 1988.
13. Hirshkowitz M, Moore CA: Sleep-related erectile activity. Neurol Clin 14:721–737, 1996.
14. Karacan I: Evaluation of nocturnal penile tumescence and impotence. In Guilleminault C (ed): Sleeping and Waking Disorders: Indications and Techniques. Menlo Park, CA, Addison-Wesley, 1982, pp 343–372.
15. Karacan I, Moore CA, Sahmay S: Measurement of pressure necessary for vaginal penetration. Sleep Res 14:269, 1985.
16. Kotagal P, Dinner DS: Usefulness of combined polysomnography and video-EEG recordings. Sleep Res 20:382, 1991.
17. Krieger J, Turlot J, Mangen P, Kurtz D: Breathing during sleep in normal young and elderly subjects: Hypopneas, apneas and correlated factors. Sleep 6:108–120, 1983.
18. Lugaresi E, Cirignotta F, Coccagna G, Montagna P: Nocturnal myoclonus and restless legs syndrome. In Fahn S (ed): Myoclonus. Adv Neurol 43:295–307, 1986.
19. Mitler MM, Gujavarty Ks, Browman CP: Maintenance of wakefulness test: A polysomnographic technique for evaluating treatment efficacy in patients with excessive somnolence. Electroencephalogr Clin Neurophysiol 53:658–661, 1982.
20. Moldfofsky H, Scarisbrick P: Induction of neurasthenic musculoskeletal pain syndrome by selective sleep stage deprivation. Psychosom Med 37:341–351, 1975.
21. Norman RG, Ahmed MM, Wlsleben JA, Rapoport DM: Detection of respiratory events during NPSG: Nasal cannula/pressure sensor versus thermistor. Sleep 20:1175–1184, 1997.
22. Radtke RA: Sleep disorders: Laboratory evaluation. In Daly DD, Pedley TA (eds): Current Practice of Clinical Electroencephalography. New York, Raven Press, 1990, pp 561–592.
23. Rechtschaffen A, Kales A: A manual for Standardized Terminology, Techniques and Scoring System for Sleep Stages of Human Subjects. Los Angeles: UCLA Brain Information Service/Brain Research Institute, 1968.
24. Saskin P, Moldofsky H, Lue FA: Sleep and posttraumatic rheumatic pain modulation disorder (fibrositis syndrome). Psychosom Med 48:319–323, 1986.
25. Wittig RM, Zorick FJ, Blumer D, et al: Disturbed sleep in patients complaining of chronic pain. J Nerv Ment Dis 170:429–431, 1982.
26. Yamashiro Y, Kryger MH: Nocturnal oximetry: Is it a screening tool for sleep disorders? Sleep 18:167–171, 1995.

INDEX

Note: Page numbers in *italics* refer to illustrations; page numbers followed by (t) refer to tables.

Absence epilepsy, in pediatric patient, 471, 472
Absence status epilepticus, 447, *448*
Absolute refractory period, 112
    determination of, 113–114, *114*
Acetylcholine, 64, 251
    receptors for, 64
        congenital deficiency of, 265, 265(t)
    release of, in Lambert-Eaton myasthenic syndrome, *253*
        in myasthenia gravis, *253*
    role of, in REM sleep, 594–595
        in wakefulness, 594
Acetylcholinesterase, congenital deficiency of, 264, 265(t)
Acoustic neuroma, abnormal brain stem AEPs associated with, 536, *537, 538*
    surgery for, monitoring of brain stem AEPs during, *566, 567, 568*
Acquired demyelinating polyneuropathies, 226(t), 226–231
    results of needle electrode examination in, 230(t)
    results of nerve conduction studies in, *226, 228,* 229(t), *230*
Acquired immunodeficiency syndrome (AIDS), dementia in, 452–453
    myopathy in, 279
Acute illness/ICU myopathy, 279
    results of needle electrode examination in, 279(t)
    results of nerve conduction studies in, 278
Acute inflammatory demyelinating polyneuropathy, electrodiagnostic findings in, 228, 230–231
Adenosine, 66
    receptors for, *60*
Adenosine triphosphate, 65–66
Adolescent(s), EEG in. See *Electroencephalography, in children and adolescents; Electroencephalography, in young adults and adolescents.*
AEPs. See *Auditory evoked potentials (AEPs).*
AIDS (acquired immunodeficiency syndrome), dementia in, 452–453
    myopathy in, 279
Alpha activity or bursts. See also *Alpha rhythm.*
    rhythmic, excessive, on EEG in newborn, 392–393, *395*
        midline frontal and midline central, on EEG in newborn, 393
        temporal, on EEG in preterm infant, 390, *391*
Alpha-delta pattern, on EEG in sleeping patient, 605
Alpha rhythm. See also *Alpha activity or bursts.*
    on electroencephalography, *414,* 414–415
        appearance of, in comatose patient, 441, *441*
        central, in children, *402*

Alpha rhythm (Continued)
        mechanisms of production of, 355
        variants of, *415,* 415–416, *416*
Alpha squeak, on electroencephalography, 415, *415*
Alternating current, 3–4
    supply of, to neurophysiology laboratory, 31, *32*
        faults in, 31, 32(t)
Alternating current signals, 3–4, *4*
Alzheimer's disease, 450–451
    cortical reflex myoclonus in, 283
Amblyopia, visual evoked potentials in, 516
Ambulatory polysomnography, 606
Amplifier(s), 20–23
    common-mode rejection ratio (CMRR) of, 8, 23
    differential, 8, *9,* 20–21, *22*
    gain of, 20, 20(t), *21*
Amplifier parameters, for recording of brain stem AEPs, 528
    for recording of SEPs, 557
    for recording of VEPs, 511
Amplitude asymmetry, on electroencephalography, 458–459
Amyloidosis, and dysautonomia, 315
    test results linking, *311, 312,* 315
Amyotrophic lateral sclerosis, 237–239
    results of needle electrode examination in, *238,* 238(t)
    results of nerve conduction studies in, 238, 238(t), *239*
    variants of, 239–241
Amyotrophy, monomelic, 240
    neuralgic, 217–218
Anesthetized cats, electroencephalography in, spindles on, 352, *353*
Anoxic encephalopathies, 446–447
Anterior horn cells, 82
Antidromic impulses, collision of orthodromic impulses with, 116
    in assessment of fast vs. slow fiber conduction, 119–120, *120*
    in assessment of refractory period in motor fibers, 119, *119*
    in blocking of unwanted nerve stimulation, 116
    in blocking of unwanted transmission through Martin-Gruber anastomosis, 116, *118*
    in production of MNSP, 297, 298, *298.* See also *Mixed nerve silent period.*
Antiepileptic drugs, dosages of, for newborn, 499(t)
    effects of, on brain neurotransmitters, 64(t)
Anxious patient, long latency electromyographic responses in, *304*
Apnea, during sleep, 603–604
    electroencephalographic findings associated with, *605*
Arousal index, myoclonic, 606
Arousal response, 354. See also *Wakefulness.*

Arousal rhythm(s), on electroencephalography, 421–422
    appearance of, in children and adolescents, 407, *409, 410*
    frontal, as benign variant, 427, *427*
Artifact(s), electroencephalographic, 435
    monitoring and, 368, *369, 383, 386, 430, 431,* 435
    physiological, 430–432, *430–432.* See also specific types, e.g., *Ocular artifacts.*
    facial motor nerve conduction study, incorrect anode placement and, *107*
    isopotential mapping, sources of, 377, *380*
Ascending reticular activating system, 592–593
Asterixis, 283
Asymmetric alpha rhythm, on electroencephalography, 415, *415*
Athetosis, 283
Atrophy, multisystem, thermoregulatory sweat test results in, *317*
    muscular/muscle fiber, 86
        progressive, 241
        spinal, 242–243
            single-fiber electromyographic findings in, *145*
Attenuation of activity, episodic, as reactive phenomenon on EEG in preterm infant, 391
    transient unilateral, on EEG in sleeping newborn, 395–396, *397*
Auditory evoked potentials (AEPs), 525–539, 565–568
    brain stem, 525
        generators of, 529, *530*
            waveform-specific, 530
        non-neurological factors affecting, 531–534
        recording of, 527
            in brain death, 534, *535*
            in central pontine myelinolysis, *536, 537*
            in cerebrovascular disease, 537, *539*
            in coma, 534
            in degenerative diseases, 537–538
            in leukodystrophies, 536, *537*
            in metabolic disorders, 538–539
            in multiple sclerosis, 534–535, *535, 536*
            in neoplastic brain disease, 536–537, *537, 538*
            intraoperative, 566, *566, 567, 568*
            parameters for, 527–528
            rationale for masking in, 527, *527*
            simultaneous with electrocochleography, *526*
        variations in, with age, *531,* 531–532, *532*
            with body temperature, 532, *532*
            with gender, 532
            with stimulus frequency, 533, *533*
            with stimulus intensity, 532–533, *533*

Auditory evoked potentials (AEPs)
(*Continued*)
  with stimulus polarity, 534, *534*
  waveform components of, *528*, 528(t),
    528–529, *529*, 530(t)
    specific generators of, 530
  long latency, 526
    in demented patients, 526, *527*
    in elderly patients, 526, *527*
  middle latency, 526, *526*
Autonomic failure, pure, thermoregulatory sweat test results in, *317*
Autonomic nervous system, 307–309, *308*
  functions of, 309(t)
  testing of, 309–321
    cardiovascular responses in, 309–315
    pharmacologic approach to, 320–321
    pupillometry in, 320
    sudomotor function in, 315–320
Averaging (signal averaging), 11
  conditions for, 16–17
  filter effect of, 17–18
    at settings obscuring variations in
      VEPs, *17*
  noise reduction achieved by, *12*, 16(t)
Axillary neuropathies, 184
Axon(s), effects of loss of. See *Axon loss*
  entries.
  refractory period of, 112
    absolute, 112
      determination of, 113–114, *114*
      effects of temperature changes on, 115
      paired stimuli determining, 113–115,
        114
        combined with single distal stimu-
          lus, in assessment of motor
          fibers, 119, *119*
    relative, 112
      clinical utility of, 115–116
      determination of, *114*, 114–115
      supernormal and subnormal peri-
        ods following, 113
        demonstration of, *115*
      trains of stimuli determining, 115
Axon loss, 168–170
Axon loss brachial plexopathy, results of
  needle electrode examination in,
    205(t), 207(t)
  results of nerve conduction studies in,
    205(t), 207(t)
Axon loss polyneuropathies, 222–225,
  223(t)
  "pure" sensory, 223, 225, 225(t)
  results of needle electrode examination
    in, 224(t)
  results of nerve conduction studies in,
    224(t), 225(t)
  sensitivity of needle electrode examina-
    tion in, 171(t)
  sensitivity of nerve conduction studies
    in, 171(t)
Axon loss radiculopathy, 192, *193*
  results of needle electrode examination
    in, 195(t)
Axon reflex test, sudomotor, 316, *317, 318*

Background activity, low-voltage, on EEG
  in newborn, 392
Background slowing, on EEG in patient
  with cerebral dysfunction, 439, *439*
Ballism, 283
Bancaud's phenomenon, 415
Band-pass filter, *23, 24*. See also *Passband*
  entries.
Band-stop filter, *23, 24*

Barbiturate-induced anesthesia,
  electroencephalographic findings
    associated with, in cats, 352, *353*
Baroreceptor reflex, 309
Bell's palsy, 176
Benign childhood epilepsy with centrotemporal spikes, 468–470, *469, 470*
  dipole in, 377, *381*, 469, *469*
Benign fasciculation syndrome, 244
Benign variant findings, on electroencephalography, 424–430
Benzodiazepines, effects of, on brain
  neurotransmitters, 64(t)
Beta activity/beta rhythm, on
  electroencephalography, 416, *417*
  appearance of, in children and adoles-
    cents, 401, *402*
    during sleep, 406, *408*
  excessive, in patients with cerebral
    dysfunction, 440, *441*
  mechanisms of production of, 355
Beta delta complexes, on EEG in preterm
  infant, 388, 390, *390*
Bifid visual evoked potential waveforms,
  509, *509*
Bifrontal rhythmic delta activity, on EEG
  in newborn, 395, *396*
Bilaterally synchronous activity,
  development of, on EEG in preterm
    infant, 388, *390*
Bipolar montages, 360, *361*
  left frontal sharp wave in, *362, 364, 365,
    372, 374, 376*
  left temporal sharp wave in, *371*
  localization of potentials in, 361–367
  longitudinal, *361*
  ocular artifacts in, *379, 380, 383*
  phase reversal in, 361–362, *362*, 362(t),
    363, 365
    negative, 361, 362, *362*
    positive, 362, *362*
    triple, 365, *366*
  right occipital sharp wave in, *364*
  right temporal sharp wave in, *363*
  transverse, *361*
Bite(s), tick, weakness following, 266
Bleeding risk, vs. safe performance of
  needle electrode examination, 138
Blind subjects, response of, to repetitive
  TMS, 340, *340*
Blinking artifacts, on electroencephalography, 377, *379, 382*
Blink reflex, 107–108, *108*
  central nervous system lesions affecting,
    109
  facial nerve lesions affecting, 108–109,
    109
  trigeminal nerve lesions affecting, 109,
    109
Blood flow, cutaneous, measured for
  response to autonomic maneuver or
    stimulus, 314
Blood pressure response, to standing,
  313–314
  to sustained hand grip, 314
  to tilt, 314
  to Valsalva maneuver, 311, *312*, 312, *313*
Body temperature, lowering of, and
  changes in brain stem AEPs, 532, *532*
Botulinum toxin, 263
  electromyographically guided injection
    of, for involuntary muscle spasms,
      292–293
Botulism, 263–264
Brachial plexopathy, 205, 206, 209. See
  also *Brachial plexus*.

Brachial plexopathy (*Continued*)
  confirmation of preserved conduction
    in, by eliciting cutaneous nerve
      silent period, 301, *303*
  results of needle electrode examination
    in, 205(t), 207(t), 209(t)
  results of nerve conduction studies in,
    205(t), 207(t), 208(t)
Brachial plexus, 202, 202–203. See also
  Brachial plexopathy.
  compound muscle action potential
    (CMAP) domains of, 204(t)
  needle electrode examination of, 204,
    206, 207, 208
    muscles frequently assessed in, 204(t),
      206(t)
    results of, in cancer involving lower
      trunk axon loss, 207(t)
      in radiation injury involving infra-
        clavicular region, 209(t)
      in trauma involving upper trunk
        axon loss, 205(t)
  nerve conduction studies of, 204, 206,
    207, 208
    results of, in cancer involving lower
      trunk axon loss, 207(t)
      in radiation injury involving infra-
        clavicular region, 208(t)
      in trauma involving upper trunk
        axon loss, 205(t)
  sensory nerve action potential (SNAP)
    domains of, 203(t)
    incidence of involvement of, in cord
      lesions, 203(t)
    in trunk lesions, 203(t), 204
Brachial plexus volley component, of
  median nerve SEP, 544–545
Brain. See also *Electroencephalography*.
  death of, 444–445, 446, 448–450
    AEPs in, 534, *535*
    vs. brain stem death, 450
  degenerative diseases of, 450–453
    AEPs in, 537–538
  dysfunction of, 445–447
    electroencephalographic findings asso-
      ciated with, 438–445, 446, 447,
        457–466. See also under specific
        conditions, e.g., *Creutzfeldt-Jakob
        disease*.
  hypoperfusion of, and syncope, 448
    electroencephalographic findings
      associated with, 448, *449*
  imaging of, results of, in epileptic
    patients, 492, *492*, 493, *493, 494*
  packing of cells in, configuration of
    scalp-recorded potentials in relation
      to, 349, *350*. See also *Electroencepha-
      lography*.
  posterior fossa tumors of, and abnormal
    AEPs, 536–537, *537, 538*
  vascular disease of, and abnormal
    AEPs, 537, *539*
  white matter lesions of, and abnormal
    AEPs, 537, *539*
    and abnormal VEPs, 514
Brain stem AEPS. See *Auditory evoked
  potentials (AEPs), brain stem*.
Brain stem death, 450. See also *Brain,
  death of*.
Brain stem gliomas, abnormal AEPs
  associated with, 536, *538*
Breach rhythm, on electroencephalography, 417, *418*
  in patient with skull defects, 459, *461,
    462, 464, 465*
Breast cancer, radiotherapy for,
  electrodiagnosis of brachial plexus
    injury from, 208(t), 209(t)

Breathing, cessation of, during sleep, 603, 606
  electroencephalographic findings associated with, 605
  deep, heart rate response to, 310–311, 311
  monitoring of, in sleep laboratory, 598–599
Bulbospinal neuronopathy, X-linked recessive, 241
Burst-suppression, on EEG in patients with cerebral dysfunction, 441, 443, 447

Cancer. See also Tumor(s).
  brachial plexopathy associated with, electrodiagnostic findings in, 207(t)
  Lambert-Eaton myasthenic syndrome associated with, 261
  motor neuron disorder associated with, 243
  radiotherapy for, electrodiagnosis of brachial plexus injury from, 208(t), 209(t)
    lower motor neuron disorder following, 243–244
  sacral plexopathy associated with, electrodiagnostic findings in, 213(t)
Capacitors, 4, 4–5
  resistors combined with. See Filter(s).
Capnography, in sleep laboratory, 600
Carbamazepine, effects of, on brain neurotransmitters, 64(t)
Cardiac artifacts, on electroencephalography, 430, 431
Cardiac tissue, passage of current through, 32, 33
  fibrillation risk associated with, 32, 34, 37
  hazards leading to, 34, 36, 36, 37, 39
Cardiovascular autonomic testing, 310–315
Carpal tunnel syndrome, 177–180
  nerve conduction studies in, 100–101, 101, 178(t), 179
    grading of lesion severity based on, 179(t)
    sources of error in, 116, 117, 118
Cat(s), barbiturate-induced anesthesia in, electroencephalographic findings associated with, 352, 353
Catecholamines, 63
  role of, in wakefulness, 594
Cellular bases, of memory, 55–56
Cellular bursting, and epileptiform discharges, 355. See also Epileptiform discharges.
Central alpha rhythm, on EEG in children, 402
Central motor conduction time (CMCT), calculation of, 332–333
Central pontine myelinolysis, brain stem AEPs in, 536, 537
Central rapid spikes, on EEG in newborn, 396–397, 398
Central rhythmic alpha/theta bursts, midline, on EEG in newborn, 393, 395
Centrotemporal spikes, benign childhood epilepsy with, 468–470, 469, 470
  dipole in, 377, 381, 469, 469
Cerebellar tremor, 287
Cerebrum. See Brain.
Cervical myelopathy, spinal SEPs in, 74
Cervical radiculopathy(ies), 194–196
  failure to elicit cutaneous nerve silent period in, 301, 303

Cervical radiculopathy(ies) (Continued)
  results of needle electrode examination in, 194
Cervical spondyloradiculomyelopathy, 244
Cervical syringomyelia, 245
  median nerve SEPs in, 548, 550
CERV N13/P13 waveform component, of median nerve SEP, 546, 547
  absence of, in cervical syringomyelia, 548, 550
  dissociation of P14 complex from, 548, 550
Chebyshev filter, 28, 29(t)
Check size parameters, for recording of VEPs, 510
Chemical synapses, 57–58
Child(ren). See also Newborn.
  electroencephalography in. See Electroencephalography, in children and adolescents.
  epileptic seizures in. See Child(ren), seizures in.
  seizures in, 468–479
    absence epilepsy and, 471, 472
    benign, with centrotemporal spikes, 377, 381, 468–470, 469, 470
    cryptogenic/symptomatic epilepsy and, 473–478
    generalized epilepsy and, idiopathic, 470–473
      symptomatic/cryptogenic, 473–477
    hemispheric epileptic syndromes and, 477–478, 478
    idiopathic epilepsy and, 468–473
    Lennox-Gastaut syndrome and, 474, 476, 476
    localization-related epilepsy and, idiopathic, 468–470
      symptomatic/cryptogenic, 477–478
    mesial temporal lobe epilepsy and, 477, 477
    myoclonic epilepsy and, 471, 473
    neocortical epilepsy and, 477
    supplementary sensorimotor area epilepsy and, 477, 478
    symptomatic/cryptogenic epilepsy and, 473–478
    temporal lobe epilepsy and, 477
    West syndrome and, 474
Chronic inflammatory demyelinating polyneuropathy, electrodiagnostic findings in, 228, 231
Circuit(s), 2
  alternating current, 3–4. See also Current, alternating.
  capacitors in, 4, 4–5
    resistors combined with. See Filter(s).
  direct current, 1, 1–2
  high-pass. See under Filter(s).
  low-pass. See under Filter(s).
  resistors in, capacitors combined with. See Filter(s).
    connection of, in parallel, 3, 3
    in series, 2, 3
    voltage divider produced by, 3, 3
Classic slow channel syndrome, 265, 265(t)
CMAP (compound muscle action potential), 97
  elicitation of, by magnetic coil, 329
CMAP (compound muscle action potential) domains, of brachial plexus, 204(t)
CMCT (central motor conduction time), calculation of, 332–333
CMRR (common-mode rejection ratio), of amplifiers, 8, 23

Cold face test, 314
Cold pressor test, 314
Collision, of antidromic and orthodromic impulses, 116
  in assessment of fast vs. slow fiber conduction, 119–120, 120
  in assessment of refractory period in motor fibers, 119, 119
  in blocking of unwanted impulse transmission through Martin-Gruber anastomosis, 116, 118
  in blocking of unwanted nerve stimulation, 116
  in production of MNSP, 297, 298, 298. See also Mixed nerve silent period.
Color parameters, for recording of VEPs, 511
Coma, brain stem AEPs in, 534
  electroencephalographic findings associated with, 440–441, 441, 442
Common-mode rejection ratio (CMRR), of amplifiers, 8, 23
Compartment syndrome, risk of, with bleeding complicating NEE, 138
Compound muscle action potential (CMAP), 97
  elicitation of, by magnetic coil, 329
Compound muscle action potential (CMAP) domains, of brachial plexus, 204(t)
Compressive polyradiculopathies, 198
Concentration gradients, production of, diffusion through nongated channels in, 45
Conductance(s), ion channel, participating in cortical "pacing," 350
Conduction studies, in electromyography. See Nerve conduction studies.
Congenital acetylcholine receptor deficiency, 264, 265(t)
Congenital acetylcholinesterase deficiency, 264, 265(t)
Congenital myasthenia gravis, 264–265, 265(t)
Congenital myopathies, structural fiber changes in, 87
Continuity of activity, development of, on EEG in preterm infant, 388, 389
Continuous slowing, on EEG in patients with cerebral dysfunction, 439–440, 440
Contractile proteins, in myofilaments, 85
Contrast parameters, for recording of VEPs, 511
Cortical reflex myoclonus, 282
  Alzheimer's disease and, 283
  electroencephalographic findings in, 283, 286
  electromyographic findings in, 283, 285, 304
Coupling, excitation-contraction, 85–86
Cranial mononeuropathies, 176–177
Cranial nerve conduction studies, 107–110
  blink reflex on, 107–108, 108
    central nervous system lesions affecting, 109
    facial nerve lesions affecting, 108–109, 109
    trigeminal nerve lesions affecting, 109, 109
  jaw jerk (masseter reflex) on, 109–110
Creutzfeldt-Jakob disease, 451
  electroencephalographic findings associated with, 444, 451, 452, 462
Cryptogenic/symptomatic epilepsy, in pediatric patient, 473–478
Current, alternating, 3–4

Current *(Continued)*
supply of, to neurophysiology laboratory, 31, *32*
faults in, 31, 32(t)
direct, *1*, 1–2
passage of, through heart, 32, *33*
fibrillation risk associated with, 32, *34*, 37
hazards leading to, *34*, 36, *36*, 37, *39*
physiological sources of, 70
Current pathways, 69
Cutaneous nerve silent period, 298–301
delayed response to attempted elicitation of, 299, *299*
lack of response to attempted elicitation of, in cervical radiculopathy, 301, 303
in conduction block involving ulnar nerve, *302*
positive response to attempted elicitation of, 299–300, *300*
confirming preserved conduction in brachial plexopathy, 301, *303*

DCV (dorsal column volley) component, of median nerve SEP, 546, *547*
Death, of brain, 445–446, *446*, 448–450
AEPs in, 534, *535*
vs. brain stem death, 450
Decay (impulse decay), time constant of, *28*
Decibel(s), *2*, *2*, 526
Deep breathing, heart rate response to, 310–311, *311*
Degeneration, of brain, 450–453
AEPs in, 537–538
of muscle fibers, 87
Delta activity, during sleep, on EEG in child, 406, *408*
polymorphic, on EEG, 465, *466*
interpretation of, 464
rhythmic, bifrontal, on EEG in newborn, 395, *396*
intermittent, on EEG in epileptic patient, 484, *486*
on EEG in patient with cerebral dysfunction, 439, *440*
vs. eye movement artifact on EEG, *380*
Delta wave, alpha "intrusion" on, on EEG in sleeping patient, 605
pathological, on electroencephalography, 355
Dementia, 450–453
dialysis and, 446
long latency auditory evoked potentials in, 526, *527*
Demyelinating polyneuropathies, 225–231, 226(t)
results of needle electrode examination in, 227(t), 230(t)
results of nerve conduction studies in, 226, *226*, 227(t), 228, *228*, 229(t), *230*
slowing of conduction in, 219, 226, *226*, 228, *228*, 230
Demyelination, focal, 170–171
Dendrites, active properties of, 55
Dental fillings, as source of artifacts on electroencephalography, 431–432
Depolarization, space constant of, 51, *53*
time constant of, 51, *52*
Depressed and undifferentiated activity, on EEG in newborn, 392
Depth electrodes, 435

Diabetic polyneuropathy, 232
cardiovascular function test in, heart rate response to standing as, *313*
pupillometric findings associated with, 320, *320*, *321*
sweat test results in, *316*, *319*
Diabetic polyradiculopathies, 198–199
Dialysis, neurologic complications associated with, 446
Differential amplifier, 8, *9*, 20–21, *22*
Diffusion, through nongated channels, in production of concentration gradients, 45
Digital filters, 25, *25*, 28–30
Digitization, 11–13
Diphasic action potential, production of, *69*, *70*
Dipole detection, in electroencephalography, *352*, 377, *379*, 381, *382*, 383, *385*, *386*
orientation maximizing, 351
Direct current, *1*, 1–2
Disproportion, muscle fiber type, 87
Distance-to-pattern parameters, for recording of VEPs, 511
Diving response, to cold stimulus at first divisions of trigeminal nerve, 314
Dopamine, 63
Dorsal column volley (DCV) component, of median nerve SEP, 546, *547*
Drug intoxication, encephalopathies due to, 446
Duchenne muscular dystrophy, 278–279
D waves, electrical stimulation eliciting, 329, *330*
magnetic stimulation eliciting, 330
Dysautonomia, amyloidosis and, 315
test results linking, *311*, *312*, 315
Dyskinesia, 283
Dystonia, 283
Dystrophin, 83
Dystrophin-glycoprotein complex, 84
Dystrophy, limb girdle, 279
muscular, 278–279
single-fiber electromyographic findings in, 145
myotonic, 279

EEG. See *Electroencephalography*.
Elbow, ulnar neuropathy at, 181–182
results of needle electrode examination in, 181(t)
results of nerve conduction studies in, 181(t)
Elderly, electroencephalography in, temporal theta activity on, 417, *419*
long latency auditory evoked potentials in, 526, *527*
Electrical artifacts, as source of problems with isopotential mapping, 377, *380*
Electrical circuit(s), 2
alternating current, 3–4. See also *Current, alternating*.
capacitors in, 4, 4–5
resistors combined with. See *Filter(s)*.
direct current, *1*, 1–2
high-pass. See under *Filter(s)*.
low-pass. See under *Filter(s)*.
resistors in, capacitors combined with. See *Filter(s)*.
connection of, in parallel, *3*, 3
in series, *2*, 3
voltage divider produced by, *3*, 3
Electric shock, sources of, in electrodiagnostic laboratory, 31, *34*, *35*, *36*, 37, *39*

Electrocardiographic artifact, on electroencephalography, 383, *386*, 430, *431*
Electrocardiography, in sleep laboratory, 600
Electrocerebral inactivity, 444–445, *446*, 448–450
Electrocochleography, 525, *526*, 566
Electrode(s), 166
depth, 435
needle, 123, *123*, 125. See also *Needle electrode examination*.
re-use of, 138
nerve conduction studies via, 92, *92*. See also *Nerve conduction studies*.
phase reversal occurring at, 361, *362*. See also *Phase reversal*.
scalp. See also *Electroencephalography*.
management of, in epilepsy monitoring unit, 434–435
system for placement of, 359, *360*. See also *Electrode montages*.
subdural, 435
Electrode montages, 359–360
bipolar, 360, *361*
left frontal sharp wave in, *362*, *364*, *365*, *372*, *374*, *376*
left temporal sharp wave in, *371*
localization of potentials in, 361–367
longitudinal, *361*
ocular artifacts in, *379*, *380*, *383*
phase reversal in, 361–362, *362*, 362(t), *363*, *365*
negative, 361, *362*, *362*
positive, 362, *362*
triple, *365*, *366*
right occipital sharp wave in, *364*
right temporal sharp wave in, *363*
transverse, *361*
false assumptions about, affecting isopotential maps, *374*, *379*
recommendations regarding, in set-up for recording of brain stem AEPs, 528
in set-up for sleep laboratory, 598
referential, 360, *361*
electrocardiographic artifact in, *431*
left temporal sharp wave in, *371*
localization of potentials in, 367–370
phase reversal in, 362(t), *367*, *368*
right temporal sharp wave in, *363*, *368*
seizure patterns in, *374*, *378*, *381*
Electrodiagnostic examination, 162(t), 162–173. See also *Needle electrode examination; Nerve conduction studies*.
accessibility of body regions to, 173(t)
Electrodiagnostic laboratory, safety hazards in, 31–39
sources of electric shock in, 31, *34*, *35*, *36*, 37, *39*
Electroencephalography, 387–432
activity of uncertain significance on, in newborn, 392–397
alpha activity or bursts on. See also *Electroencephalography, alpha rhythm on*.
rhythmic, excessive, in newborn, 392–393, *395*
midline frontal and midline central, in newborn, 393
temporal, in preterm infant, 390, *391*
alpha-delta pattern on, in sleeping patient, 606
alpha rhythm on, *414*, 414–415. See also *Electroencephalography, alpha activity or bursts on*.
appearance of, in comatose patient, 441, *441*

Electroencephalography (Continued)
  central, in children, 402
    mechanisms of production of, 355
    variants of, 415, 415–416, 416
  alpha squeak on, 415, 415
  amplitude asymmetry on, 458–459
  arousal rhythm(s) on, 421–422
    appearance of, in children and adolescents, 407, 409, 409, 410
    frontal, 427, 427
  artifacts on, 435
    monitoring and, 368, 369, 383, 386, 430, 431, 435
    physiological, 430–432, 430–432. See also specific types, e.g., Electroencephalography, ocular artifacts on.
  asymmetric alpha rhythm on, 415, 415
  attenuation of activity on, as reactive phenomenon in preterm infant, 391
    in sleeping newborn, 395–396, 397
  background activity on, low-voltage, in newborn, 392
  background slowing on, in patients with cerebral dysfunction, 439, 439
  benign variants on, 424–430
  beta activity/beta rhythm on, 416, 417
    appearance of, in children and adolescents, 401, 402
    during sleep, 406, 408
    excessive, in patients with cerebral dysfunction, 440, 441
    mechanisms of production of, 355
  beta delta complexes on, in preterm infant, 388, 390, 390
  bifrontal delta activity on, rhythmic, in newborn, 395, 396
  bilaterally synchronous activity on, development of, in preterm infant, 388, 390
  blinking artifacts on, 377, 379, 382
  breach rhythm on, 417, 418
    in patient with skull defects, 459, 461, 462, 464, 465
  burst-suppression on, in patients with cerebral dysfunction, 441, 443, 447
  cardiac artifacts on, 430, 431
  central alpha rhythm on, in children, 402
  central rapid spikes on, in newborn, 396–397, 398
  centrotemporal spikes on, in child with benign epilepsy, 377, 381, 468–470, 469, 470
  continuity of activity on, development of, in preterm infant, 388, 389
  continuous slowing on, in patients with cerebral dysfunction, 439–440, 440
  delta activity on, during sleep, in children, 406, 408
    polymorphic, 465, 466
      interpretation of, 464
    rhythmic, bifrontal, in newborn, 395, 396
    intermittent, in epileptic patient, 484, 486
      in patient with cerebral dysfunction, 439, 440
    vs. eye movement artifact, 380
  delta wave on, alpha "intrusion" mixed with, in sleeping patient, 605
    pathologic, 355
  dental fillings as source of artifacts on, 431–432
  depressed and undifferentiated activity on, in newborn, 392
  dipole detection in, 352, 377, 379, 381, 382, 383, 385, 386

Electroencephalography (Continued)
  orientation maximizing, 351
  electrocardiographic artifact on, 383, 386, 430, 431
  electromyographic artifact on, 368, 369
  EMG correlated with, in evaluation of myoclonus, 282, 302–303, 304
  epileptiform discharges on. See Epilepsy; Epileptiform discharges; Seizures, neonatal.
  episodic attenuation of activity on, as reactive phenomenon in preterm infant, 391
  excessive beta activity on, in patients with cerebral dysfunction, 440, 441
  excessive rhythmic alpha or theta activity on, in newborn, 392–393, 395
  fast alpha variant on, 415, 416
  focal slowing on. See also Electroencephalography, slow waves/slow wave activity on.
    in patients with cerebral dysfunction, 440, 457
  fourteen- and 6-hertz positive bursts on, 406, 428, 428
    in sleeping children and adolescents, 406–407, 409
  frontal arousal rhythm on, 427, 427
  frontal intermittent rhythmic delta activity on, vs. eye movement artifact, 380
  frontal sharp wave on, left, in bipolar montage, 362, 364, 365, 372, 374, 376
  frontal sharp wave transients on, in preterm infant, 390–391, 391
  fronto-central theta activity on, in children and adolescents, 401–402, 403
  generalized slowing on. See also Electroencephalography, slow waves/slow wave activity on.
    in patients with cerebral dysfunction, 438–440, 439, 440
  glossokinetic artifact on, 431, 432
  hyperventilation-induced stimulation of activity on, 422–423, 423
    in children and adolescents, 402, 404
  hypnagogic hypersynchrony on, 417, 418
    in children and adolescents, 402–403, 404, 405
  ictal activity on. See Epilepsy; Epileptiform discharges; Seizures, neonatal.
  in adolescents. See Electroencephalography, in children and adolescents; Electroencephalography, in young adults and adolescents.
  in children and adolescents, 398–409
    abnormal occipital slow activity on, 400
      vs. posterior slow waves of youth, 401(t)
    arousal rhythms on, 407, 409, 409, 410
    beta activity on, 401, 402
      during sleep, 406, 408
    central alpha rhythm on, 402
    delta activity on, during sleep, 406, 408
    fourteen- and 6-hertz positive bursts on, during sleep, 406–407, 409
    fronto-central theta activity on, 401–402, 403
    hyperventilation-induced stimulation of activity on, 402, 404
    hypnagogic hypersynchrony on, 402–403, 404, 405
    ictal activity on. See Child(ren), seizures in.

Electroencephalography (Continued)
  K complexes on, 407, 409, 409
  lambda waves on, 400, 401
  mu rhythm on, 400–401, 402
  occipital slow activity on, abnormal, 400
    vs. posterior slow waves of youth, 401(t)
  occipital slow transients on, 406, 408
  positive occipital sharp transients of sleep on, 405–406, 407
  sleep rhythms on, 402–407, 404–409
  sleep spindles on, 405, 406, 407
  vertex transients on, during sleep, 405, 405
  in comatose patients, results of, 440–441, 441, 442
  induction of activity on, by hyperventilation, 422–423, 423
    in children and adolescents, 402, 404
    by intermittent photic stimulation, 423, 424
  in elderly, temporal theta activity on, 417, 419
  in epileptic patient. See Epilepsy; Epileptiform discharges; Seizures, neonatal.
  in newborn, 387–388. See also Electroencephalography, in preterm infant.
    activity of uncertain significance on, 392–397
    central rapid spikes on, 396–397, 398
    depressed and undifferentiated activity on, 392
    ictal activity on. See Seizures, neonatal.
    low-voltage background activity on, 392
    rhythmic alpha activity or bursts on, excessive, 392–393, 395
      midline frontal and midline central, 393, 395
    rhythmic bifrontal delta activity on, 395, 396
    rhythmic theta activity or bursts on, excessive, 392–393, 395
      midline frontal and midline central, 393, 395
      occipital, 395, 396
    sleep rhythms on, 391–392, 393, 394
    temporal sharp waves on, 396, 397(t), 398
    transient unilateral attenuation of activity on, during sleep, 395–396, 397
  in patients with cerebral dysfunction, 438–445, 446, 447, 457–466. See also under specific conditions, e.g., Creutzfeldt-Jakob disease.
  in preterm infant, 388. See also Electroencephalography, in newborn.
    beta delta complexes on, 388, 390, 390
    bilaterally synchronous activity on, development of, 388, 390
    continuity of activity on, development of, 388, 389
    episodic attenuation of activity on, as reactive phenomenon, 391
    frontal sharp wave transients on, 390–391, 391, 392
    sleep rhythms on, 391
    temporal alpha or theta bursts on, 390, 391
  intermittent rhythmic delta activity on, in epileptic patient, 484, 486
    in patient with cerebral dysfunction, 439, 440
    vs. eye movement artifact, 380

Electroencephalography (Continued)
  intermittent slowing on, in patients
    with cerebral dysfunction, 439,
    439
    rhythmic. See Electroencephalography,
      intermittent rhythmic delta activity
      on.
  intracranial, in epileptic patients, 491,
    491
  in young adults and adolescents. See
    also Electroencephalography, in chil-
    dren and adolescents.
    posterior slow waves on, 399–400,
      400, 418, 419
      vs. abnormal occipital slow activity,
        401(t)
  K complexes on, 420, 422
    in children and adolescents, 407, 409,
      409
  lambda waves on, 418, 420
    in children and adolescents, 400, 401
  lateralized epileptiform discharges on,
    periodic, 442–444, 444, 449, 459,
    462
  "lateral rectus spike" on, 383, 384
  left frontal sharp wave on, in bipolar
    montage, 362, 364, 365, 372, 374,
    376
  left temporal sharp wave on, in bipolar
    montage, 371
    in referential montage, 371
  localization of potentials in, 361–374.
    See also Electroencephalography,
    dipole detection in.
    on bipolar montages, 361–367. See
      also Electrode montages, bipolar.
    on graphs. See Isopotential maps; Volt-
      age/electrode position graphs.
    on referential montages, 367–370. See
      also Electrode montages, referential.
  low-pass filtering effects in, 26
  low-voltage activity on, 418–419
    in newborn, 392
  midtemporal ictal activity on. See Elec-
    troencephalography, centrotemporal
    spikes on; Epilepsy, mesial temporal
    lobe.
  midtemporal rhythmic discharges on,
    424–425, 426
  mu rhythm on, 416–417, 417
    in children and adolescents, 400–401,
      402
  myogenic artifact on, 430–431
  NREM sleep rhythms on. See Non-rapid
    eye movement (NREM) sleep.
  occipital sharp transients of sleep on,
    positive, 420, 421
    in children and adolescents, 405–
      406, 407
  occipital sharp wave on, right, in bipo-
    lar montage, 364
  occipital slow activity on, abnormal, in
    children and adolescents, 400
    vs. posterior slow waves of
      youth, 401(t)
  occipital slow transients on, in sleeping
    child, 406, 408
  occipital theta bursts on, rhythmic, in
    newborn, 395, 396
  ocular artifacts on, 377, 379, 382–385,
    383, 430, 431
    in bipolar montage, 379, 380, 383
  overlapping dipoles in, 383
  paradoxical alpha rhythm on, 415
  pen deflections in, 358–359, 359
  periodic complexes on, in patients with
    cerebral dysfunction, 442–444, 444,
    445

Electroencephalography (Continued)
  periodic lateralized epileptiform dis-
    charges on, 442–444, 444, 449, 459,
    462
  periodic sharp wave complexes on, in
    patients with cerebral dysfunction,
    443–444, 451, 452, 452, 453
  perspiration artifact on, 431
  photic stimulation of activity on, 423,
    424
    possible abnormalities in, 462, 464,
      464
  photoconvulsive (photoparoxysmal)
    response on, 424, 425, 473
  photoelectric artifact on, 423, 424
  photomyogenic (photomyoclonic) arti-
    fact on, 423–424, 425
  photoparoxysmal (photoconvulsive)
    response on, 424, 425, 473
  physiological artifacts on, 430–432, 430–
    432. See also specific types, e.g.,
    Electroencephalography, ocular arti-
    facts on.
  polymorphic delta activity on, 465, 466
    interpretation of, 464
  positive-negative transient on, synchro-
    nous with onset of myoclonus, 283,
    286. See also Electroencephalography,
    EMG correlated with, in evaluation of
    myoclonus.
  positive occipital sharp transients of
    sleep on, 420, 421
    in children and adolescents, 405–406,
      407
  posterior slow waves on, in young
    adults and adolescents, 399–400,
    400, 418, 419
      vs. abnormal occipital slow activity,
        401(t)
  pulse artifact on, 430, 431
  REM sleep rhythms on. See Rapid eye
    movement (REM) sleep.
  respiration artifact on, 431
  rhythmic alpha activity or bursts on.
    See also Electroencephalography, alpha
    rhythm on.
    excessive, in newborn, 392–393, 395
    midline frontal and midline central,
      in newborn, 393
  rhythmic delta activity on, bifrontal, in
    newborn, 395, 396
    intermittent, in epileptic patient, 484,
      486
      in patient with cerebral dysfunc-
        tion, 439, 440
      vs. eye movement artifact, 380
  rhythmic midtemporal discharges on,
    424–425, 426
  rhythmic slowing on, intermittent. See
    Electroencephalography, rhythmic delta
    activity on, intermittent.
  rhythmic theta activity/theta bursts/
    theta discharge on. See also Electro-
    encephalography, theta rhythm on.
    excessive, in newborn, 392–393, 395
    mesiotemporal, in epileptic patient,
      483
    midline frontal and midline central,
      in newborn, 393, 395
    occipital, in newborn, 395, 396
    subclinical, in adults, 425, 426, 427
  right occipital sharp wave on, in bipolar
    montage, 364
  right temporal sharp wave on, in bipo-
    lar montage, 363
    in referential montage, 363, 368
  sampling in, placement of scalp elec-
    trodes and, 359, 360. See also Elec-
    trode montages.

Electroencephalography (Continued)
  seizure activity on. See Epilepsy; Epilepti-
    form discharges; Seizures, neonatal.
  sharp spikes on, small, 428, 428–429
  sharp wave(s) on, 481. See also Electroen-
    cephalography, sharp wave transients
    on.
    appearance of, in bipolar montage,
      362–365, 371, 372, 374, 376
      in referential montage, 363, 368, 371
    mesiotemporal, in epileptic patient,
      482
    periodic, complexes of, in patients
      with cerebral dysfunction, 443–
      444, 451, 452, 452, 453
    temporal, in newborn, 396
      normal vs. abnormal, 397(t), 398
  sharp wave transients on. See also Elec-
    troencephalography, sharp wave(s) on.
    frontal, in preterm infant, 390–391,
      391
    occipital, positive. See Electroencepha-
      lography, positive occipital sharp
      transients of sleep on.
  six-hertz spike and wave on, 429, 429
  sleep rhythms and sleep spindles on.
    See Sleep; Sleep rhythms; Sleep spin-
    dles.
  slow alpha variant on, 415, 416
  slow waves/slow wave activity on, ab-
    normal, 355–356
    appearance of, in patients with cere-
      bral dysfunction, 438–440, 439,
      440, 457, 458, 459, 460
    occipital, in children and adoles-
      cents, 400
      vs. posterior slow waves of youth,
        401(t)
    posterior, in young adults and adoles-
      cents, 399–400, 400, 418, 419
      vs. abnormal occipital slow activ-
        ity, 401(t)
  small sharp spikes on, 428–429, 429
  spikes on, 481
    centrotemporal, in child with benign
      epilepsy, 377, 381, 468–470, 469,
      470
    lateral rectus, 383, 384
    rapid, central, in newborn, 396–397,
      398
    sharp, small, 428–429, 429
    six-hertz, 429, 429
    wicket, 429, 429–430
  spindles on, 352
    during anesthesia induced by barbitu-
      rates, in cats, 352, 353
    during coma, 441, 442
    during sleep, 420, 422, 589, 593, 602.
      See also Sleep rhythms.
    in children and adolescents, 405,
      406, 407
  squeak effect on, 415, 415
  subclinical rhythmic theta discharge on,
    in adults, 425, 426, 427
  temporal alpha bursts on, in preterm
    infant, 390, 391
  temporal ictal activity on, mesial. See
    Electroencephalography, centrotem-
    poral spikes on; Epilepsy, mesial tempo-
    ral lobe.
  temporal intermittent rhythmic delta
    activity on, in epileptic patient, 484,
    486
  temporal rhythmic discharges on,
    mesial, 424–425, 426
  temporal sharp wave(s) on, appearance
    of, in bipolar montage, 363, 371

Electroencephalography (Continued)
  in referential montage, 363, 368, 371
  mesial, in epileptic patient, 482
  normal vs. abnormal, in newborn,
    396, 397(t), 398
  temporal theta activity or bursts on, in
    elderly, 417, 419
    in preterm infant, 390, 391
  theta activity/theta bursts/theta dis-
    charge on. See also Electroencephalog-
    raphy, theta rhythm on.
    fronto-central, in children and adoles-
      cents, 401–402, 403
    rhythmic. See Electroencephalography,
      rhythmic theta activity/theta bursts/
      theta discharge on.
    temporal, in elderly, 417, 419
      in preterm infant, 390, 391
  theta rhythm on, 417. See also Electroen-
    cephalography, theta activity/theta
    bursts/theta discharge on.
  alpha rhythm combined with, in
    comatose patients, 441
  hippocampal, in animals, 355
  midline, 427, 427
  theta waves on, pathologic, 355
  transient unilateral attenuation of activ-
    ity on, in sleeping newborn, 395–
    396, 397
  triphasic waves on, in patients with
    cerebral dysfunction, 441–442, 443
  vertex transients on, during sleep, 420,
    421
    in children and adolescents, 405,
      405
  video recording combined with, in mon-
    itoring of epileptic patient, 434
  wicket spikes on, 429, 429–430
Electromagnetic stimulation. See Magnetic
  coil; Transcranial magnetic stimulation.
Electromyography. See also Needle electrode
  examination; Nerve conduction studies.
  artifact from, on electroencephalogra-
    phy, 368, 369
  Botox injection guided by, 292–293
  conduction studies in. See Nerve conduc-
    tion studies.
  electroencephalography correlated with,
    in evaluation of myoclonus, 282,
    302–303, 304
  intraoperative, 572–574, 573
  LLRs on, 301–305. See also Long latency
    electromyographic responses.
  macro-, 150, 150–151
  movement disorders evaluated via, 281,
    282, 284, 289, 290, 290
  myoclonus assessment via, 283, 285,
    303, 304
    correlated with electroencephalogra-
      phy, 282, 302–303, 304
  needle electrode examination in. See
    Needle electrode examination.
  quantitative, 140
    estimation of motor unit numbers in,
      155–157, 156, 157
    interference pattern analysis in, 153–
      155
    measurement of motor unit action
      potential in, 152–153
  scanning, 151, 151–152
  silent period in, 295–301, 335. See also
    Cutaneous nerve silent period; Mixed
    nerve silent period.
  single-fiber, 140–150
    jitter in, 142–146, 143, 144, 145, 146(t),
      149
    results of, in muscular dystrophy, 145

Electromyography (Continued)
  in myasthenia gravis, 145, 257(t)
  in spinal muscular atrophy, 145
  stimulation in, 146–147, 147
  sleep laboratory applications of, 598
Electro-oculography, in sleep laboratory,
  598
Elliptical filters, 28, 29(t)
EMG. See Electromyography.
Encephalitis. See also Encephalopathy(ies).
  following measles, 452
    electroencephalographic findings asso-
      ciated with, 444, 452, 453
Encephalopathy(ies), 445–447. See also
  Brain.
  electroencephalographic findings associ-
    ated with, 438–445, 446, 447. See
    also under specific conditions, e.g.,
    Creutzfeldt-Jakob disease.
Epilepsy, 468–494, 482(t). See also
  Epileptiform discharges; Seizures,
  neonatal.
  absence, 447, 448
    in pediatric patient, 471, 472
  benign focal, with centrotemporal
    spikes, in pediatric patient, 377,
    381, 468–470, 469, 470
  cryptogenic/symptomatic, in pediatric
    patient, 473–478
  drugs for, dosages of, in newborn,
    499(t)
    effects of, on brain neurotransmitters,
      64(t)
  evaluation of, 490–491
    referential montages in, 374, 378, 381
    surgical, 490–491
  focal. See Epilepsy, localization-related.
  frontal lobe, 485–487, 492–494
    brain scan results consistent with,
      493, 494
    electroencephalographic findings asso-
      ciated with, 486, 487, 488, 493,
      493
  generalized, idiopathic, in pediatric
    patient, 470–473
    symptomatic/cryptogenic, in pediat-
      ric patient, 473–477
  hemispheric lesions and, in pediatric
    patient, 477–478, 478
  idiopathic, in pediatric patient, 468–473
  localization-related, 481–490. See also
    site-specific sub-entries, e.g., Epi-
    lepsy, mesial temporal lobe.
    benign. See Epilepsy, benign focal.
    idiopathic, in pediatric patient, 468–
      470
    symptomatic/cryptogenic, in pediat-
      ric patient, 477–478
  mesial temporal lobe, 491–492
    brain scan results consistent with,
      492, 492, 493
    electroencephalographic findings asso-
      ciated with, 477, 482, 483, 484,
      485, 486, 491
    in pediatric patient, 477
  myoclonic, 282, 447, 447
    juvenile, 471, 473
  neocortical, electroencephalographic
    findings associated with, 477, 484,
    486
  occipital lobe, 489–490, 494
    electroencephalographic findings asso-
      ciated with, 489–490, 490, 494
  parietal lobe, 487, 489
    electroencephalographic findings asso-
      ciated with, 487, 489, 489
  partial, simple, 482–483

Epilepsy (Continued)
  supplementary sensorimotor area, 477
    electroencephalographic findings asso-
      ciated with, 477, 484
    in pediatric patient, 478
  symptomatic/cryptogenic, in pediatric
    patient, 473–478
  temporal lobe, 483–485. See also Epi-
    lepsy, mesial temporal lobe.
  neocortical, electroencephalographic
    findings associated with, 477,
    484, 486
Epilepsy monitoring unit, 433–436
Epileptic myoclonus, 282
  juvenile, 471, 473
Epileptiform discharges, 444, 481. See also
  Epilepsy; Seizures, neonatal.
  cellular bursting and, 355
  focal, benign. See Epilepsy, benign focal.
  lateralized, periodic, 442–444, 444, 447,
    459, 462
    focal seizure risk associated with,
      462, 463
Equilibrium potential, in glial cells, 46
Esophageal pH, monitoring of, in sleep
  laboratory, 601
Essential tremor, 286–287
Ethosuximide, effects of, on brain
  neurotransmitters, 64(t)
Exaggerated physiological tremor, 286
Excessive beta activity, on EEG in patient
  with cerebral dysfunction, 440, 441
Excessive rhythmic alpha or theta activity,
  on EEG in newborn, 392–393, 395
Excitation-contraction coupling, 85–86
Excitatory postsynaptic potentials, fields
  generated by, 350, 351
Expected value, of random variable, 13, 15
Eye movements, artifacts due to, on
  electroencephalography, 377, 379,
  382–385, 383, 430, 430
    in bipolar montage, 379, 380, 383
  nonrapid, in sleep. See under Sleep
    rhythms.
  posterior slow transients associated
    with, on EEG in children, 400, 401
  rapid, in sleep. See under Sleep rhythms.

Facial motor conduction studies, 107
  incorrect anode placement in, artifact
    due to, 107
Facial nerve, lesions of, 176
  effect of, on blink reflex, 108–109, 109
Familial demyelinating polyneuropathies,
  225–226, 226(t)
  results of needle electrode examination
    in, 227(t)
  results of nerve conduction studies in,
    226, 226, 227(t)
Familial infantile myasthenia gravis, 264,
  265(t)
Far-field recording, 75–76, 78
Fast alpha variant, on electroencephalogra-
  phy, 415, 416
Feedback inhibition, 54, 54
Feedback system, of muscle contraction,
  81–82
Feed-forward inhibition, 54, 54
Femoral neuropathies, 188
Fibrillation, ventricular, electrical current
  producing, 32, 34, 37
Fibrillation potentials, 130
  as findings in myopathies, 270(t)
Field size parameters, for recording of
  VEPs, 510

Fillings (dental fillings), as source of artifacts on electroencephalography, 431–432
Filter(s), 6, 23–30
band-pass, 23, 24. See also *Passband* entries.
band-stop, 23, 24
Chebyshev, 28, 29(t)
design of, and rolloff, 27
tradeoffs in, 29, 30(t)
digital, 25, 25, 28–30
elliptical, 28, 29(t)
finite impulse response, 30, 30(t)
high-order, 28
ringing by, 28, 29
high-pass, 5, 23, 24, 25
frequency response of, 5
time response in relation to, 7
transient response of, 6, 6
infinite impulse response, 30, 30(t)
inverted Chebyshev, 29(t)
low-pass, 5, 23, 24
effect of, on electroencephalographic signal, 26
frequency response of, 5
time response in relation to, 7
transient response of, 6, 6
signal averaging as form of, 17–18
at settings obscuring variations in VEPs, 17
time constant of, 6–7, 7
Finite impulse response filters, 30, 30(t)
Flash visual evoked potentials, 519–520, 520
Focal epilepsy, 481–490. See also *Frontal lobe epilepsy; Occipital lobe epilepsy; Parietal lobe epilepsy; Temporal lobe epilepsy.*
childhood cases of, benign, with centro-temporal spikes, 377, 381, 468–470, 469, 470
idiopathic, 468–470
symptomatic/cryptogenic, 477–478
Focal slowing. See also *Electroencephalography, slow waves/slow wave activity on, abnormal.*
on EEG in patients with cerebral dysfunction, 440, 457
Fourteen- and 6-hertz positive bursts, on electroencephalography, 406, 428, 428
in sleeping children and adolescents, 406–407, 409
Frequency-dependent changes, in synaptic transmission, 62–63
Frequency response, of high-pass filter, 5
time response in relation to, 7
of low-pass filter, 5
time response in relation to, 7
Frontal arousal rhythm, as benign variant on EEG, 427, 427
Frontal intermittent rhythmic delta activity, vs. eye movement artifact on EEG, 380
Frontal lobe epilepsy, 485–487, 492–494
brain scan results consistent with, 493, 494
electroencephalographic findings associated with, 486, 487, 488, 493, 493
Frontal rhythmic alpha/theta bursts, midline, on EEG in newborn, 393, 395
Frontal sharp wave, left, in bipolar electrode montage, 362, 364, 365, 372, 374, 376
Frontal sharp wave transients, on EEG in preterm infant, 390–391, 391, 392
Fronto-central theta activity, on EEG in children and adolescents, 401–402, 403

F-wave, 102, 103, 575

GABA (gamma-aminobutyric acid), 64
receptors for, 64–65
Gain, of amplifiers, 20, 20(t), 21
Gamma-aminobutyric acid (GABA), 64
receptors for, 64–65
Gap junctions, 57
Generalized epilepsy, idiopathic, in pediatric patient, 470–473
symptomatic/cryptogenic, in pediatric patient, 473–477
Generalized slowing. See also *Electroencephalography, slow waves/slow wave activity on, abnormal.*
on EEG in patients with cerebral dysfunction, 438–440, 439, 440
Giant SEPs, in myoclonus, 303, 304, 305
Glaucoma, visual evoked potentials in, 516–517
Glial cells, changes in membrane potential of, in relation to extracellular potassium concentration, 46, 46, 351, 353
Glioma(s), brain stem, abnormal AEPs associated with, 536, 538
Glossokinetic artifact, on electroencephalography, 431, 432
Glutamate, 65
Glutamate receptors, 65
G-protein–coupled receptors, 59–60, 60

Hand, ulnar neuropathies in, 182
Hand grip, sustained, blood pressure response to, 314
Heart, passage of current through, 32, 33
fibrillation risk associated with, 32, 34, 37
hazards leading to, 34, 36, 36, 37, 39
Heart rate, spectral analysis of, 314–315
Heart rate response, to deep breathing, 310–311, 311
to standing, 312–313, 313
to Valsalva maneuver, 311–312
Heavy metal exposure, manifestations of, consistent with motor neuron disorders, 244
Hemifield stimulation, of pattern-reversal visual evoked potentials, 517
Hemispheric seizure syndromes, in pediatric patient, 477–478, 478
Hepatic encephalopathies, 445–446
Hereditary spinal muscular atrophy, 242–243
Hexosaminidase deficiency, manifestations of, resembling motor neuron disorders, 243
High-order filters, 28
ringing by, 28, 29
High-pass filter, 5, 23, 24, 25
frequency response of, 5
time response in relation to, 7
transient response of, 6, 6
High sciatic neuropathies, pattern of muscle involvement in, 187, 187(t)
Hippocampal theta rhythms, on EEG in animals, 355
HIV (human immunodeficiency virus) infection, and dementia, 452–453
and myopathy, 279
H-reflex, 101–102, 102, 575
Human immunodeficiency virus (HIV) infection, and dementia, 452–453
and myopathy, 279

Hypersynchrony, hypnagogic, on electroencephalography, 417, 418
in children and adolescents, 402–403, 404, 405
Hypertrophy, of muscle fibers, 87
Hyperventilation-induced stimulation, of electroencephalographic activity, 422–423, 423
in children and adolescents, 402, 404
Hypnagogic hypersynchrony, on electroencephalography, 417, 418
in children and adolescents, 402–403, 404, 405
Hypoglossal neuropathies, 177
Hypoperfusion, cerebral, and syncope, 448
electroencephalographic findings associated with, 448, 449
Hypothalamus, sleep and wake centers in, 593
Hysterical tremor, 287

Ictal activity, on electroencephalography. See *Epilepsy; Epileptiform discharges; Seizures, neonatal.*
ICU patient, myopathy in, 279
electrodiagnostic approach to, 278(t), 279(t)
Idiopathic epilepsy, in pediatric patient, 468–473
Idiopathic recurring stupor, 447
Imprint method (Silastic imprint method), of sweat testing, 316, 318, 319
Impulse decay, time constant of, 28
Impulse response filters, finite vs. infinite, 30, 30(t)
Inactivity, electrocerebral, 444–445, 446, 448–450
Inclusion body myopathy, 278
Infant(s). See *Child(ren); Newborn.*
Infantile myasthenia gravis, familial, 264, 265(t)
Infantile spasms, West syndrome and, 474, 475
Infinite impulse response filters, 30, 30(t)
Inflammatory demyelinating polyneuropathy, acute, electrodiagnostic findings in, 228, 230–231
chronic, electrodiagnostic findings in, 228, 231
Infraclavicular plexus, 207. See also *Brachial plexus.*
lesions involving, 209
radiation-induced, electrodiagnostic findings associated with, 208(t), 209(t)
Inhibition, feedback, 54, 54
feed-forward, 54, 54
Inhibitory postsynaptic potentials, fields generated by, 351
Intensive care unit patient, myopathy in, 279
electrodiagnostic approach to, 278(t), 279(t)
Interference pattern analysis, 153–155
Intermittent photic stimulation, of electroencephalographic activity, 423, 424
Intermittent rhythmic delta activity, on EEG in epileptic patient, 484, 486
on EEG in patient with cerebral dysfunction, 439, 440
vs. eye movement artifact on EEG, 380
Intermittent rhythmic slowing. See *Intermittent rhythmic delta activity.*
Intermittent slowing, on EEG in patient with cerebral dysfunction, 439, 439

Intoxication/poisoning, drug, encephalopathies due to, 446
magnesium, 266
metal, manifestations of, consistent with motor neuron disorders, 244
organophosphate, 265–266
Intracranial electroencephalography, in epileptic patients, 491, *491*
Intraoperative monitoring, 565–584, 566(t)
Inverted Chebyshev filter, 29(t)
Involuntary movements, evaluation of, 281–283, *284*
electromyographic patterns in, 281, 282, *284*
Involuntary muscle spasms. See also *Involuntary movements.*
electromyographically guided Botox injection for, 292–293
Ion(s), differences in concentrations of, at each side of cell membrane, 45, 45(t)
in production of voltage gradients, 45–46
diffusion of, 45, *45*
multiple contributions of, to resting potential, 46–47
Ion channel(s), 60–62, 84
abnormal, and disease, 84(t)
ligand-gated, 60–62
selective permeability of, 43, 45
voltage-activated, 62
Ion channel conductances, participation of, in cortical "pacing," 350
Ionic currents, neuronal, 62(t)
Ionotropic receptors, 59
Ischemic monomelic neuropathy, 217
Isopotential maps, *366, 370,* 370–371
problems with, due to electrical artifacts, 377, *380*
due to false assumptions about montages, 374, *379*
seizure patterns on, *378, 381*
voltage/electrode position graphs corresponding to, 370, *370, 373, 375, 377, 381*
I waves, electrical stimulation eliciting, 329, *330*
magnetic stimulation eliciting, 330, 337, *337*

Jaw jerk (masseter reflex), 109–110
Jerk, myoclonic. See *Myoclonus.*
Jitter, in single-fiber electromyography, 142–146, *143, 144, 145,* 146(t), 149
Juvenile absence epilepsy, 471
Juvenile myoclonic epilepsy, 471, *473*

K complexes, on electroencephalography, 420, *422*
in children and adolescents, 407, 409, *409*
Kennedy's disease, 241
Kidney disease, encephalopathy associated with, 446
Kinetic tremor, 287

Lambda waves, on electroencephalography, 418, *420*
in children and adolescents, 400, *401*
Lambert-Eaton myasthenic syndrome, 245, 260–262, 261(t)

Lambert-Eaton myasthenic syndrome (*Continued*)
acetylcholine release in, *253*
repetitive stimulation studies in, 260–261, *261*
Lateralized epileptiform discharges, periodic, 442–444, *444, 449, 459, 462*
focal seizure risk associated with, 462, *463*
"Lateral rectus spike," on electroencephalography, 383, *384*
Lateral sclerosis, amyotrophic, 237–239
results of needle electrode examination in, 238, 238(t)
results of nerve conduction studies in, 238, 238(t), *239*
variants of, 239–241
primary, 240
Left frontal sharp wave, in bipolar electrode montage, *362, 364, 365, 372, 374, 376*
Left temporal sharp wave, in bipolar electrode montage, *371*
in referential electrode montage, *371*
Leg movements, periodic, during sleep, 604–605
Lennox-Gastaut syndrome, 474, 476, *476*
Leukodystrophies, brain stem auditory evoked potentials in, 536, *537*
visual evoked potentials in, 514
Lhermitte's syndrome, visual evoked potentials in, 514
Ligand-gated ion channels, 60–62
Light flash-induced stimulation, of electroencephalographic activity, 423, *424*
possible abnormalities in, 462, 464, *464*
Limb girdle dystrophy, 279
Limb mononeuropathies, 177–188
Liver disease, encephalopathies due to, 445–446
LLRs, in electromyography. See *Long latency electromyographic responses.*
Localization-related epilepsy. See *Focal epilepsy.*
Longitudinal bipolar montage, *361.* See also *Montages (electrode montages), bipolar.*
Long latency auditory evoked potentials, 526
in demented patients, 526, *527*
in elderly patients, 526, *527*
Long latency electromyographic responses, 301–305, *302*
in anxious patient, *304*
in patient with myoclonus, 303, *304*
Long thoracic neuropathies, 184–185
Low-pass filter, *5,* 23, *24*
effect of, on electroencephalographic signal, *26*
frequency response of, *5*
time response in relation to, *7*
transient response of, 6, *6*
Low-voltage activity, on electroencephalography, 418–419
in newborn, 392
Lumbar plexus, 210, *211*
needle electrode examination of, 212
nerve conduction studies of, 211–212
Lumbar spinal cord, radiation of, lower motor neuron syndrome following, 243–244
Lumbosacral myotomes, 212(t)
Lumbosacral plexopathy, 212

Lumbosacral plexopathy (*Continued*)
limitations of nerve conduction studies in, 211
Lumbosacral radiculopathies, 196–197
Luminance parameters, for recording of VEPs, 510–511
Lung cancer, axon loss brachial plexopathy associated with, electrodiagnostic findings in, 207(t)
Lymphoma, motor neuron disorder associated with, 243

Macro-electromyography, *150,* 150–151
Magnesium intoxication, 266
Magnetic coil, electric fields induced by, 325–326, *326*
transneuroforaminal, *327*
motor cortex stimulation by, 329–330, *333.* See also *Transcranial magnetic stimulation.*
and elicitation of D waves, 330
and elicitation of I waves, 330, 337, *337*
and elicitation of silent period, *335*
intraoperative, 571
paired-pulse stimulation by, 336
phrenic nerve stimulation by, *327, 328*
Martin-Gruber anastomosis, blocking of unwanted impulse transmission through, 116, *118*
Masking, in recording of brain stem AEPs, 527, *527*
Masseter reflex (jaw jerk), 109–110
Masseter silent period, 299, *299*
Measles, subacute sclerosing panencephalitis following, 452
electroencephalographic findings associated with, 444, 452, *453*
Median nerve compression, at wrist. See *Carpal tunnel syndrome.*
Median nerve conduction studies, 97–101
applications of, in carpal tunnel syndrome. See *Carpal tunnel syndrome.*
motor, 97–98
arrangement of equipment for, *97*
sensorimotor, 100–101, *101*
sensory, 98–100
arrangement of equipment for, *99*
Median nerve somatosensory evoked potential(s), 544–553
recording of, 557, 559
waveform component(s) of, 544–553, 545(t)
brachial plexus volley, 544–545
CERV N13/P13, 546, *547*
absence of, in cervical syringomyelia, 548, *550*
dissociation of P14 complex from, 548, *550*
dorsal column volley, 546, *547*
N9, 545, *546*
N10, 545, *546*
N12, 546, *547*
N18, 548, 550–551
loss of, 551, *551*
preservation of, despite loss of N20 component, 550, *551*
N20, 551–552, *552, 553*
loss of, 550, *551, 552, 554*
preservation of, despite loss of P22 component, 552, *555*
P9, 545, *546*
P11, 546, *547*
P14 complex, 546, 548, *548, 549*
dissociation of CERV N13/P13 component from, 548, *550*

Median nerve somatosensory evoked
    potential(s) *(Continued)*
    P15, 552, *553*
    P22, 552
        loss of, 552, *555*
        preservation of, despite loss of N20
            component, 552, *554*
    proximal plexus volley, 545, *546*
Membrane potential(s), 43
    generation of, nongated ion channels in,
        43
    glial cell, changes in, with altered extra-
        cellular potassium concentrations,
        46, *46*, 351, *353*
    measurement of, 43, *44*
Memory, cellular bases of, 55–56
Meningioma, abnormal brain stem AEPs
    associated with, *538*
    surgery for, recording of SEPs during,
        571
Mental arithmetic test, pressor response to
    stress of, 314
MEPs (motor evoked potentials), 570–572
    spinal cord stimulation producing, 571–
        572, *573*
Mesial temporal lobe epilepsy, 491–492
    brain scan results consistent with, 492,
        492, 493
    electroencephalographic findings associ-
        ated with, 477, *482, 483, 484, 485,*
        486, 491
    in pediatric patient, 477
Mesiotemporal/midtemporal electroen-
    cephalographic activity, ictal. See
    Centrotemporal spikes; Mesial temporal
    lobe epilepsy.
    rhythmic discharges as, benign, 424–
        425, *426*
Metabiotropic glutamate receptors, 65
Metabolic disorders, brain stem AEPs in,
    538–539
Metabolic encephalopathies, 445
Metal exposure, sequelae of, motor
    neuron disorder–like features of, 244
Middle latency auditory evoked
    potentials, 526, *526*
Midline frontal and midline central
    rhythmic alpha/theta bursts, on EEG
    in newborn, 393, *395*
Midline theta rhythm, as benign variant
    on EEG, 427, *427*
Midtemporal/mesiotemporal
    electroencephalographic activity, ictal.
    See *Centrotemporal spikes; Mesial*
    temporal lobe epilepsy.
    rhythmic discharges as, benign, 424–
        425, *426*
Mitochondrial abnormalities, in muscle
    fibers, 87
Mixed nerve silent period, 295–298, *296,*
    296(t), *297*
    collision of antidromic and orthodromic
        impulses producing, 297, 298, *298*
    stages of, 297
        abnormally distinct separations
            between, in polyneuropathy, 297,
            297
Mixed sensorimotor nerve conduction
    studies, 100–101, *101.* See also *Nerve*
    conduction studies.
MNSP. See *Mixed nerve silent period.*
Monomelic amyotrophy, 240
Monomelic neuropathy, ischemic, 217
Mononeuropathy(ies), 174–188
    cranial, 176–177
    facial, 176
        effect of, on blink reflex, 108–109, *109*

Mononeuropathy(ies) *(Continued)*
    limb, 177–188
    multiple, 215–219
Mononeuropathy multiplex, 216–217
Montages (electrode montages), 359–360
    bipolar, 360, *361*
        left frontal sharp wave in, *362, 364,*
            365, 372, 374, 376
        left temporal sharp wave in, *371*
        localization of potentials in, 361–367
        longitudinal, *361*
        ocular artifacts in, *379, 380, 383*
        phase reversal in, 361–362, *362,* 362(t),
            363, 365
            negative, 361, 362, *362*
            positive, 362, *362*
            triple, 365, *366*
        right occipital sharp wave in, *364*
        right temporal sharp wave in, *363*
        transverse, *361*
    false assumptions about, affecting isopo-
        tential maps, 374, *379*
    recommendations regarding, in set-up
        for recording of brain stem AEPs,
        528
    in set-up for sleep laboratory, 598
    referential, 360, *361*
        electrocardiographic artifact in, *431*
        left temporal sharp wave in, *371*
        localization of potentials in, 367–370
        phase reversal in, 362(t), *367, 368*
        right temporal sharp wave in, *363,*
            368
    seizure patterns in, *374, 378, 381*
Motor conduction time, central,
    calculation of, 332–333
Motor control system, 81
Motor cortex, electrical stimulation of,
    329, *330*
    and elicitation of D waves, 329, *330*
    and elicitation of I waves, 329, *330*
    intraoperative, *571*
    magnetic stimulation of, 329–330, *333.*
        See also *Transcranial magnetic stimu-*
        lation.
        and elicitation of D waves, 330
        and elicitation of I waves, 330, 337,
            337
        and elicitation of silent period, *335*
        intraoperative, *571*
    mapping of, training effects revealed by,
        339, *339*
Motor evoked potentials (MEPs), 570–572
    spinal cord stimulation producing, 571–
        572, *573*
Motor nerve conduction studies, 97–98.
    See also *Nerve conduction studies.*
    facial, 107
        incorrect anode placement in, artifact
            due to, *107*
    median, 97–98
        arrangement of equipment for, *97*
    ulnar, 98
Motor neuron disorders, 235–248
    needle electrode examination in, 236–
        237
        muscles routinely assessed with,
            237(t)
    nerve conduction studies in, 235–236
        peripheral nerves routinely assessed
            with, 236(t)
Motor neuropathy, multifocal, 218–219,
    244
Motor unit(s), 82–83, 122–123, 268, *268*
    estimation of number(s) of, in muscle,
        155–158, *156, 157*
    recruitment of, 127–128

Motor unit action potential, 123–125, *124,*
    127(t)
    quantitative methods of measurement
        of, 152–153
Movement disorders, 281–293. See also
    specific disorders, e.g., *Tremor.*
    dementia in, 451
    electrodiagnostic approach to, 281–293
Multifocal motor neuropathy, 218–219, 244
Multiple mononeuropathies, 215–219
Multiple sclerosis, brain stem AEPs in,
    534–535, *535, 536*
    visual evoked potentials in, 513, *513,*
        514
Multiple sleep latency test, 606–607,
    609
Multisystem atrophy, thermoregulatory
    sweat test results in, *317*
Mu rhythm, on electroencephalography,
    416–417, *417*
    in children and adolescents, 400–401,
        402
Muscle contraction, 85–86
    feedback system of, 81–82
Muscle fiber(s), 83–86
    atrophy of. See *Muscular atrophy/muscle*
        fiber atrophy.
    degeneration of, 87
        vacuolar, 87
    disproportion of types of, 87
    histochemical classification of, 83, 83(t)
    hypertrophy of, 87
    mitochondrial abnormalities in, 87
    necrosis of, 87
        electrodiagnostic findings associated
            with, 275(t)
    structural abnormalities of, 86–87
        congenital, 87
Muscle metabolism, 86
Muscle spasms. See *Spasm(s); Spasticity.*
Muscular atrophy/muscle fiber atrophy,
    86
    progressive, 241
    spinal, 242–243
        single-fiber electromyographic find-
            ings in, *145*
Muscular dystrophy, 278–279
    single-fiber electromyographic findings
        in, *145*
Musculocutaneous neuropathies, 184
Myasthenia gravis, 245, 253–259
    acetylcholine release in, 253
    congenital, 264–265, 265(t)
    differential diagnosis of, 259(t)
    infantile, familial, 264, 265(t)
    needle electrode examination in, 257
    neonatal, 259
    nerve conduction studies in, 256(t)
    repetitive stimulation studies in, 256(t)
    single-fiber electromyographic findings
        in, *145,* 257(t)
Myasthenic syndrome. See *Lambert-Eaton*
    myasthenic syndrome.
Myelinolysis, central pontine, brain stem
    AEPs in, *536, 537*
Myelopathy, cervical, spinal SEPs in, *74*
Myoclonic arousal index, 606
Myoclonic epilepsy, 282, 447, *447*
    juvenile, 471, *473*
Myoclonic status epilepticus, 447, *447*
Myoclonus, 282
    cortical reflex, 282
        Alzheimer's disease and, 283
        electroencephalographic findings in,
            283, *286*
        electromyographic findings in, 283,
            285, 304

Myoclonus *(Continued)*
electroencephalographic findings in, 283, *286*
correlated with results of EMG, 282, 302–303, *304*
electromyographic findings in, 283, *285*, 303, 304
correlated with results of EEG, 282, 302–303, *304*
epileptic, 282, 447, *447*
juvenile, 471, *473*
giant SEPs in, 303, 304, *305*
long latency electromyographic responses in, 303, 304
nonepileptic, 282
pyramidal. See *Myoclonus, cortical reflex.*
reticular reflex, 282
somatosensory evoked potentials in, 303, 304, *305*
Myofibrils, 85, *85*. See also *Muscle fiber(s).*
Myofilaments, 85. See also *Muscle fiber(s).*
contractile proteins in, 85
Myogenic artifact, on electroencephalography, 430–431
Myopathy(ies), 268–279
congenital, structural fiber changes in, 87
fibrillation potentials occurring with, 270(t)
inclusion body, 278
necrotizing, 87
electrodiagnostic findings associated with, 275(t)
needle electrode examination in, 269–273
results of, 269–273, *274*, 275(t), 277(t)
in acutely ill or intensive care unit patient, 279(t)
nerve conduction studies in, 269
results of, 269, 275(t)
in acutely ill or intensive care unit patient, 278(t)
Myotomes, lumbosacral, 212(t)
Myotonic dystrophy, 279

N9 waveform component, of median nerve SEP, 545, *546*
N10 waveform component, of median nerve SEP, 545, *546*
N12 waveform component, of median nerve SEP, 546, *547*
N18 waveform component, of median nerve SEP, 548, *550–551*
loss of, 551, *551*
preservation of, despite loss of N20 component, 550, *551*
N20 waveform component, of median nerve SEP, 551–552, *552, 553*
loss of, 552, *554*
preservation of, despite loss of P22 component, 552, *555*
Near-field recording, 71
Necrosis, muscular/muscle fiber, 87
electrodiagnostic findings associated with, 275(t)
NEE. See *Needle electrode examination.*
Needle electrode(s), 123, *123*, 125. See also *Needle electrode examination.*
re-use of, 138
Needle electrode examination, 122–138
bleeding as complication of, 138
and risk of compartment syndrome, 138
muscle(s) assessed via, 168(t)
in brachial plexus domains, 204(t), 206(t)

Needle electrode examination *(Continued)*
in sites of possible myopathy, 274(t)
results of, 166(t)
in acquired demyelinating polyneuropathy, 230(t)
in amyotrophic lateral sclerosis, 238, 238(t)
in axon loss brachial plexopathy, 205(t), 207(t)
in axon loss polyneuropathy, 224(t)
in axon loss radiculopathy, 195(t). See also *Radiculopathy(ies), needle electrode examination in.*
in botulism, 263
in brachial plexopathy, 205(t), 207(t), 209(t)
in cancer-related brachial plexopathy, 207(t)
in cancer-related sacral plexopathy, 213(t)
in cervical radiculopathies, *194*. See also *Radiculopathy(ies), needle electrode examination in.*
in demyelinating polyneuropathies, 227(t), 230(t)
in familial demyelinating polyneuropathy, 227(t)
in Lambert-Eaton myasthenic syndrome, 261
in lumbar plexopathy, 212
in motor neuron disorders. See *Motor neuron disorders, needle electrode examination in.*
in myasthenia gravis, 257
in myopathies. See *Myopathy(ies), needle electrode examination in.*
in polymyositis, 275(t)
in polyneuropathies, 222, 224(t), 227(t), 230(t)
in radiation-induced brachial plexopathy, 209(t)
in radiculopathy. See *Radiculopathy(ies), needle electrode examination in.*
in sacral plexopathy, 212, 213(t)
in traumatic brachial plexopathy, 205(t)
in ulnar neuropathy at elbow, 181(t)
safe performance of, 137–138
steps for clinician to follow in, 137–138
vs. risk of excessive bleeding, 138
sensitivity of, in axon loss polyneuropathy, 171(t)
vs. nerve conduction studies, 122
Negative phase reversal, 361, 362, *362*
Neocortical epilepsy, electroencephalographic findings associated with, 477, 484, *486*
Neoplastic brain disease, and abnormal AEPs, 536–537, *537, 538*
Nerve block, physiological, collision-based production of. See *Collision, of antidromic and orthodromic impulses.*
ulnar, failure to elicit cutaneous nerve silent period in, 302
Nerve conduction studies, 89–110
colliding impulses in. See *Collision, of antidromic and orthodromic impulses.*
cranial, 107–110
blink reflex on, 107–108, *108*
central nervous system lesions affecting, 109
facial nerve lesions affecting, 108–109, *109*
trigeminal nerve lesions affecting, 109, *109*

Nerve conduction studies *(Continued)*
jaw jerk (masseter reflex) on, 109–110
determination of refractory period in, 113–115, *114*. See also *Refractory period.*
electrodes used in, 92, *92*
facial motor, 107
incorrect anode placement in, artifact due to, *107*
limitations of, in lumbosacral region, 211
median, 97–101
applications of, in carpal tunnel syndrome. See *Carpal tunnel syndrome.*
motor, 97–98
arrangement of equipment for, *97*
sensorimotor, 100–101, *101*
sensory, 98–100
arrangement of equipment for, *99*
mixed sensorimotor, 100–101, *101*
motor, 97–98
facial, 107
incorrect anode placement in, artifact due to, *107*
median, 97–98
arrangement of equipment for, *97*
ulnar, 98
results of, 166(t)
in acquired demyelinating polyneuropathies, *226, 228,* 229(t), *230*
in acute inflammatory demyelinating polyneuropathy, 228, 230–231
in amyotrophic lateral sclerosis, 238, 238(t), *239*
in axon loss brachial plexopathy, 205(t), 207(t)
in axon loss polyneuropathies, 224(t), 225(t)
in brachial plexopathy, 205(t), 207(t), 209(t)
in cancer-related brachial plexopathy, 207(t)
in cancer-related sacral plexopathy, 213(t)
in carpal tunnel syndrome. See *Carpal tunnel syndrome, nerve conduction studies in.*
in chronic inflammatory demyelinating polyneuropathy, 228
in demyelinating polyneuropathies, *226, 226,* 227(t), *228, 228,* 229(t), *230*
in familial demyelinating polyneuropathies, *226, 226,* 227(t)
in lumbar plexopathy, 211, 212
in motor neuron disorders. See *Motor neuron disorders, nerve conduction studies in.*
in myopathies. See *Myopathy(ies), nerve conduction studies in.*
in polymyositis, 275(t)
in polyneuropathies, 221–231, 224(t)–229(t), *226, 228, 230*
factors confounding interpretation of, 231(t), 231–232
"rules" regarding, 221, 221(t)
in "pure" sensory axon loss polyneuropathy, 225(t)
in radiation-induced brachial plexopathy, 208(t)
in sacral plexopathy, 211, 213(t)
in traumatic brachial plexopathy, 205(t)
in ulnar neuropathy at elbow, 181(t)
RS studies accompanying, in evaluation of myasthenia gravis, 256(t)

Nerve conduction studies (Continued)
  sensitivity of, in axon loss polyneuropa-
    thy, 171(t)
  sensorimotor, 100–101, 101
  sensory, 98–100
    median, 98–100
      arrangement of equipment for, 99
    ulnar, motor, 98
      sensory, 100
  vs. needle electrode examination, 122
Nerve roots, spinal, 189, 189. See also
    Radiculopathy(ies).
Neuralgic amyotrophy, 217–218
Neuritis, optic, visual evoked potentials
    in, 513, 514, 514
Neurodegenerative diseases, 450–453
Neuroimaging, results of, in epileptic
    patients, 492, 492, 493, 493, 494
Neuroma, acoustic, abnormal brain stem
    AEPs associated with, 536, 537, 538
  surgery for, monitoring of brain stem
    AEPs during, 566, 567, 568
Neuromuscular junction, 62, 84, 251,
    252
  transmission across, 251–253
    disorders of, 253–256. See also spe-
      cific conditions, e.g., Myasthenia
      gravis.
    drug interference with, 259
Neuronal ionic currents, 62(t)
Neuronal mechanisms, of synaptic
    plasticity, 55–56
Neuronal signal amplification, 52–53
Neuronopathy, bulbospinal, X-linked
    recessive, 241
Neuropathy(ies). See also Mononeurop-
    athy(ies); Polyneuropathy(ies).
  axillary, 184
  cranial, 176–177
  facial, 176
    effect of, on blink reflex, 108–109, 109
  femoral, 188
  hypoglossal, 177
  limb, 177–178
  long thoracic, 184–185
  median, at wrist. See Carpal tunnel syn-
    drome.
  monomelic, ischemic, 217
  motor, multifocal, 218–219, 244
  musculocutaneous, 184
  obturator, 188
  peroneal, 185–187
  polyradiculopathies accompanying,
    199–200
  sciatic, 187–188
    high, pattern of muscle involvement
      in, 187, 187(t)
  spinal accessory, 176–177
  suprascapular, 184
  tibial, 187
  trigeminal, 177
    effect of, on blink reflex, 109, 109
  ulnar, 180–182
    at elbow, 181–182
      results of needle electrode examina-
        tion in, 181(t)
      results of nerve conduction studies
        in, 181(t)
    in hand, 182
Neuropeptides, 66
Neurotonic discharges, 572, 573, 573
Neurotransmitters, 63–66
  actions of, 61(t)
  effects of antiepileptic drugs on, 64(t)
  receptors for, 59–60, 63–66
Newborn. See also Child(ren).
  brain stem AEPs in, 531, 531

Newborn (Continued)
  electroencephalography in. See Electroen-
    cephalography, in newborn; Electroen-
    cephalography, in preterm infant.
  myasthenia gravis in, 259
  seizure activity in. See Seizures, neonatal.
Nitric oxide, 66
NMDA (N-methyl-D-aspartate) glutamate
    receptors, 65
Nocturnal penile tumescence, measure-
    ment of, in sleep laboratory, 600–601,
    601, 602(t)
Noise, 22
  reduction of, via signal averaging, 12,
    16(t)
  separation of signal from, filtering in,
    20. See also Filter(s).
Nonconvulsive status epilepticus, 447–448,
    448, 449
Nonepileptic myoclonus, 282
Nongated channels, diffusion through, in
    production of concentration
    gradients, 45
Nonlinear input-output relationships, 23
Non-NMDA glutamate receptors, 65
Non-rapid eye movement (NREM) sleep,
    419–420, 420–422, 589
  brain stem center involved in, 593
  role of serotonin in, 594
  stage I, 589, 590, 602, 603
  stage II, 589, 590, 602, 603
  stage III, 589, 591, 602
  stage IV, 589, 591, 602, 604
Norepinephrine, 63
NREM sleep. See Non-rapid eye movement
    (NREM) sleep.

Obturator neuropathies, 188
Occipital lobe epilepsy, 489–490, 494
  electroencephalographic findings associ-
    ated with, 489–490, 490, 494
Occipital sharp transients of sleep, posi-
    tive, on electroencephalography,
    420, 421
    in children and adolescents, 405–
      406, 407
Occipital sharp wave, right, in bipolar
    electrode montage, 364
Occipital slow activity, abnormal, on EEG
    in children and adolescents, 400
    vs. posterior slow waves of youth,
      401(t)
Occipital slow transients, on EEG in
    sleeping child, 406, 408
Occipital theta bursts, rhythmic, on EEG
    in newborn, 395, 396
Ocular artifacts, on electroencephalogra-
    phy, 377, 379, 382–385, 383, 430, 430
    in bipolar montage, 379, 380, 383
Optic nerve diseases, visual evoked
    potentials in, 513, 514, 514
Optic neuritis, visual evoked potentials in,
    513, 514, 514
Organophosphate poisoning, 265–266
Orthodromic impulses, collision of
    antidromic impulses with, 116
    in assessment of fast vs. slow fiber
      conduction, 119–120, 120
    in assessment of refractory period in
      motor fibers, 119, 119
    in blockage of unwanted nerve stimu-
      lation, 116
    in blockage of unwanted transmission
      through Martin-Gruber anasto-
      mosis, 116, 118

Orthodromic impulses (Continued)
    in production of MNSP, 297, 298, 298.
      See also Mixed nerve silent period.
Orthostatic tremor, 287
Overdose. See Poisoning/intoxication.
Oximetry, in sleep laboratory, 600

P9 waveform component, of median
    nerve SEP, 545, 546
P11 waveform component, of median
    nerve SEP, 546, 547
P14 complex waveform component, of
    median nerve SEP, 546, 548, 548,
    549
  dissociation of CERV N13/P13 compo-
    nent from, 548, 550
P15 waveform component, of median
    nerve SEP, 552, 553
P22 waveform component, of median
    nerve SEP, 552
  loss of, 552, 555
  preservation of, despite loss of N20
    component, 552, 554
Paired-pulse magnetic stimulation, 336
Paired stimuli, in determination of
    refractory period, 113–115, 114
Palatal tremor, 287
Palsy, Bell's, 176
Panencephalitis, sclerosing, subacute, 452
  electroencephalographic findings
    associated with, 444, 452, 453
Paradoxical alpha rhythm, on
    electroencephalography, 415
Parallel resistance, 3, 3
Paralysis, facial, 176
  tick bite, 266
  tourniquet, 216
Parasomnias, during polysomnography,
    606, 608
Parietal lobe epilepsy, 487, 489
  electroencephalographic findings associ-
    ated with, 487, 489, 489
Parkinsonian tremor, at rest, 286
Parsonage-Turner syndrome, 217–218
Partial field stimulation, of pattern-
    reversal visual evoked potentials, 517
Partial seizures, simple, 482–483
Passband, 23
  rolloff outside, 28
Passband parameters, for recording of
    brain stem AEPs, 528
  for recording of SEPs, 557
  for recording of VEPs, 511
Pattern onset parameters, for recording of
    VEPs, 510
Pattern reversal parameters, for recording
    of VEPs, 509–510
Pattern-reversal visual evoked potentials
    (PVEPs), 507, 507, 508, 508
  hemifield stimulation of, 517
  normal variability in, 509
  recording of. See Visual evoked potentials
    (VEPs), recording of.
Pen deflections, in electroencephalogra-
    phy, 358–359, 359
Penile tumescence, measurement of, in
    sleep laboratory, 600–601, 601, 602(t)
Periodic complexes, on EEG in patients
    with cerebral dysfunction, 442–444,
    444, 445
Periodic lateralized epileptiform
    discharges, 442–444, 444, 449, 459, 462
  focal seizure risk associated with, 462,
    463
Periodic leg movements, during sleep,
    604–605

Periodic sharp wave complexes, on EEG
  in patients with cerebral dysfunction,
  443–444, 451, 452, *452*, *453*
Permeability, selective, of ion channels, 43,
  45
Peroneal neuropathies, 185–187
Perspiration artifact, on electroencephalography, 431
Phase cancellation, between fast and slow
  conducting sensory fibers, 73
Phase reversal, in bipolar montages,
  361–362, *362*, 362(t), *363*, *365*
  negative, 361, 362, *362*
  positive, 362, *362*
  triple, 365, *366*
  in referential montages, 362(t), *367*, *368*
Phase shift, 27–28, *28*
Phenobarbital, effects of, on brain
  neurotransmitters, 64(t)
Phenytoin, effects of, on brain
  neurotransmitters, 64(t)
Photic stimulation, of electroencephalographic activity, 423, *424*
  possible abnormalities in, 462, 464,
  464
Photoconvulsive (photoparoxysmal)
  response, on electroencephalography,
  424, *425*, 473
Photoelectric artifact, on electroencephalography, 423, *424*
Photomyogenic (photomyoclonic) artifact,
  on electroencephalography, 423–424,
  425
Photoparoxysmal (photoconvulsive)
  response, on electroencephalography,
  424, *425*, 473
Phrenic nerve stimulation, by magnetic
  coil, *327*, *328*
Physiological artifacts, on electroencephalography, 430–432, *430*–*432*. See also
  specific types, e.g., *Ocular artifacts*.
Physiological nerve block, collision-based
  production of. See *Collision, of
  antidromic and orthodromic impulses*.
Physiological tremor, 286
  exaggerated, 286
Plasticity, synaptic, neuronal mechanisms
  of, 55–56
Plexopathy(ies), 201–214
  brachial, 205, *206*, *209*. See also *Brachial
  plexus*.
    confirmation of preserved conduction
    in, by eliciting cutaneous nerve
    silent period, 301, *303*
    results of needle electrode examina-
    tion in, 205(t), 207(t), 209(t)
    results of nerve conduction studies in,
    205(t), 207(t), 208(t)
    infraclavicular, 209. See also *Plexop-
    athy(ies), brachial*.
    radiation-induced, electrodiagnostic
    findings in, 208(t), 209(t)
  lumbar, results of needle electrode examination in, 212
    results of nerve conduction studies
    in, 211–212
  lumbosacral, 212
    limitations of nerve conduction stud-
    ies in, 211
  sacral, results of needle electrode examination in, 212, 213(t)
    results of nerve conduction studies in,
    211, 213(t)
  supraclavicular, 205, *206*. See also *Plexop-
  athy(ies), brachial*.
    cancer-related, electrodiagnostic find-
    ings in, 207(t)

Plexopathy(ies) *(Continued)*
    traumatic, electrodiagnostic findings
    in, 205(t)
Poisoning/intoxication, drug,
  encephalopathies due to, 446
  magnesium, 266
  metal, manifestations of, consistent with
  motor neuron disorders, 244
  organophosphate, 265–266
Poliomyelitis, 242
Polymorphic delta activity, on electroencephalography, *465*, *466*
  interpretation of, 464
Polymyositis, 276, 278
  results of needle electrode examination
  in, 275(t)
  results of nerve conduction studies in,
  275(t)
Polyneuropathy(ies), 219–232
  axon loss, 222–225, 223(t)
    "pure" sensory, 223, 225, 225(t)
    results of needle electrode examina-
    tion in, 224(t)
    results of nerve conduction studies in,
    224(t), 225(t)
    sensitivity of needle electrode exami-
    nation in, 171(t)
    sensitivity of nerve conduction stud-
    ies in, 171(t)
  classification of, 220
  demyelinating, 225–231, 226(t)
    results of needle electrode examina-
    tion in, 227(t), 230(t)
    results of nerve conduction studies in,
    226, *226*, 227(t), *228*, 229(t),
    230
    slowing of conduction in, 219, 226,
    226, *228*, *228*, *230*
  diabetic, 232
    cardiovascular function test in, heart
    rate response to standing as, *313*
    pupillometric findings associated
    with, 320, *320*, *321*
    sweat test results in, *316*, *319*
  electrophysiological "rules" regarding,
  221, 221(t)
  mixed, 171–172
  mixed nerve silent period in, abnormally distinct stages of, 297, *297*
  results of needle electrode examination
  in, 222, 224(t), 227(t), 230(t)
  results of nerve conduction studies in,
  221–231, 224(t)–229(t), *226*, *228*, *230*
  factors confounding interpretation of,
  231(t), 231–232
  "rules" regarding, 221, 221(t)
Polyradiculoneuropathy(ies), 199–200. See
  also *Polyradiculopathy(ies)*.
Polyradiculopathy(ies), 198–199. See also
  Radiculopathy(ies).
  causes of, 198(t), 198–199
  compressive, 198
  diabetic, 198–199
  differential diagnosis of, 198(t)
  neuropathy accompanying, 199–200
Polysomnography, 597–606. See also *Sleep
  laboratory*.
  parasomnias during, 606, *608*
Pontine myelinolysis, central, brain stem
  AEPs in, *536*, *537*
Portal-systemic encephalopathy, 445
Positive-negative transient, on EEG in
  patient with myoclonus, 283, *286*. See
  also *Myoclonus, electromyographic
  findings in, correlated with results of
  EEG*.
Positive occipital sharp transients of sleep,
  on electroencephalography, 420, *421*

Positive occipital sharp transients of sleep
  (Continued)
    in children and adolescents, 405–406,
    407
Positive phase reversal, 362, *362*
Posterior fossa tumors, and abnormal
  brain stem AEPs, 536–537, *537*, *538*
Posterior slow transients, associated with
  eye movements, on EEG in children,
  400, *401*
Posterior slow waves, on EEG in young
  adults and adolescents, 399–400,
  400, 418, *419*
  vs. abnormal occipital slow activity,
  401(t)
Posterior tibial nerve somatosensory
  evoked potential(s), 553–555
  recording of, 557, 559, 561
  waveform components of, 553(t), 553–
  555
Postpolio syndrome, 242
Postsynaptic potentials, 349. See also
  Synaptic entries.
  excitatory, fields generated by, 350, *351*
  inhibitory, fields generated by, *351*
Postural tremor, 287
Potassium channels, voltage-gated, 48
  contribution of, to waveform of
  action potential, 48, *49*
PPV (proximal plexus volley) component,
  of median nerve SEP, 545, *546*
Pressor response, to stress of mental
  arithmetic test, 314
Preterm infant, electroencephalography in.
  See *Electroencephalography, in newborn;
  Electroencephalography, in preterm
  infant*.
Primary lateral sclerosis, 240
Prion(s), 451
Prion disease, 451. See also *Creutzfeldt-
  Jakob disease*.
Progressive muscular atrophy, 241
Proximal plexus volley (PPV) component,
  of median nerve SEP, 545, *546*
Pseudodementia, 453
Pulse artifact, on electroencephalography,
  430, *431*
Pupillometry, 320–321
  results of, in patients with diabetic neu-
  ropathy, 320, *320*, *321*
Pure autonomic failure, thermoregulatory
  sweat test results in, *317*
"Pure" sensory axon loss
  polyneuropathies, 223, 225, 225(t). See
  also *Polyneuropathy(ies), axon loss*.
PVEPs (pattern-reversal visual evoked
  potentials), 507, *507*, *508*, *508*
  hemifield stimulation of, 517
  normal variability in, *509*
  recording of. See *Visual evoked potentials
  (VEPs), recording of*.
Pyramidal myoclonus. See *Reflex
  myoclonus, cortical*.

Quantitative electromyography, 140
  estimation of motor unit numbers in,
  155–157, *156*, *157*
  interference pattern analysis in, 153–155
  measurement of motor unit action
  potential in, 152–153
Quantitative sudomotor axon reflex test,
  316, *317*, *318*

Radiation therapy, electrodiagnosis of
  brachial plexus injury from, 208(t),
  209(t)

Radiation therapy (Continued)
  lower motor neuron disorder following, 243–244
Radiculopathy(ies), 189–200. See also Polyradiculopathy(ies).
  axon loss, 192, 193
    results of needle electrode examination in, 195(t)
  cervical, 194–196
    failure to elicit cutaneous nerve silent period in, 301, 303
    results of needle electrode examination in, 194
  lumbosacral, 196–197
  needle electrode examination in, 191–194
    results of, in cases of axon loss, 195(t)
    in cases of cervical involvement, 194
  screening via, 191
    in cases of arm pain, 192(t)
    in cases of leg pain, 192(t)
Radiotherapy. See Radiation therapy.
Random variable, expected value of, 13, 15
Rapid eye movement (REM) sleep, 419, 421, 423, 589, 590, 591, 602, 604
  neural structures generating, 593–594
  role of acetylcholine in, 594–595
Rapid spikes, central, on EEG in newborn, 396–397, 398
RC (resistor-capacitor) filters. See Filter(s).
Reactive attenuation, episodic, of electroencephalographic activity in preterm infant, 391
Receptor(s), acetylcholine, 64
  congenital deficiency of, 265, 265(t)
  adenosine, 60
  gamma-aminobutyric acid, 64–65
  glutamate, 65
  G-protein–coupled, 59–60, 60
  ionotropic, 59
  neurotransmitter, 59–60, 63–66
  serotonin, 63–64
Rectal cancer, sacral plexopathy associated with, electrodiagnostic findings in, 213(t)
Recurring stupor, idiopathic, 447
Reference-site parameters, for recording of VEPs, 511, 511–512
Referential montages, 360, 361
  electrocardiographic artifact in, 431
  left temporal sharp wave in, 371
  localization of potentials in, 367–370
  phase reversal in, 362(t), 367, 368
  right temporal sharp wave in, 363, 368
  seizure patterns in, 374, 378, 381
Reflex myoclonus, cortical, 282
  Alzheimer's disease and, 283
  electroencephalographic findings in, 283, 286
  electromyographic findings in, 283, 285, 304
  reticular, 282
Refractory period, 112
  absolute, 112
    determination of, 113–114, 114
  effects of temperature changes on, 115
  paired stimuli determining, 113–115, 114
    combined with single distal stimulus, in assessment of motor fibers, 119, 119
  relative, 112
    clinical utility of, 115–116
    determination of, 114, 114–115
    supernormal and subnormal periods following, 113

Refractory period (Continued)
    demonstration of, 115
    trains of stimuli determining, 115
Relative refractory period, 112
  clinical utility of, 115–116
  determination of, 114, 114–115
  supernormal and subnormal periods following, 113
    demonstration of, 115
Relaxed wakefulness, electroencephalographic findings associated with, 602, 602
REM (rapid eye movement) sleep, 419, 421, 423, 589, 590, 591, 602, 604
  neural structures generating, 593–594
  role of acetylcholine in, 594–595
Renal disease, encephalopathy associated with, 446
Repetitive stimulation (RS) studies, 254–256
  in botulism, 263, 263
  in Lambert-Eaton myasthenic syndrome, 260–261, 261
  in myasthenia gravis, 256(t)
  in myopathies, 269
Repetitive transcranial magnetic stimulation, 340
  response of blind subjects to, 340, 340
  seizure risk associated with, 340
  therapeutic, 341, 341
Resistance, parallel, 3, 3
  series, 2, 3
Resistor-capacitor (RC) filters. See Filter(s).
Respiration artifact, on electroencephalography, 431
Respiratory monitoring, in sleep laboratory, 598, 600
Resting potential, 43
  contribution of multiple ions to, 46–47
Reticular nucleus of thalamus, 352, 593
Reticular reflex myoclonus, 282
Retinopathy(ies), visual evoked potentials in, 517
Reversal frequency parameters, for recording of VEPs, 511
Rhythmic alpha activity or bursts. See also Alpha rhythm.
  excessive, on EEG in newborn, 392–393, 395
  midline frontal and midline central, on EEG in newborn, 393, 395
Rhythmic delta activity, bifrontal, on EEG in newborn, 395, 396
  intermittent, on EEG in epileptic patient, 484, 486
    on EEG in patient with cerebral dysfunction, 439, 440
    vs. eye movement artifact on EEG, 380
Rhythmic electrographic discharge, subclinical, in adults, 425, 426, 427
Rhythmic intermittent slowing. See Rhythmic delta activity, intermittent.
Rhythmic midtemporal discharges, as benign variant on EEG, 424–425, 425, 426
Rhythmic theta activity/theta bursts/theta discharge. See also Theta rhythm.
  excessive, on EEG in newborn, 392–393, 395
  mesiotemporal, on EEG in epileptic patient, 483
  midline frontal and midline central, on EEG in newborn, 393, 395
  occipital, on EEG in newborn, 395, 396
  subclinical, on EEG in adult, 425, 426, 427
Right occipital sharp wave, in bipolar electrode montage, 364

Right temporal sharp wave, in bipolar electrode montage, 363
  in referential electrode montage, 363, 368
Rigidity, 291
Ringbinden, 87
Ringing, by high-order filters, 28, 29
Rolloff, filter design and, 27
  outside passband, 28
RS (repetitive stimulation) studies, 254–256
  in botulism, 263, 263
  in Lambert-Eaton myasthenic syndrome, 260–261, 261
  in myasthenia gravis, 256(t)
  in myopathies, 269

Sacral plexus, 210, 211
  needle electrode examination of, 212
    results of, in cancer involving right lumbosacral region, 213(t)
  nerve conduction studies of, 211, 212
    limitations of, 211
    results of, in cancer involving right lumbosacral region, 213(t)
Safe performance, of needle electrode examination, 137–138
  steps for clinician to follow in, 137–138
  vs. risk of excessive bleeding, 138
Safety hazards, in electrodiagnostic laboratory, 31–39
Sampling rate parameters, for recording of VEPs, 511
Sarcoplasmic reticulum, 84–85
Scalp electrodes. See also Electroencephalography.
  management of, in epilepsy monitoring unit, 434–435
  system for placement of, 359, 360. See also Montages (electrode montages).
Scanning electromyography, 151, 151–152
Sciatic neuropathies, 187–188
  high, pattern of muscle involvement in, 187, 187(t)
Sclerosing panencephalitis, subacute, 452
  electroencephalographic findings associated with, 444, 452, 453
Sclerosis, lateral, amyotrophic, 237–239
  results of needle electrode examination in, 238, 238(t)
  results of nerve conduction studies in, 238, 238(t), 239
  variants of, 239–241
  primary, 240
  multiple, brain stem AEPs in, 534–535, 535, 536
  visual evoked potentials in, 513, 513, 514
Seizures, epileptic. See Epilepsy.
  medications for, dosages of, in newborn, 499(t)
  effects of, on brain neurotransmitters, 64(t)
  neonatal, 497–505, 498(t). See also Child(ren), seizures in.
    electroencephalographic findings associated with, 500–505, 501–504
    treatment of, 499, 499(t)
  risk of induction of, via transcranial magnetic stimulation, 340
Selective permeability, of ion channels, 43, 45
Sensorimotor nerve conduction studies, 100–101, 101. See also Nerve conduction studies.

Sensory axon loss polyneuropathies, "pure," 223, 225, 225(t). See also *Polyneuropathy(ies), axon loss.*
Sensory nerve action potential (SNAP), *99*
Sensory nerve action potential (SNAP) domains, of brachial plexus, 203(t)
  incidence of involvement of, in cord lesions, 203(t)
  in trunk lesions, 203(t), *204*
Sensory nerve conduction studies, 98–100. See also *Nerve conduction studies.*
  median, 98–100
    arrangement of equipment for, *99*
SEPs. See *Somatosensory evoked potentials (SEPs).*
Series resistance, 2, 3
Serotonin, 63–64
  role of, in NREM sleep, 594
Sharp spikes, small, as benign variant on EEG, *428,* 428–429
Sharp wave(s), electroencephalographic, 481. See also *Sharp wave transients.*
  appearance of, in bipolar montage, *362–365, 371, 372, 374, 376*
    in referential montage, *363, 368, 371*
  mesiotemporal, in epileptic patient, *482*
  periodic, complexes of, in patient with cerebral dysfunction, 443–444, 451, 452, *452, 453*
  temporal, in newborn, 396
    normal vs. abnormal, 397(t), *398*
Sharp wave transients. See also *Sharp wave(s).*
  frontal, on EEG in preterm infant, 390–391, *391*
  occipital, positive. See *Positive occipital sharp transients of sleep.*
Shock, sources of, in electrodiagnostic laboratory, 31, *34, 35, 36, 37, 39*
Signal(s), accuracy in measurement of, 8, *8,* 21–23
  alternating current, 3–4, *4*
  averaging of, 11
    conditions for, 16–17
    filter effect of, 17–18
      at settings obscuring variations in VEPs, *17*
    noise reduction achieved by, *12,* 16(t)
  mechanical amplification of. See *Amplifier(s).*
  neuronal amplification of, 52–53
  separation of noise from, filtering in, *20.* See also *Filter(s).*
Signal-conditioning operations, 19, *19.* See also *Amplifier(s); Filter(s).*
Silastic imprint method, of sweat testing, 316, *318, 319*
Silence, electrocerebral, 444–445, *446,* 448–450
Silent period (electromyographically silent period), 295–301, 335–336
  cutaneous nerve, 298–301
    delayed response to attempted elicitation of, 299, *299*
    lack of response to attempted elicitation of, in cervical radiculopathy, 301, *303*
      in conduction block involving ulnar nerve, *302*
    positive response to attempted elicitation of, 299–300, *300*
      confirming preserved conduction in brachial plexopathy, 301, *303*
  masseter, 299, *299*
  mixed nerve, 295–298, *296,* 296(t), *297*
    collision of antidromic and orthodromic impulses producing, 297, *298, 298*

Silent period (electromyographically silent period) *(Continued)*
  stages of, 297
    abnormally distinct separations between, in polyneuropathy, 297, *297*
Simple partial seizures, 482–483
Single-fiber electromyography, 140–150
  jitter in, 142–146, *143, 144, 145,* 146(t), 149
  results of, in muscular dystrophy, *145*
    in myasthenia gravis, *145,* 257(t)
    in spinal muscular atrophy, *145*
  stimulation in, 146–147, *147*
Single-pulse transcranial magnetic stimulation, 340
  seizure risk associated with, 340
Six-hertz spike and wave, as benign variant on EEG, 429, *429*
Skin blood flow measurements, to record response to autonomic maneuver or stimulus, 314
Skin response, sympathetic, in sudomotor function testing, 318, *319,* 320
Skull defects, electroencephalographic findings associated with, 459, *461, 462, 464, 465*
Sleep, 419–421, 589–595. See also *Sleep rhythms.*
  apnea during, 603–604
    electroencephalographic findings associated with, *605*
  arousal from, 354. See also *Wakefulness.*
    electroencephalographic findings associated with, 421–422
      in children and adolescents, 407, *409, 409, 410*
  attenuation of electroencephalographic activity during, in newborn, 395–396, *397*
  duration of, age-related variations in, 592, *592*
  hypothalamic "center" for, 593
  laboratory studies of. See *Sleep laboratory.*
  periodic leg movements during, 604–605
Sleep laboratory, 597
  capnography in, 600
  electrocardiography in, 600
  electrode montage in, 598
  electromyography in, 598
  electro-oculography in, 598
  filter settings used in, 597(t)
  measurement of nocturnal penile tumescence in, 600–601, *601,* 602(t)
  monitoring of esophageal pH in, 601
  multiple sleep latency test performed in, 606–607, *608*
  oximetry in, 600
  parasomnias in, 606, *608*
  polysomnography in, 597–606
  respiratory monitoring in, 598, 600
  video taping in, 601
Sleep latency test, 606–607, *608*
Sleep rhythms, 419–421, 589–590. See also *Sleep; Sleep spindles.*
  appearance of, in children and adolescents, 402–407, *404–409*
    in newborn, 391–392, *393, 394*
    in preterm infant, 391
  non-rapid eye movement, 419–420, *420–422,* 589
    brain stem center involved in, 593
    role of serotonin in, 594
    stage I, 589, *590,* 602, *603*
    stage II, 589, *590,* 602, *603*

Sleep rhythms *(Continued)*
    stage III, 589, *591,* 602
    stage IV, 589, *591,* 602, *604*
  rapid eye movement, 419, 421, *423,* 589, 590, *591,* 602, *604*
    neural structures generating, 593–594
    role of acetylcholine in, 594–595
Sleep spindles. See also *Sleep rhythms.*
  on electroencephalography, 420, *422,* 589, 593, 602
    in children and adolescents, 405, *406, 407*
Slow alpha variant, on electroencephalography, 415, *416*
Slow channel syndrome, 265, 265(t)
Slow transients, occipital, on EEG in sleeping child, 406, *408*
  posterior, associated with eye movements, on EEG in children, 400, *401*
Slow waves/slow wave activity, pathological. See *Electroencephalography, slow waves/slow wave activity on, abnormal.*
  posterior, on EEG in young adults and adolescents, 399–400, *400,* 418, *419*
    vs. abnormal occipital slow activity, 401(t)
Small sharp spikes, as benign variant on EEG, *428,* 428–429
SNAP. See *Sensory nerve action potential (SNAP)* entries.
Sodium channels, voltage-gated, 48, *48*
  contribution of, to waveform of action potential, 48, *49*
Somatosensory evoked potentials (SEPs), 543–562
  giant, in patient with myoclonus, 303, 304, 305
  median nerve, 544–553
    recording of, 557, 559
    waveform components of, 544–553, 545(t). See also specific components cited under *Median nerve somatosensory evoked potential(s).*
  posterior tibial nerve, 553(t), 553–555
    recording of, 557, 559, 561
    waveform components of, 553(t), 553–555
  recording of, 557–561
    in patient with cervical myelopathy, 74
    in patient with myoclonus, 303, 304, 305
    in surgical patient, 568–570, *569, 571*
      interpretation of results of, 570(t)
    spinal, in patient with cervical myelopathy, 74
Space constant, of depolarization, 51, *53*
Spasm(s). See also *Spasticity.*
  electromyographically guided Botox injection for, 292–293
  infantile, West syndrome and, 474, *475*
Spasticity, 291. See also *Spasm(s).*
Spectral analysis, of heart rate, 314–315
Spikes, electroencephalographic, 481
  centrotemporal, in child with benign epilepsy, 377, *381,* 468–470, *469, 470*
  lateral rectus muscle activity producing, 383, *384*
  rapid, central, in newborn, 396–397, *398*
  sharp, small, 428–429, *429*
  six-hertz, 429, *429*
  wicket, *429,* 429–430
Spinal accessory neuropathies, 176–177
Spinal cord, electrical stimulation of, MEPs produced by, 571–572, *573*

Spinal cord *(Continued)*
  irradiation of, lower motor neuron syndrome following, 243–244
Spinal muscular atrophy, 242–243
  single-fiber electromyographic findings in, *145*
Spinal nerve roots, 189, *189*. See also *Radiculopathy(ies)*.
Spinal SEPs, in cervical myelopathy, *74*
Spindles, electroencephalographic, 352
  in cats anesthetized with barbiturates, 352, *353*
  in comatose patients, 441, *442*
  in sleeping subjects. See *Sleep spindles*.
Spine surgery, recording of SEPs during, *569, 570, 571*
Spinocerebellar disorders, visual evoked potentials in, 514
Spondyloradiculomyelopathy, cervical, 244
Squeak effect, on electroencephalography, 415, *415*
SREDA (subclinical rhythmic electrographic discharge in adults), 425, *426, 427*
Standing, blood pressure response to, 313–314
  heart rate response to, 312–313, *313*
Startle reflex, 282
Status epilepticus. See also *Epilepsy*.
  absence, 447, *448*
  myoclonic, 447, *447*
  nonconvulsive, 447–448, *448, 449*
Stiff man syndrome, 291
Stimulus-specific parameters, for recording of SEPs, 557
  for recording of VEPs, 509–511
Stop-band filter, 23; *24*
Stupor, recurring, idiopathic, 447
Subacute sclerosing panencephalitis, 452
  electroencephalographic findings associated with, 444, 452, *453*
Subclinical rhythmic electrographic discharge in adults (SREDA), 425, *426, 427*
Subcortical dementias, 451
Subdural electrodes, 435
Subnormal period, subsequent to supernormal period following relative refractory period, 113
  demonstration of, 115
Sudomotor function, testing of, 315–320
  methods used in. See specific approaches cited under *Sweat testing*.
Supernormal period, following relative refractory period, 113
  demonstration of, 115
  subnormal period subsequent to, 113
  demonstration of, 115
Supplementary sensorimotor area epilepsy, 477
  electroencephalographic findings associated with, 477, 484
  in pediatric patient, *478*
Supraclavicular plexus, 203. See also *Brachial plexus*.
  lesions involving, 205, 206
    cancer-related, electrodiagnostic findings associated with, 207(t)
    traumatic, electrodiagnostic findings associated with, 205(t)
Suprascapular neuropathies, 184
Surgery, monitoring during, 565–584, 566(t)
Surgical evaluation, of epileptic patient, 490–491
Sustained hand grip, blood pressure response to, 314

Sweat testing, 315–320
  quantitative axon reflex studies in, 316, *317, 318*
  Silastic imprint method of, 316, 318, *319*
  sympathetic skin response in, 318, *319, 320*
  thermoregulatory, 315
    results of, in diabetic patients, *316*
      in patients with multisystem atrophy, *317*
      in patients with pure autonomic failure, *317*
Sympathetic skin response, in sudomotor function testing, 318, *319, 320*
Symptomatic/cryptogenic epilepsy, in pediatric patient, 473–478
Synapse(s), 57. See also *Synaptic* entries; *Postsynaptic potentials*.
  chemical, 57–58
Synaptic plasticity, neuronal mechanisms of, 55–56
Synaptic transmission, 57–63
  frequency-dependent changes in, 62–63
  physiology of, 57–63
Synaptic vesicles, 58–59
Synchronous activity, development of, on EEG in preterm infant, 388, *390*
Syncope, cerebral hypoperfusion and, 448
  electroencephalographic findings associated with, 448, *449*
Syringomyelia, 245–246
  cervical, 245
    median nerve SEPs in, 548, *550*

Temperature changes, effects of, on refractory period, 115
Temporal alpha bursts, on EEG in preterm infant, 390, *391*
Temporal dispersion, 71–72
Temporal ictal activity, on electroencephalography. See *Spikes, encephalographic, centrotemporal*; *Temporal lobe epilepsy*.
Temporal intermittent rhythmic delta activity, on EEG in epileptic patient, 484, *486*
Temporal lobe epilepsy, 483–485
  mesial, 491–492
    brain scan results consistent with, 492, *492, 493*
    electroencephalographic findings associated with, 477, 482, 483, 484, 485, 486, 491
    in pediatric patient, *477*
  neocortical, electroencephalographic findings associated with, 477, 484, *486*
Temporal rhythmic discharges, mesial, as benign variant on EEG, 424–425, *426*
Temporal sharp wave(s), appearance of, in bipolar electrode montage, *363, 371*
  in referential electrode montage, *363, 368, 371*
  mesial, on EEG in epileptic patient, *482*
  normal vs. abnormal, on EEG in newborn, 396, 397(t), *398*
Temporal theta activity or bursts, on EEG in elderly, 417, *419*
  on EEG in preterm infant, 390, *391*
Thalamus, reticular nucleus of, 352, *593*
Therapeutic transcranial magnetic stimulation, 341, *341*
Thermoregulatory sweat test, 315
  results of, in diabetic patients, *316*
    in patients with multisystem atrophy, *317*

Thermoregulatory sweat test *(Continued)*
    in patients with pure autonomic failure, *317*
Theta activity/theta bursts/theta discharge. See also *Theta rhythm*.
  fronto-central, on EEG in children and adolescents, 401–402, *403*
  rhythmic, excessive, on EEG in newborn, 392–393, *395*
    mesiotemporal, on EEG in epileptic patient, *483*
    midline frontal/midline central, on EEG in newborn, 393, *395*
    occipital, on EEG in newborn, 395, *396*
    subclinical, on EEG in adult, 425, *426, 427*
  temporal, on EEG in elderly, 417, *419*
    on EEG in preterm infant, 390, *391*
Theta rhythm, on electroencephalography, 417. See also *Theta activity/theta bursts/theta discharge*.
  alpha rhythm combined with, in comatose patients, 441
  hippocampal, in animals, 355
  midline, as benign variant, 427, *427*
Theta waves, pathological, on electroencephalography, 355
Tibial nerve lesions, 187
Tic(s), 283
Tick bite paralysis, 266
Tilt, blood pressure response to, 314
Time constant, of depolarization, 51, *52*
  of impulse decay, *28*
  of RC circuits, 6–7, *7*
TMS. See *Transcranial magnetic stimulation*.
Tourniquet paralysis, 216
Toxic encephalopathies, 446
Toxin exposure. See *Poisoning/intoxication*.
Transcranial magnetic stimulation, 340–341
  repetitive, 340
    response of blind subjects to, 340, *340*
    seizure risk associated with, 340
    therapeutic, 341, *341*
  single-pulse, 340
    seizure risk associated with, 340
Transient response, of high-pass filter, 6, *6*
  of low-pass filter, 6, *6*
Transient unilateral attenuation, during sleep, on EEG in newborn, 395–396, *397*
Transmission, synaptic, 57–63
  frequency-dependent changes in, 62–63
  physiology of, 57–63
Transneuroforaminal electric field induction, by magnetic coil, 327
Transverse bipolar montage, *361*. See also *Montages (electrode montages), bipolar*.
Transverse tubules, 84
Trauma, axon loss brachial plexopathy due to, electrodiagnostic findings in, 205(t)
  multiple mononeuropathies due to, electrodiagnostic findings in, 215–216
Tremor, 283–287
  cerebellar, 287
  essential, 286–287
  evaluation of, 283–287, *288*
  hysterical, 287
  kinetic, 287
  orthostatic, 287
  palatal, 287
  Parkinsonian, at rest, 286
  physiological, 286

Tremor (Continued)
exaggerated, 286
postural, 287
Wilson's disease and, 287
writing accompanied by, 287
Trigeminal nerve, first divisions of, effects of cold stimulus applied to, 314
lesions of, 177
effect of, on blink reflex, 109, 109
Triphasic action potential, production of, 71, 72
Triphasic waves, on EEG in patients with cerebral dysfunction, 441–442, 443
Triple phase reversal, 365, 366
Tumescence, measurement of, in sleep laboratory, 600–601, 601, 602(t)
Tumor(s). See also Cancer.
posterior fossa, and abnormal brain stem AEPs, 536–537, 537, 538

Ulnar nerve, conduction block involving, failure to elicit cutaneous nerve silent period in, 302
Ulnar nerve conduction studies, motor, 98
sensory, 100
Ulnar neuropathy(ies), 180–182
at elbow, 181–182
results of needle electrode examination in, 181(t)
results of nerve conduction studies in, 181(t)
in hand, 182
Undifferentiated and depressed activity, on EEG in newborn, 392
Unilateral attenuation, transient, on EEG in sleeping newborn, 395–396, 397
Unresponsive patient, with normal electroencephalographic findings, 444, 445
Uremic encephalopathy, 446

Vacuolar degeneration, of muscle fibers, 87
Valproic acid, effects of, on brain neurotransmitters, 64(t)
Valsalva maneuver, 311
blood pressure response to, 311, 312, 312, 313
heart rate response to, 311–312

Variant findings, benign, on electroencephalography, 424–430
Vascular disease, of brain, AEPs in, 537, 539
Ventricular fibrillation, electrical current producing, 32, 34, 37
VEPs. See Visual evoked potentials (VEPs).
Vertex transients, on EEG during sleep, 420, 421
in children and adolescents, 405, 405
Vesicles, synaptic, 58–59
Video recording, in monitoring of epileptic patient, 434
in monitoring of sleeping patient, 601
Visual evoked potentials (VEPs), 507–520
flash, 519–520, 520
measurement of amplitude and latency of, 508–509. See also Visual evoked potentials (VEPs), recording of.
neuroanatomic aspects of, 508, 508
pattern-reversal, 507, 507, 508, 508
hemifield stimulation of, 517
normal variability in, 509
recording of. See Visual evoked potentials (VEPs), recording of.
recording of, 511–512
in amblyopia, 516
in glaucoma, 516–517
in leukodystrophies, 514
in Lhermitte's syndrome, 514
in multiple sclerosis, 513, 513, 514
in optic nerve diseases, 513, 514, 514
in retinopathies, 517
in spinocerebellar disorders, 514
parameters for, 509–512
check size, 510
color, 511
contrast, 511
distance-to-pattern, 511
field size, 510
luminance, 510–511
passband, 511
pattern onset, 509
pattern reversal, 509–510
reference-site, 511, 511–512
reversal frequency, 511
sampling rate, 511
stimulus-specific, 509–511
variations of, filter settings obscuring, 17
waveform components of, 507, 507
aberrant, 509
bifid, 509, 509

Voltage-activated ion channels, 62
Voltage divider, resistors producing, 3, 3
Voltage/electrode position graphs, 367, 367, 368, 368, 383
"chains" on, 371, 372, 376
linkage of, 371, 373
isopotential maps corresponding to, 370, 370, 373, 375, 377, 381
representations of phase reversals on, 365, 365, 366
Voltage-gated channels, and action potential, 47–49, 49
Voltage-gated potassium channels, 48
contribution of, to waveform of action potential, 48, 49
Voltage-gated sodium channels, 48, 48
contribution of, to waveform of action potential, 48, 49
Voltage gradients, production of, ionic concentration differences in, 45–46
Voluntary movement, evaluation of disorders of, 288–290
electromyography in, 289, 290, 290

Wakefulness. See also Sleep, arousal from.
electroencephalographic findings associated with, 602, 602
hypothalamic "center" for, 593
West syndrome, 474
White matter lesions, of brain, and abnormal AEPs, 536, 537
and abnormal VEPs, 514
Wicket spikes, as benign variant on EEG, 429, 429–430
Wilson's disease, tremor in, 287
Wrist, median nerve compression at. See Carpal tunnel syndrome.
Writing, tremor accompanying, 287

X-linked recessive bulbospinal neuronopathy, 241

Young adults, electroencephalography in, posterior slow waves on, 399–400, 400, 418, 419
vs. abnormal occipital slow activity, 401(t)

ISBN 0-7216-7656-1

90038

9 780721 676562